High-Performance
Capillary Electrophoresis

CHEMICAL ANALYSIS

A SERIES OF MONOGRAPHS ON
ANALYTICAL CHEMISTRY AND ITS APPLICATIONS

Editor
J. D. WINEFORDNER

VOLUME 146

A Wiley-Interscience Publication

JOHN WILEY & SONS, INC.

New York / Chichester / Weinheim / Brisbane / Singapore / Toronto

High-Performance Capillary Electrophoresis
Theory, Techniques, and Applications

Edited by
MORTEZA G. KHALEDI

Department of Chemistry
North Carolina State University
Raleigh, North Carolina

A Wiley-Interscience Publication

JOHN WILEY & SONS, INC.

New York / Chichester / Weinheim / Brisbane / Singapore / Toronto

This book is printed on acid-free paper. ∞

Copyright © 1998 by John Wiley & Sons, Inc.

All rights reserved. Published simultaneously in Canada.

No part of this publication may be reproduced, stored in a retrieval system or transmitted in any form or by any means, electronic, mechanical, photocopying, recording, scanning or otherwise, except as permitted under Sections 107 or 108 of the 1976 United States Copyright Act, without either the prior written permission of the Publisher, or authorization through payment of the appropriate per-copy fee to the Copyright Clearance Center, 222 Rosewood Drive, Danvers, MA 01923, (508) 750-8400, fax (508) 750-4744. Requests to the Publisher for permission should be addressed to the Permissions Department, John Wiley & Sons, Inc., 605 Third Avenue, New York, NY 10158-0012, (212) 850-6011, fax (212) 850-6008, E-Mail: PERMREQ@WILEY.COM.

Library of Congress Cataloging in Publication Data:

High-performance capillary electrophoresis : theory, techniques, and
  applications / edited by Morteza G. Khaledi.
    p. cm.—(Chemical analysis ; v. 146)
  Includes index.
  ISBN 0-471-14851-2 (cloth : alk. paper)
  1. Capillary electrophoresis. I. Khaledi, Morteza Gholi, 1956–
II. Series.
QP519.9.C36H54 1998
543'.0871—dc21   97-35765
                                                          CIP

Printed in the United States of America

10 9 8 7 6 5 4 3 2 1

CONTENTS

CONTRIBUTORS	xix
PREFACE	xxiii
CUMULATIVE LISTING OF VOLUMES IN SERIES	xxvii

PART I THEORY AND MODES OF HPCE

CHAPTER 1 CAPILLARY ELECTROPHORESIS: OVERVIEW AND PERSPECTIVE — 3
Barry L. Karger

1.1. Introduction	3
1.2. Modes of Operation	5
1.3. Capillary Electrophoresis–Mass Spectrometry	16
1.4. Electric Field Manipulation of Bulk Liquid Flow	18
1.5. Conclusions	22
List of Acronyms and Abbreviations	22
References	23

CHAPTER 2 THEORY OF CAPILLARY ZONE ELECTROPHORESIS — 25
Ernst Kenndler

2.1. Capillary Zone Electrophoresis in the Absence of Electroosmotic Flow	27
2.2. Capillary Zone Electrophoresis in the Presence of Electroosmotic Flow	58
List of Symbols	70
List of Acronyms	73
References	73

CHAPTER 3 MICELLAR ELECTROKINETIC CHROMATOGRAPHY 77
Morteza G. Khaledi

3.1.	Introduction	77
3.2.	Pseudostationary Phases	79
3.3.	Migration in Micellar Electrokinetic Chromatography	83
3.4.	Migration Parameters	84
3.5.	Resolution	85
3.6.	Structure–Retention Relationships in Micellar Electrokinetic Chromatography	87
3.7.	Characterization of the Chemical Selectivity of Pseudostationary Phases	89
3.8.	Effects of Chemical Composition of Micellar Solutions	95
3.9.	Multiparameter Optimization	117
3.10.	Conclusions and Future Trends	127
	List of Acronyms	130
	References	131

CHAPTER 4 BAND BROADENING IN MICELLAR ELECTROKINETIC CHROMATOGRAPHY 141
Joe M. Davis

4.1.	Introduction	142
4.2.	Measurements of Efficiency	143
4.3.	Plug Size	143
4.4.	The Detection Window and Time Constants	144
4.5.	Micellar Overload	145
4.6.	Longitudinal Diffusion	146
4.7.	Nonequilibrium Dispersion in Micellar Systems	148
4.8.	Electromigrative Dispersion	158
4.9.	Dependence of N on Concentration of Organized Media	158
4.10.	Dependence of N on Electrolyte Concentration and Composition	164
4.11.	Dependence of N on Organic Solvents	165

	4.12. Efficiency of Pseudostationary Phases Other Than Simple Micelles	171
	4.13. General Observations	174
	4.14. Conclusions	175
	Addendum	176
	List of Acronyms and Abbreviations	178
	References	179
CHAPTER 5	**CAPILLARY GEL ELECTROPHORESIS** *Paul Shieh, Nelson Cooke, and Andras Guttman*	**185**
	5.1. Introduction	185
	5.2. Theory and Operation Parameters	187
	5.3. Applications	196
	5.4. Conclusion	217
	List of Acronyms and Abbreviations	218
	References	218
CHAPTER 6	**CAPILLARY ISOELECTRIC FOCUSING** *John E. Wiktorowicz*	**223**
	6.1. Introduction	223
	6.2. Capillary Isoelectric Focusing	226
	6.3. Modes of Isoelectric Point Determination	241
	6.4. Future Considerations and Direction	245
	Addendum	247
	List of Acronyms	247
	References	248
CHAPTER 7	**CAPILLARY ISOTACHOPHORESIS** *Ludmila Křivánková and Petr Boček*	**251**
	7.1. Introduction	251
	7.2. Principles of Isotachophoresis	252
	7.3. Electrolyte Systems	256
	7.4. Sample Stacking	258
	7.5. The Isotachophoresis–Capillary Zone Electrophoresis Combination	260
	7.6. Dynamics of Destacking	265

	7.7. Sample-Induced Isotachophoresis (Stacking)	267
	7.8. Conclusion	273
	List of Acronyms and Abbreviations	274
	References	275
CHAPTER 8	**CAPILLARY ELECTROCHROMATOGRAPHY**	**277**
	Kathleen A. Kelly and Morteza G. Khaledi	
	8.1. Introduction	277
	8.2. Column Preparation	279
	8.3. Detection	283
	8.4. Theory	285
	8.5. Stationary Phase Considerations	287
	8.6. Mobile Phase Considerations	289
	8.7. Applications	290
	8.8. Conclusions and Future Trends	295
	Addendum	296
	List of Acronyms	297
	References	297

PART II   DETECTION SYSTEMS IN HPCE

CHAPTER 9	**CAPILLARY ELECTROPHORETIC DETECTORS BASED ON LIGHT**	**303**
	Louann Cruz, Scott A. Shippy, and Jonathan V. Sweedler	
	9.1. Introduction	304
	9.2. Requirements of Capillary Electrophoretic Detectors	305
	9.3. Absorbance	308
	9.4. Fluorescence	321
	9.5. Other Capillary Electrophoretic Optical Detectors	331
	9.6. Recent Developments	341
	9.7. Conclusions	347
	List of Acronyms and Abbreviations	347
	References	348

CHAPTER 10 ELECTROCHEMICAL DETECTION IN HIGH-PERFORMANCE CAPILLARY ELECTROPHORESIS 355
Barbara Rhoden Bryant, Franklin D. Swanek, and Andrew G. Ewing

    10.1. Introduction   355
    10.2. Methods Used to Electrically Isolate the Electrochemical Detector   356
    10.3. Modes of Electrochemical Detection   359
    10.4. Applications of High-Performance Capillary Electrophoresis with Electrochemical Detection   363
    10.5. Future Directions   369
    List of Acronyms   371
    References   371

CHAPTER 11 INDIRECT DETECTION IN CAPILLARY ELECTROPHORESIS 375
Hans Poppe and Xiaoma Xu

    11.1. Introduction   375
    11.2. Measurement Considerations   377
    11.3. The Transfer Ratio and System Zones   383
    11.4. Overload: Electromigration Dispersion   390
    11.5. Noise Induced by the Electrophoretic Process and System Peaks   393
    11.6. Choice of the Background Electrolyte   395
    11.7. Instrumental Improvements   397
    11.8. Applications and Some Special Aspects   398
    11.9. Summary   400
    List of Acronyms   401
    References   401

CHAPTER 12 HIGH-PERFORMANCE CAPILLARY ELECTROPHORESIS–MASS SPECTROMETRY 405
Kenneth B. Tomer, Leesa J. Deterding, and Carol E. Parker

    12.1. Introduction   406

	12.2.	Advantages of Mass Spectrometric Detection	407
	12.3.	Instrumentation	407
	12.4.	Disadvantages of Mass Spectrometric Detection	414
	12.5.	Potential Solutions to Disadvantages of Mass Spectrometric Detection	415
	12.6.	Applications	429
	12.7.	Summary	435
		Addendum No. 1	436
		Addendum No. 2	438
		List of Acronyms	440
		References	441

PART III OPERATIONAL ASPECTS AND SPECIAL TECHNIQUES IN HPCE

CHAPTER 13 SAMPLE INTRODUCTION AND STACKING 449
Ring-Ling Chien

	13.1.	Introduction	449
	13.2.	Sample Introduction in Capillary Zone Electrophoresis	451
	13.3.	On-Column Sample Stacking	456
	13.4.	Stacking in Sample Introduction	464
	13.5.	Applications of Sample Stacking	471
	13.6.	Conclusions	477
		Addendum	477
		List of Acronyms and Abbreviations	478
		References	478

CHAPTER 14 COATED CAPILLARIES IN HIGH-PERFORMANCE CAPILLARY ELECTROPHORESIS 481
Gerhard Schomburg

	14.1.	Introduction	481
	14.2.	The Status of Capillary Surfaces in Capillary Electrophoresis: Influence on Electroosmotic Flow	483

14.3. Influence of Electroosmotic Flow on Efficiency and Resolution of Capillary Electrophoretic Separations — 484
14.4. Analyte–Wall Interaction and the Performance of Analytical Capillary Electrophoretic Separations — 486
14.5. Surface-Modification Procedures by Coating in Fused-Silica Capillaries — 488
14.6. Methods and Applications of Dynamic Surface Modifications — 492
14.7. Methods and Applications of Permanent Surface Modifications — 507
14.8. Recent Developments in the Dynamic and Permanent Modification of Capillary Surfaces — 516
List of Acronyms and Abbreviations — 519
References — 520

CHAPTER 15 NONAQUEOUS CAPILLARY ELECTROPHORESIS — 525
Joseph L. Miller and Morteza G. Khaledi

15.1. Introduction — 525
15.2. Influence of Nonaqueous Solvents in Capillary Electrophoresis — 527
15.3. Applications — 540
15.4. Conclusions and Future Trends — 553
List of Acronyms and Abbreviations — 553
References — 554

CHAPTER 16 METHOD VALIDATION IN CAPILLARY ELECTROPHORESIS — 557
K. D. Altria

16.1. Introduction — 557
16.2. Specific Validation Aspects — 559
16.3. Conclusions — 577
List of Acronyms — 577
References — 578

CHAPTER 17	**TWO-DIMENSIONAL SEPARATIONS IN HIGH-PERFORMANCE CAPILLARY ELECTROPHORESIS**	**581**

Thomas F. Hooker, Dorothea J. Jeffrey, and James W. Jorgenson

17.1. Introduction	582
17.2. Two-Dimensional Separation Theory	583
17.3. Comprehensive Two-Dimensional Liquid Chromatography–Capillary Electrophoresis	588
17.4. Three-Dimensional Size Exclusion Chromatography–Reversed-Phase Liquid Chromatography–High-Speed Capillary Zone Electrophoresis	605
17.5. Future Directions	607
List of Acronyms	610
References	610

CHAPTER 18	**MICROFABRICATED CHEMICAL SEPARATION DEVICES**	**613**

Stephen C. Jacobson and J. Michael Ramsey

18.1. Introduction	613
18.2. Modular System Design	614
18.3. Fabrication Techniques	615
18.4. Fluid Manipulation and Injection	617
18.5. Microchip Electrophoresis	620
18.6. DNA Separations	623
18.7. Chromatographic Separations	626
18.8. Integrated Structures for Chemical and Biochemical Reactions and Analysis	627
18.9. The Future	629
Addendum	631
List of Acronyms	631
References	632

PART IV APPLICATIONS OF HPCE

CHAPTER 19 PEPTIDE ANALYSIS BY CAPILLARY ELECTROPHORESIS: METHODS DEVELOPMENT AND OPTIMIZATION, SENSITIVITY ENHANCEMENT STRATEGIES, AND APPLICATIONS   637
Gregory M. McLaughlin, Kenneth W. Anderson, and Dietrich K. Hauffe

19.1.	Introduction	638
19.2.	Effects of Capillary Dimensions, Applied Voltage, and Temperature	639
19.3.	Effects of pH and Ionic Strength	643
19.4.	Buffer Additives	647
19.5.	Use of Ion-Pairing Reagents	650
19.6.	Use of Micellar Electrokinetic Chromatography for Hydrophobic and Neutral Species	651
19.7.	Chiral Selectors	653
19.8.	Sensitivity-Enhancement Strategies	654
19.9.	Use of Coated Capillaries and Wall Coatings	656
19.10.	Methods Development Strategy	657
19.11.	Use of Capillary Electrophoresis and High-Performance Liquid Chromatography as Complementary Techniques	659
19.12.	Enhanced Detection Methods	662
19.13.	Physicochemical Measurements	663
19.14.	Concluding Remarks and Future Trends	666
	Addenda	666
	List of Acronyms, Abbreviations, and Symbols	670
	References	671

CHAPTER 20 CAPILLARY ELECTROPHORESIS OF PROTEINS   683
Fred E. Regnier and Shen Lin

20.1.	Conventional Electrophoresis of Proteins	684

20.2.	Electrophoresis of Proteins in Fused-Silica Capillaries	685
20.3.	Surface Modification of Capillaries	686
20.4.	Zone Electrophoresis in Open-Tubular Capillaries	703
20.5.	Capillary Isoelectric Focusing	712
20.6.	Capillary Gel Electrophoresis	713
20.7.	Micellar Electrokinetic Chromatography	715
20.8.	Selectivity	715
20.9.	Mass Spectrometry	716
20.10.	The Future	719
	List of Acronyms and Abbreviations	720
	References	722

CHAPTER 21 CAPILLARY ELECTROPHORESIS OF CARBOHYDRATES 729

Milos V. Novotny

21.1.	Introduction	729
21.2.	The Goals of Analytical Glycobiology	730
21.3.	Instrumental Aspects	732
21.4.	Sample Derivatization	736
21.5.	Electromigration Mechanisms	744
21.6.	Selected Applications	753
	List of Acronyms and Abbreviations	761
	References	761

CHAPTER 22 DNA SEQUENCING BY MULTIPLEXED CAPILLARY ELECTROPHORESIS 767

Edward S. Yeung and Qingbo Li

22.1.	Highly Multiplexed DNA Sequencing	768
22.2.	Acceleration of Electrophoretic Runs	774
22.3.	Base Calling and Data Handling	780
22.4.	System Integration	783
22.5.	Future Prospects	785
	List of Acronyms and Symbols	786
	References	787

CHAPTER 23 CHIRAL SEPARATIONS BY CAPILLARY ELECTROPHORESIS — 791
Fang Wang and Morteza G. Khaledi

- 23.1. Introduction — 791
- 23.2. Chiral Resolution — 793
- 23.3. Types of Chiral Selector — 795
- 23.4. Effect of the Chiral Selector Concentration — 806
- 23.5. pH and Ionic Strength — 807
- 23.6. Organic Solvents — 808
- 23.7. The Counter–Electroosmotic Flow Scheme — 811
- 23.8. Capillary Electrochromatography — 812
- 23.9. Other Methods — 813
- 23.10. Chiral Separation Efficiency — 814
- 23.11. Conclusions and Future Trends — 816
- Addendum — 817
- List of Acronyms — 818
- References — 819

CHAPTER 24 CAPILLARY ELECTROPHORESIS OF INORGANIC IONS — 825
Jeffrey R. Mazzeo

- 24.1. Introduction — 825
- 24.2. Indirect-Ultraviolet Detection: General Principles — 826
- 24.3. Direct-Ultraviolet Detection: Anion Determinations — 841
- 24.4. Direct-Ultraviolet Detection: Cation Determinations — 843
- 24.5. Other Detection Modes — 844
- 24.6. Concluding Remarks — 847
- List of Acronyms and Symbols — 849
- References — 849

CHAPTER 25 THE ANALYSIS OF PHARMACEUTICALS BY CAPILLARY ELECTROPHORESIS — 853
K. D. Altria

- 25.1. Introduction — 853

25.2. Application Areas of Capillary
Electrophoresis in Pharmaceutical Analysis 854
25.3. Future Directions 873
25.4. Conclusions 875
List of Acronyms 875
References 875

CHAPTER 26 ON-LINE IMMUNOAFFINITY CAPILLARY ELECTROPHORESIS FOR THE DETERMINATION OF ANALYTES DERIVED FROM BIOLOGICAL FLUIDS 879

Norberto A. Guzman, Andy J. Tomlinson, and Stephen Naylor

26.1. Introduction 879
26.2. Nonspecific On-Line Preconcentration
Capillary Electrophoresis 881
26.3. Specific On-Line Preconcentration Capillary
Electrophoresis 883
26.4. Conclusions 894
List of Acronyms 894
References 895

CHAPTER 27 MICROBIOANALYSIS USING ON-LINE MICROREACTORS–CAPILLARY ELECTROPHORESIS SYSTEMS 899

Larry Licklider and Werner G. Kuhr

27.1. Introduction 899
27.2. Sampling Single Biological Cells 900
27.3. On-Capillary Assays 904
27.4. On-Line Capillary Microreactors–Capillary
Electrophoresis Systems 909
27.5. Conclusions 919
List of Acronyms and Symbols 920
References 920

CHAPTER 28 ELECTROPHORETICALLY MEDIATED MICROANALYSIS 925

Bryan J. Harmon and Fred E. Regnier

28.1. Coupling On-Line Chemical Reactions to
Capillary Electrophoresis 926
28.2. Determinations of Enzymes by
Electrophoretically Mediated Microanalysis 928

28.3. Enzymatic Determinations of Substrates by Electrophoretically Mediated Microanalysis 938
28.4. Complexometric Determinations of Inorganic Ions by Electrophoretically Mediated Microanalysis 941
28.5. Determinations of Single Cells by Electrophoretically Mediated Microanalysis 941
List of Acronyms 942
References 942

PART V PHYSICOCHEMICAL STUDIES

CHAPTER 29 AFFINITY CAPILLARY ELECTROPHORESIS: USING CAPILLARY ELECTROPHORESIS TO STUDY THE INTERACTIONS OF PROTEINS WITH LIGANDS 947
Jinming Gao, Milan Mrksich, Mathai Mammen, and George M. Whitesides

29.1. Background 948
29.2. Principles of Affinity Capillary Electrophoresis 950
29.3. Technical Issues 952
29.4. Carbonic Anhydrase as a Model Protein 953
29.5. Determination of Binding Affinity 953
29.6. Determination of Kinetic Parameters for Binding 963
29.7. Using Affinity Capillary Electrophoresis to Measure the Effective Charge of a Protein 963
29.8. Determination of Stoichiometry of Binding 964
29.9. Prospects and Limitations of Affinity Capillary Electrophoresis 966
Addendum 968
List of Acronyms or Abbreviations 969
References 970

CHAPTER 30 DETERMINATION OF PHYSICOCHEMICAL PARAMETERS BY CAPILLARY ELECTROPHORESIS 973
Pier Giorgio Righetti

30.1. Introduction 973
30.2. Determination of pK Values of Weak Electrolytes 975

	30.3.	Determination of pK Values of Amphoteric Compounds	980
	30.4.	Assessment of pK Values of Silanols	981
	30.5.	Determination of pI Values of Proteins	982
	30.6.	Determination of Absolute Mobility and Its Relation to the Charge/Mass Ratio in Peptides (and Proteins)	984
	30.7.	Determination of Binding Constants	988
	30.8.	Determination of Diffusion Constants	991
	30.9.	Viscosity Measurements	992
	30.10.	Determination of T_m Values of Nucleic Acids	992
	30.11.	Conclusions	995
	List of Acronyms and Symbols		995
	References		996

CHAPTER 31 APPLICATIONS OF MICELLAR ELECTROKINETIC CHROMATOGRAPHY IN QUANTITATIVE STRUCTURE–ACTIVITY RELATIONSHIP STUDIES: ESTIMATION OF LOG P_{ow} AND BIOACTIVITY 999

Morteza G. Khaledi

	31.1.	Introduction	999
	31.2.	Solute–Micelle Interactions and Hydrophobicity	1001
	31.3.	Relationships Between Micellar Electrokinetic Chromatography Retention and log P_{ow}	1002
	31.4.	Role of the Pseudostationary Phase	1005
	31.5.	Prediction of Retention in Micellar Electrokinetic Chromatography from Solute Hydrophobicity	1011
	31.6.	Quantitative Retention–Activity Relationships in Micellar Electrokinetic Chromatography	1011
	31.7.	Conclusions	1012
	List of Acronyms		1013
	References		1014

INDEX 1015

CONTRIBUTORS

K. D. Altria, Analytical Sciences, Glaxo Wellcome Research and Development, Ware, Herts., United Kingdom

Kenneth W. Anderson, Dionex Corporation, Sunnyvale, California

Petr Boček, Institute of Analytical Chemistry, Academy of Sciences of the Czech Republic, Brno, Czech Republic

Barbara Rhoden Bryant, Department of Chemistry, Penn State University, University Park, Pennsylvania

Ring-Ling Chien, Varian Associates, Inc., Edward L. Grinzton Research Center, Palo Alto, California

Nelson Cooke, Supelco Inc., Supelco Park, Bellefonte, Pennsylvania

Louann Cruz, Department of Chemistry, Oklahoma State University, Stillwater, Oklahoma

Joe M. Davis, Department of Chemistry and Biochemistry, Southern Illinois, University at Carbondale, Carbondale, Illinois

Leesa J. Deterding, Laboratory of Molecular Biophysics, National Institute of Environmental Health Sciences, Research Triangle Park, North Carolina

Andrew G. Ewing, Department of Chemistry, Penn State University, University Park, Pennsylvania

Jinming Gao, Department of Chemistry, Harvard University, Cambridge, Massachusetts

Andras Guttman, Genetic Biosystems, Inc., San Diego, California

Norberto A. Guzman, The R. W. Johnson Pharmaceutical Research Institute, Raritan, New Jersey

Bryan J. Harmon, Biotechnology Process Engineering Center, Massachusetts Institute of Technology, Cambridge, Massachusetts

Dietrich K. Hauffe, Dionex GmbH, Idstein, Germany

Thomas F. Hooker, Department of Chemistry, The University of North Carolina at Chapel Hill, Chapel Hill, North Carolina

Stephen C. Jacobson, Oak Ridge National Laboratory, Oak Ridge, Tennessee

Dorothea J. Jeffrey, Department of Chemistry, The University of North Carolina at Chapel Hill, Chapel Hill, North Carolina

James W. Jorgenson, Department of Chemistry, The University of North Carolina at Chapel Hill, Chapel Hill, North Carolina

Barry L. Karger, Barnett Institute, Northeastern University, Boston, Massachusetts

Kathleen A. Kelly, Department of Chemistry, North Carolina State University, Raleigh, North Carolina

Ernst Kenndler, Institute for Analytical Chemistry, University of Vienna, Vienna, Austria

Morteza G. Khaledi, Department of Chemistry, North Carolina State University, Raleigh, North Carolina

Ludmila Křivánkova, Institute of Analytical Chemistry, Academy of Sciences of the Czech Republic, Brno, Czech Republic

Werner G. Kuhr, Department of Chemistry, University of California, Riverside, Riverside, California

Qingbo Li, Ames Laboratory–USDOE and Department of Chemistry, Iowa State University, Ames, Iowa

Larry Licklider, Department of Chemistry, University of California, Riverside, Riverside, California

Shen Lin, Department of Chemistry, Purdue University, West Lafayette, Indiana

Mathai Mammen, Department of Chemistry, Harvard University, Cambridge, Massachusetts

Jeffrey R. Mazzeo, GelTex Pharmaceuticals, Waltham, Massachusetts

Gregory M. McLaughlin,* Dionex Corporation, Sunnyvale, California

Joseph L. Miller, Department of Chemistry, North Carolina State University, Raleigh, North Carolina

Milan Mrksich, Department of Chemistry, Harvard University, Cambridge, Massachusetts

* Present address: Rheodyne L.P., 6815 Redwood Drive, Cotati, California 94931.

Stephen Naylor, Biomedical Mass Spectrometry Facility, Department of Biochemistry and Molecular Biology, Mayo Clinic, Rochester, Minnesota

Milos V. Novotny, Department of Chemistry, Indiana University, Bloomington, Indiana

Carol E. Parker, Laboratory of Molecular Biophysics, National Institute of Environmental Health Sciences, Research Triangle Park, North Carolina

Hans Poppe, Laboratory for Analytical Chemistry, Amsterdam Institute for Molecular Studies (AIMS), University of Amsterdam, Amsterdam, The Netherlands

J. Michael Ramsey, Oak Ridge National Laboratory, Oak Ridge, Tennessee

Fred E. Regnier, Department of Chemistry, Purdue University, West Lafayette, Indiana

Pier Giorgio Righetti, Department of Agricultural and Industrial Biotechnologies, University of Verona, Strada Le Grazie, Verona, Italy

Gerhard Schomburg, Stiftstr. 39, D-45470 Mülheim an der Ruhr, Germany

Paul Shieh, Supelco Inc., Supelco Park, Bellefonte, Pennsylvania

Scott A. Shippy, Department of Chemistry, University of Illinois, Urbana, Illinois

Franklin D. Swanek, Department of Chemistry, Penn State University, University Park, Pennsylvania

Jonathan V. Sweedler, Department of Chemistry, University of Illinois, Urbana, Illinois

Kenneth B. Tomer, Laboratory of Molecular Biophysics, National Institute of Environmental Health Sciences, Research Triangle Park, North Carolina

Andy J. Tomlinson, Biomedical Mass Spectrometry Facility, Department of Biochemistry and Molecular Biology, Mayo Clinic, Rochester, Minnesota

Fang Wang, Department of Chemistry, North Carolina State University, Raleigh, North Carolina

George M. Whitesides, Department of Chemistry, Harvard University, Cambridge, Massachusetts

John E. Wiktorowicz, Lynx Therapeutics, Hayward, California

Xiaoma Xu, Laboratory for Analytical Chemistry, Amsterdam Institute for Molecular Studies (AIMS), University of Amsterdam, Amsterdam, The Netherlands

Edward S. Yeung, Ames Laboratory–USDOE and Department of Chemistry, Iowa State University, Ames, Iowa

PREFACE

For several decades, conventional electrophoresis has been the method of choice for solving numerous biochemical problems. However, it never found a place in analytical laboratories due to problems with automation and quantitation. In fact, very few publications with a main focus on the technique could be found in analytical journals prior to 1981, when Jorgenson and Lukacs demonstrated the usefulness of performing electrophoresis in glass capillaries. Since then, and particularly over the past decade, capillary electrophoresis (CE) has undergone a phenomenal growth. Scientists from diverse fields like chemical separation, spectroscopy, mass spectrometry, biochemistry, molecular biology, pharmaceutical and clinical chemistry, and environmental sciences have contributed to the developments and exploring applications of the technique.

The tremendous interest in CE stems from a combination of unique features and capabilities that resemble and often surpass those of conventional electrophoresis and high-performance liquid chromatography (HPLC). During the early years, CE's advantages of high resolving power and speed over HPLC, as well as on-line detection and feasibility for quantitation and automation as compared to conventional electrophoresis, were demonstrated. Since the first report, it became quickly evident that, unlike HPLC, the slow diffusion of large molecules would not impose a problem in achieving CE's high performance in their separations. As a result, CE separations of biomacromolecules like proteins and DNA have been the focus of much intense research over the past decade. The microsize nature of the technique has lead to unique applications such as analysis of single cells as well as significant developments in interfacing CE with mass spectrometry for achieving on-line structural information, unprecedented low-mass detection limits with laser-induced fluorescence, and performing electrophoretic separations on microfabricated devices. Incorporation of the high resolving power of CE in two-dimensional separations holds great promise for resolving very complex mixtures. One of the unique features of the technique is the ease of method development. This is due to a combination of speed, high efficiency, and flexibility of manipulating the chemical composition of the system in a short time period. The possibility of incorporating various chemistries in free-solution CE along with additional separation mechanisms has greatly expanded the scope of applications of the technique that encompass a wide range of analytes, from small metal ions to very large molecules.

In spite of these outstanding developments and many others, there has been some concern about the future of CE, mainly from the financial aspect. Surprisingly, there are a few people who still wonder whether the technique will survive! The great achievements in the scientific arena have not instantly translated into a big commercial success. The first automated commercial instruments were introduced in the late 1980s; however, CE did not financially become the next HPLC. Considering the enormous power of the technique and its advantages over HPLC, it was anticipated that many HPLC applications would be replaced by CE. When comparing CE with HPLC, one should remember that HPLC filled a real gap for separation and analysis of nonvolatile compounds, especially biomolecules. Therefore, its rapid progress in research and development was quickly followed by substantial financial returns. Replacing and/or complementing a well-established technique such as HPLC in analytical laboratories in industrial setups would involve a great deal of investment in new equipment, personnel training, and in some cases additional time and effort to meet governmental regulations—and all this at a time of budgetary constraints and corporate downsizing! Perhaps the early expectations about a "billion dollar" market were a bit hasty and overoptimistic.

In spite of all this, the future of CE still seems bright, and interest in the technique rapidly continues to grow. This is clearly evident from the increasing number of meeting presentations and journal publications. A recent survey of readers of *LC–GC* magazine in the United States and Europe also indicates this (*LC–GC*, 15, 448, 1997). Over the past decade, the number of papers on CE has increased at a dizzying pace. It is nearly impossible to stay up to date with the developments in all areas of CE. The main objective of this book is to provide an updated comprehensive reference on various aspects of the technique. There are 31 review chapters, organized into five sections on theory; HPCE detection systems; various HPCE techniques, applications of HPCE in chemical analysis, and determination of physicochemical parameters. This monograph is intended for beginners as well as experienced users.

I am indebted to a number of people who made this project possible. First and foremost, I thank the authors of the chapters in this volume who, despite their busy schedules, accepted my invitation and contributed to it. I would also like to thank Jim Winefordner, the Series Editor, for inviting me to undertake this project. I am grateful to the graduate students in my research group, Armel Agbodjan, Jeff Bumgarner, Leesa Deterding, Kathleen Kelly, Mike Leonard, Amir Malek, Joe Miller, Mark Trone, Fang Wang, Vicki Ward, Jim Yeattes, and Wei Zhu. Some coauthored or helped me in writing some chapters; some assisted in indexing the volume; and all were patient and understanding with their "hard-to-find" advisor during several busy months. Many thanks to Bert N. Zelman and Harriet Damon Shields for their wonderful job in copyediting. Finally, I express my appreciation to my wife, Shahrzad, and my son, Arras, for their love and support, and for "putting up" with my absences (and occasionally with my presence).

Raleigh, North Carolina MORTEZA G. KHALEDI

CHEMICAL ANALYSIS

A SERIES OF MONOGRAPHS ON ANALYTICAL CHEMISTRY AND ITS APPLICATIONS

J. D. Winefordner, *Series Editor*

Vol. 1. **The Analytical Chemistry of Industrial Poisons, Hazards, and Solvents.** *Second Edition.* By the late Morris B. Jacobs
Vol. 2. **Chromatographic Adsorption Analysis.** By Harold H. Strain (*out of print*)
Vol. 3. **Photometric Determination of Traces of Metals.** *Fourth Edition*
Part I: General Aspects. By E. B. Sandell and Hiroshi Onishi
Part IIA: Individual Metals, Aluminum to Lithium. By Hiroshi Onishi
Part IIB: Individual Metals, Magnesium to Zirconium. By Hiroshi Onishi
Vol. 4. **Organic Reagents Used in Gravimetric and Volumetric Analysis.** By John F. Flagg (*out of print*)
Vol. 5. **Aquametry: A Treatise on Methods for the Determination of Water.** *Second Edition (in three parts).* By John Mitchell, Jr. and Donald Milton Smith
Vol. 6. **Analysis of Insecticides and Acaricides.** By Francis A. Gunther and Roger C. Blinn (*out of print*)
Vol. 7. **Chemical Analysis of Industrial Solvents.** By the late Morris B. Jacobs and Leopold Schetlan
Vol. 8. **Colorimetric Determination of Nonmetals.** *Second Edition.* Edited by the late David F. Boltz and James A. Howell
Vol. 9. **Analytical Chemistry of Titanium Metals and Compounds.** By Maurice Codell
Vol. 10. **The Chemical Analysis of Air Pollutants.** By the late Morris B. Jacobs
Vol. 11. **X-Ray Spectrochemical Analysis.** *Second Edition.* By L. S. Birks
Vol. 12. **Systematic Analysis of Surface-Active Agents.** *Second Edition.* By Milton J. Rosen and Henry A. Goldsmith
Vol. 13. **Alternating Current Polarography and Tensammetry.** By B. Breyer and H. H. Bauer
Vol. 14. **Flame Photometry.** By R. Herrmann and J. Alkemade
Vol. 15. **The Titration of Organic Compounds** (*in two parts*). By M. R. F. Ashworth
Vol. 16. **Complexation in Analytical Chemistry: A Guide for the Critical Selection of Analytical Methods Based on Complexation Reactions.** By the late Anders Ringbom
Vol. 17. **Electron Probe Microanalysis.** *Second Edition.* By L. S. Birks

- Vol. 18. **Organic Complexing Reagents: Structure, Behavior, and Application to Inorganic Analysis.** By D. D. Perrin
- Vol. 19. **Thermal Analysis.** *Third Edition.* By Wesley Wm. Wendlandt
- Vol. 20. **Amperometric Titrations.** By John T. Stock
- Vol. 21. **Reflectance Spectroscopy.** By Wesley Wm. Wendlandt and Harry G. Hecht
- Vol. 22. **The Analytical Toxicology of Industrial Inorganic Poisons.** By the late Morris B. Jacobs
- Vol. 23. **The formation and Properties of Precipitates.** By Alan G. Walton
- Vol. 24. **Kinetics in Analytical Chemistry.** By Harry B. Mark, Jr. and Garry A. Rechnitz
- Vol. 25. **Atomic Absorption Spectroscopy.** *Second Edition.* By Morris Slavin
- Vol. 26. **Characterization of Organometallic Compounds** (*in two parts*). Edited by Minoru Tsutsui
- Vol. 27. **Rock and Mineral Analysis.** *Second Edition.* By Wesley M. Johnson and John A. Maxwell
- Vol. 28. **The Analytical Chemistry of Nitrogen and Its Compounds** (*in two parts*). Edited by C. A. Streuli and Philip R. Averell
- Vol. 29. **The Analytical Chemistry of Sulfur and Its Compounds** (*in three parts*). By J. H. Karchmer
- Vol. 30. **Ultramicro Elemental Analysis.** By Günther Tölg
- Vol. 31. **Photometric Organic Analysis** (*in two parts*). By Eugene Sawicki
- Vol. 32. **Determination of Organic Compounds: Methods and Procedures.** By Frederick T. Weiss
- Vol. 33. **Masking and Demasking of Chemical Reactions.** By D. D. Perrin
- Vol. 34. **Neutron Activation Analysis.** By D. De Soete, R. Gijbels, and J. Hoste
- Vol. 35. **Laser Raman Spectroscopy.** By Marvins C. Tobin
- Vol. 36. **Emission Spectrochemical Analysis.** By Morris Slavin
- Vol. 37. **Analytical Chemistry of Phosphorus Compounds.** Edited by M. Halmann
- Vol. 38. **Luminescence Spectroscopy in Analytical Chemistry.** By J. D. Winefordner, S. G. Schulman, and T. C. O'Haver
- Vol. 39. **Activation Analysis with Neutron Generators.** By Sam S. Nargolwalla and Edwin P. Przybylowicz
- Vol. 40. **Determination of Gaseous Elements in Metals.** Edited by Lynn L. Lewis, Laben M. Melnick, and Ben D. Holt
- Vol. 41. **Analysis of Silicones.** Edited by A. Lee Smith
- Vol. 42. **Foundations of Ultracentrifugal Analysis.** By H. Fujita
- Vol. 43. **Chemical Infrared Fourier Transform Spectroscopy.** By Peter R. Griffiths
- Vol. 44. **Microscale Manipulations in Chemistry.** By T. S. Ma and V. Horak
- Vol. 45. **Thermometric Titrations.** By J. Barthel
- Vol. 46. **Trace Analysis: Spectroscopic Methods for Elements.** Edited by J. D. Winefordner
- Vol. 47. **Contamination Control in Trace Element Analysis.** By Morris Zief and James W. Mitchell
- Vol. 48. **Analytical Applications of NMR.** By D. E. Leyden and R. H. Cox
- Vol. 49. **Measurement of Dissolved Oxygen.** By Michael L. Hitchman
- Vol. 50. **Analytical Laser Spectroscopy.** Edited by Nicolo Omenetto

CHEMICAL ANALYSIS: A SERIES OF MONOGRAPHS

Vol. 51. **Trace Element Analysis of Geological Materials**. By Roger D. Reeves and Robert R. Brooks
Vol. 52. **Chemical Analysis by Microwave Rotational Spectroscopy**. By Ravi Varma and Lawrence W. Hrubesh
Vol. 53. **Information Theory As Applied to Chemical Analysis**. By Karel Eckschlager and Vladimir Štěpánek
Vol. 54. **Applied Infrared Spectroscopy: Fundamentals, Techniques, and Analytical Problem-Solving**. By A. Lee Smith
Vol. 55. **Archaeological Chemistry**. By Zvi Goffer
Vol. 56. **Immobilized Enzymes in Analytical and Clinical Chemistry**. By P. W. Carr and L. D. Bowers
Vol. 57. **Photoacoustics and Photoacoustic Spectroscopy**. By Allan Rosencwaig
Vol. 58. **Analysis of Pesticide Residues**. Edited by H. Anson Moye
Vol. 59. **Affinity Chromatography**. By William H. Scouten
Vol. 60. **Quality Control in Analytical Chemistry**. *Second Edition*. By G. Kateman and L. Buydens
Vol. 61. **Direct Characterization of Fineparticles**. By Brian H. Kaye
Vol. 62. **Flow Injection Analysis**. By J. Ruzicka and E. H. Hansen
Vol. 63. **Applied Electron Spectroscopy for Chemical Analysis**. Edited by Hassan Windawi and Floyd Ho
Vol. 64. **Analytical Aspects of Environmental Chemistry**. Edited by David F. S. Natusch and Philip K. Hopke
Vol. 65. **The Interpretation of Analytical Chemical Data by the Use of Cluster Analysis**. By D. Luc Massart and Leonard Kaufman
Vol. 66. **Solid Phase Biochemistry: Analytical and Synthetic Aspects**. Edited by William H. Scouten
Vol. 67. **An Introduction to Photoelectron Spectroscopy**. By Pradip K. Ghosh
Vol. 68. **Room Temperature Phosphorimetry for Chemical Analysis**. By Tuan Vo-Dinh
Vol. 69. **Potentiometry and Potentiometric Titrations**. By E. P. Serjeant
Vol. 70. **Design and Application of Process Analyzer Systems**. By Paul E. Mix
Vol. 71. **Analysis of Organic and Biological Surfaces**. Edited by Patrick Echlin
Vol. 72. **Small Bore Liquid Chromatography Columns: Their Properties and Uses**. Edited by Raymond P. W. Scott
Vol. 73. **Modern Methods of Particle Size Analysis**. Edited by Howard G. Barth
Vol. 74. **Auger Electron Spectroscopy**. By Michael Thompson, M. D. Baker, Alec Christie, and J. F. Tyson
Vol. 75. **Spot Test Analysis: Clinical, Environmental, Forensic and Geochemical Applications**. By Ervin Jungreis
Vol. 76. **Receptor Modeling in Environmental Chemistry**. By Philip K. Hopke
Vol. 77. **Molecular Luminescence Spectroscopy: Methods and Applications** (*in three parts*). Edited by Stephen G. Schulman
Vol. 78. **Inorganic Chromatographic Analysis**. By John C. McDonald
Vol. 79. **Analytical Solution Calorimetry**. Edited by J. K. Grime
Vol. 80. **Selected Methods of Trace Metal Analysis: Biological and Environmental Samples**. By Jon C. VanLoon

CHEMICAL ANALYSIS: A SERIES OF MONOGRAPHS

Vol. 81. **The Analysis of Extraterrestrial Materials.** By Isidore Adler
Vol. 82. **Chemometrics.** By Muhammad A. Sharaf. Deborah L. Illman, and Bruce Kowalski
Vol. 83. **Fourier Transform Infrared Spectrometry.** By Peter R. Griffiths and James A. de Haseth
Vol. 84. **Trace Analysis: Spectroscopic Methods for Molecules.** Edited by Gary Christian and James B. Callis
Vol. 85. **Ultratrace Analysis of Pharmaceuticals and Other Compounds of Interest.** Edited by S. Ahuja
Vol. 86. **Secondary Ion Mass Spectrometry: Basic Concepts, Instrumental Aspects, Applications and Trends.** By A. Benninghoven, F. G. Rüdenauer, and H. W. Werner
Vol. 87. **Analytical Applications of Lasers.** Edited by Edward H. Piepmeier
Vol. 88. **Applied Geochemical Analysis** By C. O. Ingamells and F. F. Pitard
Vol. 89. **Detectors for Liquid Chromatography.** Edited by Edward S. Yeung
Vol. 90. **Inductively Coupled Plasma Emission Spectroscopy: Part I: Methodology, Instrumentation, and Performance; Part II: Applications and Fundamentals.** Edited by J. M. Boumans
Vol. 91. **Applications of New Mass Spectrometry Techniques in Pesticide Chemistry.** Edited by Joseph Rosen
Vol. 92. **X-Ray Absorption: Principles, Applications, Techniques of EXAFS, SEXAFS, and XANES.** Edited by D. C. Konnigsberger
Vol. 93. **Quantitative Structure-Chromatographic Retention Relationships.** By Roman Kaliszan
Vol. 94. **Laser Remote Chemical Analysis.** Edited by Raymond M. Measures
Vol. 95. **Inorganic Mass Spectrometry.** Edited by F. Adams, R. Gijbels, and R. Van Grieken
Vol. 96. **Kinetic Aspects of Analytical Chemistry.** By Horacio A. Mottola
Vol. 97. **Two-Dimensional NMR Spectroscopy.** By Jan Schraml and Jon M. Bellama
Vol. 98. **High Performance Liquid Chromatography.** Edited by Phyllis R. Brown and Richard A. Hartwick
Vol. 99. **X-Ray Fluorescence Spectrometry.** By Ron Jenkins
Vol. 100. **Analytical Aspects of Drug Testing.** Edited by Dale G. Deutsch
Vol. 101. **Chemical Analysis of Polycyclic Aromatic Compounds.** Edited by Tuan Vo-Dinh
Vol. 102. **Quadrupole Storage Mass Spectrometry.** By Raymond E. March and Richard J. Hughes
Vol. 103. **Determination of Molecular Weight.** Edited by Anthony R. Cooper
Vol. 104. **Selectivity and Detectability Optimizations in HPLC.** By Satinder Ahuja
Vol. 105. **Laser Microanalysis.** By Lieselotte Moenke-Blankenburg
Vol. 106. **Clinical Chemistry.** Edited by E. Howard Taylor
Vol. 107. **Multielement Detection Systems for Spectrochemical Analysis.** By Kenneth W. Busch and Marianna A. Busch
Vol. 108. **Planer Chromatography in the Life Science.** Edited by Joseph C. Touchstone
Vol. 109. **Fluorometric Analysis in Biomedical Chemistry: Trends and Techniques Including HPLC Applications.** By Norio Ichinose, George Schwedt, Frank Michael Schnepel, and Kyoko Adochi
Vol. 110. **An Introduction to Laboratory Automation.** By Victor Cerdá and Guillermo Ramis

CHEMICAL ANALYSIS: A SERIES OF MONOGRAPHS

Vol. 111. **Gas Chromatography: Biochemical, Biomedical, and Clinical Applications.** Edited by Ray E. Clement
Vol. 112. **The Analytical Chemistry of Silicones.** Edited by A. Lee Smith
Vol. 113. **Modern Methods of Polymer Characterization.** Edited by Howard G. Barth and Jimmy W. Mays
Vol. 114. **Analytical Raman Spectroscopy.** Edited by Jeannette Graselli and Bernard J. Bulkin
Vol. 115. **Trace and Ultratrace Analysis by HPLC.** By Satinder Ahuja
Vol. 116. **Radiochemistry and Nuclear Methods of Analysis.** By William D. Ehmann and Diane E. Vance
Vol. 117. **Applications of Fluorescence in Immunoassays.** By Ilkka Hemmila
Vol. 118. **Principles and Practice of Spectroscopic Calibration.** By Howard Mark
Vol. 119. **Activation Spectrometry in Chemical Analysis.** By S. J. Parry
Vol. 120. **Remote Sensing by Fourier Transform Spectrometry.** By Reinhard Beer
Vol. 121. **Detectors for Capillary Chromatography.** Edited by Herbert H. Hill and Dennis McMinn
Vol. 122. **Photochemical Vapor Deposition.** By J. G. Eden
Vol. 123. **Statistical Methods in Analytical Chemistry.** By Peter C. Meier and Richard Zund
Vol. 124. **Laser Ionization Mass Analysis.** Edited by Akos Vertes, Renaat Gijbels, and Fred Adams
Vol. 125. **Physics and Chemistry of Solid State Sensor Devices.** By Andreas Mandelis and Constantinos Christofides
Vol. 126. **Electroanalytical Stripping Methods.** By khjena Z. Brainina and E. Neyman
Vol. 127. **Air Monitoring by Spectroscopic Techniques.** Edited by Markus W. Sigrist
Vol. 128. **Information Theory in Analytical Chemistry.** By Karel Eckschlager and Klaus Danzer
Vol. 129. **Flame Chemiluminescence Analysis by Molecular Emission Cavity Detection.** Edited by David Stiles, Anthony Calokerinos, and Alan Townshend
Vol. 130. **Hydride Generation Atomic Absorption Spectrometry.** Edited by Jiri Dedina and Dimiter L. Tsalev
Vol. 131. **Selective Detectors: Environmental, Industrial, and Biomedical Applications.** Edited by Robert E. Sievers
Vol. 132. **High-Speed Countercurrent Chromatography.** Edited by Yoichiro Ito and Walter D. Conway
Vol. 133. **Particle-Induced X-Ray Emission Spectrometry.** By Sven A. E. Johansson, John L. Campbell, and Klas G. Malmqvist
Vol. 134. **Photothermal Spectroscopy Methods for Chemical Analysis.** By Stephen Bialkowski
Vol. 135. **Element Speciation in Bioinorganic Chemistry.** Edited by Sergio Caroli
Vol. 136. **Laser-Enhanced Ionization and Spectrometry.** Edited by John C. Travis and Gregory C. Turk
Vol. 137. **Fluorescence Imaging Spectroscopy and Microscopy.** Edited by Xue Feng Wang and Brian Herman
Vol. 138. **Introduction to X-Ray Powder Diffractometry.** By Ron Jenkins and Robert L. Snyder
Vol. 139. **Modern Techniques in Electroanalysis.** Edited by Peter Vanýsek
Vol. 140. **Total-Reflection X-Ray Fluorescence Analysis.** By Reinhold Klockenkämper

Vol. 141. **Spot Test Analysis: Clinical, Enviromental, Forensic, and Geochemical Applications**, Second Edition. By Ervin Jungreis

Vol. 142. **The Impact of Stereochemistry on Drug Development and Use**. Edited by Hassan Y. Aboul-Enein and Irving W. Wainer

Vol. 143. **Macrocyclic Compounds in Analytical Chemistry**. Edited by Yury A. Zolotov

Vol. 144. **Surface-Launched Acoustic Wave Sensors: Chemical Sensing and Thin-Film Characterization**. By Micheal Thompson and David Stone

Vol. 145. **Modern Isotope Ratio Mass Spectrometry**. By I. T. Platzner

Vol. 146. **High-Performance Capillary Electrophoresis: Theory, Techniques, and Applications**. Edited by Morteza G. Khaledi

PART
I

THEORY AND MODES OF HPCE

CHAPTER
1

CAPILLARY ELECTROPHORESIS: OVERVIEW AND PERSPECTIVE

BARRY L. KARGER

Barnett Institute, Northeastern University, Boston, Massachusetts 02115

1.1.	Introduction	3
1.2.	Modes of Operation	5
	1.2.1. Open-Tube Capillary Zone Electrophoresis	5
	1.2.2. Capillary Isoelectric Focusing	10
	1.2.3. Polymer-Filled Capillaries	12
	1.2.4. Proteins	16
1.3.	Capillary Electrophoresis–Mass Spectrometry	16
1.4.	Electric Field Manipulation of Bulk Liquid Flow	18
	1.4.1. Capillary Electrochromatography	18
	1.4.2. Microchip Devices	20
1.5.	Conclusions	22
	List of Acronyms and Abbreviations	22
	References	23

1.1. INTRODUCTION

Electrophoresis, one of the most widely used separation techniques, is currently practiced primarily in a slab gel rather than the column format of chromatography. The reasons for this are manifold. Slab gel electrophoresis has a history of more than 25 years of development and has become a familiar tool to even the most junior biochemist. It is an inexpensive method that can be used by all laboratories, and sample cleanup prior to spotting samples on a gel is often minimal. The slab gel approach, while slow by virtue of the use of low electric fields, can be made to achieve reasonable sample analysis throughput by the use of parallel processing (i.e., multiple lanes). Additionally, two-

---

High Performance Capillary Electrophoresis, edited by Morteza G. Khaledi. Chemical Analysis Series, Vol. 146.
ISBN 0-471-14851-2 © 1998 John Wiley & Sons, Inc.

dimensional (2-D) approaches are available for resolving more than 1000 compounds of proteins from complex matrices. Indeed, 2-D gel electrophoresis is among the most powerful separation techniques available today and is finding significant applicability in functional genomics.

Slab gel electrophoresis is not a stagnant field. Today, automated systems are available for many of the steps in the process, including spotting, image detection, data analysis, and even in certain cases gel preparation. This automation necessarily leads to increases in expense. Moreover, the number of lanes that can be accommodated by such approaches is increasing rapidly, up to as many as 96, to reach compatibility with a 96 microtiter well. Some increase in separation speed per lane has occurred through reduction in gel thickness and the concomitant increase in electric field. Furthermore, it is possible today to remove spots from a gel and determine protein structure by mass spectrometry, even down to silver stain levels (1).

Having said this, it is nevertheless worth pointing out that high-performance liquid chromatography (HPLC), the premier separation method of today, is primarily operated in a column format. Columns offer the possibility of ease of quantitation, automation, fraction collection, and on-line coupling to structure specific detectors, including mass spectrometry (MS) and more recently nuclear magnetic resonance (NMR) spectroscopy. Hence, it would seem logical that column operation in electrophoresis should be a significant tool. This was recognized in the early days of electrophoresis (2); however, it is generally accepted that Jorgenson and Lukacs in 1981 were the first to produce an operational capillary electrophoresis unit and to demonstrate its high resolving power (3).

A problem in electrophoresis, especially column operation, is the need to maintain low electric power to prevent Joule heating from elevating column temperature. The problem is not so much that the whole column temperature rises but that there is a temperature difference ΔT between the walls and the center of the capillary:

$$\Delta T \sim E^2 \kappa r^2 \qquad (1.1)$$

where E = electric field; κ = conductivity of the solution; and r = capillary radius. A temperature difference leads to different migration rates at the hotter center and cooler regions near the column wall, creating band broadening (4). Active cooling, i.e., the rapid removal of the generated heat from the exterior walls of the column, can alleviate this effect to some extent. Indeed, most commercial units today utilize liquid-cartridge cooling for this purpose as well as to maintain a reproducible column temperature, a necessary ingredient to achieve reproducible migration times.

The fundamental migration equation in electrophoresis is

$$v = \mu E \qquad (1.2)$$

where v = velocity of the band and μ = electrophoretic mobility. Thus, in order to achieve high velocities and thus rapid separations, it is necessary to operate at high electric fields. Such fields, however, will increase Joule heating, as seen in Eq. (1.1). The challenge of workers in electrophoresis is to find ways to increase E while maintaining the generated power within reasonable limits, e.g., less than 1 W/m.

The current (I) and electric field are proportional to each other

$$I \sim Er^2\kappa \qquad (1.3)$$

Equations (1.1) and (1.3) strongly point to the need to miniaturize the gel thickness or the radius of the column in order to utilize high fields without significant current. Normal gel thicknesses of 1–2 mm permit fields of only roughly 5–10 V/cm before excessive heating sets in. Thin (down to $\sim 100\,\mu m$) and ultrathin (lower than 100 μm) films allow electric fields of more than 100 V/cm, a 10-fold increase in sample band velocity over normal slab gel operation. Capillaries of 25–100 μm i.d. are employed in the column format for the same reason, i.e., to maintain low current. Typically, currents range from 5 to 30 μA with electric fields of 100 to 1000 V/cm in capillary electrophoresis.

The other main factor controlling current is the conductivity of the liquid medium through which the sample ions electrophorese. Thus, low conductivity buffers are to be recommended, e.g., zwitterions or Good buffers. When proper care is taken, high-performance separations in rapid analysis time are possible (see Figure 1.1). In this case, peak efficiencies of a few hundred thousand plates in open tube operation can be routinely achieved.

1.2. MODES OF OPERATION

1.2.1. Open-Tube Capillary Zone Electrophoresis

Figure 1.1 illustrates a high-resolution separation with open-tube operation. It is interesting to note that the liquid is maintained in the capillary by virtue of the anticonvective nature of the walls. On slabs or flat plates, of course, the liquid would flow off; hence, a gel is necessary to maintain the liquid on the plate even if the porous network is not participating in the separation. It is worth noting that the liquid becomes more difficult to maintain in the column as the surface area to volume ratio decreases, i.e., wider tubes. Thus, even small height differences of the two ends of a 200 μm i.d. capillary could cause bulk liquid flow due to the small pressure drop. To operate open-tube columns of wider diameter, special precautions are necessary, such as closing one end of the capillary with either a valve or membrane.

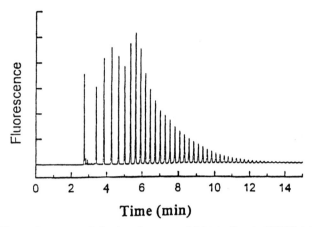

Figure 1.1. Electropherogram of the 1-aminopyrene-3,6,8-trisulfonate (APTS) labeled malto-oligosaccharide mixture: 75 μm i.d. × 37 cm (effective length, 30 cm). Buffer: 10 mM ammonium acetate buffer, pH 4.75; electric fields, $E = 333$ V/cm. Detection: LIF; excitation, 488 nm; emission, 520 nm. Injection: 5 s pressure. [From H. Suzuki, O. Müller, A. Guttman, and B. L. Karger, *Anal Chem.*, in press (1997).]

Since many aspects of open-tube operation are covered in subsequent chapters of this book, in this section only some important characteristics are outlined. First, it is possible to operate with or without bulk flow in the open tube. As long as there is a zeta potential at the walls of the capillary and the resistance to flow is not significant, bulk flow will occur when an electric field is applied, called electroosmotic flow. The velocity of this bulk flow can be expressed as

$$v = \frac{\varepsilon \zeta E}{\eta} \quad (1.4)$$

where ε = permittivity of the solution; ζ = zeta potential at the walls; and η = viscosity of the solution. Thus, as long as the temperature is maintained constant, the bulk flow increases in direct proportion to the applied field. The importance of the viscosity of the solution is emphasized in Eq. (1.4) and, since viscosity changes at roughly 2% per °C, the need for temperature control of the column to achieve reproducibility is evident. Bulk flow also depends on the ionic strength of the medium through its effect on the zeta potential. Higher ionic strengths lead to lower electroosmotic flows. The extent of bulk flow under fixed conditions depends on the charge at the capillary wall.

An important characteristic of electroosmotic flow is that it is plug-like rather than parobolic, as occurs with pressure driven flow. Consequently, solute band broadening due to differences in bulk velocity in the cross-section of the tube are considerably less with electroosmotic driven flow than pressure-driven flow. This results in the important point that wider diameter columns can be utilized in capillary electrophoresis than would be possible in capillary liquid chromatography, without the concomitant loss in efficiency. Moreover, because band broadening in the open tube is predominantly diffusion controlled, efficiencies 1–2 orders of magnitude higher than those typical for HPLC can be achieved by capillary electrophoresis (CE). High resolution is one of the most important characteristics of CE.

In the case of "bare" fused silica capillaries, bulk flow will occur toward the cathode since the wall surface will be negatively charged. This means that positively charged species migrate faster than the bulk flow, whereas negatively charged species move slower than the bulk flow. Thus, both positive and negatively charged species move in the same direction past the detector. Ultimately, however, if the electrophoretic mobility of a negatively charged species is greater than the mobility of the electroosmotic flow, the species will migrate out of the injection side. It is, of course, possible to manipulate the extent and even the direction of flow by coating the walls of the capillary (with or without charge), either dynamically or by covalent attachment.

Throughout this book there are many examples of open tube capillary electrophoretic separations, and it is not necessary to repeat them here. It is perhaps better to discuss significant areas of applicability of such systems. One important field that is gaining interest is clinical analysis (5). Here, one desires to detect normal and abnormal compounds in serum and urine. It has been shown that such complex matrices as serum can often be directly injected into the capillary with only minor cleanup, e.g., filtering or dilution. This is to be contrasted with packed bed operation where column clogging of untreated samples can be a serious problem. A second significant advantage in the clinical area is that the miniaturization afforded by CE leads to the use of quite small volumes of liquids. This is to be contrasted with a slab gel-staining operation where large solvent volume removal can be a serious problem. Clinical applications are already available for serum proteins and abnormal metabolite analysis. Immunoassay analysis has also been shown to operate successfully in research papers (6), and the future of this approach for trace level analysis would appear bright. Here, antibodies or competitors can be fluorescently labeled and detection achieved by laser-induced fluorescence (LIF). Exquisitely low detection levels are possible, for example, 10^{-11} to 10^{-12} M, or attomole to zeptomole mass levels. Indeed, single molecule detection by LIF is possible (7).

A second significant area of open tube operation is in peptide/protein analysis. Charge-based separation is obviously quite powerful, as resolution can be readily manipulated by pH changes in the running buffer. Figure 1.2 shows the separation of a mixture of peptides, e.g., a peptide map, a significant application in the biotechnology industry. Peptide mapping by CE is a complementary tool to HPLC, as compounds that do not resolve by LC are often separated by CE. Thus, very hydrophilic substances, e.g., sialylated glycopeptides, are close to being unretained in LC, whereas they separate nicely by CE (8).

Additionally, since open tube CE provides a different principle of separation than LC, the two methods are often used together to provide an assessment of compound purity. Thus, drug or protein analysis might be reported using both approaches. To be usable, it is necessary that the methods be validatable. The good news is that, whereas in the past it was a struggle to achieve validation for CE, workers are currently able to produce validated CE methods for the U.S. Food and Drug Administration (FDA) and other agencies (9).

For protein separation and analysis, a critical problem is adsorption of the substance to the capillary walls, with a concomitant loss in peak efficiency and a lack of full sample recovery from the column. From the earliest days of CE, workers have looked for ways to overcome this problem. Three approaches have been used: (a) additives to the buffer, e.g., zwitterionic buffers, nonionic detergents, or amines; (b) hydrophilic, neutral-coated capillaries, either dynamically adsorbed from the buffer or covalently attached; or (c) a combination of approaches (a) and (b). While these approaches are fully discussed in this book, we would simply like to note that today most adsorption problems are solvable. Coatings are far better than in the past—less adsorptive and more stable. Operation at pH's above 10 for long periods of time can still be a problem; however, very basic proteins can be run near neutral pH with neutral coatings and base additives, e.g., morpholine. Alternatively, positively charged coatings can be used to provide a repulsive layer to the positively charged proteins such that even highly basic proteins, e.g., histones, can be electrophoresed (10).

Figure 1.2. High-performance capillary electrophoresis (HPCE) profile of trypsin-digested recombinant human erythropoietin (rHuEPO) (3.75 mg/mL), vacuum injected for 2 s into a 50 μm i.d., 360 μm o.d. capillary of total length 75 cm and effective length 50 cm respectively. The sample was dissolved in 10% glacial acetic acid for electrophoresis. Electrophoresis was conducted at 16,000 V at 30 °C (electric field, $E = 213$ V/cm) in 40 mM sodium phosphate buffer, pH 2.5, containing 100 mM heptanesulfonic acid ion pairing agent; the profile was monitored on line at 200 nm, 0.025 AUFS (absorbance units full scale) at a data collection rate of 15 Hz for 95 min. The current level was about 110 μA for a power load of about 2.35 W/m. The nonglycosylated peptide and the glycopeptide sections of the map are designated [and the corresponding HPCE peaks are numbered according to migration time.] [From R. S. Rush, P. L Derby, T. W. Strickland, and M. F. Rohde, *Anal. Chem.* **65**, 1834 (1993).]

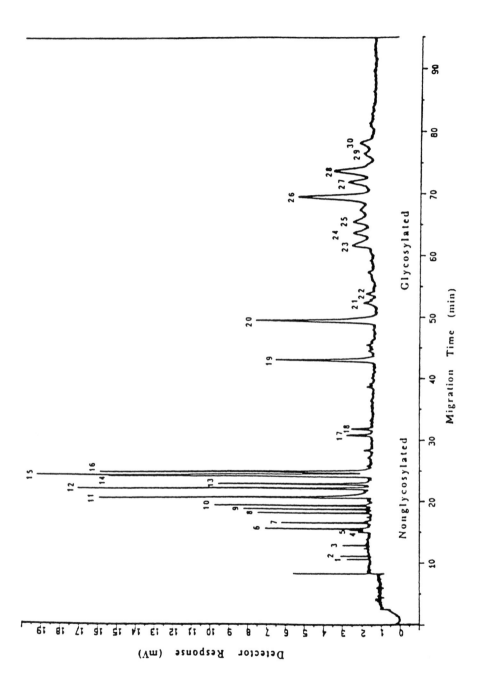

With respect to small molecules, one can operate the CE system in a pseudochromatographic mode. Through the use of detergents such as sodium dodecyl sulfate (SDS) and operation above the critical micelle concentration, micellar electrokinetic chromatography (MEKC) is achievable (11). Here, over and above any electrophoretic migration, partition in and out of the micelle is possible. In many cases, the extent of partition is hydrophobically based. Indeed, neutral substances can be resolved by MEKC based on hydrophobic differences. Of importance is that the lifetime of the micelle is very short, so the mass transfer kinetics of partition are quite rapid, leading to high performance. MEKC is today a standard tool for small molecule separation and is discussed later in this book.

Another popular application is inorganic ion analysis by CE using indirect detection (12). In this case, an absorbing substance is added to the buffer to raise the baseline substantially. When substances migrate by the detector, the added background component is partially displaced, leading to a lower concentration. This vacancy allows an indirect means of universal detection and can therefore be used in cases where no chromophore exists, e.g., inorganic ions. An example of the power of this approach is shown in Figure 1.3.

Other open-tube applications include carbohydrate analysis, chiral separations, and pharmacokinetics. For each of these cases, the high efficiency of the columns permits rapid separation of complex mixtures. Moreover, separation is often accomplished in one or only several runs. It can easily be concluded that open-tube CE analysis has a broadly based applicability for separation and analysis.

1.2.2. Capillary Isoelectric Focusing

Isoelectric focusing (IEF) is an important tool in the analysis and purification of protein mixtures. Separation is based on the positioning of molecules at their pI (i.e., the pH at which the number of exposed positive and negative charges on the surface of the protein are equal). The positioning or focusing results from the pH gradient that is created by a complex mixture of buffering components called ampholytes. IEF, typically operated on a slab gel, is an important means of protein characterization, as it leads to separation related to charge differences. In this sense, it provides somewhat comparable information to capillary zone electrophoresis (CZE).

IEF is generally the first dimension in 2-D slab gel electrophoresis, the second dimension being molecular weight separation by means of SDS–PAGE (PAGE is polyacrylamide gel electrophoresis). These two approaches provide fundamental differences of separation, leading to a powerful means of separating 1000 or more components. The method is very important in cell component analysis and protein expression (13). Until recently, one of the

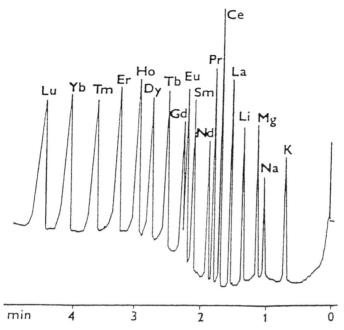

Figure 1.3. Separation of a model mixture in 0.03 M creatinine–acetate buffer, pH 4.8, containing 0.004 M hydroxybutyric acid (HIBA). Separation voltage: 12 kV, 3.6 µA. Injection: 8 s, 1.5 kV. Sample concentration: 10^{-3} M K, Na, Mg, and Li, 5×10^{-4} M La-Lu; A_{220} = change of absorbance monitored at 220 nm. [From F. Foret, S. Fanali, A. Nardi, and P. Bocek, *Electrophoresis* **11**, 780 (1990).]

difficulties with 2-D gels has been the identification of protein spots at the silver stained level. By using small-volume tryptic digestion followed by preconcentration in CE–MS, protein identification has recently become possible from 2-D gels (1). This advance allows correlation of gene expression and protein expression, the latter including posttranslational modification.

Capillary IEF (cIEF), the column format of this approach, was recognized from the early days of CE to be an important method (14); however, for some time there was concern with respect to issues of reproducibility, quantitation, and general column coating lifetime. The method of detection of the focused bands is a significant factor in achieving reproducible results. While imaging of the whole column after focusing has been shown to be possible (15), the more general approach involves mobilization of the bands by the detector after focusing. Several approaches to mobilizing the focused bands have been developed, and the one that seems to lead to the best results is the use of low-pressure mobilization while the field is in operation. This method provides linear pH vs. time plots and uniform resolution of the bands across the pH gradient (16). Obviously, it is possible to elongate the gradient in a particular pH region for improved separation, analogous to isocratic holds in gradient elution HPLC.

An important issue when one is using cIEF is that of protein precipitation. As the bands become focused and highly concentrated at the pI (solubility is often the lowest at this pH), the issue of maintaining the protein in solution is important. Several approaches have been used to improve solubility, including the addition of detergents, particularly non ionic detergents, and even the use of certain denaturants/solubilizers such as urea. Figure 1.4, for example, shows the cIEF separation of recombinant tissue plasminogen activator (rtPA) run in 36% urea. It should be pointed out that when such a method is being developed, protein recovery from the column should be determined. Since protein precipitation is an issue in cIEF, recovery should be routinely assessed when one is using this method. Even in CZE, the extent of protein losses in the column should be determined. Recovery can be measured either by using radiolabeled proteins or collecting the injected sample in one fraction and determining the amount present by a suitable method such as ELISA (enzyme-linked immunosorbent assay).

cIEF provides a powerful means of determining isoforms of complex glycoproteins. Since sialic acids will be negatively charged, a glycoprotein will separate into various bands in cIEF based on the number of sialic acids present (along with the surface charges resulting from the primary structure of the protein). IEF has been routinely used in the slab gel format to characterize glycoproteins for a number of years. Now cIEF automation, quantitative ability, and elimination of gel pouring make it a potentially important tool in protein chemistry for this application. Additionally, cIEF has recently been coupled to MS. This approach offers significant potential.

1.2.3. Polymer-Filled Capillaries

As practiced in slab gel operation, electrophoresis is predominantly a size-based separation method. This can be clearly seen by the two major applications of the technique: (1) DNA separation and analysis, and (2) SDS–PAGE separation of proteins. In the case of DNA, the electrophoretic mobility has been shown to be independent of base number for both single- and double-stranded DNA. This is a general consequence of the compensation of additional charge from the phosphate group by the mass of each nucleotide added. In the case of proteins, it has been found that fully denatured proteins will add a fixed amount of SDS per unit mass of protein. As a consequence, the charge-to-mass ratio is constant and the electrophoretic mobility will be constant in open solution.

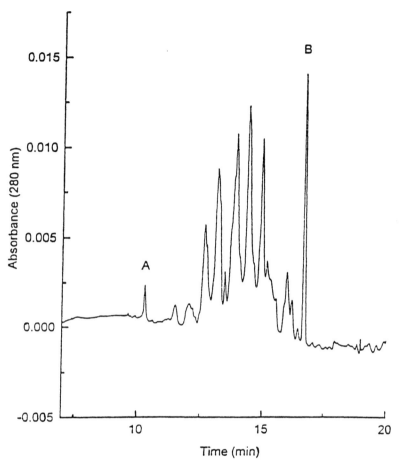

Figure 1.4. cIEF separation of rtPA with (A) human β-endorphin (pI 9.2) and (B) Gln^{11} amyloid B-protein fragment 1–28 (pI 6.4) as pI markers. Sample: 0.16 mg/mL rtPA, 0.04 mg/mL human β-endorphin, 0.04 mg/mL Gln^{11} amyloid B-protein fragment 1–28, 4.8 M urea, and 1.3% Ampholine 3.5–10. [From J. M. Thorne, W. K. Goetzinger, A. B. Chen, K. G. Moorhouse, and B. L. Karger, *J. Chromatogr.* **744**, 155 (1996).]

Separation based on size and not charge is accomplished in slab gels by using a cross-linked gel matrix. The larger molecules are retarded relative to the smaller ones by virtue of the former's greater interaction with the gel matrix. This sieving mechanism has played a significant role in such diverse areas as DNA sequencing, mutation analysis, and protein purity assessment.

Initially, it seemed logical to incorporate cross-linked gel-filled matrices in capillary electrophoresis (17). Although this approach is possible and indeed has been commercialized, such columns are not very rugged and can only operate over a narrow window of conditions. For example, column temperatures above approximately 30°C often lead to collapse of the gel and a complete loss of current. Cross-linked gels are too rigid and cannot respond effectively to stress caused by temperature, ionic strength, or pressure. It was therefore logical to examine polymers in a less rigid, non-cross-linked format (18), particularly since earlier work on slab gels had shown that sieving was possible with entangled polymers (19).

The next development was the recognition that the sieving polymer solution could be of sufficiently low viscosity that it could be automatically replaced after one or more runs. The amount of pressure required for this replacement was a function of both the molecular weight of the polymer and the concentration of the solution. Generally speaking, separation of lower-molecular-weight substances requires a high concentration of polymer, e.g., 10% concentration or higher, and therefore, in order to replace the material, low-molecular-weight polymer is required. Furthermore, in order to minimize electroosmotic flow and thus polymer solution distortion, it is typical to use a neutral-coated capillary. Figure 1.5 shows a high-speed separation of

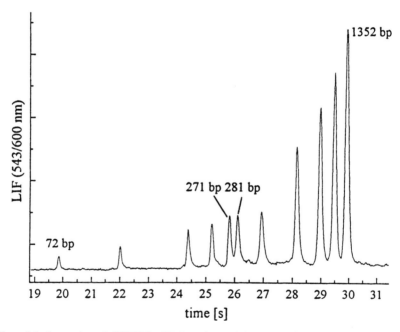

Figure 1.5. Separation of ϕX174/HaeIII. Experimental conditions: 50 μm i.d. capillary (J&W, DB-1); L= 8 cm; L= 3 cm; 1% methyl cellulose; 40 mM Tris/N-[tris(hydroxymethyl)methyl]-3-aminopropanesulfonic acid (TAPS), 1 μg/mL EtBr, E = 800 V/cm; LIF detection at 543 nm/600 nm. [From O. Müller, M. Minarik, F. Foret, and B. L. Karger, unpublished results.]

a standard restriction fragment digest using a sieving polymer solution with a high electric field and short column length.

For higher-molecular-weight substances, a dilute solution of a high-molecular-weight polymer seems to work best. For example, we have found that a 2% concentration of linear polyacrylamide with a molecular weight of roughly 10 million daltons works quite nicely for sequencing DNA fragments out to roughly 1000 bases (20). The viscosity is still sufficiently low that replacement can occur with only a few hundred psi.

Capillary electrophoresis is widely used for mutation analysis, assessment of purity of synthesized primers and antisense DNA, DNA sequencing, and DNA fingerprinting (21). For mutation studies, all the tools used in the slab gel operation have been transferred to the capillary approach, including single-strand conformational polymorphism (SSCP), ligase chain reaction (LCR), and denaturing gradient gel electrophoresis (DGGE). In last case the method of constant denaturant capillary electrophoresis (CDCE) has been utilized as a powerful means of determining rare mutations over a 100 base pair stretch in the presence of a large excess of the wild type (22). This arises from the very high separating power of this approach and thus the ability to enrich the mutant fraction relative to the wild type.

An important aspect of the success of DNA analysis has been the use of LIF detection. Fluorescent dyes can be either attached to primers, as typically occurs in DNA sequencing, or simply added as an intercalator for double-stranded DNA, e.g., ethidium bromide or thiazole orange. All dye chemistries utilized in the slab gel format are directly translatable to CE, e.g., energy transfer dyes (23). By using LIF exquisitely low detection limits are possible, down to the subattomole mass detection level. The replacement of the sieving matrices can be easily automated (as discussed Chapter 5 in this book). To enhance sample throughout, a significant effort has been underway over the past few years to develop multiple parallel capillary arrays (24). Such approaches have recently been commercialized with 96 capillaries for parallel operation.

It should be emphasized that, in the capillary format, operation at high fields is desirable in order to achieve rapid separations. However, DNA separation of large base numbers becomes difficult at high fields as a result of the influence of the electric field strength on the conformation of single- or double-stranded DNA (25). At high fields, depending on the base number, the long DNA molecules are more or less aligned axially with the field; such stretched molecules migrate with the same velocity in the sieving matrix, resulting in loss of separation. As a consequence of stretching, the speed at which separation can be achieved is related to the size of the DNA molecules. Shorter DNA molecules, e.g., 100 base pairs or less, can be separated very fast with single base pair resolution at fields of 500–1000 V/cm. Longer DNA molecules, e.g., 500–1000 base pairs require lower fields, 100–300 V/cm, and even longer species need lower fields. The success shown in Figure 1.5 at high fields is in part a result of the large differences in base numbers of the last three adjacent bands.

In order to overcome the loss in separation due to stretching, pulsed fields can be used in the capillary format (26). Especially up to 100,000 base pairs, this seems to be a useful tool, i.e., simply to invert the field periodically. However, above this base size, while some reports have appeared, the method is still in its infancy. A significant problem is the shearing of sample molecules when one is dealing with very large DNA.

1.2.4. Proteins

In a similar fashion to DNA analysis, SDS protein complexes can be separated by capillary electrophoresis (27). If ultraviolet (UV) detection is utilized, a non-absorbing polymer must be used, such as dextran or poly(ethylene oxide). Acrylamides do not work well because of the background absorbance at 210 nm. SDS protein separations have been achieved, and commercial separations kits are available from several companies. As in the case of DNA, a coating on the fused silica capillary wall seems to work best, particularly in the analysis of complex samples. Molecular weight of the protein can be estimated within 10% using this approach; however, MS is clearly the method of choice when one needs accurate molecular weights. SDS protein separations by CE would be more widely adopted today if the sensitivity of UV detection could be increased 5- to 10-fold. Such an increase would reach the silver stain level on slab gels. Recently, fluorescent labeling of the protein followed by laser-based detection achieved levels of detection of SDS proteins even lower than silver stain gels. This holds great promise. Details on all CE seiving methods can be found later in this book.

1.3. CAPILLARY ELECTROPHORESIS–MASS SPECTROMETRY

At present, the coupling of separation techniques to MS is a significant and powerful means of chemical analysis. The two methods complement each other quite well. Separation, either by HPLC or CE, often requires MS in order to be able to identify individual peaks. Furthermore, MS is fundamentally a separation technique, and in this sense, LC–MS or CE–MS represent 2-D separation processes. At the same time, MS often requires separation prior to detection in order to enable individual substances to be identified and not interferred with by other compounds.

Given this state of affairs, it is easy to conclude that in order for CE to be fully utilized, it must be coupled to MS. There are presently two approaches for this coupling. In one, an off-line approach, samples are first collected and then injected into the mass spectrometer. In the other, an on-line approach, the bulk liquid (or ions if the electroosmotic flow is very small) eluting from the capillary are electrosprayed directly into the mass spectrometer. Alternatively, material from the capillary can be deposited either on a target or membrane and matrix-assisted laser desorption time-of-flight mass spectrometry (MALDI–TOF–MS) conducted. For the most part, workers have focused on electrospray quadrupole mass spectrometry or more recently electrospray ion trap mass spectrometry for CE–MS operation.

There are two general approaches to CE coupling to electrospray MS: sheath flow and a sheathless system. In the sheath flow approach (28), a solution such as acetic acid buffer in 50/50 methanol/water flows coaxially with the capillary effluent or ions. This sheath flow is required to increase the actual flow rate from the column and to provide for a stable electrospray. A related method to this approach is the use of a liquid junction where the mixing of the sheath liquid and the effluent from the column occurs upstream from the electrospray tip (29). The sheath flow approach is predominantly used today because of its ease of operation. Although it leads to some dilution of the bands, this can be controlled to some extent through the relative flow rates in the sheath liquid and the capillary. The alternative approach is to use a sheathless system in which, as frequently constructed, the tip of the capillary is coated with a conductive metal such as gold (30). This approach provides for increased sensitivity, as much as an order of magnitude, over the sheath flow system. However, while more sensitive, the tip of the capillary is often quite small and can lead to clogging. Additionally, coating the metal on the tip can be problematic.

CE–MS is currently being incorporated as a tool for microscale analysis where the amount of sample is limited. For example, in coupling with Fourier transform mass spectrometry, attomole levels of protein could be sequenced by on-line CE–MS (31). Nevertheless, from a practical point of view some improvements in the operation and design of the interface are required. There is a need to improve concentration sensitivity from the normal micromolar range and to achieve greater compatibility with buffers required for separation. One way to overcome the low sensitivity is to perform on-column preconcentration, an example being the use of a hydrophobic membrane (32). In this manner, nanomolar concentrations can be detected and utilized. It is easy to predict that CE–MS will ultimately play a significant role, as the high resolution from the CE along with its microscale operation should provide researchers with powerful analytical tools. We anticipate significant progress for CE–MS in the next few years.

1.4. ELECTRIC FIELD MANIPULATION OF BULK LIQUID FLOW

The use of electric fields to manipulate bulk flow is not only an important characteristic of CE but has led to important developments in two related areas that have potential for significant impact on separation science and bioanalysis in the near future. In this section we briefly comment on these two areas—capillary electrochromatography and microfabricated devices. It is noteworthy that both areas have arisen from advances in CE.

1.4.1. Capillary Electrochromatography

Capillary electrochromatography (CEC) was initially introduced by Pretorius et al. in 1974 (33); however, as proper instrumentation was not then available, the initial work lay dormant for approximately 15 years. It was not until the late 1980s that researchers recognized the potential of this mode of chromatography (34). Since that time, a number of workers have become involved in the field, and the technique is at present probably one of the most active under study.

In contrast to pressurized flow, conventionally used in HPLC, in CEC the electric field is employed to drive the bulk flow through the column. Thus, electrosomotic flow resulting from the zeta potential at the column walls and at the packed particles creates the flow of mobile phase. Since this flow is close to plug-like vs. parabolic in pressurized systems, column efficiencies in CEC are found to be at least an order of magnitude higher than in HPLC. High resolution can be translated ultimately into fast analysis; alternatively, there will not be great demands on mobile phase adjustment with high plate counts (10^5 or greater plates per meter).

Both porous and nonporous particles can be used in the packed bed. The particle diameters in CEC are generally similar to those in HPLC or somewhat smaller ($\sim 1-2\,\mu m$). While such small particle sizes would require high inlet pressures for HPLC, the use of electric fields removes this restriction. Furthermore, the synthesis of packings with narrow particle size distributions is now possible, leading to such high-performance packed columns. Figure 1.6 illustrates an example of the power of separation in CEC.

An important aspect of CEC is that CE instrumentation can be directly utilized. One modification sometimes employed is to pressurize the buffer reservoirs to minimize bubble formation, particularly at the frit holding the packing in the column (35). With respect to on-column UV detection, the detector region is in an open tube section immediately downstream from the packed bed. Today, commercial columns are available for UV detection.

CEC provides the surface-mediated selectivity potential of HPLC and the high efficiency of CE. For this reason, there is a great deal of interest in the

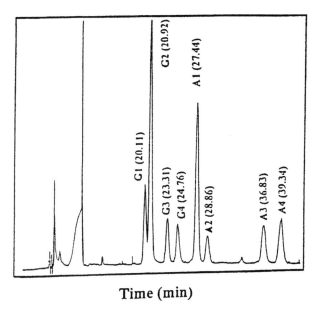

Figure 1.6. CEC–UV analysis of mixtures of diastereoisomeric deoxyguanosine and deoxyadenosine of *anti*-benzo[*g*]chrysene. [From P. Vouros and J. Ding, *Anal. Chem.* **69**, 379 (1997).]

method. However, there are a number of advances that need to occur before CEC is widely adopted. Among these is a much firmer understanding of the migration process in CEC. In addition, new stationary phases will likely be necessary for the method to be successful. For example, exposed silanol groups in reversed-phase packings aid electroosmotic flow; however, in HPLC one desires to deactivate these groups as much as possible in order to minimize adsorption of basic groups.

The coupling of CEC to MS should be an important development (36). CEC potentially can provide a higher capacity than CE (without preconcentration). Furthermore, separation based on hydrophobicity may be easier to conduct by CEC–MS than by CE–MS. In the latter, MEKC is necessary and SDS will reduce the ionization of the sample bands in the electrospray. Thus, CEC may prove especially useful for separation of neutral substances based on hydrophobic differences.

Finally, we can anticipate gradient elution playing a significant role, as is the case in HPLC. While it is possible to create a gradient based on electric field manipulation alone, many workers use HPLC pumps for this purpose. In this regard, a combination of pressure and electric field is simultaneously employed to drive the mobile phase flow. Some have predicted that we can look forward to the day when the liquid phase separation instrument will have three possible modes—HPLC, CE, and CEC—or combinations thereof, as both pressure and electric field will be available.

It is hard to predict the ultimate impact of CEC on the field of separations. As we have noted, a real potential exists; however, there are still significant barriers to wide adoption. Perhaps the greatest barrier is HPLC itself, which today has annual sales of $1.6 billion. In the final analysis, CEC must demonstrate unique applications if the method is to gain wide acceptance.

1.4.2. Microchip Devices

CE has also greatly aided the development of a second area—microfabricated or microchip devices. By means of photolithography and wet chemical etching procedures, standard methods in the semiconductor field, planar structures have been constructed with channels of dimensions comparable to those of capillary columns ($\sim 50\,\mu m$) (37). These chips have been generally made out of glass; however, commercial devices will undoubtedly be made from plastic because of cost. Plastic chips will be constructed from processes such as molding, casting, and stamping.

Fluid flow in microchips is mainly based on the use of electric fields; however, pressurized flow can be developed as well. The problem of precise sample injection on the chip has been solved, and indeed CE separations can be conducted on the chip. Figure 1.7a shows the design of a chip for CE, and the resulting separation is presented in Figure 1.7b. Note the rapid time of separation, a consequence of the short dimensions of the channel.

Microchips are expected to play a significant role in the years ahead for a number of reasons. First, once the chip has been designed, it will be possible to manufacture inexpensively a large number of identical devices that will permit disposal after use. Such an approach eliminates the problem of sample carryover, as a microchannel can be used only once. In the clinical field, this one-time use would be quite significant (38). Secondly, it should be possible to integrate a number of chemical steps on a chip, e.g., polymerase chain reaction (PCR) followed by CE separation (39). Indeed, the PCR reaction, where temperature changes play an important role, should be cycled rapidly, as the volume that needs to be heated or cooled is small. Thirdly, many channels or wells can be positioned close to one another on a chip. This means that many reactions or channels can be run in parallel, leading to significant throughput for screening purposes. Such multichannel chips can also be used as microfluidic devices to enhance throughput in MS (40). Of course, all of the above is accomplished with low amounts of sample. The future is bright for this technology, and it seems clear that laboratory experiments may well be conducted using microchips in the future.

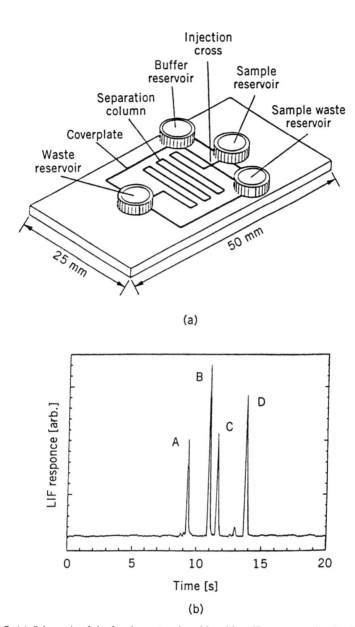

Figure 1.7. (a) Schematic of the fused quartz microchip with a 67 mm separation length. The reservoirs are configured for sample stacking injections and affixed on the microchip via epoxy. (b) Electropherograms using a pinched injection of (A) dansyl-lysine, (B) didansyl-lysine, (C) dansyl-isoleucine, and (D) didansyl-cystine. The separation field strength is 1100 V/cm, and separation length is 67 mm. The amino acid concentrations are listed in the text; for the stacked injection, the separation and sample buffer are 5 and 0.5 mM, respectively. [From S. C. Jacobson and J. M. Ramsey, *Electrophoresis* **16**, 481 (1995).]

1.5. CONCLUSIONS

The goal of this overview has been to set the stage for the forthcoming chapters. CE is a strong, growing field with many applications. It is a microscale, solution-based method with the potential for high-resolution separations. While not discussed here, the solution-based approach provides an excellent means to explore protein binding, e.g., measure dissociation constraints or combinatorial library screening. Furthermore, with proper care, the analytical procedures are validatable, a necessary ingredient for CE to be widely adopted. It is therefore highly probable that CE will continue to grow in the years ahead.

We have also included in this overview a brief discussion of CEC and microchip devices. CE is really a part of these fields, and advances in one area will impact on all. As recognition of this relationship, the major international meeting in CE, [International Symposium on High Performance Capillary Electrophoresis (HPCE)], has recently expanded its scope to include CEC and microchip devices. In total, the field of electroseparations should be exciting to follow in the coming years, as we can anticipate rapid advancement. I am sure the reader will sense this excitement from the ensuing chapters in this book.

LIST OF ACRONYMS AND ABBREVIATIONS

Acronym Abbreviation	Definition
APTS	1-aminopyrene-3,6,8-trisulfonate
AUFS	absorbance units full scale
CDCE	constant denaturant capillary electrophoresis
CE	capillary electrophoresis
CEC	capillary electrochromatography
cIEF	capillary isoelectric focusing
CZE	capillary zone electrophoresis
DGGE	denaturing gradient gel electrophoresis
ELISA	enzyme-linked immunosorbent assay
FDA	(U.S.) Food and Drug Administration
HIBA	hydroxyisobutyric acid
HPCE	high-performance capillary electrophoresis
HPLC	high-performance liquid chromatography
IEF	isoelectric focusing
LCR	ligase chain reaction
LIF	laser-induced fluorescence
MALDI–TOF–MS	matrix-assisted laser desorption ionization time-of-flight mass spectrometry
MEKC	micellar electrokinetic chromatography
MS	mass spectrometry

NMR	nuclear magnetic resonance
PAGE	polyacrylamide gel electrophoresis
PCR	polymerase chain reaction
ϕX174	recombinant human erythropoietin
rHuEPO	polymerase chain reaction
rtPA	recombinant tissue lasminogen activator
SDS	sodium dodecyl sulfate
SSCP	single-strand conformational polymorphism
TAPS	tris-aminopropanesulfonic acid
Tris	tris(hydroxymethyl) aminomethane
2-D	two-dimensional
UV	ultraviolet

ACKNOWLEDGMENT

The author thanks the National Institute of Health under GM15847 for support of this work, which is Contribution No. 684 from the Barnett Institute.

REFERENCES

1. D. Figeys, A. Ducret, J. R. Yates, and R. Aebersold, *Nat. Biotechnol.* **14**, 1579 (1996).
2. S. Hjertén, *Chromatogr. Rev.* **9**, 122 (1967).
3. J. W. Jorgenson and K. D. Lukacs, *Anal. Chem.* **53**, 1298 (1981).
4. F. Foret, L. Krivankova, and P. Bocek, *Capillary Zone Electrophoresis.* VCH, Weinheim, 1993.
5. E. Jellum, H. Dollekamp, and C. Blessum, *J. Chromatogr. B* **683**, 55 (1996).
6. L. Tao and R. T. Kennedy, *Anal. Chem.* **68**, 3899 (1996).
7. D. Y. Chen and N. J. Dovichi, *Anal. Chem.* **68**, 690 (1996).
8. R. S. Rush, P. L. Derby, T. W. Strickland, and M. F. Rohde, *Anal. Chem.* **65**, 1834 (1993).
9. K. D. Altria, J. Elgey, and J. S. Howells, *J. Chromatogr. B* **686**, 111 (1996).
10. J. E. Wiktorowicz and J. C. Colburn, *Electrophoresis* **11**, 769 (1990).
11. S. Terabe, K. Otsuka, K. Ichikawa, A. Tsuchiya, and T. Ando, *Anal. Chem.* **56**, 111 (1984).
12. F. Foret, S. Fanali, A. Nardi, and P. Bocek, *Electrophoresis* **11**, 780 (1990).
13. P. Jungblut, B. Thiede, V. Zimny-Ardnt, E. C. Muller, C. Scheler, B. Wittmann-Liebold, and A. Otto, *Electrophoresis* **17**, 839 (1996).
14. S. Hjerten, K. Elenbring, F. Kilar, J. L. Liao, A. J. C. Chen, C. J. Siebert, and M. D. Zhu, *J. Chromatogr.* **403**, 47 (1987).
15. J. Q. Wu and J. Pawliszyn, *J. Liq. Chromatogr.* **16**, 1891 (1993).

16. T. L. Huang, P. C. H. Shieh, and N. Cooke, *Chromatographia* **39**, 543 (1994).
17. A. S. Cohen, D. R. Najarian, A. Paulus, A. Guttman, J. A. Smith, and B. L. Karger, *Proc. Natl. Acad. Sci.* **84**, 9660 (1988).
18. D. N. Heiger, A. S. Cohen, and B. L. Karger, *J. Chromatogr.* **516**, 33 (1990).
19. H. J. Bode, *Electrophoresis 79*, p. 39. Gruyter, New York, 1980.
20. E. Carrilho, M. C. Ruiz-Martinez, J. Berka, I. Smirnov, W. Goetzinger, A. W. Miller, D. Brady, and B. L. Karger, *Anal. Chem.* **68**, 3305 (1996).
21. A. R. Isenberg, B. R. McCord, B. W. Koons, B. Budowle, and R. O. Allen, *Electrophoresis* **17**, 1505 (1996).
22. K. Khrapko, J. H. Hanekamp, W. G. Thilly, A. Belinkii, F. Foret, and B. L. Karger, *Nucleic Acids Res.* **22**, 364 (1994).
23. Y. Wang, J. M. Wallin, J. Y. Ju, G. F. Sensabaugh, and R. A. Mathies, *Electrophoresis* **17**, 1485 (1996).
24. E. S. Mansfield, M. Vainer, S. Enad, D. L. Barker, D. Harris, E. Rappaport, and P. Fortina, *Genome Res* **6**, 893 (1996).
25. L. Mitnik, L. Salome, J. L. Viovy, and C. Heller, *J. Chromatogr. A* **710**, 309 (1995).
26. J. Sudor and M. Novotny, *Nucleic Acids Res.* **23**, 2538 (1995).
27. K. Ganzler, K. S. Greve, A. S. Cohen, B. L. Karger, A. Guttman, and N. C. Cooke, *Anal. Chem.* **64**, 2665 (1992).
28. R. D. Smith, J. A. Olivares, N. T. Nguyen, and H. R. Udseth, *Anal. Chem.* **60**, 436 (1988). D. P. Kirby, J. M. Thorne, W. K. Goetzinger, and B. L. Karger, **68**, 4451 (1996).
29. E. D. Lee, W. Muck, J. D. Henion, and T. R. Covey, *Biomed Environ. Mass Spectrom.* **18**, 844 (1989).
30. J. C. Severs, A. C. Harms, and R. D. Smith, *Rapid Commun. Mass Spectrome.* **10**, 1175 (1996); J. F. Kelly, L. Ramaley, and P. Thibault, *Anal. Chem.* **69**, 51 (1997).
31. G. A. Valaskovic, N. L. Kelleher, and F. W. McLafferty, *Science* **273**, 1199 (1996).
32. A. J. Tomlinson, L. M. Benson, S. Jameson, D. H. Johnson, and S. Naylor, *J. Am. Soc. Mass Spectrom.* **8**, 15 (1997).
33. V. Pretorius, B. J. Hopkins, and J. D. Shieke, *J. Chromatogr.* **99**, 23 (1974).
34. J. H. Knox and I. H. Grant, *Chromatographia* **24**, 135 (1987).
35. M. M. Dittmann and G. P. Rozing, *J. Chromatogr. A* **744**, 63 (1996).
36. K. Schmeer, B. Behnke, and E. Bayer, *Anal. Chem.* **67**, 3656 (1995).
37. G. T. A. Kovacs, K. Petersen, and M. Albin, *Anal. Chem.* **68**, 407A (1996).
38. J. Cheng, M. A. Shoffner, K. R. Mitchelson, L. J. Kricka, and P. Wilding, *J. Chromatogr. A* **732**, 151 (1996). K. Fluri, G. Fitzpatrick, N. Chiem, and D. J. Harrison, *Anal. Chem.* **68**, 4285 (1996).
39. A. T. Woolley, D. Hadley, P. Landre, A. J. Demello, R. A. Mathies, and M. A. Northrup, *Anal. Chem.* **68**, 4081 (1996).
40. Q. Xue, F. Foret, Y. M. Dunayevskiy, P. M. Zavracky, N. E. McGruer, and B. L. Karger, Anal. Chem. **69**, 426 (1997).

CHAPTER

2

THEORY OF CAPILLARY ZONE ELECTROPHORESIS

ERNST KENNDLER

Institute for Analytical Chemistry, University of Vienna, 1090 Vienna, Austria

2.1.	Capillary Zone Electrophoresis in the Absence of Electroosmotic Flow		27
	2.1.1. Migration		27
		2.1.1.1. The Concept of Effective Mobility	27
		2.1.1.2. Affecting Mobility by Charge	29
		2.1.1.3. Affecting Mobility by Variation of pK_a	32
		2.1.1.4. Migration Time	35
	2.1.2. Dispersion		35
		2.1.2.1. Extracolumn Effects	37
		2.1.2.2. Dispersion Due to Capillary Coiling	38
		2.1.2.3. Longitudinal Diffusion	39
		2.1.2.4. Thermal Contribution to Dispersion	45
		2.1.2.5. Concentration Overload (Electromigration Dispersion)	48
		2.1.2.6. Wall Adsorption	50
	2.1.3. Resolution		53
2.2.	Capillary Zone Electrophoresis in the Presence of Electroosmotic Flow		58
	2.2.1. Migration		59
		2.2.1.1. Electrically Neutral Compounds	59
		2.2.1.2. Charged Solutes	60
	2.2.2. Dispersion and Dimensionless Reduced Mobility (Electromigration Factor)		63
		2.2.2.1. Contribution of the Electroosmotic Flow Profile to Peak Dispersion	63
		2.2.2.2. Dispersion by Influencing Residence Time	64
		2.2.2.3. Selectivity, Efficiency, and Resolution	67
List of Symbols			70
List of Acronyms			73
References			73

High Performance Capillary Electrophoresis, edited by Morteza G. Khaledi. Chemical Analysis Series, Vol. 146.
ISBN 0-471-14851-2 © 1998 John Wiley & Sons, Inc.

In electrophoresis ionic separands are migrating in solution under the influence of an electric field. Differences in migration velocity or migration distance may cause separation of the solutes. Three electrophoretic techniques are used in capillaries for analytical purposes: capillary zone electrophoresis (CZE), capillary isoelectrical focusing (CIEF), and capillary isotachophoresis (CITP). Two techniques, CIEF and CITP, apply buffers with chemical gradients—in a linear pH gradient and a stepwise mobility gradient, respectively.

In contrast, CZE, the topic of this chapter, has the most simple arrangement of the background electrolyte (BGE): the separation capillary is filled with a uniform buffer solution. By application of a voltage, V, in the longitudinal direction an electric field with constant strength, E, is established. The solutes are separated due to their different charge and/or size. The resulting concentration distribution of two separands, i and j, is schematically shown in Figure 2.1. Functions of the residence time are obtained that are characterized by two parameters: (i) the position of the concentration maximum of the distribution functions, which represents the migration time, t_m, of the separand when the function is depicted in the time domain; and (ii) the width of the zone, e.g., expressed by the standard deviation, σ, caused by the initial zone width and by different dispersion processes (1–5).

An appropriate theory of CZE must be able to offer a quantitative description of these two phenomena: migration and dispersion. A theoretical approach will be described in this chapter first for systems where the separands are electrophoretically moving accelerated only by an electrical force acting on the charged particle. i.e., for systems where no electroosmosis occurs. In

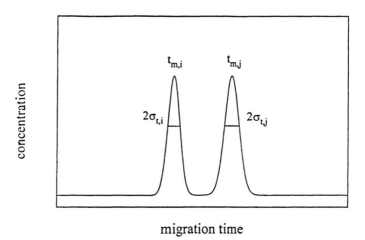

Figure 2.1. Schematic zone electropherogram of two separands, i and j, with migration times, t_m, and peak widths, expressed by the standard deviation, σ_t.

a second step migration and dispersion are described for those systems where electroosmotic flow (EOF) is present and contributes to overall electrically driven migration.

This chapter is not intended to be a complete review of the theory of CZE; rather, it deals with those important aspects of CZE that are decisive for the interpretation and improvement of practical results.

2.1. CAPILLARY ZONE ELECTROPHORESIS IN THE ABSENCE OF ELECTROOSMOTIC FLOW

2.1.1. Migration

2.1.1.1. The Concept of Effective Mobility

Charged solutes migrate under the influence of an electric field with an electrophoretic velocity, $v_{ep,i}$, which is constant within few milliseconds after application of the field. This velocity is proportional to the field strength, E, and is given for solute, i, by

$$v_{ep,i} = \mu_i E \tag{2.1}$$

The proportionality factor, μ_i, the ionic mobility, depends on the size and charge of the separand, on the kind of solvent, and on the temperature. For the case of infinitely diluted solutions (at a given temperature) it is a characteristic, constant parameter for a particular ionic species, i, and is called the absolute mobility, $\mu_{0,i}$. It can be seen from Eq. (2.1) that the ionic mobility expresses a normalized velocity; it is the migration velocity of the ionic species at unit field strength. Its dimension is m^2/Vs, with typical values in the range of $20-80 \times 10^{-9}$ for small organic and inorganic ions. It is convenient to consider the mobilities in electrophoresis as signed quantities, which is not the case in the theory of conductance. By convention, the mobility of cations has a positive sign and that of anions has a negative sign.

Obviously in practice electrophoretic separation systems operate in finite concentration ranges. In such cases the ionic mobility is no longer a constant but depends on the concentration of the electrolyte solution. The ions of a strong electrolyte are still migrating with a velocity that is proportional to the field strength, but this velocity is lower than that at infinite dilution and is given by

$$v_{ep,i} = \mu_{act,i} E \tag{2.2}$$

with the proportionality factor, $\mu_{act,i}$, termed the actual mobility.

The concentration dependence of the mobility is described by established concepts following the theories of P. J. W. Debye, E. Hückel, and L. Onsager, based on the assumption of the existence of a cloud of counterions. Readers are referred to standard textbooks on the topic (6,7). Deviations from the ideal case of infinite dilution are expressed by a correction factor, f, depending on various physicochemical properties like ionic strength, by increasing complexity with increasing ionic concentration, linking the actual mobility to the absolute mobility according to

$$\mu_{act,i} = f_i \mu_{0,i} \quad (2.3)$$

These theories have some limitations in electrophoretic practice, because they are restricted to relatively low concentration ranges, and mainly to ions in their own solutions of different concentrations rather than to solutions of ions in different background electrolytes (8).

In practice the actual mobilities can vary strongly with ionic strength. For example, this variation is depicted in Figure 2.2 for small organic acids with

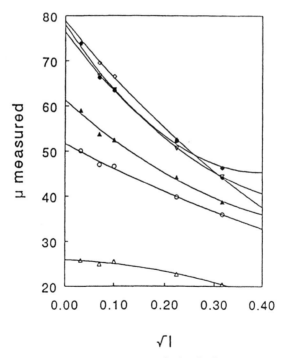

Figure 2.2. Dependence of the actual mobility μ [10^{-9} m^{2}V^{-1}s^{-1}], on the square root of the ionic strength, I [mol/L], of six aromatic sulfonates measured with CZE. The solutes have different charge numbers, z: 1, △; 2, ○; 3, ▲; 4, ●; 5, ▽; 6, ◇. (From ref. 9 with permission.)

different charge numbers, z_i, between 1 and 6. Higher charged ions at higher ionic strengths of the BGE on the order of 0.1 mol/L can exhibit an actual mobility that is only one-half of the value of the absolute mobility.

Supplementing theory, an empirical expression for the factor f_i in Eq. (2.3) for the actual mobility was derived for organic anions as a function of the charge number, z_i, and the buffer ionic strength, I, in the range between 10^{-1} and 10^{-3} mol/L (9).

$$f_i = \exp(-0.77\sqrt{z_i I}) \qquad (2.4)$$

The ionic strength is defined as

$$I = \tfrac{1}{2} \sum_{k}^{k} c_k z_k^2 \qquad (2.5)$$

where c_k is the molal concentration of ionic species k of the buffer. The sample ions must be present in such low concentrations that they do not contribute significantly to the ionic strength. From Eq. (2.4) it can be deduced that (at least for this set of organic anions described in ref. 9) the logarithm of the actual mobility decreases with the square root of the ionic strength of the background electrolyte and the square root of the charge number of the separand. For ions with the same valence a roughly parallel change of the actual mobility with ion strength will be observed; consequently, a variation of the ionic strength can significantly improve the separation only for ions with different charge numbers. Its importance as a means of adjusting mobilities in order to resolve separands therefore seems rather limited.

2.1.1.2. Affecting Mobility by Charge

There are many more weak electrolytes than strong ones. In all cases of the former type a pH can be found where the separands are not fully charged but are only partially ionized due to acid–base equilibria. Then the degree of ionization, α, determines the effective mobility, $\mu_{\text{eff},i}$, of the separand, which is given for a monovalent solute by

$$\mu_{\text{eff},i} = \mu_{\text{act},i}\,\alpha = \mu_{0,i}\,f_i\,\alpha \qquad (2.6)$$

Protolysis is one of the most important equilibria to establish separation in CZE for weakly acidic and basic separands. For the aforementioned case of a monovalent weak acid with a certain pK_a the degree of dissociation, α, is connected to the pH of the buffering background electrolyte by

$$\alpha = \frac{1}{1 + 10^{(pK_a - pH)}} \quad (2.7)$$

For weak bases $(pK_a - pH)$ is replaced by $(pH - pK_a)$.

The effective mobility of weak monovalent acids thus depends on the pH of the buffering background electrolyte by

$$\mu_{eff,i} = \mu_{act,i}\alpha = \frac{\mu_{act,i}}{1 + 10^{(pK_{a,i} - pH)}} \quad (2.8)$$

If the solute is a multivalent weak acid, i, it consists of a number of subspecies, j, with a different valence, each with a certain $pK_{a,j}$ and actual mobility, $\mu_{act,j}$. In this case the effective mobility can be described by the following equation, where the individual contributions of the subspecies to the total effective mobility are summarized (5; for examples of another approach, cf. refs. (10–12):

$$\mu_{eff,i} = \sum^{j} \frac{\mu_{act,j} - \mu_{act,(j-1)}}{1 + 10^{(pK_{a,j} - pH)}} \quad (2.9)$$

An example of the dependence of the effective mobility of solutes on the pH of the buffering background electrolyte is shown in Figure 2.3. Measured values of the effective mobility match well the curve fitted according to Eq. (2.9), at least for the mono- (13,14), di-, and trivalent acids; for the hexabasic acid shown in the example only poor accordance is found (15).

A major problem exists, however, in calculating the effective mobilities of separands as a function of the pH of the background electrolyte. It is necessary to have accurate data of the pK_a values of all subspecies (which are rare in the literature, and which are given only as limiting values for infinite dilution), and to have data of actual mobilities of the particular subspecies (which are also rare in the literature for oligo- or multivalent solutes). Useful compilations of data for common inorganic and organic ions have been published by Hirokawa et al. (16) and Pospichal et al. (17).

Protolysis and complexation (the latter being especially useful for the separation of enantiomers, e.g., as inclusion complexes) (18,19) are the most important tools for the adjustment of the effective mobility in CZE practice. It can be seen, however, from Eq. (2.9) that not only the pH but also changes in the actual mobility and the pK_a value can influence the effective mobility. This occurs either by changing the ionic strength, as mentioned above, or by replacing water as the solvent either partially or totally by organic solvents.

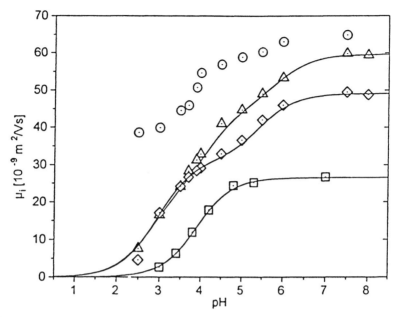

Figure 2.3. Dependence of the effective mobility, μ_i, experimentally determined by CZE, on the pH of the buffering electrolyte: ▫, 3,5-dimethoxybenzoic acid; ◇, 1,2-benzenedicarboxylic acid; △, 1,2,3-benzenetricarboxylic acid, ⊙, benzenehexacarboxylic acid, Solid lines: calculated curve resulting from a least squares fit according to Eq. (2.9) (Reprinted from ref. 15 with permission.)

The change of the actual mobilities follows roughly the viscosity of the solvent but can still differ for particular pairs of solutes. In other words, the Walden rule

$$\mu_0 \eta = \text{const} \qquad (2.10)$$

is not fully obeyed; moreover, the generalized Walden rule

$$\mu_0 r \eta = \text{const} \qquad (2.11)$$

is also seldom fulfilled. In Eqs. (2.10) and (2.11), η is the viscosity of the solvent and r is the radius of the solvated ion in the particular solvent. A more pronounced effect of the solvent on the pK values is observed in many cases, on the other hand, and may lead to selective changes of the effective mobilities.

Table 2.1. Thermodynamic pK_a Values of Simple Organic Acids and Bases in Pure Organic Solvents[a] at 25 °C

	pK_a			
Solute	Water	MeOH	DMSO	ACN
Acetic acid	4.76	9.52	12.6	22.4
Chloroacetic acid	2.86	7.81	8.9	18.8
Benzoic acid	4.20	9.28	11.0	20.7
Anilinium	4.6	—	3.8	10.6

Sources: Data from sources cited in ref. 20.
[a] MeOH, methanol; DMSO, dimethylsulfoxide; ACN, acetonitrile.

2.1.1.3. Affecting Mobility by Variation of pK_a

The pK values depend strongly on the nature of the solvent. It is often found that solvents increase (in many cases drastically) the pK_a values of acids and change to a smaller degree that of bases. In Table 2.1 examples are given of some typical solutes and solvents applicable to capillary electrophoresis (20).

The thermodynamic pK_a values of the neutral acids increase in pure methanol compared to water by about 5 units, in dimethylsulfoxide (DMSO) by 7–8 units, and in acetonitrile (ACN) by a remarkable 16–18 units. For a base like aniline (or its corresponding cation acid, anilinium) the change is much lower. The problem of defining pH and pK scales in nonaqueous media will not be discussed here (see, e.g., ref. 21).

The effect of the solvent on the pK_a value is best described by the concept of the transfer activity coefficient (21–24), which is applied to all species involved in the protolysis equilibrium. This concept takes into consideration the chemical potential, $\bar{\omega}_i$, of species i in a particular solvent, S,

$$\bar{\omega}_i^S = \bar{\omega}_{0,i}^S + RT \ln c_i^S + RT \ln \gamma_i^S \qquad (2.12)$$

and in water, W,

$$\bar{\omega}_i^W = \bar{\omega}_{0,i}^W + RT \ln c_i^W + RT \ln \gamma_i^W \qquad (2.13)$$

where $\bar{\omega}_{0,i}$ is the chemical potential in the standard state, taken for the hypothetical molality of 1; c_i is the molal concentration of the species i; γ_i is the activity coefficient; and R and T are the gas constant and the absolute temperature, respectively.

Combining Eqs. (2.12) and (2.13) leads to the expression of the medium effect, $\ln {}_m\gamma_i$, for species i,

$$\ln {}_m\gamma_i = \ln \gamma_i^W - \ln \gamma_i^S = \frac{\bar{\omega}_{0,i}^S - \bar{\omega}_{0,i}^W}{RT} \tag{2.14}$$

Here ${}_m\gamma_i = \gamma_i^W/\gamma_i^S$ is the transfer activity coefficient or medium activity coefficient. Its (natural or decadic) logarithm, the medium effect, is proportional to the reversible work of transferring 1 mol of species i from infinite dilution in water (W) to solvent (S). It has a negative value when the species is more stable in solvent S, and a positive value when it is more stable in water.

In the following discussion of the variation of the pK_a value of a certain acid, HA, or base, B, the concept of the transfer activity coefficient is applied to all particles involved in the protolysis of either a neutral acid, HA

$$HA = H^+ + A^- \tag{2.15}$$

or a base B. For formal reasons the protolysis of the corresponding cation acid, HB^+, is considered here:

$$HB^+ = H^+ + B \tag{2.16}$$

The reversible work of transferring 1 mol of H^+ and A^- (or H^+ and B) from infinite dilution in water to infinite dilution in solvent S, and transferring 1 mol of HA (or HB^+) from solvent S to water, is equal to the standard free energy of transfer, ΔG_t^0:

$$\Delta G_t^0 = RT \ln \left[\frac{({}_m\gamma_{H^+})({}_m\gamma_{A^-})}{({}_m\gamma_{HA})} \right] \tag{2.17}$$

and

$$\Delta G_t^0 = RT \ln \left[\frac{({}_m\gamma_{H^+})({}_m\gamma_{B})}{({}_m\gamma_{HB^+})} \right] \tag{2.18}$$

where ${}_m\gamma_{H^+}$, ${}_m\gamma_{A^-}$, ${}_m\gamma_{HA}$, ${}_m\gamma_{HB^+}$, and ${}_m\gamma_{B}$ are the transfer activity coefficients of the particular species.

The standard free energy of transfer is directly related to the change in the protolysis constants, K_a:

$$\Delta pK_{a,\text{HA}} = pK^S_{a,\text{HA}} - pK^W_{a,\text{HA}} = \log\left[\frac{(_m\gamma_{\text{H}^+})(_m\gamma_{\text{A}})}{(_m\gamma_{\text{HA}})}\right] \quad (2.19)$$

$$\Delta pK_{a,\text{HB}^+} = pK^S_{a,\text{HB}^+} - pK^W_{a,\text{HB}^+} = \log\left[\frac{(_m\gamma_{\text{H}^+})(_m\gamma_{\text{B}})}{(_m\gamma_{\text{HB}^+})}\right] \quad (2.20)$$

Equations (2.19) and (2.20) relate the effect of stabilization of the particular species in the protolysis equilibrium in water and organic solvent (or solvent mixture) to the shift of the pK_a value. The contribution of the medium effect on the proton, $\log {}_m\gamma_{\text{H}^+}$, reflects the basicity of the solvent, compared to water. The contributions of the neutral species, HA and B, are normally negligible compared to the ionic species.

The consequence for the pK_a shift for cation acid (i.e., HB^+) is different from that for neutral acids like HA. It can be assumed that a solvent that stabilizes (or destabilizes) the proton (H^+) will also stabilize (or destabilize) the cation (HB^+) to some extent. The effect of the solvent on these two cations is then nearly canceled in Eq. (2.20) for cation acids (where we have the ratio of the particular transfer activity coefficients of the two cations), and the shift in pK will not be large. This is what in fact is found, e.g., for substituted organic ammonium compounds of type R_3NH^+ in aqueous solutions of common solvents like alcohols or DMSO (25).

The effect is significantly different for neutral acids of type HA. Organic solvents normally have poorer solvation properties for anions than does water. The resulting destabilization of the anion often leads to large positive values for the transfer activity coefficient for A^- in Eq. (2.19), and even a higher basicity, e.g., of DMSO cannot compensate for this influence: most solvents therefore produce a shift of the pK_a of neutral acids to higher values as seen in Table 2.1. The drastic changes in ACN are caused not only by its extremely poor solvating ability for anions but also by its being a very weak base ($\log {}_m\gamma_{\text{H}^+}$ is about 4) (21).

As this shift strongly depends on the transfer activity coefficient of the particular anion, it may be different for different neutral acids (26, 27). This makes organic solvents useful as tools with which we can adjust the separability of solutes with only slightly differing pK_a values in water. Mixed aqueous–organic solvents rather than pure solvents seem to be preferable for this purpose because water present even in trace concentrations in organic solvents can markedly affect the pK_a and can be hard to control in normal practice. Therefore controlling water content will lead to higher accuracy and greater precision of the established conditions, i.e., of the pK_a and thus of the effective mobility adjusted at a certain pH for better selectivity of capillary electrophoretic separation.

2.1.1.4. Migration Time

It is obvious that analytes must have different electrophoretic velocities to be separated by CZE; in other words, they must exhibit different residence time, t_m, after migration across the distance, L_d, from the point of injection to that of detection. This distance is normally not identical with the total length of the capillary, L_t, which determines the field strength for a certain applied voltage, V:

$$E = V/L_t \tag{2.21}$$

The migration or residence time is given for solute i by (28–30)

$$t_{m,i} = \frac{L_d}{v_{ep,i}} = \frac{L_d}{\mu_{eff,i} E} = \frac{L_d L_t}{\mu_{eff,i} V} \tag{2.22}$$

The magnitude of the difference of migration times necessary for sufficient resolution is determined by those dispersive processes that counteract the separation due to sample zone broadening effects.

2.1.2. Dispersion

A number of processes lead to peak broadening during the migration of a sample zone through the capillary. One of them is inevitable in CZE, namely, longitudinal diffusion. It occurs because the sample exhibits a concentration gradient in the direction of the (longitudinal) electric field, which causes diffusional mass flux. Other contributions to peak dispersion are caused by the following:

1. Joule heating of the electrolyte solution
2. Conductivity differences between the sample and the background electrolyte solution, leading to electrophoretic dispersion, or dispersion due to concentration overload
3. Wall adsorption
4. Different migration distances due to coiling of the capillary

The EOF normally has no significant effect on peak dispersion due to its plug flow profile, at least when its velocity is constant in the axial direction. Its potential contribution to peak broadening will, however, be discussed below.

Besides the effects listed above, which arise during the electrophoretic process, extracolumn peak broadening may occur, e.g., due to the finite plug length of the injection or due to the width of the detector aperture.

The particular peak broadening can be expressed by the individual second moment of the concentration distribution, the variance, σ_{ind}^2. If one assumes that all these processes are independent of each other (more precisely, if it is assumed that the system is linear), the additivity of the variances enables us to sum up the single contributions, giving the total peak variance, σ_{tot}^2, as

$$\sigma_{tot}^2 = \sum \sigma_{ind}^2 = \sum \sigma_{extr}^2 + \sigma_{dif}^2 + \sigma_{joule}^2 + \sigma_{conc}^2 + \sigma_{ads}^2 + \sigma_{coil}^2 \qquad (2.23)$$

In fact, this assumption is not fully valid because, e.g., the electromigration dispersion counteracts longitudinal diffusion on one edge of the sample zone.

A way to describe the individual contributions to peak broadening is based on the plate height model, which is commonly employed in isocratic chromatography. Also in CZE a fundamental relation exists between the peak variance in the length domain, σ_x^2, and the migration distance, x, given by

$$\sigma_x^2 = Hx \qquad (2.24)$$

The factor of proportionality, H, is the theoretical plate height. It is a measure of the mixing properties of the separation system. It has a misleading (historical) name, because it stems from the theory of multiple extraction, where it indeed refers to a real plate. In CZE (as in chromatography) it must not be considered as an elementary plate of a certain height (although it has the dimension of length) where dispersion takes place step by step: both separation methods are based on continuous processes.

It is clear from Eqs. (2.23) and (2.24) that the total plate height of the electrophoretic peak is the sum of the individual plate height contributions, caused by the particular broadening effects after migration across the distance L_d. Equation (2.24) can be rearranged to

$$H = \frac{\sigma_x^2}{x} = \frac{\sigma_x^2}{L_d} \qquad (2.25)$$

In CZE the peak is normally registered not in the space domain but in the time domain, with standard deviation, σ_t. Transformation of σ_t^2 in σ_x^2 or vice versa is carried out with the aid of the migration velocity, v_i, of the separand:

$$\sigma_{t,i}^2 = \frac{\sigma_{x,i}^2}{v_t^2} \qquad (2.26)$$

The plate height can be determined from the electropherogram by using the standard deviation of the peak (half of the width of a Gaussian peak in 0.63 of the height), as follows:

$$H = L_d \left(\frac{\sigma_{t,i}}{t_m}\right)^2 \tag{2.27}$$

2.1.2.1. Extracolumn Effects

There are several types of sample injection in CZE: (i) electrokinetic injection (the sample ions are inserted into the capillary by application of an electric field prior to separation); (ii) by electroosmosis (described later); or (iii) pneumatically by application of a pressure or vacuum. In the first two cases the sample plug has a rectangular concentration distribution (when thermal effects are excluded); in the third case it is inserted by a laminar flow, which generates a parabolic profile.

For a rectangular distribution function of rectangle width τ, the peak variance is given in general by (31)

$$\sigma_x^2 = \tau^2/12 \tag{2.28}$$

The corresponding plate height for a rectangular injection plug with width τ_{inj} is

$$H_{inj} = \frac{\tau_{inj}^2}{12L_d} \tag{2.29}$$

For pressurized injection the contribution to peak broadening in an open tube is more complicated. Given that the sample plug would be inserted symmetrically by a laminar flow into the cylindrical tube, the profile would be deformed due to Taylor dispersion (32), which is expressed by

$$\sigma_{lam}^2 = \frac{R_c^2 \hat{v}^2 t}{24D} \tag{2.30}$$

where R_c is the capillary inner radius; D is the diffusion coefficient of solute; and $\hat{v}$ is the average linear velocity. The corresponding plate height is

$$H_{lam} = \frac{R_c^2}{24D} \hat{v} \tag{2.31}$$

The average velocity when applying a pressure difference of ΔP is given by

$$\hat{v} = \frac{\Delta P R_c^2}{8 L_t \eta} \tag{2.32}$$

Combining Eqs. 2.30 and 2.32 yields the expression for the plate height contribution due to pressurized injection of the total symmetrical sample plug into the capillary:

$$H_{\text{lam},i} = \frac{R_c^6 \Delta P^2 t_{\text{inj}}}{1536 L_t^2 \eta^2 D_i L_d} \tag{2.33}$$

Thus the plate height contribution depends on the capillary radius by the power of 6, which means that it is larger by a factor of 64 in a capillary with 100 μm i.d. compared to one with 50 μm i.d.

The situation is more complicated in practice when the sample is injected pneumatically, because the sample plug is not symmetrical but is deformed only at the front end by Taylor dispersion. This is because after application of the pressure the sample reservoir is replaced by the buffer reservoir. This manipulation retains the curved profile of the sample in the capillary but produces a rectangular profile of the sample plug at the open end of the capillary. Therefore the resulting extracolumn zone broadening of the sample will lay between the two extreme cases: the "Taylor-dispersed" zone on the one hand, and a rectangular profile on the other.

The detector cell is usually of rectangular geometry in the axial direction, with aperture width τ_{det}. Its contribution to peak broadening can be described in the same way as that caused by the rectangular injection plug, equivalent to Eq. (2.29):

$$H_{\text{det}} = \frac{\tau_{\text{det}}^2}{12 L_d} \tag{2.34}$$

2.1.2.2. Dispersion Due to Capillary Coiling

Zone broadening can occur in coiled capillaries because the migration distance of the solutes moving at the outer circumference differs from that at the inner circumference (33, 34). This difference, s_d, depends on the number of coils, n_c, and the radius of the capillary, R_c (35):

$$s_d = (2n_c + 1) 2\pi R_c \tag{2.35}$$

The differences in length of the migration paths result in peak broadening expressed by the spacial variance, σ^2 coil, given by

$$\sigma^2_{coil} = \frac{S_d^2}{16} \tag{2.36}$$

The corresponding plate height, H_{coil}, is then

$$H_{coil} = \frac{R_c^2 L_d}{4r_{in}^2} \tag{2.37}$$

where r_{in} is the internal radius of the capillary coil. The effect of capillary coiling on peak broadening can be relevant when capillaries with small coil radii are used, e.g., for CZE on microchips (36). The effect may also become visible when the inevitable diffusion is negligible, which is the case for very large biomolecules with low diffusion coefficients.

Peak broadening due to the finite length of the injected plug and detector aperture and peak broadening due to capillary coiling are not dependent on the migration time of the solutes because they are not caused by processes occurring during electrophoretic movement. In contrast, a number of dispersion effects occur during the migration of the solutes. They will be discussed in the following subsection first for conditions where EOF is absent. It was mentioned above that the EOF will mostly not contribute directly to peak broadening; however, it will indirectly influence the magnitude of the particular dispersion effects, because it determines the residence time of the solutes in the capillary and thus governs the time that these processes will continue.

2.1.2.3. Longitudinal Diffusion

As the sample inserted at one end of the capillary forms a concentration gradient in the axial direction, a gradient of the chemical potential exists that causes a diffusional mass flux in this direction according to Fick's second law, which describes the change of the concentration as a function of time and space:

$$\frac{\partial c}{\partial t} = D \frac{\partial^2 c}{\partial x^2} \tag{2.38}$$

When the ionic sample pulse is also transported by the electric force, the corresponding transport equation describing the mass flux is

$$\frac{\partial c}{\partial t} = D\frac{\partial^2 c}{\partial x^2} - v\frac{\partial c}{\partial x} = D\frac{\partial^2 c}{\partial x^2} - \mu E\frac{\partial c}{\partial x} \qquad (2.39)$$

where v is the (net) migration velocity of the sample.

Supposing that the initial sample pulse is a Dirac (or δ) function, which is infinitely narrow, then the solution of the differential equation at the interval $x \in (-\infty/\infty)$ for $t \geq 0$ is

$$c(x,t) = \frac{Q}{(4\pi Dt)^{1/2}} \exp\left(-\frac{x^2}{4Dt}\right) \qquad (2.40)$$

where Q is the amount of sample. This is the formula for a Gaussian bell–shaped curve with the spacial variance $2Dt$, in accordance with the random walk theory and given by the Einstein–Smoluchowski equation:

$$\sigma_x^2 = 2Dt \qquad (2.41)$$

Here t is the time available for the solute particles in the sample zone to move and is thus the residence time, t_m, in CZE.

The concentration as a function of x and t can be numerically calculated under various much more complicated conditions, using the equations given above, as carried out by a number of authors (37–46).

The other, static approach is based on the plate height model, which is applied here throughout. Introduction of the residence time [Eq. (2.22)] into Eq. (2.41) leads to

$$\sigma_{x,i}^2 = \frac{2D_i L_t L_d}{\mu_{\text{eff},i} V} \qquad (2.42)$$

The corresponding plate height expressing the contribution of longitudinal diffusion to zone broadening (47–50) is

$$H_{\text{dif},i} = \frac{2D_i L_t}{\mu_{\text{eff},i} V} \qquad (2.43)$$

The plate number, N which is more decisive for separation in such systems than the plate height, is defined as

$$N = \frac{L}{H} \qquad (2.44)$$

and is thus given for the case of longitudinal diffusion by

$$N_{\text{dif},i} = \frac{\mu_{\text{eff},i} V L_d}{2 D_i L_t} = \frac{\mu_{\text{eff},i} V_d}{2 D_i} \qquad (2.45)$$

where

$$V_d = V(L_d/L_t) \qquad (2.46)$$

is the voltage drop from the point of injection to that of detection. This is the electric potential difference across which the ionic solutes migrate. This difference must not be confused with the total applied voltage, V.

Equation (2.45) enables us to explain the high plate numbers that can be reached with CZE: the very high voltage that can be applied allows separation to occur in the rather short time remaining for diffusional peak broadening.

Note that the ratio μ/D in Eq. (2.45) appears in a number of relations relevant to the efficiency of CZE (51). Its significance will be discussed in detail below.

Equation (2.45) determines that the efficiency attainable for "ultimate" performance—when no other peak broadening contributions than longitudinal diffusion takes place (52)—is explicitly not dependent on the length of the separation capillary; this is in contrast to what happens in chromatography. Therefore it is questionable whether the efficiency of CZE should be characterized by plates per meter, as is done in chromatography as well as by many authors in CZE. It is more suitable to characterize the separation efficiency of the system by plates per volt (thus making the deviation from the ultimate case clear):

$$\frac{N_{\text{dif},i}}{V_d} = \frac{\mu_{\text{eff},i}}{2 D_i} \qquad (2.47)$$

The plate number per volt depends only on two substance-specific properties: the effective mobility and the diffusion coefficient. Both parameters characterize transport processes: the effective mobility, the mass flow across the gradient of the electric potential; the diffusion coefficient, the mass flow across the gradient of the chemical potential. Note that in diffusion no real force acts on the particles, in contrast to what occurs in conductance, where a real electric force acts.

The two parameters, the effective mobility and the diffusion coefficient, are correlated; for ions at infinite dilution their relation is described by the Nernst–Einstein equation, which for an ion with charge number z_i is as follows:

$$D_i = \frac{\mu_{0,i} RT}{z_i F} \tag{2.48}$$

Here F is the Faraday constant. This relation can be used to derive the diffusion coefficient of a solute from its ionic mobility, because the latter can easily be measured by capillary electrophoresis.

In a strict sense the Nernst–Einstein relation is a limiting law and is valid only for infinite dilution. If the ion charge and the ionic strength are not too high, the deviation from ideality is in an acceptable range for analytical purposes. In this case Eq. (2.48) can be approximated by

$$D_i = \frac{\mu_{\text{eff},i} RT}{z_i F} \tag{2.49}$$

where z_i is the overall effective charge number of the solute. Rearrangement of Eq. (2.49) yields

$$\frac{\mu_{\text{eff},i}}{D_i} = z_i \frac{F}{RT} \tag{2.50}$$

Note again that this ratio is of key importance for the description of the efficiency in the diffusional controlled case of peak dispersion. Substitution of this ratio in Eqs. (2.43) and (2.45) yields the expression for the plate height and the plate number, respectively, given by (47, 48, 53, 54)

$$H_{\text{dif},i} = \frac{2RT}{z_i EF} \tag{2.51}$$

and

$$N_{\text{dif},i} = \frac{L_d}{H_{\text{dif},i}} = \frac{z_i V_d F}{2RT} \tag{2.52}$$

Substitution of the ratio μ/D by the application of the Nernst–Einstein relation produces the remarkable result that the plate height or plate number due to longitudinal diffusion is not dependent on the diffusion coefficient of the solute (at least within the limitations mentioned above, namely, small ions at low ionic strength of the background electrolyte, so that the deviations from ideality are not too large). It can be seen from Eq. (2.52) that there are only two parameters that control diffusional band broadening (besides the temperature): the effective charge of the solute, and the voltage. Both determine the time available for diffusion (note that no EOF is considered here). The higher the voltage, the shorter is this time. The same is true for the charge. Again it is not the total voltage applied to the entire capillary but only the voltage drop between the point of injection and the detector; the difference between these two voltages can reach 30% and even more in practice.

Insertion of the appropriate values for the constants F and R in Eq. (2.52) gives (for a temperature of 25 °C) the following approximate expression for the plate number:

$$N_{\text{dif},i} \approx 20 z_i V_d \qquad (2.53)$$

From this equation it can be seen again why CZE shows enormously high plate numbers especially for large biomolecules like proteins or DNA fragments: besides the high voltage applicable, these solutes are often highly charged. Although especially for large and highly charged solutes the experimental conditions do not allow the correct application of the Nernst–Einstein relation, Eq. (2.53) is still generally practicable for our purposes (see below).

This discussion also offers a tool whereby we can characterize the deviation of the efficiency measured under practical conditions from that which is theoretically reachable when only longitudinal diffusion takes place [the "ultimate" performance pointed out by Knox (52)]. In practice the plate number is measured in the time domain from the migration time, related to the width of the (Gaussian) peak, either to the standard deviation, $\sigma_{t,i}$, or to the peak's full width at half-maximum height, $w_{1/2}$, all given in the same units, according to

$$N_{i,\text{meas}} = \left(\frac{t_{m,i}}{\sigma_{t,i}}\right)^2 = 5.54 \left(\frac{t_{m,i}}{w_{1/2,i}}\right)^2 \qquad (2.54)$$

By comparing the plate number measured in this way with that given in Eq. (2.53), the deviation from the ultimate case can be determined.

An example will now be given demonstrating that the conditions of the Nernst–Einstein relation are fairly well fulfilled for practical purposes, e.g., for partially charged small solutes. Obviously the agreement with theory will be better the smaller the difference between the diffusion coefficients of the neutral and ionic species of the solute involved.

For a monovalent weak acid the effective charge number is equal to the degree of dissociation, α. Then Eq. (2.53) can be rewritten as

$$N_{\text{dif},i} \approx 20 \alpha V_d = \frac{20 V_d}{1 + 10^{\text{p}K_a - \text{pH}}} \qquad (2.55)$$

This means that the plate number for a given solute is dependent on the pH of the background electrolyte. The shape of the N vs. pH curve should approximate that of a titration curve, with a plateau corresponding to the value of N given in Eq. (2.55) at high pH (namely, 20 V_d), with half of this plateau value (10 V_d) at the pH equal to the pK_a of the particular acid, and with decreasing plate numbers for lower pH. In Figure 2.4 such a dependence is shown for three benzoic acids. Indeed the curves support the theoretical predictions; so we can conclude that the pH of the background electrolyte influences not only the selectivity of CZE but also its efficiency within the limitations already mentioned).

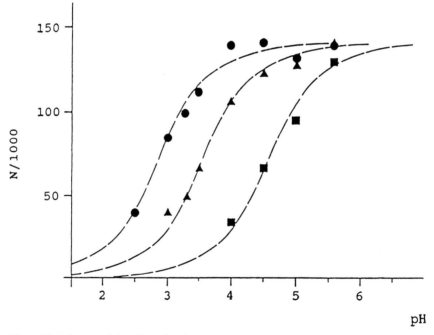

Figure 2.4. Influence of the pH on the plate number, N, of different monovalent acids based on the dependence of N on the charge number, z, and the degree of dissociation, α, respectively. The dashed curves give the theoretical dependence of N on pH, according to Eq. (2.55). Experimental values: ●, 2,3-dihydroxybenzoic acid, pK 2.38; ▲, 2,3-dimethoxybenzoic acid, pK 3.53; ■, 4-dimethoxybenzoic acid, pK 4.58. (Reprinted with permission from ref 53.)

2.1.2.4. Thermal Contribution to Dispersion

During ion migration heat is generated that causes a parabolic temperature gradient in the radial direction inside the capillary schematically depicted in Figure 2.5. Although the temperature increase can be substantial, the temperature difference between the center of the capillary and the wall remains comparable low when advanced instruments are used (with forced cooling of the capillary outside) (52,55–72). The main part of the gradient falls off between the outside wall and the surrounding. The discussion below will concentrate on the effects inside the capillary, because these contribute to zone broadening.

Under the conditions of forced cooling the temperature, T_w, at the inner wall of the capillary with inner radius R_c and outer radius R_o is

$$T_w = T_o + \frac{\kappa E^2 R_c^2}{2\lambda_c} \ln\left(\frac{R_o}{R_c}\right) \tag{2.56}$$

where T_o is the cooling temperature at the outside of the capillary; κ is the specific electric conductivity of the buffer; and λ_c is the thermal conductivity of the capillary material. Here κE^2 is the electric power generated.

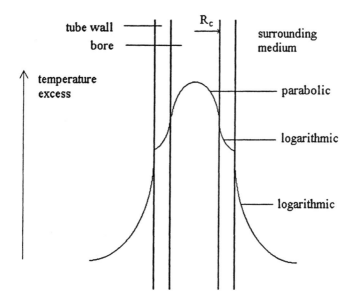

Figure 2.5. Schematic temperature profiles inside and outside of the capillary due to Joule heating, according to ref. 52; R_c, inner capillary radius.

Inside the capillary a parabolic radial temperature profile, $T(r)$, is established that is expressed as a function of the radial coordinate, r, by

$$T(r) = T_w + \frac{\kappa E^2 R_c^2}{4\lambda_s}\left(1 - \frac{r^2}{R_c^2}\right) \tag{2.57}$$

where λ_s is the thermal conductivity of the BGE solution. This profile is responsible for the thermal contribution to peak broadening; r varies from zero at the center of the capillary to R_c at the wall.

The average temperature, $\bar{T}$, of the BGE solution is

$$\bar{T} = T_o + \kappa E^2 R_c^2 \left[\frac{1}{2\lambda_c}\ln\left(\frac{R_o}{R_c}\right) + \frac{1}{8\lambda_s}\right] \tag{2.58}$$

This equation is an approximation, because it is assumed that the specific conductivity is constant in the radial direction (which is not fulfilled in principle), so it is acceptable only for small temperature changes within the capillary.

In fact the temperature dependence of the mobility is quite pronounced. The mobility increases by about 2–3% per °C, or—in other words—the temperature factor of the mobility

$$f_T = \frac{1}{\mu}\frac{d\mu}{dT} \tag{2.59}$$

is about $+0.02$ to $+0.03$. Therefore ions in the center of the capillary will migrate faster than those closer to the wall.

The parabolic temperature gradient in the radial direction leads to peak broadening, and its contribution can be derived (73) in a similar way as that caused by the laminar flow in a noncompressible fluid in a cylindrical tube, namely, the Taylor dispersion.

Assuming a constant temperature factor, f_T, we can rewrite Eq. (2.59) thus:

$$f_T = \frac{\mu_T - \mu_{T_w}}{\mu_{T_w}(T - T_w)} \tag{2.60}$$

where μ_T is the mobility at temperature T, and μ_{T_w} is the mobility at the temperature directly at wall. Equation (2.60) can be rearranged, leading to

$$\mu_T - \mu_{T_w} = f_T \mu_{T_w}(T - T_w) \tag{2.61}$$

Substitution of $(T - T_w)$ by Eq. (2.57) and multiplication with the field strength, E, to replace the mobility by the velocity, leads to the change of the electrophoretic velocity, $\Delta v_{ep,i}$, as a function of the radial coordinate

$$\Delta v_{ep,i}(r) = \frac{f_T \kappa E^3 R_c^2 \mu_{Tw}}{4\lambda_s}\left(1 - \frac{r^2}{R_c^2}\right) \tag{2.62}$$

which is the analogue of the Hagen–Poiseuille equation for laminar flow.

The average linear velocity is then

$$\hat{v}_{ep,i} = \frac{f_T \kappa E^3 R_c^2 \mu_{eff,i}}{8\lambda_s} \tag{2.63}$$

which can be inserted in the variance describing Taylor dispersion [Eq. (2.30)]; this leads to the variance, σ_{th}^2, caused by the self-heating

$$\sigma_{th,i}^2 = \frac{f_T^2 \kappa^2 E^6 R_c^6 \mu_{eff,i}^2}{1536 \lambda_c^2 D_i} t_{m,i} \tag{2.64}$$

and the corresponding plate height, $H_{th,i}$

$$H_{th,i} = \frac{f_T^2 \kappa^2 E^6 R_c^6 \mu_{eff,i}^2}{1536 \lambda_c^2 D_i} \frac{t_{m,i}}{L_d} \tag{2.65}$$

or

$$H_{th,i} = \frac{f_T^2 \kappa^2 E^5 R_c^6 \mu_{eff,i}}{1536 \lambda_c^2 D_i} \tag{2.66}$$

which is, after substitution of μ/D according to the Nernst–Einstein relation,

$$H_{th,i} = \frac{f_T^2 \kappa^2 E^5 R_c^6 F z_i}{1536 RT \lambda_s^2} \tag{2.67}$$

Thermal peak broadening can be formally treated by the plate height model like the longitudinal diffusion by replacing the diffusion coefficient, D_i, by the thermal dispersion coefficient, $D_{th,i}$. Peak broadening is then expressed analogously to the Einstein–Smoluchowsky equation for one-dimensional diffusion [Eq. (2.41)] by

$$\sigma_{th,i}^2 = 2 D_{th,i} t_{m,i} \tag{2.68}$$

with the thermal dispersion coefficient

$$D_{th,i} = \frac{f_T^2 \kappa^2 \mu_{eff,i}^2 E^6 R_c^6}{3072 D_i \lambda_s^2} \qquad (2.69)$$

The plate height contribution due to self-heating is strongly dependent on the radius, which should therefore be minimized. It also depends on the fifth power of the field strength, which, in contrast, allows rapid separations with low diffusional dispersion. It depends further on the second power of the specific conductivity of the buffer, which is directly proportional to its ionic concentration and the sum of the effective mobilities of the cation and anion involved.

Hence, to achieve low thermal dispersion, it follows that buffers should be used with low conductivity, on the one hand, and with low effective mobilities, on the other. This implies the favorable use of buffering zwitterions. Buffers with low ionic concentration, however, often produce another undesirable peak-broadening phenomenon, namely, peak triangulation due to sample concentration overload (discussed next).

2.1.2.5. *Concentration Overload (Electromigration Dispersion)*

A special type of peak deformation is triangulation, an example of which is shown in Figure 2.6. It occurs when the sample concentration is excessively high compared to the BGE concentration and when the mobilities of the sample ion and the co-ion of the BGE (that with the same sign) differ. This effect is not accurately described by the plate height model used here, because the formation of stepwise field gradients at the edges of the sample zone (formed due to stepwise conductivity gradients according to the Ohm law, equal current density presumed) counteracts diffusion on the one edge as in isotachophoresis; therefore the additivity of the variances is not strictly applicable (the system is not linear). Thus a more rigorous treatment of the transport of the ions and the shaping of the zones is that based on dynamical numerical simulations of the concentration as a function of time and position in the capillary, as carried out by several authors (37–46,74).

Here the contribution of concentration overload to the plate height is conceptualized in a way that is based on the Kohlrausch regulation function and formulated by Everaerts group [see Mikkers et al. (75)]. It has the precondition that the effective mobilities of the sample ion and the co-ion, A, of the buffering BGE differ, which means that the ratio of these mobilities (which is the definition of the selectivity coefficient, r_{ji}) must be unequal to 1:

$$r_{iA} = \frac{\mu_{eff,i}}{\mu_{eff,A}} \neq 1 \qquad (2.70)$$

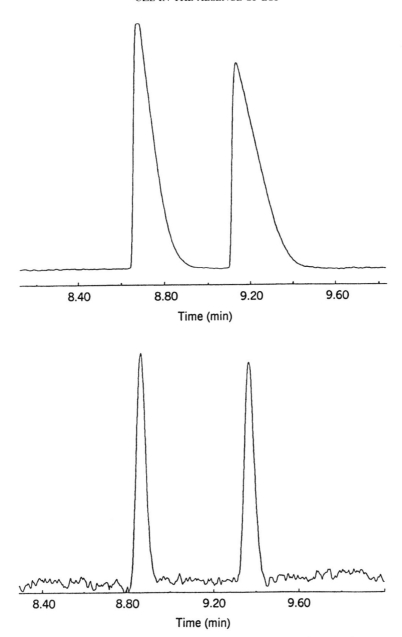

Figure 2.6. Electropherograms demonstrating qualitatively the effect of the sample concentration on peak triangulation due to concentration overload. Electropherogram (bottom) obtained after 50-fold dilution of the initial sample (top).

Then the base width, $\delta_{tr,i}$, of the triangular concentration partition function with initial width δ_{inj} is given after migration by

$$\delta_{tr,i} = 2\sqrt{|L_d \, a_i \, c_i \, \delta_{inj}|} \qquad (2.71)$$

with

$$a_i = \frac{(r_{iA} - r_{BA})}{(1 - r_{BA})r_{iA}} \frac{(1 - r_{iA})}{I} \qquad (2.72)$$

where subscript B indicates the counterion; consequently r_{BA} is always negative; c_i is the (nonadjusted) initial solute concentration in the sample injected. The sign of a_i designates the form of the triangle: when the mobility of the solute ion is higher than the co-ion mobility, the analyte peak shows a leading edge; in the reverse case, the triangle has a trailing edge—both recorded in the time domain.

The plate height stemming from the concentration overload is given by (5)

$$H_{conc,i} = \left| \frac{2 a_i c_i \delta_{inj}}{9} \right| \qquad (2.73)$$

The triangulation depends on the mobilities related to the mobility of the co-ion in a rather complex way: peak deformation will be smaller, the smaller is the concentration of the sample (see Figure 2.6). It must be emphasized again that this approach is an oversimplification and the real peak shape is better derived using numerical simulations.

2.1.2.6. Wall Adsorption

The effect of wall adsorption on peak shape can be treated either dynamically by computer simulation methods using numerical calculations (76–79) or by the static concept of the plate height. As done above, the discussion presented here follows the latter approach.

If slow linear adsorption at the wall of the cylindrical capillary takes place, the resulting contribution to the plate height H_{ads} is given for the case of electrophoretic migration (80–82) by

$$H_{ads} = \frac{(1-\phi)^2}{4D} R_c^2 v + \frac{2\phi(1-\phi)}{k_d} v \qquad (2.74)$$

Here k_d is the rate constant of desorption; v is the velocity of an unretained solute; ϕ is the fraction of the analyte in solution, defined as

$$\phi = \frac{n_l}{n_l + n_s} \tag{2.75}$$

where n_l and n_s represent the number of moles of the free sample in solution and adsorbed on the wall, respectively. Equation (2.74) is also valid for the case in which electroosmosis with the typical plug flow profile occurs; in this case the velocity is composed of the electrophoretic and the electroosmotic velocity (see below).

The following expression relating the fraction ϕ to the distribution coefficient K can be derived (83):

$$\phi = \frac{\pi R_c^2 c_{eq}}{\pi R_c^2 c_{eq} + 2\pi R_c a_{eq}} = \frac{R_c}{R_c + 2K} \tag{2.76}$$

where the distribution coefficient is defined as $K = a_{eq}/c_{eq}$; here c_{eq} and a_{eq} are the equilibrium concentrations of the analyte in the liquid solution and the solid surface, respectively. It should be noted that due to adsorption the mean velocity of the sample, v_m, is given by $v_m = v \cdot \phi$.

Substituting ϕ by $R_c/(R_c + 2K)$ and v by v_m/ϕ in Eq. (2.74) gives

$$H_{ads} = \left[\frac{K^2}{D} \frac{R_c}{R_c + 2K} + \frac{4K}{(R_c + 2K)k_d} \right] v_m \tag{2.77}$$

Equation (2.77) expresses the contribution of slow linear adsorption to the plate height for an analyte moving in a flat radial velocity profile. The two terms in the brackets in this equation indicate that both radial diffusion and kinetics of adsorption contribute to the dispersion of analyte peaks.

Equation (2.77) is equivalent to the C-term of the Golay equation for chromatography in open capillaries, given as

$$H_{tot} = \frac{B}{v} + Cv \tag{2.78}$$

where the B-term expresses the effect of longitudinal diffusion (it is inversely proportional to the solute's migration velocity) and the C-term expresses the effects of the mass transfer resistance in the mobile phase and stationary phase, respectively (increasing linearly with the migration velocity). In fact, the plate height contribution due to wall adsorption should also increase in CZE with increasing migration velocity, whereby the steepness of the H vs. v curve is determined by the magnitude of the C-term.

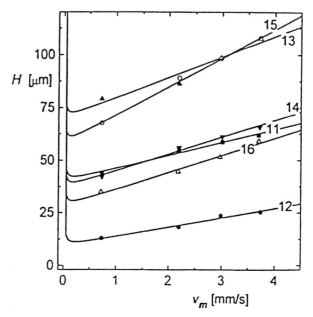

Figure 2.7. Dependence of the plate height, H, on the migration velocity, v_m, for a protein (myoglobin) in different fused silica capillaries with 75 μm i.d., indicating peak dispersion due to adsorptive interaction with the wall. The different migration velocities in the uncoated capillaries (indicated by the numbers) were adjusted by application of different voltages. Solid lines are fitting functions to Eq. (2.78). Borate buffer, pH 8.5 (From ref. 83 with permission.)

This increase has been confirmed experimentally, as shown for example in Figure 2.7 for the adsorption of a protein onto the surface of a fused silica separation capillary (83). Values of the distribution coefficient K of the protein derived from the slope of the curve according to Eq. (2.78) were found to range between 33×10^{-8} and 79×10^{-8} [m], in good agreement with those determined by static adsorption experiments with 2 μm diameter nonporous silica beads (with a partition coefficient of 63×10^{-8} m). Values of the corresponding capacity factor, k', given by

$$k' = Kq = \frac{2K}{R_c} \tag{2.79}$$

are between 0.018 and 0.042 for the foregoing example. In Eq. (2.79) q is the phase ratio, as in chromatography, i.e., the ratio of the volume of liquid phase and the area of the inner capillary wall.

Table 2.2. Expressions for the Plate Height from the Individual Contributions of the Different Peak-Broadening Effects

Source of Zone Broadening	Plate Height Expression
Rectangular injection plug	$H_{inj} = \dfrac{\tau_{inj}^2}{12 L_d}$
Pressurized injection[a]	$H_{lam} = \dfrac{R_c^6 \Delta P^2 t_{inj}}{1536 L_t^2 \eta^2 D L_d}$
Capillary coiling	$H_{coil} = \dfrac{R_c^2 L_d}{4 r_{in}^2}$
Detection (rectangular aperture)	$H_{det} = \dfrac{\tau_{det}^2}{12 L_d}$
Longitudinal diffusion	$H_{dif} = \dfrac{2 D L_t}{\mu_{eff} V}$
Longitudinal diffusion (Nernst–Einstein)	$H_{dif} = \dfrac{2 R T}{z E F}$
Self-heating	$H_{th} = \dfrac{f_T^2 \kappa^2 E^5 R_c^6 \mu_{eff}}{1536 \lambda_c^2 D}$
Self-heating (Nernst–Einstein)	$H_{th} = \dfrac{f_T^2 \kappa^2 E^5 R_c^6 F z}{1536 R T \lambda_s^2}$
Concentration overload	$H_{conc} = \left| \dfrac{2 a_i c_i \delta_{inj}}{9} \right|$
Wall adsorption	$H_{ads} = \left[\dfrac{K^2}{D} \dfrac{R_c}{R_c + 2K} + \dfrac{4K}{(R_c + 2K) k_d} \right] v_m$

[a] Maximum value; in practice it is not reached (see text).

The contributions of the individual peak-broadening effects are summarized in Table 2.2.

2.1.3. Resolution

The measure that describes the degree of separation of a pair of subsequently migrating analytes, i and j, is the resolution, R_{ji}, a dimensionless number. It combines both effects, migration and dispersion. There are several definitions of the resolution stemming from chromatography; all have in common that the difference of the migration times is related to the peak width. The definition used in most cases is (see Figure 2.1)

$$R_{ji} = \frac{t_{m,j} - t_{m,i}}{2(\sigma_{t,i} + \sigma_{t,j})} \qquad (2.80)$$

It is obvious that this expression does not easily enable us to manipulate the resolution experimentally; therefore it is transformed into a more operative form by substituting t_m and σ_t according to Eqs. (2.22) and (2.27), and introducing the arithmetic means of the mobilities and the plate heights of the two separands, $\bar{\mu}$ and $\bar{H}$, resulting in

$$R_{ji} = \frac{1}{4} \frac{\mu_i - \mu_j}{\bar{\mu}} \sqrt{\frac{L}{\bar{H}}} = \frac{1}{4} \frac{\Delta\mu}{\bar{\mu}} \sqrt{\bar{N}} \qquad (2.81)$$

The resolution consists of two terms, the selectivity term, $\Delta\mu/\bar{\mu}$, with the relative difference of the mobilities, and the efficiency term with the square root of the mean plate number. Note that the plate height, $\bar{H}$, on which this plate number is based consists of all the contributions to peak broadening discussed above. Although very high plate numbers can be reached in CZE, it can be seen from Eq. (2.8) that this high plate number does not linearly affect the reachable resolution because it is only in a square root. This means that an enhancement of the plate number, say, by a factor of 2 brings only a 40% higher resolution.

The selectivity term is much more sensitive to minute experimental changes leading to a drastic increase in resolution. Very small changes in the pH of the buffering background electrolyte in the range around the pK of the analytes can lead to such changes in the difference of the separands' effective mobilities that resolution is achieved (84,85). This has been demonstrated, e.g., for benzoic acids that differ only in the oxygen isotope composition of their carboxylic groups and thus very slightly in their dissociation constants (about 2%), as shown in Figure 2.8a,b (86).

It was pointed out earlier that of all peak-broadening effects only one is inevitable, namely, longitudinal diffusion. For this reason the expression for the resolution can be transformed for this special case in which all other peak-broadening effects are negligible. Then Eq. (2.51) for the plate height can be introduced, and the resolution can be reformulated for closely migrating peaks (87,88) as

$$R_{ji} = (r_{ij} - 1)\left[\frac{(z_i z_j)^{1/2}}{z_i^{1/2} + z_j^{1/2}}\left(\frac{FV_d}{8RT}\right)^{1/2}\right] \qquad (2.82)$$

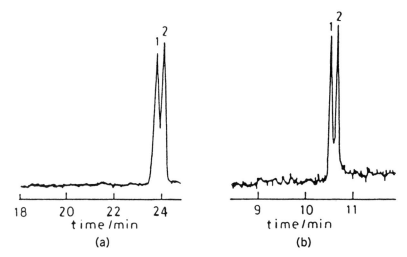

Figure 2.8. Separation of benzoic acids differing in their oxygen isotope content and therefore slightly in their pK. Acid with (1) $C^{16}O^{16}OH$ and (2) $C^{18}O^{18}OH$. Capillary 50 µm i.d., 0.750 (0.500) m length. Voltage: (a) 20 kV; (b) 40 kV. Acetate buffer, pH 4.0, 0.05 mol L^{-1}, 0.1% hydroxypropyl cellulose. UV absorbance at 210 nm. (From ref. 86 with permission.)

Here r_{ij} is the selectivity coefficient as given above:

$$r_{ij} = \frac{\mu_{\text{eff},i}}{\mu_{\text{eff},j}} \qquad (2.83)$$

This selectivity coefficient is the ratio of the effective mobilities of the two separands of interest; it is equivalent to the selectivity coefficient in chromatography, where it is defined as the ratio of the capacity factors. It is normally symbolized by α_{ij} in chromatography (89), but this symbol is commonly used in electrophoresis for the degree of dissociation. The analogy between chromatography and electrophoresis is noteworthy, however, as the separation selectivity is determined in both high-performance methods by the selectivity coefficient, defined as the ratio of those parameters that are responsible for migration—the capacity factor and the effective mobility.

The resolution can be manipulated by two main terms, the selectivity term, given by $(r_{ij} - 1)$, and the efficiency term, given in brackets in Eq. (2.82). It follows that the efficiency term is affected not only by the voltage, which is well known (90), but also by the charge number of the analytes, and thus by the pH.

The resolution therefore depends on the following.

1. Physical constants like R and F
2. Instrumental parameters: V (and T)
3. Analyte parameters: μ_{act}, pK_a, α, and z

In electrophoretic separation, however, the only analyte parameters that are at our disposal (i.e., are alterble experimentally) are the degree of dissociation and (connected with it) the charge number. They are determined by the selection of the pH of the buffer. The actual mobility and the pK value are more or less invariable (see, however, the discussion on the use of organic solvents).

Mainly two parameters can be utilized to optimize the resolution: (i) increasing voltage increases the resolution (only by one-half power) but reaches a plateau, and a reversal may even take place due to increasing Joule heat; (ii) the pH of the buffer influences the selectivity by adjusting the effective mobilities of the separands, and the pH also influences the efficiency of the system by adjusting the charge numbers of the separands.

The pH of the buffer thus plays two important roles when affecting the resolution—both selectivity and efficiency are influenced by it. Under favorable conditions, namely, when other dispersion effects besides longitudinal diffusion are negligible and when the data for the actual mobilities and the pK values of the subspecies of the separand pair are known, Eq. (2.82) is highly predictive because in principle it enables the resolution to be calculated at any pH of the buffer. An example of the high predictability of the resolution is shown in Figure 2.9. The baseline resolution was calculated for the separands based on the actual mobilities and pK_a values according to Eq. (2.82) at a pH of exactly 4.33 and lower. This is in accord with what can be seen in the experiment (88). For a pH only 0.04 units higher the resolution was entirely lost for two separands (2C and 3A), as predicted.

Figure 2.9. Resolution as a function of the pH of the buffer calculated by Eq. (2.82) for di- and tribasic benzoic acids. As a result of the calculation, the baseline resolution is predicted for pH values of 4.33 and lower; this was confirmed by experiment (right electropherogram at pH 4.33). Resolution between solutes 3A and 2C is lost at a pH only 0.04 units higher (left electropherogram at pH 4.37). Acetate buffer, 0.01 mol/L. Coated capillary, 75 μm i.d., 0.269 (0.202) m length. Field strength, $18.600 \, \mathrm{V\,m^{-1}}$. Detection at 214 nm. (From ref. 88 with permission.)

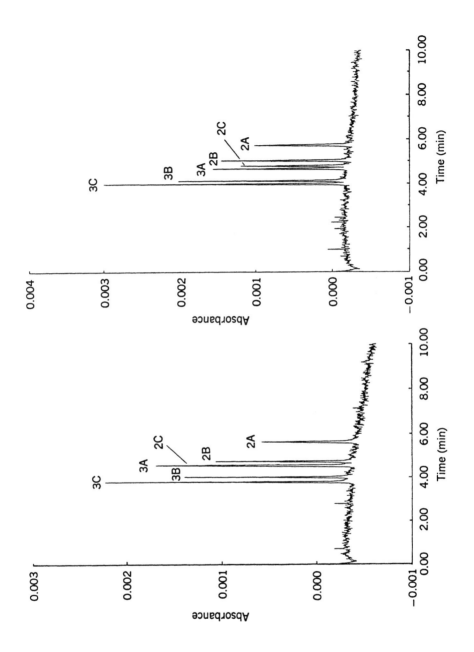

2.2. CAPILLARY ZONE ELECTROPHORESIS IN THE PRESENCE OF ELECTROOSMOTIC FLOW

EOF, an important phenomenon in CZE, is caused by the fact that at the interface between a solid and a liquid an electric double layer is formed. The structure of the double layer is described by the Stern–Gouy–Chapman model, which will not be discussed here. Readers are referred to textbooks on this topic (91–94).

In fused silica capillary material, which is often used in CZE, the solid surface is normally negatively charged due to protolysis of the silanol groups and due to specific adsorption of ions. The negative charge of the surface is counterbalanced by positive ions in the diffuse layer of the liquid. Consequently, the electroneutrality condition, which prevails in the bulk electrolyte solution, does not obtain in the diffuse layer.

When a longitudinal electric field of strength E is applied, a volume force $E \cdot \rho$ acts upon the diffusion layer, which exhibits the electric charge density ρ. The diffusion layer therefore moves, and that motion is propagated through the bulk liquid due to viscous forces. In practice, where narrow capillaries with up to 100 μm i.d. are used, a constant velocity is attained within milliseconds after application of the voltage and constant electroosmotic, v_{eo}, of the bulk liquid is reached. This velocity is proportional to the electric field strength applied. For the case under consideration it is directed toward the cathode.

The electric potential at the shear plane of the double layer is called the *zeta* (ζ) *potential*. Obviously for fused silica material it is dependent on the pH of the buffer, because the silanol groups at the solid surface are only weakly acidic. At high pH, where the silanol groups are fully dissociated, the ζ potential reaches about -120 mV. As the electroosmotic velocity is proportional to the ζ potential, it is also pH dependent. It is given by the Helmholtz–Smoluchowski equation:

$$v_{eo} = -\frac{\varepsilon \zeta E}{\eta} \qquad (2.84)$$

where ε is the permittivity and η is the dynamic viscosity of the liquid. Strictly speaking, ε and η in Eq. (2.84) are those in the double layer. The values of these parameters may deviate significantly from those in the bulk liquid due to the orientation of dipolar molecules of the solvent near the charged surface.

From Eq. (2.84) it can be seen that the electroosmotic velocity is not a function of the capillary radius. Outside the diffuse region (whose thickness depends on the ionic strength of the solution) v_{eo} has a constant value throughout the liquid. Furthermore, it has the same value for capillaries with different diameters, given that they have the same ζ potential.

2.2.1. Migration

2.2.1.1. Electrically Neutral Compounds

The electroosmotic velocity vector acts on each particle in the separation capillary. The electrically neutral components move toward the cathode at exactly this velocity under the experimental conditions obtaining in fused silica capillaries. In this way the EOF is used to transport such neutral components in micellar electrokinetic chromatography (MEKC). To determine the electroosmotic mobility in CZE, neutral marker substances are injected. However, in many cases baseline irregularities or system peaks are observed by using a UV absorbance detector most, a UV active marker, and a buffer, which is also UV absorbing. Such irregularities are often ignored, as their occurrence is attributed to inappropriate experimental or instrumental conditions. They may nevertheless have other well-defined reasons: they may result from pH or mobility gradients that are characteristic of isotachophoretic conditions (95–99); or they may stem from the geometry of the capillary end (100,101), or be caused by thermal fluctuations (42,102), or occur as the result of concentration boundaries between the sample and the background electrolyte (103). The following discussion will concentrate on the last case.

In CZE the neutral marker is often not inserted into a solution with the same ionic composition as the BGE but is dissolved either in water or in the buffer, which results in the marker sample solution having a lower ionic concentration than the BGE. When the sample zone is injected into the capillary, concentration boundaries are established between the lower-concentrated electrolyte solution containing the neutral marker, on the one hand, and the BGE, on the other.

When an electric field is applied, these zones and boundaries migrate by electroosmosis and electrophoresis. The neutral marker zone moves at exactly the electroosmotic velocity. The initial sample zone containing the ions at lower concentration also migrates at the velocity of the EOF, but the concentration boundaries between the ionic sample zone and the BGE may undergo an additional migration. This may be the case when there is a difference in the concentration dependence of the transference numbers of the ions involved. The transference number, τ^+, is defined for the cation as

$$\tau^+ = \frac{\mu^+}{\mu^+ + \mu^-} \qquad (2.85)$$

and is that fraction of current which the particular ionic species transports.

Even in the absence of electroosmosis the concentration boundary in such a system may migrate after application of an electrical field. The velocity of the concentration boundary in the absence of an electroosmotic flow is given by (104,105)

$$v = \frac{J}{F}\frac{\tau' - \tau''}{c' - c''} \tag{2.86}$$

where τ' and τ'' are the transference numbers of the ion in the two solutions of different ionic concentrations, c' and c'', respectively, and J is the current density.

The migration of the concentration boundaries is superimposed on the electroosmotic migration of the bulk phase. What results is a different net migration velocity of the neutral marker, on the one hand, and the sampled electrolyte zone, on the other, producing a kind of separation or *demixing* of the considered zones. This demixing is recorded by the UV absorbance detector, and a baseline irregularity is observed (103), as shown in Figure 2.10. The deviation of the migration times of the boundaries of the lower-concentrated electrolyte sample (Figure 2.10c) from those of the neutral marker (dissolved in a solution with the exact ionic concentration of the BGE in the example shown and moving with the velocity of the EOF; Figure 2.10b) reaches about 10s this study. In Figure 2.10a the baseline irregularities become visible when the neutral marker is dissolved in lower-concentrated BGE and run electrophoretically as the sample. As just discussed, the demixing of the neutral marker zone and that between the ion concentration boundaries results in a positive and a negative jump in the UV absorbance record. A calculated signal of the same shape is obtained by numerical simulation of the electropherogram based on the aforementioned model of migrating concentration boundaries (103,107).

2.2.1.2. *Charged Solutes*

In analogy with electrophoresis, the electroosmotic mobility can be defined as

$$\mu_{eo} = \frac{v_{eo}}{E} = -\frac{\varepsilon \zeta}{\eta} \tag{2.87}$$

and is the velocity of the bulk liquid at unit field strength. The electroosmotic mobility mostly has a positive sign for untreated fused silica, because the ζ potential is negative there; only at very low pH of the BGE is its sign reversed. Treatment of the capillary wall by cationic modifiers can also reverse its charge.

Again the mobility is considered as a signed quantity here for convenience, which is not the case in the theory of conductance. Operating with signed

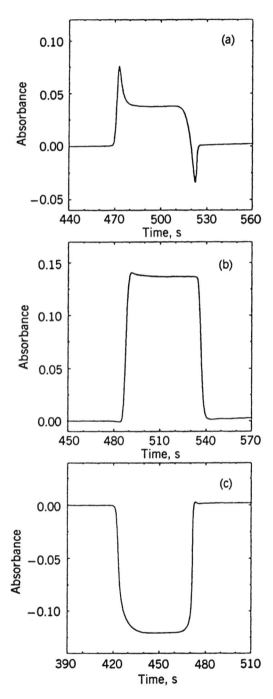

Figure 2.10. Electropherogram with baseline irregularity, with LiI as the BGE and mesityloxide as the neutral marker. Uncoated capillary, 50 μm i.d., 0.560 (0.500) m length. BGE LiI 0.010 mol/L. Pressurized injection for 120 s. UV absorption at 214 nm (cathodic side). Voltage, 10 kV. (a) Sample mesityloxide (0.020 mol/L) dissolved in LiI solution (0.0025 mol/L). (b) Sample mesityloxide (0.020 mol/L) dissolved in LiI solution (0.010 mol/L). (c) Sample LiI (0.0025 mol/L). (From ref. 103 with permission.)

mobilities facilitates our consideration of the mutual effects of electrophoresis and electrosomosis on the anionic and cationic separands.

The EOF is superimposed onto the electrophoretic velocity of ionic separands, resulting in the total (or net or apparent) velocity $v_{tot,i}$, which consists of a specific electrophoretic contribution and an unspecific electroosmotic contribution:

$$v_{tot,i} = v_{ep,i} + v_{eo} \tag{2.88}$$

In bare fused silica capillaries, cationic separands will move faster in the presence of an EOF under most circumstances, in contrast to anions. If the elctroosmotic velocity of the liquid is larger than the electrophoretic velocity of an anion, it will be swept to the cathodic side of the capillary.

The ionic migration velocity of the separands is expressed by

$$v_{tot,i} = (\mu_{eff,i} + \mu_{eo}) E \tag{2.89}$$

in which the sign of the mobilities must be taken into consideration: cations have positive signs and anions negative signs. For obvious reasons in all equations the sign of the voltage and of the field strength must be taken into account as well.

The total migration time is given as in Eq. (2.22) by

$$t_{m,i}^{tot} = \frac{L_t L_d}{(\mu_{eff,i} + \mu_{eo}) V} \tag{2.90}$$

The effect of the EOF on the migration time, compared with the migration time of a system with EOF excluded but otherwise with the same conditions (indicated by a superscript "0") can be expressed by the ratio of the particular migration times [Eqs. (2.22) and (2.90)]:

$$t_{m,i}^{rel} = \frac{t_{m,i}^{tot}}{t_{m,i}^0} = \frac{\mu_{eff,i}}{\mu_{eff,i} + \mu_{eo}} \tag{2.91}$$

It can be seen that the effect of the EOF on the migration time is given by the reduced mobility, μ_i^* (108), or electromigration factor, f_{em} (5), a dimensionless number defined as

$$\mu_i^* = \frac{\mu_{eff,i}}{\mu_{eff,i} + \mu_{eo}} \tag{2.92}$$

The reduced mobility describes in different forms all the variations that the EOF causes on separation parameters. Note that the migration time always has a positive sign, because a negative sign of the reduced mobility is compensated by negative signs of either the voltage or the field strength under appropriate conditions. Inclusion of the sign of the charge number of the separand always produces positive values for the plate heights or plate numbers.

2.2.2. Dispersion and Dimensionless Reduced Mobility (Electromigration Factor)

All contributions to zone broadening that have already been discussed for the case without electroosmosis are affected by the EOF except extracolumn contributions and those caused by the coiling of the capillary. Thus the plate height due to the finite injection zone length as well as the length of the detector cell is the same as described above.

It will be seen that the reduced mobility, defined by Eq. (2.92), plays an essential role in a description of the plate height contributions from the other individual processes: diffusion, thermal dispersion, concentration overload, and wall adsorption. However, before we go into more detail regarding these factors, the potential dispersion due to the radial velocity profile of the EOF is treated.

2.2.2.1. Contribution of the Electroosmotic Flow Profile to Peak Dispersion

The excess net charge in the diffuse part of the double layer, in the case of untreated fused silica consisting of cations, is electrically driven in the longitudinal direction after the application of an electric field in this direction. Due to the viscosity of the solvent this movement is propagated into the bulk solution, which moves at a constant velocity outside the diffuse layer within a short time; then the entire liquid is pumped by the EOF toward the cathode.

The stationary velocity profile of the EOF was derived by Rice and Whitehead (109) for cylindrical geometry using a linear approximation of the exponential terms of the Poisson–Boltzman equation, which was also used by Martin and Guiochon (110,111) to determine the contribution of the EOF profile to peak dispersion. Numerical calculations of the radial velocity profile were undertaken by Andreev and Lisin (112) and by Gas et al. (113). The latter authors calculated the effective thickness of the diffuse layer, β, which is the thickness of a cylindrical sheath of liquid that sticks to the column wall without moving thereby contributing to peak dispersion equivalent to wall adsorption. Under normal conditions this effective thickness has the dimension of the Debye length, well known in electrochemistry. For the case where the effective thickness (which is in the low nm range) is much smaller than the diameter of the capillary, a simplification of the approach of Martin and Guiochon leads to the following expression for peak broadening due to the radial EOF profile:

$$\sigma_{eo,i}^2 = \frac{\beta^2}{D_i} v_{eo}^2 t \qquad (2.93)$$

The corresponding contribution to the plate height can be quantified by

$$H_{eo,i} = \frac{\beta^2}{D_i} v_{eo} \qquad (2.94)$$

Under usual conditions the dimension of the plate height is much smaller than that in the other dispersion processes and can be neglected even in relation to the inevitable diffusion. For some special cases, e.g., for very large separands like viruses that exhibit very small diffusion coefficients, and for small concentrations of the BGE (with resulting large values of the effective thickness, β, in the 10^2 nm range), peak dispersion due to the profile of the EOF may, however, be potentially worthy of notice. An extreme example is a large protein with a diffusion coefficient in the range of $10^{-11}\,m^2s^{-1}$ at a concentration of $0.01\,mmol/L^{-1}$ of the uni-univalent BGE. At a ζ potential of the capillary of $20\,mV$ (and thus an effective thickness, β, of about 95 nm) the plate height reaches the magnitude of the contribution of longitudinal diffusion.

2.2.2.2. Dispersion by Influencing Residence Time

Diffusion, thermal effects, and concentration overload all are causes of dispersion during the migration of the sample zone in the capillary. As the EOF may vary the migration velocity, it has a direct influence on the magnitude of dispersion due to these processes.

The Einstein–Smoluchowski equation for the length-based variance, given in Eq. (2.41), for longitudinal diffusion, its modification for thermal dispersion, and the equation for the base width of the triangle due to concentration overload are valid also for the case with the EOF (5), where we substitute for the initial migration time as shown in Eq. (2.90). This yields the expression for the diffusional plate height contribution:

$$H_{dif,i} = \frac{2RT}{z_i EF} \mu_i^* \qquad (2.95)$$

where μ_i^* is the reduced mobility given in Eq. (2.92). In a similar way the thermal dispersion introduced into the modified Einstein–Smoluchowski equation leads to the plate height contribution from self-heating in the presence of an EOF:

$$H_{\text{ther}} = \frac{f_T^2 \kappa^2 E^5 R_c^6 z_i F}{1536 R T \lambda_s^2} \mu_i^* \qquad (2.96)$$

The appropriate replacement of the migration time gives the plate height contribution due to concentration overload:

$$H_{\text{conc}} = \left| \frac{2 a_i c_i \delta_{\text{inj}}}{9} \mu^* \right| \qquad (2.97)$$

Interestingly all three expressions for the plate height differ only by the reduced mobility or electromigration factor defined by Eq. (2.92) in comparison to the case without the EOF. Its variation with the EOF is directly expressed by μ_i^*, so it is worth considering in more detail the dependence of this reduced mobility on the magnitude of the EOF. The corresponding plate height due to wall adsorption is inversely proportional to μ_i^*. The following discussion will concentrate on the cases of diffusion, self-heating, and concentration overload.

The dependence of μ_i^* on the electroosmotic mobility is shown for cations in Figure 2.11a (under normal conditions, where the EOF is directed toward the cathode). It can be seen that, starting from a value of 1 in the absence of electroosmosis, μ_i^* decreases with increasing EOF. For cations with relatively small mobilities of, e.g., 20×10^{-9} m^2/Vs, this decrease is drastic even with an EOF of medium mobility between 40×10^{-9} and 60×10^{-9} m^2/Vs; here μ_i^* and therefore the values for all three plate heights are reduced by a factor of about 3. This means that the separation efficiency, expressed by the plate number, is enhanced by this factor as well. The result is clear-cut because, due to the larger total mobility of the cations, the residence time of the sample zone in the capillary is shortened and therefore less time is available for zone broadening. From the point of view of efficiency it is preferable to apply EOF for CZE of cations (the opposite is the case when adsorptional peak dispersion predominates).

For anions the situation is more complex, e.g., for an anion with an effective mobility of 20×10^{-9} m^2/Vs (Figure 2.11b). Again the reduced mobility has the value of 1 when no EOF occurs. The values of μ_i^*, and therefore the respective plate heights, then increase with increasing EOF because the residence time of the anion, which is migrating against the EOF to the anodic side of the capillary, increases too, and the dispersion processes last longer as compared with the EOF-free situation. If the electroosmotic mobility approaches the value of the effective mobility of the solute, μ_i^* and the plate height increase drastically and approach assymptotic infinity (due to the

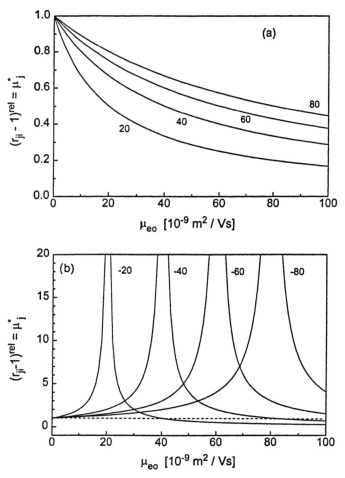

Figure 2.11. Dependence of the reduced mobility, μ_j^*, defined in Eq. (2.92), on the mobility of the EOF, μ_{eo}, for (a) cations and (b) anions with different mobilities between 20×10^{-9} and 80×10^{-9} m^{2}V^{-1}s^{-1}. EOF toward the cathode is assumed. (From ref. 108 with permission.)

infinitely long residence time that would be established when the two mobilities have the same value but opposite signs). A further increase of the electroosmotic mobility, which is then larger than the effective anion mobility, leads to the migration of the anion to the cathodic side of the capillary (obviously after reversing the polarity of the voltage). The anions migrate faster to the cathode as the EOF is increased further, and the migration time decreases accordingly (μ_i^* and therefore the plate heights also become proportionally smaller). Finally, when the electroosmotic mobility has double the value of effective mobility of the anion, it migrates exactly with the same velocity as when there is no EOF, but in the opposite direction. Here the value of μ_i^* is 1 again and the plate heights have the same values as in the case without electroosmosis. A further increase of the EOF leads to a faster migration to the cathode as to the anode in the case without EOF: μ_i^* and thus the plate heights now are smaller than without EOF; the efficiency, given by the plate number, is higher than under any comparable conditions without EOF. Under these circumstances the efficiency can be enhanced by the application of a fast EOF, especially for anions with low mobility.

2.2.2.3. Selectivity, Efficiency, and Resolution

Efficiency as discussed above is only one side of the coin when solutes have to be separated. The other side is the *separation selectivity*, which is also influenced by the EOF, because selectivity is defined as the difference of the mobilities related to the average mobility; it is clear that the difference remains the same with and without the EOF, but the average mobility changes. For cations the mean mobility will always increase with the mobility of the EOF. For cations selectivity will always be reduced when EOF is applied (and efficiency gained; see above). For anions the situation will again be more complicated.

For a simpler approach the electrophoretic resolution is expressed by a less complex but still appropriate definition compared to that given above [Eq. (2.80)]. Again the migration difference will be taken into account; as the time scale whereby this difference is expressed, not both peak widths are used, but only that of one peak (114) according to

$$R_{ji} = \frac{t_{m,j} - t_{m,i}}{\sigma_{t,i}} \qquad (2.98)$$

Substitution of t and σ leads to the following simpler expressions for the resolution, which has the disadvantage that the width of the second peak is not taken into account, but has the advantage that it does not contain the averages of diffusion coefficients, mobilities, or plate numbers:

$$R_{ij} = (r_{ij} - 1)\sqrt{N_i} \qquad (2.99)$$

where r_{ij} is the selectivity coefficient as introduced above.

The influence of the EOF on the different separation parameters can be seen in Table 2.3. All variations of the relative parameters caused by the EOF (which are compared with the corresponding EOF-free systems) can be described by different forms of μ_i^*, namely, the reciprocal, the square root, the square root of the reciprocal, etc. Most pertinent for the case under discussion is the dependence of the efficiency term (the square root of the relative plate

Table 2.3. Expressions of the Change of the Various Separation Parameters Due to the EOF by the Reduced Mobility, μ^*

Parameter Changed by EOF	Definition	Different Forms of μ_i^*
Migration time	$t_m^{\text{rel}} = t_m^{\text{eo}}/t_m^0$	μ_i^*
Plate number	$N^{\text{rel}} = N^{\text{eo}}/N^0$	$1/\mu_i^*$
Efficiency/time	$(N/t_m)^{\text{rel}} = (N^{\text{eo}}/t_m^{\text{eo}})/(N^0/t_m^0)$	$1/\mu_i^{*2}$
Efficiency term	$(\sqrt{N^{\text{rel}}}) = \sqrt{N^{\text{eo}}}/\sqrt{N^0}$	$1/\sqrt{\mu_i^*}$
Selectivity coefficient	$r_{ji}^{\text{rel}} = r_{ji}^{\text{eo}}/r_{ji}^0$	μ_j^*/μ_i^*
Selectivity term	$(r_{ji} - 1)^{\text{rel}} = (r_{ji}^{\text{eo}} - 1)/(r_{ji}^o - 1)$	μ_j^*
Resolution	$R_{ji}^{\text{rel}} = R_{ji}^{\text{eo}}/R_{ji}^o$	$\mu_j^*/\sqrt{\mu_i^*}$
Resolution/time	$(R_{ji}/t_{m,j})^{\text{rel}} = (R_{ji}^{\text{eo}}/t_{m,j}^{\text{eo}})/(R_{ji}^0/t_{m,j}^0)$	$1/\sqrt{\mu_i^*}$

number) and the selectivity term in the equation for the (relative) resolution. The concept of reduced mobility allows us to discuss this quantitatively.

The two terms can be expressed by $1/\sqrt{\mu_i^*}$ and μ_j^*. The resulting (relative) resolution as combination of these two terms is expressed by $\mu_j^*/\sqrt{\mu_i^*}$, which can be approximated for similar effective mobilities by $\sqrt{\mu_j^*}$.

The effect of the EOF on the resolution can be seen from Figure 2.12a. For cations, where the value of the reduced mobility is always smaller than unity when EOF occurs, the resolution always decreases despite the fact that the efficiency increases. The loss in selectivity overcompensates for the gain in efficiency. Application of an EOF (toward the cathode) always leads to a decline in resolution, compared to the case without EOF. The loss of resolution is higher, the lower is the effective mobility of the separands and the higher is that of the EOF. Its exact magnitude can be quantified for any effective mobility of the cation and electroosmotic mobility by the relations containing the reduced mobility given above.

For anions a gain in resolution can be achieved, as can be seen from the example shown in Figure 2.12b of two anions—one with mobility of 40×10^{-9} m^2/Vs; the second differing in mobility by 1, 5, or 10%, respectively. The application of an EOF increases the resolution of the anionic separands migrating to the anode at first slightly, then steeply when the mobility of the EOF approaches that of the separands. In this region every resolution value can be reached, in principle, but enhanced resolution must be paid for by a simultaneous prolongation of the migration time. At electroosmotic mobilities higher than those of both separands, the resolution decreases sharply

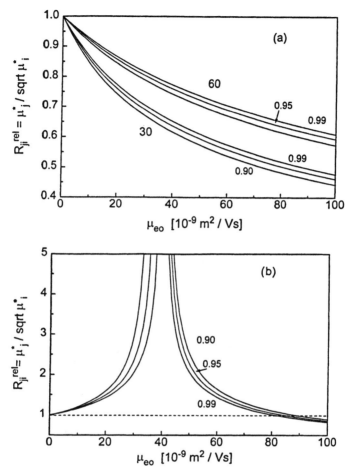

Figure 2.12. Dependence of the relative resolution, R_{ji}^{rel} expressed by the reduced mobility (Table 2.3) as a function of the electroosmotic mobility, μ_{eo}, for (a) cations with effective mobility of 30 or $60 \times 10^{-9}\,m^2V^{-1}s^{-1}$ (mobility of separand j differs by 1, 5, or 10% from that of separand i, (b) anions with effective mobility of $40 \times 10^{-9}\,m^2V^{-1}s^{-1}$ and the same percentage differences in mobility as for case (a) EOF is assumed toward the cathode. (From ref. 108 with permission.)

(when the potential is reversed and the anions migrate toward the cathode). After the EOF reaches twice the value of the effective mobility, the same resolution is established than in the case without EOF; any further increase of the EOF leads to only a slight decrease of the relative resolution below the value of 1. In that range anions with very low mobility can be separated without a drastic loss of resolution but within much shorter separation times compared with the conditions without EOF.

LIST OF SYMBOLS

Symbol	Parameter	Dimension
D	diffusion coefficient	$m^2 s^{-1}$
D_i	diffusion coefficient of compound i	$m^2 s^{-1}$
D_{th}	thermal dispersion coefficient	$m^2 s^{-1}$
$\bar{D}$	mean diffusion coefficient	$m^2 s^{-1}$
E	electric field strength	$V m^{-1}$
F	Faraday constant	$C mol^{-1}$
ΔG_t^0	standard free energy of transfer	$J mol^{-1}$
H	theoretical plate height	m
H_{ads}	plate height due to wall adsorption	m
H_{coil}	plate height due to capillary coiling	m
H_{conc}	plate height due to concentration overload	m
H_{det}	plate height from detector aperture width	m
H_{dif}	plate height from longitudinal diffusion	m
H_{eo}	plate height due to EOF profile	m
H_{inj}	plate height due to injection zone length	m
H_{lam}	plate height due to laminar flow	m
H_{th}	plate height due to thermal selfheating effects	m
H_{tot}	total plate height	m
$\bar{H}$	mean plate height	m
I	ionic strength	$mol L^{-1}$
J	current density	$A m^{-2}$
K	distribution coefficient for wall adsorption	m
K_a	ionization constant	—[a]
L_d	length of capillary to detector	m
L_t	overall capillary length	m
N	plate number	—
N_{dif}	plate number from longitudinal diffusion	—
$N_{meas,i}$	measured plate number for separand i	—
N_i^{eo}	plate number for separand i for case of EOF	—
N_i^{rel}	relative plate number for separand i	—
N_i^0	plate number for separand i in absence of EOF	—
ΔP	pressure difference	$N m^{-2}$
Q	sample amount	mol
R	gas constant	$J mol^{-1} K^{-1}$
R_c	inner capillary radius	m
R_0	outer capillary radius	m
R_{ji}	resolution of separands i and j	—
R_{ji}^{eo}	resolution in presence of EOF	—
R_{ji}^{rel}	relative resolution	—
R_{ji}^0	resolution in absence of EOF	—
T	absolute temperature	K
T_w	temperature at inner capillary wall	K
T_0	temperature at outer capillary wall	K
$T(r)$	temperature as function of the radial coordinate	K

LIST OF SYMBOLS (*Continued*)

Symbol	Parameter	Dimension
$\bar{T}$	mean temperature	K
V	total voltage	V
V_d	(effective) voltage from injector to detector	V
a_i	overload parameter	L mol^{-1}
a_{eq}	equilibrium concentration on solid surface	mol m^{-2}
c	molal concentration	mol kg^{-1}
c_i	molal concentration of compound i	mol kg^{-1}
c_{eq}	equilibrium concentration in solution	mol m^{-3}
f_{em}	electromigration factor	—
f_i	correction factor for actual mobility	—
f_T	temperature factor	K^{-1}
i,j	indices of separands	—
k_d	rate constant of desorption	s^{-1}
k'	capacity factor	—
n_c	number of capillary coils	—
n_l	number of moles in liquid phase	mol
n_s	number of moles adsorbed on solid surface	mol
q	phase ratio	—
r	radial coordinate; radius of solvated ion	m
r_{ij}	selectivity coefficient (ratio of effective mobilities)	—
r_{in}	radius of inner circumference of coiled capillary	m
r_{ij}^{eo}	selectivity coefficient in presence of EOF	—
r_{ij}^{rel}	relative selectivity coefficient	—
r_{ij}^{0}	selectivity coefficient in absence of EOF	—
s_d	different migration distance in coiled capillary	m
t	time	s
t_{inj}	time of injection	s
t_m	migration time	s
$t_{m,i}$	migration time of separand i	s
$t_{m,i}^{eo}$	migration time of separand i presence of EOF	s
$t_{m,i}^{rel}$	relative migration time of separand i	s
$t_{m,i}^{tot}$	total migration time of separand i	s
$t_{m,i}^{0}$	migration time of separand i in absence of EOF	s
v	migration velocity	ms^{-1}
v_m	average velocity in case of wall adsorption	ms^{-1}
v_{eo}	electroosmotic velocity	ms^{-1}
$v_{ep,i}$	electrophoretic velocity of solute i	ms^{-1}
v_i	migration velocity of solute i	ms^{-1}
$v_{tot,i}$	total migration velocity of separand i	ms^{-1}
$\hat{v}$	mean linear velocity at laminar flow	ms^{-1}
$\hat{v}_{ep,i}$	mean electrophoretic velocity of solute i	ms^{-1}
$w_{1/2}$	full peak width at half maximum height	m, s
x	spacial direction	m
z_i	charge number of solute i	—

LIST OF SYMBOLS (*Continued*)

Symbol	Parameter	Dimension
α	degree of dissociation	—
β	thickness of diffuse layer	m
$\delta_{tr,i}$	triangle base width at concentration overload	m
δ_{inj}	length of injection zone	m
ε	permittivity	$F\,m^{-1}$
ϕ	mass ratio	—
γ_i	activity coefficient	—
$_m\gamma_i$	medium or transfer activity coefficient of solute i	—
η	dynamic viscosity	$N\,s\,m^{-2}$
κ	specific electric conductivity	$S\,m^{-1}$
λ_c	thermal conductivity of capillary	$W\,m^{-1}\,K^{-1}$
λ_s	thermal conductivity of solution	$W\,m^{-1}\,K^{-1}$
μ_i	ionic mobility of solute i	$m^2\,V^{-1}\,s^{-1}$
$\mu_{0,i}$	absolute ionic mobility of solute i	$m^2\,V^{-1}\,s^{-1}$
$\mu_{act,i}$	actual mobility of solute i	$m^2\,V^{-1}\,s^{-1}$
$\mu_{eff,i}$	effective mobility of solute i	$m^2\,V^{-1}\,s^{-1}$
μ_{eo}	mobility of electroosmotic flow	$m^2\,V^{-1}\,s^{-1}$
μ_T	mobility at temperature T	$m^2\,V^{-1}\,s^{-1}$
μ_{T_w}	mobility at temperature of inner capillary wall	$m^2\,V^{-1}\,s^{-1}$
μ_i^*	reduced mobility	—
ρ	electric charge density	$C\,m^{-2}$
σ	standard deviation	m,s
σ_t^2	variance in time domain	s^2
σ_x^2	spacial variance	m^2
σ_{ads}^2	variance due to wall adsorption	m^2, s^2
σ_{coil}^2	variance due to capillary coiling	m^2, s^2
σ_{conc}^2	variance due to concentration overload	m^2, s^2
σ_{dif}^2	variance due to longitudinal diffusion	m^2, s^2
σ_{eo}^2	variance due to the profile of the EOF	m^2, s^2
σ_{extr}^2	variance due to extracolumn effects	m^2, s^2
σ_{lam}^2	variance due to laminar flow	m^2, s^2
σ_{th}^2	variance due to Joule selfheating	m^2, s^2
σ_{tot}^2	total variance	m^2, s^2
τ	width of rectangular function	m
τ_{inj}	width of rectangular injection zone	m
τ_{det}	width of rectangular detector aperture	m
τ^+	transference number of cation	—
$\bar{\omega}_i$	chemical potential	$J\,mol^{-1}$
$\bar{\omega}_{0,i}$	chemical potential at standard state	$J\,mol^{-1}$
ζ	zeta potential	V

[a] — dimensionless.

LIST OF ACRONYMS

Acronym	Definition
ACN	acetonitrile
BGE	background electrolyte
CIEF	capillary isoelectric focusing
CITF	capillary isotachophoresis
DMSO	dimethylsulfoxide
EOF	electroosmotic flow
MEKC	micellar electrokinetic chromatography

REFERENCES

1. P. D. Grossman, in *Capillary Electrophoresis, Theory and Practice* (P. D. Grossman and J. C. Colburn, eds.), pp. 1–43. Academic Press, San Diego, CA, 1992.
2. R. A. Wallingford and A. G. Ewing, *Adv. Chromatogr.* **29**, 1 (1989).
3. F. Foret, L. Krivankova, and P. Bocek, in *Capillary Zone Electrophoresis* (B. J. Radola, ed.), VCH, Weinheim, 1993.
4. S. F. Y. Li, *Capillary Electrophoresis. Principles, Practice and Applications.* Elsevier, Amsterdam, 1992.
5. J. C. Reijenga and E. Kenndler, *J. Chromatogr.* **659**, 403 (1994).
6. R. A. Robinson and R. H. Stokes, *Electrolyte Solutions*, 2nd ed. Butterworth, London, 1970.
7. J. O. M. Bockris and A. K. N. Reddy, *Modern Electrochemistry.* Plenum/Rosetta, New York, 1977.
8. T. Erdey-Gruz, *Transport Phenomena in Aqueous Solutions.* Akadémiai Kiadó, Budapest, 1974.
9. W. Friedl, J. C. Reijenga and E. Kenndler, *J. Chromatogr. A* **709**, 163 (1995).
10. A. Tiselius, *Nova Acta Regiae Soc. Sci. Ups.* **7**, 1 (1930).
11. V. Kasicka and Z. Prusik, *J. Chromatogr.* **470**, 209 (1989).
12. A. Cifuentes and H. Poppe, *Electrophoresis* **16**, 516 (1995).
13. M. G. Khaledi, S. C. Smith, and J. K. Strasters, *Anal. Chem.* **63**, 1820 (1991).
14. S. C. Smith and M. G. Khaledi, *Anal. Chem.* **65**, 193 (1993).
15. W. Friedl and E. Kenndler, *Fresenius J. Anal. Chem.* **348**, 576 (1994).
16. T. Hirokawa, N. Nishino, N. Aoki, Y. Kiso, Y. Sawamoto, T. Yagi, and J. I. Akiyama, *J. Chromatogr.* **271**, DI (1983).
17. J. Pospichal, P. Gebauer, and P. Bocek, *Chem. Rev.* **89**, 419 (1989).
18. A. Guttman, A. Paulus, A. Cohen, N. Grinberg, and B. Karger, *J. Chromatogr.* **448**, 41 (1988).
19. Y. Y. Rawjee, R. L. Williams, and Gy. Vigh, *J. Chromatogr. A* **680**, 599 (1994).

20. E. Kenndler, in *Capillary Electrophoresis Technology* (N. Guzman, ed.), pp. 161–186. Dekker, New York, 1993.
21. R. G. Bates, in *Solute–Solvent Interactions*, (J. F. Coetzee and C. D. Ritchie, eds.), pp. 45–96, Dekker, New York, 1969.
22. E. J. King, in *Physical Chemistry of Organic Solvent Systems* (A. K. Covington and T. Dickinson, eds.), pp. 331–404. Plenum Press, London, 1973.
23. I. M. Kolthoff and M. K. Chantooni, in *Treatise on Analytical Chemistry* (I. M. Kolthoff and P. J. Elving, eds.), Part I, Vol. 2, Sect. D, pp. 239–301. Wiley, New York, 1979.
24. A. P. Popov and H. Caruso, in *Treatise on Analytical Chemistry* (I. M. Kolthoff and P. J. Elving, eds.), Part I, Vol. 2, Sect. D, pp. 303–348. Wiley, New York, 1979.
25. E. Kenndler and P. Jenner, *J. Chromatogr.* **390**, 185 (1987).
26. E. Kenndler and P. Jenner, *J. Chromatogr.* **390**, 169 (1987).
27. M. Chiari and E. Kenndler, *J. Chromatogr. A* **716**, 303 (1995).
28. J. W. Jorgenson and K. D. Lukacs, *Anal. Chem.* **53**, 1298 (1981).
29. J. W. Jorgenson and K. D. Lukacs, *J. High Resolut. Chromatogr.* **4**, 202 (1981).
30. J. W. Jorgenson and K. D. Lukacs, *J. Chromatogr.* **218**, 209 (1981).
31. J. Sternberg, *Adv. Chromatogr.* **2**, 205 (1966).
32. G. I. Taylor, *Proc. R. Soc. London, Ser A*, **219**, 186 (1953).
33. T. Srichaiyo and S. Hjertén, *J. Chromatogr.* **604**, 85 (1992).
34. S. Wicar, M. Vilenchik, A. Belenkii, A. S. Cohen, and B. L. Karger, *J. Microcolumn Sep.* **4**, 339 (1992).
35. V. Kasicka, Z. Prusik, B. Gas, and M. Stedry, *Electrophoresis* **16**, 2034 (1995).
36. S. C. Jakobson, R. Hergenröder, L. B. Koutny, R. J. Warmack, and J. M. Ramsey, *Anal. Chem.* **66**, 1107 (1994).
37. R. A. Mosher, D. Dewey, W. Thormann, D. A. Saville, and M. Bier, *Anal. Chem.* **61**, 362 (1989).
38. R. A. Mosher, D. A. Saville, and W. Thormann, *The Dynamics of Electrophoresis.* VCH, Weinheim, 1992.
39. H. Poppe, *J. Chromatogr.* **506**, 45 (1990).
40. H. Poppe, *Anal. Chem.* **64**, 1908 (1992).
41. B. Gas, J. Vacik, and I. Zelensky, *J. Chromatogr.* **545**, 225 (1991).
42. B. Gas, *J. Chromatogr.* **644**, 161 (1993).
43. Ch. Schwer, B. Gas, W. Lottspeich, and E. Kenndler, *Anal. Chem.* **65**, 2108 (1993).
44. E. V. Dose and G. A. Guiochon, *Anal. Chem.* **63**, 1063 (1991).
45. M. S. Bello, M. Y. Zhukov, and P.-G. Righetti, *J. Chromatogr. A* **693**, 113 (1995).
46. G. O. Roberts, P. H. Rhodes, and R. S. Snyder, *J. Chromatogr.* **480**, 35 (1989).
47. J. C. Giddings, *Sep. Sci.* **4**, 181 (1969).
48. J. C. Giddings, in *Treatise on Analytical Chemistry* (I. M. Kolthoff and P. J. Elving, eds.), Part 1, Vol. 5, Chapter 3, p. 63. Wiley, New York, 1981.
49. J. C. Giddings, *J. Chromatogr.* **480**, 21 (1989).

50. J. W. Jorgenson and K. D. Lucas, *Science* **222**, 266 (1983).
51. S. Hjertén, *Electrophoresis* **11**, 665 (1990).
52. J. H. Knox, *Chromatographia* **26**, 329 (1988).
53. E. Kenndler and Ch. Schwer, *Anal. Chem.* **63**, 2499 (1991).
54. Ch. Schwer and E. Kenndler, *Chromatographia* **33**, 135 (1992).
55. J. H. Knox and I. H. Grant, *Chromatographia* **24**, 135 (1987).
56. S. Hjertén, *Chromatogr. Rev.* **9**, 122 (1967).
57. J. H. Knox and K. A. McCormack, *Chromatographia* **38**, 207 (1994).
58. J. H. Knox and K. A. McCormack, *Chromatographia* **38**, 215 (1994).
59. F. Foret, M. Deml, and P. Bocek, *J. Chromatogr.* **452**, 601 (1988).
60. E. Grushka, R. M. McCormack, and J. J. Kirkland, *Anal. Chem.* **61**, 241 (1989).
61. A. E. Jones, and E. Grushka, *J. Chromatogr.* **466**, 219 (1989).
62. R. J. Nelson, A. Paulus, A. S. Cohen, A. Guttman, and B. L. Karger, *J. Chromatogr.* **480**, 111 (1989).
63. G. J. M. Bruin, P. P. H. Tock, J. H. Kraak, and H. Poppe, *J. Chromatogr.* **517**, 557 (1990).
64. W. A. Gobie and C. F. Ivory, *J. Chromatogr.* **516**, 191 (1990).
65. A. Vinther and H. Soberg, *J. Chromatogr.* **559**, 27 (1991).
66. D. S. Burgi, K. Salomon, and R. L. Chien, *J. Liq. Chromatogr.* **14**, 847 (1991).
67. M. S. Bello and P.-G. Righetti, *J. Chromatogr.* **606**, 95 (1992).
68. M. S. Bello, M. Chiari, M. Nesi, and P.-G. Righetti, *J. Chromatogr.* **625**, 323 (1992).
69. J. P. Landers, R. P. Oda, B. Madden, T. P. Sismelich, and T. C. Spelsberg, *J. High Resolut. Chromatogr.* **15**, 517 (1992).
70. E. V. Dose and G. Guiochon, *J. Chromatogr.* **652**, 263 (1993).
71. M. S. Bello, E. I. Levin, and P.-G. Righetti, *J. Chromatogr.* **652**, 329 (1993).
72. A. Cifuentes, W. T. Kok, and H. Poppe, *J. Microcolumn Sep.* **7**, 365 (1995).
73. R. Virtanen, *Acta Polytech. Scand.* **123**, 1 (1967).
74. W. Thormann, *Electrophoresis* **4**, 383 (1983).
75. F. E. P. Mikkers, F. M. Everaerts, and Th. P. E. M. Verheggen, *J. Chromatogr.* **169**, 1 (1979).
76. B. Gas, M. Stedry, A. Rizzi, and E. Kenndler, *Electrophoresis* **16**, 958 (1995).
77. M. Stedry, B. Gas, and E. Kenndler, *Electrophoresis* **16**, 2027 (1995).
78. S. V. Ermakov, M. Y. Zhukov, L. Capelli, and P.-G. Righetti, *J. Chromatogr.* **699**, 297 (1995).
79. I. Paganobarraga, J. Bafaluy, and J. M. Ruby, *Phys. Rev. Lett.* **75**, 461 (1995).
80. J. C. Giddings, *Dynamics of Chromatography*. Dekker, New York, 1965.
81. J. C. Stegman, H. Poppe, and J. C. Kraak, *J. Chromatogr.* **634**, 149 (1993).
82. R. J. Wieme, in *Chromatography, A Laboratory Handbook of Chromatographic and Electrophoretic Methods* (E. Heftman, ed.), pp. 228–281. Van Nostrand-Reinhold, New York, 1975.

83. M. Minarik, B. Gas, and E. Kenndler, *J. Capillary Electrophor.* **2**, 89 (1995).
84. W. G. Kuhr and E. S. Yeung, *Anal. Chem.* **60**, 2642 (1988).
85. S. Wren, *J. Microcolumn Sep.* **3**, 147 (1991).
86. S. Terabe, T. Yashima, Y. Tanaka, and M. Araki, *Anal. Chem.* **60**, 1673 (1988).
87. E. Kenndler and W. Friedl, *J. Chromatogr.* **608**, 161 (1993).
88. W. Friedl and E. Kenndler, *Anal. Chem.* **65**, 2003 (1993).
89. B. L. Karger, L. R. Snyder, and C. Horvath, eds., *An Introduction to Separation Science.* Wiley, New York, 1973.
90. S. L. Delinger and J. M. Davis, *Anal. Chem.* **64**, 1947 (1992).
91. R. J. Hunter, *Zeta Potential in Colloid Science.* Academic Press, London, 1981.
92. P. Delahay, *Double Layer and Electrode Kinetics.* Interscience, New York, 1965.
93. E. J. W. Verwey and J. Th. G. Overbeek, *Theory of the Stability of Lyophobic Colloids.* Elsevier, New York, 1948.
94. O. F. Devereux and P. L. De Bruyn, *Interaction of Plane Parallel Double Layers.* MIT Press, Cambridge, MA, 1963.
95. J. L. Beckers and M. T. Ackermans, *J. Chromatogr.* **629**, 371 (1993).
96. J. L. Beckers, *J. Chromatogr. A* **662**, 153 (1994).
97. J. L. Beckers, *J. Chromatogr. A* **693**, 347 (1995).
98. J. L. Beckers, *J. Chromatogr. A* **696**, 285 (1995).
99. F. A. Vinther, F. M. Everaerts, and H. Soberg, *J. High Resolut. Chromatogr.* **13**, 639 (1990).
100. C. L. Colyer and K. B. Oldham, *J. Chromatogr. A* **716**, 3 (1995).
101. C. L. Colyer and K. B. Oldham, *Anal. Chem.* **67**, 3234 (1995).
102. M. S. Bello, P. de Besi, and P.-G. Righetti, *J. Chromatogr.* **652**, 317 (1993).
103. K. Kenndler-Blachkolm, S. Popelka, B. Gas, and E. Kenndler, *J. Chromatogr. A* **734**, 351 (1996).
104. L. G. Longsworth, *J. Am. Chem. Soc.* **65**, 1755 (1943).
105. W. H. Stockmayer, *Trans. N. Y. Acad. Sci.* **13**, 226 (1951).
106. J. L. Beckers, F. M. Everaerts, and M. T. Ackermans, *J. Chromatogr.* **537**, 407 (1991).
107. M. Stedry, S. Popelka, B. Gas, and E. Kenndler, *Electrophoresis* **17**, 1121 (1996).
108. E. Kenndler, *J. Capillary Electrophor.* **4**, 191 (1996).
109. C. L. Rice and R. Whitehead, *J. Phys. Chem.* **69**, 4017 (1965).
110. M. Martin and G. Guiochon, *Anal. Chem.* **56**, 614 (1984).
111. M. Martin, G. Guiochon, Y. Walbroehl, and J. W. Jorgenson, *Anal. Chem.* **57**, 559 (1985).
112. V. P. Andreev and E. E. Lisin, *Electrophoresis* **13**, 832 (1992).
113. B. Gas, M. Stedry, and E. Kenndler, *J. Chromatogr. A* **709**, 63 (1995).
114. J. F. K. Huber, *Fresenius Z. Anal. Chem.* **277**, 341 (1975).

CHAPTER 3

MICELLAR ELECTROKINETIC CHROMATOGRAPHY

MORTEZA G. KHALEDI

Department of Chemistry, North Carolina State University, Raleigh, North Carolina 27695-8204

3.1.	Introduction	77
3.2.	Pseudostationary Phases	79
	3.2.1. Micelles	80
	3.2.2. Polymeric Phases	83
3.3.	Migration in Micellar Electrokinetic Chromatography	83
3.4.	Migration Parameters	84
	3.4.1. Uncharged Solutes	84
	3.4.2. Charged Solutes	85
3.5.	Resolution	85
3.6.	Structure–Retention Relationships in Micellar Electrokinetic Chromatography	87
3.7.	Characterization of the Chemical Selectivity of Pseudostationary Phases	89
3.8.	Effects of Chemical Composition of Micellar Solutions	95
	3.8.1. Surfactant Concentration	96
	3.8.2. Types of Pseudostationary Phase	97
	3.8.3. Modifiers	113
3.9.	Multiparameter Optimization	117
3.10.	Conclusions and Future Trends	127
List of Acronyms		130
References		131

3.1. INTRODUCTION

Micellar electokinetic chromatography (MEKC) is a mode of capillary electrophoresis (CE) that is capable of separating uncharged compounds. The

---

High Performance Capillary Electrophoresis, edited by Morteza G. Khaledi. Chemical Analysis Series, Vol. 146.
ISBN 0-471-14851-2 © 1998 John Wiley & Sons, Inc.

technique has also been referred to in the literature as micellar electrokinetic capillary chromatography (MECC). Since its introduction by Terabe and coworkers (1,2) over 10 years ago. MEKC has become widely popular as hundreds of studies and applications have been reported. MEKC uses the same instrumental setup as CE; however, charged organized media such as micelles are incorporated that act as the separation medium for uncharged

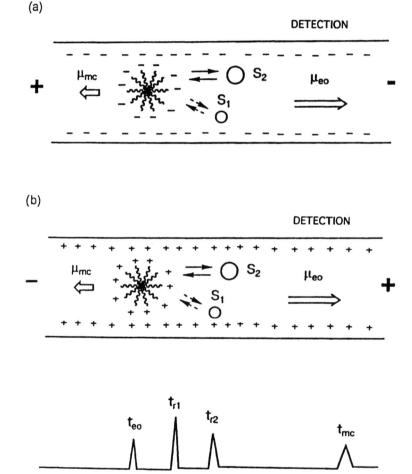

Figure 3.1. Migration of uncharged compounds in MEKC using (a) anionic and (b) cationic pseudostationary phases. Separation of solutes S_1 and S_2 is achieved due to their differential partitioning into the pseudostationary phase. The uncharged solutes are eluted within an elution window (t_{mc}/t_{eo}). From M. G. Khaledi, *J. Chromotogr.* A (1997) in press, with permission.

solutes. Charged micelles migrate in the electric field at an electrophoretic velocity that is proportional to their charge-to-size ratio. Uncharged solutes with different micelle–water partition coefficients, P_{mw}, can then be separated. Since the separation mechanism is based on differential partitioning, MEKC is viewed as a chromatographic technique with migrating charged micelles (or other types of organized media) acting as pseudostationary phases. Consequently, there exists a limited elution window in MEKC. All uncharged solutes are to be separated between the elution time of an unretained solute, t_{eo}, and the migration time of micelles, t_{mc} (Figure 3.1). MEKC can be viewed as a hybrid of reversed-phase liquid chromatography (RPLC) and capillary zone electrophoresis (CZE), as the separation process incorporates hydrophobic and polar interactions, a partitioning mechanism, and electromigration. However, MEKC offers a combination of unique features of CZE and RPLC such as high efficiencies, rapid analysis, small sample size, small solvent consumption, and versatility of incorporating chemical selectivity in the separation process. MEKC is the only CE technique that is capable of separating mixtures of charged and uncharged molecules. It offers higher efficiency than RPLC and capillary electrochromatography (CEC). The number of theoretical plates in a MEKC separation can be 10 times or higher that in RPLC.

Figure 3.2 represents a comparison between typical MEKC and high-performance liquid chromatography (HPLC) separations that demonstrates the superior speed and efficiency of MEKC over RPLC. Another major advantage of MEKC over conventional chromatographic techniques as well as CEC is the flexibility and ease of changing the chemical composition of the pseudostationary phases. For example, the type and/or composition of the micellar solution can be easily modified or replaced by simply rinsing the capillary with a new type of pseudostationary phase. The equilibrium times are often very rapid. This inherent flexibility of controlling key parameters leads to enhanced separations and greatly facilitates the process of method development.

This chapter provides an overview of MEKC with an emphasis on recent developments (since 1993). The literature prior to that year was previously reviewed elsewhere (4). Special attention has been given here to the characterization of selectivity and selection of pseudostationary phases on a rational basis in MEKC.

3.2. PSEUDOSTATIONARY PHASES

Various types of pseudostationary phases have been used in MEKC over the past 10 years. They can be categorized into two general groups: the first and

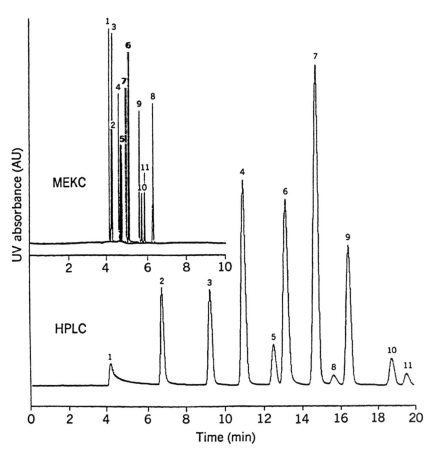

Figure 3.2. Comparison of typical MEKC and RPLC separations. MEKC conditions: applied voltage = +20 kV; λ = 230 nm; total length = 57 cm; phosphate/borate buffer containing 25 mM SDS. HPLC conditions: column = Supelcosil LC-18 column (25 cm × 4.6 mm i.d.); flow rate = 1.0 mL/min; solvent system = water/methanol with 20 min gradient from 40 to 60% methanol. Compounds are (1) 1,3,5,7-tetranitro-1,3,5,7-tetraazacyclooctane; (2) 1,3,5-trinitro-1,3,5-triazacyclohexane; (3) 2,4,6-trinitrobenzene; (4) 2,4-dinitrobenzene; (5) nitrobenzene; (6) 2,4,6-N-tetranitro-N-methylaniline; (7) 2,4,6,-trinitrotoluene; (8) amino-2,6-dinitrotoluene; (9) 2,6-dinitrobenzene; (10) 2-nitrotoluene; (11) 4-nitrotoluene. (From ref. 3 with permission).

most widely used are charged micelles (i.e., dynamic aggregates of charged surfactants); the other group consists of covalently bonded or polymerized charged organized assemblies.

3.2.1. Micelles

Surfactants are amphiphilic molecules that comprise a hydrophobic moiety and a polar or ionic head group. They can be recognized by the charge of the head group (as nonionic, anionic, cationic, and zwitterionic surfactants) or by the variations in the nature of hydrophobic moiety (as hydrocarbon, bile salts, and fluorocarbon surfactants). A list of typical surfactants is given in Table 3.1. Above a critical micelle concentration (CMC), surfactants begin to form aggregates that are in dynamic equilibrium with the monomers in the bulk

Table 3.1. Long-Chain Surfactants and Bile Salts

Type	Name	CMC (mM)
Long-chain surfactants:		
Anionic	Sodium dodecyl sulfate (SDS)	8.1[a]
	Sodium tetradecyl sulfate	2.1[b]
	Sodium dodecyl sulfonate	9.3[b]
	Lithium perfluorooctane sulfonate (LiPFOS)	6.72[c]
Cationic	Cetyltrimethylammonium bromide (CTAB)	0.92[d]
	Cetyltrimethylammonium chloride	1.3[d]
	Dodecyltrimethylammonium bromide (DTAB)	15[a]
Nonionic	Polyoxyethylene (23) dodecanol (Brij 35)	0.1[a]
	Polyoxyethylene (20) sorbitane monooleate (Tween 80)	0.01[d]
	Polyoxyethylene (20) sorbitane monolaurate (Tween 20)	0.059[d]
Zwitterionic	N-dodecyl-N,N-dimethylammonio-3-propanesulfonate (Sulfobetain, SB-12)	3.3[d]
	3-[(3-Cholamidopropyl)dimethylammonio]-1-propane-sulfonate (CHAPS)	4.2–6.3[d]

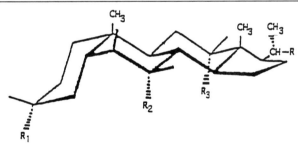

Type	R_1	R_2	R_3	R	CMC
Bile sodium salts:					
Cholate	OH	OH	OH	$-CH_2CH_2COO^-$	12.5[a]
Deoxycholate	OH	H	OH	$-CH_2CH_2COO^-$	6.4[a]
Taurocholate	OH	OH	OH	$-CH_2CH_2CONHCH_2SO_3^-$	—
Glycodeoxycholate	OH	H	OH	$-CH_2CH_2CONHCH_2COO^-$	—
Taurodeoxycholate	OH	H	OH	$-CH_2CH_2CONHCH_2CH_2SO_3^-$	—

Source: M. G. Khaledi, *J. Chromatogr.* (1997) in press, with permission.
[a] Ref. 5.
[b] Ref. 7.
[c] Ref. 6.
[d] Ref. 8.

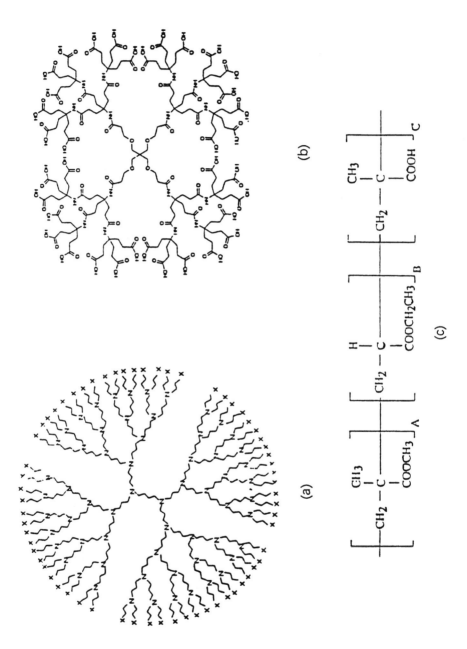

aqueous solution. The number of monomer surfactants in the aggregate form (called the aggregation number) and size of micelles vary greatly between surfactants. For example, surfactants with alkyl chains form roughly spherical micelles with a diameter between 3 and 6 nm and aggregation number of between 30 and 100 (5–8). On the other hand, micelles of bile salts have much smaller aggregation numbers (typically 2–10 for primary micelles), presumably with a helical structure. Bile salts are biological surfactants that have a hydrophobic steroidal backbone with substituted hydrophilic groups (such as hydroxyl or carbonyl) (9–11). For the MEKC separation of uncharged solutes, the pseudostationary phase must be charged; thus nonionic and zwitterionic surfactants can be used only in mixed micelles with a charged surfactant.

3.2.2. Polymeric Phases

This group includes pseudostationary phases that have widely different structural properties but share one common feature: they are covalently bonded organized assemblies. Examples of three types that have been used in MEKC are polymerized micelles (where monomer surfactants are covalently bonded together through a polymerization process) (12,13), cascade macromolecules (dendrimers) (14–17), and ionic polymers (18–20) (see Figure 3.3). These charged organized media provide sites of interactions for solutes and can effectively play the role of pseudostationary phases in MEKC. They can be used in organic-rich solvents, and their primary application has been for the separation of highly hydrophobic compounds in MEKC, where addition of organic cosolvent is often necessary (*vide infra*). Structures of conventional micelles are disrupted at high concentrations of organic solvents; thus their capabilities for separating hydrophobic molecules are limited.

3.3. MIGRATION IN MICELLAR ELECTROKINETIC CHROMATOGRAPHY

Figure 3.1 illustrates the typical migration scheme for uncharged compounds in MEKC using an anionic surfactant and a cationic surfactant in an uncoated fused silica capillary.

Anionic micelles migrate in the opposite direction of the electroosmotic flow (EOF) in an uncoated capillary (Figure 3.1). Typically, the EOF velocity

---

Figure 3.3. Structures of typical polymeric pseudostationary phases in MEKC: (a) diaminobutane-based poly(propyleneamine)dendrimer; (b) general structure of amide-based dendrimers; (c) Elvacite 2669 [poly(methyl methacrylate–ethyl acrylate–methacrylic acid)]. (Part a from ref. 15; part b, from ref. 20; part c, from ref. 17— with permission.)

is stronger than the electrophoretic velocity of anionic micelles under "typical" conditions (e.g., an uncoated capillary and pH greater than 6). As a result, the anionic micelles are carried toward the cathode. When cationic micelles are used, the capillary wall is coated with the positively charged surfactants which oftentimes leads to a reversal in the direction of the EOF. It is therefore necessary to reverse the polarity of the electrodes in the CE setup to ensure the elution of the cationic micelles and consequently the uncharged solutes through the detection window (Figure 3.1).

There exists two extremes that define an elution window in MEKC. Analytes that do not interact with micelles ($P_{mw} \sim 0$) spend all of their migration times in the bulk aqueous phase and migrate at the electroosmotic mobility. These are typically uncharged polar molecules like methanol or acetonitrile that are EOF markers and elute at t_{eo}. The other end is defined by the elution of analytes that interact so strongly with the micelles ($P_{mw} \sim \infty$) that they spend all of their migration time with micelles. The t_{mc} markers are typically very hydrophobic compounds that are sparingly soluble in the aqueous media—reported examples being Sudan III and dodecanophenone. The elution times for these analytes coincide with the micellar migration time, t_{mc}. The existence of an elution window limits peak capacity in MEKC, as all uncharged solutes are to be separated between the migration time of an unretained solute, t_{eo}, and a fully retained solute, t_{mc}. The size of the elution window can be enhanced by using organic modifiers or mixed micelles, or by modifying the capillary walls (*vide infra*).

3.4. MIGRATION PARAMETERS

3.4.1. Uncharged Solutes

As in chromatography, the retention factor in MEKC is defined as the ratio of the number of moles of solute in the micellar pseudostationary phase, n_{mc}, and that in the bulk aqueous phase, n_{aq}. The retention factor is directly proportional to the micelle–water partition coefficient, P_{mw}, and the phase ratio, Φ, as

$$k' = \frac{n_{mc}}{n_{aq}} = P_{mw}\Phi \tag{3.1}$$

The retention factor in MEKC can be determined from migration time data using Eq. (3.2)

$$k' = \frac{t_r - t_{eo}}{t_{eo}\left[1 - \left(\frac{t_r}{t_{mc}}\right)\right]} \tag{3.2}$$

This is very similar to the equation for the retention factor in conventional chromatography, with the exception of the additional term $(1 - t_r/t_{mc})$ in the denominator. This term indicates the existence of an elution window, because the "stationary" phase in MEKC is actually mobile. If t_{mc} approaches infinity (i.e., stationary micelles), the extra term in the denominator is omitted and the retention factor equation becomes the same as that in conventional chromatography.

Other retention parameters such as mobility (4), retention indices (21,22), migration indices (23), and modified capacity factors (24) for MEKC have recently been introduced. Retention indices in MEKC are based on the same concept as that for gas chromatography (GC), introduced by Kovats in 1958. These parameters provide methods of normalizing retention in MEKC. They should be independent of surfactant concentration and are less sensitive to variations in t_{eo} and t_{mc} values than is the retention factor, k'.

3.4.2. Charged Solutes

In addition to partitioning into micelles and migrating at the micellar mobility, charged compounds possess electrophoretic mobilities of their own in the bulk aqueous solvent (25,26). As a result, the observed retention time also includes the time that solute migrated electrophoretically in the bulk aqueous phase, t_o. In calculating the retention factor, this electromigration time should be taken into account:

$$k' = \frac{t_r - t_0}{t_0[1 - (t_r/t_{mc})]} \qquad (3.3)$$

3.5. RESOLUTION

The fundamental resolution equation for uncharged solutes in MEKC, Eq. (3.4), has the same format as that for conventional chromatography, as it indicates that resolution depends on three terms related to efficiency, selectivity, and retention (2); the fourth term is unique to MEKC, representing the existence of an elution window:

$$R = \left(\frac{N^{1/2}}{4}\right)\left(\frac{\alpha - 1}{\alpha}\right)\left(\frac{k'_2}{1 + k'_2}\right)\left[\frac{1 - (t_{eo}/t_{mc})}{1 + (t_{eo}/t_{mc})k'_1}\right] \qquad (3.4)$$

Again, if micelles were truly stationary (i.e., if $t_{mc} \sim \infty$), the fourth term would drop out and the equation would be identical to that in conventional chromatography. The size of the elution window has a significant impact on MEKC separations. The difference between conventional liquid chromatography (LC) and MEKC as well as the influence of the elution window is illustrated in Figure 3.4, where the product of the last two terms in Eq. (3.4) [retention function, $f(k)$] is plotted against k' at various t_{eo}/t_{mc} ratios (i.e., inverse of the size of the elution window, defined as t_{mc}/t_{eo}). The $t_{eo}/t_{mc} = 0$ curve shows the well-known behavior in LC where resolution first increases rapidly before reaching a plateau. The curve's behavior is strikingly different in

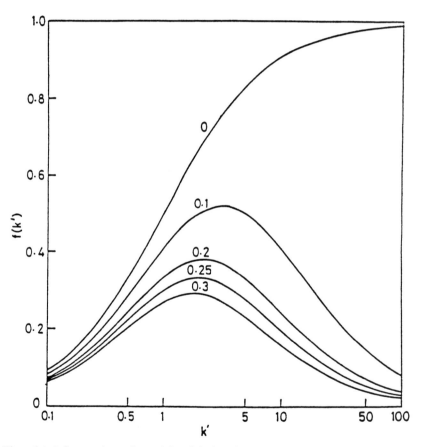

Figure 3.4. Influence of retention and size of elution window on resolution in MEKC. Resolution is directly related to $f(k')$, a product of the last two terms in the resolution equation (3.4). (From ref. 2 with permission.)

MEKC, where resolution reaches a maximum over an optimum range of k'. The price of excessive retention in MEKC is a loss of resolution. This behavior is different from that in LC, where the cost of large retention factors is prolonged analysis times without much gain in resolution. It is therefore crucial to operate within or close to the optimum k' range in MEKC separations. Figure 3.4 also shows that better resolution is achieved as the size of the elution window is increased (or t_{eo}/t_{mc} is reduced).

Likewise, peak capacity, n, can be increased with wider elution windows:

$$n = 1 + \frac{\sqrt{N}}{4}\ln\frac{t_{mc}}{t_{eo}} \qquad (3.5)$$

The optimum k' for achieving maximum resolution is approximately related to the square root of the size of the elution window (27):

$$k'_{opt} = \sqrt{\frac{t_{mc}}{t_{eo}}} \qquad (3.6)$$

As in chromatography, better resolution is achieved at higher efficiency. This disadvantage of the limited elution window is mostly compensated by the large number of theoretical plates that are routinely achieved in MEKC. Unfortunately, the exact sources for band broadening in MEKC have yet to be determined. A detailed discussion of the band dispersion in electrokinetic chromatography (EKC) is given in Chapter 4. In the following sections, the main emphasis is placed on the influence of the composition of micellar solution on retention and selectivity in MEKC.

3.6. STRUCTURE–RETENTION RELATIONSHIPS IN MICELLAR ELECTROKINETIC CHROMATOGRAPHY

Since the introduction of the technique, it has been widely believed that retention in MEKC is due to the hydrophobicity of the solute. There is no doubt that hydrophobic interaction plays a major role in solute–micelle interactions and consequently retention in MEKC. However, the large variations in migration patterns that are observed for different types of pseudo-stationary phases indicate the existence of other types of interactions. Yang and Khaledi (28,29) reported the use of Kamlet and Taft's linear solvation energy relationships (LSER) (30–34) to quantitatively describe retention in MEKC in terms of the following structural properties of the solute:

$$\log k' = SP_0 + mV_1/100 + s\pi^* + b\beta + a\alpha \qquad (3.7)$$

Table 3.2. LSER Modeling of Retention in MEKC with Different Pseudostationary Phases [Eq. (3.7)][a]

MEKC System[b]	SP_0	m	s	b	a
0.04 M LiPFOS	−1.51	2.44	−0.25	0.16[d]	−0.98
2% Elvacite 2669[c]	−1.55	3.00	0.09[a]	−2.33	0.09
0.06 M SC	−1.62	3.89	−0.27	−2.88	0.23
0.08 M SC	−1.38	3.82	−0.32	−2.85	0.18
0.01 M C_{14}TAB	−1.78	3.96	−0.26	−2.75	0.99
0.02 M SDS	−1.87	4.00	−0.25	−1.79	0.16[d]
0.04 M SDS	−1.49	3.95	−0.26	−1.80	−0.18[d]
$\log P_{ow}$	0.17	5.62	−0.66	−3.90	0.14

Source: Data from refs. 29 and 35.
[a] In all cases, $n = 60$ and correlation coefficients for the regression better than 0.95 were observed.
[b] Buffer: 50 mM phosphate, pH = 7.0.
[c] Buffer: 100 mM (3-cyclohexylamino)-1-propanesulfonic acid (CAPS), pH = 10.00.
[d] Values are not significant at the 95% confidence level.

where SP_0 is the regression constant and contains information about the chromatographic phase ratio; V is solute molar volume; π^* is the measure of dipolarity–polarizability of the solute; β is solute hydrogen bond acceptor strength (basicity); and α is solute hydrogen bond donor strength (acidity). The three parameters π^*, β, and α have been derived based on solvatochromic comparison methods that were first developed by Kamlet, Taft, and coworkers.

Using a group of aromatic test solutes, the LSER modeling of retention in MEKC for various micellar pseudostationary phases was investigated. The coefficients m, s, b, and a in Eq. (3.7) represent the properties of the micellar systems, and the values for representative systems are listed in Table 3.2. The data suggest that solute size plays the most important role in retention, as m is the largest coefficient in the LSER models. The positive sign for m indicates that bulkier molecules are retained longer in MEKC (i.e., due to a stronger solute–micelle interaction). The mV/100 term in the LSER model is related to hydrophobic interaction, as it represents an unfavorable energy term for formation of a properly sized cavity in the solvent system for solute accommodation. The next largest coefficient in most systems is b (with the exception of the fluorocarbon micelles of LiPFOS), indicating that type A hydrogen bonding (solute acceptor–solvent donor) is the second most important type of interaction that influences retention in nearly all MEKC systems (Table 3.2). The negative sign for b indicates that stronger hydrogen bond acceptor solutes (i.e., larger β) would have less interaction with the micelles and are retained less. This is reasonable since water is a stronger hydrogen bond donor than are micelles; thus more basic solutes would have a stronger interaction with the bulk aqueous media than with micelles. A combined effect of these two factors mainly determines solute retention in MEKC. For example, as compared to benzene, a substituted aromatic compound such as nitrobenzene is bulkier and is a stronger hydrogen bond acceptor. The larger size of the substituted aromatic compound causes stronger interaction with micelles than that experienced by the parent benzene molecule; however, the increase in the substituted compound's basicity reduces its interaction with the micelle. Subsequently, the relative retention of these two solutes is mainly determined by the net effect of these two opposing factors and can vary greatly among various micellar systems according to their interactive properties (*vide infra*).

3.7. CHARACTERIZATION OF CHEMICAL SELECTIVITY OF PSEUDOSTATIONARY PHASES

A great majority of MEKC separations have been performed using SDS micelles. Over the past few years, there have been more and more investigations of the usefulness of other types of micellar, polymeric, and mixed pseudostationary phases. On the one hand, the availability of a great variety of pseudostationary phases with different selectivities is quite advantageous in method development. On the other hand, the numerous choices tend to make the process of selecting the appropriate type of pseudostationary phase difficult. This problem is especially pronounced for the separation of complex mixtures where operating at optimum conditions is crucial.

Snyder's selectivity triangle in LC and the Rohrschneider-McReynolds scale in GC have greatly facilitated the selection of high-performance liquid chromatography (HPLC) mobile phases and GC stationary phases. Presently, due to a lack of knowledge about the exact chemical nature of solute–micelle interactions, the selection of the chemical composition of pseudostationary phases in MEKC is based on trial and error or the experience of the analyst. As a result, a general understanding of properties of various surfactants would be useful in selecting the optimum conditions for MEKC separations.

The chemical nature of the pseudostationary phase has a tremendous influence on selectivity, the overall elution pattern, and ultimately on resolution in MEKC. Figure 3.5a–f (pp. 90–93) illustrate examples of the large variations in selectivity among four micellar and one polymeric pseudostationary phases, a triblock copolymer known as Elvacite 2669 (Figure 3.3). A comparison of Figures 3.5a and 3.5b shows that micelle concentration has little influence on selectivity of uncharged solutes in MEKC. This can also be

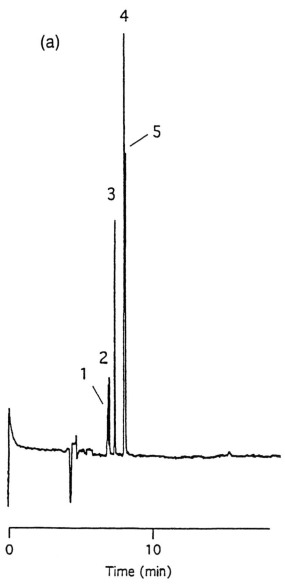

Figure 3.5. Influence of surfactant type on elution pattern and selectivity in MEKC: (a) 40 mM SDS; (b) 60 mM SDS; (c) 40 mM LiPFOS; (d) 100 mM SC; (e) 4% Elvacite 2669; (f) 25 mM C_{14}TAB, Peak identifications are given in Table 3.3. (From ref. 35 with permission.)

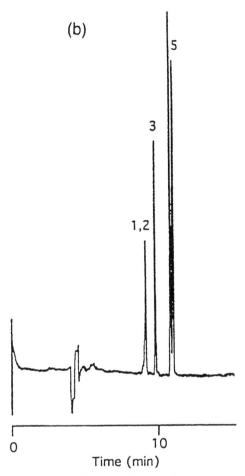

Figure 3.5. (*Continued*)

seen from the data in Table 3.2, where the LSER coefficients (with the exception of the constant) are independent of micelle concentration. The structural properties of the five test compounds in terms of solvatochromic parameters, molecular volume, and a measure of their hydrophobicity, the logarithm of the partition coefficient between *n*-octanol and water ($\log P_{ow}$), are listed in Table 3.3. The chromatograms shown in Figures 3.5a–f clearly indicate two important points. First, the overall separation patterns in MEKC strongly depend on the type of surfactant used. Second, solutes do not necessarily elute according to their hydrophobicity. In fact, in systems such as

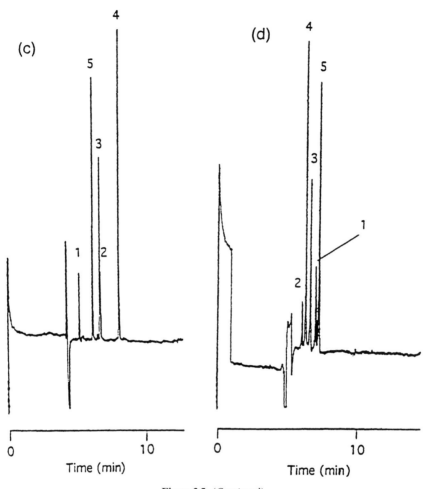

Figure 3.5. (*Continued*)

SDS and LiPFOS, the two most hydrophobic solutes, benzene (peak 1) and anisole (peak 5), are the least retained. Note that surfactant and polymer concentrations were selected to provide equivalent elution times for all five systems; thus there are some overlapping peaks. However, the main purpose was to demonstrate selectivity changes rather than resolving the five test solutes. The LSER methodology can provide valuable information about the nature of the underlying forces that lead to different selective interactions between solute and micelles (35).

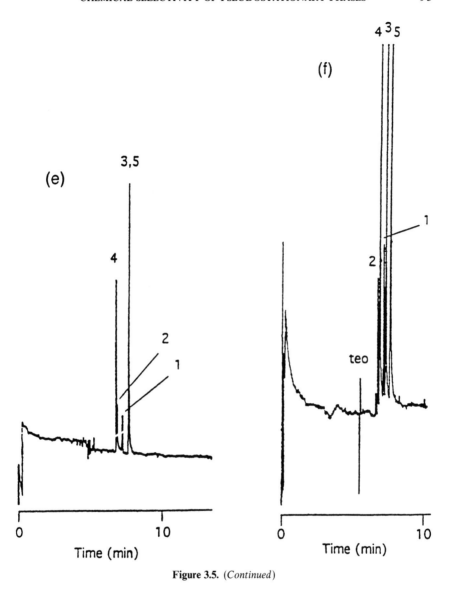

Figure 3.5. (*Continued*)

The *m* coefficient is related to the cohesiveness of the micellar phase. The LiPFOS micelles are the most cohesive (smallest *m* coefficient), whereas the hydrocarbon micelles of SDS, $C_{14}TAB$, and SC (sodium cholate) are the least cohesive phases (large *m* coefficients) (29). As a point of reference, recall that water is the most cohesive solvent and hydrocarbon liquids are among the

Table 3.3. Structural Descriptors for Representative Test Solutes Used in LSER Modeling of Retention in MEKC

Compound (type)	$V_1/100$	π^*	β	α	$\log P_{ow}$
1. Benzene (NHB[a])	0.491	0.59	0.10	0.00	2.13
2. Benzonitrile (HBA)	0.590	0.90	0.37	0.00	1.56
3. Nitrobenzene (HBA)	0.631	1.01	0.30	0.00	1.86
4. Acetophenone (HBA)	0.690	0.90	0.49	0.04	1.58
5. Anisole (HBA)	0.639	0.73	0.32	0.00	2.11
6. Phenol (HBD)	0.536	0.72	0.33	0.61	1.49

Source: Ref. 29.
[a]NHB = non-hydrogen bonding

least cohesive phases. According to Table 3.2, the *m* values are very similar for the hydrocarbon micelles (i.e., C_{14}TAB, SDS, and SC). This suggests that cavity formation energy has a minor effect on the difference in chemical selectivity between hydrocarbonaceous surfactants in MEKC. However, differences in selectivity between LiPFOS or Elvacite 2669 and the hydrocarbonaceous surfactants can partly be due to the cavity formation term.

The coefficient *b* represents the relative strength of the micellar phases as hydrogen bond donors (HBDs i.e., acidity). The larger *b* value means that the micellar phase is the stronger HBD (29). Based on Table 3.2, the relative HBD strength of the micellar systems can be ranked as follows: LiPFOS > SDS > Elvacite 2669 > C_{14}TAB > SC > 1-octanol. This suggests that a LiPFOS–MEKC system is the strongest HBD system, followed by the SDS–MEKC system. The acidities of Elvacite 2669 and C_{14}TAB are between SDS and SC. The values of coefficient *b* are very different in these five MEKC systems, ranging from -2.88 to $+0.16$. Therefore, type A hydrogen bonding interaction contributes significantly to the chemical selectivity differences among these five systems. The negative *b* values indicate that the micelles are weaker HBDs than the bulk aqueous phase.

On the other hand, the term ($a\alpha$) corresponds to type B hydrogen bond interaction, which involves solutes acting as HBDs (acids) and solvents as HBAs (bases); the coefficient *a* is a measure of the relative strength of micellar phases as hydrogen bond acceptors (HBAs, i.e., basicity). The larger value for coefficient *a* refers to higher HBA strength of the micellar phase. According to Table 3.2, the HBA strength of the micellar systems can be ranked as follows: C_{14}TAB > Elvacite 2669 > SC > 1-octanol > SDS > LiPFOS. This means that the C_{14}TAB MEKC system is the strongest HBA (basic) system among these five MEKC systems, followed by Elvacite 2669. The values of the coefficient a are also very different for these five MEKC systems (−0.98 to +0.99), which suggests that type B hydrogen bonding interaction also contributes significantly to the chemical selectivity among these five MEKC systems.

The term $s\pi^*$ represents the dipolar interactions between solutes and solvents. Because s values are small in magnitude and are similar for the systems studied, it can be concluded that dipolar interactions have little or no effect on retention and selectivity in these MEKC systems.

In summary, the LSER results indicate that hydrogen bonding is a major reason for the selectivity differences among different surfactants. In general, the fluorocarbon micelles of LiPFOS have the strongest HBD strength among the five MEKC systems. This conclusion is consistent with the observed elution behavior as HBA solutes interact strongly with the LiPFOS micelles. One such example is given in Figure 3.5c, where the five test solutes elute according to their basicity (i.e., β values) (35). On the other hand, SC and $C_{14}TAB$, which are viewed as HBA micelles, have a much stronger interactions with the HBD solutes such as phenols and alcohols, whereas they exhibit weaker affinities for solutes with HBA functionalities. The elution pattern in the SDS system is mainly according to the solute size and HBA strength of solute. For instance, the two early eluting peaks in the SDS system, benzene (peak 1) and benzonitrile (peak 2), have the smallest sizes. Acetophenone (peak 4) is bulkier than anisole, which favors more retention. However, the effect of the larger size of acetophenone on retention is somewhat offset by its having greater HBA strength than anisole. As a result, it elutes slightly earlier than anisole. According to the LSER model, retention in the SDS system decreases as the basicity of solute increases due to the negative b coefficient. Based on the LSER results, SDS has intermediate properties in terms of hydrogen bonding as compared to LiPFOS, on the one hand, and SC, Elvacite 2669, and $C_{14}TAB$, on the other. It is a stronger HBD than SC, $C_{14}TAB$, and polymeric Elvacite 2669 but weaker than LiPFOS, while as a HBA, it is weaker than SC, CTAB, and Elvacite 2669, but stronger than LiPFOS.

3.8. EFFECTS OF CHEMICAL COMPOSITION OF MICELLAR SOLUTIONS

In MEKC, resolution is a function of retention (k'), selectivity (α), efficiency (N), and the size of the elution window (t_{mc}/t_{eo}). In turn, these four terms are influenced by experimental parameters such as surfactant type and concentration, type and percentage of modifiers, pH, temperature, ionic strength, and applied field strength. In the following subsections the roles of various parameters in MEKC separations are discussed.

3.8.1. Surfactant Concentration

The primary role of surfactant concentration is to adjust the retention factor to within the optimum range in order to achieve better resolution. According to Terabe et al. (2), the relationship between retention factor k' and surfactant concentration can be described as follows:

$$k' = \frac{v(C_{sf} - \text{CMC})}{1 - v(C_{sf} - \text{CMC})} P_{mw} \quad (3.8)$$

where v is surfactant molar volume; C_{sf} is the total surfactant concentration; CMC is the critical micelle concentration; and P_{mw} is the partition coefficient of a solute between an aqueous phase and micelles. At low micelle concentrations the second term in the denominator becomes negligible and a linear relationship between the retention factor and surfactant concentration can be described as follows (2):

$$k' = P_{mw} v(C_{sf} - \text{CMC}) \quad (3.9)$$

Figure 3.6 demonstrates the linear relationship between k' and surfactant concentration for a group of uncharged solutes. This relationship has also

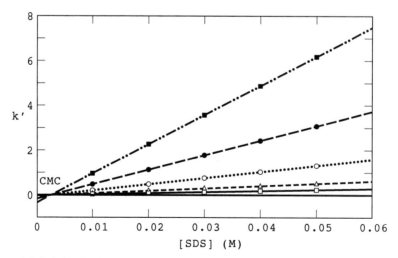

Figure 3.6. Relationship between k' and [SDS] for neutral solutes. Compounds are 2-naphthol (solid squares); toluene (solid circles); nitrobenzene (open circles); phenol (open triangles); and resorcinol (open squares). (From ref. 25 with permission.)

been shown for groups of charged solutes (25,26). Partition coefficient P_{mw} and CMC can be determined from the slope and the x-intercept of Eq. (3.9), respectively. Recently, the use of polyacrylamide-coated capillaries with negligible electroosmotic flow have been reported for the determination of micelle–water partition coefficients of hydrophobic and slightly polar molecules (36). In adddition to unmodified and dynamically coated fused silica capillaries, hollow polymeric fibers of pH-stable polypropylene have been used as separation columns in MEKC (37). The surfactant concentration may also alter the size of the elution window and efficiency; however, it has little if any effect on selectivity (compare Figures 3.5a and 3.5b). Note that selectivity between a pair of solutes is simply defined as the ratio of the retention factors k'_2 and k'_1, which is approximately equal to the ratio of the micelle–water partition coefficients at low surfactant concentration:

$$\alpha = k'_2/k'_1 \cong P_{mw,2}/P_{mw,1} \quad (3.10)$$

3.8.2. Types of Pseudostationary Phase

Variations in the hydrophobic moiety, the ionic head group, or the type of counterion can influence retention, selectivity, the size of elution window, and efficiency in MEKC.

3.8.2.1. *Anionic Alkyl Chain Surfactants*

Anionic alkyl chain surfactants, especially SDS, have been the most widely used surfactant type (38–68). The popularity of SDS can be attributed to its high aqueous solubility, low CMC, low Kraft point, small ultraviolet (UV) molar absorptivity even at low wavelengths, availability, and cost. Serendipitously, SDS might provide the "right" type of selectivity for many solute mixture. The LSER studies indicate that SDS is a stronger HBD as compared to most other surfactant systems studied thus far such as bile salts, cationic C_{14}TAB surfactants, and methacrylate-based copolymers. Consequently SDS should be a better surfactant type in many situations considering that the great majority of small solutes that are separated by MEKC contain a HBA functional group such as nitro, carbonyl, or cyano. Even HBD solutes such as phenols or alcohols have hydrogen-bond-accepting characteristic (see Table 3.3). As shown in Figure 3.7, a mixture of 24 explosive chemicals—mostly nitroaromatic molecules—is separated in approximately 10 minutes using an SDS buffer in MEKC (63). Ahuja and Foley reported that the type of counterion (e.g., Na^+, Li^+, or K^+) for dodecyl sulfate micelles can greatly influence efficiency, elution pattern, retention, and the size of the elution window in MEKC (69).

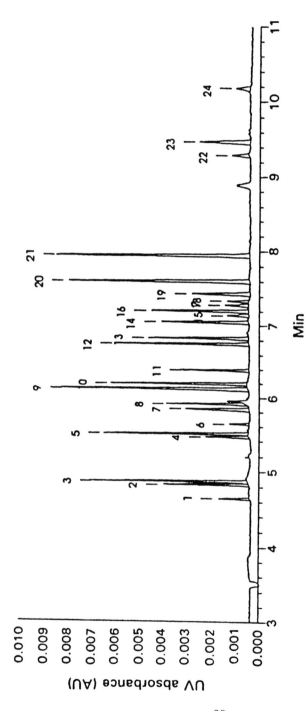

Figure 3.7. MEKC separation of a mixture of 24 nitroaromatic compounds. MEKC conditions: 50 mM SDS micelles in borate buffer; applied voltage = +25 kV; λ = 230 nm; total length = 60 cm. Peak identifications: (1) 1,3,5,7-tetranitro-1,3,5,7-tetraazacyclooctane; (2) 1,3,5-trinitro-1,3,5-triazacyclohexane; (3) 1,3,5-trinitrobenzene; (4) 1,4-dinitrobenzene; (5) 1,3-dinitrobenzene; (6) nitrobenzene; (7) 2,4-dinitro-6-methylphenol; (8) 1,2-dinitrobenzene; (9) 3-methyl-4-nitrophenol and 2,4,6-trinitrotoluene; (10) 2-methyl-3-nitroaniline; (11) 2,4,6-N-tetranitro-N-methylaniline; (12) 2-methyl-5-nitroaniline; (13) 2,4-dinitrotoluene; (14) 2,6-dinitrotoluene; (15) 2-nitrotoluene (16) 3,4-dinitrotoluene; (17) 4-nitrotoluene; (18) 3-nitrotoluene; (19) 2,3-dinitrotoluene; (20) 2-amino-4,6-dinitrotoluene; (21) 4-amino-2,6-dinitrotoluene; (22) 1,5-dinitronapthalene; (23) diphenylamine; (24) Sudan III (t_{mc}). (From ref. 63 with permission.)

Anionic alkyl chain surfactants, especially SDS, have been used in a variety of applications such as MEKC separations of phenols and other acidic solutes (25,58), pharmaceutical amines (26), organic solvents (38), keto compounds and carbohydrate derivatives (40,43,61), fungicide and phytotoxin (41), caffeine matabolites in human urine (42), DNA adducts (44), pharmaceuticals such as antibiotics, vitamins, sulfonamides and xanthines (45,46,52,60), peptides and proteins (47–51), organometallic compounds and ligands (39,53), amino acids (54,57,67,68), sulfur and nitro compounds in environmental applications (55,63), flavonoids in foods (56), and nucleic acid constituents in body fluids (59).

The resolving power of the SDS–MEKC systems, however, has been somewhat limited for certain groups of solutes such as polycyclic aromatic hydrocarbons (PAHs) and steroids (64–66). For example, separation of corticosteroids has been a challenging problem due to their high structural similarities, as oftentimes two corticosteroids are different by only a double bond or a small functional group. Corticosteroids are bulky molecules with high structural similarities. They have a steroidal backbone with polar substituents such as hydroxyl or carbonyl groups. As discussed previously, bulky molecules such as these have a stronger interaction with the micelles, which results in longer retention (recall the large positive m coefficient in LSER; Table 3.2). Figure 3.8a shows an example of a MEKC separation of 17 structually similar corticosteroids with SDS micelles (66). The majority of the compounds eluted near or with SDS micelles. Addition of a small percentage of an organic cosolvent or urea and inclusion of cyclodextrins in the SDS solution are possible options for enhancing the resolution (*vide infra*). The use of modifiers like acetonitrile, urea, and cyclodextrin improved the separation; however, it did not result in the complete separation of all compounds. The other alternatives are changing the type of surfactant and/or use of mixed micelles.

3.8.2.2. Bile Salts

Changing the type of surfactant to a bile salt dramatically improves the separation of corticosteroids, as shown in Figure 3.8b (p. 101) and 3.8c (p. 102). The two bile salt surfactants of sodium glycodeoxycholate and sodium taurocholate provided different selectivities from each other in the corticosteroids separation. As an alternative to SDS, the use of bile salt surfactants in MEKC separations has become popular (66–85). Various types of bile salts have been used as pseudostationary phases in MEKC. They provide different selectivities as compared to SDS. As mentioned previously, bile salt surfactants also have very different aggregation properties and structures from those of SDS micelles. They can tolerate relatively higher concentrations of organic

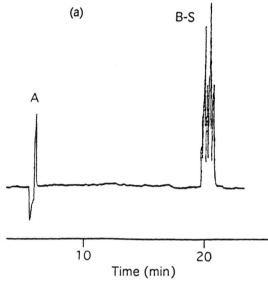

Figure 3.8. Influence of surfactant type on MEKC separation of a mixture of 17 corticosteroids: (a) 100 mM SDS; (b) 100 mM glycodeoxycholate; (c) 100 mM taurocholate; (d) 40 mM LiPFOS. MEKC conditions: for parts (a)–(c), applied voltage = +20 kV, $\lambda = 254$, total length = 62 cm, phosphate buffer; for part (d), applied voltage = +15 kV, phosphate/borate buffer. Peak identifications: (A) t_{eo}; (B) triamcinolone; (C) prednisone; (D) cortisone; (E) fludrocortisone; (F) hydrocortisone; (G) prednisolone; (H) prednisone acetate; (I) fludrocortisone acetate; (J) cortisone acetate; (K) prednisolone acetate; (L) hydrocortisone 21-acetate; (M) corticosteronic; (N) triamcinolone acetonide; (O) fluocinolone acetonide; (P) 6α-methyl prednisolone; (Q) deoxycorticosterone; (R) progesterone; (S) t_{mc}. (Parts a–c from ref. 66 with permission; part d from Jefferson G. Bumgarner, Ph. D. dissertation, North Carolina State University, Raleigh, 1996.)

modifier (30% organic solvent) than do SDS micelles without a disruption of their structural integrity. Bile salt micelles are generally considered to be more "polar" than the SDS micelles. This arises from the general observation that most solutes have a stronger interaction with the SDS micelles. This perception is somewhat misleading as "polarity" is defined within a broad context. According to the LSER results, however, both SDS and SC micelles have nearly identical m (i.e., cohesiveness) and s (dipolarity–polarizability) coefficients. SDS micelles are stronger HBDs whereas SC are stronger HBAs. This trend was also consistent for another bile salt surfactant, sodium deoxycholate (SDC). Therefore, one can expect that HBA solutes would have stronger interaction with the SDS micelles whereas HBDs have a greater affinity for the bile salt micelles. As mentioned earlier, the majority of solutes bear a HBA

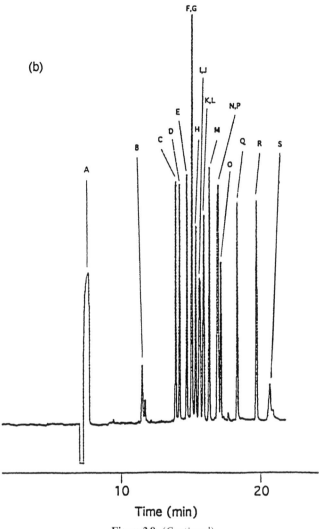

Figure 3.8. (*Continued*)

functional group, which has lead to the general notion of stronger interaction with the SDS micelles. In fact, one such example is the corticosteroids mixture. Since these compounds contain several HBA functional groups, they do not interact as strongly with the bile salts micelles as they do with the SDS micelles [Figure 3.8 (pp. 100–103)]. Therefore, the problem of coelution with the

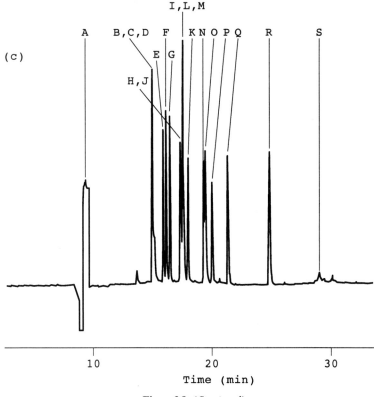

Figure 3.8. (*Continued*)

micelle is resolved. Bile salts have been used in a variety of MEKC separations such as steroids (66), amino acids (67,68), benzodiazepines (70), synthetic colors (71), bioactive compounds (72), environmental analysis (73,76,78), proteins and peptides (74,85), bilirubin (75), organic acids (76,77,81), hydrophobic compounds including polycyclic aromatic hydrocarbons (79,82,84), chiral and diastereoisomers (80), and anti-HIV agents (83).

3.8.2.3. *Fluorocarbon Surfactants*

In general, the use of fluorinated surfactants in MEKC separations has been very limited. This has been mostly due to the lack of availability of MEKC-compatible fluorocarbon surfactants in high purity. Ye et al. (86) reported the first application of the fluorocarbon micelles of LiPFOS for the MEKC separations of a group of small peptides. They observed that the selectivity and

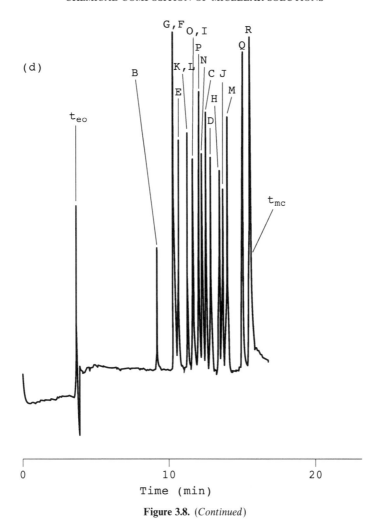

Figure 3.8. (*Continued*)

elution pattern for LiPFOS differ dramatically from that of the typical anionic alkyl chain surfactants like SDS or LiDS. As mentioned earlier, LiPFOS micelles are more cohesive than SDS, as is evident from the smaller m coefficient. This results in smaller solute interaction with LiPFOS micelles. On the other hand, LiPFOS is a stronger HBD than are other micellar systems that have been studied thus far. Consequently, a stronger interaction between solutes with HBA functional groups and LiPFOS micelles is observed. In principle, a HBD micelle should provide better selectivity for a mixture of HBA-solutes such as corticosteroids. In the MEKC separation of the corticosteroids mixture mentioned earlier, the use of LiPFOS micelles resulted in at least partial resolution of 16 of the 17 peaks [Figure 3.8d (p. 103)]. Bulky molecules like corticosteroids do not have as much interaction with the LiPFOS micelles as do SDS or other micelles due to the smaller m coefficient, but corticosteroids experience stronger and more selective hydrogen bond interaction. Bear in mind that the surfactant concentration in Figure 3.8d is not an optimum value and a small elution window is observed. Adjusting these two parameters should lead to even better separations.

3.8.2.4. Cationic Surfactants

Cationic surfactants generally interact with the negatively charged silica capillary wall and reverse the direction of the electroosmotic flow. As a result of the reversed EOF, the polarity of the electrodes would have to be reversed in order to elute solutes through the detector window (Figure 3.1b).

According to the LSER analysis, the cationic micelles of CTAB have a very different interactive behavior than SDS micelles. Figure 3.9 presents a comparison of retention behavior in the two micellar systems for a group of 60 aromatic compounds. The existence of three distinct lines is indicative of different selectivity in the two systems. Interestingly, the patterns of grouping of solutes into three lines can be explained in terms of the interactive properties of the two micelles as concluded from the LSER analysis. The CTAB micelles are stronger hydrogen bond acceptors. The first line (top line with solid triangles) includes phenols, which are HBDs that have a stronger affinity for the hydrogen-bond-accepting CTAB micelles. The second line (solid squares) consists of some HBDs such as alkyl benzyl alcohols, and weak HBAs such as alkyl benzenes, PAHs, or benzonitrile. The third line (open squares) contains strong HBAs such as alkyl phenones, and aromatic esters that have the lowest interactions with the HBA micelle.

The usefulness of cationic surfactants has been explored in a number of MEKC applications. A few examples are phenolic carboxylic acids (87), charged molecules with nearly identical electrophoretic mobilities like bis(amidinohydrazones) (88), nucleic acid constituents (89,90), cocaine and illicit drugs (91), adrenergic blocking agents (92), glycosylated compounds (93–95), and aromatic hydrocarbons (96).

3.8.2.5. Novel and Chiral Surfactants

Other anionic surfactants that have been examined are *in situ* borate complexed surfactants such as the N-D-gluco-N-methylalkanamide (MEGA)

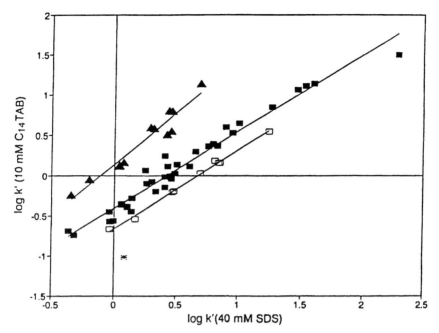

Figure 3.9. Relationships between retention in the cationic micelles of CTAB and the anionic micelles of SDS for 60 aromatic uncharged solutes. See the text for a further explanation. (From ref. 35, with permission.)

series and N,N-bis-(3-D-gluconamidopropyl)cholamide and -deoxycholamide (Big CHAP and Deoxy Big CHAP). Charge density on the micelles and consequently the size of the elution window can be controlled through proper adjustment of the pH and borate concentration. These surfactants have been used for the MEKC separations of compounds such as herbicides, barbiturates, and amino acids (97–99).

Due to their chiral functionalities, Big CHAP and Deoxy Big CHAP have also been applied for the chiral separations of compounds such as Troger's base and Silvex herbicide in MEKC. Other novel anionic chiral surfactants such as (R) and (S)-N-dodecoxycarbonylvaline and sodium N-dodecanoyl-L-valinate, as well as polymerized chiral micelles, have been reported for the separation of chiral molecules (100–104). Many of these chiral surfactants are synthetic and consist of a hydrophobic alkyl chain tail and an amino acid or a carbohydrate head group with a chiral center. More information on chiral separations can be obtained from Chapter 23.

3.8.2.6. Nonionic and Zwitterionic Surfactants

In order to separate uncharged molecules in MEKC, charged surfactants must be used. Surfactants with a zero net charge such as nonionic and zwitterionic surfactants can be used along with an ionic surfactant as mixed micelles. They can also have a great influence on the MEKC separation of charged molecules. Both nonionic and zwitterionic surfactants have been used alone for the separations of charged compounds such as amino acids and polypeptides (105,106). Nonionic surfactants and mixtures of anionic and nonionic surfactants have been shown to provide different selectivities, reduced currents, and changes in elution window sizes than those of anionic surfactant systems. Zwitterionic surfactants have been used alone and in conjuction with SDS in MEKC. Because these surfactants do not increase the conductivity of the buffer, they can be used at high concentrations while still allowing the use of high voltages and large i.d. capillaries.

3.8.2.7. Mixed Micelles

As mentioned above, different retention behavior and selectivity can be observed for various types of surfactants. For certain complex mixtures of structurally similar solutes, however, one might not be able to find a suitable surfactant type that provides adequate resolution. This is typically due to a lack of selectivity and/or a narrow elution window. In these situations, the use of mixed micelles can lead to enhanced separations. Mixing surfactants with different interactive properties can lead to great changes in selectivity for a given mixture. Selection of optimum composition would then be crucial in improving the quality of separation. The size of the elution window for certain mixed micellar systems is often larger than that for the individual constituent surfactants. The retention factor and selectivity change with the mole fraction of surfactants in the mixed micelles. Efficiency in the mixed micellar systems is often not much different from those of the single surfactant systems; however, in a few cases, loss of efficiency has been observed.

A good example of the usefulness of mixed micelles is the separation of a mixture of 17 corticosteroids with high structural similarities (see Figure 3.8). As mentioned earlier, this complex mixture cannot be separated by SDS micelles due to strong solute interaction with the micelles and elution of the majority of corticosteroids near or with SDS micelles (Figure 3.8a). Much better separations have been achieved using bile salts such as glycodeoxycholate (GDC; Figure 3.8b) and taurocholate (TC; Figure 3.8c). Note the different separation patterns between GDC and TC. For example, a number of peak pairs that have been resolved in GDC are coeluted with the TC system and vice versa. Thus these two systems might be viewed as complementary to

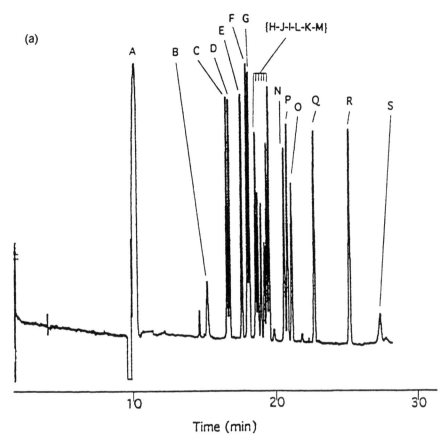

Figure 3.10. Mixed micellar electrokinetic chromatography of a mixture of 17 corticorsteroids. (a) Binary mixed micelles of two bile salts: 50 mM glycodeoxycholate and 50 mM taurocholate: (b) Ternary mixed micelles of the two bile salts and SDS: 33 mM glycodeoxycholate, 33 mM taurocholate, and 33 mM SDS, (c) Mixed micelles of the two bile salts and a short-chain alkyl surfactant: 33 mM glycodeoxycholate, 33 mM taurocholate, 70 mM butane sulfonate. Peak identifications the same as in Figure 3.8. Other conditions: applied voltage = +15 kV; $\lambda = 254$; total length = 62 cm; phosphate/borate buffer. (From ref. 107 with permission.)

one another and should be suitable for a mixed micellar system. Figure 3.10a shows the separation of the mixture using an equimolar mixture of the two surfactants GDC and TC, at the same total concentration. Although the resolution between the previously coeluted peaks has improved, a number of peaks still overlap and are not fully resolved. Note that the size of the elution window in the single and mixed bile salt systems is small, which might be due to the typically small aggregation number of these systems and subsequent

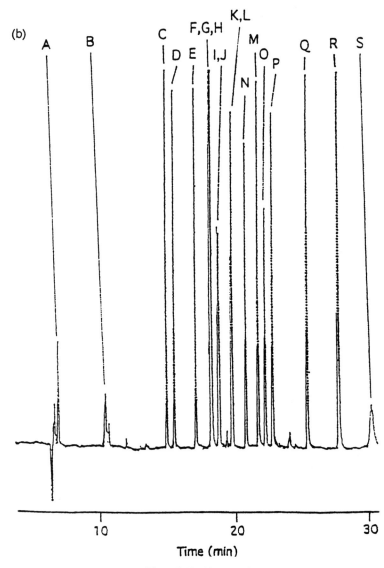

Figure 3.10. (*Continued*)

small values of the charge-to-mass ratio. Extending the size of the elution window should lead to increased separation. This can be achieved by incorporating an alkyl chain ionic surfactant with the bile salts. As illustrated in Figure 3.10b, incorporation of the SDS to the mixed bile salts systems has resulted in a significant increase in the size of the elution window (from around

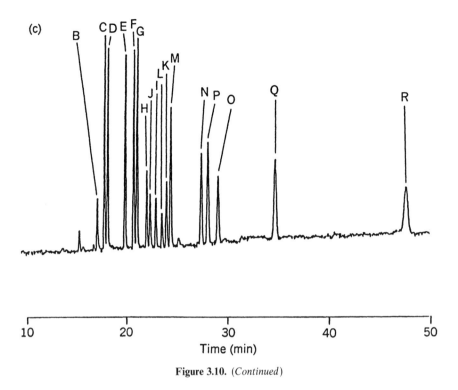

Figure 3.10. (*Continued*)

3 in the bile salts to about 4.4 in the ternary system of the two bile salts and SDS). However, several peaks still overlap and/or coelute, apparently due to the loss of selectivity and/or strong solute interaction with the SDS in the ternary mixed micellar system (i.e., k' too large). In order to maintain the wide elution window without sacrificing selectivity, an anionic surfactant with a short alkyl chain length, such as butane sulfonate (BuS) can be added to the bile salts mixture. BuS does not form micelles and therefore has little effect on the overall retention behavior and selectivity. As shown in Figure 3.10c, the ternary system of the GDC–TC–BuS has a very wide elution window (~ 7) and baseline separation of all peaks has been achieved. This increase in the size of the elution window is mainly due to the increase in the migration time of the micelles (i.e., t_{eo} remained nearly constant). Bile salts monomers have bulky steroid backbones with only one charged head group. The aggregation number of the bile salt micelles is between 2 and 10, which means that their charge-to-mass ratio is small. The anionic alkyl surfactants are attracted to the hydrophobic portion of the bile salt micelles, thus increasing the charge-to-

mass ratio of the mixed micelles. This increase in the negative charge density would enhance the mobility of the mixed micelles in the opposite direction of the EOF and causes the micelles to elute for a longer time. The size of the elution window increases with the concentration of the alkyl surfactant initially before reaching a plateau. For BuS, this is achieved at 70mM. It is worth noting that the efficiency for the GDC–TC–BuS system (Figure 3.10c) is considerably lower (by a factor of 4–5) than that achieved with the GDC–TC–SDS system (Figure 3.10b). This is probably due to the Joule heating that occurs as a result of the use of high concentrations of BuS. In spite of the lower efficiency, however, much better resolution is achieved in the BuS system due to a wider elution window, higher overall selectivity, and perhaps near-optimal retention.

The popularity of mixed micelles has dramatically increased over the past few years (107–118). Some examples of the groups of compounds that have been separated by mixed micelles in MEKC include nitroaromatic explosive compounds (63), corticosteroids (66,107), amino acids (109,112), organic anionic species (111), herbicides (112), fatty acids (113), analgesic compounds in pharmaceutical samples (114), tetracycline antibiotics (116), and drugs in body fluids (118). Various combinations of different surfactants such as bile salt/anionic (66,107), anionic/zwitterionic (63,108), cationic/zwitterionic (109), fluorocarbon/anionic (86), anionic/nonionic (108,113,115,117,118), cationic/cationic (110,111), anionic/cationic (112), bile salt/bile salt (66,114), and nonionic/nonionic (for charged solutes) (116) have been used in MEKC. The use of mixed micelles of the nonionic Brij 35 and anionic SDS has resulted in an infinite elution window at a specific Brij 35 concentration (108). Typically, addition of Brij 35 yields narrower elution window size and large changes in selectivity. However, at a specific composition, the mobility of anionic micelles equals the electroosmotic mobility, thus making the charged micelles effectively stationary.

3.8.2.8. Polymeric Pseudostationary Phases

In recent years, there has been a great deal of interest in the use of polymeric pseudostationary phases. Three different types of polymeric phases have been reported: (i) polymer micelles where surfactant monomers are polymerized and covalently bonded together (12,13); (ii) cascade macromolecules like starburst dendrimers (14–17); and (iii) ionic block copolymers (18–20). As mentioned earlier, the primary advantage in using polymeric phases is the stability of their structure in the presence of large concentrations of organic modifiers. Conventional micelle-forming surfactants such as SDS and bile salts can tolerate up to 20–30% organic modifier before micelle formation is inhibited. The use of a high concentration of organic modifier is necessary for

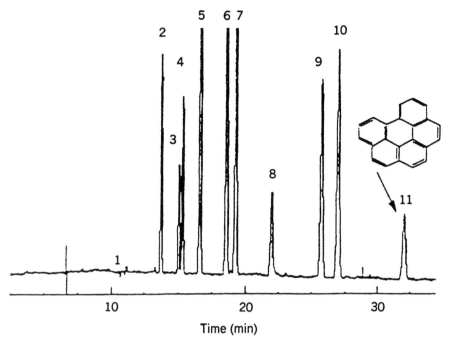

Figure 3.11. Separation of a polyaromatic hydrocarbon (PAH) mixture using the polymerized micelles of sodium 10-undecylenate in 35% acetonitrile at pH 8.2. Peak identifications: (1) methanol; (2) naphthalene; (3) acenaphthene; (4) fluorene; (5) phenanthrene; (7) pyrene; (8) chrysene; (9) benzo[b]fluoranthene; (10) benzo[a]pyrene; (11) benzo[g,h,i]perylene. (From ref. 12 with permission.)

the separation of highly hydrophobic solutes that interact strongly with the micelles. Therefore, most reported applications for the polymeric pseudo-stationary types include hydrophobic solutes. For example, Figure 3.11 shows the separation of a mixture of PAHs using a polymer micelle (12). Separation of PAHs has also been achieved using other polymeric phases such as starburst dendrimers with a 90% methanol solvent (15) and ionic block copolymers with a 50% methanol in the aqueous buffer (20). Recently, the usefulness of these polymer micelles with a chiral head group has been demonstrated for the separations of optical isomers (104); see Chapter 23. Figure 3.12 illustrates the first reported separation of fullerenes by a CE method using a triblock copolymer of poly(methyl methacrylate–ethyl acrylate–methacrylic acid), commercially known as Elvacite 2669. As shown the C_{60} and C_{70} peaks are easily separated due to sufficient difference in the size of the two molecules by use of an organic-rich buffer. Figure 3.13 shows the

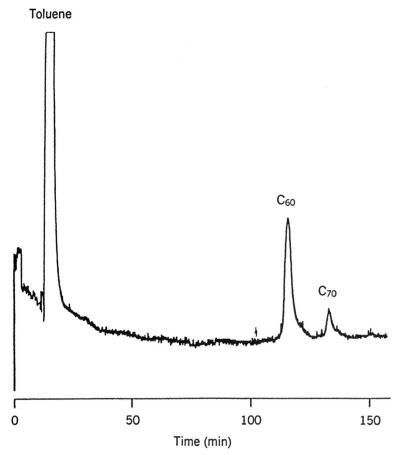

Figure 3.12. Separation of fullerenes in a fullerite sample using a polymeric pseudostationary phase, 2% Elvacite 2669 polymer (see Figure 3.3 for the structure), 84% methanol in water. The fullerite sample was dissolved in toluene. MEKC conditions: applied voltage = +28 kV; $\lambda = 260$; total length = 62 cm; temperature = 40°C; CAPS buffer. (From Jefferson G. Bumgarner, Ph. D. dissertation, North Carolina State University, Raleigh, 1996.)

separation of a mixture of homologous series of n-alkylphenones, from acetophenone to the very hydrophobic octanodecanophenone (20). Note the existence of a general elution problem, as early eluting peaks are somewhat overlapping while the late eluting peaks are broad with long retention times. Lower concentration of the organic solvent would improve the separation of the early peaks, and operating at a higher percentage of organic modifier

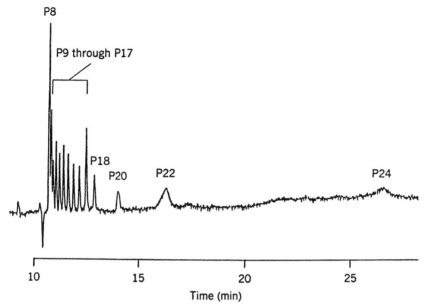

Figure 3.13. Separation of *n*-alkylphenones using a polymeric pseudostationary phase, 2% Elvacite 2669 triblock copolymer (see Figure 3.3 for the structure) and 60% methanol in water. Other conditions: applied voltage = 30 kV, $\lambda = 248$ nm; ambient temperature. Peak identifications: (P8) acetophenone; (P9) propiophenone; (P10) butyrophenone; (P11) valerophenone; (P12) hexanophenone; (P13) heptanophenone; (P14) octanophenone; (P15) nonanophenone; (P16) decanophenone (P17) undecanophenone; (P18) dodecanophenone; (P20) tetradecanophenone; (P22) hexadecanophenone; (P24) octanodecanophenone. (From ref. 20 with permission.)

would reduce the analysis time of the later-eluting compounds while improving their peak shape. As in conventional LC, there is a need for gradient elution to improve the overall quality of the separation and analysis time.

3.8.3. Modifiers

Modifiers such as organic solvents, cyclodextrins, and urea are typically incorporated into the aqueous buffers of MEKC in order to reduce the retention factors of strongly bound solutes to micelles. Their presence can also lead to wider elution ranges and/or higher selectivity.

3.8.3.1. *Organic Solvents*

Organic modifiers such as methanol and acetonitrile have been extensively utilized for improving resolution in MEKC (118–141). In RPLC, organic modifiers play an important role in controlling retention and selectivity for a wide variety of compounds. Their use in MEKC separations, however, has been mostly for improving the separations of hydrophobic compounds that interact strongly with micelles and elute at or near migration time of micelles. The main role of organic modifiers in MEKC has been to reduce retention factors of highly hydrophobic solutes to within or near the optimum range. Typically inclusion of organic solvents leads to an increase in the size of the elution window. As a result, resolution in MEKC can be enhanced at the expense of longer analysis times. The influence of organic solvents on selectivity of partitioning into micelles for a wide range of molecules with polar functional groups is not clear. The concentration of organic solvents would have to be limited (typically $\leq 20\%$) in order to maintain the integrity of micelles. The use of polymeric phases will provide an opportunity to investigate the role of organic solvents over a wider range of concentrations.

Chromatographic behavior with more hydrophobic modifiers can be different from that with polar solvents. This is because polar modifiers like methanol or acetonitrile have little or no interaction with the micelles whereas the more hydrophobic ones such as longer-chain alcohols are incorporated into micelles. Aiken and Huie examined the effects of addition of 1-alkanols (C_4–C_8) in MEKC (126). In general, the size of the elution window increased upon addition of the long-chain alcohols, which was attributed to the increase in mobility of the SDS micelles as the EOF remained relatively constant. Interestingly, higher retention factors were observed with use of the longer-chain modifiers such as octanol. This increase was more pronounced for the hydrophobic solutes. For example, the retention factors of alkyl phenols in a 100 mM SDS increased upon addition of 50 mM long-chain alcohols from butanol to octanol. Inclusion of the long-chain alcohol, however, only caused a small increase in retention of more polar solutes. On the other hand, retention factors for more hydrophobic solutes were considerably higher when the micellar solution with the longer-chain alcohols such as octanol was used rather than that for butanol. This was attributed to the incorporation of the long-chain alcohols into the micelles, which could increase the micelle–water partition coefficient and phase ratio due to an increase in hydrophobicity and volume of the micelle (126). Aiken and Huie also reported that selectivity was changed with the inclusion of alcohol modifiers for a group of alkyl phenol test solutes. This behavior is different from that typically observed with polar modifiers like methanol or acetonitrile, where retention factor is smaller in the presence of the modifier. Several other reports (138–141) have focused on the effects of organic modifiers such as 1-hexanol, methanol, acetonitrile, and dimethylformamide on SDS and bile salts micelles. Katsuta and coworkers (138) demonstrated that selectivity can be altered by the inclusion of organic modifiers in MEKC systems. It was determined that this is primarily due to the saturation of the micellar palisade layer with the modifier and the hydrogen bonding interaction between the modifier and analyte molecules.

Organic modifiers have been used to improve the MEKC separations of cardiovascular drugs (119), herbicides (120), amino acids (121), chlorophylls (122,130), taxol and related antitumor compounds (123), benzodiazepines (124), vitamins (125), illicit drugs (127,132), porphyrin compounds (128), coumarins (129), natural compounds with anticancer activity (131), nucleosides (133), hydrophobic compounds including PAHs, alkyl benzenes, alkyl phenones, alkyl parabenes, and fullerenes (12–15,18–20,135), and phthalate esters (136).

Figure 3.14 shows the MEKC separation of a group of 10 estrogens. Figure 3.14a shows no separation in the 50 mM SDS buffer solution. Addition of 15% acetonitrile improves the separation (Figure 3.14b), with peaks 5 and 8 coeluting. These peaks correspond to 4-hydroxyesteroidal (peak 5) and 4-hydroxyesterone (peak 8). Upon changing the type and concentration of the organic modifier to 20% methanol (Figure 3.14c), the two peaks are resolved and the elution order for peaks 4,6, and 7 has been changed.

3.8.3.2. Glucose and Urea

In addition to typical organic solvents, the use of other modifiers like urea and glucose has also been reported. Urea reduces the interactions of hydrophobic compounds with micelles by increasing their solubility in the aqueous solutions (142–146). Retention factors of hydrophobic compounds in micellar solutions is therefore decreased dramatically. The size of the elution window has also been shown to increase with the addition of urea to MEKC systems.

Kaneta and coworkers have reported the effects of adding glucose as a modifier to enhance resolution in MEKC. In the separation of nine nucleosides, inclusion of 1 M glucose resulted in variations in selectivity as well as an increase in electrophoretic mobility of SDS micelles and consequently in a wider elution window. For this mixture, glucose was reported to be even more effective than methanol in improving the separation (133).

3.8.3.3. Cyclodextrins

Another type of modifier that has been used in MEKC are the cyclodextrins (CDs) (147–153). The first report of cyclodextrin-modified MEKC (CD–MEKC) with SDS micelles was by Terabe et al. (147) for the separation of PAHs. The hydrophobic cavity of cyclodextrins provides an alternative site of interaction to micelles for the hydrophobic solutes. Since uncharged cyclodextrins migrate at the EOF velocity and in the opposite direction of anionic SDS micelles, the net retention factor of solutes decreases in the presence of cyclo-

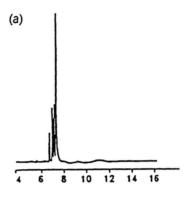
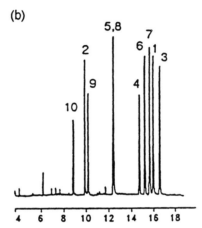
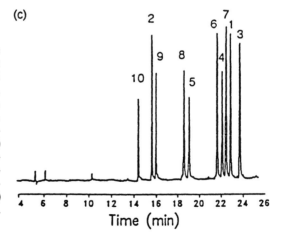

Figure 3.14. Effect of organic modifier in MEKC separation of a group of estrogens: 50 mM SDS in phosphate buffer, pH 7.0. (a) No organic solvent; (b) 15% acetonitrile; (c) 20% methanol. Peak identifications: (1) 17β-estradiol; (2) 16-keto-17β-estradiol; (3) 2-methoxyestradiol; (4) 2-hydroxyestradiol; (5) 4-hydroxyestradiol; (6) estrone; (7) 2-methoxyestrone; (8) 4-hydroxyestrone; (9) 16α-hydroxyestrone; (10) estriol. (From ref. 149 with permission.)

dextrins. As a result, hydrophobic solutes that would otherwise elute with micelles, can be better separated. In addition, cyclodextrins introduce a shape selectivity effect that is beneficial for the separation of structural isomers.

Cyclodextrin-modified MEKC has been used in the separation of various classes of compounds such as PAHs (147,150,152), chlorinated benzens and trichlorobiphenyl isomers (147), enantiomeric mixtures (148,151), steroids (149), and mycotoxins (153). Figure 3.15 presents an example of the use of two different types of cyclodextrins for the separation of a group of 10 estrogens. Note that the same mixture has also been separated using organic modifiers (see Figure 3.14). γ-CD has a larger cavity, which provides a better selectivity for the mixture than does β-CD. Most of the reported applications of CD–MEKC have involved the use of γ-CD, which apparently provides better results than do other types of CD. Note that charged CDs can also be used either in conjuction with the micelles or alone for the separation of uncharged solutes.

3.9. MULTIPARAMETER OPTIMIZATION

A number of papers have addressed optimization of composition of micellar solutions for achieving better MEKC separations (4,154). Optimum micelle concentration (which results in optimum k' range) for achieving best resolution is a function of the size of the elution window (27). In many MEKC separations, more than one solution parameter should be optimized in order to achieve the desired results. This typically involves optimization of other parameters like pH or concentration of a modifier in addition to micelle concentration. Interactive parameters would have to be optimized simultaneously (as compared to sequentially or one at a time). One such example arises in the separations of ionizable solutes in MEKC, where optimization of pH and micelle concentration is of great significance in achieving successful MEKC separations. Interpretive optimization methodologies are quite useful as a chromatographic parameter (typically retention or resolution) is modeled as a function of experimental parameters. This can be quite effective since based on a minimum number of experiments the chromatographic behavior can be predicted and various experimental parameters can be optimized simultaneously.

In addition to the primary partitioning mechanism into micelles, one can incorporate secondary chemical equilibria (SCE) such as protropic, ion pairing, or metal ligand complexation in the bulk aqueous solution. The use of acid–base equilibria is particularly important, as it involves the separation of ionizable compounds. A great majority of small molecules of pharmaceutical and clinical significance have acidic or basic functional groups. For separation

118   MICELLAR ELECTROKINETIC CHROMATOGAPHY

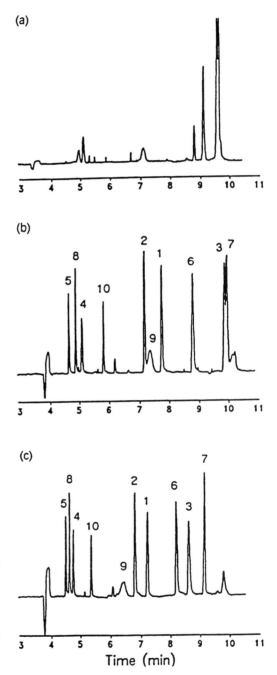

Figure 3.15. Effect of cyclodextrin (CD) on MEKC separation of a group of estrogens. MEKC conditions: 50 mM SDS in 10 mM sodium borate pH 9.2 with (a) no CD; (b) 20 mM β-CD; (c) 20 mM γ-CD. For peak identifications see Figure 3.14. (From ref. 149 with permission.)

of compounds with very similar electrophoretic mobilities, incorporation of micelles can result in better selectivity. For mixtures of charged and uncharged compounds, MEKC is the method of choice. Retention behavior of ionizable compounds is much more complicated than that of uncharged solutes. The migration schemes for acidic and basic compounds are shown in Figures 3.16 and for cationic and anionic micelles. Both charged and uncharged forms of the solutes can interact and migrate with the micelles. The charged fraction of the solutes would also migrate in the bulk aqueous media at its own electrophoretic mobility. For these groups of compounds, controlling the pH is of

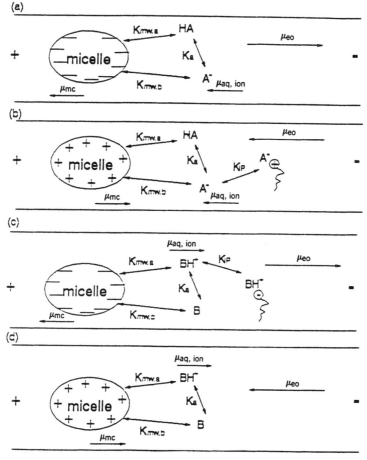

Figure 3.16. Schematic diagrams for the migration of ionizable solutes in MEKC: (a) acidic solute, HA, with anionic micelles; (b) acidic solute with cationic micelles; (c) basic solute, B, with anionic micelle; (d) basic solute with cationic micelle. (From ref. 155 with permission.)

great significance as pH determines the position of the acid–base equilibrium and consequently the net charge on the molecule. Migration behavior of ionizable compounds in MEKC has been quantitatively described through simple mathematical models that were developed on the basis of the acid–base equilibrium and micelle–water partitioning equilibria for the charged and uncharged forms of solute. The equations describe retention factor or net mobility as a function of two experimental variables, pH and surfactant concentration, as well as the dissociation constant and micelle–water partition coefficients (25,26,154,155). If the equilibrium constant values are known for a given compound, one can easily predict the coresponding migration behavior over a wide range of pH and micelle concentration. In a majority of cases, however, the constant values are unknown. One can achieve an estimate of the values by measuring retention at a few initial pH and micelle concentration and by fitting the experimental values to the models through nonlinear regression. The retention behavior of a group of 17 amines was successfully predicted on the basis of only five initial experiments at different pH and SDS concentrations. High correlation was observed between the predicted and observed migration parameters. Figures 3.17a and b (pp. 121 and 122) show examples of predicted chromatograms at various pH and micelle concentrations for a group of amines. Figures 3.18a,b illustrates high correlations between the predicted and observed chromatograms at the predicted optimum pH and SDS concentration. However, the separation is not satisfactory due to overlapping peaks and low efficiency. In order to improve the separation, one should incorporate another parameter. Upon addition of 10% acetonitrile, baseline separation of all 17 compounds was achieved in less than 15 minutes. Higher micelle concentration was used in order to achieve better efficiency, even though it was a nonoptimum value from the retention point of view (Figure 3.19).

The usefulness of iterative regression strategy to optimize selectivity in CZE and MEKC has been reported by Corstjens and coworkers (156,157). Pyell and Butehorn have examined the role of temperature in the light of minimizing analysis time and optimizing resolution (158). Temperature is not a widely used parameter, as its effect on selectivity is not pronounced. It has a great effect on viscosity (and therefore EOF), and this can be significant (as shown in Figure 3.20) when the separation of a group of amino acids at two different temperatures are compared (54).

Figure 3.17. Computer assisted optimization in MEKC. Predicted retention behavior of 17 aromatic amines as a function of (a) pH and (b) micelle concentration. Peak identifications: (1) nicotine; (2) 4-nitrobenzylamine; (3) benzylamine; (4) 3-methyldopamine; (5) norephedrine; (6) ephedrine; (7) phenethylamine; (8) 4-nitrophenethylamine; (9) N-methylphenethylamine; (10) 4-chlorobenzylamine; (11) 2-methylphenethylamine; (12) phenylpropylamine; (13) 4-bromobenzylamine; (14) 2-tolylethylamine; (15) 4-chlorophenethylamine; (16) phenyl-n-butylamine; (17) 1-(methylphenyl)propylamine. (From ref. 155 with permission.)

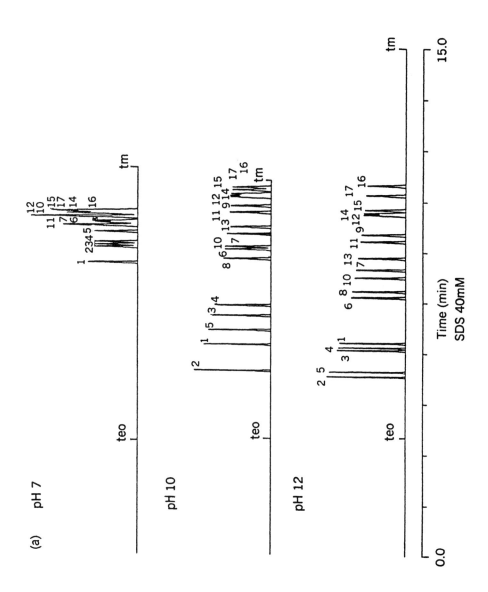

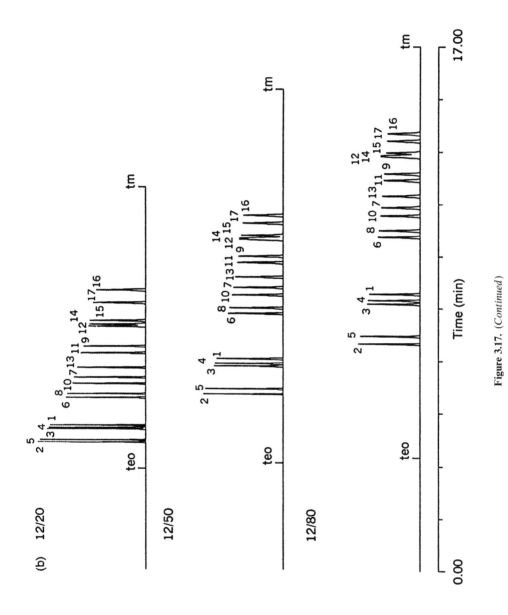

Figure 3.17. (*Continued*)

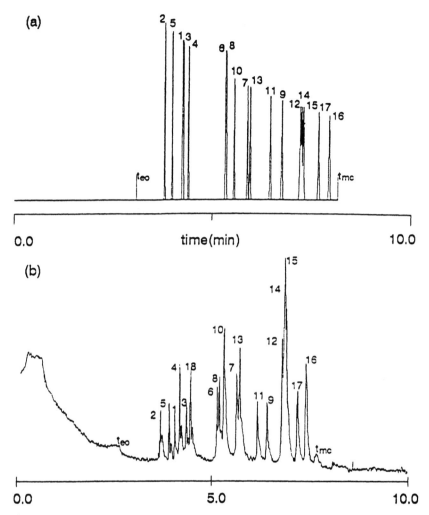

Figure 3.18. Comparison of the predicted (a) and the observed (b) separation of 17 aromatic amines at one of the optimum conditions (pH 11.0, 20 mM SDS). See Figure 3.17 for peak identifications. (From ref. 155 with permission.)

Pyell and Butehorn discussed the strategies for rapid one-parameter optimization of concentrations of SDS and modifiers like urea and glucose (159). They also developed a computer-aided method for simultaneous optimization concentrations of SDS and urea for the separation of various mixtures like nitroaromatic compounds, urea pesticides, and amines (160). Bretnall and

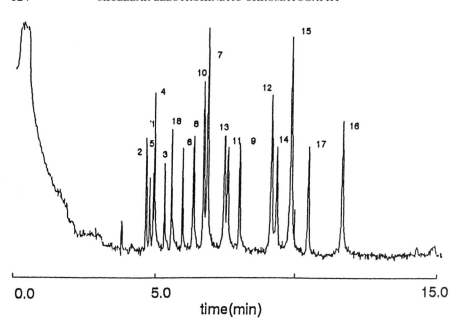

Figure 3.19. MEKC separation of 17 aromatic amines using 40 mM SDS, pH 11.0, and 10% acetonitrile. Peak identifications the same as in Figure 3.17. (From ref. 155 with permission.)

Clarke also investigated the optimum modifier composition for the separation of several cardiovascular drugs. They examined different modifiers such as various alcohols and ketones as well as acetonitrile (161). Ng et al. used an interpretive optimization scheme based on overlapping resolution mapping for the separation of flavonoids (162). Beattie and Richards investigated the optimization of separation conditions such as different electrolyte compositions and capillary wall modifications for the separation of proteins (163). Wan et al. used a full factorial design to optimize SDS concentration and pH for the separation of diastereomers amino acids (164).

In addition to optimizing MEKC separations, several reports have focused on the enhancement of detectability and efficiency of uncharged (109,144) as well as charged (58) compounds in MEKC through zone sharpening. Nielsen

Figure 3.20. Influence of temperature on MEKC separation of dansylated amino acids: (a) 20 °C, 100 mM SDS and (b) 10 °C, 102.5 mM SDS. Buffer: 20 mM borax, pH 9.2. (From ref. 54 with permission.)

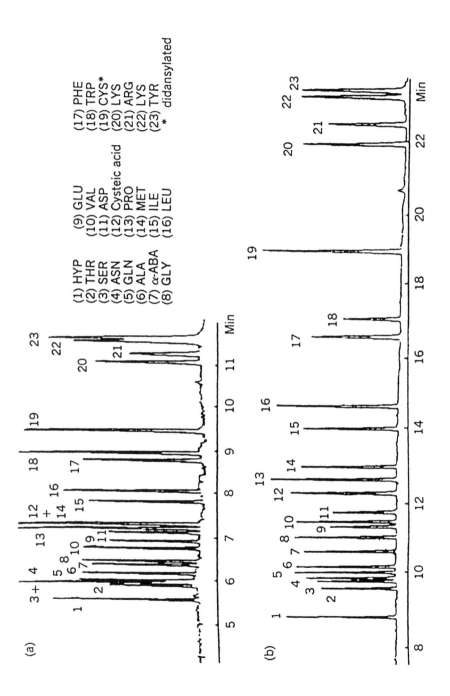

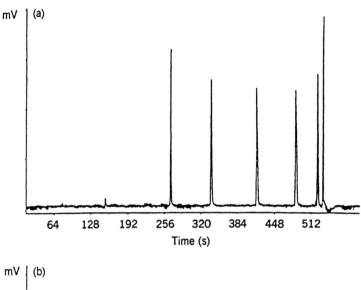
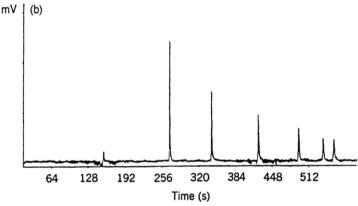
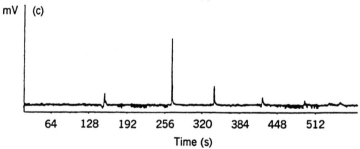

Figure 3.21. Zone sharpening of uncharged solutes in MEKC. Comparison of the MEKC separation of alkylphenones (acetophenone through heptaphenone) after 10 s: 2.5 kV electrokinetic injection with (a) appropriate zone sharpening, injection buffer containing mixed micelles of CTAB and SB-12; (b) no zone sharpening, injection buffer same as running buffer; (c) inappropriate zone sharpening, injection buffer mixed micelles of SDS and SB-12. The running buffer was 50 mM SDS and 20 mM phosphate buffer, pH 7.2. (From ref. 109 with permission.)

and Foley (109) reported on the zone sharpening of neutral solutes by sharpening the zones of charged micelles that serve as carriers for neutral (and charged) molecules. By means of electrokinetic injection, zone sharpening can be accomplished only when the effective electrophoretic mobility of the solute and the electroosmotic velocity migrate in the same direction during the injection process. A cationic–zwitterionic mixed micellar system had to be used in the sample buffer, with the running buffer containing SDS micelles, in order to achieve zone sharpening for homologous series of alkyl phenones. As a result, limits of detection for the solutes were reduced by as much as an order of magnitude. Efficiencies for heptaphenone exceeding a million theoretical plates were generated in under 10 minutes on a 50 cm capillary. Figure 3.21 illustrates the dramatic effects of zone sharpening (109).

Liu et al. achieved on-column concentration of neutral molecules by field-amplified sample stacking (144). The neutral analytes were dissolved in a low-concentration micellar solution (above CMC) with a lower ionic strength than that of the running micellar buffer. The negatively charged micelles migrate rapidly into the boundary between the sample and the running buffer, where they stack up. A 75-fold to 80-fold increase in sensitivity was observed for dioxins (144).

3.10. CONCLUSIONS AND FUTURE TRENDS

Over the past 13 years, MEKC has been utilized in hundreds of applications. A primary reason for the exploding interest in MEKC separations is a combination of high efficiency, versatility, speed, and ease of use, as well as the feasibility of manipulating selectivity through changes in chemical composition of the micellar solutions and incorporation of secondary equilibria. The applicability of this HPCE-based technique, which was primarily developed for the separation of uncharged solutes, has grown far beyond its initial intent. The range of applications cover wide groups of organic, inorganic, and biochemical compounds that are of interest in various disciplines such as the pharmaceutical, clinical, biotechnology, food, and environmental fields. Some examples of compounds and samples that have been analyzed by MEKC are given in Table 3.4 (165–234). This list is by no means exhaustive, and only selected references are listed that have been mostly published over the past three or four years. However, even this partial list provides evidence of the versatility of the technique.

As mentioned earlier in this chapter, the chemical composition of the pseudostationary phase plays a major role in the separation process. An area of research that has attracted interest and deserves even more attention is the use of pseudostationary phases with new and different chemistries. This would greatly enhance control over chemical selectivity, should extend the range of applications, and should lead to better separations. Various groups of pseudostationary phases such as micelle-forming synthetic and biological surfactants, polymerized micelles, ionic polymer, dendrimers, and other types of organized media have already been examined. In addition, through incorporation of various types of modifiers, especially organic solvents, and mixed pseudostationary phases, one can induce dramatic effects in MEKC separations. This combination of various types of organized assemblies and modifiers provides an unprecedentedly wide range of choices for manipulation of selectivity that can easily be incorporated into the separation process. In

Table 3.4. Recent Representative Applications of MEKC

Compounds/samples/applications	Reference(s)
Adrenaline and precursors	165
Aflatoxins	82, 153
Aldehydes	61
Alkaloids	166
Alkyl aromatics/aromatic hydrocarbons/PAH's	18–20,64,79,82,96,98,134,135,144,147,148, 152,167
Alkylphenones/alkylbenzenes	20,98,109,111,118
Amines and analines	26,148,155,168
Amino acids	16,19,54,57,67,68,98,103,105,112,121,164,169
Anticancer-related drugs	123,131,170,171
Anti-HIV drugs	83
Antibiotics	45,116,172
Antibodies	48
Antipyretic analgesics	173
Aromatic choline esters	77
Beer components/hop bitter acids	58,81,174
Benzodiazapines	70,124,175,176
Bilirubin	75
Bis(amidinohydrazones)	88
β-Blockers	23,92,148,177
Body fluids/biological samples	59,92,117,124,143,170,178–188
Caffeine	42,59,62
Carbohydrates/degradation products	40,43
Cardiovascular drugs	119,177,191
Catechols/amines	25,48,192
Cell extracts	96,193
Chiral separation	61,102–104,194–201
Chlorophylls	122,130

CMC determination	189,190
Colors and dyes	71,73,76
Coumarins	129
Diconazole and uniconazole	145
Diuretics	202
DNA adducts	203
Drug formulations	15,62,72,114,176,181,196,204
Drugs of abuse	91,127,132,205
Enalapril maleate	49
Environmental pollutants: pesticides, herbicides, fungicides	3,76,78,97,98,111,120,150,206–211
Enzymes	47,90
Fatty acids	113
Flavonoids	56,131,162,212
Food analysis	213–217
Fullerenes	20
Glucosinolates	89,95
Immunoassays	218,219
Indoles	169
Industrial chemicals	206
Inorganic ions	110
Isotopes	220
Metallocompounds	39,53,163
Microcystins	193,221
Nitroaromatics/warfare explosives	55,63,222
Nucleotides/nucleic acids	90,146
Oligo/disaccharides	93,151
Organic solvents	38
Peptides/proteins	50,51,74,85,106,163,223,224
Phenols	25,126,148,212,225–228
Phenothiazines	229
Phospholipids	230
Phthalates	76,77,87,136
Porphyrins	128
Steroids	64–66,149,231–233
Sulfonamides	60
Taxol and analogs	182
Vitamins and metabolites	52,76,125,234

order to fully take advantage of this great flexibility, however, a better characterization of these pseudostationary phases and roles of various modifiers is necessary. The application of the LSER methodology has provided promising preliminary results leading toward a better understanding of the separation mechanism in MEKC. Through a better grasp of the nature of solute interactions with individual and mixed pseudostationary phases as well as the effects of modifiers, investigators can select the composition of the MEKC buffer on a rational basis. Such a capability will result in better separations of mixtures of increasing complexity and a broadening of the scope of MEKC applications to wider groups of compounds.

Finally, one area of research that deserves more attention is on-line coupling of MEKC with mass spectrometry (MEKC–MS). The major obstacle with MEKC–MS is the contamination of the ion source by micelle-forming surfactants. Two successful attempts in on-line coupling of MEKC with electrospray MS have been reported (235,236). Lamoree et al. (235) made some modifications of the interface to minimize contamination of the ion source, while Ozaki et al. (236) demonstrated the possibility of using a high-molecular-mass surfactant in MEKC with on-line MS detection.

Foley and Masucci used a semipermeable membrane that allowed selective permeation of small molecules to the MS, while retaining the larger surfactant molecules (237). Terabe's group (238,239) as well as Lee's group (240–242) investigated the coupling of partial filling MEKC with electrospray ionization MS (ESI–MS) detection. In this aproach, only a portion of the capillary is filled with the micellar solution. The uncharged solutes are separated in the micellar segment before entering the last part of the capillary that is filled with electrophoresis buffer. Micelles never enter the ionization source of the MS in partial filling arrangement.

LIST OF ACRONYMS

Acronym	Definition
Big CHAP	$N,N,$-bis-(3-D-gluconamidopropyl) cholamide
BuS	butanesulfonate
CAPS	(3-cyclohexylamino)-1-propanesulfonic acid
CD	cyclodextrin
CE	capillary electrophoresis
CEC	capillary electrochromatography
CHAPS	3-[(3-chlamidopropyl)dimethylammonio]-1-propane sulfonate
CMC	critical micelle concentration
CTAB	cetyltrimethylammonium bromide
CZE	capillary zone electrophoresis
Deoxy Big CHAP	N,N-bis-(3-D-gluconamidopropyl)deoxycholamide
DTAB	dodecyltrimethylammonium bromide
EKC	electrokinetic chromatography
EOF	electroosmotic flow
GC	gas chromatography
GDC	glycodeoxycholate
HBA	hydrogen bond acceptor

HBD	hydrogen bond donor
HPLC	high-performance liquid chromatography
LC	liquid chromatography
LiPFOS	lithium perfluorooctane sulfonate
LSER	linear solvation energy relationships
MEGA	N-D-gluco-N-methyalkanamide
MECC	micellar electrokinetic capillary chromatography
MEKC	micellar electrokinetic chromatography
MS	mass spectrometry
PAHs	polycyclic aromatic hydrocarbons
RPLC	reversed-phase liquid chromatography
SB-12	Sulfobetain or N-dodecyl-N,N-dimethylammonio-3-propanesulfonate
SC	sodium cholate
SCE	secondary chemical equilibria
SDC	sodium deoxycholate
SDS	sodium dodecyl sulfate
TC	taurocholate
UV	ultraviolet

ACKNOWLEDGMENTS

The author thanks Dr. Jefferson G. Bumgarner for his assistance in literature search and for help in preparation of figures and the application table (3.4). Funding from the National Institutes of Health is gratefully acknowledged.

REFERENCES

1. S. Terabe, K. Otsuka, K. Ichikawa, A. Tsuchiya, and T. Ando, *Anal. Chem.* **56**, 111 (1984).
2. S. Terabe, K. Otsuka, and T. Ando, *Anal. Chem.* **57**, 834 (1985).
3. W. Kleibohmer, K. Cammann, J. Robert, and E. Mussenbrock, *J. Chromatogr.* **638**, 349 (1993).
4. M. G. Khaledi, in *Handbook of Capillary Electrophoresis*, (J. P. Landers, ed.), pp. 43–93. CRC Press; Boca Raton, FL, 1994.
5. W. L. Hinze, in *Ordered Media in Chemical Separations* (W. L. Hinze and D. W. Armstrong, eds.), pp. 2–82. American Chemical Society, Washington, DC, 1987.
6. K. Yoda, K. Tamori, K. Esumi, and K. Meguro, *J. Colloid Interface Sci.* **131**, 1 (1989).
7. P. Mukerjee and K. J. Mysels, *Natl. Stand. Ref. Data Ser.* **36**, 227 (1971).

8. R. Kuhn and S. Hoffstetter-Kuhn, *Capillary Electrophoresis: Principles and Practice*, p.192, Springer-Verlag, Berlin, 1993.
9. G. Conte, R. DiBlase, E. Giglio, A. Parretta, and N. Pavel, *J. Phys. Chem.* **88**, 5720 (1984).
10. G. Esposito, E. Giglio, N. Pavel, and A. Zanobi, *J. Phys. Chem.* **91**, 356 (1987)
11. A. Campanelli, S. De Sanctis, E. Chiessi, M. D'Alagni, E. Giglio, and L. Scaramuzza, *J. Phys. Chem.* **93**, 1536 (1989).
12. C. P. Palmer, M. Y. Khaled, and H. M. McNair, *J. High Resolut. Chromatogr.* **15**, 756 (1992).
13. C. P. Palmer and S. Terabe, *J. Microcolumn Sep.* **8**(2), 115 (1996).
14. N. Tanaka, T. Fukutome, K. Hosoya, K. Kimata, and T. Araki, *J. Chromatogr. A* **716**, 57 (1995).
15. S. A. Kuzdzal, C. A. Monnig, G. R. Newkome, and C. N. Moorefield, *J. Chem. Soc., Chem. Commun.*, 2139 (1994).
16. M. Castagnola, L. Cassiano, A. Lupi, I. Messana, M. Patamia, R. Rabino, D. V. Rossetti, and B. Giardina, *J. Chromatogr. A* **694**, 463 (1995).
17. P. G. H. M. Muijselaar, H. A. Claessens, C. A. Cramers, J. F. G. A. Jansen, E. W. Meijer, E. M. M. de Brabander van den Berg, and S. van der Wal, *J. High Resolut. Chromatogr.* **18**, 121 (1995).
18. H. Ozaki, S. Terabe, and A. Ichihara, *J. Chromatogr.* **680**, 117 (1994).
19. H. Ozaki, A. Ichihara, and S. Terabe, *J. Chromatogr. A* **709**, 3 (1995).
20. S. Yang, J. G. Bumgarner, and M. G. Khaledi, *J. High Resolut. Chromatogr.* **18**, 443 (1995).
21. E. S. Ahuja and J. P. Foley, *Analyst (London)* **119**, 353 (1994).
22. P. G. H. M. Muijselaar, H. A. Claessens, and C. A. Cramers, *Anal. Chem.* **66**, 635 (1994).
23. H. Siren, J. H. Jumppanen, K. Manninen, and M. L. Riekkola, *Electrophoresis* **15**, 779 (1994).
24. M. Idei, E. Kiss, and G. Keri, *Electrophoresis* **17**, 762 (1996).
25. M. G. Khaledi, S. C. Smith, and J. K. Strasters, *Anal. Chem.* **63**, 1820 (1991).
26. J. K. Strasters and M. G. Khaledi, *Anal. Chem.* **63**, 2503 (1991).
27. J. P. Foley, *Anal. Chem.* **62**, 1302 (1990).
28. S. Yang and M. G. Khaledi, *J. Chromatogr. A* **692**, 301 (1995).
29. S. Yang and M. G. Khaledi, *Anal. Chem.* **67**, 499 (1995).
30. M. J. Kamlet, R. M. Doherty, M. H. Abraham, Y. Marcus, and R. W. Taft, *J. Phys. Chem.* **92**, 5244 (1988).
31. R. W. Taft, M. H. Abraham, G. R. Famini, R. M. Doherty, J. M. Abboud, and M. J. Kamlet, *J. Pharm. Sci.* **74**, 807 (1985).
32. P. C. Sadek, P. W. Carr, R. W. Doherty, M. J. Kamlet, R. W. Taft, and M. H. Abraham, *Anal. Chem.* **57**, 2971 (1985).
33. P. W. Carr, *Microchem. J.* **48**, 4 (1993).

34. F. H. Quina, E. O. Alonso, and J. P. S. Farah, *J. Phys. Chem.* **99**, 11708 (1995).
35. S. Yang, J. G. Bumgarner, and M. G. Khaledi, *J. Chromatogr. A.* **738**, 265 (1996).
36. G. M. Janini, G. M. Muschik, and H. J. Issaq, *J. High Resolut. Chromatogr.* **18**, 171 (1995).
37. A. Fridstrom, K. E. Markides, and M. L. Lee, *Chromatographia* **41**(5/6), 295 (1995).
38. K. D. Altria and J. S. Howells, *J. Chromatogr. A* **696**, 341 (1995).
39. M. Macka, P. R. Haddad, and W. Buchberger, *J. Chromatogr. A* **706**, 493 (1995).
40. H. Schwaiger, P. J. Oefner, C. Huber, E. Grill, and G. K. Bonn, *Electrophoresis* **15**, 941 (1994).
41. G. Lai and J. P. Pachlatko, *J. High Resolut. Chromatogr.* **17**, 565 (1994).
42. R. Guo and W. Thormann, *Electrophoresis* **14**, 547 (1993).
43. C. Corradini and D. Corradini, *J. Microcolumn. Sep.* **6**, 19 (1994).
44. S. D. Harvey, R. M. Bean, and H. R. Udseth, *J. Microcolumn. Sep.* **4**, 191 (1992).
45. C. L. Flurer and K. A. Wolnik, *J. Chromatogr. A* **674**, 153 (1994).
46. Q. X. Dang, L. X. Yan, Z. P. Sun, and D. K. Ling, *J. Chromatogr.* **630**, 363 (1993).
47. J. Pedersen, M. Pedersen, H. Soeberg, and K. Biedermann, *J. Chromatogr.* **645**, 353 (1993).
48. D. E. Hughes and P. Richberg, *J. Chromatogr.* **635**, 313 (1993).
49. X. Z. Qin, D. P. Ip, and E. W. Tsai, *J. Chromatogr.* **626**, 251 (1992).
50. D. C. James, R. B. Freedman, M. Hoare, and N. Jenkins, *Anal. Biochem.* **222**, 315 (1994).
51. V. E. Klyushnichenko, D. M. Koulich, S. A. Yakimov, K. V. Maltsev, G. A. Grishina, I. V. Nazimov, and A. N. Wulfson, *J. Chromatogr. A* **661**, 83 (1994).
52. S. Boonkerd, M. R. Detaevernier, and Y. Michotte, *J. Chromatogr. A* **670**, 209 (1994).
53. K. Li and S. F. Y. Li, *J. Chromatogr. Sci.* **33**, 309 (1995).
54. E. Skocir, J. Vindevogel, and P. Sandra, *Chromatographia* **39** (1/2), 7 (1994).
55. R. L. Cheicante, J. R. Stuff, and H. D. Durst, *J. Chromatogr. A* **711**, 347 (1995).
56. C. Delgado, F. A. Tomas-Barberan, T. Talou, and A. Gaset, *Chromatographia* **38** (1/2), 71 (1994).
57. Y. F. Yik and S. F. Y. Li, *Chromatographia* **35** (9–12), 560 (1993).
58. R. Szucs, J. Vindevogel, P. Sandra, and L. C. Verhagen, *Chromatographia* **36**, 323 (1993).
59. W. Thormann, A. Minger, S. Molteni, J. Caslavska, and P. Gebauer, *J. Chromatogr.* **593**, 275 (1992).
60. Q. X. Dang, Z. P. Sun, and D. K. Ling, *J. Chromatogr.* **603**, 259 (1992).
61. S. Takeda, S. I. Wakida, M. Yamane, and K. Higashi, *Electrophoresis* **15**, 1332 (1994).
62. M. Korman, J. Vindevogel, and P. Sandra, *Electrophoresis* **15**, 1304 (1994).

63. E. Mussenbrock and W. Kleibohmer, *J. Microcolumn Sep.* **7**, 107 (1995).
64. H. Nishi and M. Matsuo, *J. Liq. Chromatogr.* **14**(5), 973 (1991).
65. J. H. Jumppanen, S. K. Wiedmer, H. Siren, M. L. Riekkola, and H. Haario, *Electrophoresis.* **15**, 1267 (1994).
66. J. G. Bumgarner and M. G. Khaledi, *Electrophoresis* **15**, 1260 (1994).
67. C. Bjergegaard, H. Simonsen, and H. Sorensen, *J. Chromatogr. A* **680**, 561 (1994).
68. S. Michaelsen, P. Moller, and H. Sorensen, *J. Chromatogr. A* **680**, 299 (1994).
69. E. S. Ahuja and J. P. Foley, *Anal. Chem.* **67**, 2315 (1995).
70. S. Boonkerd, M. R. Detaevernier, Y. Michotte, and J. Vindevogel, *J. Chromatogr. A* **704**, 238 (1995).
71. C. O. Thompson and V. C. Trenerry, *J. Chromatogr. A* **704**, 195 (1995).
72. S. J. Sheu and C. F. Lu, *J. High Resolut. Chromatogr.* **18**, 269 (1995).
73. W. C. Brumley, C. M. Brownrigg, and A. H. Grange, *J. Chromatogr. A* **680**, 635 (1994).
74. H. Frokiaer, P. Moller, H. Sorensen, and S. Sorenser, *J. Chromatogr. A* **680**, 437 (1994).
75. A. D. Harman, R. G. Kibbey, M. A. Sablik, Y. Fintschenko, W. E. Kurtin, and M. M. Bushey, *J. Chromatogr. A* **652**, 525 (1993).
76. W. C. Brumley and C. M. Brownrigg, *J. Chromatogr.* **646**, 377 (1993).
77. C. Bjergegaard, L. Ingvardsen, and H. Sorensen, *J. Chromatogr. A* **653**, 99 (1993).
78. W. C. Brumley and C. M. Brownrigg, *J. Chromatogr. Sci.* **32**, 69 (1994).
79. T. Kaneta, T. Yamashita, and T. Imasaka, *Anal. Chem. Acta* **299**, 371 (1995).
80. R. C. Williams, J. F. Edwards, and C. R. Ainsworth, *Chromatographia* **38** (7/8), 441 (1994).
81. P. A. Marshall, V. C. Trenerry, and C. O. Thompson, *J. Chromatogr. Sci.* **33**, 426 (1995).
82. R. O. Cole, M. J. Sepaniak, W. L. Hinze, J. Gorse, and K. Oldiges, *J. Chromatogr.* **557**, 113 (1991).
83. K. C. Chan, F. Majadly, T. G. McCloud, G. M. Muschik, H. J. Issaq, and K. M. Snader, *Electrophoresis* **15**, 1310 (1994).
84. T. Kaneta, T. Yamashita, and T. Imasaka, *Electrophoresis* **15**, 1276 (1994).
85. I. Beijersten and D. Westerlund, *Anal. Chem.* **65**, 3484 (1993).
86. B. Ye, M. Hadjmohammadi and M. G. Khaledi, *J. Chromatogr. A* **692**, 291 (1995).
87. C. Bjergegaard, S. Michaelsen, and H. Sorensen, *J. Chromatogr.* **608**, 403 (1992).
88. P. Lukkari, J. Jumppanen, T. Holma, H. Siren, K. Jinno, H. Elo, and M. L. Riekkola, *J. Chromatogr.* **608**, 317 (1992).
89. R. S. Ramsey, G. A. Kerchner, and J. Cadet, *J. High Resolut. Chromatogr.* **17**, 4 (1994).
90. A. Loregian, C. Scremin, M. Schiavon, A. Marcello, and G. Palu, *Anal. Chem.* **66**, 2981 (1994).

91. V. C. Trenerry, J. Robertson, and R. J. Wells, *Electrophoresis* **15**, 103 (1994).
92. P. Lukkari, T. Nyman, and M. L. Riekkola, *J. Chromatogr. A* **674**, 241 (1994).
93. S. Michaelsen, P. Moller, and H. Sorensen, *J. Chromatogr.* **608**, 363 (1992).
94. S. Michaelsen, M. B. Schroder, and H. Sorensen, *J. Chromatogr. A* **652**, 503 (1993).
95. C. Feldl, P. Moller, J. Otte, and H. Sorensen, *Anal. Biochem.* **217**, 62 (1994).
96. T. Kaneta and T. Imasaka, *Anal. Chem.* **67**, 829 (1995).
97. J. T. Smith and Z. El Rassi, *J. Chromatogr. A* **685**, 131 (1994).
98. J. T. Smith, W. Nashabeh, and Z. El Rassi, *Anal. Chem.* **66**, 1119 (1994).
99. Y. Mechref and Z. El Rassi, *J. Chromatogr. A* **724**, 285 (1996).
100. K. Otsuka, J. Kawahara, K. Tatekawa, and S. Terabe, *J. Chromatogr.* **559**, 209 (1991).
101. J. R. Mazzeo, E. R. Grover, M. E. Swartz, and J. S. Petersen, *J. Chromatogr. A* **680**, 125 (1994).
102. D. D. Dalton, D. R. Taylor, and D. G. Waters, *J. Chromatogr. A* **712**, 365 (1995).
103. D. C. Tickle, G. N. Okafo, P. Camilleri, R. F. D. Jones, and A. J. Kirby, *Anal. Chem.* **66**, 4121 (1994).
104. J. Wang and I. M. Warner, *Anal. Chem.* **66**, 3773 (1994).
105. N. Matsubara and S. Terabe, *J. Chromatogr. A* **680**, 311 (1994).
106. H. K. Kristensen and S. H. Hansen, *J. Chromatogr.* **628**, 309 (1993).
107. J. G. Bumgarner and M. G. Khaledi, *J. Chromatogr. A* **738**, 275 (1996).
108. E. S. Ahuja, E. L. Little, and J. P. Foley, *J. Microcolumn Sep.* **4**, 145 (1992).
109. K. R. Nielsen and J. P. Foley, *J. Chromatogr. A* **686**, 283 (1994).
110. P. R. Haddad, A. H. Harakuwe, and W. Buchberger, *J. Chromatogr. A* **706**, 571 (1995).
111. D. Crosby and Z. El Rassi, *J. Liq. Chromatogr.* **16**(9/10), 2161 (1993).
112. C. P. Ong, C. L. Ng, H. K. Lee, and S. F. Y. Li, *Electrophoresis* **15**, 1273 (1994).
113. F. B. Erim, X. Xu, and J. C. Kraak, *J. Chromatogr. A* **694**, 471 (1995).
114. S. Boonkerd, M. Lauwers, M. R. Detaevernier, and Y. Michotte, *J. Chromatogr. A* **695**, 97 (1995).
115. H. T. Rasmussen, L. K. Goebel, and H. M. McNair, *J. Chromatogr.* **517**, 549 (1990).
116. S. Croubels, W. Baeyens, C. Dewaele, and C. V. Peteghem, *J. Chromatogr. A* **673**, 267 (1994).
117. W. Thormann, S. Lienhard, and P. Wernly, *J. Chromatogr.* **636**, 137 (1993).
118. P. G. H. M. Muijselaar, H. A. Claessens, and C. A. Cramers, *J. Chromatogr. A* **696**, 273 (1995).
119. A. E. Bretnall and G. S. Clarke, *J. Chromatogr. A* **700**, 173 (1995).
120. G. Dinelli, A. Vicari, and V. Brandolini, *J. Chromatogr. A* **700**, 201 (1995).
121. A. Tivesten and S. Folestad, *J. Chromatogr. A* **708**, 323 (1995).
122. K. Saitoh, H. Kato, and N. Teramae, *J. Chromatogr. A* **687**, 149 (1994).

123. K. C. Chan, G. M. Muschik, H. J. Issaq, and K. M. Snader, *J. High Resolut. Chromatogr.* **17**, 51 (1994).
124. M. Schafroth, W. Thormann, and D. Allemann, *Electrophoresis* **15**, 72 (1994).
125. G. Dinelli and A. Bonetti, *Electrophoresis* **15**, 1147 (1994).
126. J. H. Aiken and C. W. Huie, *J. Mlicrocolumn Sep.* **5**, 95 (1993).
127. M. Krogh, S. Brekke, F. Tonnesen, and K. E. Rasmussen, *J. Chromatogr. A* **674**, 235 (1994).
128. C. Kiyohara, K. Saitoh, and N. Suzuki, *J. Chromatogr.* **646**, 397 (1993).
129. C. T. Chen and S. J. Sheu, *J. Chromatogr. A* **710**, 323 (1995).
130. H. Kato, K. Saitoh, and N. Teramae, *Chem. Lett.* 547 (1995).
131. L. G. Song, S. M. Zhang, Q. Y. Ou, and W. L. Yu, *Chromatographia* **39** (11/12), 682 (1994).
132. R. Weinberger and I. S. Lurie, *Anal. Chem.* **63**, 823 (1991).
133. T. Kaneta, S. Tanaka, M. Taga, and H. Yoshida, *J. Chromatogr.* **609**, 369 (1992).
134. M. Greenaway, G. Okafo, D. Manallack, and P. Camilleri, *Electrophoresis* **15**, 1284 (1994).
135. K. Otsuka, M. Higashimori, R. Koike, K. Karuhaka, Y. Okada, and S. Terabe, *Electrophoresis* **15**, 1280 (1994).
136. S. Takeda, S. I. Wakida, M. Yamane, A. Kawahara, and K. Higashi, *Anal. Chem.* **65**, 2489 (1993).
137. C. Bjergegaard, S. Michaelsen, K. Mortensen, and H. Sorensen, *J. Chromatogr. A* **652**, 477 (1993).
138. S. Katsuta, T. Tsumura, K. Saitoh, and N. Teramae, *J. Chromatogr. A* **705**, 319 (1995).
139. N. Chen, S. Terabe, and T. Nakagawa, *Electrophoresis* **16**, 1457 (1995).
140. N. Chen and S. Terabe, *Electrophoresis* **16**, 2100 (1995).
141. T. Kaneta, T. Yamashita and T. Imasaka, *Electrophoresis* **15**, 1276 (1994).
142. S. Terabe, Y. Ishihama, H. Nishi, T. Fukuyama, and K. Otsuka, *J. Chromatogr.* **545**, 359 (1991).
143. K. J. Lee, J. J. Lee, and D. C. Moon, *Electrophoresis* **15**, 98 (1994).
144. Z. Liu, P. Sam, S. R. Sirimanne, P. C. McClure, J. Grainger, and D. G. Patterson, *J. Chromatogr. A* **673**, 125 (1994).
145. R. Furuta and T. Doi, *J. Chromatogr. A* **676**, 431 (1994).
146. C. D. Bevan, I. M. Mutton, and A. J. Pipe, *J. Chromatogr.* **636**, 113 (1993).
147. S. Terabe, Y. Miyashita, O. Shibata, E. R. Barnhart, L. R. Alexander, D. G. Patterson, B. L. Karger, K. Hosoya, and N. Tanaka, *J. Chromatogr.* **516**, 23 (1990).
148. A. Aumatell and R. J. Wells, *J. Chromatogr. A* **688**, 329 (1994).
149. K. C. Chan, G. M. Muschik, H. J. Issaq, and P. K. Siiteri, *J. Chromatogr. A* **690**, 149 (1995).
150. W. C. Brumley and W. J. Jones, *J. Chromatogr. A* **680**, 163 (1994).

151. M. Greenaway, G. N. Okafo, P. Camilleri, and D. Dhanak, *J. Chem. Soc., Chem. Commun.,* 1691 (1994).
152. C. L. Copper and M. J. Sepaniak, *Anal. Chem.* **66**, 147 (1994).
153. R. D. Holland and M. J. Sepaniak, *Anal. Chem.* **65**, 1140 (1993).
154. M. G. Khaledi, R. P. Sahota, J. K. Strasters, C. Quang, and S. C. Smith in *Capillary Electrophoresis Technology* (N. Guzman, ed.), p. 187. Dekker, New York, 1993.
155. C. Quang, J. K. Strasters, and M. G. Khaledi, *Anal. Chem.* **66**, 1646 (1994).
156. H. Crostjens, A. E. E. Oord, H. A. H. Billiet, J. Frank, and K. Ch. A. M. Luyben, *J. High Resolut. Chromatogr.* **18**, 551 (1995).
157. H. Corstjens, H. A. H. Billiet, J. Frank, and K. Ch. A. M. Luyben, *J. Chromatogr. A* **715**, 1 (1995).
158. U. Pyell and U. Butehorn, *Chromatographia* **40** (1/2), 69 (1995).
159. U. Pyell and U. Butehorn, *Chromatographia* **40** (3/4), 175 (1995).
160. U. Pyell and U. Butehorn, *J. Chromatographia. A* **716**, 81 (1995).
161. A. E. Bretnall and G. S. Clarke, *J. Chromatogr. A* **716**, 49 (1995).
162. C. L. Ng, C. P. Ong, H. K. Lee, and S. F. Y. Li, *Chromatographia* **34**(3/4), 166 (1992).
163. J. H. Beattie and M. P. Richards, *J. Chromatogr. A* **700**, 95 (1995).
164. H. Wan, P. E. Andersson, A. Engstrom, and L. G. Blomberg, *J. Chromatogr. A* **704**, 179 (1995).
165. Y. Esaka, K. Tanaka, B. Uno, M. Goto, and K. Kano, *Anal. Chem.* **69**, 1332 (1997).
166. S. S. Yang, I. Smetena, and A. I. Goldsmith, *J. Chromatogr.* **746**, 131 (1996).
167. W. L. Ding and J. S. Fritz, *Anal. Chem.* **69**, 1593 (1997).
168. S. Takeda, S. I. Wakida, M. Yamane, A. Kawahara, and K. Higashi, *J. Chromatogr. A.* **653**, 109 (1993).
169. K. C. Chan, G. M. Muschik, and H. J. Issaq, *J. Chromatogr. A* **718**, 203 (1995).
170. H. J. Issaq, K. C. Chan, G. M. Muschik, and G. M. Janini, *J. Liq. Chromatogr.* **18**(7), 1273 (1995).
171. B. W. Wenclawiak and M. Wollmann, J. Chromatogr. A **724**, 317 (1996).
172. S. H. Chen, H. L. Wu, S. M. Wu, H. S. Kou, and S. J. Lin, *J. Chin. Chem. Soc.-TAIP* **43**, 393 (1996).
173. X. Y. Fu, J. D. Lu, and A. Zhu, *J. Chromatogr. A* **735**, 353 (1996).
174. R. Szucs, E. VanHove, and P. Sandra, *J. High Resolut. Chromatogr.* **19**, 189 (1996).
175. I. Bechet, M. Fillet, P. Hubert, and J. Crommen, *Electrophoresis* **15**, 1316 (1994).
176. M. F. RenouGonnord and K. David, *J. Chromatogr. A* **735**, 249 (1996).
177. A. E. Bretnall and G. S. Clarke, *J. Chromatogr. A* **745**, 145 (1996).
178. Q. Y. Chu, B. T. Evans, and M. G. Zeece, *J. Chromatogr. B* **692**, 293 (1997).
179. Z. K. Shihabi and M. E. Hinsdale, *J. Chromatogr. B* **669**, 75 (1995).

180. F. von Heeren, E. Verpoorte, A. Manz, and W. Thormann, *Anal. Chem.* **68**, 2044 (1996).
181. H. Nishi, S. Terabe, *J. Chromatogr. A* **735**, 3 (1996).
182. G. Hempel, D. Lehmkuhl, S. Krumpelmann, G. Blaschke, and J. Boos, *J. Chromatogr. A* **745**, 173 (1996).
183. K. C. Panak, O. A. Ruiz, S. A. Giorgieri, and L. E. Diaz, *Electrophoresis* **17**, 1613 (1996).
184. W. Thormann, C. X. Zhang, and A. Schmutz, *Ther. Drug Monit.* **18**, 506 (1996).
185. D. K. Lloyd, *J. Chromatogr. A* **735**, 29 (1996).
186. S. Mayer and M. Schleimer, *J. Chromatogr A* **730**, 297 (1996).
187. F. vonHeeren, E. Verpoorte, A. Manz, and W. Thormann, *Anal. Chem.* **68**, 2044 (1996).
188. G. C. Penalvo, M. Kelly, H. Maillols, and H. Fabre, *Anal. Chem.* **69**, 1364 (1997).
189. J. C. Jacquier and P. L. Desbene, *J. Chromatogr. A* **718**, 167, (1995).
190. J. C. Jacquier and P. L. Desbene *J. Chromatogr. A* **743**, 307, (1996).
191. N. T. Nguyen and R. W. Siegler, *J. Chromatogr. A* **735**, 123 (1996).
192. I. Rodriguez, H. K. Lee, and S. F. Y. Li, *J. Chromatogr. A* **745**, 255 (1996).
193. N. Bouaicha, C. Rivasseau, M. C. Hennion, and P. Sandra, *J. Chromatogr. B* **685**, 53 (1996).
194. J. Wang and I. M. Warner, *J. Chromatogr. A* **711**, 297 (1995).
195. H. Nishi, *J. Chromatogr. A* **735**, 57 (1996).
196. M. E. Swartz, J. R. Mazzeo, E. R. Grover, and P. R. Brown, *J. Chromatogr. A* **735**, 303 (1996).
197. A. Amini, I. Beijersten, C. Pettersson, and D. Westerlund, *J. Chromatogr. A* **737**, 301 (1996).
198. M. L. Marina, I. Benito, J. C. DiezMasa, and M. J. Gonzalez, *J. Chromatogr. A* **752**, 265 (1996).
199. P. L. Desbene and C. E. Fulchic, *J. Chromatogr. A* **749**, 257 (1996).
200. A. G. Peterson and J. P. Foley *J. Chromatogr. B* **752**, 683, 15 (1996).
201. J. G. Clothier and S. A. Tomellini, *J. Chromatogr. A* **723**, 179 (1996).
202. M. L. Riekkola and J. H. Jumppanen, *J. Chromatogr. A* **735**, 151 (1996)
203. K. V. Penmetsa, D. Shea, and R. B. Leidy, *J. High Resolut. Chromatogr.* **18**, 90 (1995).
204. S. D. Fazio, *LC GC—Mag. Sep. Sci.* **13**, 338 (1995).
205. O. Naess and K. E. Rasmussen, *J. Chromatogr. A* **760**, 245 (1996).
206. G. M. McLaughlin, A. Weston, and K. D. Hauffe, *J. Chromatogr. A* **744**, 123 (1996).
207. K. V. Penmetsa, R. B. Leidy, and D. Shea, *J. Chromatogr.* 000 **745**, 201–208, (1996).
208. K. V. Penmetsa, R. B. Leidy, and D. Shea, *Electrophoresis* **18**, 235 (1997).

209. M. Rossi and D. Rotilio, *J. High Resolut. Chromatogr.* **20**, 265 (1997).
210. R. C. Martinez, E. R. Gonzalo, A. I. M. Dominguez, J. D. Alvarez, and J. H. Mendez, *J. Chromatogr. A* **733**, 349 (1996).
211. W. C. Brumley, *LC GC—Mag. Sep. Sci.* **13**, 556 (1995).
212. H. R. Liang, H. Siren, M. L. Riekkola, P. Vuorela, H. Vuorela, and R. Hiltunen, *J. Chromatogr. A* **746**, 123 (1996).
213. S. Abrantes, M. R. Philo, A. P. Damant, and L. Castle, *J. High Resolut. Chromatogr.* **20**, 270 (1997).
214. U. Butehorn and U. Pyell, *J. Chromatogr. A* **736**, 321 (1996).
215. M. Strickland, B. C. Weimer, and J. R. Broadbent, *J. Chromatogr. A* **731**, 305 (1996).
216. E. Skocir and M. Prosek, *Chromatographia* **41**, 638 (1995).
217. J. Lindeberg, *Food Chem.* **55**, 73 (1996).
218. M. Kats, P. C. Richberg, and D. E. Hughes, *Anal. Chem.* **69**, 338 (1997).
219. L. Steinmann and W. Thormann, *Electrophoresis* **17**, 1348 (1996).
220. Chiari, M. Nesi, G. Ottolina, and P. G. Righetti, *J. Chromatogr. A* **680**, 571 (1994).
221. N. Onyewuenyi and P. Hawkins, *J. Chromatogr. A* **749**, 271 (1996).
222. S. Kennedy, B. Caddy, and J. M. F. Douse, *J. Chromatogr. A* **726**, 211 (1996).
223. M. Petersson, K. Walhagen, A. Nilsson, K. G. Wahlund, and S. Nilsson, *J. Chromatogr. A* **769**, 301 (1997).
224. I. Beijersten and D. Westerlund, *Electrophoresis* **17**, 161 (1996).
225. Y. He and H. K. Lee, *J. Chromatogr. A* **749**, 227 (1996).
226. S. Kar and P. K. Dasgupta, *J. Chromatogr. A* **739**, 379 (1996).
227. M. C. Boyce and I. J. Bennett, *Anal. Lett.* **29**, 1805 (1996).
228. G. Li and D. C. Locke, *J. Chromatogr. A* **734**, 357 (1996).
229. P. G. H. M. Muijselaar, H. A. Claessens, and C. A. Cramers, *J. Chromatogr. A* **735**, 395 (1996).
230. R. Szucs, K. Verleysen, G. Duchateau, P. Sandra, and B. Vandeginste, *J. Chromatogr. A* **738**, 25 (1996).
231. R. E. Milofsky, M. G. Malberg, and J. M. Smith, *J. High Resolut. Chromatogr.* **17**, 731 (1994).
232. L. Vomastova, I. Miksik, and Z. Deyl, *J. Chromatogr. B* **681**, 107 (1996).
233. S. K. Poole and C. F. Poole, *J. Chromatogr. A* **749**, 247 (1996).
234. A. Profumo, V. Profumo, and G. Vidali, *Electrophoresis* **17**, 1617 (1996).
235. M. H. Lamoree, U. R. Tjaden, and J. van der Greef, *J. Chromatogr. A* **712**, 219 (1995).
236. H. Ozaki, N. Itou, S. Terabe, Y. Takada, M. Sakairi, and H. Koizumi, *J. Chromatogr. A* **716**, 69 (1995).
237. J. P. Foley and J. A. Masucci, *Int. Symp. Capillary Chromatography Electrophor., 17th* Wintergreen, VA, 1995, Abstr., p. 278 (1995).

238. S. Terabe, H. Ozaki, Y. Takada, M. Sakairi, H. Koizuni, *Int. Symp. High Perform. Capillary Electrophor., 7th* Wurtzberg, 1995, Abstr., p. 53 (1995).
239. P. G. Muijselaar, M. Yokoi, K. Otsuka, and S. Terabe, *Int. Symp. High Perform. Capillary Electrophor., 10th* Kyoto, 1997, Abstr., p. 48 (1997).
240. W. M. Nelson and C. S. Lee, *Anal. Chem.* **68**, 3265 (1996).
241. W. M. Nelson, Q. Tang, A. K. Harrata, and C. S. Lee, *J. Chromatogr. A* **749**, 219 (1996).
242. L. Yang, A. K. Harrata, and C. S. Lee, *Anal. Chem.* **69**, 1820 (1997).

CHAPTER

4

BAND BROADENING IN MICELLAR ELECTROKINETIC CHROMATOGRAPHY

JOE M. DAVIS

Department of Chemistry and Biochemistry, Southern Illinois University at Carbondale, Carbondale, Illinois 62901-4409

4.1.	Introduction	142
4.2.	Measurements of Efficiency	143
4.3.	Plug Size	143
4.4.	The Detection Window and Time Constants	144
4.5.	Micellar Overload	145
4.6.	Longitudinal Diffusion	146
4.7.	Nonequilibrium Dispersion in Micellar Systems	148
	4.7.1. Dispersions Not Attributed to Joule Heating or Wall Adsorption	148
	4.7.2. Dispersions Attributed to Joule Heating	154
	4.7.3. Dispersions Attributed to Wall Adsorption	155
	4.7.4. Other Temperature Effects	157
4.8.	Electromigrative Dispersion	158
4.9.	Dependence of N on Concentration of Organized Media	158
	4.9.1. Single Micellar Systems	159
	4.9.2. Mixed Media	163
4.10.	Dependence of N on Electrolyte Concentration and Composition	164
4.11.	Dependence of N on Organic Solvents	165
	4.11.1. Single Micellar Systems	166
	4.11.2. Mixed Micellar Systems	169
4.12.	Efficiency of Pseudostationary Phases Other Than Simple Micelles	171
	4.12.1. Charged Cyclodextrins	171
	4.12.2. Microemulsions	171
	4.12.3. Bile Salts	172
	4.12.4. Mixed Media	172
	4.12.5. Oligomerized and Polymerized Micelles	173
	4.12.6. Ion-Exchange Polyelectrolytes	173
	4.12.7. Dendrimers	174

High Performance Capillary Electrophoresis, edited by Morteza G. Khaledi. Chemical Analysis Series, Vol. 146.
ISBN 0-471-14851-2 © 1998 John Wiley & Sons, Inc.

4.13. General Observations	174
4.14. Conclusions	175
Addendum	176
List of Acronyms and Abbreviations	178
References	179

4.1. INTRODUCTION

Few developments in separation science over the past decade have had the impact of electrokinetic separations. For small (and especially neutral) molecules, micellar electrokinetic chromatography (MEKC) and related methods have had a phenomenal impact. Dozens of publications have addressed the fundamentals of MEKC, and hundreds of applications have been reported.

This chapter addresses dispersion or band broadening in MEKC. To keep the chapter at a reasonable length, the reviewer was selective. For example, separations based on neutral organized media (e.g., cyclodextrins or crown ethers) are not considered, although charged and charged/neutral media other than micelles are covered. In addition, efficiency losses due to coupling MEKC capillaries to other systems, e.g., detection or reaction capillaries, are not addressed.

Here studies have been classified as primary and secondary. In the former, efficiency was addressed in either extensive or fair detail; in the latter, efficiency was of secondary concern and was reported in only a sentence or two. The latter usually are summarized in separate subsections under the heading *Secondary Studies*.

A word of caution is relevant to one's reading of this review, which attempts to be fairly comprehensive. Because of MEKC's high efficiency, researchers have not always made efforts to develop experiments in which dispersion is carefully controlled. This assessment is not a criticism but simply a fact. The concerns of most researchers in MEKC are practical, and efficiency losses rarely are important to them as long as sufficient efficiency exists for their purposes. In contrast, plate numbers and heights are easy quantities to measure, and researchers are expected to report them as figures of merit. The consequence is that not all reports of efficiency are equally meaningful. Because many variables affect dispersion in MEKC, it is possible (indeed, probable) that some results and interpretations reviewed here are artifactual and specious, respectively, because of poor variable control. The reader is encouraged to bear this possibility in mind.

In most cases only brief summaries of issues affecting dispersion in MEKC, *as interpreted by the authors of the studies*, are presented. The most thorough papers are examined more carefully and, occasionally the reviewer ventures to offer some comments.

4.2. MEASUREMENTS OF EFFICIENCY

The two measures of efficiency used here are plate number N and plate height H:

$$N = L_d^2 \bigg/ \sum_i \sigma_i^2 = t_r^2 \bigg/ \sum_i \sigma_{t_i}^2; \quad H = L_d/N \qquad (4.1)$$

where L_d is the length of the capillary between the detector window and injection end; t_r is the analyte retention time; and σ_i^2 ($\sigma_{t_i}^2$) is the spatial (temporal) variance due to the ith independent source of dispersion. Various approximations were used in calculating these measures from experimental data, the most common being the approximation of peak shapes by Gaussians. In some cases, N and H were calculated by moments analysis.

4.3. PLUG SIZE

Because plate numbers N in MEKC usually exceed 10^5, the analyte plug size strongly affects efficiency, as it does in capillary electrophoresis (CE) (1,2), and sets an upper limit to efficiency. Most users of MEKC introduce short unfocused hydrodynamic plugs of 1 to 2 mm. To a good approximation, the plug is a rectangular concentration pulse and generates a variance σ_{inj}^2 equal to $\ell^2/12$, where ℓ is the plug length. For a capillary with $L_d = 0.4$ m, Eq. (4.1) predicts that such plugs set upper limits to N of 4.8 to 19.2×10^5 plates.

Plug size ℓ can be estimated from commonly reported equations for hydrodynamic and electrokinetic transport. Some commercial instruments use a split, by which a fraction of a large-volume injection passes into the capillary; the injected volume is determined by the split ratio. In our lab, plug size is determined from peak area A^* (3):

$$\ell = \mu A^* E / A_0 \qquad (4.2)$$

where μ is the apparent (or measured) analyte mobility; $E = V/L_t$ is the electric field strength generated by the potential difference V between the ends of a capillary having length L_t; and A_0 is the maximum analyte response determined by introducing analyte continuously until a breakthrough is achieved. In other words, A_0 is the response difference between the plateaus of the sigmoidal signal generated by frontal chromatography. For mixtures of analytes, breakthroughs must be generated electrokinetically.

Severe efficiency losses result from large plugs. Burton et al. (4) demonstrated significant losses in N by electrokinetically injecting caffeine and related compounds at low voltages for long times (e.g., 5 kV for 60 s) and at high voltages for short times (e.g., 30 kV for 30 s). Row et al. (5) and Northrup et al. (6) also measured efficiency losses for nucleotides and explosive constituents, respectively, under similar conditions. Terabe et al. verified that $\sigma_{inj}^2 = \ell^2/12$ by siphoning phenol for different time intervals; for 4.5 mm plugs, the plate height increased by 2.5, relative to $\ell = 0$ (7). Vindevogel and Sandra demonstrated efficiency losses for iso-α-acids by hydrodynamically injecting them into a capillary for 1–8 s; N decreased nearly 10-fold in some cases (8).

Stacking and field-amplification techniques are used routinely in CE to focus analyte plugs to acceptable sizes (9,10). In both techniques, a large conductivity difference between buffer (high conductivity) and sample (low conductivity) causes a sharp localized gradient in field strength, toward which sample ions rapidly migrate and focus. Their use in MEKC, however, has been limited. Weinberger et al. obtained substantial improvements in N's of urinary porphyrins by dissolving them in 20 mM (3-cyclohexylamino)-1-propanesulfonic acid (CAPS) buffer but no micelles [the working buffer contained both 20 mM CAPS and 100 mM sodium dodecyl sulfate (SDS)]; N's were nearly constant for up to 5 s for vacuum injections. The authors argued, however, that most porphyrins were electrostatically repelled from the micelles, which were added to reduce wall adsorption (11). Szücs et al. focused iso-α-acids by hydrodynamic injection but failed to focus them by electrokinetic injection, because the micelles moved faster and opposite to the electroosmotic flow (EOF). They also determined the ratio of sample-to-working buffer conductivity necessary for adequate focusing and determined N at various injection times for a 10-fold diluted sample; N varied little for up to 20 s injections under these conditions (12). Nielsen and Foley showed that surfactants in the sample migrating with the EOF avoided the problem of Szücs et al. when surfactants in the working buffer migrated against the EOF; relative to unfocused zones, the N's of alkylphenones and simple aromatics having capacity factors $k > 30$ were increased by two to five times. For less retained analytes, however, N decreased by roughly twofold (13). Liu et al. used field-amplified injection to preconcentrate tetrachlorodibenzodioxins; they also determined N as a function of injection volume for up to 200 nL injections and found the optimal SDS concentration for stacking was about 8 mM, or about the tabulated critical micelle concentration (CMC) in the absence of electrolyte salts (14).

4.4. THE DETECTION WINDOW AND TIME CONSTANTS

A rectangular window for on-line optical detection typically is generated by removing a small section of polyimide from the capillary. If w is the width of the illuminated region (which is not necessarily the window width), the detected peak can be described by the convolution of the on-column peak with a rectangular response function of unit area and temporal width $T = w/v_s$, where v_s is the analyte velocity. The resultant temporal variance σ_{det}^2 equals $T^2/12$.

The reviewer notes that this dispersion is apparent (i.e., mesurable) but not real. The actual on-column spatial peak width is unaffected by w, but because the peak is detected over the time required to move distance w, the *observed* temporal width is affected.

Terabe et al. expressed the convolution integral for an on-column peak represented by a Gaussian and concluded that T should be less than 0.7 times the temporal standard deviation of the on-column peak if the observed peak width is to be no greater than 2.4% of the on-column peak width (7). [*Note*: This profile can be described analytically (15).] Huang et al. developed a simple algebraic conversion from temporal to spatial peak widths; the conversion was verified by analysis of CE peaks of dansylated lysine at different voltages (1). Luckey et al. approximated the spatial variance for detection of DNA bands by laser-induced fluorescence in CE as the square of the beam waist, divided by 4; this expression accounted for the Gaussian beam shape (16).

The time constants of detectors and recorders also affect the observed peak profile. As observed by Terabe et al., the former contributes a temporal variance equal to the square of the time constant; the latter, the square of the time required for full-scale deflection, divided by 18 (7). In our lab, we typically filter noise with time constants of 0.1 s or less and have observed distortion of high-efficiency peaks (e.g., $N \approx 6 \times 10^5$) smoothed by 0.3 s filters.

4.5. MICELLAR OVERLOAD

Efficiency can be lost if insufficient organized media are present to solubilize analytes. The origin of the loss is not always clear. It is not necessary that one micelle exist for each analyte molecule or ion; indeed, theories for the spectroscopic behavior of fluorescent micellar probes often postulate a Poisson distribution of occupancy numbers (17). In general, however, efficiency is highest at low analyte concentration, and overload is most severe for highly retained analytes.

Some studies reflecting media overload have been reported. Row et al. reported significant decreases in N with increasing analyte concentration and noted that the maximum N was reached near the detection limit; N's of nucleotides decreased by 20–50% after a 10-fold concentration increase (5). Terabe et al. observed that the commonly reported decrease in N with decreasing media concentration (which also has been explained by resistance to mass transfer due to intermicellar diffusion; see below) could result from micellar overload (7). Little and Foley observed efficiency losses and peak fronting in low concentrations of mixed polyoxyethylene (23) dodecanol (Brij 35)/SDS micelles; these losses were less severe than for low concentrations of SDS micelles. The efficiency losses in SDS were attributed to overload; this explanation seemed less likely for the mixed micelles (18). Song et al. found that *n*-butanol/*n*-octane/SDS microemulsions, SDS micelles, and mixed SDS/poly(ethyleneglycol 400 monolaurate) micelles generated N's less than 10^5, possibly because analyte concentrations were roughly 1–3 mM (the authors noted that overload was a possibility) (19).

In our lab, we observe no evidence of overload in 50–100 mM SDS, regardless of analyte capacity factor, when 40 µM analyte concentrations are used. However, in 15 mM SDS, 40 µm concentrations of highly retained analytes *occasionally* generate broadened peaks (e.g., decreases in N by factors of 2 or so, relative to other times). Efficiency has been restored by washing the capillary with 0.1 M NaOH, suggesting (but not proving) that analyte was adsorbed by the capillary wall (20). We also have observed the skewing of peaks (i.e., peaks shaped like right triangles) for highly retained compounds at roughly 300 µM in 10 mM SDS solutions; dilution with working buffer caused the peaks to return to Gaussian shapes (21).

4.6. LONGITUDINAL DIFFUSION

Unless plug size or overload limits efficiency, the dominant source of dispersion in MEKC at low field strengths is longitudinal diffusion. The near-linear increase in N with voltage V at low voltage is attributable to longitudinal diffusion; the more rapidly analytes elute, the less time they diffuse. The dispersion's origin is an axial gradient in analyte concentration, which generates a diffusive flux along the separation axis.

The variance σ^2_{diff} of this dispersion is given at time t by the Einstein diffusion equation $\sigma^2_{\text{diff}} = 2Dt$, where D is the weighted analyte diffusion coefficient (see below). At the time the analyte passes the detection window, the plate height contribution H_{diff} is

$$H_{\text{diff}} = 2D/(\mu E) \quad (4.3a)$$

where μ is

$$\mu = \mu_{\text{eo}} + (1 - R)\mu_{\text{mc}} \quad (4.3b)$$

Here μ_{eo}, μ_{mc}, and R are respectively the electroosmotic flow coefficient, the micellar electrophoretic mobility, and the analyte retardation factor (or retention ratio), i.e., the fraction of analyte in the mobile phase. For typical values of D and μ (e.g., $3 \times 10^{-10}\,m^2/s$ and $2.5 \times 10^{-8}\,m^2/Vs$), H_{diff} is on the order of μm for typical field strengths (e.g., 1.2 μm for the above numbers and $E = 20\,kV/m$). Plate heights larger than a few micrometers indicate that other dispersion sources are present.

Sepaniak and Cole suggested that longitudinal diffusion in MEKC occurs only in the mobile phase (22). The use of a diffusion coefficient weighted by the equilibrium distribution of analyte between aqueous and micellar phases goes back to at least 1955 (23), however, and it is generally accepted that the axial diffusion coefficient D is so weighted. It can be expressed as (7,23,24)

$$D = RD_m + (1 - R)D_{mc} \qquad (4.3c)$$

where D_m is the mobile-phase analyte diffusion coefficient and D_{mc} is the diffusion coefficient of the analyte–micelle adduct, which almost equals the micellar diffusion coefficient. Equation (4.3a–c) was derived independently by Terabe et al. (7) and Davis (24).

Terabe et al. showed by van Deemter-like graphs of plate height H vs. EOF velocity that H's of a large number of simple aromatics were limited by longitudinal diffusion at EOFs up to 2 mm/s (7). In a recent study, Mazzeo et al. showed that plate numbers of N-methylpseudoephedrine enantiomers resolved by a chiral surfactant were large when the pH favored large interfacial charges and EOFs (e.g., 263,000 at pH = 6) but were small when it did not (e.g., 82,000 at pH = 4.5) (25); hence, dispersion was related directly to time in the capillary. Yu and Davis showed that N's of hydrophilic compounds could be described fairly quantitatively by theory for longitudinal diffusion and plug size, when $E \leq 30\,kV/m$ (3,20).

In an extensive study, Nielsen found that the H's for focused plugs of alkylketones, nitroalkanes, aromatic acids, and other pure compounds were governed by longitudinal diffusion in 50 μm capillaries for EOFs less than roughly 3.5 mm/s. Several media/buffer systems were examined, including SDS/borate, SDS/butanol/borate, and cetyltrimethylammonium bromide (CTAB)/Tris (26). This is the most detailed study of longitudinal diffusion known to the reviewer. Part of the study's purpose was to find nonequilibrium dispersion of the types proposed by Terabe (7) and Davis (24); one closely related case (discussed below) was found.

Secondary Studies. Smith and El Rassi explained the small N's of alkylphenylketones resolved by n-octanoylsucrose–borate micelles by low analyte retention and extensive diffusion in the mobile phase (27). Smith et al. showed that

N's of herbicides increased with solubilization by N-D-gluco-N-methylalkanamide/borate surfactants, which reduced diffusion coefficient D (28). Feldl et al. found that N's of indolyl glucosinolates and related compounds decreased with increasing temperature, especially at 50–60 °C, possibly because of increases in diffusion with temperature (29); similar observations were reported by Castagnola et al. for dansylated amino acids resolved by dendrimers and dodecyltrimethylammonium bromide (DTAB) micelles (30) and were modeled in CE by Knox and McCormack (31) and in MEKC by Yu and Davis (3).

4.7. NONEQUILIBRIUM DISPERSION IN MICELLAR SYSTEMS

Nonequilibrium dispersion is often observed in MEKC. As in CE, this dispersion may arise from mobility differences due to Joule heating or analyte adsorption to the capillary wall. In other cases, the dispersion's origin may be unique to MEKC itself. In all cases, the dispersion results from analyte molecules or ions moving through the capillary with a distribution of velocities. Regardless of origin, the result is an increase in plate height with increasing E. It is particularly troublesome, because rapid separations require high E's and thus cannot be carried out with optimal efficiency.

This section is divided into four categories: the first addresses dispersions not attributed to Joule heating or wall adsorption; the second addresses dispersions attributed to Joule heating; the third addresses dispersions attributed to wall adsorption; and the last addresses dispersions attributed to changes in buffer temperature. The inclusion here of the last category is not very logical, but Joule heating causes changes in buffer temperature, so a connection does exist. Some topical overlap is inevitable.

4.7.1. Dispersion Not Attributed to Joule Heating or Wall Adsorption

In a thorough experimental study, Sepaniak and Cole found that nonequilibrium dispersion for amines derivatized by 7-chloro-4-nitrobenz-2-oxa-1,3-diazole [i.e., NBD-derivatized amines] depended on capacity factors, capillary diameter, and electrolyte concentration (see Figure 4.1a). Micellar, intermicellar, and intercolumn mass-transfer resistances were considered, in analogies with regular chromatography; the latter two were judged the more important. Dispersion also was increased by increasing electrolyte concentration; this increase was attributed mainly to Joule heating (22). In a related study, Cole et al. showed that high-voltage H's of aflatoxins in 50 mM SDS exhibited a radius-dependent dispersion at high E's, i.e., H was less for 25 μm than for 50 μm capillaries. The dispersion was explained by thermal loading, mass-

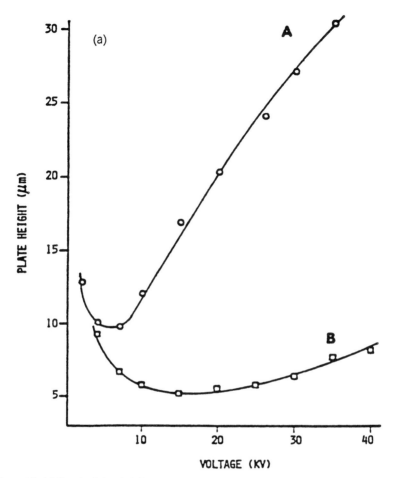

Figure 4.1. (a) Graph of plate height H vs. voltage V for NBD–cyclohexylamine in (A) 75 µm and (B) 25 µm capillaries. [Reprinted with permission from M. J. Sepaniak and R. O. Cole, *Anal. Chem.* **59**, 472 (1987). Copyright 1987 American Chemical Society.] (b) Graph of H vs. EOF velocity, v_{eo}, for (1) resorcinol; (2) phenol; (3) p-nitroaniline; (4) nitrobenzene; (5) toluene; (6) 2-naphthol; and (7) Sudan III. [Reprinted with permission from S. Terabe, K. Otsuka, and T. Ando, *Anal. Chem.* **61**, 251 (1989). Copyright 1989 American Chemical Society.] (c) Graph of plate number N vs. field strength E for highly retained aromatics. [Reprinted with permission from L. Yu and J. M. Davis, *Electrophoresis* **16**, 2104 (1995); published by VCH Verlagsgesellschaft.] (d) Graphs of H vs. V for (A) microemulsion and (B) SDS micelles. Symbols: □, resorcinol; ■, phenol; ○, p-nitroaniline, ●, nitrobenzene; ▼, toluene; ▽, 2-naphthol. [Reprinted with permission from S. Terabe, N. Matsubara, Y. Ishinama, and Y. Okado, *J. Chromatogr.* **608**, 23 (1992).]

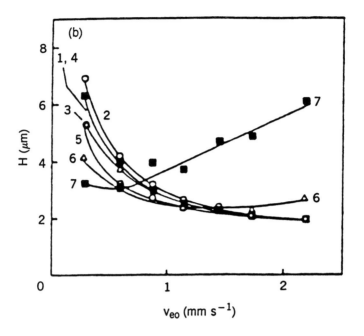

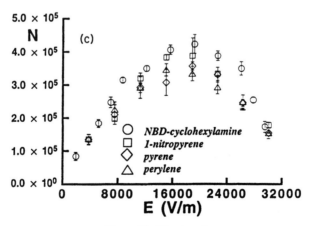

Figure 4.1. (*Continued*)

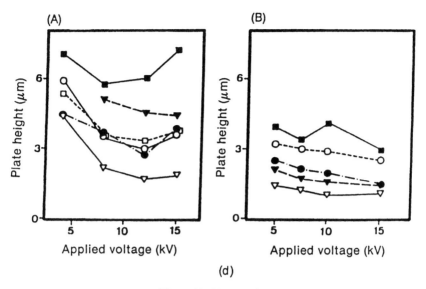

Figure 4.1. (*Continued*)

transfer kinetics, and radial variations in partition coefficient and phase ratio (32).

The conclusions deduced from these two references were justified qualitatively. In perhaps the most quantitative study of dispersion in MEKC, Terabe et al. examined the effects of extracolumn dispersion (specifically, effects of injection, detection, and time constants), adsorption–desorption (or mass-transfer) kinetics, intermicellar diffusion, radial temperature gradients, and electrophoretic dispersion due to microheterogeneity. Analytical expressions of plate height were derived for most of these sources. The H's of several simple aromatics, as generated in a 50 μm capillary containing a 0.1 M phosphate, 50 mM SDS buffer, were limited by diffusion as noted earlier; however, the micellar marker Sudan III exhibited a nonequilibrium dispersion that was attributed mainly to micellar polydispersity (see Figure 4.1b). To explain this behavior, a theory was proposed in which micellar mobility varied with aggregation number (7). However, a large numerical error in this theory was noted later by Vindevogel and Sandra (33) and a minor one by Yu and Davis (3).

Because of the importance of this work, some details are presented here. The consequences of extracolumn dispersion were addressed earlier. By applying nonequilibrium theory, Terabe et al. showed that the plate height contribution H_{ad} from adsorption–desorption kinetics is

$$H_{ad} = 2R^2(1-R)\mu_{mc}^2 t_{mc}^* E/\mu \tag{4.4}$$

where the notation of the original paper has been changed to μ, μ_{mc}, and R and where t_{mc}^* is the mean lifetime of analyte in the micelle, which can vary from nanoseconds to milliseconds (17) or more. The authors concluded that this dispersion is small unless there are electrostatic forces or ionic interactions between analyte and micelles. The reason is that t_{mc}^* and R offset each other; the former cannot become "large" without the latter approaching zero.

As discussed later, Sepaniak and Cole (22) rationalized the dependence of N on surfactant concentration by proposing a dispersion mechanism in which analyte diffuses among micelles. They argued that resistance to mass transfer increases with increasing distances between micelles and that plate height should increase with decreasing surfactant concentration. By a random-walk calculation, Terabe et al. determined the plate height H_{imd} for this intermicellar diffusion to be

$$H_{imd} = \frac{R(1-R)^2\mu_{mc}^2}{\mu} \frac{d^{*2}}{4D_m} E \tag{4.5}$$

where d^* is the mean distance between micelles. Because d^* is very small (e.g., 100 Å for a 0.1 M phosphate, 50 mM SDS buffer), $H_{imd} \approx 10^{-5}$ μm and was judged by Terabe et al. to be negligibly small.

To explain the anomalous dispersion of Sudan III in Figure 4.1b, Terabe et al. developed a random-walk calculation for a polydispersity contribution H_{pd} to plate height resulting from the distribution $\sigma_{\mu mc}$ of mobilities of micelles having different aggregation numbers

$$H_{pd} = \left[\frac{\sigma_{\mu mc}}{\mu_{mc}}\right]^2 \mu_{mc}^2(1-R)t_{mc}^* E \Big/ \mu \tag{4.6}$$

where $\sigma_{\mu mc}$ was estimated from the variation of Henry's correction factor with the cube root of the aggregation number. After correcting for the numerical errors noted earlier, one finds that $[\sigma_{\mu mc}/\mu_{mc}]^2 = 3.33 \times 10^{-5}$ for a 0.1 M phosphate, 50 mM SDS buffer (3,33). This coefficient was originally reported as 0.026, which made H_{pd} a plausible explanation for the dispersion; the corrected expression for H_{pd} seems too small.

Another possible explanation for the dispersion of Sudan III was reported later by Terabe et al., who concluded from twin-detector experiments that the high-field dispersion of Sudan IV (not III) could arise from mobility heterogeneity due to sample impurity. Sudan III also is impure. In contrast, pure timpidium bromide generated a smaller dispersion as a micellar marker; more than half the band broadening for Sudan IV was attributed to mobility heterogeneity (34). A similar behavior was reported by Nielsen, who found that the 5% impurities in dodecylbenzenesulfonic acid caused a nonequilibrium dispersion, in contrast to all other analytes examined (26). In both cases, partial resolution of the components of these impure mixtures was realized with increasing E, but the separation was insufficient to generate distinct peaks; rather, one peak was generated, whose breadth increased with increasing resolution (i.e., increasing E).

Davis also concluded from quantitative considerations that adsorption–desorption kinetics were not important in MEKC with SDS micelles and suggested that radial temperature gradients caused a radial variation in partition coefficient, leading to dispersion. The calculation agreed roughly with experiment (24); the theory also was used by Smith and El Rassi to rationalize the dependence of N's of alkylphenylketones on the electrophoretic mobilities of alkylglycoside–borate surfactants (27) and by Vindevogel and Sandra to rationalize the high-field dispersion of hop bitter acids in a Tris buffer but not a borate one (35). However, more rigorous calculations by Yu and Davis demonstrated the theory's shortcoming, at least when simple heat theory was considered (the original equation had the correct functional dependence on μ, μ_{mc}, D, and R, but the coefficient was wrong as is common for random-walk calculations) (3).

Yu and Davis also showed that highly retained hydrophobic aromatics (e.g., pyrene and perylene) exhibited nonequilibrium dispersion (see Figure 4.1c) and concluded from theory and experiment that mass-transfer kinetics, micellar polydispersity (as expressed by theory in ref. 7), hydrodynamic flow, radial temperature gradients calculated from simple heat theory, wall adsorption, and sample impurity were unlikely explanations of the dispersion (3). Yu and Davis also reduced analyte concentration by three-fold and showed that micellar overload did not cause the dispersion [see Yu (36)].

Some additional details of the published study by Yu and Davis may be useful. The compounds pyrene and perylene have a 10-fold difference in t_{mc}^*, and their identical N's ruled out dispersion mechanisms depending on t_{mc}^*, e.g., adsorption–desorption kinetics and micellar polydispersity. The effect of hydrodynamic flow was assessed by generalizing the plate height expression of Grushka to MEKC (37); to explain the dispersion by these means, buffer levels would have had to differ by 10–15 mm, which they did not. The adsorption of hydrophobics by the capillary wall was judged unlikely, because the pressure-induced transport times of a hydrophobic compound were statistically equal to those of the EOF markers acetone and methanol. Finally, chemical impurity and mobility heterogeneity largely were ruled out by chemical analysis.

Several other cases of dispersions attributed to nonequilibrium exist, and the origins of the efficiency losses are not always clear. For example, Terabe et al. found that a heptane/SDS/butanol microemulsion exhibited nonequilibrium dispersion at high voltages; plate heights (e.g., 2–7 µm at 30 kV/m) were about twice that of a comparable SDS system and were attributed to slow adsorption–desorption kinetics (see Figure 4.1d) (38). Dang et al. (39) and Lecoq et al. (40) reported N's for active ingredients in theophylline tablets, and for nucleosides and nucleotides, respectively, that increased and then decreased as E was increased from roughly 24 to 36 kV/m and from 18 to 37 kV/m, respectively. Castagnola et al. reported increases in H for high-voltage separations of dansylated amino acids by DTAB micelles that could not be explained only by Joule heating, because the use of starburst dendrimers as a pseudostationary phase caused H to decrease over the same E range (30). More recently, Bächmann et al. showed that H's of polynuclear aromatics (PNAs) partitioned among macrocyclic resorcarenes yielded minimum H's that varied between 3.6 and 5.4 µm (41).

The conclusions reported in the above paragraph were not supported by theory.

4.7.2. Dispersions Attributed to Joule Heating

Radial temperature gradients in electrokinetic capillaries cause radial gradients in the electrophoretic mobilities of micelles and analytes if the latter are charged. Regardless of whether the analyte is neutral or charged, it is subject to mobility dispersion when solvated by charged micelles. The severity of the dispersion increases with field strength. Several theories for the dispersion have been reported (3,7,42–46); however, none of the equations is cited here for reasons noted below.

All conclusions of the following studies were reached qualitatively unless otherwise noted. In a thorough study, Vindevogel and Sandra showed that N's of ionic hop bitter acids maximized at 20–30 kV/m in a 75 µm capillary and then decreased slowly with increasing E for a borate buffer. In contrast, N decreased more rapidly with E for a Tris buffer; the rate of decrease was related quantitatively to the amount of dissipated power, increase in capillary-buffer temperature, and temperature-induced shifts in pH (47). Row et al. observed decreases of N with V in a 60 µm capillary for some (but not all) neutral and charged nucleotides and attributed them to thermal and concentration overloading (5). Griest et al. showed that N's of dAMP oligomers in a 60 µm capillary decreased systematically with increasing E (48) for reasons discussed by Row et al. (5). Smith and El Rassi showed, for alkyldisaccharide-borate micelles, that N's of herbicides usually decreased as borate concentration was increased from 100 to 400 mM; the decrease was attributed to Joule heating (49).

4.7.2.1. Problems with Current Theories of Joule-Heating Dispersion

The heat-transport theories used in predicting Joule-heating dispersion show that it is fairly small, in marked contrast with experimental observations. This issue has puzzled many people, including the reviewer. In late 1994, however, Liu et al. reported the first measurements by Raman spectroscopy of temperature profiles in 75 μm CE capillaries; these measurements exceeded some of the best theoretical predictions by up to fivefold. Apparently, temperature differences ΔT between the center and wall from 2 to 4 °C are possible when power levels approach 3 W/m in 75 μm capillaries (50). These measurements show that temperature gradients developed by Joule heating are more serious than calculated by theory but, fortunately, can be removed by efficient convection. The most probable reason for error is the inapplicability of simple boundary conditions on the differential equations governing heat transfer (51). Consequently, the reviewer is skeptical of current Joule-heating dispersion calculations, including his own, based on simple boundary conditions.

Nevertheless, the theoretical framework of Taylor (52,53) and Aris (54) for estimating nonequilibrium dispersion should still be valid if the proper ΔT is used to estimate viscosity variations. Using the framework of Taylor, Yu et al. recently approximated the expected Joule-heating variance for a ΔT of 4 °C and the experimental results in ref. 3 (corresponding to roughly 3 W/m in a 75 μm capillary, as in Liu and colleagues' work); the variance, equal to 6.6×10^{-7} m^2 at 29.5 kV/m, agrees fairly well with the residual variance, 9.1×10^{-7} m^2, determined at this E (20). Consequently, Joule heating may be sufficient to explain most of the dispersion of highly retained compounds reported in ref. 3. However, less retained neutral compounds are not subject to Joule-heating dispersion, even at these power densities, because they must be in the micelles to be dispersed.

4.7.3. Dispersions Attributed to Wall Adsorption

As in CE, wall adsorption can occur in MEKC and is severe when analytes and capillary walls carry opposite charges. The behavior in CE has been modeled thoroughly (55), but a comparable study in MEKC appears to be lacking. The CE model probably could be generalized to MEKC if the analyte–micelle mass-transfer rate greatly exceeds the analyte–wall mass-transfer rate.

A brief summary of some studies is presented here; all conclusions were reached qualitatively. Cohen et al. added divalent (e.g., Mg^{2+}) cations to generate selective metal–ion complexations with oligonucleotides; however, efficiency was lost, because the cations adsorbed to the wall and caused analyte–wall interactions. Efficiency was restored by adding SDS micelles, which complexed a fraction of these cations (56). Wallingford and Ewing found that cationic norepinephine produced tailing peaks having $N = 10,000$ due to secondary and tertiary equilibria and wall interactions; tailing was worse with micelles and was attributed to analyte–micelle interaction (57). Balchunas and Sepaniak attributed to weak wall interactions the efficiency losses ($N \approx 12,000$) of NBD derivatives of amines separated by 25 mM SDS in silanized capillaries [similar efficiency losses in silanized capillaries were reported by Ahuja et al. (58)]; efficiency was increased to $N \approx 65,000$ by adding 10% 2-propanol. Similar efficiency increases also were found in underivatized capillaries; in both cases, the behavior was attributed to improved mass-transfer kinetics (59). Wallingford and Ewing found that N's of cationic catechols in pH = 7 buffers increased by one- to two-fold, and that tailing decreased, as SDS concentration was increased from 10 to 20 mM. The behavior was attributed to multiple equilibria and increased competition between micelle solubilization and wall interaction (60). Ishihama et al. reported efficiency losses in the optical resolution of drugs at pH = 5 due to adsorption of the pseudostationary phase, ovomucoid, to the capillary wall, as well as due to slow mass-transfer kinetics and heterogeneity. The efficiency was almost doubled by adding the zwitterion, o-phosphorylethanolamine, which probably reduced the wall adsorption of ovomucoid (61).

In a letter to the reviewer, Terabe commented that wall adsorption is a problem in MEKC and requires either rinsing or replacing the capillary (62). In our lab, we have had similar experiences; occasionally, wall adsorption becomes serious, even if a series of previous injections exhibits little or no adsorption. Usually, but not always, flushing the capillary with 0.1 M NaOH for brief periods eliminates the problem.

Secondary Studies. Nishi et al. found that peak tailing due to interaction of cationic β-lactam antibiotics with negative capillary walls was reduced by raising pH; the effect was probably due to ion suppression of cationic solute groups (63). Otsuka et al. found that addition of urea to N-dodecanoyl-L-valinate micelles improved peaks of chiral phenylthiohydantoin (PTH) amino acids, possibly because adsorption of urea by the capillary wall reduced their adsorption (64,65). Yik and Li observed that subsequent analyses of di-nitrophenyl (DNP)-derivatized amino acids produced increasingly broadened peaks because of wall interactions (66); similar results were reported by Otsuka and Terabe for PTH amino acids (67). Beijersten and Westerlund resolved peptides in polyacrylamide-coated capillaries with taurodeoxycholic acid micelles; N's were 2–4 times smaller than their CE equivalents, because of interactions of the cationic solutes with wall-adsorbed surfactant monomers (68). Pedersen et al. found that 20 mM NaCl slightly reduced interactions of basic nuclease isoforms with capillary walls but that higher concentrations of NaCl, or Mg^{2+} ion, broadened peaks, possibly due to analyte interactions with wall-adsorbed metal ions (69).

4.7.4. Other Temperature Effects

Increases in buffer temperature usually degrade separation efficiency in MEKC. Most behaviors probably originate in Joule heating; the reduction of viscosity at elevated temperatures favors dissipation of power as heat. They have not been included with the others above, however, because researchers did not propose Joule heating as the behaviors' origin.

It may be useful to note that the commonly reported 2.2% increase in mobility per °C is correct only near room temperature. The rate of change in the viscosity of water decreases with increasing temperature; it is about 1.8 and 1.5% at 40 and 60 °C, respectively.

Balchunas and Sepaniak found that N's of NBD-derivatized amines decreased 5- to 6-fold as the temperature of a thermostatted capillary containing 100 mM SDS was increased from 28 to 57 °C. However, in 15 mM SDS, a 6- to 7-fold increase was first observed, followed by a 9- to 10-fold decrease (relative to the maximum N). The behaviors qualitatively were attributed to changes in solute diffusivity, mass-transfer resistances, Joule heating, and micellar polydispersity (70). In a similar study, Lecoq et al. showed that N's of nucleotides and nucleosides decreased by roughly 1.5-fold as the capillary temperature was increased from 20 to 45 °C (40).

Secondary Studies. Bjergegaard et al. noted a slight reduction in N's of benzoic and cinnamic acid derivatives resolved by CTAB micelles as the capillary temperature was increased from 30 to 60 °C, possibly due to changes in CMC, aggregation number, and micellar size and shape (71). Michaelsen et al. found that N's of dansylated amino acids decreased by roughly 1.5-fold when the thermostatting temperature was changed from only 22.4 to 29 °C (72). Bjegegaard et al. determined that N's of aromatic choline esters decreased slightly for most compounds over a 27–50 °C range but that N of one compound (sinapine) increased above 30 °C (73). Michaelsen et al. found, for glycosaminoglycan (GAG) separations with CTAB micelles spanning 25–60 °C, that N's were largest for two compounds at high temperatures but for two others at 30 °C (74).

4.8. ELECTROMIGRATIVE DISPERSION

Electromigrative dispersion is common in CE but rarely observed in MEKC. The dispersion arises from conductivity differences between the analyte and buffer co-ion, resulting in axial gradients in field strength and skewed peaks (i.e., peaks shaped like right triangles). A few reports of skewed peaks in MEKC have appeared.

Corran and Sutcliffe observed that peaks generated by derivatives of gonadorelin that were highly retained by CTAB micelles were slightly asymmetric; the asymmetry was much worse in zwitterionic 3-[(3-cholamidopropyl)dimethylammonio]-1-propanesulfonate (CHAPS) and neutral Triton X-100 micelles (75). Chiari et al. reported skewed peaks for benzene-h_6 and d_6 resolved by capillaries having suppressed EOF and containing 50 mM SDS in 50 mM 3-(N-morpholino)propanesulfonic acid (MOPS) buffer (76). However, neither group attributed these behaviors to electromigrative dispersion, and therefore the inclusion of these papers here may be inappropriate.

In contrast, Qin et al. found that enantiomeric separations of lisinopril from its RSS diastereomer by sodium cholate micelles was more efficient at 30 °C than at 45 °C; at the higher temperature, one enantiomer generated a sharp peak but the other produced a skewed one. The behavior was attributed to greater mobility differences between analyte and buffer at higher temperatures than at lower ones. A similar behavior was observed in separations with SDS, except that both peaks were broadened (77).

Stathakis and Cassidy reported simple-ion separations using chromate and benzoate derivatives of the polyelectrolyte poly-(1,1-dimethyl-3,5-dimethylenepiperidinium chloride); N's varied from 35,500 to 398,000 for the chromate derivative and 282,000 to 906,000 for the benzoate one. The relative inefficiency of the chromate polyelectrolyte was attributed to high chromate mobility, which caused peak asymmetry (78).

As noted earlier, in our lab, we have observed skewed peaks at roughly 300 μM concentrations of highly retained analytes in 10 mM SDS (21). The peaks returned to a Gaussian shape on dilution, and the mobilities of the Gaussian and skewed peaks differed.

4.9. DEPENDENCE OF N ON CONCENTRATION OF ORGANIZED MEDIA

The dependence of N on organized-media concentration has been studied extensively. In general, as organized-media concentration increases, N first increases in a near-linear manner and then either approaches a maximum or decreases. Although several explanations for this behavior have been suggested, a quantitative understanding still is lacking. Because of this, all conclusions reported under this heading are qualitative unless otherwise noted. Surfactant concentrations are represented by brackets, e.g., [SDS].

4.9.1. Single Micellar Systems

Sepaniak and Cole attributed increases in N's of NBD-derivatived amines with increasing [SDS] to reduction of distances between micelles, which favored rapid mass transfer. Efficiencies increased by 5- to 10-fold as [SDS] was raised from 5 to 50 mM (22). It should be noted that the N's of this study were unduly small (e.g., $\sim$ 40,000), however, in part because a 100 µm capillary was used. Row et al. similarly explained the increase in N of nucleotides with increasing [SDS] (5). This explanation is accepted by many and will be reported again. As noted earlier, however, a random-walk calculation by Terabe et al. suggested that this behavior was insufficient to cause dispersion; instead, the authors suggested that overloading of the micellar phase was responsible for small N's at low surfactant concentrations (7).

Because the reviewer has a current interest in this subject, he has examined it in some detail. Some reported behaviors may be due to wall adsorption. For example, Wallingford and Ewing found that N's of cationic catechols increased but that those of nonionic catechols decreased, as [SDS] was raised from 10 to 20 mM. Also, borate–catechol complexes had N's that increased, then decreased slightly or leveled off, as [SDS] was increased from 7.5 to 100 mM; the optimal concentration was about 10 mM for net-neutral complexes. The variations were attributed to surfactant–solute–micelle equilibria (60). However, the apparent mobilities of some cationic catechols were unusually small (e.g., the μ for dopamine in Figure 1 of ref. 60 was less than 10^{-8} m^2/Vs) and may have been biased by wall adsorption, as noted by the authors.

Other studies may be biased by large plug sizes. Dang et al. reported that N's of active ingredients of theophylline tablets increased, then usually decreased, as [SDS] increased; [SDS] was optimal near 40–50 mM for the largest N's and somewhat higher for smaller N's (see Figure 4.2a) (39). Some of these N's appear to the reviewer to be plug-size limited; by estimating mobilities in ref. 39, one concludes that ℓ is about 4.5 mm in unfavorable cases. When longitudinal diffusion is taken into account, N's of 60,000 or so are expected, in agreement with experiment. The small N's, and small variations in N with media concentration, in studies by Shihabi and Hinsdale (79) and by Perrett and Ross (80), may be similarly explained. Of course, the reviewer could be wrong in these assessments.

Other studies are not justified so simply. For example, Sepaniak et al. attributed the roughly 45-fold increase in N's of NBD-n-butylamine, as [SDS] was increased from 15 to 100 mM, to increased monomer–micelle exchange

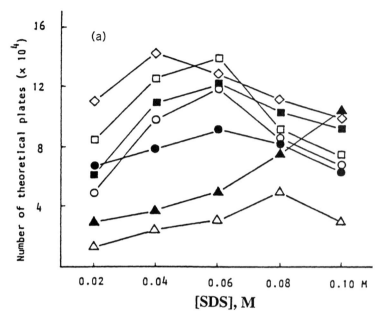

Figure 4.2. (a) Graph of N vs. [SDS] for active ingredients of theophylline tablets. Symbols: ◇, ephedrine HCl; □, theophylline; ●, caffeine; ■, phenobarbital; ○, theobromine; ▲, phenacetin; △, amidopyrine. [Reprinted with permission from Q. Dang, L. Yan, Z. Sun, and D. Ling, *J. Chromatogr.* **630**, 363 (1993).] (b) Graph of average N's of nitrobenzene, prometon, and prometryn vs. [octylglucoside]. Buffer: 200 mM borate, pH = 10. [Reprinted with permission from J. Cai and Z. El Rassi, *J. Chromatogr.* **608**, 31 (1992).] (c) Graph of N vs. [SB-12] in 20 mM SDS. Symbols: ●, acetophenone; □, propiophenone; ■, butyrophenone; ○, valerophenone; ▲, hexanophenone. [Reprinted with permission from E. S. Ahuja, B. P. Preston, and J. P. Foley, *J. Chromatogr. B* **657**, 271 (1994). Graph regenerated by the reviewer using digitization.] (d) Graph of N vs. carboxymethyl-β-CD (CM-β-CD) concentration in 30 mM hydroxypropyl-β-CD for various aromatics. [Reprinted with permission from O. H. J. Szolar, R. S. Brown, and J. H. T. Luong, *Anal. Chem.* **67**, 3004 (1995). Copyright 1995 American Chemical Society. Graph generated by the reviewer from tabular data.]

rates that reduced micellar polydispersity. However, the compound's capacity factor actually was less in 75 mM SDS (1.02) than in 15 mM SDS (1.20). The authors stated that unspecified differences in room temperature caused the behavior (81). Cole et al. found that the efficiency of binaphthyl enantiomers separated by sodium deoxycholate bile salts increased as surfactant concentration was increased from 5 to 10 mM; however, efficiency decreased for 50 mM surfactant. The behavior was attributed to either a reduction in micellar polydispersity or increases in micelle–monomer exchange rates due to increasing surfactant concentration (82). Using octylglucoside–borate micelles, Cai

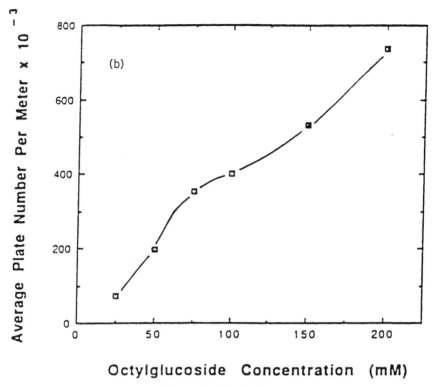

Figure 4.2. (*Continued*)

and El Rassi obtained a nearly *10-fold increase* in N's of nitrobenzene and herbicides as octylglucoside concentration was increased from 25 to 200 mM in 200 mM borate buffer; the behavior was attributed to reduction of distances between micelles (see Figure 4.2b) (83).

Similar behaviors are reported elsewhere. Bevan et al. found that N's of one unnatural oligonucleotide systematically increased over a 50–100 mM SDS range when 7 M urea was added; efficiencies also were about six- to seven-fold higher with the urea than without it (84). By reducing surfactant concentration from 50 to 25 mM, Cole et al. decreased the generation rate of plate numbers, N/time, of aflatoxins by 70% for sodium deoxycholate micelles and 20% for SDS micelles; the behavior was attributed to micellar polydispersity at low surfactant concentrations (32). Crosby and El Rassi found that peaks of herbicides broadened when [CTAB] was reduced below roughly 31 mM; the behavior was attributed to increased intermicellar distances. Similarly, increases in concentrations of tetradecyltrimethylammonium chloride (TTAC)

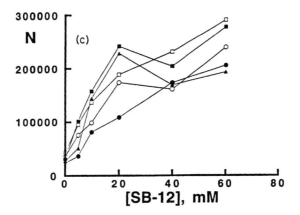

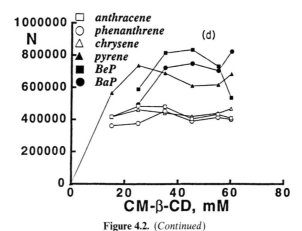

Figure 4.2. (*Continued*)

from 27 to 40 mM increased the number of plates per meter, N/m, by roughly 2.5-fold (85). Smith and El Rassi showed that N/m for herbicides increased by roughly 1.5-fold as alkyldisaccharide–borate surfactant concentration was increased from 25 to 100 mM in 200 mM borate, in part due to increases in capacity factor and decreased diffusion (49). In contrast to most studies, Lecoq et al. showed that N's of nucleotides and nucleosides *decreased* as [SDS] was increased from 75 to 150 mM; they attributed the behavior to greater analyte retention (40). These experiments were implemented in a constant-current mode, however, and voltage decreased with increasing [SDS]. Most likely increases in longitudinal diffusion, due to lower voltages and longer elution times, at least partly explain these results.

Yu et al. recently made a reassessment of the dependence of N on [SDS]. For a 10 mM phosphate/6 mM borate buffer at pH = 9 and E's < 15 kV/m in a 50 μm capillary, they found *no* variation of N over the [SDS] range, 15–100 mM, for each of the following compounds: 2'-deoxyadenosine (0.70 < R < 0.94 over this [SDS] range); nitrobenzene (0.26 < R < 0.74); NBD-cyclohexylamine ($\approx 0 < R < 0.072$), and 1-nitropyrene ($R \approx 0$). In effect, μ and D decreased at equal rates with increasing [SDS], such that their ratio remained constant in the longitudinal-diffusion domain. These findings are supported by theory. Between 15 and 30 kV/m, efficiency losses were observed for the two most retained compounds, probably due to Joule heating (as suggested by calculations reported above). The high-field dispersion was most severe for 100 mM SDS and least severe for 15 mM SDS, in contrast to the commonly reported decrease in N with decreasing media concentration (20).

Secondary Studies. Bjergegaard et al. found that N's of benzoic and cinnamic acid derivatives resolved by CTAB micelles increased as the surfactant concentration was increased from 20 to 50 mM, possibly due to changes in micellar size, shape, and aggregation number (71). Pedersen et al. found that the optimal separation for nuclease isoforms was obtained with 25–50 mM SDS; peaks broadened with 75 mM SDS, possibly because of thermal effects (69). Michaelsen et al. found that the largest N's for GAGs separated by CTAB micelles spanning a 10–60 mM concentration range were obtained at 50 mM (74).

4.9.2. Mixed Media

4.9.2.1. Micellar Systems

Wallingford et al. observed a roughly 1.5- to 3.5-fold increase in the N's of borate-complexed catechols, as sodium octyl sulfate (SOS) concentration was increased from 5 to 40 mM in 20 mM SDS. They concluded that SOS "loosened" the mixed micellar structure, resulting in increased mass-transfer kinetics (86). Ahuja et al. observed two- to five-fold increases in N's for alkylphenones and simple aromatics as zwitterionic N-dodecyl-N,N-dimethylammonio-3-propanesulfonate (SB-12) concentration was increased from 0 to 60 mM in 20 mM SDS; the increase was largest for highly retained species. The increase was attributed to increased mass-transfer kinetics and reduction of the weighted diffusion coefficient, with the former behavior dominating (see Figure 4.2c) (87). In contrast to many studies, Crosby and El Rassi found that N/m for herbicides systematically *decreased* by roughly 4.5-fold, when octyltrimethylammonium chloride surfactant concentration was increased from 0 to 100 mM in 40 mM TTAC (85). This decrease cannot be explained by the commonly suggested reduction of distances between micelles, which would favor faster mass transfer.

4.9.2.2. *Micellar/Cyclodextrin Systems*

Sepaniak et al. found that N's of PNAs resolved by mixed β-CD/carboxymethyl-β-CD (CM-β-CD) media were comparable in efficiency to pure SDS micelles. Efficiency improved markedly by increasing the 1:1 CD (cyclodextrin) mixture from 1 to 10 mM, possibly due to increased temperature and improved solubility but more likely due to reduction of distances between CDs (88). Szolar et al. showed that N's of PNAs in 30 mM hydroxypropyl-β-CD increased, then slightly decreased, as the concentration of anionic CM-β-CD was varied from 0 to 60 mM; the least efficient separations were associated with highly retained species, and N varied only slightly with [CM-β-CD]. The authors attributed the efficiency loss to heterogeneity of CM-β-CD due to multiple charge-site distribution (see Figure 4.2d). A similar behavior was observed when the concentration of anionic sulfobutyl ether-β-CD was varied from 0 to 25 mM in 20 mM methyl-β-CD (89).

4.10. DEPENDENCE OF N ON ELECTROLYTE CONCENTRATION AND COMPOSITION

The dependences of N on electrolyte concentration and composition are subtle. A high concentration causes efficiency losses, due to Joule heating, but N often increases with modest increases in concentration. As with surfactant concentration, a quantitative understanding of this behavior often is lacking. In addition, efficiency can vary substantially among different buffers.

Ishihama et al. found that N's of drugs optically resolved by ovomucoid increased more than three-fold as phosphate buffer concentration was increased from 10 to 50 mM. The buffer choice also was crucial, e.g., for a 50 mM phosphate buffer, N equaled only 10,800, but for a 50 mM phosphate/100 mM borate buffer, N increased by almost 20-fold (61). In contrast to other studies, Cole et al. reported systematic *decreases* in the rate of plate generation, N/time, for aflatoxins solvated by SDS or sodium deoxycholate micelles as phosphate/borate buffer concentration was increased by four-fold (32). Field strengths of 60 kV/m were used here, however, and Joule-heating dispersion undoubtedly was important.

A few studies on the effect of dodecyl sulfate and buffer counterions have been reported. Lecoq et al. show that N's of nucleotides and nucleosides were two times larger under constant-current conditions with Li-buffer salts than with Na-buffer salts; N's for K-buffer salts were intermediate. The behavior in part was due to greater EOF for the lithium cation and less longitudinal diffusion. In contrast to many findings, they also determined that N's of nucleotide and nucleosides *decreased* by roughly 1.5-fold as Li-buffer concentration was increased from 10 to 40 mM in 100 mM SDS (40); however, these were constant-current experiments accompanied by decreases in EOF and increases in diffusion. Ahuja and Foley showed that simple aromatics had larger N's when retained by dodecylsulfate micelles having Li^+, instead of Na^+, counterions; for hydrophobic analytes, N was 2.5 times larger for the Li-surfactant (90).

Buffer composition and pH may be more crucial than commonly considered. For example, Balchunas and Sepaniak reported an N of only 6500 for NBD-n-hexylamine in a 15 mM SDS, 10 mM Na_2HPO_4/5 mM $Na_2B_4O_7$ buffer at pH = 7 (70); in contrast, over the same E range, Yu et al. determined N's for NBD-cyclohexylamine (almost the same compound) in excess of 3.5×10^5 in a 15 mM SDS, 10 mM Na_2HPO_4/6 mM $Na_2B_4O_7$ buffer at pH = 9 (20). The compounds' capacity factors were identical; also, SDS concentrations and capillary diameters were identical. Only the buffer composition slightly differed (actually, the capillary and chemical sources also differed).

Secondary Studies. Bjergegaard et al. found that N's of most aromatic choline esters increased as phosphate buffer concentration was increased from 50 to 150 mM (73). For a phosphate/borate buffer spanning 16–48 mM concentrations at 40 °C and $E \approx 27$ kV/m, Michaelsen et al. found that N's generally were largest for the larger buffer concentrations (74). For 50 mM octylglucoside surfactant, Cai and El Rassi found that N's of simple aromatics and herbicides were almost independent of borate buffer concentrations spanning 25–400 mM (83); this study should be compared to that by Smith and El Rassi, which showed that N's of herbicides generated by alkyldisaccharide–borate micelles usually *decreased* as borate concentration was increased from 100 to 400 mM (49).

4.11. DEPENDENCE OF N ON ORGANIC SOLVENTS

Various organic solvents, e.g., acetonitrile (ACN), methanol (MeOH), isopropanol (IPA), and urea, have been added to MEKC buffers as modifiers to reduce EOF and capacity factors and to increase resolution. The modifiers can either enhance or degrade efficiency, although the latter effect is more common. The modifiers cause several changes, including increased viscosity, reduced zeta potentials, and modification of the micellar structure (the latter especially by the more hydrophobic modifiers, e.g., cyclohexanol). In most cases, simple explanations for changes in N are not clear. In a few cases, generalizations from studies in micellar liquid chromatography have been made. Consequently, only observations are reported here; all conclusions were reached qualitatively.

4.11.1. Single Micellar Systems

Balchunas and Sepaniak reported two- to fivefold *increases* in N's of NBD-derivatized amines in silanized and unsilanized capillaries after addition of 10% IPA to 25 mM SDS; the efficiency increase was larger in the unsilanized capillary and was attributed to "loosening" the micellar structure (59). These observations should be contrasted against subsequent ones of Balchunas and Sepaniak, who reported roughly three- to fourfold *decreases* in N for NBD derivatives of n-butylamine and n-hexylamine in 75 mM SDS as IPA concentration was increased from 0 to 30%. The efficiency then decreased rapidly with 35% IPA, possibly due to dissolution of micelles by the organic (see Figure 4.3a) (70). Similarly, Sepaniak et al. reported a 20% reduction in N's of NBD-n-butylamine after addition of 10% IPA (the fractional decrease was comparable to that in ref. 70) and attributed the decrease to micellar polydispersity brought about by partial micellar solubilization (81).

Cole et al. found that N's of a binaphthyl compound retained by sodium cholate bile salts were decreased by roughly 1.5-fold after addition of 6% MeOH; however, subsequent additions of MeOH caused little variation in N over the range 6–18% MeOH (82). Vindevogel and Sandra found that N's of testosterone esters decreased with increasing [ACN] and noise levels; the nucleation of SDS and formation of turbid solutions by ACN were proposed as explanations. The efficiency seemed more correlated with noise than with elution time (91). Cole et al. found that dispersion of aflatoxins was slightly increased, and then slightly decreased, in the longitudinal-diffusion and nonequilibrium domains, respectively, of van Deemter graphs after addition of 10% ACN to 50 mM SDS (32). In contrast to the findings of Balchunas and Sepaniak (70), Ishihama et al. observed a near-linear *increase* in N's of eperisone HCl resolved by 250 μM ovomucoid as a pseudostationary phase when IPA was increased from 0 to 15% (see Figure 4.3b); the behavior was attributed to reduction of the hydrophobic interaction between the ovomucoid and analyte (61). Michaelsen et al. found that N/m of dansylated amino acids decreased almost linearly by 2.5-fold as [MeOH] in the sample (not the working buffer) was increased in 150 mM SDS from 5 to 50%; they

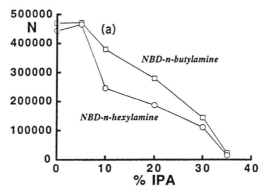

Figure 4.3. (a) Graph of N vs. IPA concentration for NBD-n-butylamine and -n-hexylamine. [Reprinted with permission from A. T. Balchunas and M. J. Sepaniak, *Anal. Chem.* **60**, 617 (1988). Copyright 1988 American Chemical Society. Graph generated by the reviewer from tabular data.] (b) Graph of N vs. IPA concentration for eperisone HCl in 250 μM ovomucoid. [Reprinted with permission from Y. Ishihama, Y. Oda, N. Asakawa, Y. Yoshida, and T. Sato, *J. Chromatogr. A* **666**, 193 (1994).] (c) Graphs of N vs. ACN concentration in (A) NaDS and (B) Mg(DS)$_2$ micelles (DS = dodecyl sulfate). Symbols: ▲, nitrobenzene; □, anisole; ●, benzophenone; ■, biphenyl. [Reprinted from K. R. Nielsen and J. P. Foley, *J. Microcolumn Sep.* **5**, 347 (1993). Copyright 1993 John Wiley & Sons, Inc; reprinted by permission of John Wiley & Sons, Inc.] (d) Graphs of N for simple aromatics in different concentrations [ACN (acetonitrile) = MeCN (methyl cyanide)] for DS micelles having Li (top), Na (middle), and K (bottom) counterions. [Reprinted with permission from E. S. Ahuja and J. P. Foley, *Anal. Chem.* **67**, 2315 (1995). Copyright 1995 American Chemical Society. Graph regenerated by the reviewer using digitization.]

attributed the behavior to reduced analyte interaction with SDS or altered stacking conditions during the first minutes of separation (72).

The use of different buffer and micellar counterions alters the effects of organics on efficiency. In contrast to many studies, Lecoq et al. showed that N's of nucleotides and nucleosides *increased* by roughly 1.5-fold after addition of 10% ACN to 100 mM SDS. Although these were constant-current experiments, voltage varied very little with [ACN] (40). Nielsen and Foley showed that N's of simple aromatics also *increased* by one- to two-fold, when [ACN] was varied from 4 to 15% and when Mg^{2+} replaced Na$^+$ as the dodecyl sulfate counterion. The behavior possibly was due to improved mass transfer, reduced wall effects, or reduction in micellar heterogeneity. In agreement with most other experiments, however, N usually decreased by one- to two-fold in SDS/ACN systems (see Figure 4.3c) (92). However, these N's are very small, even for the sodium ion, and probably are not representative. Ahuja and Foley showed that increases in [ACN] usually, but not always, decreased N's of simple aromatics partitioned into dodecyl sulfate surfactants having Li$^+$, Na$^+$, and K$^+$ counterions; on average, N was greater for the Li-surfactant

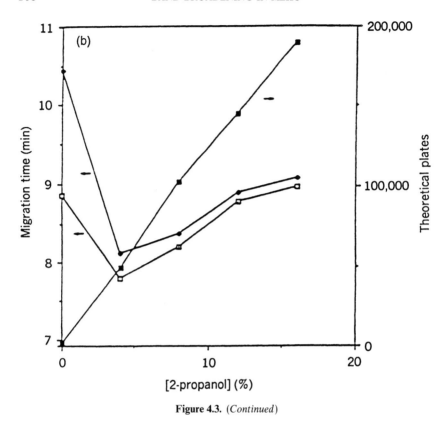

Figure 4.3. (*Continued*)

than for the Na- or (especially) K-surfactant, although the trend of N's for the Li- and Na-surfactants occasionally reversed (see Figure 4.3d) (90).

Secondary Studies. Aiken and Huie reported that addition of 0.5% 1-alkyl alcohols to SDS improved resolution but caused N to decrease when the alcohol/SDS ratio was greater than 3, possibly because of poor mass transfer, wall interactions, or viscosity increases (93). Pedersen et al. showed that addition of up to 25% MeOH decreased the efficiency of isoform separations, probably because of MeOH-induced micellar polydispersity (69). Bjergegaard et al. determined that N's of aromatic choline esters *increased* as up to 12% IPA was added, probably due to increased micellar stability and increased rates of mass transfer between Na-cholate micelles and analytes (73). Katsuta et al. showed that 0.10 M cyclohexanol in SDS buffers *increased* the N's of acetophenone and 2-naphthol by 25–30% (94); here, modification of the micellar structure is probable.

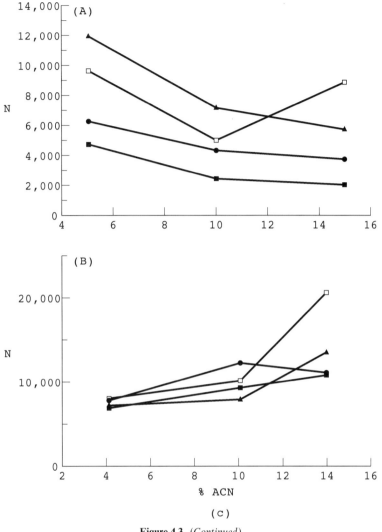

Figure 4.3. (*Continued*)

4.11.2. Mixed Micellar Systems

The efficiency of mixed-media separations also is altered by organic modifiers. Wan et al. showed that some N/m of dansylated amino acids in 15 mM SDS and 12 mM β-CD usually increased, then decreased, as [IPA] was increased from 5 to 20%; in contrast, other N's were almost invariant. High concentrations of ACN (30–50%) caused N to decrease in 10 mM SDS; the decreases

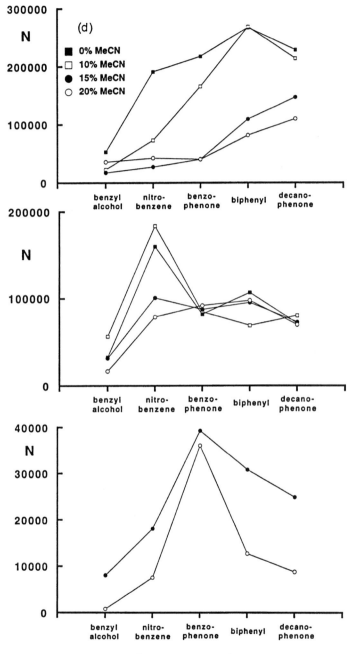

Figure 4.3. (*Continued*)

were attributed to using SDS concentrations near the CMC (95). Ng et al. increased the efficiency of impurity peaks in the drug DS1 (a basic drug closely related to ondansetron), as separated with 30 mM sodium cholate/15 mM hydroxypropyl-β-CD, by dissolving the sample in 50% MeOH/50% working buffer instead of 100% MeOH; the efficiency increase was attributed to reduced analyte solubilization by MeOH, increased analyte solubilization by micelles, and possibly stacking (96).

4.12. EFFICIENCY OF PSEUDOSTATIONARY PHASES OTHER THAN SIMPLE MICELLES

The selectivity required for separation is not always obtainable with alkylsulfate micelles. For example, highly hydrophobic compounds elute closely to the micellar marker and are not separated well. A variety of organized media, including CDs (both neutral and charged), microemulsions, mixed media (e.g., zwitterionic/charged micelles, charged/neutral micelles, and micelles/CDs), bile salts, dendrimers, and fluorocarbon micelles, has been used to obtain better selectivity. The subject are quite diverse, and so only brief summaries are reported here; all conclusions were reached qualitatively.

4.12.1. Charged Cyclodextrins

CDs were introduced in 1985 as alternatives to micelles as organized media for chromatography. Selective inclusion depends on hydrophobicity, size, and hydrogen bonding. The efficiencies of several charged CD systems have been discussed earlier. In addition, Terabe et al. demonstrated the resolution of cresol isomers with the Na salt of 2-O-carboxymethyl-β-cyclodextrin (CM-β-CD), with N equal to 1.2–1.3×10^5 plates. Plate heights for buffers containing 25 mM CM-β-CD were one to two times larger than comparable SDS systems, for reasons that are not clear (97).

4.12.2. Microemulsions

Watarai first reported the use of a water/SDS/1-butanol/heptane microemulsion as a pseudostationary phase and noted the absence of serious dispersion for ketones and β-diketones due to rapid aggregate formation and breakdown on a timescale comparable to that for micelles (98,99). As noted earlier, Terabe et al. found that a heptane/SDS/butanol microemulsion exhibited nonequilibrium dispersion at E's greater than $20\,\mathrm{kV/m}$ or so (see Figure 4.1d) (38). In contrast, Song et al. found that an n-butanol/n-octane/SDS microemulsion separated herbicides with N's 1–3 times greater than that for SDS alone;

however, as noted by the authors, these results may be biased by concentration overload (19). Boso et al. obtained very large variations of N for n-butanol-based microemulsion separations of water- and lipid-soluble vitamins by varying an organic solvent ($<1\%$) among SDS and tetradecyltrimethyl-ammonium bromide (TTAB) micelles (e.g., for the analyte thiamine and SDS, N was 13,500 with diethyl ether as the solvent but 505,600 with n-hexane) (100).

4.12.3. Bile Salts

The efficiencies of several bile salt separations have been discussed previously. In addition, Cole et al. found that N's of NBD-derivatized amines were larger for 50 mM sodium cholate micelles in 20% MeOH than for 50 mM SDS micelles in 20% MeOH, possibly due to a reduced tendency of the bile salt to become polydisperse in MeOH (101). Bjergegaard et al. compared N's of heterocyclic compounds separated by 50 mM SDS containing 5% 1-propanol to those separated by 15 mM sodium cholate micelles containing 10% 1-propanol; N's varied by as much as fourfold and, for the same compound, were sometimes larger for the SDS system and other times for the sodium cholate system (102).

Secondary Studies. Terabe et al. introduced chiral MEKC separations by using 50 mM taurodeoxycholate chiral micelles to resolve dansylated amino acids, with N of 70,000 or so (103). Nishi et al. separated corticosteroids and diltiazem-related compounds by sodium cholate, deoxycholate, and taurocholate micelles, with N's of $2-3.5 \times 10^5$ (104). Beijersten and Westerlund resolved peptides by CE and MEKC with taurodeoxycholic acid micelles; N's for MEKC were two to three times smaller than their CE counterparts because of interactions of the cationic solutes with wall-adsorbed surfactant (68). Michaelsen et al. obtained values of N/m of dansylated amino acids equaling 228,000–428,000 with 50 and 80 mM sodium cholate micelles (72).

4.12.4. Mixed Media

Mixed micellar systems offer advantages in changes in selectivity and hydrophobicity over alkylsulfate micelles. Those containing zwitterionic or neutral surfactants also can tolerate higher E's before thermal effects increase dispersion.

Little and Foley showed that N's of PTH amino acids were 2.5–3.5 times larger for Brij 35/SDS micelles than for pure SDS micelles, possibly because of rapid mass-transfer kinetics. The N's of weakly retained analytes also were more reproducible, when developed by mixed micelles than by SDS alone (18). Ahuja et al. used mixed Brij 35/SDS micelles to obtain an infinite elution window in which the EOF and micellar velocities were equal but opposite; highly retained n-alkylphenones (e.g., hexanophenone) generated very broad peaks requiring corrections to the plate number due to slow migration velocities (58). Song et al. showed that herbicides separated by mixed SDS/poly(ethylene glycol 400 monolaurate) micelles had N's one to two times greater than for pure SDS alone (19); however, as noted by the authors, this study might be biased by concentration overload. Rundlett and Armstrong reported a large increase in efficiency when the antibiotic vancomycin was complexed by SDS micelles to enable chiral separations of amino acids; the mobility of vancomycin was reversed by the SDS (105). Copper and Sepaniak attributed the low N/m (e.g., 80,000) of "infinitely" retained benzo[a]pyrene (BaP) in 20 mM SDS to polydispersity; the addition of 10–15 mM γ-CD, however, increased N/m to 2×10^5. The efficiency suggested a direct transfer of analyte from micelle to γ-CD, as opposed to an intermediate analyte-transfer step back to the mobile phase (106).

4.12.5. Oligomerized and Polymerized Micelles

Polymerized micelles offer several advantages over dynamic ones, including avoidance of monomer–micelle exchange kinetics, possession of CMCs equaling zero, and stability in organic solvents. Palmer and McNair oligomerized sodium undecylenate into a polymeric pseudostationry phase whose efficiency, as measured by benzyl alcohol, benzene, and acetophenone in 100 μm capillaries, was comparable to that of both sodium undecylenate and SDS (107). The N's varied from 44,000 to 69,000; these N's possibly were limited by Joule heating in these large capillaries. Wang and Warner synthesized the chiral micellar polymer poly(sodium N-undecylenyl-L-valinate); a 0.5% solution of it generated N's for ($\pm$)-1,1'-bi-2-naphthol that were about 3.5 times greater than a 1% micellar solution of the monomer. The efficiency increase was attributed to fast mass transfer resulting from compactness of the polymerized micelle, which does not allow deep penetration of analytes (108). Ozaki et al. developed high-efficiency separations using Na salts of butyl acrylate–butyl methacrylate–methacrylic acid copolymers. The efficiency for benzene and naphthalene derivatives was slightly less than for low-mass surfactants, but no serious efficiency loss resulted from the large polymer-mass distribution (109).

4.12.6. Ion-Exchange Polyelectrolytes

Polyelectrolytes can act as organized media by virtue of ion-exchange reactions with charged analytes. Terabe and Isemura reported N's of 250,000 for monobasic acids separated by 0.3% poly(diallyldimethylammonium chloride), or PDDAC; dibasic acids also were separated by 0.01% PDDAC, but naphthalenedisulfonates generated broad tailing peaks (110). Stathakis and Cassidy reported separations of simple anions (e.g., Cl^- and Br^-) using four different cationic polyelectrolytes (see ref. 111 for their identities); N's varied from 258,000 to 780,000. Two of the four polyelectrolytes consistently generated N's about 40% larger than the other two, possibly due to more rapid mass transfer (111). As observed earlier, Stathakis and Cassidy also reported simple-ion separations using several polyelectrolytes, including chromate and benzoate derivatives of poly-(1,1-dimethyl-3,5-dimethylenepyrrolidinium chloride); N's varied from 35,500 to 398,000 for the chromate derivative and 282,000 to 906,000 for the benzoate derivative (78).

4.12.7. Dendrimers

Dendrimers are pseudostationary phases developed by "cascade" syntheses starting from an initiator core of a branched building block. Castagnola et al. used starburst dendrimers (SBDs) to obtain H's for dansylated amino acids that were two to three times smaller than for DTAB micelles at $E \approx 35\,kV/m$; they attributed the higher efficiency of the dendrimers to their microhomogeneity, which was greater than that of the micelles. Efficiency also decreased with the generation number (e.g., for the first generation, $N = 614,000$, but for the third, $N = 463,000$) but was still larger than that for DTAB (e.g., $N = 277,000$) (30). Tanaka et al. noted that the efficiency of a SBD was slightly lower than for SDS micelles, possibly due to polydispersity or slow equilibration. An alkylated SBD, however, generated better efficiencies for aromatic compounds than did either the parent SBD or SDS (112).

4.13. GENERAL OBSERVATIONS

Below are some general observations and findings that were not easily integrated into the above topics. Lux et al. observed that rinsing the external capillary body after its removal from sample vials reduced peak tailing by eliminating a second analyte injection (113). Bevan et al. noted a 35% increase in efficiency in the separation of unnatural oligonucleotides by using D_2O instead of H_2O, but this gain was offset by a reduction in the capacity factor (84). Smith and El Rassi observed that "bubble" capillaries from a commercial vendor caused a 25% reduction in N's of herbicides (49); such efficiency losses were examined theoretically by Xue and Yeung (114). Crabtree et al. found that acetone-2,4-dinitrophenylhydrazine could be split into two peaks at high ACN concentrations; this artifactual behavior was not observed at low ACN concentrations and was attributed to interaction of ACN with SDS micelles (115). Reijenga and Hutta developed and discussed training software for MEKC that described many dispersion sources (116).

4.14. CONCLUSIONS

The reviewer will now conclude this chapter with some personal comments and observations. First, it is clear that dispersion processes in MEKC have been studied extensively and that much empirical evidence exists to guide our choice of practical experimental conditions. Secondly, it is apparent that some dispersion processes are clearly understood. Thirdly, it also is clear that a general theoretical understanding of dispersion in MEKC is incomplete.

The difference between theoreticians and experimentalists on the significance of Joule heating may be approaching its end. The former (including me, but also much better scientists than me) have argued that heat-transport theory shows dispersion by Joule heating is small; the latter have demonstrated clearly that dispersion is reduced by convective cooling. Liu and associates' measurements of electrokinetic temperature profiles show that these profiles are much greater than their calculated counterparts. The preliminary calculations by Yu et al. of the expected Joule-heating dispersion due to these profiles agree fairly well with experimental observations. Further work is required for a total reconciliation.

Unfortunately, theories for many other important dispersion processes in MEKC are lacking. For example, the dependence of N on micellar concentration has been attributed to micellar polydispersity, reduction in axial diffusion coefficient, reduction of intermicellar distances, increases in mass-transfer rates, and micellar overload. However, no theory developed to date completely justifies these explanations, and experimental studies have been, at best, consistent with the explanations but have not proven them. Similarly, the finding that an organic modifier can increase N for one set of analytes and buffer but simultaneously decrease it for another precludes a simple explanation. It seems to me that all too often we invoke plausible but not proven explanations for these behaviors, and while we qualify the explanations (i.e., by using "might," "could," etc.), it is disturbing to see how readily we invoke them as citations in subsequent publications in the absence of proof. From the perspective of analysis, these invocations probably are immaterial; we have a large empirical body of data to guide experimental development. From the perspective of understanding, however, they are not.

Perhaps I think (or try to think!) too much like a physical chemist, but I don't think we really understand something unless we have a quantitative theory for it that agrees with experimental observation within a reasonable tolerance. In some cases, perhaps a definitive understanding of factors affecting efficiency in MEKC is not possible. But then we should acknowledge this more often and use fewer "mights" and "coulds" in our papers.

ADDENDUM

Polymeric micelles. Palmer and Terabe found that the mean N's of substituted aromatic compounds, as separated by 50 mM SDS and 1% solutions of the polymerized micelles, sodium polyundecenyl sulfate (SUS) and sodium polyundecylenate (SUA), varied negligibly. In contrast, the polydispersities of SUS (1.71) and SUA (1.09) varied substantially (117). Although results are inconclusive, the author believes this study is a challenge to suggestions that polydispersity is a major cause of dispersion in MEKC. The similar efficiencies of the polymerized micelles also were discussed elsewhere (118).

MEKC on a chip. Microchip-based separations, first used with CE and liquid chromatography, are now being used with MEKC. Moore et al. found that coumarin dyes having different capacity factors generated constant plate heights of 5–6 µm in 60- to 80-µm-wide channels and also generated N's varying linearly with L_d at constant E. Van Deemter-like graphs were developed from a well-retained dye (119). Von Heeren et al. separated derivatized amino acids in a synchronized cyclic channel having a 40-µm width. The channel design allowed recycling; plate heights of tagged serine increased nonlinearly from 0.5 µm after 1/4 cycle to 1.66 µm after 3 1/4 cycles. A countervoltage limited the injection volume, reducing dispersion due to plug size (120).

Mixed media. Whitaker et al. reported van Deemter-like graphs, and graphs of H vs [CD], for several PAHs separated by CM-β-CD and β-CD. Efficiency losses at high E's were attributed to thermal loading; those at low E's were attributed to mass transfer. Because of mobile-phase insolubility, solute transfer of PAHs requires the proper orientation between CDs; Whitaker et al. calculated from two expressions of exchange rate that <1% of properly oriented collisions had sufficient energy for solute exchange (121). While the calculation was not substantiated by independent work, it is a refreshing contrast to the speculation so common in this field.

Wan and Blomberg showed that N's of derivatized dipeptides resolved by vancomycin generally improved with addition of SDS and that N's of the later-eluting isomers increased with increasing [SDS] (122). They also showed that N's of derivatized amino acids developed with 12 mM γ-CD/50 mM SDS usually were greater than N's developed with 12 mM β-CD/50 mM SDS; in addition, N usually increased and decreased, respectively, when 15% IPA was added. This behavior may have resulted from mass-transfer kinetics due to cavity size and differential interactions of IPA with the CDs (123). Related studies containing similar N's were reported by the same group (124,125).

Brown et al. resolved PNAs by mixtures of sulfobutyl ether-β-CD (SB-β-CD) and methyl-β-CD (M-β-CD); N's were comparable or slightly greater when the SB-β-CD to M-β-CD ratio was 25 mM/20 mM instead of 35 mM/15 mM. Early eluters had the largest N's. Improvements in N were attributed to lower media and analyte concentrations and to injection from a more aqueous solvent; dispersion might have arisen from inclusion kinetics, injection dispersion and heterogeneity in SB-β-CD (126). Nguyen and Luong added 4 mM α-CD to 35 mM SB-β-CD and 10 mM M-β-CD to improve selectivity; PNAs that eluted more rapidly in the presence of α-CD had larger N's than in its absence, whereas others had comparable or lower ones (127). Liu et al. showed that the average N's of norgestrel enantiomers maximized as [γ-CD] and [SDS] were both increased in the ratio, [γ-CD]/[SDS] = 0.3. The reduced N's at low and high media concentrations were attributed to slow mass-transfer kinetics and long elution times, respectively (128).

Charge-transfer additives. Charge-transfer additives recently attracted interest as organized media. Using π-electron rich dye anions, Welsch et al. developed N's of dinitrotoluene isomers equaling 85,000; efficiency was lowered relative to SDS systems by counterpressure, mass-transfer kinetics, and boundaries between buffer zones caused by partial capillary filling (129). Miller et al. used planar organic cations in nonaqueous ACN electrolytes to develop peaks from PAHs having N's of 150,000–250,000; triphenylpyrylium ion generated larger N's than tropylium ion (130).

Miscellaneous. Tivesten et al. reported 2- to 20-fold losses in N's of on-column derivatized amino acids in SDS buffers; the losses, most severe for hydrophobic analytes, were partially offset by stacking and were attributed to on-column mixing and mobility differences between the analytes and their derivatives (131). Sun et al. separated PAHs using p-(carboxyethyl) calix[n]arenes; the analytes 2-aminoanthracene and pyrene generated van Deemter curves having minima of about 5 and 10 µm, respectively. The large minimum for pyrene was attributed to low solubility affecting diffusion and mass-transfer kinetics (132). Li and Locke reported distortion of pentachlorophenol peaks near the CMC of nonionic Tween 40; peaks were near-Gaussian above and below the CMC but triangular near the CMC (133). Kar and Dasgupta found that films of 0.5 M NaOH offered the best compromise between MEKC efficiency and extraction of acidic phenols from air when looped wires were used for injection; higher concentrations favored recovery but decreased efficiency by destacking (134). Muijselaar et al. found that alkylbenzenes generated larger N's from dodecylsulfate anions having tris(hydroxymethyl)aminomethane counterions, instead of Na ones; the increase was attributed to rigidity differences in the polar headgroups of the micelles and resulting differences in mass-transfer kinetics (135).

Models for N. Davis used an analytical method differing from that in Yu et al. (20) to determine from N's of micellar markers both diffusion coefficients of SDS micelles and instrumental contributions to dispersion; the values so determined enabled prediction of the N's in Yu et al. (20) with greater accuracy than before (136). Seals and Davis developed N's of analytes having various capacity factors in 50 mM SDS buffers containing 0–15% IPA; the N's largely were consistent with theory for plug size and longitudinal diffusion (137).

Measurements of diffusion coefficients. Muijselaar et al. used MEKC to determine the diffusion coefficients of alkylaryl ketones and SDS micelles by longitudinal diffusion (138). Davis used a different method to determine the micellar diffusion coefficients of SDS and showed the values were nearly identical to ones determined by pulsed-field gradient NMR (139).

Other. The reader is directed to various tables or graphs illustrating variations of N's of benzene derivatives with capacity factor (140), increases in N's of derivatized dipeptides with increasing [Brig 35] (141), variation of N's of antipyretic analgesics determined by microemulsion chromatography using different core phases (142), decreases in N's of vanilla flavorings with increasing volume fractions of ethanol and acetone (143), variations of N's of local anaestic drugs resolved by different concentrations of taurodeoxycholate micelles and Brig 35 (144), variations of N's of methylnitroanilines (145) and derivatized amino acids (146) with temperature, and variations of N's of antiviral drugs with additives like polyethylene glycol and dextran (147).

LIST OF ACRONYMS AND ABBREVIATIONS

Acronym or Abbreviation	Definition
ACN	acetonitrile (same as MeCN, methyl cyanide)
BaP	benzo[*a*]pyrene
CAPS	(3-cyclohexylamino)-1-propanesulfonic acid
CD	cyclodextrin
CE	capillary electrophoresis
CHAPS	3-[(3-cholamidopropyl)dimethylammonio]-1-propanesulfonate
CM	carboxymethyl
CM-β-CD	2-*O*-carboxymethyl-β-cyclodextrin
CMC	critical micelle concentration
CTAB	cetyltrimethylammonium bromide
dAMP	2'-deoxyadenosine 5'-phosphate
DNP	dinitrophenyl

DS	dodecyl sulfate
EOF	electroosmotic flow
GAG	glycosaminoglycan
IPA	isopropanol
MeCN	methyl cyanide (same as ACN, acetonitrile)
MEKC	micellar electrokinetic chromatography
MeOH	methanol
MOPS	3-(*N*-morpholino)propanesulfonic acid
NBD	4-nitrobenz-2-oxa-1,3-diazole
PDDAC	poly(diallyldimethylammonium chloride)
PNAs	polynuclear aromatics
PTH	phenylthiohydantoin
SBDs	starburst dendrimers
SB-12	Sulfobetain or *N*-dodecyl-*N*,*N*-dimethyl-ammonio-3-propanesulfonate
SDS	sodium dodecyl sulfate
SOS	sodium octyl sulfate
Tris	tris(hydroxymethyl)aminomethane
TTAB	tetradecyltrimethylammonium bromide
TTAC	tetradecyltrimethylammonium chloride

ACKNOWLEDGMENTS

The author thanks Professor Shigeru Terabe (Himeji Institute of Technology) for helpful comments on this chapter. This review was supported in part by the National Institutes of Health (1 R15 GM/OD55894-01).

REFERENCES

1. X. Huang, W. F. Coleman, and R. N. Zare, *J. Chromatogr.* **480**, 95 (1989).
2. S. L. Delinger and J. M. Davis, *Anal. Chem.* **64**, 1947 (1992).
3. L. Yu and J. M. Davis, *Electrophoresis* **16**, 2104 (1995).
4. D. E. Burton, M. J. Sepaniak, and M. P. Maskarinec, *Chromatographia* **21**, 583 (1986).
5. K. H. Row, W. H. Griest, and M. P. Maskarinec, *J. Chromatogr.* **409**, 193 (1987).
6. D. M. Northrup, D. E. Martire, and W. A. MacCrehan, *Anal. Chem.* **63**, 1038 (1991).
7. S. Terabe, K. Otsuka, and T. Ando, *Anal. Chem.* **61**, 251 (1989).
8. J. Vindevogel and P. Sandra, *Introduction to Micellar Electrokinetic Chromatography*, p. 100. Hüthig Buch Verlag, Heidelberg, 1992.
9. R. L. Chien and D. S. Burgi, *Anal. Chem.* **64**, 489A (1992).
10. D. S. Burgi and R. L. Chien, *J. Microcolumn Sep.* **3**, 199 (1991).

11. R. Weinberger, E. Sapp, and S. Moring, *J. Chromatogr.* **516**, 271 (1990).
12. R. Szücs, J. Vindevogel, P. Sandra, and L. C. Verhagen, *Chromatographia* **36**, 323 (1993).
13. K. R. Nielsen and J. P. Foley, *J. Chromatogr. A* **686**, 283 (1994).
14. Z. Liu, P. Sam, S. R. Sirimanne, P. C. McClure, J. Grainger, and D. G. Patterson, Jr., *J. Chromatogr. A* **673**, 125 (1994).
15. J. Crank, *The Mathematics of Diffusion*, 2nd ed., p. 15. Clarendon Press, Oxford, 1975.
16. J. A. Luckey, T. B. Norris, and L. M. Smith, *J. Phys. Chem.* **97**, 3067 (1993).
17. M. Almgren, F. Grieser, and J. K. Thomas, *J. Am. Chem. Soc.* **101**, 279 (1979).
18. E. L. Little and J. P. Foley, *J. Microcolumn Sep.* **4**, 145 (1992).
19. L. Song, Q. Ou, W. Yu, and G. Li, *J. Chromatogr. A* **699**, 371 (1995).
20. L. Yu, T. H. Seals, and J. M. Davis, *Anal. Chem.* **68**, 4270 (1996).
21. T. H. Seals and J. M. Davis, unpublished results.
22. M. J. Sepaniak and R. O. Cole, *Anal. Chem.* **59**, 472 (1987).
23. D. Stigter, R. J. Williams, and K. J. Mysels, *J. Phys. Chem.* **59**, 330 (1955).
24. J. M. Davis, *Anal. Chem.* **61**, 2455 (1989).
25. J. R. Mazzeo, M. E. Swartz, and E. R. Grover, *Anal. Chem.* **67**, 2966 (1995).
26. K. R. Nielsen, Ph.D. Dissertation, Villanova University, Villanova, PA (1994).
27. J. T. Smith and Z. El Rassi, *J. Chromatogr. A* **685**, 131 (1994).
28. J. T. Smith, W. Nashabeh, and Z. El Rassi, *Anal. Chem.* **66**, 1119 (1994).
29. C. Feldl, P. Møller, J. Otte, and H. Sørensen, *Anal. Biochem.* **217**, 62 (1994).
30. M. Castagnola, L. Cassiano, A. Lupi, I. Messana, M. Patamia, R. Rabino, D. V. Rossetti, and B. Giardina, *J. Chromatogr. A* **694**, 463 (1995).
31. J. H. Knox and K. A. McCormack, *Chromatographia* **38**, 207 (1994).
32. R. O. Cole, R. D. Holland, and M. J. Sepaniak, *Talanta* **39**, 1139 (1992).
33. J. Vindevogel and P. Sandra, *Introduction to Micellar Electrokinetic Chromatography*, p. 136 Hüthig Buch Verlag, Heidelberg, 1992.
34. S. Terabe, O. Shibata, and T. Isemura, *J. High Resolut. Chromatogr.* **14**, 52 (1991).
35. J. Vindevogel and P. Sandra, *Introduction to Micellar Electrokinetic Chromatography*, p. 138. Hüthig Buch Verlag, Heidelberg, 1992.
36. L. Yu, Masters' Thesis, Southern Illinois University, Carbondale (1996).
37. E. Grushka, *J. Chromatogr.* **559**, 81 (1991).
38. S. Terabe, N. Matsubara, Y. Ishinama, and Y. Okado, *J. Chromatogr.* **608**, 23 (1992).
39. Q. Dang, L. Yan, Z. Sun, and D. Ling, *J. Chromatogr.* **630**, 363 (1993).
40. A.-F Lecoq, L. Montanarella, and S. Di Biase, *J. Microcolumn Sep.* **5**, 105 (1993).

41. K. Bächmann, A. Bazzanella, I. Haag, K.-Y. Han, R. Arnecke, V. Böhmer, and W. Vogt, *Anal. Chem.* **67**, 1722 (1995).
42. J. H. Knox and I. H. Grant, *Chromatographia* **24**, 135 (1987).
43. J. H. Knox, *Chromatographia* **26**, 329 (1988).
44. E. Grushka, R. M. McCormick, and J. J. Kirkland, *Anal. Chem.* **61**, 241 (1989).
45. J. M. Davis, *J. Chromatogr.* **517**, 521 (1990).
46. J. H. Knox and K. A. McCormack, *Chromatographia* **38**, 215 (1994).
47. J. Vindevogel and P. Sandra, *J. High Resolut. Chromatogr.* **14**, 795 (1991).
48. W. H. Griest, M. P. Maskarinec, and K. H. Row, *Sep. Sci. Technol.* **23**, 1905 (1988).
49. J. T. Smith and Z. El Rassi, *J. Microcolumn Sep.* **6**, 127 (1994).
50. K.-L. Liu, K. L. Davis, and M. D. Morris, *Anal. Chem.* **66**, 3744 (1994).
51. M. D. Morris, personal communication (1995).
52. G. Taylor, *Proc. R. Soc. London, Ser. A* **219**, 186 (1953).
53. G. Taylor, *Proc. R. Soc. London, Ser. A* **225**, 473 (1954).
54. R. Aris, *Proc. R. Soc. London, Ser. A* **235**, 67 (1956).
55. M. R. Schure and A. M. Lenhoff, *Anal. Chem.* **65**, 3024 (1993).
56. A. S. Cohen, S. Terabe, J. A. Smith, and B. L. Karger, *Anal. Chem.* **59**, 1021 (1987).
57. R. A. Wallingford and A. G. Ewing, *Anal. Chem.* **60**, 258 (1988).
58. E. S. Ahuja, E. L. Little, K. R. Nielsen, and J. P. Foley, *Anal. Chem.* **67**, 26 (1995).
59. A. T. Balchunas and M. J. Sepaniak, *Anal. Chem.* **59**, 1466 (1987).
60. R. A. Wallingford and A. G. Ewing, *J. Chromatogr.* **441**, 299 (1988).
61. Y. Ishihama, Y. Oda, N. Asakawa, Y. Yoshida, and T. Sato, *J. Chromatogr. A* **666**, 193 (1994).
62. S. Terabe, personal communication (1995).
63. H. Nishi, N. Tsumagari, T. Kakimoto, and S. Terabe, *J. Chromatogr.* **477**, 259 (1989).
64. K. Otsuka, J. Kawahara, K. Tatekawa, and S. Terabe, *J. Chromatogr.* **559**, 209 (1991).
65. K. Otsuka, M. Kashihara, Y. Kawaguchi, R. Koike, T. Hisamitsu, and S. Terabe, *J. Chromatogr. A* **652**, 253 (1993).
66. Y. F. Yik and S. F. Y. Li, *Chromatographia* **35**, 560 (1993).
67. K. Otsuka and S. Terabe, *Electrophoresis* **11**, 982 (1990).
68. I. Beijersten and D. Westerlund, *Anal. Chem.* **65**, 3484 (1993).
69. J. Pedersen, M. Pedersen, H. Søeberg, and K. Biedermann, *J. Chromatogr.* **645**, 353 (1993).
70. A. T. Balchunas and M. J. Sepaniak, *Anal. Chem.* **60**, 617 (1988).
71. C. Bjergegaard, S. Michaelsen, and H. Sørenson, *J. Chromatogr.* **608**, 403 (1992).
72. S. Michaelsen, P. Møller, and H. Sørenson, *J. Chromatogr. A* **680**, 299 (1994).
73. C. Bjergegaard, L. Ingvardesen, and H. Sørensen, *J. Chromatogr. A* **653**, 99 (1993).

74. S. Michaelsen, M.-B. Schrøder, and H. Sørenson, *J. Chromatogr. A* **652**, 503 (1993).
75. P. H. Corran and N. Sutcliffe, *J. Chromatogr.* **636**, 87 (1993).
76. M. Chiari, M. Nesi, G. Ottolina, and P. G. Righetti, *J. Chromatogr. A* **680**, 571 (1994).
77. X.-Z. Qin, D.-S. T. Nguyen, and D. P. Ip, *J. Liq. Chromatogr.* **16**, 3713 (1993).
78. C. Stathakis and R. M. Cassidy, *J. Chromatogr. A* **699**, 353 (1995).
79. Z. K. Shihabi and M. E. Hinsdale, *J. Chromatogr. B* **669**, 75 (1995).
80. D. Perrett and G. A. Ross, *J. Chromatogr. A* **700**, 179 (1995).
81. M. J. Sepaniak, D. F. Swaile, A. G. Powell, and R. O. Cole, *J. High Resolut. Chromatogr.* **13**, 679 (1990).
82. R. O. Cole, M. J. Sepaniak, and W. L. Hinze, *J. High Resolut. Chromatogr.* **13**, 579 (1990).
83. J. Cai and Z. El Rassi, *J. Chromatogr.* **608**, 31 (1992).
84. C. D. Bevan, I. M. Mutten, and A. J. Pipe, *J. Chromatogr.* **636**, 113 (1993).
85. D. Crosby and Z. El Rassi, *J. Liq. Chromatogr.* **16**, 2161 (1993).
86. R. A. Wallingford, P. D. Curry, Jr., and A. G. Ewing, *J. Microcolumn Sep.* **1**, 23 (1989).
87. E. S. Ahuja, B. P. Preston, and J. P. Foley, *J. Chromatogr. B* **657**, 271 (1994).
88. M. J. Sepaniak, C. L. Copper, K. W. Whitaker, and V. C. Anigbogu, *Anal. Chem.* **67**, 2037 (1995).
89. O. H. J. Szolar, R. S. Brown, and J. H. T. Luong, *Anal. Chem.* **67**, 3004 (1995).
90. E. S. Ahuja and J. P. Foley, *Anal. Chem.* **67**, 2315 (1995).
91. J. Vindevogel and P. Sandra, *Anal. Chem.* **63**, 1530 (1991).
92. K. R. Nielsen and J. P. Foley, *J. Microcolumn Sep.* **5**, 347 (1993).
93. J. H. Aiken and C. W. Huie, *J. Microcolumn Sep.* **5**, 95 (1993).
94. S. Katsuta, T. Tsumura, K. Saitoh, and N. Teramae, *J. Chromatogr. A* **705**, 319 (1995).
95. H. Wan, P. E. Andersson, A. Engström, and L. G. Blomberg, *J. Chromatogr. A* **704**, 179 (1995).
96. C. L. Ng, C. P. Ong, H. K. Lee, and S. F. Y. Li, *J. Chromatogr. A* **680**, 579 (1994).
97. S. Terabe, H. Ozaki, K. Otsuka, and T. Ando, *J. Chromatogr.* **332**, 211 (1985).
98. H. Watarai, *Chem. Lett.*, 391 (1991).
99. H. Watarai, *Anal. Sci.* **7**, Suppl., 245 (1991).
100. R. L. Boso, M. S. Bellini, I. Miksik, and Z. Deyl, *J. Chromatogr. A* **709**, 11 (1995).
101. R. O. Cole, M. J. Sepaniak, W. L. Hinze, J. Gorse, and K. Oldiges, *J. Chromatogr.* **557**, 113 (1991).
102. C. Bjergegaard, H. Simonsen, and H. Sørensen, *J. Chromatogr. A* **680**, 561 (1994).
103. S. Terabe, M. Shibtata, and Y. Miyashita, *J. Chromatogr.* **480**, 403 (1989).
104. H. Nishi, T. Fukyyama, M. Matsuo, and S. Terabe, *J. Chromatogr.* **513**, 279 (1990).

105. K. L. Rundlett and D. Armstrong, *Anal. Chem.* **67**, 2088 (1995).
106. C. L. Copper and M. J. Sepaniak, *Anal. Chem.* **66**, 147 (1994).
107. C. P. Palmer and H. M. McNair, *J. Microcolumn Sep.* **4**, 509 (1992).
108. J. Wang and I. M. Warner, *Anal. Chem.* **66**, 3773 (1994).
109. H. Ozaki, S. Terabe, and A. Ichihara, *J. Chromatogr. A* **680**, 117 (1994).
110. S. Terabe and T. Isemura, *J. Chromatogr.* **515**, 667 (1990).
111. C. Stathakis and R. M. Cassidy, *Anal. Chem.* **66**, 2110 (1994).
112. N. Tanaka, T. Fukutome, T. Tanigawa, K. Hosoya, K. Kimata, T. Araki, and K. K. Unger, *J. Chromatogr. A* **699**, 331 (1995).
113. J. A. Lux, H.-F. Yin, and G. Schomburg, *Chromatographia* **30**, 7 (1990).
114. Y. Xue and E. S. Yeung, *Anal. Chem.* **66**, 3575 (1994).
115. H. J. Crabtree, I. D. Ireland, and N. J. Dovichi, *J. Chromatogr. A* **669**, 263 (1994).
116. J. C. Reijenga and M. Hutta, *J. Chromatogr. A* **709**, 21 (1995).
117. C. P. Palmer and S. Terabe, *Anal. Chem.* **69**, 1852 (1997).
118. C. P. Palmer and S. Terabe, *J. Microcolumn Sep.* **8**, 115 (1996).
119. A. W. Moore, Jr., S. C. Jacobson, and M. J. Ramsey, *Anal. Chem.* **67**, 4184 (1995).
120. F. von Heeren, E. Verpoorte, A. Manz, and W. Thormann, *Anal. Chem.* **68**, 2044 (1996).
121. K. W. Whitaker, C. L. Copper, and M. J. Sepaniak, *J. Microcolumn Sep.* **8**, 461 (1996).
122. H. Wan and L. G. Blomberg, *J. Microcolumn Sep.* **8**, 339 (1996).
123. H. Wan and L. G. Blomberg, *J. Chromatogr. Sci.* **34**, 540 (1996).
124. H. Wan, A. Engström, and L. G. Blomberg, *J. Chromatogr. A* **731**, 283 (1996).
125. H. Wan and L. G. Blomberg, *J. Chromatogr. A* **758**, 303 (1997).
126. R. S. Brown, J. H. T. Luong, O. H. J. Szolar, A. Halasz, and J. Hawari, *Anal. Chem.* **68**, 287 (1996).
127. A.-L. Nguyen and J. H. T. Luong, *Anal. Chem.* **69**, 1726 (1997).
128. Y. Liu, J. Gu, and R. Fu, *J. High Resolut. Chromatogr.* **20**, 159 (1997).
129. T. Welsch, S. Kolb, and J. P. Kutter, *J. Microcolumn Sep.* **9**, 15 (1997).
130. J. L. Miller, M. G. Khaledi, and D. Shea, *Anal. Chem.* **69**, 1223 (1997).
131. A. Tivesten, E. Örnskov, and S. Folestad, *J. High Resolut. Chromatogr.* **19**, 229 (1996).
132. S. Sun, M. J. Sepaniak, J.-S. Wang, and C. D. Gutsche, *Anal. Chem.* **69**, 344 (1997).
133. G. Li and D. C. Locke, *J. Chromatogr. A* **734**, 357 (1996).
134. S. Kar and P. K. Dasgupta, *J. Chromatogr. A* **739**, 379 (1996).
135. P. G. Muijselaar, H. A. Claessens, and C. A. Cramers, *J. Chromatogr. A* **764**, 127 (1997).
136. J. M. Davis, *J. Microcolumn Sep.* In press.
137. T. H. Seals and J. M. Davis, unpublished results.

138. P. G. Muijselaar, M. A. van Straten, H. A. Claessens, and C. A. Cramers, *J. Chromatogr. A* **766**, 187 (1997).
139. J. M. Davis, *The Analyst*. In press.
140. P. G. H. M. Muijselaar, H. A. Claessens, and C. A. Cramers, *J. Chromatogr. A* **696**, 273 (1995).
141. I. Beijersten and D. Westerlund, *J. Chromatogr. A* **716**, 389 (1995).
142. X. Fu, J. Lu, and A. Zhu, *J. Chromatogr. A* **735**, 353 (1996).
143. U. Bütehorn and U. Pyell, *J. Chromatogr. A* **736**, 321 (1996).
144. A. Amini, I. Beijersten, C. Pettersson, and D. Westerlund, *J. Chromatogr. A* **737**, 301 (1996).
145. U. Pyell and U. Bütehorn, *Chromatographia* **40**, 69 (1995).
146. P. L. Desbène and C. E. Fulchic, *J. Chromatogr. A* **749**, 257 (1996).
147. R. Singhal, J. Xian, and O. Otim, *J. Chromatogr. A* **756**, 263 (1996).

CHAPTER

5

CAPILLARY GEL ELECTROPHORESIS

PAUL SHIEH AND NELSON COOKE

Supelco Inc., Supelco Park, Bellefonte, Pennsylvania

ANDRAS GUTTMAN

Genetic Biosystems, Inc., San Diego, California

5.1.	Introduction	185
5.2.	Theory and Operation Parameters	187
	5.2.1. Theory	187
	5.2.2. Chemical vs. Physical Gels	189
	5.2.3. Operation Variables	190
5.3.	Applications	196
	5.3.1. Separation of Single-Stranded Oligonucleotides	196
	5.3.2. Separation of Double-Stranded DNA	204
	5.3.3. Separation of Proteins and Peptides	210
	5.3.4. Capillary Gel Electrophoresis of Complex Carbohydrates	214
5.4.	Conclusion	217
List of Acronyms and Abbreviations		218
References		218

5.1. INTRODUCTION

The principle of electrophoresis (from the Greek words *elektron* = electron and *phoresis* = carrying) is that molecules migrate under the influence of an electric field at various rates depending on their charge-to-mass ratio. By controlling the applied electric field strength, the medium through which the molecules move, and other separation variables such as temperature, one can separate various molecules from complex biological mixtures into discrete peaks. In the mid-1930s, in Uppsala, Sweden, Arne Tiselius demonstrated the

High Performance Capillary Electrophoresis, edited by Morteza G. Khaledi. Chemical Analysis Series, Vol. 146.
ISBN 0-471-14851-2 © 1998 John Wiley & Sons, Inc.

potential of electrophorsis as a new research tool; he was awarded a Nobel Prize ten years later for his pioneering work in this area. He used the so-called moving boundary method to separate human serum into albumin, α-globulin, β-globulin, and γ-globulin (1). Since then, electrophoresis has become a basic separation method employed to analyze biopolymers such as proteins and nucleic acids in the pharmaceutical, biotechnology, molecular biology, forensics, and immunology fields. Traditional electrophoresis methodology consists of gel casting equipment, separation chambers, power supplies, and the gels (2,3). Supporting materials such as polyacrylamide and agarose are normally used in these convential electrophoresis methods. Here the gel acts as an anticonvective medium, eliminating conductive transport and reducing diffusion, so the various components of the analyte mixture migrate in sharp zones during the electrophoretic separation. The gel also acts as a molecular sieving matrix leading to separation based on analyte size. Classical electrophoresis is still performed today mainly by manual methods that require a lot of skill in gel pouring, sampling, separation, and staining/destaining processes for visualization of the separated bands. Separations can be archieved by either scanning the stained gels or using photographic methods. All of the above show that classical gel electrophoresis is a very labor-intensive and time-consuming technique. The only slab-gel-type systems automated so far are the different DNA sequencers. Recently, the gel casting equipment has become a very significant component of slab-gel-based separation systems due to the increasing interest in ultrathin gels and capillary casting (4).

In its short history of a little more than a decade, capillary electrophoresis (CE) has proven its applicability in modern separation science (5,6). As a relatively new, powerful separation technique CE is ideally suitable for handling minute amounts of samples. There is an increasing demand for CE in bioanalytical research, e.g., in biotechnology and in various clinical, diagnostic, genetic, and forensic applications. CE is quite similar to high-performance liquid chromatography (HPLC) in its ease of use, high resolution, speed, on-line detection, and full automation capability and is actually viewed as an automated and instrumental approach to classical electrophoresis. Several recent books have covered issues of CE theory, instrumentation, and applications (7–10). (See Chapters 1–8, 19–31 in this volume.)

The first papers on the use of capillary gel electrophoresis were published in the late 1980s by Kasper et al. (11) and Cohen et al. (12). These publications addressed the application areas of DNA and sodium dodecyl sulfate (SDS)–protein separation, respectively. Since then, it has become more and more evident that almost all methods developed for slab gel electrophoresis can easily be transferred to a capillary format, with the advantages of fast analysis with high resolution, full automation, and on-line data acquisition and storage capability. Additionally, multiple injections can be made from only a few microliters of sample, enabling easy validation of analytical methods. High-sensitivity detection systems such a laser-induced fluorescence (LIF) have opened the way to new, until now unimaginable low detection limits in the zeptomole range (13).

This chapter will describe the main principles of capillary gel electrophoresis and its main applications to the separation of biopolymers such as DNA, proteins, and complex carbohydrate molecules. The main focus will be on applications that can be easily performed with commercially available instrumentation.

5.2. THEORY AND OPERATION PARAMETERS

5.2.1. Theory

In capillary gel electrophoresis (CGE), the size-dependent retardation of the solute is a primary function of the concentration of the separation polymer ($P^\%$):

$$v = v_0 \exp(-K_R P^\%) \qquad (5.1)$$

where v is the apparent velocity; v_0 is the free solution velocity of the solute; and K_R is the retardation coefficient (14–17).

If the solute is a polyion holding the same amount of charge per unit, such as DNA molecules or SDS–protein complexes, the total electrostatic force on the migrating ion can be assumed to be constant per molecular weight unit. Actually, the amount of SDS molecules surrounding the protein depends on the protein's molecular weight (MW), so the higher the MW of the protein, the more SDS is attached. Thus, in these cases the charge (Q) is proportional to the MW of the biopolymer (18), and therefore one can consider using the MW instead of the net charge:

$$Q \sim (\text{MW})^k \qquad (5.2)$$

where k, the exponent of (MW), represents information about the apparent shape of the polyionic molecule under the electric field used, i.e., random coil, oriented, stretched, etc. (19–21).

Incorporating all the constants together in one (Const), the electrophoretic velocity of the migrating polyion can be described as

$$v = \text{Const}\, E(\text{MW})^k \exp(1/T) \exp(-K_R/P^\%) \qquad (5.3)$$

where $E(MW)^k \exp(1/T)$ corresponds to the free solution velocity (v_0) at zero polymer concentration, with E as the field strength and T as the temperature and with $\exp(-K_R P^{\%})$ responsible for the sieving.

When effective molecular sieving exists and the Ogston theory applies (22), the average pore size of the matrix should be in the same range as that of the hydrodynamic radius of the migrating solute. In this instance the logarithm of the velocity of the migrating solute is proportional to the size of the solute:

$$v \sim E \exp[-(MW)] \qquad (5.4)$$

This theory also assumes that the migrating particles behave like unperturbed spherical objects. However, large biopolymer molecules with flexible chains, e.g., DNA or protein molecules, still migrate through polymer networks that have a pore size significantly smaller than the size of the solute. This phenomenon is explained by the reptation migration model. The reptation model describes the migration of the polyion, suggesting a "head first" motion through the pores of the polymer network (19,23–25). According to the reptation model (19), the velocity of the migrating solute is inversely proportional to the size of the solute; thus, in this case of large molecules with flexible, chains

$$v \sim E/(MW) \qquad (5.5)$$

When higher electric field strengths are used, biased reptation may occur, in which case the velocity of the solute can be described as

$$v \sim E/(MW) + bE^2 \qquad (5.6)$$

where b is a function of the mesh size of the polymer network as well as the charge and segment length (26) of the migrating biopolymer molecule.

If there is any special additive in the gel, such as a complexing agent, the solute will have a distribution between the complex and the electrolyte:

$$v = \text{Const } E(MW)^k \exp(1/T) \exp(-K_R P^{\%}) R_p (1 + K_M)^{-1} \qquad (5.7)$$

where K_M is the complex formation constant of the M ion, and R_p is the molar ratio of the free solute (27). Depending on the charge and size of the complexing agent, the complex may migrate faster or slower than the free solute. When the complexing agent has a charge opposite to that of the solute, an uncharged complex may be formed so that the higher the concentration of the complexing agent, the slower will be the migration velocity of the solute.

5.2.2. Chemical vs. Physical Gels

The primary difference between classical gel electrophoresis and CE is the use of the fused silica capillary column, which can be filled with any separation medium. Gel-filled capillaries were at first filled with cross-linked polyacrylamide gel similar to that which was used in traditional gel electrophoresis and have proven to be particularly useful in the analysis of DNA and SDS–protein molecules (32). Although these columns were originally filled by the users, nowadays many manufacturers are offering prefilled gel capillary columns (33). However, bubble formation is quite common in cross-linked polymer matrix filled capillaries, as will be discussed later.

Gels may vary from viscous fluids to solids. In CE two types of sieving matrices are used: the chemically cross-linked ("chemical") gels, with relatively high viscosity, and the non-cross-linked low-viscosity ("physical") gels.

The so-called chemical, or cross-linked, gels have a well-defined pore structure due to the cross-linker. They are usually chemically bound to the capillary wall, to prevent electroosmotic-flow (EOF)-mediated disruption of their structure. The pore size of the cross-linked gels can not be varied after polymerization. These gels are heat sensitive, and any slight precipitation—originated, e.g., from the injected samples—can damage the gels in the capillary (bubble formation). On the other hand chemical gels provide very high resolution for low-molecular-weight biopolymers such as single-stranded oligonucleotides and peptides. Due to the extremely high viscosity of the cross-linked gels, they are not replaceable in the capillary; actually they should be polymerized within the column (as they cannot be pushed into or out of the capillary). Almost exclusively, acrylamide is used as monomer cross-linked with N,N'-methylenebisacrylamide (BIS) for chemical gels in capillary gel electrophoresis.

The other type of gel matrices that have recently become very popular in CGE are the so-called physical gels, or polymer networks. These non-cross-linked linear polymers have a very flexible dynamic pore structure, and they are not attached to the inside wall of the capillary. Their pore size can be varied any time, even during the separation, by simply changing physical parameters such as the column temperature. Also, non-crosslinked polymer networks are heat insensitive, and any particulate originated from injection can be easily removed by simply replacing the matrix in the capillary. These so-called replaceable gels are very popular for the separation of high-molecular-weight double-stranded DNA (dsDNA) molecules and SDS–protein complexes (34). In the beginning, linear, non-crosslinked polyacrylamides and various cellulose derivatives, such as methyl, ethyl, or hydroxypropyl cellulose, were used to separate DNA restriction fragments. Agarose, which is well known from conventional slab gel separations, can also be filled into narrow bore capillaries and used as separation media (35). Recently, other polymers such as slightly branched dextran and linear poly(ethylene oxide) (PEO) were applied to SDS-mediated protein separations in CGE (36), and PEO was later also applied to DNA sequencing (37). Very dilute polymer solutions (0.001%) were also successfully applied to dsDNA separations, suggesting that a fully entangled polymer network is not a prerequisite for separaton of biopolymers (38). Another remarkable advantage of using non-cross-linked hydrophilic polymer solutions is their applicability to the quantitative pressure injection method. In contrast, the use of cross-linked gels, due to their high viscosity, is only possible with the electrokinetic injection mode.

In the application of both chemical and physical gels, denaturing and nondenaturing conditions can be employed. Denaturants such as urea or formamide generally are used for single-stranded DNA (ssDNA) separations, in DNA sequencing gels, and in some cases in special dsDNA separations (39). Nondenaturing conditions are used in most dsDNA separations [DNA restriction fragments; polymerase chain reaction (PCR) products] and in cases when point mutations should be detected in ssDNA molecules (40). In this latter case differences based on the shape, size, and charge of the analytes are exploited. SDS is frequently used as a denaturant in protein molecular weight determination.

The quality of the capillary wall coating is very important if one wishes to obtain good reproducible separations with gel- or polymer-network-filled capillary columns. In fact, a good coating is essential with the use of replaceable polymer matrices. Good coatings must be stable over a wide range of pH and be usable for a reasonable operating lifetime. The coating must also minimize EOF and prevent sample interaction with wall surface silanols. Several different types of coatings have been used in CGE, including cross-linked and non-cross-linked polyacrylamide, dextran, poly(ethylene glycol), poly(vinyl alcohol) and carbohydrate derivatives (41–45). Many of them are commercially available and can be used for hundreds of runs.

5.2.3. Operation Variables

Separations in gel-filled capillary columns can be influenced by alteration of the separation parameters, such as the applied electric field strength, temperature, and capillary dimensions (diameter and length). Note that DNA and SDS–protein complexes are negatively charged; therefore, their separations with gel-filled capillaries require the polarity of the CE instrument's power supply to be reversed (i.e., the cathode must be at the injection side). The separation is monitored in real time by UV (254 nm for DNA or 214 nm for proteins) or LIF detection.

5.2.3.1. Voltage

As Eq. (5.1) indicates, the migration velocity of a given solute is proportional to the electric field strength (E). In cases of the separation of DNA, SDS–protein, and complex carbohydrate molecules, this simple linear relationship has been experimentally verified (46). Besides the application of a constant electric field (isoelectric mode), other separation modes of constant current, constant power, as well as gradient modes of the above have proved useful. Separation of dsDNA fragments, for example, was significantly improved by employing stepwise and continuous field strength gradients (47). Guttman et al. investigated the effect of electric field strength gradients on the separation of DNA restriction fragments. They demonstrated that in CGE improved separation of DNA restriction fragments, up to the kilobase-pair (kbp) range, can be achieved by applying some kind of field strength gradient method (Figure 5.1). Note that with the increase of the applied field strength, the peak efficiency and the concomitant resolution of the separated peaks are increased while separation time is decresed. The use of increasing, decreasing, or stepwise voltage gradient techniques showed that the resolving power can be optimized for a given DNA chain length range and separation time can be significantly reduced. In Guttman and colleagues' study (47) on the separation of the ϕX174 DNA restriction fragments by capillary polyacrylamide gel electrophoresis, the best separation with minimum time requirement was achieved by using a continuously decreasing applied electric field. With the use of field strength gradient methods, the apparent peak efficiency and resolution may be misleading since the different size components migrate past the detector window with a velocity that is determined by the voltage in use at that point in time. The shape of the gradient can be continuous or stepwise over time. Both methods can be used to increase separation efficiency and resolution in CGE dsDNA molecules.

5.2.3.2. Temperature

Like voltage gradients, temperature gradients can be applied during CGE. However, elevated temperatures can be used only with non-cross-linked physical gels where the temperature changes do not initiate bubble formation and concomitant discontinuity of the gel matrix. The increase of temperature decreases the viscosity of the polymer matrix (14,48) and increases the average mesh size of the dynamic pore structure. In this way, if necessary, the actual sieving effect can be modified anytime during the separation. Guttman and Cooke (14) studied two different separation modes in high-performance CGE to determine the effect of temperature on the separation of DNA restriction fragments. In the instances of isoelectrostatic (use of a constant applied electric

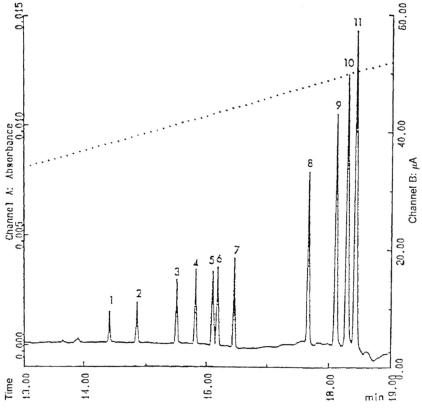

Figure 5.1. Separation of the φX174 DNA restriction fragment mixture by capillary polyacrylamide gel electrophoresis using an increasing voltage gradient. Conditions: eCAP DNA capillary, total length 47 cm with 40 cm effective length. Buffer: 0.1 M Tris–borate, 2 mM EDTA (pH 8.35). Voltage: field gradient from 0 to 400 V/cm in 20 min. Peaks: (1) 72; (2) 118; (3) 194; (4) 234; (5) 271; (6) 281; (7) 310; (8) 603; (9) 872; (10) 1078; (11) 1353 bp. (From ref. 47 with permission.)

field) and isorheic (use of a constant current) separation modes, migration properties and resolution of the DNA molecules were studied as a function of column temperature between 20 and 50 °C. In the isoelectrostatic separation mode, migration time and resolution decreased as temperature increased (Figure 5.2). In the isorheic separation mode, increasing column temperature resulted in a maximum in migration time for all of the DNA fragments (Figure 5.3) and showed maximum resolution for the lower-molecular-weight fragments [$< 2 \times 10^5$ Da (daltons)]. Guttman et al. found dramatic differences in the effect of temperature on the electrophoretic separation of SDS–protein

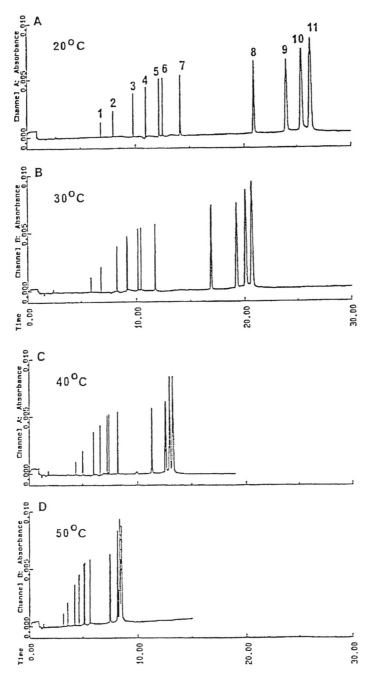

Figure 5.2. Separation of the φX174 DNA restriction fragment mixture by capillary polyacrylamide gel electrophoresis at different temperatures in isoelectrostatic separation mode. Conditions: capillary, 27 cm total length with 20 cm effective length. Buffer: 0.1 M Tris–boric acid–2 mM EDTA (pH 8.5). Sample: 50 μg/mL of φX174 DNA/*Hae*III digest. Peaks: same as Figure 5.1. (From ref. 14 with permission.)

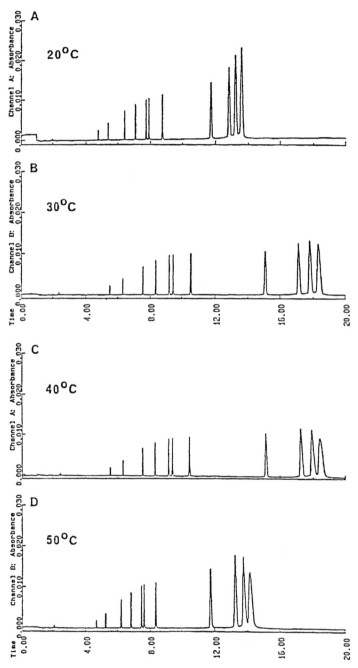

Figure 5.3. Separation of the ɸX174 DNA restriction fragment mixture by CGE at different temperatures in isorheic mode. Conditions: constant current mode, 20 μA; otherwise as in Figure 5.2. (From ref. 14 with permission.)

molecules using PEO and dextran-gel-based capillary columns (48). Separation behavior could not be rationalized merely by the viscosity–temperature relationship of the two polymer network systems. Indeed, the activation energy data for the polymer network formation suggested that a dynamic temperature-dependent formation and dissociation of oriented conformers occurred in both polymer solutions (48). These results suggested that the more or less oriented arrangements, i.e., channel-like structures, in concentrated polymeric solutions play a very important role in CGE separations. All the parameters helping to arrange the organized structures tend to enhance the separation efficiency in high-performance CGE.

The other important feature of the applicability of higher temperature of up to 70 °C is very useful in point mutation studies with temperature-dependent duplex formation (49). Gelfi et al. reported a continuous denaturing capillary gel electrophoresis (CDCGE) for the mutation analysis of DNA samples from four cystic fibrosis patients (50). The wild-type and mutant DNA were separated on a capillary filled with a poly(N-acryloylaminoethoxyethanol) (PAAEE) gel.

5.2.3.3. Capillary Dimensions

The capillary dimensions have a significant effect on the achievable separation; for example, longer capillaries can provide higher resolving power. However, increases in the applied electric field strength increase the separation efficiency; and when the selectivity is constant the resolution is also increased. Therefore, one has to find the optimal combination of column length and field strength to achieve the desired efficiency and resolution. The effect of the internal diameter (i.d.) of the capillary is very important for detection purposes. The higher the i.d., the more sample can be injected, resulting in better detection limits. On the other hand, Joule heating, and thus column temperature, is exponentially increased with the increasing column diameter. Even with one of the most effective heat-removal (liquid cooling) systems (30,51), the maximum applicable power to a capillary is limited to approximately 5 W. This suggests that the i.d. of the separation column should also be optimized for a given separation problem in such a way that the detection is sensitive enough and the heat dissipation does not decrease separation efficiency.

5.2.3.4. Injection

There are two basic injection techniques used in CE. One is called hydrodynamic injection, and it can be mediated by pressure, vacuum, or buffer vial level differences. This injection mode is easily applied to low-viscosity "physical" polymer-network-filled capillary columns but cannot be applied to high-viscosity polymer or cross-linked ("chemical") gel-filled capillary columns. In hydrodynamic injection, sample preparation is simplified, as no sample prepurification such as desalting is necessary. With the use of an internal standard, migration time precision of less than 0.1% RSD can be achieved (52). The other injection mode in CGE is *electrokinetic injection*, which is accomplished by starting the electrophoresis process from the sample vial. Electrokinetic injection often yields superior peak efficiency compared to hydrodynamic injection, especially when the injection is made from low-ionic-strength solutions or from water since the sample components are stacked against the relatively viscous gel matrix (concentration effect). In the case of separation of small molecules this injection method can result in a sampling bias because different mobility sample components move into the capillary with different speeds. No such bias occurs in the separations of species with a similar charge-to-mass ratio such as biopolymers (DNA or SDS–protein complexes). However, quantification of the actual injected amount is very difficult during electrokinetic injections.

Separation performance in CGE is often highly dependent on the sample matrix, especially in the case of electrokinetic injection when high amount of salt is present (53). Desalting or ultrafiltration of the sample usually helps, but it is labor intensive and can be avoided by special sampling techniques. In both hydrodynamic and electrokinetic injection modes, sampling can be made by the simple injection of the analyte into the capillary, preceded by preinjection of a different buffer system or water to increase injected amount and precision (54). With cross-linked high-viscosity gels, preinjection of water greatly increased peak area reproducibility and even increased the injected amount. Preinjection of a low-viscosity buffer plug helped to obtain high-efficiency separation with sharp peaks from high-salt-containing PCR products (55).

5.3. APPLICATIONS

5.3.1. Separation of Single-Stranded Oligonucleotides

CGE has become a powerful tool for the analysis of natural and synthetic oligonucleotides because of its high efficiency and great resolving power (56). Methods have been developed such as purity check of primers for DNA sequencing, point mutation analysis, and DNA diagnostics (57–59). Similar to slab gel electrophoresis, cross-linked polyacrylamide gels were first used in DNA separation by CGE (60). The gel capillary contains denaturants such as 7 M urea or formamide to denature DNA molecules. Figure 5.4 shows a separation of a $pd(A)_{40-60}$ mer mixture by a cross-linked polyacrylamide gel (7.5% T, 3.3% C) containing 7 M urea at pH 8.3. The 20 single-stranded

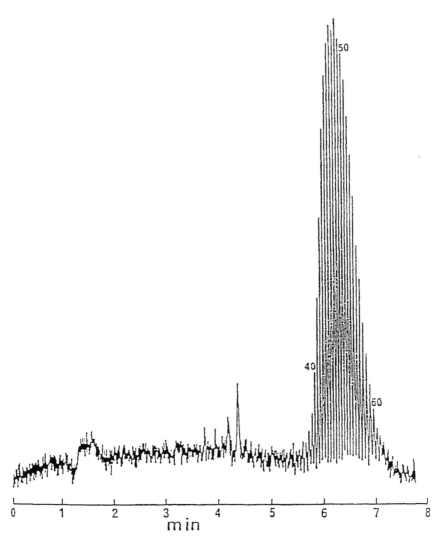

Figure 5.4. CGE separation of oligonucleotides. Conditions: capillary, 27 cm total length with 13 cm effective length. Buffer: 0.1 M Tris–0.25 M borate, 7 M urea, pH 8.3. Gel: 7.5% T, 3.3% C. $E = 400$ V/cm. Sample: pd(A)$_{40-60}$ mers. (From ref. 60 with permission.)

oligonucleotides were baseline separated in less than 8 min. Compared to HPLC, CGE offers higher speed and resolution for this type of analysis. Warren and Vella investigated the difference between CGE and anion exchange chromatography in the separation of oligonucleotides (61). In CGE, the mobility of a molecule through the gel matrix is determined by two factors:

one is the charge-to-mass ratio of the DNA molecule, which influences DNA electrophoretic mobility, and the other is the sieving characteristics imposed by the gel matrix, which separate DNA by size. By comparison, the separations of DNA by anion exchange chromatography are dominated by the total number of negative charges on the oligonucleotides. The authors concluded that both HPLC and HPCE can be used for the separation of small oligonucleotides. However, for the analysis of large oligodeoxyribonuclotides (> 50 bases), CGE provides higher resolution than HPLC (Figure 5.5).

Although cross-linked polyacrylamide gels offer high resolution with rapid analysis, the method suffers from poor stability and short shelf life due to bubble formation during gel polymerization or during a run. The cross-linked gel is also not stable either when exposed to elevated temperature or high electric fields. Recently, Guttman et al. reported the use of linear polyacrylamide gels for the separation of DNA. This type of gel has the advantage of a much longer shelf life while still providing high resolution similar to cross-linked polyacrylamide gels (62). The resolving power of linear polyacrylamide was demonstrated by the baseline separation of three human K-*ras* oncogene mixtures with only one base difference in less than 13 min (Figure 5.6). The authors studied the migration behavior of oligonucleotides in the gel-filled capillary with and without denaturants (7 M urea). When nondenaturing gel is used; the relative migration order is not constant for homooligonucleotides of the same chain length, but is dependent on the base number. For a base number less than 14, the order is $A > C > G > T$, and for base numbers larger than 18, it is $G > A > C > T$. The discrepancy is probably caused by the secondary structure of the oligonucleotides. When denaturing gels are employed; the migration order of the homooligonucleotides is the same for the entire chain-length range examined, normally $A > C > G > T$.

Capillary affinity gel electrophoresis (CAGE) has also been used for the separation of oligonucleotides by specific base recognition. Baba et al. synthesized poly(9-vinyladenine)(PVAD) as a water-soluble polynucleotide analogue that has a polyvinyl group instead of a sugar-phosphate backbone in the polynucleotides and employed it as an affinity macroligand (63). They intercalated PVAD into polyacrylamide gel by copolymerization with acrylamide and BIS–acrylamide inside the capillary. PVAD–polyacrylamide gel decreases the electrophoretic mobility of oligo(dT), by forming hydrogen bonds with DNA samples. Figure 5.7 shows a separation of $pd(A)_{12-18}$mers and $pd(T)_{15}$ using polyacrylamide gel with and without PVAD. In this case, $pd(T)_{15}$ was coeluted with the pd(A) mixture on a 8%T, 5%C polyacrylamide gel. However, with PVAD–polyacrylamide gel, the d(T) was retained and separated from the pd(A) mixture.

Phosphothioate oligonucleotides, which have the substitution of a sulfur for an oxygen atom on each phosphorus within the oligonucleotide back-

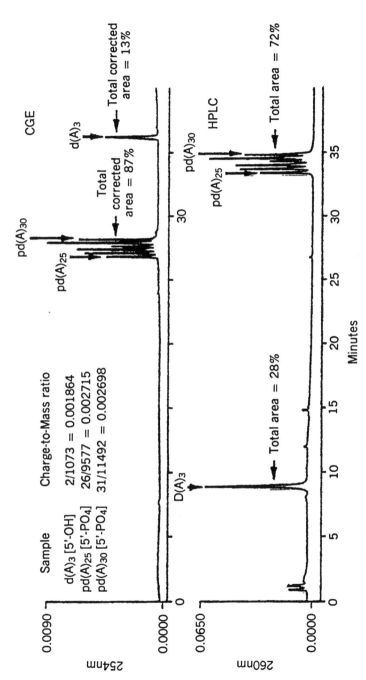

Figure 5.5. Separation of oligonucleotides by CGE and anion exchange HPLC. Conditions: CGE–capillary, μPAGS-5 (5% T, 5% C), 60 cm total length with 51 cm effective length. Voltage: 13 kV. Ambient temperature. HPLC–linear gradient from 0–0.6 M NaCl in 30 min. Flow rate: 0.75 mL/min. Sample: d(A)$_3$ and pd(A)$_{25-30}$. (From ref. 61 with permission.)

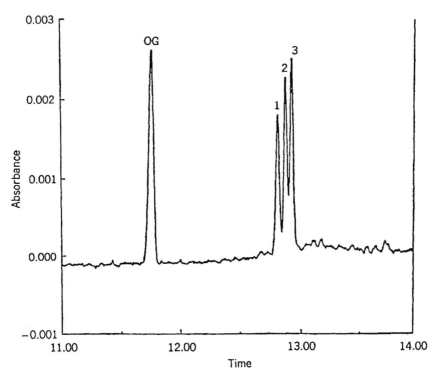

Figure 5.6. Nondenaturing capillary polyacrylamide gel electrophoresis separation of a human K-*ras* oncogene (OG) mixture. Conditions: capillary, 47 cm total length with 40 cm effective length. $E = 400$ V/cm. Buffer: 0.1 M Tris–borate, 2 mM EDTA, (pH 8.5). Gel: linear polyacrylamide gel. Peaks: (1) dGTTGGAGCT-G-GTGGCGTAG; (2) dGTTGGAGCT-C-GTGGCGTAG; (3) dGTTGGAGCT-T-GTGGCGTAG. (From ref. 62 with permission.)

bone, are used for antisense therapeutic investigations. Srivatsa et al. have used a commercially available gel-filled capillary for quantitative CGE assay of phosphorothioate oligonucleotides in pharmaceutical formulations (64). They reported good day-to-day ($<0.43\%$ RSD), column-to-column ($<0.092\%$ RSD) migration time reproducibility. The gel capillary also offers a good separation of phosphorothioate oligonucleotides (21 mers) and its $(n-1)$ deletion sequence (Figure 5.8). However, "thioation failures" (mono or higher-order partial phosphodiesters) could not be detected by CGE. It is primarily a length-based separation method for DNA molecules, yielding excellent resolution of the deletion sequences of phosphorothioate oligonucleotides.

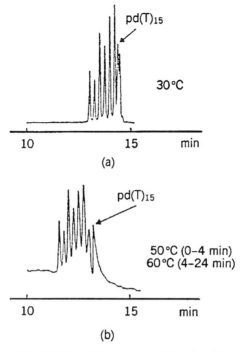

Figure 5.7. (a) CGE and (b) CAGE of pd(A)$_{12-18}$ and pd(T)$_{15}$. Conditions: capillary, 42 cm total length with 22 cm effective length. Buffer: 0.1 M Tris–borate, 7 M urea, pH 8.6. Gel: (a) 8% T, 5% C; (b) 8% T, 5% C and 0.1% (PVAD). Samples: pd(A)$_{12-18}$ and pd(T)$_{15}$. (From ref. 63 with permission.)

DNA sequencing is one of the faster growing areas in the field of molecular biology. CGE has resolving power similar to that of slab gels but offers very rapid separation with the possibility of automation. Figure 5.9 shows the separation of DNA sequencing fragments by CGE using a capillary column filled with linear polyacrylamide gel (65). The capillary used for this application was coated with a cross-linked polyacrylamide and was stable for multiple injections. The viscosity of this linear polyacrylamide gel was low, which allowed it to be replaced in the capillary after each analysis. Fast separation with good read length (> 500 bases) was obtained with CGE of DNA sequencing fragments.

In order to achieve single base resolution of oligonucleotides, most polyacrylamide gels used possess very high viscosity and are not easy to replace. The gel can degrade with use, decreasing its performance and shortening its

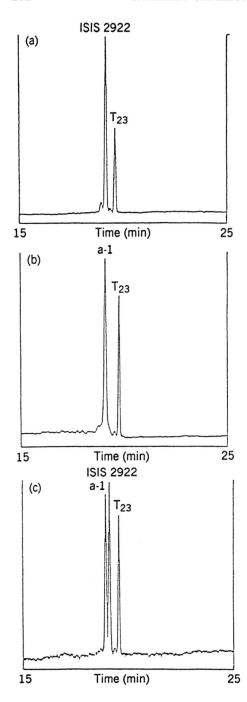

Figure 5.8. Separation of a phosphothioates mixture by CGE: (a) ISIS 2922; (b) an $n-1$ deletion sequence of ISIS 2922; (c) a mixture of (a) and (b). Conditions: eCAp ssDNA 100 capillary, 47 total length with 40 cm effective length. $E = 300$ V/cm. $T = 30$ °C. (From ref. 64 with permission.)

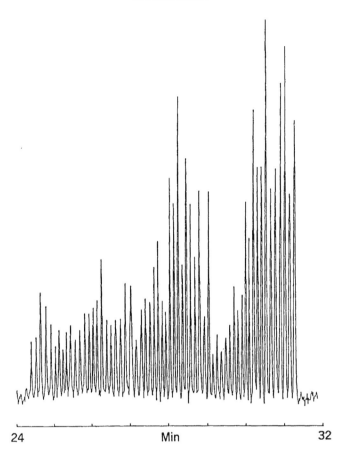

Figure 5.9. Electropherogram of chain-termination sequencing reaction products. Conditions: capillary, 68 cm total length with 54 cm effective length. $E = 300$ V/cm. Buffer: 0.1 M Tris–borate, 2.5 mM EDTA, 7 M urea, pH 7.6. (From ref. 65 with permission.)

lifetime. Recently, a new replaceable polyacrylamide gel matrix has been reported specifically for the separation of short oligonucleotides (66). The gel matrix has low viscosity and can be replaced by many commercial instruments. In combination with a coated capillary, the gel-filled capillary yields high resolution of DNA molecules with good stability. Figure 5.10 shows the first and 324th injection of $pd(A)_{40-60}$ mers on a replaceable-gel-filled capillary. Good resolution and long operating life were observed with the replaceable gel.

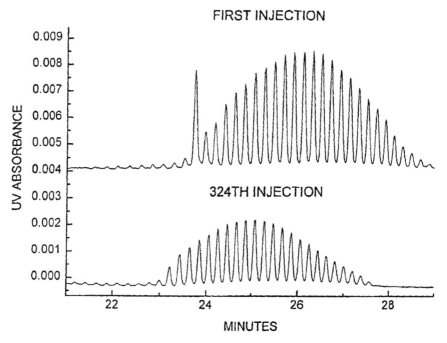

Figure 5.10. Stability study of the separation of oligonucleotides by replaceable polyacrylamide gel. Conditions: eCAP DNA capillary, total length 27 cm with 20 cm effective length. Buffer: 0.1 M Tris–borate, 7 M urea, 2 mM EDTA, pH 8.5. $E = 300$ V/cm. $T = 30\,°C$. Sample; $pd(A)_{40-60}$. (From ref. 66 with permission.)

5.3.2. Separation of Double-Stranded DNA

The ability to analyze dsDNA, such as PCR products or restriction enzyme digest fragments, is a fundamental method in molecular biology. The separation and identification of individual fragments can provide important information regarding the structure of DNA. Restriction mapping of DNA fragments is currently a major application of slab gel electrophoresis, where separation is based on the difference of their size. CGE, which offers the benefits of rapid separation, automation, and convenience over slab gel electrophoresis, has become a powerful tool for the separation of dsDNA fragments. Since all the DNA polymers have a very similar charge-to-mass ratio, their electrophoretic mobility is constant regardless of the differences in their molecular weight. The separation is based upon the differences in size, as DNA molecules migrate through pores of the gel inside the capillary.

5.3.2.1. Cross-Linked or Linear Polyacrylamide Gel

Heiger et al. have studied the separation of dsDNA molecules by low or zero cross-linked polyacrylamide gel (67). A 1 kbp DNA ladder, fragment length varying from 75 to 12,216 base pairs (bp), was successfully separated on a slightly cross-linked polyacrylamide gel/capillary (3%T, 0.5%C) in less than 20 min at 250 V/cm (Figure 5.11). The authors also studied different linear polyacrylamide gel compositions for the separation of ϕX174/*Hae*III digest DNA fragments and demonstrated that size selectivity is a function of gel concentration. Shorter columns can be used with increased gel concentrations; or, alternatively lower polyacrylamide concentrations can be compensated for the increase of capillary length.

Large DNA molecules can also be separated by more diluted polyacrylamide gels. Koh et al. (68) reported the separation of a 1 kbp DNA ladder, ranging from 75 to 12,216 bp, using a commercially available chemistry kit (Figure 5.12). The gel/buffer contains a low concentration of polyacrylamide gel that can be replaced after each injection. A 47 cm × 100 µm i.d. coated capillary consisting of two polymer layers was used: the first layer is hydrophobic and prevents hydrolysis of Si-O-Si linkage, whereas the second layer is hydrophilic and prevents sample interaction with the first layer. In this instance, the gel-filled capillary could be used for more than 200 runs with good stability and reproducibility. The RSD of run-to-run and day-to-day reproducibility is below 2%.

CAGE also has been used to expand the separation potential of DNA restriction fragments. Figure 5.13 shows the separation of ϕX174/*Hae*III digest DNA fragments with and without ethidium bromide in the gel buffer (68). When ethidium bromide is incorporated into the polyacrylamide gel, high resolution of closely migrating DNA fragments is achieved. However, this type of affinity ligand only effects dsDNA and shows no interaction with single-stranded oligonucleotides.

5.3.2.2. Separation by Entangled Polymer Solutions

Like polyacrylamide gels, linear polymers such as polyethylene glycols, dextrans, and methyl cellulose derivatives have been used for the separation of DNA restriction fragments. These linear polymers, known as entangled polymer solutions, simulate pore structure and have sieving effects similar to those of polyacrylamide gel. The degree of sieving can be controlled by the concentration of polymer and or the type of polymer used for the separation. A coated capillary is usually required in conjunction with polymer additives. A comparison of separation of DNA restriction fragments by a coated and an uncoated capillary with 0.5% methyl cellulose additive was reported by Strege

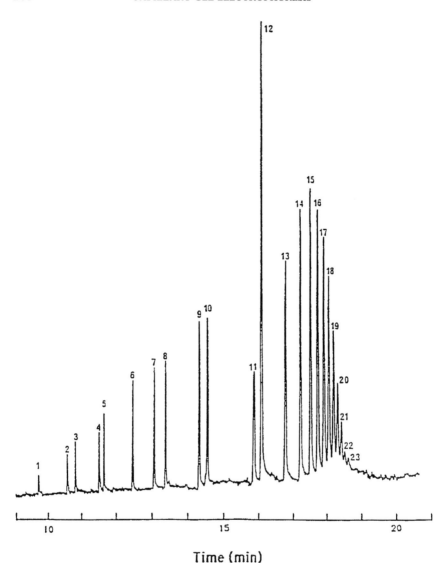

Figure 5.11. Separation of a 1 kbp DNA ladder by CGE. Conditions: capillary, 40 cm total length with 30 cm effective length. $E = 250$ V/cm. Buffer: 0.1 M Tris–borate, 2 mM EDTA, pH 8.3. Gel: 3% T, 0.5% C. Peaks: (1) 75; (2) 142; (3) 154; (4) 200; (5) 220; (6) 298; (7) 344; (8) 394; (9) 506; (10) 516; (11) 1018; (12) 1635; (13) 2036; (14) 3054; (15) 4072; (16) 5090; (17) 6108; (18) 7126; (19) 8144; (20) 9162; (21) 10,180; (22) 11,198; (23) 12,216 bp. (From ref. 67 with permission.)

Figure 5.12. Separation of large DNA restriction fragments. Conditions: eCAP DNA capillary, 47 cm × 100 μm i.d. with 40 cm effective length. Gel buffer: eCAP dsDNA 20,000 gel buffer. Voltage: 9 kV. $T = 20$ °C. Sample: 1 kbp DNA ladder. (From ref. 14 with permission.)

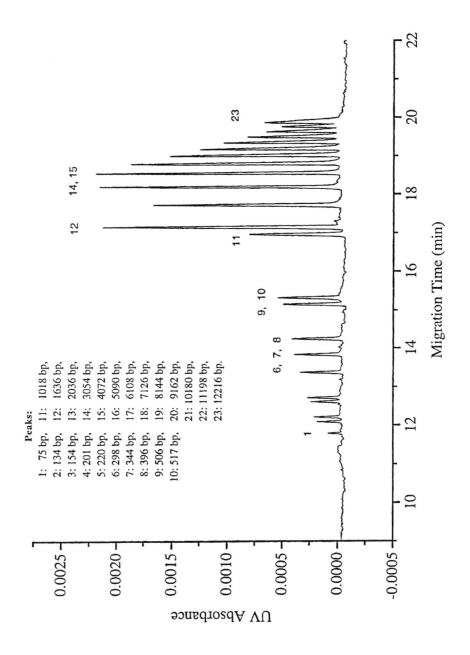

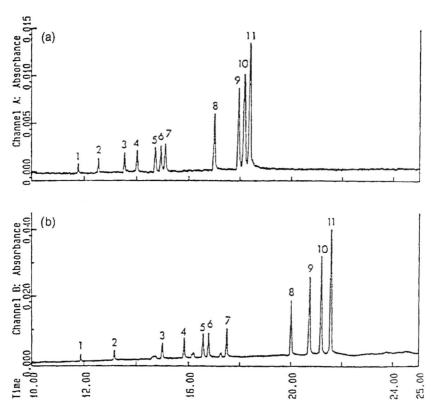

Figure 5.13. CGE separation of φX174 DNA/*Hae*III digest restriction fragment mixture in the absence (a) and in the presence (b) of 10 μM ethidium bromide in the gel buffer. Conditions: capillary, 47 cm total length with 40 cm effective length. Buffer: 0.1 M Tris–borate, 2 mM EDTA, pH 8.35. $E = 200$ v/cm. (From ref. 68 with permission.)

and Lagu (69). Methods employing EOF in an untreated capillary resulted in only partial separation of 23 fragments of a 1 kbp DNA ladder. However, all 23 fragments can be separated in less than 18 min by using a polyacrylamide-coated capillary.

Ulfelder et al. reported restriction fragments length polymorphism (RFLP) analysis of ERBB2 oncogene using hydroxymethyl cellulose (70). A surface modified fused capillary (DB-1, J&W, Folsom, CA) filled with 0.5% hydroxypropyl methylcellulose in 89 mM Tris–borate buffer, pH 8.5, was used for the separation of DNA fragments. Figure 5.14 shows the separation of three selected samples: one homozygous for the 500 bp allele, one homozygous for the 520 bp allele, and one heterozygous having both 500 and 520 bp fragments. All three samples show nonpolymorphic fragments at 220 and 330 bp. The bp

APPLICATIONS 209

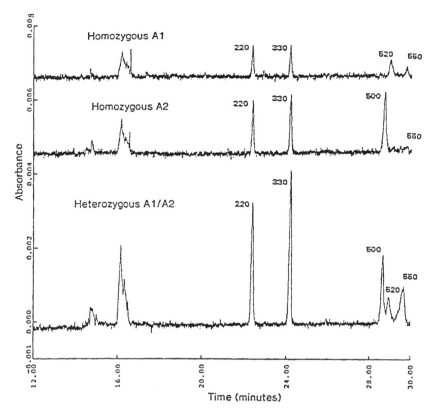

Figure 5.14. CE separation of PCR-derived RFLP smaples demonstrating *Mbo*I polymorphism: top, homozygous for allele A1 (520 bp fragment); middle, homozygous for allele A2 (500 bp fragment); bottom, heterozygous for A1 and A2. Constant fragments of 220, 330 and 550 bp can be seen in all three runs as a result of incomplete *pvu*II digestion. Conditions; DB-1 coated capillary, total length 57 cm with 50 cm effective length. Buffer: 0.5% hydroxypropyl methylcellulose in 89 mM Tris–borate, 2 mM EDTA, pH 8.5, with 10 mM ethidium bromide. $E = 175$ V/cm. (From ref. 70 with permission.)

number was compared to a standard calibration curve obtained by the separation of known bp number DNA fragments of the ϕX174/*Hae*III digest. The study demonstrated the applicability of CE, with fast separation time, as an alternate method for RFLP analysis.

Kleparnik et al. used a low-melting-point agarose solution for the separation of DNA fragments by CE (71). Figure 5.15 shows the separation of a 123 bp DNA ladder in 1% SeaPrep agarose at 40°C. Depending on DNA size, the separation mechanism can be divided into three regimes. Molecules in

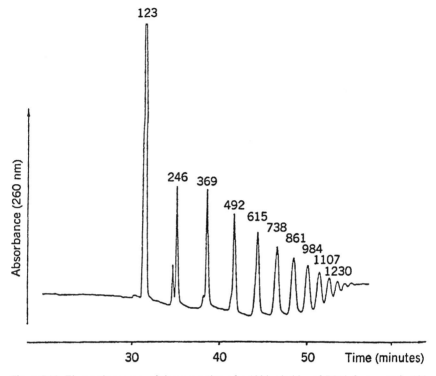

Figure 5.15. Electropherogram of the separation of a 123 bp ladder of DNA fragments in 1% SeaPrep agarose. Conditions: capillary, total length 41 cm with 30 cm effective length. $E = 490$ V/cm. $T = 40$ °C. (From ref. 71 with permission.)

the size range up to ca. 400 bp migrate in the sieving mode. Under optimum selectivity conditions, molecules in the size range ca. 400–800 bp are separated in the regime of entropically regulated transport. The transport of molecules larger than 800 bp is driven by a reptation mechanism.

5.3.3. Separation of Proteins and Peptides

5.3.3.1. Cross-Linked and Non-Cross-Linked Polyacrylamide Gels

Both cross-linked and non-cross-linked polyacrylamide gels have been used for conventional SDS–PAGE separation of proteins (72). These matrices can also be used in capillary SDS gel electrophoresis. Hjertén first reported SDS–PAGE separation of membrane proteins by a polyacrylamide-gel-filled capillary (73). Cohen and Karger provided a detailed investigation of a crosslinked polyacrylamide gel for SDS protein separation by CGE (74). Figure 5.16 shows capillary SDS gel electrophoretic separation of four standard proteins with 12.5% T, 3.3%C polyacrylamide gel in 100 mM Tris–phosphate, pH 8.3, 0.1 M SDS and 8 M urea. The proteins migrated according to their molecular weight. The mobility of SDS–protein complexes is a function of the composition of polyacrylamide gel used for the separation.

Tsuji used a linear polyacrylamide gel filled capillary for the separation of proteins by SDS–CGE (75). The polyacrylamide gel capillary was prepared according to the method described by Cohen and Karger (74). A protein mixture whose molecular weight ranged from 14.4 to 97.4 kDa was separated by capillary SDS gel electrophoresis (Figure 5.17). The plot of protein migration time vs. protein molecular weight yielded a linear relationship ($r > 0.999$). The presence of ethylene glycol in SDS–PAGE at a level of 1.8–2.7 M significantly improves the longevity of the capillary to more than 300 injections; otherwise it lasts for only a few runs.

5.3.3.2. *Entangled Polymers, Linear Polyacrylamide, Dextrans, Poly(ethylene glycol), and Poly(ethylene oxide)*

Although polyacrylamide gel provides good separation of SDS–protein mixtures and can be used for protein molecular weight determination, the gel matrix suffers from poor stability particularly at the high fields employed. In addition, the strong UV absorption of polyacrylamide gel at 214 nm significantly reduces the linear dynamic range and detection limits. To overcome these problems, several UV-transparent polymer matrices have been used for the protein separation by capillary SDS gel electrophoresis. Ganzler et al. described the separation of SDS–protein complexes using a UV transparent polymer matrix (dextran or PEO) of low-to-moderate viscosity (36). Six standard protein mixtures whose molecular weights range from 17 to 206 kDa were separated on a 18 cm dextran-coated capillary filled with 10% w/v dextran (MW 2,000,000) in 0.06 M 2-amino-2-methyl-1,3-propanediol (AMPO)–cacodylic acid (CACO) buffer, pH 8.8, and 0.1% SDS (Figure 5.18). In a given set of separation conditions, the relative mobility of the SDS–protein mixture also depends on the molecular weight of dextran used for the separation. At constant concentration, higher-molecular-weight dextran provides higher viscosity and slower mobility of samples.

Guttman et al. studied the temperature effect of dextran and PEO polymer networks for the separation of SDS–protein mixtures (48). At elevated temperature, dextran chains are able to form oriented channel-like structures by polymer–polymer interactions and yield better resolution compared to room temperature. On other hand, the PEO chains at room temperature are

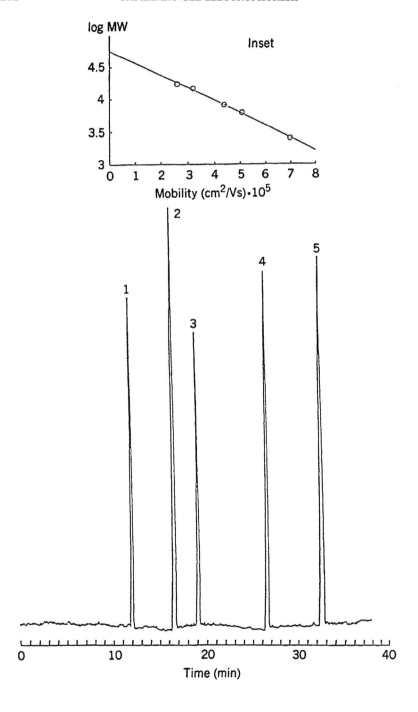

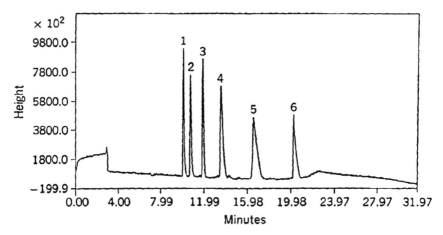

Figure 5.17. High performance capillary SDS–PAGE separation of myoglobin and several of its fragments. Conditions: capillary, 7 cm × 75 μm i.d. $E = 83$ V/cm. $T = 25\,°C$. Buffer: 375 mM Tris (pH 8.8)–0.1% SDS–ethylene glycol. Gel: 10% T, 3.3% C. Peaks: (1) lysozyme, MW 14,400; (2) trypsin inhibitor, MW 21,500; (3) carbonic anhydrase, MW 31,000; (4) ovalbumin, MW 45,000; (5) serum albumin, bovine, MW 66,200; (6) phosphorylase B, MW 99,400. (From ref. 75 with permission.)

flexible enough to form polymer–polymer interactions and favor the size separation of the SDS–protein mixture. At elevated temperatures, the mobility of the chains increased and counteracted the orienting force of the applied electric field. The whole PEO polymer solution has an almost random, irregular structure over $40\,°C$.

Shieh et al. reported the stability and reproducibility of SDS–protein separations by capillary SDS gel electrophoresis. Using a commercially available coated capillary, the gel/capillary can be used for more than 200 injections with acid washes between runs (76). Seven protein mixtures were separated by this polymer network in less than 15 min (Figure 5.19). The linearity of the plot of log MW vs. SDS–protein mobility is greater than $r = 0.99$, which demonstrates the pure sieving mechanism of the polymer

←

Figure 5.16. High-performance capillary SDS–PAGE separation of myoglobin and several of its fragments. Conditions: capillary, 20 cm × 75 μm i.d. $E = 400$ V/cm. $T = 27\,°C$. Buffer: 0.1 M Tris–H_3PO_4 (pH = 8.6), 8 M urea, 0.1% SDS. Gel: 12.5% T, 3.3% C. Peaks: (1) fragment III, MW 2510; (2) fragment II, MW 6210; (3) fragment I, MW 8160; (4) fragments I & II, MW 14,400; (5) myoglobin, MW 17,000. Inset: calibration plot of log MW vs. mobility for the above species. (From ref. 74 with permission.)

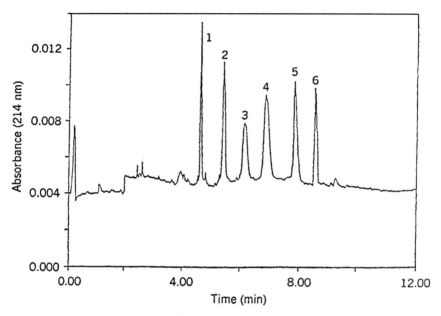

Figure 5.18. Separation of standard SDS–protein complexes in dextran polymer network. Conditions: capillary, effective length 18 cm. Buffer: 0.06 M AMPD–CACO, pH 8.8, 0.1% SDS. Polymer: 10% dextran (MW 2,000,000). $E = 400$ V/cm. Peaks: (1) myoglobin; (2) carbonic anhydrase; (3) ovalbumin; (4) bovine serum albumin; (5) β-galactosidase; (6) myosin. (From ref. 36 with permission.)

network used for the separation (Figure 5.20). The coated capillaries also prevent sample interaction with the capillary surface. Crude samples such as chicken egg white or milk can be directly injected into the capillary without any sample purification. The method also exhibits good run-to-run and day-to-day reproducibility (corrected migration time $< 2\%$ RSD). Thus, the method can be used for a rapid protein molecular weight estimation or a purity check.

5.3.4. Capillary Gel Electrophoresis of Complex Carbohydrates

The recent increase in the use of glycoproteins as pharmaceutical products has necessitated the development of rapid, high-resolution, and reproducible methods for the analysis of carbohydrates (77,78). Understanding the role of glycosylated proteins in cell function is becoming more and more important, as there is increasing evidence that carbohydrate moieties of glycoproteins are

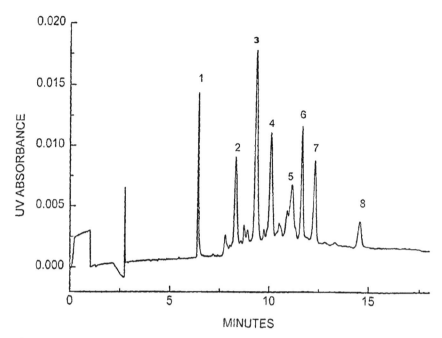

Figure 5.19. Separation of seven SDS–protein complexes by the eCAP SDS 14-200 kit. Conditions: eCAP SDS capillary, 27 cm total length with 20 cm effective length. Buffer: eCAP SDS buffer, pH 8.8. $E = 300$ V/cm. $T = 20\,°C$. Peaks: (1) orange G; (2) α-lactalbumin; (3) carbonic anhydrase; (4) ovalbumin; (5) bovine serum albumin; (6) phosphorylase b; (7) β-galactosidase; (8) myosin. (From ref. 76 with permission.)

key recognition factors in receptor–ligand or cell–cell interactions, in the modulation of immunogenicity, in the folding/unfolding process of protein molecules, and in the regulation of their bioactivity (79). Minor changes in such carbohydrate structures may have a great influence on the biological activity of glycoprotein. CGE combined with LIF detection have proved to be excellent means of analyzing complex carbohydrates (80–84). Complex carbohydrates released from glycoproteins are labeled with a charged fluorescent label of 8-aminopyrene-1,3,6-trisulfonate (APTS) for CGE analysis, using simple reductive amination chemistry. The labeled oligosaccharides can then be rapidly profiled. Individual carbohydrates can be detected at the low femtomole range by the LIF detection system. Figure 5.21 exhibits the separation of APTS-labeled glycans of bovine fetuin (middle trace) and bovine ribonuclease B (lower trace) compared to the maltooligosaccharide ladder standard (upper trace) (85). As Figure 5.21 demonstrates, CGE provides an excellent carbohydrate profiling tool that is able to resolve differences in oligosaccharide size and net charge, as well as differences in linkage positions of the individual monosaccharide units.

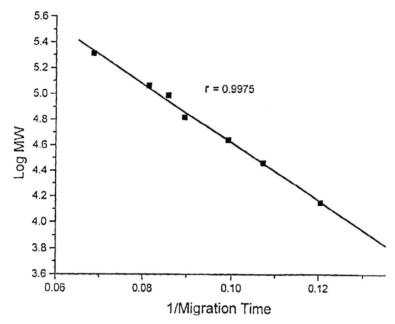

Figure 5.20. Linearity plot of logarithm of protein molecular weight vs. mobility as seen in Figure 5.18. (From ref. 76 with permission.)

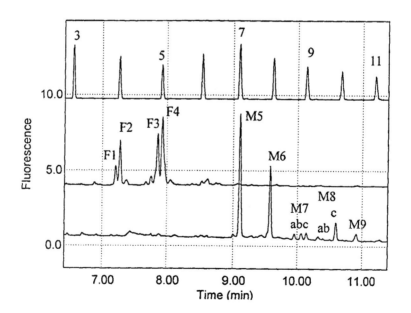

5.4. CONCLUSION

Capillary gel electrophoresis offers high-speed separations with automation and can be used as an improved method over slab gel electrophoresis. Single-stranded oligonucleotides, dsDNA, and protein–SDS mixtures can be readily analyzed by CGE. The separation mechanism is based on the sieving effect when the sample passes through a gel or polymer network. Very fast separation of synthetic oligonucleotides and DNA sequencing products with single-base resolution can easily be achieved by cross-linked or linear polyacrylamide gel. DNA restriction digest fragments and PCR products can also be analyzed with lower-concentration gels or polymer networks. Careful pretreatment and surface modification of the capillary wall can significantly reduce EOF and prevent any sample interaction with the capillary surface, thus increasing gel/capillary stability and reproducibility. By using coated capillaries in conjunction with gel or polymer networks, SDS–protein mixtures can be separated according to their molecular weight. The methods can be used for protein molecular weight estimation and purity checking. CGE will undoubtedly become an important tool for the separation of biopolymers.

Cross-linked polyacrylamide gel, which provides good resolution for the separation of biopolymers, was first used in CGE. Linear polyacrylamide gel, on the other hand, offers better stability and reproducibility than cross-linked gel and is becoming more widely used. Coated capillaries are commonly used in CGE. However, most commercially available coated capillaries are expensive and not stable when exposed to extreme conditions such as elevated temperatures, high pH, and high electric fields. Dynamically coated capillaries, which can be regenerated, have recently gained more attention and may replace capillaries with permanent coatings. The combination of using dynamically coated capillaries with polymer solutions provides users with a stable, reproducible, less toxic method that can be automated for the analysis of biopolymers by CGE.

←

Figure 5.21. Electropherograms of APTS-labeled glycans of bovine fetuin (middle trace) and bovine ribonuclease B (lower trace) compared to the maltooligosaccharide ladder standard (upper trace). Numbers on the upper trace correspond to the degree of polymerization of the glucose oligomers. Peaks: (F1) tetrasialo-triantennary-2 × 2,6; (F2) tetrasialo-triantennary-2 × 2,3; (F3) trisialo-triantennary-2 × 2, 6; (Fu) trisialo-triantennary-2 × 2, 3; (M5–M9) mannose 5–mannose 9; (M7a,b,c) 3 position isomers of manose 7, (M8a,b,c) 3 position isomers of manose 8. Conditions: 40 cm (effective) neutrally coated capillary (eCAP N-CHO) with 50 μm i.d. Buffer: 25 mM acetate buffer, pH 4.75. Detection: LIF–excitation: 488 nm, emission: 520 nm; E: 500 V/cm; T: 20 °C. (Ref. 85).

LIST OF ACRONYMS AND ABBREVIATIONS

Acronym or Abbreviation	Definition
AMPO	2-amino-2-methyl-1,3-propanediol
APTS	8-aminopyrene-1,3,6-trisulfonate
BIS	N,N'-methylenebisacrylamide
bp	base pairs
CACO	cacodylic acid
CAGE	capillary affinity gel electrophoresis
CDCGE	continuous denaturing capillary gel electrophoresis
CE	capillary electrophoresis
CGE	capillary gel electrophoresis
dsDNA	double-stranded deoxyribonucleic acid
EDTA	edetic acid or ethylenediaminetetraacetic acid
EOF	electroosmotic flow
HPCE	high-performance capillary electrophoresis
HPLC	high-performance liquid chromatography
kbp	kilobase-pairs
LIF	laser-induced fluorescence
PAAEE	poly(N-acryloylaminoethoxyethoxyethanol)
PAGE	polyacrylamide gel electrophoresis
PCR	polymerase chain reaction
PEG	poly(ethylene glycol)
PEO	poly(ethylene oxide)
PVAD	poly(9-vinyladenine)
RFLP	restriction fragments length polymorphism
RSD	relative standard deviation
SDS	sodium dodecyl sulfate
UV	ultraviolet

REFERENCES

1. A. Tiselius, *Trans. Faraday. Soc.* **33**, 524 (1937).
2. A. Chrambach, *The Practice of Quantitative Gel Electrophoresis*. VCH, Deerfield Beach, FL, 1985.
3. A. T. Andrews, *Electrophoresis*, 2nd ed., Clarendon Press, Oxford 1986.
4. A. J. Kostichka, M. L. Marchbanks, L. Robert, J. Brumley, H. Drossman, and M. L. Smith, *Bio Technology* **10**, 78 Clarendon (1992).
5. J. W. Jorgensen and K. D. Lukacs, *Science* **222**, 266 (1983).
6. B. L. Karger, *Nature (London)* **339**, 641 (1989).
7. S. F. Y. Li, *Capillary Electrophoresis*. Elsevier, Amsterdam, 1993.

8. J. P. Landers, *Handbook of Capillary Electrophoresis.* CRC Press, Boca Raton, FL, 1994.
9. R. Kuhn and S. Hofstetter-Kuhn, *Capillary Electrophoresis.* Springler-Verlag, New York, 1993.
10. P. Camilleri, ed., *Capillary Electrophoresis.* CRC Press, Boca Raton, F., 1993.
11. T. J. Kasper, M. Melera, P. Gozel, and R. G. Brownlee, *J. Chromatogr.* **458**, 303 (1988).
12. A. S. Cohen, A. Paulus, and B. L. Karger, *Chromatographia* **24**, 15 (1987).
13. S. L. Pentoney and J. V. Sweedler, in *Handbook of Capillary Electrophoresis*, (J.P. Landers, ed.), p. 147. CRC Press, Boca Raton, FL, 1994.
14. A. Guttman and N. Cooke, *J. Chromatogr.* **559**, 285 (1991).
15. A. Guttman, P. Shieh, D. Hoang, J. Horvath, and N. Cooke, *Electrophoresis* **15**, 221 (1994).
16. B. D. Hames and D. Rickwood, *Gel Electrophoresis of Proteins.* IRL Press, London, 1981.
17. D. Rickwood and B. D. Hames, *Gel Electrophoresis of Nucleic Acids.* IRL Press, London, 1982.
18. R. C. Cantor, C. L. Smith, and M. K. Mathew, *Annu. Rev. Biophys. Biophys. Chem.* **17**, 287 (1988).
19. S. L. Lerman and H. L. Frisch, *Biopolymers* **21**, 995 (1982).
20. G. W. Slater and J. Noolandi, Polym. Prepr. Am. Chem. Soc. Div. Polym. Chem. **29**, 416 (1988).
21. N. C. Stellwagen, *Electrophoresis* **10**, 332 (1989).
22. A. G. Ogston, *Trans. Faraday Soc.* **54**, 1754 (1958).
23. P. G. De Gennes, *Scaling Concept in Polymer Physics.* Cornell University Press, Ithaca, NY, 1979.
24. O. J. Lumpkin, P. Dejardin, and B. H. Zimm, *Biopolymers* **24**, 1573 (1985).
25. J. L. Viovy and T. Duke, *Electrophoresis* **14**, 332 (1993).
26. P. D. Grossman, S. Menchen, and D. Hersey, *GATA* **9**, 9 (1992).
27. A. Guttman and N. Cooke, *Anal. Chem.* **63**, 2038 (1991).
28. S. Terabe, K. Otsuka, and T. Ando, *Anal. Chem.* **61**, 251 (1989).
29. B. L. Karger, A. S. Cohen, and A. Guttman, *J. Chromatogr.* **492**, 585 (1989).
30. R. J. Nelson, A. Paulus, A. S. Cohen, A. Guttman, and B. L. Karger, *J. Chromatogr.* **480**, 111 (1989).
31. R. J. Wieme, *Chromatography: A Laboratory Handbook of Chromatography and Electrophoretic Methods.* Van Nostrand, New York, 1975.
32. D. M. Freifelder, *Physical Biochemistry.* Freeman, San Francisco, 1982.
33. D. N. Heiger and R. E. Majors, *LG-GC* **13**, 12 (1995).
34. A. Guttman, *U.S. Pat.* 5,332,481 (1994).
35. P. Bocek and A. Chrambach, *Electrophoresis* **12**, 1059 (1991).

36. K. Ganzler, K. S. Grene, A. S. Cohen, B. L. Karger, A. Guttman, and N. Cooke, *Anal. Chem.* **64**, 2665 (1992).
37. E. Fung and E. S. Yeung, *Anal. Chem.* **67**, 13, 1913 (1995).
38. A. Barron, H. W. Blanch, and D. S. Soane, *Electrophoresis* **15**, 597 (1994).
39. A. Guttman, in *Handbook of Capillary Electrophoresis* (J. P. Landers, ed.), Chapter 6, p. 129. CRC Press, Boca Raton, FL, 1994.
40. B. K. Clark, C. L. Nickles, K. C. Morton, J. Kovak, and M. Sepaniak, *J. Microcolumn Sep.* **6**, 503 (1994).
41. D. K. Schmalzing, C. A. Piggee, F. Foret, E. Carriho, and B. L. Karger, *J. Chromatogr. A* **652**, 149 (1993).
42. S. Hjertén and K. Kubo, *Electrophoresis* **14**, 390 (1993).
43. J. Townes, J. Bao, and F. Regnier, *J. Chromatogr.* **599**, 227 (1992).
44. P. Shieh, A. Hedeyati, N. Cooke, W. Goetzinger, and B. L. Karger, *Symp. High Perform. Capillary Electrophor.*, 8th, Orlando, FL, 1996.
45. J. L. Liao, J. Abramson, and S. Hjertén, *J. Capillary Electrophoresis* **2**, 191 (1995).
46. A. Guttman, *Appl. Theor. Electrophor.* **3**, 91 (1992).
47. A. Guttman, B. Wanders, and N. Cooke, *Anal. Chem.* **64**, 2348 (1992).
48. A. Guttman, J. Horvath, and N. Cooke, *Anal. Chem.* **65**, 199 (1993).
49. K. Khrapko, J. S. Hanekamp, W. G. Thilly, F. Foret, and B. L. Karger, *Nucleic Acids Res.* **22**, 364 (1994).
50. C. Gelfi, P. G. Righetti, L. Cremonesi, and M. Ferrari, *Electrophoresis* **15**, 1506 (1994).
51. A. Cifuentes, W. T. Kok, and H. Poppe, *J. Microcolumn Sep.* **7**(4), 365 (1995).
52. J. M. Butler, B. R. McCord, J. M. Jung, R. Wilson, B. Budowle, and R. O. Allen, *J. Chromatogr.* **658**, 271 (1994).
53. H. E. Schwartz and A. Guttman, *Primer: Separation of DNA*, Beckman Instruments, Fullerton, 1995.
54. A. Guttman and H. E. Schwartz, *Anal. Chem.* **67**, 2279 (1995).
55. M. J. Van der Schans, J. K. Allen, B. J. Wanders, and A. Guttman, *J. Chromatogr.* **680**, 511 (1994).
56. B. A. Johnson, S. G. McClain, E. R. Doran, G. Tice, and M. A. Kirsh, *BioTechniques* **8**(4), 424 (1990).
57. X. C. Huang, S. G. Stuart, P. F. Bente, III, and T. M. Brennan, *J. Chromatogr.* **600**, 289 (1992).
58. K. Hebenbrock, P. M. Williams, and B. L. Karger, *Electrophoresis* **16**, 1429 (1995).
59. A. Ganguly and D. J. Prockop, *Electrophoresis* **16**, 1830 (1995).
60. A. Cohen, D. R. Najarian, A. Paulus, A. Guttman, J. A. Smith, and B. L. Karger, *Proc. Natl. Acad. Sci. U.S.A.* **85**, 9660 (1988).
61. W. Warren and G. Vella *BioTechniques* **14**, (4), 598 (1993).
62. A. Guttman, R. Nelson, and N. Cooke, *J. Chromatogr.* **593**, 297 (1992).

63. Y. Baba, M. Tsuhako, T. Sawa, M. Akashi, and E. Yashima, *Anal. Chem.* **64**, 1920 (1992).
64. C. S. Srivatsa, M. Batt, J. Schuette, R. Carlson, J. Fitchett, C. Lee, and D. Cole, *J. Chromatogr. A* **680**, 469 (1994).
65. A. S. Cohen, D. R. Najarian, and B. L. Karger, *J. Chromatogr.* **516**, 49 (1990).
66. W. Goetzinger, B. L. Karger, P. Shieh, A. Hedayati, and N. Cooke, *Symp. High Perform. Capillary Electrophor.*, 5th, Orlando, FL, 1996 Poster No. 226 (1996).
67. D. Heiger, A. Cohen, and B. L. Karger, *J. Chromatogr.* **516**, 33 (1990).
68. E. Koh, P. Shieh, and N. Cooke, *Symp. High Perform. Capillary Electrophor.*, 7th, Würzburg, 1995, Poster No. 230 (1995).
69. M. Strege and A. Lagu, *Anal. Chem.* **63**, 1233 (1991).
70. K. Ulfelder, H. Schwartz, J. Hall, and F. Sunzeri, *Anal. Biochem.* **200**, 260 (1992).
71. K. Kleparnik, M. M. Garner, and P. Bocek, *J. Chromatogr. A* **698**, 375 (1995).
72. K. Weber and M. Osborn, *J. Biol. Chem.* **244**, 4406 (1969).
73. S. Hjertén, in *Electrophoresis' 83* (H. Hirai, ed.), pp. 71–79. de Gruyter, New York, 1984.
74. A. S. Cohen, and B. L. Karger, *J. Chromatogr.* **397**, 409 (1987).
75. K. Tsuji, *J. Chromatogr.* **550**, 823 (1991).
76. P. Shieh, D. Hoang, A. Guttman, and N. Cooke, *J. Chromatogr.* **676**, 219 (1994).
77. R. R. Hardy and R.R. Townsend, *Proc. Natl. Acad. Sci. U.S.A.* **85**, 3289 (1988).
78. Z. El Rassi, ed., *Carbohydrate Analysis.* Elsevier, Amsterdam, 1995.
79. A. Varki, *Glycobiology* **3**, 97 (1993).
80. S. Honda, S. Iwase, A. Makino, and S. Fujiwara, *Anal. Biochem.* **176**, 72 (1989).
81. J. Liu, O. Shirota, D. Wiesler, and M. Novotny, *Proc. Natl. Acad. Sci. U.S.A.* **88**, 2302 (1991).
82. C. Chiesa and Cs. Horvath, *J. Chromatogr.* **645**, 337 (1993).
83. R. J. Linhardt, *Methods Enzymol.* **230**, 265 (1994).
84. A. Klockow, R. Amado, H. M. Widmer, and A. Paulus, *J. Chromatogr.* **716**, 241 (1995).
85. A. Guttman, *Nature (London)* **380**, 461 (1996).

CHAPTER

6

CAPILLARY ISOELECTRIC FOCUSING

JOHN E. WIKTOROWICZ

Lynx Therapeutics, Hayward, California 94545

6.1.	Introduction	223
6.2.	Capillary Isoelectric Focusing	226
	6.2.1. Capillary Coating and Control of Electroosmotic Flow	228
	6.2.2. Detection and Gradient Mobilization	232
6.3.	Modes of Isoelectric Point Determination	241
6.4.	Future Considerations and Direction	245
Addendum		247
List of Acronyms		247
References		248

6.1. INTRODUCTION

Isoelectric focusing (IEF) is a technique by which a charged molecule is electrophoretically driven through a pH gradient until it encounters a pH at which it no longer carries a charge. At the point of zero net charge, it experiences zero electromotive force and stops migrating. For zwitterions undergoing IEF, this point of zero net charge is called the isoelectric point (pI) and species separate according to their pI's. The pH gradient is established by synthetic compounds called ampholytes—small amphoteric compounds (containing both acidic and basic functional groups) exhibiting slightly different pI's. These compounds act as strong buffers at their pI's and thus are capable of maintaining a stable pH gradient. The theoretical limit for this technique is therefore dictated by the resolution of the pH gradient (differences in pI's of adjacent ampholyte species), and control of their concentrations. In fact it was not until Vesterberg (1) succeeded in synthesizing and purifying carrier ampholytes exhibiting properties identified as ideal by Svensson (2) that the

---

High Performance Capillary Electrophoresis, edited by Morteza G. Khaledi. Chemical Analysis Series, Vol. 146.
ISBN 0-471-14851-2 © 1998 John Wiley & Sons, Inc.

224 CAPILLARY ISOELECTRIC FOCUSING

technique received attention from the analytical community (particularly biochemists).

In its earliest commercial manifestation, IEF was performed in a vertical column in free solution (Figure 6.1). Samples were dissolved in ampholytes, and a gradient of glycerol, sucrose, sorbitol, polyvinylpyrrolidone, or other

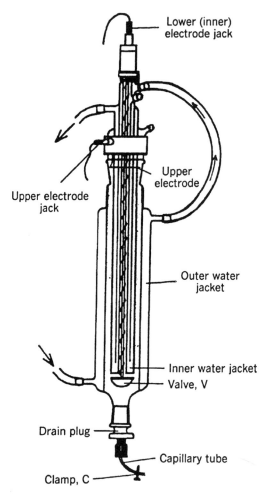

Figure 6.1. Isoelectric focusing column. Arrows show the direction of flow of the cooling water through the outer and then the inner water jackets. The shaded area is the region occupied by the sample and pH gradient; below this and surrounding the inner electrode is the dense electrode solution, and above it is the light (upper) electrode solution. Release of valve V permits removal of the sample through the capillary tube. (Reproduced with kind permission of Pharmacia.)

density-imparting material was created in the column in order to minimize convective mixing (3). Generally, the cathode is placed in a strong base (typically NaOH), the anode in a strong acid (typically H_3PO_4), and the region in between then filled with ampholyte-sample solution. In a mixture, ampholytes exhibit a pH of about 7.8, so near the cathode (high pH) the ampholyte species will carry a net negative charge and near the anode (low pH) they will carry a net positive charge. When the voltage is applied, they move toward the appropriate electrode and will continue to migrate until they experience a pH imparted by the surrounding ampholytes equal to their pI's. A steady state is achieved since any disturbance (such as diffusion) will result in the acquisition of a charge, causing the ampholyte to move back to its focused pI position under the influence of an electric field.

Samples dissolved in ampholytes are normally focused slowly until zero (or near-zero) current is measured. This signifies completion of the focusing procedure, as all of the current-carrying species stop migrating due to their zero net charge. In order to minimize loss of resolution due to Joule heating, low voltages may be used and the column jacketed. Very long focusing times—up to 72 hours for preparative isoelectric focusing, can be typical due to the low voltages. The pH gradient containing separated analytes may then be pumped from the column through an on-line UV detector and into a fraction collector for further analysis. While a very high degree of resolution is theoretically possible, the requirement for hydrodynamic mobilization for detection and collection, as well as the unavoidable convection due to Joule heating, limits its attainment.

IEF became a practical *analytical* tool with the use of polyacrylamide gel matrices and better control of ampholyte synthesis. Standard detection strategies and gel-handling mechanisms were quickly applied to IEF analysis of proteins and peptides. The higher degree of resolution, brought about by the anticonvective properties of rigid gels, led to its popularity as an analytical technique for the measurement of pI's, as well as another method for gauging purity and establishing identity. Commercial systems designed to partially automate the separation and detection became available. Although rigid gels permitted higher resolution, this was achieved at the expense of quantifiability, as chemical staining (inherently semiquantitative) substituted for on-line detection. However, the other limitations of gel-based separations, in addition to the semiquantitative nature of current staining strategies, hinder the realization of the full potential of IEF. Some loss of resolution due to local Joule heating remains, analysis time is lengthened due to labor-intensive preparation and postseparation handling, and large sample mass requirements still serve as impediments to the full implementation of IEF as a widely used analytical technique. The most recent developments in gel-based IEF address some of these limitations, such as the application of silver staining to decrease mass loading requirements, thinner gels to permit more efficient heat dissipation, or shorter gels to decrease analysis time, but all achieve their benefit through some sacrifice, either in expense, resolution, or gel fragility.

6.2. CAPILLARY ISOELECTRIC FOCUSING

With the increased interest in the application of capillary electrophoresis (CE) to the analysis of proteins in the early to mid-1980s, it was natural for IEF to be among the first specialized techniques to find use (4–6). This method rediscovers the merits of free-solution electrophoresis by performing separations in 50 μm diameter fused-silica capillary, thereby minimizing the detrimental effects created by convection. The high efficiency of heat dissipation of the smaller diameter tube permits high field strengths (volts per unit length) and therefore higher resolution and shorter analysis times. The narrow-diameter capillary permits analysis with minute mass requirements (pg), while the UV transparency of the fused silica permits on-line detection with precise and accurate quantification without the need for staining/destaining. Finally the free-solution format eliminates the need for polymerization steps in preparation for electrophoresis and permits complete control over the composition of the separation buffer. While CE contributed high resolution, speed of analysis, and true quantification to the technique, several difficulties had to be overcome before IEF could be implemented as an application of the technology.

Generally, application of IEF to CE (cIEF) poses two challenges: the first arises from the fact that focusing must be completed before samples pass the detector; the second comes about because detection requires either mobilization of the inherently neutral species or movement of the detector, so that focused species may be detected and quantified. Either detection scheme is accompanied by unique problems.

Three general chemistry approaches have been developed to address the focusing/detection challenge (Figure 6.2). Two of these approaches address the need for separate and mobilization steps, and the third attempts to use electroosmotic flow (EOF) as the mobilizing agent. Obviating the chemistry issues, others have modified the detection hardware to accomplish detection without mobilization of the gradient.

In order to achieve true quantification, detection in CE is on-line; i.e., light is shone through the separation capillary, perpendicular to the electrophoretic axis, at a fixed point. More generally, either the capillary and detector are fixed with respect to each other or one is drawn past the other. In either scheme, light-absorbing species pass in front of the detector and register as peaks.

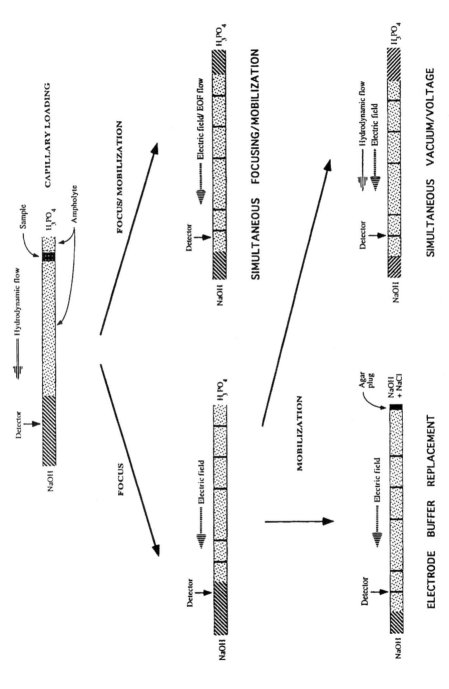

Figure 6.2. Mobilization schemes in cIEF.

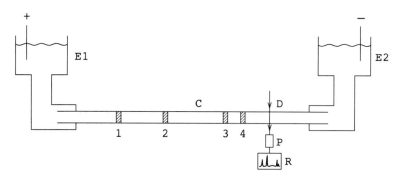

Figure 6.3. The principle of capillary electrophoresis: C, capillary; E1, E2, electrode vessels; D, detecting light beam; P, photodetector; R, recorder; 1–4, focused zones. (Reproduced from ref. 19, with kind permission of Academic Press.)

Thus, focusing must be completed before the gradient (and samples) pass the detector (Figure 6.3).

In order to address the mobilization issue in cIEF, the complication presented by EOF in fused-silica capillaries at pH values typically used in cIEF schemes must be discussed. Silanols created upon hydration of silica deprotonate at pH's above 4, leaving a negatively charged surface. The charged capillary surface causes a bulk migration of fluid toward the cathode, with the magnitude of the flow dependent upon voltage, temperature (viscosity), pH, and ionic strength. In cIEF, uncontrolled EOF will cause peaks to appear prematurely or not at all (in the case of flow away from the detector), depending upon the polarity of the system. Thus, as is often the case in CE, the capillary wall plays a major role in defining the maximum performance of the system. Careful control of the capillary surface is critical in achieving the optimum performance in cIEF.

6.2.1. Capillary Coating and Control of Electroosmotic Flow

Several strategies have been utilized in order to control the capillary surface for cIEF, including masking the charged surface with covalent coatings, adding dynamic coating agents to the focusing solution, or adjusting the EOF to control the elution time of the peaks. The most popular approach is to covalently modify the surface with neutral ligands in order to abolish EOF. The greatest difficulties with covalent coatings in general, however, are related to the difficulty in obtaining complete coverage, as well as maintaining the coatings intact in face of aqueous solutions of varying pH.

The early developers of cIEF relied on neutral coatings (e.g., methylcellulose or non-cross-linked polyacrylamide) designed to abolish EOF and decrease wall interactions due to coulombic attraction. These succeeded in diminishing EOF, although it was found necessary to plug one end of the capillary with agarose, suggesting that the coatings were insufficient to completely abolish it.

In general covalent coatings operating through the siloxy linkage are susceptible to hydrolysis, even at slightly alkaline pH's. Since most cIEF schemes expose the capillary to pH extremes (either due to the mobilization scheme or during focusing or both), these coatings are very labile. The loss of coating creates ionized silanols at pH's above 4–6. These silanols contribute to EOF to a degree dependent upon the part of the pH gradient they happen to be experiencing at any given time. The net result is that over the focusing and elution time, EOF is variable. If the flow is present during the focusing phase, it may interfere by driving incompletely focused sample components past the detector prematurely or, in the case of partially coated capillaries, it may be variable due to the variable pH in the capillary during the mobilization phase. The net result of sample and standard peaks eluting nonlinearly is that the gradient appears nonlinear and so the pI vs. mobilization time is nonlinear. Continued exposure to pH extremes yields increasingly denuded capillaries. Because of the cumulative nature of the coating damage, reproducibility of these systems is generally also poor. When deterioration of the coating reaches a point at which the separation is severely impacted, capillary replacement is the only corrective action that can be taken. Thus a good indication of capillary performance and coating stability is to establish the linearity (by calculation of the least squares correlation coefficient) of the pH gradient with standards included at regular run intervals (7) (Figure 6.4). Any loss in linearity, or departure from an acceptable correlation coefficient, suggests that mobilization is variable and coating has been lost. This degradation generally occurs over 5–10 successive cIEF runs with conventional capillary coatings (8–10).

Attempts to address the lability of siloxane-derived covalent coatings have yielded some success by creating a covalent linkage directly between silicon and carbon (8,11,12) (Figure 6.5). IEF using acrylamide-coated capillaries coated by this technique yielded stable coating beyond 30 runs (9). Similar results have been achieved through the use of extremely hydrophobic coatings with the addition of methylcellulose in the focusing solution (7,13). Although the latter strategy achieves coating via conventional siloxane chemistry, the inclusion of methylcellulose in the ampholyte mixture provides a means of masking the presumed loss of covalent coating (7). This resulted in stable pI gradients measured with internal standards for over 400 runs (7) (Figure 6.4).

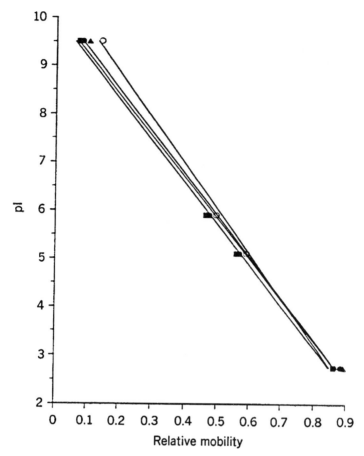

Figure 6.4. Capillary coating stability. IEF was performed over a period of more than 2 months. Standard curves were obtained at regular intervals during this period. Shown are the curves obtained from days 1 (●, $R^2 = 0.997$), 19 (○, $R^2 = 0.998$), 37 (■, $R^2 = 0.998$), and 64 (▲, $R^2 = 0.993$). (Reproduced from ref. 7 with kind permission of Academic Press.)

The extent to which wall interactions were diminished by any of these coating strategies was unclear, however, as electrostatic interactions account for only a part of the capillary wall–protein problem. Wall interactions mediated by weak, nonionic forces are unaffected (and may even be maximized) by neutral coatings. In addition, any focused protein will be specially disadvantaged due to the locally high concentration created by focusing into a narrow zone and by its neutrality. Thus proteins at their isoelectric points will have a tendency to precipitate and interact with neutral wall coatings to a greater extent than when they are charged (Figure 6.6). Nevertheless, as

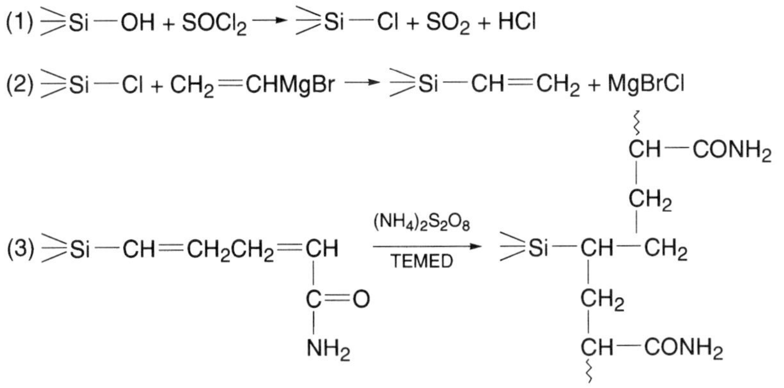

(1) ≥Si—OH + SOCl2 → ≥Si—Cl + SO2 + HCl
(2) ≥Si—Cl + CH2=CHMgBr → ≥Si—CH=CH2 + MgBrCl
(3) ≥Si—CH=CH2CH2=CH
C=O
NH2
(NH4)2S2O8
TEMED
CH—CONH2
CH2
≥Si—CH—CH2
CH2
CH—CONH2

Figure 6.5. Reaction scheme for the preparation of vinyl-bound polyacrylamide-coated capillaries. (Reproduced from ref. 8 with kind permission of American Chemical Society. Copyright 1992 American Chemical Society.)

A 280
0.14
0.12
0.10
0.08
0.06
0.04
0.02
0
45.9
46.0
46.1
Time (min)

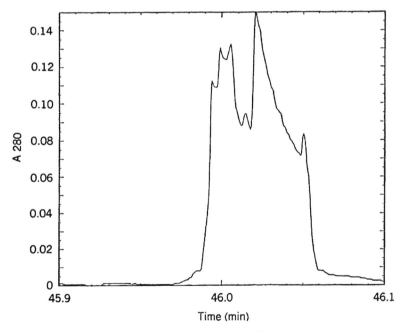

Figure 6.6. Focusing of a mixture of bovine hemoglobin and bovine serum albumin in 2% ampholytes. Focusing was performed in a polytetrafluoroethylene (PTFE) capillary, demonstrating severe sample–wall interactions. Very little focusing is evident, even after 45 min at 4 kV. (Reproduced from ref. 9 with kind permission of Elsevier.)

mentioned above, neutral wall coatings generally decrease EOF, and they can by made more reliable by the addition of dynamic coatings in the focusing solution. Moreover, the likelihood of protein precipitation can be diminished by the addition of neutral, solubilizing agents (such as deionized urea or detergents) in order to interfere in the precipitation process (13) (Figure 6.7). Complete elimination of EOF, however, requires a mobilization step, assuming that the detector remains fixed.

An alternate approach that has achieved a certain level of success utilizes uncoated capillaries as the separation column (14–18). In this strategy, EOF is diminished but not abolished. The magnitude of EOF is controlled by the addition of methylcellulose derivatives. These act both as coating agents to mask the charged silanols and as a viscous medium to slow EOF. As a result, EOF is controlled to a low enough level to permit focusing coincidental with mobilization.

6.2.2. Detection and Gradient Mobilization

Specifically, three strategies have emerged since the introduction of cIEF to accomplish detection: (i) focusing in the absence of EOF (necessitating a separate mobilization step); (ii) simultaneous focusing and mobilization through EOF control; and (iii) mobilization of the detector or capillary in order to determine the separation profile.

The original cIEF methods utilized anode or cathode buffer replacement after the focusing step to mobilize the gradient past a stationary detector. Typically, in its most recent version, this method involves filling a fused-silica capillary, coated with a uniform, neutral coating (e.g., polyacrylamide), with a 1–2% solution of carrier ampholytes. The sample may be dissolved in this solution or injected separately. The cathode and anode ends of the capillary are immersed in 0.02 M sodium hydroxide and 0.02 M phosphoric acid, respectively, and voltage (4000–6000 V) is applied until the current has fallen to 10–25% of its original value (19). As the steady state is achieved, the electroneutrality can be represented as follows (6):

$$C_{H^+} + \Sigma C_{NH_3^+} = C_{OH^-} + \Sigma C_{COO^-} \qquad (6.1)$$

Figure 6.7. cIEF in the presence of additives. Linearity of standards in variable urea in electrode buffers and ampholyte solutions. Absorbance trace demonstrates the focusing in the presence of 4 M urea. No effect of the urea on the formation or recovery of the gradient is observed. (Reproduced from ref. 13 with kind permission of Academic Press.) (b) Separation of two isoforms (Types I and II) of t-PA (tissue plasminogen activator). Samples were dialyzed, concentrated, and reconstituted in 1.0 M sodium bicarbonate. cIEF was performed with 0.2% reduced Triton X-100 in all solutions. Omitting detergent yields an inseparable mix (data not shown).

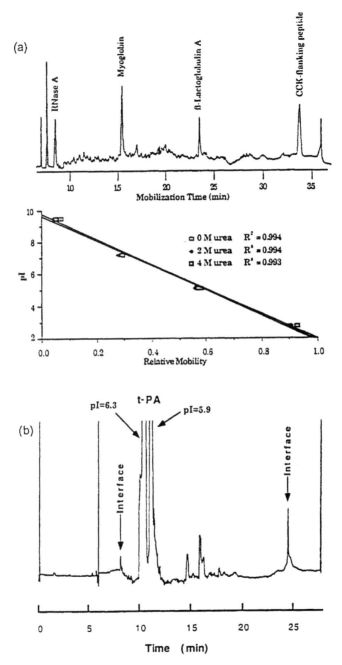

Figure 6.7

where C_{H^+}, $\Sigma C_{NH_3^+}$, C_{OH^-}, and ΣC_{COO^-} are the concentrations in equivalents per liter of the respective species. At this point the voltage is switched off and either the anode buffer is replaced with 0.02 M sodium hydroxide containing 0.02–0.04 M NaCl (anodic mobilization) or the cathode is replaced with 0.02 M phosphoric acid containing 0.02–0.04 M NaCl (cathodic mobilization). By arranging the electrodes and buffer solutions appropriately before focusing, upon the application of voltage, the gradient is mobilized toward the detector. This occurs because the additional Na^+ ions or Cl^- ions disturb the steady state. The addition of migrating ions to the focused zone causes the pH of the zone to shift as the ampholytes become oppositely charged (electrical neutrality must prevail). The new steady state can be represented as follows:

$$C_{Na^+} + C_{H^+} + \Sigma C_{NH_3^+} = C_{OH^-} + \Sigma C_{COO^-} \tag{6.2}$$

in the case of anodic mobilization with a sodium salt. The entry of a sodium ion (under the influence of an electric field) into a focused zone results in the displacement of a proton and movement of the proton into the adjacent zone. This has a double effect: the first zone experiences an increase in C_{COO^-} and the depletion of a proton, while the adjacent zone is disturbed by the appearance of a proton. The net result is that the pH of each zone changes and the ampholytes and sample molecules reacquire charge, causing them to move. They never again reach a neutral steady state in the capillary, and thus the entire gradient moves in the desired direction.

In the case of cathodic mobilization with a chloride salt, the following describes the steady-state condition:

$$C_{H^+} + \Sigma C_{NH_3^+} = C_{OH^-} + \Sigma C_{COO^-} + C_{Cl^-} \tag{6.3}$$

Either mobilization scheme causes movement of the gradient in the appropriate direction as the additional ions continue to migrate through the capillary, disturbing the steady state as discussed.

A major difficulty with salt (or pH) mobilization is the loss of gradient at the extremes of pH. Generally, most separations performed by this technique result in linearity only between pH's 5 and 9, even though ampholytes between pI's 3 and 10 were used. The loss (or lack) of separation at the higher pI's may be explained by the low concentrations of ampholytes at the higher end, as supplementing the ampholytes and mobilization buffer with amines [e.g., N,N,N',N'-tetramethylethylenediamine (TEMED)] (20,21) seems to permit focusing of high pI standard proteins such as horse heart cytochrome c. This is achieved at the expense of resolution, since resolution is somewhat lost due to the increased slope of the pH gradient (21). Very little discussion surrounds the

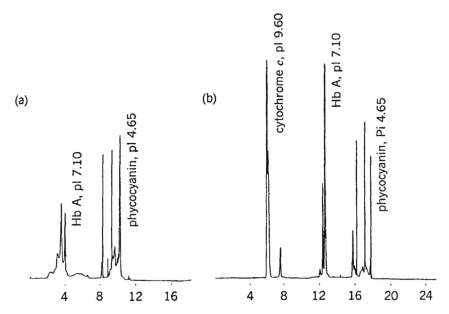

Figure 6.8. Comparison of mobilization agents. Mobilization with pI 3.22 zwitterion in basic solution. (b) Mobilization with pI 6.90 zwitterion. Note in part b increased mobilization time, loss of phycocyanin, and loss in resolution in the middle of the gradient. Hb A = hemoglobin A. (Reproduced from ref. 21, with kind permission of Elsevier.)

poor performance at low pH's, although the addition of a zwitterion with a pI of 3.2 to the cathode buffer seems to have permitted the detection of phycocyanin (pI 4.65) (Figure 6.8). The suggestion has been made that, in fact, the phenomenon known in gel IEF as cathodic and anodic drift (23) is the cause of a diminished focusing range (pI 5–9) in cIEF (19). More recent evidence has shown, however, that a linear pH gradient (pH 3–10) can be achieved by an alternative mobilization scheme not relying on electrode buffer replacement (see the discussion of Figure 6.9 below).

This strongly suggests the following points: (1) the full, *linear* pH range is defined by the ampholyte distribution and is established during focusing, and (2) any loss in the linearity and/or observed focusing range is caused by the mobilization scheme chosen for detection. It cannot be ruled out, however, that anodic or cathodic buffer replacement is specially sensitive to pH drift. If so, the net result is that it may be physically impossible to extend the pH range much beyond the observed values. In general, then, the buffer replacement mode of mobilization is useful only for proteins or peptides whose pI's range from $5 < pI < 9$.

A modification (7,13) of the electrode buffer replacement scheme described above was first suggested early in the development of cIEF (4), i.e., mobilization by simultaneous voltage and vacuum (Figure 6.2). In this early work, no benefit was observed by this mobilization scheme when compared to electrode buffer replacement. It was suggested that further performance benefit might be realized with more precise control of vacuum. With the introduction of commercial systems that precisely control vacuum, it became feasible to return to this scheme. Although it was unrecognized in the early work, the pH gradient generated by focusing ampholytes in a capillary can be exquisitely linear, with focusing power extending through the whole ampholyte pH range (Figure 6.9). Critical to the performance of this scheme is the simultaneous application of vacuum and voltage. Mobilization by the application of only vacuum (pressure) results in parabolic flow profile that will destroy the resolution achieved in focusing (S.-M. Chen, unpublished observations 1991). Increasing the viscosity of the medium as well as utilizing a low vacuum permits control of the flow so that dispersion due to the parabolic profile is minimized. Moreover, voltage serves to maintain the proteins within their respective pH zones, thereby preventing diffusion from destroying the resolution. The efficacy of this technique is demonstrated by its performance (7,13, 23) and the wide range of proteins and peptides separated. At the extremes of pH this approach is unchallenged, permitting the estimation of pI's of proteins and peptides never before focused (either in gels or capillaries) due to their low or high pI's (e.g., RNase T_1, pI = 2.9; RNase ba (*Bacillus amylofaciens*), pI = 9.0; CCK flanking peptide, pI = 2.75).

It should be emphasized that the measurement of the linearity of the pH gradient, as well as a discussion of its reproducibility with a variety of proteins, is of critical importance in gauging the efficacy of a given cIEF technique. Without it, the reader is faced with the prospect of a determining whether the mix of proteins presented in a separation was selected for their ability to favorably separate under the conditions presented. In addition, the linearity demonstrates the condition of the capillary surface, and serves as a baseline of performance by which all customizations can be compared. In short, it establishes a standard of performance, providing the user with a basis of comparison among competing techniques (gel- or capillary-based).

Addressing the disadvantages surrounding a second, mobilization step necessitated by the elimination of EOF, several individuals have reported the

---

Figure 6.9. cIEF calibration and focusing performance. Unknowns (RNase T_1 mutants and RNase ba) and standards were injected and focusing performed for 6 min at 30 kV in ampholyte pH (3–10) and methylcellulose. The linearity of the pH gradient was measured by plotting the pI of the standards against relative mobility. (Reproduced from ref. 7 with kind permission of Academic Press.)

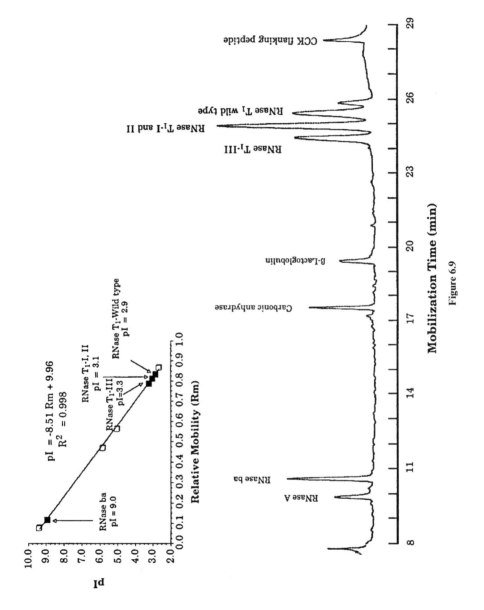

Figure 6.9

use of controlled EOF to simultaneously focus and mobilize peaks past the detector (Figure 6.2) (14–18). This strategy involves slowing EOF with a flow modifier, such as methylcellulose, to the extent that focusing occurs before the first peaks pass the detector. Theoretically the advantages are obvious. (i) no covalent capillary coatings are required—flow modifiers often double as dynamic coating agents; (ii) no loss in resolution occurs due to hydrodynamic flow; and (iii) analysis times are dramatically shortened.

In practical use, however, the major difficulty lies in the loss in gradient pH range as well as reproducibility. Because the magnitude of EOF in uncoated silica capillaries is dependent upon the pH and ionic strength of the buffer in the capillary, the changing nature of the capillary buffer (EOF draws anode buffer, $\sim$pH 1, or cathode buffer, $\sim$pH 12, into the capillary), results in changing EOF. Thus the elution times of the peaks will vary during the run. Plots of elution time vs. pH will therefore be nonlinear, making the analysis and identification of unknowns difficult. Nonlinear gradients make the use of the typical analytical tools for gauging and assuring performance arduous. In addition, the variable EOF limits the ability to perform analysis of acidic proteins or peptides with pI's below 5.3 [see Mazzeo et al. (16)] (Figure 6.10). Since the weakness in the system is due to the variable EOF, the use of a modified capillary that exhibits a more stable EOF as a function of pH permits linear gradients over selected, smaller regions (16). Thus the use of a C_8-coated capillary with sufficient residual EOF to permit simultaneous focusing mobilization yields linear pH gradients from 7 to 9.3 and a linear gradient from 5.3 to 4.7 (although these lines exhibit different slopes) (Figure 6.11). The use of internal standards permits the determination of pI of unknowns in such a system. Under these conditions, the authors report capillary stability of about 2 weeks (16). Nevertheless, understanding the constraints of the system, it is possible to use this cIEF scheme to qualitatively analyze complex samples in complex matrices (24,25).

A second approach that eliminates the need for gradient mobilization is based upon whole-column detection. Several schemes have been published that describe these approaches. One of the first is a linear potential gradient array detector (26), consisting of a discrete number of sensing electrodes. This system may have limited sensitivity, while resolution is dictated by the placement and spacing of sensing electrodes.

Another whole-column detection system utilizes a moving capillary past a fixed absorbance detector to scan for peaks (27) (Figure 6.12). The advantage lies in the ability to average the signal of multiple scans in order to minimize the noise created as a consequence of the motion. This is limited, of course, by the diffusion that occurs during the scan(s), as well as the nonlaminar flow induced by the motion of the capillary. Multiple scans would result in increased broadening of the peaks (although not mentioned, presumably

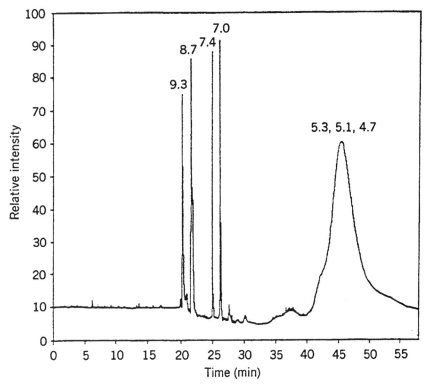

Figure 6.10. cIEF of standard proteins in an uncoated capillary. Poor resolution is evident below pI ~ 5.3. (Reproduced from ref. 16 with kind permission of Academic Press.)

a short voltage pulse would refocus the diffused zones). Since the whole length of the capillary must be imaged, the need to strip the capillary of all external coating diminishes the ruggedness of the capillary and leads to a shortened life span.

A third detection scheme, utilizing an imaging refractive index detector (28), demonstrates the advantages of whole-column detection as well as the disadvantages (Figure 6.13). This system utilizes a laser and various lenses to create a broad probe beam. A photodiode mounted on a single-axis stage scans in the detector plane over the separation length of the capillary. The relatively high concentration of focused proteins causes a change in intensity of the probe beam that is proportional to the second derivative of the refractive index change. The magnitude of the change in refractive index is nearly linearly dependent upon the concentration of sample. Although the use of a mechanical stage introduced noise into the system, use of a photodiode

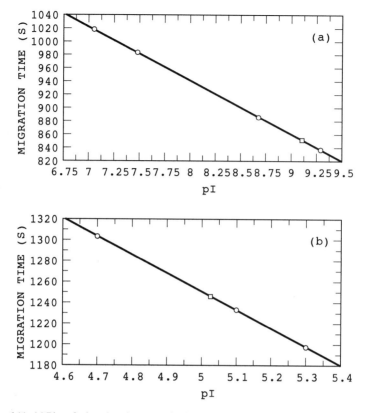

Figure 6.11. (a) Plot of migration time vs. pI for basic and neutral standards. (b) Plot of migration time vs. pI for acidic standards. Linearity is obvious for segments of the separation. (Reproduced from ref. 16 with kind permission of Academic Press.)

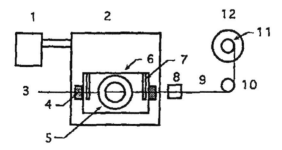

Figure 6.12. Apparatus for whole-column scanning detection: (1) integrator; (2) detector; (3) capillary; (4) wiper; (5) foam ring; (6) adjustable aperture cell body; (7) spring wire; (8) connector; (9) thread; (10) guide wheel; (11) driving wheel; (12) motor. (Reproduced from ref. 27, with kind permission of American Chemical Society. Copyright 1992 American Chemical Society.)

MODES OF ISOELECTRIC POINT DETERMINATION 241

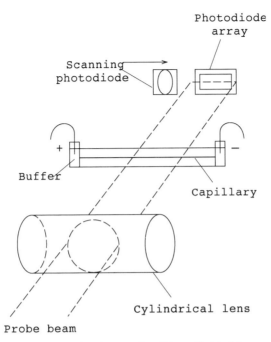

Figure 6.13. Imaging refractive index detector describing cylindrical lens, capillary cartridge, scanning photodiode, and diode array. (Reproduced from ref. 28 with kind permission of American Chemical Society. Copyright 1992 American Chemical Society.)

array eliminated this noise. However, the capillary lengths, and therefore resolution and peak capacity, are limited by the size of the array, with the expense of the array governing the practical limit. In addition, although this detection scheme is suited to the high concentration gradients (and therefore refractive index gradients) of focused proteins in cIEF, it suffers from the limited sensitivity (10^{-6} to 10^{-7} M) of more conventional refractive index detectors (28,29) but has less susceptibility to thermal influences. Nevertheless these factors diminish the acceptance of these detectors in cIEF.

6.3. MODES OF ISOELECTRIC POINT DETERMINATION

The free-solution format of conventional CE permits enormous flexibility in analytical design. The ability to control the separation matrix and separation conditions through all degrees of freedom (viscosity, pH, ionic strength, additives, voltage, temperature, etc.) permits the transfer of many (if not all) forms of conventional electrophoresis to the capillary format. An example of this ability is the determination of pI's without the need for IEF.

In this technique, first described by Rohde et al. (30) and Werner and Wiktorowicz (31), and later by Rush et al. (32), Kleparnik et al. (33) and Yao et al. (34), EOF and the neutral marker that permits its measurement are used to establish the pH at which proteins or peptides have net zero charge. The strategy is to start the analysis of an unknown at a low enough pH so that the most proteins/peptides are cationic ($\sim$ pH 2). By using a coated capillary that prevents wall interactions (35), the mobility of the protein can be compared to an internal neutral marker as the pH of the separation buffer is increased in successive runs. The pH at which the protein mobility can be interpolated to zero defines its isoelectric point (Figure 6.14). This can be directly compared to the pI's estimated from IEF even though the mobility of many proteins departs from linear dependence on pH (particularly at extremes). This departure, most likely due to pH-dependent conformational change, further demonstrates the power of this technique.

The free solution mode of pI estimation represents a fundamental departure from IEF. In IEF, proteins are driven to their pI's by the electrophoretic process. The physical attainment of net zero charge is necessary for the separation and the measurement. The alternate mode of free solution analysis does not require that the proteins actually experience either the range of pH's to which IEF exposes them or the actual point of neutrality.

Thus, IEF represents less the elecronic state of the native protein than the electronic state of the protein in the unnatural state as it aproaches zero net charge. In other words, the changes in conformation that the protein undergoes during focusing as it migrates toward its pI may, in fact, alter its pI. In addition, the conformational changes it undergoes as it migrates through the extremes of the pH gradient may be different in one direction than the other. That is to say, proteins dissolved in the ampholyte solution near one electrode experience a different pH series than the same proteins near the other electrode; or protein samples injected at one end experience a different series of pH's than when they are injected at the other end. To generalize, the pI that a given protein exhibits during IEF reflects its lowest energy state attained as a result of the conformational changes accumulated during its travel through the gradient to its final state of zero net charge. The pI that the protein achieves may be dependent upon the final configuration it assumes, and thus may be dependent upon the path taken. It can be stated that IEF produces an estimate of the "true" pI of a protein, i.e., the pI that describes more closely the pI obtainable from the fully denatured protein. This is particularly true if the pI measured by IEF differs substantially from the protein's native pH.

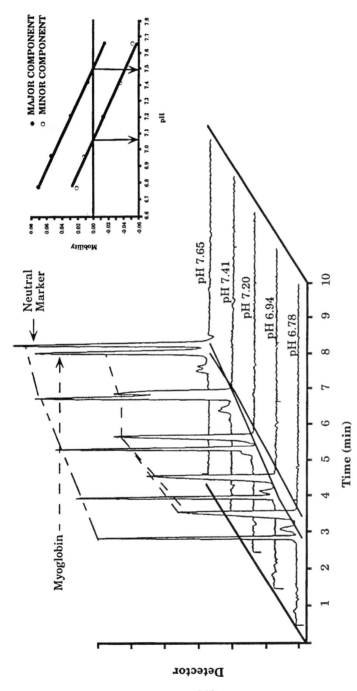

Figure 6.14. Free solution estimation of pI's of myoglobin and minor contaminant; pI's can be estimated graphically as shown in the inset. (Modified from ref. 30, reproduced with kind permission of Academic Press.)

This is in contrast to the pI estimate obtained from free-solution electrophoresis. This technique permits interpolation to zero net charge from the behavior on either side of the putative pI. If the putative pI is far removed from the "native" pH range, then this value can be estimated from extrapolation, and it becomes descriptive of the native conformation of the protein if it is estimated from the mobility observed in or near the native pH range. While the estimates of pI of many proteins show agreement between techniques, differences have been observed between apparent and "true" pI's for certain proteins and peptides (30,31). This suggests that pI's derived from IEF may not be quite so absolute as has been assumed.

The free-solution approach permits testing of the hypothesis by analysis in the presence and absence of urea (Figure 6.15). For soybean trypsin inhibitor, the "apparent" pI estimated in the absence of urea is 0.6 pH units lower than the "true" pI, estimated in the presence of urea by the free-solution approach. Interestingly, the pI as measured by cIEF is closer to the value obtained by free solution in urea. It should be noted, however, that urea treatment alone is insufficient to fully denature the protein, as soybean trypsin inhibitor is known to have multiple disulfide bridges that will remain oxidized. Other proteins have been studied similarly, and their behavior in the presence and absence of urea and comparison to cIEF is shown in Table 6.1. In general most of the

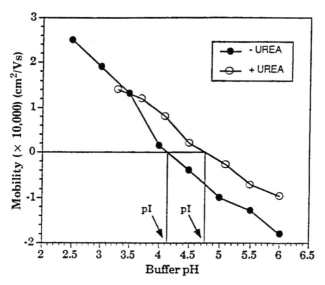

Figure 6.15. Effect of urea on free solution estimate of pI. The mobility of soybean trypsin inhibitor as a function of buffer pH in the absence and presence of 8 M urea. (Reproduced from ref. 31.)

Table 6.1. Comparison of Isoelectric Points Obtained from cIEF and FSCE

Proteins	cIEF[a]	FSCE[b]	FSCE + Urea[c]
α-Lactalbumin	5.1	4.7	5.1
β-Lactalbumin	5.3	4.7	5.1
Myoglobin	7.4	6.8	6.8
Conalbumin	6.4	4.8	6.0
Ovalbumin	4.8	4.3	5.2
Bovine serum albumin	5.4	4.2	5.5
Soybean trypsin inhibitor	4.5	4.1	4.7

[a]Capillary isoelectric focusing.
[b]Free-solution capillary electrophoresis (20 mM sodium phosphate or sodium citrate, pH 2.5–7.5).
[c]8 M urea.

proteins analyzed differ in their pI's estimated in free solution in the presence and absence of urea. In almost all cases, the values derived in urea are closer to the values estimated via cIEF. It should not be surprising that this is so, given the assumption that proteins must undergo conformational changes during their path to the isoelectric point. This "path dependence," measurable only by comparison to the free-solution approach, may have significance beyond the simple assumptions that have lent significance to the pI measurement in the past. Together, FSCE and cIEF can be utilized to characterize the process of pH-dependent denaturation and permit the generation of structural inferences relevant to the process of denaturation.

6.4. FUTURE CONSIDERATIONS AND DIRECTION

Currently, conventional gel-based IEF holds a specific niche in the analytical pallet of the protein/analytical chemist. In general, it is to establish quantitative purity, purity from degradation products, determination of multiple isoforms, and to a certain extent identity of a given protein or group of proteins [see Jones (36) for a review]. It is selected as the analytical method to use where there is a need for gentle, nondenaturing separations (the above discussion notwithstanding); it is selected where there is a need for high resolution between similarly charged molecules; and it is selected where no other separation scheme can yield information for a small amount of material. It is, as of yet, unchallenged in this niche.

This picture is less than clear where peptides are concerned. Fixing and staining of gel-based IEF causes the loss of peptides, such that IEF is not a method that a peptide chemist would use, for example, to establish the identity or purity of his peptide. Moreover, as with all gel-based electrophoretic systems that rely on chemical staining, true quantification is not attainable, particularly in the presence of ampholytes that may produce a high background. A particular difficulty with gel-based IEF occurs when the analytes precipitate at their pI's or under the conditions of focusing (experiencing a pH during focusing at which the protein is insoluble). Finally, gel-based IEF suffers from cathodic and anodic drift that severely limits the effective focusing range to pH values between 4 and 9.

Capillary-based IEF extends the utility of IEF by providing the means by which most of these limitations can be overcome. Thus, on-line detection (no fixing/staining/destaining) permits the analysis and quantification of peptides and other small molecules, such that chemists may now exploit the resolving power of cIEF in their analytical programs. The benefits of solution-based IEF can be rediscovered in that agents that interfere with or prevent precipitation of complex analytes (e.g., urea or nonionic detergents) at their pI's or in the neutral state can be included as a diluent of the ampholytes. The same advantage permits the analysis of comparatively dilute solutions of analytes. The capillary can be filled with comparatively large volumes of samples diluted in the ampholyte solution. Focusing will cause the entire mass of analyte species in the capillary to focus to its appropriate pI, thus concentrating the species and permitting detection. Finally the pH range limitations in gel-based IEF are abolished by the selection of the appropriate mobilization scheme in cIEF. This results in pI values being measured for proteins and peptides for which no value exists in the literature due to their extreme nature.

This is not to say that no problems exist with cIEF. Indeed there is considerable room for improvement in capillary coatings. Regardless of the mobilization scheme selected, all schemes would benefit from the development of stable coatings. For example the method of dynamic cIEF could benefit from charged coatings that remain ionized through much of the pH gradient. Coatings such as those containing sulfonic acid or quaternary ammonium would maintain their charge in all but the most extreme pH's and presumably yield constant EOF's.

Additionally, great benefit could be gained by the development of ampholytes that permit detection at low wavelengths. Current ampholytes absorb quite strongly at wavelengths < 250 nm, thereby limiting sensitive detection to high concentrations of peptides and proteins containing aromatic residues (280 nm). In addition, because of this absorbance and the variable amount of each ampholyte species in the mixture, baselines are somewhat unstable. Development of nonabsorbing ampholytes, as well as closer control of relative ampholyte concentrations, would go a long way toward minimizing these problems.

Future developments will most likely parallel the development of CE in general. On the horizon are truly effective CE–MS interfaces, i.e., an interface that is more in tune with the advantages of both techniques and does not penalize one for the sake of the other. Since no hyphenation of cIEF and MS has ever been performed, such a match could be of great benefit. The development of transfer systems tailored to the minuscule volumes inherent in CE might permit a CE analogue to the O'Farrell two-dimensional eletcrophoresis system (37) to become a reality. These systems need not only utilize IEF and sodium dodecyl sulfate (SDS). With appropriate buffer systems, capillary coatings, etc., other electrophoretic separation strategies could (and no doubt will) be combined. Thus, the future of cIEF and CE in general is at this time bounded only by the imagination of the reader.

ADDENDUM

Since the completion of this manuscript, recent attention has indeed been paid to the concept of cIEF–MS by several investigators. In particular are two publications that report analyses of model proteins sequentially by cIEF and MS. The first describes cIEF with off-line matrix-assisted laser desorption ionization time-of-flight mass spectrometry (MALDI–TOF–MS) (38). This system uses a collection interface that transfers peaks of interest into collection capillaries. The proteins were then mixed with the appropriate matrix solution and analyzed via MALDI–MS. The second report describes an on-line cIEF–electrospray mass spectrometry system (39), utilizing a capillary positioning mechanism that permits adjustment of the separation capillary to an optimum position for cIEF followed by adjustment for MS. While both of these hyphenated system strategies are in their infancy, the credible demonstration of the necessary interfacing components and a frank discussion of the difficulties of the systems will certainly stimulate further work into this new and promising arena.

LIST OF ACRONYMS

Acronym	Definition
CE	capillary electrophoresis
cIEF	capillary isoelectric focusing
EOF	electroosmotic flow
FSCE	free-solution capillary electrophoresis
Hb A	hemoglobin A
IEF	isoelectric focusing
MALDI–TOF–MS	matrix-assisted laser desorption ionization time-of-flight mass spectrometry

LIST OF ACRONYMS (*Continued*)

Acronym	Definition
MS	mass spectrometry
PTFE	polytetrafluoroethylene (or Teflon)
SDS	sodium dodecyl sulfate
TEMED	N,N,N',N'-tetramethylethylenediamine
t-PA	tissue plasminogen activator
UV	ultraviolet

REFERENCES

1. O. Vesterberg, *Acta. Chem. Scand.* **23**, 2653 (1969).
2. H. Svensson, *Acta Chem. Scand.* **15**, 325 (1961).
3. A. Winter, in *Electrofocusing and Isotachophoresis* (B. J. Radola and D. Graesslin, eds.), p. 433. de Gruyter, Berlin, 1977.
4. S. Hjertén, *J. Chromatogr.* **347**, 191 (1985).
5. S. Hjertén and M. Zhu, *J. Chromatogr.* **346**, 265 (1985).
6. S. Hjertén, J. Liao, and K. Yao, *J. Chromatogr.* **387**, 127 (1987).
7. S.-M. Chen, and J. E. Wiktorowicz, *Anal. Biochem.* **206**, 84 (1992).
8. K. A. Cobb, V. Dolnik, and M. Novotny, *Anal. Chem.* **62**, 2478 (1990).
9. T. J. Nelson, *J. Chromatog.* **623**, 357 (1992).
10. C. Silverman, M. Komar, K. Shields, G. Diegnan, and J. Adamovics, *J. Liq. Chromatogr.* **15**, 207 (1992).
11. J. J. Pesek and S. A. Swedberg, *J. Chromatogr.* **361**, 83 (1986).
12. J. J. Pesek and S. A. Swedberg, U. S. Pat. 4,904,632 (1992).
13. S.-M. Chen and J. E. Wiktorowicz, in Techniques in Protein Chemistry IV (R. H. Angeletti, ed.), p. 333. Academic Press, San Diego, CA, 1992.
14. J. R. Mazzeo and I. S. Krull, *Anal. Chem.* **63**, 2852 (1991).
15. J. R. Mazzeo and I. S. Krull, *J. Chromatog.* **606**, 291 (1992).
16. J. R. Mazzeo, J. A. Martineau, and I. S. Krull, *Methods* **4**, 205 (1992).
17. J. Chmelik and W. Thormann, *J. Chromatog.* **589**, 321 (1992).
18. W. Thormann, S. Caslavska, and J. Chemelik, *J. Chromatog.* **589**, 321 (1992).
19. S. Hjertén, in *Capillary Electrophoresis: Theory and Practice* (P. D. Grossman and J. C. Colburn, eds.), p. 191. Academic Press, San Diego, CA, 1992.
20. G. Yao-Jun and R. Bishop, *J. Chromatogr.* **234**, 459 (1982).
21. M. Zhu, R. Rodriguez, and T. J. Wehr, *J. Chromatogr.* **559**, 479 (1991).
22. A. Chrambach, P. Doerr, G. R. Finlayson, L. E. M. Miles, R. Shierin, and D. Rodbard, *Ann. N. Y. Acad. Sci.* **209**, 44 (1973).
23. J. M. Hempe and R. D. Craver, *Clin. Chem.* **40**, 2288 (1994).

24. S. Molteni, H. Frischknecht, and W. Thormann, *Electrophoresis* **15**, 22 (1994).
25. J. R. Mazzeo, J. A. Martineau, I. S. Krull, *Anal. Biochem.* **208**, 323 (1993).
26. W. Thormann, A. Tsai, J. Michaud, R. A. Mosher, and M. Bier, *J. Chromatogr.* **389**, 75 (1987).
27. T. Wang and R. A. Hartwick, *Anal. Chem.* **64**, 1745 (1992).
28. J. Wu and J. Pawliszyn, *Anal. Chem.* **64**, 224 (1992).
29. J. Wu and J. Pawliszyn, *J. Chromatogr.* **608**, 121 (1992).
30. M F. Rohde, K. S. Stoney, and J. E. Wiktorowicz, in Techniques in Protein Chemistry III (R. H. Angelleti, ed.), p. 121. Academic Press, San Diego, CA, 1992.
31. W. A. Werner and J. E. Wiktorowicz, Int. Soc. Peptides, Protides, Polynucleotides, San Francisco, 1993 (1993).
32. R. S. Rush, M. D. McGinley, K. A. Stoney, and M. F. Rohde, *Methods* **4**, 191 (1992).
33. K. Kleparnik, K. Slais, and P. Bocek, *Electrophoresis* **14**, 475 (1993).
34. Y. J. Yao, K. S. Khoo, M. C. M. Chung, and S. F. Y. Li, *J. Chromatog. A* **680**, 431 (1994).
35. J. E. Wiktorowicz and J. C. Colburn, *Electrophoresis* **11**, 7669 (1990).
36. A. J. S. Jones, *Adv. Drug Delivery Rev.* **10**, 29 (1993).
37. P. H. O'Farrell, *J. Biol. Chem.* **250**, 4007 (1975).
38. F. Foret, O. Müller, J. Thorne, W. Götzinger, and B. L. Karger, *J. Chromatogr. A.* **716**, 157 (1995).
39. D. P, Kirby, J. M. Thorne, W. K. Götzinger and B. L. Karger, *Anal. Chem.* **68**, 4451 (1996).

CHAPTER 7

CAPILLARY ISOTACHOPHORESIS

LUDMILA KŘIVÁNKOVÁ AND PETR BOČEK

Institute of Analytical Chemistry, Academy of Sciences of the Czech Republic, 611 42 Brno, Czech Republic

7.1.	Introduction	251
7.2.	Principles of Isotachophoresis	252
7.3.	Electrolyte Systems	256
7.4.	Sample Stacking	258
7.5.	The Isotachophoresis–Capillary Zone Electrophoresis Combination	260
7.6.	Dynamics of Destacking	265
7.7.	Sample-Induced Isotachophoresis (Stacking)	267
7.8.	Conclusion	273
	List of Acronyms and Abbreviations	274
	References	275

7.1. INTRODUCTION

The goal of this chapter is to provide a brief description of isotachophoresis (ITP) in such a way that it may be helpful to users of capillary zone electrophoresis (CZE). It is well known that sample stacking is frequently required at the beginning of CZE, as it is very advantageous for efficient separations. The principles of stacking are in fact the principles of ITP, with concentrating effects of zones and self-sharpening effects of their boundaries. Hence we present in the following sections a consideration of the principles of ITP, the characteristic features of ITP, the stacking of samples, the combination of ITP with CZE, the destacking of ITP zones in CZE systems, and sample-induced stacking in CZE. Those more interested in ITP itself, including the relevant instrumentation and qualitative and quantitative analysis by ITP, are referred to a monograph (1) and a recent brief overview (2).

High Performance Capillary Electrophoresis, edited by Morteza G. Khaledi. Chemical Analysis Series, Vol. 146.
ISBN 0-471-14851-2 © 1998 John Wiley & Sons, Inc.

7.2. PRINCIPLES OF ISOTACHOPHORESIS

In ITP a sample is separated in a discontinuous electrolyte system formed by the leading and terminating electrolytes. If a certain degree of simplification is employed, i.e., only strong electrolytes are considered, the leading electrolyte contains the ion with the highest mobility, whereas the terminating electrolyte contains the ion with the lowest mobility. When a sample is injected in between the leading and terminating electrolytes, the ions with mobilities that are between those of the leading and terminating ions will migrate isotachophoretically and create typical stacked isotachophoretic zones with sharp boundaries. For weak electrolytes, one must employ effective mobilities and rather complicated criteria must be fulfilled to obtain ITP (see the discussion on separability in Section 7.3, below). However, the following phenomenological description of the process is simple and straightforward. Figure 7.1 shows an isotachophoretic anionic separation of a mixture of acetate and formate. The initial state, where the sample of acetate and formate is injected between the leading and terminating electrolytes, is depicted in Figure 7.1a. The leading electrolyte consists of an equimolar mixture of

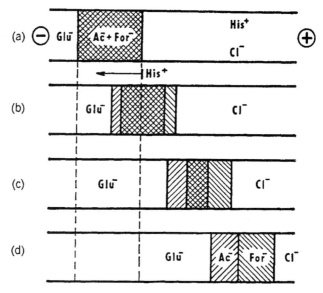

Figure 7.1. Anionic ITP of a mixture of acetate and formate. Chloride is the leading ion; glutamate is the terminating ion: (a) initial conditions; (b,c) the separation process; (d) the final steady state. (Reprinted form L. Křivánková and P. Boček, Isotachophoresis, in *Encyclopedia of Analytical Science*, Academic Press, London, 1995, with kind permission of Academic Press, Ltd.)

histidine (His) and histidine hydrochloride. Thus, it contains the leading anion Cl^- (later denoted as L) and the buffering counterionic system His^+ (later denoted as R). The terminating electrolyte consists of a solution of glutamic acid; thus it provides the terminating anion Glu^- (later denoted as T). After the current is switched on, the counterion of the leading electrolyte (protonated histidine) moves toward the cathode and becomes the common counterion for the whole system. The mixed zone of anions in the sample moves toward the anode, and the separation occurs. The original mixed zone becomes narrower (Figure 7.1b,c) until it disappears and the steady state is attained where all the anions move toward the anode in separate zones (Figure 7.1d).

The fundamental properties of an isotachophoretic system can be described as follows:

1. The zone of any ion of a substance A, which migrates in an isotachophoretic way behind the zone of L, migrates with the same velocity as the leading zone L:

$$v_{iso} = v_T = v_X = v_L \tag{7.1}$$

$$v_{iso} = \bar{\mu}_T E_T = \bar{\mu}_X E_X = \bar{\mu}_L E_L \tag{7.2}$$

where v_{iso} is the isotachophoretic migration velocity; v_X is the migration velocity of a zone X; μ_X is the effective mobility of a substance X; and E_X is the electric field strength in a zone X. The effective mobility of a given substance is of key importance here and is defined as follows. A substance X present in a solution as the ionic species or neutral molecules $X_0, X_1, X_2, \ldots, X_n$, with ionic mobilities $\mu_0, \mu_1, \mu_2, \ldots \mu_n$, migrates in the electric field as a single substance with the effective mobility μ_X given by

$$\bar{\mu}_X = \frac{1}{\bar{c}_X} \sum_{i=0}^{n} c_i \mu_i \tag{7.3}$$

where $\bar{c}_X$ is the total (analytical) concentration of the substance:

$$\bar{c}_X = \sum_{i=0}^{n} c_i \tag{7.4}$$

The equation defining $\bar{\mu}_X$ can be rewritten in the form

$$\bar{\mu}_X = \sum_{i=0}^{n} \alpha_i \mu_i \tag{7.5}$$

where α_i is the mole fraction of species X_i.

For practical calculations of effective mobilities, the following useful formulas are available providing that the dissociation constants of the species as well as the pH of the zone are known:

$$\bar{\mu}_A = \frac{K_{HA}}{K_{HA} + [H^+]} \mu_A \quad (7.6)$$

$$\bar{\mu}_B = \frac{[H^+]}{K_{BH} + [H^+]} \mu_{BH} \quad (7.7)$$

where K_{HA} and K_{BH} are dissociation constants of an acid and base respectively; μ_A and μ_{BH} are ionic mobilities of ions A^- and BH^+, respectively; and $[H^+]$ is the concentration of hydrogen ions.

A practical example of the effective mobility in solution is provided by the effective mobility of acetate, $\bar{\mu}_{Ac}$:

$$\bar{\mu}_{Ac} = \alpha_{Ac}\mu_{Ac} + \alpha_{HAc}\mu_{HAc} \quad (7.8)$$

where it is obvious that the neutral molecule HAc has zero mobility, $\mu_{HAc} = 0$, and thus

$$\bar{\mu}_{Ac} = \alpha_{Ac}\mu_{Ac} \quad (7.9)$$

2. When the ITP proceeds in a channel of a constant cross section, the driving current density i is constant along the whole path:

$$i = E_T \cdot \kappa_T = E_A \cdot \kappa_A = E_L \cdot \kappa_L \quad (7.10)$$

where κ is the specific conductivity.

3. The boundaries between isotachophoretic zones are sharp, and they exhibit self-sharpening effect. For any moving boundary $2 \rightarrow 1$, for which $|\mu_2| < |\mu_1|$, it also holds that $\kappa_2 < \kappa_1$, and form the relation stated in Eq. (7.2) it follows that $E_2 > E_1$. This means that if an ion of lower mobility, say, μ_{Ac}, enters the zone of ions of higher mobility μ_{For} due to diffusion, its migration velocity $v = E_{Ac} \cdot \bar{\mu}_{Ac}$ decreases to $v = E_{For} \cdot \bar{\mu}_{Ac}$, and as a result the acetate ion is again caught up by its own zone. Thus, the step change of the electric field strength between the two zones causes a permanently sharp boundary between the acetate and formate.

Steady-state boundaries exhibiting the self-sharpening effect also exist in systems containing weak electrolytes, but the mathematical description is more complex. Relationships of the type $|\bar{\mu}_{2,2}| < |\bar{\mu}_{1,2}|$ must be used, indicating that the substance 1 in the zone 2 has higher mobility than the substance 2 in its own zone. Thus the 2→1 boundary exhibits the self-sharpening effect.

4. The concentrations of substances in their own isotachophoretic zones are not arbitrary but are adjusted to the composition of the leading zone. For uni-univalent electrolytes, the situation may be advantageously described with the help of the Kohlrausch regulating function. This is a well-known function frequently employed in simple calculations; however, it is not an independent equation but is derived as the sum of the fundamental balance equations (either in differential form or in the rational form of the moving boundary equation). The function was derived by Friedrich Kohlrausch in 1897 for strong electrolytes by assuming constant mobilities. Later, the Kohlrausch regulating function was redefined so as to apply even to uni-univalent weak electrolytes. The function is not defined for the general case of multivalent weak acids or bases, complexes, or zwitterions. The importance of the Kohlrausch function stems from its highly illustrative physicochemical connotation. In general, it can be stated that, prior to the beginning of electromigration, the value of the Kohlrausch regulating function, $\omega(x)$, is given by the composition of the original electrolyte along the migration path:

$$\sum_{j=1}^{s} \frac{c_j z_j}{\mu_j} = \omega(x) \tag{7.11}$$

where z is the charge of an ion j. This value is retained even during electromigration at a certain point x. If step changes of $\omega(x)$ occur in the electrolyte composition along the migration path prior to electromigration, then the concentrations of the migrating substances vary during electromigration so that the value of $\omega(x)$ is retained. The adjusted concentrations of the substances in the zones copy the state prior to the start of the experiment. The number of values of the Kohlrausch regulating function along the migration axis corresponds to the number of phases (zones) that were in the system before the electric current started to flow. The composition of migrating zones is then automatically adjusted to $\omega = f(x)$. For strong uni-univalent electrolytes, where the electroneutrality condition in the zones is valid in the form $c_{R,L} = c_L$ and $c_{R,A} = c_A$ (here A is an analyte and R is the counterion), application of the Kohlrausch regulating function to zones L and A gives

$$c_A = c_L \frac{\mu_A}{\mu_L} \frac{\mu_L + \mu_R}{\mu_A + \mu_R} \tag{7.12}$$

This well-known equation, commonly also termed the Kohlrausch regulating function, explicitly expresses the concentration of a substance A in its zone after the isotachophoretic steady state has been reached. The concentration of the ion A in its zone is thus adjusted for a given leading electrolyte, independent of the concentration of this ion in the original sample. By this mechanism the concentrations of all the ions in their zones are adjusted. A diluted sample is concentrated and a concentrated sample is diluted when passing through the stationary boundary.

7.3. ELECTROLYTE SYSTEMS

A key problem in obtaining the required ITP is the selection of a suitable electrolyte system in which the analytes are completely separated and form stable isotachophoretic zones. Electrolytes are selected on the basis of data on the effective mobilities of the relevant substances and on ways of changing these effective mobilities by employing variable factors, especially the pH. This variation is based on the definition of the effective mobility $\bar{\mu}$, which permits calculation of the effective mobility of the substance under the given conditions in the presence of acid–base and complexation equilibria. The tabulated values of the effective mobilities for series of tested electrolyte systems over a wide range of pH_L values are very useful (3).

Basic rules for the selection of the electrolyte system are listed in Table 7.1. The leading ion is usually a suitable fast ion of a completely dissociated substance. The terminating ion is usually H^+ for cations or OH^- for anions, which must always form the last zone in the system (at least theoretically). If the terminating ion consisting of H^+ or OH^- is too slow (the electric potential gradient is too high), it is preferable to use another faster substance as the terminating ion. If the separated substances are weak bases or acids, sufficient

Table 7.1. Characteristics of Electrolyte Systems in ITP

	ITP	
	Cationic	Anionic
Suitable leading ion	K^+, NH_4^+, Na^+	Cl^-
Terminating ion	H^+	OH^-
	Weak base	Weak acid
Counterion	Weak acid	Weak base
Condition for ionization of analytes	$pH \leq pK_{BH} + 1$	$pH \geq pK_{HA} - 1$

Source: After ref. 1 with kind permission of VCH Verlagsgesellschaft.

ionization should be ensured by proper selection of the pH of the leading electrolyte and the pK of the counterion such that the condition given in the Table 7.1 is approximately fulfilled.

Another problem is to determine whether the substance of interest forms a stable individual zone in the selected system, i.e., that the composition and volume of a zone do not change with time. Experimental verification of zone stability consists in testing whether the calibration graph constructed for the substance, i.e., the dependence of the amount of the analyte detected in its isotachophoretic zone on the injected amount, passes through the origin. Theoretical considerations are based on effective mobilities; for the stable zone of a substance X we must have

$$|\bar{\mu}_{T,T}| < |\bar{\mu}_{X,T}| \quad \text{and} \quad |\bar{\mu}_{X,X}| < |\bar{\mu}_{L,X}| \quad (7.13)$$

which means that the effective mobility of the substance X in the terminating zone, $|\bar{\mu}_{X,T}|$, is higher than the effective mobility of the terminator in its own zone, $|\bar{\mu}_{T,T}|$, and the effective mobility of the leading ion in the zone of the substance X, $|\bar{\mu}_{L,X}|$, is higher than the effective mobility of the substance X in its own zone, $|\bar{\mu}_{X,X}|$. When we consider acid–base equilibria only, the above parameters of $|\bar{\mu}_{i,j}|$ (the mobility of a substance i in a zone j) mean

$$\bar{\mu}_{i,j} \equiv \bar{\mu}_i \quad (\text{pH}_j) \quad (7.14)$$

that is, $\bar{\mu}_{i,j}$ is given by the effective mobility of a substance i at pH equal to that of the zone of a substance j.

A third task is to verify whether the substances of interest are separable, i.e., whether they form their individual zones separated from each other. It is not sufficient to compare the tabulated or calculated effective mobilities of the substances in their own zones (parameters of the $\mu_{i,j}$ type) since some pairs of substances of the same effective mobilities in their own zones can be completely separated and migrate separately while other pairs of different effective mobilities in their own zones do not separate. Experimental verification of the separability is simple and is based on measuring calibration graphs for standard solutions of individual substances and for their mixtures of varying mutual ratio. Theoretical evaluation of the separability of substances on the basis of parameters of $\bar{\mu}_{i,j}$ type is identical with the principle of unambiguous determination of the migration order of substances. If the migration order can be determined unambiguously for a pair of substances i and j in a given electrolyte system, then substances i and j can be separated. A simple criterion can be employed to evaluate separability based on the migration order. Two cases representing separation are possible:

(i) the zone of a substance i migrates ahead of the zone of a substance j if

$$|\bar{\mu}_{i,j}| > |\bar{\mu}_{j,j}| \quad \text{and} \quad |\bar{\mu}_{j,i}| < |\bar{\mu}_{i,i}| \tag{7.15}$$

(ii) the zone of a substance j migrates ahead of the zone of a substance i if

$$|\bar{\mu}_{i,j}| < |\bar{\mu}_{j,j}| \quad \text{and} \quad |\bar{\mu}_{j,i}| > |\bar{\mu}_{i,i}| \tag{7.16}$$

If conditions (i) and/or (ii) are not valid for a given pair of substances, then a third case takes place where

$$|\bar{\mu}_{i,j}| < |\bar{\mu}_{j,j}| \quad \text{and} \quad |\bar{\mu}_{j,i}| < |\bar{\mu}_{i,i}|. \tag{7.17}$$

Here, the migration order cannot be determined unambiguously and the two substances form a stable mixed zone and cannot be separated.

When the selected electrolyte system does not meet all the criteria demanded (e.g., analytes are insufficiently ionized, form unstable zones, or their separability is low), another electrolyte is searched for. The most common way is to influence the effective mobilities of analytes by changing the pH of the separation medium or by incorporating complex formation during the separation process. Here, a very important part is played by the counterion of the leading electrolyte, which can have a buffering effect and/or interact with the analytes on forming complexes or ion pairs. Usually, buffered electrolyte systems are used where the counterion is an anion of a weak acid or a weak base. In such a system, the migration of fast H^+ or OH^- ions is controlled due to their combination with the counterion according to the dissociation balance so that they create a real or potential terminating zone and do not disturb the separation process. For the pH range required for the analyte to be sufficiently ionized, suitable counterions are chosen according to their pK values. Several of the commonly used and recommended counterions are presented in Table 7.2. In some special cases the selection of buffer-free or nonbuffered electrolyte systems is advantageous on condition that correct stable zones are formed (4,5).

7.4. SAMPLE STACKING

From the principle of the separation process in ITP it follows that the concentration of the analytes in the sample changes during separation until it

Table 7.2. Recommended Counterions for ITP Separations in Buffered Electrolyte Systems[a]

Anionic ITP		Cationic ITP	
Counterion	pH_L	Counterion	pH_L
β-Alanine	3.1–4.1	Formate	3.2–4.2
EACA	4.1–5.1	Acetate	4.2–5.2
Histidine	5.5–6.5	MES	5.7–6.7
Imidazole	6.6–7.6	Veronal	6.9–7.9
Tris	7.6–8.6	$AspNH_2$	8.3–9.3
Ethanolamine	9.0–10.0	Glycine	9.1–10.1

[a] EACA = ε-amino-n-caproic acid; MES = 2-(N-morpholino)ethanesulfonic acid; Tris = tris-(hydroxymethyl)aminomethane.

Source: After ref. 1 with kind permission of VCH Verlagsgesellschaft.

is adjusted to a defined ratio in respect to the concentration of the leading ion (Figure 7.2). Thus the concentration of an analyte in its zone is constant for a given leading electrolyte regardless of whether a diluted or a concentrated sample has been injected, with the only variation being in the length of the adjusted zone. In commercial instruments equipped with capillaries of 0.2–0.8 mm i.d. supplied with conductivity or potential gradient detectors usually up to 30 µL of 10^{-6} M solution of an analyte can be injected. The concentration of the analyte is finally adjusted so as to be close to that of the leading ion (10^{-2}–10^{-3} M), and the created zones can be detected unless they are shorter than 0.2 mm, which is the detection limit resulting from the length of the detection cell. Shorter zones, though separated, cannot be distinguished by the detector. Some improvement in such differentiation can be achieved by an optical detector on which individual zones not long enough to form a rectangular trace on the record with a plateau can still be detected as sharp peaks called spikes. The linear dependence of spike height on the amount of the analyte can also be used for quantitative analysis. However, if a series of such short zones migrate one after another, spikes are impossible to distinguish and nonabsorbing spacers must be included in the sample to isolate spikes from one another. Boundaries of isotachophoretic zones are very sharp, as are the boundaries of spikes. This follows from the principle of ITP, and ensures that the analyte, once separated, is present only in its own zone. The whole system of zones in ITP once the steady state has been reached, moves with a constant velocity, and nothing else but a stack of zones fills the space between the leading and terminating ions.

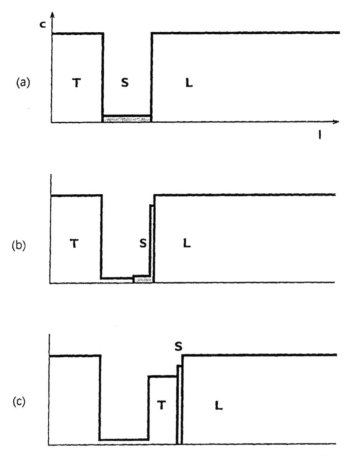

Figure 7.2. Adjustment of the sample concentration to the concentration of leading electrolyte during the ITP separation process: (a) initial conditions; (b) the separation process; (c) the final steady state. Key: L, leading electrolyte; S, sample; T, terminating electrolyte; c, concentration; l, length of the migration path.

7.5. THE ISOTACHOPHORESIS–CAPILLARY ZONE ELECTROPHORESIS COMBINATION

From the description above it is obvious that to distinguish individual analytes in stack is a difficult task. There is an elegant solution to this problem, however, and it is based on the combination ITP–CZE, where the ITP stack of

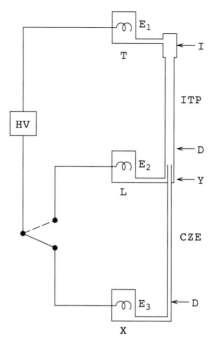

Figure 7.3. Scheme of a column-switching system for the on-line combination of ITP and CZE. Key: L, leading electrolyte; T, terminating electrolyte; X, background electrolyte; I, injection point; D, detector; HV, high-voltage power supply; E_1, E_2, E_3, electrode chambers; Y, bifurcation point.

analytes serves as the ideal narrow sample pulse for CZE separation and detection. The combination ITP–CZE requires a special instrumental arrangement and operating procedure. A schematic diagram of such an arrangement is shown in Figure 7.3. In the first stage, the sample is injected in between the leading and terminating electrolyte, current is switched on and passes between the terminating and auxiliary electrodes only through the preseparation capillary. The components of the sample form isotachophoretic consecutive zones with sharp boundaries, arrange themselves in the order of decreasing mobilities, and change their concentrations in dependence on the concentration of the leading zone. All the zones move with the same and constant velocity along the detector toward the bifurcation block, and macrocomponents, i.e., the components present at high concentrations, are driven out of the system to the auxiliary electrode. Before the stack of analytes of interest reaches the bifurcation point between the preseparation and analytical capillaries, the current is switched to the analytical capillary. The ITP stack now becomes the sample for the CZE analysis that proceeds in the second capillary.

In principle there are three ways of performing an ITP–CZE combination technique as far as the electrolyte system is concerned. The simplest way is to use the terminating electrolyte (TE) as the background electrolyte (BGE) for CZE, and the wet work with the preparation of the system is done before the analysis starts. When the separated zones reach the inlet of the analytical capillary, current is switched on across the system of both capillaries and the separation proceeds continuously. In the second approach the leading electrolyte (LE) is employed as the BGE. In this case, when the stack of isotachophoretically arranged zones of analytes reaches the bifurcation point, the current is switched on and the migration of analytes proceeds continuously into the analytical capillary. After the last analyte of the stack has entered the analytical capillary, the current is switched off and the TE within the preseparation capillary is flushed out and replaced with the LE. Then the current is applied again. The third possibility is to use a BGE different from the LE or TE. The procedure is similar to the L–S–L system: after the current is switched off, the TE is replaced by the BGE.

The utility of the ITP–CZE method derives from complementary advantages of ITP and CZE. Isotachophoresis ensures accurate dosage of relatively large volumes of the sample, separation and concentration adjustment of individual components, determination of macrocomponents followed by their removal from the system, and ideal sampling for CZE separation in the form of short zones with sharp boundaries. During the CZE step, individual components of the sample stack leave the stack and migrate independently with different velocities in the BGE. Zones are detected with highly sensitive detectors and identified according to their mobility.

A comparison of ITP and ITP–CZE analyses has been carried out using practical examples. In Figure 7.4, a typical UV record of an ITP separation performed in a simply designed device is presented (6) and can be used for both qualitative and a quantitative evaluation if the concentration of analytes is above the order of 10^{-6} M (Figure 7.4a). Below this concentration level, the zones formed are shorter than the slit width. As they migrate together in a stack, they cannot be distinguished and only one spike is recorded (Figure 7.4b). When the same quantity of the sample is analyzed by ITP–CZE concentration, independent migration in the ZE (zone electrophoretic) mode results in formation of individual zones isolated from one another by the nonabsorbing background electrolyte so that their identification and quantification are possible (Figure 7.4c).

Another practical application of ITP–CZE is demonstrated in Figure 7.5. The determination of a coccidiocidic drug halofuginone (HFG) by ITP directly in feedstuff extracts is complicated by its high content of sodium and

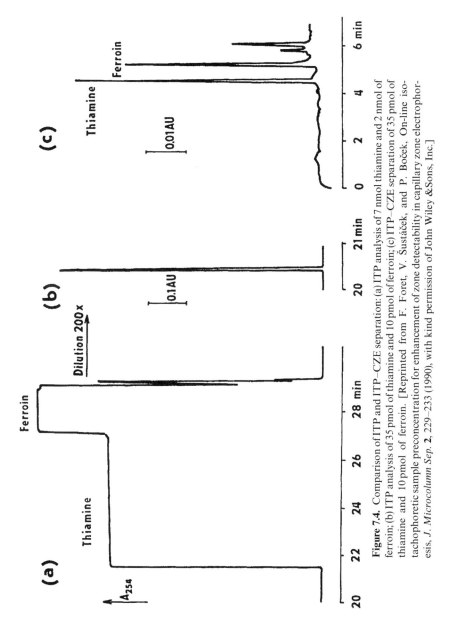

Figure 7.4. Comparison of ITP and ITP–CZE separation: (a) ITP analysis of 7 nmol of thiamine and 2 nmol of ferroin; (b) ITP analysis of 35 pmol of thiamine and 10 pmol of ferroin; (c) ITP–CZE separation of 35 pmol of thiamine and 10 pmol of ferroin. [Reprinted from F. Foret, V. Šustáček, and P. Boček. On-line isotachophoretic sample preconcentration for enhancement of zone detectability in capillary zone electrophoresis. *J. Microcolumn Sep.* **2**, 229–233 (1990), with kind permission of John Wiley &Sons, Inc.]

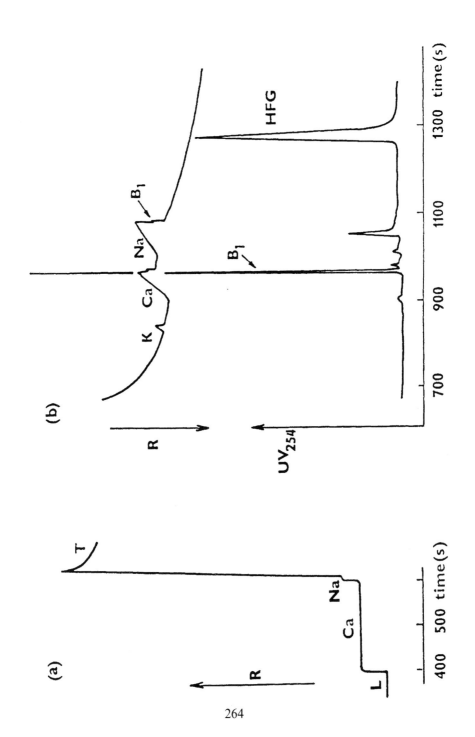

calcium, which means that even if the full-load capacity of the isotachophoretic system and the column-coupling system are employed, only short zones of microcomponents including halofuginone are created. Neither the step height on the conductivity record nor the migration order on the UV record ensures univocal evaluation of the analysis, as the composition of individual feedstuff batches differs (Figure 7.5a). In comparison to this, the trace of the CZE separation performed as the second step of the analysis was easy to evaluate both qualitatively and quantitatively (Figure 7.5b). The surplus of sodium and calcium was removed from this system by driving it to the auxiliary electrode in the first step; however, residues of these zones can still be seen on the conductivity record. No response of the conductivity to HFG is seen, as the conductivities of HFG and the co-ion of the BGE (here T) are very close (7).

7.6. DYNAMICS OF DESTACKING

Correct performance and evaluation of ITP–CZE analyses requires a detailed understanding of the separation process, particularly the transition from ITP to CZE (8,9). The ITP migration in the first capillary proceeds until the rear boundary of the leading electrolyte appears in front of the entrance to the second capillary. In practice, however, it is necessary to perform this switching-over step earlier in order not to lose a part of the sample due to outflow to the auxiliary electrode. Thus, some amount of the leading ion is always introduced into the analytical capillary together with the sample. When any other background electrolyte including the leading electrolyte is used for the CZE step, a segment of the TE is cut together with the sample stack and accompanies it into the second capillary. This means that in any type of electrolyte combination the sample entering the second capillary is sandwiched between the LE and TE and conditions are created for a temporary continuation of ITP migration. Though the front boundary of the leading zone penetrating into the terminating zone, which serves as the BGE in the case of the T–S–T system, or the rear boundary of the TE entering the BGE formed by LE in the case of the L–S–L system, or both front and rear boundaries of the leading and terminator zones, respectively, entering a BGE in the case of the BGE–S–BGE system, do not have self-sharpening properties and have

---

Figure 7.5. Practical application of ITP–CZE combination. Analysis of halofuginone (HFG) in a feedstuff supplement: (a) trace from the ITP preseparation; (b) trace from follow-up CZE analysis. Key: B_1 = vitamin B_1. [Reprinted from L. Křivánková, F. Foret, and P. Boček, Determination of halofuginone in feedstuffs by the combination of capillary isotachophoresis and capillary zone electrophoresis in a column-switching system, *J. Chromatogr.* **545**, 307–313 (1991), with kind permission of Elsevier Science Publisher, Amsterdam, The Netherlands.]

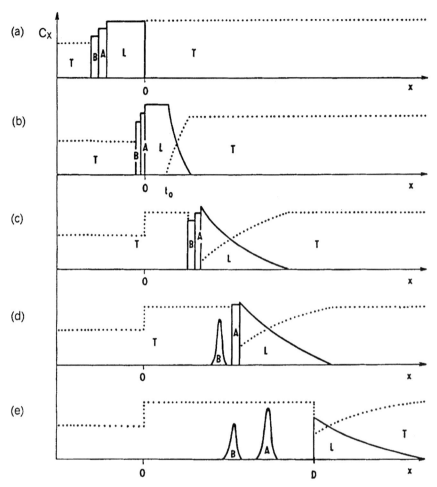

Figure 7.6. Scheme of transition of the stack of zones from isotachophoretic migration to ZE movement in the T–S–T electrolyte system. The situations are shown at the following times: (a) when the current is switched to the analytical capillary; (b) when the entire zone of the leader electrolyte has just migrated into the analytical capillary; (c) when the ITP concentration plateau of the leading zone has just disappeared; (d) when zone B has destacked while zone A is still migrating in the stack; (e) when all the sample zones migrate in the ZE mode and the rear boundary of the leading zone just passes the detector, thus opening the time window for detection. Key: L, leading electrolyte; T, terminating electrolyte; A & B, sample zones; x, longitudinal coordinate; c_X, concentration of a component in the separation system; 0, interface between the analytical and preseparation capillaries; D, position of the detector. [Reprinted from L. Křivánková, P. Gebauer, W. Thormann, R. A. Mosher, and P. Boček, Options in electrolyte systems for on-line combined capillary isotachophoresis and capillary zone electrophoresis, *J. Chromatogr.* **638**, 119–135 (1993), with kind permission of Elsevier Science Publishers, Amsterdam, The Netherlands.]

velocities different from the velocity of ITP migration, conditions for ITP migration of the sample stack persist till the isotachophoretic concentration plateau disappears. After that, the sample components gradually leave the stack, lose their contact with each other, and migrate independently in the ZE mode. The zone whose mobility is closest to that of L (in the T–S–T system), or T (in the L–S–L system), or both L and T (in the BGE–S–BGE system) is destacked last. The process of destacking is illustrated in Figure 7.6 for the T–S–T system (8). It is evident that the ZE migration starts at a different time and at a different column position. Thus, the methods using the time when the zone passes through a detector, i.e., the migration or detection time, directly for qualitative evaluation cannot be applied here. The detection time and variance of a zone change in different ways for the same analytes migrating in the T–S–T, L–S–L or BGE–S–BGE systems and depend on the amount of the accompanying segments of the leading and/or terminating zones (for T–S–T, L–S–L and BGE–S–BGE system, respectively), on the ratio of analyte/stacker mobilities, and on the concentration of the BGE.

In T–S–T systems, the detection times of the analytes linearly increase with the amount of the introduced leader. In L–S–L systems, the detection time linearly decreases with the amount of the introduced terminator. These effects relate only to sample zones that are already destacked when passing the detector; to ensure this, the amount of the introduced leader/terminator must not exceed a certain value that differs from analyte to analyte. The qualitative identification of the analytes based on their variable detection times is possible using a standardization procedure requiring that the sample be run with one internal standard substance. Figure 7.7 (9) illustrates the construction of the calibration curve used for standardization.

The temporal stacking also affects the final dispersion of the analyte zones at the point of detection since the dispersion of zones is frozen as long as they are in stack. The longer the analyte zone remains stacked, the sharper is its peak when detected. The sharpest zones therefore provide high-mobility analytes in T–S–T systems and low-mobility analytes in L–S–L systems. BGE–S–BGE systems show anomalous zone-dispersion distribution, with analytes of intermediate mobilities showing the most dispersed zones (Figure 7.8).

7.7. SAMPLE-INDUCED ISOTACHOPHORESIS (STACKING)

The effect of sample stacking by transient ITP migration in zone electrophoresis can be induced by a proper composition of the original sample or

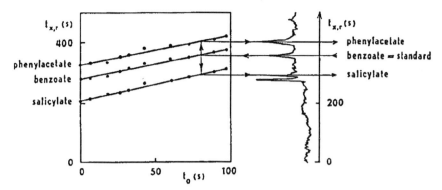

Figure 7.7. Principle of the standardization method for ITP–CZE combination. To be practicable for identification of analytes in the sample, the calibration curve for the dependence of the detection time of analytes and of an internal standard on the length of the leading electrolyte comigrating with the sample into the CZE step in the case of the T–S–T system (or terminator in the case of the L–S–L system) should be drawn and used as shown here for the T–S–T system. [Reprinted from L. Křivánková, P. Gebauer, and P. Boček, Some practical aspects of utilizing the on-line combination of isotachophoresis and capillary zone electrophoresis, *J. Chromatogr. A* **716**, 35–48 (1995), with kind permission of Elsevier Science Publisher, Amsterdam, The Netherlands.]

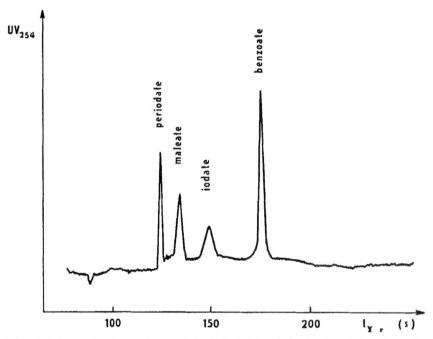

Figure 7.8. Dispersion of zones characteristic of ITP–CZE analysis when the mobility of the BGE co-ion used for the CZE stage lies between the mobility of the leading and terminating ions used for the ITP stage. [Reprinted from L. Křivánková, P. Gebauer, and P. Boček, Some practical aspects of utilizing the on-line ombination of isotachophoresis and capillary zone electrophoresis, *J. Chromatogr. A* **716**, 35–48 (1995), with kind permission of Elsevier Science Publishers, Amsterdam, The Netherlands.]

artificially after the sample composition is changed so that necessary conditions for stacking are fulfilled. In any case, three stages of migration can be distinguished: first analytes are arranged in the order of their effective mobilities and create a stack with sharp mutual boundaries; then they all migrate in this stack with the same velocities in the ITP mode; finally they gradually destack, and each starts migrating in a single-zone electrophoresis mode at different distances from the injection point. It is evident that the existence of transient ITP migration affects the basic electrophoretic parameters, migration time, and zone variation (10).

In practice, inherent self-stacking can be expected in samples of biological origin containing, e.g., a large amount of NaCl next to the analyzed microcomponents. The shift of migration time and change of peak shape is demonstrated in Figure 7.9a–c. It is clear that for the ratio of chloride to analytes (1000:1) the migration times of analytes whose mobility was higher than that of the BGE co-ion (analytes 1–9) increased in comparison with the migration velocity in the absence of chlorides (Figure 7.9b). At a higher chloride concentration (Figure 7.9c) the fastest analytes were not yet destacked when they passed through the detector and sharp isotachophoretic spikes were registered (11).

From the example given above it follows that for the correct evaluation of CZE analysis involving transient ITP stacking the mechanism of this process should be understood and parameters affecting it should be described so that both the sharpening effect can be employed and the analytes can be destacked in time prior to detection (12). Generally, minor analytes can be involved in the sample self-stacking mechanism only if at least one macrocomponent of the same charge is also present in the injected sample mixture. The macrocomponent for some time plays the role of the leading or terminating zone, with the BGE ion being the terminator or leader, respectively. Analytes create a train of stacked zones with sharp boundaries migrating isotachophoretically between the macrocomponent (stacker) and BGE. This means that the second basic condition for sample-induced stacking is that the analyte mobility must lie between those of the stacker and BGE co-ion. Concentrations of the analyte, BGE co-ion, and stacker as well as their mobilities determine whether the stacker creates the leader or the terminator for the stack. For

$$v_{X,S} > v_{S/B} > v_{X,B} \qquad (7.18)$$

the stacker is the transient terminator, whereas for

$$v_{X,S} < v_{S/B} < v_{X,B} \qquad (7.19)$$

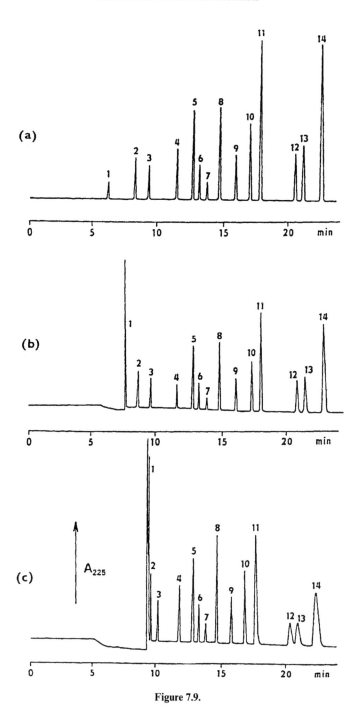

Figure 7.9.

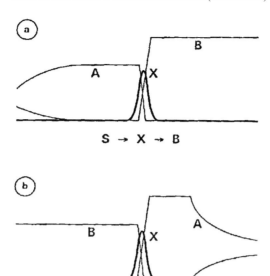

Figure 7.10. Concentration profile of sample self-stacking: (a) the sample major component (A) acts as the terminator, the BGE co-ion (B) as the leading ion; (b) the sample major component (A) is a temporal leader, the BGE co-ion (B) is the terminator. Key: X, analyte; S, sample zone. [Reprinted from P. Gebauer, W. Thormann, and P. Boček, Sample self-stacking and sample stacking in zone electrophoresis with major sample components of like charge: General model and scheme of possible modes, *Electrophoresis* **16**, 2039–2050 (1995), with kind permission of VCH Verlagsgesellschaft.]

the stacker is the transient leading zone. Here $v_{X,S}$ is the migration velocity of an analyte X in the sample zone; $v_{S,B}$ is the migration velocity of an analyte X in the BGE; and $v_{X/B}$ is the migration velocity of the sharp boundary between the sample zone and BGE (Figure 7.10) (12).

To decide whether the stacking is or is not effective in an actual case, one can carry out the following calculations. Two explicit requirements must be fulfilled, namely, the mobility and concentration criteria. Concerning the

---

Figure 7.9. The shift of migration times and the change of peak shapes caused by sample-induced stacking. Concentration of analytes: 0.1 mM. Concentration of NaCl in the sample: (a) 0; (b) 100 mM; (c) 300 mM. Peaks: 1–14, components of anionic model mixture. [Reprinted from R. Vespalec, P. Gebauer, and P. Boček, Identification of peaks in capillary zone electrophoresis based on actual mobilities, *Electrophoresis* **13**, 677–682, with kind permission of VCH Verlagsgesellschaft.]

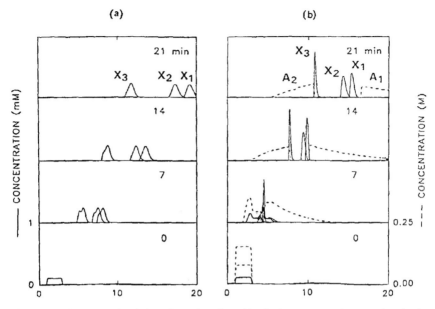

Figure 7.11. Computer simulation of sample self-stacking in the presence of two stackers in the sample with mobilities higher, A_1, and lower A_2, than the mobility of the BGE co-ion. Mobilities of analytes X_1, X_2 and X_3 are $\mu_{A_1} > \mu_{X_1} > \mu_{X_2} > \mu_{X_3} > \mu_{A_2}$. Analytes X_1 and X_2 are stacked by A_1; X_3 is stacked by A_2. [Reprinted from P. Gebauer, W. Thormann, and P. Boček, Sample self-stacking and sample stacking in zone electrophoresis with major sample components of like charge: General model and scheme of possible modes, *Electrophoresis* **16**, 2039–2050 (1995), by kind permission of VCH Verlagsgesellschaft.]

mobility criterion for the terminating stacker, it is necessary that

$$\bar{\mu}_{B,B} \frac{\bar{\mu}_{A,S}}{\bar{\mu}_{B,S}} < \bar{\mu}_{X,B} < \bar{\mu}_{B,B} \qquad (7.20)$$

For the leading stacker, it is necessary that

$$\bar{\mu}_{B,B} < \bar{\mu}_{X,B} < \bar{\mu}_{B,B} \frac{\bar{\mu}_{A,S}}{\bar{\mu}_{B,S}} \qquad (7.21)$$

Moreover, the concentration of the stacker must be sufficiently high to bring effective stacking. As for the concentration ratio of the stacker and the BGE co-ion in the sample, $\bar{c}_{A,S}/\bar{c}_{B,S}$, it was shown for strong ions (12) that it must be higher than a certain break value:

$$\frac{c_{A,S}}{c_{B,S}} > \left(\frac{c_{A,S}}{c_{B,S}}\right)_{br} = \frac{\mu_A - \mu_X}{\mu_X - \mu_B} \frac{\mu_B + \mu_R}{\mu_A + \mu_R} \qquad (7.22)$$

This means that for a given electrolyte system and for a given analyte a concentration of a component destined to be a stacker can be calculated and artificial sample self-stacking effect induced. Obviously, stacking of all components of a complex sample, including analytes whose mobilities are both higher and lower than the mobility of the BGE co-ion, can be achieved when stackers of both the leading and terminating types are present in the injected sample (Figure 7.11).

Though relationships have been derived enabling calculation of the detection time and variance of individual zones in a model system analyzed by zone electrophoresis involving sample self-stacking (10), evaluation of complicated samples of natural origin with varying and unknown types and amounts of stackers is difficult. In any case, to avoid incorrect evaluation of the analytic trace, only destacked zones should be evaluated and two internal standards should be analyzed simultaneously with the sample and used for elimination of any transient stacking effect (11).

7.8. CONCLUSION

Capillary isotachophoresis is one of the principal techniques of analytical electrophoresis. It has been employed both in a single-stage setup and in various sophisticated combinations in various fields of analytical chemistry for more than 20 years (1). The unique and characteristic features of ITP are the self-sharpening properties of the zone boundaries and adjustment of the concentration of species in their zones (stacking). In this chapter, the aforementioned characteristic features have been described and a selection of recommended electrolyte systems providing good isotachophoretic results have been presented. Moreover, it has been shown that the combination of ITP with CZE is beneficial to both techniques. Especially when on-line coupling of two capillaries in series, one for ITP and one for CZE, is utilized, sensitivity and resolution are enhanced relative to ITP analysis alone and detectability and reproducibility are enhanced relative to CZE analysis alone. This combination always starts with ITP when a large volume of a diluted sample can be injected and the stacking power of ITP converts it into sharp concentrated zones. These stacked zones then serve as the initial pulse for CZE analysis carried out in the second capillary. The transition of the ITP migration mode into the CZE mode proceeds progressively and is affected by various parameters, and therefore the simple rules pertaining to the theory of migration in the CZE mode cannot be applied directly. The effect of individual parameters on the transformational process have been studied and now, based on the results, a suitable composition of the electrolyte system and separation conditions can be selected for the required separation, and a reliable qualitative and quantitative evaluation of the analysis can be performed.

The isotachophoretic mode of migration, nonetheless, can also proceed transiently in a single-column CZE separation when conditions for ITP are induced either by the sample composition itself or by intentional involvement of a zone possessing features of the leading or terminating zone with respect to the co-ion of the BGE. Such a transient ITP may also produce the benefit of stacking of diluted analytes. Conditions were described above that must be fulfilled for a component to act as a stacker or to be stacked and the subsequent destacking process to be understood. Therefore the stacking and destacking process can be controlled, ensuring that an optimal sample concentration is attained during the stacking phase as well as enabling a correct evaluation of the results.

Finally, the analytical utility of ITP and of its combination with CZE may be summarized as follows: The application of capillary isotachophoresis both in combination with CZE or involved transiently in CZE is especially useful in biology and biochemistry, where the composition of the samples is very complex and variable and affects the course of the separation in CZE alone. The use of ITP alone is advantageous in determination of macrocomponents, and ITP coupled with CZE allows for separation of analyzed samples. In such a case the composition of the analyzed sample is modified so that it does not affect the separation process in the subsequent CZE step. Reproducible results of a highly sensitive analysis are thus obtained.

LIST OF ACRONYMS AND ABBREVIATIONS

Acronym or Abbreviation	Definition
BGE	background electrolyte
CZE	capillary zone electrophoresis
EACA	ε-amino-n-caproic acid
HFG	halofuginone
His	histidine
ITP	isotachophoresis
LE	leading electrolyte

MES	2-(N-morpholino)ethanesulfonic acid
TE	terminating electrolyte
Tris	tris(hydroxymethyl)aminomethane
UV	ultraviolet
ZE	zone electrophoretic (mode)

REFERENCES

1. P. Boček, M. Deml, P. Gebauer, and V. Dolnik, *Analytical Isotachophoresis.* VCH, Weinheim, 1988.
2. L. Křivánková and P. Boček, Isotachophoresis, In *Encyclopedia of Analytical Science*, pp. 1075–1082. Academic Press, London, 1995.
3. T. Hirokawa, M. Nishino, Y. Aoki, Y. Kiso, Y. Sawamoto, T. Yagi, and J. Akiyama, *J. Chromatogr.* **271**, D1 (1983).
4. L. Křivánková, F. Foret, P. Gebauer, and P. Boček, *J. Chromatogr.* **390**, 3 (1987).
5. P. Gebauer, L. Křivánková, and P. Boček, *J. Chromatogr.* **470**, 3 (1989).
6. F. Foret, V. Šustáček, and P. Boček, *J. Microcolumn Sep.* **2**, 229 (1990).
7. L. Křivánková, F. Foret, and P. Boček, *J. Chromatogr.* **545**, 307 (1991).
8. L. Křivánková, P. Gebauer, W. Thormann, R. A. Mosher, and P. Boček, *J. Chromatogr.* **638**, 119 (1993).
9. L. Křivánková, P. Gebauer, and P. Boček, *J. Chromatogr. A* **716**, 35 (1995).
10. P. Gebauer, W. Thormann, and P. Boček, *J. Chromatogr.* **608**, 47 (1992).
11. R. Vespalec, P. Gebauer, and P. Boček, *Electrophoresis* **13**, 677 (1992).
12. P. Gebauer, W. Thormann, and P. Boček, *Electrophoresis* **16**, 2039 (1995).

CHAPTER 8

CAPILLARY ELECTROCHROMATOGRAPHY

KATHLEEN A. KELLY and MORTEZA G. KHALEDI

Department of Chemistry, North Carolina State University, Raleigh, North Carolina 27695

8.1.	Introduction	277
8.2.	Column Preparation	279
	8.2.1. Frits	279
	8.2.2. Slurry-Packed Columns	280
	8.2.3. Drawn Packed Columns	282
	8.2.4. Electrokinetically Packed Columns	282
	8.2.5. Bubble Formation	282
8.3.	Detection	283
	8.3.1. Formation of the Detection Window	283
	8.3.2. "On-Column" vs. "In-Column" Detection	283
8.4.	Theory	285
	8.4.1. Electroosmotic Flow and the Electrical Double Layer (δ)	285
	8.4.2. Band Broadening	286
8.5.	Stationary Phase Considerations	287
8.6.	Mobile Phase Considerations	289
8.7.	Applications	290
	8.7.1. Pharmaceutical Compounds	290
	8.7.2. Polycyclic Aromatic Hydrocarbons	290
	8.7.3. Chiral Compounds	295
8.8.	Conclusions and Future Trends	295
Addendum		296
List of Acronyms		297
References		297

8.1. INTRODUCTION

The great success of open-tubular gas chromatography (OTGC) prompted much intense effort to achieve equivalent efficiency in high-performance liquid

---

High Performance Capillary Electrophoresis, edited by Morteza G. Khaledi. Chemical Analysis Series, Vol. 146.
ISBN 0-471-14851-2 © 1998 John Wiley & Sons, Inc.

chromatography (HPLC). The technological problems associated with open-tubular liquid chromatography (OTLC) that require column diameters on the order of a few micrometers have precluded its widespread use in analytical laboratories. Another approach for achieving high efficiencies in a short time period is to minimize stationary phase particle size for packed columns, as chromatographic theory shows that it has significant implications in terms of achieving low plate heights and high separation speed (1, 2). Tremendous improvements have been made in the performance of HPLC columns over the past two decades. However, the trend in reducing particle size faces a major obstacle in HPLC—that is, pressure drop is the main limitation in achieving higher column efficiency and faster analysis time.

An alternative is to simply replace the pressure-driven pumping mechanism with the one that is electrically driven. In 1974, Pretorius et al. (3) compared the plate heights between pressure-driven (laminar flow) systems and electro-osmotically driven (plug flow) systems. They studied the flow of acetone (unretained) in both open tubes and in packed tubes. Although capillaries with an internal diameter of less than 1 mm were not studied, this investigation is one of the first examples of using the electroosmotic pumping action to move solvent through packed tubes. In their landmark paper in 1981, Jorgenson and Lukacs showed the first electromigration separation using a 10 µm ODS stationary phase packed into a 170 µm i.d. capillary (4). The technique now known as capillary electrochromatography (CEC) is a hybrid of HPLC and capillary electrophoresis (CE). In CEC, separation of uncharged molecules is achieved on the basis of differential partitioning into a stationary phase. The mobile phase is pumped electrically, and thus analytes are carried through the column by electroosmotic flow (EOF). Basically CEC has a similar experimental setup to that of CE except that the capillary is packed with stationary phase particles. In a sense, CEC is an electrically driven HPLC. Figure 8.1 shows

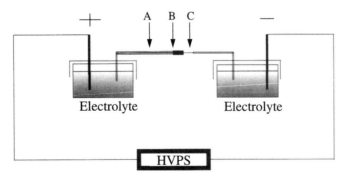

Figure 8.1. Schematic of CEC equipment: A = packed capillary; B = frit; C = detection window; HVPS = high-voltage power supply.

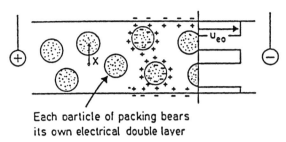

Each particle of packing bears its own electrical double layer

Figure 8.2 Plug flow profile through a packed capillary: u_{eo} = electroosmotic velocity; [From J.H. Knox, *Chromatographia* **26**, 329–335 (1988), with permission.]

a block diagram of CEC. In order to eliminate the formation of bubbles in the column, the buffer reservoirs may be slightly pressurized.

Due to the lack of pressure limitations in CEC, the stationary phase particle diameter can be reduced to the submicrometer level. In addition, larger number of theoretical plates can be generated by increasing the column length to values that are not practical in HPLC separations. This is achieved, however, at the expense of longer analysis time. It is known that the laminar flow profile in pressure-driven HPLC contributes to zone dispersion. In electrically driven CEC, one expects to achieve higher efficiencies due to the plug profile (Figure 8.2). Some of the most prevalent problems with CEC involve column preparation. This includes the preparation of the frits (which keep the packing material in the column), the method used to pack the column, reproducibility between columns, and Joule heating (which may also contribute to bubble formation in the column as well as band broadening).

8.2. COLUMN PREPARATION

8.2.1. Frits

Frits are necessary in CEC in order to keep the stationary phase migrating out of the column. The silica-based stationary phase is negatively charged, thus would be drawn toward the positive electrode when the voltage is applied. The production of mechanically stable frits is important both for column packing (so the frit is not blown out of the column when packing pressure is applied) and for reduction of the band broadening associated with frits being present in the column.

Boughtflower et al. (5) examined two methods of frit preparation in CEC. One method involves wetting the silica stationary phase and tapping it into the capillary. The frit is then produced by heating this material (sintering) until the silica melts slightly. This type of frit is not very porous due to large particles of silica that can block the capillary. The second method is to produce the frit by sintering the stationary phase material itself. This method yields more uniform and porous frits. Care must be taken, however, when the stationary phase is used to produce frits. If the stationary phase is heated too quickly or if the temperature is too high, the reversed-phase material bonded to the silica support will become pyrolized, turning black (6). This can result in peak tailing, as the solutes might interact with the frit (7).

Benhke et al. (8) examined other methods of preparing frits. One involves *in situ* polymerization of formamide and potassium silicate solution heated at 120 °C for 1 h. Another method is tapping of silica gel wetted with the potassium silicate solution. A third method uses bare silica gel. The latter two methods involve sintering the frit into place. It was found that the first two methods resulted in frits that have essentially fused together into an amorphous frit with irregular channels. The third method resulted in distinct silica particles fusing together at the points of contact between the particles. However, this third method is not as mechanically stable as the first two. The frit prepared with the pure silica gel cannot withstand the high pressures necessary to pack the column. If the inner diameter of the column is less then 50 μm, the frit will then break. Above this threshold, columns can be packed with pressures up to 400 bar.

8.2.2. Slurry-Packed Columns

There are two general methods for packing capillaries, slurry packing and drawn packed columns. Slurry-packed columns are formed by making a slurry with the packing mobile phase and the stationary phase. To pack a capillary, the most commonly used technique is the procedure developed by Boughtflower and coworkers (5). In step 1, a packing end frit is sintered (Figure 8.3). The capillary is packed with a slurry (step 2) consisting of the stationary phase and the packing mobile phase. In step 3, the packing mobile phase is flushed out of the capillary with water. A second end frit (a few centimeters from the packing end frit) is sintered. This second end frit is made of the stationary phase, and the packing end frit is cut off. A retaining frit (step 4) is then sintered close to where the detection window will be formed. The retaining frit prevents the stationary phase from migrating out of the capillary (due to its own electrophoretic mobility) when the electric field is applied. Low pressure is applied to remove any remaining stationary phase between the retaining frit and the end of the capillary (step 5). A detection

COLUMN PREPARATION   281

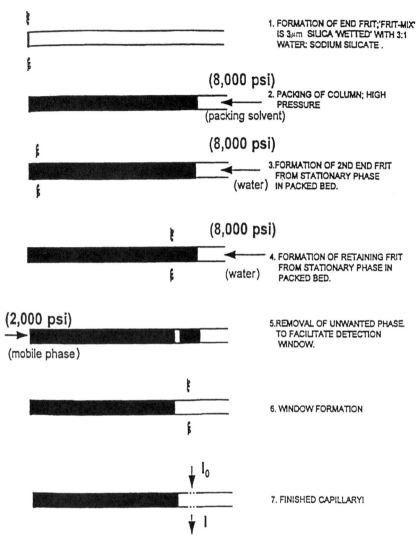

Figure 8.3 A technique used to slurry pack capillaries: I and I_o are the intensity of light through the detection window. [From R. J. Boughtflower, T. Underwood, and C. J. Paterson, *Chromatogr.* **40** 329–335 (1995), with permission.]

window is then formed by gently burning off and removing the polyimide coating (step 6).

It was found that more uniform columns could be produced if the slurry used for packing the column was sonicated for a short time, both prior to packing and during the packing procedure. The slurry reservoir and the capillary were placed in an ultrasonic bath to prevent the slurry from settling. The longer the slurry is maintained, the longer the column packing may proceed. Boughtflower et al. (9) designed a new ultrasound slurry chamber to which the slurry reservoir, capillary, ultrasound probe, and solvent inlet (from the packing pump) are all immersed. With this new apparatus, they were able to routinely produce high-performance columns (200,000 plates/m).

8.2.3. Drawn Packed Columns

Drawn packed capillaries are prepared by packing a large-bore capillary (1–2 mm i.d.) with a dry stationary phase. A glass-drawing machine then draws the capillary to the desired inner diameter. The stationary phase can become partially imbedded in the walls of the capillary, which helps to maintain the stability of the capillary. The production of drawn packed capillaries is laborious because any moisture in the stationary phase makes the drawing process difficult. Therefore heating the stationary phase under a vacuum is done both before packing and during the drawing of the capillary.

8.2.4. Electrokinetically Packed Columns

The common material used to pack columns, silica gel, has multiple charges on its surface. As a result, the stationary phase particles possess electrophoretic mobility (this is one of the reasons that retaining frits must be present on a column). By ultrasonically maintaining a slurry, a voltage can be applied to the system so that silica particles migrate into the column and a packed column is produced. Inagaki and coworkers (10) and Yan (11) have both packed columns using this electrokinetic technique.

8.2.5. Bubble Formation

In CEC, there is speculation about the cause of bubble formation in the column. One theory is that Joule heating (12,13) may specifically promote the formation of bubbles in the CEC system. This is due to the stationary phase, which could act as nucleation sites ("boiling chips") for the formation of gas bubbles (12). Some studies have found that either pressurizing the buffer reservoirs (8,13,14) or degassing the solvent prior to using the column (5,8) helps to reduce bubbles. A different theory suggested by Rebscher and Pyell is that bubbles form where unpacked and packed regions meet (7). This is believed to result from differing electroosmotic mobilities due to inhomogeneous packing, and not due to Joule heating. As of yet, there is no definitive theory as to what the cause of the bubble formation is, or whether there are several contributing factors rather than just one. As mentioned above, it is best to keep the system pressurized to reduce the risk of bubble formation.

8.3. DETECTION

8.3.1. Formation of the Detection Window

There are two methods of forming the detection window. The first is similar to the method used in CE today. In CEC, the polyimide coating is burned off after the outlet frit, and on-column detection is performed (Figure 8.4). The disadvantage to this method is that when the polyimide coating is removed, that region of the capillary becomes very fragile. If the capillary breaks at the window, the entire column is rendered useless. The second method (Figure 8.4) involves attaching a second unpacked capillary to the capillary that contains the packed column by means of Teflon tubing (8). This attachment is made adjoining the outlet frit. The unpacked capillary contains a detection window. This method allows replacement of the detection window if it should break. However, this method requires the use of lower electrical field strengths (less than 10 kV) due to an increase in noise.

8.3.2. "On-Column" vs. "In-Column" Detection

On-column detection refers to a detection window that is placed after the outlet frit. This is the most common method of detection for capillary electroseparation methods. However, detection after the frit results in peaks

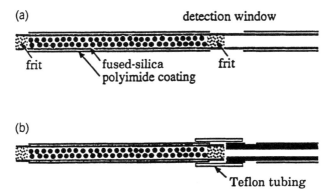

Figure 8.4. Two different types of detection windows. [From B. Behnke, E. Grom, and E. Bayer, *J. Chromatogr. A* **716**, 207–213 (1995), with permission.]

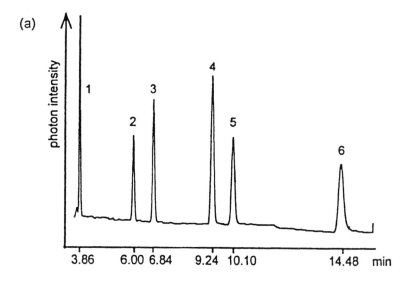

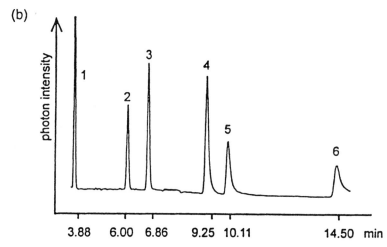

Figure 8.5. Comparison between in-column (a) and on-column (b) detection. Peaks: (1) resorcinol; (2) naphthalene; (3) fluorene; (4) pyrene; (5) crysene; (6) (1,2:5,6)-dibenzanthracene. [From H. Retscher and U. Pyell, *J. Chromatogr. A* **737**, 171–180 (1996), with permission.]

that have been subjected to band broadening due to the presence of the frit (Figure 8.5) (15). Yan et al. (16) found that the number of theoretical plates were significantly reduced when detection was made using this method. Detection can be made by detecting solutes while they are passing through the stationary phase. This is referred to as in-column detection. Detection may be made with either UV or fluorescence detectors. Although light scattering limits the linear working range for quantitative determinations, this is advantageous, as a source of band broadening is eliminated. Guthrie and Jorgenson calculated that in-column detection should be more sensitive than on-column detection for OTLC system (17). Rebscher and Pyell have predicted that this signal enhancement should also apply to the CEC system (15).

8.4. THEORY

8.4.1. Electroosmotic Flow and the Electrical Double Layer (δ)

In CE, the electrical double layer only exists around the walls of the capillary. In CEC, each particle possesses its own electrical double layer. When the electric field is applied, the solvent moves through the capillary around each particle (Figure 8.2). If the particle diameter becomes too small, the electrical double layers that surround the particles will overlap and the EOF will stop. Also within each particle, channels or pores exist. However, the electrical double layers overlap within these channels, and therefore there is no flow of the mobile phase within a particle, only in between particles. The electrical double layer is also influenced by the column preparation method used (i.e., drawn packed or slurry-packed columns). It was found that electrical double layers of adjacent particles overlap if the particle diameter, d_p, is less than 20 times the electrical double layer thickness, δ (i.e., $d_p \leq 20\delta$) for drawn-packed columns, and δ is reduced when $d_p \leq 40\delta$ for slurry-packed columns (1).

In accord with the Stern–Gouy–Chapman theory, δ is given by the following equation:

$$\delta = \{\varepsilon_0 \varepsilon_r RT/2cF^2\}^{1/2} \tag{8.1}$$

where ε_0 is the permittivity of a vacuum; ε_r is the dielectric constant; R is the universal gas constant; T is the temperature; c is the molar concentration of the electrolyte; and F is the Faraday constant. Knox (18) examined the experimental conditions under which no loss of EOF is observed. On the basis of the criterion $d_p \leq 40\delta$, he estimated (Table 8.1) the minimum particle diameter, $d_{p,\text{min}}$, where no loss of EOF is at different concentrations of background electrolyte. As the concentration of electrolyte increases from 0.001 to 0.1 M, the ratio of δ/d_p remains constant (δ and d_p change with respect to each other). Table 8.1 shows that packing submicrometer particles for use in the CEC system is theoretically possible.

Table 8.1. Relationship Between Concentration of Electrolyte (c), the Electrical Double Layer (δ), and the Minimum Particle Diameter (d_p)

Concentration of electrolyte (mol/L)	0.001	0.01	0.1
δ (nm)	10	3	1.0
Minimum d_p (μm)	0.4	0.12	0.04

8.4.2. Band Broadening

Band broadening in CE has been shown to be the sum of various contributions (7,19,20). The mechanisms that contribute to band broadening in CEC can originate from both its electromigration and chromatographic characteristics. This results in potentially complicated electroseparation technique. The chromatographic plate height, H, includes contributions from longitudinal diffusion (the B term), eddy diffusion (the A term), and resistance to mass transfer (the C term), as described by a van Deemter equation (18):

$$H = 2\lambda d_p + \{(2\gamma D_m)/u\} + \{(1/30)(k'/(1+k')^2)(d_p^2 u/D_{sz})\} \quad (8.2)$$

$$H = A + B/u + Cu \quad (8.3)$$

$A = 2\lambda d_p$, where λ is the packing constant and d_p is the particle diameter;
$B = \{(2\gamma D_m)/u\}$, where γ is the obstruction factor for the stationary phase, D_m is the diffusion coefficient for the solute, and u is the linear velocity;
$C = \{(1/30)(k'/(1+k')^2)(d_p^2 u/D_{sz})\}$, where k' is the capacity ratio and D_{sz} is the diffusion coefficient for the solute in the stagnant mobile phase.

The A and C terms are dependent upon the particle diameter of the stationary phase (d_p); and the B term is unaffected by the particle size. Theoretically, as the particle diameter decreases to the submicrometer level, the A and C terms would become insignificant and the contribution due to axial diffusion (the B term) would dominate (18). In other words, the plate height would be equal to $2\gamma D_m/u$; assuming that band dispersion mechanisms due to electromigration, such as Joule heating and electrodispersion, as well as other sources like sample plug size, detection window size, and broadening due to the presence of frits in CEC, have negligible effects and/or could be eliminated.

Like other electroseparation methods, miniaturization is necessary for CEC in order to alleviate Joule heating effects (1,21,22). Knox and McCormack (23) theoretically examined the problem of Joule heating for electroseparation methods that also applies to CEC. The temperature at the center of a CEC column can be much higher than that at the walls. This leads to a laminar flow profile and contributes to zone dispersion.

Band broadening is also due to instrumental factors specific to CEC. Instrumental band broadening is dependent upon the preparation of the column (i.e., packing technique used, frit production, etc.) for the system being examined. One of the differences is the method used to pack the capillary (slurry-packed vs. drawn packed capillaries). To study instrumental band broadening, Rebscher and Pyell (7) switched the voltage back and forth, thereby making the peak pass through the detection area several times. This also allowed the peak to pass through the frit several times. Their work agrees with Knox and Grant's conclusion (1) that the drawn packed capillaries perform better than the slurry packed capillaries (less band broadening). This may be due to the absence of a retaining frit before the detection window on the drawn packed capillaries, since detection is generally made within the capillary (through the packing material). However, due to the difficulties in preparation, drawn packed columns are not widely used.

Lelièvre et al. (24) examined the extracolumn effects unique to the CEC system. Their study involved the frit and the detection zone after the frit. The number of plates decreased due to the presence of the frit by approximately 28% of the column variance. The frit is essentially a turbulent area in which partitioning does not occur. The presence of the detection zone also reduced the number of theoretical plates by approximately 6%. The combined total was approximately 34%, which significantly reduced the number of theoretical plates. Yan et al. (16) experienced an even more drastic drop in theoretical plates when detection was made after the frit. The number of theoretical plates obtained for the retained solutes (in the beginning of the electropherogram) were calculated to be 110,000–150,000 when detection was made 1–2 mm after the outlet frit. However, when a detection window was made before the outlet frit, theoretical plates of 400,000 were obtained.

8.5. STATIONARY PHASE CONSIDERATIONS

The most commonly used stationary phase in CEC is octadecylsilane (ODS). This C_{18} stationary phase is commercially available in various particle diameters (the size most used is 3 μm). The use of open-tubular columns has also been reported. Guo and Colón (25) prepared a solgel that coated a thin glass film onto the capillary walls. Their glass film is porous, and incorporates a C_8 moiety into the glass substrate. By controlling the processing parameters, they can control the pore sizes and surface area. Pesek and Matyska (26) etched the interior of a capillary to increase the surfce area of the capillary. They then modified the walls of the capillary by attaching a C_{18} bonded phase.

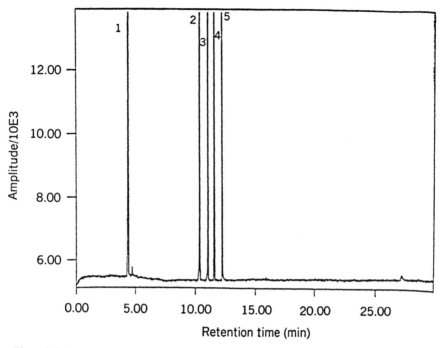

Figure 8.6. Separation of basic antidepressants. Peaks: (1) bendroflumethaiazide; (2) nortriptyline, (3) clomipramine; (4) imipramine; (5) methdilazine. [From N.W. Smith and M.B. Evans *Chromatographia* **41** 197–203 (1995) with permission.]

Although the capillary is not physically packed with stationary-phase materials, the separation process is similar. Chiral stationary phases have also been packed into capillaries. The same stationary phases that have been used in HPLC columns can be used to pack capillaries. Li and Lloyd (27) packed their column with α_1-acid glycoprotein (AGP) bound on silica as the stationary phase. Zare et al. (24) packed a column with hydroxypropyl-β-cyclodextrin (HPβCD). Both groups used their respective stationary phases for enantiomeric separations (see Section 8.7, below). Smith and Evans (28) used an ion-exchange material (1.8 μm Zorbax SB C_8) as a stationary phase to separate very basic antidepressants (Figure 8.6). The separation of these compounds on a conventional ODS stationary phase is difficult due to interactions between the solutes and the unprotected silanol groups. The authors reported an enormously large number of theoretical plates (8 million). They did not elaborate on the unusually large plate number. The standard equation for efficiency was used. However, Liao et al. (29) recently indicated that this equation is not applicable where band stacking occurs due to a solvent gradient.

Recently, the use of monolithic columns (or continuous-bed columns) in CEC has been reported in the literature (29). A monolithic column is a continuous stationary phase that is prepared within the capillary. This eliminates problems of uniformly packing a capillary with individual stationary phase particles and also problems of producing frits (which are not necessary for this column). The continuous column is prepared by using a polymerization chain reaction. By changing the monomers used for polymerization, as well as the polymerization conditions, the pore sizes within a column can be controlled. There can be several advantages of a continuous column over silica-based stationary-phase columns, one of these being that the electroosmotic flow can be independent of the pH (29).

8.6. MOBILE PHASE CONSIDERATIONS

In CE, the choice of mobile phases extends from a purely aqueous phase to a totally nonaqueous mobile phase. In CEC, the mobile phases from which one can choose are determined by the type of stationary phase employed. For a column packed with ODS silica gel, the most commonly used mobile phase is an aqueous mobile phase with a volume fraction of acetonitrile in the running buffer. Rebscher and Pyell (7) were unable to obtain a stable system on an ODS column when the fraction of acetonitrile was below 0.50. They concluded that this was possibly due to wetting difficulties of the ODS by the mobile phase when there was a high percentage of aqueous phase. They also found that the electroosmotic velocity increased when the volume fraction of acetonitrile increased. Yamamoto et al. (30) found, however, that the EOF decreased upon increasing acetonitrile percentages in the mobile phase. There are alternatives to using the ODS/acetonitrile combination for a separation. An ODS column with methanol/water (70:30 v/v) with 2 mM sodium dihydrogen phosphate buffer was used to separate polycyclic aromatic hydrocarbons (PAHs) (1). However, acetonitrile has been the most widely used organic modifier with the ODS stationary phases.

Various modifiers have been used with other (i.e., non-ODS) stationary phases. Guo and Colón (25) used a methanol/water mobile phase to separate PAHs on a solgel-based column. Smith and Evans (28) used an acetonitrile/water mobile phase with an ion-exchange column. Li and Lloyd (27) examined the effects of five different organic modifiers on the separation of chiral compounds using an AGP stationary phase. They found that the best separations were obtained by using 2-propanol.

8.7. APPLICATIONS

8.7.1. Pharmaceutical Compounds

Smith and Evans (14) have separated steroids from their respective impurities using CEC. Using the same packing technique as that of Boughtflower et al. (5), they pressurized the buffer reservoirs to reduce bubble formation and were able to obtain more than 300,000 theoretical plates/m. They have also used an ion-exchange material to separate very basic antidepressants (Figure 8.6), (28). Tetracyclines (Figure 8.7) were separated by Pesek and Matyska (31) using a wall-modified CEC system. They etched the inner capillary wall (referred to as a hydride-modified capillary) and then modified the wall with a C_{18} moiety (a C_{18} capillary). They were able to separate the tetracyclines from their degradation products using both capillaries; however, the C_{18} capillary resulted in the best separation. In a comparison with capillary zone electrophoresis (CZE) separation, they reported better resolution between doxycycline and methacycline (a challenging separation) when using CEC with the modified capillary with a citric acid/β-alanine buffer system.

8.7.2. Polycyclic Aromatic Hydrocarbons

Yan et al. used CEC to separate 16 PAHs (16). They used two detection methods: in-column through the stationary phase, and on-column after the retaining frit. The separation obtained showed a general elution problem. However, they were able to resolve 15 of the 16 PAHs with an isocratic mobile phase (Figure 8.8a). The number of theoretical plates obtained for the retained solutes was calculated to be some 110,000–150,000 when detection was made after the outlet frit; however, when a detection window was made before the outlet frit, theoretical plates of 400,000 were obtained. The limit of detection increased when detection was made in the column due to a high background (light scattering due to the stationary phase). The efficiency of the CEC system was higher than that of a micro-HPLC (75 µm i.d. column, 33 cm packed length), e.g., $N = 102,000$ for CEC vs. $N = 67,000$ for micro-HPCL for naphthalene. Yan and coworkers (32) recently reported a gradient elution separation of PAHs (Figure 8.8b). The schematic of the gradient elution system is shown in Figure 8.9. Various gradient profiles can be created through careful control and voltage programming of the two power supplies.

Figure 8.7. Separation of some likely compounds in commercial tetracycline: (a) bare capillary pH 3.0 buffer (30 mM citric acid and 24.5 mM β-alanine, $l = 50$ cm; (b) bare capillary same as (a) but with buffer-methanol, 60:40; (c) hydride capillary, conditions same as for (b) except $l = 25$ cm; (d) C_{18} capillary, conditions same as for (c). Peak: (e) 4-epianhydrotetracycline. [From J.J. Pesek and M. T. Matyska, *J. Chromatogr. A* **736**, 313–320 (1996), with permission.]

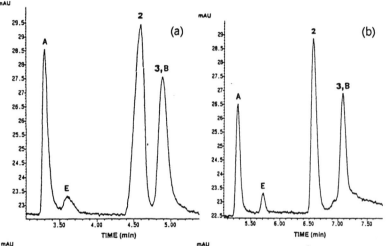

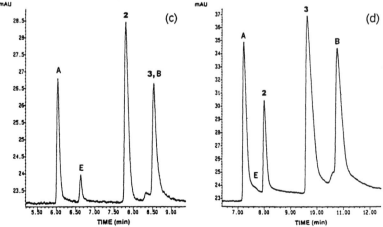

Figure 8.7

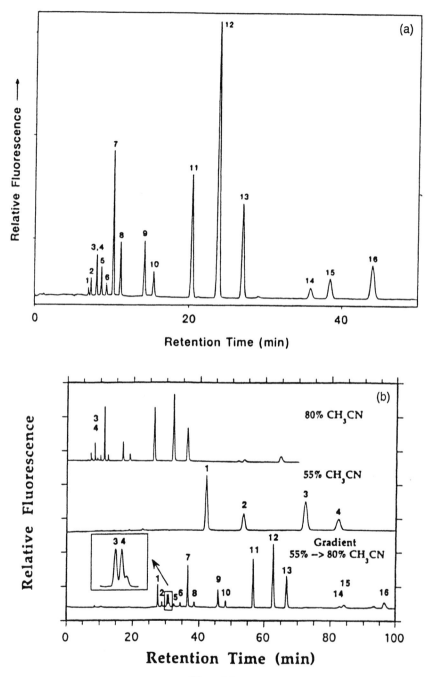

Figure 8.8

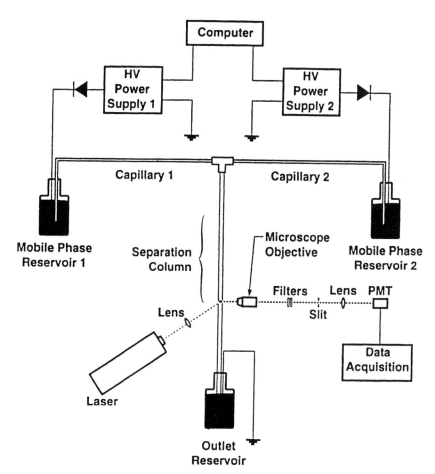

Figure 8.9. Schematic of equipment used for gradient elution in CEC: PMT = photomultiplier tube. [From C. Yan, R. Dadoo, R. Zare, D. Rakestraw, and D. Anex, *Anal. Chem.* **68**, 2726–2730 (1996), with permission.]

Guo and Colón (25) separated PAHs using a solgel-based column. They coated the capillary with a porous glass film and examined the separations obtained with these capillaries using both OTLC and CEC. Figure 8.10

Figure 8.8. Separation of 16 PAHs with isocratic (a) and gradient elution (b) solvent systems. Peaks: (1) naphthalene; (2) acenaphtylene; (3) acenaphthene; (4) fluorene; (5) phenanthrene; (6) anthracene; (7) floranthene; (8) pyrene; (9) benz[*a*]anthracene; (10) chrysene; (11) benzo[*b*] fluoranthene; (12) benzo[*k*] fluoranthene; (13) benzo[*a*] pyrene; (14) dibenz [*a,h*] anthracene; (15) benzo[*g,h,i*] perylene; and (16) indeno[1,2,3-*cd*] pyrene. [(a) [From C. Yan, R. Dadoo, H. Zhao, R. Zare, and D. Rakestraw, *Anal. Chem.* **67**, 2726–2729 (1995).]

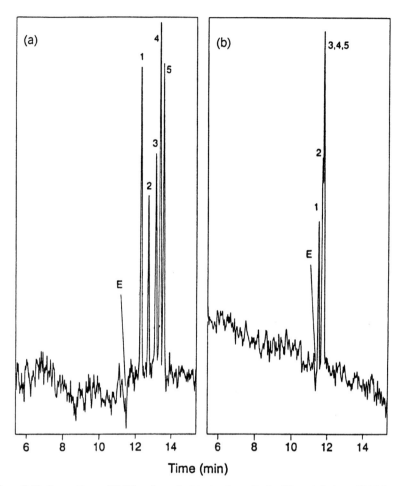

Figure 8.10. Separations of PAHs using solgel technology. Peaks: (1) naphthalene; (2) biphenyl; (3) fluorene; (4) 2-ethylnaphthalene; (5) 2,6-dimethylnaphthalene. E indicates electroosmotic mobility. [From Y. Guo and L. Colón, *Anal. Chem.* **67**, 2511–2515 (1995), with permission.]

illustrates the separation of 5 PAHs using two different stationary solgel-based phases (Figure 8.10a) and an ODS column (Figure 8.10b). Although they were able to separate the five PAHs within 14 min, the retention factors of the solutes were small on the solgel-based column ($k' < 1$). The longest retained solute, 2,6-dimethylnaphthalene, has a k' value of 0.179 and 382,000 theoretical plates. Note that for longer-retained solutes, resistance to mass transfer can be the predominant source of band broadening, thus reducing the plate number.

8.7.3. Chiral Compounds

The chiral selector in CEC separations can be either incorporated in the mobile phase or immobilized on the stationary phase. Lelièvre et al. (24) reported that an achiral stationary phase (3 μm ODS) with HPβCD in the mobile phase resulted in baseline resolution of the enantiomers of chlorthalidone and mianserin. They also reported the use of a HPβCD stationary phase. This method required a long equilibration time (15), but separation of the enantiomers of mianserin was also obtained. In order to achieve baseline resolution of the enantiomers the amount of organic modifier necessary was different for each method.

Li and Lloyd (27) used a chiral stationary phase packing AGP with a propanol mobile phase to separate 19 enantiomeric compounds. They were able to separate 10 of the 19 isomers. These compounds were generally β-blockers (i.e., oxprenolol). The remaining unresolved compounds were negatively charged and were not eluted from the capillary. These compounds were mainly orgainc acids (such as ketoprofen) and a few β-blockers (such as atenolol). According to the authors, there was a possibility that none of the anionic compounds (or a small amount) may have actually been injected electrokinetically.

8.8. CONCLUSIONS AND FUTURE TRENDS

The plug flow profile in electroseparation methods, including CEC, results in higher efficiency than that of the pressure–driven HPLC system. In addition, the lack of pressure limitation in CEC allows the use of smaller stationary-phase particle sizes and/or longer columns that should lead to even higher numbers of theoretical plates. In theory, by using submicrometer particle sizes in CEC, it should be possible to virtually eliminate the two main sources of band broadening in chromatography: eddy diffusion and resistance to mass transfer. Although the reported efficiency values in CEC are not close to the theoretical limits, it provides higher efficiency than HPLC. The majority of CEC reports have used 3 μm particles, even though the theory shows that particles as small as 0.4 μm can be used without a loss in EOF. Obviously, the use of submicrometer particles will greatly enhance efficiency; however, making suitable frits and packing them into a column are major challenges. Likewise, packing very long columns would be a difficult task. Obviously, the main focus of future research will be on solving current problems and improving the column technology.

Other electroseparation techniques that are solution based such as micellar electrokinetic chromatography (MEKC) or CZE offer clear advantages over

CEC where separation occurs at solid interfaces. In addition to offering higher efficiency, MEKC and CZE provide the feasibility and versatility of incorporating different chemistries to manipulate selectivity. The possibility of using different HPLC stationary phases is often viewed as "higher selectivity" for CEC. In the best scenario, CEC would provide selectivity equivalent to that of reversed-phase liquid chromatography (RPLC) for separation of small molecules. The use of normal-phase LC type mobile phases is not feasible in electroseparation techniques. Ion-exchange columns would not provide any advantage since one can simply separate charged molecules by CZE. The possibility of on-line coupling of CEC with mass spectrometry for analysis of uncharged solutes is a promising application of CEC, considering the difficulties associated with MEKC–MS.

ADDENDUM

Since this chapter was first submitted, some important research has been presented at two recent conferences: The 9th International Symposium on High Performance Capillary Electrophoresis and Related Microscale Techniques on January 26–30, 1997, in Anaheim, California (HPCE '97), and the 19th International Symposium on Capillary Chromatography and Electrophoresis on May 18–22, 1997, in Wintergreen, Virginia (CC&E '97). The following is a summary of the important developments reported in the abstracts of the papers at these two meetings.

Work has been presented by several groups in which submicron particles were used as the stationary phase for CEC. Bayer (33) investigated modified submicron silica particles packed in columns with different lengths. Ludthe and coworkers (34) examined the efficiencies obtained with 500 nm silica that was derivatized with ODS. Colón and co-workers (35) synthesized 560 nm silica particles with C_8 functional group. This group has devised a new method for packing capillaries (36). With the use centripetal force, they were able to pack multiple capillaries at a time. They have also investigated both open tubular solgel columns and submicron packed capillary columns (37). Stobaugh et al. (38) examined stationary phases ranging from 500 nm to 5 μm. These stationary phases also were derivatized with various functional groups, from bare silica to C_{18}, phenyl, SCX (strong cation exchange), mixed mode (SCX/C_{18}), and C_8. Fujimoto and coworkers (39) showed a separation using 14.2 nm colloidal silica. Venema et al. (40) used 3 and 5 μm size-exclusion stationary phases for nonaqueous size-exclusion chromatography (SEC) separation of polystyrene standards. They used dimethylformamide as the solvent, and compared pressure-driven (μHPLC) and electric-driven (CEC) systems.

Tan and coworkers (41) have made several different molecular imprinted polymer stationary phases that were prepared for the separation of D,L-phenylalanines. Hobo et al. (42) have synthesized a molecular imprinted polymer to separate dansyl amino acids. Nilsson and co-workers (43) have prepared a molecularly imprinted monolithic bed for chiral separations. Horváth and coworkers (44) coated the inside of a capillary with a highly cross-linked polymer. This would be useful to decrease protein adsorption, and would also reverse the EOF. Dorsey and co-workers (45) prepared various monolithic columns of differing diameter, lengths, and pore diameters. They examined the effects of aqueous/organic binary solvent mixtures, as well as pure solvents on the EOF and currents produced. Malik and Hayes (46) have studied monolithic porous beds that do not shrink upon the evaporation of solvents during preparation. Palm and Novotny (47) have prepared a monolithic column of a macroporous polyacrylamide–polythylene glycol gel as the stationary phase for CEC.

LIST OF ACRONYMS

Acronym	Definition
AGP	α_1-acid glycoprotein
CE	capillary electrophoresis
CEC	capillary electrochromatography
CZE	capillary zone electrophoresis
EOF	electroosmotic flow
HPCD	hydroxypropyl-β-cyclodextrin
HPLC	high-performance liquid chromatography
HVPS	high-voltage power supply
MEKC	micellar electrokinetic chromatography
MS	mass spectrometry
ODS	octadecylsilane
OTGC	open-tubular gas chromatography
OTLC	open-tubular liquid chromatography
PMT	photomultiplier tube
RPLC	reversed-phase liquid chromatography
SCX	strong cation exchange
SEC	size-exclusion chromatography

REFERENCES

1. J. H. Knox and I. H. Grant, *Chromatographia* **24**, 135–143 (1987).
2. J. C. Giddings, *Unified Separation Science.* Wiley, New York, 1991.

3. V. Pretorius, B. J. Hopkins, and J.D. Schieke, *J. Chromatogr.* **99**, 23–30 (1974).
4. J. W. Jorgenson and K.D. Lukacs, *J. Chromatogr. A* **218** 209–216 (1981).
5. R. J. Boughtflower, T. Underwood, and C. J. Paterson, *Chromatographia* **40**, 329–335 (1995).
6. H. Rebscher and U. Pyell, *Chromatographia* **42**, 171–176 (1996).
7. H. Rebscher and U. Pyell, *Chromatographia* **38**, 737–743 (1994).
8. B. Behnke, E. Grom, and E. Bayer, *J. Chromatogr. A* **716**, 207–213 (1995).
9. R. J. Boughtflower, T. Underwood, and J. Maddin, *Chromatographia* **41**, 398–402 (1995).
10. M. Inagaki, S. Kitagawa, and T. Tsuda, *Chromatography* 55R–60R (1993).
11. C. Yan, Electrokinetic Packing of Capillary Columns, U.S. Pat. 5,453, 163 (1993).
12. J. H. Knox and I. H. Grant, *Chromatographia* **32** 317–328 (1991).
13. T. Tsuda, *Anal. Chem.* **59** 521–523 (1987).
14. N. W. Smith and M. B. Evans, *Chromatographia* **38**, 649–657 (1994).
15. H. Rebscher and U. Pyell, *J. Chromatogr. A.* **737**, 171–180 (1996).
16. C. Yan, R. Dadoo, H. Zhao, R. Zare, and D. Rakestraw, *Anal. Chem.* **67** 2026–2029 (1995).
17. E. Guthrie and J. Jorgenson, *Anal. Chem.* **56**, 483 (1984)
18. J. H. Knox, *Chromatographia* **26**, 329–335 (1988).
19. F. Foret, M. Deml and P. Bocek, *J. Chromatogr.* **452** 601–613 (1988).
20. X. Huang, W. Coleman and R. Zare, *J. Chromatogr.* **480** 95–110 (1989).
21. J. H. Knox and K. A. McCormark, *Chromatographia* **38**, 207–214 (1994).
22. J. H. Knox and K. A. McCormack, *Chromatographia* **38**, 215–221 (1994).
23. J. H. Knox and K. A. McCormack, *Chromatographia* **38**, 279–282 (1994).
24. F. Lelièvere, C. Yan, R. Zare, and P. Gareil, *J. Chromatogr. A* **723**, 145–156 (1996).
25. Y. Guo and L. Colón, *Anal. Chem.* **67**, 2511–2515 (1995).
26. J.J. Pesek and M.T. Matyska, *J. Chromatogr. A* **736**, 225–264 (1996).
27. S. Li and D. Lloyd, *Anal. Chem.* **65**, 3684–3690 (1995).
28. N.W. Smith and M.B. Evans, *Chromatographia* **41**, 197–203 (1995).
29. J. Liao, N. Chen, C. Ericson, and S. Hjertén, *Anal. Chem.* **68**, 3468–3472 (1996).
30. H. Yamamoto, J. Baumann, and F. Enri, *J. Chromatogr.* **593**, 313–319 (1992).
31. J. J. Pesek and M.T. Matyska, *J. Chromatogr. A* **736**, 313–320 (1996).
32. C. Yan, R. Dadoo, R. Zare, D. Rakestraw, and D. Anex, *Anal. Chem.* **68**, 2726–2730 (1996).
33. E. Bayer, *Proc. Int. Symp. High Perform. Capillary Electrophoresis (HPCE 97) Relat. Microscale Tech., 9th*, Anaheim, California, 1997, Abstr., p. 62 (1997).
34. S. Ludthe, T. Adam, and K. K. Unger, *Proc. Int. Symp. High Perform. Capillary Electrophoresis (HPCE' 97) Relat. Microscale Tech., 9th*, Anaheim, California, 1997, Abstr., p. 62 (1997).

35. L. A. Colón, Y. Guo, A. M. Fermier, K. J. Reynolds, and S. Rodriguez, *Proc. Int. Symp. Capillary Chromatogr. Electrophoresis (CC&E '97), 19th*, Wintergreen, Virginia, *1997*, Abstr., p. 62 (1997).
36. A. M. Fermier, L. A. Colón, and K. J. Reynolds, *Proc. Int. Symp. High Perform. Capillary Electrophoresis (HPCE '97) Relat. Microscale Tech., 9th*, Anaheim, California, 1997, Abstr., p. 62. (1997).
37. L.A. Colón, A.M. Fermier, G. Young, and K. Reynolds, *Proc. Int. Symp. High Perform. Capillary Electrophoresis (HPCE 97) Relat. Microscale Tech., 9th*, Anaheim, California, 1997, Abstr., p. 80. (1997).
38. J. Stobaugh, P. Angus, C. Demarest, K. Payne, K. Sedo, and L. Kwok, *Proc. Int. Symp. Capillary Chromatogr. Electrophoresis (CC&E '97), 19th*, Wintergreen, Virginia, *1997*, Abstr., p. 88 (1997).
39. C. Fujimoto, M. Sakurai, and Y. Fujise, *Proc. Int. Symp. Capillary Chromatogr. Electrophoresis (CC&E '97), 19th*, Wintergreen, Virginia, *1997*, Abstr., p. 70 (1997).
40. E. Venema, J. Kraak, H. Poppe and R. Tijssen, *Proc. Int. Symp. Capillary Chromatogr. Electrophoresis (CC&E '97), 19th*, Wintergreen, Virginia, *1997*, Abstr., p. **92** (1997).
41. Z. J. Tan, M. Cipoletti, and V. T. Remcho, *Proc. Int. Symp. Capillary Chromatogr. Electrophoresis (CC&E '97), 19th*, Wintergreen, Virginia, *1997*, Abstr., p. 466 (1997).
42. T. Hobo, J-M. Lin, Y. Gao, T. Nakagama, and K. Uchiyama, *Proc. Int. Symp. Capillary Chromatogr. Electrophoresis (CC&E '97), 19th*, Wintergreen, Virginia, *1997*, Abstr., p. 454 (1997).
43. S. Nilsson, L. Schweitz, and L. Andersson, *Proc. Int. Symp. Capillary Chromatogr. Electrophoresis (CC&E '97), 19th*, Wintergreen, Virginia, *1997*, Abstr., p. 82 (1997).
44. C. Horváth, G. Choudhary and X. Huang, *Proc. Int. Symp. Capillary Chromatogr. Electrophoresis (CC&E '97), 19th*, Wintergreen, Virginia, *1997*, Abstr., p. 60 (1997).
45. J. Dorsey, A. Lister, and P. Wright, *Proc. Int. Symp. Capillary Chromatogr. Electrophoresis (CC&E '97), 19th*, Wintergreen, Virginia, *1997*, Abstr., p. 62 (1997).
46. A. Malik and J. Hayes, *Proc. Int. Symp. High Perform. Capillary Electrophoresis (HPCE 97) Relat. Microscale Tech., 9th*, Anaheim, California, 1997, Abstr., p. 80. (1997).
47. A. Palm and M. Novotny, *Proc. Int. Symp. High Perform. Capillary Electrophoresis (HPCE 97) Relat. Microscale Tech., 9th*, Anaheim, California, 1997, Abstr., p. 125 (1997).

PART II

DETECTION SYSTEMS IN HPCE

CHAPTER

9

CAPILLARY ELECTROPHORETIC DETECTORS BASED ON LIGHT

LOUANN CRUZ

Department of Chemistry, Oklahoma State University, Stillwater, Oklahoma 74078

SCOTT A. SHIPPY and JONATHAN V. SWEEDLER

Department of Chemistry, University of Illinois, Urbana, Illinois 61801

9.1.	Introduction	304
9.2.	**Requirements of Capillary Electrophoretic Detectors**	305
	9.2.1. The Detector Cell	305
	9.2.2. Response Time	306
	9.2.3. Qualitative and Structural Information	307
	9.2.4. Figures of Merit	307
9.3.	**Absorbance**	308
	9.3.1. Introductory Remarks	308
	9.3.2. Theory	310
	9.3.3. Instrumentation	311
	9.3.4. Universal Ultraviolet–Visible Absorbance Detection	316
	9.3.5. Qualitative Ultraviolet–Visible Absorbance Detection	317
9.4.	**Fluorescence**	321
	9.4.1. Introductory Remarks	321
	9.4.2. Theory	321
	9.4.3. Intrumentation	322
	9.4.4. Multidimensional Fluorescence	327
	9.4.5. Native Fluorescence	329
	9.4.6. Derivatization	330
9.5.	**Other Capillary Electrophoretic Optical Detectors**	331
	9.5.1. Chemiluminescence Detectors	331
	9.5.2. Refractive Index Detectors	333

High Performance Capillary Electrophoresis, edited by Morteza G. Khaledi. Chemical Analysis Series, Vol. 146.
ISBN 0-471-14851-2 © 1998 John Wiley & Sons, Inc.

	9.5.3.	Thermooptical Detection	336
	9.5.4.	Radionuclide Detection	338
	9.5.5.	Hyphenated Detection	341
9.6.	**Recent Developments**		341
	9.6.1.	Raman Detection	341
	9.6.2.	Nuclear Magnetic Resonance Detection	342
	9.6.3.	Light-Scattering Detection	343
	9.6.4.	Living Cells as Detectors	346
9.7.	**Conclusions**		347
List of Acronyms and Abbreviations			347
References			348

9.1. INTRODUCTION

Detection in capillary electrophoresis (CE) is a challenge due to the high peak efficiencies, small peak volumes, and limited time available to observe analytes. The wide scale use of CE has required innovative methods to detect low concentrations of nanoliter (or smaller) volume analytes. Many of the detection schemes developed for conventional liquid chromatography (LC) have been extensively modified to adapt them to the more stringent requirements of CE. As one example, commercial high-performance liquid chromatography (HPLC) fluorescence detectors use lamp sources; in CE, almost all fluorescence detection is based on laser excitation due to the difficulty in focusing incoherent sources into the capillary. To date, a large number of detector schemes have been demonstrated in CE.

This chapter describes those methods that involve probing the sample with electromagnetic radiation (light), as opposed to electrochemistry and mass spectrometry (covered in Chapters 10 and 12, respectively). The greatest emphasis is on ultraviolet–visible (UV–Vis) absorbance and fluorescence detection, as these are by far the most common optical detection modes used today. In addition, newer and less widely used methods are described such as chemiluminescence and refractive index. We have interpreted optical detection to include methods such as nuclear magnetic resonance (NMR, which probes a sample using radio frequency photons) and many forms of radioactivity detection (where β particles are converted to photons).

In view of the large number of detection schemes available for CE, careful choice of the detector is one of the first steps to make for a particular analysis. Many different types of chemical compounds, both ionic and nonionic, have now successfully been analyzed by CE. These range in size from small molecules such as inorganic ions, amino acids, and peptides to large biomolecules such as proteins and DNA, and to even larger moieties such as particles. Many different and complex sample matrices such as river water, blood, urine, tissue, and foodstuffs are analyzable with CE; potential interferences from matrix components in the detection process must be accounted for in the selection of the detection mode. Thus, the most important considerations for detection in CE are the properties of the analyte, the sample matrix composition, and the required detection limits. The separation mode as well as the type and size of the capillary are additional points. Other considerations are whether qualitative and structural information are required in addition to quantitative information and whether all analytes or just a select type need to be measured. If an analyte is not easily detected by any of the available optical detection schemes, it may be possible to convert it to a molecule that is detectable through precolumn, on-column, and postcolumn derivatization techniques (1). Further considerations are whether the detection scheme is available commercially or whether it will have to be constructed in-house, as well as the net cost and ease of operation. Obviously, a serious consideration of all of the above factors restricts the type of detector amenable to each application.

The purpose of this chapter is to provide the reader with a detailed description of the more commonly used optical detection schemes in CE, as well as current performance specifications for those that are actively being developed. Each discussion includes basic theory, instrumentation design, and a listing of advantages and disadvantages. Prior to a discussion of the individual schemes, a review of the basic requirements of CE optical detectors and a summary of the optical detectors that have been used sucessfully for particular compounds are presented. As other reviews of optical detection are available (2–6), this chapter emphasizes more recent developments.

9.2. REQUIREMENTS OF CAPILLARY ELECTROPHORETIC DETECTORS

9.2.1. The Detector Cell

The capillary columns used for CE separations are typically fused silica with internal diameters of 10–100 µm, outer diameter of 100–375 µm, and polyimide coatings anywhere from 10–30 µm thick. Total capillary volumes range from tens of nanoliters to a few microliters. There are different locations to detect analytes: on-column, end-column, postcolumn, and even entire column. In on-column detection, such as that used in commercial UV–Vis absorbance detectors, a small section of the outer polyimide coating is removed near the outlet end of the capillary to form a cylindrical window that serves as the detector cell. The window is positioned in the path of the light beam in the detector assembly, and analytes are detected as they flow past the window. The advantages of on-column methods are simplicity and alleviation of any postcolumn band spreading. Because the optical pathlength is restricted to the internal diameter of the capillary, optical detection schemes that are pathlength dependent will exhibit relatively poor detection limits. In end-column detection, the detector is placed just after the capillary outlet. An important example is the sheath-flow detector cell used in laser-induced fluorescence (LIF) detection described later (7). Because of the detector design and proximity of the detection region to the capillary outlet, minimal band spreading is observed. In postcolumn detection, detection of analytes occurs at some position remote to the capillary outlet. One example is the postcolumn radionuclide detection scheme developed by Tracht et al. (8), where the capillary effluent is continuously collected onto a moving membrane that has been coated with a solid scintillator. Such approaches have the advantage of providing a means to independently optimize the separation conditions and detection conditions (in this case, by increasing the exposure time of analytes to enhance sensitivity without sacrificing electrophoretic resolution). Finally, in entire-column detection, such as that used in UV–Vis or LIF capillary scanning systems (9), the entire length of the separation capillary is monitored. This type of detection is useful when it is desirable to view the separation process while it is developing.

The volume of the detector cell should be approximately one-tenth of the peak volume to ensure that losses in separation efficiency do not occur. The analyte peak volume is no less than the injected volume (neglecting stacking or other concentration methods); typical sample injection volumes in CE range from 0.1 to 10 nL, with the detector cells in the picoliter range. In contrast to LC with microliter volume analyte bands, the nanoliter analyte bands in CE are particularly susceptible to excessive spreading, making the use of column connections and postcolumn detection cells difficult. Thus, most systems use either on-column or end-column detection so that the detector does not degrade the separation efficiency.

9.2.2. Response Time

For most optical detection methods the longer one observes the analytes, the better precision one obtains. Typical flow rates in CE are on the order of 10–100 nL/min, resulting in peaks that migrate through the detector in times as short as 1–2 s. To describe such a peak profile adequately, 10 data points over the width of the entire peak are desired. This translates into a time measurement for each data point of approximately 100 ms. For this reason, detector response times in CE are faster than for most other separations. If the response of the detector is too slow or if too much time is needed to obtain a desired level of sensitivity, then separation efficiency and resolution are lost. For those optical detection schemes where more than 0.2 s are needed to obtain a desired sensitivity, several options are available. For example, fractions can be collected and then analyzed later (8), or the voltage can be lowered so that analyte zones remain in the detector window for a longer period of time (10).

9.2.3. Qualitative and Structural Information

If a mathematical relationship can be derived that establishes how the interaction of light with an analyte of interest is related to its concentration, then quantitative information can be obtained. In many situations, it is also desirable to identify the peaks of an electropherogram by using detection schemes that provide additional information. This information can confirm the presence or absence of specific functional groups, elucidate stereochemistry, provide structure, or allow determination of size and mass. The optical detection schemes for CE that are capable of providing limited qualitative information are multiwavelength UV–Vis absorbance, multiwavelength fluorescence, optical activity, and laser light scattering. NMR and Raman spectroscopy (and the nonoptical mass spectrometric methods) provide much more detailed information for unknown identification.

9.2.4. Figures of Merit

In order to evaluate the performance of various optical detectors, several standards of comparison are required. Four of the most important figures of merit are dynamic range, detection limit, sensitivity, and selectivity. Table 9.1 lists the major optical detection schemes and compares their detection limits. The comparison is made in several ways: as the typical range of minimum detectable quantities, as minimum detectable concentrations, and as the best reported detection limit for a specific analyte. From Table 9.1, it is clear that LIF is the most sensitive optical detection scheme available for CE, currently approaching single-molecule detection.

As mentioned previously, many kinds of molecules are now analyzable by CE, ranging from small ions to large biopolymers. Table 9.2 groups these molecules into broad categories and lists the types of optical detection schemes that have been used in their analysis. This table is meant to provide an overview of the diferent detection schemes that are discussed in the remainder of this chapter and is not comprehensive. Even though many approaches are listed, UV–Vis absorbance and fluorescence are the most commonly employed methods and are available from a number of commercial sources. Thus, an extensive coverage of these two methods is presented.

Table 9.1. Summary of Detection Limit Performance of CE Optical Detectors

Detector	MDQ (mol)	MDC (M)	Best Reported Detection Limit/ Molecule	Ref.
Direct absorbance	10^{-13}–10^{-16}	10^{-5}–10^{-7}	2×10^{-8} M Malachite Green	19
Indirect absorbance	10^{-12}–10^{-15}	10^{-4}–10^{-6}	1 fmol (1×10^{-7} M) pyruvate	33
Laser-induced fluorescence (LIF)	10^{-18}–10^{-21}	10^{-9}–10^{-12}	6 molecules sulforhodamine 101	41
Indirect fluorescence	10^{-14}–10^{-16}	10^{-6}–10^{-8}	70 amol 5'-monophosphate nucleotides	10a
Chemiluminescence (CL)	10^{-14}–10^{-16}	10^{-7}–10^{-9}	100 amol (5×10^{-9} M) ATP	96
Refractive index (RI)	10^{-13}–10^{-15}	10^{-5}–10^{-7}	7 fmol (2×10^{-7} M) ovalbumin	115
Thermooptical absorbance	10^{-15}–10^{-18}	10^{-5}–10^{-7}	8 amol (1×10^{-5} M) DABSYL–amino acids	10b
Radioactivity	10^{-14}–10^{-18}	10^{-6}–10^{-10}	88 zmol ^{32}P	8
Raman	10^{-12}–10^{-15}	10^{-3}–10^{-5}	500 amol (1×10^{-7} M) Methyl Orange	133

9.3. ABSORBANCE

9.3.1. Introductory Remarks

There are many reasons for the popularity of UV–Vis detection. Aside from its simplicity, ease of use, and relatively low cost, UV–Vis absorbance is an extremely versatile detection technique. A large fraction of the analytes separated by CE can be detected by using UV–Vis absorbance in its various forms, although some of the molecules require derivatization in order to be detected. In addition, UV–Vis absorbance is compatible with all of the modes of CE, i.e., capillary zone electrophoresis (CZE), capillary isotachophoresis (CITP), capillary isoelectric focusing (CIEF), capillary gel electrophoresis (CGE), capillary electrochromatography (CEC) and micellar electrokinetic chromatography (MEKC) though some restrictions in the running buffers exist due to the optical properties of the buffers themselves. On-column detection can be carried out with capillaries made out of fused silica as well as other materials (11), and with capillaries that are internally coated (12).

Table 9.2. Uses of Optical Detectors in CE

Detector	Dyes	Small Organic and Inorganic Ions	Amino Acids	Peptides	Proteins	Sugars, Oligosaccharides, and Carbohydrates	Fatty Acids and Lipids	Nucleotides, Oligonucleotides, and Nucleic Acids	Vitamins	Pharmaceutical and Forensic Drugs	Pesticides and Pollutants	Surfactants and Polymers	Particles	Single Cells, Single-Cell Contents, and Cell Homogenates
Direct absorbance Low-wavelength UV absorbance	×	×	×	×	×	×	×	×	×	×	×	×	×	×
Indirect absorbance	—	×	×	×	×	×	×	×	×	×	×	×	—	×
Incoherent-fluorescence	×	—	×	—	—	×	×	—	—	—	—	×	—	—
Indirect fluorescence	—	×	×	×	×	×	—	×	×	—	×	—	—	—
Native fluorescence	×	—	×	×	—	—	—	×	—	—	×	—	—	×
Laser-induced fluorescence (LIF)	×	×	×	×	×	×	—	×	×	×	×	—	—	×
Chemiluminescence (CL)	×	—	×	×	×	—	—	×	—	×	—	—	—	—
Radioactivity	—	×	×	×	×	—	—	×	—	×	—	—	—	×
Refractive index (RI)	×	×	×	×	×	×	×	—	—	—	—	—	—	—
Thermooptical absorbance	—	—	×	×	×	—	—	—	×	—	—	—	—	—
Raman	×	×	×	—	—	—	—	×	×	—	—	—	—	—
Optical Activity	×	×	—	—	—	—	—	×	×	—	—	—	—	—
Nuclear magnetic resonance (NMR)	—	—	×	—	—	—	—	—	—	—	—	—	—	—
Light-scattering	—	—	×	×	×	—	—	—	—	—	—	—	×	—

Absorbance detectors may function as universal detectors by using low-UV wavelengths or through operation in an indirect mode. Quantitative data is easily extractable from UV-Vis absorbance electropherograms. Qualitative information to aid in peak purity assessment and peak identification also can be obtained from multiwavelength spectra acquired by fast-scanning spectrometers and photodiode array detectors. Lastly, UV–Vis absorbance detection is nondestructive of the analyte, so further analysis may be achieved by connecting other detectors downstream from the UV–Vis detector. UV–Vis absorbance can be used to measure physicochemical properties of analytes, such as electrophoretic mobilities (13), binding constants (14) and viscosity (15). The largest drawback to UV–Vis absorbance detection is its relatively poor concentration sensitivity relative to that attainable by conventional UV–Vis in HPLC. Recent advances in instrument design by companies and research laboratories have led to improved concentration detection limits.

9.3.2. Theory

Any molecule that possesses a chromophore (a functional group that absorbs radiation in the UV or visible region of the electromagnetic spectrum) can be detected by UV–Vis absorbance. Detection of analytes that do not contain a chromophore or that absorb only weakly can be accomplished through derivatization with a molecule that has a strongly absorbing transition. In the basic CE UV–Vis absorbance instrument, there is a decrease in the intensity of light transmitted through the detection cell to the detector as analytes pass through the detector region. Beer's law relates this absorbance to the concentration of an analyte. The magnitude of the absorbance signal is analyte dependent; analytes with molar absorptivities of 10^4 to 10^5 are strong absorbers, and those with molar absorptivities of $\leq 10^3$ are weak absorbers. Obviously, detection limits are better for analytes that have larger molar absorptivity values. Additionally, the pH, running buffer composition, or degree of ionization is also important. This can result in changes in the molar absorptivity value and a shift in the wavelength of maximum absorption.

Because the absorbance signal is proportional to the optical pathlength, the sensitivity of the UV–Vis absorbance detector decreases as the inner diameter of the capillary is reduced. Also, due to the circular cross section of the capillary, not all of the light rays are able to pass through the capillary center and the actual optical pathlength is less than the inner diameter of the capillary by a factor of $\pi d/4$ (16), where d is the inner diameter of the capillary. In CE, one important limitation in the linear dynamic range arises from the presence of stray light (i.e., light from the source that reaches the detector without interacting with the analyte). Stray light becomes more problematic as narrower-i.d. capillaries are used because more light misses the lumen of the capillary; it is reduced through the use of optimized optics.

9.3.3. Instrumentation

The basic CE absorbance detector is composed of a light source, focusing optic that direct the light onto the detection cell, a wavelength selector, and a photodetector. The photodetector can be a photomultiplier tube, a photodiode, or a photodiode array. The output of the photodetector is converted to an electrical signal, amplified, electronically filtered to reduce noise, and recorded or digitized.

Because the detection limit in absorbance is dictated by the ability to see a small decrease in the light intensity of a high-intensity background, source stability becomes one of the most important criteria of an illumination system. The effect of source intensity instabilities is reduced in many commercial systems by using a double-beam system where the second optical path serves as a reference to correct for source fluctuations. The best systems approach shot-noise-limited performance. As the arrival rate of photons follows Poisson statistics, the minimum statistical fluctuation in the number of photons arriving at the photodetector is shot noise, equal to the square root of the number of photons incident on the detector. The shot-noise-limit for most UV–Vis absorbance detectors in CE is about 10^{-6} AU (absorbance units). The following subsections describe the more important components of CE absorbance detectors.

9.3.3.1. The Light Source

Several types of light sources are used in UV–Vis absorbance detectors (17). The simplest UV light sources are atomic lamps that produce strong emission lines at well-defined wavelengths. Examples are the low-presssure Hg lamp, the Cd lamp, and the Zn lamp that have sharp lines at 254, 229, and 214 nm, respectively. A medium-pressure Hg lamp is also available that has useful emission lines at 254, 280, 313, 365, 405, 436, and 546 nm. The low-pressure deuterium arc lamp is a continuous source of UV light and has fairly stable energy output in the range 160–400 nm. The tungsten–halogen lamp provides a continuum of energy from about 350 to 2500 nm and is a useful source of visible and near-infrared (near-IR) light. These light sources are incoherent, and so it is difficult to focus a significant fraction of the light into the smallest capillaries. The low luminosity and the inefficient coupling of the light source into the capillary (low intensity and high stray light) result in a decrease in detector performance (16).

Spatially coherent lasers are easily focused into the smallest i.d. capillaries. The higher intensity of a focused laser implies better shot-noise performance. However, absorbance systems based on lasers rarely obtain shot-noise-limited performance due to intensity instability. Even with a power stabilizer, stability is rarely better than $\pm 0.01\%$. Xue and Yeung (18) report the development of an on-column double-beam laser absorbance detector for CE that utilizes a He–Ne laser operating at 632.8 nm that gives a 25-fold improvement in detection limit over that achievable by a commercial UV–Vis detector with malachite green as the test analyte. More recently, they report the use of an argon ion laser operating at 305 nm for UV detection of hemoglobin A_0 in CE (19). However, only a fourfold improvement in the detection limit over that of a commercial detector is reported. Thus, given the high cost and complexity associated with UV lasers, the limited-excitation wavelengths, and the small improvement in concentration performance, we feel it is unlikely that laser-based absorbance detectors will become common for commercial CE instruments.

Laser diodes (LDs) and light-emitting diodes (LEDs) are solid-state light sources that are miniature, have a stable output and long lifetime, and are relatively inexpensive. Unfortunately, current LDs and LEDs operate in the Vis–near-IR region (400–1000 nm). The design of a LED-based absorbance detector for CE has been described by Bruno et al. (20). If increased power output and shorter emission wavelengths can be developed for LDs and LEDs, these devices will prove to be useful light sources for CE absorbance detectors.

9.3.3.2. The Wavelength Selector

Wavelength selection is achieved in commercial UV–Vis absorbance detectors in CE in numerous ways. Selection of a particular wavelength of light prior to its passage through the capillary is referred to as a "forward optics" design. Most single-wavelength detectors employ this design. For instruments that use atomic lamps as the source, only an appropriate filter is needed for wavelength selection. When a broadband source is used such as a deuterium lamp, the line of interest can be isolated either by a filter or by a grating monochromator. Through the rotation of the grating, the desired wavelength passes through a slit while other wavelengths remain blocked. The fast-scanning UV–Vis detector (21) enables the grating monochromator to be moved very rapidly so that a series of measurements at different wavelengths can be taken on a single analyte zone as it traverses the detector region.

Another multiwavelength detector, the diode array detector (DAD) (21), employs the "reverse optics" design in which the broadband illumination (white light) first passes through the capillary. The light is then dispersed into its component wavelengths by a diffraction grating and focused onto a solid-state photodiode array detector. The number of individual diode elements in the array often determines the wavelength resolution of the detector. Continuous monitoring of absorbance measurement at all wavelengths generates a large amount of data; to reduce this, multiwavelength detectors can also be operated as single-wavelength detectors.

9.3.3.3. *Focusing Optics and the Detector Viewing Region*

In CE, the light source intensity and the effective coupling of the light source into the capillary are important factors that determine the sensitivity and linearity of the absorbance detector. To efficiently image the light source into the capillary, CE instruments use a series of conventional lenses, ball lenses, or fiber-optic cables (22). Three dimensions specify the illuminated volume of the capillary: the inner diameter and geometry of the aqueous core; the length; and the width of the illuminated region. For optimized performance, each of these dimensions should be maximized. However, the separation places constraints on the maximum values of these dimensions. For instance, increasing the inner diameter of the capillary may improve concentration sensitivity (longer optical pathlength) but increases Joule heating and so often degrades separation efficiency. Similarly, increasing the length of the detector viewing region may increase light throughput but can reduce efficiency.

For instruments that use sapphire or quartz ball lenses for focusing, performance depends on the distance between the lens and the capillary and on the ratio of the inner and outer diameters of the capillary. With fiber optics, the fiber dimensions must also be optimized to the inner and outer diameters of the capillary. In some CE instruments, lenses or fiber optics are used both to focus the light source onto the capillary and to collect the light and focus it onto the photodetector. From the standpoint of the user, easy placement and alignment of the capillary with the detector optics are large advantages.

The easiest method of improving detection limits is to extend the pathlength. Several approaches to extending pathlengths in CE absorbance detection have been demonstrated, including the multireflection absorbance cell (23), axial capillary illumination (24), the rectangular capillary (25), the Z-cell (26), and the "bubble" cell (27). The last three approaches, depicted in Figure 9.1, extend the pathlength by using detection cells differing in their geometry and hence are discussed in greater detail below.

In the rectangular capillary cell, a rectangular (e.g., 50 μm × 1000 μm) cross-section capillary is used (25). The width of the capillary is much greater than the height (20:1) in order to maximize pathlength (better absorbance performance) and heat dissipation (higher surface-to-volume ratio). Such capillaries are not readily available because of manufacturing difficulties, especially from

CE DETECTORS BASED ON LIGHT

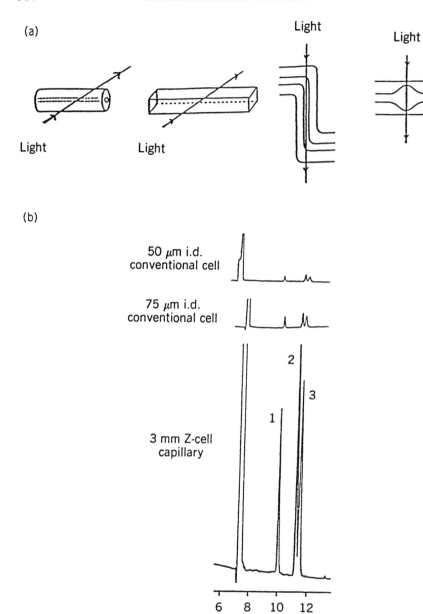

Figure 9.1

fused silica. Although much of the original isotachophoresis work used rectangular geometry capillaries (28), this type of pathlength extension is not in widespread use.

The Z-cell is a method of bending a section of capillary so that illumination occurs down the bore of a section of capillary (see Figure 9.1), thereby offering greatly increased opical pathlength. For example, a Z-cell length of 600 μm length in a 75 μm i.d. capillary offers an eightfold increase in pathlength without a loss of separation efficiency. However, the smallest Z-length that is commercially available is 3000 μm; in many cases, separation efficiency is reduced along with the large gain in sensitivity. Because the Z-cell has a reduced optical throughput as compared to a conventional capillary, an increase in baseline noise partly offsets the signal gain from the pathlength extension. For example, a 3000 μm long Z-cell with a 75 μm i.d. yields a 12-fold increase in sensitivity rather than the theoretical value of 40. Overall, the Z-cell provides more than an order of magnitude improvement in S/N ratio [MDC (minimum detectable concentration) to 10^{-8} M] over that of a conventional capillary, with an increase in linearity from 3.5 to more than 4 orders of magnitude and only a 14% loss in resolution (26).

The "bubble" cell (27), so called because of its shape, is an enlarged section of capillary with a diameter three to five times larger than that of the rest of the capillary and leads to a three to fivefold increase in the signal-to-noise (S/N) ratio. Because the diameter is essentially independent of the capillary i.d., sensitivity and linearity are not compromised, even for capillaries of 25 μm i.d. or less. The volume of the bubble is considerably larger than that found in the same section of nonenlarged capillary. Separation efficiency remains unaffected due to compression of the analyte band in the axial direction arising from the radial expansion of the analyte zone. For example, in a bubble with a threefold enlarged diameter, the zone is reported to be nine times narrower in the axial direction and three times longer in the radial dimension, although the zone has the same concentration it had prior to entering the bubble. Whereas the detector viewing region of a conventional 50 μm i.d. capillary is rectangular (40 × 600 μm), the length of the detector viewing region in a bubble cell is reduced in order to preserve electrophoretic resolution because of the zone compression. For this reason, the detector viewing region of the bubble cell

---

Figure 9.1. Different detector cells used to increase the optical pathlength. (a) From left to right: conventional capillary, rectangular capillary, Z-cell capillary, and "bubble" cell capillary. (Adapted from ref. 27.) (b) Sensitivity enhancement of Z-cell capillary in the separation of pyrrolidinones. Conditions: buffer, 20 mM sodium phosphate; 90 mM hexanesulfonic acid; 10 mM tetramethylammonium phosphate; pH 6.5. Sample: 3.3 μg/mL each. Injection: 1s vaccum. Detection: 242 nm. Separation: 15 kV, 45°C. [Reprinted with permission from S. E. Moring, C. Pairaud, M. Albin, S. Locke, P. Thibault, and G. W. Tindall, *Am. Lab.* **25**, 32–39 (1993).]

typically has square dimensions (145 × 145 µm). The maximum diameter of the bubble is limited to five to six times the i.d. of the capillary. Bubble cells larger than this require increasingly smaller detector viewing regions to maintain resolution. The decreased light throughput leads to a higher baseline noise that negates the gain from the increased optical pathlength.

9.3.4. Universal Ultraviolet–Visible Absorbance Detection

9.3.4.1. Low-Wavelength Ultraviolet

Absorbance is not normally considered a universal detection scheme. However, an absorbance detector begins to take on the attributes of a universal detector when operated at far-UV wavelength (180–200 nm), as most analytes have some absorbance at these wavelengths (as do the buffers and many solvents). Thus, at far-UV wavelengths there is a large increase in the number of detectable peaks and so detection of molecules without obvious chromophores becomes possible.

Universal low-wavelength UV detection is an example of where detection in CE outperforms that of HPLC. Low UV wavelengths are difficult to use in LC for both isocratic and gradient elution separations due to impurities in the mobile phases, flow-induced noise, and refractive index effects (29). In addition, absorption of the < 200 nm radiation by the fused-silica walls is tolerable in CE as compared to LC, due to the thinner wall thickness.

One complication of universal detection is that there are significant restrictions in the selection of running buffers and additives as many absorb in this region. Although many difficult-to-detect compounds can be sensed, the sensitivity can be low because of the low molar absorptivity values of such compounds. However, when an analyte is relatively concentrated, this type of detection scheme is an attractive and inexpensive approach. For example, Kakehi et al. (30) used low-wavelength UV absorbance at 185 nm to detect several O-linked sialooligosaccharide alditols at the 10^{-4} M level. These alditols cannot be sensitively detected by direct UV–Vis absorbance or fluorescence. Conversion to absorbing or fluorescing derivatives is also hampered because of lack of an aldehyde group.

9.3.4.2 Indirect Detection

Indirect detection involves placing an easily detected molecule into the running buffer and then looking at the effect of the analyte on the detectable molecule. Indirect detection schemes approach universal detection and are used in the analysis of compounds that do not possess the necessary characteristic for direct detection. Indirect detection does not require pre- or postcolumn derivatization to convert the analyte of interest into a species that can give an acceptable detector response. Although any detection method can be used in an indirect mode, UV–Vis absorbance is one of the most commonly employed. One reason for this is that the same instrumentation used for direct absorbance can also be used for indirect absorbance. For both methods a very stable background is required, as the detection limit involves measuring two relatively large signals and looking for a small difference between them. The principles of indirect detection are thoroughly described by Yeung and Kuhr (31) and are also discussed in detail in Chapter 11 of this book, and so this discussion is brief.

Basically, in indirect absorbance a UV–Vis absorbing chromophore with a high molar absorptivity is added to the running buffer to provide a constant background signal. The analyte displaces the chromophore, and so as the analyte passes through the detector a decrease in the background signal occurs. Therefore, the peak that is generated in indirect detection arises not from the analyte but from a displacement of the detectable chromophore. A well-characterized displacement mechanism, a large transfer ratio, and a stable background are all important in obtaining the highest performance. It is important for the detectable chromophore to have the same electrophoretic mobility as the analytes of interest, for otherwise nonsymmerical peaks are obtained (32).

Requirements of indirect absorbance detection are that the pH and composition of the buffer have to be carefully chosen in order to achieve optimum resolution and sensitivity. Because lower buffer concentrations (2–10 mM) are typically used to improve performance, buffering capacity, and mass sample overloading become more of a concern, especially when one is analyzing a series of widely varying samples. When smaller-diameter capillaries are used, laser sources can be employed to increase the mass sensitivity in indirect absorbance detection just as is done in direct detection. Xue and Yeung (33) report the use of a laser-based double-beam absorbance detector in the indirect analysis of pyruvate that gives a 15-fold improvement in detection limit over that of a commercial system. Indirect absorbance detection can sometimes be more sensitive than low-wavelegth UV detection.

9.3.5. Qualitative Ultraviolet–Visible Absorbance Detection

Because of overlapping electronic transitions from multiple functional groups, the superposition of numerous vibrational transitions upon each electronic transition, and the distortion by various solvent interactions, absorbance spectra usually lack fine detail and are charcterized by broad bands. In fact, UV–Vis spectra of chemically related species can be very similar. Therefore, UV–Vis absorbance detection appears to have limited utility in qualitative analysis because CE separations are often used to separate classes of structurally similar compounds. Multiwavelength absorbance detectors (both fast-scanning UV-Vis and photodiode array) can acquire high-resolution UV-Vis spectra of migrating analytes as they traverse the detector region. When use in conjunction with sophisticated mathematical algorithms, these detectors provide some qualitative information about an analyte. The ways that multiwavelength UV-Vis spectra are used in qualitative analysis are discussed below (27).

9.3.5.1. *Wavelength Selection*

Selecting a particular wavelength can be problematic when one is working with samples that contain either unknown compounds or a wide variety of different types of molecules. Because the absorbance of an analyte at several different wavelengths is measured simultaneously during a single electrophoretic run by multiwavelength detectors, the selection of the optimum wavelength is simplified. In fact, different single wavelength electropherograms can be constructed at specific wavelengths after the run. The optimum wavelength can be the one that gives the maximum S/N ratio for the analyte or the one where the analyte absorbs but where interfering compounds do not absorb.

9.3.5.2. *Peak Tracking*

During method development, different analytical conditions are applied to a sample to determine the best set of separation parameters; peaks can change both migration time and order. In single-wavelength detection, tracking the peak under the changing conditions requires sample spiking or rerunning of standards. With multiwavelength detection, the spectral features of the peaks allow compounds to be tracked as the peaks change position. One of the electropherograms obtained during a method development run can be saved and stored in the computer as a reference spectrum. A second method development run can then be made, and the multiwavelength spectra of a particular peak from the second run is compared to each peak in the reference spectra until a match is found. Thus, peak tracking between different runs is accomplished by spectral matching, which is a numerical interpretation of the similarity of spectra. Peak tracking in this way does not require any prior knowledge about the analyte or the use of pure standards. Also, spectral fidelity appears to be maintained even as capillary i.d. or the separation efficiency are changed. A complication is that changing separation media can have a profound effect on spectral bandshape and thus the ability to track a peak.

9.3.5.3. Peak Purity

One of the more difficult tasks in a separation is confirming that each individual peak corresponds to a single component. In single-wavelength detection, this is traditionally done by careful inspection of peak shape. However, skewed peaks are not necessarily impure; non-Gaussian peaks result from mass overloading or electrodispersion. Multiwavelength absorbance detection can aid in peak purity assessments. For example, UV–Vis spectra obtained at the upslope, apex, and downslope of a single peak can be normalized and overlaid. If any of the spectra differ, heterogeneity is indicated. Spectral analysis by itself cannot establish peak purity; it can only increase confidence that a peak is pure. This is because the UV–Vis spectra of two comigrating species can be too similar to differentiate them or an impurity may not be in high enough concentration to be detected. As the linear dynamic range of most DADs is 4 orders of magnitude, they comply with the current FDA requirement of being able to detect impurities down to the 0.1% level. Giuffre and Sievert (34) describe the use of software tools to establish the validity of the qualitative and quantitative results obtained from peak purity data.

9.3.5.4. Peak Identification

With single-wavelength absorbance detection, separated components can only be identified by migration time, problematic in complicated sample matrices. The presence of interferents in the matrix can cause migration time shifts, loss of resolution, or comigration of peaks relative to runs made with pure standards. Although multiwavelength absorbance spectra do not normally contain sufficient information to enable a priori identification of compounds, spectra of unknown peaks can be matched to those of reference compounds in a standard library. There are many factors that can affect the ability to match spectra for identification purposes, including separation conditions, acquisition parameters, and even use of different instruments. Normally, peak identification is performed on peaks that appear to be homogeneous after peak purity analysis. However, Wheat et al. (35) demonstrated that multiwavelength absorbance detection in conjunction with chemometric tools can aid in the identification and quantitation of non-homogeneous CE peaks. They used principle component analysis to deconvolute heterogeneous peaks and then reconstructed pure spectra using iterative target transform factor analysis.

As shown in Figure 9.2, Grimm et al. (36) have provided an example of multiwavelength absorbance detection for compound identification in CE by specifically identifying peptides containing either tryptophan or tyrosine out of a complex pool of peptides generated by enzymatic digests. While the

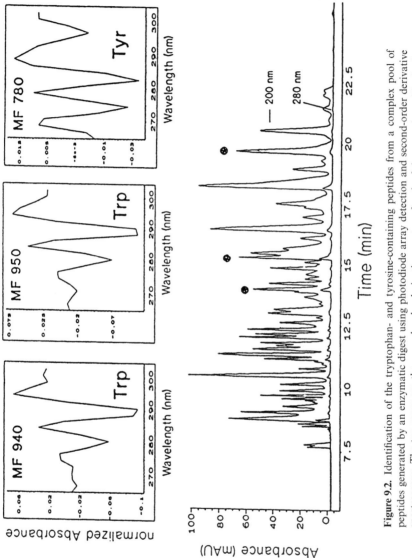

Figure 9.2. Identification of the tryptophan- and tyrosine-containing peptides from a complex pool of peptides generated by an enzymatic digest using photodiode array detection and second-order derivative spectroscopy. The top scans are the second-order derivative spectra of each of the starred peptides in the lower scan. [Reprinted from R. Grimm, A. Graf, and D. N. Heiger, *J. Chromatogr.* **679**, 173–180 (1994), with kind permission of Elsevier Science-NL, Amsterdam, The Netherlands.]

zero-order UV spectra of tryptophan and tyrosine overlap with an absorption maximum of about 278 nm for both amino acids, the second-order derivative spectra show significant differences. In the second-order derivative spectra, tryptophan shows a main minimum of 290 ± 2 nm and a first side minimum at 280 ± 2 nm, whereas tyrosine shows a main minimum at 282 ± 2 nm and a first side minimum at 274 ± 2 nm. Thus, peptides that contain tryptophan can be differentiated from peptides that contain tyrosine using high-resolution diode array detection in the region 250–320 nm in conjunction with second-order derivative spectroscopy. Peptides that contain neither tryptophan nor tyrosine exhibit no clear minimum. Peptides that contain both tryptophan and tyrosine have spectra like that of tryptophan; tyrosine is only detected in the presence of tryptophan if the ratio Tyr/Trp is 3:1 or greater. Wavelength-resolved absorbance provides limited structural information and moderate detection limits; it is commercially available from several vendors.

9.4. FLUORESCENCE

9.4.1. Introductory Remarks

The most sensitive and second most common method of detection in CE is fluorescence. There have been many papers reporting a wide variety of designs for detection of analytes ranging from biopolymers such as proteins and DNA to inorganic ion complexes. The popularity of this method is due to the extremely high sensitivity obtained by using native fluorescence or after labeling with a fluorescent moiety. Fluorescence detection has the lowest reported detection limits to date, with single-molecule detection achieved in the Vis (37) and near-IR (38,39) spectral regions. In CE, the best concentration detection limits are in the 100 fM range (40), with mass detection limits of less than 10 molecules (41). Furthermore, fluoresence detection provides wide linear dynamic ranges. However, the detection limits for the same fluorescent molecule may vary by more than 4 orders of magnitude depending on the particular configuration used, and so instrumentation plays an important role in method selection. Just as for wavelength-resolved absorbance, wavelength-resolved fluorescence detection provides information regarding analyte identity and separation conditions. As many molecules are not fluorescent, a large body of literature describes methods of derivatizing or converting nonfluorescent analytes into detectable molecules.

9.4.2. Theory

In fluorescence, the analyte molecule absorbs a photon, and then some (a given fraction) of the electronically excited molecules emit a fluorescent photon upon returning to the ground state. To describe it more empirically, a fraction of the incident photon of a particular wavelength is absorbed by a molecule to become excited. The detected fluorescence signal is the product of the fraction of excited molecules emitting photons on the return to the ground state by fluorescence, the fluorescent quantum efficiency, and a function describing the instrumental efficiency (light collection and detection efficiency).

Thus, the fluorescence signal is linearly dependent on the concentration for analytes below $\sim 10^{-4}$ M. The most important molecular properties that determine the utility of a fluorophore are the molecule's absorptivity, fluorescence quantum yield, and photostability. Fluorescence detection is capable of low detection limits for two main reasons: the fluorescence emission is at a different wavelength than excitation yielding a low background, and a single fluorescent analyte can fluoresce repeatedly (up to 10^5 cycles in aqueous systems). Although the fluorescence signal is pathlength dependent, the S/N ratio is not strictly pathlength dependent. Background fluorescence and solvent Raman scattering are the major contributors to the background signal that increase the detection limit. As a smaller-diameter capillary is used, the fluorescence signal from a particular concentration of analyte decreases but so does the spectral background. Thus, the best reported fluorescence detection limits are in the 10–100 fM range for a wide range of pathlengths.

9.4.3. Instrumentation

There have been a variety of optical configurations for fluorescence detection with vastly differing performance specifications. The basic fluorescence detection system in CE consists of several component groups including an excitation source, a detection cell, collection optics, and a detector. Chen et al. (42) and Dovichi and Wu (43) have described the design constraints in developing an LIF system, and Yeung and coworkers have outlined the principles of an easily constructed LIF system that achieves 3 pM limits of detection for fluorescein (44). The major components of a fluorescent detector and the various innovations in different instruments are now discussed in turn.

9.4.3.1. The Detector Cell

There have been three detector cell designs used with fluorescence: on-, post-, and end-column. By far, on-column detection is the most widely employed. A small window is made on the capillary by burning off a section of the polyimide coating. The excitation light is focused and the fluorescence emission collected through this window. This offers high performance and simplicity. However, a broad spectral background is associated with the capillary walls and there is significant scattering of the excitation light because of the cylindrical capillary geometry, both of which degrade S/N. The scattering can be reduced by focusing a laser through the capillary center at Brewster's angle. Alternately, the capillary can be immersed in index-matching fluid to reduce scattering from the outside walls. Neither of these methods reduces the capillary spectral background.

One example of a postcolumn design (45) involves etching the outlet end of the capillary and inserting it into a larger, secondary capillary. Fluorescent reagents are pumped into a "tee" connector for reaction and makeup flow (Figure 9.3a). Subsequently, detection is performed on the reaction column. Although detection does not take place on the separation capillary, postcolumn detection takes place on a capillary, so there are similar scattering and background effects. Also, the postcolumn connectors introduce band broadening (peak efficiencies of 50,000 theoretical plates.) The reason for such postcolumn methods and advantages of them are discussed later in Section 9.4.6.

The background sources can be greatly reduced by eliminating the capillary walls. Cheng and Dovichi adapted the sheath flow cell used in flow cytometry for CE (7). In this design, the outlet end of the capillary is placed in a specially designed quartz cuvette (Figure 9.3b). A sheath fluid flows past the outlet end of the capillary under conditions of laminar flow so that the eluent remains constrained in a core region where detection occurs. The excitation light is focused into the core stream in the sheath just downstream from the outlet. The fluorescence background from the capillary is eliminated, and the background and scattering from the cuvette is spatially removed from the fluorescence of the core stream and can be easily spatially filtered before detection. Thus, this design allows for extremely low mass detection limits (41).

9.4.3.2. Excitation Sources

The first step of any fluorescence experiment is to excite the analyte molecules. Important characteristics of fluorescence sources include emission wavelength, intensity, and stability. Sources can be conveniently divided into incoherent and coherent (lasers). In the notable work of Jorgenson and Lukacs (11,46,47), a high-pressure mercury lamp was used for illumination. Combined with high-quality filters or monochromators, arc lamps provide flexibility in wavelength selection to match the excitation to an analyte absorption maxima; however, stability can be poor, leading to difficulties for trace-level detection. For example, Green and Jorgenson report (48) that focusing the light of a high-pressure Hg–Xe arc lamp into a slit on a double monochromator and then into a small capillary window is a major challenge. They reported detection limits in the range of 0.1–1 µM. The spatial instability of arc lamps is a significant problem when one is focusing onto a small area such as

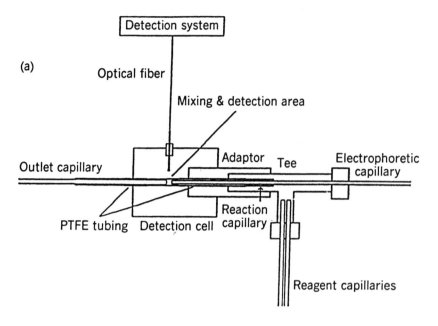

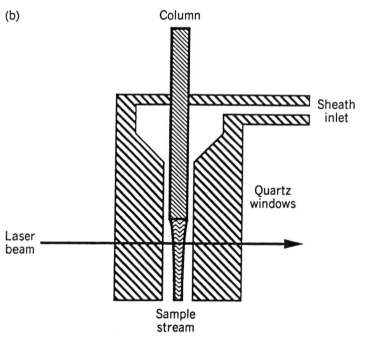

Figure 9.3

a capillary inner lumen. By using a stabilized arc lamp, spatial filtering and optimized optics, Arriaga et al. (49) demonstrated a high-efficiency filter fluorimeter that uses a Xe arc lamp source (75 W) and a sheath flow detector that obtains a concentration detection limit of 10 pM for florescein isothiocyanate (FITC)–labeled amino acids (corresponding to 0.2 amol injected onto the column).

Lasers have a number of advantages as excitation sources for CE, including the ability to be focused near the diffraction limit of light, simpler spectral backgrounds, and often an excitation power that can be varied over a wide range to optimize it for the particular analytes being studied. LIF was first demonstrated by Gassmann et al. (50). Although most lasers have only one emission line or relatively few, a variety of lasers are available with emission lines spanning the UV to IR regions. The most common lasers are the Ar ion (457,488,514 nm), He–Cd (325,354,442 nm), and the green He–Ne (543.5 nm). These lasers are in common use, as they are dependable, relatively inexpensive, and have emission lines that match commonly used fluorescent reagents.

Although most LIF in CE has used Vis laser excitation, near IR excitation is now attracting considerable interest because of the availability of inexpensive diode lasers and low near-IR spectral background (51,52). Excellent detection limits are obtainable because the $1/\lambda^4$ dependence of Raman scattering reduces the Raman background, few materials fluoresce in the region so the spectral background is low, and the lower energy of the photons increase the photostability of many fluorophores. Although these advantages are important, there are few derivatizing chemistries available; recently, there has been substantial progress in developing new tagging chemistries and more is expected (53,54).

There has also been considerable effort in extending LIF to UV regions, as several important classes of biomolecules absorb in the UV and floresce. Thus, UV excitation is an alternative to derivatizing these compounds. The biggest hindrance to UV excitation is the lack of UV lasers and the high cost of those that are available. Starting with the work of Swaile and Sepaniak (55), several groups have investigated the detection of a variety of compounds (56–60), with detection limits in the high pM concentration range (high zmol masses) in the best cases. Despite significant advantages in a number of applications, the widespread use of UV–LIF will depend on the introduction of an easy-to-use inexpensive laser.

Figure 9.3. (a) Schematic of postcolumn "tee" detector cell. [Reprinted from N. Wu and C. W. Huie, *J. Chromatogr.* **634**, 309–315 (1993), with kind permission of Elsevier Science-NL, Amsterdam, The Netherlands.] (b) Schematic of sheath flow detection design. [Reprinted from M. Yu and N. J. Dovichi, *Appl. Spectrosc.* **43**(2), 196–201 (1989), with permission.]

9.4.3.3. Collection Optics and Detectors

The function of the collection optics and detector are to collect a large fraction of the fluorescent emission, spectrally filter it to pass the fluorescence emission while discriminating against various background sources, and detect it as sensitively as possible. In nearly all the fluorescence detection systems the signal is collected at 90° relative to excitation, as the scattering is lower at this angle. Collection optics can be as simple as a single plano-convex lens and spectral filter to as complex as a custom series of optics and a monochromator. Almost all reported systems use single-channel detecton, with a photomultiplier tube as the detector. Two of the most common collection arrangements are the microscope objective (Figure 9.4) and fiber optics.

A number of systems use a high-numerical-aperture microscope objective because of its ability to collect a large fraction of the light from a small region and focus this light onto the detector. The microscope objective can be used alone or as part of a complete microscope. Hernandez et al. (61) reported the use of an epillumination microscope for LIF. The excitation wavelength is

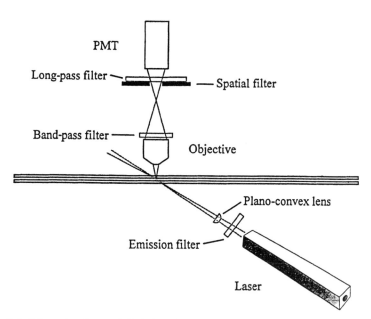

Figure 9.4. Schematic of general LIF detection system. [Reprinted from S. A. Shippy, J. A. Jankowski, and J. V. Sweedler, *Analytica Chim. Acta* **307**, 163–171 (1995), with kind permission of Elsevier Science-NL, Amsterdam, The Netherlands.]

reflected into an objective lens, and the fluorescence emission is collected with the same objective but directed toward the detector using a dichroic filter. Chen et al. have reported detection limits of less than 10 molecules using a high-collection-efficiency microscope objective with a sheath flow cell (41), and Henandez et al. have reported detection limits below pM concentration (100 molecules) (61,62). Most microscope objective systems include a spatial filter located at the image plane of the objective that consists of an aperture just smaller than the capillary lumen image. Thus, only light originating from the aqueous core reaches the detector. In addition, a series of spectral filters are used to reject as much of the Rayleigh and Raman scattering as possible. Careful selection of the filter wavelength range and rejection characteristics are important to achieve highest performance.

9.4.4. Multidimensional Fluorescence

Just as in multiwavelength absorbance detection, multidimensional fluorescence can be interfaced with CE to provide qualitative information. In wavelength-resolved fluorescence, fluorescence excitation and emission spectra are obtained at every data acquisition. An example of a wavelength-resolved emission electropherogram can be seen in Figure 9.5b; a complete fluorescence emission spectrum is obtained for each analyte band. There have been a number of studies on wavelength-resolved fluorescence emission since the first report from Swaile and Sepaniak (63). They used a photodiode array and a spectrograph to achieve 4 nm per channel resolution with a detection limit of 20 µM (60 fg) for fluorescein. Until the advent of low-noise detector arrays such as charge-coupled devices (CCDs) and intensified photodiode arrays (64), the acquisition of spectral information also entailed a large decrease in sensitivity compared to single channel detection. There have been several more recent reports of wavelength-resolved fluorescence detection with improved figures of merit (40,65–67). The lowest reported detection limits are 5×10^{-14} M for sulforhodamine 101 (~ 80 molecules) (40), indicating that for well-designed systems there is no longer a sensitivity decrease associated with obtaining spectral information. In Figure 9.5a, a schematic diagram of a wavelength-resolved system is shown.

Applications for wavelength-resolved CE have centered around analyte identification based on fluorescence wavelength and bandshape. The ability to spectrally differentiate the four fluorescently labeled dideoxy bases in the Sanger DNA sequencing method has driven the development of much of the wavelength-resolved fluorescence instrumentation to date (66–68). A second application involves using UV excitation to detect the native fluorescence of the tyrosine and tryptophan residues in peptides and proteins. The emission spectra can be used to categorize unknown peptides into several classes: those

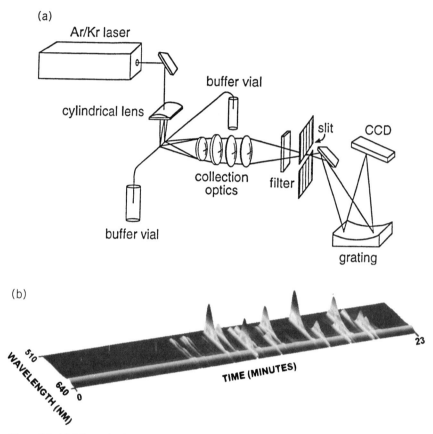

Figure 9.5. (a) Schematic of a wavelength-resolved LIF detection system. (b) Electropherogram of amino acids labeled with Bodipy 503/512 C_3 and Bodipy 576/589C_3. The Bodipy 503/512 C_3 features are those at shorter wavelengths than those of the Bodipy 576/589C_3. The continuous feature at ~ 583 nm is the major Raman band of water. [Both reprinted from A. T. Timperman, K. Khatib, and J. V. Sweedler, *Anal. Chem.* **67**, 139–144 (1995), with permission. Copyright 1995 American Chemical Society.]

containing tyrosine, those containing tryptophan, and those containing both, with 3σ detection limits of 200 pM (800 zmol) for tryptophan (59). By use of excitation in the 250–260 nm range, phenylalanine-containing proteins can also be characterized. The methods of peak purity and peak tracking developed for absorbance detection can be adapted to fluorescence. In addition, the optimum wavelength range can be determined after the separation to take into account spectral background features such as Raman bands from water and buffer additives (40).

Fluorescence emission depends on the immediate chemical and physical environment of the fluorescent probe, and large changes in fluorescence emission can occur depending on factors such as the ionic strength, pH, chelation, and temperaure (68). One of the most important separation parameters is electrolyte pH, as small changes in pH can cause large changes in the resulting separation. By means of wavelength-resolved fluorescence detection and carboxy-SNARF-1 {2 (or 4)-[10-dimethylamino-3-oxo-$3H$-benzo[c]xanthene-7-yl]-benzenedicarboxylic acid} the pH at a fixed point in the capillary can be monitored on-line (68). Large pH gradients (more than 3 units) are reported when reduced volume outlet vials are used. Many other separation and analyte parameters can be followed on-line using the information in the fluorescence spectrum.

Other forms of multidimensional fluorescence have been reported. For example, time-resolved fluorescence has a number of advantages and has been applied to CE (69). In this method, fluorescence is excited with a pulsed laser and the time course of the fluorescence signal is measured. Time resolution provides a reduction in spectral background, as Rayleigh and Raman scattering occur on a much faster timescale than that of fluorescence and so can be discriminated against. The fluorescence lifetime of an analyte can aid in peak identification. In addition to fluorescence, time-resolved luminescence detection of europium(III) and terbium(III) chelates have been studied in CE. As luminescence from these molecules involves a long-lived $D \to F$ electronic transition, the instrumentation is much simpler than for other forms of time-resolved luminescence (70,71). In addition to excitation and emission wavelength and time resolution, polarization can yield important information concerning chiral analytes. Christensen and Yeung demonstrated fluorescence-detected circular dichroism for chiral fluorescent compounds such as riboflavin, where they obtained 3σ mass detection limits of 0.2 fmol (72). This system allows optical activity measurements from picoliter volumes and can be carried out with a system that can also be used for conventional LIF measurements.

9.4.5. Native Fluorescence

The most straightforward use of fluorescence is detecting analytes that are fluorescent without having to modify (derivatize) them; such a process has been termed native fluorescence. Few molecules fluoresce significantly when excited with visible excitation. Examples of native fluorescence in CE include detecting vitamin B_6 metabolites (73), pharmaceutical compounds (74–76) bilirubins (77), and porphyrins (78). The fluorescent properties of these analytes can vary greatly depending on the separation pH and ionic strength. At basic pH's, for example, the porphyrin δ absorption band absorbs in the

490–510 nm range and so matches the emission of an Ar ion laser. Thus, separation pH is chosen to be compatible with both the separation and the detection. Several important classes of molecules exhibit native fluorescence using UV excitation below 300 nm. These include proteins and peptides that contain the Tyr, Phe, or Trp residues, polycylic aromatic hydrocarbons (PAH), and DNA (56–59,79,80)

9.4.6. Derivatization

Unfortunately most analytes are not fluorescent, and so a significant part of fluorescent detection involves attaching fluorescent probe molecules to the analyte to make them detectable. Often, the characteristics of a particular tagging chemisty limit analytical performance more than instrumental figures of merit. This is also true in fluorescence microscopy and fluorescence detection in LC, and there has been an intense effort to develop chemistries to tag particular classes of molecules (81). The aim of this subsection is not to describe derivatizing chemistries, as the chemistry depends on the specific applications (which are outlined elsewhere in this book in the appropriate chapters). Instead, this discussion focuses on the methods developed to tag small volume samples in CE.

In most cases derivatization has been done precolumn and often with a large sample volume that avoids losses due to sample handling and dilution. This is generally done for instrument characterization or to prove the feasibility of a particular chemistry. However, there are instances when a large volume tagging reaction is not possible. Kennedy and Jorgenson (82) have shown that the amino acids and peptides in a single *Helix aspersa* cell can be derivatized with naphthalenedicarboxaldehyde (NDA) in a 25 nL volume. Reaction vials were made by drilling small holes into a quartz block. By means of this methodology amino acids and peptides have been detected in single adrenal medullary cells (83). Alternately, derivatizaion of intracellular compounds can be accomplished by using the cell as the reaction chamber. Hogan and Yeung (84) describe selective derivatization of thiol containing compounds in a single red blood cell by monobromobimane and Orwar et al. (85) have shown glutathione derivatization by NDA in single mammalian neuroblastoma cells. Significantly, in the absence of CN^- the NDA reacts highly selectively with the glutathione, thereby allowing for this analysis.

Alternately, analyte derivatization can be accomplished on-column to preclude sample handling difficulties. Gilman and Ewing injected entire PC12 cells followed by an injection of the derivatizaion reagent, NDA (86). Sample dilution is greatly reduced and high peak efficiencies are obtained (340,000–540,000). Additionally, enzymes can be used to generate multiple fluorescent products from nonfluorescent reagents. The methodology developed by

Regnier et al. (87) and used by others (88–91) enables single enzyme molecules to be assayed on-column.

Not all pre- and on-column methods of derivatization are suitable for a given analyte. Tagged analytes become more similar chemically, which often makes the separation more difficult. If a number of labeling sites are available, multiple side products may be produced for each analyte, further complicating the separaton. As the derivatization chemistry changes, the separation often needs to be reoptimized as a different set of analytes needs to be separated. Postcolumn reaction schemes avoid these problems with derivatization just after the separation. Employing a set up similar to HPLC designs, Tsuda et al. (92) used three pumps and a four-way connector; however, lower separation efficiencies and a nonlinear calibration were obtained. Rose and Jorgenson (93) demonstrated a post-column reactor with a reaction capillary and reagent addition via a connecting "tee" to the outlet end of the capillary (Figure 9.3a). Alternately, Albin et al. (94) demonstrated a gap reactor, and this idea has been further developed by Gilman et al. (95). Analytes are derivatized as they flow through a 4–160 µm gap between two ends of a broken capillary without large losses in resolution (theoretical plates as high as 230,000.)

Regardless of the method of derivatization applied, fluorescence detection achieves low detection limits ($< 10^{-21}$ mol). While there are a large number of tagging chemistries, choosing the optimum chemistry and derivatization scheme is crucial. Potential problems include fluorescent side reactions (extraneous peaks), the requirement that some fluorescent reagents be in high concentrations for appreciable yield, and the poor fluorescent properties of some labeled analytes. We expect that new and suitable derivatizing reagents will improve the utility of fluorescence detection.

9.5. OTHER CAPILLARY ELECTROPHORETIC OPTICAL DETECTORS

9.5.1. Chemiluminescence Detectors

Another means of obtaining luminescence from an analyte is to electronically excite the molecules of interest by using a chemical reaction instead of light, a process known as chemiluminescence. A molecule excited by chemical reaction emits a photon that can be detected and quantified just as in fluorescence. Chemiluminescence (CL) instrumentation is conceptually simple due to the lack of excitation source and emission filtering. In fluorescence, detection is limited by spectral background. CL reactions can also be limited by background luminescence, but the CL background can be much lower than that in fluorescence. Unlike most fluorophores, which can undergo thousands of fluorescence cycles (and emit thousands of photons), most CL reactions destroy the analyte and so can only be cycled once. Thus CL does not normally have the senstivity of fluorescence, but detection limits can be several orders of magnitude better than absorbance detection (10^{-8} M) (96). The biggest limitation of chemiluminescence is that there are relatively few CL reactions available. Most analytes need to be derivatized with the chemiluminescent reagent just as in fluorescence detection.

A large advantage is that CL detectors are simple in design. They involve mixing the CL reagents to the CE eluent in a reaction capillary and detecting the emitted photons. Typically a smaller analytical capillary is inserted into a reaction capillary with the CL reagents pumped in as makeup flow (Figure 9.3a). Light emitting species that are generated flow by a detection window to a photomultiplier tube (PMT) for photon counting. While CL detection systems are usually simple, inexpensive, and sensitive, there are a number of challenges in utilizing this technique in CE. Because the CL detection requires postcolumn addition of reagents, the peak efficiencies from CL tend to be significantly lower than efficiencies obtained with other detection techniques. Secondly, there can be complications due to a difference between the optimal conditions for the separation and the CL reaction. Factors such as pH, temperature, ionic strength, and the type of running buffer may affect both the separation and CL reaction differently. Also, as the time course of the reaction and, in turn, the light emission varies for each analyte, specific reaction times and flow rates may not be optimal for every analyte. Finally, quantitation of light emission depends on complex reaction kinetics. Despite these concerns there are a number of CL reactions and designs that are effective.

In HPLC, the CL reaction most commonly used involves that of peroxyoxalate with hydrogen peroxide in the presence of a fluorophore. In CE, Wu and Huie (97) describe a detection system for dansylated amino acids that yields detection limits more than 30 times better than those of absorbance detection. Unless the analyte is natively fluorescent, derivatization with a suitable tag is required. The ability to combine chemical excitation with a selected fluorophore offers advantages, as was demonstrated by Hara et al. (98–100) in protein separations with peroxyoxalate detection. Sensitivity and limits of detection were improved by optimizing the fluorophore used for labeling proteins. The use of bis(2,4,6-trichlorophenyl)oxalate, commonly known as TCPO, requires the addition of a significant amount of an organic solvent, which complicates the CE separation. As another CL reaction, Ruberto and Grayeski describe acridinium chemiluminescence applied to peptide separations (101,102). Precolumn derivatization of peptides was done at picomole level and used without purification. An improvement over peroxyoxalate detection limits was shown, but the peak efficiencies were low (4000 theoretical plates). Further improvements in detection limits were seen with another common CL reaction involving luminol derivatives with hydrogen peroxide in the presence of a catalyst. Moreover all reagents are soluble in an aqueous buffer, allowing for a simplified separation. In the detection scheme described by Dadoo et al. (103), a detection window was placed in the focal point of a parabolic mirror for light collection. Improvements in detection limits for luminol derivatives in this system vs. absorbance techniques are approximately 2–3 orders of magnitude. At present luminol appears to be the most sensitive CL reaction available, although the working range is somewhat limited.

In order to improve separation efficiencies, alternate methods for CL reagent introduction have been explored recently. Dadoo et al. (96) developed an end-column method of detection whereby analytes react upon entering the outlet vial. A fiber-optic probe collects the luminescence near the end of the capillary. Both luminol and firefly luciferase CL were characterized with this method; the detection limits were improved with increased separation efficiencies. Zhao et al. (104) demonstrated a sheath flow reactor with two-PMT coincidence detection that obtained 40 amol detection limits with 100,000–200,000 theoretical plates for a series of isoluminol-labeled amino acids. We expect to see additional reports showing optimized separation efficiencies and detectabilities in this rapidly evolving detection area.

9.5.2. Refractive Index Detectors

Several types of refractive index (RI) detectors have been developed for use in CE, although none are yet available on a commercial basis. The RI detector is a universal detection scheme that gives a uniform response for most analytes. It is based upon the measurement of a change in the refractive index, Δn, of the capillary effluent. It is most useful for analytes such as carbohydrates that do not possess strong UV chromophores, fluorophores, electrochemical activity, or ionic conductivity. A light source, usually a laser, is focused onto the buffer inside the capillary. When an analyte passes through the detector window (as long as the analyte has a RI that is different from that of the running buffer), the light beam is deflected by the change in the RI and the deflection is detected by a position-sensitive detector. The RI detector is similar to an absorbance detector in that it measures a small change in a large background signal. It is also similar to indirect detection, where the proper choice of a running buffer leads to higher sensitivities. The best buffer is one with a low conductivity and a large difference in molar refractivity from that of the analyte.

For each particular analyte, there is a characteristic dependence of the RI on concentration. For example, Δn for aqueous solutions of most proteins is $1-2 \times 10^{-4}$ RI unit per 1% (w/v) of protein (105). Therefore, the concentration sensitivity of RI measurements is inherently low. Additionally, many external factors provide sources of background fluctuations or drift that limit the performance of RI detectors. These include laser beam intensity and position instabilities as well as small capillary position movements (106). Joule heat poses a major challenge in the development of RI detectors suitable for CE; the Δn of the running buffer is on the order of 10^{-4} RI units/°C (3). Three main types of RI detectors for CE are described in the literature. With addition of a second laser to the instrumental design, an RI detector can be used as a thermooptical absorbance detector. This will be addressed below in Section 9.5.3.

9.5.2.1. "Off-Axis" Laser Beam Deflection

This method was originally proposed for microbore and capillary separations by Bornhop and Dovichi (107,108) and by Synovec (109). An "off-axis" light beam (a beam that is offset from the capillary center axis), usually from a He–Ne laser or a LD, illuminates the capillary. The beam is split by the inner interface of the capillary into a reflected beam fan and a refracted beam fan that overlap in the far field, yielding a characteristic fringe pattern. A change in the RI of the liquid in the capillary shifts the positions of the fringes, whose lateral displacements are monitored by a position-sensitive photodiode; upon processing, that displaement constitutes the output signal. Bruno et al. (110) optimized the geometric arrangement of the beam to the internal diameter by selecting the most sensitive fringe to Δn, and by surrounding the capillary with RI-matching fluids to stabilize the temperature inside the capillary in the detection region and to eliminate the reflections and refractions at the outer capillary wall. With these improvements, an rms (root mean square) noise level of 3×10^{-8} RI units is achieved with baseline drifts of 2×10^{-6} RI units/h at 1 Hz and stabilities of angular deflection on the order of 100 nrad. These noise levels are only about an order of magnitude above the calculated shot noise limit when a 50 μm capillary is used. The lowest measured detection limit at $S/N = 2$ is 1×10^{-5} M for sucrose. The linear dynamic range extends over about 3 decades. This method works well for capillaries having ≥ 15 μm i.d.

To eliminate the baseline drift that occurs during use of the aforementioned off-axis RI detector, Chen et al. (105) applied a 400 Hz ac modulation to the dc voltage across the capillary. The modulation in the separation voltage produces a modulation in the analyte velocity and, as a result, in the RI signal. During demodulation of the signal, it is possible to reject the low-frequency drift in the measurement. With this technique, a detection limit of 0.5×10^{-6} M ($S/N = 3$) is obtained for the protein, carbonic anhydrase.

9.5.2.2. "On-Axis" Hologram-Based Detection

This detector uses a holographic element to perform the optical functions that are performed by the capillary in the off-axis method. In that method, the laser beam does not probe the maximum optical path available, resulting in a loss of sensitivity. The probing arm of the interferometer in the hologram-based RI detector crosses the capillary center, where the optical path and phase changes are larger and diffraction effects smaller. Also, in the holographic version, the observed fringes are equally spaced, allowing the use of multiple fringe detection with diode arrays. Krattiger et al. (111) report the CE separation of small cations in a 10 μm i.d. capillary using this detector. A separation of a mixture of underivatized saccharides in a 25 μm i.d. capillary has also been reported by Bruno et al. (20). While the concentration detection limits and linear dynamic range rendered by this RI detector are comparable to those of the off-axis version, the hologram-based RI detector allows smaller-i.d. capillaries to be used, increasing the mass detection limits.

9.5.2.3. The Concentration Gradient

The narrow peaks produced in CE separations generate high concentration gradients in the detection volume that can effectively be monitored with a concentration gradient RI detector based on Schlieren optics (106,112). In this method, a laser beam is focused directly into the separation capillary and is deflected when it encounters the refractive index gradient ($\partial n/\partial x$) that is associated with the concentration gradient ($\partial C/\partial x$) of an eluting analyte. The deflection is monitored using a photodiode light beam position sensor. The signal produced by the concentration gradient detector has the shape of a Gaussian derivative rather than a Gaussian peak. The derivative or differential nature of the detector response is effective at removing the low-frequency noise associated with the drifts normally seen in RI detection and produces a flat baseline.

Obviously, the sharper the electrophoretic zones, the sharper the concentration gradients at the boundaries, resulting in a higher-sensitivity detection. This is why the concentration gradient detector is ideally suited for use in CE modes where highly focused zones are achieved, such as moving boundary electrophoresis (113), isotachophoresis (114), and isoelectric focusing (115–117). Detection limits are usually expressed in the units of the concentration gradient, i.e., mol/Lm. However, they can also be expressed in units of concentration because the concentration gradient at the inflection point of the Gaussian peak is proportional to its height. Linearity between the peak height and analyte concentration extends over 3 orders of magnitude. The best reported detection limit of the detector is in the CIEF mode (115) (on the order of 10^{-6} to 10^{-7} M for proteins) and is of the same order of magnitude as that of a UV absorbance or a diode-laser-based fluorescence detector. However, an advantage to the concentration gradient RI detector is its universal nature, making it possible to detect all eluted analytes without derivatization.

9.5.3. Thermooptical Detection

Thermooptical detection is an absorbance technique that uses the refraction of the probe beam as a means of detection. Basically, a pump beam is focused onto the center of the capillary and modulated at 10–120 Hz. The molecules in the region of the pump beam that absorb the light are heated, creating a "thermal lens" due to the different refractive indices of the heated and nonheated regions. A probe beam is then focused on the thermal lens (Figure 9.6a). Changes in the demodulated signal yield information about molecules in the stream that affect the thermal lens. The benefits of using such a technique are that many more molecules have suitable molar absorptivities compared to those that are fluorescent. It is interesting to note that most reported studies with this detector use fluorescently labeled compounds due to their high absorptivities. The best reported detection limit of 50 amol for 4-dimethylaminoazobenzene-4′-sulfonyl (DABSYL)-glutamic acid is significantly better than that of conventional UV–Vis techniques (118). Thermooptical detectors have not become routine instruments, and there are no commercial models available. In part this is due to the technical expertise required for aligning lasers to the center of the capillary, especially when the beam diameter is on the order of the capillary diameter. With optimized detectors, pump laser beam instability is the limiting source of noise in the system. Dissipation of the thermal lens in the flowing streams lowers the signal. Pulsed lasers, high modulation of the cotinuous wave, and focusing the laser to small probe volumes helps to counteract this problem.

With a pump He–Cd laser of 4 mW at 442 nm and a probe He–Ne laser at 632.8 nm, Yu and Dovichi (119) demonstrated a separation for DABSYL–amino acids with thermooptical detection. Because absorbance sensitivity is proportional to pump laser power, a second generation instrument (118) used the 130 mW, 457 nm line from an Ar ion laser as the pump laser. Detection limits were improved an order of magnitude to $5–50 \times 10^{-8}$ M (50–500 amol). A later improvement was the development of a two-color thermooptical system that used the 40 mW, 442 line of a He–Cd laser and the 5 mW, 632.8 line of a He–Ne laser as pump and probe beams for detection (120). Mixtures of amaranth and Fast Green were separated and detected in this two-channel system without any significant cross talk between channels. Saz et al. (121) have applied thermooptical detection for native proteins. With a pump beam at 257 nm and a He–Ne probe, detection limits of 7×10^{-9} M for bovine serum albumin were obtained. Significantly, proteins not containing an appreciable amount of surface Tyr, Phe, or Trp were detected, complementary to fluorescence techniques.

In addition to continuous-wave lasers, pulsed lasers can be used in thermooptical detection and provide an improvement in the signal due to the high

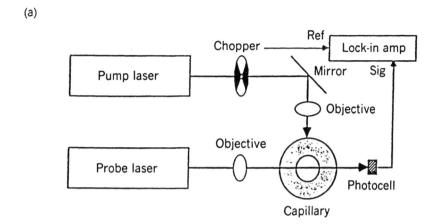

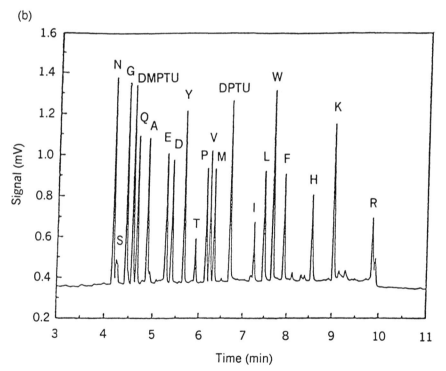

Figure 9.6. (a) Schematic of thermooptical detection design. (Reprinted from ref. 118 with permission.) (b) Electropherogram of 19 phenylthohydantoin (PTH)–amino acids and diphenylthiourea (DPTU) and dimethylphenylthiourea (DMPTU) as purchased from Applied Biosystems. Running buffer; 10.7 mM phosphate, 1.8 mM borate, and 25 nM sodium dodecyl sulfate (SDS), pH = 6.7. [Reprinted from M. Chen, K. C. Waldron, Y. Zhao, and N. J. Dovichi, *Electrophoresis* **15**, 1290–1294 (1994), with permission.]

repetition rates that are possible. Waldron and Dovichi demonstrated the use of a pulsed excimer laser pump at 248 nm with 50 ns pulses at 610 Hz with an average power of 10 mJ (200 mW maximum power) (122). The detection limits for phenylthiohydantoin–amino acids (200–5000 amol) are more than 2 orders of magnitude better than the same molecules in UV absorbance in LC and other UV thermooptical absorbance detectors for CE. From a manually run Edman degradation, the separation of 19 labeled amino acids was demonstrated (as shown in Figure 9.6b) using this excimer laser thermooptical system (123).

9.5.4. Radionuclide Detection

Radionuclide detection is employed extensively in many biological applications because of its extremely high sensitivity and outstanding selectivity. Multiple CE radionuclide detection schemes have been reported, as well as several that are commercially available. The sensitivity arises because the energetic decay events characteristic of most nuclear processes are easy to detect and there is a very low natural background; the selectivity occurs because only radiolabeled compounds produce a response at the detector. For example, it is possible to perform experiments where a radiolabeled material is introduced into a living organism and the metabolic pathways are investigated by following the movement of the material and chemical changes in it throughout the organism without interferences from unlabeled compounds. Radionuclide detection has the advantage that the decay process is independent of the chemical and physical environment of the analyte and that detection is truly mass sensitive.

The detection limits of radionuclide detection depend on the energetics of decay and the number of events that are observed. As radioactivity follows first-order rate law, the number of detected events depends on the observation time of the analyte and the fraction of events detected. The rate of decay for an isotope is expressed in terms of the time for half the nuclei in a particular sample to decay. Rate constants vary widely, depending on the nucleus; for example, ^{32}P has a half-life of 14.2 days, whereas that of ^{14}C is 2.03×10^6 days. Because total peak times in CE are limited to several seconds, observation time fundamentally limits detection sensitivity. For example, 3 amol of a ^{32}P labeled analyte will produce 1 decay/s (1 Bq) and, as several events are needed at the detection limit, it is difficult to improve on-line ^{32}P detection limits to less than 1 amol. In addition to the half-life, the higher-energy decay events are easier to detect than lower-energy events. Commonly used isotopes have energies ranging from 0.019 MeV for ^3H to 1.71 MeV for ^{32}P. Besides the magnitude of the energy, there are different forms of decay, with the commonly employed isotopes involving β^- emission (beta particles or negative electrons), β^+ (positrons or positive electrons), and γ-rays (neutral photons).

In CE, a scintillator is normally used to convert some of the energy of the radiation to photons that are detected. Typically, there is a region in the capillary where the radioactive events interact with a scintillator to generate light for detection. Because shielding on either side of the detector is necessary to control detection zone width, a balance must be found between larger signals from wide detection windows and lower separation efficiencies. As in other detection modes, on-column detection is the most common. With larger samples separated in a 300 μm i.d. fluorinated ethylene–propylene capillary, Kaniansky et al. (124,125) presented a postcolumn flow cell detection scheme for isotachophoresis. The separation eluent flows directly into a plastic scintillator cell (210 nL) that is between two PMTs for coincidence detection. A limit of detection of 16 Bq was obtained for ^{14}C analytes; the design enabled detection of radionuclides with β^- energies in the 0.4–0.7 MeV range. Sufficient separation efficiencies were obtained for acetate and glutamate ions. Altria et al. (126) developed an on-column scheme for γ-ray detection of ^{99}Tc in radiopharmaceuticals using a smaller capillary. Pentoney et al (127) reported a similar geometrical design for the detection of the biologically important nuclide ^{32}P using a γ and β^--sensitive Cd–Te detector with detection limits of ~ 20 Bq with a collection efficiency of $\sim 26\%$. An improved design with a significantly higher collection efficiency ($\sim 65\%$) was presented by Pentoney et al. (10). In this configuration, the separation capillary is inserted through a solid-parabolic plastic scintillator on which the curved surfaces are coated with an aluminum film to reflect light toward a cooled PMT. Westerberg et al. (128) report a similar 4π setup with a 85% detector efficiency and detection limits for ^{11}C and ^{18}F in the single-Bq range. The short half-lives of these radionuclides (< 2 h) contribute to excellent sensitivity. To increase the signal from a particular band and improve sensitivity, Pentoney et al.[10] used flow programming—reducing the voltage while the analyte is at the detection window—and increased the time that a particular band was in the detection window. This resulted in an order of magnitude improvement in detection limits for preselected analytes.

Two of the most important radionuclides, ^{3}H and ^{35}S, lack sufficient energy to penetrate the capillary wall. Thus, Tracht et al. (8) developed a postcolumn method to detect these low-energy β^- emitters. As shown in Figure 9.7, the end of the capillary is painted with a conductive paint to complete the electrical circuit and the capillary eluent is deposited directly onto an analyte binding membrane. Scintillation compounds deposited on the membrane luminesce from the radioactive decay of adsorbed radionuclide analytes and the luminescence is detected with a cooled CCD. As the membrane can be viewed for hours, even with relatively poor efficiency, large improvements in detection

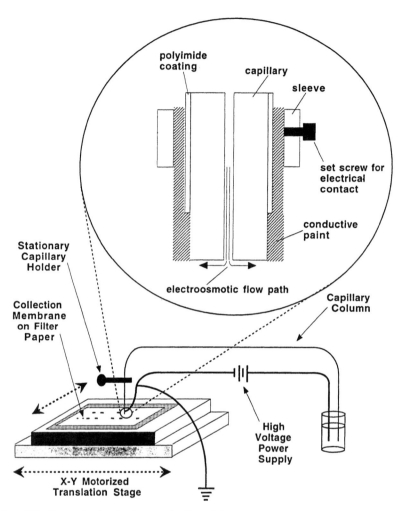

Figure 9.7. Schematic of end-column collection of the electrophoretic eluent. Electrical connection at the end of the capillary is shown in the detail. [Reprinted from S. Tracht, V. Toma, and J. V. Sweedler, *Anal. Chem.* **66**, 2382–2389 (1994), with permission. Copyright 1994 American Chemical Society.]

limits are obtainable compared to those obtained with on-line methods. Recently, improvements in detection linearity and lower LODs have been obtained when membrane scintillation is replaced with a solid-state phosphoimaging plate that is read in a phosphoimaging system (129).

Although radionuclide detection involves working with potentially hazardous materials, the unique characteristics are such that many applications in the biological sciences and environmental monitoring will continue to use this method. An advantage to CE is the large reduction in the volume of difficult-to-dispose-of waste compared to that with other separation methods.

9.5.5. Hyphenated Detection

The use of two different detection schemes to gain information about an analyte is referred to as hyphenated detection. Detectors can be connected in series or in parallel. When linked in series, the first detector must be nondestructive of the analyte. In most applications of hyphenated detection in CE, one detector is selective and the other is universal, with UV–Vis absorbance most commonly used as the first detector. The second detector can be any other optical or nonoptical detection scheme. For instance, Caslavska et al. (130) describe simultaneous absorbance and fluorescence detectors capable of profiling drugs and metabolites in bodily fluids. The absorbance electropherogram is rather complicated and detects the presence of many compounds, whereas the fluorescence detector is sensitive only to the presence of underivatized drugs, drug metabolites, or endogenous compounds that are fluorescent.

Another useful application of hyphenated detection in CE is peptide mapping by isoelectric focusing. In CIEF, a conventional absorbance detector must operate at ~ 280 nm because of the UV absorbance of the carrier ampholytes. At this wavelength, only peptides containing tryptophan and tyrosine residues are detected. Wu and Pawliszyn (131) describe the development of a simultaneous, dual-detection scheme for CIEF consisting of an optical absorption imaging detector and a concentration gradient RI detector. All peptides are detected by the universal RI detector, and only tryptophan- and tyrosine-containing peptides are detected by the absorbance detector.

In addition to these examples, many commercial CE–MS instruments include an absorbance detector to indicate the presence of an eluting compound. Combining different detectors with complementary sensitivity and selectivity characteristics aids in evaluating complex samples.

9.6. RECENT DEVELOPMENTS

9.6.1. Raman Detection

Although most detectors signal the presence of a compound, relatively few techniques provide information about chemical structure or composition. Raman detection in CE is one means of obtaining qualitative information about the analytes but is complicated because the Raman process is inherently insensitive (132). Chen and Morris (133) demonstrated Raman detection in CE using a remote probe design in the separation of methyl orange and methyl red dyes. The capillary window with polyimide removed was placed in a holder where the 532 nm line of a frequency-doubled Nd–YAG laser operating at about 300 mW was used for excitation. At 15° to the plane of polarization of the excitation beam, two fiber optics collect the resonant light which is collimated, focused to a slit of a 0.64 m spectrograph, and dispersed across a CCD. Detection limits are relatively poor (low mM); however, they can be improved by employing voltage programming where a lower potential is applied while analytes are resident in the detection window. Recently, improvements were demonstrated by Walker et al. and Kowalchyk et al. with a Raman detection scheme mounted in a microscope with an objective for light collection. In isotachophoresis they report minimum detection concentrations of 5×10^{-6} M for adenosine nucleotides (134) and 1×10^{-5} M for nitrates and perchlorates (135).

9.6.2. Nuclear Magnetic Resonance Detection

NMR involves absorption of radio frequency photons by a sample when it is placed into a strong magnetic field. The NMR spectrum that results gives valuable information that can be used for structure elucidation and the determinations of the chemical environment of the molecule. NMR detectors for conventional HPLC have previously been described (136). The first example of an on-line NMR detector for CE has recently been reported by Wu et al. (137). The technique involves wrapping a 50 μm wire around a fused-silica capillary to form a 1 mm long detector cell. The CE setup is housed in an NMR probe that fits into the magnet. A 75 μm capillary yields a sample chamber within the microcoil of just 5 nL. They report an on-column free-solution separation in a deuterated solvent/buffer system of three underivatized amino acids with a mass detection limit ($S/N = 3$) for arginine of 0.2 nmol and a concentration detection limit of 35 mM. Initial line widths obtained in the CE experiment were on the order of 10 Hz and prevented observation of proton scalar coupling. Subsequent improvements made in the microcoil fabrication (138) now enable the achievement of line widths of ~ 0.6 Hz, which is similar to the resolution obtained with conventional 300 MHz NMR.

Because NMR is an insensitive detection technique, the most suitable application of proton NMR detection in CE may be for modes where the analytes are concentrated into sharp zones, such as in CIEF. The sensitivity is enhanced by methods where the sample remains in the detection chamber for a longer period of time, allowing longer signal acquisitions. Proton NMR can be considered to be a universal detection scheme for CE, as most organic molecules possess hydrogen atoms. Careful comparisons of the microcoil approach to conventional NMR show that the S/N per micromole of analyte is enhanced 130-fold compared to traditional 5 mm spinning-tube probes. However, CE–NMR still has a poor detection limit compared to that of other CE detection modes and so will be most useful in applications where high concentration samples are available and the chemical information provided by NMR is important.

9.6.3. Light-Scattering Detection

Light-scattering detection in CE utilizes a wavelength of light that is not absorbed by the capillary effluent. Instead, when light comes in contact with the effluent, it is scattered and the angular dependence of the scattered light is measured. Three different types of light-scattering detectors are being developed for use in CE: the condensation nucleation light-scattering detector (CNLSD), the laser-light-scattering (LLS) detector, and the laser-based particle-counting microimmunoassay.

9.6.3.1. Condensation Nucleation Light-Scattering Detector

The CNLSD belongs to the class of universal detectors where any analyte that is less volatile than the mobile phase can potentially be detected. It resembles the evaporative light-scattering detector, a commercially available detector for HPLC separations (139). However, the CNLSD uses a particle growth process that substantially increases the sensitivity. The CNLSD itself has recently been interfaced to HPLC by Allen et al. (140) and Szostek and Koropchak (141) are currently working on interfacing the CNLSD to CE.

In this post-column detection technique (illustrated in Figure 9.8), effluent from the capillary is nebulized into an aerosol. Under the assumption that the running buffer contains only volatile components, the aerosol particles are desolvated, leaving behind the dry particles of nonvolatile analyte. The analyte particles, some as small as 2–3 nm, are then "grown" to tens of micrometers, a size that is much more effective at scattering light. The growth process is achieved by saturating the gas surrounding the particles with the vapors of a condensible fluid such as butanol, then passing the mixture through a condenser, where the butanol vapor condenses onto the particles. All particles above a selected threshold size grow to the same size and are detected; those below the threshold are discriminated against. Therefore, the detector response is a function of both the size and number distribution of input particles. An increase or decrease in analyte concentration changes the number of particles available to condensation nucleation and thus alters the level of response. The detector can be operated either in a continuous mode, where the average intensity of light is measured at a given angle from the incident light,

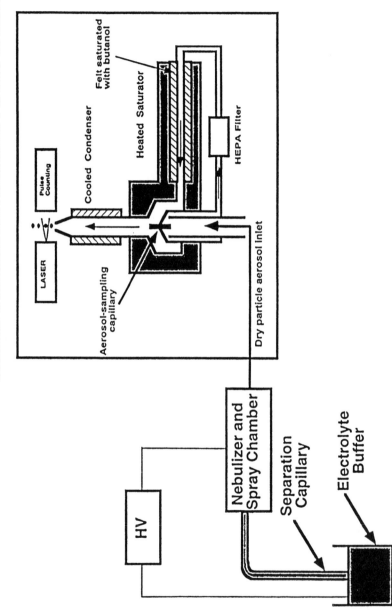

Figure 9.8. Schematic of the condensation nucleation light-scattering detector. (Provided by B. Szostek and J. A. Koropchak.)

or in a counting mode, where individual droplets are counted by a particle counter. Separations of underivatized amino acids and peptides are shown with detection limits below the 1 ppm level. The major limitation to achieving ppb detection limits is the background observed from nonvolatile components in the running buffer.

One advantage to this type of detection scheme is the ability to detect all analyte types without using pre- or postcolumn derivatization. Another advantage is the potentially low detection limits. Lewis and Jorgenson (142) are currently evaluating the utility of an electrospray–condensation particle counter as a detector for CE. The difference between this detector and the CNLSD is that an electrospray needle rather than a nebulizer is used for aerosol generation. The proposed theoretical detection limit is 1 molecule injected.

9.6.3.2. The Laser-Light-Scattering Detector

Currently, Wyatt Technology is working on interfacing a LLS detector to CE (143) and have already commercialized a LLS detector for HPLC that is primarily used for absolute macromolecular characterization of analytes separated by size exclusion chromatography (144). Such a system generally assumes Rayleigh scattering for calculations using the intensity of scattered light as a function of scattering angle. For the HPLC detector, the absolute molecular weights (10^3–10^7 g/mol) and sizes (10–70 nm) of polymers, biopolymers, and particles have been determined. Successful interfacing of a LLS detector to CE provides for a detector that can classify separated species on the basis of their mass, and for larger molecules, their size or conformation.

In this technique, light-scattering data are collected at three angles (37°, 90°, and 143°) and at various protein concentrations using 690 nm light from a 20 mW semiconductor laser. The average molecular weight (1000 to more than 1,000,000 Da) and the mean square radius of the protein are derived. When the LLS detector is used in conjunction with an on-line concentration-sensitive detector such as UV–Vis absorbance, masses of separated proteins are easily determined, as the light-scattering signal is directly proportional to the product of concentration and mass. Wyatt has successfully made on-line-scattering measurements of the CE-separated proteins β-lactoglobulin and myoglobin.

9.6.3.3. Laser-Based Particle-Counting Microimmunoassay

A laser-based particle-counting microimmunoassay system using light-scattering detection allows ultrasensitive and selective analysis of proteins (145).

Polyclonal antibody-coated latex particles are used in the running buffer. After the analyte (a protein) is injected onto the capillary, the electrophoresis is stopped to allow incubation of the analyte with the antibody-coated particles. The particles are then electrophoretically moved past the detection window. Light scattering of the laser beam is more intense for larger agglutinated particles than for unreacted particles. Therefore, counts from unreacted particles are rejected and only light scattering from agglutinated particles measured. A detection limit of 1 zmol (10^{-21} mol) of the enzyme glucose-6-phosphate dehydrogenase, corresponding to only 620 molecules, is reported. The detection method is useful in monitoring the glucose-6-phosphate dehydrogenase levels of single human erythrocytes. An advantage of the particle-counting microimmunoassay is that the enzyme mass rather than enzymatic activity is measured. As in other immunochemical methods, the assay is highly selective for a particular analyte, but it is necessary to produce a different antibody for many analytes. The overall assay should be applicable to a large variety of analytes, as antibodies can be produced for many compounds, but the compounds of interest must be known in advance.

9.6.4. Living Cells as Detectors

A novel end-column detection scheme for CE is the single-cell biosensor (SCB) recently described by Shear et al. (146). The outlet of a fused-silica capillary is held directly above a single cultured cell, and separated components are directed onto the surface of the cell. If a component binds to receptors on the cell membrane, a response is elicited that causes an increase in the free calcium ion concentration within the cell. The cell is previously loaded with a fluorescent intracellular calcium indicator and the increase in calcium ion concentration is detected by fluorescence microscopy, making this approach an indirect optical detection scheme.

A major challenge in the use of living cells as detectors is the poor selectivity. Cells have numerous surface receptors and can respond to many chemical species simultaneously. Through genetic engineering, it is possible to design a living cell biosensor that predominantly possesses a single receptor expressed in a particular cell type. The sensor will respond to a family of structurally related molecules. An enhancement, as described by Fishman et al. (147) is the use of known receptor antagonists. An electrophoretic run is first made without the presence of the antagonist. The cell is then incubated with a known antagonist that binds to a specific cell receptor, and a second electrophoretic run is made. The disappearance of a particular peak identifies that peak as the ligand that binds to the specific receptor. Using the living cell detector in conjunction with a UV–Vis absorbance detector, the researchers show that the sensor responds to bradykinin in a cell lysate but not to bradykinin degradation products. Thus, the SCB represents a CE detector that discriminates between separated components based upon their biological function rather than the chemical or optical properties.

9.7. CONCLUSIONS

A number of optical detection schemes have been described with a wide range of performance specifications. Although detection in CE is a challenge, detection schemes are available commercially that provide figures of merit adequate for many applications, and research instruments have demonstrated the high sensitivity, selectivity and applicability required in most analytical separations. Although the vast majority of CE systems are equipped with UV–Vis detectors, we expect additional detection modes to become more widely available over the next decade, as well as enhanced performance from existing methods. As demonstrated in other chapters, nonoptical detection modes based on electrochemical methods and mass spectrometry have complementary characteristics to the methods described in this chapter. Further details on a number of methods mentioned here appear in the chapters dealing with these specific applications.

ACKNOWLEDGMENTS

The support of the National Institutes of Health through Grant NS31609, the Searle Foundations Chicago Community Trust, and an Alfred P. Solan Fellowship are gratefully acknowledged.

LIST OF ACRONYMS AND ABBREVIATIONS

Acronym or Abbreviation	Definition
AU	absorbance units
Bq	becquerel (unit: 1 Bq = 1 decay/s)
CCD	charge-coupled device
CE	capillary electrophoresis
CEC	capillary electrochromatography
CGE	capillary gel electrophoresis
CIEF	capillary isoelectric focusing
CITP	capillary isotachophoresis
CL	chemiluminescence (detector)
CNLSD	condensation nucleation light-scattering detector
CZE	capillary zone electrophoresis

LIST OF ACRONYMS AND ABBREVIATIONS (Continued)

Acronym or Abbreviation	Definition
DABSYL	4-dimethylaminoazobenzene-4'-sulfonyl
DAD	diode array detector
DMPTU	dimethylphenylthiourea
DPTU	diphenylthiourea
FITC	fluorescein isothiocyanate
HPLC	high-performance liquid chromatography
IR	infrared
LD	laser diode
LED	light-emitting diode
LIF	laser-induced fluorescence
LLS	laser-light-scattering (detector)
MDC	minimum detectable concentration
MECC	micellar electrokinetic capillary chromatography
MEKC	micellar electrokinetic chromatography (same as MECC)
NDA	naphthalenedicarboxaldehyde
NMR	nuclear magnetic resonance
PAHs	polycyclic aromatic hydrocarbons
PMT	photomultiplier tube
PTH	phenylthiohydantoin
RI	refractive index (detector)
rms	root mean square
SCB	single-cell biosensor
SDS	sodium dodecyl sulfate
S/N	signal-to-noise ratio
SNARF-1	2 (or 4)-[10-dimethylamino)-3-oxo-3H-benzo-[c]xanthene-7-yl]-benzenedicarboxylic acid
TCPO	bis (2,4,6-trichlorophenyl) oxalate
UV–Vis	ultraviolet–visible (range)

REFERENCES

1. M. E. Szulc and I. S. Krull, *J. Chromatogr.* **659**, 231–245 (1994).
2. E. S. Yeung, in *Capillary Electrophoresis Technology Chromatogr. Sci. Ser.* (N. A. Guzman, ed.), pp. 587–603, Vol. 64. Dekker, New York, 1993.
3. R. Kuhn and S. Hoffstetter-Kuhn, *Capillary Electrophoresis: Principles and Practice*, pp. 109–151. Springer-Verlag, Berlin, 1993.
4. S. L. Pentoney, Jr. and J. V. Sweedler, in *Handbook of Capillary Electrophoresis*, (J. P. Landers, ed.), pp. 147–183. CRC Press, Boca Raton, FL, 1994.

5. E. S. Yeung in *Adv. Chromatogr.* **35**, 1–51 (1995).
6. D. R. Baker, *Capillary Electrophoresis*, pp. 113–141. Wiley, New York, 1995.
7. Y. F. Cheng and N. J. Dovichi, *Science* **242**, 562–564 (1988).
8. S. Tracht, V. Toma, and J. V. Sweedler, *Anal. Chem.* **66**, 2382–2389 (1994)
9. S. C. Beale and S. J. Sudmeier, *Anal. Chem.* **67**, 3367–3371 (1995).
10. S. L. Pentoney, Jr., R. N. Zare, and J. F. Quint, *Anal. Chem.* **61**, 1642–1647 (1989).
10a. W. G. Kuhr and E. S. Yeung, *Anal. Chem.* **60**, 2642–2646 (1988).
10b. T. Odake, T. Kitamori, and T. Sawada, *Anal. Chem.* **67**, 145–148 (1995).
11. J. W. Jorgenson and K. D. Lukacs, *Science* **222**, 266–272 (1983).
12. J. K. Towns and F. E. Regnier, *Anal. Chem.* **63**, 1126–1132 (1991).
13. V. Kašička, Z. Prusik, P. Mudra, and J. Štěpánek, *J. Chromatogr.* **709**, 31–38 (1995).
14. J. L. Carpenter, P. Camilleri, D. Dhanak, and D. Goodall, *J. Chem. Soc., Chem. Commun.* **11**, 804–806 (1992).
15. M. S. Bello, R. Rezzonico, and P. G. Righetti, *J. Chromatogr.* **659**, 199–204 (1994).
16. G. J. M. Bruin, G. Stegeman, A. C. van Asten, X. Xu, J. C. Kraak, and H. Poppe, *J. Chromatogr.* **559**, 163–182 (1991).
17. D. A. Skoog and J. J. Leary, *Principles of Instrumental Analysis*, 4th ed., pp. 123–140. Harcourt Brace Jovanovich, Orlando, FL, 1992.
18. Y. Xue and E. S. Yeung, *Anal. Chem.* **65**, 1988–1993 (1993).
19. Y. Xue and E. S. Yeung, *Appl. Spectrosc.* **48**, 502–506 (1994).
20. A. E. Bruno, F. Maystre, B. Krattiger, P. Nussbaum, and E. Gassmann, *Trends Anal. Chem.* **13**, 190–198 (1994).
21. M. Albin, P. D. Grossman, and S. E. Moring, *Anal. Chem.* **65**, 489A–497A (1993).
22. W. Beck, R. vanHoek, and H. Engelhardt, *Electrophoresis* **14**, 540–546 (1993).
23. T. Wang, J. H. Aiken, C. W. Huie, and R. A. Hartwick, *Anal. Chem.* **63**, 1372–1376 (1991).
24. J. A. Taylor and E. S. Yeung, *J. Chromatogr.* **550**, 831–837 (1991).
25. T. Tsuda, J. V. Sweedler, and R. N. Zare, *Anal. Chem.* **62**, 2149–2152 (1990).
26. S. E. Moring, R. T. Reel, and R. E. J. van Soest, *Anal. Chem.* **65**, 3454–3459 (1993).
27. D. N. Heiger, P. Kaltenbach, and H. J. P. Sievert, *Electrophoresis* **15**, 1234–1247 (1994).
28. P. Boček, M. Deml, and J. Janák, *J. Chromatogr.* **106**, 283–290 (1975).
29. M. Munk, in *A Practical Guide to HPLC Detection* (D. Parriott, ed.), pp. 26–28. Academic Press, San Diego, 1993.
30. K. Kakehi, A. Susami, A. Taga, S. Suzuki, and S. Honda, *J. Chromatogr.* **680**, 209–215 (1994).
31. E. S. Yeung and W. G. Kuhr, *Anal. Chem.* **63**, 275A–282A (1991).

32. W. Buchberger, S. M. Cousins, and P. R. Haddad, *Trends Aanl. Chem.* **13**, 313–319 (1994).
33. Y. Xue and E. S. Yeung, *Anal. Chem.* **65**, 2923–2927 (1993).
34. R. W. Giuffre and H. J. P. Sievert, *Pittsburgh Conf.*, New Orleans, LA, *1995*, Abstr. No. 891, (1995).
35. T. E. Wheat, F. M. Chiklis, and K. A. Lilley, *J. Liq. Chromatogr.* **18**, 3643–3657 (1995).
36. R. Grimm, A. Graf, and D. N. Heiger, *J. Chromatogr.* **679**, 173–180 (1994).
37. D. C. Nguyen, R. A. Keller, J. H. Jett, and J. C. Martin, *Anal. Chem.* **59**, 2158–2161 (1987).
38. Y. H. Lee, R. G. Maus, B. W. Smith, and J. D. Winefordner, *Anal. Chem.* **66**, 4142–4149 (1994).
39. Q. L. Mattingly, P. Vegunta, and S. A. Soper, *Anal. Chem.* **65**, 740–747 (1993).
40. A. T. Timperman, K. Khatib, and J. V. Sweedler, *Anal. Chem.* **67**, 139–144 (1995).
41. D. Y. Chen, X. L. Chen-Aldelhem, and N. J. Dovichi, *Analyst (London)* **119**, 349–352 (1994).
42. D. Y. Chen, H. P. Swerdlow, H. R. Harke, J. Z. Zhang, and N. J. Dovichi, *J. Chromatogr.* **559**, 238–246 (1991).
43. S. Wu and N. J. Dovichi, *J. Chromatogr.* **480**, 141–155 (1990).
44. E. S. Yeung, P. Wang, W. Li, and R. W. Griese, *J. Chromatogr.* **608**, 73–77 (1993).
45. B. Nickerson and J. W. Jorgenson, *J. Chromatogr.* **480**, 157–168 (1989).
46. J. W. Jorgenson and K. D. Lukacs, *Anal. Chem.* **53**, 1298–1302 (1981).
47. J. W. Jorgenson and K. D. Lukacs, *J. Chromatogr.* **218**, 209–216 (1981).
48. J. S. Green and J. W. Jorgenson, *J. Chromatogr.* **352**, 337–343 (1986).
49. E. Arriaga, D. Y. Chen, X. L. Cheng, and N. J. Dovichi, *J. Chromatogr.* **652**, 347–53 (1993).
50. E. Gassmann, J. E. Kuo, and R. N. Zare, *Science* **230**, 813–814 (1985).
51. T. Higashijima, T. Fuchigami, T. Imasaka, and N. Ishibashi, *Anal. Chem.* **64**, 711–714 (1992).
52. D. C. Williams and S. A. Soper, *Anal. Chem.* **67**, 3427–3432 (1995).
53. R. J. Williams, M. Lipowska, G. Patonay, and L. Strekowski, *Anal. Chem.* **65**, 601–605 (1993).
54. D. B. Shealy, M. Lipowska, J. Lipowski, N. Narayanan, S. Sutter, L. Strekowski, and G. Patonay, *Anal. Chem.* **67**, 247–251 (1995).
55. D. F. Swaile and M. J. Sepaniak, *J. Liq. Chromatogr.* **14**, 869–893 (1991).
56. T. Lee and E. S. Yeung, *J. Chromatogr.* **595**, 319–325 (1992).
57. K. C. Chan, G. M. Janini, G. M. Muschik, and K. L. Issaq, *J. Liq. Chromatogr.* **16**, 1877–1890 (1993).
58. H. T. Chang and E. S. Yeung, *Anal. Chem.* **67**, 1079–1083 (1995).

59. A. T. Timperman, K. E. Oldenburg, and J. V. Sweedler, *Anal. Chem.* **67**, 3421–3426 (1995).
60. S. A. Shippy, J. A. Jankowski, and J. V. Sweedler, *Anal. Chim. Acta* **307**, 163–171 (1995).
61. L. Hernandez, R. Marquina, J. Escalona, and N. A. Guzman, *J. Chromatogr.* **502**, 257–255 (1990).
62. L. Hernandez, N. Joshi, J. Escalona, and N. Guzman, *Pittsburgh Conf.*, Chicago, 1991, Abst. No. 15 (1991).
63. D. F. Swaile and M. J. Sepaniak, *J. Microcolumn Sep.* **1**, 155 (1989).
64. J. V. Sweedler, *Crit. Rev. Anal. Chem.* **24**, 59–98 (1993).
65. J. V. Sweedler, J. B. Shear, H. A. Fishman, and R. N. Zare, *Anal. Chem.* **63**, 496–502 (1991).
66. S. Carson, A. S. Chohen, A. Belenii, M. C. Ruiz-Martinez, J. Berka, and B. L. Karger, *Anal. Chem.* **65**, 3219–3226 (1991).
67. A. E. Karger, J. M. Harris, and R. F. Gestland, *Nucleic Acids Res.* **19**, 4955–4962 (1991).
68. A. T. Timperman and J. V. Sweedler, *Analyst (London)* **121**, 45R–52R (1996).
69. K. Miller and F. E. Lytle, *J. Chromatogr.* **648**, 245–250 (1993).
70. M. Latva, T. Ala-Kleme, H. Bjennes, J. Kankare, and K. Haapakka, *Analyst (London)* **120**, 367–372 (1995).
71. M. Latva, T. Ala-Kleme, H. Bjennes, J. Kankare, and K. Haapakka, *J. Chromatogr.* **608**, 85–92 (1992).
72. P. L. Christensen and E. S. Yeung, *Anal. Chem.* **61**, 1344–1347 (1989).
73. D. E. Burton, M. J. Sepaniak, and M. P. Maskarinec, *J. Chromatogr. Sci.* **24**, 347–351 (1986).
74. M. C. Roach, P. Gozel, and R. N. Zare, *J. Chromatogr.* **426**, 129–140 (1988).
75. N. J. Reinhoud, U. R. Tjaden, H. Irth, and J. van der Greef, *J. Chromatogr.* **574**, 327–334 (1992).
76. H. Soini and M. V. Novotny, *J. Microcolumn Sep.* **4**, 313–318 (1992).
77. N. Wu, J. V. Sweedler, and M. Lin, *J. Chromatogr.* **654**, 185–191 (1994).
78. N. Wu, B. Li, and J. V. Sweedler, *J. Liq. Chromatogr.* **17**, 1917–1927 (1994).
79. S. Nie, R. Dadoo, and R. N. Zare, *Anal. Chem.* **65**, 3571–3575 (1995).
80. R. E. Milofsky and E. S. Yeung, *Anal. Chem.* **65**, 153–157 (1993).
81. R. P. Haugland, *Molecular Probes*. Molecular Probes, Eugene OR, 1992–1994.
82. R. T. Kennedy and J. W. Jorgenson, *Anal. Chem.* **61**, 436–441 (1989).
83. B. R. Cooper, R. M. Wightman, and J. W. Jorgenson, *J. Chromatogr.* **653**, 25–34 (1994).
84. B. L. Hogan and E. S. Yeung, *Anal. Chem.* **64**, 2841–2845 (1992).
85. O. Orwar, H. A. Fishman, N. E. Ziv, R. H. Scheller, and R. N. Zare, *Anal. Chem.* **67**, 4261–4268 (1995).
86. S. D. Gilman and A. G. Ewing, *Anal. Chem.* **67**, 58–64 (1995).

87. F. E. Regnier, D. H. Patterson, and B. J. Harmon, *Trends Aanl. Chem.* **14**, 177–181 (1995).
88. Q. F. Xue and E. S. Yeung, *Anal. Chem.* **66**, 1175–1178 (1994).
89. Q. F. Xue and E. S. Yeung, *Nature (London)* **373**, 681–683 (1995).
90. J. Y. Zhao, N. J. Dovichi, O. Hindsgaul, S. Gosselin, and M. M. Palcic, *Glycobiology* **4**, 239–242 (1994).
91. Y. N. Zhang, X. C. Le, N. J. Dovichi, C. A. Compston, M. M. Palcic, P. Diedrich, and O. Hindsgaul, *Anal. Biochem.* **227**, 368–376 (1995).
92. T. Tsuda, T. Mizuno, and J. Akiyama, *Anal. Chem.* **59**, 799–800 (1987).
93. D. J. Rose and J. W. Jorgenson, *Anal. Chem.* **60**, 642–648 (1988).
94. M. Albin, R. Weinberger, and E. Sapp, *Anal. Chem.* **63**, 417–422 (1991).
95. S. D. Gilman, J. J. Pietron, and A. G. Ewing, *J. Microcolumn Sep.* **6**, 373–384 (1994).
96. R. Dadoo, A. G. Seto, L. A. Colon, and R. N. Zare, *Anal. Chem.* **66**, 303–306 (1994).
97. N. Wu, and C. W. Huie, *J. Chromatogr.* **634**, 309–315 (1993).
98. T. Hara, H. Nishida, and R. Nakajima, *Anal. Sci.* **10**, 823–825 (1994).
99. T. Hara, J. Yokogi, S. Okamura, S. Kato, and R. Nakajima, *J. Chromatogr.* **652**, 361–367 (1993).
100. T. Hara, H. Nishida, S. Kayama, and R. Nakijima, *Bull. Chem. Soc. Jpn.* **67**, 1193–1195 (1994).
101. M. A. Ruberto and M. L. Grayeski, *Anal. Chem.* **64**, 2758–2762 (1992).
102. M. A. Ruberto and M. L. Grayeski, *J. Microcolumn Sep.* **6**, 545–550 (1994).
103. R. Dadoo, L. A. Colón, and R. N. Zare, *J. High Res. Chromatogr.* **15**, 133–135 (1992).
104. J. Y. Zhao, J. Labbe, and N. J. Dovichi, *J. Microcolumn Sep.* **5**, 331–339 (1993).
105. C. Y. Chen, T. Demana, S. D. Huang, and M. D. Morris, *Anal. Chem.* **61**, 1590–1593 (1989).
106. J. Pawliszyn, *Anal. Chem.* **60**, 2796–2801 (1988).
107. D. J. Bornhop and N. J. Dovichi, *Anal. Chem.* **58**, 504–505 (1986).
108. D. J. Bornhop and N. J. Dovichi, *Anal. Chem.* **59**, 1632–1636 (1987).
109. R. E. Synovec, *Anal. Chem.* **59**, 2877–2884 (1987).
110. A. E. Bruno, B. Krattiger, F. Maystre, and H. M. Widmer, *Anal. Chem.* **63**, 2689–2697 (1991).
111. B. Krattiger, G. J. M. Bruin, and A. E. Bruno, *Anal. Chem.* **66**, 1–8 (1994).
112. J. Pawliszyn, *J. Liq. Chromatogr.* **10**, 3377–3392 (1987).
113. J. Pawliszyn and J. Wu, *J. Chromatogr.* **559**, 111–118 (1991).
114. J. Pawliszyn and J. Wu, *Anal. Chem.* **63**, 1884–1889 (1991).
115. J. Wu, P. Frank, and J. Pawliszyn, *Appl. Spectrosc.* **46**, 1837–1840 (1992).
116. J. Wu and P. Pawliszyn, *Anal. Chem.* **64**, 219–224 (1992).
117. J. Wu and J. Pawliszyn, *Anal. Chem.* **64**, 224–227 (1992).

118. M. Yu and N. J. Dovichi, *Appl. Spectrosc.* **43**(2), 196–201 (1989).
119. M. Yu and N. J. Dovichi, *Mikrochim. Acta* **3**, 27–40 (1989).
120. C. W. Earle and N. J. Dovichi, *J. Liq. Chromatogr.* **12**, 2574–2585 (1989).
121. J. M. Saz, B. Krattiger, A. E. Bruno, J. C. Diez-Masa, and H. M. Widmer, *J. Chromatogr.* **699**, 315–322 (1995).
122. K. C. Waldron and N. J. Dovichi, *Anal. Chem.* **64**, 1396–1399 (1992).
123. M. Chen, K. C. Waldron, Y. Zhao, and N. J. Zovichi, *Electrophoresis* **15**, 1290–1294 (1994).
124. D. Kaniansky, P. Rajec, A. Švec, J. Marák, M. Koval, M. Lúčka, Š. Franko, and G. Sabanoš, *J. Radioanal. Nucl. Chem.* **129**, 305–325 (1989).
125. D. Kaniansky, P. Rajec, A. Švec, P. Havaši, and F. Macášek, *J. Chromatogr.* **258**, 238–243 (1983).
126. K. D. Altria, C. F. Simpson, A. K. Bharij, and A. E. Theobald, *Electrophoresis* **11**, 732–734 (1990).
127. S. L. Pentoney, Jr., R. N. Zare, and J. F. Quint, *J. Chromatogr.* **480**, 259–270 (1989).
128. G. Westerberg, H. Lundqvist, F. Kilár, and B. Långström, *J. Chromatogr.* **645**, 319–325 (1993).
129. J. A. Jankowski, S. Tracht, and J. V. Sweedler, *Trends Anal. Chem.* **14**, 170–176 (1995).
130. J. Caslavska, E. Gassmann, and W. Thormann, *J. Chromatogr.* **709**, 147–156 (1995).
131. J. Wu and J. Pawliszyn, *Anal. Chem.* **66**, 867–873 (1994).
132. R. L. McCreery, in *Charge Transfer Devices in Spectroscopy* (J. V. Sweedler, K. L. Ratzlaff, and M. B. Denton, eds.), pp. 227–280. VCH, New York, 1994.
133. C. Y. Chen and M. D. Morris, *J. Chromatogr.* **540**, 355–363 (1991).
134. P. A. Walker, III, W. K. Kowalchyk, and M. D. Morris, *Anal. Chem.* **67**, 4255–4260 (1995).
135. W. K. Kowalchyk, P. A. Walker, III, and M. D. Morris, *Appl. Spectrosc.* **49**, 1183–1188 (1995).
136. K. Albert and E. Bayer, in *HPLC Detection: Newer Methods* (G. Patoney, ed.), pp. 197–229. VCH Publishers, New York, 1992.
137. N. Wu, T. L. Peck, A. G. Webb, R. L. Magin, and J. V. Sweedler, *J. Am. Chem. Soc.* **116**, 7929–7930 (1994).
138. D. L. Olson, T. L. Peck, A. G. Webb, and J. V. Sweedler, *Science* **270**, 1967–1970 (1995).
139. J. V. Amari, P. R. Brown, and J. G. Turcotte, *Am. Lab.* **24**, 26–36 (1992).
140. L. B. Allen, J. A. Koropchak, and B. Szostek, *Anal. Chem.* **67**, 659–666 (1995).
141. B. Szostek and J. A. Koropchak, *Fed. Anal. Chem. Spectrosc. Soc. (FACSS) Conf.*, Cincinnati, OH, 1995, Abstr. No. 536 (1995).

142. K. C. Lewis and J. W. Jorgenson, *High Perform. Capillary Electrophor.*, 7th, Wurzburg, *1995*, Poster No. 410 (1995).
143. P. J. Wyatt and G. R. Janik, *Pittsburgh Conf.*, New Orleans, LA, 1995, Abst. No. 802 (1995).
144. P. J. Wyatt, *Anal. Chim. Acta* **272**, 1–40 (1993).
145. Z. Rosenzweig and E. S. Yeung, *Anal. Chem.* **66**, 1771–1776 (1994).
146. J. B. Shear, H. A. Fishman, N. L. Allbritton, P. Garigan, R. N. Zare, and R. H. Scheller, *Science* **267**, 74–77 (1995).
147. H. A. Fishman, O. Orwar, R. H. Scheller, and R. N. Zare, *Proc. Natl. Acad. Sci. U.S.A.* **92**, 7877–7881 (1995).

CHAPTER
10

ELECTROCHEMICAL DETECTION IN HIGH-PERFORMANCE CAPILLARY ELECTROPHORESIS

BARBARA RHODEN BRYANT, FRANKLIN D. SWANEK,
and ANDREW G. EWING

Department of Chemistry, Penn State University, University Park, Pennsylvania 16802

10.1.	Introduction	355
10.2.	Methods Used to Electrically Isolate the Electrochemical Detector	356
10.3.	Modes of Electrochemical Detection	359
	10.3.1. Amperometry	359
	10.3.2. Scanning Electrochemistry	361
	10.3.3. Potentiometry	361
	10.3.4. Conductivity	361
	10.3.5. Indirect Electrochemical Detection	363
10.4.	Applications of High-Performance Capillary Electrophoresis with Electrochemical Detection	363
10.5.	Future Directions	369
List of Acronyms		371
References		371

10.1. INTRODUCTION

High-performance capillary electrophoresis (HPCE) has proven to be a powerful method of separation since Jorgenson and Lukacs demonstrated its potential as an analytical technique (1–4). The versatility of HPCE is partially derived from its numerous modes of operation: capillary zone electrophoresis (CZE), micellar electrokinetic chromatography (MEKC), capillary gel electrophoresis (CGE), capillary isoelectric focusing (CIEF), and capillary isotachophoresis (CITP). Each mode offers extremely low mass detection limits, high separation efficiency, and small sample volumes. This places great demands on the detection schemes used. The most commonly

High Performance Capillary Electrophoresis, edited by Morteza G. Khaledi. Chemical Analysis Series, Vol. 146.
ISBN 0-471-14851-2 © 1998 John Wiley & Sons, Inc.

used detection schemes include ultraviolet–visible (UV–Vis) absorption (5,6), fluorescence (1,7–9), mass spectrometry (MS) (10–13), and electrochemical (EC) detection (14–16).

UV–Vis absorption, the most widely used method, is pathlength dependent and is therefore limited when small-diameter capillaries are used. Typical concentration detection limits for UV–Vis absorption range from 10^{-5} to 10^{-8} M (17–23). Laser-induced fluorescence detection schemes are more sensitive but are often limited by the range of excitation wavelengths provided by the laser and the need for sample derivatization. Attomole detection limits are common, and zeptomole detection limits have been obtained (8,9). MS is ideal for elucidating the structure of very small quantities of analytes, but there is difficulty in coupling it to HPCE since on-column schemes are not yet feasible. This results in a loss of the high efficiency associated with HPCE. Mass detection limits have been reported in the attomole range (24).

EC methods are highly selective and useful for very sensitive detection of many electroactive species without prior sample derivatization. Typical detection limits range from 10^{-17} to 10^{-19} mol (17). Detection limits are not compromised by the small dimensions inherent in HPCE since EC detection is not pathlength dependent. An additional advantage of EC methods is that there is no requirement for an optical carrier, which results in less costly instrumentation. The major limitation of EC detection is related to the noise from the high separation voltage interfering with the detection. Current as high as 100 µA may pass through the capillary whereas the current detected may be as small as 100 fA. Several methods have therefore been developed to isolate the electrophoretic current from the EC cell.

10.2. METHODS USED TO ELECTRICALLY ISOLATE THE ELECTROCHEMICAL DETECTOR

Off-column and end-column detection schemes are used to decrease the effect of the separation potential on the electrode. In order to perform off-column detection it is necessary to electrically isolate or decouple the separation capillary from the EC cell. To do this, the system is generally split into two segments: a separation capillary and a detection capillary. At the point where the two capillaries meet, a crack or joint is used to electrically ground the CE system. Separation only occurs up to the point in the capillary where the joint is placed. Electroosmotic flow from the separation capillary pushes solutes from the joint to the detector. The distance from the joint to the detector must therefore be kept short to avoid dispersion. The use of off-column detectors involves electrically isolating a separate detection capillary from the sepa-

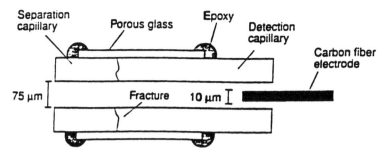

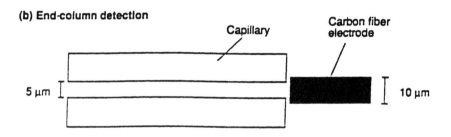

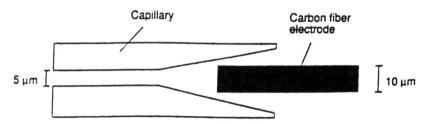

Figure 10.1. Schematic of EC detection schemes for CZE; (a) Porous glass coupler; (b) End-column detection; (c) optimized end-column detection. [From J. M. Mesaros, A. G. Ewing, P. F. Gavin, *Anal. Chem.* **66**, 527A–537A (1994), with permission.]

ration capillary by a grounded crack or gap. The first amperometric detection schemes applied to HPCE employed a porous glass coupler (16). As shown in Figure 10.1, a porous glass tube is positioned into place over a cracked portion of the capillary and sealed with epoxy. The porous glass allows the flow of current to ground the capillary while ions continue to the detector. An electrode is then inserted into the detection capillary, where it is isolated from the high potential field. Detection limits of 300 zmol have been achieved with this technique (25). Disadvantages of this scheme include difficulty in construction and limited availability of porous glass. Yik et al. have demonstrated a similar decoupler with porous graphite tubing, which is more available (26). Huang and Zare have demonstrated a decoupler using an on-column frit (27). A laser is used to drill a hole near the end of the capillary. The hole is then filled with a slurry of solder glass and fused silica. This decoupler has good grounding characteristics but is difficult to construct. O'Shea et al. have demonstrated a decoupler made by covering a crack in a capillary with Nafion tubing (28). The advantages of this system are that Nafion is readily available and fairly simple to construct into a decoupler. Detection limits of 6×10^{-9} M have been reported for hydroquinone. Park et al. have demonstrated that noise due to separation current decreased even further when a gap rather than a crack is used between the separation and detection capillaries (29). By casting a Nafion membrane across a break in the capillary, gaps as long as 1 mm can be constructed. Detection limits of 5×10^{-9} M have been achieved for hydroquinone and 2×10^{-9} M for phenolic acids. However, since Nafion is a cation-exchange membrane, separation efficiency for cations degrades with increasing gap length. Linhares and Kissinger have utilized a simple fracture in the capillary to decouple the separation current (30). This allows electrokinetic injections to be carried out without the sample bias based on charge. Electroosmotic flow is initiated at the fracture and pulls analyte into the end of the capillary without sample bias. Application of this technique to EC detection is possible, but proper capillary alignment is necessary since it lacks the external support present in the other schemes described.

Another approach to isolation of the electrode in HPCE is end-column detection (31). As shown in Figure 10.1b, the electrode is placed just outside the exit of the separation capillary. End-column detection is most effectively used with low-conductivity buffers and small i.d., capillaries since virtually all of the high-potential field associated with the separation potential is dropped across the capillary. The potential drop across the EC cell is on the order of microvolts, allowing for the placement of electrodes at the end of the capillaries without couplers. Detection limits are not as low as off-column methods: 56 amol (5×10^{-7} M) for dopamine. The advantage of this approach partially lies in its ease of application—a micromanipulator and a microscope are needed to place the electrode at the exit of the capillary. Capillaries with 5 μm i.d. have been used here. However, Lu et al. have shown that capillaries as large as 25 μm i.d. can be used without significant interference from background noise (32). Detection limits of 5×10^{-10} M (1.1 amol) for catechol have been achieved with these capillaries. In previously described examples of end-column detection, the electrode has generally had the same or smaller diameter than the i.d. of the capillary. In contrast, Ye and Baldwin have introduced an end-column detector based on a macrodisk (> 100 µm) electrode (33). This technique boasts easier electrode construction, a more rugged electrode, and simpler capillary-electrode alignment. Copper electrodes have been used for the detection of carbohydrates and detection limits of 1×10^{-6} M (1 fmol) have been obtained for ribose.

The difficulties associated with placement of the electrode in the end-column format have led to low precision. Sloss and Ewing (34) circumvented these difficulties by chemically etching the end of capillaries for electrode placement. Figure 10.1c shows the capillary–electrode alignment. Convection currents and analyte diffusion outside the capillary are avoided, so greater coulometric efficiencies of 91% can be achieved. Detection limits of 10 amol for dopamine and 11 amol for catechol have been reported.

10.3. MODES OF ELECTROCHEMICAL DETECTION

The EC methods employed as detectors for HPCE include amperometry, conductivity, potentiometry, scanning electrochemistry, and indirect electrochemistry.

10.3.1. Amperometry

Wallingford and Ewing first demonstrated the use of amperometric detection for HPCE (16). In amperometry, current is measured at a working electrode as analytes undergo oxidation or reduction reactions by the loss or gain of electrons at the electrode surface. The electrode is held at a constant potential relative to the reference electrode, and current, which is proportional to the concentration of analyte present, is measured as a function of time as analytes elute from the capillary. Typical detection limits range from 10^{-18} to 10^{-19} mol.

The off-column and end-column schemes previously described utilize the amperometric mode of detection. Amperometric detection of catecholamines has also been demonstrated with MEKC (35). Although the presence of high concentrations of sodium dodecyl sulfate appeared to interfere with the carbon electrode surface, detection limits of 0.2 fmol were achieved. Surface modification of the electrode with Nafion allows lower detection limits to be achieved (36).

HPCE with amperometric detection has been used in the analysis of single cells (16, 36–38) and microdialysis (39–41). Detection limits of 10^{-7} to 10^{-9} M have been reported for several neurotransmitters. For derivatized amino acids 10^{-7} to 10^{-8} M detection limits have been reported with HPCE–EC–microdialysis. The sensitivity achieved with amperometric detection results in it being the most commonly used EC detection mode. However, slow electron transfer kinetics at the electrode surface can result in oxidation occurring at potentials greater than the thermodynamic potential. The use of chemically modified electrodes has circumvented this problem, since surface-bound mediators catalyze the redox reaction of analytes at reduced potentials. Enzymes and electrocatalysts have been immobilized on carbon paste to detect glucose and cysteine, respectively (42,43). Detection limits in the 10^{-8} M range have been reported. Ru-CN-based modified electrodes have been used in conjunction with HPCE to detect thiols and disulfides (44).

One drawback of amperometric detection is that only easily oxidized or reduced species can be detected. Metal electrodes can sometimes be used to monitor compounds such as glycols, alcohols, and carbohydrates that are not electroactive or not detectable at carbon fibers. These molecules will, however, interact with empty *d*-orbitals on the surface of metal electrodes to form an electrochemically active species. The use of an amalgamated gold wire microelectrode for the detection of nanomolar levels of biological thiols through the catalytic oxidation of mercury-thiol species after CZE analysis has been reported (45). Lin et al. improved this technique and demonstrated the use of a dual gold–mercury amalgam detector for the analysis of disulfides and free thiols (46). The determination of 14 metal ions has been achieved by a gold microdisk electrode with a mercury film (47). Engstrom-Silverman and Ewing developed a 10 μm copper wire microelectrode detector for the analysis of amino acids and catechols (48). The detection principle for this detection scheme is based on the complexation of copper ions with specific analytes at the electrode surface, which results in a change in the solubility of the copper oxide film on the surface of the electrode. This dissolution of the copper oxide film results in an oxidation current that is dependent on the analyte concentration. Subfemtomole detection limits have been achieved. These electrodes have also been used to detect carbohydrates at high pH (49).

A limitation of metal electrodes is the tendency of electrochemically generated products to accumulate on the surface of the electrode. This results in fouling, or loss of the electrode's ability to detect analytes. Fouling may be minimized by the use of pulse amperometric detection (PAD). PAD involves rapid continuous application of a three-step potential waveform to the electrode. The first step, which is used to detect the analytes, is followed by a more positive potential that cleans the electrode surface (the second step). The third step is at a potential that reconditions the electrode for further detection. PAD was first used with HPCE in the analysis of carbohydrates with detection limits of 22.5 fmol (50). In addition Lu and Cassidy have detected Tl^+, Pb^{2+}, and Cu^{2+} with PAD (47).

10.3.2. Scanning Electrochemistry

Although PAD aids in the expansion of the range of analytes detected, no additional information is gained. HPCE with scanning EC detection can be used to provide complete chromatographic and voltammetric information. The technique involves application of a rapid potential sweep to the working electrode in the range in which the analyte is oxidized or reduced. Ferris et al. have achieved detection limits in the range of 360–740 amol for catechols by using this technique (51). Resolution of two neutral solutes, catechol and tryptophan, has also been achieved. Swanek et al. (52) have used this method to identify dopamine in two vesicular compartments of the giant dopamine neuron in the pond snail, *Planorbis corneus*.

10.3.3. Potentiometry

Potentiometric detectors can be coupled to HPCE and used for the simultaneous analysis of organic and inorganic ions with high resolution and sensitivity. These detectors or ion-selective microelectrodes (ISMEs) were first shown in the off-column position (53). Their use on-column has attained precision better than 5% in the range of 10^{-5} to 10^{-3} M for histamine and alkali metals (54). An anion-selective microelectrode has been recently shown to be capable of detecting perchlorate at 5×10^{-8} M (55). In these experiments, an Ag/AgCl-coated Pt wire electrode encased in a 1 μm diameter glass micropipette has been used. The tip is filled with the anion-selective membrane, and the electrode is used on-column with HPCE. Potential differences between the ISME and the reference electrode are monitored. The data obtained are shown in Figure 10.2. This system demonstrates that anion-selective microelectrodes can be used as powerful detectors in HPCE.

10.3.4. Conductivity

Conductivity is the earliest form of EC detection used in HPCE (56). It is a nonselective, bulk-property detection scheme that is a relatively simple means of detecting ionic species in solution and thus is ideally suited for HPCE. In this detection mode, solution flows between two inert indicator electrodes across which a small constant current is applied. The potential across the electrodes is measured when analytes with different conductivity than that of the buffer pass between these electrodes. The signal arises from the difference in the conductance of the electrolyte and analyte ions.

Conductivity detection has been explored since many inorganic ions and aliphatic carboxylic acids display very low optical absorption, in contrast to most biomolecules. Similar systems have been used to check the purity of

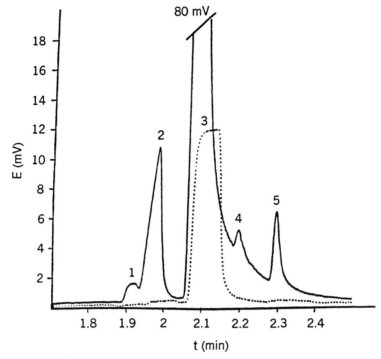

Figure 10.2. A CZE–ISME analysis of 10^{-8} M ClO_4^- in tap water. Conditions were as follows: Buffer; 20 mM Na_2SO_4, adjusted to pH 2.5 with H_2SO_4; separation voltage; 30 kV; electrokinetic injection, 10 kV for 10 s. Uncoated capillary. Solid line = 10^{-8} M $NaClO_4$ in tap water; dotted line = 10^{-8} M ClO_4^- + 20 mM Na_2SO_4 in tap water. Peaks; (1) Br^-; (2) Cl^-; (3) NO_3^-; (4) ClO_4^-; (5) unknown. [From A. Nann and E. Pertsch, *J. Chromatogr.* **676**, 437–442 (1994), with permission.]

synthetic peptides and to determine the amount of chloride, nitrate, and sulfate in drinking water (57). Beckers et al. have described a method for determination of the electrophoretic mobilities of very small inorganic ions and carboxylic acids (58). A dual-conductivity detector system is used to determine their electrophoretic mobilities. Foret and coworkers have adapted a commercial conductivity detector for use off-column with CE and have achieved detection limits of 10^{-6} M for Cl^-, SO_4^{2-}, and NO^{3-} (59).

Huang et al. have described an on-column detector for CE that has been used to detect alkali metal ions and small ionic components in human serum (15). This detector is constructed by epoxying Pt electrodes in 40 μm diameter holes formed by use of a CO_2 laser. Huang et al. have also developed a conductivity detector that detects changes in conductivity outside the end of the capillary (31). This end-column detection scheme is easily implemented.

Unlike other detection methods, conductivity detection shows a direct relationship between effective mobilities and peak area, independent of the kind of ionic species. Hence conductivity detection in HPCE has the unique advantage of quantitating the components in a mixture on an absolute basis with the use of an internal standard. Nonsuppressed conductivity detection in HPCE is limited when there is a large difference in the mobility of the electrolyte and analyte ion, causing excessive peak tailing or fronting. This results in a conflict between optimum sensitivity and separation efficiency. Typical detection limits are comparable to those of UV detection (10^{-6} M). Background noise from the high voltage and the fact that the sensitivity of the method relies on the difference in equivalent conductivity of the analyte and electrolyte ion are other problems associated with the application of nonsuppressed conductivity to the detection of ions. Use of suppressed conductimetric detection provides a low detection background and high sensitivity, therby enhancing the detection limit (10–20 µg/L) (60). Avdalovic et al. have obtained detection limits of 1–10 ppb for common organic and inorganic ions by using suppressed conductivity detection (61).

10.3.5. Indirect Electrochemical Detection

Detection of nonelectroactive compounds can be achieved via the indirect detection scheme (62). This technique is based on displacement of the run buffer by migrating analyte zones. The displacement of deliberately added detectable buffer ions causes a decrease in the signal because the concentration of the additive is lower than steady-state concentrations. Detection of nonelectroactive amino acids and dipeptides has been accomplished by addition of a cationic electrophore, dihydroxybenzylamine (DHBA), to the electrophoretic buffer. As zones of nonelectroactive analyte pass through the detector, a decrease in the current for DHBA oxidation is observed as a negative peak. Figure 10.3 demonstrates schematically the displacement of the additive by the analyte. The lowest detection limit achieved to date is 380 amol for arginine. Indirect amperometry combines the advantages of sensitive detection, instrumental simplicity, and the capability to detect electroactive and nonelectroactive analytes.

10.4. APPLICATIONS OF HIGH-PERFORMANCE CAPILLARY ELECTROPHORESIS WITH ELECTROCHEMICAL DETECTION

HPCE–EC has carved its niche in the area of single-cell analysis (63–65). CE is ideally suited for single-cell analysis owing to its ability to provide rapid,

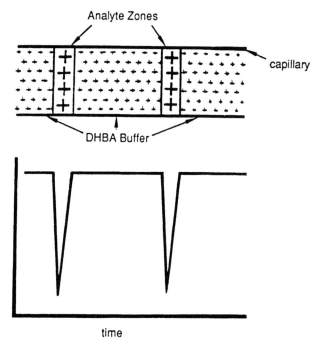

Figure 10.3. Schematic diagram of cation displacement zones of cationic solutes in indirect amperometric detection. [From T. M. Olefirowicz and A. G. Ewing, *Anal. Chem.* **62**, 1872–1876 (1990), with permission.]

high-efficiency separations of ionic species from extremely small volume samples. EC detection provides a sensitive and selective technique for single-cell analysis. The selectivity of EC detection allows analysis of complex biological systems. Many neurotransmitters found in the brain are electroactive and are therefore ideally suited to EC detection. These include catecholamines such as dopamine, epinephrine, and norepinephrine, as well as indolamines such as serotonin.

A number of studies have been performed utilizing CE with EC for single-cell analysis. Analysis of single lymphocytes from human cerebrospinal fluid by CE has led to the discovery of endogenous catecholamines in these cells (25). Figure 10.4 shows an electropherogram of a single human lymphocyte (cell diameter $\approx 7\,\mu m$) analyzed by CE with EC after injection of the whole cell onto an etched 10 μm i.d. capillary by electroosmotic flow. The cationic peak at 10.8 min (peak 1) has been tentatively identified as dopamine based on its electrophoretic mobility, and an average dopamine content of 2.3 ± 1.7 amol ($n = 3$) is found in cerebrospinal fluid lymphocytes. In addition,

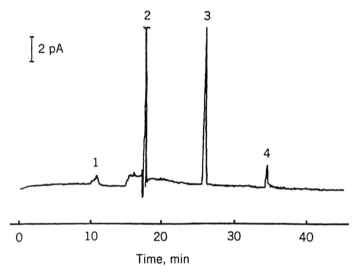

Figure 10.4. Capillary electropherogram of a single human cerebrospinal fluid lymphocyte. Conditions were as follows: separation capillary, 10 μm i.d., 80 cm length, buffer, 25 mM MES (pH 5.65); injection 60 s at 1 kV to inject the cell followed by a 15 s injection of digitonin to permeabilize the cell; separation potential, 25 kV; amperometric detection potential, 0.8 V vs. sodium-saturated calomel reference electrode. Electrophoretic mobilities of the major peaks correspond to the calculated electrophoretic mobilities of dopamine (peak 1), a neutral species (peak 2), uric acid (peak 3), and DOPAC (peak 4). [From J. Bergquist, A. Tarkowski, R. Ekman, A. G. Ewing, *Proc. Natl. Acad. Sci. U.S.A.* **91**, 12912–12916 (1995), with permission.]

the electrophoretic mobility of peak 4 corresponds to dihydroxyphenylacetic acid (DOPAC), a metabolite of dopamine, which further suggests that peak 1 is due to dopamine. Cloned $CD4^+$ T lymphocytes and B lymphocytes have also been found to contain catecholamines. Analysis of three $CD4^+$ T lymphocytes gives a value of 31 ± 29 amol of catecholamine, and one $CD4^+$ B lymphocyte has been found to contain 310 zmol (3.1×10^{-19} mol) of catecholamine. The impressive detection limits in these experiments have been achieved through the use of a porous glass coupler to isolate the detector.

In another application of CE to study of single cells, scanning electrochemical detection has been employed to conclusively identify two vesicular compartments of dopamine from a single snail neuron. Kristensen et al. first reported the observation of two vesicular compartments of dopamine in the giant dopamine neuron of *Planorbis corneus* (66). Injection of an intact cell into a CE capillary, followed by lysing the plasma membrane on-column and separation of the cellular components, resulted in several distinct peaks. A cation corresponding to the electrophoretic mobility of dopamine, a neutral species, and the anionic metabolites uric acid and DOPAC have been detected. A second cation has also been observed. This peak is believed to be due to a second compartment of dopamine that can be observed when the cell is not completely lysed before beginning the separation. The first dopamine peak represents dopamine in the functional compartment, whereas the second represents that in a nonfunctional compartment. When long lyse times are employed, the second peak disappears while the first peak increases. This evidence strongly suggests that the two cations observed in these separations are in fact both dopamine. This hypothesis has been tested using a more qualitative detection scheme.

Scannning EC detection with CE has been used to confirm the identity of both cations in the separation as dopamine (52). Figure 10.5 illustrates the data obtained using scanning detection on the separation of the components of the snail neuron injected onto the capillary and allowed to lyse for 1 min. Figure 10.5a shows a three-dimensional surface plot of data obtained through scanning electrochemical detection. The current response along the time axis shows that three peaks are clearly resolved. This is more clearly seen in Figure 10.5b, which shows the electropherogram at a 0.7 V electrode potential. If the current is observed along the potential axis at a time corresponding to the elution of a peak, the voltammogram for that peak can be seen. This has been plotted in Figure 10.5c. The voltammetry of peaks 1 and 2 have been extracted and plotted together. Shown is the forward wave of a linear sweep voltammogram from 0.0 to 0.6 V vs. a Ag/AgCl electrode. The resulting voltammograms are almost identical. This strongly suggests that both peaks detected from the *Planorbis* neuron are indeed dopamine.

In addition to single-cell analysis, CE–EC has been used to study brain dialysates through coupling with microdialysis. Here, CE–EC has been used to analyze derivatized dialysates from the brain of a living animal (28). Amino acids such as alanine, glutamate, and aspartate are detected electrochemically by derivatizing them with the electroactive tag, napthalene-2,3-dicarboxaldehyde (NDA). The microdialysis probe is inserted into the rat frontoparietal cortex and perfused at 1 µL/min, and samples are collected every 5 min for over an hour following administration of K^+-elevated perfusate. The dialysate samples are derivatized with NDA, and 8 nL injections are used for analysis by CE–EC. Detection limits (including probe recoveries) of 8.2×10^{-8} M,

---

Figure 10.5. (a) Three-dimensional surface plot of electrophoretic separation of the contents of a dopamine cell with scanning EC detection; separation capillary, 80.5 cm long, 30 µm i.d.; buffer, 25 mM MES at pH 5.65; separation potential, 25 kV; detection, carbon fibre electrode scanned from -0.2 V to 0.8 V vs. Ag/AgCl; injection, inject cell at 10 kV for 10 s; lyse in capillary tip for 40 s. (b) Single voltage electropherogram extracted from part (a) at 0.8 V. (c) Background subtracted voltammograms of peaks 1 and 2.

APPLICATIONS OF HPCE–CE DETECTION 367

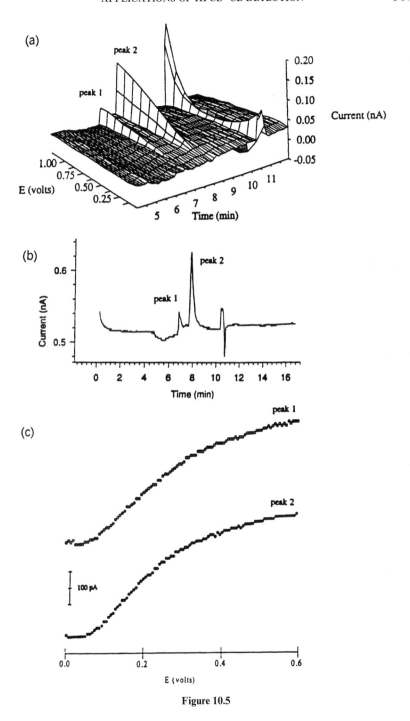

Figure 10.5

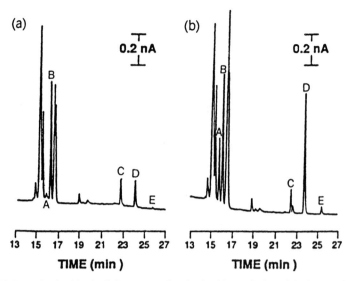

Figure 10.6. Derivatized brain dialysate samples obtained by perfusion with (a) normal Ringer's solution and (b) high K^+ Ringer's solution. Peaks: (a) GABA; (b) alanine; (c) internal standard; (d) glutamate; (e) aspartate. Separation conditions: 0.02 M borate buffer (pH 9.0); voltage, 30 kV; column length, 1 m; detection potential, +800 mV vs. Ag/AgCl. [From T. J. O'Shea, P. L. Weber, B. P. Bammel, C. E. Lunte, and S. M. Lunte, *J. Chromatogr.* **608**, 189–195 (1995), with permission.]

4.8×10^{-7} M, and 3.7×10^{-7} M have been obtained *in vivo* for alanine, glutamate, and aspartate, respectively. Figure 10.6 shows electropherograms from the control (part a) and K^+-elevated (part b) experiments. The excitatory amino acids, glutamate and aspartate, are clearly elevated in the electropherogram in Figure 10.6b as well as an impure peak that contains γ-aminobutyric acid (GABA). A plot of concentration vs. time as determined from dialysate samples shows that aspartate and glutamate levels increase by nearly a factor of 4 over basal levels after K^+ stimulation while alanine levels remain constant.

In an application of CE to ion analysis, Huang et al. have performed HPCE with on-column conductivity detection to quantitate Li^+ in human serum (67). Lithium treatment for acute mania is routinely performed. It is therefore necessary to monitor Li^+ levels in serum, since levels above 3.5 mmol/L are fatal. Quantitation of Li^+ in serum by flame emission photometry and atomic absorption spectroscopy is difficult due to spectral interference from Na^+ and K^+. Huang et al. have used the CZE–conductivity cell system to separate and detect Li^+, K^+, and Na^+. A comparison of normal human serum and serum

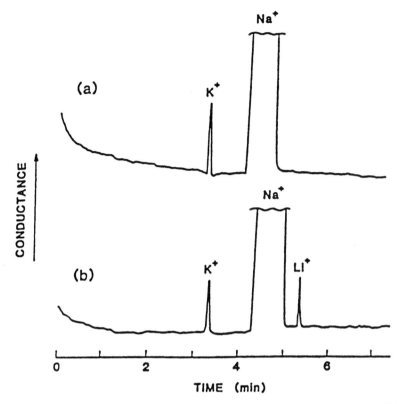

Figure 10.7. Electropherograms of human serum: (a) normal subject; (b) patient on lithium therapy. Dilution is 1:19 with 20 mM 2-(N-morpholino)ethanesulfonic acid (MES)–His buffer, pH 6.1. Conditions: capillary, 75 μm i.d., 70 cm length; gravity injection 10 cm for 30 s; applied voltage, 25 kV. The Na$^+$ peak is off scale. [From X. Huang, M. J. Gordon, R. N. Zare, *J. Chromatogr.* **425**, 385–390 (1988), with permission.]

samples from a patient on Li$^+$ therapy shows that 1.0 mM Li$^+$ is completely resolved from the Na$^+$ peak (Figure 10.7). CZE–conductivity detection has proven to be a viable method for the quantitation of Li$^+$ in human serum.

10.5. FUTURE DIRECTIONS

Current research conducted in the area of EC detection is aimed at improving the detection limits and discovering new applications for the technique. Sinusoidal voltammetry has recently been described (68) and could be applied to CZE with a possibility of achieving nanomolar to picomolar detection limits. In addition, the development of a continuous microseparation technique based on CZE has been reported (65). Continuous separation methods allow dynamic events to be investigated and implementation of EC detection for this continuous separation technique has revealed detection limits in the femtomole region for dopamine. Figure 10.8 shows a schematic of the channel

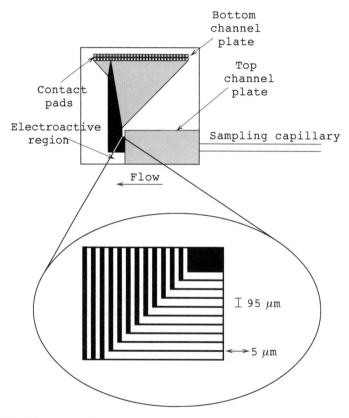

Figure 10.8. Schematic of electrode array detector for continuous electrophoresis in narrow channels. Shaded rectangular portion represents the top channel plate (quartz, 2.5 × 5.0 × 0.23 cm). Electroactive region is composed of 100 individually spaced gold microelectrodes (each 95 μm wide; 5 μm interelectrode spacing). The electrodes are fabricated directly onto the bottom of the channel plate (quartz, 7.5 × 7.5 × 0.23 cm) by standard microlithographic techniques. An individual address to each electrode is accomplished through the termination at the appropriate contact pad. The inset micrograph shows a 10 electrode portion of the array: the white horizontal portions are electroactive regions; the black portions are quartz substrate. [From J. M. Mesaros, P. F. Gavin and A. G. Ewing, *Anal. Chem.* **66**, 527A–537A (1994), with permission.]

with the EC array detector that is used. Application of this technique in the areas of dynamic single-cell analysis, DNA sequencing, and microdialysis should be a major future direction for the development of capillary electrophoresis with electrochemical detection.

ACKNOWLEDGMENTS

The efforts of our colleagues, past and present, in development of HPCE with electrochemical detection and financial support from the National Institutes of Health and the National Science Foundation are gratefully acknowledged.

LIST OF ACRONYMS

Acronyms	Definitions
CE	capilllary electrophoresis
CGE	capillary gel electrophoresis
CIEF	capillary isoelectric focusing
CITP	capillary isotachophoresis
CZE	capillary zone electrophoresis
DHBA	dihydroxybenzylamine
DOPAC	dihydroxyphenylacetate
EC	electrochemical (detection)
GABA	γ-aminobutyric acid
HPCE	high-performance capillary electrophoresis
ISME	ion-selective microelectrode
MEKC	micellar electrokinetic chromatography
MES	2-(N-morpholino)ethanesulfonic acid
MS	mass spectrometry
NDA	naphthalene-2,3-dicarboxaldehyde
PAD	pulse amperometric detection
UV–Vis	ultraviolet–visual (range)

REFERENCES

1. J. W. Jorgenson and K. D. Lukacs, *Anal. Chem.* **53**, 1298–1302 (1981).
2. J. W. Jorgenson and K. D. Lukacs, *Clin. Chem.* **27**, 1551–1553 (1981).
3. J. W. Jorgenson and K. D. Lukacs, *J. Chromatogr.* **218**, 209–216 (1981).
4. J. W. Jorgenson and K. D. Lukacs *Science* **222**, 266–272 (1983).
5. S. Hjertén, *J. Chromatogr.* **347**, 191–198 (1985).

6. S. Hjertén, K. Elenbring, F. Kilar, J. Liao, A. J. Chen, C. J. Siebert, and M. Zhu, *J. Chromatogr.* **403**, 47–61 (1987).
7. W. G. Kuhr and E. S. Yeung, *Anal. Chem.* **60**, 2642–2644 (1989).
8. S. Wu and N. Dovichi, *J. Chromatogr.* **480**, 141–155 (1989).
9. J. V. Sweedler, J. B. Shear, H. A. Fishman, R. N. Zare, and R. H. Scheller, *Anal. Chem.* **63**, 496–502 (1991).
10. R. A. Smith, J. A. Olivares, N. T. Nguyen, and H. R. Udseth, *Anal. Chem.* **60**, 436–441 (1988).
11. E. D. Lee, W. Muck, J. D. Henion, and T. R. Covey, *Biomed. Environ. Mass Spectrom.*, **18**, 844–850 (1989).
12. M. Moseley, L. Deterding, K. Tomer, and J. Jorgenson, *J. Chromatogr.* **480**, 197–210 (1989).
13. R. Caprioli, W. Moore, M. Martin, B. DaGue, K. Wilson, and S. Moring, *J. Chromatogr.* **480**, 247–258 (1989).
14. F. Mikkers, F. Everaerts, and T. Verheggen, *J. Chromatogr.* **169**, 11–20 (1979).
15. X. Huang, T. K. Pang, M. J. Gordon, and R. N. Zare, *Anal. Chem.* **59**, 2747–2749 (1987).
16. R. N. Wallingford and A. G. Ewing, *Anal. Chem.* **59**, 1762–1766 (1987).
17. A. G. Ewing, R. A. Wallingford, and T. M. Olefirowicz, *Anal. Chem.* **61**, 292A–303A (1989).
18. M. Yu and N. J. Dovichi, *Anal. Chem.* **61**, 37–40 (1989).
19. T. Tsuda, J. V. Sweedler, and R. N. Zare, *Anal. Chem.* **62**, 2149–2152 (1990).
20. J. P. Chervet, R. E. J. Van Soest, and M. Ursem, *J. Chromatogr.* **543**, 439–449 (1991).
21. J. A. Taylor and E. S. Yeung, *J. Chromatogr.* **550**, 831–837 (1991).
22. T. Wang, J. H. Aiken, C. W. Huie, and R. A. Hartwick, *Anal. Chem.* **63**, 1372–1376 (1991).
23. Y. Xue and E. S. Yeung, *Anal. Chem.* **65**, 1988–1993 (1993).
24. R. N. Smith, J. H. Whale, D. R. Goodlet, and S. A. Hofstadler, *Anal. Chem.* **65**, 547A–584A (1993).
25. J. Bergquist, A. Tarkowski, R. Ekman, and A. Ewing, *Proc. Natl. Acad. Sci. U.S.A.* **91**, 12912–12916 (1994).
26. Y. F. Yik, H. K. Lee, S. Y. Li, and S. B. Khoo, *J. Chromatogr.* **585**, 139–144 (1991).
27. X. Huang and R. N. Zare, *Anal. Chem.* **63**, 2193–2196 (1991).
28. T. J. O'Shea, R. D. Greenhangen, S. M. Lunte, C. E. Lunte, M. R. Smyth, D. M. Radzik, and N. Watanabe, *J. Chromatogr.* **593**, 305–312 (1992).
29. S. Park, S. M. Lunte, and C. E. Lunte, *Anal. Chem.* **67**, 911–918 (1995).
30. M. C. Linhares and P. T. Kissinger, *Anal. Chem.* **63**, 2076–2078 (1991).
31. X. Huang, R. N. Zare, S. Sloss, and A. G. Ewing, *Anal. Chem.* **63**, 189–192 (1991).

32. W. Lu, R. M. Cassidy, and A. S. Baranski, *J. Chromatogr.* **640**, 433–440 (1993).
33. J. Ye and R. P. Baldwin, *Anal. Chem.* **65**, 3525–3527 (1993).
34. S. Sloss and A. G. Ewing, *Anal. Chem.* **65**, 577–581 (1993).
35. R. A. Wallingford and A. G. Ewing, *Anal. Chem.* **60**, 258–263 (1988).
36. R. A. Wallingford and A. G. Ewing, *Anal. Chem.* **61**, 98–100 (1989).
37. R. A. Wallingford and A. G. Ewing, *Anal. Chem.* **60**, 1972–1975 (1988).
38. T. M. Olefirowicz and A. G. Ewing, *Anal. Chem.* **62**, 1872–1876 (1990).
39. T. J. O'Shea, M. W. Telting-Diaz, S. M. Lunte, C. E. Lunte, and M. R. Smyth, *Electroanalysis* **4**, 463–468 (1992).
40. T. J. O'Shea, P. L. Weber, B. P. Bammel, C. E. Lunte, S. M. Lunte, and M. R. Smyth, *J. Chromatogr.* **608**, 189–195 (1992).
41. S. M. Lunte, M. A. Malone, H. Zuo, and M. R. Smyth, *Curr. Sep.* **13(3)**, 75–78 (1994).
42. T. J. O'Shea and S. M. Lunte, *Anal. Chem.* **66**, 307–311 (1994).
43. J. Zhou, T. J. O'Shea, and S. M. Lunte, *J. Chromatogr.* **680**, 271–277 (1994).
44. J. Zhou and S. M. Lunte, *Anal. Chem.* **67**, 13–18 (1995).
45. T. J. O'Shea and S. M. Lunte, *Anal. Chem.* **65**, 247–250 (1993).
46. B. L. Lin, L. A. Colón, and R. N. Zare, *J. Chromatogr.* **680**, 263–270 (1994).
47. W. Lu and R. M. Cassidy, *Anal. Chem.* **65**, 1649–1653 (1993).
48. C. E. Engstrom-Silverman, and A. G. Ewing, *J. Microcolumn Sep.* **3**, 141–145 (1991).
49. L. A. Colón, R. Dadoo, and R. N. Zare, *Anal. Chem.* **65**, 476–481 (1993).
50. T. J. O'Shea, S. M. Lunte, and W. R. LaCourse, *Anal. Chem.* **65**, 948–951 (1993).
51. S. S. Ferris, G. Lou, and A. G. Ewing, *J. Microcolumn Sep.* **6**, 263–268 (1994).
52. F. D. Swanek, G. Chen, and A. G. Ewing, *Anal. Chem.* **68**, 3912–3916 (1996).
53. R. Virtanen, *Acta Polytech. Scand.* **123**, 1–67 (1974).
54. A. Nann, I. Silvestri, and W. Simon, *Anal. Chem.* **65**, 1662–1667 (1993).
55. A. Nann and E. Pretsch, *J. Chromatogr.* **676**, 437–442 (1994).
56. F. E. P. Mikkers, F. M. Everaerts, and Th. Verheggen, *J. Chromatogr.* **169**, 11–20 (1979).
57. P. Gebauer, M. Deml, P. Boček, and J. Janak, *J. Chromatogr.* **267**, 455–457 (1983).
58. J. Beckers, T. Verheggen, and F. Everaerts, *J. Chromatogr.* **452**, 591–600 (1988).
59. F. Foret, M. Demyl, V. Khale, and P. Boček, *Electrophoresis* **7**, 430–432 (1986).
60. P. K. Dasgupta and L. Bao, *Anal. Chem.* **65**, 1003–1011 (1993).
61. N. Avdalovic, C. A. Pohl, R. D. Rocklin, and J. R. Stillian, *Anal. Chem.* **65**, 1470–1475 (1993).

62. T. M. Olefirowicz and A. G. Ewing, *J. Chromatogr.* **499**, 713–719 (1990).
63. J. A. Jankowski, S. Tracht, and J. V. Sweedler, *Trends Anal. Chem.* **14**, 170–176 (1995).
64. A. G. Ewing, *J. Neurosci. Methods* **48**, 215–224 (1993).
65. J. M. Mesaros, A. G. Ewing, and P. F. Gavin, *Anal. Chem.* **66**, 527A–537A (1994).
66. H. K. Kristensen, Y. Y. Lau, and A. G. Ewing, *J. Neurosci. Methods* **51**, 183–188 (1994).
67. X. Huang, M. J. Gordon, and R. N. Zare, *J. Chromatogr.* **425**, 385–390 (1988).
68. P. Singhal, K. T. Kawagoe, C. N. Christian, and W. G. Kuhr, *Anal. Chem.* **69**, 1662–1668 (1997).

CHAPTER

11

INDIRECT DETECTION IN CAPILLARY ELECTROPHORESIS

HANS POPPE and XIAOMA XU

Laboratory for Analytical Chemistry, Amsterdam Institute for Molecular Studies (AIMS), University of Amsterdam, 1018 WV Amsterdam, The Netherlands

11.1.	Introduction	375
11.2.	Measurement Considerations	377
11.3.	The Transfer Ratio and System Zones	383
	11.3.1. Transfer Ratios	383
	11.3.2. System Zones	386
	11.3.3. The Physical Nature of Eigenvectors	388
	11.3.4. Transfer Ratios in Complex Background Electrolytes	388
	11.3.5. Universal Calibration	389
11.4.	Overload: Electromigration Dispersion	390
11.5.	Noise Induced by the Electrophoretic Process and System Peaks	393
11.6.	Choice of the Background Electrolyte	395
11.7.	Instrumental Improvements	397
11.8.	Applications and Some Special Aspects	398
11.9.	Summary	400
List of Acronyms		401
References		401

11.1. INTRODUCTION

In the early days of high-performance liquid chromatography (HPLC) it was said of this potentially powerful technique, often compared to gas chromatography (GC) in those days, that detection was the main bottleneck. Today, with capillary zone electrophoresis, history appears to repeat itself. The comparison is now to HPLC, and in this comparison detection problems are again a major issue. They are indeed greater than in HPLC; detection has to be

---

High Performance Capillary Electrophoresis, edited by Morteza G. Khaledi. Chemical Analysis Series, Vol. 146.
ISBN 0-471-14851-2 © 1998 John Wiley & Sons, Inc.

performed in the same phase but always in the presence of many ions, and in a few nanoliters of volume rather than in 10–100 µL, as in HPLC with 4–5 mm columns.

For that reason, and because interfacing of a CE separation system to a stand-alone detector is problematic, on-column ultraviolet (UV) detection is the only available method in many cases.

With that—again a situation similar to that in the early days of HPLC—one would be limited to the analysis of UV-absorbing compounds. This is a severe limitation considering the fact that CE is an attractive method for the separation of all kinds of biogenic and bioactive compounds, many of which have little or negligible UV absorption.

Indirect detection, using UV absorption, or using another technique such as fluorometry as the parent detection technique, partly ameliorates this limitation. With its application every ion can be detected, albeit at an often less impressive detection limit than is possible with the same method in the direct mode.

The possibility of indirect detection in capillary electrophoresis (CE) was apparently first proposed by Hjertén et al. (1) and then demonstrated in early studies by Foret et al. (2) and Kuhr and Yeung (3). It stems from the displacement effect in a zone electrophoretic experiment: where the analyte ion is, the concentrations of the background electrolyte (BGE) ions are different from that in the undisturbed BGE. In the simplest model it is assumed that each analyte ion "displaces" its electrochemical equivalent of BGE ions of the same charge sign, the co-ion, thus maintaining electrical neutrality in the zone. For example; in a zone of a three-valent positive ion in a BGE with Na^+ as the only positive ion, the Na^+ concentration in the zone would be diminished by three times the analyte concentration. In a later section this model will be further refined, and it will be shown that the concentration of all ions of the BGE, including those of opposite charge (counterions), are affected. See Figure 11.1 for an illustration of the principle.

Indirect detection becomes possible in the above situation when instead of Na^+ a UV-absorbing ion is chosen, such as a benzylammonium ion (which is generally "detector responsive"). The presence of the analyte in its zone is revealed by the decrease in the concentration of that carefully chosen BGE ion. The French call this latter ion the *ion revélateur*. In English we will have to make do with the less beautiful term "visualizing ion." Although we actually cannot *see* anything, we prefer this to the term "monitoring ion," which is also in use.

As is immediately clear from this mechanism, the method is expected to be universal; all ions should show this effect. The degree of displacement, indicated by the transfer ratio (or displacement ratio) is given in the above model by the ratio of the solute charge to the charge of the visualizing ion. This

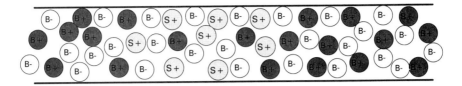

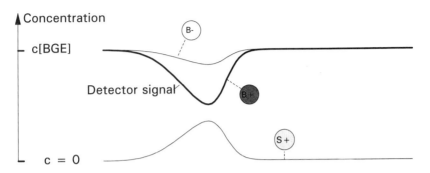

Figure 11.1. General schematic diagram illustrating the overall process of indirect detection for the case where the analyte ion (S^+), moving to the right, is positively chraged, and the background electrolyte consists of a detector-responsive positive ion B^+ (co-ion of S^+) and a nonresponsive negative ion B^-. The horizontal axis is position in capillary. Electroneutrality is maintained, mainly by opposite excursions of S^+ and B^- concentrations, but a small excursion of the B^- concentration also occurs. The asymmetry corresponds to the case where the mobility of S^+ is smaller than that of B^+.

value determines the sensitivity (sensitivity meant here as defined by IUPAC, i.e., the slope of the calibration line, rather than something like the detection limit). Following Yeung, one of the pioneers in this field, and his group (3–6), we indicate the transfer ratio by TR_i:

$$TR_i = -\frac{dc_{vis}}{dc_i} \qquad (11.1)$$

with c_{vis} being the concentration of the visualizing ion.

11.2. MEASUREMENT CONSIDERATIONS

Also clear from the principle described above is the fact that in the measurement two peculiarities occur (see Figure 11.2). In the first place most peaks give a negative signal, something that can cause trouble in some data systems

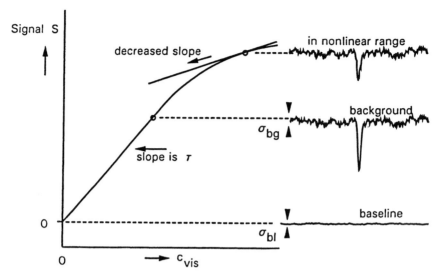

Figure 11.2. Signal S as a function, partly nonlinear, of the concentration of the visualizing ion, c_{vis}, the baseline trace, the background trace with an indirect detection peak, and the background trace with an indirect detection peak observed in the nonlinear range of the detector. Standard deviations in the noise in the traces σ_{bl} and σ_{bg} are indicated by arrows, that are too wide apart (for graphic reasons).

(integrators). We note in passing that switching the two connecting wires may not be such a good remedy: usually one of the electrical connections is kept at some "ground" potential in one or both of the apparatuses, and connecting the "hot" wire to a grounded terminal may cause trouble. It is better to acquire data systems able to handle negative peaks.

The other, more important, point is that the peaks ride on (or rather *hang down from*, because the signal is negative in sign) a high background signal; when UV is used this background signal corresponds to the absorbance of the undisturbed visualizing ion concentration.

Detector noise and drift may be substantially larger due to the presence of the visualizing ion. It is therefore useful to make a clear distinction here. On the one hand, there is noise and drift as observed with what will be called a "non-responsive" buffer (producing no signal whatsoever from any fluorescence or absorption present, the situation that prevails with regular direct detection). This condition will be indicated in the following discussion (rather arbitrarily) by the term "baseline." On the other hand, the condition when the visualizing ion is present will be indicated by the term "background." Thus, *baseline noise* is defined here as what the detector gives when no responding component is present, whereas *background noise*, nearly always larger, is what it gives when the visualizing ion is present. [The terms baseline and background are ambiguous; the choice made here is mainly based on the useful association between "background" and "background electrolyte" (BGE).]

The high background signal can have two detrimental effects. First, the linearity of the detection method can be degraded at the high signal value. When peaks are visualized in a CE indirect detection experiment, the calibration line for the visualizing ion—and its curvature—is followed backward, as the calibration for the sample starts at the high signal level (see Figure 11.3). At low analyte concentrations we still may have acceptable linearity, but there is less response (most calibration lines curve downward!) than was expected on the basis of the transfer ratio. The second detrimental effect is that the noise level of the detector at the high concentration may be larger than that with a nonresponsive buffer. In earlier work on indirect detection in ion exchange chromatography (7) it was even assumed that noise is strictly proportional to the signal level. This point appeared so important that Yeung [see Pfeffer et al.

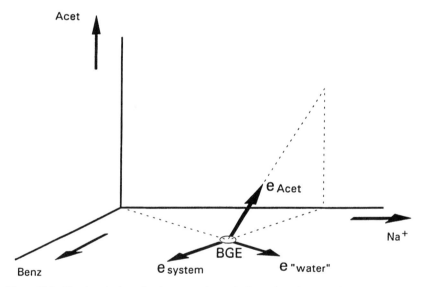

Figure 11.3. The description of a three-constituent indirect detection experiment, with a BGE consisting of benzoic acid/benzoate (Benz) as the buffer and sodium ion. In the axes labeled "Acet" and "Benz" the charges are omitted, as the plotted concentrations stand for the total (dissociated as well as undissociated) amount of material. In the ground plane, formed by "Benz" and Na^+, the BGE constituents, the eigenvectors of two system zones are plotted: $e_{"water"}$ is one, corresponding to the zone with zero mobility; e_{system} corresponds to a zone with a nonzero mobility, and it satisfies the Kohlrausch equation. The vector marked e_{Acet} is the one corresponding to the analyte (acetate) zone; its slope in the plane formed by "Acet" and "Benz" equals the negative of the replacement ratio, $-TR$.

(8)] proposed a new term to describe it: the "dynamic reserve" DR. Formally,

$$\mathrm{DR} = \frac{\text{signal level}}{\text{noise level}} = \frac{S}{\sigma} \qquad (11.2)$$

where S stands for the signal and σ for the standard deviation in the signal.

Purists will remark that this is (as indicated by the original authors) nothing other than the generally used definition of the signal-to-noise ratio, in which, contrary to what is often assumed, there is no need whatever to take the baseline noise for σ. Despite this caveat as to the nomenclature, the discussion of DR is useful for the following reason:

Assuming for the moment linearity in the detector response, we define the signal, S_{vis}, evoked by the visualizing ion, i, in concentration c_{vis}:

$$S_{vis} = S_{BGE} = c_{vis} \cdot \tau_{vis} \qquad (11.3)$$

where τ_{vis} is the instrumental concentration sensitivity, or calibration slope for the visualizing ion, i.e.,

$$\tau_i = \frac{dS}{dc_{vis}} \quad \{ = \varepsilon_{vis} \cdot b \text{ for UV absorption,}$$
$$\text{with } S \text{ in absorbance units (AU)} \} \qquad (11.4)$$

The background noise level, σ_{bg}, equals [rearranging Eq. (11.2)]:

$$\sigma_{bg} = 1/(\mathrm{DR}) \cdot S = 1/(\mathrm{DR}) \cdot c_{vis} \cdot \tau_{vis} \qquad (11.5)$$

The signal S_i brought about by the presence of an analyte ion i, in concentration c_i is found by considering the change in c_{vis} via the transfer ratio, and equals (assuming that the response of i itself is zero):

$$S_i = (\mathrm{TR}_i) \cdot c_i \cdot \tau_{vis} \qquad (11.6)$$

The signal-to-noise ratio $(S/N)_i$ for a zone of i is found by dividing Eqs. (11.6) and (11.5):

$$\left(\frac{S}{N}\right)_i = \frac{S_i}{\sigma_{bg}} (\mathrm{TR}_i) \cdot (\mathrm{DR}) \cdot \frac{c_i}{c_{vis}} \qquad (11.7)$$

and the detection limit in concentration of the detector (i.e., not in the sample, the latter being higher by a factor equal to the dilution during separation),

c_{LOD}, is found by setting $(S/N)_i$ equal to k, where $k = 2-3$ depending on the significance level required:

$$c_{LOD} = k/[(DR)\cdot(TR_i)]\cdot c_{vis} \qquad (11.8)$$

This clarifying (but under some conditions also deceptive) equation was used by Yeung, and later by many others, as a guide to the choice of conditions. Favorable conditions for good detection limits (at given DR) are high transfer ratios TR_i and a small BGE concentration c_{vis}. A lot more should be said about this, but that will be done in Section 11.6 on the choice of the BGE. The parameter DR describes the suitability of the particular detection system at hand for indirect detection (but see our remarks below).

Surprisingly, the instrumental sensitivity, τ_{vis}, does not occur in Eq. (11.8). This can be explained by considering that having a larger τ_{vis} increases the signal but with a constant DR the noise level increases in the same proportion.

As DR can be quite important, it is useful to consider the factors that determine it and the question of whether it can indeed be treated as a constant for a particular technique. Declaring DR a constant essentially means that we are assuming that we have only proportional noise in the system, i.e., proportional to the signal level itself. Generally, it would be more appropriate to assume that the total noise, σ_{bg}, in the background signal, S_{bg}, also contains contributions from baseline noise, σ_{bl}. Using the rule of addition of variances, one can write

$$\sigma_{bg}^2 = \sigma_{bl}^2 + (\sigma_{prop}\cdot S)^2 \qquad (11.9)$$

where σ_{prop}, the relative standard deviation in the signal, is equal to $1/(DR)$, as follows from the definition of the latter, Eq. (11.2).

Note that σ_{bg} and σ_{bl} have the dimension corresponding to that of the signal scale (millivolts, absorbance units, etc.) but σ_{prop} is dimensionless. A few rearrangements show that DR can only be constant when σ_{bl} is zero; in that case DR equals $1/\sigma_{prop}$.

A condition σ_{bl} equal to zero cannot be generally true since were it so; Eq. (11.2) would predict a zero noise level for a zero c_{vis}. The point is, however, that sometimes the proportional contribution form the high background signal vastly predominates at any reasonable value for c_{vis}.

We shall now discuss this point for fluorescence first (because that is easier) and next for UV absorption (which is more often used).

Fluorescence, especially with laser excitation, is typically a technique where proportional noise is relatively large. In direct detection it is a method with a very low detection limit. When measuring a compound having a reasonable fluorescence yield at concentrations such as those occurring in a BGE, the signal is very large compared to the baseline noise, so that the latter often can be neglected.

The remaining proportional noise nearly always stems from the instability of the excitation source. When the intensity of the excitation source varies with a standard deviation of 1%, each signal generated will have at least the same standard deviation. Thus, for fluorescence the assumption of constant DR is often quite appropriate.

Noise in UV absorption is slightly harder to describe. One part stems from the shot noise in the streams of photons (9,10), having a standard deviation equal to the square root of the count, N, of photons observable in one measuring time constant. In terms of transmission T this contributes a standard deviation equal to $T/\sqrt{N}$. Another part stems from the instability of the source or (having the same effect on the end result) from effects proportional to the light level in the detection path. This part can be described by the relative standard deviation, σ_{source}. Finally, there may be a contribution from the dark-current noise in the photoreceptor that can also be expressed in a standard deviation in photon counts, $\sigma_{N\text{dc}}$. If we assume that only these three effects are present, the noise level in the measured transmission, T, would be

$$\sigma_T^2 = T/N + (\sigma_{\text{source}} T)^2 + \sigma_{N\text{dc}}^2/N^2 \tag{11.10}$$

Translating the transmission noise into absorbance (A) units, with $\sigma_A = 0.43 \, \sigma_T/T$ and $T = 10^{-A}$, we have

$$\sigma_A = 0.4343[1/N \cdot 10^{-A} + (\sigma_{\text{source}})^2 + (\sigma_{N\text{dc}}/N)^2 \, 10^{2A}]^{1/2} \tag{11.11}$$

This expression *never* gives a constant value for DR ($= A/\sigma_A$). The second term is constant. For low values of A the powers of 10 in Eq. (11.11) are close to 1. It is then found that DR $\approx A$/constant, with the constant equal to the normally observed noise level (in AU) of the detector. In other words, contrary to what Eq. (11.5) suggests, the background noise is equal to the baseline noise. When, on the other hand, A is large, i.e., when either the concentration or the absorptivity of the visualizing ion is large, the dark-current contribution [the third term in Eq. (11.11)] will increase and decrease the DR value. As a result, absorbance values in excess of 1 or 2 can hardly be used. There is another, and often even more compelling, reason for that: at such high values the optical purity of the radiation will be impaired; stray-light and bandwidth effects will give a serious downward curvature to the calibration plot.

The above discussion shows that for UV absorption the concept of dynamic reserve is not that useful when only the instrumental effects detection are considered. Often, it is appropriate to substitute Eqs. (11.6) and (11.7) for the signal-to-noise ratio and the detection limit, respectively:

$$(S/N)_i = S_i/\sigma_{\text{bg}} = (TR_i) \cdot c_i/\sigma_{\text{bl}} \tag{11.12}$$

$$c_{\text{LOD}} = k\sigma_{\text{bl}}/(TR_i) \tag{11.13}$$

Wang and Hartwick (11) demonstrated experimentally, however, that additional noise is generated within the capillary and that this typically behaves like proportional noise, i.e., its magnitude increases with the concentration of the visualizing ion. They were convinced that this effect is due to real concentration fluctuations in c_{vis}. These noise contributions again lead to a rather constant value of DR, and in such a situation this term again becomes useful.

Note, however, that its physical meaning is now different from that in the original concept of Yeung, who explicitly referred to the "dynamic reserve of the detector": the value of DR now does not depend on the detection instrument but rather is a powerful function of the chemistry (buffer composition, concentration) and conditions (capillary diameter, field strength) under which CE is carried out. In retrospect we ascribe this type of noise to thermal disturbances (12).

The concept of dynamic reserve is useful for another reason: it indicates how well it is possible to avoid electromigration dispersion. We will come back to that in Section 11.4 on overload.

Other detection devices might be considered for indirect detection, e.g., electrochemical detectors (13,14). Unfortunately, such detectors are not commercially available for CE. Apart from that, we believe that in HPLC as well as in CE electrochemical detection is in general less attractive than fluorescence. The two methods share a number of characteristics: both are capable of extremely low detection limits in terms of concentration (for both this applies only for selected compounds, which is an advantage in indirect detection; both can be carried out in very small volumes, allowing detection of attomole or smaller quantities.

However, the compounds suitable for electrochemical detection (i.e., mostly oxidizable ones) tend to be less stable than compounds that fluoresce well. Also, the response factors in electrochemical detection are often affected by the history of the electrode, and less stable background signals may occur. Thus, at present, UV absorbance and, to a lesser extent, fluorescence are the only alternatives available for the indirect detection technique to be considered in an application laboratory.

11.3. THE TRANSFER RATIO AND SYSTEM ZONES

11.3.1. Transfer Ratios

In the above discussion it was assumed that the transfer ratio, TR_i, equals

$$TR_i = \frac{z_i}{z_{vis}} \tag{11.14}$$

where the z_i and z_{vis} are the charges of the analyte and visualizing ion, respectively. However, a more careful consideration of zone-electrophoretic behavior is needed.

Lets us now consider a strong ion i and a 1:1 BGE consisting of the strong ions "vis" and "count." This case was first treated in 1979 by Mikkers, Everaerts, and Verheggen (15). The composition of the sample zone of i in terms of the respective concentrations c_i, c_{vis}, and c_{count} is uniquely determined by c_i. This dependence of c_{vis} and c_{count} on c_i can be derived from two relations:

1. Electroneutrality must be maintained. This gives the following relation between c_i, c_{vis}, and c_{count}:

$$c_i z_i + c_{vis} z_{vis} + c_{count} z_{count} = 0$$

2. A constant value for the Kohlrausch regulating function (KRF) is necessary [see, e.g., Everaerts et al. (16)]. The function KRF is generally given by:

$$\text{KRF} = \sum_{\text{all ions } k} c_k z_k / \mu_k \quad (11.15)$$

where the z's are the charges of the ions and the μ's the mobilities. Both types of parameters are signed, i.e., negative ions have negative z's and μ's. The expression KRF must be constant in time and, as the zone is supposed to move, has to be the same within and outside the zone containing i. This gives a second relation between c_i, c_{vis}, and c_{count} in the zone:

$$c_i z_i / \mu_i + c_{vis} z_{vis} / \mu_{vis} + c_{count} z_{count} \mu_{count} = \text{constant}$$

where the constant is given by the value for the BGE.

Having two relations pertaining to the three concentrations means that with a given c_i it is possible to find c_{vis} and c_{count}. In this way the transfer ratio TR can be derived.

The result for the foregoing simple case, as has been propounded by, e.g., Nielen (17) and Buchberger et al. (18), is (taking the z's to be unity, i.e., considering only monovalent ions)

$$\text{TR}_{i,vis} = -\frac{\mu_{vis}(\mu_i - \mu_{count})}{(\mu_{count} - \mu_{vis})\mu_i} \quad (11.16)$$

$$\text{TR}_{i,count} = -\frac{\mu_{count}(\mu_i - \mu_{vis})}{(\mu_{count} - \mu_{vis})\mu_i} \quad (11.17)$$

in which two TR's have been used—the first being the one considered so far, toward the visualizing ion; the second, the one towards the counterion.

For values z_i, z_{vis}, and z_{count} different from 1, the two expressions (11.16) and (11.17) have to be multiplied on the right-hand side by z_i/z_{vis} and $-z_i/z_{count}$, respectively. These equations have some important features:

1. They appear entirely symmetrical and indeed are so, formally. However, when one is inserting the μ_{count} value, opposite in sign to the analyte μ_i, it turns out that the transfer ratio for the counterion, $TR_{i,count}$, is much smaller than the transfer ratio for the co-ion, $TR_{i,vis}$. One could use the counterion as the visualizing ion, except that the response would be much smaller, and even zero at a particular position in the electropherogram.
2. When μ_{vis} equals μ_i, the simple result is $TR_{i,count} = 0$ and $TR_{i,vis} = z_i/z_{vis}$, which are the values assumed in the simplified treatment in Section 11.2.
3. In that case, the solute zone has the same conductivity as the BGE, as the only change is the replacement of an ion with another having the same mobility. Thus, this cause (the only one for strong ions) of peak distortion (see below) is absent, and peaks for which $\mu_i = \mu_{vis}$ are symmetrical (15).

In practice more complicated BGE systems have to be used. They nearly always have to also serve as a pH buffer and thus must contain at least one weak acid or base. Many analytes are also weak acids or bases; they may have several dissociation stages, with increasing or decreasing charges (z's). Considering even the least of these complications rapidly leads to very complex expressions. One example, generated with a symbolic mathematical program, can be found in Xu et al. (19). Although such approaches may be precise, they do not provide much intuitive understanding.

Therefore it is useful to think about these phenomena in a more general way. Figure 11.3 serves to illustrate the approach taken by Poppe and colleagues (20–22). The idea of this figure is that the composition of the electrolyte, in the BGE as well as in the zone, can be described as a point in an N-dimensional space, each axis corresponding to a concentration. The concentration of an analyte ion corresponds to one of the axes; other axes (those in the base plane of Figure 11.3) correspond to the BGE constituents. In the example used for Figure 11.3, the analyte is an acetate ion, the BGE consists of benzoate (visualizing ion and co-ion) and a sodium ion (counterion). Although part of two organic compounds may be present in the form of the corresponding acid, there is no need to have more than one axis for each, as the proportion of the ion to the associate free acid is fixed by the remaining composition of the electrolyte. In other words, when the three total concentrations of acetic acid/acetate, benzoic acid/benzoate, and sodium are given, the electrolyte is exhaustively described and with that the pH (and the conductivity) is also known. Thus $N = 3$ suffices. The BGE is described by a point indicated by the flat-lying circle in the figure, on the base plane.

Having a transfer ratio, $TR_{Acet,Benz}$, in Figure 11.3 means that a change, Δ, in c_{Acet}, from the value 0 in the (base plane) BGE, is accompanied by a change in c_{Benz} equal to $-(TR_{Act,Benz})\Delta$. Likewise, c_{Na} changes by $(TR_{Acet,Na})\Delta$. Thus, the replacement process can be visualized in the N-dimensional space by a vector, protruding from the BGE, with components Δ, $(TR_{Acet,Benz})\Delta$, and $(TR_{Acet,Na})\Delta$. How far this vector is followed in the solute zone is determined by the analyte concentration.

Extension of this concept to systems with more components just requires an increase in the number of dimensions (axes), but with two complications. First, graphic representation in a figure becomes impossible. Second, and more importantly, the vectors now have more than three components, $N > 3$. With a given c_i, we have $N - 1$ unknowns. These $N - 1$ (> 2) unknowns cannot be found by employing the two relations electroneutrality and KRF. Instead, one has to use, e.g., the moving-boundary equations (19,20,23) for all components in order to solve the system mathematically.

11.3.2. System Zones

As is indicated in Figure 11.3, there are two other vectors possible in the given system. These are confined to the BGE plane, i.e., the analyte concentration stays at the value zero. These vectors correspond to system zones. An understanding of these is important for achieving an insight into various effects observed in indirect detection.

As has been reported by papers from our laboratory (20,21), as well as other papers on liquid chromatography (22,24), system zones are disturbances in the composition of the BGE or mobile phase, respectively. They migrate through the capillary or column as if they were analyte zones, but they do not contain any analyte. Rather, the concentrations of the components of the BGE or mobile phase are disturbed, the disturbances in each zone being in a fixed proportion (given by the vectors such as $e_{"water"}$ and e_{system} in Figure 11.3), quite similar to the vector formed by the transfer ratios for the analyte. The system zones can be assigned a mobility, or a capacity factor in LC, both being an eigenvalue. Thus, for each system zone there is an eigenvector and a mobility, the latter being an eigenvalue. Also, the system zones are transported in the same manner as are the analytes; on migration they can undergo diffusion and electromigration dispersion, etc.

In a stationary situation system zones are absent or, rather, latent; before they can actually occur they have to be activated in some way or other. In a regular CE experiment, this takes place at the injection. Any injection plug that is not exactly equal in composition to the BGE will activate system zones.

It is not necessary that the injected solution contains "foreign" (analyte) ions; injecting an amount of a slightly modified BGE solution (e.g., diluted, pH-shifted) will generate system peaks. In a nonresponsive BGE (i.e., without using indirect detection) these peaks migrate through the capillary without ever being noticed (although sometimes one sees "differentiated" peak shapes as a result of the refraction effects in a UV detector). With a visualizing agent in the BGE, they all become—to our dismay—visible.

The number of possible system zones depends on the number of constituents in the BGE, N_{BGE}, i.e., the number of components that uniquely determine the composition of the BGE ($N_{BGE} = 2$ in Figure 11.3, as explained above). In pure NaCl, and not taking the H^+ and OH^- ions into account, N_{BGE} would be 1, as electroneutrality requires that $c_{Na^+} = c_{Cl^-}$.

Theory (20,22) predicts that in a BGE with N_{BGE} constituents one can have N_{BGE} system zones. The following rules may be helpful in the interpretation of observed system peaks:

1. One system zone has zero mobility. In it the concentrations in the BGE go up (or down) in proportion to the original concentration, i.e. when one increases by 1%, the others do the same. Application of the Kohlrausch regulating function (KRF; see Section 11.3.1, above) shows that the value of the KRF differs within this zone from that in the undisturbed BGE. It follows that this zone cannot move (relative to the liquid; it does move with the electroosmotic flow when this is present), it must have zero mobility. This is often (and also in Figure 11.3), indicated as the "water peak." Thus, in pure NaCl there is only one system zone, with zero mobility.
2. Other system zones occur only when N_{BGE} is larger than 1. The mobilities of these are a complicated function of the mobilities, pK_a's, and concentrations of the constituents in the BGE.
3. When there are two strong ions of the same sign, there is a system zone having a mobility in between those of the two ions. When the concentration of one ion becomes large compared to the other, the system zone mobility approaches the mobility of the latter ion.

For the simple case of a BGE consisting of ions A and B of the same sign and charge 1, and a counter ion "count" with an opposite single charge, it can be derived, e.g., from the KRF, or from consideration of the moving boundary equations (25), that apart from the system peak with zero mobility ("water") there is a system peak with mobility μ_{system}:

$$\mu_{system} = \frac{c_A \mu_A \mu_B + c_B \mu_A \mu_B - c_B \mu_A \mu_{count} - c_A \mu_B \mu_{count}}{c_A \mu_A + c_B \mu_B - c_A \mu_{count} - c_B \mu_{count}} \quad (11.18)$$

11.3.3. The Physical Nature of Eigenvectors

Many people, including ourselves, find it difficult to get an intuitive understanding of system zones or eigenzones. The question is often asked: why is it that such rules as Eqs. (11.16) and (11.17), or more complex ones for more complex systems, apply so strictly? There seems not to be a straightforward cause-and-effect reasoning that explains these rules; they just appear to be the result of mathematical manipulation of some kind or the other. Also the discussion of the constant value for KRF is not very satisfactory in this respect.

The point has been discussed by Poppe et al. (20,22), with an only partly satisfactory result. The main idea is that disturbances not having the prescribed vector composition will be resolved by the electrophoretic process into several zones that do have this prescribed composition.

11.3.4. Transfer Ratios in Complex Background Electrolytes

A few further rules that can be derived (23,26) for the TRs of an analyte ion may be useful:

1. The TR for weak monovalent solutes is BGE of a weak monovalent buffer constituent and one counterion (such a that in Figure 11.3 can be predicted by means of Eq. (11.16) also when the variation in the degree of dissociation is taken into account. Thus, when calculations are performed with the total concentrations of both weak acid or base constituents, the TR is again given by Eq. (11.16). However, for the prediction of response in practice this is only of use when the visualizing ion is strong. In the opposite case the degree of dissociation as well as the effective absorptivity of the visualizing ion would vary. For instance, when benzoic acid is used, the UV absorption of that acid is different from that of the benzoate ion. When one is operating within the buffering pH range, the proportion of benzoic acid to benzoate in the zone differs from that in the BGE (19).
2. When one ionic constituent matches the analyte in charge and mobility, its concentration is the only one affected in the zone, with TR $= z_i/z_{vis}$; all other concentrations remain constant.
3. When an analyte peak sits close to a system peak, the TR may reach very high values. However, this is not analytically useful, for at the same time peaks are usually strongly deformed.
4. When several rather than one co-ion are present, the TR with respect to each of them is usually smaller than with a single co-ion. It is as if the

response is distributed between the two co-ions. An exception is when rule 2 above, applies and only one co-ion has nonzero TR. It follows that the detector response usually will be decreased, unless both co-ions evoke a response in the detector. Exceptions can occur near system peaks, but those areas in the electropherogram are generally useless.
5. Despite all this, making a first rough guess that TR = z_i/z_{vis} is often the only sensible thing to do, especially when no data on mobilities and the like are yet available.

The use of very complicated expressions for the TR (that have to be evaluated by computer anyway before they can be of practical use) can be avoided by applying a full numerical method, based upon the concept of Figure 11.3. This method has been described by Poppe (20). It enables prediction of the entire electropherogram, including overloading effects (*vide infra*) and TRs, on the basis of pK_a values and mobilities, even for rather complicated mixtures. This is done by solving the moving boundary equations according to a numerical eigenvalue/eigenvector scheme.

The approach has been criticized on various grounds by those who have derived more explicit expressions. However, in all cases where the conditions for such expressions are met [e.g., Eqs. (11.16) and (11.17), or those given in several studies (15, 17–19)], the numerical method yields identical results, while it also reproduces other numerical work (23,26) that has a much narrower range of applicability.

11.3.5. Universal Calibration

Williams et al. (27) have called attention to the fact that relations such as Eqs. (11.16) and (11.17) enable us to perform universal calibration. With direct detection, sensitivies for the analytes, as given by their UV or fluorescence spectra, are in general entirely unrelated and individual calibration for each analyte is necessary. In contrast, when indirect detection is used, the calibration constant of an ion for which no calibration has been carried out can be calculated from that of any other ion or from that of the visualizing ion. The mobilities are needed to do this [e.g., using equations such as (11.16) and (11.17)], but they can be derived from the observed electropherogram.

It appears that this idea has not yet been tested in quantitative terms; indeed, we are not aware of any research where the accuracy and precision of such a procedure have been compared to those of the more elaborate comprehensive calibration procedure.

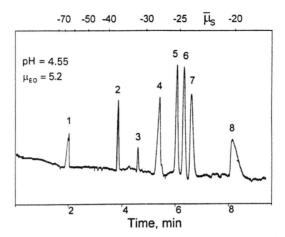

Figure 11.4. CE separation of a mixture of acids; the concentration of each in the sample is about 0.1 mol/L, determined by using indirect detection. Capillary: 75 μm i.d. coated with acrylamide, length 49.5 cm, to detector 30.5 cm. Voltage: -20 kV. BGE: 6-amino-n-hexanoic acid and dimethoxyanthracenesulfonate ($\mu = 25.2 \times 10^{-9}$ m^{2}V^{-1}s^{-1}), 0.002 mol/L, as the visualizing ion, pH 4.55; μ_{eo} is the electroosmotic mobility; μ_S is detection: UV, 268 nm. Peaks: (1) chloride; (2) azide; (3) an impurity; (4) octyl sulfate; (5) undecanesulfonate; (6) decanesulfonate; (7) dodecylsulfonate; (8) acetate. Injection: hydrodynamic, 167 mbar for 0.5 s.

11.4. OVERLOAD: ELECTROMIGRATION DISPERSION

Quite often electropherograms made with indirect detection look like Figure 11.4. Peaks with small and large migration times have a more or less triangular shape: fronting for the early peaks, tailing for the late peaks. This is caused by electromigration dispersion (EMD), an important phenomenon in CE that was treated as early as 1979 by the Everaerts group [see Mikkers et al. (15)], already cited in the foregoing (section 11.3.1), as this discussion produced the TRs as a kind of by-product.

EMD results when the analyte migration rate becomes a function of the analyte concentration, a situation quite analogous to peak deformation in chromatography at high concentration. As long as the analyte concentrations are far below that of the BGE, the effect is negligible.

The migration rate in CE can vary with the concentration as a result of two effects. In the first place the electric field E_α within the analyte zone, α, is usually different from that in the BGE because the conductivity in the zone differs from that in the BGE. As current (density) is constant across the CE tube, and everywhere it holds that $I = \kappa E$, the E_α must satisfy the following equation:

$$E_\alpha = E_\beta \cdot \frac{\kappa_\beta}{\kappa_\alpha} \quad (11.19)$$

As is shown in Mikkers et al. (15), the replacements of BGE ions by analyte ions lead to a change in conductivity that becomes more pronounced as the difference between the mobilities of analyte ions and the co-ion increases. For more complicated cases [e.g., see Poppe (20)], the extent of the change given by Eq. (11.19) can be derived.

A second reason why migration rates at finite analyte concentrations differ is that the degree of dissociation (possibly in several steps) can vary within the zone. Also this effect can be calculated explicitly for some cases (19,23,26), and all effects are numerically accessible with the method described in Poppe (20).

It is beyond the scope of this chapter to treat all the details of EMD. However, in order to asses the possibilities and peculiarities of various BGE systems, a few general rules are needed:

1. As stated earlier, when charges, pK_a, s, and mobilities of the analyte ion match those of some BGE ion, EMD is absent. Usually this means that near one particular position in the electropherogram the peaks are symmetrical.
2. If that is not the case, the larger the ratio $c_i/c_{BGE} = c_i/c_{vis}$ is, the stronger is the effect. Usually, therefore, it is considered good practice to keep c_i below 1–10% of c_{BGE}.
3. It is the *end* concentration (the one recorded by the detector) that matters. Thus, when a highly concentrated sample is injected in such a small volume that the resulting zone is still dilute when reaching the detector, there is little EMD. In other words, for a given CE capillary under given conditions (BGE, voltage), it is the amount of analyte injected that matters.
4. In some special cases the effects brought about by conductivity change and pH shift can have opposite signs, with the result that the peak deformation is less than would be expected on the basis of each of these effects separately (or it may be even absent).

The consequences of all this for indirect detection must now be stated, especially as they constitute some bad news:

From the point of view of peak shape one needs a high BGE concentration. From the point of view of detection, however, low c_{vis} values are nearly always favorable; this is so in the case of constant DR (fluorescence), since according to Eq. (11.7) a low c_{vis} leads to low detection limits. In the case of UV detection, where DR is generally not constant, one is also limited to low c_{vis} values, since the background absorbance should not be too high. Note, however, that favorable c_{vis} concentrations in UV detection are usually higher than those for indirect fluorescence detection. The reason lies in the smaller contribution from proportional noise.

As a result of this compromise, one often sees electropherograms like Figure 11.4, where appreciable EMD is visible.

This state of affairs lends a new significance to the concept of dynamic reserve, DR, discussed earlier as being a means to judge the detection performance of the setup used. When DR is too small it becomes impossible to perform any reasonable quantitative analysis, irrespective of the sample concentrations available. This can be made clear as follows:

In order to avoid EMD, one has to see to it that c_{vis} (assuming it to be the only co-ion) within the zone does not deviate too much from the BGE value. Depending on the relative values of the mobilities, the pH effect, and the desired peak sharpness ("plate number"), c_{vis} may not change more than by 1–10%. Let us assume that the fractional change in c_{vis}, acceptable from the point of view of EMD, equals α_{max} ($0 < \alpha_{max} < 1$). The signal in the peak crest is then

$$S_{max} = \alpha_{max} \tau_{vis} c_{vis} \qquad (11.20)$$

With the background noise σ_{bg} equal to

$$\sigma_{bg} = \tau_{vis} c_{vis}/(DR) \qquad (11.21)$$

one obtains the maximum value of the signal-to-noise ratio that can be obtained:

$$(S/N)_i = \alpha_{max}/(DR) \qquad (11.22)$$

With a typical value of 0.1 for α_{max}, one needs a DR value of at least 100 to obtain an even rather unsatisfactory $(S/N)_i$ value of 10.

Thus, the DR describes how well one can avoid both EMD and degradation of the signal-to-noise ratio at the same time.

The Hartwick group has made an attempt [see Wang and Hartwick (28)] to get around the limitation set by EMD by using a BGE with two visualization ions that have differing mobilities. The idea is (qualitatively) as follows: Each of the mobilities of the two ions correspond to a position in the electropherogram. Analytes at or close to these positions will have zero or negligible EMD, as stated above in rule 1. Having two positions in the electropherogram with zero EMD is preferable is itself, and one might also expect a better overall behavior. Although the equations used by the aforementioned authors are partially incorrect, they were able to show convincingly that this approach works. However, some serious drawbacks should be mentioned: The EMD in between the two "symmetrical peak positions" is actually not better but worse than in the single co-ion case. Also, in this part of the electropherogram system (called a "disturbance" in the paper cited above) peaks occur, effectively blocking a significant part of the electropherogram from interpretation.

11.5. NOISE INDUCED BY THE ELECTROPHORETIC PROCESS AND SYSTEM PEAKS

Virtually all reports on indirect UV detection show, sometimes implicitly, that the noise level observed is higher than would be expected on the basis of the performance of the instrument, as observed in nonresponsive BGE. The most detailed experimental study of this has been published by the Hartwick group [see Wang and Hartwick (11)]. Recent work in our laboratory (12) has clarified some important issues in this area.

The observed additional noise can have various shapes. Sometimes one observes random fluctuations, possibly slower than the instrumental detector noise, and often of considerably larger amplitude. Apart from that, nonrandom excursions of the background signal are seen, some having the width and shape of regular sample peaks, others being even slower and having odd shapes.

Wang and Hartwick in 1992 (11) indicated that all these fluctuations in the background signal most probably reflect actual changes in the concentration of the visualizing ion, c_{vis}. However, they could not ascertain what phenomenon caused these changes. They mentioned in addition the occurrence of system peaks, as noted in the previous section. They showed that the effect increases with increasing driving voltage and conductivity of the BGE. Also, Ma and Zhang (29) observed that smaller-i.d. capillaries can lead to better DR values, that is, to a significant decrease in noise level.

All this suggests that the effect has to do with the heat dissipation in the capillary. This conjecture is in line with the fact that the effect is observed nearly exclusively in UV absorbance work, where usually capillary diameters are in the 50–100 μm range and BGE concentrations are high, as opposed to fluorescence detection, in which the use of smaller diameters and lower concentrations are the rule.

We believe that examining the system zones is essential for a better understanding of these observations (12). The occurrence of system zones explains the presence of undesired disturbances, shaped as peaks (Hartwick's "disturbance peaks") in the electropherogram. They are simply activated in the injection process. With a numerical program (20) based on the knowledge of mobilities and pK_a values, the observed disturbance peaks can be tracked down to system zones in nearly all cases we encountered. This explains a part of the undesired responses in indirect detection, but certain responses and additional random noise remain unexplained.

However, more mechanisms at other positions in the capillary than just the injection process may activate system zones. All that is needed for this is that at these positions some kind of disturbance occurs. The aforementioned experimental evidence suggests that these disturbances are thermal in nature.

It is known that under CE conditions applied in indirect UV detection (i.e., c_{BGE} on the order of 0.01 mol/L; diameter on the order of 100 μm) the thermal dissipation especially in air-cooled system leads to a significant temperature rise in the liquid lumen (30–32); it can amount to a few to tens of degrees Kelvin.

In themselves these temperature deviations do not necessarily lead to the undesired effects under discussion. For instance, when each part of the system is at an elevated temperature—even at differing temperatures in different positions—but stationary (independent of time), no disturbances will be generated. Each ion [or liquid element in the case of electroosmotic flow (EOF)] that passes the detection window has experienced various temperatures, but all ions (or liquid elements) have had the same history and so a stationary concentration of all ions in the detection window must be the result.

However, thermal changes that are nonstationary, i.e., time dependent *and* nonuniform across the length, will generate concentration disturbances (12). Such situations occur when the voltage is switched on, when all parts of the capillary are heated up, simultaneously but with differing rates and with differing final temperatures. Especially important in this respect are the parts where the heat transfer differs strongly from the average value. This is the case in the buffer vials, at the detector position, and at points where the capillary is fastened. At each of these points the electrophoretic transport is disturbed temporarily, the disturbance will activate system peaks, and these system peaks may eventually pass the detection window.

Connecting such observed disturbances, especially their "migration times," to those specific points along the tube requires some care: what is "behind" the detector (i.e., at the side of the end vial) matters too: System peaks generated there, e.g., in the end vial (a "cold spot" in air-cooled systems) with a mobility opposite in sign to regular mobilities of the analytes, will move to the detector window and become visible. Often these "cold spots" or "hot spots" are wider than a normal injection plug, with the result that the generated signal peaks are relatively broad.

Many experimentally observed disturbances can be tracked down to such inadvertent disturbances, while artificially induced thermal disturbances never fail to give the expected responses. Without this insight, the excursions appear to be more or less random.

We assume further that continuous thermal variations are responsible for the remaining effects, and especially for the remaining random noise.

For optimizing indirect detection in CE analytical experiments, these observations are extremely important. First, thermal effects should be kept to a minimum as noise is generated by them. Liquid thermostatting appears preferable, but it is often seen as inconvenient or may not be available. Also, at the position where the capillary leaves the thermostatting bath and enters the buffer vials, less favorable heat transfer is possible; i.e., a "hot spot" may still develop. Generally, therfore, electrical conductivity of the BGE should be kept down, more than in direct detection CE, and as much as possible without degrading resolution (EMD). The alternatives—decreasing the tube diameter or diminishing the electric field strength—are unattractive because of degradation of speed and detection sensitivity, respectively. Second, the more complicated is the BGE, the more system peaks can be expected. It is therefore advisable to keep the BGE composition as simple as possible.

11.6. CHOICE OF THE BACKGROUND ELECTROLYTE

It no doubt has become clear from the preceding discussion that the choice of the BGE is of paramount importance in indirect detection applications. Of course, apart from the requirements imposed by the indirect detection scheme, it is also necessary to take more general requirements of the CE separation into account.

Thus, often the pH in the BGE is dictated by the need to have the solutes ionized; e.g., in the case of sugars (33,34), one needs a pH near 12 to obtain sufficient mobilities. In many other cases it is desired to work at a pH value close to the pK_a's of a group of similar compounds in order to obtain sufficiently large differences in the migration rate.

In the choice of the type of BGE ions one first needs to consider the pH-buffering capacity and the desire to match the mobility of the most important and/or most difficult-to-separate compounds. Taking these aspects into account will lead to symmetrical and narrow peaks where they are most needed (35).

With regard to the BGE concentration, the electrical conductivity is important. It determines power dissipation and the upper limit of the field strength. For that reason generally concentrations above 0.01 mol/L are not useful. On the other hand a too low concentration will result in excessive EMD, even when other choices have been made judiciously. Finally, sometimes special additives to the BGE have to be used, e.g., to induce enantiomeric separation, diminish wall adsorption, or cope with undesired activities of impurities like metal ions [ethylenediaminetetraacetic acid (EDTA)].

In the case of indirect detection this already long list is enlarged by a number of other requirements, which are best itemized (as a preliminary summary of this chapter):

1. In general the visualizing ion should be the co-ion and the only one. Were other nonresponsive co-ions present, the transfer ratio and with that the sensitivity would be affected. Thus, the BGE concentration normally equals c_{vis}.

2. The co-ion should have a stable response in the detector.
3. The detector response factor (fluorescence calibration slope, UV-molar absorptivity) should of course be sufficiently large to have a reasonably large response in the detector.
4. The signal value at the preferred BGE concentration, c_{vis}, should be high enough to reach the maximum DR value. On the other hand, it should not be so high as to bring the detector into the nonlinear range (e.g., AU > 1 in most UV detectors) or bring about excessive proportional noise. When Eq. (11.5)—with constant DR—holds, the smallest concentration that still gives acceptable peak shapes and widths should be chosen.
5. The need to keep thermal dissipation low is even stronger than in other forms of CE, because thermal dissipation generates noise. Thus, low-conductivity BGEs satisfying the above requirements should be preferred to high-conductivity BGEs.
6. System peaks are best avoided by choosing the BGE composition as simple as possible. Especially there should be no ions with the same sign as the visualizing ion, because injection-generated system peaks may emerge in the middle of the electropherogram. With high pH's one should be aware that carbon dioxide from ambient air can gradually turn a simple BGE into one with carbonate, which may generate additional system peaks and additional noise (apart from the shift in pH resulting from the absorption of carbon dioxide, undesirable for other reasons).

As we have shown in earlier sections, fluorescence and UV absorption have somewhat different requirements:

In fluorescence, with constant DR, the magnitude of the detector response factor (τ_{vis} in Section 11.2) does not matter as long as DR indeed remains constant. That is, once a good visualizing ion is available for which DR reflects the capabilities of the instrument, it is pointless to search for another ion having a larger response. Using such a new ion would increase the signal [equal to $(TR)\cdot\tau_{vis}$] and noise [$c_{vis}\cdot\tau_{vis}/(DR)$] in the same proportion, and no improvement would be obtained.

In contrast, for UV absorption the background noise is constant over a large signal range. Here it is worthwhile to search for an ion with a particularly high molar absorptivity, thereby enhancing the signal [$(TR)\cdot\tau_{vis}$] while the instrumental noise is not affected. This is meaningful as long as the absorbance does not become excessive, say, 0.5–1.0 AU (absorbance units). In that case one might well consider choosing a lower value for c_{vis}. The associated smaller conductivity will often decrease thermally induced noise, as discussed in the previous section, while the increased absorptivity will lead to a stronger signal. However, one should be aware of possible negative effects as well. For one thing the decreased BGE concentration could lead to EMD and peak deformation for the major constituents in the sample. Another result could be that sample stacking at the injection, often depending on the conductivity difference between sample and BGE, could be less successful.

As regards the charge of the visualizing ion, we would be inclined to prefer singly charged ions because choice yields the highest TR value. However, because the choice is already constrained by the many other limitations noted above, and because quite successful systems have been developed with double charged ions, we offer this advice only with some reservations.

11.7. INSTRUMENTAL IMPROVEMENTS

In Section 11.2 on measuring considerations it was pointed out that baseline noise works out the same way as in direct detection, but too much proportional noise, leading to what was called background noise, can make indirect detection practically impossible. Especially for fluorescence it is imperative to make proportional noise as small as possible. That amounts mainly to improving on the short-term stability of the excitation source.

Yeung and his group have been active in exploring the limits of indirect detection in terms of detection limits, in concentration as well as in amount, using laser fluorescence. The use of lasers (e.g., He–Ne, emitting at $\lambda = 632.8$ nm; He–Cd, emitting at $\lambda = 442$ or 325 nm; or Ar ion, emitting at $\lambda = 488$ nm and a number of other nearby wavelengths) obviously has the advantage that large amounts of power can be concentrated in a small, well-defined volume element within the liquid lumen in the CE capillary. This allows very narrow capillaries to be used, with numerous advantages. For example, separation speed is much higher. The limit of detection (in terms of the amount) is bettter due to the absence of thermally induced noise.

However, the stability of most other lasers is not as good as, say, a deuterium UV source, being in the percent-to-subpercent range rather than in the low-ppm range. As explained in Section 11.2, above, any signal instability constitutes a major drawback when one is using indirect detection. Decreasing the detrimental effects of instability has been a main issue in the work by the Yeung group. Several instrumental measures have been explored [part of this work has been carried out with ion chromatography and open-tubular liquid chromatography; however, the instrumental aspects are largely independent of the separation mode chosen (36, 37)]: choice of the type of laser; installing stabilizing devices in the laser; and compensation in the measurement path by rationing the emission signal to the laser intensity tapped off by a beam splitter, using some kind of rationing amplifier.

In attempts to make analysis in single cells (extracts) possible, usually rather low ($c_{BGE} = c_{vis}$) concentrations down to 10^{-4} mol/L were used. Obviously, under such conditions the peaks suffer from EMD. Still, the goal was reached; impressively low detection limits have been achieved (38).

The Yeung group also succeeded in improving the noise level in laser-based absorption detection, by suitable electronic compensation of source instability (39). Using laser radiation is practically a prerequisite for working with capillaries of 25 μm or less, but without such stabilization absorption measurements suffer from high noise levels.

Williams et al. (27) have devised an absorption detector for 675 nm based on a solid-state diode laser emitting at this wavelength. This wavelength is in an unattractive spectral range for direct CE detection. However, with indirect detection it can be extremely useful, as one can select the visualizing ion so as to have favorable absorptivity at the wavelength used and there will be little chance that the native absorption by the analyte will degrade the performance. The main advantage of the diode laser as a radiation source is the extreme stability at which it can operate. As discussed in Section 11.2, above, the instability in the source, amounting to, say, P ppm, leads to a corresponding noise in the transmission, and to a noise level of $0.43\ P\ 10^{-6}$ in the absorbance. In many commercial UV–Vis absorption instruments the source instability is a major contributor to the instrumental noise. With the diode laser the forementioned authors could attain a noise level 2×10^{-6} AU, measured with a time constant of 100 ms, indeed more than a factor of 10 better than values obtained with commercial UV–Vis detectors equipped with more conventional radiation sources.

Nielen (40) used time-resolved luminescence of the terbium ion and its acetylacetonate complex for indirect detection proper, as well as in a dyamic quenching or ligand exchange mode. The advantage expected for time-resolved measurement was that relatively simple HPLC equipment of this type can serve the purpose in CE. The nonuniversal dynamic quenching mode delivered by far the best detection limit.

11.8. APPLICATIONS AND SOME SPECIAL ASPECTS

Amino acids (5), proteins, nucleotides (3), and peptides from tryptic digests (41) have been studied by the Yeung group using laser fluorescence, with salicylate as the visualizing ion. As indicated above, in this technique it is often favorable to use low c_{vis} values—in this case often as low as 0.5 mmol/L.

Indirect detection has been applied by many workers to analysis of carbohydrates (42–45), which should come as no surprise, as there are hardly any alternatives for sensitive detection of this group of compounds. A special aspect here is that ionization of sugars and the like occurs only at pH 11 or higher. In such a solution the concentration of hydroxide ion becomes comparable in magnitude to that of the added visualizing ion and will "absorb" some of the replacement effect; therefore at too high a pH a deteriorated response can be expected. The choice of pH therefore involves a careful compromise between sensitivity, on the one hand, and sufficient ionization, on the other. This is further complicated by the desire to "tune" pH in such a way that good separation is obtained for carbohydrates of similar size, where enough selectivity can be accomplished only by exploiting pK_a differences (43). Heat production, sharply rising with pH because of the high mobility of the OH^- ion, is also of importance. A pH value of 11.9–12.3 appears to be a favorable choice with, e.g., sorbate as the visualizing ion. Klockow et al. (45) compared the CE method to HPLC with pulsed amperometric detection. The improved resolution in CE is obtained at the expense of the detection limit, which is 2–3 orders of magnitude worse. Nevertheless, the method was suitable for routine analysis in fruit juices and the like.

Another area where there has been much activity is in the analysis of inorganic anions such as chloride, nitrate, phosphate, or sulfate in water (18, 29, 46–49) and pharmaceutical preparations (50). Such ions were traditionally determined by "ion chromatography," with rather low efficiency and separation speed. The most successful approach appears to be that using dichromate at a pH around 11 (dichromate, by the way, does not exist as such in the BGE; in the pH = 11 solution it forms chromate, CrO_4^{2-}). In one example (46), 36 anions separated, nearly all with baseline resolution, in less than 3 min, with about 1 ppm sample concentration.

Short-chain organic acids and fatty acids have also been analyzed via indirect detection techniques (51, 52, 54).

Amino acids have attracted the attention of other workers (5,21,44,53). Chiral separation of amino alcohols has been accomplished by Armstrong et al. (55) using indirect detection. This work shows an interesting feature: the visualizing ion also performs the role of a chiral selector. This appears to be a good approach; having separate ions for chiral selection and indirect detection may deteriorate sensitivity (TR) and lead to complications due to the occurrence of system peaks.

On the other hand, neutrals or counterions as separate chiral selectors would do little harm to the detection process, as demonstrated by the work on γ-butyric acid derivatives by Walbroehl and Wagner (56), who used a crown ether for chiral selection and benzyltrimethylammonium as the visualizer. We note in passing that the complexation of the chiral selector with the analytes will lead to a more complicated dependence of the TR on conditions.

Indirect detection in CE has been applied to numerous other interesting analytical challenges. Without trying to be exhaustive, we should mention the following: anionic 17,57–59, and cationic (60) surfactants; various phosphates and phosphonates (61–63); organic acids (51), also in single cells (63); inorganic cations (49,50), also in difficult matrices such as hydrogen peroxide (64) and eye lens tissue (65), as well as in single cells (39,66); drugs in biological fluids (67); heparin fragments (68); and polyamines in biological samples (53).

Organic solvents were determined by micellar electrokinetic chromatography using indirect detection by Altria and Howell (69), with excellent analytical performance. In this variety of CE the normal ionic displacement mechanism is not in operation for the simple reason that the analytes are neutral. Nevertheless, the method works quite well, as had been demonstrated earlier by Szucs et al. (70) and Amankwa and Kuhr (71).

It is unlikely that c_{vis} in such experiments is changing as a result of the conductivity and pH effects. Therefore, another mechanism for mutual interaction has to be proposed. By analogy with work in HPCL, with indirect detection (24) one may safely assume that the increasing concentration of the analyte in the micelle leads to a changing distribution of the Veronal, which in turn leads to a concentration modulation. The description of such a system might likewise be performed by means of an eigenvector treatment (22), in quite general terms, which would make this method more accessible to numerical predictions; unfortunately this is not yet practicable, as the necessary data on competitive isotherms are hardly ever available.

11.9. SUMMARY

Indirect detection is an important technique in those applications of capillary zone electrophoresis (CZE) where no suitable detection device can be found for the analytes at hand. This applies to virtually all ions with no or negligible fluorescence or UV absorption. The effective use of indirect detection requires some insight into the underlying electrophoretic phenomena and the instrumental limitations.

In the choice of the combination of detection technique and visualizing ion, points to consider are the response factor, the baseline noise level, and the dynamic reserve (proportional noise). The choice of the concentration of the visualizing ion in the BGE involves a compromise between avoiding EMD, on the one hand, and degradation of detection limits due to detector proportional noise (low dynamic reserve) and/or nonlinearity of detection, on the other.

Under certain conditions, the occurrence of system peaks can degrade the applicability of indirect detection schemes. It is therefore often best to keep the composition of the BGE as simple as possible, as the number of systems peaks goes up with the number of components in the BGE.

All of these considerations have been dealt with in this chapter.

The most important limitation upon indirect detection is the restricted concentration range available between the detection limit and the onset of inacceptable electromigration dispersion. This limitation is mainly instrumental in nature, but thermal dissipation in the capillary also plays a role.

LIST OF ACRONYMS

Acronyms	Definitions
AU	absorbance units
BGE	background electrolyte
CE	capillary electrophoresis
CZE	capillary zone electrophoresis
DR	dynamic reserve
EDTA	ethylenediaminetetraacetic acid
EMD	electromigration dispersion
EOF	electroosmotic flow
GC	gas chromatography
HPLC	high-performance liquid chromatography
IUPAC	International Union of Pure and Applied Chemistry
KRF	Kohlrausch regulating function
TR	transfer ratio
UV–Vis	ultraviolet–visual (range)

REFERENCES

1. S. Hjertén, K. Elenbring, F. Kilar, J. L. Liao, A. J. C. Chen, C. J. Siebert, and M. D. Zhu, *J. Chromatogr.* **403**, 47 (1987).
2. F. Foret, S. Fanali, L. Ossicini, and P. Bocek, *J. Chromatogr.* **470**, 299 (1989).
3. W.G. Kuhr and E.S. Yeung, *Anal. Chem.* **60**, 1832–1834 (1988).
4. E.S. Yeung and W.G. Kuhr, *Anal. Chem.* **63**, 275A (1991).
5. W.G. Kuhr and E.S. Yeung, *Anal. Chem.* **60**, 2642–2648 (1988).
6. L. Gross and E.S. Yeung, *Anal. Chem.* **62**, 427–431 (1990).
7. H. Small and T.E. Miller, *Anal. Chem.* **54**, 463 (1982).
8. W.D. Pfeffer, T. Takeuchi, and E.S. Yeung, *Chromatographia* **24**, 123–126 (1987).
9. W. Baumann, *Fresenius'z. Anal. Chem.* **284**, 31 (1977).
10. H. Poppe, *Anal. Chim. Acta* **145**, 17 (1983).
11. T. Wang and R.A. Hartwick, *J. Chromatogr.* **607**, 119 (1992).
12. X. Xu, W. Th. Kok, and H. Poppe, *J. Chromatogr. A* **786**, 333 (1997).

13. T. Olefirowitz and A.G. Ewing, *J. Chromatogr.* **499**, 713–719 (1990).
14. A. Manz and W. Simon, *Anal. Chem.* **59**, 74–79 (1987).
15. F.E.P. Mikkers, F.M. Everaerts, and Th. P.E.M. Verheggen, *J. Chromatogr.* **169**, 11 (1979).
16. F.M. Everaerts, J.L. Beckers, and Th. P.E.M. Verheggen, *Isotachophoresis: Theory, Instrumentation and Practice,* J. Chromatogr. Libr., Vol. 6 Elsevier, Amsterdam, 1976.
17. M.W.F. Nielen, *J. Chromatogr.* **588**, 321 (1991).
18. W. Buchberger, S.M. Cousins, and P.R. Haddad, *Trends Anal. Chem.* **13**, 313 (1994).
19. X. Xu, W. Th. Kok, and H. Poppe, *J. Chromatogr. A* **742**, 211 (1996).
20. H. Poppe, *Anal. Chem.* **64**, 1908–1919 (1992).
21. G.J.M. Bruin, A.C. van Asten, X. Xu, and H. Poppe, *J. Chromatogr. A* **608**, 97–107 (1992).
22. H. Poppe, *J. Chromatogr.* **506**, 45–60 (1990).
23. J.L. Beckers, *J. Chromatogr. A* **693**, 347–357 (1995).
24. E. Arvidsson, J. Crommen, G. Schill, and D. Westerlund, *Chromatographia* **26**, 45 (1988).
25. H. Poppe, *Advances in Chromatography.* Marcel Dekker, New York, in press.
26. J.L. Beckers, *J. Chromatogr. A* **696**, 285–294 (1995).
27. S.J. Williams, E.T. Bergstrom, and D.M. Goodall, *J. Chromatogr. A* **636**, 39–45 (1993).
28. T. Wang and R.A. Hartwick, *J. Chromatogr.* **589**, 307–313 (1992).
29. Y. Ma and R. Zhang, *J. Chromatogr. A* **625**, 341–348 (1992).
30. J.H. Knox, *Chromatographia* **26**, 329–337 (1988).
31. M.S. Bello and P.G. Righetti, *J. Chromatogr.* **606**, 95 (1992).
32. A. Cifuentes, W. Th. Kok, and H. Poppe, *J. Microcolumn Sep.* **7**, 365 (1995).
33. A.E. Vorndran, P.O. Oefner, H. Scherz, and G.K. Bonn, *Chromatographia* **33**, 163 (1992).
34. X. Xu. W. Th. Kok, and H. Poppe, *J. Chromatogr. A* **716**, 231 (1995).
35. X. Xu, W. Th. Kok, J.C. Krak, and H. Poppe, *J. Chromatogr. B. Biomed, Appl.* **661**, 35–45 (1994).
36. S.-I. Mho and E.S. Yeung, *Anal. Chem.* **57**, 2253–2256 (1985).
37. W.D. Pfeffer and E.S. Yeung, *J. Chromatogr.* **506**, 401 (1990).
38. B.L. Hogan and E.S. Yeung, *Anal. Chem.* **64**, 2841–2845 (1992).
39. Y. Xue and E.S. Yeung, *Anal. Chem.* **65**, 2923–2927 (1993).
40. M.W.F. Nielen, *J. Chromatogr. A* **608**, 85–92 (1992).
41. B.L. Hogan and E.S. Yeung, *J. Chromatogr. Sci.* **28**, 15 (1990).
42. A.E. Vorndran, P.J. Oefner, H. Scherz, and G.K. Bonn, *Chromatographia* **33**, 163–168 (1992).

43. P.P.J. Oefner, A. Vorndran, E. Grill, C. Huber, and G.K. Bonn, *Chromatographia* **34**, 308–316 (1992).
44. T.W. Garner and E.S. Yeung, *J. Chromatogr. A* 515, 639 (1990).
45. A. Klockow, A. Paulus, V. Figueiredo, R. Amado, and H.M. Widmer, *J. Chromatogr. A* **680**, 187 (1994).
46. W.R. Jones and P. Jandik, *J. Chromatogr.* **608**, 385–393 (1992).
47. L. Kelly and D.S. Burgi, *Res. Discl.* **340**, 597 (1992).
48. M.M. Rhemrev-Boom, *J. Chromatogr.* **680**, 675–84 (1994).
49. K. Bächmann, K.-H. Steeg, T. Groh, I. Haumann, J. Boden, and H. Holthues, *J. Mocrocolumn Sep.* **4**, 431–438 (1992).
50. J.B. Nair and C.G. Izzo, *J. Chromatogr. A* **640**, 445–461 (1993).
51. G. Gutnikov, W. Beck and H. Engelhardt, *J. Microcolumn Sep.* **6**, 565–570 (1994).
52. F.B. Erim, X. Xu, and J.C. Kraak, *J. Chromatogr. A* **694**, 471–479 (1955).
53. Y. Ma, R. Zhang, and C.L. Cooper, *J. Chromatogr. A* **608**, 93–96 (1992).
54. L. Kelly and R.J. Nelson, *J. Liq. Chromatogr.* **16**, 2103–2112 (1993).
55. D.W. Armstrong, K. Rindlett, and G.L. Reid, III, *Anal. Chem.* **66**, 1690 (1994).
56. Y. Walbroehl and J. Wagner, *J. Chromatogr. A* **685**, 321–329 (1994).
57. S. Chen and D.J. Pietrzyk, *Anal. Chem.* **65**, 2770–2775 (1993).
58. M.P. Harrold, M.J. Wojtusik, J. Riviello, and P. Henson, *J. Chromatogr. A* **640**, 643 (1993).
59. L.K. Goebel and H.M. McNair, *J. Microcolumn* Sep. **5**, 47–50 (1993).
60. C.S. Weiss, J.S. Hazlett, M.H. Datta, and M.H. Danzer, *J. Chromatogr.* **608**, 325 (1992).
61. S.A. Shamsi and N.D. Danielson, *Anal. Chem.* **67**, 1845–1852 (1995).
62. A. Henshall, M.P. Harrold, and J.M.Y. Tso, *J. Chromatogr.* **608**, 413–419 (1992).
63. Q. Xue and E.S. Yeung, *J. Chromatogr. A* **661**, 287–295 (1994).
64. R.A. Carpio, P. Jandik, and E. Fallon, *J. Chromatogr. A* **657**, 185–191 (1993).
65. H. Shi, R. Zhang, G. Chandrasekher, and Y. Ma, *J. Chromatogr. A* **680**, 653 (1994).
66. Q. Li and E.S. Yeung, *J. Capillary Electrophor.* **1**, 55–61 (1994).
67. D. Leveque, C. Gallion, E. Tarral, H. Monteil, and F. Jehl, *J. Chromatogr. B. Bioned. Appl.* **665**, 320–324 (1994).
68. J.B.L. Damm and G.T. Overklift, *J. Chromatogr. A* **678**, 151–166 (1994).
69. K.D. Altria and J. S. Howell, *J. Chromatogr. A* **696**, 341–348 (1995).
70. R.K. Szucs, J. Vindevogel and P. Sandra, *J. High Resolut. Chromatogr.* **14**, 692 (1991).
71. L.N. Amankwa and W.G. Kuhr, *Anal. Chem.* **63**, 692 (1991).

CHAPTER

12

HIGH-PERFORMANCE CAPILLARY ELECTROPHORESIS–MASS SPECTROMETRY

KENNETH B. TOMER, LEESA J. DETERDING, AND CAROL E. PARKER

Laboratory of Molecular Biophysics, National Institute of Environmental Health Sciences, Research Triangle Park, North Carolina 27709

12.1.	Introduction	406
12.2.	Advantages of Mass Spectrometric Detection	407
12.3.	Instrumentation	407
	12.3.1. Mass Spectrometric Ionization Techniques	407
	12.3.1.1. Electrospray Ionization	408
	12.3.1.2. Continuous-Flow Fast Atom Bombardment	408
	12.3.2. Interface Designs	408
	12.3.2.1. The Sheath-Flow Interface	409
	12.3.2.2. The Liquid Junction Interface	409
	12.3.2.3. The Sheathless Interface	412
12.4.	Disadvantages of Mass Spectrometric Detection	414
	12.4.1. Sensitivity	414
	12.4.2. Buffer Choice	414
	12.4.2.1. Nonvolatile Buffers	414
	12.4.2.2. High-Conductivity Buffers	415
12.5.	Potential Solutions to Disadvantages of Mass Spectrometric Detection	415
	12.5.1. Sensitivity	415
	12.5.1.1. Sample Stacking	415
	12.5.1.2. Isotachophoretic Approaches	416
	12.5.1.3. Precolumns and Membranes	416
	12.5.1.4. Column Technology	418
	12.5.1.5. Mass Spectrometric Improvements	422
	12.5.1.5.1. Sheathless Interfaces	422
	12.5.1.5.2. Enhanced Detector Sensitivity	422

High Performance Capillary Electrophoresis, edited by Morteza G. Khaledi. Chemical Analysis Series, Vol. 146.
ISBN 0-471-14851-2 © 1998 John Wiley & Sons, Inc.

		12.5.1.5.3.	Selected Ion Monitoring	423
		12.5.1.5.4.	Alternative Analyzer Designs	423
	12.5.2.	Increasing Buffer Choices		428
		12.5.2.1.	Nonvolatile Buffers	428
		12.5.2.2.	Micellar Electrokinetic Chromatography Surfactants	428
		12.5.2.3.	Organic Modifiers	429
12.6.	Applications			429
	12.6.1.	Peptides and Proteins		430
	12.6.2.	Pharmaceutical Applications		430
	12.6.3.	Oligonucleotides		432
	12.6.4.	Affinity Capillary Electrophoresis		432
12.7.	Summary			435
Addendum No. 1				436
Addendum No. 2				438
List of Acronyms				440
References				441

12.1. INTRODUCTION

Of the numerous detection techniques that are used in conjunction with capillary electrophoresis (CE), mass spectrometry (MS) provides the best combination of sensitivity and structural information, especially for complex mixtures. Thus, it is logical that a significant amount of effort has gone into interfacing CE and MS. What is somewhat surprising, however, is that 6 years passed between the report of the high separation efficiencies of CE (1) and the first report of the coupling of CE and MS (2). Part of this delay was due to the profound changes that were occurring within MS. In 1980, mass spectrometrists were hard pressed to analyze polar compounds, such as peptides, or compounds with molecular weights significantly above 1000 Da. Introduction of a liquid stream, such as the eluent from an liquid chromatography (LC) separation, was difficult and tricky. The advent of fast atom bombardment (FAB) in 1981 (3) increased the range of compounds amenable to MS determination to include polar, thermally labile compounds with masses up to 5000–10,000 Da. The development of continuous flow (CF) introduction for FAB in the middle of the decade allowed the direct coupling of liquid phase separation techniques with FAB. At the same time as CF–FAB was being developed, another ionization technique, electrospray ionization (ESI), was beginning to be explored. ESI, like FAB, is suitable for the study of large polar and thermally labile molecules and has an even higher mass range than does FAB, i.e., more than 100,000 Da. As ESI is based on a flowing liquid stream, it turned out to be even easier to couple with liquid phase separation techniques than was CF–FAB.

12.2. ADVANTAGES OF MASS SPECTROMETRIC DETECTION

Once the new developments in MS ionization techniques began to be assimilated, reports on the coupling of CE and MS began to appear (2, 4–8). Although the early interest in coupling CE and MS came from MS laboratories, there was interest also from the growing CE community. The interest of the MS community was due to the growing need for handling complex samples containing low levels of analytes, especially biological samples. CE provided a highly efficient separation technique that was well suited for handling low level samples. The CE community noted that MS could provide a high level of structural information about analytes, a highly specific analysis using selected ion-monitoring techniques, and a method of component identification more reliable than comparing reproducible migration times.

At the present time, however, only about two dozen groups worldwide are actively engaged in the development and application of CE combined with MS. With the introduction of most combinations of analytical techniques, the number of practitioners increases rapidly unless significant problems with the combination are noted or assumed to be present by the scientific community. So the immediate questions become: What factors are retarding the rapid growth of CE–MS? Are these factors still operative? What efforts are being made to overcome any problems, and how effective are such efforts? These questions must be addressed in a critical review of the state of CE–MS.

12.3. INSTRUMENTATION

Before the foregoing questions can be answered in detail, an overview of the MS ionization techniques with which CE has been interfaced and the general approaches to interfacing CE and MS is appropriate.

12.3.1. Mass Spectrometric Ionization Techniques

CE has been interfaced with three main ionization techniques: ESI, CF–FAB, and matrix-assisted laser desorption ionization (MALDI) (9). On-line interfacing techniques have been described for all three ionization types, but off-line techniques are more common with MALDI and so will not be discussed further here.

12.3.1.1. Electrospray Ionization

ESI is based on a liquid stream at atmospheric pressure flowing through a needle electrode that is held at approximately a 3 kV potential. The flowing stream becomes charged as it passes through the needle. Evaporation of the solvent leads to highly charged droplets. Ions are thought to be either directly desorbed from the droplets or the droplets undergo coulombic explosion leading to smaller droplets that eventually can lead to single ions, or both. The recent application of this technique in MS (10,11) [and the closely related ionization process, ion spray (12)] has revolutionized the field, for ESI produces multiply-charged ions. Because mass spectrometers measure the mass-to-charge ratio (m/z) of a compound, not the mass directly, an increase in the number of charges on a molecule will decrease the observed m/z of the compound. If a sufficient number of charges are placed on a very large molecule, it is possible to analyze that molecule on a relatively limited mass range mass spectrometer such as a quadrupole mass spectrometer. ESI is typically used for compounds below 100,000 Da, although its upper mass limit is substantially higher. The multiple charging effect observed in ESI also does not produce ions with only one multiple charge state but yields a population of ions having a number of different charges. This permits the calculation of the molecular weight of the compound.

12.3.1.2. Continuous-Flow Fast Atom Bombardment

In FAB, the analyte is dissolved in a liquid matrix—glycerol and dithiothreitol/dithioerythritol being the most common—on the probe tip and is then bombarded with fast atoms (8–10 keV xenon) or ions (20–30 keV cesium). This bombardment desorbs the upper layers of the analyte/matrix solution into the gas phase. Ionization of the analyte occurs either in the matrix or in the high-pressure selvage above the matrix. Continuous liquid introduction systems were introduced by Ito et al. in 1985 (13) and by Caprioli et al. in 1986 (14). In these designs, the analyte is introduced into the system as an aqueous matrix solution. The typical upper mass limit for CF–FAB is about 3000 Da.

Although CF–FAB has been superseded to some extent by ESI, it is still a viable ionization technique that continues to solve problems in the hands of many practitioners. It is, however, the experience of the present authors that CE is somewhat easier to interface with ESI than with CF–FAB.

12.3.2. Interface Designs

Although there are significant differences in the two ionization techniques, most of the on-line interfaces that have been described can be divided into two general types: the sheath-flow interface and liquid junction interface. A third type primarily applicable to ESI, the sheathless interface, has recently been explored. The main purpose of these interfaces is to deliver any makeup flow and/or makeup solvent appropriate to the ionization technique.

12.3.2.1. The Sheath-Flow Interface

The main feature of the sheath-flow interface is that the capillary column extends all the way to the ionization source with no transfer lines involved (Figure 12.1). Because both the ESI and CF–FAB sources are designed for flow rates of ca. 2–10 μL/min, a makeup flow must be added to the nL/min electroosmotic flow (EOF) rates associated with CE. This makeup flow is delivered through the sheath column. Schematics of the sheath-flow interfaces for an ESI and for a CF—FAB interface are shown in Figures 12.1a and 12.1b, respectively. Note that a transfer line can be substituted for the CE capillary without changing the design.

Although the sheath-flow interfaces for ESI and CF–FAB are conceptually similar, there are differences in details due to the differences in the ionization techniques. In the ESI interface, the sheath-flow capillary is stainless steel and provides electrical contact for the ESI voltage (Figure 12.1a) (15). In the CF–FAB system, the sheath-flow capillary is fused silica and terminates at the probe tip, which is at source potential (6–8 keV) (Figure 12.1b). Contact between the source potential and the CE column is provided by the sheath fluid. A further difference is in the inner diameters of the CE columns, which are compatible with the interface. In the CF–FAB interface, the probe tip is under high vacuum (10^{-5} to 10^{-6} torr) whereas the end of the CE capillary in ESI is at atmospheric pressure. Because the CF–FAB capillary is exposed to high vacuum, it is susceptible to vacuum-induced flow, which can significantly reduce separation efficiency. It has been found that inner diameters of up to only 13 μm can be used without significant vacuum-induced flow being observed. The tip of the CE column is withdrawn slightly into the sheath column so that the sheath fluid helps protect the end of the column from the high vacuum (16,17). On the other hand, there is no vacuum constraint on the ESI capillary diameters.

12.3.2.2. The Liquid Junction Interface

Liquid junction interfaces have been designed for both ESI and CF–FAB sources, primarily by Henion's, group for ESI [see Lee et al. (18,19)] and by Caprioli and co-workers for CF–FAB (7). In the liquid junction interface the separation column ends outside of the mass spectrometer with a transfer line

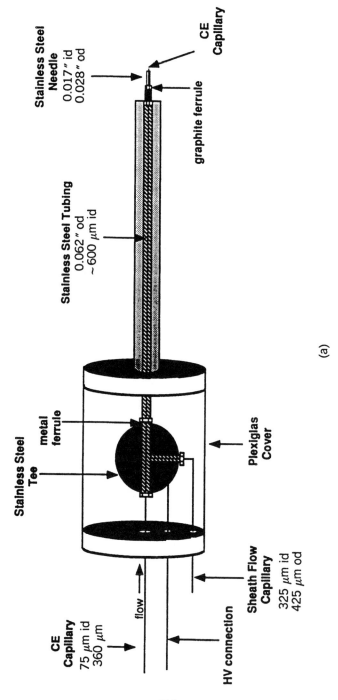

(a)

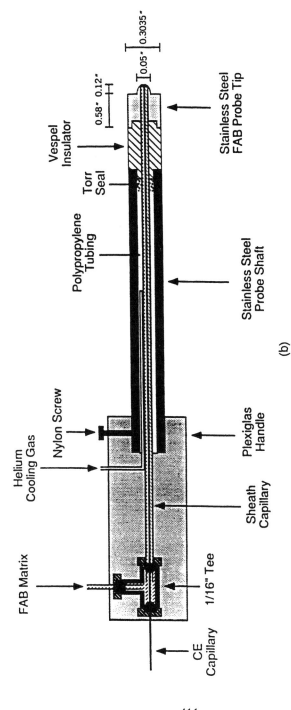

Figure 12.1. (a) Sheath-flow interface for ESI. [Reprinted with permission of Elsevier Science Inc. from C. E. Parker, J. R. Perkins, K. B. Tomer, Y. Shida, K. O'Hara, and M. Kono, *J. Am. Soc. Spectrom.* **3**, 563 (1992). Copyright 1992 by the American Society for Mass Spectrometry.] (b) Sheath-flow interface for CF–FAB. [Adapted with permission from M. A. Moseley, L. J. Deterding, K. B. Tomer, and J. W. Jorgenson, *Anal. Chem.* **63**, 109 (1991). Copyright 1991 American Chemical Society.]

being positioned opposite it with a very narrow gap between them (10–23 µm). Surrounding the CE column and the transfer line is a reservoir of makeup fluid. The flow of makeup fluid into the source helps constrain and carry the column effluent into the transfer line. Schematics of the ESI and CF–FAB liquid junction interfaces are shown in Figures 12.2a and 12.2b, respectively.

Both Pleasance et al. (20) and Suter and Caprioli (21) have compared liquid junction and coaxial sheath-flow interfaces on an ESI and CF–FAB source, respectively. Both groups found that some band broadening occurred in the liquid junction interfaces relative to the coaxial sheath-flow interfaces, resulting in a 1.67 to 10-fold loss in the observed theoretical plates. Pleasance et al. also reported that the coaxial design was more rugged and reproducible in conjunction with ESI (20), whereas Suter and Caprioli favored their modified liquid junction interface (21). It is quite possible that robustness and ease of use have as much to do with familiarity with the design as with actual interface characteristics.

12.3.2.3. *The Sheathless Interface*

Recently, several groups have investigated ESI sources operating at low flow rates (22–28). In all of these designs, the spray is maintained by the flow of effluent from a small column with flow rates ranging from 10–800 nL/min. Generally, the latter designs incorporate an etched and tapered fused-silica column with either a metallic-coated tip or a gold-wire connection through the fused-silica capillary that carries the voltage. The source design by Caprioli's group has electrical connection outside of the source [see Emmett and Caprioli (24); Andren et al. (25)]. A generalized schematic is shown in Figure 12.3.

A major advantage of the sheathless source design is a reduction in the level of background ions. The lack of sheath fluid also reduces competition between the analyte and electrolyte for the charge. The net result is an increase in sensitivity, with sensitivities in the low femtomole region being reported (24,26,27). A second advantage of the sheathless interface is its utility with a wide range of solutions from high-performance liquid chromatography (HPLC), water through 50 mM NaH_2PO_4 (28). Due to the lower flow rates, these buffers may be tolerated by the mass spectrometer, and thus this design may circumvent the problem of nonvolatile buffers (see below).

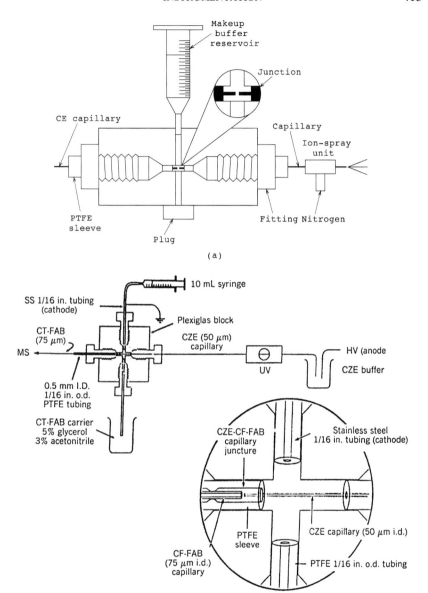

Figure 12.2. (a) Liquid junction interface for ESI (PTFE is polytetrafluoroethylene.) [Reprinted from E. D. Lee, *Biomed. Environ. Mass Spectrom.* **18**, 844 (1989), with permission of John Wiley & Sons, Ltd. Copyright 1989.] (b) Liquid junction interface for CF–FAB. [Reprinted with permission of Elsevier Science Inc. from R. M. Caprioli, W. T. Moore, M. Martin, B. B. DaGue, K. Wilson, and S. Moring, *J. Chromatogr.* **480**, 247 (1989). Copyright 1989.]

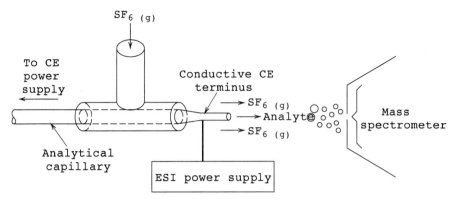

Figure 12.3. Sheathless interface for ESI. [Reprinted with permission from J. H. Wahl, D. C. Gale, and R. D. Smith, *J. Chromatogr. A* **659**, 217 (1994).

12.4. DISADVANTAGES OF MASS SPECTROMETRIC DETECTION

12.4.1. Sensitivity

As pointed out in Section 12.1, the combination of CE and MS has not been routinely exploited. The primary reason is the limited dynamic range of CE–MS. The injection volumes in CE are typically very small, which leads to a high concentration detection limit even though the absolute amount detected is quite low. For example, Moseley et al. demonstrated detection levels in the low-femtomole range for chemotactic peptides using a coaxial CF–FAB interface (17). The concentrations of the peptides used in this analysis were on the order of 10^{-4} to 10^{-5} molar, however. With routine MS detection limits under full-scan conditions being in the low-picomole to high-femtomole range, there is clearly a limited concentration range encompassing the amount injected and the amount detectable.

12.4.2. Buffer Choice

The second major disadvantage of combining CE with MS detection stems from the limitations placed on the choice of buffers.

12.4.2.1. Nonvolatile Buffers

As in LC–MS, nonvolatile buffers present problems. The use of nonvolatile buffers can lead to plugging the interface and/or source contamination.

Additionally, nonvolatile buffers are usually alkali metal salts, e.g., sodium phosphate. The presence of the alkali metal often leads to cationized analytes in addition to protonated analytes. This splits the analyte into several distinct ions, reducing the overall sensitivity (29). Other nonvolatile buffers such as those used in micellar electrokinetic chromatography (MEKC) or sieving buffers also present problems. Because there is typically a limited number of ions that can be formed at a given time, the ionic buffers also compete with the analyte in ion formation. Thus, many separations that have been developed for other methods of detection are incompatible with the mass spectrometer.

12.4.2.2. High-Conductivity Buffers

Moseley et al. investigated the effects of a number of volatile buffers and sheath fluids on sensitivity using a sheath-flow ESI interface (30). They found that acidic buffers of low ionic strength (e.g., 0.01 M acetic acid) gave the best results. Recently Wahl and Smith reported on a comparison of buffer systems for both a sheath-flow and sheathless ESI interface (31). This study also showed that signal intensity was reduced with increasing ionic strength of the buffer and also with increasing buffer concentration. An additional problem with high-conductivity/high-concentration buffers when used with ESI–MS is the possibility that the voltage at the electrospray needle will be affected by the voltage applied to the CE column. We have observed this to occur on two very different instruments with concomitant changes in tuning (32).

12.5. POTENTIAL SOLUTIONS TO DISADVANTAGES OF MASS SPECTROMETRIC DETECTION

12.5.1. Sensitivity

Potential solutions to the problem of sensitivity have involved developments on both the CE end and the MS end.

12.5.1.1. Sample Stacking

On the CE side, substantial advances have been made in the area of sample pre-concentration. The simplest approach to sample preconcentration is sample stacking where the sample is injected in water rather than in buffer. Some concentration, up to a factor 10, occurs at the front of the sample plug (33).

12.5.1.2. Isotachophoretic Approaches

Isotachophoresis (ITP) and transient isotachophoresis (tIPT) have been examined by several groups as a means of concentrating the sample. In ITP analytes are stacked in bands that have the same concentration as the leading electrolyte. A disadvantage of ITP is that low-level analytes form very narrow bands that may be too narrow to produce a good mass spectrum. Van der Greef and coworkers have reported the use of a dual-column ITP–CE system in which ITP occurs in the first capillary (34). The focused analytes are then transferred to the CE capillary. Improvements in concentration by a factor of 200 have been reported when this approach was used.

An attractive alternative is to use transient isotachophoretic focusing as described by Thompson et al. (35a). These authors demonstrated a 24-fold increase in sensitivity of a mixture of proteins by filling more than one-third of the CE column with the analyte solution (Figure 12.4). Separation efficiencies are reduced but are still better than LC. Improvements of 2 orders of magnitude in concentration detection limits by using transient ITP–CE have been reported (35a,b). Lamoree et al. demonstrated trace analysis of β-agonists spiked into calf urine at the 2 ng/mL level using a similar approach (36). In this study, they preconcentrated 290 nL of extract by means of a single-column ITP–CE system. Concentration detection limits in the range of 10^{-7} were observed. One *caveat* that should be mentioned is that the mobility of the background electrolyte used must be lower than the mobilities of any of the analytes. If it is not less, focusing will not occur.

12.5.1.3. Precolumns and Membranes

Naylor's group have explored two other approaches to on-line preconcentration for CE–MS (PC–CE–MS) [see Tomlinson et al. (37a,b)]. Initially, they used a small precolumn packed with C18 packing material with an approach similar to that of Morita and Sawada and (38a) and of Swartz and Merion (38b).

Figure 12.4. (a) CE–MS full-scan (m/z 600–2000) reconstructed ion electropherogram of a 150 nL injection of 12 µM each of lysozyme (1), cytochrome *c* (2), ribonuclease (3), myoglobin (4), β-lactoglobulin A (5), β-lactoglobulin B (6), and carbonic anhydrase (7) dissolved in water. BGE: 0.02 M 6-aminohexanoic acid + acetic acid to pH 4.4. (b) Transient CITP–MS full-scan (m/z 600–11,850) reconstructed ion electropherogram of a 750 nL injection of ∼500 nM each of lysozyme (1), cytochrome *c* (2), ribonuclease (3), myoglobin (4), β-lactoglobulin A (5), β-lactoglobulin B (6), and carbonic anhydrase (not detected) in 0.005 M ammonium acetate buffer. The peak marked with an asterisk (*) is from the rear boundary of the ammonium zone. BGE: 0.02 M 6-aminohexanoic acid + acetic acid, pH 4.5 [Reprinted with permission from T. J. Thomson, F. Foret, P. Vouros, and B. L. Karger, *Anal Chem.* **65**, 900 (1993). Copyright 1993 American Chemical Society.]

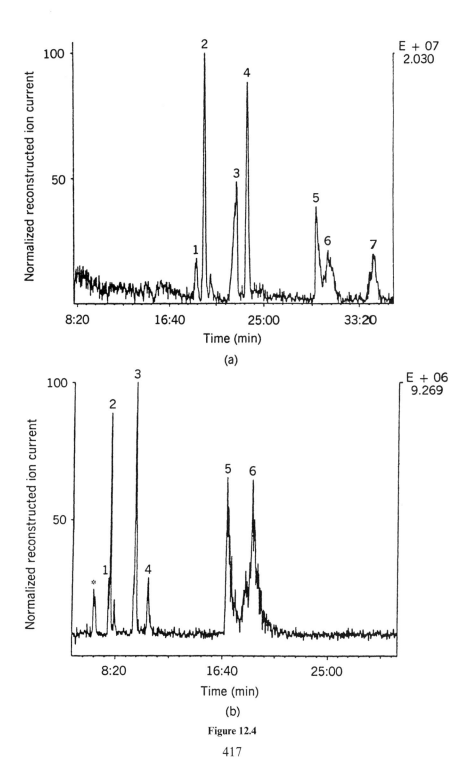

Figure 12.4

Nylor's group, however, found that CE performance could be compromised. Analyte peaks were broader and often tailed, and the electroosmotic flow (EOF) was often reduced due to the use of an organic solvent to elute the analytes from the preconcentration column and to the presence of the solid phase [see Tomlinson et al. (39)]. They then turned to adsorption on a C18 or polymeric membrane in a metal cartridge (40). They observed that this reduced the volume of organic phase needed to elute the absorbed analytes and that the affect of the adsorbent on the EOF was also reduced. An example of a separation using this system is shown in Figure 12.5, where a sample spiked with insulin-like growth factors I and II (IGF-I and -II) at the 20 nM level was analyzed.

12.5.1.4. Column Technology

Significant improvements in sensitivity have also been made through modifications of the CE column. Some of these, such as use of derivatized columns for basic peptides and proteins, are applicable to CE separations in general (41) and will not be specifically detailed here other than to note that any improvements in column technology that leads to reduced adsorption and increased separation efficiency will yield a corresponding improvement in CE–MS sensitivity.

Several groups, however, have focused on features of the CE column that are particularly relevant to their interfacing with MS (42,43). Tetler et al. investigated the effect of CE capillary diameter in a coaxial sheath-flow ESI interface on the sensitivity observed in the analysis of a mixture of angiotensins I and II, bradykinin, and substance P. These workers varied the dimensions of the CE column, the sheath column, and the outer nebulizing gas column. In these sets of experiments the i.d. of the CE column was maintained at 75 μm and the o.d. was varied from 170 to 375 μm. The (stainless steel) sheath column i.d. was varied from 228 to 406 μm, while the o.d. was varied from 406 to 710 μm. The nebulizing gas capillary (also stainless steel) was varied from 520 to 850 μm i.d. and from 805 to 1240 μm o.d.. The net result was a postulate that an arrangement with a CE capillary of 75 μm i.d. × 375 μm o.d., a sheath-flow capillary of 406 μm i.d. × 635 μm o.d., and a nebulizing gas capillary of 750 μm i.d. × 1067 μm o.d. would provide optimum performance in terms of sensitivity and spray stability. The aforementioned authors also noted that the sheath-flow capillary dimensions were nonstandard and that they were unable to construct such an interface.

Wahl et al. approached the problem of CE capillary dimensions and sensitivity by changing the i.d. of the capillaries (Figure 12.6) (42). Using aminopropyltrimethoxysilane-derivatized columns with an ESI interface and a mixture of proteins—aprotinin, carbonic anhydrase, cytochrome *c*, and

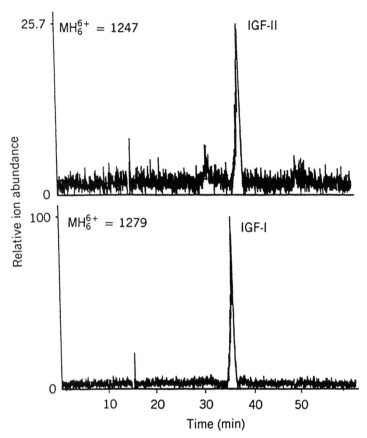

Figure 12.5. Preconcentration(PC)–CE–MS analysis of a condition medium spiked with IGF-I and -II to give a final concentration of 20 nM of each; 1 μL of this solution was applied to a PC–CE capillary (50 μm × 80 cm polybrene coated) by hdrodynamic pressure. Sample cleanup was effected by washing the adsorbed sample with 1 % formic acid, and analyte elution was with ~ 80 nL of methanol/water/acetic acid (80:20:1 v:v:v). Separation voltage −20kV. [Reprinted with permission from L. M. Benson, R. P. Oda, W. D. Braddock, B. L. Riggs, J. A. Katzmann, and S. Naylor, A. J. Tomlinson, *J. Capillary Electrophor.* **2**, 97 (1995).]

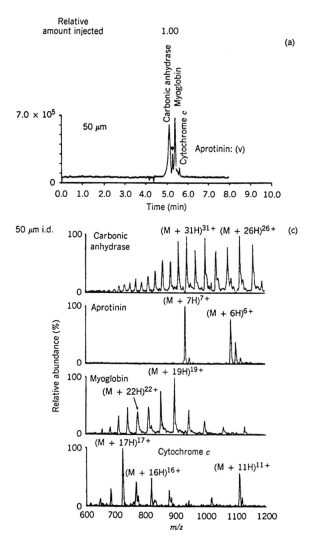

Figure 12.6. (a) Total ion current electropherogram obtained for the 30 μM protein mixture containing carbonic anydrase, aprotinin, myoglobin, and cytochrome c using 50 μm i.d. capillaries. Injected amount 60 fmol; voltage drop = 300 V/cm (100 cm column); 0.01 M acetic acid buffer solution, pH = 3.4. (b) Total ion current electropherogram of 600 amol injection of the mixture in part a using a 5 μm i.d. column—other conditions remaining the same. Comparison ot parts a and b shows only a two-to four-fold difference in signal intensity. (c) Mass spectra of components obtained from the separation in part a. (d) Mass spectra of components obtained from the separation in part b. [Reprinted from J. H. Wahl, D. R. Goodlett, H. R. Udseth, and R. D. Smith, *Electrophoresis* **14**, 448 (1993), with permission.

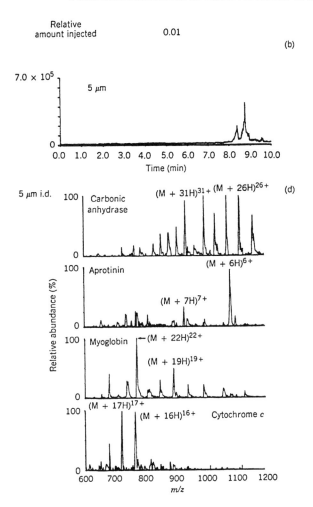

Figure 12.6. (*Continued*)

myoglobin—they observed that there was only a two- to four-fold decrease in signal intensity when 100 times less sample was injected and the column size was decreased from 50 to 5 μm. The quality of the spectra obtained from 600 amol injected onto the 5 μm i.d. column was quite similar to that obtained from 60 fmol injected onto the 50 μm i.d. column.

12.5.1.5. *Mass Spectrometer Improvements*

A number of developments in ion sources, detectors, and analyzers show promise of increasing sensitivity in CE–MS experiments.

12.5.1.5.1. Sheathless Interfaces. Smith's group combined their observations about increased sensitivity with decreased column diameter [see Wahl et al. (42)] and the increased sensitivity of the sheathless electrospray source described above [see Gale and Smith (23)] to investigate CE interfaced with the sheathless ESI source [see Wahl et al. (27)]. In this 1994 paper, the authors showed the analyses of digests of proteins in the 30 fmol range using a 10 μm i.d. aminopropylsilylated column. The concentration of the digests, however, were approximately 50 μm.

There are several other observations in the paper by Wahl et al. (27) that are also worthy of note. One is that the use of a Ag-coated capillary leads to formation of Ag adduct ions. These ions can be useful for determination of the charge state of the ions. Secondly the authors noted that the reconstructed ion electropherograms showed additional solute zones with the same mass as the tryptic fragment. Wahl et al. attributed these zones to collisional activation of other tryptic fragments in the source. We have noted the same feature in our laboratory using a coaxial CF–FAB interface and 10 μm i.d. columns and at high concentrations on 75 μm i.d. columns in ESI. When identical solutions were analyzed by packed nanoscale capillary LC, these extraneous peaks were absent. We have attributed this multiple-peak effect to a micellar effect at the concentrations used, with one component carrying along a small amount of another component. Thirdly, Wahl et al. noted that a relatively conductive buffer system was used. Although not explicitly stated in their paper, there is an implication that the buffers compatible with operation of this source may be more limited than even with CE–MS.

12.5.1.5.2. Enhanced Detector Sensitivity. Improvements in detector sensitivity will also facilitate the interfacing of CE with MS. The most promising detector improvements are in the use of array detectors. With array detectors, a range of ions are detected at one time rather than scanning from ion to ion. The most common array detection system that has been applied to CE–MS detection is a scanning array detector. Reinhoud et al. evaluated an array detector system for CE–MS using a liquid junction interface with CF–FAB (44). An improvement in sensitivity of 100 to 1000-fold over conventional detection was demonstrated in the scanning and static modes, with absolute detection limits achievable in the 1–5 fmol range. Tomlinson et al., using a static array detector with an 8% mass window, obtained detection levels of approximately 100 amol of haloperidol standard and metabolites of haloperidol in urine (45). An array detector that can cover 40% of the mass range has been used in the acquisition of tandem MS data from on-line CE-separated peptides with coaxial CF–FAB ionization (46). Although this detector was not used after MS-I for CE–MS acquisitions, it would be expected to enhance the sensitivity of CE–MS analyses, especially if used with ESI, where there is less chemical background.

12.5.1.5.3. Selected Ion Monitoring. The most common method of improving detector sensitivity is the use of selected ion monitoring (SIM). In SIM, the mass spectrometer only detects selected masses rather than a complete spectrum. This eliminates the time wasted on a scanning instrument while it is scanning mass ranges that are not of interest. This also means, however, that the masses to be detected must be known before the analysis. Thus, SIM is primarily used for targeted analyses. If a survey scan of the mixture is obtained prior to separation, ions can be selected for a CE–SIM–MS analysis, but one must keep in mind that suppression effects and dynamic range effects may limit observation of ions of some analytes in the survey scan. In our laboratory, we have successfully used this approach in the analysis of polypeptides in the venom from several species of snakes where over 70 components were separated and detected (Figure 12.7) (47–49).

12.5.1.5.4. Alternative Analyzer Designs. Although the majority of mass spectrometers interfaced with CE have been based on either quadrupole or magnetic sector mass analyzers, several other designs have either been interfaced with CE or have the potential to offer improvements in sensitivity if interfaced to CE. These alternative analyzer designs fall broadly into two categories, ion-storage analyzers or time-of-flight (TOF) analyzers.

Ion-storage analyzers have the capability of storing and accumulating ions prior to detection. The most common types are ion trap mass spectrometers and Fourier transform mass spectrometers. ESI is the most common liquid introduction ionization technique associated with these analyzers.

The first reports of the coupling of CE with an ion trap mass spectrometer appeared in conference proceedings in 1992 (50,51). In late 1993, Ramsey et al. reported the use of a combination of broadband collisional activation and

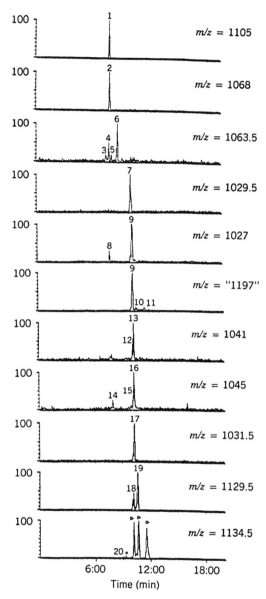

Figure 12.7. Selected ion electropherograms of the peptides observed by the CE–ESI–MS of 26 of the most abundant ions from the fullscan spectrum of the venom from *Haemachatus haemachatus*. The triangular symbol ▷ indicates a peak that could not be characterized using these data but was identified following further experiments. [Reprinted with permission from J. R. Perkins and Tomer, *J. Capillary Electrophor.* **1**, 231 (1994).]

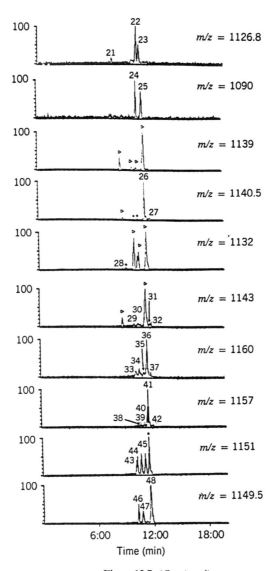

Figure 12.7. (Continued)

resonance ejection to reduce background and noise signal levels in an ion trap mass spectrometer with CE–ESI introduction of the analytes (52). The total ion electropherogram of a mixture of bradykinin (180 fmol), xenopsin (200 fmol), and neurotensin (220 fmol) showed improvement in the signal-to-noise ratio by a factor 3. Estimated limits of detection of the components from the total ion electropherogram (not extracted-ion electropherograms) were on the order of 10 fmol. A more recent publication by this group demonstrated mid-attomole detection level for a full-scan spectrum of leu-enkephalin using a sheathless interface and a 20 μm i.d. column (53). Henion et al. demonstrated the acquisition of full-scan spectra from a CE separation of isoquinoline alkaloids at the 370 amol level (Figure 12.8) (54). Detection levels of 90–130 amol were reported.

The capabilities of the second ion storage analyzer, the Fourier transform ion cyclotron resonance mass spectrometer (FTMS), has been explored by Smith's group [see Hofstadler et al. (55,56) and Wahl et al. (57)] and by Johnson et al. (58). A major advantage claimed for FTMS over other MS

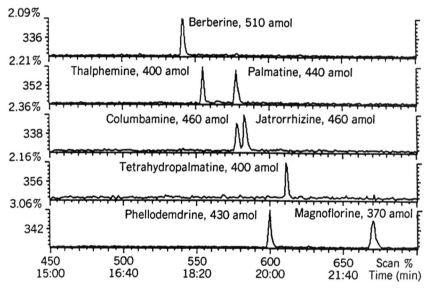

Figure 12.8. Extracted ion current electropherograms of a synthetic mixture of isoquinoline alkaloids. Capillary: 25 μm i.d.. Buffer 40% methanol/60 mM ammonium acetate, pH 4.5. [Reprinted with permission from J. D. Henion, A. V. Mordehai, and J. Cai, *Anal. Chem.* **66**, 2103 (1994). Copyright 1994 American Chemical Society.]

analyzers is its high-resolution capabilites. High sensitivity and its capabilities for multiple-tandem MS experiments are also characteristics of FTMS. FTMS instrumentation relies on a very high vacuum (10^{-9} torr range for high-resolution detection), which makes coupling with on-line separation techniques not as straightforward as with quadrupole-based instruments. The instrumentation reported by Johnson et al. (58) utilized a sheath flow interface. Due to the pumping speed of their vacuum system, spectra were obtained every 15s (a 15s ion accumulation). They demonstrated the separation and detection of dopamine, bradykinin/substance P (not separated) and leuenkephalin at the 5–9 pmol level. Smith's group combined modifications of the vacuum system that permit very rapid manipulation of the pressure in the instrument with a sheathless electrospray source to enable data acquisition with a duty cycle of 4 s. Although this is significantly slower than that of, say, a quadrupole, it is fast enough to be useful for many CE analyses. Separations of a mixture of proteins and of a tryptic digest were demonstrated.

In a TOF mass spectrometer, ions are separated in time, with the flight time being proportional to the mass, and all ions reaching the end of the flight tube are detected. Acquisition time is fast with a short duty cycle. TOF therefore offers the same sensitivity enhancements as do other multichannel analyzers such as array detection. TOF–MS, like FTMS, requires a relatively high vacuum, 10^{-7} torr. Zare's group recently demonstrated the coupling of CE with a TOF mass spectrometer [see Faug et al. (59)]. Using a sheathless electrospray source, they acquired a separation of bag cell peptides as well as standard mixtures at the low-femtomole level. Complete mass spectra were recorded every 100 μs. Andrien et al. have also recently shown rapid CE separations detected on a TOF mass spectrometer using a sheath-flow ESI interface (60). Eight spectra were acquired per second with a mass range of 100–1700 Da.

At present, the routine limits of detection for these alternative analyzer designs in combination with CE appear to be in the low-femtomole regime. Although this may not be a significant improvement over detection limits observed using the more traditional analyzers, the fact that in most cases the instrument designs are still very much in the development stage holds the promise of even lower detection limits in the near future. As an example, McLafferty's group recently reported a new ESI–FTMS design based on a sheathless interface and narrow capillary tip (~ 2 μm i.d.). Spectra of proteins at the mid-attomole level and of a DNA 50-mer at the mid-femtomole level and an estimated limit of detection of 10 amol were reported [see Valaskovic et al. (61)].

12.5.2. Increasing Buffer Choices

12.5.2.1. Nonvolatile Buffers

Wahl and Smith have also observed that the use of nonvolatile buffer systems led to decreased signal intensity (31). But they found that, using the sheathless ESI interface, a CE–MS analysis of a complex sample (tryptic digest of *Candida krusei* cytochrome c) could be performed at the low-femtomole range (18 fmol) using a nonvolatile buffer system (1 mM Na_2HPO_4) (Figure 12.9) (31).

12.5.2.2. Micellar Electrokinetic Chromatography Surfactants

Because of the use of nonvolatile surfactants in the running buffers, there have been few attempts at interfacing MEKC with MS. Surfactants, in addition to being relatively nonvolatile, compete with the analyte for ionization and significantly reduce sensitivity. Varghese and Cole used cetyltrimethylammonium chloride (CTAC) in the buffer solution in the separation of peptides and of cationic dyes. The surfactant was used primarily to inhibit interactions of the analytes with the column walls. One example, however, was shown of a separation of peptides using the CTAC above its critical micellar concentration (62). These researchers also noted that, in addition to sensitivity limitations due to the presence of the surfactant, the ion source rapidly became dirty and needed cleaning after each full day of use.

Ozaki et al. reported the use of a high-molecular-weight surfactant, butyl acrylate/butyl methacrylate/methacrylic acid copolymer sodium salt (BBMA), as a pseudostationary phase for MEKC–ESI–MS (63). The average

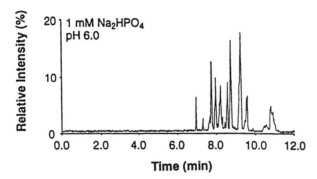

Figure 12.9. Total ion electropherogram of a tryptic digest of 18 fmol *Candida krusei* cytochrome c using a 1 mM Na_2HPO_4 buffer system, pH 6.0, and a 20 μm i.d. column. [Reprinted with permission from J. H. Wahl and R. D. Smith, *J. Capillary Electrophor.* **1**, 62 (1994).]

molecular mass of BBMA was determined to be about 40,000 Da. The critical micellar concentration of BBMA was effectively zero, which makes it suitable for use with ESI–MS. Separations of solutions of phenyltrimethylammonium chloride, and 1-naphthylamine, quinine sulfate, tetraphenylphosphonium chloride, and octaoxyethylenedodecanol using 2% BBMA in ammonium acetate and of a solution of sulfamethazine, sulfisomidine, sulfadiazine, and sulfisoxazole using 1% BBMA in 10% methanol in 100 mM borate–50 mM phosphate buffer were shown. Because of the high mass of the BBMA, there was little background associated with BBMA.

Nashabeh et al. used a combination of a coated capillary (cross-linked polyvinylmethylsiloxane sublayer and a polyacrylamide layer) with 20 mM ε-amino-n-caproic acid (EACA)/acetic acid, pH 4.4, 5 mM N-dodecyl-N,N-dimethyl-3-amino-1-propanesulfonate (DAPS) and 10% acetonitrile to analyze IGF-I variants at about the 100 fmol level (64). Under these conditions there was no EOF. The critical micellar concentration of DAPS in 10% acetonitrile is 5 mM, but the authors noted that the interaction of the IGF-I with free DAPS molecules may have been more selective (i.e., hydrophobic interaction) than that with the micelles.

Thus, although MEKC–ESI–MS is recognized to present problems with competitive ionization and source contamination, the utility of MEKC has resulted in several groups working to overcome the problems. The success of the aforementioned reports indicates that the barriers are not insurmountable.

12.5.2.3. Organic Modifiers

Another type of buffer modification that has drawn attention is the use of organic modifiers. Organic solvents can change the solubilities of analytes in the buffer solution, the migration time, and [as illustrated by Nashabeh et al. (64)] the interactions between analytes and buffer components (65). There have not been extensive studies of the effects of organic cosolvents on CE–ESI, but Naylor, Tomlinson, and coworkers have published extensively on the CE–MS determination of drugs and drug metabolites in which organic cosolvents were incorporated into the buffer system (37–40,45,66–71).

12.6. APPLICATIONS

CE–MS has been used for a number of different types of analyses, with applications to protein mixtures and protein digests being most common. Unfortunately, the majority of these have been with artificial mixtures and/or known compounds. Due to space limitations and to the intention that this be a critical review, there will not be an attempt to mention all CE–MS applications. Instead, we hope to highlight those that involve "real" samples and the more novel applications.

12.6.1. Peptides and Proteins

CE–ESI–MS has been used to detect and identify recombinant and synthetic proteins. Minor components in recombinant bovine and porcine somatotropins were analyzed by Tsuji et al. (72), and five impurities in the synthesis of a synthetic 37-residue fragment of an antiherpes monoclonal antibody were separated and identified by Kostiainen et al. (73).

As part of their characterization of preconcentration using membranes (see above), Tomlinson, Naylor, and colleagues demonstrated the analysis of Bence Jones proteins in 1 µL of a 25 h patient urine collection without pretreatment (Figure 12.10) (40). The presence of these proteins is indicative of multiple myeloma.

Perkins and associates have demonstrated the utility of CE–ESI–SIM–MS and CE–ESI–MS for the analysis of complex mixtures of proteins in a series of papers characterizing snake venoms (47–49,74). In this series, there were several examples of the detection of more than 70 proteins in a single venom (Figure 12.7). Full-scan spectra of several of the components were obtained at the low-femtomole level.

Hofstadler et al. have recently reported one of the more intriguing applications of CE–MS. They report the CE–ESI–MS analysis of hemoglobin from 20 intact human erythrocyte cells (Figure 12.11) (75). This corresponds to 10 fmol of hemoglobin. When 10 cells were used, the separation of the α- and β-chains was nearly complete. Intact cells were drawn into the CE column by EOF at a -4 kV potential. A micromanipulator incorporating a stereomicroscope was used to manipulate and observe the column. The cells underwent lysis when exposed to the running buffer, thus releasing the cellular contents.

12.6.2. Pharmaceutical Applications

Although there have been a number of papers illustrating the separation of drugs and drug metabolites, most have been proof of concept papers. Naylor, Tomlinson, and their coworkers at the Mayo Clinic have perhaps the most extensive series of reports of the application of CE–MS to clinical and related problems, especially of neuroleptic drugs. They have investigated rat microsomal metabolites of haloperidol and mifentidine (71,76), as well as performing an analysis of haloperidol and its metabolites from the urine of a patient on haloperidol treatment (40,45). Using membrane preconcentration, they were

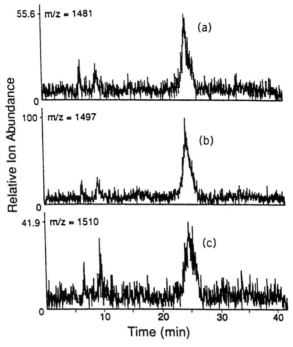

Figure 12.10. PC–CE–MS analysis of 1 μL of patient urine showing the detection of Bence Jones proteins. Conditions as in Figure 12.5 except that the CE voltage was −30 kV with a simultaneous application of low pressure. [Reprinted with permission from L. M. Benson, R. P. Oda, W. D. Braddock, B. L. Riggs, J. A. Katzmann, and S. Naylor, A. J. Tomlinson, *J. Capillary Electrophor.* **2**, 97 (1995).]

able to obtain full-scan spectra of three phase I metabolites and the parent compound (Figure 12.12) (12). In previous reports with no preconcentration, only the parent drug and its reduced form were detected by means of SIM and static array detection (45). If the membrane preconcentration techniques combined with MS becomes widely available, the number of applications to clinical samples is likely to increase significantly due to the increased sensitivity and specificity. Henion's group has also published a number of papers demonstrating CE–MS for drug analysis. In addition to the identification of flurazepam metabolites in urine [see Johansson et al. (77)], this group has recently demonstrated the separation and analysis of enantiomers of terbutaline and ephedrine using an electrolyte solution of heptakis (2,6-di-*O*-methyl)-*β*-cyclodextrin and SIM detection (78). Chiral purity at the 99% level was determinable based on a 1% spike of (1*R*,2*S*)-(−)-ephedrine in (1*S*,2*R*)-(+)-ephedrine.

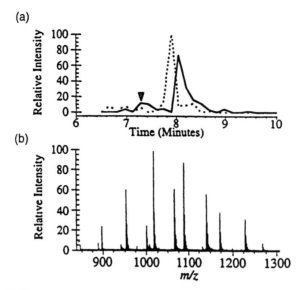

Figure 12.11. (a) The reconstructed ion electropherogram of the $(M + 17H)^{17+}$ species of the hemoglobin α-chain (solid line) and the $(M + 16H)^{16+}$ species of the hemoglobin β-chain (dashed line). (b) A mass spectrum acquired 7.3 min into the run showing the presence of both chains. [Reprinted with permission from S. A. Hofstadler, F. D. Swanek, D. C. Gale, A. G. Ewing, and R. D. Smith, *Anal. Chem.* **67**, 1477 (1995). Copyright 1995 American Chemical Society]

12.6.3. Oligonucleotides

Although CE is widely applied in the separation and sequencing of DNA, there have been very few reports of the CE–MS analysis of DNA. The most probable reason for this has been the use of sieving buffers in the separations. The extrusion of these buffers into the mass spectrometer is viewed as most likely causing serious and rapid source contamination, and so has been avoided by mass spectrometrists. Zhao et al. have recently reported the separation of several dinucleotides at the 200 pg level using ITP–ESI–MS under positive ion conditions (79).

12.6.4. Affinity Capillary Electrophoresis

An emerging area that we expect will become of major interest in the near future is affinity capillary electrophoresis (ACE). Essentially, either the ligand or the receptor can be dissolved in the running buffer, with the other half of the interaction being injected. Differences in migration time are related to the

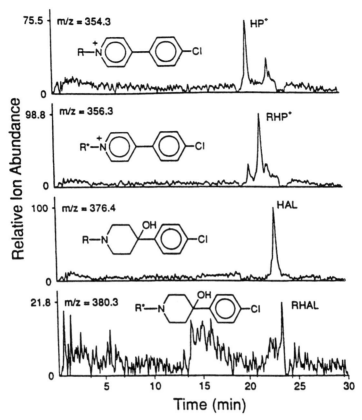

Figure 12.12. PC–CE–MS analysis of 10 μL of the supernatant solution of a patient urine after protein removal by $ZnSO_4$ precipitation and centrifugation. 4-(4-Chlorophenyl)-1-(4-fluorophenyl)-4-oxobutyl-pyridinium (HP^+), reduced HP^+ (RHP^+), reduced haloperidol (RHAL) are metabolites of haloperidol (HAL). [Reprinted with permission from L. M. Benson, R. P. Oda, W. D. Braddock, B. L. Riggs, J. A. Katzmann, and S. Naylor, A. J. Tomlinson, *J. Capillary Electrophor.* **2**, 97 (1995).]

binding constant. Multiple interactions can be compared by varying the concentrations of the ligands being studied. (A detailed description of the application of ACE to the determination of protein binding constants is given in Chapter 29 by Gao and colleagues in Part II of this monograph.) Karger's group developed an ACE–ESI–MS methodology for the identification of candidate peptides from combinatorial libraries (Figure 12.13) [see Chu et al. (80)]. In their example, they showed the identification of ligands to vancomycin from a peptide combinatorial library. Vancomycin was dissolved in

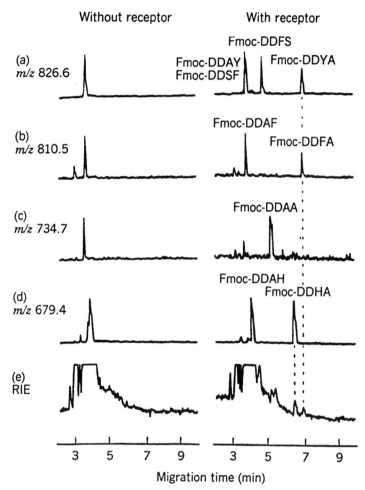

Figure 12.13. ACE–MS of synthetic all-D, Fmoc-DDXX library of 100 tetrapeptides using vancomycin as the receptor: a–d selected ion electropherograms for the masses indicated; (e) reconstructed ion electropherogram (RIE) for runs without (left) and with (right) vancomycin in the electrophoresis buffer. [Reprinted with permission from Y.-H. Chu, D. P. Kirby, and B. L. Karger, *J. Am. Chem. Soc.* **117**, 5419 (1995). Copyright 1995 American Chemical Society.]

the running buffer, and the combinatorial library (100 components) was injected into the column. Comparison of data from control runs without receptor and analysis with receptor showed that four peptides bound to the receptor. Flow of the vancomycin into the mass spectrometer was eliminated by use of a CE column coated with a neutral hydrophilic polymer that minimized the EOF. MS detection was important because it allowed identification of the ligands that bound to the vancomycin, and it readily uncovered comigrating components. With the rapidly expanding interest in the use of combinatorial libraries for screening molecular diversity for lead drug candidates and the necessity of identifying the structure of the active components, ACE–MS should become an area of increasing interest and utility.

Goodlett et al. have reported the CE–ESI–MS analysis of RNase S (18). RNase S is a noncovalent complex of a 2166 Da peptide (S-peptide) and a 11,534 Da protein (S-protein). When they analyzed the complex using a pH 3.4 buffer, only peaks due to S-peptide and S-protein were observed. Under more physiological pH conditions (pH 7.9), a broad peak was observed whose mass spectrum showed ions due to the intact RNase S complex, indicating that the complex had remained intact during the CE analysis. Similarly, Henion's group have investigated noncovalent complexes between the immunophilin FKBP (FK binding protein) and the macrolide immunosuppressive drug FK506 and with the related macrolide ligand rapamycin by CE–ESI–MS [see Hsieh et al. (82)]. Relative binding affinities consistent with values reported in the literature were determined from the data. Hamdan et al. have investigated the complexation between vancomycin, ristocetin A, teicoplanin, and two bacterial cell wall analogs, a di-and a tripeptide (83). The incorporation of CE verified that the complexes were present in solution prior to MS analysis.

12.7. SUMMARY

CE has been successfully interfaced on-line with continuous-flow fast atom bombardment and ESI. Although the combination of CE and MS offers high separation efficiency and high analyte flux for the mass spectrometrist and high structural information for the separations scientist, the combination suffers from a low dynamic range, poor concentration detection limits, and limited buffer selection. Recent advances in methods for loading higher sample amounts from dilute solutions, such as tITP and membrane preconcentration, along with newer developments in MS detectors (array detectors) and analyzers (sheathless electrospray and ion storage instruments), indicates that the sensitivity issues are being dealt with successfully.

Although heretofore most examples of CE–MS have been proof-of-concept experiments, more recent examples are coming to involve "real" problems in biological and clinical areas. The appearance of these types of applications indicates that, even though still not routine, CE–MS is beginning to be more widely accepted and applied.

ADDENDUM NO. 1

When an author is given the opportunity to update a book chapter after submission, it is difficult not to end up with a hodgepodge of disconnected thoughts. We found it not only difficult but impossible, so please accept our apologies. The following discussion does follow roughly the outline of the body of the text.

Although there have not been major advances in MS interface designs, there have been several reports that are of significance to improvements in CE–MS data. Aebi and Henion improved sensitivity in CE–ESI by using a sheath-flow interface and by incorporating a cover to the heated shield used for desolvation (reduces airborne contaminants) in an (API) interface, removing all plastic components near the heated zone, and using clean bath and curtain gas (84). The net result was an enhancement to the signal by a factor of 2 and a reduction in noise by a factor of 5. Johnson et al. reported a detailed procedure for capillary conditioning and ESI configuration for optimization of CE–MS performance (85).

Banks and Dresch reported on the use of a TOF detector with ESI for fast-scanning capabilities that more closely match the peak widths observed with optimal CE separations (86). Full-scan spectra were recorded every 0.125 s, with each spectrum being the signal average obtained from 1024 complete scans when an integrated transient recorder operating at 200 MHz was used. The authors noted that a sampling rate slower than 0.5 s/scan distorted the reconstructed total ion electropherogram. Sample sensitivities in the 250 fmol per component range were easily achieved. It is rumored that ESI–TOF mass spectrometers are to become commercially available in the near future. Banks has also reported on an improved liquid-sheath probe in which the position of the CE capillary within the ESI needle can be easily manipulated for optimum sensitivity (87).

The absence of new applications of membrane preconcentration for CE–MS, especially by groups other than the developers, is disappointing.

A number of exciting "real" applications of CE–MS have appeared recently. Hofstadler et al. have obtained spectra of hemoglobin from a single human erythrocyte using CE introduction and a sheathless electrospray interface to an FTMS (88). Levels of hemoglobin in a single cell are on the order of 450 amol. Tang et al. have reported the use of capillary isoelectric focusing with ESI–MS for the analysis of hemoglobin variants with a concentration detection limit of 10^{-8} M when SIM was employed (89). Because the ampholyte can suppress hemoglobin ion intensity, a 0.5% pharmalyte 5–8 solution was used as the carrier ampholyte. A good separation efficiency was achieved at this concentration without unacceptable loss in sensitivity (ion intensity suppression of up to 65% was noted at this concentration). Variants with a p*I* difference of 0.05 pH units were almost baseline resolved. Tang et al. noted that due to the high current generated during the mobilization step, a resistor ladder was used in parallel with electrospray high-voltage supply.

Kelly et al. have reported on the analysis of protein glycoforms and glycopeptides from digests of glycoproteins using Polybrene capillary and acidic buffers (90). Incorporation of skimmer-induced fragmentations provide structural information about the sugar moieties. Applications to proteins such as horseradish peroxidase were shown. Kelly et al. also developed methodologies for the separation and characterization of *O*-deacylated lipooligosaccharides and glycans using CE–ESI–MS with applications to lipooligosaccharides from *Moraxella catarrhalis* (91).

Novel uses of CE–MS to probe biological system have also been reported. Takada et al. have combined CE–ESI with *in vivo* microdialysis to detect, although not as yet quantify, γ-aminobutyric acid in a rat brain (92). Zhao et al. have used CITP–ESI–MS to monitor the liver alcohol dehydrogenase (LADH)–catalyzed peroxidation of nicotinamide adenine dinucleotide (NAD^+)(93). The latter authors were able to distinguish between covalent and noncovalent associations in the reaction and to provide structural information about intermediates in the reaction. Lu et al. employed 7 mM sodium dodecyl sulfate (SDS) and methanol to separate tamoxifen metabolites with on-line detection by ESI–MS (94).

There have also been several reports on the analysis of oligonucleotides and modified oligonucleotides. Barry et al. used an aqueous solution of polyvinylpyrrolidone to separate modified oligonucleotides four to six bases in length with detection by negative ion ESI–MS (95). Barry et al. investigated the CE–ESI–MS analysis of benzo[*a*]pyrene diol epoxide adducts with oligonucleotides (as the modified mononucleotide) (96). With use of sample stacking, detection limits of 10^{-8} M were achieved. Adducts at the level of four adducts in 10^7 unmodified bases were detected. Schrader and Linscheid investigated the products of the reaction of styrene oxide with DNA components (97). Tjaden's group have separated and detected inositol phosphates (mono through hexakis) at the high femtomole injected level, and analyzed inositol triphosphate spiked into a plasma sample [see Buscher et al. (98)].

Finally, Vouros and coworkers have applied CE–ESI–MS to the determination of a combinatorial library of 171 xanthene derivatives [see Dunayevskiy et al. (99)]. Linking the high separation efficiency of CE and the high structural information that can be obtained by MS promises to be invaluable in screening (cataloging?) combinatorial libraries.

ADDENDUM NO. 2

Sensitivity Enhancement

- The Mayo group has continued to explore the applicability of membrane preconcentration (mPC) with CE to biological problems (100,101). The sequence of a naturally occurring peptide ligand for thymic positive selection was determined using mPC–CE–MS/MS (102). Transient isotachophoresis (tITP) was combined with mPC–CE and was demonstrated by the analysis of major histocompatibility complex class I peptides; and their structural characterization by mPC–CE–MS/MS was reported (103). The direct analysis of haloperidol metabolites in urine, the analysis of urine from a patient suffering from multiple myeloma, 3-phenylamino-1,2-propanediol metabolites, and the analysis of MHC class I peptides from EG-7 and EL-4 cells by mPC–CE–MS was also reported (104–106).

- Koezuka et al. and Nelson et al. employed a partial filling technique combined with MEKC. In this approach, the capillary is not filled with the micellar solution, and the micellar zone migrates slower than the analytes and was not introduced into the analyzer (107,108).

- Tang et al. compared capillary isoelectric focusing–ESI/MS with CE–ESI/MS (109). In the analysis of model proteins, detection limits in the region of 10^{-7} M were achieved.

- Rundlett and Armstrong discussed the mechanism of signal suppression by anionic surfactants in CE–ESI/MS (110). They postulate that analyte suppression is caused by Coulombic interactions between oppositely charged solute and surfactant ions.

Instrumentation

- There have been a number of reports of improvements to CE–MS instrumentation. Chang and Yeung have reported on the development of an on-line laser vaporization/ionization interface for use with a TOF–MS (111). Concentration detection levels at the 10^{-7} M level for amines and peptides were reported.

- Severs and Smith have characterized a microdialysis junction for interfacing CE with microelectrospray. In this design electrical contact is made at the CE column terminus via polysulfone dialysis tubing (112,113). Cao and Moini have reported the use of a small Pt wire inserted into the CE capillary to provide electrical contact for use with

sheathless electrospray (114). Bateman et al. described a new method for preparing disposable microsprayers for use with microelectrospray (115). Kelly et al. have reported on practical considerations and performance of CE–ESI/MS at a submicroliter flow rate (116). Detection levels in the order of 0.1–5 fmol for peptides were reported.
- Ramsey and Ramsey reported a method for generating electrospray from microchips (117).
- Li et al. and Wu et al. described the coupling of CE–ESI with an ion trap storage/reflectron TOF–MS for rapid scanning of CE peaks with peak widths of less than 1 s at half-height (118,119). The application of the combination to the analysis of mutant and normal hemoglobins and protein digests were presented.
- Foret et al. have developed a UV detection system based on fiber optics that permits UV detection close to the electrospray tip (120).
- Smith's group and McLafferty's group reported further developments of the combination of CE with FT–MS (12–123).
- Wachs et al. have designed a self-aligning liquid-junction interface with improved stability for CE–MS, even with nonvolatile buffers (124).
- Kirby et al. reported the development of a CE–ESI/MS interface that can be actively positioned during the electrospray operation and can be used with liquid sheath flows as low as 250 nL/min (125).

Applications

CE–MS applications have been reported for the analysis of the following:
- High-mannose glycoproteins (126)
- *Rauwolfia* alkaloids (127)
- Proteins, including identification of unknown proteins by digestion, CE separation of the proteolytic peptides, MS/MS analysis of the peptides, and sequence database searching (128–133)
- Drugs and metabolites (134–136)
- Shellfish toxins (137)
- Oligosaccharides from *Yersinia reckeri* (138) and lipopolysaccharides from *Pseudomonas aeruginosa* (139)
- Chiral drugs (140)
- Herbicides (141,142)
- Priority phenol pollutants with a detection limit of ca. 50 ppb (143)
- DNA adducts with phenyl glycidyl ethers (144).
- Combinatorial libraries (145)

- Cellulose chemical modifications as a basis for a new and efficient dating technique for archaeological textiles taken from the same site (146).

LIST OF ACRONYMS

Acronym	Definition
ACE	affinity capillary electrophoresis
API	atmospheric pressure ionization
BBMA	butyl acrylate/butyl methacrylate/methacrylic acid copolymer sodium salt
CE	capillary electrophoresis
CF	continuous flow
CITP	capillary isotachophoresis
CTAC	cetyltrimethylammonium chloride
DAPS	N-dodecyl-N,N-dimethyl-3-amino-1-propanesulfonate
EACA	ε-amino-n-caproic acid
EOF	electroosmotic flow
ESI	electrospray ionization
FAB	fast atom bombardment
FTMS	Fourier transform (ion cyclotron resonance) mass spectrometer
HP$^+$	4-(4-chlorophenyl)-1-(4-fluorophenyl)-4-oxobutyl-pyridinium
HPCE	high-performance capillary electrophoresis
HPLC	high-performance liquid chromatography
IGF-I & -II	insulin-like growth factors I and II
ITP	isotachophoresis
LADH	liver alcohol dehydrogenase
LC	liquid chromatography
MALDI	matrix-assisted laser desorption ionization
MEKC	micellar electrokinetic chromatography
mPC	membrane preconcentration
MS	mass spectrometry
NAD$^+$	nicotinamide adenine dinucleotide
PC	preconcentration
PTFE	polytetrafluoroethylene (or Teflon)
RHP$^+$	reduced HP$^+$
RHAL	reduced haloperidol
SDS	sodium dodecyl sulfate
SIM	selected ion monitoring
tITP	transient isotachophoresis
TOF	time-of-flight (analyzer)

REFERENCES

1. J. W. Jorgenson, and K. D. Lukacs, *Anal Chem.* **53,** 1298 (1981).
2. J. A. Olivares, N. T. Nguyen, C. R. Yonker, and R. D. Smith, *Anal. Chem.* **59**, 1230 (1987).
3. M. Barber, R. S. Bordoli, R. D. Sedwick, and A. N. Tyler, *J. Chem. Soc., Chem. Commun.*, 325 (1981).
4. R. D. Smith, C. J. Barinaga, and H. R. Udseth, *Anal. Chem.* **60**, 1948 (1988).
5. M. A. Moseley, L. J. Deterding, K. B. Tomer, and J. W. Jorgenson, *Rapid Commun. Mass Spectrom.* **3**, 87 (1989).
6. E. D. Lee, W. Muck, J. D. Henion, and T. R. Covey, *J. Chromatogr.* **458**, 313 (1988).
7. R. M. Caprioli, W. T. Moore, M. Martin, B. B. DaGue, K. Wilson, and S. Moring, *J. Chromatogr.,* **480**, 247 (1989).
8. R. D. Minard, D. Chinn-Fatt, P. Curry, Jr., and A. G. Ewing, *ASMS Confe. Mass. Spectrom. and Allied Top. 36th,* p. 950 (1988).
9. F. Hillenkamp and M. Karas, *Methods Enzymol.* **193**, 280 (1990).
10. C. M. Whitehouse, R. N. Dreyer, M. Yamashita, and J. B. Fenn, *Anal. Chem.* **57**, 675 (1985).
11. M. Yamashita and J. B. Fenn, *J. Phys. Chem.* **88**, 4451 (1984).
12. A. P. Bruins, T. R. Covey, and J. D. Henion, *Anal. Chem.* **59**, 2642 (1987).
13. Y. Ito, T. Takeuchi, D. Ishi, and M. Goto, *J. Chromatogr.* **346**, 161 (1985).
14. R. M. Caprioli, T. Fan, and J. S. Cottrell, *Anal. Chem.* **58**, 2949 (1986).
15. C. E. Parker, J. R. Perkins, K. B. Tomer, Y. Shida, K. O'Hara, and M. Kono, *J. Am. Soc. Mass Spectrom.* **3**, 563 (1992).
16. J. S. M. deWit, L. J. Deterding, M. A. Moseley, K. B. Tomer, and J. W. Jorgenson, *Rapid Commun. Mass Spectrom.* **2**, 100 (1988).
17. M. A. Moseley, L. J. Deterding, K. B. Tomer, and J. W. Jorgenson, *Anal. Chem.* **63**, 109 (1991).
18. E. D. Lee, W. Muck, J. D. Henion, and T. R. Covey, *J. Chromatogr.* **458**, 313 (1988).
19. E. D. Lee, W. Muck, J. D. Henion, and T. R. Covey, *Biomed. Environ. Mass Spectrom.* **18**, 844 (1989).
20. S. Pleasance, P. Thibault, and J. Kelly, *J. Chromatogr.* **591**, 325 (1992).
21. M. J.-F. Suter, and R. M. Caprioli, *J. Am. Soc. Mass Spectrom.* **3**, 198 (1992).
22. S. K. Chowdhury, and B. T. Chait, *Anal. Chem.* **63**, 1660 (1991).
23. D. C. Gale and R. D. Smith, *Rapid Commun. Mass Soectrom.* **7**, 1017 (1993).
24. M. R. Emmett and R. M. Caprioli, *J. Am. Soc. Mass Spectrom.* **5**, 605 (1994).
25. P. E. Andren, M. R. Emmett, and R. M. Caprioli, *J. Am. Soc. Mass Spectrom.* **5**, 867 (1994).
26. M. S. Wilm and M. Mann, *Int. J. Mass Spectrom. Ion Processes* **136**, 167 (1994).
27. J. H. Wahl, D. C. Gale and R. D. Smith, *J. Chromatogr. A* **659**, 217 (1994).

28. M. S. Kriger, K. D. Cook, and R. S. Ramsey, *Anal. Chem.* **67**, 385 (1995).
29. M. A. Moseley, L. J. Deterding, K. B. Tomer, and J. W. Jorgenson, *J. Chromatogr.* **480**, 197 (1989).
30. M. A. Moseley, J. W. Jorgenson, J. Shabanowitz, D. F. Hunt, and K. B. Tomer, *J. Am. Soc. Mass Spectrom.* **3**, 389 (1992).
31. J. H. Wahl and R. D. Smith, *J. Capillary Electrophor.* **1**, 62 (1994).
32. J. R. Perkins, C. E. Parker, and K. B. Tomer, *J. Am. Soc. Mass Spectrom.* **3**, 139 (1992).
33. D. S. Burgi and R.-L. Dhien, *J. Microcolumn Sep.* **3**, 199 (1991).
34. A. P. Tinke, J. J. Reinhoud, W. M. A. Niessen, U. R. Tjaden, and J. van der Greef, *Rapid Commun. Mass Spectrom.* **6**, 560 (1992)
35a. T. J. Thompson, F. Foret, P. Vouros, and B. L. Karger, *Anal. Chem.* **65**, 900 (1993).
35b. K. B. Tomer, C. E. Parker, and L. J. Deterding, in *Capillary Electrophoresis: An Analytical Tool in Biotechnology* (P. G. Righetti, ed.), p. 121. CRC Press, Boca Raton, 1995.
36. M. H. Lamoree, N. J. Reinhoud, U. R. Tjaden, W. M. A, Niessen, and J. van der Greef, *Biol. Mass Spectrom.* **23**, 339 (1994).
37a. A. J. Tomlinson, L. M. Benson, W. D. Braddock, R. P. Oda, and S. Naylor, *J. High Resolut. Chromatogr.* **17**, 729 (1994).
37b. A. J. Tomlinson, W. D. Braddock, L. M. Benson, R. P. Oda, and S. Naylor, *J. Chromatogr. B: Biomed. Appl.* **669**, 67 (1995).
38a. I. Morita and J. I. Sawada, *J. Chromatogr.* **641**, 375 (1993).
38b. M. E. Swatz and M. Merion, *J. Chromatogr.* **632**, 209 (1993).
39. A. J. Tomlinson, L. M. Benson, R. P. Oda, and S. Naylor, *J. High Resolut. Chromatogr.* **17**, 669 (1994).
40. A. J. Tomlinson, L. M. Benson, R. P. Oda, W. D. Braddock, B. L. Riggs, J. A. Katzmann, and S. Naylor, *J. Capillary Electrophor.* **2**, 97 (1995).
41. N.A. Guzman, ed., *Capillary Electrophoresis Technology*. Dekker, New York, 1993.
42. J. H. Wahl, D. R. Goodlett, H. R. Udseth, and R. D. Smith, *Electrophoresis* **14**, 448 (1993).
43. L. W. Tetler, P. A. Cooper, and B. Powell, *J. Chromatogr. A* **700**, 21 (1995).
44. N. J. Reinhoud, E. Schroeder, U. R. Tjaden, W. M. A. Niessen, M. C. Ten Noever de Brauw, J. van der Greef, *J. Chromatogr.* **516**, 147 (1990).
45. A. J. Tomlinson, L. M. Benson, K. L. Johnson, and S. Naylor, *Electrophoresis* **15**, 62 (1994).
46. R. L. Cerny, J. M. Y. Wellemans, M. L. Gross, L. J. Deterding, and K. B. Tomer, *Proc. ASMS Conf. Mass Spectrom. Allied Top. 40th*, p. 51 (1992).
47. J. R. Perkins and K. B. Tomer, *Eur. J. Biochem.* **233**, 815 (1995).
48. J. R. Perkins and K. B. Tomer, *J. Capillary Electrophor.* **1**, 231 (1994).
49. J. R. Perkins and C. E. Parker and K. B. Tomer, *Electrophoresis* **14**, 458–68 (1993).
50. A. Mordehai and Henion, *Proc. ASMS Conf. Mass Spectrom. Allied Top. 40th*, p. 197 (1992).

51. J. C. Schwartz and I. Jardine, *Proc. ASMS Conf. Mass Spectrom. Allied Top. 40th,* p. 707 (1992).
52. R. S. Ramsey, D. E. Goeringer, and S. A. McLuckey, *Anal. Chem.* **65**, 3521 (1993).
53. R. S. Ramsey, and S. A. McLuckey, *J. Microcolumn Sep.* **7**, 461 (1995).
54. J. D. Henion, A. V. Mordehai, and J. Cai, *Anal. Chem.* **66**, 2103 (1994).
55. S. A. Hofstadler, J. H. Wahl, J. E. Bruce, and R. D. Smith, *J. Am. Chem. Soc.* **115**, 6983 (1993).
56. S. A. Hofstadler, J. H. Wahl, R. Bakhtiar, G. A. Anderson, J. E. Bruce, and R. D. Smith *J. Am. Soc. Mass Spectrom.* **5**, 894 (1994).
57. J. H. Wahl, S. A. Hofstadler, Z. Shao, D. C. Gale and R. D. Smith, *Proc. ASMS Conf. Mass Spectrom. Allied Top., 42nd,* p. 1019 (1994).
58. P. J. Johnson, D. S. Gross, P. D. Schnier, and E. R. Williams, *Proc. ASMS Conf. Mass Spectrom. Allied Top., 42nd,* p. 235 (1994).
59. I. Fang, R. Zhang, E. R. Williams, and R. N. Zare, *Anal. Chem.* **66**, 3696 (1994).
60. B. A. Andrien, J. F. Banks, J. G. Boyle, T. Dresch, P. F. Haren, and C. M. Whitehouse, *Proc. ASMS Conf. Mass Spectrom. Allied Top. 43rd,* p. 999 (1995).
61. G. A. Valaskoviec, N. L. Kelleher, D. P. Little, D. J. Aaserud, and F. W. McLafferty, *Anal. Chem.* **67**, 3802 (1995).
62. J. Varghese and R. B. Cole, *J. Chromatogr. A* **652**, 369 (1993).
63. H. Ozali, N. Itou, S. Terabe, Y. Takada, M. Sakairi, and H. Koizumi, *J. Chromatogr. A* **716**, 69 (1995).
64. W. Nashabeh, K. F. Greve, D. Kirby, F. Foret, B. L. Karger, D. H. Reifsnyder, and S. E. Builder, *Anal. Chem.* **66**, 2148 (1994).
65. E. Kenndler, in *Capillary Electrophoresis Technology* (N. A. Guzman, ed), Chromatogr. Sci. Ser., Vol. 64, p. 161. Dekker, New York, 1993.
66. A. J. Tomlinson, L. M. Benson, J. W. Gorrod, and S. Naylor, *J. Chromatogr. B: Biomed. Appl.* **657**, 373 (1994).
67. S. Nakylor, A. J. Tomlinson, L. M. Benson, and J. W. Gorrod, *Eur. J. Drug Metab. Pharmacokinet.* **19**, 235 (1994).
68. A. J. Tomlinson, L. M. Benson, and K. L. Johnson, *J. Chromatogr. Biomed. Appl.* **621**, 239 (1993).
69. A. J. Tomlinson, L. M. Benson, and S. Naylor, *J. Capillary Electrophor.* **1**, 127 (1994).
70. A. J. Tomlinson, L. M. Benson, J. W. Gorrod, and S. Naylor, *J. Chromatogr. B: Biomed. Appl.* **657**, 373 (1994).
71. A. J. Tomlinson, L. M. Benson, and S. Naylor, *J. High Resolut. Chromatogr.* **17**, 175 (1994).
72. K. Tsuji, L. Baczynskyj, and G. E. Bronson, *Anal. Chem.* **64**, 1864 (1992).
73. R. Kostiainen, E. Lasonder, W. Bloemhoff, P. A. van Veelen, G. W. Welling, and A. P. Bruins, *Biol. Mass Spectrom.* **23**, 346 (1994).
74. J. R. Perkins, and K. B. Tomer, *Anal. Chem.* **66**, 2835 (1994).
75. S. A. Hofstadler, F. D. Swanek, D. C. Gale, A. G. Ewing, and R. D. Smith, *Anal. Chem.* **67**, 1477 (1995).

76. A. J. Tomlinson, L. M. Benson, K. L. Johnson, and S. Naylor, *J. Chromatogr.* **621**, 239 (1993).
77. I. M. Johansson, R. Pavelka, and J. D. Henion, *J. Chromatogr.* **559**, 515 (1991).
78. R. L. Sheppard, X. Tong, J. Cai, and J. D. Henion, *Anal. Chem.* **67**, 2054 (1995).
79. Z. Zhao, J. H. Wahl, H. R. Udseth, S. A. Hofstadler, A. F. Fuciarelli, and R. D. Smith, *Electrophoresis* **16**, 389 (1995).
80. Y.-H. Chu, D. P. Kirby, and B. L. Karger, *J. Am. Chem. Soc.* **117**, 5419 (1995).
81. D. R. Goodlett, R. R. Ogorzalek Loo, J. A. Loo, J. H. Wahl, H. R. Udseth, and R. D. Smith, *J. Am. Soc. Mass Spectrom.* **5**, 614 (1994).
82. Y.-L. Hsieh, J. Cai, Y.-T. Li, J. D. Henion, and B. Ganem, *J. Am. Soc. Mass Spectrom.* **6**, 85 (1995).
83. M. Hamdan, O. Curcuruto, and E. Di Modugno, *Rapid Commun. Mass Spectrom.* **9**, 883 (1995).
84. B. Aebi, and J. Henion, *Rapid Commun. Mass Spectrom.* **10**, 947 (1996).
85. K. L. Johnoson, A. J. Tomlinson, and S. Naylor, *Rapid Commun. Mass Spectrom.* **10**, 1159 (1996).
86. J. F. Banks, Jr. and T. Dresch, *Anal. Chem.* **68**, 1480 (1996).
87. J. F. Banks, Jr. *J. Chromatogr. A.* **712**, 245 (1995).
88. S. A. Hofstadler, J. C. Severs, R. D. Smith, F. D. Swanek, and A. G. Ewing, *Rapid Commun. Mass Spectrom.* **10**, 919 (1996).
89. Q. Tang, A. K. Harrata, and C. S. Lee, *Anal. Chem.* **68**, 2482 (1996).
90. J. F. Kelly, S. J. Locke, L. Ramaley, and P. Thibault, *J. Chromatogr. A.* **720**, 409 (1996).
91. J. Kelly, H. Masoud, M. B. Perry, J. C. Richards, and P. Thibault, *Anal. Biochem.* **233**, 15 (1996).
92. Y. Takada, M. Yoshida, M. Sakairi and H. Koizumi, *Rapid Commun. Mass Spectrom.* **9**, 895 (1995).
93. Z. Zhao, H. R. Udseth and R. D. Smith, *J. Mass Spectrom.* **31**, 193 (1996).
94. W. Lu, G. K. Poon, P. L. Carmichael and R. B. Cole, *Anal. Chem.* **68**, 668 (1996).
95. J. P. Barry, J. Muth, S.-J. Law, B. L. Karger and P. Vouros, *J. Chromatogr. A.* **732**, 159 (1996).
96. J. P. Barry, C. Norwood and P. Vouros, *Anal. Chem.* **68**, 1432 (1966).
97. W. Schrader and M. Linscheid, *J. Chromatogr. A.* **717**, 117)1995).
98. B. A. P. Buscher, R. A. M. van der Hoeven, U. R. Tjaden, E. Andersson, and J. van der Greef, *J. Cromatogr. A.* **712**, 235 (1995).
99. Y. M. Dunayevskiy, P. Vouros, E. A. Wintner, G. W. Shipps, T. Carell, and J. Rebek, Jr. *Proc. Natl. Acad. Sci. U. S. A.* **93**, 6152 (1996).
100. E. Kurian, F. G. Prendergast, A. J. Tomlinson, J. W. Holmes, and S. Naylor, *J. Am. Soc. Mass Spectrom.* **8**, 8 (1997).
101. S. Naylor and A. J. Tomlinson, *Biomed. Chromatogr.* **10**, 325 (1996).
102. K. A. Hogquist, A. J. Tomlinson, W. C. Kieper, M. A. McGargill, M. C. Hart, S. Naylor, and S. C. Jameson, *Immunity* **6**, 389 (1977).
103. A. J. Tomlinson, L. M. Benson, S. Jameson, and S. Naylor, *Electrophoresis* **17**, 1801 (1996).

104. A. J. Tomlinson, L. M. Benson, S. Jameson, D. H. Johnson, and S. Naylor, *J. Am. Soc. Mass Spectrom.* **8**, 15 (1997).
105. A. J. Tomlinson, S. Jameson, and S. Naylor, *J. Chromatogr. A.* **744**, 273 (1996).
106. L. M. Benson, A. J. Tomlinson, A. N. Mayeno, G. J. Gleich, D. Wells, and S. Naylor, *J. High Resolut. Chromatogr.* **19**, 291 (1996).
107. K. Koezuka, H. Ozaki, N. Matsubara, and S. Terabe, *J. Chromatogr. B: Biomed. Appl.* **689**, 3 (1997).
108. W. M. Nelson, Q. Tang, A. K. Harrata, and C. S. Lee, *J. Chromatogr. A* **749**, 219 (1996).
109. Q. Tang, A. K. Harrata, and C. S. Lee, *J. Mass Spectrom.* **31**, 1284 (1996).
110. K. L. Rundlett and D. W. Armstrong, *Anal. Chem.* **68**, 3493 (1996).
111. S. Y. Chang and E. S. Yeung, *Anal. Chem.* **69**, 2251, (1997).
112. J. C. Severs and R. D. Smith, *Anal. Chem.* **69**, 2154 (1997).
113. J. C. Severs, A. C. Harms, and R. D. Smith, *Rapid Commun. Mass Spectrom.* **10**, 1175 (1996).
114. P. Cao and M. Moini, *J. Am. Soc. Mass Spectrom.* **8**, 561 (1997).
115. K. P. Bateman, R. L. White, and P. Thibault, *Rapid Commun. Mass Spectrom.* **11**, 307 (1997).
116. J. F. Kelley, L. Ramaley, and P. Thibault, *Anal. Chem.* **69**, 51 (1997).
117. R. S. Ramsey and J. M. Ramsey, *Anal. Chem.* **69**, 1174 (1997).
118. M. X. Li, J.-T. Wu, L. Liu, and D. M. Lubman, *Rapid Commun. Mass Spectrom.* **11**, 99 (1997).
119. J.-T. Wu, M. G. Qian, M. X. Li, L. Liu, and D. M. Lubman, *Anal. Chem.* **68**, 3388 (1996).
120. F. Foret, D. P. Kirby, P. Vouros, and B. L. Karger, *Electrophoresis* **17**, 1829 (1996).
121. J. C. Severs, S. A. Hofstadler, Z. Zhao, R. T. Senh, and R. D. Smith, *Electrophoresis* **17**, 1808 (1996).
122. S. A. Hofstadler, J. C. Severs, R. D. Smith, F. D. Swanek, and A. G. Ewing, *J. High Resolut. Chromatogr.* **19**, 617 (1996).
123. G. A. Valaskovic, N. L. Kelleher, and F. W. McLafferty, *Science* **273**, 1199 (1996).
124. T. Wachs, R. L. Sheppard, and J. Henion, *J. Chromatogr. B* **685**, 335 (1996).
125. D. P. Kirby, J. M. Thorne, W. K. Goetzinger, and B. L. Karger, *Anal. Chem.* **68**, 4451 (1996).
126. B. Yeung, T. J. Porter, and J. E. Vath, *Anal. Chem.* **69**, 2510 (1997).
127. D. Stockigt, M. Unger, D. Belder, and J. Stockigt, *Nat. Prod. Lett.* **9**, 265 (1997).
128. D. Figeys and R. Aebersold, *Electrophoresis* **18**, 360 (1997).
129. D. Figeys, A. Ducret, and R. Aebersold, *J. Chromatogr. A.* **763**, 295 (1997).
130. I. D. Cruzado, S. Song, S. F. Crouse, B. C. O'Brien, and R. D. Macfarlane, *Anal. Biochem.* **243**, 100 (1996).
131. D. Figeys, A. Ducret, J. R. Yates III, and R. Aebersold, *Nat. Biotechnol.* **14**, 1579 (1996).

132. D. Figeys, I. van Oostveen, A. Ducret, and R. Aebersold, *Anal. Chem.* **68**, 1822 (1996).
133. E. Forest, Y. Petillot, J. Gagnon, J. Cotton, and C. Vita, *Electrophoresis,* **17**, 962 (1996).
134. A. E. Ashcroft, J. J. Major, I. D. Wilson, A. Nicholls, and J. K. Nicholson, *Anal. Commun.* **34**, 41 (1997).
135. J. Cai and J. Henion, *J. Anal. Toxicol.* **20**, 27 (1996).
136. W. Lu, G. K. Poon, P. L. Carmichael, and R. B. Cole, *Anal. Chem.* **68**, 668 (1996).
137. S. Gallacher, K. J. Flynn, J. M. Franco, E. E. Brueggemann, and J. B. Hines, *Appl. Environ. Microbiol.* **63**, 239 (1997).
138. K. P. Bateman, J. H. Banoub, and P. Thibault, *Electrophoresis* **17**, 1818 (1996).
139. S. Auriola, P. Thibault, I. Sadovskaya, E. Altman, H. Masoud, and J. C. Richards, *ACS Symp. Ser.* **619**, 149 (1996).
140. M. H. Lamoree, A. F. H. Sprang, U. R. Tjaden, and J. van der Greef, *J. Chromatog. A* **742**, 235 (1996).
141. E. Moyano, D. E. Games, and M. T. Galceran, *Rapid Commun. Mass Spectrom.* **10**, 1379 (1996).
142. D. Wycherley, M. E. Rose, K. Giles, T. M. Hutton, D. A. Rimmer, *J. Chromatogr. A* **734**, 339 (1996).
143. C.-Y. Tsai and G.-R. Her, *J. Chromatogr. A* **743**, 315 (1996).
144. D. L. D. DeForce, F. P. K. Ryniers, E. G. van den Eeckhout, F. Lemiere, and E. L. Esmans, *Anal. Chem.* **68**, 3575 (1996).
145. Y.-H. Chu, Y. M. Dunayevskiy, D. P. Kirby, P. Vouros, and B. L. Karger, *J. Am. Chem. Soc.* **118**, 7827 (1996).
146. D. A. Kouznetsov, A. A. Ivanov, and P. R. Veletsky, *ACS Symp. Ser* **625**, 254 (1996).

PART III

OPERATIONAL ASPECTS AND SPECIAL TECHNIQUES IN HPCE

CHAPTER

13

SAMPLE INTRODUCTION AND STACKING

RING-LING CHIEN

Edward L. Ginzton Research Center, Varian Associates, Inc., Palo Alto, California 94304

13.1.	**Introduction**	449
13.2.	**Sample Introduction in Capillary Zone Electrophoresis**	451
	13.2.1. Hydrodynamic Injection	451
	13.2.2. Electrokinetic Injection	453
	13.2.3. Spontaneous Injection	455
	13.2.4. Artifacts in Sample Introduction	456
13.3.	**On-Column Sample Stacking**	456
	13.3.1. Isotachophoresis–Capillary Zone Electrophoresis	457
	13.3.2. Field-Amplified Sample Stacking	460
13.4.	**Stacking in Sample Introduction**	464
	13.4.1. Field-Amplified Sample Injection	464
	13.4.2. Pseudoisotachophoretic Sample Injection	467
	13.4.3. Polarity-Switching Sample Injection	467
13.5.	**Applications of Sample Stacking**	471
	13.5.1. Stacking of Neutral Moleules in Micellar Electrokinetic Chromatography	471
	13.5.2. Trace Enrichment of Environmental Samples	474
	13.5.3. Sample Stacking in Mass Spectrometric Analysis	474
13.6.	**Conclusions**	477
	Addendum	477
	List of Acronyms and Abbreviations	478
	References	478

13.1. INTRODUCTION

One of the major advantages of capillary zone electrophoresis (CZE) compared with other separation techniques is its high resolution. To obtain high

High Performance Capillary Electrophoresis, edited by Morteza G. Khaledi. Chemical Analysis Series, Vol. 146.
ISBN 0-471-14851-2 © 1998 John Wiley & Sons, Inc.

resolution in CZE, the zone length or volume of the sample has to be much smaller than the widths generated from either diffusion broadening or the detector window. However, detection of a small volume using a conventional ultraviolet (UV) detection is hindered by the short optical path defined by the diameter of the capillary. Although the mass limit in CZE can be very low because of the small volume, the concentration limit is usually on the order of 10^{-6} M. A major challenge in CZE is how to increase sample loading while maintaining its high resolution.

Ideally, one would like to introduce the maximum amount of sample as an infinitely narrow zone and using a zero volume detector. Theoretically, the dispersion will then be limited only by the longitudinal diffusion process and the number of theoretical plates is (1)

$$N_D = \mu V_d / 2D \tag{13.1}$$

where μ is the apparent mobility of the analyte, equal to the sum of the electrophoretic mobility μ_{ep} and electroosmotic mobility μ_{eo}; V_d is the voltage normalized to the detector position; and D is the diffusion coefficient. For example, if $\mu = 4 \times 10^{-8}\,m^2/V \cdot s$, $D = 6 \times 10^{-10}\,m^2/s$, and $V = 20\,kV$, the theoretical maximum number of plates is 6.67×10^5.

On the other hand, if the resolution is limited by the injection length, then the number of theoretical plates is (2)

$$N_{inj} = 12(L_d/l_{inj})^2 \tag{13.2}$$

or

$$N_{inj} = 12(1/x_{inj})^2 \tag{13.3}$$

where L_d is the length to the detector window; l_{inj} is the length of the injected sample zone; and $x_{inj} = (l_{inj}/L_d)$ is the normalized injection length. For a sample zone length that is 1% of the separation column, i.e., $x_{inj} = 0.01$, the plate number is 1.2×10^5, smaller than the diffusion limited plate number. To achieve diffusion limited dispersion, the injection length has to be kept short enough so that N_{inj} will be much larger than N_D.

Following Grushka and McCormick (3), we can express the injection length as a function of the decrease in the theoretical plates or the loss of resolution, ε, by the equation

$$l_{inj} = (24D\varepsilon t)^{1/2} \tag{13.4}$$

$$= (24D\varepsilon/\mu V_d)^{1/2} L_d \tag{13.5}$$

where t is the sample injection time and ε is the loss of resolution. Since this expected injection length will depend on the separation column length, it is more rational to rearrange it to the normalized sample length as

$$x_{inj} \equiv l_{inj}/L_d = (24D\varepsilon/\mu V_d)^{1/2} \qquad (13.6)$$

Using the parameters given in the previous paragraph, the normalized injection zone length is 0.0013 for $\varepsilon = 0.1$. This means that if the decrease in N is expected to be within 10% from the diffusion limited case, the sample zone length should be less than 1.3 mm for a 1 m long separation column. For a capillary with 100 μm i.d., this zone length is equivalent to a sample volume of 10 nL. Such a small zone volume generates a host of difficulties and challenges for analytical chemists, who not only require reproducible and precise analysis but also sensitive detection.

To increase sample loading and minimize the impact of zone length on the sensitivity and resolution of detection, one has to control and understand the basic principles of various injection mechanisms. Many efforts have been undertaken to improve repetition accuracy by using automated injectors. An alternative means of reducing injection effects is to perform on-column sample stacking to further narrow the sample zone. Sample stacking can provide better detection sensitivity while maintaining high resolution. Several different stacking techniques have been reported in the literature (4–7). Each technique has its advantages and drawbacks. This chapter will review various injection and stacking methods and describe some applications involving sample stacking.

13.2. SAMPLE INTRODUCTION IN CAPILLARY ZONE ELECTROPHORESIS

13.2.1. Hydrodynamic Injection

There are basically two different methods of sample introduction into the capillary—either hydrodynamic injection or electrokinetic injection (8). The hydrodynamic method is based on pressure differences between the inlet and outlet ends of the capillary. This pressure difference can be achieved by various methods such as gravimetric, overpressure, and vacuum. For example, in gravimetric injection, the buffer reservoir at the injection end of the capillary is replaced with the sample vial. The sample vial is then raised vertically to a predetermined height for a fixed interval of time. This height difference between the liquid levels of the inlet and outlet end of the capillary creates hydrostatic pressure that forces the sample into the capillary. After sample introduction, the sample vial is lowered to ground level and replaced by the buffer reservoir. A high voltage is then applied to the capillary and starts the separation.

The length of the sample zone injected into the capillary by the hydrodynamic method is given by the time integral of the flow velocity curve:

$$l_{inj} = \int_0^{t_{inj}} v_{hf}\, dt \tag{13.7}$$

where t_{inj} is the total injection time and v_{hf} is the hydrodynamic flow velocity, which is a function of injection time. As a first-order approximation, the pressure and hence the flow velocity can be assumed to be constant, so that

$$l_{inj} = t_{inj} v_{hf} \tag{13.8}$$

In a capillary tube with a fixed length and diameter, the hydrodynamic flow is proportional to the pressure difference and inversely proportional to the viscosity of the liquid. The flow velocity is described by the Poisseuille equation:

$$v_{hf} = \frac{\Delta P d^2}{32 \eta L_t} \tag{13.9}$$

where ΔP is the pressure difference between the inlet and outlet ends of the capillary; d is the capillary inner diameter; η is the liquid viscosity; and L_t is the capillary length. For injection by gravity, the pressure difference is simply proportional to the height difference Δh between the liquid levels of the sample vial and the buffer vial, or

$$\Delta P = \rho g \Delta h \tag{13.10}$$

where ρ is the density of the sample solution and g is the acceleration of gravity, 9.8 m/s².

From Eqs. (13.8) and (13.9), we can express the pressure required as a function of normalized injection length:

$$\Delta P = 8\eta \frac{x_{inj}}{t} \left(\frac{L_t}{r}\right)^2 \tag{13.11}$$

where we have assumed that the detector is located at the end of the capillary. For a 1 m capillary with a 100 μm i.d. and $\eta = 0.001$ kg/m³, Eq. (13.11) shows that the pressure must be smaller than 0.32 kPa for 10 s if one is to achieve a normalized injection length of 0.001. This pressure is roughly equivalent to a height difference of 3 cm. One could also increase the pressure and shorten the injection time to achieve the same zone length.

In general, hydrodynamic injection has better reproducibility and greater control over the amount of sample injected into the capillary. Since the injection is based on the pressure difference, it is universally applied to all kinds of sample matrices without any bias on the sample components. However, the implementation of the pressure control is rather complicated and involves extra equipment. In addition, the response time (or "travel time") is rather slow and has a major impact during the short injection time (8).

13.2.2. Electrokinetic Injection

In electrokinetic injection, one end of the capillary and the injecting electrode are removed from the separation buffer reservoir and placed in the sample vial. High voltage is then applied for a short period of time. Analyte ions then migrate into the capillary due to the combination of electrophoretic migration of the ions and electroosmotic flow of the sample solution. The length of sample zone introduced into the capillary during electrokinetic injection can be expressed in a manner analogous to Eq. (13.7) for hydrodynamic injection:

$$l_{inj} = \int_0^{t_{inj}} (v_{eo} + v_{epi}) \, dt \tag{13.12}$$

where v_{eo} is the electroosmotic velocity of the bulk solution and v_{epi} is the electrophoretic velocity of ion species i at the injection interface. The electrophoretic velocity and the electroosmotic velocity may or may not be in the same direction and are assumed to be positive if migrating toward the output end of the capillary and to be negative otherwise. In order to inject sample ions into the capillary, the combination of the electrophoretic velocity and the bulk electroosmotic velocity must be positive.

There is a major difference between electrophoresis of a charge paticle and electroosmosis of a bulk solution. Electrophoresis is the movement of charged particles in response to a local electric field, and the electrophoretic velocity is simply the product of electrophoretic mobility and the local field strength. On the other hand, electroosmosis refers to the displacement of the bulk solution and has to be averaged over the whole length of capillary (9). If the sample buffer is the same as the separation buffer, a uniform electric field strength will be established across the length of the capillary. For a short injection time, one could simplify Eq. (13.12) as

$$l_{inj} = (v_{eo} + v_{epi})t \tag{13.13}$$

and

$$v_{eo} = \mu_{eo}E \qquad (13.14)$$

$$v_{epi} = \mu_{epi}E \qquad (13.15)$$

where E ($= V/L$) is the electric field strength across the capillary; μ_{eo} is the electroosmotic mobility; and μ_{ep} is the electrophoretic mobility of ion species i.

The amount of sample injected into the capillary is simply proportional to the injection length. For ion species i, the injected amount is

$$m_i = AC_i l_{inj} = AC_i(\mu_{eo} + \mu_{epi})Et \qquad (13.16)$$

where A is the cross-section area of the capillary and C_i is the concentration of ion species i. We have assumed that the electrophoretic mobility remains the same as the ions cross the boundary between the capillary and the sample vial. Equations (13.13) and (13.16) predict a difference in the sample zone length and hence an injection bias determined by the difference in the electrophoretic mobilities of analytes (10).

In deriving the foregoing equations, we have assumed that v_{epi} remains the same as the ion across the injection interface into the capillary. Under certain operating conditions, the electrophoretic mobility of the analyte ions inside the capillary may be different from their mobility in the sample solution. Bocek's group recently studied the injection bias of DNA fragments in free solution by electromigration into a capillary filled with the sieving medium [see Kleparnik et al. (11)]. They found injection bias according to the sample size even for fragments having the same electrophoretic mobilities. This size bias originated from the difference in mobilities in the sieving medium and is affected by electroosmosis. Figure 13.1 schematically shows electrokinetic injection under these circumstances. Following the derivation by these authors, we can express the amount injected as

$$m_i = AC_i Et(\mu_{eo} + \mu_{epi2})\frac{\mu_{epi1}}{\mu_{epi2}} \qquad (13.17)$$

where μ_{epi1} is the electrophoretic mobility of the ion species i in the free solution in the sample vial and μ_{epi2} is its electrophoretic mobility in the sieving medium inside the capillary.

Although there is a strong bias operating on the injected quantity, electrokinetic injection has much wider and faster controlling capabilities than does the hydrodynamic process; moreover, it can inject reproducibly a much smaller sample volume than can hydrodynamic injection (12). In addition, the injection apparatus has basically the same arrangement as the separation

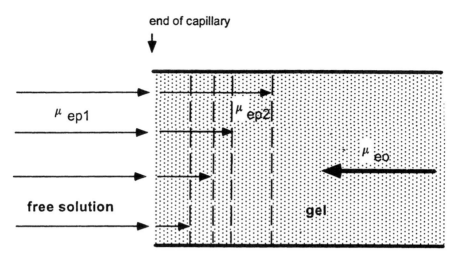

Figure 13.1. Schematic representation of electrokinetic injection in capillary electrophoresis with sieving. The electrophoretic mobility of analytes in free solution is independent of size. The size-selective mobility in a sieving medium is affected by electroosmosis mobility, μ_{eo}. [Adapted from K. Kleparnik, M. Garner, and P. Bocek, *J. Chromatogr. A* **698**, 375 (1995), with permission.]

process, with the exception that the injecting electrode is moved into the sample vial. Ease of operation makes electrokinetic injection the preferred technique in many CE applications.

13.2.3. Spontaneous Injection

The insertion of the capillary into the sample solution and withdrawal of the capillary from it have long been recognized as causing extraneous sample injection. Grushka and McCormick were the first to report this "ubiquitous injection" phenomeon in capillary electrophoresis (3). They showed that the sample solution can penetrate into the capillary after simple contact. The length of the sample soluton in the capillary can reach about 700 μm, a normalized length of 0.014 for a 50 cm long column. This extra length causes a serious deterioration of system performance and separation efficiency.

Fishman et al. have recently characterized the mechanism of this extraneous injection in capillary electrophoresis (13). Using a fluorescence imaging system equipped with a charge-coupled device, they identified that spontaneous fluid displacement resulting from an interfacial pressure difference across a droplet at the inlet of the capillary is a main cause for extraneous injection. They also proposed several approaches to minimize this extraneous injection. Minimizing the volume of the droplet, lowering the interfacial pressure difference by decreasing the curved surface, providing a counteracting back pressure, or reducing the outer diameter of the capillary to reduce the drop volume can all act to decrease sample penetration into the capillary.

Although spontaneous fluid displacement will cause injection error in normal capillary electrophoresis, this mechanism can also be used as an ultrasmall sample injection technique if controlled properly. These controlling parmeters include precise timing when one is withdrawing the capillary from the sample and reinserting it into the inlet reservoir, constant vapor pressure surrounding the capillary during the transfer time, and surface wettability and morphology of the capillary walls at the inlet. Fishman *et al.* found that a 3.5 nL sample could be injected by hand using spontaneous injection with a reproducibility of 6.0% RSD. Significant improvement can be expected with an automated system.

13.2.4. Artifacts in Sample Introduction

Recently, Guttman and Schwartz reported two injection-related artifacts in capillary gel electrophoresis that may affect precision and accuracy as well as separation performance (14). The first problem occurs with consecutive electrokinetic injections from small sample volume and results in progressively smaller analyte peak heights and areas. This effect was also mentioned earlier by Rose and Jorgenson (8), resulting from an electrochemical reaction as current flows through the sample solution. The solution to this problem is to perform an electrokinetic injection of water just prior to the sample injection to create an ion-depleted zone at the end of the capillary. With this "preinjection," Guttman and Schwartz found only 2% increase in the pH of the sample vial after five consecutive injection, as opposed to a 20% increase without the preinjection of water.

The second problem involves the physical shape of the capillary inlet. An oblique-edge capillary or poor cutting at the inlet results in substantial peak distortion and resolution loss. Thus, careful cutting and examination of the capillary inlet are essential for achieving optimum performance in capillary electrophoresis.

13.3. ON-COLUMN SAMPLE STACKING

One of the main drawbacks of CZE is its poor concentration detection limit, resulting from the small dimension of the separation column. There are two obvious ways to achieve a lower concentration detection limit in CZE. One method is to improve the detection system such as the use of laser-induced fluorescence, the installation of a Z-shaped cell, or bubble cell, etc. A much simpler and better way to increase CE sensitivity is to use the sample stacking technique.

The sample stacking approach involves loading a large volume of low-concentration analyte into a sharper zone of higher concentration. Many different techniques for on-column sample stacking have been discussed in the literature. They can generally be grouped into either of two conditions according to the movement of the boundary between buffers: moving boundary stacking and stationary boundary stacking. The best example of moving boundary stacking is isotachophoresis (ITP), where all sample zones ultimately migrate at the same velocity. An example of stationary boundary stacking is field-amplified sample stacking, where the boundary between a lower concentration sample buffer and higher concentration support buffer remains steady during the electrophoresis process.

13.3.1. Isotachophoresis–Capillary Zone Electrophoreis

ITP is a well-known technique that provides a very high sample loading into a sharp zone (15,16). In conventional ITP, the analyte ions are loaded into a capillary filled with a leading electrolyte that has a higher electrophoretic mobility than that of the analyte ions. The injecion end is then placed into the buffer vial filled with a trailing electrolytes that has a lower mobility than that of the analyte ions. After the electric field is applied, all electrolytes will eventually reach a steady-state condition and separate into different zones according to their electrophoretic mobilities. The concentrations of analyte ions in each zone are adjusted to the concentration of the leading ions according to the Kohlraush regulating function (described in detail in Chapter 11).

Although ITP has an enormous concentrating power and has extremely narrow zones of analyte ions, it has unfortunately never been widely accepted by the analytical community. The technique was thought to be too complicated and mathematics intensive. In addition, since all sample zones stacked continuously according to their mobilities, the detection and identification of the analytes was considered a major challenge for separation scientists. Nevertheless, the concentration capability of ITP offers a good starting point for CZE. There are many different ITP–CZE configurations. It can be arranged by directly coupling ITP into CZE, or a transient ITP–CZE with single capillary. In direct-coupled ITP–CZE, as shown in Figure 13.2, the isophoretic concentration stage is performed in the first capillary and followed by on-line transfer of the sample into the second capillary, where analytical zone electrophoresis is performed. The on-line transfer between two capillaries is always accompanied by the segments of some additional amount of the leading and terminating electrolytes from the ITP step. Bocek's group recently did a detailed study of this process [see Krivankova et al. (17)]. They found

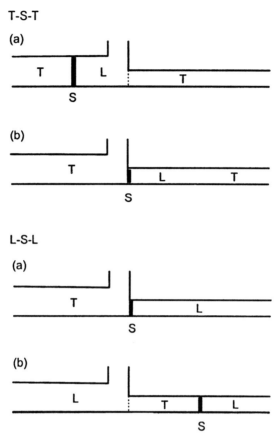

Figure 13.2. Scheme of the direct-coupled ITP–CZE arrangements for the T–S–T and L–S–L systems. The concentration (wide) capillary is located on the left, and the analytical (narrow) capillary on the right. The leading, sample and terminating zones are marked L, S and T, respectively [Adapted from L. Krivankova, P. Gebauer, and P. Bocek, *J. Chromatogr. A* **716**, 35 (1995), with permission.]

that both the detection time and variance of the analyte zone depend strongly on the amount of the accompanying segments during ITP–CZE transfer.

In transient ITP–CZE, the isophoretic step is induced by the composition of the sample containing a major component temporarily playing the role of the leading or terminating ion (7). Figure 13.3 provides a schematic overview

Figure 13.3. Schematic diagrams of the two single-capillary transient ITP–CZE procedures: A, $\mu_{Sample} < \mu_{BGE}$; B, $\mu_{Sample} > \mu_{BGE}$. BGE: background electrolyte. LE: leading electrolyte. TE: terminating electrolyte. [Adapted from M. Mazereeuw, U. R. Tjaden, and N. J. Reinhoud, *J. Chromatogr. Sci.* **33**, 686 (1995), with permission.]

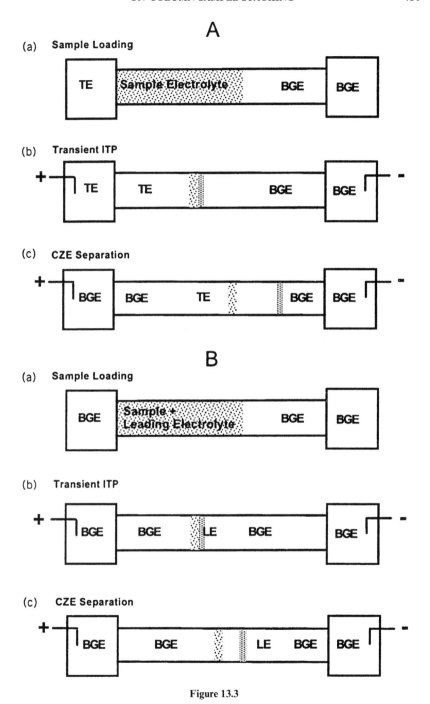

Figure 13.3

of two transient ITP–CZE procedures. In procedure A, the co-ion of the background electrolyte has a higher electrophoretic mobility than that of all sample ions. A transient ITP can be created by inserting a short zone of the terminating electrolyte after the sample zone. After the ITP focusing step, the terminating buffer vial is replaced by the leading electrolyte buffer vial. The normal zone electrophoretic separation is then started. In procedure B, the co-ions of the background electrolyte has a lower mobility than that of the sample ions. To perform ITP, one has to add fast-moving ions into the sample matrix to function as the leading electrolyte.

Reinhoud et al. recently developed an automated single-column ITP–CZE system using hydrodynamic back-pressure programming as shown in Figure 13.4 (18). The back pressure reduced the effects of electroosmotic flow during the ITP step and allowed the removal of terminating buffer before the CZE run was started. The authors demonstrated an improvement in the detection limit of more than 100 as compared to conventional CZE.

13.3.2. Field-Amplified Sample Stacking

In the simplest form of field-amplified sample stacking (19), a long zone of sample containing ions prepared in a lower-concentration buffer is injected hydrodynamically into a capillary of the higher-concentration buffer. High voltage is then applied across the capillary, causing electrophoresis to occur. Because of the difference in the conductivity between different buffers inside the capillary, the electric field strength will be much larger in the sample buffer region than in the rest of the capillary. The ions inside the lower-concentration region experience a higher electric field strength and move faster than the ions inside the higher concentration region, as shown in Figure 13.5. Once they pass the concentration boundary, they experience a lower electric field strength and slow down.

Since the flux of the ions across the concentration boundary has to be conserved, we have

$$C_{i1} v_{epi1} = C_{i2} v_{epi2} \tag{13.18}$$

where C_{i1}, C_{i2} and v_{epi1}, v_{epi2} are the concentrations of sample ions and their electrophoretic velocities inside the sample zone and the separation buffer, respectively. The electrophoretic velocity of the ions in their respective regions is simply proportional to the local electric field strength:

$$v_{epi1} = \mu_{epi1} E_1 \tag{13.19}$$

$$v_{epi2} = \mu_{epi2} E_2 \tag{13.20}$$

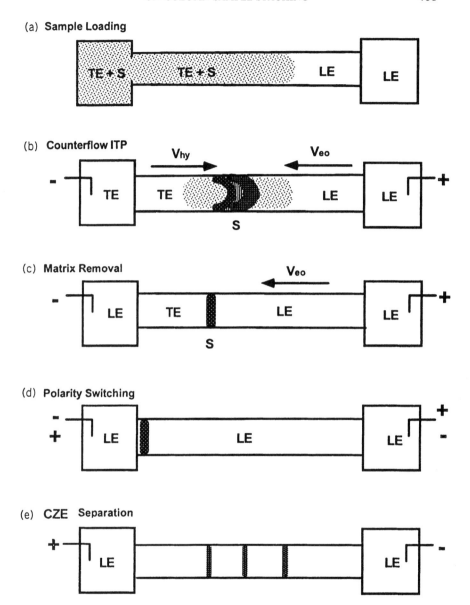

Figure 13.4. Schematic overview of ITP–CZE with a counterflow. In step a, the sample in the terminating electrolyte is injected hydrodynamically into the capiillary filled with the leading electrolyte. In step b, electrophoretic flow is counterbalanced by the hydrodynamic flow. The terminating electrolyte is then removed in step c. [Adapted from N. J. Reinhoud, U. R. Tjaden, and J. van der Greef, *J. Chromatogr.* **641**, 155 (1993), with permission.]

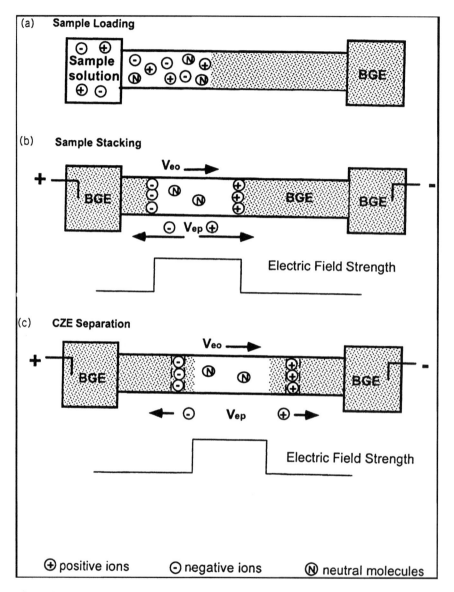

Figure 13.5. Sample in a diluted BGE is injected into the capillary filled with the same electrolyte of higher concentration. Stacking is caused by the change in the electrophoretic velocity as the ions cross the concentration boundary.

where μ_{epi1} and μ_{epi2} are the electrophoretic mobilities of the sample ions. If the mobility remains the same across the concentration boundary, i.e., $\mu_{\text{epi1}} = \mu_{\text{epi2}}$, Eqs. (13.18–13.20) yield

$$C_{i2} = \gamma C_{i1} \tag{13.21}$$

where $\gamma (= E_1/E_2)$ is the field-enhancement factor. Equation (13.21) shows that the concentration of sample ions that migrated into the high-concentration buffer region will be enhanced by the field ratio γ. Since the total number of sample ions has to be conserved, the length of the sample zone in the high-concentration buffer region also has to be reduced by the same factor γ, as shown schematically in Figure 13.5. If we neglect the diffusion effect, the effective sample zone length after stacking is

$$l_i = l_{\text{inj}}/\gamma \tag{13.22}$$

This thin zone of ions then moves through the support buffer and separates into individual zones by conventional free-zone electrophoresis. The stacking mechanism occurs for both positively and negatively charged species. The positive species stack up in front of the sample zone, and the negative species stack up in back of the sample zone. The neutral compounds are left in the sample zone and coelute out.

The major advantage of field-amplified CZE is its simplicity; it can be applied universally to any electrolyte system. One simply dissolves the sample in a lower-concentration background electrolyte and inject it into the capillary. Theoretically, the amount of stacking is simply propotional to the field-enhancement factor. The larger the difference in concentrations, the narrower the peak. An extrapolation can be made that a rather long sample zone prepared in water or very-low-concentration buffer should be stacked into a very thin zone in the high-cocentration support buffer. However, a laminar flow is generated inside the capillary due to the mismatch between local electroosmatic velocities and the bulk velocity. This laminar flow will broaden the sharp zone generated by the stacking process. The larger difference in concentration will result in a greater laminar flow. These two effects, stacking and broadening, work against each other, so there is an optimal point as to the length of water zone one can introduce into the capillary and still achieve the same high resolution. A general rule of thumb is to use a separation buffer that is about 10 times the concentration of the sample buffer (19).

For larger sample volume, the nonuniform distribution of the electric field along the capillary will lead to a disturbance in the electroosmotic flow and will result in extra peak broadening. Another problem with introducing a long zone of low conductivity sample buffer into the capillary is the redistribution of the electric field strength. In the case of a sample prepared in water, almost all electric field strength will drop across the sample buffer due to its low conductivity. The field strength in the supporting buffer region where the separation actually occurs approaches zero. Therefore, once the sample ions migrate into the separation buffer their velocity drops to zero. All ions will then stack up at the concentration boundary and will only move through the capillary with the electroosmotic flow, and then no further separation can be achieved.

To stack an extremely long sample zone while retaining high resolution and separation, it is necessary that the sample buffer be removed after the stacking process to eliminate the nonuniform distributions of both the field strength and the electroosmotic velocity. One way is to physically remove the sample buffer by using the switching-column technique. A more elegant method of removing the sample buffer is to use the electroosmotic pump itself (20). This pumping technique involves the following steps, as shown in Figure 13.6:

a. Introduction hydrodynamically of a zone of the analyte in a low conductivity buffer
b. Immediately after sample injection, the polarity of the electrodes is reversed from the normal separation configuration; with a negatively charged silica wall, the negative sample ions are stacked at the back end of the zone of sample buffer and will follow the zone as it moves out from the injection end under the reversed electric field
c. When the sample buffer is almost completely out of the capillary, which can be monitored by the measuring the electric current, the polarity is reversed again to the normal separation configuration
d. Separation and detection of the stacked negative analytes

Since the technique relies on pushing out of sample buffer, only ions having negative mobility with respect to the bulk electroosmotic flow will remain inside the capillary. To stack and separate positive species, the charge on the silica wall of the capillary can be made positive by adding cetyltrimethylammonium bromide (CTAB) to the buffer, thus changing the direction of the electroosmotic flow. Another scheme for large-volume sample stacking of negative species without the need to switch polarity is to use an electroosmotic flow modifier as described by Burgi (21).

13.4. STACKING IN SAMPLE INTRODUCTION

13.4.1. Field-Amplified Sample Injection

It is well known that the conductivity of the sample matrix has a strong effect on the efficiency of electrokinetic injection (22). When a sample vial containing

STACKING IN SAMPLE INTRODUCTION 465

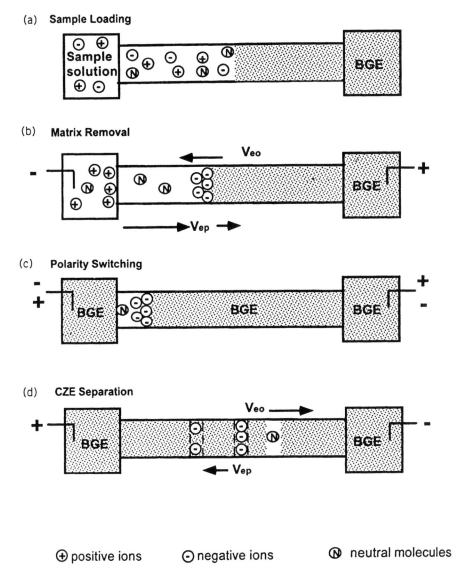

Figure 13.6. Scheme for removal of a large volume of sample buffer using electroosmotic flow after sample stacking is completed. (See the text.)

a low-conductivity buffer is in contact with a capillary filled with a high-conductivity buffer, an enhanced electric field strength will arise at the injection interface. The analyte ions in the sample vial will obtain a very high electrophoretic velocity under this amplified field at the injection point. On the other hand, the velocity of the bulk solution during injection only changes a relatively small amount from the electroosmotic velocity of the pure buffer system even under this enhanced field. Thus the injection of analyte ions into the capillary will be more rapid than the movement of bulk solution.

As more and more of the low-conductivity sample solution migrates into the capillary by electroosmotic flow, the characteristics of resistivities distribution across the length of capillary will change continuously. As mentioned earlier, the length of the sample zone and the amount injected is proportional to the time integral of the flow velocity curve. Consequently, detailed knowledge is required of the time dependence of the electroosmotic velocity, conductivity profile, and electric field strength distribution with the injection time. In a first-order approximation, we assume a steady-state condition at the sample interface during the injection of the low-conductivity sample buffer. For a short injection time, the amount injected can be simplified as

$$m_i = AC_i(\mu_{eo} + \gamma\mu_{epi})E_0 t \tag{13.23}$$

where we have used the approximation $v_{epi} = \gamma\mu_{epi}E_0$ and $v_{eo} = \mu_{eo}E_0$. Here $E_0 (\equiv V/L)$ is the average field strength.

In the special case where the enhanced electrophoretic velocity is much higher than the bulk electroosmotic velocity, we have

$$m_i = AC_i\gamma\mu_{epi}E_0 t \tag{13.24}$$

This relationship is different from conventional electrokinetic injection shown in Eq. (13.16), where the amount is proportional to the combination of the electrophoretic and the electroosmotic mobilities. In field-amplified sample injection, the amount of injection is proportional to only their electrophoretic mobility.

Once the ions migrate into the region of the high-conductivity buffer inside the capillary, they experience a lower field strength and stack together in a narrow zone of enhanced concentration. The end of the capillary is then returned to the separation buffer reservoir and a high voltage is applied to perform separation. The length of this effective sample zone after stacking is

$$l_i = (\mu_{eo}/\gamma + \mu_{epi})E_0 t \tag{13.25}$$

For large γ, the contribution of the electroosmotic flow can be neglected and the effective zone length is again simply proportional to the electrophoretic mobility. Equations (13.13) and (13.25) also show that because of the stacking effect the effective sample zone length when field-amplified sample injection is used will be smaller than when conventional electroinjection is used.

One problem in injecting the low-conductivity sample buffer directly into the capillary is that the sample ions will stack at the injection point and cause a degradation of the field enhancement. Since the stacked sample ions have either reduced or normal electrophoretic velocity, it is also possible for them to be moved out of the capillary if the electroosmotic flow has an opposite direction. A simple way to solve this problem is to preinject a short plug of low-concentration buffer before sample introduction, as shown in Figure 13.7. This short plug of buffer will provide a zone of high electric field strength at the injection point. The analyte ions will then migrate rapidly into the capillary and stack at the other end of the plug regardless of the direction of electroosmotic flow.

13.4.2. Pseudoisotachophoretic Sample Injection

In field-amplified sample injection just described, the sample is prepared in a diluted electrolyte having the same composition as the running background electrolyte in the capillary. If the low-concentration sample solution is enriched with ions having a lower mobility instead of using diluted background electrolyte, a pseudo-ITP condition can be set up. The procedure is illustrated in Figure 13.8. The capillary is filled with a background electrolyte that has a higher mobility and plays the role of the leading electrolyte. After high voltage is applied the ions migrate into the capillary, creating a frontal electrophoresis state. Since the sample ions have higher mobility, they will migrate deeper into the capillary. After sample injection, the sample vial is replaced by the background electrolyte. Pseudo-ITP stacking will take place at the concentration boundary and focus all sample ions into a narrow zone. Finally, the background electrolyte from the injecting end of the capillary will surpass the sample ions and zone electrophoresis begins.

Jandik and Jones performed this pseudo-ITP sample injection to concentrate the inorganic anions (23). In their experiment, the chromate electrolyte is used as the leading electrolyte while the sample is made into the terminating electrolyte through the addition of 40–80 µM sodium octane sulfonate. They obtained detection limits in the nanomolar range for several inorganic ions, an improvement of several orders of magnitude compared with hydrodynamic injections in an aqueous solution.

13.4.3. Polarity-Switching Sample Injection

In the absence of electroosmotic flow, field-amplified sample injection technique can be applied to either positive or negative ions by choosing a proper

(a) **Dilute Buffer Pre-Loading**

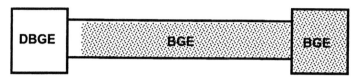

(b) **Field-Amplified Sample Injection**

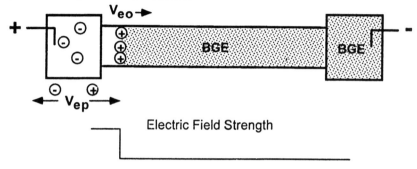

(c) **CZE separation**

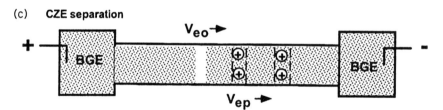

Figure 13.7. Schematic overview of field-amplified sample injection for positive ion injection. (a) A short plug of diluted sample buffer is hydrodynamically introduced into the capillary. (b) The analyte prepared in the diluted buffer is then injected electrokinetically into the capillary. (c) After sample injection, the inlet end of the capillary is transferred into the high-concentration run buffer to start the separation process.

polarity of the electrodes. However, if the electroosmotic flow (EOF) is larger than the electrophoretic flow, field-amplified sample injection will inject only ions that migrate in the same direction as the bulk flow. With the normal polarity arrangement, where the EOF is toward the outlet of the capillary, the high field strength at the injection point will push away ions having negative

STACKING IN SAMPLE INTRODUCTION

(a) **Pseudo-ITP Sample Injection**

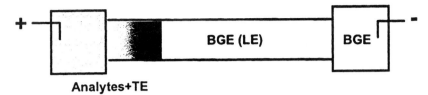

Analytes+TE

(b) **Replaced with Run Buffer**

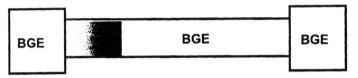

(c) **Pseudo-ITP Sample Stacking**

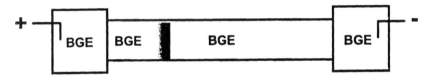

(d) **CZE separation**

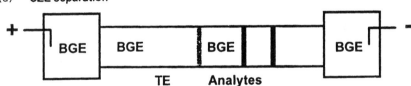

Figure 13.8. Schematic overview of pseudo-ITP sample injection.

mobility with respect to the EOF. With reversed polarity, although the high field strength will pull the negative ions into the capillary, they will be carried out immediately from the injection end by the EOF during sample injection (as shown in Figure 13.9).

Under the enhanced field stength, however, the bulk electroosmotic velocity of the solution will be much slower than the electrophoretic velocity of

(a) **normal-polarity injection of sample in diluted buffer**

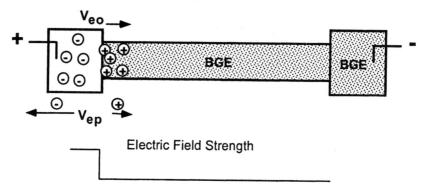

(b) **reversed-polarity injection of sample in diluted buffer**

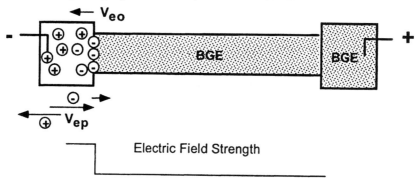

Figure 13.9. Schematic diagram of sample rejection under reversed polarity.

sample ions at the sample interface, even if the EOF might be larger than the electrophoretic velocity of the sample ions in the rest of the column. Thus, by introducing a short zone of diluted buffer prior to sample injection and reversing the polarity of the electrodes, one can inject and stack negative ions inside the capillary. Since the diluted sample buffer is now moving out of the capillary from the injection end, the length of the diluted buffer should be long enough for the whole injection process. The positive ions will now be pushed away from the interface during injection under the reversed polarity. A very effective charge selective and signal enhanced injection is achieved (24).

After sample injection, the capillary is switched back to the high-concentration support buffer and the separation voltage with the normal polarity is applied to the capillary. The bulk solution and the sample zone now migrate toward the detector, and the negative ions migrate back toward the injection end. Once the negative ions reach the high concentration region, they experience the normal field strength and stack into a narrow band between the concentration boundary. If the electrophoretic mobility of the negative ions is smaller than the bulk electroosmotic mobility, the negative ions in the low-electric-field region will eventually migrate toward the detector at the outlet end of the capillary.

By arranging the electrode configuration, one can perform field-amplified sample injection of ions having either positive or negative mobilities with respect to the EOF. Thus, it is straightforward to extend the setup so as to inject and stack both positive and negative ions into the capillary. Physically this means switching the polarity of the electrodes at proper times during the injection. A schematic diagram showing this polarity-switching sample injection is shown in Figure 13.10. Different time-programming injection schemes can be used to inject both positive and negative ions. All switching and timing of the applied high voltage obviously can be controlled automatically.

13.5. APPLICATIONS OF SAMPLE STACKING

13.5.1. Stacking of Neutral Molecules in Micellar Electrokinetic Chromatography

Micellar electrokinetic chromatography (MEKC) has been shown to be an excellent tool for separation of neutral molecules. In MEKC, ionic surfactants such as sodium dodecyl sulfate (SDS) are introduced into the buffer solution at concentrations above their critical micelle concentration. The separation of neutral molecules is achieved due to the differential partitioning of solutes into the micellar pseudophase.

In an untreated fused-silica capillary, the bulk EOF is toward the cathode whereas the anionic micelles migrate toward the anode. Generally, the strong EOF carries the negatively charged micelles toward the cathode (e.g., at pH > 5). The dominant electroosmotic velocity of the buffer solution and the opposite electrophoretic movement of the micelles provide an ideal situation for stacking neutral molecules. An example of sample stacking of neutral molecules in MEKC is reported by Liu et al. (25). In their work, a sample zone containing neutral molecules in a low-conductivity micellar solution was introduced into a capillary filled with micellar solution containing the same surfactants. Under high voltages, the neutral molecules that partition into the

(a) Dilute Buffer Pre-Loading

(b) Field-Amplified Cation Injection

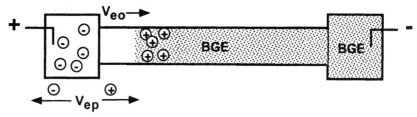

(c) Field-Amplified Anion Injection

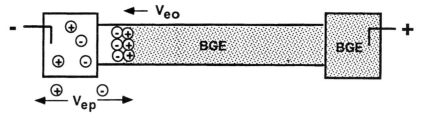

(d) CZE separation

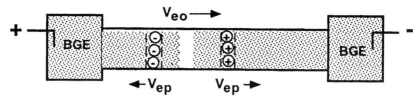

Figure 13.10. Schematic overview of field-amplified polarity-switching injection. (a) A short plug of diluted buffer is introduced hydrodynamically into the capillary. (b) Positive ions are injected electrokinetically into the column first using a normal polarity configuration. (c) Negative ions are then injected using a reversed polarity. (d) After sample introduction, the polarity of the electrodes is switched back to the normal setting to start the separation process.

charged micelles inside the sample zone experienced a much higher electric field and migrated rapidly toward the anode. The charged micelles slowed down and stacked together once they crossed the boundary between two buffers of different concentrations. Using reversed electrode polarity during the sample stacking process, Liu el al. were also able to remove the low-concentration buffer from the capillary. Figure 13.11 shows the separation of a mixture of 16 polycyclic aromatic hydrocarbons (PAHs) using cyclodextrin-

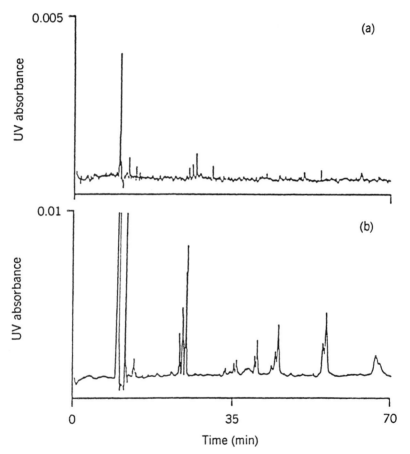

Figure 13.11. MEKC chromatograms of a mixture of PAHs with and without field-amplified sample stacking: (a) 2 nL injection (0.005 absorbance range setting); (b) 54 nL injection (0.01 absorbance range setting) with normal polarity sample stacking. Conditions: The running buffer consists of 100 mM sodium borate, 100 mM SDS, 5 M urea, and 10 mM γ-cyclodextrin, pH 9.0; the sample buffer, 9 mM SDS; applied voltage, 27 kV. [From Z. Liu, P. Sam, S. R. Sirimanne, P. C. McClure, J. Grainger, and D. G. Patterson, Jr., *J. Chromatogr. A* **673**, 125 (1994), with permission.]

modified MEKC: a normal 2 nL injection of the standard solution of PAHs is shown in Figure 13.11a; a 54 nL injection of the same standard using sample stacking produced a significant sensitivity enhancement while maintaining high resolution Figure 13.11b.

13.5.2. Trace Enrichment of Environment Samples

Despite its great resolving power, applications of capillary electrophoresis (CE) have mostly been limited to the field of biochemical analysis. A major obstacle to applying CE to other fields of chemical analysis, such as work with environmental samples, is its poor sensitivity. However, using sample stacking and suitable field-amplified injection, it is possible to perform ultra-trace analyses of environmental sample interest. Nielen has applied field-amplification technique for trace enrichment of environmental samples (26). The separation of a 20 ppm standard solution containing phenoxy acid herbicides MCPP, 2,4-DP, MCPA, and 2,4-D, was first performed using normal hydrodynamic injection. Next, the 20 ppm standard solution was diluted 1000-fold with water and a large volume (about 35% of the entire capillary) was injected and preconcentrated with matrix removal under reversed polarity. The resulting electropherogram shows that the resolution has been fully maintained even with large volume sample loading. Nielen also investigated another similar field-amplified injection technique in which the buffer vial was replaced by the sample vial during the reversed-polarity step.

To further enhance the detection sensitivity and minimize the sample pretreatment, Nielen used C18 membrane disks for simultaneous filtering and solid phase extraction. Desorption from the disks was then carried out using an acetonitrile/water mixture and with a sample matrix having a low conductivity. The general procedure for environmental trace analysis of aqueous samples using CE is summarized in Table 13.1.

13.5.3. Sample Stacking in Mass Spectrometric Analysis

Wolf and Vouros have used sample stacking to improve the detection limits of CE coupled to continuous-flow fast atom bombardment mass spectrometry (CF–FAB–MS) for the analysis of DNA adducts (27). They are interested in the analysis of adducts formed by the covalent bonding of PAH metabolites with DNA. Since PAH–deoxynucleoside adducts show dissociation behavior similar to that of the parent base, CE should be able to perform analysis for various adduct applications. The authors' initial results showed that the analysis of PAH–deoxynucleoside adducts CE–CF–FAB–MS has mass detection limits in the range of 50–100 fmol, limited by the low sampling volume

Table 13.1. General Procedure for Environmental Trace Analysis of Aqueous Samples Using CZE

1. Prewet extraction disk with 4 mL methanol.
2. Condition extraction disk with 10 mL water.
3. Apply 20 or 40 mL water sample (acidified with HCl to pH = 1).
4. Wash extraction disk with 5 mL water.
5. Apply vacuum for 1 min.
6. Elute the herbicide from the disk into the CZE sample vial with 2 mL of 50 μM lithium acetate, pH 4.8, in acetonitrile/water (1:1).
7. Put sample vial into CZE apparatus and inject the large volume hydrodynamically (1 min at 500 mbar).
8. Apply field amplification and perform matrix removal at -5 kV until current reaches 95% of its normal value at -5 kV.
9. Perform CZE at $+30$ kV with 50 mM lithium acetate buffer, pH 4.8, in water.

Adapted from Nielen (26) with permission.

of CE. To solve the sample-loading problem, they redesigned their CE–MS interface to perform large-volume sample injection and stacking as described earlier (Section 13.3.2).

The mechanics of stacking with the CE–MS interface is shown in Figure 13.12. In large-volume stacking, it is necessary to include a step in which the sample solvent is pumped out under a reversed polarity of high voltage. During this step, the buffer solution flows into the capillary from the cathode reservoir. In the typical CE–CF–FAB–MS interface, the cathode reservoir has been replaced by the CF–FAB solution. Nevertheless, the sample stacking technique requires that the cathode buffer be the same as the separation buffer, so an off-line stacking prior to CE–MS analysis is employed by Wolf and Vouros. In step a, the sample solution in water was hydrodynamically injected into the anode end of the capillary. In step b, a high voltage of opposite polarity to that employed in separation was applied to cause stacking and pumping out of the sample solution. Finally, the cathode end of the capillary was transferred to the liquid junction interface and the electrodes were reversed back to their normal configuration (as shown in step c) to perform separation and mass analysis.

The results showed that the concentration detection limit can be improved by as much as 3 orders of magnitude with stacking. Using a standard acetylaminofluorene deoxyguanosine 5'-monophosphate adduct (AAF–dGMP), Wolf and Vouros found the mass detection limits to be in the low-picomole range for full scanning and the low-femtomole range for multiple reaction monitoring of a selected fragmentation.

(a) Injection

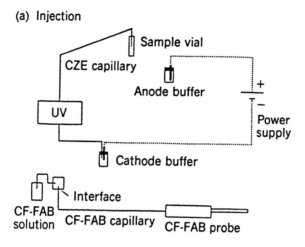

(b) Stacking

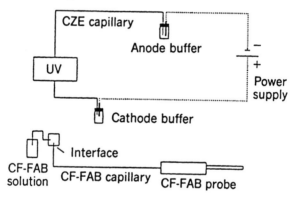

(c) Analysis

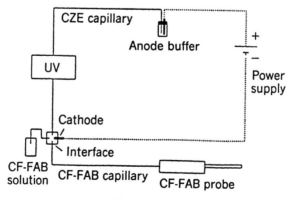

Figure 13.12

13.6. CONCLUSIONS

The goal of sample introduction in any separation technology is to inject the largest possible sample into the narrowest band to achieve ultimate detection sensitivity with highest resolution. A particular problem related to CE is its poor detection sensitivity due to the small volume of sample involved. Therefore sample stacking techniques, either on-column or during sample introduction, are of tremendous importance for the applicability of CE, especially in trace analysis. Many sample introduction and stacking methods have been described in this chapter. Generally speaking, large-volume stacking using hydrodynamic injection has the best reproducibility. However, it can achieve only limited stacking due to the extra peak broadening from laminar flow. On-column ITP–CE can achieve highest stacking capability, the popularity of the technique has unfortunately been limited by its complexity. Field-amplified sample injection is easy to implement and automate. Nevertheless, one should be aware of the sample matrix effect that can cause bias in the injected amount. Depending on the user's specific application, proper injection and stacking methods should be chosen. In addition, one can also combine any injection technique with the removal of a large sample matrix prior to the stacking process to obtain both high detection sensitivity and high resolution.

ADDENDUM

Many recent articles on the subject sample treatment have been published since the submission of this manuscript. Palmarsdottir has investigated the procedures for sample concentration in CZE for determination of drugs in biosamples in her thesis (28). Turnes et al. combined the sample stacking techniques with the use of an extended light path to allow the derermination of pentachlorophenol in the ppt range in water samples (29). Large sample volume stacking was also used by Albert et al. for quantitative analysis of anions (30). A robust, reproducible procedure of using field-amplified sample injection with a short plug of water at the capillary inlet was demonstrated by Zhang and Thormann (31). A >1000-fold sensitivity enhancement was

---

Figure 13.12. Utilization of the large-volume sample stacking technique involving the following steps: (a) hydrodynamic sample injection of the desired sample volume (high voltage off); (b) application of a high voltage opposite in polarity to that normally employed for analysis in order to effect off-line sample stacking; and (c) transfer of the CZE capillary from the cathode buffer to the liquid junction interface for CZE–UV–MS analysis. [From S. M. Wolf and P. Vouros, *Anal. Chem.* **67**, 891 (1995), with permission.]

achieved for the determination of positively charged hydrophobic compounds. A variant of large sample injection, partial-filling micellar electrokinetic chromatography (MEKC), was also developed by Lee's group (32). Finally, several articles on ITP and other fundamental processes in capillary electrophoresis are also recommended (33–36).

LIST OF ACRONYMS AND ABBREVIATIONS

Acronym or Abbreviation	Definition
AAF–dGMP	acetylaminofluorene deoxyguanosine 5′-monophosphate
BGE	background electrolyte
CF–FAB–MS	continuous-flow fast atom bombardment mass spectrometry
CTAB	cetyltrimethylammonium bromide
CZE	capillary zone electrophoresis
2,4-D	trade name for (2,4-dichlorophenoxy)acetic acid, a herbicide
2,4-DP	trade name for (2,4-dichlorophenoxy)propanoic acid or dichlorprop, a herbicide
EOF	electroosmotic flow
ITP	isotachophoresis
LE	leading electrolyte
MCPA	trade name for (4-chloro-2-methylphenoxy)acetic acid, a herbicide
MCPP	trade name for ($\pm$)-2-(4-chloromethylphenoxy) propanoic acid or mecoprop, a herbicide
MEKC	micellar electrokinetic chromatography
PAH	polycyclic aromatic hydrocarbon
RSD	relative standard deviation
TE	terminating electrolyte
UV	ultraviolet

REFERENCES

1. J. W. Jorgenson and K. D. Lukacs, *Anal. Chem.* **53**, 1298 (1981).
2. J. C. Sterberg, *Adv. Chromatogr.* **2**, 205 (1966).
3. E. Grushka and R. M. McCormick, *J. Chromatogr.* **471**, 421 (1989).
4. S. Hjerten, J. L. Liao, and Z. Zhang, *J. Chromatogr.* **676**, 409 (1994).

5. P. Gebauer, W. Thormann, and Pl Bocek, *J. Chromatogr.* **608**, 47 (1992).
6. R. L. Chien and D. S. Burgi, *Anal. Chem.* **64**, 489A (1992).
7. M. Mazereeuw, U. R. Tjaden, and N. J. Reinhoud, *J. Chromatogr. Sci.* **33**, 686 (1995).
8. D. J. Rose and J. W. Jorgenson, *Anal. Chem.* **60**, 642 (1988).
9. R.-L. Chien and J. C. Helmer, *Anal. Chem.* **63**, 1354 (1991).
10. X. Huang, M. J. Gordon, and R. N. Zare, *Anal. Chem.* **60**, 375 (1988).
11. K. Kleparnik, M. Garner, and P. Bocek, *J. Chromatogr. A* **698**, 375 (1995).
12. S. Delinger and J. M. Davis, *Anal. Chem.* **64**, 1947 (1992).
13. H. A. Fishman, N. M. Amudl, T. T. Lee, R. H. Scheller, and R. N. Zare, *Anal. Chem.* **66**, 2318 (1994).
14. A. Guttman and H. E. Schwartz, *Anal. Chem.* **67** 2279 (1995).
15. F. M. Everats, J. L. Beckers, and Th. P. E. M. Verheggen, *Isotachophoresis: Theory, Instrumentation and Practice,* J. Chromatogr. Libr., Vol. 6. Elsevier, Amsterdam, 1976.
16. P. Bocek, M. Deml, P. Gebauer, and V. Dolnik, *Analytical Isotachophoresis.* VCH, Weinheim, 1988.
17. L. Krivankova, P. Gebauer, and P. Bocek, *J. Chromatogr. A* **716**, 35 (1995).
18. N. J. Reinhoud, U. R. Tjaden, and J. van der Greef, *J. Chromatogr.* **641**, 155 (1993).
19. D. S. Burgi and R.-L. Chien, *Anal. Chem.* **63**, 2042 (1991).
20. D. S. Burgi and R.-L. Chien, *Anal. Chem.* **64**, 1048 (1992).
21. D. S. Burgi, *Anal. Chem.* **65**, 3726 (1993).
22. L. Gross and E. S. Yeung, *J. Chromatogr.* **480**, 169 (1989).
23. P. Jandik and W. R. Jones, *J. Chromatogr.* **546**, 431 (1991).
24. R.-L. Chien and D. S. Burgi, *J. Chromatogr.* **559**, 153 (1991).
25. Z. Liu, P. Sam, S. R. Sirimanne, P. C. McClure, J. Grainger, and D. G. Patterson, Jr., *J. Chromatogr. A* **673**, 125 (1994).
26. M. W. F. Nielen, *Trends Anal. Chem.* **12**, 345 (1993).
27. S. M. Wolf and P. Vouros, *Anal. Chem.* **67**, 891 (1995).
28. S. Palmarsdottir, Doctoral Dissertation. Lund University, Lund, Sweden, 1996.
29. M. I. Turnes, M. C. Mejuto, and R. Cela, *J. Chromatogr. A* **733**, 395 (1996).
30. M. Albert, L. Debusschere, C. Demesmay, and J. L. Rocca, *J. Chromatogr. A* **757**, 281 (1997).
31. C. Zhang and W. Thormann, *Anal. Chem.* **68**, 2523 (1996).
32. W. M. Nelson and C. S. Lee, *Anal. Chem.* **68**, 3265 (1996).
33. M. J. van der Schans, J. L. Beckers, M. C. Molling, and F. M. Everaerts, *J. Chromatogr. A* **717**, 139 (1995).
34. J. L. Beckers, *J. Chromatogr. A* **741**, 265 (1996).
35. F. E. P. Mikkers, *Anal. Chem.* **69**, 333 (1997).
36. S. Auriola, I. Jaaskelainen, M. Regina, and A. Urtti, *Anal. Chem.* **68**, 3907 (1996).

CHAPTER

14

COATED CAPILLARIES IN HIGH-PERFORMANCE CAPILLARY ELECTROPHORESIS

GERHARD SCHOMBURG

Stiftstr. 39, D-45470 Mülheim an der Ruhr, Germany

14.1.	Introduction	481
14.2.	The Status of Capillary Surfaces in Capillary Electrophoresis: Influence on Electroosmotic Flow	483
14.3.	Influence of Electroosmotic Flow on Efficiency and Resolution of Capillary Electrophoretic Separations	484
14.4.	Analyte–Wall Interaction and the Performance of Analytical Capillary Electrophoretic Separations	486
14.5.	Surface-Modification Procedures by Coating in Fused-Silica Capillaries	488
	14.5.1. Dynamic Coating	489
	14.5.2. Permanent Coating	491
14.6.	Methods and Applications of Dynamic Surface Modifications	492
14.7.	Methods and Applications of Permanent Surface Modifications	507
14.8.	Recent Developments in the Dynamic and Permanent Modification of Capillary Surfaces	516
	14.8.1. Separation of Small Molecules: Pharmaceutical Compounds	517
	14.8.2. Separation of Proteins	517
	14.8.3. Application of Various Kinds of Permanent Polymer Coatings; Miscellaneous	518
List of Acronyms and Abbreviations		519
References		520

14.1. INTRODUCTION

Separations in electrophoresis are achieved by differences in the electromobilities of the charged separands (analytes) in *parallel* with the direction of the electric field. No flow of the liquid medium nor any interaction of the analyte

---

High Performance Capillary Electrophoresis, edited by Morteza G. Khaledi. Chemical Analysis Series, Vol. 146.
ISBN 0-471-14851-2 © 1998 John Wiley & Sons, Inc.

with a specially modified capillary wall or surfaces of a support is necessary to effect a separation. In addition to the electrophoretic migration of charged analytes a flow of the buffer medium is produced by electroosmotic effects as a consequence of the charging of the capillary surfaces, but such a flow may also be created hydrodynamically in capillary electrophoresis (CE) for sample introduction, in capillary electrochromatography (CEC), or in capillary isoelectric focusing (CIEF). The following discussion considers separation mechanisms in different analytical CE systems performed in narrow-bore fused-silica capillaries with < 100 μm i.d.

Separations in systems of free capillary zone electrophoresis (CZE) can be faster and much more efficient than chromatographic separations, especially with large molecules. The vertical equilibration of the analytes between the buffer and system walls by slow diffusion in liquids during the course of separation is not essential for common CE separations.

In fused-silica capillaries of CZE systems the inner walls of the capillary are in contact with the aqueous buffer medium and are negatively charged in their native state because of the presence of silanols on the surfaces. Positively charged ions of the buffer may become attached to such negatively charged capillary surfaces by ionic and other types of intermolecular interaction. Due to the strong electrostatic interaction of cations with the negatively charged capillary surface, a "fixed" layer of positive ions is formed. The concentration of positive ions in a "diffuse" layer above this fixed (Stern) layer is still higher, as is the case in the bulk of the buffer medium. This gradient in the concentration of positive ions between the surface and the center of the capillary leads to a radial (ζ) potential, which is the reason for the cathodic flow of the buffer medium if an electric field is applied to the capillary ends. The resulting electroosmotic flow (EOF) depends on the pH of the buffer, which changes the dissociation of the surface silanols, and also depends on the presence of different kinds of interacting molecules in the buffer, which may be adsorbed at the capillary surface. The EOF can be modified by coatings of polymeric molecules that are permanently adsorbed or bonded to the surface.

In CEC, which can be considered a hybrid of chromatography and electrophoresis, interaction of the analytes with a phase takes place in a way that is really stationary, but the mobile phase is driven in the electroosmotic mode. CEC systems have the advantage of higher separation efficiency as a consequence of the EOF profiles, which are very flat compared to the parabolic Hagen–Poiseuille profiles arising in pressure-driven systems of liquid chromatography by hydrodynamic flow. In CEC the selectivities of separations are determined by equilibria between a mobile and a stationary phase but can simultaneously be influenced by electromigration in the electric field in the presence of charged analytes. When the "chromatographic" mobile phase flow is to be achieved by electroosmotic effects, it may be difficult to modify the surface of the support or the capillary walls by a stationary phase coating that is suited to supply the desired chromatographic selectivities and also gives rise to an optimal EOF that moves the mobile phase.

Especially in CZE and for the larger biomolecules, maximum separation efficiencies can only be achieved if analyte–wall interaction is carefully suppressed through changes of the charging and adsorptivity of the fused-silica surface by different methods of modification or "coating". Changes of the chemical status of the capillary surface can be done either by "permanent coatings" or simply by interaction with buffer additives. In the latter case special small and preferably polymeric additive molecules form a kind of coating the stability and configuration of which depends on an equilibrium between the buffer and the surface; therefore such coatings are called "dynamic." The interaction of the silica surfaces with the analyte molecules that leads to a decrease of separation efficiency can also be influenced or altered in this manner, but the various mechanisms of analyte–wall interaction and its suppression are complex and not yet well understood. For the analytical application of CE methods at the maximum or optimum performance with regard to efficiency, resolution and reproducibility, the avoidance or suppression of analyte–wall interaction is as essential as the manipulation of the EOF. Any separation systems being developed and optimized for analytical applications should admit stable and reproducible, absolute and relative migration times. It is also necessary to elucidate and optimize the chemistry in the separation capillaries with special regard to both the manipulation and variation of EOF and the suppression of analyte–wall interaction. In CEC, mobile–stationary phase equilibria must be considered as well.

14.2. THE STATUS OF CAPILLARY SURFACES IN CAPILLARY ELECTROPHORESIS: INFLUENCE ON ELECTROOSMOTIC FLOW

In CZE and EKC (electrokinetic chromatography) the negatively charged silica surface attracts the hydrated positive ions from the buffer to form an electric double layer according to the theory developed by H. L. M. von Helmholtz, Marian Smoluchowski, Otto Stern, and others. In free zone electrophoresis, performed in uncoated, "bare" capillaries, the EOF depends on the number of silanol groups on the surface of the silica tubing material, the dissociation of the silanols at the given pH of the buffer, the ionic strength of the buffer, the strength of the electric field, the viscosity of the buffer, and the dielectric constants of the buffer solvents. In the presence of charged or uncharged buffer constituents, which modify the surface in the dynamic mode by ionic or other types of intermolecular interaction, the EOF can be retarded, accelerated, suppressed, or even reversed. In the last case the negative charging of the silica surface has to become positive in order to generate an anodic EOF.

The manipulation of the EOF in CE systems for analytical application is important to:

- Accelerate or retard the separation of analytes, depending on their direction of electrophoretic migration, with influence on the migration times (speed of analysis), separation efficiency, and resolution
- Transfer cations or anions against the direction of their electrophoretic migration to the detector side of the instrumental setup to allow for the analysis of cationic and anionic species in the same analysis
- Transfer uncharged species to the detection side of the instrumental setup
- Transfer analytes into coupled instrumentation such as fraction collectors, or the interfaces or ion sources of mass spectrometers, and in general if an electric field cannot be applied in the section where the capillary is coupled to such devices

In capillary gel electrophoresis (CGE), the size-selective separation of biomolecules such as oligonucleotides in sieving matrices such as gels [cross-linked polyacrylamide (PAA)] or non-cross-linked, entangled polymer solutions (physical gels), the EOF is usually suppressed by covalently bonded coatings with polymers such as PAA (1). Especially with highly diluted polymer solutions, the EOF would disturb the separations and hence must be suppressed by a prior polymer coating of the surface. Hydroxylic polymers, i.e., HEC (hydroxyethyl cellulose), HPMC (hydroxypropyl methylcellulose), dextran, or PVA (polyvinyl alcohol), as constituents of such polymer solutions, for size-selective separations of oligonucleotides and DNA fragments suppress the EOF, so that a coating with PAA or other polar permanent coatings before the introduction of the separation medium may not be required. Suppression of the EOF can be enhanced by rinsing steps with solutions of these polymers, before the actual diluted polymer solution is introduced (2–4).

14.3. INFLUENCE OF ELECTROOSMOTIC FLOW ON EFFICIENCY AND RESOLUTION OF CAPILLARY ELECTROPHORETIC SEPARATIONS

The efficiency of CE separations can, in analogy to chromatography, be described by plate heights H or the number of theoretical plates N as follows:

$$N = \frac{1}{H} = \frac{(\mu_{eo} + \mu_{ep}) \cdot V \cdot l}{2D \cdot L} \tag{14.1}$$

where μ_{eo} and μ_{ep} are the electroosmotic mobility (m^2/Vs) and electrophoretic mobility; respectively; V is applied voltage; l is the capillary length to the detector; D is the diffusion coefficient; and L is the total capillary length. The resolution R can be described by the equation

$$R = \frac{1}{4} \frac{\Delta \mu}{\mu_{ep} \pm \mu_{eo}} \cdot \sqrt{N} \tag{14.2}$$

where $\Delta \mu$ is mobility difference. These equations indicate what influence changes in the EOF that occur with surface modifications for the suppression of analyte–wall interactions may also have on the efficiency and resolution of an electromigrative separation with a separation factor α:

$$\alpha = \frac{(\mu_{ep})_2}{(\mu_{ep})_1} \tag{14.3}$$

In the resolution equation (14.2) the electroosmotic mobility μ_{eo} appears twice, in the denominator and also in the dispersion term under the square root sign. Therefore faster execution of a separation by increasing the apparent electromobility with a stronger (cathodic) EOF, which can be achieved, e.g., by increasing number of negative charges on the surface, would lead to a higher number of theoretical plates in the dispersion term of Eq. (14.2), which influences the resolution only by the square root. Simultaneously the "selectivity" term is proportionally decreased even more. The EOF in the direction of the electrophoretic migration decreases the resolution and vice versa.

The manipulation of EOF in order to optimize efficiency N, speed of analyses, and the resolution R is one of the aims of procedures of chemical surface modification. If surface modifications are performed for suppression of analyte–wall interactions and if the influence of such procedures on the separation efficiency or resolution is to be investigated, it must be taken into consideration that the apparent electromobilities depending on the EOF are also changed and have a significant influence on efficiency and resolution.

Secondary contributions to band broadening may originate from the following: sample introduction; influence of sample load and volume, stacking effects, etc.; inhomogeneity of field strength within the peak volume, especially at high solute concentrations; analyte–wall interaction; thermoconvection by Joule heating and related temperature inhomogeneities; detection; and data handling. These effects produce band broadening and distortion of peak symmetry, which may further diminish the resolution necessary for certain analytical applications. For investigations of the influence of EOF and analyte–wall interaction on efficiency and resolution it is necessary to eliminate these secondary contributions to band broadening.

14.4. ANALYTE–WALL INTERACTION AND THE PERFORMANCE OF ANALYTICAL CAPILLARY ELECTROPHORETIC SEPARATIONS

High performance of analytical CE separations can only be achieved in systems that require the operation of a thermostated, window-equipped capillary with defined chemistry of the uncoated or coated inner surface and a specially composed buffer medium. Various aqueous and nonaqueous buffers modify the electromobilities of the different analytes, the EOF, and the adsorptivity of the surface by special additives. In CZE and EKC, separations with optimal efficiency should be performed without reversible or even irreversible interaction of the analyte molecules with the walls of the capillary. If molecules collide statistically with the capillary walls they may weakly or strongly interact with the surface and are retarded in their migration through the capillary under band broadening and symmetry distortion, depending on the type of interaction and the velocity of diffusion back from the wall into the bulk of the separation medium. This retentive type of band broadening by analyte interaction is, according to chromatographic theory, a consequence of nonequilibrium zone broadening and resistance to mass transfer in the stationary phase, which strongly decreases the separation efficiency especially in the case of large molecules such as proteins and peptides. Otherwise such large analytes could be separated in CE with very high efficiency because of their very slow axial diffusion. These phenomena have been treated theoretically by Schure and Lenhoff (6). They concluded:

- Small ions with relatively large diffusivity should be separated in lower-diameter capillaries (< 50 μm i.d.) to avoid potential loss of resolution by analyte–wall interaction.
- The presence of adsorption is particularly devastating for medium-to-large macromolecules. Even with a minor amount of adsorption (small k'), the normally high efficiency that can be achieved with CE in open tubes is diminished.

In routine CE analyses series performed with validated and rugged separation methods, the analytical performance may not be decreased by contamination of the surface by matrix components that change the adsorptivity of the surface, its charging, and the related EOF. It may be more difficult with certain types of "real" samples to maintain the performance of CE separation systems in longer series of measurements due to contamination from the sample matrix and instabilities of the many previously discussed parameters that influence the performance of a CE analysis. One of the procedures in routine CE analysis is regeneration of the system by the refreshment of the buffer medium. In analytical CZE, EKC, and CGE, i.e., when entangled polymer solutions are used, the refreshment of the buffer medium after each or a few separations has proved a suitable approach to achieve reproducible analytical data. Such procedures consist of several steps that can easily be executed automatically and preprogrammed in modern CE instrumentation, provided that the separation medium is not too viscous and higher pressure drops between the capillary ends can be applied in the instrumentation; then these rinsing or refreshment steps can be accomplished in a short time. These steps comprise the displacement of the old buffer, rinsing of the capillary by various solvents, by aqueous solutions of variable (even extreme) pH, and with the separation buffer itself before the introduction of fresh buffer for the actual separation. This approach can only be useful with regard to high reproducibility of the analyses if the capillary surfaces are indeed regenerable by rinsing procedures, e.g., by removal of adsorbed contaminants from the sample matrix.

A simple way to set up CZE and EKC systems in practice might be to apply fused-silica capillary tubing as delivered from the manufacturer, i.e., without any special permanent coating. In many cases this can be achieved by introduction of a buffer that contains specially selected "surface modifiers" into capillaries from such tubing material in order to establish the separation system. For such an approach, the tubing material should have a defined surface with regard to the concentration of silanols and the presence or absence of contaminants of the silica that may change the EOF and the adsorptivity of different types of charged or uncharged small and large polar molecules. Tubing material from different manufacturers and even from the same lot of a single manufacturer may have different surface properties, probably because of varying SiOH concentration on the surface. Many applications require a preliminary step of etching or rinsing with strongly acidic or basic aqueous solutions by which the coverage of the surface by silanol groups is changed or (advantageously) homogenized in order to achieve reproducible conditions of the EOF in different capillaries; adsorbed contaminants may also be removed by such a procedure. An extensive investigation on the influence of pretreatment with water and of acidic or basic etching of fused-silica tubing originating from four different manufacturers was performed by Coufal et al. (7). In aqueous systems the Si-O-Si groups at the surface may be hydrated to form Si(OH) groups. Repeated (up to 10) measurements with the pretreated capillaries were executed using five different buffers to evaluate the EOF magnitude and reproducibility. EOF values between 6 and 8×10^{-4} cm^2/V·s were found. Even without acidic or basic treatment very similar EOF values were found at mean RSD between 0.5 and 1.5% with improvement to lower values after a period of conditioning with the buffer. Coufal et al. (7) state that the nature of the running buffer plays the most significant role in determining the repeatability and magnitude of the electroosmotic mobility; the effects of the manufacturing process of the tubing and its pretreatment are less important. The application of buffers consisting of Tris [tris(hydroxymethyl)aminomethane] and ethylenediamine resulted in an EOF with better characteristics than the widely used PHOS buffers. Tris or PHOSAM (phosphates + ethylenediamine) buffers proved to be better with respect to stability and repeatability of EOF.

Rinsing procedures with organic solvents and water are combined with such pretreatment or etching steps, as detailed in many published reports. The aqueous buffers for CZE separations contain the co-ion and counterion of the buffer salts at a defined pH and an ionic strength that is not too high with regard to Joule heat formation or not too low with regard to suppression of adsorption of large biomolecules, for example. The buffer also contains additives ("modifiers" or "selectors") for modification of the EOF and especially the electromigration of the analytes by selective intermolecular interaction or complexation. For regeneration of the separation systems, refreshment of the separation buffer by removal from the capillary and refining was first proposed by Karger et al. (8) for CGE separations of oligonucleotides in non-cross-linked polyacrylamide solutions. This method is now applied to most CZE and EKC separations, combined with rinsing steps with water, buffers, and organic solvents. Lauer and McManigill (9) have described the regeneration of uncoated capillaries for protein separations by rinsing with NaOH solutions of very high pH. Whether basic or acidic rinsing solutions should be applied in such a procedure depends on the pI value of the adsorbed material.

14.5. SURFACE-MODIFICATION PROCEDURES BY COATING IN FUSED-SILICA CAPILLARIES

Coating or modification of the surfaces in separation capillaries is performed with the aim of manipulation of the EOF and suppression of analyte–wall interaction, mainly to achieve fast separations at the adequate or optimum resolution of the species of analytical relevance. High-performance separation systems can thus be established that can be validated for application, e.g., in pharmaceutical and biochemical analytical laboratories. They should not be too sophisticated with regard to the procedures of chemical modification of the silica surfaces, nor to the composition of the optimal buffer, e.g., background electrolyte (BGE). In the context of this chapter the term *coating* of surfaces in CE capillaries is applied to different modes of modification of the capillary surface. Coatings may be generated in a pretreatment procedure, i.e., before the running buffer is introduced, by chemical reactions with the silanols as anchor groups or by strongly chemi- or physisorbed layers of polymers, which may also be cross-linked. Silanization procedures lead to chemical bonding of small and oligomeric molecules that are covalently bonded to the silicon atom of the silanization reagent. The coatings obtained by silanization or with involvement of silanization are considered to be "permanent." It is generally assumed that such coatings have reasonably good stability with aggressive buffers.

Coatings of polymeric and surfactant molecules may also be formed by "dynamic" interaction of modifiers that are included in the separation buffer as additives in order to establish high enough concentrations of the additive (modifier) on the surface to form a homogeneous coating layer.

14.5.1. Dynamic Coating

The formation of common "dynamic" coatings occurs in equilibria between the buffer and the capillary surface. Such coatings cover the surface and the charges on it and prevent or change the formation of the double layer that gives rise to the EOF. The effectivity and stability of such coatings depend mainly on the energy of intermolecular interaction or adsorption of the modifying molecule at the surface and on the concentration of the modifying additive in the buffer. Such dynamic coating layers are to prevent analyte molecules, also in competition with the positive-ions in the buffer, to become attached to the silica surface, and to undergo the aforementioned disturbing analyte–wall interaction. Special problems of optimization arise with the extreme pH of the buffers that must be applied to ionize the analytes of interest and achieve highly effective electromobilities. The optimal pH of buffers applied to separations of acidic or basic biomolecules may range from 2 to 12. The types of interaction forces (ionic, hydrophilic, hydrophobic) involved determine the fixation of the different surface-modifying molecules that may be charged or uncharged. Negatively charged additives increase the concentration of negative charge on the surface if they are adsorbed. Additives with positive charge, i.e., the opposite charge to that of silica surface, are able to reverse the polarity of the charges at the surface. Neutral polymeric additives, when they are adsorbed at the surface, shield the negatively charged silanol groups and prevent the formation of the positively charged Stern double layers from which the EOF evolves. Shielding of the silanols against the adhesion of positive buffer ions and simultaneously against interaction or adsorption of strongly basic compounds such as positive with high pK was a topic of extensive investigations on the dynamic and especially the permanent surface coating.

Another topic of great interest, notably in pharmaceutical analysis, is the suppression of analyte–wall interaction for enhancement of efficiency of separations of basic and acidic drugs and their enantiomers in fused-silica capillaries. At present it is preferable to achieve this by dynamic coatings, but in the future it will probably be advantageous in commercial capillaries to use permanent coatings.

Dynamic or permanent deactivation of the surface against adsorption of cationic or acidic analytes and the simultaneous suppression or variation of the EOF are expected to lead to significant improvements in efficiency (avoidance of band broadening) and peak symmetry, and consequently much higher resolution. Surface modification by coatings is even more important for the separation of various types of large biomolecules such as peptides or proteins and for enantiomeric separations that require optimum efficiencies at sometimes lower enantioselectivities.

Dynamic coatings are successfully applied in the fast separation of inorganic and small organic anions for which cationic surfactants or polymeric compounds are added to the buffer to reverse the EOF, i.e., to operate the separation with co-EOF, which increases the apparent electromobility of the anions (10–16). Variation of the EOF in such a manner that the apparent electromobilities are strongly increased and the migration times are shortened leads to an increase of efficiency—but to a decrease in resolution, according to Eq. (14.1) and (14.2).

From the published separations of basic proteins or of basic compounds of pharmaceutical interest it can be concluded that especially polymers, either the neutral hydroxyalkyl celluloses, PVA, and dextrans (17–24) or cationic polymers such as Polybrene (25,26) can be used as dynamic modifiers when contained in the buffers in low concentrations. In other studies small diamines (27) and cationic surfactants, also with a fluorinated carbon skeleton (28,29), have been applied for separation of proteins. Excellent results with dynamic modification by hydroxylic surfactants such as Tween were obtained in combination with a hydrophobic permanent coating (see below) produced by silanization of the surface silanols with C_{18}-substituted silanization reagents (30). By prior permanent coating with a hydrophobic layer, the interaction of the hydrophobic part of the nonionic hydrophilic surfactant with the hydrophobized silica surface was strong and effective. A mechanism for reversal of the surface charge by the interaction of a cationic surfactant with a bare fused-silica surface was proposed by Towns and Regnier (31) and Snopek et al. (23), see Figure 14.1. Cationic surfactants such as tetradecyltrimethylammonium bromide (TTAB) adhere to the negatively charged wall probably by strong involvement of Coulomb interaction (32). Then the alkyl groups of the

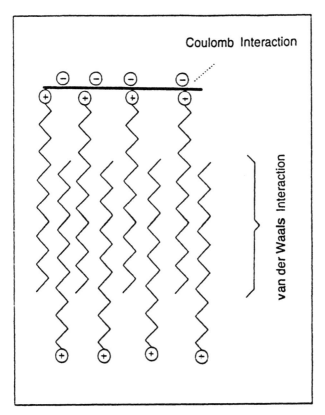

Figure 14.1. Reversal of the negative charge of fused-silica surfaces by a cationic surfactant such as TTAB. The TTAB forms a first layer by Coulomb interaction of the cationic ammonium group with the negative charge on the surface. Another TTAB layer is fixed tail to tail by van der Waals interaction between the alkyl groups of the two layers.

monomolecular layer of the surfactant protrude into the buffer and further surfactant molecules interact by van der Waals interaction to this surface layer between the hydrophobic alkyl chains. A biomolecular layer may be formed that bears positive charges at the interface to the buffer while the hydrophilic and/or charged groups of the second layer protrude into the electrolyte.

14.5.2. Permanent Coating

These modification methods are characterized by various types of coating layers generated either by silanization of the SiOH as anchor groups for the fixation of polar, neutral, or ionic small and larger oligomers or by multipoint adsorption of neutral and charged polymers that can also be cross-linked. Ionic interaction of cationic polymers or surfactant molecules with the negative charged surface may also be involved for fixation.

The practical aim in generating such coatings must be to form surface layers that are stable and not disrupted during long series of measurements even with chemically aggressive buffers, i.e., in a wide range of buffer pH's. Moreover, permanent coatings may not be destroyed by rinsing procedures with any of the aggressive rinsing solutions that might be applied for regeneration of the separation performance as determined by the status of the capillary surfaces.

At present different permanently coated CE capillaries with reproducible properties have become commercially available, with either hydrophobic (alkylpolysiloxanes) or hydrophilic [poly(ethylene glycols) or PEGs, and PVAs] coatings and also charged coatings that are fixed by covalent bonding or by strong adsorption and cross-linking of polymeric layers. It would be analytically advantageous if only a few types of coatings were developed that had more or less universal application in the various separation systems with different types of buffers. For such commercial products it is desirable that manufacturers supply sufficient information about the chemistry of their coatings to facilitate the development of separation methods for analytes of special chemical properties.

14.6. METHODS AND APPLICATIONS OF DYNAMIC SURFACE MODIFICATIONS

Because of the negatively charged surfaces in fused-silica capillaries, both small and large cationic modifiers have been applied as dynamic coatings, as well as the nonionic hydroxylic modifiers preferred in the more recent work of several groups. Everaerts et al. (33) first investigated the influence of polymeric buffer additives on bandshapes in isotachophoresis (ITP), using polymers such as PVA, HEC, and HPMC, which all suppress the EOF by shielding. For the structures of these polymers, see Figure 14.2.

It was also found (33) that Priminox 32 (Röhm & Haas, Frankfurt, Germany), a tertiary amine with longer chains of -CH_2CH_2O units, and cetyltrimethylammonium bromide [$(C_{16}H_{33})(CH_3)_3N$]Br, or CTAB, reverse the EOF in polytetrafluoroethylene (PTFE) and fused-silica capillaries. CTAB is a surfactant whose influence on EOF was investigated by Tsuda (10). Altria

---

Figure 14.2. Polar polymers used for adsorptive modification of fused-silica surfaces, showing structures of hydrophilic polymers that are neutral (hydroxylic, ethylene oxide type), cationic, or anionic: (1) PEG; (2) PVA; (3) HEC and HPMC; (4) poly(galacturonic acid), PGA; (5) sodium alginate; (6) ι-carrageenan; (7) κ-carrageenan; (8) Polybrene; (9) Praestol.

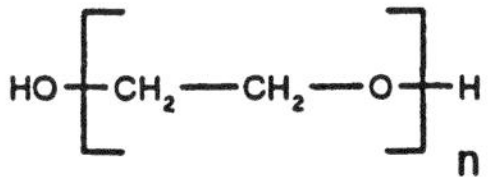

Figure 14.2

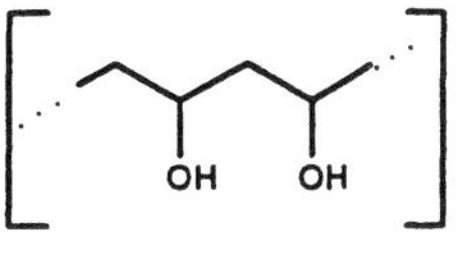

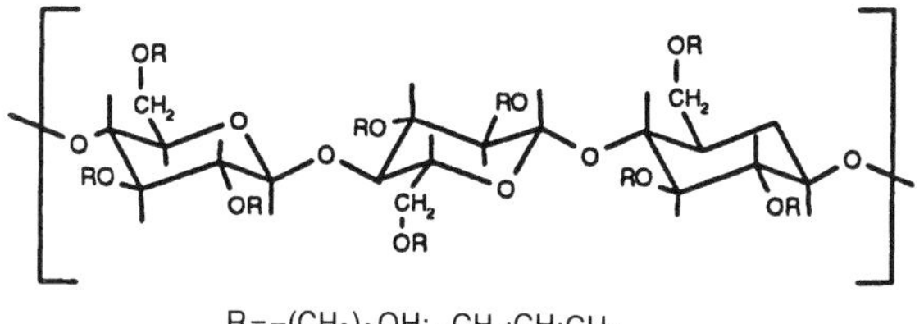

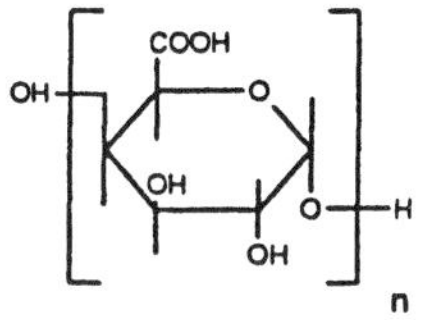

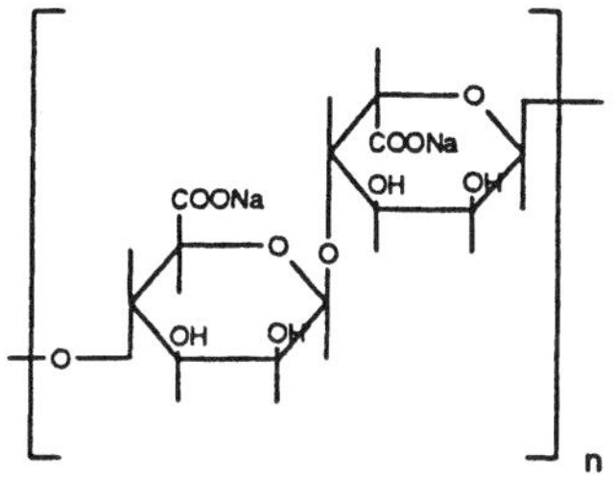

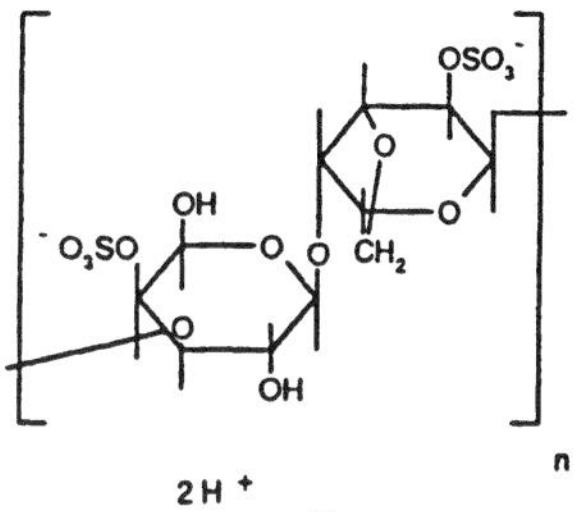

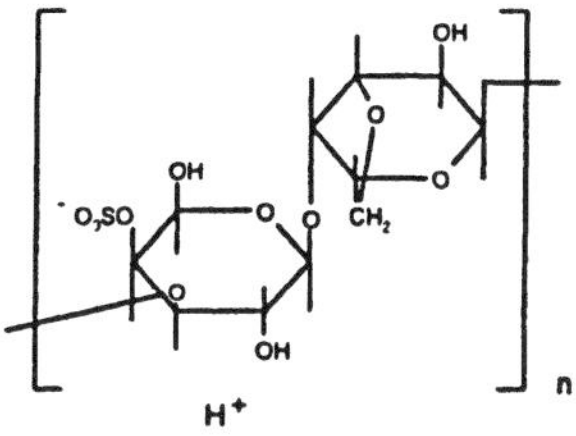

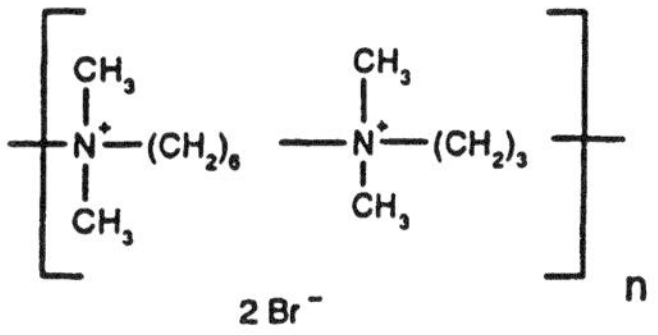

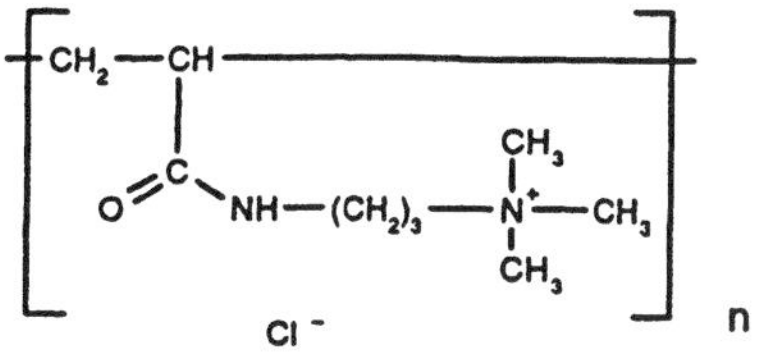

and Simpson (34) studied the influence of alkyl chain length in alkyltrimethylammonium bromide compounds. TTAB was first used by Huang et al. (35) for separations of inorganic anions with reversal of EOF and was extensively applied in the work of Jandik et al. (11–13) for the separation of anions. Smaller cationic compounds such as diethylenetriamine (DETA) or hexamethonium bromide $[(CH_3)_3N(CH_2)_6N(CH_3)_3]^{2+}\ 2Br^-$ have also been applied for anion separations. A method of anion separation (36) makes use of quaternary amines that are covalently bonded to the fused-silica surface.

In many papers on peptide and protein separation, dynamic and semipermanent surface modification by neutral and cationic surface coatings have been described. The extensive research work in the area of separation of proteins in the CZE mode with dynamic surface modification and also on the investigation of alternative permanent coatings gives the impression that particular difficulties have arisen with these classes of compounds in CZE systems. The charging of proteins is characterized by their pI and depends on the pH of the separation medium. Due to multimodal interaction, the large basic or acidic protein molecules tend to be strongly adsorbed on the capillary surfaces. Size-selective separations with proteins can be achieved with sodium dodecyl sulfate (SDS) in buffers that equalize the charge density of the different protein species. The effectiveness of dynamic modification methods in separations of proteins is not always sufficient, probably because of the especially strong affinity of basic proteins toward the acidic surface caused by simultaneous involvement of other electrostatic and hydrophobic types of intermolecular interaction. Figure 14.3 shows four separations of a test mixture of basic proteins with uncharged "hydroxylic" polymers as dynamic modifiers. In this series of experiments the best separation was achieved with HPMC. The capillary tubing applied for these separations was from Polymicro Technologies (Phoenix, Arizona). It is noteworthy that only for this special tubing material was PVA not an optimal *dynamic* modifier; however, it was found that this might well be different for tubing material from other sources. After thermal immobilization (probably effected by cross-linking between the polymer chains) PVA is also a perfect modifier for this tubing material, as shown in Figure 14.4a (see p. 497). The effective cationic polymeric modifier Telec DX [poly(diethylaminoethyl) methacrylate] (Du Pont Co., Wilmington, Delaware), described by Green (37), is also strongly adsorbed to the fused-silica surface. Surface modification procedures that involve rinsing prior to the introduction of the running buffer, which can be applied without the presence of the modifying molecule in the buffer at appreciable concentration, can be considered semipermanent. Similar observations were made by Motsch (16) and Schomburg et al. (20), using the strong base Praestol as the modifier for separation of inorganic anions. (For the structure of this polymer, see Figure 14.2.) Surface modification by a rinsing procedure with strongly chemi-

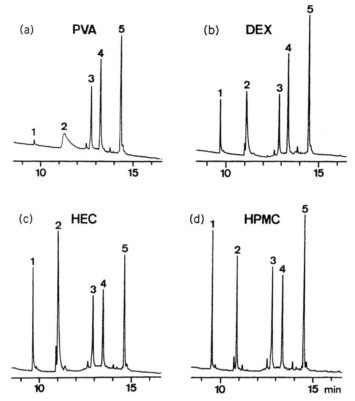

Figure 14.3. Efficiency, peak symmetry, and resolution in separation of basic proteins with dynamic coating by hydroxylic polymers. The optimal deactivation of the fused-silica surface in capillary tubing (originating from Polymicro Technologies, Phoenix, Arizona) was achieved by dynamic modification with HPMC. Capillary: 57 cm effective length; 70 cm total length; 50 μm i.d. Sample: (1) cytochrome c; (2) lysozyme; (3) trypsin; (4) trypsinogen; (5) α-chymotrypsinogen A—each 0.2 mg/mL. Buffers: 50 mM Na phosphate, pH 3.0; 0.05% (w/w) polymer additive. (DEX = dextran.) Conditions: 30 kV (429 V/cm), 28 μA; 20°C. Injection: 10 kV, 5 s. Detection: UV, 214 nm.

and physisorbed modifiers before the introduction of the run buffer for the separation of proteins were proposed earlier by Wiktorowicz and Colburn (25,26) using Polybrene (see Figure 14.2), which is hexadimethrine bromide and originates from Aldrich-Chemie (Steinheim, Germany). These authors observed that the biogenic amines spermine (Aldrich-Chemie) and dodecyltrimethylammonium bromide (DTAB) proved not to be as effective for such rinsing procedures because of rather weak adsorption and corresponding easy desorption, consequently resulting in unstable EOF and protein adsorption. Towns and Regnier (32,38) applied cross-linked polyethyleneimine (PEI), but later rinsed the capillaries with PEI only. Recently (39) a new method has been reported for the application of PEI to protein separations.

The practical application of even strongly adsorbed polymeric hydroxylic and cationic modifiers requires regeneration after a number of runs by rinsing with a coating solution in order to achieve reproducible performance in a long series of analyses. Insufficiently adsorbed dynamic coating layers necessitate the presence of the modifying polymer in the buffer. This was shown with the hydroxyalkyl celluloses, which were very successfully applied for separation of histones by Lindner et al. (17) and for basic proteins in general (see Figure 14.3). PVA (for the structure, see Figure 14.2) is a large linear molecule with a high content of hydroxyl groups for strong multipoint interaction via hydrogen bridge bonding with silica surfaces that is capable of strong self-association between the molecular chains of the polymer. The effectiveness of PVA as a *dynamic* modifier is more or less similar to that of the hydroxyalkyl celluloses but may sometimes depend on the properties of the tubing material. The immobilization of PVA is improved by a thermal treatment (see Figure 14.4a below), which probably leads to cross-linking of the polymer chains and to a more effective suppression of analyte–wall interaction by more stable coating layers. These immobilized PVA coatings are just as effective as those obtainable with the cationic polymers, which are assumed to be strongly attached to the surface by Coulomb interaction (19). PVA coatings are neutral and not dependent on pH with regard to changes of EOF and the adsorptive properties of the coating. Hydroxylic coatings, such as PVA, and hydrophilic coatings with PEGs, which are less effective as dynamic modifiers (compared to PVA), seem to be generally characterized by weak hydrophobic interaction with many proteins. The PEG molecule does not contain hydroxyl groups, but it can be assumed that the ether O atoms of the PEG will be hydrated by interaction with the aqueous buffer. It has been shown that PVA as a dynamic modifier and as a kind of permanent modifier was of varying effectivity for the separation of basic proteins [see Figures 14.3 and 14.4 (19)] depending on the origin and properties of the capillary tubing applied (see Figure 14.4b). Figure 14.5 (p. 499) shows a separation of chicken egg white proteins and illustrates the resolution of the glycosilated derivatives that can be achieved with PVA as the thermo-immobilized cross-linked polymeric modifier (19).

Enantiomeric separations of basic and acidic compounds of pharmaceutical interest require careful optimization of both the enantioselectivity (relative electrophoretic migration times) and efficiency. The final aim of such separations is the complete resolution of the enantiomers even in cases of extreme ratios of the concentrations of the enantiomers in the sample to be analyzed. In a basic study of the mechanisms of surface modifications in CZE and EKC with cyclodextrin derivatives as selectors, neutral (hydroxylic), cationic, and anionic small molecules and preferably such polymeric compounds as shown

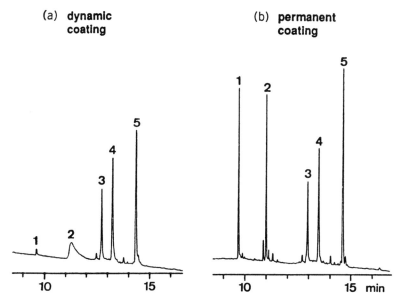

Figure 14.4. Separation of basic proteins with (a) dynamic and (b) permanent PVA coating, showing the decrease of analyte–wall interaction by the immobilized coating layer of PVA. Capillaries: Polymicro Technologies (Phoenix, Arizona) tubing, 57 cm effective length; 70 cm total length; 50 μm i.d. Coating: (a) 0.05% (w/w) PVA in buffer; (b) thermo-immobilized PVA. Buffer: 50mM Na phosphate, pH 3.0. Sample: (1) cytochrome c; (2) lysozyme; (3) trypsin; (4) trypsinogen; (5) α-chymotrypsinogen A—each 0.2 mg/mL. Conditions: 30 kV (429 V/cm), 28 μA; 20°C. Injection: 10 kV, 5 s. Detection: UV, 214 nm. (c, d) Influence of th fused-silica material on dynamic coating with PVA. Sample: same as in parts a and b; above. Capillary: 57 cm effective length; 70 cm total length; 50 μm i.d. (c) MicroQuartz, Munich; (d) Polymicro Technologies, Phoenix, Arizona. Buffer: 50 mM Na phosphate, pH 3.0; 0.05% (w/v) PVA (50,000 Da, Aldrich). Conditions: 30 kV (429 V/cm); (c) 32 μA, (d) 261A; 20°C. Injection and detection: same as in parts a and b, above.

in Figure 14.2 were applied as modifiers by Belder and Schomburg (21,22,40). The major aim of this study was to apply different types of surface modifiers so as to achieve the suppression, increase, or reversal of the EOF, thereby influencing the efficiency and resolution of separations in accord with Eq. (14.1) and (14.2). Simultaneously it was sought to suppress analyte–wall interaction by dynamic or permanent coatings. Polymers as dynamic modifiers can be affixed to the fused-silica surface by various chemi- or physisorption mechanisms of these molecules: by hydrogen bridge bonding, by dipole–dipole interactions; and by ionic (Coulomb) interactions. Polymers that give rise to multipoint interactions and formation of strongly fixed multimolecular layers with effective shielding of the negative charges at the surface are beneficial. These coatings may have an undesirable effect on the attraction and interaction of certain classes of analyte molecules that are present in a sample

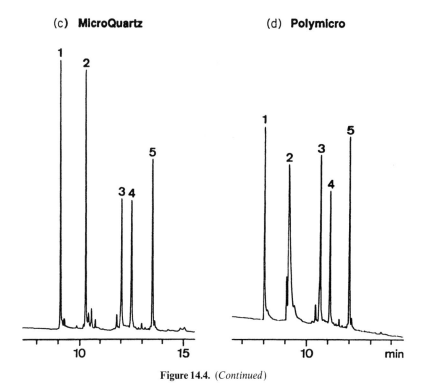

Figure 14.4. (*Continued*)

in addition to those analytes whose wall interaction has been suppressed. Of course, the coating layer itself should have negligible interaction with the analyte molecules. The cationic polymers can be expected to undergo stronger ionic interaction with the negatively charged silica surface than the neutral hydroxylic polymers do, but can effect the positive charging of the surface and may cause reversal of the EOF. Negatively charged, i.e., anionic polymers such as PGA or the carrageenans (see formulas 4, 6, and 7 in Figure 14.2) are also adsorbed on the negatively charged silica surfaces in spite of electrostatic repulsion via the charged groups in these polymers that is usually thought to be effective. Belder (40) has proved that these negatively charged polymers are attached to the surface by adsorption, possibly by other than ionic types of intermolecular interaction. The EOF in capillaries modified by anionic polymers is increased due to an increase in the number of negative charges at the surface. Correspondingly higher plate numbers but a decrease of resolution were observed, as Eqs. (14.1) and (14.2) predict (see Figure 14.6). The resolution in separation a without the PGA coating was even slightly better. The peak symmetries in both separations are excellent for these basic drugs,

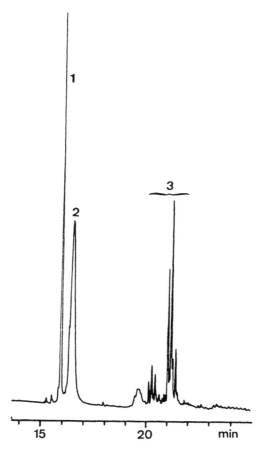

Figure 14.5. Separation of chicken egg white proteins with thermo-immobilized PVA. Sample: Chicken egg white, diluted 10-fold with buffer, filtered through 5 μm membrane. Peaks: (1) lysozyme; (2) conalbumin; (3) ovalbumin. Capillary: 57 cm effective length; 70 cm total length; 50 μm i.d. (Polymicro Technologies, Phoenix, Arizona). Coating: thermo-immobilized PVA. Buffer: 150 mM Na phosphate, pH 3.0. Conditions: 30 kV (429 V/cm); 78 μA; 20°C. Injection: 15 kV, 5 s. Detection: UV, 214 nm.

although the surface had a high concentration of negative charges, as can be concluded from the enhanced EOF.

All the polymers shown in Figure 14.2 can be applied as additives for dynamic modification in very low concentrations (50 mM, 0.05% w/w of polymers) that do not change the viscosity of the bulk buffer significantly. In surface adsorption the viscosity of the buffer close to the surface may be much higher, however, and could influence the flat flow profile of the buffer. The

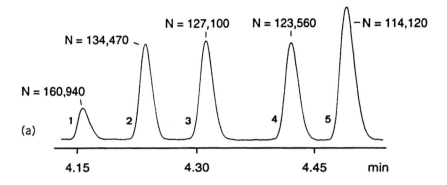

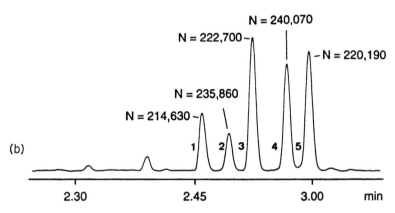

Figure 14.6. Separation of basic drugs (tocainides): influence of dynamic coating with κ-carrageenan on migration times, efficiency, and resolution. Note the increase of EOF with higher concentrations of negative charges on the surface in separation b. With the κ-carrageenan coating the efficiencies in separation b become twice as high as in the bare capillary, but the resolution in separation a is about 10% higher [the ratio of resolutions ($\Delta t/t$) is 1.0] as a consequence of the influence of EOF. Capillary: 43.5 cm effective length; 59.6 cm total length; 50 μm i.d. Buffer: (a) 50 mM Na phosphate, pH 3; (b) 50 mM Na phosphate, pH 3; 0.05% (w/w) κ-carrageenan. Conditions: 30 kV; 20°C.

nonionic hydroxyalkyl methylcelluloses have proved to be very effective as dynamic surface modifiers for the separation of cationic and anionic small (e.g., drug) molecules. They suppress the EOF effectively within a pH range between 4 and 10, depending on the immobilization of the coating layer (see Figures 14.7a,b); here PVA was used as the dynamic and immobilized modifier. In these EOF measurements PVA was an additive in the buffer for curve B in diagram a and was thermally immobilized for curve B of diagram b. The thermal immobilization of several PVA layers at temperatures of about 130–140°C leads to stable "permanent" layers of the PVA without covalent bonding to the surface. The layers shield the negative charges and prevent the formation of the cationic double layer. Without a shielding PVA layer, the EOF increases with increasing pH because of stronger dissociation of the SiOH groups. The shielding of the negative surface charges prevents the formation of the Stern double layer by the positive buffer ions. With cationic polymers as modifiers (see Figure 14.8), negative, i.e., reversed, EOF data are observed after prior rinsing with a 0.1% w/w aqueous solution of these polymers. The polymers do not need to be contained in the buffer at the time of actual measurement. Also in this case the EOF depends on the buffer pH. The EOF data in Figure 14.8 are plotted together with those obtained with a bare fused-silica capillary and with those obtained after EOF modification by a vertical electric field (see the *curve). The modification by hydroxylic neutral polymers such as HPMC or PVA and similar compounds leads to separations with higher efficiency and resolution by suppression of analyte–wall interaction and simultaneously of the EOF (see separations a and c in both Figures 14.9 and 14.10). In the faster separations (a) the EOF of a bare fused-silica capillary was effective in the same direction as the co-electroosmotic migration of the solutes to the cathodic detector side of the capillary. Only about 60,000 theoretical plates were achieved in a bare capillary without modification in spite of the co-electroosmotic mode of separation and the high apparent electromobility compared to that of separations c, for which suppression of the EOF by PVA or HPMC, respectively, as additive was achieved. The apparent electromobilities in separations c are much lower and the migration times longer because only electrophoretic migration is effective. Surprisingly the efficiency is higher (213,000 in Figure 14.9 and 245,000 theoretical plates in Figure 14.10) than in the bare capillary despite the longer residence times of the solute molecules in the capillary and the related higher number of capillary wall collisions. With EOF suppression the efficiency N depends [according to Eq. (14.1)] on only the electrophoretic mobility, which is assumed to stay approximately the same in the modified system, and lower than the sum $\mu_{eo} + \mu_{ep}$ in the unmodified system. In view of the influence of the apparent electromobility on N much higher plate numbers ought to be achieved without suppression of the EOF; however, in reality the plate counts achieved are higher when the EOF is suppressed and the apparent μ_{ep} is lower. The reason for this is the suppression of analyte–wall interaction by the dynamic coating with a neutral modifier. The resolution in separations c of Figures 14.9 and 14.10 is higher not only because of the increase of the "dispersion" term (square root of N) of Eq. (14.2) but also because of the smaller denominator in the "retardation and selectivity" term of this equation. For the assessment of separations in CZE and EKC it is important to consider the influence of the

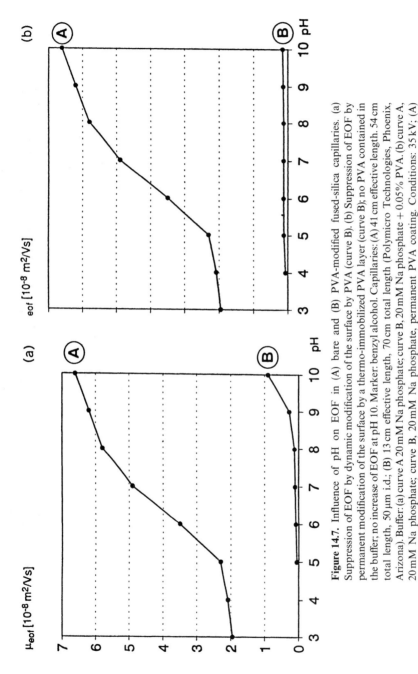

Figure 14.7. Influence of pH on EOF in (A) bare and (B) PVA-modified fused-silica capillaries. (a) Suppression of EOF by dynamic modification of the surface by PVA (curve B). (b) Suppression of EOF by permanent modification of the surface by a thermo-immobilized PVA layer (curve B); no PVA contained in the buffer; no increase of EOF at pH 10. Marker: benzyl alcohol. Capillaries: (A) 41 cm effective length. 54 cm total length, 50 μm i.d.; (B) 13 cm effective length, 70 cm total length (Polymicro Technologies, Phoenix, Arizona). Buffer: (a) curve A 20 mM Na phosphate; curve B, 20 mM Na phosphate + 0.05% PVA. (b) curve A, 20 mM Na phosphate; curve B, 20 mM Na phosphate, permanent PVA coating. Conditions: 35 kV; (A) 648 V/cm, 19–45 μA; (B) 500 V/cm, 15–35 μA; 20°C. Detection: UV, 214 nm.

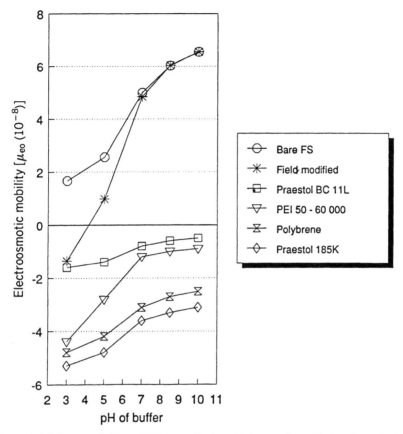

Figure 14.8. Influence of pH on EOF in capillaries with bare and modified surfaces. Surface modification by four cationic polymers; Praestol BC 11L; PEI 50–60,000; Polybrene; Praestol 185K. They were contained in rinsing solutions (0.1% w/w) in water applied prior to the introduction of the actual buffer, which did not contain these polymers. For comparison, the pH dependency of EOF is shown after modification by a vertical electric field. Capillary: 55 cm effective length; 70 cm total length; 75 μm i.d. Field strength (E): 500 V/cm. Buffer: 20 mM phosphate. Marker: dimethyl sulfoxide (DMSO).

electroosmotic mobility on efficiency and resolution, which is a vector as the electrophoretic mobility. A small vector sum $\mu_{eo} + \mu_{ep}$ leads to lower efficiencies but higher resolution.

It has been shown that the enhancement of efficiency by the suppression of analyte–wall interaction and when μ_{eo} in Eq. (14.2) becomes zero is much more significant than when the EOF is suppressed by a vertical electric field (see separation b of Figure 14.9). This electrical modification (suppression) of the EOF does not lead to an improvement of separation efficiency and

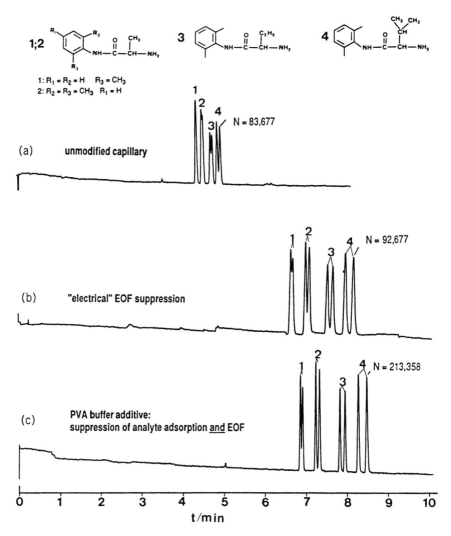

Figure 14.9. Influence of EOF suppression by a vertical electric field in comparison to chemical suppression with PVA as the buffer additive. EOF suppression by the vertical electric field increases migration times at slightly increased efficiency. Suppression of EOF by PVA increases migration times, efficiency (theoretical plate numbers), and resolution. Sample: (0.2 mg/mL of each compound dissolved in water). Capillary: 43.2 cm effective length; 56 cm total length; 50 μm i.d., 0.36 mm o.d. Conditions; (a,c) 35 kV (617 V/cm); (b) HV_1, +17.5 kV; Hv_2–17.5 kV; 53 μA; 20°C. Buffers: (a,b) 40 mM Na phosphate, pH 3.0, and 50 mM γ-CD; 0.05% (w/w) PVA 77,000–79,000 Da. Injection: hydrodynamic, Δp: 85 mbar, 2 s. Detection: UV, 210 nm.

resolution at the same migration times as are achieved with the chemisorption of the hydroxylic polymers [see separation c of Figures 14.9 and 14.10 (21)]. In the separations of both figures, the enantiomeric pairs are nearly baseline resolved with either PVA or HPMC surface modification as a consequence of reduced analyte–wall interaction and because the EOF suppression as well as the longer migration times increases the resolution in accord with Eq. (14.2). It was observed by Belder and Schomburg (21) that the efficiency of separations of basic compounds in bare fused-silica columns decreases strongly toward high electric field strengths without dynamic modification but that the efficiency N increases proportionally to field strength with PVA or HEC modification. Separation b in Figure 14.10 was achieved with a buffer that contained the negatively charged PGA (see formula 4 in Figure 14.2) in low concentration as the modifier, the electroosmotic mobility was higher and the migration times shorter due to the higher concentration of negative charges on the surface originating from the adsorbed anionic polymer. It should be mentioned here that higher plate numbers than 300,000 have generally not been observed in EKC with separation of small molecules with higher diffusion coefficients. The increase of the apparent electromobility and the shorter migration times indicate that the anionic PGA molecules adhere to the negatively charged surface, although one might have assumed that they should have been repulsed because of their negative charging were Coulomb interaction assumed to be solely effective.

For separation d in Figure 14.10 the capillary was rinsed with the cationic Polybrene before the run buffer was introduced. The chemisorption of Polybrene causes reversal of the EOF and of the elution order of the enantiomeric pairs of the chiral drugs. Longer migration times are achieved because the EOF is effective against the direction of electromigration. Efficiency and resolution are as low as in the unmodified capillary. Before the execution of these separations it was expected that with a cationic surface the cationic analytes would undergo weaker interaction with the surface because of ionic repulsion. It is possible that a stable, immobilized, and more homogeneous cationic polymer coating might lead to other results in this regard. Another explanation for this phenomenon could be that the cationic analyte ions are not repulsed from the coating layer because the buffer anions form a stable layer of reverse charge polarity above this coating. Moreover, not only ionic (Coulomb) interaction should be considered for the interaction of the analytes. Organic molecules and especially proteins undergo more complex multimodal interaction. Similar oberservations with regard to the influence of charged dynamic coating layers were also performed with anionic drugs (NSAID = non-steroidal anti-inflammatory drugs) as analytes and neutral, cationic, or anionic polymers for dynamic coatings (20–22, 41). If only Coulomb interaction between the ionic analytes and the surface modified by

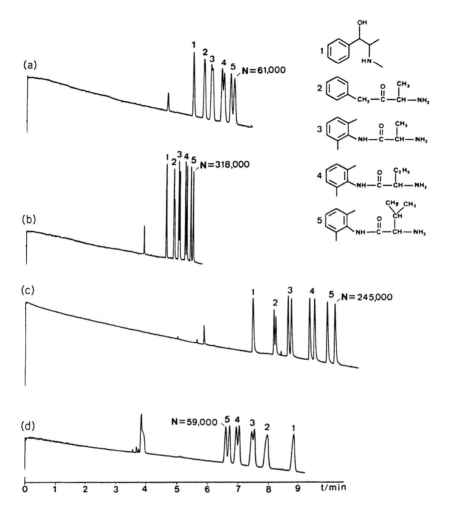

Figure 14.10. Chiral separations of basic drugs (tocainides) with γ-cyclodextrin (γ-CD) in fused-silica dynamically modified by neutral, cationic, and anionic polymers. (a) Higher efficiencies are achieved with neutral or anionic polymers. (b) With PGA, high efficiency is attained because of higher EOF, and better resolution because of suppression of analyte–wall interaction. (c) With HPMC, even higher efficiency is achieved because of suppression of analyte–wall interaction; also high resolution, because of suppression of EOF and analyte–wall interaction. (d) Polybrene reverses EOF and the retention sequence of analytes 1–5: separation occurs with electromigration against the EOF; there is poor efficiency and resolution despite slow separation and a cationic surface. Sample: 0.02 mg/mL of each compound dissolved in water. Capillaries: (a,b) 45.7 cm effective length, 59.5 total length; (c) 46.2 cm effective length, 59.3 cm total length; (d) 45.7 cm effective length, 59.3 cm total length; 50 μm i.d. Conditions: 35 kV; 29–32 mA; 20°C. Buffers: (a) 40 mM Na phosphate, pH 3.0, and 50 mM γ-CD; (b) + 0.05% (w/w) PGA; (c) 0.05% (w/w) HPMC; (d) preconditioned with Polybrene. Injection: vaccum, $\Delta p - 85$ mbar, 2 s. Detection: UV, 210 nm.

coatings with different charge were effective, a decrease in efficiency should arise, but also in this case the opposite was observed. High resolution can only be achieved in slow separations, which is unfortunate for routine analyses. Separations of racemic acidic drugs were performed in a bare fused-silica capillary and also with cationic surface modification achieved by rinsing with Polybrene. In spite of the cationic surface, higher efficiencies and resolution were observed with anodic EOF at reversed polarity of the electric field (22).

The interpretation of theoretical plate counts and the assessment of peak symmetries as indicators of the contribution of analyte–wall interaction to band broadening is not without problems because of the influence of secondary contributions originating from sample introduction (sample load of the system, sample volume, stacking effects), thermoconvection within the capillary, and detection. Such secondary contributions have to be eliminated or kept constant in any series of comparative measurements. The influence of surface coatings on EOF and analyte–wall interaction can only be studied when the necessary measurements are executed in systems with active cooling of the capillary in which all other parameters are kept constant when the modifier is added to the buffer.

14.7. METHODS AND APPLICATIONS OF PERMANENT SURFACE MODIFICATIONS

The generation, testing, and application of permanent coatings of fused-silica surfaces in capillary tubing have been treated in many publications since the development of modern CE. In Section 14.2 it was pointed out that the status of the surfaces of CE capillaries, either unmodified or modified, has a strong influence on the performance of analytical separations as regards their efficiency, absolute and relative migration times, and especially resolution. The major thrust of research work on permanent coatings has been directed to the separation of large biomolecules such as oligonucleotides, proteins, and polysaccharides, although the same modification methods are effective for the separation of small molecules, especially those of cationic drug compounds. In permanently coated capillaries separation systems can be established with buffers of less sophisticated composition. This approach of permanent coating may also be important for CE–MS work to prevent the large molecules of polymeric modifiers from entering interfaces and ion sources. High-performance liquid chromatographic (HPLC) separations of large molecules are performed in different aqueous phase systems but mainly with selectivity changes during the separation by gradient elution in order to achieve good resolution and higher peak capacities. The achievable efficiencies of separations of large molecules are quite limited in HPLC, however; but this lack of efficiency in HPLC is compensated by gradient elution. The various CE methods can be performed with high efficiency and sufficiently high resolution at high speed over a wide range of molecular weights without a procedure that is analogous to gradient elution in HPLC. In this context CIEF methods should be mentioned that are performed with a constant pH gradient but without movement of the buffer at the actual separation. The early work on electrophoretic separations of biopolymers in capillaries, including the chemistry of surfaces, was influenced by the extensive experience with the application of electrophoretic separations in flatbeds.

Coatings for reduction or suppression of the EOF must be "polar" and/or optimally hydroxylic, such as PAAs, hydroxyalkyl celluloses (HACs), dextran, PVAs, and PEG. Permanent coatings with the desired properties for universal CE application, especially those with a silanization sublayer, are rather difficult to generate with high reproducibility for commercial manufacture; they require a sequence of steps of chemical reactions at the surface in the miniaturized space of a narrow-bore capillary. Single-step procedures are described by Malik et al. (42) and Zhao et al. (43) that begin with static coating of the capillary (as applied for GC capillary columns) with a solution containing PEG, plus a bifunctional silanization reagent like HMDS (hexamethyldisilazane) and dicumylperoxide for radical cross-linking. After static coating of the surfaces the volatile solvent is evaporated and the cross-linking or chemical bonding reactions are initiated by heating. Afterward the capillaries are freed from nonvolatile reaction by-products by rinsing. The aforementioned authors also describe a similar method without silanization that is merely based on immobilization by cross-linking with radical catalysts. The capillaries are characterized as stable at high pH and allow for good separations of basic proteins. The simplest procedure that leads to stable coatings for proteins even at high pH is the cross-linking of "living" PEG polymers by either thermal or catalytic treatment. Such polymers can be purchased from Innophase Corp. (44)

Characteristics and criteria of commercial coated capillaries that provide a suitable level of analytical performance are as follows:

- Coatings of defined polarity and thickness for the shielding of the silanol charges and weak interaction with the analytes
- Stability of the coatings over a wide range of pH of the buffers
- Defined and reproducible EOF, i.e., reproducible migration behavior in use with standard buffers and in a wide range of pH.
- Long-term stability of the coating for the analysis of real samples with complex matrices when rinsing by special solutions for regeneration becomes necessary

Hydrophobic alkylpolysiloxane coatings may be of interest for CZE and MEKC (micellar electrokinetic chromatography). The procedures of generation of such coatings on fused-silica surfaces are well known from capillary GC (although much thinner films are applied in CE). At present alkylpolysiloxanes that are to be immobilized can be "living" and contain, e.g., vinyl or other end groups by which coatings can be cross-linked or chemically bonded by thermal or catalytic treatment after coating of the capillary walls. Living polymers of the alkylpolysiloxane and PEG type can be purchased from several companies [e.g., Innophase Corp. (44)]. Details on the formation and chemical properties of such coatings developed by manufacturers are unfortunately often proprietary.

Three typical applications of either polysiloxane-coated or silanized capillaries are discussed next.

Lux et al. (45) coated fused-silica capillaries for EOF modification in MEKC with very thin films (0.04 μm) thickness of polymethylsiloxane OVI [and with PEG (Carbowax 20 M, 0.09 μm thickness)] to study the EOF that occurs after dynamic formation of an SDS layer. Thicker films effect no further change of the EOF in both cases, as can be seen in Figure 14.11. The influence of the EOF on the separation was studied with a test mixture containing theobromine, theophylline, caffeine, and ureic acid and with a test mixture of nucleobases. The EOF increases with the OV1 coating and decreases with the PEG coating in the presence of SDS in the buffer. With OV1 a faster separation was obtained with relatively low resolution; with PEG a slower separation, with much higher resolution (see Figure 14.12). The separation efficiencies were as high as they generally are with MEKC, even without the presence of polymeric modifiers; here the surfaces seem to be behaving as if modified by adsorption of the SDS, in spite of its negative charge. By adsorption of the SDS via the hydrophobic alkyl group of this surfactant, the hydrophobized surface becomes more negatively charged and a higher EOF occurs than with a bare fused-silica capillary. In contrast, with the polar PEG the SDS is chemisorbed via the polar charged sulfate group, forming a less negative and more hydrophobic surface; consequently a lower EOF occurs than with the bare fused-silica surface. The influence of surface treatments on the EOF in MEKC was also treated by Wu et al. (46).

Regnier and Wu have discussed the principles of several modification techniques described in the literature that can provide surfaces suitable for the CZE of proteins by reducing surface adsorption, providing good protein recovery, and maintaining EOF. These authors proposed modification of fused-silica capillary surfaces with a C_{18}-substituted silanization reagent and nonionic surfactants such as Brij, Tween, and Pluronic, which are adsorbed to the hydrophobized surface by van der Waals interaction via the hydrophobic part of the molecule (30).

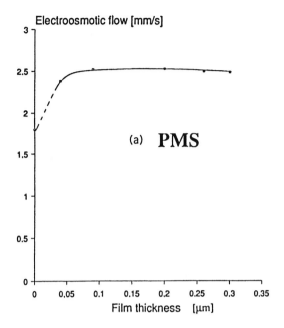
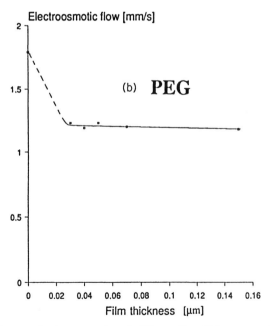

Figure 14.11. Influence of polymer coating of different film thicknesses on EOF: (a) PMSC polymethylsiloxane OV1); (b) PEG (or Carbowax 20M).

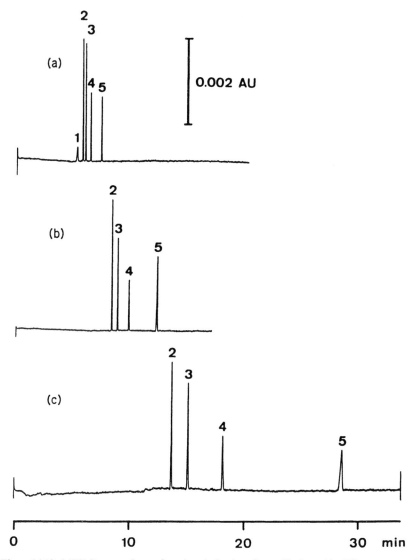

Figure 14.12. MEKC separations of purine derivatives in capillaries with different types of polymer coating showing the influence of modification of EOF on migration time and resolution. Sample: 0.2 mg/mL each. Peaks: (1) water; (2) theobromine; (3) theophylline; (4) caffeine; (5) uric acid. Capillaries: 80 cm effective length, 120 cm total length; (a) PMS, 50 μm i.d., with film thickness of 0.09 μm; (b) uncoated capillay, 50 μm i.d., (c) PEG (Carbowax 20M), 50 μm i.d., with film thickness 0.04 μm. Buffer: 0.02 mM phosphate, + 0.05 M SDS. Injection: electrokinetic, 7500 V; (a) 2 s; (b) 5 s; (c) 8 s. Conditions: 35 kV; 298 K (temp.). Detection: UV, 245 nm.

Heiger (48) modified fused-silica capillary surfaces with *n*-octadecyl-triethoxysilane, which then adsorbed the hydroxylic surfactant dodecyl-D-mannoside; on this coating a layer of acrylamide–bisacrylamide was generated by copolymerization with the dodecyl-D-mannoside. The author claimed that such capillaries, modified in this somewhat sophisticated triple-step procedure, were useful for size-selective separations with linear polyacrylamide gels.

Other permanent coatings to be discussed in this last section consist of polymeric layers that are covalently bonded to the fused-silica surface at multiple sites. It is generally expected that such bonded coating layers are more stable against chemical aggression. They can be generated by various methods. The polymer (e.g., a copolymer of an organosilane and a hydrophilic monomer such as ethyleneoxide), which is externally synthesized, contains additional chemically reactive groups that can bond the copolymer to the surface at multiple sites or can be cross-linked with adjacent polymer chains through siloxane bonding (47–49). Another approach is to bond to the surface a silane that contains a further functional group suited for copolymerization. This functional group is then incorporated into a layer of a polymer that is formed *in situ*, i.e., within the capillary. For such procedures silanes are applied that bear two types of reactive groups: one for the covalent bonding to the surface via the silanol anchor group; the other for the initiation of an *in situ* polymerization or the fixation of an externally synthesized oligomer or polymer. Hjertén (1) described a polyacrylamide coating generated *in situ* by copolymerization on a surface previously silanized with the methacryloylpropyl–silylation reagent. The reactions that have to be performed to achieve such a polyacrylamide coating are shown in Figure 14.13. Since the introduction of CE, this coating has been widely used in the work of many investigators on the separation of proteins and oligonucleotides as well as in CZE or MEKC for the separation of small molecules. Other reaction schemes for the production of sublayers to which polymer layers are bonded are based on the vinylation of the silica surface by reaction with highly reactive vinyldimethylalkoxysilane or on the formation of a vinylated silica surface by a Grignard reaction of the $SOCl_2$ clorinated silanols with $BrMgCHCH_2$; the vinyl group is bonded via SiC (50). Such silanization and copolymerization reactions for the generation of hydrophilic polymer coatings are well known from the aforementioned research work, reported by Van Alstine's group (51–53), where the authors described a variety of polymer coatings based on the formation of sublayers by silanization that are suited to control protein adsorption and related phenomena of biotechnical interest. This group applied aminopropyltrialkoxysilane and mercaptopropyltrialkoxysilane to form sublayers for covalent attachment of hydrophilic polymers to quartz glass surfaces. Polymer coatings based on PEG and three polysac-

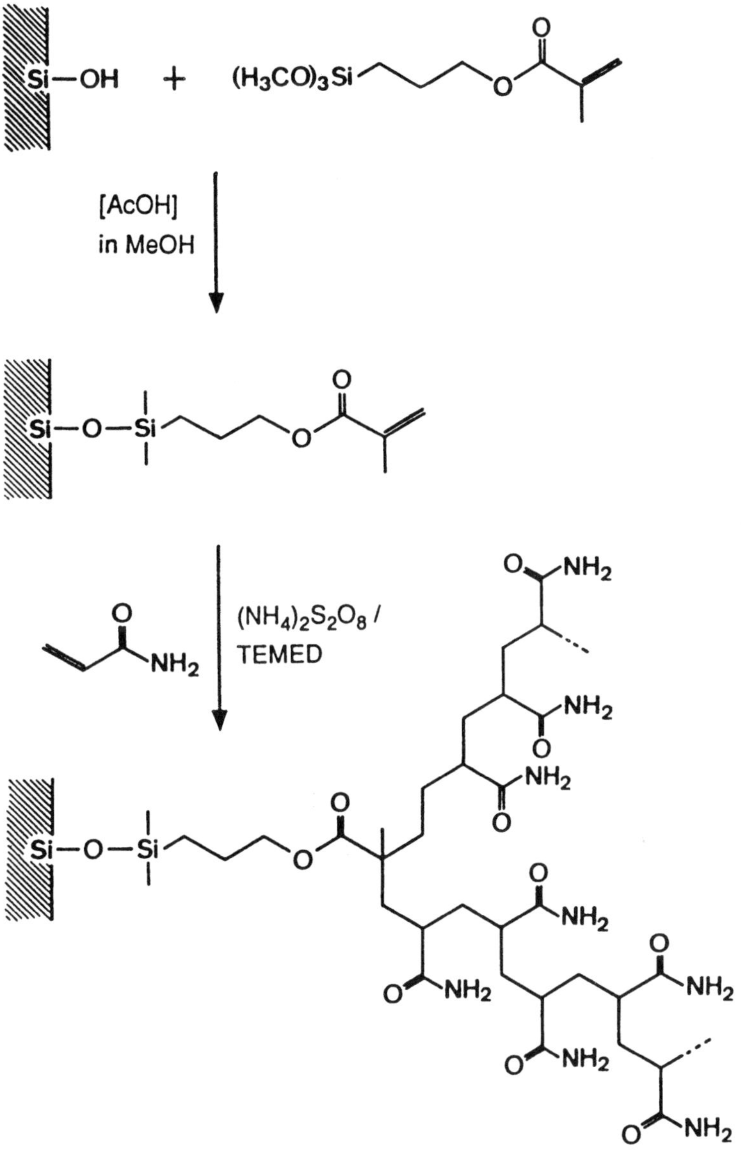

Figure 14.13. Generation of polyacrylamide coating according to Hjertén (1). *In situ* (within-capillary) copolymerization of methacrylamide at a sublayer is obtained by silanization with methacryloxytrialkoxysilane. (TEMED = N,N,N',N'-tetramethylethylenediamine).

charides (dextran, hydroxyethyl cellulose, and HPMC) were found to reduce capillary electroosmosis and microsphere electrophoretic mobility over a broad pH range. This reduction of EOF was stated to correspond to the ability of these coatings to reduce protein adsorption and control surface wetting by aqueous polymer two-phase systems. In their conclusion the authors noted the ability of covalent PEG coatings of 1000–20,000 Da to control the quartz zeta potential expression and verified similar capabilities for covalent dextran coatings of much higher molecular weight. Similar PEG polymer coatings for protein separations were investigated by Bruin et al. (54). In another report (55) from the same laboratory, a sequence of reactions is described including silanization by a glycidoxypropyltrialkoxysilane and a subsequent reaction with oligoethylene glycols under BF_3Et_2O catalysis to form covalently bonded PEG coatings for protein separations (see Figure 14.14). Recently a procedure for the generation of covalently bonded PVA has been proposed by Götzinger (56) that is based on the fixation of a sublayer containing a vinyl group to the fused-silica surface by means of Cl_3Si—$CH = CH_2$. The vinyl group is subsequently incorporated in a copolymerization with vinyl acetate to form a poly(vinyl acetate) layer and can be hydrolyzed to form a covalently bonded PVA layer. This author reported that highly efficient protein separations were obtained with such capillaries, which he claimed were stable after long-term usage even at high pH.

In many recent publications permanent coatings for CZE separations of proteins have been described. Most of these coatings were obtained by coupling of presynthesized hydroxylic oligomers with reactive sublayers obtained by silanization with reagents of the aforementioned type (γ-glycidoxypropyltriethoxysilane, 3-aminopropyltriethoxysilane, and 3-mercaptopropyltrimethoxysilane, etc.).

Liu et al. (57) describe a combination of silanization with γ-glycidoxypropyltrimethoxysilane and of coating with a thin film of diglycidyl ether of bisphenol A (DEBA) in methylene chloride and cross-linking with ethylenediamine; this combination was applied to achieve coated capillaries for the separation of basic proteins. The capillaries proved to be stable for more than 350 measurements in a pH range of 2–10. Merchref and El Rassi (58) describe the generation of a surface-bound dextran layer cross-linked with diepoxy-PEG after the capillaries were silanized with γ-glycidoxypropyltrimethoxysilane. The authors found that the capillaries were stable and suited for separations of acidic and basic proteins and other biomolecules. Smith and El Rassi (59) silanized fused-silica tubing with γ-glycidoxypropyltrimethoxysilane and treated the coating layer obtained with methylated and hydroxyethylated polyethyleneimine. Then a third layer was formed by reaction of diepoxy-PEG 200 and diepoxy-PEG 600 and 30% triethylamine above the previous layers. The capillaries were used for the separation of acidic

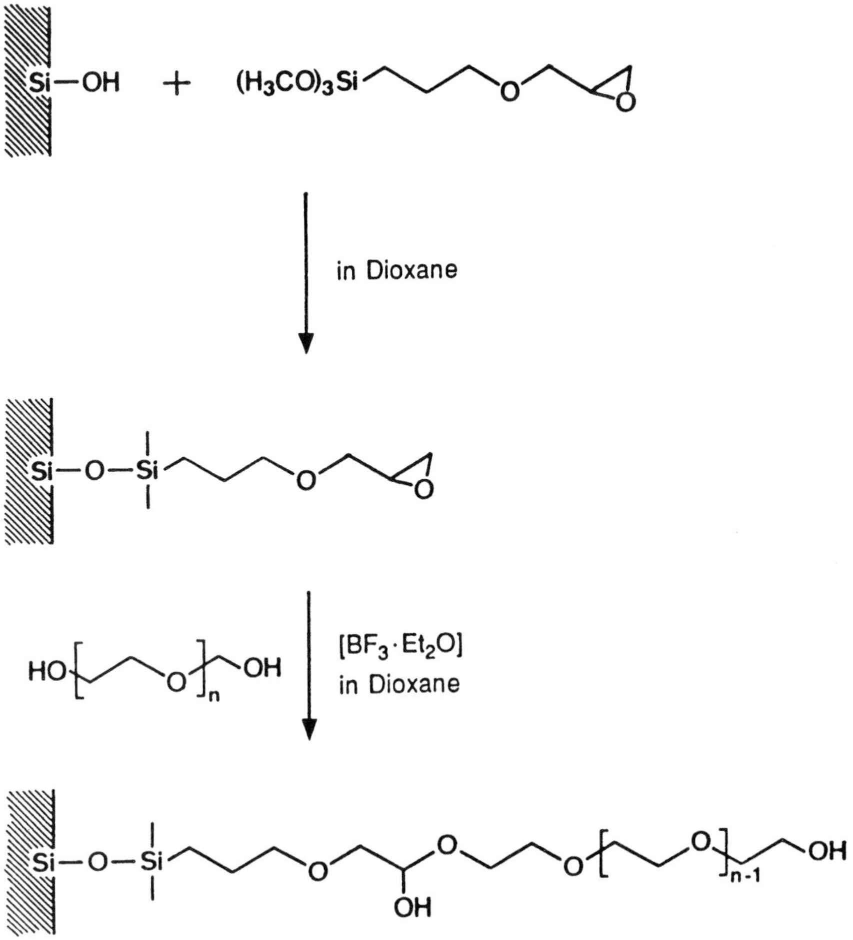

Figure 14.14. Surface modification by PEG, covalently bonded to a sublayer achieved by silanization with glycidoxypropyltrimethoxysilane. Injection: hydrodynamic, Δp: 85 mbar, 2 s. Detection: UV, 210 nm. [From Bruin et al. (54)].

peptides. Sun et al. (60) describe a chitosan-coated capillary with reversed EOF for the separation of basic drugs and proteins.

Fused-silica capillaries coated by cellulose acetate without a sublayer obtained by silanization for protein separations were used by Busch et al. (61). This was achieved by coating the internal wall with a 5% solution of cellulose acetate in acetone and vaporization of the solvent; it was not stable at pH higher than 7.5. Liao et al. (62) report on a stable methylcellulose coating by covalent bonding of a hydroxyl group of methylcellulose to the epoxy group of γ-glycidoxypropyltrimethoxysilane attached to the fused-silica walls. The coating was stated to be stable for 135 consecutive runs under extreme conditions (pH 2–12). The EOF remained suppressed after storage of the capillary with 0.01 M NaOH for 30 days. From this work and from that of other authors it can be concluded that hydroxylic and cationic polymeric coatings are more stable against aggressive buffers than are polyacrylamide coatings, for example. It is not yet clear if cross-linked coatings of such polymers are sufficiently stable to avoid the covalent bonding to sublayers created by silanization.

In most of the publications on covalently bonded coatings of hydrophilic and hydroxylic polymers, separations of test mixtures preferably of basic proteins with $pI > 7$ are shown that are separated at pH < 6 at high efficiency, resolution, and peak symmetry. According to the experience from HPLC, problems with the stability of the covalent SiOSi bond may be expected to arise at pH > 8, which would have a bearing on the separation of acidic proteins and NSAID for example. Further investigations and studies are necessary on the chemistry to be performed within the capillaries and on the application of such capillaries in routine analytical work. It must be ascertained whether the different polymer-coated capillaries discussed above can be reproducibly manufactured as commercial products and have sufficient long-term stability of performance in universal application with aggressive buffers in a wide pH range. Effective dynamic modification procedures that are comparable in performance with permanent coated capillaries would probably be preferred in practical analyses for their simplicity of application. They can be applied in modern instrumentation by a sequence of automated rinsing and filling steps.

In a recently published review on "Capillaries and Chemistries for Capillary Electrophoresis" (63) it was concluded that currently various types of coated capillaries are commercially available together with a specially composed buffer in a kit. The foregoing discussion of the diverse coating procedures being developed, as well as the special buffers containing many different modifiers, may well suggest that such a commercial approach to optimization of the various CE separation systems is decidedly promising.

14.8. RECENT DEVELOPMENTS IN THE DYNAMIC AND PERMANENT MODIFICATION OF CAPILLARY SURFACES

The recent progress in the field of coating of fused-silica capillaries may be illustrated by reference to some publications of the years 1996 and 1997 that were not covered in the foregoing sections. Fused silica is still the preferred tubing material for capillaries used in HPCE.

The generation and application of immobilized capillary coatings for separations in the CZE, CGE, and CEC (capillary electrochromatography) modes was achieved by either adsorption or covalent bonding or both. Surfaces in fused-silica and organic polymer capillaries were modified, as well as those of microparticles for packings used in electrochromatography. The coatings for fused-silica and organic polymer capillaries were mainly executed to suppress the EOF and especially analyte–wall interactions (particularly of basic compounds such as peptides and proteins, as well as pharmaceuticals) in CZE. In electrochromatography with microparticle-packed capillaries chemical modification of the packing surfaces is intended to create a high EOF for the mobile phase and simultaneously to avoid strong analyte–wall interaction.

14.8.1. Separation of Small Molecules: Pharmaceutical Compounds

Fillet et al. (64) have recently compared methods of dynamic coating with charged and uncharged, small and large polar additives with methods of permanent PEG and PVA coatings of fused-silica capillaries for the separation of basic drugs. A similar comparison of CZE methods for the separation of small chiral compounds has been carried out by Cunat-Walter et al. (65). Burt and coworkers (66) have separated inorganic anions by CZE with capillaries coated by an azetidinium-ion-based reactive polyamide resin using a buffer also containing triethanolamine [which can be considered an adsorbable hydroxylic surface modifier according to Fillet et al. (64)]. Burt and coworkers have also separated nine organic cations using different buffers for comparison of separations in uncoated and coated capillaries (67). Guo et al. have separated small molecules of similar structure, e.g., adrenaline (epinephrine) enantiomers (68).

14.8.2. Separation of Proteins

Elsewhere in this monograph (Chapters 19 and 20) it has been shown that the performance of CZE separations of differently charged large molecules, e.g., peptides and especially cationic proteins, can be seriously impaired by analyte–wall interactions. Hydroxylic polymer coatings of various kinds have proved to be advantageous for protein separations, probably because the hydrophobic interaction of such large molecules is also suppressed. Another method of generation of the most hydroxylic PVA layer is proposed by the

U.S. patent of Karger and Götzinger (69). According to the method described therein the PVA polymer is formed by a special process within the capillary and is then chemically bonded to the silica surface. The general application of PVA coatings was investigated by Jegle and coworkers (70). A nonbonded neutral poly(ethylene oxide) coating was applied for protein separations by Iki et al. (71) "Noncovalent" polycationic coatings of the acryloylaminoethoxyethanol type were used by Cordova et al. (72). For their work on highly efficient protein separations, Huang et al. used hydrophilic coatings (e.g., Brij 35) on C_{18} polysiloxane-bonded columns (73). Other investigations by Huang et al. deal with the separation of peptide drugs in differently coated capillaries (74) or apply hydrolytically stable cellulose-derivative coatings for CZE of peptides, proteins, and glycoconjugates (75). PVA was applied for surface modification in separations of physiologically derived proteins with CIEF by Clarke et al. (76). Epoxy resin coatings under a hydroxylic poly(ethylene glycol–propylene glycol) coating for CZE separations of proteins were described by Ren (79). Liu et al. also generated epoxy resin coatings for the CZE of proteins (78).

14.8.3. Application of Various Kinds of Permanent Polymer Coatings; Miscellaneous

Although fused silica is the favored tubing material for CE separations, several groups have investigated the hydrophilic coating of polypropylene hollow fibers (79–81). Other types of coatings were formed by sol–gel technology from aminoprop 20, ethanol, and HCl in capillaries that were previously modified by aminosilylation (68). Some basic compounds including the enantiomers of adrenaline (epinephrine) could be separated with the capillaries achieved. The work of Huang and Horvath (82) deals with generation of a fluid-impervious polymer tubing (wall thickness 3 to 5 µm) inside the fused-silica capillary. These authors characterize such capillaries as being of the tube-in-tube type. They state that such a thick wall coating will protect the silica surface from the effects of a strongly basic buffer that would destroy the silica structure and disrupt chemical bonding to the polymeric coating layer. The capillaries achieved were applied to protein separations. For the separation of oligonucleotides by capillary polymer sieving electrophoresis, acryloylaminoethoxyethanol-coated fused-silica capillaries have been utilized (83). Investigations on the adsorption of proteins on fused-silica surfaces were performed by Bonvent et al. using atomic force microscopy (84). Preisler and Yeung have characterized nonbonded poly(ethylene oxide) coatings via continuous monitoring of EOF (85).

LIST OF ACRONYMS AND ABBREVIATIONS

Acronym or Abbreviation	Definition
BGE	background electrolyte
γ-CD	γ-cyclodextrin
CE	capillary electrophoresis
CEC	capillary electrochromatography
CGE	capillary gel electrophoresis
CIEF	capillary isoelectric focusing
CTAB	cetyltrimethylammonium bromide
CZE	capillary zone electrophoresis
DEBA	diglycidyl ether of bisphenol A
DETA	diethylenetriamine
DMSO	dimethyl sulfoxide
DTAB	dodecyltrimethylammonium bromide
EKC	electrokinetic chromatography
EOF	electroosmotic flow
GC	gas chromatography
HACs	hydroxyalkyl celluloses
HEC	hydroxyethyl cellulose
HMDS	hexamethyldisilazane
HPC	hydroxypropyl cellulose
HPLC	high-performance liquid chromatography
HPMC	hydroxypropyl methylcellulose
ITP	isotachophoresis
MEKC	micellar electrokinetic chromatography
MS	mass spectrometry
NSAID	non-steroidal anti-inflammatory drugs
PAA	polyacrylamide
PEG	poly(ethylene glycol)
PEI	polyethyleneimine
PGA	poly(galacturonic acid)
PHOS	phosphate (buffers)
PHOSAM	phosphates + ethylenediamine (buffers)
PMS	polymethylsiloxane OV1
PTFE	polytetrafluoroethylene (or Teflon)
PVA	poly(vinyl alcohol)
SDS	sodium dodecyl sulfate
TEMED	N,N,N',N',-tetramethylenediamine
Tris	tris(hydroxymethyl)aminomethane
TTAB	tetradecyltrimethylammonium bromide
UV	ultraviolet

REFERENCES

1. S. Hjertén, *J. Chromatogr.* **347**, 191 (1985).
2. M. H. Kleemiss, Ph. D. Thesis, p. 74. University of Marburg, Germany (1993).
3. M. H. Kleemiss, M. Gilges, and G. Schomburg, *Electrophoresis* **14**, 515 (1993).
4. G. Schomburg, in *Capillary Electrophoresis Technology* (N. A. Guzman, ed.), p. 311. Dekker, New York (1993).
5. G. Schomburg, D. Belder, M. Gilges, M. H. Kleemiss, J. A. Lux, and St. R. Motsch, *GIT Spez. Chromatogr.* **1**, 7–17 (1993).
6. M. R. Schure and A. M. Lenhoff, *Anal. Chem.* **65**, 3024 (1993).
7. P. Coufal, K. Stulik, H. A. Claessens, and C. Cramers, *J. High Resolut. Chromatogr.* **17**, 325 (1994).
8. B. L. Karger, A. S. Cohen, D. N. Heiger, and K. Ganzler, *High Perform. Capillary Electrophor.*, 3rd, San Diego (1991).
9. H. Lauer and D. McManigill, *Anal. Chem.* **587**, 166 (1986).
10. T. Tsuda, *HRC & CC, J. High Resolut. Chromatogr., Chromatogr. Commun.* **10**, 622 (1987).
11. W. R. Jones and P. Jandik, *Am. Lab.* **22**, 51 (1990).
12. W. R. Jones and P. Jandik, *J. Chromatogr.* **546**, 445 (1991).
13. P. Jandik and G. Bonn, *Capillary Electrophoresis of Small Molecules and Ions*. VCH Publishers, Weinheim, 1993.
14. A. Weston, P. R. Brown, P. Jandik, W. R. Jones, and A. L. Heckenberg, *J. Chromatogr.* **593**, 289 (1992).
15. W. Beck and H. Engelhardt, *Chromatographia* **33**, 313 (1991).
16. St. R. Motsch, Ph. D. Thesis, University of Saarbrücken, Germany (1994).
17. H. Lindner, W. Heiliger, A. Dirschlmayer, M. Jaquemar, and B. Puschendorf, *Biochem. J.* **238**, 467 (1992).
18. M. Gilges, H. Husmann, M.-H. Kleemiss, St. R. Motsch, and G. Schomburg, *J. High Resolut. Chromatogr.* **15**, 452 (1992).
19. M. Gilges, M.-H. Kleemiss and G. Schomburg, *Anal. Chem.* **66**, 2038 (1994).
20. G. Schomburg, D. Belder, M. Gilges, and St. R. Motsch, *J. Capillary Electrophor.* **1**, 219 (1994).
21. D. Belder and G. Schomburg, *J. High Resolut. Chromator.* **15**, 686 (1992).
22. D. Belder and G. Schomburg, *J. Chromatogr. A* **666**, 351 (1994).
23. J. Snopek, H. Soini, M. Novotny, E. Smolkova-Keulemansova, and I. Jelinek, *J. Chromatogr.* **559**, 215 (1991).
24. M. W. F. Nielen, *Anal. Chem.* **65**, 885 (1993).
25. J. E. Wiktorowicz and J. C. Colburn, *Electrophoresis* **11**, 769 (1990).
26. J. E. Wiktorowicz, U. S. Pat. 5,015,350 (1991).

27. J. A. Bullock and L.-C. Yuan, *J. Microcolumn Sep.* **3**, 241 (1991).
28. A. Emmer, M. Hanson, and J. Roeraade, *J. High Resolut. Chromatogr.* **14**, 738 (1991).
29. S. A. Swedberg, *Anal. Biochem.* **185**, 51 (1990).
30. J. K. Towns and F. E. Regnier, *Anal. Chem.* **63**, 1126 (1991).
31. J. K. Towns and F. E. Regnier, *Anal. Chem.* **64**, 2473 (1992).
32. B. T. Ingram and R. H. Ottwil, in *Cationic Surfactants: Physical Chemistry* (D. N. Rubing and P. M. Holland, eds.). New York, Dekker, 1991.
33. F. M. Everaerts, J. L. Beckers, and Th. P. E. M. Verheggen, *Isotachophoresis: Theory, Instrumentation and Practice*, J. Chromatogr. Lib. Vol. 6. Elsevier, Amsterdam, 1976.
34. K. D. Altria and C. F. Simpson, *Anal. Proc.* **25**, 85 (1988).
35. H. Huang, J. A. Luckey, M. J. Gordon, and R. N. Zare, *Anal. Chem.* **61**, 766 (1989).
36. R. Kerr and L. Jung, Fr. Pat. PCT/FR91/00734 (1990).
37. J. St. Green, Ph. D. Thesis, University of North Carolina, Chapel Hill (1986).
38. J. K. Towns and F. E. Regnier, *J. Chromatogr.* **516**, 69 (1990).
39. F. B. Erim, A. Cifuentes, H. Poppe, and J. C. Kraak, *J. Chromatogr. A* **708**, 356 (1995).
40. D. Belder, Thesis, University of Marburg, Germany (1994).
41. F. Kobor, D. Belder, and G. Schomburg, *Int. Symp. Column Liq. Chromatogr.*, 19th, Innsbruck, Poster (1995).
42. A. Malik, Z. Zhao, and M. L. Lee, *J. Microcolumn Sep.* **5**, 119 (1993).
43. Z. Zhao, A. Malik, and M. L. Lee, *Anal. Chem.* **65**, 2747 (1993).
44. Innophase Corp., *Thermoset Polymer List*. Westbrook, CT (1995).
45. J. A. Lux, J. A., H. Yin, and G. Schomburg, *J. High Resolut. Cromatogr.* **13**, 145 (1990).
46. Wu, Q, H. A. Classen, and C. A. Cramers, *Chromatographia* **33**, 303 (1992).
47. F. E. Regnier and D. Wu, in *Capillary Electrophoresis* (N. A. Guzman, ed.). Chromatogr. Sci. Ser., Vol. 64, p. 287. Dekker, New York and Basel, 1993.
48. D. Heiger, Ph. D. Thesis, Northeastern University, Boston (1995).
49. A. M. Dougherty and M. R. Schure, in *Capillary Electrophoresis Technology* (N. A. Guzman, Chromatogr. Sci. Ser., Vol. 64, Part III, No. 12, p. 435. Dekker, New York, 1993.
50. K. A. Cobb, V. Dolnik, and M. Novotny, *Anal. Chem.* **62**, 2478 (1990).
51. J. M. Van Alstine, N. L. Burns, J. A. Riggs, K. Holberg, and J. M. Harris, *Colloids Surf. A: Phys. Eng. Aspects* **77**, 149 (1993).
52. B. J. Herren, S. G. Shafer, J. Van Alstine, J. M. Harris, and R. S. Snyder, *J. Colloid Interface Sci.* **115**, 46 (1986).
53. J. M. Van Alstine, J. M. Harris and S. G. Shafer, U. S. Pat. 4,690, 749 (1987).

54. G. J. M. Bruin, R. Huisden, J. C. Kraak, and H. Poppe, *J. Chromatogr.* **480**, 339 (1989).
55. G. J. M. Bruin, J. P. Chang, R. H. Kuhlmann, K. Zegers, J. C. Kraak, and H. Poppe, *J. Chromatogr.* **471**, 429 (1989).
56. B. Götzinger, *Symp. High Perform. Liq. Chromatogr.*, 15th, Würzburg, Poster (1995).
57. Y. Liu, R. N. Fu, and J. L. Gu, *J. Chromatogr.* **694**(2) 498 (1995).
58. Y. Merchref and Z. El Rassi, *Electrophoresis* **16**, 617 (1995).
59. J. T. Smith and Z. El Rassi, *Electrophoresis* **14**, 396 (1993).
60. P. Sun, A. Landmann, and R. A. Hartwick, *J. Microcolumn Sep.* **6**, 403 (1994).
61. M. H. A. Busch, J. C. Kraak, and H. Poppe, *J. Chromatogr.* **695**(2), 187 (1995).
62. J. L. Liao, J. Abramson, and S. Hjertén, *J. Capillary Electrophor.* **2**, 191 (1995).
63. D. N. Heiger and R. E. Majors, *LC–GC* **13**, 12 (1995).
64. M. Fillet, Ph. Hubert, G. Schomburg, and J. Crommen, *J. Chromatogr. A* To be published.
65. M.-A. Cunat-Walter, E. Bossu and H. Engelhardt, *J. Capillary Electrophor.* 3,275 (1997).
66. H. Burt, D. M. Lewis, and K. N. Tapley, *J. Chromatogr. A* **739**, 367 (1996).
67. H. Burt, D. M. Lewis, and K. N. Tapley, *J. Chromatogr. A* **736**, 265 (1996).
68. Y. Guo, G. A. Imahori, and L. A. Colon, *J. Chromatogr. A* **744**, 17 (1996).
69. B. L. Karger and W. Götzinger, PCT Int. Apl. WO 96; 23,220; U. S. Pat. 379, 834.
70. U. Jegle, R. Grimm, H. Godel, R. Schuster, G. Ross, and T. Soga; *LC-GC-Int.* **10**, 176–188 (1997).
71. N. Iki and E. S. Yeung, *J. Chromatogr. A* **731**, 273 (1996).
72. E. Cordova, J. Gao, and G. M. Whitesides, *Anal. Chem.* **69**, 1370 (1997).
73. M. Huang, M. Bigelow, and M. Byers, *LC-GC-Int.* **9**, 658 (1996).
74. M. Huang, D. Mitchell, and M. Bigelow, *J. Chromatogr. B* **677**, 77 (1996).
75. M. Huang, J. Plocek, and M. V. Novotny, *Electrophoresis (Weinheim)* **16**, 396 (1995).
76. N. J. Clarke, A. J. Tomlinson, G. Schomburg, and St. Naylor, *Anal. Chem.* To be published.
77. X. Ren, Y. Shen, and M. L. Lee, *J. Chromatogr. A* **741**, 115 (1996).
78. Y. Liu, R. N. Fu, and J. L. Gu, *J. Chromatogr. A* **723**, 157 (1996).
79. X. Ren, P. Z. Liu, A. Malik, and M. L. Lee, *J. Microcol. Sep.* **8**, 535 (1996).
80. A. Fridstrom, N. Lundell, B. Ekstroem, and K. E. Markides, *J. Microcol. Sep.* **9**, 1 (1997).

81. X. Ren, P. Z. Liu, and M. L. Lee, *J. Microcol. Sep.* **8**, 529 (1996).
82. X. Huang and Cs. Horvath, *Anal. Chem.* In press.
83. K. W. Talmadge, M. Zhu, L. Olech, and C. Siebert, *J. Chromatogr. A* **744**, 347 (1996).
84. J. J. Bonvent, R. Barberi, R. Bartolino, L. Capelli and P. G. Righetti, *J. Chromatogr. A* **756**, 233 (1996).
85. J. Preisler and E. S. Yeung, *Anal. Chem.* **68**, 2885 (1996).

CHAPTER

15

NONAQUEOUS CAPILLARY ELECTROPHORESIS

JOSEPH L. MILLER and MORTEZA G. KHALEDI

Department of Chemistry, North Carolina State University, Raleigh, North Carolina 27695

15.1.	Introduction	525
15.2.	Influence of Nonaqueous Solvents in Capillary Electrophoresis	527
	15.2.1. Electroosmotic Flow	527
	15.2.2. Selectivity	533
	15.2.3. Compatibility with Special Detection Techniques	538
	15.2.4. Other Considerations	539
15.3.	Applications	540
	15.3.1. Separation of Charged Solutes	540
	15.3.2. Separation of Uncharged Solutes	547
	15.3.3. Separation of Chiral Solutes	550
15.4.	Conclusions and Future Trends	553
List of Acronyms and Abbreviations		553
References		554

15.1. INTRODUCTION

Since its introduction in 1981, capillary electrophoresis (CE) has quickly become a method of providing high-efficiency separations for a wide variety of solutes. As one might expect, the type of medium employed in these separations plays an important role, and in the vast majority of cases the use of a water-based medium has yielded sufficient results. However, there are cases in which such aqueous environments are unsuitable, and as a result the use of organic solvents has recently become more commonplace in CE separations. This chapter summarizes the effects of nonaqueous solvents in CE separations and reviews the use of nonaqueous CE methods in separating both charged and uncharged species.

High Performance Capillary Electrophoresis, edited by Morteza G. Khaledi. Chemical Analysis Series, Vol. 146.
ISBN 0-471-14851-2 © 1998 John Wiley & Sons, Inc.

The popularity of water as a CE solvent stems from its unique characteristics, including low cost, low viscosity, nonvolatility, safety, and availability. Some of the other favorable characteristics of water include its well-known acid–base chemistry, its compatibility with various detection schemes, and its ability to solvate a variety of electrolytes, thus providing a high-conductivity medium. In spite of these advantages, however, the application of aqueous CE is limited to compounds that are somewhat polar. The extension of CE to nonpolar solutes necessitates the addition of modifiers such as micelles, urea, or organic cosolvents. These modifiers alter the bulk properties of the solvent and influence the solvation of the analytes, thus affecting the susceptibility of the analytes to electrophoretic migration. The addition of organic cosolvents, because of their wide range of properties, is especially advantageous when the solutes show similar electrophoretic behavior in aqueous media. Therefore, in the separation of nonpolar solutes, one may maximize the advantages of using an organic cosolvent by preparing a medium that is totally (or at least predominantly) nonaqueous.

Prior to the development of CE, several reports involving the use of nonaqueous media in classical electrophoresis were published. The first of these reports dates to 1951, when Hayek separated carbon black particles suspended in kerosene (1). Since then, nonaqueous electrophoresis has been employed in the separation of many other classes of analytes including weak acids and bases (2,3), lubricating oil additives (4–6), metal ions (7–10), anions (11,12), mineral acids (13), and biological compounds such as cholesterol, steroids, and fatty acids (14). Korchemnaya et al. published an excellent review of classical electrophoretic methods using nonaqueous media (15).

Most of the reports involving nonaqueous capillary electrophoresis (NACE) methods, on the other hand, have only recently appeared. Early reports dealing with NACE methods were published in the mid-1980s by Walbroehl and Jorgenson (16,17), but not until some 10 years later did the first comprehensive reports of NACE appear. Sahota and Khaledi demonstrated the usefulness of certain amides such as formamide and N-methylformamide as CE solvents. They compared NACE to aqueous CE by considering a number of key factors such as efficiency, analysis time, and selectivity (18). Tomlinson and coworkers, as well as Ng et al., reported the first applications of NACE in the separation of hydrophobic drugs and metabolites using pure and mixed methanolic solvents (19–21). These and other reports that have been published on the subject over the past ten years are summarized in this chapter. While some papers on organic-rich mixed solvents are included, the reports on hydro-organic (< 50% organic) media are not covered.

15.2. INFLUENCE OF NONAQUEOUS SOLVENTS IN CAPILLARY ELECTROPHORESIS

The wide variety of physical and chemical properties that characterize nonaqueous solvents allows one to optimize a number of operating parameters in CE. Among the parameters that are ultimately affected by the choice of solvent are the analysis time, selectivity, efficiency, and detectability. In the following subsections, the relationships between solvent properties and the CE conditions that may produce an optimized separation are discussed.

15.2.1. Electroosmotic Flow

The Smoluchowski equation states:

$$\mu_{eo} = \varepsilon_0 \varepsilon \zeta / 4\pi \eta \qquad (15.1)$$

where μ_{eo} is the electroosmotic mobility; ε_0 is the permittivity of free space; ε is the dielectric constant; ζ is the zeta potential; and η is the viscosity (22). Therefore, the ability of a solvent to generate electroosmotic flow is proportional to $\varepsilon \zeta / \eta$. To assess the practicality of using a given solvent in a CE application, one may often compare the ε/η ratios of solvents by assuming that ζ (which is difficult to measure accurately) remains constant. Table 15.1 lists the ε, η, and ε/η values of several NACE solvents.

Table 15.1. Physical Properties of Some Candidate Nonaqueous Solvents for CE

Solvent	η (cP, 25 °C)	ε	ε/η	bp (°C)	Polarity	pK_{auto}
Water	0.89	80	90	100	10.2	14.00
Methanol (MeOH)	0.54	32.7	61	65	5.1	17.2
Acetonitrile (ACN)	0.34	37.5	110	82	5.8	>33.3
Formamide (FA)	3.3	111	34	210	9.6	16.8
N-Methylformamide (NMF)	1.65	182	110	182	6.0	10.74
N,N-Dimethylformamide (DMF)	0.80	36.7	46	153	6.4	29.4
N,N-Dimethylacetamide (DMA)	0.78	37.8	48	166	6.5	23.95
Dimethylsulfoxide (DMSO)	1.98	46.7	24	189	7.2	33.3
Acetic acid	1.1	6.2	6	118	6.0	—
Propylene carbonate	—	64.9	—	242	5.0	—

Source: Data from Sahota and Khaledi (18) and Hansen et al. (28), and references cited therein.

For pure solvents (and some solvent mixtures), the assumption that μ_{eo} is directly proportional to ε/η appears to be somewhat valid. Several studies have shown that the choice of solvent may have a profound effect on μ_{eo} and may ultimately affect the analysis time, efficiency, and even detectability. Sahota and Khaledi, for example, showed that with formamide (FA, a solvent with a lower ε/η value than water), a higher concentration of electrolyte could be used, thereby increasing the ionic strength of the solution (18). This, in turn, increased the sample capacity (thus improving detectability), and it created a greater ionic strength difference between the buffer and sample zone, which resulted in sample stacking and ultimately higher efficiencies (Figure 15.1). Conversely, the amount of current generated using FA was lower than in water for a given ionic strength (in accordance with Ohm's law). Finally, it was found that an electric field strength of 90 kV/m in FA generated the same amount of current as 22.5 kV/m in an aqueous buffer with an identical electrolyte composition. This had the effect of shortening the analysis times (in spite of

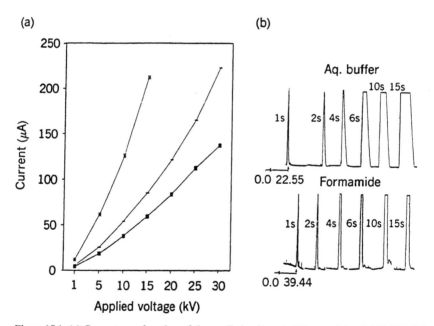

Figure 15.1. (a) Current as a function of the applied voltage in FA containing 0.1 M KH_2PO_4 (solid squares), H_2O containing 0.1 M KH_2PO_4 (asterisks), and H_2O containing 0.04 M KH_2PO_4 (dashes). (b) Peak widths of a 2.5 mg/mL Leu-Trp dipeptide sample in aqueous buffer and FA. The injection time in seconds is noted. [Reproduced from R. S. Sahota and M. G. Khaledi, *Anal. Chem.* **66**, 1141 (1994), with permission.]

FA's higher viscosity). Figure 15.1a shows the Ohm's law plots for FA and aqueous systems. In an aqueous buffer, deviations from linearity are observed at higher field strengths, which indicate that Joule heating has become significant. For the FA system, however, this relationship is more linear. Because of the overall lower currents that are produced, one may apply much higher field strengths in the FA system without encountering the problem of Joule heating (18).

Other studies involving the effects of nonaqueous solvents on electroosmosis have produced similar results. Tomlinson and coworkers found that increasing the methanol content of a 20 mM ammonium acetate/1% acetic acid buffer from 30 to 100% resulted in slower electroosmosis and improved resolution, as might be expected based on MeOH's lower ε/η value. However, they also found that increasing the MeOH content of the medium enhanced the solubility of the analytes, which gave rise to greater detectability and sample recovery (20,21). Ng and coworkers discovered that the use of ammonium acetate in MeOH provided complete separation of several drug compounds, but broad peaks and long analysis times were observed (19). However, the addition of up to 50% acetonitrile (ACN) shortened the analysis time (as would be expected from ACN's lower viscosity) while maintaining full resolution of all components (19). Tjørnelund and Hansen found that the addition of a small amount (6%) of dimethylformamide (DMF) to a mixture of methanol (MeOH) and ACN decreased μ_{eo}, which led to an increase in resolution (Figure 15.2) (23). Also increasing the MeOH/ACN ratio (while keeping the electrolyte concentration constant) increased μ_{eo} as well as efficiency, whereas decreasing the ratio resulted in lower efficiencies but higher selectivity (23). In our laboratory, Ye and Khaledi discovered that the addition of a large concentration (300 mM) of β-cyclodextrin to N-methylformamide (NMF) increased the viscosity to the point where μ_{eo} was sufficiently lowered to provide the full separation of several basic solutes (24).

The foregoing examples show that altering the ε/η ratio may affect analysis time, resolution, and detectability. There are some cases, however, where the ζ potential of the capillary wall must also be taken into account. Salimi-Moosavi and Cassidy, for instance, showed that the addition of either MeOH or DMF to an aqueous solution caused μ_{eo} to decrease, except at organic concentrations above 75%, in which case μ_{eo} increased (see Figure 15.3) (25). The occurrence of a minimum value for μ_{eo} was attributed to the changes in the ε/η ratio, which reached a minimum in the 60–80% organic range. However, changes in the ζ potential may have also contributed, as this quantity steadily decreased upon addition of organic solvent. This occurred due to the increase in the pK_a (and thus, the degree of dissociation) of the silanol groups that composed the capillary wall (25). In a subsequent work, Salimi-Moosavi and Cassidy found that μ_{eo} in an ACN/MeOH system increased as the ACN

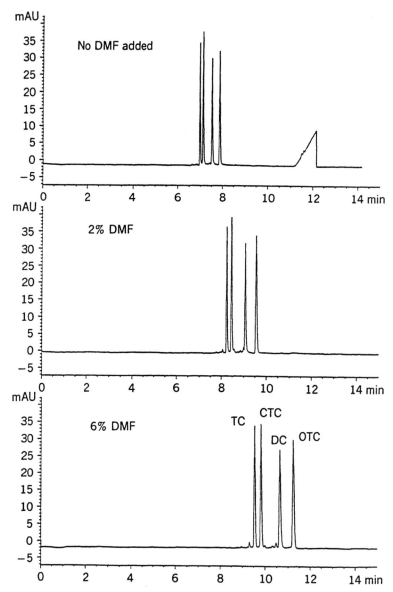

Figure 15.2. Electropherograms of a mixture of tetracycline (TC) and some related antibiotics (CTC = chlorotetracycline; DC = doxycycline; OTC = oxytetracycline); each in a concentration of 0.2 mg/mL using various mixtures of solvents. Electrophoresis medium was MeOH/ACN (48:52, v/v), with the solvent as indicated on the electropherograms and 25 mM ammonium acetate, 10 mM citric acid; and 118 mM methanesulfonic acid added. [Reproduced from J. Tjørnelund and S. H. Hansen, *J. Chromatotgr. A* **737**, 291 (1996), with permission.]

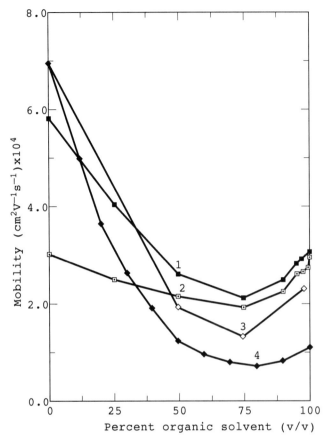

Figure 15.3. Electroosmotic mobility as a function of organic solvent: (1) separation electrolyte, 0.05 mol/L TEAP in DMF; (2) separation electrolyte, 0.10 mol/L TEAP and 0.1 mol/L n-BuNH$_2$ in DMF; (3) separation electrolyte, 0.01 mol/L KHP 0.02 mol/L n-BuNH$_2$ and 2% (v/v) H$_2$O in MeOH; (4) calculated values for MeOH. Electrolyte concentrations were constant for all H$_2$O/solvent mixtures. [Reproduced from H. Salimi-Moosavi and R. M. Cassidy, *Anal. Chem.* **67**, 1067 (1995), with permission.]

content was increased to 50%, due to the lowering of viscosity. However, μ_{eo} decreased as the percentage of ACN was further increased to 75%, most likely as a result of the increasing contribution of the ζ potential, which decreased upon addition of ACN (26).

In other studies, Hansen and coworkers discovered that μ_{eo} decreased from solvent to solvent as follows: NMF > DMF > DMA > DMSO (Figure 15.4). The viscosity and dielectric constant appeared to be the main contributing factors, but changes in the ζ potential may have played a role as well

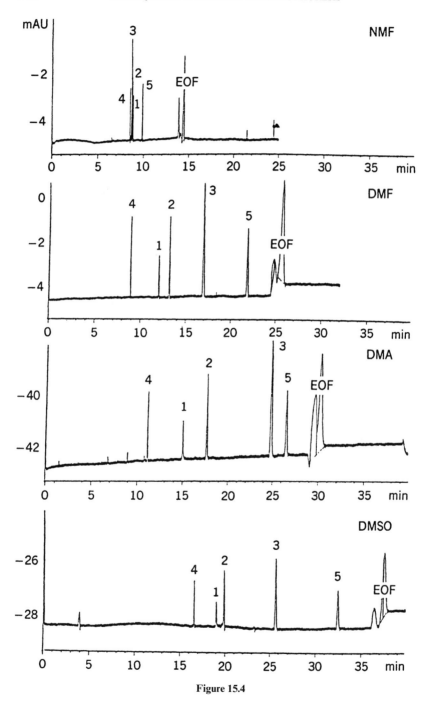

Figure 15.4

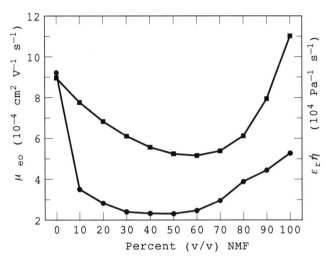

Figure 15.5. The electroosmotic mobility (●) and the relative dielectric constant/viscosity ratio (■) as a function of the percentage (v/v) of NMF. Conditions: varying amounts of pure NMF with 50 mM H_3BO_3 added, mixed with 50 mM H_3BO_3 aqueous buffer (pH 9.15). [Reproduced from M. Jansson and J. Roeraade, *Chromatographia* **40**, 163 (1995), with permission.]

(27,28). Figure 15.5 shows the calculated ε/η ratios and the measured μ_{eo} values as a function of the NMF volume fraction in water, as reported by Jansson and Roeraade (29). They found that the two curves reached minima in the 40–60% concentration region and attributed the differences in shape between the two curves to changes in the ζ potential (29).

15.2.2. Selectivity

Like electroosmosis, the electrophoretic mobility (μ_{ep}) of a solute is affected by the viscosity and dielectric constant of the solvent (29), according to the following equation:

$$\mu_{ep} = 2\varepsilon_0 \varepsilon \zeta_{ion} / 3\eta \tag{15.2}$$

Here ζ_{ion} represents the ζ potential of the ion, which is influenced by another solvent property: the autoprotolysis (or autoionization) constant. This

---

Figure 15.4. Separation of five test solutes (imipramine derivatives) using different organic solvents. Electrophoresis medium: 25 mM ammonium acetate/1 M acetic acid in the organic solvent. (EOF = electroosmotic flow.) [Reproduced from S. H. Hansen, J. Tjørnelund, and I. Bjørnsdottir, *Trends Anal. Chem.* **15**, 175 (1996), with permission.]

property in part determines the extent to which a free ion exists in solution, as opposed to remaining in an undissociated state. A separation, thus, may be optimized by selecting a solvent that can differentiate between solutes based on their degrees of ionization. Because organic solvents often differ from one another in autoprotolytic behavior, the choice of an appropriate solvent may yield separations that are not possible with aqueous systems.

It is well known that a change in pK_a occurs as an acid or base is transferred from water to another solvent. As explained by Kenndler in Chapter 2 of this volume, the change in pK_a as HA is transferred from water (W) to solvent S is

$$\Delta pK_{a,HA} = (pK_{a,HA})^S - (pK_{a,HA})^W = \log[(_m\gamma_{H^+})(_m\gamma_{A^-})/(_m\gamma_{HA})] \quad (15.3)$$

where $_m\gamma$ denotes the transfer activity coefficient of a particular species (based on medium effects derived from thermodynamic relationships). Likewise, for a neutral base:

$$\Delta pK_{a,BH^+} = (pK_{a,BH^+})^S - (pK_{a,BH^+})^W = \log[(_m\gamma_{H^+})(_m\gamma_B)/(_m\gamma_{BH^+})] \quad (15.4)$$

as BH^+ is transferred from water to solvent S. Since the transfer activity coefficients for the neutral species (HA and B) are essentially unity, the change in pK_a for HA becomes

$$\Delta pK_{a,HA} \approx \log[(_m\gamma_{H^+})(_m\gamma_{A^-})] \quad (15.5)$$

which contains the product of transfer activity coefficients for two ions of opposite charge (H^+ and A^-). On the other hand, the change in pK_a for B reduces to

$$\Delta pK_{a,BH^+} \approx \log[(_m\gamma_{H^+})/(_m\gamma_{BH^+})] \quad (15.6)$$

which contains a ratio of two terms involving ions of like charge (H^+ and BH^+). Because of the general inability of organic solvents to stabilize anions (contrasted to water's ability to stabilize both cations and anions), one might expect the change in pK_a to be greater for neutral acids than for neutral bases. This is indeed observed, as Kenndler's Table 2.1 (in this volume) shows. The increase in pK_a as several neutral acids are transferred from water to organic solvents shows the relative inability of these solvents to stabilize the dissociated species. On the other hand, the change in pK_a for the anilinium ion is much smaller as this species is transferred to an organic medium (30).

On a molecular level, the stabilities of dissociated acids and bases in a given solvent may be traced to the mechanism through which dissociation takes place. For an acidic solute, the transfer of a proton to the solvent occurs in two steps:

$$S + HA^{n+} \rightleftharpoons SH^{+}A^{(n-1)+} \rightleftharpoons SH^{+} + A^{(n-1)+} \quad (15.7)$$

where S is the solvent and HA^{n+} is the undissociated acid. The transfer of a proton to a base from the solvent proceeds through a similar sequence of reactions:

$$S + B^{n+} \rightleftharpoons S^{-}HB^{(n+1)+} \rightleftharpoons S^{-} + HB^{(n+1)+} \quad (15.8)$$

where SH is the solvent and B^{n+} is the base. The first step in each reaction involves the formation of a solvent–solute ion pair and is influenced by the solvent's ability to accept or donate a proton. On the other hand, the degree to which the second (dissociation) step of Eq. (15.7) or (15.8) goes to completion depends on the solvent's ability to solvate the separated species (especially in cases where the species bear opposite charges). The dielectric constant and the hydrogen-bonding ability of the solvent appear to be the most important factors in this step (31).

The autoprotolysis constant (K_{auto}) is influenced by the stability of the ionic species in the solvent. Thus, one would expect the dissociation of any uncharged acid to be related to the magnitude of this constant. A comparison of Tables 15.1 and 2.1 (30) in fact, shows a direct (albeit nonlinear) relationship between a solvent's pK_{auto} and the pK_a of an acidic species in that solvent. Of the solvents listed in Table 2.1, ACN (because of its relative inability to solvate ionic solutes) possesses the highest pK_{auto} and gives rise to pK_a values that are greater than in any other solvent.

In CE separations, the change in pK_a as a solute is transferred to a nonaqueous solvent is reflected by a change in its electrophoretic mobility as well. For the conjugate base of a neutral acidic solute, the effective mobility is described by the following equation:

$$\mu_{eff,A^-} = \mu_{act,A^-}\alpha = \mu_{act,A^-}[K_a/([H^+] + K_a)] \quad (15.9)$$

where μ_{act,A^-} is the mobility of A^- (upon complete dissociation of HA); α is the fraction of A^- present at a given acidity; and K_a is the dissociation constant of HA. For the conjugate acid of a neutral base, a similar relationship applies:

$$\mu_{eff,BH^+} = \mu_{act,BH^+}\alpha = \mu_{act,BH^+}[[H^+]/([H^+] + K_a)] \quad (15.10)$$

In CE separations, the significance of Eq. (15.9) and (15.10) has been demonstrated, and many researchers have found that the transfer of ionizable solutes from one solvent to another results in nonuniform changes in the solutes' pK_a values. Therefore, differences in relative mobility (i.e., selectivity) are generally observed in NACE separations. Bjørnsdottir and Hansen, for example, found that the separation of amines in nonaqueous media yielded relative migration times that corresponded with the apparent pK_a's of the solutes in the chosen solvent. The apparent pK_a's depended on the solvent's hydrogen-bonding characteristics. In MeOH, DMF, DMA, and DMSO—solvents that are capable of acting as either hydrogen-bond donors or acceptors—the primary amine eluted first, followed by the secondary and tertiary amines. However, in ACN and NMF (which behave predominantly as hydrogen-bond acceptors), the elution order was reversed (27). In another study, Thormann and coworkers (using a 60:40 organic/aqueous system) learned that the pK_a values for organic bases varied with the type of organic solvent used (Figure 15.6) (32). When the pH of the buffer was adjusted to match the pK_a's of the solutes in a particular solvent mixture, improved separations were obtained (32). Kenndler and Chiari achieved similar results as they successfully separated six acids using MeOH as the solvent (33). The same six solutes could not be separated in aqueous media, however, due mainly to the ability of MeOH to "differentiate" between solutes based on their different pK_a values. Water, on the other hand, exhibited a "leveling" effect; that is, the pK_a's of the solutes were "leveled" to similar values and adequate separations were not obtained (33).

Not all NACE separations of acidic solutes involve deprotonation. Occasionally, an undissociated weak acid (RH) may combine with the conjugate base of a strong acid (A^-) to form a "heteroconjugate"

$$RH + A^- \rightleftharpoons RH \cdots A^- \qquad (15.11)$$

rather than transferring its proton to A^-. Such formation, which depends on the hydrogen-bonding capabilities of the weak acid, can occur only in an aprotic solvent (34). The possibility of using heteroconjugation in NACE applications was recently investigated by Okada, who separated several benzoic acids and phenols in ACN through the formation of heteroconjugate anions with the perchlorate ion (35).

In the preparation of a nonaqueous buffer, one should realize that the pH (and thus the pK_a) cannot be accurately determined through the potentio-

---

Figure 15.6. pH dependence of the effective electrophoretic mobility in binary system with 0.04 M phosphate containing (a) 60% (v/v) 1-propanol and (b) 60% (v/v) THF (tetrahydrofuran). Key: (1) ketoconazole; (2) dextromethorphan; (3) methadone; (4) desethylamiodarone; (5) amiodarone. [Reproduced from C. X. Zhang, F. von Heeren, and W. Thormann, *Anal. Chem.* **67**, 2070 (1955), with permission.]

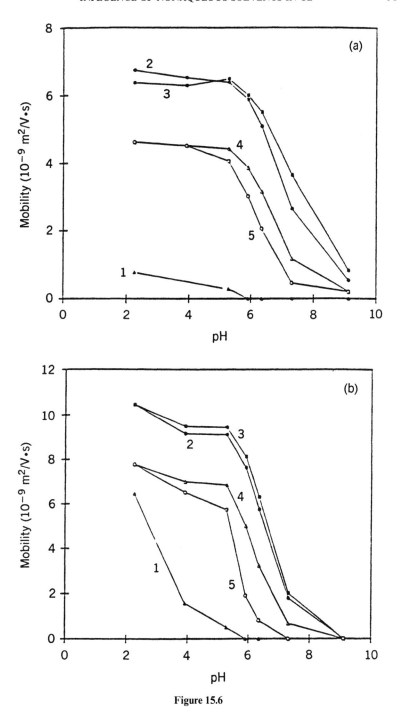

Figure 15.6

metric measurement using a standard pH-meter. Because the liquid junction potential that results from the presence of two different solvents at the glass-calomel electrode is not taken into account, the pH that is measured (commonly referred to as the "apparent pH," or pH*) differs from the actual (thermodynamic) pH. Similarly, the apparent pK_a (or pK_a^*) will also be different from the actual pK_a. This need not pose a difficulty, however, considering that the differences in solute mobility are based on the *relative* changes that occur in their pK_a values and that these changes are usually reflected in their apparent pK_a's. Ye and Khaledi, for instance, used the apparent pK_a's of several acidic solutes and constructed a model that accurately predicted the relative migration behavior of these solutes in buffered NMF (24).

15.2.3. Compatibility with Special Detection Techniques

In CE, there are a number of detection methods that are either currently used or have been recently investigated. These methods have been shown to be suitable for separations performed in both aqueous and nonaqueous media.

15.2.3.1. Ultraviolet–Visible Absorbance Detection

By far, the most commonly used detection method in CE is ultraviolet (UV) or visible absorbance. In most applications, direct UV absorbance has been the method of choice since the separated solutes usually absorb in the UV region at wavelengths above the cutoff for the medium. For analytes that have low-UV absorbances, however, indirect UV detection in NACE may be more applicable. This detection technique was investigated by Salimi-Moosavi and Cassidy, who separated and detected a series of monovalent anions using chromate, phthalate, and benzoate electrolytes in MeOH. Detection limits were as low as 3×10^{-5} M, but sensitivity plots for peak areas showed that response factors in the 8×10^{-4} to 4×10^{-5} M range varied widely among solutes (25).

15.2.3.2. Mass Spectrometry

A CE instrument may be interfaced with a mass spectrometer (MS) through an electrospray-ionization (ESI) interface. The use of MS detection in NACE was explored by Tomlinson and coworkers, who found that a nonaqueous medium (MeOH) was most suitable for this technique. Such a medium produced CE currents that were comparable to that of the ESI interface; this provided the most stable CE–MS performance and improved MS sensitivity. In this study, the researchers successfully obtained mass-spectral data following the complete separation of the H_2 antagonist mifentidine and seven standards-putative metabolites in 5 mM ammonium acetate and 100 mM acetic acid in MeOH. The data obtained from the CE–MS in this study proved to be valuable, as the authors were able to postulate the mechanism through which mifentidine is metabolized (20,21).

15.2.3.3. *Electrochemical Detection*

The possibility of using electrochemical detection in NACE applications occurs as a result of the wide working potential ranges that are possible with nonaqueous solvents. Salimi-Moosavi and Cassidy explored the use of electrochemical detection in DMF using a platinum ultramicroelectrode. They found that the linear ranges of this detection method were limited, even though low limits of detection (1×10^{-9} to 6×10^{-8} M) were measured (25).

15.2.3.4. *Laser-Induced Fluorescence Detection*

Although numerous reports have been published regarding the use of laser-induced fluorescence (LIF) detection in CE applications, very little research has been accomplished using this detection method in nonaqueous media. A common problem one encounters using LIF detection in aqueous CE is the signal quenching that occurs as the excited analyte molecules collide with the solvent molecules. The use of a solvent that is more viscous and/or less polar than water (such as DMSO, FA, NMF, DMF, or 2-propanol) would likely reduce the frequency of these collisions. Thus, in these solvents, quenching would be suppressed, which would result in enhanced fluorescence signals. A preliminary study, for example, showed the limit of detection of derivatized maltohexaose in FA to be 7.7×10^{-7} M, whereas in water a detection limit of 1.42×10^{-6} M was measured (36).

Soper and coworkers explored the possibility of using fluorescence lifetime determinations as an on-line detection method in CE separations. In this study, two near-IR dyes, DTTCI (diethylthiatricarbadyanine iodide, cationic) and IR-125 (anionic), were separated in a 20 mM borate (pH 9.5) buffer. However, because of the signal quenching that occurred in aqueous media, it was necessary to use a solvent system that was predominantly nonaqueous (95:5 MeOH/H_2O). As a result, this allowed the analytes to be detected in subattomole quantities (37).

15.2.4. Other Considerations

In addition to the factors already discussed, there are several others that one might consider in choosing a NACE solvent. An important concern is the solubility of the analytes and electrolytes in a given solvent. One of the advantages of using a nonaqueous solvent in CE stems from the increased solubility of many analytes, which can then be more easily detected. In choosing an electrolyte, on the other hand, one should realize that many of the buffer salts commonly used in aqueous CE have low solubilities in nonaqueous solvents. Some examples of electrolytes that researchers have found to be practical in NACE applications are mentioned below in Section 15.3 of this chapter.

The volatility of a solvent is another property that needs to be addressed prior to performing NACE separations. Depending on the design of the CE instrument, a solvent's volatility may or may not present difficulties during the course of a separation. For example, if the run buffers are not well sealed, the loss of solvent through evaporation may change the electrolyte concentration either during a run or between runs. This leads to a gradual increase in ionic strength, which results in poor reproducibility of migration behavior. The boiling points of some NACE solvents are listed in Table 15.1.

Among the other factors one may consider in selecting a solvent for NACE are the availability, purity, flammability, toxicity, and reactivity of the solvent.

15.3. APPLICATIONS

15.3.1. Separation of Charged Solutes

The electrophoretic mobility of an ion is determined by its ζ potential, which depends on the size and the charge of the ion. Most of the reported NACE separations of ions involve ionizable organic acids and bases, and the charge on the solute (as described by its degree of ionization) is the factor that normally varies the most among species. Thus, in NACE (as in aqueous CE), the selection of an appropriate buffer is crucial.

15.3.1.1. Cations

Early work in the separation of cations was performed by Walbroehl and Jorgenson, who found that nitrogenated polycyclic aromatic compounds could be separated in ACN using 50 mM tetraethylammonium perchlorate and 10 mM hydrochloric acid to adjust the ionic strength and the acidity of the medium (16).

In more recent reports, other researchers have used similar approaches to separate basic solutes in nonaqueous media. Tomlinson and coworkers found that in the separation of the antitumor drug pyrazoloacridine and two synthetic standards-putative metabolites, the use of a 20 mM ammonium acetate/1% acetic acid buffer in MeOH resulted in the complete separation of all three components (as well as an impurity) and provided a medium that was compatible with MS detection. The H_2 antagonist mifentidine and seven other similar compounds were also separated under the same conditions (Figure 15.7) (20,21). Hansen and coworkers successfully separated a series of cationic drugs using 25 mM ammonium acetate and 1 M acetic acid in several different solvents. They noticed that lowering the pH of a nonaqueous solution through addition of acetic acid caused changes in the elution order of these analytes (27,28). In a subsequent study, the researchers separated tetracycline hydrochloride from its major impurities using a solution containing 118 mM methanesulfonic acid (to keep the solutes ionized) as well as 25 mM ammonium acetate and 10 mM citric acid (used as supporting electrolytes) in a solvent mixture of 47:47:6 MeOH/ACN/DMF. This medium allowed the solutes to be differentiated based on their pK_a's and yielded selectivities that equaled or exceeded those obtained using high-performance liquid chromatography (HPLC) or aqueous CE (23). Also, they obtained high-efficiency separations of opium alkaloids with both aqueous (cyclodextrin-modified) and nonaqueous (ACN) systems but found that selectivities differed greatly between the two systems (Figure 15.8) (28).

In addition to separating organic cations, NACE methods have been utilized in the separation of metal ions. In our laboratory, Ye and Khaledi successfully separated a number of metal ions using 4 mM 4-methylbenzylamine and 8 mM hydrochloric acid in MeOH. The authors noticed that the mobility of each ion was determined mainly by its limiting ionic conductance in the medium but that ion–solvent interactions also played a role, as evidenced by the different elution orders that were obtained in various solvents (24).

15.3.1.2. Anions

As in the case of cations, several studies have shown that anionic solutes can be separated by NACE methods through the selection of an appropriate buffer. Most of the anions that have been separated by means of NACE involve dissociated organic acids. Sahota and Khaledi, for example, used a 250 mM phosphate buffer in FA to separate six dipeptides. Also, a mixture of 10 dipeptides was separated by using 20 mM sodium hydroxide and 180 mM boric acid in FA (Figure 15.9) (18). Chiari and Kenndler obtained full resolution of six organic acids using 100 mM Tris–acetate in MeOH. This separation was only partially successful in an aqueous medium (Figure 15.10) (33). Jansson and Roeraade separated two carboxylic acids in less than a minute using pure, unbuffered NMF (29). Similar conditions were employed in the separation of two water-insoluble drugs with similar results

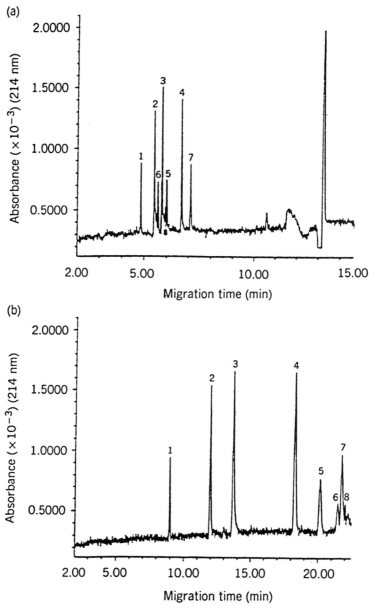

Figure 15.7. Separation of mifentidine (MIF) and eight synthetic standards by capillary zone electrophoresis (CZE) using variable concentrations of organic modifier. Peaks: (1) MIF; (2) MIF–amine; (3) MIF–amine–OH; (4) MIF–urea; (5) MIF–amide; (6) MIF–azo; (7) MIF–nitro; (8) MIF–azoxy. CZE separation buffer/solvent: (a) 20 mM NH_4OAc in 30% aqueous MeOH containing 1% acetic acid; (b) 20 mM NH_4OAc in 100% MeOH containing 1% acetic acid. [Reproduced from A. J. Tomlinson, L. M. Benson, J. W. Gorrod, and S. Naylor, *J. Chromatogr. B. Biomed. Appl.* **657**, 373 (1994), with permission.]

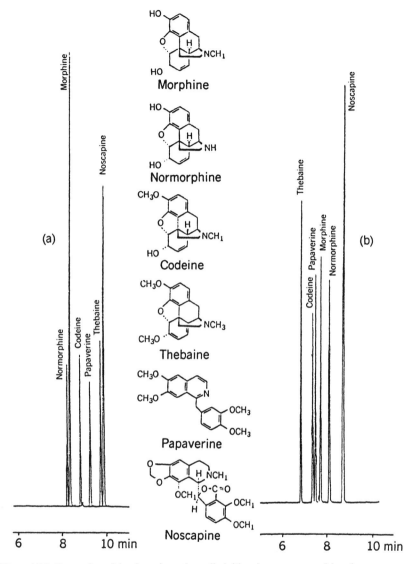

Figure 15.8. Separation of the six major opium alkaloids using an aqueous (a) and a nonaqueous (b) CE system. Electrophoresis medium: (a) 0.05 M 6-aminocaproic acid pH 4.0 with 30 mM of 2,6-di-O-methyl-β-cyclodextrin added; (b) 25 mM ammonium acetate/1 M acetic acid in ACN. [Reproduced from S. H. Hansen, J. Tjørnelund, and I. Bjørnsdottir, *Trends Anal. Chem.* **15**, 175 (1996), with permission.]

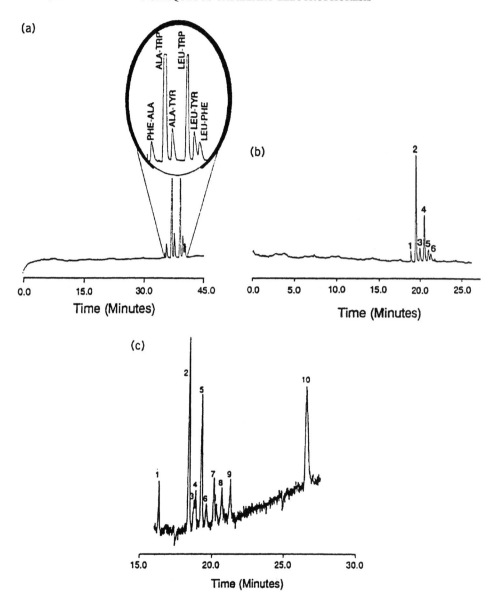

Figure 15.9. Separation of peptide mixtures. (a) Six peptides, identified in the inset. Applied voltage: 20 kV. (b) Same peptide mixture as in part a with same elution order, but under an applied voltage of 30 kV. (c) Mixture of 10 peptides: (1) Tyr-Tyr-Tyr; (2) Trp-Gly-Gly; (3) Tyr-Gly-Gly; (4) Gly-Ala-Tyr; (5) Leu-Trp; (6) Phe-Gly-Gly; (7) Phe-Phe (8) Tyr-Ala; (9) Ala-Tyr; (10) Glu-Trp. Separating buffer: 250 mM NaH_2PO_4 in FA at an apparent pH of 3.0. [Reproduced from R. S. Sahota and M. G. Khaledi, *Anal. Chem.* **66**, 1141 (1994), with permission.]

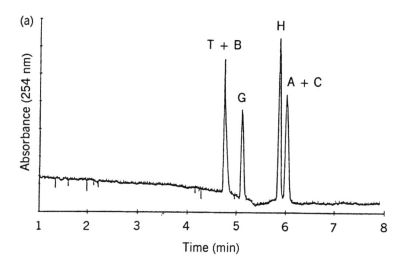

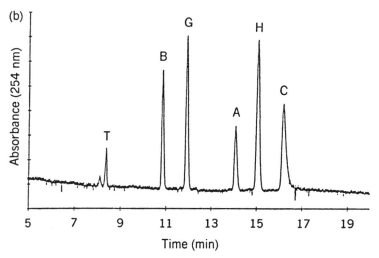

Figure 15.10. Electropherograms of anions on (a) 50 mM Tris–acetate in water at pH 8.0 and (b) 100 mM Tris–acetate in MeOH at apparent pH 8.5. Symbols: B = benzoic acid; T = *p*-toluenesulfonic acid; C = caffeic acid; H = *p*-hydroxycinnamic acid; G = *N*-acryloylglycine; A = *N*-acryloyl-γ-aminobutyric acid. [Reproduced from M. Chiari and E. Kenndler. *J. Chromatogr. A* **716**, 303 (1995), with permission.]

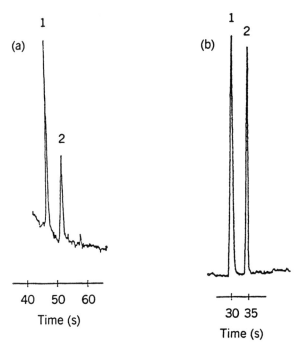

Figure 15.11. Electropherograms of (a) 2-pyridinecarboxylic acid (1) and salicyclic acid (2) and (b) propranolol HCl (1) and felodipine (2) in pure NMF. [Reproduced from M. Jansson and J. Roeraade, *Chromatographia* **40**, 163 (1995), with permission.]

(Figure 15.11) (29). Salimi-Moosavi and Cassidy separated anionic surfactants in several liquid detergents and shampoos. Nonaqueous conditions were necessary in separating long-chain (>12 carbon units) surfactants by CE. Also, these authors learned that the separation of anionic surfactants using 10 mM sodium *p*-toluenesulfonate and 5 mM *p*-toluenesulfonic acid in MeOH yielded efficiencies that far exceeded those obtained in H_2O (probably due to the solubility problems that occurred in H_2O). Finally, they separated linear alkyl benzenesulfonates using 20 mM tetramethylammonium hydroxide and 10 mM perchloric acid in ACN/MeOH mixtures (26).

The only NACE separation of inorganic anions that has been reported was performed by Salimi-Moosavi and Cassidy, who separated 11 such anions using 10 mM KHP, 20 mM *n*-butylamine, and 2% H_2O in MeOH (Figure 15.12). The role of the buffer components was to control the degree of ion association, which was shown to affect selectivity (25).

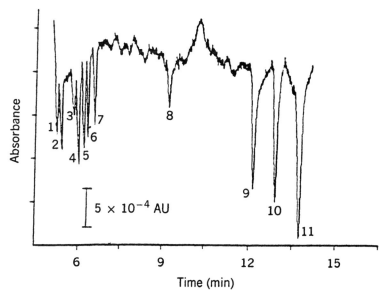

Figure 15.12. Separation and detection of inorganic anions with phthalate electrolyte: 0.01 mol/L KHP, 0.02 mol/L n-BuNH$_2$, and 2% (v/v) water in MeOH; indirect UV detection at 254 nm. Peaks (1) I^-; (2) SCN^-; (3) NO_3^-; (4) Br^-; (5) NO_2^-; (6) N_3^-; (7) Cl^-; (8) F^-; (9) $C_2O_4^{2-}$; (10) $S_2O_3^{2-}$; (11) SO_4^{2-}. [Reproduced from H. Salimi-Moosavi and R. M. Cassidy, *Anal. Chem.* **67**, 1067 (1995), with permission.]

15.3.2. Separation of Uncharged Solutes

The separation of neutral solutes by NACE presents a challenge since these solutes cannot migrate on their own in an electric field. As a result, neutral solutes must interact with charged species during the course of an electrophoretic separation. The extent to which these interactions occur determine the mobilities, and thus the separation, of a series of neutral solutes.

In some cases, the interaction mechanism between the charged and uncharged species involves the presence of hydrophobic moieties on the two species. In one study, Tanaka and coworkers investigated the use of alkylated starburst dendrimers (SBDs) in the separation of uncharged analytes by both aqueous and nonaqueous CE. As the MeOH content of the buffer containing the SBD was increased, better separations of the more hydrophobic solutes were obtained. For example, the complete resolution of a series of alkylphenones was obtained at 20% MeOH, whereas higher MeOH contents (up to 90%) were necessary for the rapid and complete separation of the more hydrophobic compounds. Figure 15.13 shows the separations that were

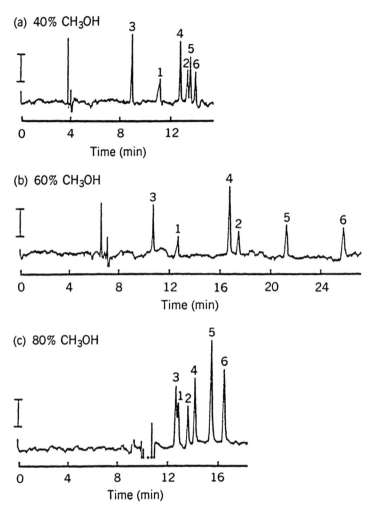

Figure 15.13. Effect of MeOH addition on the separation of aromatic hydrocarbons with 5 mM SBD (X)-C_{12} and 20 mM borate: (a) pH 10.8 in 40% MeOH; (b) pH 11.0 in 60% MeOH; (c) pH 11.2 in 80% MeOH. Solutes: (1) diphenylmethane; (2) o-terphenyl; (3) naphthalene; (4) anthracene; (5) pyrene; (6) triphenylene. [Reproduced from N. Tanaka, T. Fukutome, K. Hosoya, K. Kimata, and T. Araki, *J. Chromatogr. A* **716**, 57 (1995), with permission.]

obtained by using SBD solutions in 40%, 60%, and 80% MeOH (38). In another study, Yang and coworkers examined the use of ionic polymers in the separation of several classes of hydrophobic solutes. In 60% MeOH and 2% Elvacite 2669 (a triblock copolymer bearing negative charges), several alkylphenones were separated. The separation of the larger alkylphenones was

APPLICATIONS

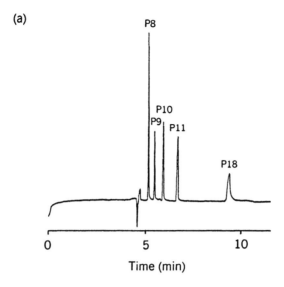

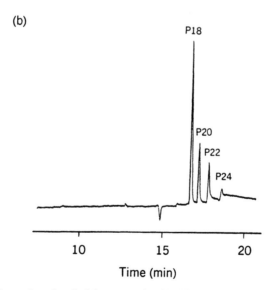

Figure 15.14. Separation of *n*-alkylphenones using 2% Elvacite 2669 in (a) 100% H$_2$O and (b) 70:30 MeOH/H$_2$O. Peaks: (P8) acetophenone; (P9) propiophenone; (P10) butyrophenone; (P11) valerophenone; (D18) dodecanophenone; (P20) tetradecanophenone; (P22) hexadecanophenone; (P24) octadecanophenone. [Reproduced from S. Yang, J. G. Bumgarner, and M. G. Khaledi, *J. High Resolut. Chromatogr.* **18**, 443 (1995), with permission.]

improved by increasing the MeOH content to 70% (Figure 15.14). Using up to 84% MeOH and 2% Elvacite 2669, these authors also obtained separations of polycyclic aromatic hydrocarbons (PAHs) and fullerenes (39).

Sometimes, solute–ion interactions may arise from the complexation of the neutral species with a cationic electrolyte. In the NACE separation of polyethers via complexation with cations in MeOH, Okada found that the ability of polyoxyethylene (POE) to form complexes depended on the nature of the cation. The ammonium ion was used to separate POEs containing at least four repeating oxyethylene units. On the other hand, the addition of potassium ion (or another alkali metal ion) in small quantities allowed smaller POEs to be separated. The stability (and thus the migration) of the POE–cation complexes was determined by the POE chain length regardless of the ion used (Figure 15.15). However, shape characteristics seemed to play a role in the separation of crown ethers, as selectivity differences were observed from cation to cation (40). In a more recent study, Miller and coworkers separated a series of PAHs using charge-transfer and induced-dipole interactions with planar organic cations (tropylium and 2,4,6-triphenylpyrylium) in pure ACN. Figure 15.16 shows the separation that was obtained using a 40 mM solution of 2,4,6-triphenylpyrylium tetrafluoroborate in ACN. It was found that the elution order of the PAHs depended mainly on the size of the solute molecule, with the largest of the PAHs (coronene) eluting first. With the tropylium ion, however, the shape of the solute molecule (as well as its size) appeared to play an important role. In addition, these authors observed that with increasing cation concentration, the electroosmotic flow decreased whereas the degree of PAH–cation binding increased. These two effects ultimately led to increased resolution (41). In a later study, they found that other phenyl-substituted pyrylium salts (tetrafluoroborates and perchlorates) could be used in the separation of PAHs. Substituted PAHs were also separated under the same conditions. It was observed that the number, position, and nature of the PAH substituent(s) affected the binding interaction with the cation (thus affecting the mobility of the solute–cation complex) (42).

15.3.3. Separation of Chiral Solutes

Chapter 23 of this volume discusses the chiral separations that have been performed using CE techniques. In this section, the NACE separations of chiral solutes reported to date are briefly mentioned.

Often, the CE resolution of chiral solutes may be facilitated by the use of a nonaqueous medium. Recently, in our laboratory, Ye and Khaledi found that a 250 mM solution of β-cyclodextrin (β-CD) in NMF in the presence of triflic acid (trifluoromethanesulfonic acid) could be used in the chiral separation of the trimipramine enantiomers (24). Because β-CD was 30 times more

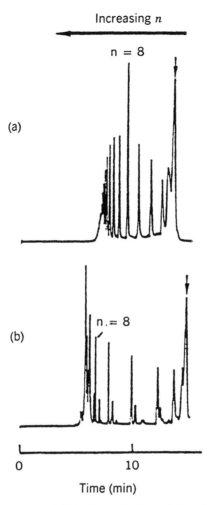

Figure 15.15. Electropherograms of dinitrobenzoylpolyoxyethylene (n = number of repeating units). Solution: (a) 50 mM NH_4Cl + 50 mM Et_3N; (b) 5 mM KCl + 45 mM NH_4Cl + 45 mM Et_3N. The peak of the EOF marker, which is a dinitrobenzoyl derivative not having complexation ability, is indicated with an arrow. [Reproduced from T. Okada, *J. Chromatogr. A* **695**, 309 (1995), with permission.]

soluble in NMF than in H_2O, the separations achieved using NMF as the CE solvent were superior to those obtained under aqueous conditions. In a subsequent study, Wang and Khaledi discovered that β- and γ-CDs could be employed in the enantiomeric separation of several tricyclic antidepressants. These separations were best achieved using a nonaqueous medium since the

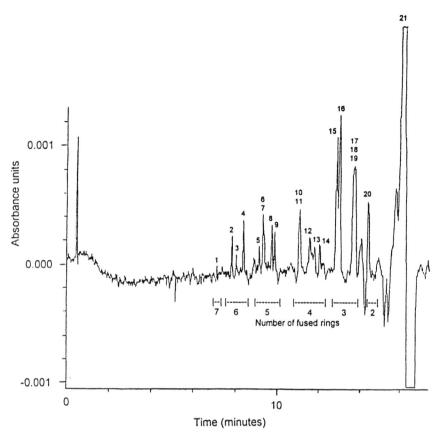

Figure 15.16. Blank-subtracted electropherogram showing the separation of PAHs using 40 mM 2,4,6-triphenylpyrylium tetrafluoroborate in acetonitrile. Solutes: (1) coronene; (2) naphtho[2,3-*a*]pyrene; (3) benzo[*ghi*]perylene; (4) indeno[1,2,3-*cd*] pyrene; (5) perylene; (6) benzo [*a*] pyrene; (7) dibenz[*ah*]anthracene; (8) benzo[*b*]fluoranthene; (9) benzo[*k*]fluoranthene; (10) benz[*a*]-anthracene; (11) chrysene; (12) pyrene; (13) fluoranthene; (14) naphthacene; (15) anthracene; (16) phenanthrene; (17) acenaphthene; (18) acenaphthylene; (19) fluorene; (20) naphthalene; (21) FA (EOF marker). [Reproduced from J. L. Miller, M. G. Khaledi, and D. Shea, *Anal. Chem.* **69**, 1223 (1997), with permission.]

degree of CD–solute binding in water was too high to allow for adequate separations (43).

Chiral CE separations may also be achieved through diastereomeric ion pairing between charged chiral solutes and an enantiomerically pure counterion. This approach was followed by Bjørnsdottir and coworkers as they used an anionic counterion [30 mM (*R*)- or (*S*)-camphorsulfonate] to separate cisapride as well as substances with a β-amino alcohol configuration. In this study, ACN was chosen as the solvent because of its moderately low dielectric constant, low-UV cutoff, and ability to dissolve sufficient amounts of the camphorsulfonate ion. Acetic acid (1 M) was added to control the acidity of medium and to further increase the solubility of the counterion. In some cases, Tween 20 (0.1–0.2 mM) was also added to decrease the electroosmotic flow (thereby enhancing resolution) (44).

15.4. CONCLUSIONS AND FUTURE TRENDS

The use of nonaqueous media broadens the scope of CE applications. Through the careful selection of an appropriate solvent, separations of a wide variety of solutes are possible. There are several nonaqueous solvents that researchers have shown to be capable of serving as CE media, and in many cases more than one solvent may be suitable for a given application. In the coming years, research in NACE separations will likely blossom in the medical, pharmaceutical, environmental, and industrial fields, as well as other fields in which chemical separations play a key role. Therefore, it will be crucial to examine more fully the cause-and-effect relationships in NACE separations and to better understand the mechanisms that govern these separations.

LIST OF ACRONYMS AND ABBREVIATIONS

Acronym or Abbreviation	Definition
ACN	acetonitrile
AU	absorbance units
n-BuNH$_2$	n-butylamine
β-CD	β-cyclodextrin
CE	capillary electrophoresis
CTC	chlorotetracycline
CZE	capillary zone electrophoresis
DC	doxycycline
DMA	dimethylacetamide
DMF	dimethylformamide
DMSO	dimethyl sulfoxide
DTTCI	diethylthiatricarbacyanine iodide
EOF	electroosmotic flow
ESI	electrospray-ionization (interface)
FA	formamide
HPLC	high-performance liquid chromatography
IR	infrared

LIST OF ACRONYMS AND ABBREVIATIONS (Continued)

Acronym or Abbreviation	Definition
KHP	potassium hydrogen phthalate
LIF	laser-induced fluorescence
MeOH	methanol
MIF	mifentidine
MS	mass spectrometry (or mass spectrometer)
NACE	nonaqueous capillary electrophoresis
NMF	N-methylformamide
OTC	oxytetracycline
PAHs	polycyclic aromatic hydrocarbons
POE	polyoxyethylene
SBD	starburst dendrimer
TC	tetracycline
TEAP	tetraethylammonium perchlorate
THF	tetrahydrofuran
Tris	tris(hydroxymethyl) aminomethane
UV	ultraviolet

REFERENCES

1. M. Hayek, *J. Phys. Colloid Chem.* **55**, 1527 (1951).
2. N. J. Parekh, A. A. Fatmi, M. A. Tshabalala, and L. B. Rogers, *J. Chromatogr.* **314**, 65 (1984).
3. N. J. Parekh, and L. B. Rogers, *J. Chromatogr.* **330**, 19 (1985).
4. R. T. Johansen, R. J. Heemstra, and H. N. Dunning, *Proc. Am. Pet. Inst., Sect. VIII* **42**, 60 (1962).
5. D. Leighton, G. J. Moody, and J. D. R. Thomas, *Analyst (London)* **99**, 442 (1974).
6. M. A. Tshabalala, S. B. Schram, F. G. Gerberich, D. W. Lowman, and L. B. Rogers, *J. Chromatogr.* **207**, 353 (1981).
7. M. M. Tuckerman and H. H. Strain, *Anal. Chem.* **32**, 695 (1960).
8. G. Marcu and A. Botar, *Anal. Abstr.* **17**, No. 786 (1969).
9. J. Becka and V. Jokl, *Collect. Czech. Chem Commun.* **36**, 2467 (1971).
10. P. B. Chakrabarti and S. K. Gupta, *Vijnana Parishad Anusandhan Patrika* **15**, 177 (1972).
11. J. L. Beckers and F. M. Everaerts, *J. Chromatogr.* **51**, 339 (1970).
12. J. L. Beckers and F. M. Everaerts, *J. Chromatogr.* **68**, 207 (1972).
13. M. Lederer, *Chem. Ind. (London)*, 1481 (1954).
14. M. H. Paul and E. L. Durrum, *J. Am. Chem. Soc.* **74**, 4721 (1952).

15. E. K. Korchemnaya, A. N. Ermakov, and L. P. Bochlova, *J. Anal. Chem. USSR (Engl. Transl.)* **33**, 635 (1978).
16. Y. Walbroehl and J. W. Jorgenson, *J. Chromatogr.* **315**, 135 (1984).
17. Y. Walbroehl and J. W. Jorgenson, *Anal. Chem.* **58**, 479 (1986).
18. R. S. Sahota and M. G. Khaledi, *Anal. Chem.* **66**, 1141 (1994).
19. C. L. Ng, H. K. Lee, and S. F. Y. Li, *J. Liq. Chromatogr.* **17**, 3847 (1994).
20. A. J. Tomlinson, L. M. Benson, and S. Naylor, *LC-GC* **12**, 122 (1994).
21. A. J. Tomlinson, L. M. Benson, J. W. Gorrod, and S. Naylor, *J. Chromatogr. B: Biomed. Appl.* **657**, 373 (1994).
22. M. Smoluchowski, *Handbuch der Elektrizität und des Magnetismus,* Vol. 2, p. 366. Leipzig, Germany (1914).
23. J. Tjørnelund, and S. H. Hansen, *J. Chromatogr. A* **737**, 291 (1996).
24. B. Ye and M. G. Khaledi, unpublished results (1995).
25. H. Salimi-Moosavi and R. M. Cassidy, *Anal. Chem.* **67**, 1067 (1995).
26. H. Salimi-Moosavi and R. M. Cassidy, *Anal. Chem.* **68**, 293 (1996).
27. I. Bjørnsdottir and S. H. Hansen, *J. Chromtogr. A* **711**, 313 (1995).
28. S. H. Hansen, J. Tjørnelund, and I. Bjørnsdottir, *Trends Anal. Chem.* **15**, 175 (1996).
29. M. Jansson and J. Roeraade, *Chromatographia* **40**, 163 (1995).
30. See Chapter 2 of this volume. See also E. Kenndler, in *Capillary Electrophoresis Technology,* (N. A. Guzman, ed.), Chromatogr. Sci. Ser., Vol. 64. Dekker, New York.
31. J. F. Coetzee and G. R. Padmanabhan, *J. Phys. Chem.* **69**, 3193 (1965).
32. C. X. Zhang, F. von Heeren, and W. Thormann, *Anal. Chem.* **67**, 2070 (1995).
33. M. Chiari and E. Kenndler, *J. Chromatogr. A* **716**, 303 (1995).
34. I. M. Kolthoff and M. K. Chantooni, Jr., *J. Am. Chem. Soc.* **91**, 4621 (1969).
35. T. Okada, *Chem. Commun.,* 1779 (1996).
36. V. Ward and M. G. Khaledi, unpublished results (1996).
37. S. A. Soper, B. L. Legendre, Jr., and D. C. Williams, *Anal. Chem.* **67**, 4358 (1995).
38. N. Tanaka, T. Fukutome, K. Hosoya, K. Kimata, and T. Araki, *J. Chromatogr. A* **716**, 57 (1995).
39. S. Yang, J. G. Bumgarner, and M. G. Khaledi, *J. High Resolut. Chromatogr.* **18**, 443 (1995).
40. T. Okada, *J. Chromatogr. A* **695**, 309 (1995).
41. J. L. Miller, M. G. Khaledi, and D. Shea, *Anal. Chem.* **69**, 1223 (1997).
42. J. L. Miller, unpublished results (1996).
43. F. Wang and M. G. Khaledi, *Anal. Chem.* **68**, 3460 (1996).
44. I. Bjørnsdottir, S. H. Hansen, and S. Terabe, *J. Chromatogr. A* **745**, 37 (1996).

CHAPTER

16

METHOD VALIDATION IN CAPILLARY ELECTROPHORESIS

K. D. ALTRIA

Analytical Sciences, GlaxoWellcome Research and Development, Ware, Herts. SG12 ODP, United Kingdom

16.1.	**Introduction**	557
16.2.	**Specific Validation Aspects**	559
	16.2.1. Accuracy (Cross-Correlation)	559
	16.2.2. Buffer Depletion Effects	561
	16.2.3. Linearity	561
	16.2.4. Peak Purity	562
	16.2.5. Reagents Supply	564
	16.2.6. Recovery	564
	16.2.7. Response Factors	565
	16.2.8. Repeatability (Precision)	566
	16.2.9. Robustness	569
	16.2.10. Selectivity	570
	16.2.11. Sample Diluent	570
	16.2.12. Sensitivity	571
	16.2.13. Solution Stability	574
	16.2.14. System Suitability	574
	16.2.15. Training	575
	16.2.16. Transfer of Methods Between Laboratories	575
16.3.	**Conclusions**	577
List of Acronyms		577
References		578

16.1. INTRODUCTION

Appropriate validation of an analytical method is essential to fully demonstrate that the method is fit for its intended use. Validation therefore

High Performance Capillary Electrophoresis, edited by Morteza G. Khaledi. Chemical Analysis Series, Vol. 146.
ISBN 0-471-14851-2 © 1998 John Wiley & Sons, Inc.

covers all analytical parameters that impact on the performance of the method.

The level and extent of the validation of a specific method is dependent upon the complexity of the separative task and the intended extent of usage. For instance, an extensive validation would be required for a method to be used in a number of laboratories worldwide to quantify trace levels of closely related impurities, whereas the validation requirements would be more limited for an identity confirmation method used on one occasion within one laboratory.

There are now many examples of validated capillary electrophoretic (CE) methods across a number of application areas. The validation aspects addressed are largely equivalent when CE and high-performance liquid chromatographic (HPLC) methods are being assessed (1–5). However, there are important differences between CE and HPLC validation exercises, the principal differences being that factors such as instrument-to-instrument transfer and long-term separation performance are of increased importance in CE.

Table 16.1 shows the range of validation testing that may be appropriate for the three main activity areas of CE: identity confirmation, purity determinations, and main-component assays.

The majority of validation reports have originated from pharmaceutical analysis laboratories, where the simplicity and cost effectiveness of CE has been widely appreciated. A number of validated CE methods have been successfully incorporated into regulatory submissions (6).

In this chapter the various validation aspects will be individually covered. A discussion of each aspect will include reference to examples in the literature illustrating the level of performance that may be achieved during validations.

Table 16.1. Validation Requirements[a]

Test	Identity Confirmation	Chiral and Achiral Purity	Main-Component Assay
Accuracy	—	Yes	Yes
Precision	—	Yes	Yes
Linearity	—	Yes	Yes
Selectivity	(Yes)	Yes	Yes
LOD	(Yes)	Yes	Yes
LOQ	(Yes)	Yes	Yes
Recovery	Yes	Yes	Yes
Solution stability	Yes	Yes	Yes
Robustness	Yes	Yes	Yes

[a](Yes) denotes that the test may be applicable for analysis of placebo (or low-level) samples.

16.2. SPECIFIC VALIDATION ASPECTS

16.2.1. Accuracy (Cross-Correlation)

It is a vital part of method validation to demonstrate that the method is generating true and accurate data. This can be demonstrated in part by recovery experiments (which are discussed later). Another, more comprehensive approach is to generate data using an alternative method/technique and to compare the data sets obtained by the two methods/techniques. The most frequently employed combination is that of HPLC and CE. This combination is especially complementary, as both techniques can be used to generate data in a similar fashion, i.e., using a common detection wavelength and quantitation approach. This is not necessarily the case with other techniques such as thin-layer chromatography (TLC), where detection principles and response factors may be somewhat different. HPLC and CE are also capable of similar performance levels in terms of key analytical figures of merit, such as precision and sensitivity, enabling equivalent data sets to be generated.

16.2.1.1. Assay

Combinations of CE with a variety of analytical techniques have been reported for main component analysis. This cross-correlation has been predominantly shown using CE and HPLC (7–9). Correlation studies between HPLC and CE results from the testing of a range of drug formulations produced (8) correlation coefficients of > 0.999 between the data sets. Table 16.2 shows comparable MEKC and HPLC results obtained for assay of components in cola drinks (10).

Table 16.2. Analysis of Various Components in Cola Drink (mg/L)

Cola Sample	Aspartame		Benzoic Acid		Caffeine	
	MECK	HPLC	MEKC	HPLC	MEKC	HPLC
1	510	530	170	165	140	130
2	440	405	170	160	100	100
3	450	430	175	175	90	95
4	470	450	170	165	85	85
5	335	335	150	150	60	60
6	N/D[a]	N/D[a]	320	320	80	80

Source: Reproduced from C. O. Thompson, V. G. Trennery, and B. Kemmery, J. Chromatogr. **694**, 507 (1995), with permission.

[a] N/D denotes "none detected"; this sample contained saccharin [MEKC (micellar electrokinetic chromatography) 75 mg/L; HPLC 80 mg/L].

An additional means of demonstrating method accuracy is to compare the data obtained with the label claim of the test material. This has been shown in many reports, including drug content in formulations (9), potassium levels in drug substance (11), vitamin content (12), and metal ion content in mineral water (13).

16.2.1.2. Related Impurities

Cross-validation of CE data with results generated by an alternative technique can constitute a powerful demonstration of the validity of a method. For example, if CE impurity data for a sample shows five impurities present with a total impurity level of 1.3% and HPLC (or TLC) analysis of the same material gives an equivalent number of impurities and similar total impurity content, then greater confidence can be placed in the data generated.

CE and HPLC, when used together (5,14), are very complementary techniques for related impurity determinations, as common wavelengths can be selected and results quoted as percent area/area. Comparisons with other separative techniques such as TLC become more complicated, as response factors have to be calculated and applied for each impurity. Statistical analysis can be applied to measure how closely the data sets are correlated. For example, a correlation coefficient of 0.995 was reported between HPLC and MEKC data for content of an impurity in drug substance batches (5). In a similar study, a statistical difference at the 5% probability level was found between impurity levels in batches of salbutamol drug substance analyzed by both CE and HPLC (4).

MEKC has been used to assess impurity levels in cephalosporin samples and the data statistically compared to those obtained using HPLC (15,16).

16.2.1.3. Chiral Purity Determinations

Several reports have shown (17) good agreement between the HPLC and CE for enantiomeric purity determinations. Cross-comparison of CE and HPLC data in enantiomeric separations has also been employed extensively (17).

16.2.1.4. Inorganic Ion Contents and Bioassays

CE and atomic absorption spectrometric results for sulfate levels in detergent powder have also been statistically demonstrated to be equivalent (18). Statistical treatment of comparative data sets is especially important in bioanalysis, and several reports (19–21) have shown statistical correlation of CE and HPLC data.

16.2.2. Buffer Depletion Effects

Deterioration of the electrolyte due to electrolysis effects may occur during extended analytical sequences (22). This should be appropriately checked, and changes of electrolyte—or switches between vials containing fresh electrolyte—should be considered if necessary. For instance, if a typical operating sequence involves 40 injections, then injection sequences of this length should be included in the validation work.

16.2.3. Linearity

It is necessary to show that the detector gives a linear response for changes in the solute concentration, over the concentration range that the method is intended to be used to measure. Generally peak areas are measured in CE, as these gives longer dynamic ranges than do peak heights (23). Peak height increases are not linear at higher sample concentrations due to band-broadening effects. This broadening results in wider, rather than taller, peaks being observed at higher concentration, which limits the peak height dynamic range. However, peak heights can be successfully used for quantitation at low solute concentrations; providing that an acceptable linearity has been demonstrated.

The use of an internal standard to improve precision is discussed later (in Section 16.2.8). Use of an internal standard also reduces the scatter on a linearity plot and can improve correlation coefficient data. For example, an internal standard was used in the determination of detector linearity with levothyroxine concentration and improved the correlation coefficient from 0.99856, when calculated using peak areas, to 0.99991, when calculated using peak area ratios (24).

16.2.3.1. Main-Component Assay and Identity Confirmation Testing

For main component analysis it is typical to assess the linear response of the detector by using calibration solutions covering a range around the concentrations to be routinely employed. Typically linearity is assessed by analyzing five standards covering concentrations over the range 50–150% of the nominal assay concentration. Several reports (7–9) have shown that linearity correlation coefficients of > 0.999 are possible with standard CE instruments.

16.2.3.2. Chiral and Achiral Impurity Determinations

When an analyst is determining impurity levels, or trace enantiomer content, it is also necessary to demonstrate detector linearity for the trace enantiomer in the presence of the main peak. This involves maintaining the main component at a constant concentration and varying the content of the minor component(s) to mimic real samples that contain variable levels of impurities. For example, if the intention of a specific method is to quote impurity levels at the 0.05–1.0% level, then it may be appropriate to prepare standards covering the range 0.025–1.5% of the concentration of the main component. These standards should be injected in duplicate and the peak area obtained plotted against the percentage of nominal concentration. Acceptable correlation coefficients (0.99 or greater) and an intercept close to the origin should be achieved. For example, levels of *p*-toluenesulfonic acid were spiked into the drug substance at levels of 0.02–5.0%, and a correlation coefficient of 0.9999 was obtained for peak area plotted against impurity concentration (5). Figure 16.1 shows separation of the drug substance spiked with 2.0% impurity.

Correlation coefficients of > 0.99 have been reported (25) for fluparoxan enantiomer levels of 1–10% of the desired enantiomer content. A further example of this validation aspect has been provided (26) in the determination of levels of the undesired enantiomer of a drug in which the main component was held constant and the trace enantiomer spiked over the range 0.1–1.5% (with a correlation coefficient of 0.9995).

16.2.4. Peak Purity

As with all separative techniques, when solutes of very similar chemical properties are separated, comigration of peaks is possible in CE. This may be shown as comigration of impurity peaks or as impurity peaks migrating under the main peak. Approaches to peak purity determination in CE are similar to those employed in HPLC, and the principal methods are fraction collection (27) and spectral characterization (28). Common approaches would be used in related impurity determinations, main-component assays, and enantiomeric separations.

Micropreparative fraction collection is possible in CE, and fractions can be collected from the CE separation and analyzed using an alternative separative technique such as HPLC (27). If a single peak is indicated by the secondary test method, this gives further confirmation of peak purity.

Many commercial CE instruments are now equipped with diode array facilities that can be used in assessing peak purity. Standards of the main component and impurity can be analyzed and their UV spectra recorded and compared (28). Figure 16.2 shows the separation of impurities in a drug substance and the spectra of one of the impurities present at the 0.1% level. If good correlation between the spectra is not achieved, this can indicate co-migration of peaks. As in HPLC, this approach is limited in that closely related impurities will have similar ultraviolet (UV) spectra to each other and the main component.

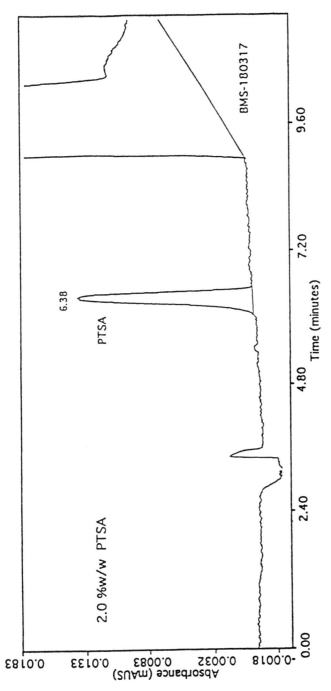

Figure 16.1. BMS-180317 drug substance spiked with 2% *p*-toluenesulfonic acid (PTSA). Separation conditions—electrolyte: 100 mM boric acid/20 mM sodium borate/20 mM sodium dedecyl sulfate (SDS); detection: 190 nm; voltage: +5 kV; temperature: 25 °C; capillary: 50 μm × 25 cm. [Reproduced from P.A. Shah. and L. Quinones, *J. Liq. Chromatogr.* **18**, 1349 (1995), with permission.]

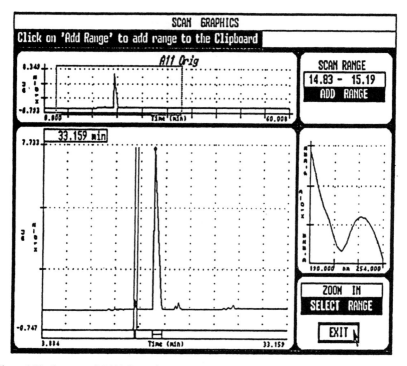

Figure 16.2. Spectra of 0.1% impurity determined using a diode array detector. Separation conditions—electrolyte: 50 mM NaH_2PO_4, pH 2.1; detection: 230 nm; voltage: + 10 kV; temperature: ambient; capillary: 50 μm × 7 cm; concentration: 1 mg/mL of a basic drug dissolved in water.

16.2.5. Reagents Supply

The purity of additives such as cyclodextrins should be carefully assessed, as changes in additive purity can alter the selectivity obtained (29). For example, chemical derivatization of cyclodextrins often results in a product containing cyclodextrins having variable degrees of substitution. In these instances there can be appreciable differences between cyclodextrin suppliers and among batches from a single supplier.

16.2.6. Recovery

It is necessary to demonstrate full extraction of components from the sample matrix. This is achieved by spiking drug and related impurities into the matrix of concern (30). For example, in the case of a pharmaceutical formulation it

Table 16.3. Recovery Results from Spiking of BMS-180431-09 with Undesired Enantiomer[a]

Spiking Level (% w/w)	Average Observed (%)	Average Recovery (%)
0.36	0.37	102.7
0.65	0.65	100.5
0.80	0.84	104.6
1.10	1.11	100.9

Source: Reproduced from J. E. Noroski, D. J. Mayo, and M. Moran, J. Pharm. Biomed. Anal. 13, 45 (1995), with permission.

[a] Three samples were prepared and analyzed at each spiking level.

would be appropriate to add known levels of drug and related impurities to a placebo mixture (30,31). Analysis of the spiked sample should confirm acceptable extraction. For example, various levels of the undesired enantiomer of chiral drug BMS-180431-09 was spiked into the pure enantiomer and analyzed by CE. Table 16.3 shows that 100% recovery was successfully obtained (32).

Levels of a synthetic impurity were added to drug substance over the range of 0.05–2.0% and the average recovery was 99% (5). Demonstration of acceptable recovery is especially important in bioanalysis (33), where sample pretreatments can be extensive.

16.2.7. Response Factors

The response factors for each component of a mixture that is quantified by a given method should be established to ensure that they are being accurately determined. This is essential if impurity data are to be calculated as percent area/area of the total peak area. Impurities may have markedly different UV characteristics from each other and from the main component and may therefore be considerably over- or underestimated. Data handling is different in CE than in HPLC, as the separated peaks in CE move through the detector at different speeds. Therefore, to compare the areas of peaks within a single separation, all peak areas must be divided by their corresponding migration times (34). The calculated areas are often referred to as "normalized areas." All impurity data (and response factors) should be calculated using these normalized areas, which compensate for the different migration speed of peaks through the detector.

Response factors obtained from HPLC measurements cannot be directly applied to peaks obtained in CE separations. The ionic form of the solute may be different in CE, compared to HPLC, as it may be charged. Changes in solute charge can cause UV absorbance shifts and alter response factors. Inclusion of solutes into commonly used CE additives such as surfactant micelles and cyclodextrins may also cause spectral shifts.

16.2.8. Repeatability (Precision)

It is important to demonstrate that the method can be repeated successfully. This may involve repeated injection of a single sample solution or may be more extensive and involve performing the method in different laboratories, on different days, using different staff, reagents, equipment, and capillaries. It is also appropriate to assess how precisely operations can be repeated; for example, a single sample solution may be injected several times to measure the operation of the CE injector system and the stability of the separation.

16.2.8.1. Method Repeatability

It is necessary to demonstrate that the method can be reproduced in a variety of situations. As with HPLC, reproducibility between capillaries (or columns), instruments, analysts, days, and laboratories is required. As the method may be employed on CE instruments from various manufacturers, it is important to demonstrate that acceptable performance can be attained using different sample introduction modes and capillary dimensions. Examples of such studies include assessments of CE methods for the determination of sumatriptan (7) and hydrochlorothiazide (31), and for the chiral separation of tryptophan (35).

16.2.8.2. Injection Precision (Migration Times and Areas)

Repeated analyses (typically 10 sequential injections) of both a sample and a calibration solution should be conducted. Acceptable precision in terms of migration times and peak areas should be demonstrated to a predetermined level. Precision in CE is typically of the order of 0.5–2% RSD relative standard deviation for main peak assay (7,9,31,36). Injection precision is generally improved by use of internal standards and high sample concentrations (37).

For trace impurities the precision obtained for the trace component would be expected to be < 10% RSD (1,38). Repeated analyses of the same sample solution should give consistent results in terms of number of impurities, levels and relative migration time for each impurity, and total level of impurities (15,16).

16.2.8.3. Sample and Calibration Preparation

Repeatable preparation of samples and standards should be demonstrated. For example, 10 individual standards should be prepared and analyzed in duplicate. The precision for the response factors obtained should be within acceptable levels. Similarly 10 individual samples should be prepared and analyzed in duplicate; the precision for the pooled assay values should be acceptable. For example, in the validation of a method for determination of potassium levels in the potassium salt of an acidic drug, 10 individual calibration solutions were prepared and analyzed in duplicate (11). The calculated response factors for the calibrations produced a precision of 0.8% RSD. Ten samples were also prepared and analyzed (Table 16.4).

Levels of a synthetic impurity were determined in six replicate samples of drug batches (5). Each sample was injected in duplicate. The mean result was 0.11% w/w, with an acceptable RSD of 7%.

16.2.8.4. Repeatability on Different Days

Repeatablility of a free solution CE method for the analysis of deamination products in insulin solutions has been assessed (39) by repeating the analysis of three batches on five separate days. Batches of antibiotics were consistently analyzed on a number of days using different instruments (40).

Table 16.4. Precision of Sample Preparation[a]

Sample No.	Injection 1	Injection 2	Average Result
1	5.69	5.71	5.70
2	5.74	5.73	5.74
3	5.80	5.78	5.79
4	5.73	5.73	5.73
5	5.72	5.72	5.72
6	5.66	5.67	5.67
7	5.74	5.75	5.75
8	5.71	5.79	5.75
9	5.70	5.72	5.71
10	5.77	5.76	5.77
Mean			5.73[b]

Source: Reproduced from K. D. Altria, T. Wood, R. Kitscha, and A. Roberts-McIntosh, J. Pharm. Biomed. Analy. 13, 33 (1995), with permission.

[a] Results as percent w/w potassium (calculated using peak area ratios).
[b] 0.65% RSD.

16.2.8.5. Repeatability When Different Analysts Are Used

Typically, analysis of a specific sample solution will be conducted by two or more analysts (31,40). Each of the analysts should conduct the analysis exactly as detailed in the analytical method. The results generated by each analyst should not differ significantly from those generated by the others. The number of analyses and reporting format [i.e., significant figures and the limit of detection (LOD) and limit of quantitation (LOQ) should be specified]. Each analyst should independently prepare her or his own sample solutions, capillaries, and reagents.

16.2.8.6. Repeatability When Different Reagents Are Used

It is necessary to demonstrate that it is possible to repeat the separation using different batches of reagents and materials (29). To achieve this it is necessary to be very specific as to how to prepare the electrolyte in terms of the electrolyte concentration, exact diluent composition, and the procedure for pH adjustment (i.e., the concentration and type of acid/base used to adjust the pH). Inadvertent variations will lead to nonrepeatable ionic strength electrolytes, leading to variability in selectivity.

16.2.8.7. Repeatability When Different Capillaries Are Used

The separation should be satisfactorily repeated on different capillaries (41), possibly from different capillary lots (and manufacturers, if appropriate). The initial preconditioning and storage conditions for capillaries should be specied, and capillaries should be dedicated to individual methods to prevent cross-contamination and nonrepeatable separations.

16.2.8.8. Repeatability on Different Instruments

This aspect is somewhat different from that encountered in HPLC, where transfer between instrument manufacturers is not considered a primary validation aspect. However, this is of great importance in CE, as instrument manufacturers have produced considerably different instrument configurations. This means that operating conditions must be modified to achieve similar separations on a second instrument type. This is necessary due to different detector positions (i.e., variations in the lengths along the capillary to the detector from the point of injection) and to different injection procedure (i.e., some instruments use vacuum, pressure- or gravity-based injections). In addition each instruent uses different presure settings, and therefore injection times cannot be directly transferred. Therefore, it is preferable to specify an injection volume (in nanoliters) in the method, which can then be considered instrument independent (36).

Transfer between instruments requires (36,42,43) different settings on the various instruments to obtain equivalent separations. There should be some limited revalidation on the additional instrumentation type as to, say, precision, sensitivity, and linearity.

16.2.9. Robustness

Robustness relates to the sensitivity of the method to small deliberate deviations from the method. For instance, if the method states an operating temperature of 30 °C, will acceptable performances be maintained at either 25 or 35 °C? Two approaches to method robustness evaluation are possible: univariate or multivariate. The univariate approach involves systematically varying each parameter sequentially. This "one-by-one" approach has been performed for the determination of enalapril content in tablets by CE (30).

Typically each parameter may be varied by 5–10% above and below the value set in the method. Parameters examined may include temperature, pH, electrolyte concentration, rinse times, additive concentration, detector wavelength, and sample loading. The responses measured may include resolutions, peak efficiencies, relative and absolute migration times, peak areas, and migration order.

In CE simultaneous changes in various operating factors cause an alteration in system performance, but the same changes alone in each factor may not unduly affect the system performance. For example, if during robustness testing both electrolyte concentration and temperature are independently increased by 10%, no effect may be recorded, although both cause an increase in the heat generated within the capillary. However, if both factors are then simultaneously increased, the increase in heat will be cumulative and may be sufficiently high to impact on separation performance. When two factors have a combined effect in this way they are said to interact. Such interactions are generally more likely in CE than in HPLC.

Interaction of factors cannot be assessed, or quantified, when factor-by-factor univariate robustness testing is performed. However, appropriately constructed experiments using statistical designs such as central composites and fractional factorials can allow both the effect of parameter variation and interactions to be measured. The use of experimental designs in robustness testing of CE methods has been shown for drug-related impurities determinations (44), levels of potassium (41), and trace levels of surfactants (45). The results from these robustness testing exercises allow the tolerance limits to be set on each parameter in the method.

16.2.10. Selectivity

This is arguably the most important parameter to evaluate in an analytical separative method as consistent selectivity is a fundamental requirement for all types of analysis undertaken. It is essential that the required selectivity be demonstrated. For example, in the case of a related impurities determination method it is necessary that all potential impurities be resolved from each other and the main component at the desired quantitation levels.

16.2.10.1. Achiral and Chiral Purity Determinations

Appropriate selectivity of the method should be demonstrated for all known likely synthetic or degradative impurities (31,46,47). This would be performed by analyzing test solutions spiked with all available impurities at their expected levels. If appropriate the method may be further challenged by analyzing stressed samples. The stressed samples may be generated by exposure to high temperatures, pH extremes, and both visible and UV light. During validation of a chiral method it is necessary to demonstrate that no synthetic or degradative impurities will interfere with the determination of either enantiomer at the required levels (32). For example, Figure 16.3 shows separation of a range of domperidone impurities. (14). The migration order of the enantiomers should be confirmed by spiking experiments with the pure enantiomer and the racemate if available (17).

16.2.10.2. Main-Component Assay and Identity Confirmation Testing

Selectivity for any likely interferents (from excipients, dissolving solvents, etc.) should be demonstrated in addition to the likely sample-related impurities. For example, when drug content in a formulation is being determined, samples of placebo formulations should be prepared and analyzed to confirm that no interfering peaks are present.

16.2.11. Sample Diluent

The ionic strength and composition of the sample-dissolving solvent can have a pronounced effect on the quality of the separation achieved (48). For example, Figure 16.4 shows separation of a range of basic drugs dissolved in either water or 100 mM NaCl solution. Other detrimental effects associated with sample diluents involve run failures due to outgassing when aqueous organic solvent mixtures are being analyzed.

It is therefore essential that the solution used to dissolve the calibration and sample be the same. In addition, the exact composition of the diluent should be

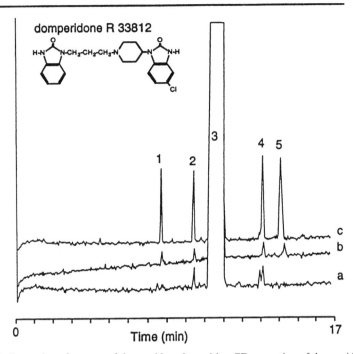

Figure 16.3. Separation of a range of domperidone impurities. CE separation of domperidone (R33812) and major known impurities: (a) batch G1A041; (b) 0.1% reference mixture; (c) 1% reference mixture. Peaks: (1) R29676: (2) R45771; (3) domperidone R33812; (4) R48557; (5) R52211. Separation conditions—electrolyte: citrate–phosphate, pH 4; voltage: +25 kV; temperature: 30 °C; detection; 250 nm; Capillary: 50 μm × 72 cm. [Reproduced from A. Pluym, W. Van Ael, and M. De Smet, *Trends Anal Chem.* **11**, 27 (1992), with permission.]

specified in the method and used throughout the validation and subsequent routine operation. In the case of a complex diluent, such as methanol/water mixtures, variations in the methanol/water ratio should be included in the robustness study.

16.2.12. Sensitivity

16.2.12.1. *The Limit of Detection*

The LOD value denotes the minimum detectable level of impurities or the lowest detectable level of the compound(s) of interest. The LOD is often defined (1,4) as the sample concentration that produces a peak with a height

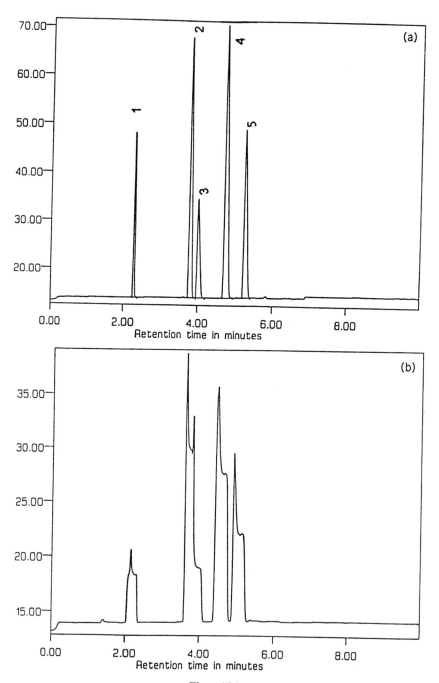

Figure 16.4

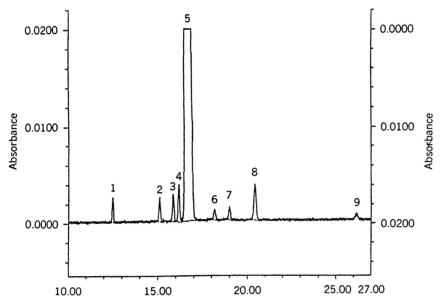

Figure 16.5. Separation of remoxipride and range of impurities at 0.1% level. Peaks: (1) FLA 708; (2) FLA 739; (3) FLA 838; (4) FLB 526; (5) FLA 731 (remoxipride); (6,7) (7) NCR 513; (8) FLA 740; (9) FLA 797. Separation conditions—electrolyte: 20 mM NaH_2PO_4, pH 3, containing methyl-β-cyclodextrin and 8/12 TPA and TBA; detection: 200 nm; voltage: + 10 kV; temperature: ambient; capillary: 75 μm × 37 cm. [Reproduced from O. Stalberg, D. Westerlund, U.-B. Rodby, and S. Schidmt, *Chromatographia* **41**, 287 (1995), with permission.]

three times the level of the baseline noise. Noise is defined as the random fluctuations of the detector output generated with time arising from variability in factors such as lamp output and spurious electronic signals.

The LOD may alternatively be expressed as the level attributable to the smallest peak that can be detected as a percent area/area of the electropherogram when determining purity. These LOD values will be dependent upon the molar absorptivity of the solute, the capillary bore, and the sample loading. LOD values of achiral impurities may be lower than 0.1%; often values of 0.01% are required. For example, levels of remoxipride impurities were determined at the 0.05% (Figure 16.5 shows separation of a test mixture

---

Figure 16.4. Separation of range of basic drugs dissolved in either (a) water or (b) 100 mM NaCl solution. Peaks: (1) imidazole; (2) aminobenzoate; (3) lamivudine; (4) salbutamol; (5) aspartame. Separation conditions—electrolyte: 25 mM NaH_2PO_4, pH 2.5; detection: 200 nm; voltage: + 10 kV; temperature: ambient; capillary: 75 μm × 37 cm; concentration: 0.1 mg/mL of basic drugs dissolved in water or 100 mM NaCl. [Reproduced from M. A. Kelly, B. J. Clark, and K. D. Altria, *J. High Resolut. Chromatogr.* in press (1998), with permission.]

containing 0.1% of each impurity) (49). A LOD of 0.1% of trace enantiomer in the presence of the main enantiomer has been reported (17,32,50).

16.2.12.2. The Limit of Quantitation

The LOQ value refers to the lowest level of impurity that can be precisely and accurately measured. The LOQ may be calculated (25) as 10 times the signal-to-noise ratio. Typically, replicate analysis of samples at the LOQ should give an RSD value of 10% or better (1,9,25,38).

16.2.13. Solution Stability

Typically in CE the volumes of electrolyte used daily may be on the order of 10–20 mL. Therefore preparation of 500 mL of electrolyte may be sufficient for several weeks of use. This is unlike HPLC, where typical daily requirements of mobile phase exceed 2000 mL, which is usually prepared fresh daily. Therefore, it is necessary to determine the shelf life of all CE reagent solutions. For example, a 3 month shelf life has been assigned for a phosphate–borate electrolyte when stored in a plastic container at room temperature, unprotected from light (45). However, some electrolytes are more sensitive to deterioration and need to be prepared daily or stored refridgerated.

Similarly, as in HPLC, storage conditions should be established for sample solutions. For example, a shelf life of 8 days (stored in a refrigerator) has been assigned for aqueous solutions of various basic drugs tested by CE (9).

16.2.14. System Suitability

The performance of the CE instrument is normally assessed prior to initiation of an analytical test sequence. As in HPLC, this testing commonly involves an assessment of parameters such as selectivity, precision, and separation performance. Table 16.5 gives some general guidelines as to the applicability of the system-suitability test to various application areas. For example, system suitability may be specified as any or all of the following parameters:

1. Resolution of > 2.0 for all components in a mixture
2. Precision of injection for the main components of < 2%
3. Peak efficiency of > 5000 for the main component

Other system-suitability parameters less commonly employed may include linearity and LOD. Linearity may be assessed by employing three calibration solutions, for example, covering 75–125% of the target sample concentration. A measurement of LOD may be significant when trace level quantitation is

Table 16.5. General Guidelines as to the System-Suitability Requirement for Application Areas[a]

System-Suitability Test	Identity Confirmation	Chiral/Achiral Impurity Test	Main-Component Assay
Precision	N/A	N/A	Yes
Linearity	N/A	N/A	Possibly
Sensitivity	Possibly	Yes	Possibly
Resolution	Yes	Yes	Yes
Migration time	Yes	Yes	Yes

Source: Reproduced from K. D. Altria and D. R. Rudd, *Chromatographia* **41**, 325 (1995), with permission.
[a]N/A denotes "not generally applicable."

performed. A maximum permissible peak tailing factor may also be specified in the method. However, this is not commonly calculated in CE as peak tailing and symmetry factors are very dependent on the sample concentration used. These aspects have been discussed by Gutmann and Cooke (50) with regard to determination of the trace enantiomer in naproxen. The main naproxen peak was severely asymmetric (peak asymmetry at 10% of peak width, $A_{10\%}$, was greater than 10); however, the system suitability criteria of (=resolution) greater than 2.0 was maintained despite the relatively poor peak shape.

Providing that the system meets these minimum requirements then the analysis may proceed. In some instances test injections are also positioned throughout the sequence and/or at the end of the sequence to demonstrate the continued suitability of the instrument through the analytical sequence.

16.2.15. Training

Suitable training is required for CE users, as such operations as installation of the capillary into the optical center of the detector are critical for obtaining optimal sensitivity.

16.2.16. Transfer of Methods Between Laboratories

Succesful formal transfer of CE methods between laboratories and sites has been reported (6). Formal method transfer between laboratories involves a considerable amount of comparative testing, and the protocols employed are established on a case-by-case basis decided between the companies and laboratories concerned. Generally this activity is concerned with the simulta-

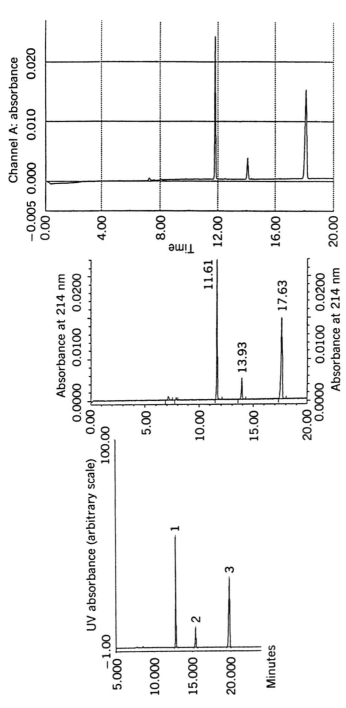

Figure 16.6. Separations of paracetamol capsules by MEKC in intercompany collaboration exercise. Peaks: (1) paracetamol; (2) caffeine; (3) 4-hydroxyacetophenone (interval standard). Conditions—electrolyte: 40 mM disodium tetraborate/125 mM SDS; detection: 210 or 214 nm; capillary: fused silica, 72 cm × 50 μm (1, 3), 57 × 50 μm (2); temperature: 40 °C. [Reproduced from K. D. Altria, N. G. Clayton, R. C. Harden, M. Hart, J. Hevizi, J. Makwana, and M. J. Portsmouth, *Chromatographia* **39**, 180 (1994), with permission.]

neous testing of identical samples. Comparisons of results generated by both laboratories should not differ significantly.

A series of intercompany cross-validation exercises have been conducted in which methods have been repeated in seven independent pharmaceutical companies. These were a chiral separation (36), an assay of the main component in a formulation (42), and a drug stoichiometry determination (43).

Figure 16.6 shows representative separations achieved by three companies for the separation of paracetamol, caffeine (another component of the paracetamol capsules tested), and acetophenone, the internal standard.

16.3. CONCLUSIONS

A general overview of the requirements for the validation of CE methods is presented, including literature examples to illustrate the applicability and suitability of the validation approach.

Basically an approach similar to that used for HPLC may be used, but there are differences between HPLC and CE that need to be recognized when one is validating CE methods. In particular, these include greater emphasis on the need to demonstrate transferability of CE methods from one manufacturer's equipment to another, the implication of different data-handling procedures, and the need to demonstrate satisfactory solution stability for electrolytes, for example, over periods of several weeks.

Within any analytical technique, validation of an analytical method should always be considered on an individual basis. Thus, while a number of official guidelines exist, often directed specifically toward certain fields of application (e.g., pharmaceuticals), such documents should be regarded as advisory only. Scientific judgment, supported by a sound rationale, should always be exercised—with the primary objective of method validation remaining uppermost: to demonstrate that the analytical method is suitable for the intended purpose.

LIST OF ACRONYMS

Acronym	Definition
CE	capillary electrophoresis
HPLC	high-performance liquid chromatography
LOD	limit of detection
LOQ	limit of qunatitation
MEKC	micellar electrokinetic chromatography
PTSA	p-toluenesulfonic acid
RSD	relative standard deviation

LIST OF ACRONYMS (Continued)

Acronym	Definition
SDS	sodium dodecyl sulfate
TBA	tetrabutylammonium
TLC	thin-layer chromatography
TPA	tetrapropylammonium
UV	ultraviolet

REFERENCES

1. K. D. Altria, and Y. L. Chanter, *J. Chromatogr.* **652**, 459 (1993).
2. B. R. Thomas, X. G. Fang, X. Chen, R. J. Tyrell, and S. Ghodbane, *J. Chromatogr.* **657**, 383 (1994).
3. K. D. Altria and D. R. Rudd, *Chromatographia* **41**, 325 (1995).
4. K. D. Altria, *J. Chromatogr.* **634**, 323 (1993).
5. P.A. Shah and L. Quinones, *J. Liq. Chromatogr.* **18**, 1349 (1995).
6. K. D. Altria and M. Kersey, *LG-CG*, January, 40 (1995).
7. K. D. Altria and S. D. Filbey, *J. Liq. Chromatogr.* **16**, 2281 (1993).
8. M. T. Ackermans, J. L. Beckers, F. M. Everaerts, and I. G. J. A. Seelen, *J. Chromatogr.* **590**, 341 (1992).
9. K. D. Altria, P. Frake, I. Gill, T. A. Hadgett, M. A. Kelly, and D. R. Rudd, *J. Pharm. Biomed. Anal.* **13**, 951 (1995).
10. C. O. Thompson, V. G. Trennery, and B. Kemmery, *J. Chromatogr.* **694**, 507 (1995).
11. K. D. Altria, T. Wood, R. Kitscha, and A. Roberts-McIntosh, *J. Pharm. Biomed. Anal.* **13**, 33 (1995).
12. S. Boonkerd, M. R. Detaevernier, and Y. Michotis, *J. Chromatogr.* **670**, 209 (1994).
13. W. Beck and H. Engelhardt, *Chromatographia* **33**, 313 (1992).
14. A. Pluym, W. Van Ael, and M. De Smet, *Trends. Anal. Chem.* **11**, 27 (1992).
15. G. C. Penalvo, E. Julien, and H. Fabre, *Chromatographia* **42**, 159 (1996).
16. P. Emaldi, S. Fapanni, and A. Baldini, *J. Chromatogr. A* **711**, 339 (1995).
17. K. D. Altria, D. M. Goodall and M. M. Rogan, *Electrophoresis* **15**, 824 (1994).
18. J. M. Jordan, R. L. Moese, R. Johnson-Watts, and D. E. Burton, *J. Chromatogr.* **640**, 445 (1994).
19. W. Thormann, S. Molteni, J. Caslavska, and T. Schmutz, *Electrophoresis* **15**, 3 (1994).
20. Z. Deyl, F. Tagliaro, and I. Miksik, *J. Chromatogr.* **656**, 3 (1994).
21. W. Thormann, S. Lienhard, and P. Wernly, *J. Chromatogr.* **636**, 137 (1993).
22. A. Shafaati and B. J. Clark, *Anal. Proc.* **30**, 481 (1993).
23. H. Watzig, *J. Chromatogr. A* **700**, 1 (1995).

24. K. D. Altria and J. Bestford, *J. Capillary Electrophor.* **3**, 13 (1996).
25. K. D. Altria, A. R. Walsh, and N. W. Smith, *J. Chromatogr.* **645**, 193 (1993).
26. A. Werner, T. Nassaure, P. Kiechle, and F. Erni, *J. Chromatogr.* **666**, 375 (1994).
27. K. D. Altria and K. Dave, *J. Chromatogr.* **633**, 221 (1993).
28. W. Beck, R. Van Hoeck, and H. Engelhardt, *Electrophoresis* **14**, 540 (1993).
29. E. C. Rickard and R. J. Bopp, *J. Chromatogr.* **680**, 609 (1994).
30. B. R. Thomas and S. Ghodbane, *J. Liq. Chromatogr.* **16**, 1983 (1993).
31. B. R. Thomas, X. G. Fang, X. Chen, R. J. Tyrell, and S. Ghodbane, *J. Chromatogr.* **657**, 383 (1994).
32. J. E. Noroski, D. J. Mayo, and M. Moran, *J. Pharm. Biomed. Analy.* **13**, 45 (1995).
33. J. Prunonosa, R. Obach, A. Diez-Cascon, and L. Gouesclou, *J. Chromatogr.* **581**, 219 (1992).
34. K. D. Altria, *Chromatographia* **35**, 177 (1993).
35. K. D. Altria, P. Harkin, and M. Hindson, *J. Chromatogr. B* **686**, 103 (1996).
36. K. D. Altria, R. C. Harden, M. Hart, J. Hevizi, P. A. Hailey, J. V. Makwana, and M. J. Portsmouth, *J. Chromatogr.* **641**, 147 (1993).
37. K. D. Altria and H. Fabre, *Chromatographia* **40**, 313 (1995).
38. M. E. Swartz, *J. Liq. Chromatogr.* **14**, 923 (1991).
39. G. Madrup, *J. Chromatogr.* **604**, 267 (1992).
40. C. L. Flurer and K. A. Wolnik, *J. Chromatogr.* **663**, 259 (1993).
41. S. D. Filbey, and K. D. Altria, *J. Capillary Electrophor.* **1**, 190 (1994).
42. K. D. Altria, N. G. Clayton, R. C. Harden, M. Hart, J. Hevizi, J. Makwana, and M. J. Portsmouth, *Chromatographia* **39**, 180 (1994).
43. K. D. Altria, N. G. Clayton, R. C. Harden, M. Hart, J. Hevizi, J. V. Makwana, and M. J. Portsmouth, *Chromatographia* **40**, 47 (1995).
44. K. D. Altria, and S. D. Filbey, *Chromatographia* **39**, 306 (1994).
45. K. D. Altria, I. Gill, J. Howells, C. N. Luscombe, and R. Z. Williams, *Chromatographia* **40**, 527 (1995).
46. A. Shafaati, and B. J. Clark, *Anal. Proc.* **30**, 481 (1993).
47. M. Korman, J. Vindevogel, and P. Sandra, *J. Chromatogr.* **645**, 366 (1993).
48. M. A. Kelly, B. J. Clark and K. D. Altria, *J. High Resolut. Chromatogr.*, in press (1998).
49. O. Stalberg, D. Westerlund, U.-B. Rodby, and S. Schidmt, *Chromatographia* **41**, 287 (1998).
50. A. Guttman and N. Cooke *J. Chromatogr. A* **685**, 155 (1995).

CHAPTER

17

TWO-DIMENSIONAL SEPARATIONS IN HIGH-PERFORMANCE CAPILLARY ELECTROPHORESIS

THOMAS F. HOOKER, DOROTHEA J. JEFFERY, and JAMES W. JORGENSON

Department of Chemistry, The University of North Carolina at Chapel Hill, Chapel Hill, North Carolina 27599-3290

17.1.	Introduction	582
17.2.	Two-Dimensional Separation Theory	583
	17.2.1. Peak Capacity	583
	17.2.2. The Statistical Model of Overlap	584
	17.2.3. Two-Dimensional Peak Capacity	586
	17.2.4. Criteria for Two-Dimensional Separations	586
17.3.	Comprehensive Two-Dimensional Liquid Chromatography–Capillary Electrophoresis	588
	17.3.1. High-Performance Reversed-Phase Liquid Chromatography–Capillary Zone Electrophoresis	589
	17.3.2. Microcolumn Size Exclusion Chromatography–Capillary Zone Electrophoresis	591
	17.3.3. Microcolumn Reversed-Phase Liquid Chromatography–Capillary Zone Electrophoresis	597
	17.3.4. High-Performance Reversed-Phase Liquid Chromatography–High-Speed Capillary Zone Electrophoresis	601
17.4.	Three-Dimensional Size Exclusion Chromatography–Reversed-Phase Liquid Chromatography–High-Speed Capillary Zone Electrophoresis	605
17.5.	Future Directions	607
	List of Acronyms	610
	References	610

High Performance Capillary Electrophoresis, edited by Morteza G. Khaledi. Chemical Analysis Series, Vol. 146.
ISBN 0-471-14851-2 © 1998 John Wiley & Sons, Inc.

17.1. INTRODUCTION

Samples of biological and environmental origin are often complex in composition. The number of potentially detectable components present in these types of mixtures can exceed a hundred and, in some cases, a thousand. Separation techniques are playing an ever-increasing role in the analysis of these samples. Capillary gas chromatography (GC), high-performance liquid chromatography (HPLC), and capillary electrophoresis (CE) are used extensively for the separation and quantitation of components in both natural and synthetic mixtures. While capillary GC is an extremely powerful technique for the analysis of volatile compounds, HPLC and CE are better suited for the analysis of nonvolatile, polar compounds since they operate with a liquid mobile phase. Because they can be operated under mild, nondenaturing conditions, they are popular techniques for use in biological analyses.

There are many cases where one-dimensional (1D) techniques are incapable of handling samples due to the sheer number of components they contain. This limitation has generated interest in two-dimensional (2D) separations in recent years due to the increased peak capacity of these systems over 1D systems. That increase is because the total peak capacity of 2D system is determined by multiplying the peak capacities of the individual constituent dimensions together. This results in an increased amount of separation space into which peaks may be distributed.

There are certain criteria that must be met for a 2D separation to be considered comprehensive 2D. This will be defined clearly below. The purpose of this chapter is to focus on comprehensive 2D separations that use CE. Many research groups have used CE in an off-line or on-line mode to further enhance a primary separation (1–7); while these separations may be 2D, they are not comprehensive 2D by definition. To our knowledge, the only comprehensive 2D work involving CE has been in the author's laboratory and at Hewlett Packard (8).

Comprehensive 2D separations were first employed in a planar format. The earliest work involved 2D paper chromatography (9); work with 2D paper chromatography–electrophoresis (10) and 2D electrophoresis (11) soon followed. Other planar 2D separations were developed (2D TLC (12,13), as well as coupled column–planar systems (GC–TLC, HPLC–TLC) (14), to extend resolving power as well as applicability of 2D systems to a wider range of samples.

The most powerful 2D planar separation was demonstrated by O'Farrell in 1975 (15). In this 2D gel electrophoresis system, isoelectric focusing (IEF) was coupled with sodium dodecyl sulfate–polyacrylamide gel electrophoresis (SDS–PAGE) for the separation of proteins. The 2D IEF/SDS–PAGE system was able to resolve more than 1000 proteins from an *Escherichia coli* sample.

A result of this work was increased interest in the field of 2D electrophoresis (16).

Work on 2D coupled-column systems grew out of the desire to automate 2D separations and make them more quantitative. Numerous 2D combinations exist due to the diversity of separation techniques available, and many have been implemented. Limits of space prevent our mentioning all but a few of them here: GC–GC, SFC–GC, SFC–SFC, LC–GC, LC–LC, LC–SFC, and LC–CE. (SFC is supercritical fluid chromatography) Interested readers are directed to reviews of these techniques for more information (17–24).

This chapter will focus on comprehensive 2D separations that combine HPLC with CZE. In comprehensive systems, all analytes in the sample mixture are subjected to separation in both dimensions. A brief look at the statistical model of overlap (SMO) and the limitations of 1D separations will be examined. The criteria by which 2D separations are defined will be discussed. Following this, the different 2D schemes developed using CZE will be presented. This includes the coupling of high-performance reversed-phase liquid chromatography (RPLC) and size exclusion chromatography (SEC) with CZE in 2D systems and tandem SEC–RPLC–CZE in a three-dimensional (3D) system. Our emphasis will be on the instrumentation involved in these systems and, in particular, the interfacing involved between the LC and CZE dimensions.

17.2. TWO-DIMENSIONAL SEPARATION THEORY

17.2.1. Peak Capacity

In 1967, Giddings developed the concept of peak capacity (n_c) for determining the maximum number of components that could be resolved within a chromatographic run (25). For isocratic or isothermal separations, the peak capacity is given as

$$n_c = \frac{\sqrt{N}}{4R_S}\ln\left(\frac{t_{max}}{t_{min}}\right) + 1 \qquad (17.1)$$

where N is the number of theoretical plates; R_S is the resolution of the separation (usually 1); and t_{min} and t_{max} are the retention times of the first and last peaks in the elution window, respectively. As the theoretical plate is less easily determined in programmed runs, such as temperature gradient GC or gradient elution LC, and peak widths throughout such runs are relatively constant, the expression for peak capacity becomes

$$n_c = (t_{max}/t_{min})/w_t \qquad (17.2)$$

where w_t is the width of a peak in units of time. A column that has an elution range of 100 minutes with eluting peaks that are 2 min wide is said to have a peak capacity of 50. A peak capacity of 50 means that 50 single-component peaks can be lined up, one after another, with a stated resolution (usually 1), into a system's elution range. However, the components in very few mixtures follow this ideal elution profile. Generally, peaks will scatter randomly and often overlap within a separation space and show no evidence of regular, even spacing. Therefore, peak capacity is to be treated as an ideal number and is to be regarded only as the *maximum* number of components that can be resolved by a particular separation system under a given set of conditions. A mixture of 50 components will not be fully resolved on a column with $n_c = 50$. The size of the peak capacity should be much larger than the number of components if one wishes to resolve them completely. This is due to the seriousness of the overlap of peaks within a chromatogram.

17.2.2. The Statistical Model of Overlap

The SMO was developed by Davis and Giddings to address the random distribution of peaks within a chromatogram and determine the extent and effect of the phenomenon (26). The theory assumes that components fall randomly across the 1D or 2D space the system provides. Theoretical reasons have been provided, and experimental data have supported the random model (26–28). Work by Guiochon's group has also provided additional evidence and support for the random hypothesis (29–32). Limitations in space prevent an in-depth discussion of SMO theory here. We shall only briefly discuss the SMO as it pertains to the material in this chapter. Readers should refer to the aforementioned articles for more detailed information.

Equation (17.3) derived from the SMO, illustrates the relationship between the number of resolved singlets (s), the number of components in the sample (m), and the peak capacity (n_c):

$$s = m \exp\left(-2\frac{m}{n_c}\right) \quad (17.3)$$

The ratio m/n_c is the saturation factor, which is indicative of how well the peak capacity is used by the sample components. Dividing each side of Eq. (17.3) by m gives the probability (P_s) that a component will be resolved as a single peak:

$$P_s = \frac{s}{m} = \exp\left(-2\frac{m}{n_c}\right) \quad (17.4)$$

The ratio of s/m is indicative of what fraction of the m components are separated as single peaks. This is illustrated in Figure 17.1, where the peak capacity is plotted against the fraction isolated for a mixture of 100 components ($m = 100$). We can see the sharp increase in peak capacity that is required if we wish to resolve a greater percentage of the components into singlets. If we want to resolve 50 of the 100 components ($s/m = 0.5$), then a peak capacity of 290 is required. This translates into 250,000 theoretical plates when we use the relationship between N and n_c in Eq. (17.1) for a system where the ratio of t_{max} to t_{min} is 10.

This is an amount of separation efficiency that some 1D techniques can provide, and yet only half of the components are resolved into singlets! This limited amount of peak capacity is a major drawback of 1D separations. In order to resolve 90% of the 100 components into singlets ($s/m = 0.9$), a peak capacity of 1900 is needed with an equivalent plate count of 10 million. The situation becomes worse with mixtures containing more than 100 components. Clearly, current 1D techniques cannot generate efficiency values of this magnitude. The fact that many techniques are already operating near maximum efficiency indicates that large gains in peak capacity will not be realized with incremental improvements in 1D efficiency. This provides the motivation for pursuit of 2D techniques.

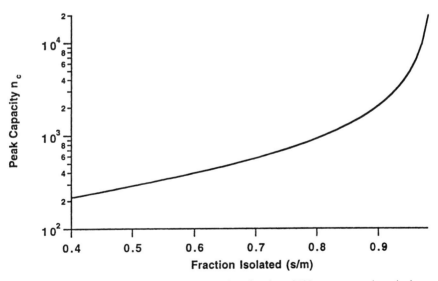

Figure 17.1. Peak capacity required to resolve a given fraction of 100 components into singlets.

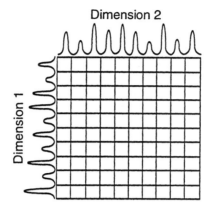

Figure 17.2. The peak capacity of a 2D system, represented by the number of boxes, is approximately equal to the product of the peak capacities n_{c1} and n_{c2} generated along the two individual axes, as represented by the number of adjacent Gaussian profiles. [Adapted from J. C. Giddings, HRC & CC, J. High Resolut. Chromatogr. Chromatogr. Commun. **10**, 319–323 (1987).]

17.2.3. Two-Dimensional Peak Capacity

Greater separation power is achieved in 2D systems because of the multiplicative nature of peak capacity. Work by Giddings as well as Guiochon and colleagues has shown that the peak capacity of a 2D system is roughly equal to the product of the peak capacities of the individual dimensions (33,34):

$$n_c = n_{c1} \times n_{c2} \tag{17.5}$$

This is illustrated in Figure 17.2 for a simple 2D system. In this hypothetical 2D separation, each 1D separation has an n_c of 10 whereas the 2D separation has a total n_c of 100. It is clear that coupling separations of even modest efficiencies can generate a peak capacity beyond what can be obtained in a 1D case. This is the advantage of 2D separations and is the reason for research in this area.

17.2.4. Criteria for Two-Dimensional Separations

17.2.4.1. Orthogonality

Giddings has defined two criteria to be met in order for a separation to be considered multidimensional (35). First, the coupled separations must be orthogonal to one another. To be orthogonal, each dimension must base its

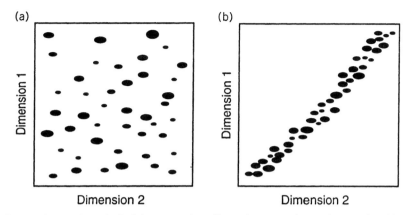

Figure 17.3. Two hypothetical 2D separations illustrating an orthogonal separation (a) and a correlated, nonorthogonal separation (b).

separation mechanism on as different molecular characteristics as possible. The amount of redundancy in the system will therefore be reduced, and the most will be made out of the particular coupling. Figure 17.3 illustrates two hypothetical 2D separations: in Figure 17.3a, the two combined techniques are orthogonal and the 2D space is used effectively; in Figure 17.3b, the two separations are correlated, as indicated by the diagonal pattern, and little improvement is realized in their coupling. Work by Freeman has indicated that coupling two similar techniques results in only minor improvement over a 1D separation (36). A particular concern arises when one is attempting to couple orthogonal separation techniques. Often, the more dissimilar they are in operation, the more difficult it may be to couple the two systems. However, this has not posed a problem thus far in the LC–CE systems discussed here, as both methods of separation employ aqueous liquid phases.

17.2.4.2. Resolution Preservation

The second criterion that must be met for a separation to be considered 2D states that any separation or resolution obtained in the first dimension must be preserved throughout the entire displacement (35). This stipulation has a number of implications for a coupled-column system. A simple tandem column arrangement, where effluent from one column is fed directly into another, is ruled out as being 2D. Components separated on the first column may recombine in the second due to a different order of migration rates. In order to preserve the resolution generated by the first dimension, there must be discrete and efficient sample transfer from one dimension to the next. When there is frequent sampling of the first dimension by the second, a representative peak profile is generated. In practice, this is achieved by operating the second dimension on a much faster time scale than the first. When improper sampling occurs there is a misrepresentation of the first dimension separation and unfortunately separated components are allowed to recombine. There are situations, however, when one has interest only in a few substances in a complex mixture.

When only a portion of a mixture is of interest, heart-cutting techniques are used in order to eliminate the part of the sample that is not relevant to the particular study. In a heart-cutting scheme, a "cut" or fraction eluting from the first dimension is shunted to a secondary column of different selectivity for additional resolution of a few components. Knowledge of the elution profile of the first column is needed in order to select the region of interest. Typically, the size of the region that is chosen is larger than the component peak widths to ensure transfer of the entire sample band. Because such a large fraction is transferred, components resolved on the first-dimension column recombine.

Comprehensive sampling is essentially a logical extension of the heart-cutting approach though the two differ in a number of ways. The sampling frequency in a comprehensive system is much greater than that in a heart-cut mode. In a comprehensive system, many more slices are taken across a peak and analyzed by the second dimension. This results in a better representation of the first-dimension separation and reduces the chances of recombination. Another important characteristic of these systems is that all sample components are subjected to separation in both dimensions. In contrast, the heart-cut mode subjects only a selected portion of the mobile phase eluted from the primary column to a secondary separation. Additionally, because all components are analyzed in a comprehensive system, a priori knowledge of the first dimension elution profile is not needed.

17.3. COMPREHENSIVE TWO-DIMENSIONAL LIQUID CHROMATOGRAPHY–CAPILLARY ELECTROPHORESIS

A general block diagram illustrating the major components of a comprehensive 2D LC–CE system is shown in Figure 17.4. The LC–CE systems that have been used were designed in this general format. The first dimension consists of an HPLC system: a pump, injection valve, and column. An interface that is responsible for quick and efficient transfer of LC effluent to the CE dimension is also required. The interfacing schemes developed to this point and their usefulness with various LC column sizes will be discussed in detail in the following subsections. Components eluted from the LC column are injected

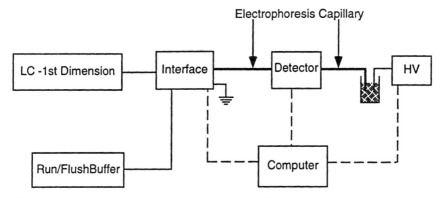

Figure 17.4. A general schematic for 2D LC–CE. The LC dimension is run on a time scale of hours, whereas CE runs are performed in seconds. Various interfacing schemes have been developed for efficient transfer of LC effluent to the electrophoresis capillary. These are described more fully later in the text.

onto a fused-silica capillary that is used for the second-dimension electrophoretic separation. Ultraviolet (UV) absorption and laser-induced fluorescence (LIF) have been the two popular detection techniques used with these systems. Automation is a key part to the operation of the LC–CE system. Computer control of the interface and the high-voltage CE power supply has been implemented so that injections of LC fractions onto the electrophoresis capillary are performed in a consistent and reproducible manner. The computer is also used for data acquisition to allow permanent storage of the data.

17.3.1. High-Performance Reversed-Phase Liquid Chromatography–Capillary Zone Electrophoresis

The first comprehensive coupling of LC with CZE was achieved by Bushey and Jorgenson in 1990 (37). The first-dimension separation utilized RPLC. The LC column was 1 mm i.d. and 25 cm long, and was operated under gradient conditions. The key part to coupling the LC to the CZE in this system was an electrically actuated six-port valve fitted with a 10 μL loop. The two positions for the valve are shown in Figure 17.5. In the "run" position, mobile phase eluted from the LC column (C1), operated at 20 μL/min, filled the 10 μL loop (L) in 30 s, while the flush pump (P2) delivered CZE buffer past the inlet of the capillary and out of the low-dead-volume tee (T) to waste (W). The run voltage was applied to the CE capillary at this time, and flush buffer migrated

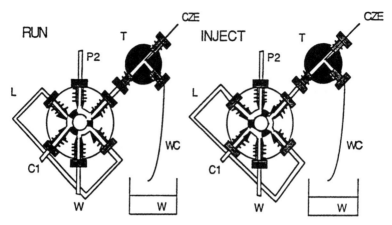

Figure 17.5. Two configurations of six-port, computer-controlled valve used in the first comprehensive coupling of LC with CZE. Key: C1, reversed-phase HPLC column; P2, flush pump 2; L, 10 μL loop, CZE, capillary electrophoresis fused-silica capillary; WC, waste capillary; T, Valco low-dead-volume tee; W, waste. [Reprinted from M. M. Bushey and J. W. Jorgenson, *J. Microcolumn Sep.* **2**, 293–299 (1990), with permission of John Wiley and Sons, Inc., New York.]

onto the capillary. The tee in this case acted as an inlet buffer vial, one in which the buffer was continuously replenished. In the "inject" position, mobile phase eluted from the RPLC column went waste while buffer from the flush pump forced the contents of the 10 μL loop past the inlet of the CE capillary. During this period, generally 5 s, the injection voltage was applied to the CZE capillary and an electromigration injection of the contents of the flowing stream was performed. The injection voltage applied to the CE capillary at this time was much less than the run voltage in order to prevent sample overload. After the contents of the loop were flushed by, the valve was returned to the run position restoring the flow of flush buffer. The run voltage was then reapplied, and data acquisition was begun for the next CZE separation. This process was continued until all sample components were eluted from the LC column.

The entire volume of eluted mobile phase from the LC column was not injected onto the CE capillary due to the volume mismatch between the RPLC and CZE systems. Peak elution volumes from the LC column were 20–40 μL, whereas injection volumes on the CZE capillary were on the order of nanoliters. Nonetheless, representative portions of the eluted LC mobile phase were transferred to the CE capillary. This was accomplished by performing the electromigration injection while the loop contents were being flushed by the CE capillary inlet. In this way, a percentage of the loop contents was injected onto the CE capillary (much like what happens with split injection in microcolumn LC).

The separation of a fluorescamine-labeled tryptic digest of horse heart cytochrome *c* by this system is shown in Figure 17.6. A surface plot and a contour map of the separation are both presented. The surface plot allows for the intensities of peaks to be seen, but not all of the peaks in the 2D separation space can be seen simultaneously. The contour plot allows the entire data set to be seen and is useful for protein fingerprinting or peptide mapping (38). The results from this separation clearly illustrate that neither technique alone has sufficient resolving power for the mixture but that when the two are combined a more complete separation results. The orthogonality of the two techniques can be easily seen as well. Peaks are randomly scattered in the 2D space, and no diagonal correlation is observed.

An improvement upon this same basic system was made by Larmann et al. (39). The system utilized a 2.1 mm i.d. RPLC column operated under gradient conditions for the first-dimension separation. The effluent from this column was collected in two 10 μL loops fitted onto an eight-port electrically actuated valve. The contents of one loop would be flushed by the CZE inlet and a representative sample would be injected onto the CZE capillary. At the same time, the effluent from the RPLC column would be filling the second loop. The peak capacity of the system was improved fivefold over the previously described system. This improvement was attributed to optimization of the LC system [water/acetonitrile/trifluoroacetic acid (TFA) eluents were used instead of the phosphate buffer] and an improvement in the voltage programming for CZE injections. The improved injection scheme resulted in narrower sample injection widths, which led to an overall improvement in the CE separation.

17.3.2. Microcolumn Size Exclusion Chromatograaphy–Capillary Zone Electrophoresis

The use of microcolumns in a 2D LC–CZE system was pursued in an effort to increase the peak capacity of the overall 2D system. Previous work has demonstrated the improvement in separation efficiency when packed capillaries are used for LC separations (40–43). Since microcolumns are more efficient per unit length and they can be packed in lengths of 1 m or more, they provide increased resolution and greater peak capacity when they are used in a 2D system. Efficiencies of well over 100,000 theoretical plates per meter have been obtained with these packed microcolumns.

Two other factors prompted the switch to microcolumns for 2D work. First, there is a closer volume match between a microcolumn and a CE capillary. Previous 2D LC–CE systems used columns with millimeter-sized bores that had peak elution volumes ranging from microliters to tens of microliters. Injection volumes in CE are typically hundreds of picoliters to

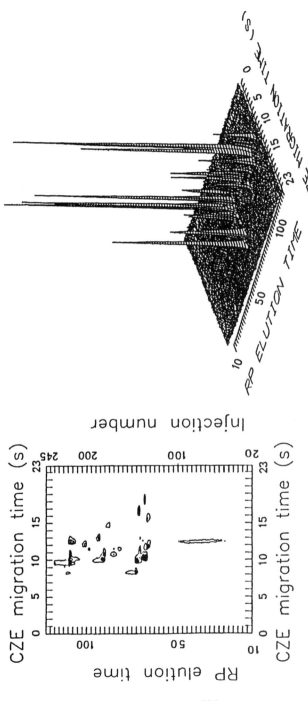

Figure 17.6. Contour and 3D surface plot of 2D LC–CE separation of fluorescamine-labeled tryptic digest of horse heart cytochrome *c*. Obtained with 15 μm i.d. CE capillary. CE injections: −1 kV, 5 s; CE runs: −28 kV, 0.5 min. Pump 1 (P1) flow rate, 20 μL/min; pump 2 (P2) flow rate, 0.3 mL/min. CE capillary: 6.5 cm to detector, 26 cm overall. Eight points per second collected. Every other point displayed for injections 20 through 245. [Reprinted from M. M. Bushey and J. W. Jorgenson, *J. Microcolumn Sep* **2**, 293–w299 (1990), with permission of John Wiley and Sons, Inc., New York.]

nanoliters. This is a 1000- to 10,000-fold mismatch in volumes between the two dimensions. Thus, a large portion of the LC column effluent is wasted, since such volumes cannot be loaded onto a CE capillary. When a microcolumn is used, peak elution volumes are in nanoliters. With proper interfacing, there is a larger percent transfer to the CE capillary. The move to microcolumns was also prompted for sensitivity reasons. As the total column volume for a microcolumn is much less than that of a conventional sized column, sample dilution is diminished. A perfect example of this is the analysis of single cell contents where the cell volume injected onto the microcolumn was hundreds of picoliters (44–47). Reduction in sample dilution in the first dimension results in improved sensitivity for the entire 2D system.

17.3.2.1. Valve–Loop Interface

The first use of microcolumns in a 2D system was demonstrated by Lemmo and Jorgenson. Microcolumn SEC was used in combination with CE for the separation of proteins in bovine, equine, and human serum samples (48). A valve–loop interface similar to the one in Figure 17.5 was used to couple the SEC column to the CE capillary. The loop in this case was a 50 μm i.d., 15 cm long fused-silica capillary that had a volume of 300 nL. The SEC microcolumn used was 250 μm i.d. and 105 cm long. The amount of SEC effluent to be collected was determined by the nonloop volumes in the valve (port and rotor volumes) in addition to the size of the loop used. This fixed volume was ~900 nL and was generated in the time it took complete one CZE run. In order to fill this volume in a few minutes the column had to be operated at a flow rate well over optimum. The result was a diminished SEC separation efficiency. Although slower SEC flow rate could be used, the time to fill the loop would then be too long and CZE runs would not be performed as frequently. This change would cause diminished sampling of the first dimension and would nullify some of the resolution obtained in the first dimension.

17.3.2.2. Internal Rotor Collection Scheme

The first attempt to overcome the problem of collecting a fixed volume can be seen in Figure 17.7 (39,49). Instead of collecting SEC effluent in a loop, an internal rotor of the valve (500 nL volume) was used. This figure shows the "run" position when electrophoresis buffer flowed past the CE inlet and SEC mobile phase was collected in the internal loop. When the valve was switched to the "inject" position (not shown), the collected fraction was flushed by the CE capillary and electroinjected. The smaller collection volume allowed the flow rate of the SEC microcolumn to be decreased closer to the optimum rate, increasing resolution. Additionally, when this system was used, the dilution of

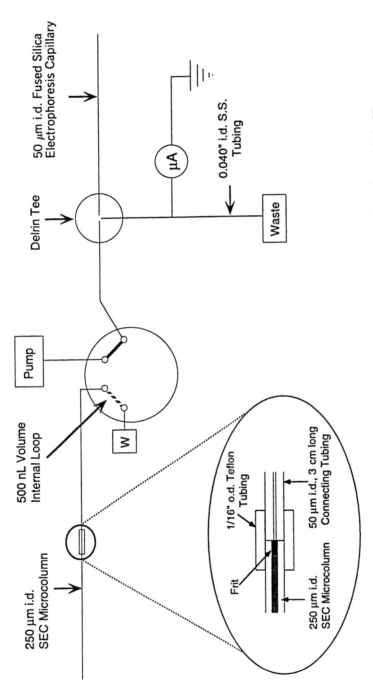

Figure 17.7. View of the internal rotor collection scheme used to interface the SEC microcolumn with the CE capillary. The inset shows the union between the SEC microcolumn and the fused-silica connecting tubing used to interface the microcolumn to the valve. [Reprinted from J. P. Larmann, Jr., A. V. Lemmo, A. W. Moore, Jr., and J. W. Jorgenson, *Electrophoresis* **14**, 439–447 (1993), with permission of VCH Verlagsgesellschaft mbH, Weinheim, Germany.]

eluted SEC peaks was reduced from 50 to 20% compared to the previously described system (49). A separation of protein standards and formamide (a low-molecular-weight marker) using this 2D system can be seen in Figure 17.8. Baseline resolution of all the components was not achieved when either separation was run separately. By combining the two techniques in a 2D system, all sample components were resolved. Although this was an improvement over the valve–loop method, the need to collect a fixed volume of sample proved to be too great a restriction on operating conditions. A new interfacing scheme was developed to address this limitation.

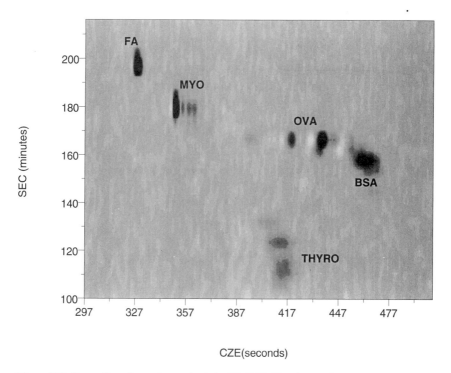

Figure 17.8. Separation of protein standards by 2D SEC–CE using the internal rotor collection scheme. SEC injection was 10 min at 7 bar (100 p.s.i.). Head pressure of 28 bar (400 p.s.i.) was applied to generate a flow rate of 180 nL/min for the chromatographic separation. Electrophoresis capillary was 58 cm overall, 40 cm to the detection window. CE conditions are 6 s electromigration injection at −5 kV and 9 min runs at −11 kV (injection made every 4.5 min). Data collection was 2 points/s. Key: BSA, bovine serum albumin; FA, formamide; MYO, horse heart myoglobin; THYRO, thyroglobulin; OVA, chicken egg albumin. [Reprinted from J. P. Larmann, Jr., A. V. Lemmo, A. W. Moore, Jr., and J. W. Jorgenson, *Electrophoresis* **14**, 439–447 (1993), with permission of VCH Verlagsgesellschaft mbH, Weinheim, Germany.]

17.3.2.3. Transverse-Flow-Gating Interface

The transverse-flow-gating interface was developed to overcome the problems and limitations encountered with previously discussed interfacing schemes (50). In Figure 17.9 a schematic for the interface is displayed. Two stainless steel plates were separated from each other by a Teflon spacer (or gasket), generally 0.005 in. (130 µm) thick. A channel cut into the Teflon allowed buffer flow between the two plates. The outlet of the LC microcolumn was positioned directly across from the inlet of the CE capillary, and they were separated by the thickness of the gasket.

A transverse flow of CE buffer (controlled by an electrically or pneumatically actuated valve) acted as a "gate" preventing eluted LC mobile phase from electromigrating onto the CE capillary. When an injection was made the transverse flow was diverted to waste, allowing sample to fill the gap between the two capillaries. An electrokinetic injection of this material was then performed. After the injection, the transverse flow was resumed while the CE voltage was kept at 0 V. This short period, called the "slew-up" time, allowed excess eluted mobile phase to be sufficiently swept out from between the capillaries; without this slew-up period, tailed peaks resulted. Following this

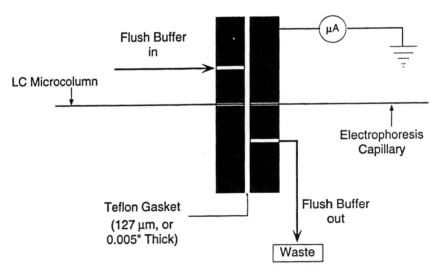

Figure 17.9. Schematic of the flow-gating interface. The interface was built in-house from two 3 in. diameter plates of stainless steel. Sandwiched between these plates is a Teflon gasket (shown here with a thickness of 127 µm) with a channel cut in it to allow liquid flow. Details of the operation of the device are given in the text. [Reprinted from A. V. Lemmo, and J. W. Jorgenson, *Anal. Chem.* **65**, 1576–1581 (1993), with permission of The American Chemical Society, Washington, DC.]

injection process, the CE run voltage was applied to the capillary. Before the next injection was made, the run voltage was decreased to 0 V; after a short period called the "slew-down" time, the transverse flow was again stopped; then the injection process just described was repeated. This slew-down time was necessary to prevent any preinjection that might occur while the high voltage was lowered from run to injection voltage. Figure 17.10 shows a timing diagram for the entire process.

Figure 17.11 shows the 2D SEC–CZE separation of a mixture of protein standards using the flow-gating interface. The SEC microcolumn used for this separation was 100 μm i.d. and was operated at a flow rate of 20 nL/min. The improvement in resolution in the SEC dimension is due to operation of the column at a flow rate close to optimal. The ability to operate at this low flow rate with the valve–loop scheme is simply not possible. It would take 45 min to collect a 900 nL LC fraction, but peaks eluting off the SEC column are narrower. Separated peaks would recombine during the transfer partially or completely, nullifying the resolution obtained in the first-dimension separation.

17.3.3. Microcolumn Reversed-Phase Liquid Chromatography–Capillary Zone Electrophoresis

Use of the flow-gating interface was also demonstrated with the coupling of μRPLC (microcolumn RPLC) with CZE for the analysis of peptides (51,52). The first-dimension separation consisted of a RPLC microcolumn that was 60 cm long, 50 μm i.d., and was packed with C_8 modified silica particles. The column was operated with an acetonitrile/water gradient with TFA added as a mobile phase modifier. In order to have CZE runs completed on a timescale of seconds, a short length of capillary [total capillary length $(L) = 24.5$ cm; length of capillary to detector $(l) = 15.5$ cm] was used while dropping a large potential (-22.5 kV) across it. In order to reduce the amount of CZE current and Joule heating, a capillary i.d. of 15 μm was used. This precluded using UV detection because of the extremely short pathlength and subsequent lack of sensitivity that would be obtained with absorption detection. LIF of the tetramethylrhodamine isothiocyanate (TRITC)–derivatized amines was the detection scheme used instead. The green helium–neon laser (543.5 nm) was an excellent match for the TRITC tag with its maximum absorbance at 548 nm. Figure 17.12 shows the 2D μRPLC–CZE separation of TRITC-derivatized human plasma. This sample contained only the low-molecular-weight fraction of the plasma. Large serum proteins were removed with a molecular-weight cutoff filter.

A tremendous number of components in the sample have been resolved with this technique. It is interesting to note how three vertical bands are seen with respect to the CZE separation. These three bands correspond to species

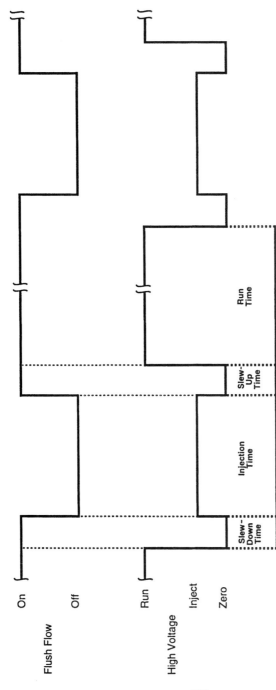

Figure 17.10. Timing sequence used to perform injections with the flow-gating interface. During the "run time" the flush buffer is on, carrying LC effluent to waste. The injection sequence begins by dropping the high voltage to 0 V. Keeping the transverse flow on during this "slew-down" period prevents any spurious injection and peak fronting while the high voltage drops from run to 0 kV. During the "injection time" the transverse flow is stopped, allowing effluent to flow across to the CZE capillary and bee lectroinjected. The injection is terminated by allowing the transverse flow to resume. The "slew-up" time allows the gap between the two capillaries to be swept clear of LC effluent before the high voltage slews to the run voltage.

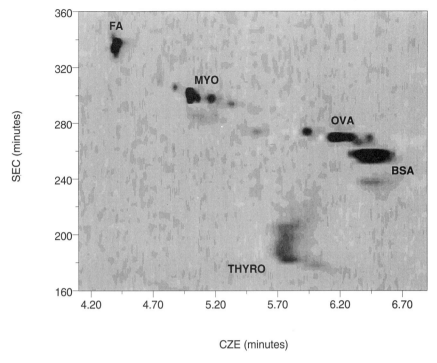

Figure 17.11. Separation of protein standards by 2D SEC–CZE with the flow-gating interface and a 100 μm i.d. microcolumn. Each protein was present at 0.5% (w/v) with 2.5% (w/v) formamide (FA). See Figure 17.8 for all explanation of the other abbreviations. SEC injection was 8 min at 7 bar (100 p.s.i.). Head pressure of 14 bar (200 p.s.i.) was applied to generate a flow rate of 20 nL/min for the SEC separation. The electrophoresis capillary was 53 cm long, 33 cm to the detection window. CZE conditions were 30 s electromigration injection at 0 kV and 4 min overlapped runs at −11 kV. The actual CZE run time was 8 min. Head pressure of 0.5 bar (7 p.s.i.) generated a flush flow rate of 100 μL/min. The Teflon gasket thickness was 76μm. Data collection was 2 points/s. [Reprinted from A. V. Lemmo, and J. W. Jorgenson, *Anal. Chem.* **65**, 1576–1581 (1993), with permission of The American Chemical Society, Washington, DC.]

with net charges of 0, −1, and −2 at migration times of 36, 41, and 47 s, respectively. It is believed that since the analytes in the sample are all relatively small, those with the same charge will have very similar mobilities. In a sample where there is a greater diversity in the size of components (and thus mobilities), there is a more effective use of the 2D separation space. In this case, it is clear that separation by CZE alone would not be sufficient for this sample. In addition there are many peaks that coelute on the RPLC column but are resolved further by CE. The peak capacity for the LC dimension has been

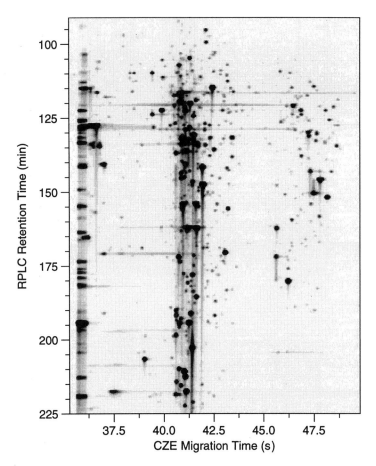

Figure 17.12. Two-dimensional μRPLC–CE separation of TRITC-derivatized human plasma. The plasma was filtered through a 10 kDa molecular-weight cutoff filter to remove large serum proteins prior to derivatization. The RPLC microcolumn was operated with a gradient from 15 to 45% acetonitrile for 240 min. The remainder of the mobile phase composition was H_2O. The H_2O and CH_3CN both contained 0.1% TFA. The flush flow of CZE buffer was 0.4 mL/min. The electrophoresis capillary was 15 μm i.d., $L = 24.5$ cm, $l = 15.5$ cm. The injection voltage was -1.1 kV for 3 s, and the run voltage was -22.5 kV. The total CZE analysis time was 29 s. [Reprinted from J. P. Larmann, Jr., Doctoral Dissertation, University of North Carolina, Chapel Hill, (1993).]

estimated at 200, whereas for the CZE it is 100. This translates into a 2D peak capacity of 20,000, which is unmatched by any coupled-column system.

The great resolving power of this system was also demonstrated with other biological samples including a urine sample, a tryptic digest of porcine thyroglobulin, and the contents of a single snail neuron from *Helix aspersa*. All of these samples contained a tremendous number of components (more than 400 in the thyroglobulin digest), and there would be no hope of resolving all of them with a 1D separation method. The ability to separate and detect the contents of a single cell by this method is a tribute to the small column volumes involved and the minimal amount of dilution and dispersion introduced by the flow-gating interface. The large number of components detected at low levels with this system also demonstrated another important aspect of complex mixtures. Although an ultrasensitive LIF detection scheme was used in these studies, it was determined that the resolving power of the separation was as important as the sensitivity of the detector in determining low levels of analytes. In order to determine low levels of endogenous compounds in biological matrices a sensitive detection scheme is needed, but unless the analytes of interest are resolved from the "chemical noise" in the sample, the increased sensitivity has only served to increase the complexity of the system and not lower detection limits. The use of selective detectors could be a solution to the dilemma if they have adequate sensitivity and selectivity for each particular application. The components of interest still need to be resolved.

17.3.4. High-Performance Reversed-Phase Liquid Chromatography–High-Speed Capillary Zone Electrophoresis

A high-speed 2D LC–CZE setup was designed by Moore and Jorgenson for the analysis of peptide samples [(53); see also Larmann et al. (39)]. The second-dimension separation used was high-speed CZE originally developed by Monnig and Jorgenson (54) and later improved upon by Moore and Jorgenson (55). Fast CZE used a unique injection scheme referred to as optical gating. Figure 17.13 illustrates the instrumental setup for high-speed CZE. The ends of the capillary rested in buffer reservoirs to which high voltage was applied. The upper buffer reservoir was also the sample reservoir. The capillary was mounted vertically, and the electroosmotic flow (EOF) was toward the more negative electrode (bottom). Gravity-induced flow was not a concern due to the small capillary internal diameters of 10 µm or less that were used. Optically gated injections use a powerful laser beam from an argon ion laser that is split and focused onto two different spots on the CE capillary. A gating beam of approximately 95% of the laser power was focused near the top of the capillary. Further down from this point (usually 1–2 cm) the other 5% of the laser power, or the probe beam, was focused. Voltage applied to the buffer reservoir continuously electromigrated sample onto the capillary. When the high-power gating beam is on and fluorescently labeled [fluorescein-5-isothiocyanate (FITC)] species migrated into the beam, they were

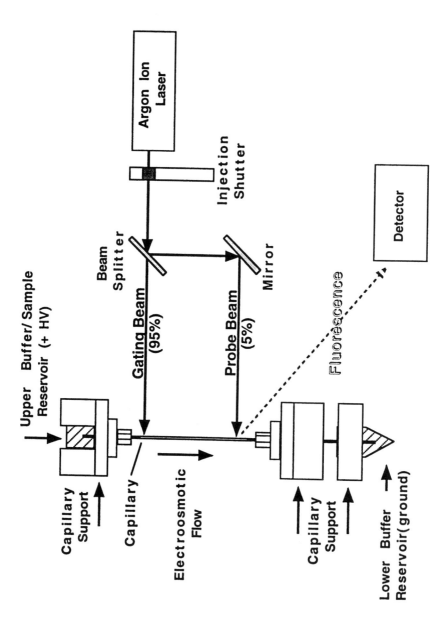

Figure 17.13. Experimental setup for high-speed CZE.

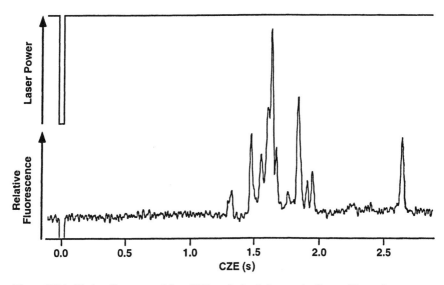

Figure 17.14. Timing diagram and fast CZE analysis of the tryptic digest of horse heart cytochrome c: injection time, 20 ms; 20 kV applied over total capillary length of 8 cm, with 2 cm between gating and probe beams. Data were smoothed with a five-point moving average filter to reduce baseline noise. [Reprinted from A. W. Moore, Jr. and J. W. Jorgenson, *Anal. Chem.* **67**, 3448 (1995), with permission of The American Chemical Society, Washington DC.]

photobleached and became nonfluorescent. No fluorescent species were detected at the probe beam as long as the gating beam was on. To perform an injection, the gating beam was blocked momentarily (10–50 ms) with a mechanical shutter and then a tiny sample of undestroyed fluorescent material passed by. This sample was then separated by the applied electric field in the region between the gating and probe beams. Figure 17.14 illustrates the correlation between the laser power intensity and the fluorescence signal seen at the probe beam. Run times less than 3 s were achieved with this experimental setup because high voltage was applied to a short length of capillary and the distance between injection and detection was very small.

Figure 17.15 illustrates the experimental setup for using high-speed CZE in a 2D LC–CZE system. A conventional-sized C_{18} reversed-phase column operated with an aqueous/acetonitrile solvent gradient was used for the first-dimension separation of fluorescently tagged peptides. In this setup, the electrophoresis capillary inlet resided in a stainless steel tee instead of in a buffer reservoir. The LC mobile phase flowed through the tee, and the applied electrophoresis voltage continuously electromigrated sample onto the

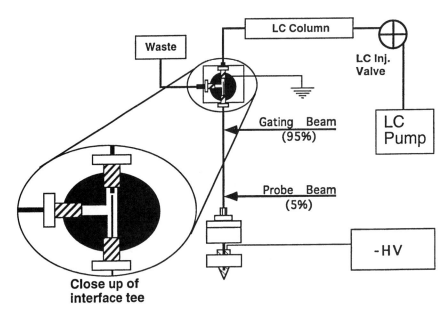

Figure 17.15. Two-dimensional RPLC–fast CZE instrumental diagram. [Reprinted from J. P. Larmann Jr., A. V. Lemmo, A. W. Moore, Jr., and J. W. Jorgenson, *Electrophoresis*. **14**, 439–447 (1993), with permission of VCH Verlagsgesellschaft mbH, Weinheim, Germany.]

capillary as previously described. Optically gated injections and high-speed CZE separations of the electromigrated LC effluent were performed. The result was a further separation of components that eluted from the LC column to give a 2D separation. Figure 17.16 contains the 2D LC–fast CZE separation of horse heart cytochrome *c*. The entire 2D separation was performed in 9 min. As was seen in previous separations, neither technique alone had sufficient resolving power for this mixture but when combined nearly all of the components were separated.

Since the CZE buffer for this analysis was made up entirely from the RPLC effluent, the RPLC mobile phase had to also act as a suitable buffer system for CZE. The traditional mobile phase of water/acetonitrile/TFA used in protein and peptide separations would not be an effective CZE buffer. Instead, a phosphate buffer was used as the aqueous mobile phase (solution A) and a 30% acetonitrile/70% phosphate solution was used as the organic modifier (solution B). A gradient with increasing percentage of solution B was run on the reversed-phase column to elute the peptides. This change in CE buffer composition due to the LC gradient had an effect on migration times. As the

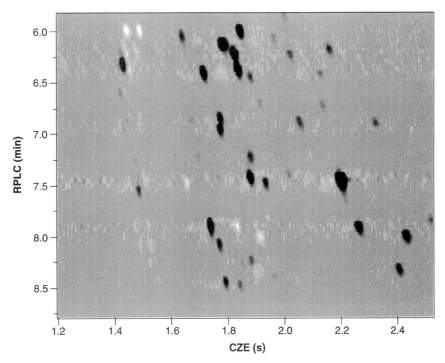

Figure 17.16. Two-dimensional at RPLC–fast CZE analysis with 5 min LC gradient, 2.5 s CZE analyses. Sample is FITC-tagged tryptic digest of horse heart cytochrome c. RPLC conditions: 5 min linear gradient from 10% B to 50% B, and hold 1 min at 50% B; return to initial conditions over 0.1 min. CZE conditions: 20 kV applied over 8 cm total capillary length, with 2 cm between gating and probe beams, CZE injection time, 10 ms; total CZE analysis time 2.5 s. [Reprinted from A. W. Moore Jr. and J. W Jorgenson, *Anal. Chem.* **67**, 3448 (1995), with permission of The American Chemical Society, Washington, DC.]

percentage of acetonitrile increased, the EOF diminished. Because of this decrease in EOF over the course of the gradient, the peaks in the 2D plot in Figure 16.16 appear to tail toward later migration times.

17.4. THREE-DIMENSIONAL SIZE EXCLUSION CHROMATOGRAPHY–REVERSED-PHASE LIQUID CHROMATOGRAPHY–HIGH-SPEED CAPILLARY ZONE ELECTROPHORESIS

The ability to perform the 2D LC–high-speed CZE separation on the timescale of minutes allowed for its incorporation into a 3D coupled-column

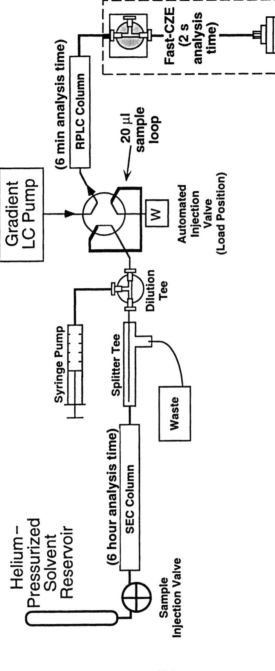

Figure 17.17. Schematic diagram of 3D SEC–RPLC–fast CZE instrument. See the text for an explanation. [Reprinted from A. W. Moore Jr. and J. W. Jorgenson, *Anal. Chem.* **67**, 3456 (1995), with permission of The American Chemical Society, Washington, DC.]

system (56). The reason for creating a 3D system is the same as for a 2D system, increased peak capacity. Ideally, the peak capacity of a 3D system will be the product of the individual peak capacities ($n_{c1} \times n_{c2} \times n_{c3}$). This will not be the case experimentally, but nonetheless there will be a substantial gain in peak capacity for the 3D system over that of the 2D. SEC was chosen for the first-dimension separation because it can be run on the timescale of hours and since its mode of separation (hydrodynamic size) differs from that of RPLC (hydrophobicity) and CZE (charge).

Figure 17.17 diagrams the experimental setup for the 3D system. The SEC separation was performed over a period of 5 h. In order to ensure that the peptides were separated based on their size, a mobile phase containing 85% methanol was used to keep the peptides from interacting with the SEC stationary phase through hydrophobic interactions. Fractions from the SEC column to be transferred to the LC column were collected in a 20 µL sample loop. The splitter tee was needed in order to reduce the amount of volume entering the RPLC column from the SEC column. A split of 10:1 was used to generate a flow rate of 1 µL/min. The dilution tee was used to mix 4 µL of aqueous diluant with 1 µL of SEC mobile phase, making the flow rate entering the sample loop 5 µL/min. This was necessary in order to reduce the high organic content of the SEC eluent so the analytes would be retained on the RPLC column. The operation of the RPLC and high-speed CZE system was identical to that described in the previous subsection. RPLC runs were performed in 6 min, and high-speed CZE runs were completed in 2 s.

The separation of a FITC-labeled tryptic digest of ovalbumin with this 3D system is shown in Figure 17.18. This image was generated by stacking the 2D RPLC–CZE runs of the SEC fractions together to yield a 3D separation "volume," where each of the three axes represents a different separation method. An improvement in the overall peak capacity of the system was realized by incorporating a third dimension, but there were drawbacks: there was an increase in the instrumental complexity with this system, and significant amounts of time were required for data processing and analysis. The increased peak capacity of this system may not be worth the extra effort and added complexity that is entailed. Instrumental miniaturization may help to make a higher-order separation such as this more realistic and achievable.

17.5. FUTURE DIRECTIONS

Although the results that have been obtained from LC–CZE systems so far are very impressive, there are many areas where improvement is desired. Microcolumns have demonstrated utility for improved separation efficiency and

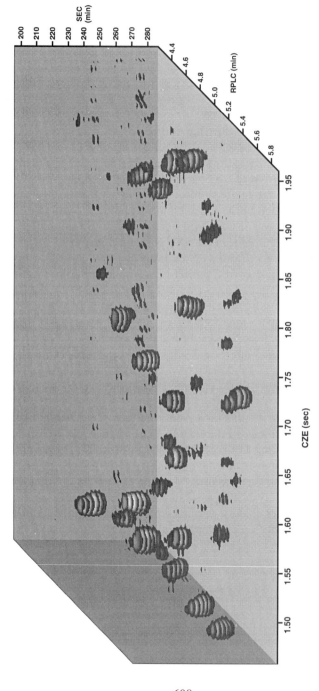

Figure 17.18. Representation of 3D data "volume" of tryptic digest of ovalbumin. Peaks are seen by making a series of planar slices through the data volume and so have the appearance of "stacks" of disks or ellipsoids. [Reprinted from A. W. Moore, Jr. and J. W. Jorgenson, *Anal. Chem.* **67**, 3456 (1995), with permission of The American Chemical Society, Washington, DC.]

analysis of limited sample volumes. A more universal adoption of microcolumns has been hampered by a slow development of pumps, injectors, and detectors developed specifically for them. Work is progressing and several companies are now solely dedicated to development of instrumentation for LC microcolumns (57–60). Work in this area must continue, as there is a growing need for pumps that can deliver very low flow rates ($< 1 \mu L/min$) in an isocratic and gradient mode for these microcolumns. Autosamplers and microinjectors are needed as well if reproducibility and automation are to become more routine with microcolumn systems.

There is much work left to be done in designing a 2D LC–CE system for proteins. Due to their less favorable chromatographic and electrophoretic behavior, designing a 2D system for their separation has been difficult. Sensitive detection is also a problem: UV detection is limited due to the short pathlength with CE capillaries. Precolumn derivatization for fluorescence is more difficult for proteins than for amino acids or peptides due to the large number of sites on most proteins that can be tagged with a fluorescent moiety. Generally, a mixture of reaction products will be generated, and these will all have slightly different electrophoretic mobilities. Native fluorescence detection of proteins is limited because it requires the protein to have tryptophan and tyrosine residues. Also, since excitation generally has to be performed in the UV region (~ 250–280 nm), there is a large background due to luminescence of the capillary, buffer components, and optics. Work on postcolumn fluorescence detection in CE has been characterized (61,62), and one attempt has been made to use it in a 2D LC–CE system for the separation of proteins (63). These postcolumn systems, however, are not easy to operate and will require more work.

To date, a majority of the 2D LC–CE work has been applied to peptide and protein mixtures. Future work must include expanding the applicability of the technique to other analyte systems and real-world samples. The technique holds great promise in the area of clinical analysis. Since hundreds of components can be separated with this instrumentation, one can examine a tremendous number of analytes in samples, such as urine and plasma, from healthy and diseased individuals. Generally, only the presence or absence of a few species can be determined in these samples when the analyst is using LC or CE alone due to the large amount of component overlap.

Further adoption of 2D LC–CE systems will occur when more researchers in the field of chromatography realize the limitations of 1D separations and the extent of component overlap in separations. The 2D separations that have been demonstrated indicate that the complexity of a sample is generally severely underestimated because many components comigrate or coelute in a run and give the appearance of a single peak when in reality there are many more.

LIST OF ACRONYMS

Acronym	Definition
CE	capillary electrophoresis
CZE	capillary zone electrophoresis
1D, 2D, 3D	one-, two-, three-dimensional
EOF	electroosmotic flow
FITC	fluorescein-5-isothiocyanate
GC	gas chromatography
HPLC	high-performance liquid chromatography
IEF	isoelectric focusing
LC	liquid chromatography
LIF	laser-induced fluorescence
PAGE	polyacrylamide gel electrophoresis
RPLC	reversed-phase liquid chromatography
SDS	sodium dodecyl sulfate
SEC	size exclusion chromatography
SFC	supercritical fluid chromatography
SMO	statistical model of overlap
TFA	trifluoroacetic acid
TLC	thin-layer chromatography
TRITC	tetramethylrhodamine-5-isothiocyanate
UV	ultraviolet

REFERENCES

1. H. Yamamoto, T. Manabe, and T. Okuyama, *J. Chromatogr.* **480**, 277–284 (1989).
2. P. S. L. Janssen, J. W. Van Nispen, M. J. M. Van Zeeland, and P. A. T. A. Melgers, *J. Chromatogr.* **470**, 171–183 (1989).
3. H. Yamamoto, T. Manabe and T. Okuyama, *J. Chromatogr.* **515**, 659–666 (1990)
4. W. Steuer, I. Grant, and F. Erni, *J. Chromatogr.* **507**, 125–140 (1990).
5. M. Castagnola, L. Cassiano, R. Rabino, D. V. Rossetti and F. A. Bassi, *J. Chromatogr.* **572**, 51–58 (1991).
6. M. Strömqvist, *J. Chromatogr.* **667**, 304–310 (1994).
7. S. Palmarsdottir and L.-E. Edholm, *J. Chromatogr.* **693**, 131–143 (1995).
8. V. K. Smith, R. R. Holloway, C. A. K. Templin, and W. D. Cole, *6th Ann. Frederick Conf. Capillary Electrophoresis,* Frederick, MD, p. 70 (1995).
9. R. Consden, A. H. Gordon, and A. J. P. Martin, *Biochem. J.* **38**, 244 (1944).
10. G. Haugaard and T. D. Kroner, *J. Am. Chem. Soc.* **70**, 2135–2137 (1948).
11. E. L. Durrum, *J. Colloid Sci.* **6**, 274–290 (1951).

12. A. Jeans, C. S. Wise, and R. J. Dimler, *Anal. Chem.* **23**, 425 (1951).
13. H. P. Lenk, *Fresenius' Z. Anal. Chem.* **184**, 107 (1961).
14. J. Janak *Thin-Layer Chromatography and Related Methods,* p. 63. Ann Arbor Sci. Publishers, Ann Arbor, MI, 1971.
15. P. H. O'Farrell, *J. Biol. Chem.* **250**, 4007–4021 (1975).
16. M. J. Dunn, *Electrophoresis* **12**, 461–606 (1991).
17. H. J. Cortes, In *Chromatographic Science: A Series of Monographs* (J. Cazes, ed.), Vol. 50, p. 378. m. Dekker; New York, 1990.
18. H. J. Cortes and L. D. Rothman, in *Multidimensional Chromatography: Techniques and Applications* (H. J. Cortes, ed.), Vol. 50, pp. 219–250. Dekker, New York, 1990.
19. H. J. Cortes, *J. Chromatogr.* **626**, 3–23 (1992).
20. K. D. Bartle, I. Davies, M. W. Raynor, A. A. Clifford, and J. P. Kithinji, *J. Microcol. Separ.* **1**, 63–70 (1989).
21. R. E. Majors, *Chromatographia* **18**, 571–579 (1980).
22. W. Bertsch, in *Multidimensional Chromatography: Techniques and Applications* (H. J. Cortes, ed.), Vol. 50, pp. 75–144. Dekker, New York, 1990.
23. I. L. Davies, K. E. Markides, M. L. Lee, and K. D. Bartle, in *Multidimensional Chromatography: Techniques and Applications* (H. J. Cortes, ed.), Vol. 50, pp. 301–330. Dekker, New York, 1990.
24. H. J. Cortes, in *Multidimensional Chromatography: Techniques and Applications* (H. J. Cortes, ed.), Vol. 50, pp. 251–300. Dekker, New York, 1990.
25. J. C. Giddings, *Anal. Chem.* **39**, 1027–1028 (1967).
26. J. M. Davis and J. C. Giddings, *Anal. Chem.* **55**, 418–424 (1983).
27. J. M. Davis and J. C. Giddings, *Anal. Chem.* **57**, 2168–2177 (1985).
28. J. M. Davis, *J. Chromatogr.* **449**, 41–52 (1988).
29. G. Guiochon, M.-F. Gonnord, M. Zakaria, L. A. Beaver, and A. M. Siouffi, *Chromatographia* **17**, 121 (1983).
30. P. Herman, M.-F. Gonnord, and G. Guiochon, *Anal. Chem.* **56**, 995 (1984).
31. M. Martin, and G. Guiochon, *Anal. Chem.* **57**, 289 (1985).
32. M. Martin, D. P. Herman, and G. Guiochon, *Anal. Chem.* **58**, 2200–2207 (1986).
33. J. C. Giddings, *Anal. Chem.* **56**, 1258A (1984).
34. G. Guiochon, L. A. Beaver, M.-F. Gonnord, A. M. Siouffi, and M. Zakaria, *J. Chromatogr.* **255**, 415–437 (1983).
35. J. C. Giddings, *HRC & CC, J. High Resolut. Chromatogr. Chromatogr. Commun.* **10**, 319–323 (1987).
36. D. H. Freeman, *Anal. Chem.* **53**, 2–5 (1981).
37. M. M. Bushey, and J. W. Jorgenson, *Anal. Chem.* **62**, 978–984 (1990).
38. M. M. Bushey, and J. W. Jorgenson, *J. Microcolumn Sep.* **2**, 293–299 (1990).
39. J. P. Larmann, Jr., A. V. Lemmo, A. W. Moore, Jr., and J. W. Jorgenson, *Electrophoresis* **14**, 439–447 (1993).

40. K.-E. Karlsson, and M. Novotny, *Anal. Chem.* **60**, 1662–1665 (1988).
41. R. T. Kennedy, and J. W. Jorgenson, *Anal. Chem.* **61**, 1128–1135 (1989).
42. R. T. Kennedy, and J. W. Jorgenson, *J. Microcolumn Sep.* **2**, 120–126 (1990).
43. S. Hsieh, and J. W. Jorgenson, *Anal. Chem.* **68**, 1212–1217 (1996).
44. R. T. Kennedy, M. D. Oates, B. R. Cooper, B. Nickerson, and J. W. Jorgenson, *Science* **246**, 57–63 (1989).
45. B. R. Cooper, J. A. Jankowski, D. J. Leszczyszyn, R. M. Wightman, and J. W. Jorgenson, *Anal. Chem.* **64**, 691–694 (1992).
46. B. R. Cooper, R. M. Wightman, and J. W. Jorgenson, *J. Chromatogr.* **653** 25–34 (1994).
47. L. A. Holland, and J. W. Jorgenson, *Anal. Chem.* **67**, 3275–3283 (1995).
48. A. V. Lemmo, and J. W. Jorgenson, *J. Chromatogr.* **633**, 213–220 (1993).
49. A. V. Lemmo, Doctoral Dissertation, University of North Carolina, Chapel Hill (1994).
50. A. V. Lemmo, and J. W. Jorgenson, *Anal. Chem.* **65**, 1576–1581 (1993).
51. J. P. Larmann, Jr., Doctoral Dissertation, University of North Carolina, Chapel Hill (1993)
52. J. P. Larmann, Jr., and J. W. Jorgenson, *Anal. Chem.* Submitted for publication.
53. A. W. Moore, Jr. and J. W. Jorgenson, *Anal. Chem.* **67**, 3448 (1995).
54. C. A. Monnig, and J. W. Jorgenson, *Anal. Chem.* **63**, 802–807 (1991).
55. A. W. Moore, Jr. and J. W. Jorgenson, *Anal. Chem.* **65**, 3550–3560 (1993).
56. A. W. Moore, Jr. and J. W. Jorgenson, *Anal. Chem.* **67**, 3456 (1995).
57. Micro-Tech Scientific, Inc.; Sunnyvale, CA, 1995.
58. J.-P. Chervet, In *LC Packings Newsletter,* Vol. 6, p. 1. LC Packings; San Francisco, 1995.
59. M. Ursem, J. P. Chervet, R. E. J. Van Soest, and J. P. Salzman, *Proc. Int. Symp. Capillary Chromatogr. Electrophor. 17th.* Wintergreen VA, *1995*; pp. 28–29 (1995).
60. F. J. Yang, P. Rippington, J. Powers, and W. S. Lai, *Proc. Int. Symp. Capillary Chromatogr. Electrophor. 17th.* Wintergreen, VA, *1995*; pp. 40–41 (1995).
61. D. J. Rose, Jr. and J. W. Jorgenson, *J. Chromatogr.* **447**, 117–131 (1988).
62. B. Nickerson and J. W. Jorgenson, *J. Chromatogr.* **480**, 157–168 (1989).
63. A. V. Lemmo, and J. W. Jorgenson, *J. Microcolumn Sep.* Submitted for publication.

CHAPTER

18

MICROFABRICATED CHEMICAL SEPARATION DEVICES

STEPHEN C. JACOBSON and J. MICHAEL RAMSEY

Oak Ridge National Laboratory, Oak Ridge, Tennessee 37831-6142

18.1.	Introduction	613
18.2.	Modular System Design	614
18.3.	Fabrication Techniques	615
18.4.	Fluid Manipulation and Injection	617
18.5.	Microchip Electrophoresis	620
18.6.	DNA Separations	623
18.7.	Chromatographic Separations	626
18.8.	Integated Structures for Chemical and Biochemical Reactions and Analysis	627
18.9.	The Future	629
Addendum		631
List of Acronyms		631
References		632

18.1. INTRODUCTION

Open-tubular separations in the liquid phase, i.e., capillary electrophoresis, liquid chromatography, and micellar electrokinetic chromatography, are conventionally practiced in capillary tubes with diameters of a few tens of micrometers, as described in this book. Structures having similar dimensions can be micromachined on planar substrates by standard photolithographic, thin-film deposition, etching, and bonding techniques. Microfabricated separation devices have been demonstrated for capillary electrophoresis (1–8), synchronized cyclic electrophoresis (9), free-flow electrophoresis (10), and capillary gel electrophoresis (11–14). Open channel electrochromatography

High Performance Capillary Electrophoresis, edited by Morteza G. Khaledi. Chemical Analysis Series, Vol. 146.
ISBN 0-471-14851-2 © 1998 John Wiley & Sons, Inc.

(15), micellar electrokinetic capillary chromatography (16), and packed bed liquid chromatography (17) have been integrated onto microchips for the separation of neutral species. These devices are able to manipulate picoliter volumes with high precision without the use of mechanical valves. Microfabricated liquid phase separation devices have demonstrated performance features equivalent to or exceeding that of conventional laboratory analogues. Much of this success for liquid phase analysis is due to the use of electrokinetic phenomena to transport buffers and samples through the nonsymmetric channel cross sections.

Microfabricated chemical analysis devices have not been limited to separation techniques, and include structures that perform chemical reactions such as arrays for solid-phase chemistry (18), reaction wells for polymerase chain reactions (19), channels with immobilized enzymes for flow injection analysis (20), and stacked modules for flow injection analysis (21). Coupling of miniature sample preparation devices with a separation technique would provide complete chemical analysis on a microchip. Some simple integrated devices have appeared in the literature, including pre- (22,23) and postcolumn (24) chemical reactions in conjunction with electrophoretic separations. Such devices have several potential benefits for the acquisition of chemical information in many different application scenarios from environmental monitoring to medical diagnostics. The dexterity of the fluidic manipulations can provide increased precision and accuracy of measurement. Miniaturization can also increase measurement speed and allow massively parallel measurement systems to become feasible. This chapter will discuss the development of electrically driven separation techniques on microchips and the coupling of chemical reactions to these miniaturized separation devices.

18.2. MODULAR SYSTEM DESIGN

Miniature chemical instruments can be assembled in much the same way that integrated circuits are in the microelectronics industry if a collection of basic components can be realized. In Figure 18.1 three basic components are depicted: a channel, a tee, and a cross. The channel is used as a separation column or reaction chamber; the tee enables the mixing of reagents or the dilution of a sample, and the cross functions as a valve to dispense or inject samples or reagents. With these three components, sophisticated analytical systems can be designed and fabricated. By combining a cross and channel (Figure 18.1d), an electrophoresis or chromatography microchip performs chemical separations. By combining a tee, cross, and channel (Figure 18.1e), a precolumn reactor microchip can perform reactions and chemical separations sequentially. Alternately, by rearranging the order

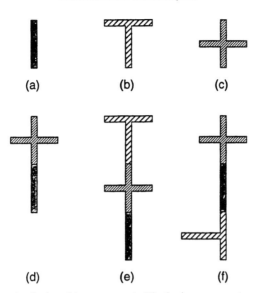

Figure 18.1. Schematic of microchip components. The basic components are (a) a channel, (b) a tee, and (c) a cross to perform chemical separations/reactions, mixing, and valving, respectively. System designs include (d) a capillary electrophoresis or chromatography microchip, (e) a precolumn reactor microchip, and (f) a postcolumn reactor microchip.

of components a cross, channel, and tee (Figure 18.1f), a postcolumn reactor microchip can carry out separations followed by postseparation reactions, e.g., derivatization. Examples of these systems are described below. Additional basic components are necessary for the realization of a complete "laboratory-on-a chip" (25). Other microfabricated components of primary interest are physical filters (26), heterogeneous partitions, and detectors. No doubt the designs will become more sophisticated as microchip development progresses, but the basic components will remain structurally simple by design.

18.3. FABRICATION TECHNIQUES

Insulating substrates, e.g., glass, BK7 (optical-grade borosilicate glass), and fused quartz, are used rather than semiconducting substrates to allow high voltages to be applied to the microchips to transport samples and reagents through the channel manifolds. Much of the technology developed for the semiconductor industry can be transferred directly to microchip fabrication using insulating substrates. Although slight variations in fabrication techniques of microchips exist, the general fabrication aspects are similar (Figure 18.2). For fused quartz or BK7, a chromium/gold film is sputtered on the substrate as an etch mask. Onto the gold film, a positive photoresist is spin-coated, and a channel design is transferred to the substrate using a photomask and ultraviolet (UV) exposure of the photoresist. Photomasks are generally manufactured by means of optical or electron beam direct-write processes. Following exposure and development of the photoresist, the metal films are etched by using KI/I_2 for Au and $K_3Fe(CN)_6/NaOH$ for Cr. The channels are then etched into the substrate in a dilute, stirred HF/NH_4F bath. To form the closed network of channels, a cover plate is bonded to the substrate over the etched channels. Two bonding techniques have been reported. The first technique is to melt the cover plate to the substrate (1). The second method is a direct bonding technique in which the surface of the substrate and cover plate are hydrolyzed, brought into contact with each other, and finally thermally processed at 500°C for soda lime glass and BK7

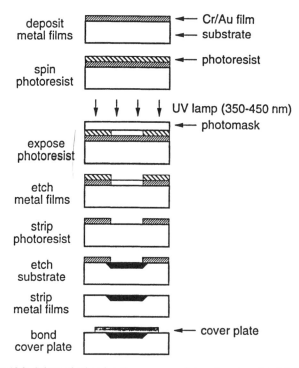

Figure 18.2. Schematic showing the sequence of steps for microchip fabrication.

substrates or 1100°C for fused quartz (6,27). Amorphous materials etch uniformly in all directions, resulting in nonrectangular channel profiles. Typically, channels range in width at half-height from 20 to 100 μm with depths from 5 to 30 μm.

Due to the relative ease of micromachining and bonding, glass substrates are used predominantly. However, quartz substrates have the benefit of superior optical properties, and a fused quartz microchip has been fabricated to perform capillary electrophoresis (27). By incorporating the direct bonding technique during the fabrication of the microchip, the substrate and cover plate can be fused together below the melting temperature for fused quartz.

18.4. FLUID MANIPULATION AND INJECTION

For the devices described here, the primary means of material transport on microchips is electrokinetic, i.e., electrophoresis and/or electroosmosis. For separation devices, precise valving is required for the injection process. In the simplest scenario (Figure 18.3a1), to perform an injection a potential is applied between the sample and sample waste reservoirs, with the potentials at the buffer and waste reservoirs removed (3). The sample moves electrokinetically through the injection cross as a frontal electropherogram. After the slowest-moving compound has moved through the channel intersection, the potentials are reconfigured for the separation mode—where a potential is applied between the buffer and waste reservoirs, with the potentials at the sample and sample waste now removed as seen in Figure 18.3a2. The reproducibility of this technique is $\sim 2\%$ RSD (relative standard deviation). In the above injection and separation procedures, potentials are not actively controlled at each reservoir. Problems with leakage of analyte into the separation channel during a run have been observed using this injection procedure, which increases detection background signals. Injection volumes are also dependent upon the duration of the sample-loading step of the injection procedure (Figure 18.3a1). These problems are alleviated with the multiport electric potential control scheme shown in Figures 18.3b1 and 18.3b2.

Spatial confinement of the sample plug, and thus injection volume control, is attained by controlling the potentials at each of the reservoirs during the injection and separation modes (6). In Figure 18.3b1, the sample flowing through the injection cross is both defined and confined by electric fields in the buffer and separation channels. To prevent bleeding of excess sample from the sample and sample waste reservoirs into the separation column during the separation, the sample and sample waste reservoirs are maintained at a fraction of the potential applied to the buffer reservoir (Figure 18.3b2). This

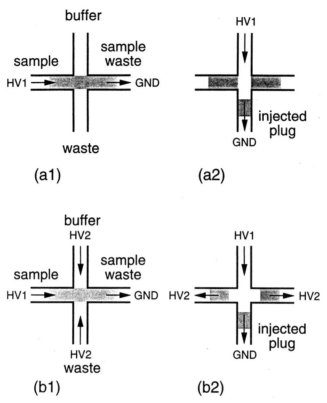

Figure 18.3. Schematic of injection procedure. (a1) The sample (shaded area) is loaded into the injection cross by applying a potential between the sample and sample waste reservoirs. (a2) The sample plug is injected onto the separation column for the separation by applying a potential between the buffer and waste reservoirs. (b1) For the fixed-volume valve, the sample plug can be spatially confined in the channel intersection by applying potentials to the buffer and waste reservoirs. (b2) During the separation, secondary flows of buffer into the sample and sample waste channels prevent bleeding of excess sample onto the separation channel. (c1) For the variable-volume valve, the sample (shaded area) is loaded into the injection cross by applying a potential between the sample and sample waste reservoirs, and the potential applied to the buffer reservoir prevents sample from migrating onto the separation column. (c2) The sample plug is migrated onto the separation column by removing the potential at the buffer reservoir. (c3) The plug is broken off by reapplying the potential at the buffer reservoir. High voltage 1 (HV1) is greater than high voltage 2 (HV2), and HV2 is greater than high voltage 3 (HV3). The arrows indicate the direction of flow for the sample and buffer streams.

scheme introduces the smallest spatial extent sample plug for a given channel dimension, providing the highest separation efficiency. This technique also provides a nonbiased, time-independent injection with high volumetric reproducibility ($<0.3\%$ RSD).

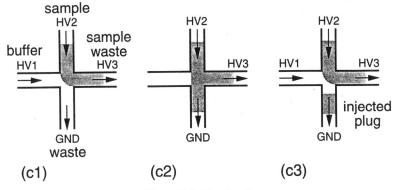

Figure 18.3. (*Continued*)

A variable-volume valve has also been developed (22,24). The positions of the sample and buffer reservoirs in Figure 18.3b are now exchanged, and the sample is migrated down through the injection cross toward the sample waste reservoir (Figure 18.3c1). The potential applied to the buffer reservoir prevents sample transport into the separation column. The sample is injected onto the separation column as follows: the potential at the buffer reservoir is removed by opening a high-voltage relay for a brief period of time (0.1 s or longer; see Figure 18.3c2), and then the sample migrates into the separation column as in an electrokinetic injection. To break off the injection plug, the potential at the buffer reservoir is reapplied (Figure 18.3c3). This valve implementation can deliver operator-selected volumes. The injection is biased by the relative electrophoretic mobilities of the analyte ions as with conventional electrokinetic injections. The variable-volume valve also provides unidirectional flow of fluids, which can be advantageous in some systems.

The variable-volume valve can also be used to enhance the sensitivity of microchip devices by performing sample stacking (28). The sample is prepared in a buffer with a lower conductivity than the separation buffer. The variable-volume valve can be used to inject a plug of the low-conductivity sample, surrounding it with high-conductivity buffer. Due to the higher field strength in the sample plug than in the separation buffer, the sample stacks at either the front or rear boundary of the sample plug depending on the net charge of the analyte. This technique requires the manipulation of just two fluid streams: the separation and sample buffers. The advantage of the microchip format is the reproducibility with which these types of injections can be made ($\sim 2\%$ RSD).

Lastly, two fluid streams can be proportionately mixed at a tee (Figure 18.1b) in any ratio from 0 to 100% for either stream simply by varying the relative field strengths in the two channels. Dilution at a tee has been demonstrated by stepping the potentials to produce a simple stair step (29). One of the two buffer streams was doped with an organic dye to enable the changing ratio of the two streams to be observed. Again, reagents can be mixed in whatever proportions are desired. This precise fluid control can also be applied to homogeneous reaction systems and solvent programming for gradient elution separations.

18.5. MICROCHIP ELECTROPHORESIS

Initial electrophoretic separations were demonstrated on a three-port device that had 30 μm channels etched into a glass slab with dimensions of 148 × 39 × 10 mm (1). A moderate separation efficiency of 35,000 theoretical plates was achieved in ~7 min. Bleeding of analyte into the separation channel from the injection channel was observed because of the injection procedure employed. Significant improvements came with a second-generation device (Figure 18.4a) that yielded more promising separation efficiencies for fluorescein-5-isothiocyanate (FITC)–labeled amino acids (3). In Figure 18.4b (p. 622), up to 160,000 plates were generated in less than 14 s with a separation length of 22 mm and separation electric field strength of 1 kV/cm. The rapid analysis time and a migration time reproducibility better than 0.1% RSD for these experiments demonstrated the potential of microchip electrophoresis. The sample was introduced onto the separation column in a nonbiased fashion with a fixed-volume "double-tee" injector (150 μm, 100 pL; see the inset in Figure 18.4a). The double-tee injector greatly reduced the valve leakage problems observed in earlier designs while still only controlling electric potentials at two reservoirs at a time.

On a microchip format, efficient high-speed separations can be accomplished by having short separation distances, extremely short injection plug lengths, and high separation field strengths. A simple chip design was used for the demonstration of high-speed separations as shown in Figure 18.5a. This chip had a channel cross section of 80 μm wide by 15 μm deep. By using this microchip, for a 0.9 mm separation length, electrophoretic separation of a binary mixture with baseline resolution was achieved in less than 150 ms with a separation field strength of 1.5 kV/cm (Figure 18.5b) (7). Active potential control at all four reservoirs was used for the injection and separation (Figure 18.3b). The resolution was limited by the spatial extent of the injection plug and observation length of the detector, 200 and 80 μm, respectively. Improvements in separation efficiency are possible by using narrower channels to reduce the injection plug length, tighter spatial filtering to minimize the detector observation region, and higher separation field strengths to decrease migration times.

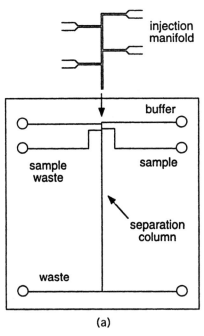

Figure 18.4. (a) Schematic of a capillary electrophoresis microchip with an offset "double-tee" injector (inset). The microchip had substrate dimensions of 80 × 70 × 3 mm and channel dimensions of 50 μm width and 12 μm depth. This design was used to obtain the data in part b. (b) Electropherogram of six FITC-labeled amino acids. The separation field strength was 1.06 kV/cm, and separation length was 22 mm. [Parts a and b reprinted with permission from C. S. Effenhauser, A. Manz, and H. M. Widmer, *Anal. Chem.* **65**, 2637 (1993), copyright 1993 American Chemical Society.]

There is considerable interest in designing capillary electrophoresis microchips that can generate high plate numbers yet have minimal dimensional extent. Two approaches to increasing the separation length are to squeeze the separation channel into a small area or to have the sample retravel a short separation channel. For compact column geometries, separations were performed on a microchip with a serpentine column geometry that had a 165 mm separation channel in an 8 × 8 mm area (6). Band-broadening effects due to the corners appear to be simply geometric in nature, and consequently, by limiting channel width, this contribution to the plate height can be sufficiently minimized. As a method for having the sample retravel the separation column, an 8 cm column was segmented into four 2 cm sections in a square format (9). Applied voltages at the corners of the structure are manipulated appropriately

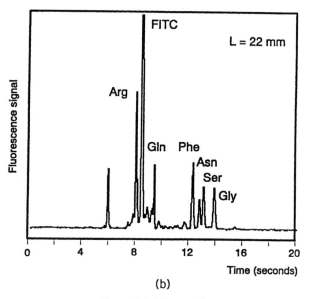

Figure 18.4. (*Continued*)

to make ions within a given mobility window cycle around the channel loop. The sample cycles the separation column as many times as needed to achieve the desired resolution. An advantage of the cyclic electrophoreis format is that lower applied potentials are required due to the short segment lengths but the effective applied potential scales with the number of cycles the ions experience. After just five cycles with fluorescein, 850,000 theoretical plates are generated with only 2.5 kV applied at any given time. As the number of cycles increases and the resolution—in principle—grows, the migration window that can be observed decreases. Ions with migration times outside of this window exit the cyclical structure. Also, a loss in separation efficiency and of sample can be observed as the band traverses the corners. The cyclic architecture is an effective approach for analyzing specific components at high resolution.

As a method for continuous separation, free-flow electrophoresis has been demonstrated on a silicon microchip (10). The stream of sample is continuously fed into a device by a syringe pump, and an electric field is applied perpendicular to the flow path as depicted schematically in Figure 18.6a. The sample is resolved spatially in the transverse direction in the separation bed, as opposed to axially as seen with the other techniques discussed herein. Figure 18.6b shows the continuous separation of rhodamine-labeled amino acids at

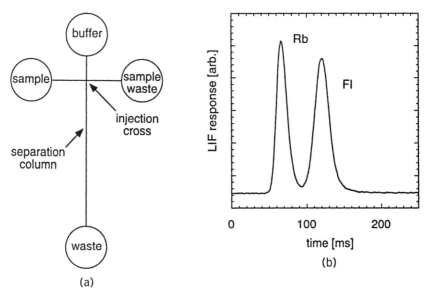

Figure 18.5. (a) Schematic of a cross microchip. This design was used to obtain the data in part b and Figures 18.7 and 18.8. (b) High-speed electrophoretic separation of rhodamine B (Rb) and disodium fluorescein (Fl). Separation field strength was 1.5 kV/cm, and separation length was 0.9 mm. The microchip depicted in part a had substrate dimensions of $75 \times 25 \times 1$ mm and channel dimensions of 80 μm width and 15 μm depth. (LIF = laser-induced fluorescence.) [Parts a and b reprinted with permission from S. C. Jacobson, R. Hergenröder, L. B. Koutny, and J. M. Ramsey, *Anal. Chem.* **66**, 1114 (1994), copyright 1994 American Chemical Society.]

24 mm downstream from the sample inlet. Free-flow electrophoresis is ideal for sample cleanup or low–resolution separations where a higher throughput is needed. For a baseline peak width of 1.2 mm for lysine, the peak capacity of the device is estimated to be 8.2 for the 10 mm wide separation bed. In the experiments conducted so far, electrolysis of water in the side channels has posed some problems for general utility, but miniaturized free-flow electrophoresis could become a valuable component in an analytical system where low-resolution purification is required.

18.6. DNA SEPARATIONS

Like capillary-based separation techniques, microchips are not limited to free-solution electrophoresis. Microchip technology has also been applied to the analysis of nucleic acids where high sample throughput methods are of

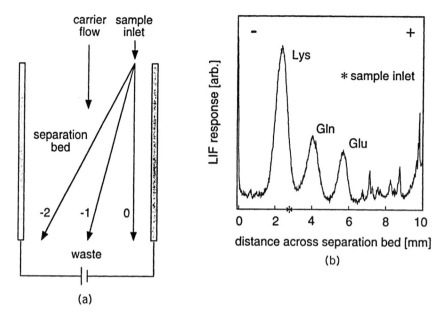

Figure 18.6. (a) Schematic of free-flow electrophoresis. The sample is introduced at the top of the separation bed in a hydrostatically pumped carrier flow. The separation field strength is applied perpendicular to the flow of the sample; 0, −1, and −2 denote the expected position following separation of a neutral, monoanion, and dianion, respectively. (b) Free-flow electrophoretic separation of rhodamine B isothiocyanate–labeled lysine (Lys), glutamine (Gln), and glutamic acid (Glu). The separation bed width was 10 mm, the separation length was 24 mm, and the voltage applied across the separation bed was 50 V. [Parts a and b reprinted with permission from D. E. Raymond, A. Manz, and H. M. Widmer, *Anal. Chem.* **66**, 2858 (1994), copyright 1994 American Chemical Society.]

interest. Sieving media have been incorporated into microchips for the separation of both single- and double-stranded DNA. The first demonstration of DNA separations using a microchip involved the analysis of restriction enzyme digests. The microchip used in this work was manufactured first by fabricating a high precision die in silicon and molding a polymer to obtain a device with a channel 250 μm wide, 50 μm deep, and 50 mm long (11). The restriction fragments of ϕX174 DNA (*Hae*III digest) were separated by use of 10% (w/v) linear polyacrylamide, and the analysis was completed in less than 30 min. The separation was later duplicated and improved upon by using a glass substrate and 0.75% (w/v) hydroxyethyl cellulose for the sieving medium (13). The analysis time was decreased by an order of magnitude to 3 min for the separation, with a field strength of 180 V/cm and separation length of 35 mm. The electroosmotic flow was minimized by covalently bonding linear polyacrylamide to the channel wall.

For single-stranded DNA, end-labeled antisense oligonucleotides ranging from 10 to 25 bases in length were resolved by using 10% (w/v) linear acrylamide (12). The most impressive feature of this study was that the antisense oligonucleotides were resolved in less than 45 s by Effenhauser et al., using a separation field strength of 2300 V/cm and a separation distance of 38 mm. The potential of using microfabricated devices for DNA sequencing is promising because the microchips can be fabricated with a highly parallel architecture, similar to capillary arrays but possibly more practical with respect to fabrication and handling. In Figure 18.7, sequencing extension fragments were resolved in less than 10 min with a resolution greater than 0.5 out to 200 bases by using 9% T, 0% C polyacrylamide and a one-color

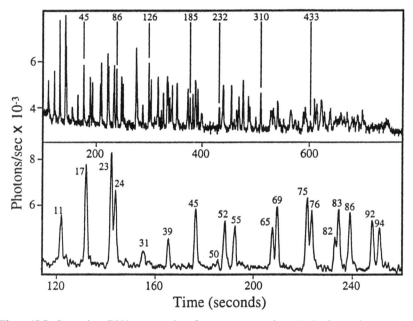

Figure 18.7. One-color DNA sequencing fragment separation: (*top*) electropherogram of M13mp18 A sequencing fragments generated with the primer F10F; (*bottom*) expanded view of the peaks corresponding to the first 100 bases, demonstrating single-base resolution. The separation field strength was 250 V/cm with a 9% T, 0% C polyacrylamide-filled channel, and the separation length was 35 mm. The microchip depicted in Figure 18.5a was used and had substrate dimensions of 75 × 50 × 2 mm and channel dimensions of 50 μm wide and 8 μm deep. [Reprinted with permission from A. T. Woolley and R. A. Mathies, *Anal. Chem.* **67**, 3676 (1995), copyright 1995 American Chemical Society.]

detection system (14). Similar resolution was obtained for a four-color detection scheme, with single-base resolution out to ~ 150 bases and 97% base-calling accuracy in 9 min. These runs, however, fall short of the resolution that has been obtained using capillary-based systems. The microchip format might be multiplexed so as to have a set of parallel channels with which to run many separations simultaneously.

18.7. CHROMATOGRAPHIC SEPARATIONS

Three methods for separating neutral species have been investigated: open-channel electrochromatography (15), micellar electrokinetic chromatography (MEKC) (16), and packed-bed chromatography (17). The first two schemes employ electroosmotic flow to pump the mobile phase through the channel manifold, whereas the packed-bed scheme uses an external syringe pump.

For the microchip electrochromatography, Jacobson et al. chemically modified the surface of the separation channel with octadecylsilane to function as the stationary phase (15). Using a separation length of 58 mm and a separation field stength of 160 V/cm, these authors resolved three coumarin dyes in less than 2.5 min. Plate heights as low as 4.1 and 5.0 µm were obtained for unretained and retained components, respectively. Improvements in efficiency would be seen with better stationary phase coating, channel geometry, and solvent programming.

In Figure 18.8, fast MEKC separations are performed in 30 s generating plate heights of 1.5 and 4.0 µm for the unretained and most retained components, respectively, using a cross microchip similar in design to Figure 18.5a (16). A loss in efficiency is observed between the unretained peak, C440, and the most retained peak, C460, at the high mobile phase velocity associated with the 500 V/cm electric field strength. In addition to having a higher efficiency than the open-channel electrochromatography, MEKC has the advantages of higher stationary phase densities in the separation channel, replaceable separation media, and fabrication simplicity with the elimination of surface modification requirements. The primary drawback is that micelles elute from the column, and consequently a finite separation window can be limiting for some separations.

A third method for the separation of neutral molecules on a microchip was demonstrated by Ocvirk et al. using a silicon microchip filled with 5 µm spherical packing material (17). On the silicon microchip, a split injector, a small-bore column, a frit, and a detection cell were fabricated, but an external syringe pump was used to deliver the mobile phase. The separation column volume was 500 nL, with all connections having a total dead volume of

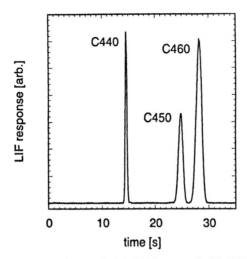

Figure 18.8. MEKC separation of coumarin 440 (C440), coumarin 450 (C450), and coumarin 460 (C460). The separation field strength was 500 V/cm, the separation length was 13 mm, and the buffer was 10 mM sodium tetraborate, 50 mM sodium dodecyl sulfate, and 10% (v/v) methanol. The microchip depicted in Figure 18.5a was used and had substrate dimensions of 75 × 25 × 1 mm and channel dimensions of 70 μm width and 10 μm depth. [Reprinted with permission from A. W. Moore, S. C. Jacobson, and J. M. Ramsey, *Anal. Chem.* **67**, 4184 (1995), copyright 1995 American Chemical Society.]

~ 3.5 nL. Fluorescein and acridine orange were resolved in 3 min and had a maximum of 200 plates. This technique demonstrated the poorest efficiency of the three methods discussed.

18.8. INTEGRATED STRUCTURES FOR CHEMICAL AND BIOCHEMICAL REACTIONS AND ANALYSIS

The separations described above establish a foundation on which liquid phase analysis using microchips can be built. One advantage of microfabrication is the relative ease in which low-dead-volume fluidic connections can be made. Additional functions such as reagent mixing or dilution can be incorporated on separations microchips to perform sample preparation procedures. Adding functionality to microchips greatly enhances their potential utility. To demonstrate this concept, the glass microchip in Figure 18.9a was constructed to perform chemical reactions and capillary electrophoresis sequentially (22). A sample and a reagent are combined by using the previously discussed mixing tee in the short reaction chamber shown in Figure 18.9a. The reaction chamber

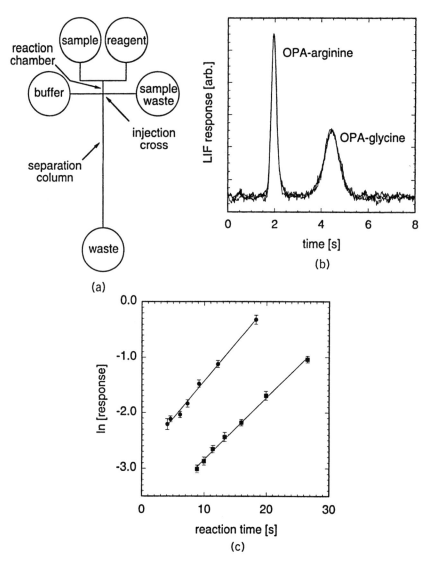

Figure 18.9. (a) Schematic of a precolumn reactor microchip with a reaction chamber, injection cross, and separation channel. The microchip had substrate dimensions of 50 × 25 × 1 mm, channel dimensions of 30 μm width and 6 μm depth, and a reaction chamber volume of 1 nL. This design was used to obtain the data in parts b and c. (b) Overlay of three electropherograms of arginine and glycine following precolumn derivatization with o-phthaldialdehyde (OPA). The separation field strength was 1.8 kV/cm, and the separation length was 10 mm. (c) Variation of product formation with reaction time for arginine (●) and glycine (■) derivatized with OPA. Lines represent linear fits. Error bars are $\pm \sigma$ for three runs. [Parts a–c reprinted with permission from S. C. Jacobson, R. Hergenröder, A. W. Moore, and J. M. Ramsey, *Anal. Chem.* **66**, 4127 (1994), copyright 1994 American Chemical Society.]

has a volume ~ 1 nL. The reagents and products flow continuously through the reactor to the sample waste reservoir, with the four-way cross operating as a variable-volume valve. Small aliquots from the reactor are periodically introduced onto the separation column to be analyzed by triggering the variable-volume valve. The reactor/capillary electrophoresis integrated microchip was evaluated by reacting amino acids with o-phthaldialdehyde (OPA) to generate a fluorescent product detected by LIF. In Figure 18.9b, three sequential runs were performed for arginine and glycine reacted with OPA prior to the separation. The reaction time corresponds to the transit time of the sample through the reaction chamber. The reaction time can be varied by changing the potential applied to the microchip, which in turn affects the residence time of the sample in the reaction chamber. Figure 18.9c shows the variation of product formation with reaction time, from which reaction kinetics can be ascertained. The measured half-times of reaction were 5.1 and 6.2 s for arginine and glycine, respectively, and are comparable to reported values.

Biochemical reactions and analysis are also possible on microchips. A restriction digestion and electrophoretic sizing experiment were performed sequentially on an integrated microchip (23). The microdevice in Figure 18.10a mixes a DNA sample with a restriction enzyme in a 0.7 nL reaction chamber and after a digestion period injects the fragments onto a 67 mm long capillary electrophoresis channel for fragment sizing. The channel walls were coated with linear polyacrylamide for passivation thus all manipulations are by electrophoresis. Figure 18.10b shows the fragments from the digestion of the plasmid pBR322 by the enzyme *Hin*f I. The digestion time was 129 s (120 s dwell time plus a 9 s transit time), and the separation was completed in less than 150 s. Due to precise manipulation of picoliter volumes, only 30 amol of DNA and 2.8×10^{-3} units of enzyme were consumed per run. Migrating fragments were interrogated using on-microchip LIF with an intercalating dye, thiazole orange dimer (TOTO-1). On-chip labeling of the DNA fragments after digestion but prior to interrogation was accomplished by migrating TOTO-1 countercurrent to the separation from the waste reservoir toward the buffer reservoir.

18.9. THE FUTURE

Efforts to miniaturize chemical measurement techniques have primarily focused on chemical separations, i.e., capillary electrophoresis and capillary chromatography. While chemical separations are a very important area of chemical analysis, the value of miniaturized chemical measurement techniques will be greatly increased by the integration of many functions to solve

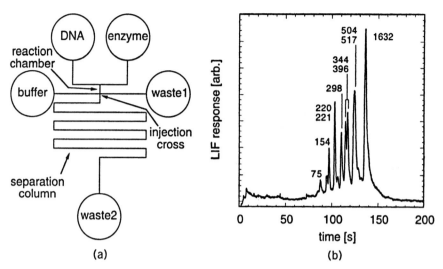

Figure 18.10. (a) Schematic of a precolumn reactor microchip with reaction chamber, injection cross, and serpentine separation channel. The microchip had substrate dimensions of 75 × 25 × 1 mm, channel dimensions of 60 μm width and 12 μm depth, separation channel length of 67 mm, and reaction chamber volume of 0.7 nL. This design was used to obtain the data in part b. (b) Electropherogram of products from the digestion of the plasmid pBR322 by the enzyme *Hin*f I. The separation field strength was 380 V/cm, and the separation length was 67 mm. The numbers correspond to the fragment lengths in base pairs. [Parts a and b reprinted with permission from S. C. Jacobson and J. M. Ramsey, *Anal. Chem.* **68**, 720 (1996), copyright 1996 American Chemical Society.]

a complete measurement problem, not just a single step in the measurement procedure. We call this form of integration, serial integration. Demonstrations to date show promise of building integrated monolithic devices that include pretreatment to reduce sample complexity, analysis by partitioning, derivatization for detection, and finally detection. Once functional components (Figure 18.1) are developed and understood under various operating conditions, it may be possible to design chemical measurement systems with computer-aided design (CAD) tools similar to microelectronics. Such a capability can potentially lead to significant reductions in cost for chemical instrumentation. An additional development that will surely follow the microelectronics paradigm is the use of parallel integration (many identical structures on the same substrate) to solve large measurement problems. The funding base for pursuing these concepts is presently increasing at a significant rate. The developments suggested here are likely to come in the near future.

ADDENDUM

The most recent advances in microfluidics include improvements in previously explored areas and the development of several unique techniques. In capillary electrophoresis, higher efficiencies were achieved in a postseparation derivatization of amino acids with o-phthaldialdehyde by empirically optimizing the reactor geometry and reagent flow rates (30). Microchip electrophoresis was also used to resolve the reaction products of homogeneous immunological reactions (31) [similar to that in Koutny et al. (8)]. For MEKC, the microchip platform enabled rapid mixing of buffer streams to produce linear, concave, and convex gradients of organic modifier content (32). For DNA analysis, up to twelve sizing ladders and products were analyzed simultaneously using a microchip with twelve parallel separation lanes (33). In other work, a hybridized device combined a microfabricated silicon thermal cycler for PCR amplification and a glass microchip for product analysis (34). In addition, a monolithic glass device combined multiplexed PCR amplification and electrophoretic product sizing (35). For detection, two groups have developed electrospray interfaces (36,37) which couple the microchip to a mass spectrometer to allow for structural analysis of molecular species manipulated on microchips. Also, confocal fluorescence detection was used to observe single-molecule bursts following electrophoretic separations (38). Less traditional techniques included continuous flow, homogeneous enzyme assays to elucidate enzyme kinetic information and inhibition constants (39) and the use of electrokinetic forces for two-dimensional confinement of both ions and fluids in order to reduce the detection probe volume (40). Lastly, electric fields were used to direct the migration and reaction of cells in the channel manifold (41).

LIST OF ACRONYMS

Acronym	Definition
CAD	computer-aided design
DOE	U.S. Department of Energy
FITC	fluorescein-5-isothiocyanate
LIF	laser-induced fluorescence
MEKC	micellar electrokinetic chromatography
OPA	o-phthaldialdehyde
PCR	polymerase chain reaction
RSD	relative standard deviation
TOTO-1	thiazole orange dimer
UV	ultraviolet

ACKNOWLEDGMENTS

This research is sponsored by U.S. Department of Energy (DOE), Office of Research and Development. Oak Ridge National Laboratory is managed by Lockheed Martin Energy Research Corporation for the DOE under contract DE-AC05-96OR22464.

REFERENCES

1. D. J. Harrison, A. Manz, Z. Fan, H. Lüdi, and H. M. Widmer, *Anal. Chem.* **64**, 1926 (1992).
2. A. Manz, D. J. Harrison, E. M. J. Verpoorte, J. C. Fettinger, A. Paulus, H. Lüdi, and H. M. Widmer, *J. Chromatogr.* **593**, 253 (1992).
3. C. S. Effenhauser, A. Manz, and H. M. Widmer, *Anal. Chem.* **65**, 2637 (1993).
4. K. Seiler, D. J. Harrison, and A. Manz, *Anal. Chem.* **65**, 1481 (1993).
5. D. J. Harrison, K. Fluri, K. Seiler, Z. Fan, C. S. Effenhauser, and A. Manz, *Science* **261**, 895 (1993).
6. S. C. Jacobson, R. Hergenröder, L. B. Koutny, R. J. Warmack, and J. M. Ramsey, *Anal. Chem.* **66**, 1107 (1994).
7. S. C. Jacobson, R. Hergenröder, L. B. Koutny, and J. M. Ramsey, *Anal. Chem.* **66**, 1114 (1994).
8. L. B. Koutny, D. Schmalzing, T. A. Taylor, and M. Fuchs, *Anal. Chem.* **68**, 18 (1996).
9. N. Burggraf, A. Manz, C. S. Effenhauser, E. Verpoorte, N. F. de Rooij, and H. M. Widmer, *J. High Resolut. Chromatogr.* **16**, 594 (1993).
10. D. E. Raymond, A. Manz, and H. M. Widmer, *Anal. Chem.* **66**, 2858 (1994).
11. B. Ekström, G. Jacobson, O. Ohman, and H. Sjödin, Int. Pat. WO 91/16966 (1990), in *Adv. Chromatogr.* **33**, 1 (1993).
12. C. S. Effenhauser, A. Paulus, A. Manz, and H. M. Widmer, *Anal. Chem.* **66**, 2949 (1994).
13. A. T. Woolley and R. A. Mathies, *Proc. Natl. Acad. Sci. U.S.A.* **91**, 11348 (1994)
14. A. T. Woolley and R. A. Mathies, *Anal. Chem.* **67**, 3676 (1995).
15. S. C. Jacobson, R. Hergenröder, L. B. Koutny, and J. M. Ramsey, *Anal. Chem.* **66**, 2369 (1994).
16. A. W. Moore, S. C. Jacobson, and J. M. Ramsey, *Anal. Chem.* **67**, 4184 (1995).
17. G. Ocvirk, E. Verpoorte, A. Manz, M. Grasserbauer, and H. M. Widmer, *Anal. Methods Instrum.* **2**, 74 (1995).
18. S. P. A. Fodor, J. L. Read, M. C. Pirrung, L. Stryer, A. T. Lu, and D. Solas, *Science* **251**, 767 (1991).
19. P. Wilding, M. A. Shoffner, and L. J. Kricka, *Clin. Chem. (Winston-Salem, N.C.)* **40**, 1815 (1994).
20. Y. Murakami, T. Takeuchi, K. Yokoyama, E. Tamiya, I. Karube, and M. Suda, *Anal. Chem.* **65**, 2731 (1993).

21. J. C. Fettinger, A. Manz, H. Lüdi, and H. M. Widmer, *Sens. Actuators B* **17**, 19 (1993).
22. S. C. Jacobson, R. Hergenröder, A. W. Moore, and J. M. Ramsey, *Anal. Chem.* **66**, 4127 (1994).
23. S. C. Jacobson and J. M. Ramsey, *Anal. Chem.* **68**, 720 (1996).
24. S. C. Jacobson, L. B. Koutny, R. Hergenröder, A. W. Moore, and J. M. Ramsey, *Anal. Chem.* **66**, 3472 (1994).
25. J. M. Ramsey, S. C. Jacobson, and M. R. Knapp, *Nat. Med.* **1**, 1093 (1995).
26. W. D. Volkmuth and R. H. Austin, *Nature (London)*, **358**, 600 (1992).
27. S. C. Jacobson, A. W. Moore, and J. M. Ramsey, *Anal. Chem.* **67**, 2059 (1995).
28. S. C. Jacobson and J. M. Ramsey, *Electrophoresis* **16**, 481 (1995).
29. K. Seiler, Z. H. Fan, K. Fluri and D. J. Harrison, *Anal. Chem.* **66**, 3485 (1994).
30. K. Fluri, G. Fitzpatrick, N. Chiem and D. J. Harrison, *Anal. Chem.* **68**, 4285 (1996).
31. N. Chiem and D. J. Harrison, *Anal. Chem.* **69**, 373 (1997).
32. J. P. Kutter, S. C. Jacobson, and J. M. Ramsey, *Anal. Chem.* **69**, 5165 (1997).
33. A. T. Woolley, G. F. Sensabaugh, and R. A. Mathies, *Anal. Chem.* **69**, 2181 (1997).
34. A. T. Woolley, D. Hadley, P. Landre, A. J. deMello, R. A. Mathies, and M. A. Northrup, *Anal. Chem.* **68**, 4081 (1996).
35. L. C. Waters, S. C. Jacobson, N. Kroutchinina, J. Khandurina, R. S. Foote, and J. M. Ramsey, *Anal. Chem.* **70** (1998).
36. R. S. Ramsey and J. M. Ramsey, *Anal. Chem.* **69**, 1174 (1997).
37. Q. Xue, F. Foret, Y. M. Dunayevshiy, P. M. Zavracky, N. E. McGruer, and B. L. Karger, *Anal. Chem.* **69**, 426 (1997).
38. J. C. Fister III, S. C. Jacobson, L. M. Davis, and J. M. Ramsey, *Anal. Chem.* **70** (1998).
39. A. G. Hadd, D. E. Raymond, J. W. Halliwell, S. C. Jacobson, and J. M. Ramsey, *Anal. Chem.* **69**, 3407 (1997).
40. S. C. Jacobson and J. M. Ramsey, *Anal. Chem.* **69**, 3212 (1997).
41. P. C. H. Li and D. J. Harrison, *Anal. Chem.* **69**, 1564 (1997).

PART IV
APPLICATIONS OF HPCE

CHAPTER
19

PEPTIDE ANALYSIS BY CAPILLARY ELECTROPHORESIS: METHODS DEVELOPMENT AND OPTIMIZATION, SENSITIVITY ENHANCEMENT STRATEGIES, AND APPLICATIONS

GREGORY M. McLAUGHLIN[*] and KENNETH W. ANDERSON

Dionex Corporation, Sunnyvale, California 94088-3603

DIETRICH K. HAUFFE

Dionex GmbH, 65510 Idstein, Germany

19.1.	Introduction	638
19.2.	Effects of Capillary Dimensions, Applied Voltage, and Temperature	639
	19.2.1. Capillary Length	639
	19.2.2. Capillary Inner Diameter	640
	19.2.3. Applied Voltage	640
	19.2.4. Capillary Temperature	642
19.3.	Effects of pH and Ionic Strength	643
	19.3.1. pH	643
	19.3.2. Ionic Strength	645
19.4.	Buffer Additives	647
	19.4.1. Organic Solvents	647
	19.4.2. Other Organic Chemical Additives and Modifiers	648
	19.4.3. Metal Additives	649
19.5.	Use of Ion-Pairing Reagents	650
19.6.	Use of Micellar Electrokinetic Chromatography for Hydrophobic and Neutral Species	651
19.7.	Chiral Selectors	653
19.8.	Sensitivity-Enhancement Strategies	654
19.9.	Use of Coated Capillaries and Wall Coatings	656
19.10.	Methods Development Strategy	657

* *Present address*: Rheodyne L.P., 6815 Redwood Drive, Cotati, California 94931.

High Performance Capillary Electrophoresis, edited by Morteza G. Khaledi. Chemical Analysis Series, Vol. 146.
ISBN 0-471-14851-2 © 1998 John Wiley & Sons, Inc.

	19.10.1.	Ensure Sample Solubility in the Separation Solution	657
	19.10.2.	Choose the Capillary Length and Diameter	657
	19.10.3.	Select Capillary Temperature	657
	19.10.4.	Optimize Buffer pH	657
	19.10.5.	Optimize Buffer Concentration	658
	19.10.6.	Optimize Separation Voltage (Ohm's Law Plot)	658
	19.10.7.	Select Additives to Maximize Differences or Mask Interactions	658
19.11.	**Use of Capillary Electrophoresis and High-Performance Liquid Chromatography as Complementary Techniques**		659
	19.11.1.	Selectivity Differences Between Capillary Electrophoresis and High-Performance Liquid Chromatography	659
	19.11.2.	Use of Capillary Electrophoresis to Check High-Performance Liquid Chromatography Fractions for Purity and as a Micropreparative Technique	660
19.12.	**Enhanced Detection Methods**		662
	19.12.1.	Capillary Electrophoresis and Mass Spectrometry	662
	19.12.2.	Capillary Electrophoresis and Laser-Induced Fluorescence Detection of Peptides	662
	19.12.3.	Capillary Electrophoresis of Peptides Combined with Other Sensitive Types of Detection	663
19.13.	**Physicochemical Measurements**		663
	19.13.1.	Electrophoretic Mobility Determination and Prediction	663
	19.13.2.	Reaction Monitoring, Kinetics, Binding Constants, and Stoichiometries	664
19.14.	**Concluding Remarks and Future Trends**		666
Addenda			666
List of Acronyms, Abbreviations, and Symbols			670
References			671

19.1. INTRODUCTION

The purpose of this chapter is to offer a practical guide to peptide analysis by capillary electrophoresis (CE). To this end the fundamental concepts of CE are presented in terms of operating conditions, buffer selection, and sensitivity enhancement strategies through chemical and physical means. Practical guidelines for methods development and optimization using a wide range of chemical additives are offered as they apply to peptide separations. Combined use of CE with other techniques, including HPLC and novel detectors, and usage to determine physicochemical parameters are described. These concepts are illustrated through a review of published reports describing methodologies and applications of CE for peptide analysis.

19.2. EFFECTS OF CAPILLARY DIMENSIONS, APPLIED VOLTAGE, AND TEMPERATURE

19.2.1. Capillary Length

In capillary electrophoresis (CE) efficiency and resolution are proportional to the capillary length, providing that the field strength is held constant. Efficiency and migration times increases linearly with length, whereas resolution depends on the square root of length. Therefore, improving efficiency and resolution by increasing capillary length occurs at the expense of increased analysis times. Figure 19.1 illustrates the effect of varying capillary length while applying a nearly constant electric field strength on the separation of a tryptic digest of bovine serum albumin (BSA) (1). The longer capillary gave greater efficiency, increased resolution, and revealed additional peaks not seen with the shorter capillary. These results agree with the theoretical relationship between length and migration time, efficiency, and resolution described by

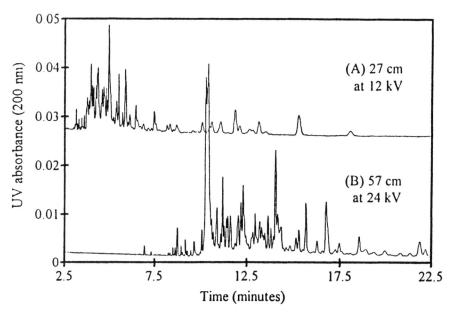

Figure 19.1. Effect of capillary length on migration time and resolution of tryptic digests of BSA. Conditions—buffer: 100 mM sodium phosphate, pH 2.50/1.5 M urea; capillary: (A) 50 μm (i.d.) × 27 cm (L_t) fused-silica (FS), (B) 50 μm (i.d.) × 57 cm (L_t) FS; voltage: (A) 12 kV or 444 V/cm, (B) 24 kV or 421 V/cm; temperature: 25 °C; detection: UV at 200 nm; injection: pressure, 10 s at 0.5 psi; concentration: 10.1 dilution of BSA tryptic digests in water. [Redrawn with permission from McLaughlin et al. (1).]

Cohen et al. (2). In general, longer capillaries are useful for analyzing complex mixtures, whereas shorter capillaries are preferable for less complex mixtures or when migration times are excessive.

The main effects of *increasing* capillary *length* (at constant field strength) are as follows (1):

- Longer migration times (directly proportional to length)
- Increased theoretical plates (directly proportional to length)
- Increased resolution (directly proportional to the square root of length)
- Less heat produced if voltage is held constant (increased surface area for heat dissipation)

19.2.2. Capillary Inner Diameter

Theory dictates that efficiency and resolution are inversely proportional to the capillary inner diameter (2). Smaller-bore capillaries give reduced sensitivity during optical detection because the pathlength is shorter. Larger-bore capillaries realize greater light throughput resulting in reduced noise and enhanced detection signal. They are also useful for micropreparative applications because they allow more mass to be loaded. The loading limit for 14 synthetic opioid peptides was investigated by Lee and Desiderio (3) as a function of capillary inner diameter (50–200 µm). In terms of resolution, time, and mass, they found that the 100 µm i.d. capillary gave the optimal loading of each peptide (39–78 pmol). When other types of detectors are used, smaller-i.d. capillaries may be preferable. Wahl et al. (4) used small-bore capillaries (10–100 µm i.d.) for high-efficiency peptide detection sensitivity in CE–MS. Smaller-i.d. capillaries also allow use of higher voltages and ionic strength buffers because less heat will be generated. Capillary i.d. has only a small effect on electroosmotic flow (EOF).

The main effects of *increasing* capillary *diameter* are as follows (1):

- Better signal-to-noise ratio (S/N)
- Better mass loading (proportional to diameter squared)
- Less efficiency and poorer resolution
- Increased heat generated (proportional to diameter squared)

19.2.3. Applied Voltage

Cohen et al. (2) showed theoretically that efficiency and resolution increase and migration time decreases with increase in voltage. However, higher voltages produce larger currents, thus generating more Joule heat. The optimum applied voltage for a given capillary and separation buffer can be determined by generating an Ohm's law plot of the current produced by step changes in the applied voltage. Wolze et al. (5) investigated the effect of increasing voltage on the separation of peptides. Figure 19.2 demonstrates that increasing applied voltage from 5 to 25 kV caused a decrease in migration times. Maximum efficiency and resolution were obtained at the voltage (20 kV) where the Ohm's law plot deviates from linearity (Figure 19.2F, inset).

Certain buffers, including the "Good" buffers [i.e., low-conductivity buffers, such as 2-morpholinoethanesulfonic acid (MES), 2-(cyclohexylamino) ethanesulfonic acid (CHES), and 4-(2-hydroxyethyl)-1-piperazineethanesulfonic acid (HEPES)], at concentrations of 50–100 mM produce relatively low currents at applied potentials of 20–30 kV. They yield linear Ohm's law plots, as demonstrated by McLaughlin et al. (1,6). Chen et al. (7) determined that the

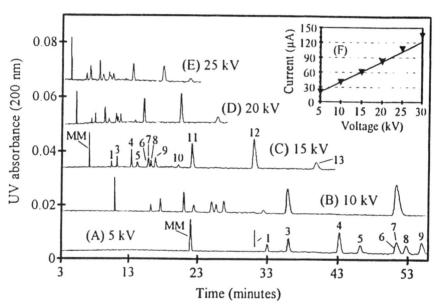

Figure 19.2. (A–E) Effect of voltage on migration time of peptide mixture. (F) Effect of voltage on current. Conditions—buffer: 50 mM sodium phosphate, pH 2.50; capillary: 75 μm (i.d.) × 50 cm (L_t) FS; voltage: (A) 5 kV, (B) 10 kV, (C) 15 kV, (D) 20 kV, (E) 25 kV; temperature: 34 °C; detection: UV at 200 nm; injection: pressure, 5 s at 0.5 psi; concentration: 2.5–17.5 μg/mL. Peaks: (MM) mobility marker; (1) angiotensin I; (3) Arg-Gly-Ala-Gly-Gly-Leu-Gly-Leu-Gly-Lys-Amide; (4) angiotensin II; (5) β-endorphin fragment 61-91; (6) Ac-Arg-Gly-Gly-Gly-Gly-Leu-Gly-Leu-Gly-Lys-Amide; (7) Ac-Arg-Gly-Ala-Gly-Gly-Leu-Gly-Leu-Gly-Lys-Amide; (8) Ac-Arg-Gly-Val-Gly-Gly-Leu-Gly-Leu-Gly-Lys-Amide; (9) Ac-Arg-Gly-Val-Val-Gly-Leu-Gly-Leu-Gly-Lys-Amide; (10) impurity (11) Des-Tyr-methionine enkephalin; (12) methionine enkephalin; (13) oxcytocin.

migration order of peptides was independent of the operating current and the capillary temperature.

The main effects of *increasing voltage* (at constant capillary length) are as follows (1):

- Increased efficiency and resolution (directly proportional to voltage)
- Shorter analysis times (inversely proportional to voltage)
- Increased heat production (current increases linearly with increasing voltage)

19.2.4. Capillary Temperature

Capillary temperature has important effects on the viscosity and the pH of the separation buffer, its current stability, analyte solubility, and migration time reproducibility. Therefore, efficient capillary temperature control allows better migration time reproducibility. Furthermore, high-concentration buffers and field strengths may be applied, resulting in higher efficiencies and shorter analysis times. The effects of temperature on migration behavior of 13 peptides were demonstrated by McLaughlin et al. (8). As shown in Figure 19.3, the peptides migrated faster as the temperature was increased from 25 to 60 °C. The decreased migration times appeared directly proportional to the decrease in buffer viscosity and EOF as the temperature was increased. Also, due to decreasing viscosity, the amount of injected sample increased as temperature increased (see relative size reference arrow). Temperature changes induced minor selectivity changes. However, some larger peptides (see peaks 11–13) migrated relatively later at 60 °C than at 51 °C. These results agree with previous reports on temperature effects (1).

Zhang et al. (9) studied the effect of capillary temperature (15–60 °C at 5 °C increments) on peptide migration time (t_m) and determined that a plot of log t_m vs. $1/T$ was linear. Changes in t_m with temperature correlated with changes in buffer viscosity, while selectivity remained constant. Issaq et al. (10) studied the effects of pH, buffer additives, and temperature on the separation of peptides. Variable temperature control can be used for studies such as determinations of thermal denaturation rates and melting points. Ma et al. (11) described gains in resolution of the cis and trans conformers of small peptides when CE was used at subzero temperatures.

The main effects of *increasing temperature* are as follows (1):

- Lower viscosity and faster analysis time (inversely proportional to temperature)
- Higher current (directly proportional to temperature)

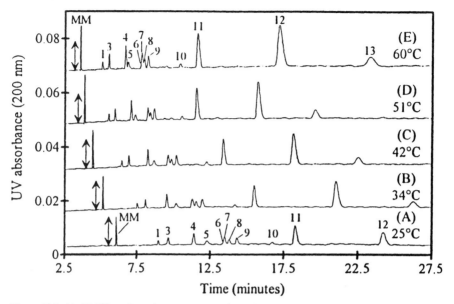

Figure 19.3. (A–E) Effect of varying temperature from 25 to 60 °C on migration time of peptide mixture. Conditions—buffer: 50 mM sodium phosphate, pH 2.50; capillary: 75 μm (i.d.) × 50 cm (L_t) FS; voltage: 20 kV; temperature: (A) 25 °C, (B) 34 °C, (C) 42 °C, (D) 51 °C, (E) 60 °C; detection: UV at 200 nm; injection: pressure, 5 s at 0.5 psi; concentration: 2.5–17.5 μg/mL. Peaks: same as in Figure 19.2 (↕ indicates peak height for mobility marker at 25 °C as a relative reference).

- Minor selectivity changes
- Changes in resolution, efficiency, peak shape, buffer pH, and analyte solubility

19.3. EFFECTS OF pH AND IONIC STRENGTH

19.3.1. pH

Buffer pH is important in CE because it affects both the net charge of the analyte and the magnitude of the EOF. Optimization of pH produces gains in efficiency, resolution, and analysis time. This applies to a wide variety of peptide forms (synthetic, deamidated, formylated, glycosylated, and digest fragments) that can be separated by CZE. Field et al. (12) described the separation of five structurally similar synthetic peptides with the same amino acid composition but different primary sequences. Figure 19.4 shows that

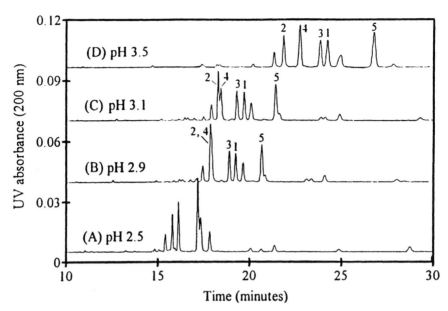

Figure 19.4. (A–D) Determination of optimum pH by titrating across the pK_a's of a closely related set of peptides. Conditions—buffer: 50 mM sodium phosphate, (A) pH 2.5, (B) pH 2.9, (C) pH 3.1, (D) pH 3.5; capillary: 75 μm (i.d.) × 57 cm (L_t) FS; voltage: 15 kV; temperature: 25 °C; detection: UV at 200 nm; injection: pressure, 5 s at 0.5 psi. Peaks: (1) Arg-His-Asp-Ala-Ala-Ala-Tyr-Leu-Leu-OH; (2) Leu-Leu-Tyr-Ala-Ala-Ala-Asp-His-Asp-OH; (3) His-Arg-Ala-Leu-Asp-Tyr-Ala-Ala-Leu-OH; (4) His-Ala-Leu-Asp-Tyr-Ala-Arg-Leu-Ala-OH; (5) Asp-His-Ala-Tyr-Leu-Ala-Ala-Arg-OH. [Redrawn with permission from Field et al. (12).]

increasing the pH from 2.5 to 3.5 allowed peptides 2 and 4 to be resolved, probably due to pK_a differences.

While peptides can be separated over a wide range of pH, two pH regions, 2–4 and 7–9, are especially useful. McCormick (13) used low pH (1.5–3.0) to achieve resolution of octapeptides. Strickland and Strickland (14) used small changes in pH from 2.45 to 2.67 to optimize the separation of a 14-component peptide mixture. The effect of pH (2.5, 5.5, and 8.5) on the separation of opioid peptides was reported by Lee and Desiderio (15). Grossman et al. (16) studied the effect of pH (2.5, 4.0, and 11.0) on the selectivity of peptide separations by CZE. Separations of sulfated and non-sulfated forms of cholecystokinin, enkephalin, and hirudin peptides were performed at pH 6.5 by Hortin et al. (17). Ye and Baldwin (18) reported the separation of amino acid and peptide samples at elevated pH. Langenhuizen and Janssen (19) observed that the best separation for basic and neutral peptides was achieved in the low pH region and for acidic peptides in the neutral pH region.

Optimization of buffer pH frequently yields good separations of peptides. Some examples include: Miller et al. (20), purity analysis; Rivier et al. (21), purity of synthetic peptides; Fukuoka et al. (22), bradykinin and its metabolites; Cifuentes and Poppe (23), peptide migration and electropherogram prediction using a computer program; Pessi et al. (24), single-residue changes in synthetic branched and multiple-antigen peptides; Chen et al. (25), deletion by-products and anticoagulant peptide separation; Yannoukakos et al. (26), phosphorylated and deamidated peptides; Ferranti et al. (27), abnormal globins; and Florance et al. (28), the influence of motilin fragment net charge on migration time.

Some conclusions regarding pH are as follows (1):

- It is the most important parameter to vary for optimizing selectivity.
- Small differences in pK_a can be the basis of separating structurally similar peptides.
- pK_a of charged groups on the peptide determine pH ranges to evaluate.
- Unknown mixtures can be optimized using survey runs over the range of pH 2–10.
- Resolution, selectivity, and peak shape can be dramatically altered.

19.3.2. Ionic Strength

In contrast to proteins, which often adsorb onto the capillary wall, peptides can generally be separated on native, fused-silica capillaries. Increases in ionic strength improve peak efficiency, resolution, sensitivity, and recovery. As ionic strength increases, mobilities decrease (mobility is directly related to zeta potential and inversely to time) (1). Increased ionic strength reduces the magnitude of the zeta potential by screening the fixed charge on the capillary wall. The result is decreased EOF and longer migration times. Ionic strength differences between sample and separation buffer can lead to sensitivity increases of 5–200 times due to sample stacking.

The effect of ionic strength (25–500 mM) on the separation of BSA tryptic peptides, as reported by McLaughlin et al. (29), is shown in Figure 19.5. The migration times of early migrating peaks increased with increased ionic strength (see the dashed line at 4–6 min). Migration times increased as buffer concentration increased from 25 to 200 mM. Most peaks sharpened, resolution improved, and peak heights increased (ionic focusing effect). Efficiency increased from 99,000 to 406,000 plates/m (the peak marked ▼), and resolution increased from 0.4 to 0.9 (the pair of peaks marked with ♦). The number of peaks and shoulders resolved increased from 58 to 99 as the ionic strength

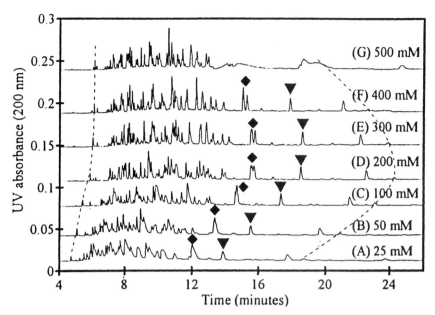

Figure 19.5. Effect of ionic strength on the separation of a tryptic digest of BSA. Conditions—buffer: sodium phosphate, pH 2.40, at (A) 25 mM, (B) 50 mM, (C) 100 mM, (D) 200 mM, (E) 300 mM, (F) 400 mM, and (G) 500 mM; capillary: 50 μm (i.d.) × 50 cm (L_t) FS; voltage: 20 kV; temperature: ambient; detection: UV at 200 nm; injection: gravity 10 s at 100 mm; concentration: 0.4 mg/mL for total digest.

was increased from 25 to 400 mM. Superimposed on these effects was an increase in current from less than 25 μA to over 200 μA. At 200 mM buffer, the ability of the CE system to remove heat effectively was exceeded and the buffer viscosity began to decrease. As a result, the later peaks started to move faster (see the dashed line between 18 and 24 min). Excessive heating increases diffusion and band dispersion, with the loss of resolution.

Nielsen and Rickard (30,31) optimized ionic strength to separate human growth hormone tryptic peptides. Chen et al. (32) demonstrated that increasing buffer concentration did not alter the electrophoretic mobilities of small peptides. Lee and Desiderio (33) studied the effects of buffer and analyte concentration on the resolution and detectability of peptides. Increasing the ionic strength led to increased peak height and decreased peak width, resulting in higher detectability and resolution. Peak area was linear with peptide concentration, even when overloading lead to peak distortion. Chen et al. (34) separated peptides with very high ionic strength buffers (500 mM). In general, larger volumes can be injected using higher ionic strength buffers.

The main effects of *increasing* the *ionic strength* of the separation buffer are as follows (1):

- Decreased EOF
- Increased migration time in direct proportion to ionic strength
- Increased efficiency and resolution and improved peak shape
- Enhanced detection sensitivity due to sample stacking
- Increased Joule heating

19.4. BUFFER ADDITIVES

19.4.1. Organic Solvents

The addition of organic solvents to the separation buffer used in CZE or MEKC has proved useful for modifying selectivity, viscosity, and zeta potential. Organic additives offer a convenient way to modify these parameters. The effect of acetonitrile and methanol on the migration behavior of a bioactive peptide mixture was examined by McLaughlin et al. (1). The results are shown in Figure 19.6 for acetonitrile and methanol at 0–50% (v/v) in a sodium phosphate buffer at pH 2.5. Addition of acetonitrile resulted in slight increases followed by steady decreases in migration times (Figure 19.6A). Increased migration times were seen over the same concentration range with methanol (Figure 19.6B). Peptide migration time behavior paralleled the changes in viscosity produced by adding the organic solvents. The viscosity of acetonitrile solutions increases slightly up to 15% (v/v) and decreases at higher percentages. The viscosity of methanol solutions increases up to 50% (v/v). In both cases small selectivity changes were observed, particularly with 0–20% (v/v) organic modifier.

Idei et al. (35) studied the effect of adding 0–30% (v/v) organic solvent (acetonitrile and alcohols) on migration behavior of synthetic peptides. The effect of increasing acetonitrile concentration on peptide migration time was opposite to that for the alcohols. Furthermore, peptide migration time increased with increasing alcohol chain length. Florance (36,37) proposed that incorporation of organic solvents improved resolution by modifying secondary structure and the hydrodynamic profile of normal and deamidated motilin peptides. Florance concluded that organic solvents may be useful for resolving structurally similar peptides with small differences in charge. Sahota and Khaledi (38) separated a mixture of 10 peptides using nonaqueous CE (NACE) with formamide containing 20 mM NaOH and 180 mM boric acid. NACE shows promise for separating hydrophobic peptides.

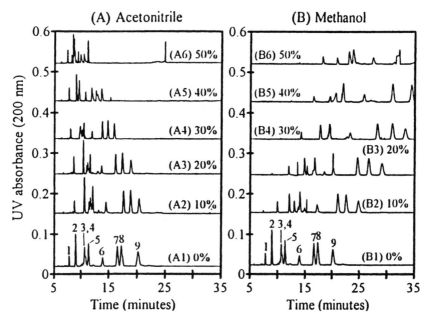

Figure 19.6. Effect of addition of acetonitrile (A1–A6) and methanol (B1–B6) on migration time of peptide mixture. Conditions—buffer: 50 mM sodium phosphate, pH 2.44, with (A1–A6) 0–50% acetonitrile, as indicated, and (B1–B6) 0–50% methanol, as indicated; capillary: 75 μm (i.d.) × 50 cm (L_t) FS; voltage: 25 kV; temperature: 30 °C; detection: UV at 200 nm; injection: pressure, 5 s at 0.5 psi. Peaks: (1) dynorphin; (2) bradykinin; (3) angiotensin II; (4) thyrotropin-releasing hormone (TRH); (5) (LH-RH); (6) bombesin; (7) leucine enkephalin; (8) methionine enkephalin; (9) oxytocin. [Redrawn with permission from McLaughlin et al. (1).]

The main *effects* of using *organic solvents* are as follows (1):

- Migration times correlate directly with the viscosity of the buffer–organic solvent mixture.
- Analysis times vary depending on the type and concentration of organic modifier.
- Both efficiency and resolution can be affected.
- Current decreases as the concentration of organic solvent is increased.
- Selectivity, solubility, and zeta potential are altered.

19.4.2. Other Organic Chemical Additives and Modifiers

Peptides that are strongly basic, hydrophilic, or hydrophobic, or that have very similar mobilities, often present separation challenges. Adsorption onto the capillary wall (particularly at low pH), poor efficiency, and asymmetric peak shape are the most common separation problems. A variety of additives have been used to coat the capillary wall in order to eliminate adsorption, modify buffer polarity, form an *in situ* ion exchanger, and introduce sieving or size-based separation mechanisms. Other additives modify or reverse the EOF.

Bullock (39) proposed the use of 1,3-diaminopropane as an EOF modifier along with moderate levels of alkali-metal salts to suppress protein–capillary wall interactions. The combined buffer additives allowed the proteins to be separated at pH values below their isoelectric points. This same technique is applicable to small basic peptides that would normally stick to the wall at low pH. Oda et al. (40) developed multiple buffer additive strategies to separate a series of six peptides. Hexanesulfonic acid (HSA), acetonitrile, and hexamethonium bromide were found to alter selectivity dramatically. Hexadecyltrimethylammonium chloride was used to improve the separation of two mixtures of basic peptides by Varghese and Cole (41). Liu et al. (42) used hexadecyltrimethylammonium bromide, sodium dodecyl sulfate (SDS), and cyclodextrins to separate derivatized hydrophobic peptides. Basic proteins and tryptic peptides of horse heart cytochrome *c* were separated using chitosan as a modifier by Yao and Li (43). The chitosan interacted with the capillary surface to reverse EOF and reduce wall interactions at pH values below the isoelectric points of the analytes. Thorsteinsdottir et al. (44) used MEKC with *N,N*-dimethyldodecylamine and Brij 35 to improve separations of enkephalin-related peptides and protein kinase A peptide substrates. Ma et al. (45) separated cis and trans conformers of small peptides using a sodium borate buffer containing 23% (w/v) glycerol at $-12\,°C$. Krueger et al. (46) added NaCl and zwitterions to separate endoproteinase Arg C digests of adrenocorticotropic hormone (ACTH).

19.4.3. Metal Additives

Metal additives can be useful in peptide separations, either by directly interacting with the peptide or indirectly through modification of the buffer system. Issaq et al. (47) studied the separation of dipeptides and oligopeptides using buffers at low and high pH. For peptides with different net charge, the best separation was achieved at pH 2.5 in phosphate buffer with 50 mM Zn^{2+}. Kajiwara (48) used glycine and Tris buffers containing 2 mM $CaCl_2$, 2 mM $ZnCl_2$, or ethylenediaminetetraacetic acid (EDTA) to separate calcium-binding proteins, including proteolytic peptides of calmodulin and zinc-binding proteins. On-column copper(II) complexation and electrochemical detection were used by Deacon et al. (49) to analyze peptides. Richards and Beattie (50) reviewed the use of CE for determining metalloproteins and metal-binding peptides. Dette and Waetzig (51) separated r-hirudin from related compounds using an acetate/ZnCl$_2$ buffer and 3% poly(ethylene glycol) (PEG) 20,000.

19.5. USE OF ION-PAIRING REAGENTS

Ion-pairing reagents (IPRs), such as tetrabutylammonium hydroxide (TBAOH), trifluoroacetic acid (TFA), and HSA are commonly used in HPLC to alter retention characteristics of ionic solutes. An early study of the use of IPR in CE employed HSA for the separation of synthetic peptides, and a tryptic map of a protein showed selectivity changes and improved resolution, particularly for the smaller, hydrophilic peptides (1).

Separation of tryptic peptides of BSA using sodium butane and pentanesulfonates (PSA) was studied by McLaughlin et al. (29). Compared to phosphate buffer (Figure 19.7A), addition of butanesulfonate (Figure 19.7B) or PSA (Figure 19.7C) improved the efficiency and increased the number of resolved

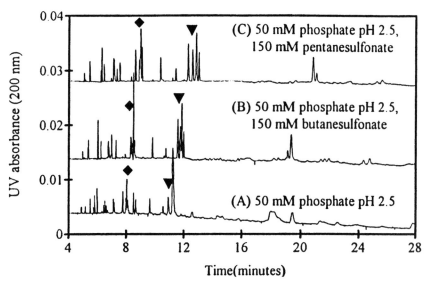

Figure 19.7. Effect of IPR on the resolution and efficiency of tryptic maps of fetuin. Conditions—buffer: (A) 50 mM sodium phosphate, pH 2.5, (B) 50 mM sodium phosphate, pH 2.5/150 mM sodium butanesulfonate, (C) 50 mM sodium phosphate, pH 2.5/150 mM sodium pentanesulfonate; capillary: 50 μm (i.d.) × 70 cm (L_t) FS; voltage: 30 kV; current: 50–160 μA; temperature: ambient; detection: UV at 200 nm; injection: gravity, 30 s at 150 mm; concentration: 0.2 mg/mL for total digest. Peaks: tryptic peptides of fetuin.

peaks. Furthermore, resolution improved with increased alkyl chain length. In particular, PSA resolved several pairs of peptides (indicated by ◆ and ▼) that were unresolved with butanesulfonate. McLaughlin et al. (29) also investigated the effects of octanesulfonic acid and increasing HSA concentration on peptide separation. As the IPR concentration increased, the resolution improved, theoretical plate number grew dramatically, and the number of peaks resolved increased.

Okafo et al. (52,53) used phytic acid to improve resolution in the separation of dipeptides, bradykinin-related peptides, cytochrome c tryptic digests, and hemoglobin variants. Rush et al. (54) used HSA to evaluate glycopeptide microheterogeneity. Hagono et al. (55) used HSA to resolve pharmaceutically important polypeptides.

Some *effects* of IPRs are as follows (1):

- Changes in migration time and zeta potential due to IPR concentration and chain length
- Improved resolution, efficiency, and number of separated components
- Selectivity changes that resolve components with similar mobility
- Increased current with increased IPR concentration

19.6. USE OF MICELLAR ELECTROKINETIC CHROMATOGRAPHY FOR HYDROPHOBIC AND NEUTRAL SPECIES

In CZE neutral solutes migrate at the speed of the EOF and do not separate. Addition of surfactants above their critical micelle concentration to the electrolyte resolves this problem, resulting in the highly selective separation of neutral molecules and hydrophobic analytes. This technique, called micellar electrokinetic chromatography (MEKC), was developed by Terabe and coworkers (56,57). MEKC has recently been used to analyze peptides and proteins, especially hydrophobic species, and to examine their chemical modification (deamidation, phosphorylation, acetylation, and formylation), McLaughlin et al. (29) compared CZE and MEKC for the separation of three hydrophobic peptides: N-acetyl, N-formyl, and unmodified Met-Leu-Phe (Figure 19.8). While the CZE separation showed good peak efficiency ($>$ 500,000 plates/m), inadequate resolution of peaks 2 and 3 was achieved. All three hydrophobic peptides, however, were easily resolved using MEKC by adding 100 mM SDS to the buffer. Although efficiency with CZE was higher, selectivity with MEKC was superior. Recently, nonionic surfactants have been used in place of SDS to improve selectivity. Terabe et al. (58) and Matsubara et al. (59) optimized electrolyte pH and surfactant concentration for peptide

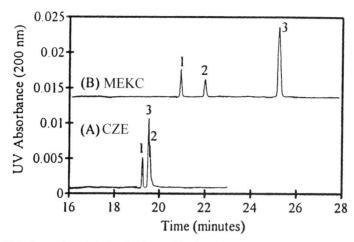

Figure 19.8. Separation of hydrophobic peptides by (A) CZE and (B) MEKC. Conditions—buffer: (A) 250 mM sodium borate, pH 8.8, and (B) 100 mM sodium borate, pH 8.8/100 mM SDS; capillary: 50 μm (i.d.) × 90 cm (L_t) FS; voltage: 30 kV; temperature: ambient; detection: UV at 200 nm; injection: gravity, 10 s at 150 mm; concentration: 50–200 μg/mL. Peaks: (1) N-acetyl-Met-Leu-Phe; (2) N-formyl-Met-Leu-Phe; (3) Met-Leu-Phe.

separations. The use of nonionic surfactants enabled the total ionic strength to be increased without increasing Joule heating. Ye et al. (60) achieved unique selectivity control of small tryptophan-containing peptides having very similar charge and mass using MEKC with mixed fluorocarbon–hydrocarbon anionic surfactants. (MEKC is described by Khaledi in Chapter 3 of this monograph.)

Further published reports on peptide separations by MEKC include: Yashima et al. (61), large peptides with the addition of organic modifiers; Gaus et al. (62), basic peptides with different charges and charge orientations; Karim et al. (63), modified BSA derivatives using coated capillaries with mobility differences of only 0.33%; Greve et al. (64), zwitterionic detergent and an alcohol or acetonitrile to probe hydrophobic selectivity differences in closely related gastrin peptides; Issaq et al. (65), cancer research peptides; Van de Goor et al. (66), modified ACTH fragments; Swedberg (67), nonionic and zwitterionic surfactants with heptapeptides; Liu et al. (42), structurally similar peptides with dodecyl- and hexadecyltrimethylammonium bromide, SDS, and cyclodextrins (CDs); Matsubara et al. (68), nonionic surfactant, Tween 20, with angiotensin II analogues; Matsubara and Terabe (69), nonionic surfactant, Polysorbate 20, with angiotensin II peptides; Beijersten and Westerlund (70), enkephalin-related peptides separated with taurodeoxycholate; and Wainright (71), pharmaceutical drug classes and peptides.

Some *key aspects* of *MEKC* are as follows:

- Neutral and charged solutes separate simultaneously.
- Micelle concentration, surfactant alkyl chain length, and mixed micelle composition alter migration times and the separation window.
- High-pH buffers maintain reasonable EOF and ensure the migration direction.
- Organic modifiers can be added to manipulate solute–micelle interaction, lower viscosity, and alter partitioning [note: organic solvents above 25% (v/v) can disrupt micelle formation].

19.7. CHIRAL SELECTORS

Chiral capillary electrophoresis (CCE) uses a buffer with chiral additives, frequently CDs, to separate enantiomers. (CE is covered more extensively by Waug and Khaledi in Chapter 23 of this monograph.) McLaughlin et al. (29) described the use of chiral selectors for the separation of a set of closely related dipeptides. Capillary zone electrophoretic (CZE) conditions produced a poor separation of the aromatic peptide pair L-Phe-Gly and Gly-L-Phe (Figure 19.9A). Addition of 10 mM hydroxyethyl-β-cyclodextrin allowed resolution of the dipeptides (Figure 19.9B). Chiral selectivity was based on interaction of the aromatic residues with the CD cavity.

Published reports of the use of chiral selectors for the separation of peptides include: Altria et al. (72,73), drugs and peptides; Terabe et al. (74), peptides; Kuhn et al. (75,76), peptides with optically active crown ether; Survay et al. (77,78), oligoglycines and -alanines for modeling peptide mobility; and Quang and Khaledi (79,80), tetraalkylammonium ions to reverse EOF and cyclodextrins to form inclusion complexes with peptides and other basic compounds.

Some *conclusions* regarding *chiral selectors* are as follows:

- CD cavity size and substituted CD R-groups (e.g., HO, CH_3) are crucial to recognition.
- Addition of solubilizers (e.g., SDS, methanol, urea) can increase the resolution of enantiomers by increasing CD solubility.
- More hydrophilic substituted CDs are more soluble in aqueous buffers.
- Lower temperatures lead to higher resolution.
- Differences in selectivity and solubility may be obtained by using substituted CDs, bile salts, metal complexation, and crown ethers.

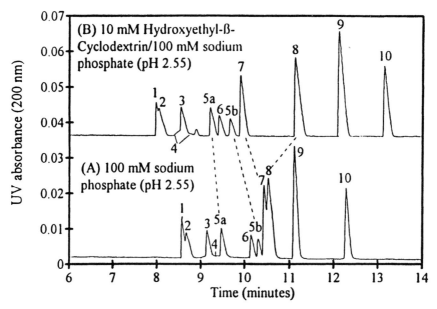

Figure 19.9. Use of hydroxyethyl-β-CD to enhance selectivity and resolution for a mixture of closely related dipeptides: (A) without and (B) with hydroxyethyl-β-CD. Conditions—buffer: (A) 100 mM sodium phosphate, (B) 100 mM sodium phosphate, pH 2.55/10 mM hydroxyethyl-β-CD; and capillary: 50 μm (i.d.) × 70 cm (L_t) FS; voltage: 30 kV; temperature: ambient; detection: UV at 200 nm; injection: gravity, 10 s at 100 mm; concentration: 0.5 mg/mL each component. Peaks: (1) Gly-L-Ala; (2) L-Ala-Gly; (3) L-Ala-L-Ala; (4) Gly-L-Pro; (5a,5b) Gly-L-Val; (6) L-Leu-Gly; (7) L-Phe-Gly; (8) Gly-L-Phe; (9) Gly-L-Tyr; (10) L-Leu-Ala.

19.8. SENSITIVITY-ENHANCEMENT STRATEGIES

Typical detection limits for peptides using UV absorbance detection are on the order of 50–100 ng/mL. This means that many types of peptide samples are too dilute to analyze by standard CZE. A number of strategies have been developed to increase detection sensitivity. One approach is known as ionic focusing or stacking. The sample is injected in a low-conductivity (high-resistance) solution and is separated in a higher-conductivity (low-resistance) buffer. This results in a higher field strength in the sample injection plug than in the run buffer when voltage is applied. The analytes migrate quickly in the sample zone because of the high electric field. When they reach the lower field in the separation buffer, they concentrate or stack. Figure 19.10 (29) compares separation of a tryptic digest under nonstacking [diluted with separation buffer (Figure 19.10A)] and stacking conditions [diluted with water (Fig-

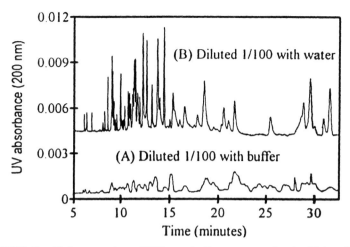

Figure 19.10. Sensitivity comparison of BSA tryptic digests separated under (A) focusing conditions and (B) nonfocusing conditions. Conditions—buffer: 100 mM sodium phosphate, pH 2.40, capillary: 50 μm (i.d.) × 50 cm (L_t) FS; voltage: 20 kV; temperature: ambient; detection: UV at 200 nm; injection: gravity, 40 s at 100 mm; concentration: (A) digest diluted 1:100 with water (0.02 mg/mL), and (B) digest diluted 1:100 with separation buffer (0.02 mg/mL).

ure 19.10B)]. Ionic focusing can increase sensitivity up to 50 times while improving peak shape and efficiency.

Aebersold and Morrison (81) injected an ammonium hydroxide plug prior to the sample, which produced stacking at the separation buffer interface. Schwer and Lottspeich (82) [see also Schwer et al. (83)] described analytical and micropreparative peptide separations using one to three discontinuous buffer systems that allowed a 30- to 50-fold increase in sample load to increase sensitivity. Sensitivity enhancement of greater than 100 times for dilute peptide and protein mixtures was achieved by Foret et al. (84) using "on-column" transient isotachophoretic (ITP) preconcentration. Liao et al. (85) demonstrated automated, 1000-fold concentration of peptides combining two electrophoresis modes and CZE. Other concentration strategies include: Kasicka et al. (86), ITP applied to a synthetic C-terminal tetrapeptide fragment of growth-hormone-releasing peptide; McLaughlin et al. (29), MEKC with stacking to detect less than 0.1% by weight acetylated and formylated impurities of a peptide; Witte et al. (87), on-line ITP preconcentration of heterocyclic peptides in plasma; and Satow et al. (88), low salt concentration in the sample zone to improve peptide resolution.

Alternative methods to enhance sensitivity have been developing rapidly. A preconcentrator was used for low-level detection of cytochrome c tryptic digests by Hoyt et al. (89). A replaceable concentrator tip and capillary hybrid was used by Strausbauch et al. (90) to achieve low detection limits (1–10 ng/mL) for BSA tryptic peptides. Strausbauch et al. (91) described the mechanism of peptide separation using on-line solid phase extraction. Tomlinson et al. (92,93) used improved on–line membrane preconcentration combined with ITP for peptide detection at low level (50 fmol/nL and 200 amol/nL for different mixtures).

Some *conclusions* regarding *sensitivity enhancement strategies* are as follows:

- Sensitivity improves by stacking when the sample is in a low-conductance solution.
- ITP and preconcentrators can increase sensitivity 50- to 1000-fold.

19.9. USE OF COATED CAPILLARIES AND WALL COATINGS

Coated or surface-modified capillaries can eliminate wall interactions, modify EOF, alter separation selectivity, and change the hydrophobicity and hydrophilicity of the capillary wall. The capillaries can separate a variety of neutral and basic proteins and peptides in buffers at acidic, neutral, and basic pH. A variety of wall coatings have been reported, and several of these are commercialy available. Thorsteinsdottir et al. (94) used amino-silylated, fused-silica capillaries for fast and highly efficient separation of enkephalin-related peptides. A pH stable cellulose coating for separations of peptides and glycoconjugates was described by Huang et al. (95). Phillips and Kimmel (96) used PEG-coated capillaries for the analysis of small peptides and inflammatory cytokines. Nashabeh and El Rassi (97) separated glycopeptides with hydrophilic coated capillaries. Smith and El Rassi (98) used methylated hydroxyethylated polyethyleneimine-coated capillaries to separate a mixture of acidic peptides. Huang et al. (99) developed a stable charged surface coating that allows switchable EOF. Other separations using coated capillaries include: Piccoli and coworkers (100), cytochrome *c* and BSA digests on a hydrophilic phase; Castagnola et al. (101), myoglobin tryptic peptides with acrylamide-coated capillaries; and McLaughlin et al. (102), use of a hydrophilic capillary to separate a peptide mixture.

The main *advantages* to use of a *coated capillary* for peptide analysis are as follows:

- Eliminate adsorption to the capillary wall.
- Modify, stabilize, eliminate, or reverse EOF.

- Manipulate the hydrophobicity or hydrophilicity of the capillary inner surface.
- Alter separation selectivity (e.g., affinity capillary electrophoresis).
- Produce a permanent, stable, and reproducible surface.

19.10. METHODS DEVELOPMENT STRATEGY

19.10.1. Ensure Sample Solubility in the Separation Solution

- Most peptides are soluble in aqueous buffers.
- Hydophobic solutes may require the use of additives.
- Need for an organic solvent may indicate use of MEKC.

19.10.2. Choose the Capillary Length and Diameter

- A good first choice is a 50 µm i.d. × 50 cm fused-silica capillary.
- Separation compexity dictates capillary length (35–40 cm for 2–10 analytes, 50–60 cm for 11–50 analytes, 70–80 cm for 50–80 analytes, and 90–100 cm for > 80 analytes).
- Efficiency, resolution, detection limits, and mass loading requirements dictate the capillary diameter: for best efficiency use 25 or 50 µm i.d. capillaries; for best UV detection limits use 50–100 µm i.d. capillaries; and for best mass loading use 100–200 µm i.d. capillaries.
- Coated capillaries are recommended for difficult separations and better reproducibility.

19.10.3. Select Capillary Temperature

- A good first choice temperature is 20–25 °C.
- For bioactivity recovery, chiral separations, and high concentration buffers use 15–20 °C.
- For faster separations use 30–60 °C.
- Vary the set temperature from 20–60 °C in 5 °C increments to optimize solubility, selectivity, or conformational stability.

19.10.4. Optimize Buffer pH

- Select a buffer that gives good pH control in the region of interest.
- Use pH 2.5–3.5 (C-terminal pK_a) or pH 8–9 (N-terminal pK_a) for most peptides.

- Use small pH changes (i.e., 0.1–0.5 pH units) to optimize the separation.
- If the pK_a values for a peptide sample are unknown, conduct initial separations in appropriate buffers at or near pH 2.5, 4.0, 5.5, 7.0, 8.5, and 10 to determine a promising pH range.

19.10.5. Optimize Buffer Concentration

- A good first choice for buffer concentration is 50–100 mM with a 50 μm i.d. capillary or 25–50 mM with a 75 μm i.d. capillary.
- Select buffer concentration depending on separation requirements. Use lower ionic strengths for speed, relatively few analytes, or when separation selectivity is high. Use higher ionic strengths for closely related analytes, numerous analytes, or micropreparative scale-up.
- To stack samples, maximize ionic strength differences between the sample and the buffer.
- Zwitterionic and denaturing agents (e.g., urea) can be used at high concentrations (> 1 M).

19.10.6. Optimize Separation Voltage (Ohm's Law Plot)

- Rinse and fill the capillary with the desired buffer and apply voltage from 0 to 30 kV in 2.5 kV increments while monitoring the current.
- For a given buffer and capillary the Ohm's law plot indicates the voltage that will give the fastest separation with optimum efficiency and resolution.

19.10.7. Select Additives to Maximize Differences or Mask Interactions

- Ion pair reagents (10–100 mM) are effective in promoting the separation of short, hydrophilic peptides and tryptic digests.
- Use MEKC with 25–200 mM SDS for hydrophobic and neutral peptides. SDS works well at neutral and basic pH, whereas bile salts such as cholates are effective at acidic pH.
- Add nonionic surfactants (up to 50 mM) or organic modifiers (1–25% by volume) to modify partitioning and change selectivity.
- Use dibasic amines (0.1–5 mM) to modify EOF and reduce wall interactions at low pH.
- Use zwitterions (25 mM to 1.5 M), denaturants (50 mM to 8 M), or ethylene glycol to minimize hydrophobic interactions among peptides and with the capillary wall.
- Use sieving buffers [e.g., 1–15% (w/v) of celluloses or dextrans] for size-based separations of peptides with large differences in size.

19.11. USE OF CAPILLARY ELECTROPHORESIS AND HIGH-PERFORMANCE LIQUID CHROMATOGRAPHY AS COMPLEMENTARY TECHNIQUES

19.11.1. Selectivity Differences Between Capillary Electrophoresis and High-Performance Liquid Chromatography

CE and HPLC are ideal complementary techniques for peptide analysis. They offer efficient, high-resolution, and rapid analyses based on different separation mechanisms. Since they have different selectivities, the techniques can be utilized to check the purity of peptides and proteins, compare tryptic maps of proteins to verify the identity of the correct product, and satisfy cross-method confirmation requirements. Grossman et al. (103) demonstrated this orthogonal use of HPLC and CE in the analysis peptides and protein tryptic fragments.

Selectivity differences of CE and HPLC are illustrated in the peptide separation shown in Figure 19.11. Peptides varying from 4 to 31 amino acids in

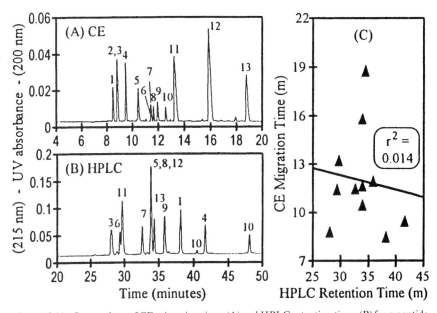

Figure 19.11. Comparison of CE migration times (A) and HPLC retention times (B) for a peptide mixture. Technique selectivity difference plots, correlation of CE migration times and HPLC retention times (C). (A) CE Conditions—buffer: 100 mM sodium phosphate, pH 2.55; capillary: 50 μm (i.d.) × 50 cm (L_t) FS; voltage: 30 kV; temperature: ambient; detection: UV at 200 nm; injection: gravity, 10 s at 100 mm. (B) HPLC conditions—column, Vydac 218TP51 (300 Å C_{18}, 250 mm × 1 mm); flow rate: 50 μL/min; mobile phase: (Solvent A) 0.1% TFA in water, (Solvent B) 0.085% TFA in 10% water/90% acetonitrile; gradient, 5–60% B over 60 min; concentration: (A) 50–350 μg each in water, (B) 10–70 μg each in water. Peaks: same as in Figure 19.2.

length were separated by CZE (Figure 19.11A) and HPLC (Figure 19.11B). The results clearly reveal different orders of elution. The selectivity difference plot of CE migration time vs. HPLC retention time (Figure 19.11C) shows almost no correlation ($r^2 = 0.014$), demonstrating that the techniques are orthogonal.

Guarino and Phillips (104) described the use of CE and HPLC for peptide purity analysis. Janini et al. (105) used HPLC, CE, and MS to determine the purity of charged peptides. Other published reports of the combined use of CE and HPLC for peptide analysis include: Strickland (106), screening narrow-bore HPLC fractions; Rudnick et al. (107), peptide mapping of recombinant human insulin-like growth factor; Stromqvist (108), combining SEC, HPLC, and CE for peptide mapping; Idei et al. (109), somatostatin analogue peptides; Yildiz et al. (110), peptide research; Sutcliffe and Corran (111), neurohypophyseal peptides and analogues by CZE and MEKC; Zhang et al. (112), small hydrophilic peptides differing in one glycine residue; Young et al. (113), effects of pH and buffer composition on peptides; Rudnick et al. (114), peptide mapping; Kruegar et al. (115), specificity and rate of cleavage of ACTH peptide bonds; Cui et al. (116), derivatization reagents for selectively determining the content of specific amino acid residues; and Buoen et al. (117), proline-rich peptides. Larmann et al. (118) used on-line LC–CE to produce two-dimensional, high-resolution peptide separations.

19.11.2. Use of Capillary Electrophoresis to Check High-Performance Liquid Chromatography Fractions for Purity and as a Micropreparative Technique

Another important use of CE is for checking the purity of peptide fractions collected from HPLC separations prior to sequencing. Because of selectivity differences, it is highly unlikely that analytes which coelute by HPLC will also comigrate by CE. HPLC fractions were collected from a semipreparative scale-up of fetuin tryptic peptides (102). Some of the collected fractions were analyzed by CE using a neutral, hydrophilic coated capillary. Figure 19.12 shows overlays of the CE analysis of HPLC fractions 1–13. The separation of the complete tryptic digest by CE is shown on the overlay for reference. The neutral, hydrophilic coated capillary produced rapid separations with high efficiencies (160,000 to over 650,000 plates/m). Many HPLC fractions contained multiple peaks when analyzed by CE. Ward and coworkers developed sequencing strategies for purified proteins and peptide fragments separated by micropreparative CE and HPLC (119). Lee and Desiderio (120) used preparative CZE for recovery of a synthetic peptide. Sufficient material for direct analysis of peptide sequence of carboxymethylated insulin B chain was obtained by Bergman et al. (121) using five repetitive fraction collection runs.

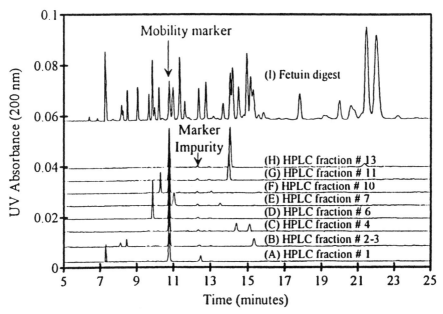

Figure 19.12. Purity checks of HPLC fractions of fetuin tryptic digest using CE: (A–H) CE of fractions 1–13 collected from HPLC; (I) entire fetuin digest. Conditions—buffer: 100 mM sodium phosphate, pH 2.55; capillary: 50 μm (i.d.) × 50 cm (L_t) neutral, hydrophilic coated; voltage: 30 kV; temperature: ambient; detection: UV at 200 nm; injection: gravity, 30 s at 150 mm: Peaks: fractions (A) 1; (B) 2–3; (C) 4; (D) 6; (E) 7; (F) 10; (G) 11; (H) 13; (I) entire fetuin digest. (Note: All peaks in this figure were mobility corrected using a mobility marker at times indicated above and overlaid as migration-time-corrected electropherograms.)

In the case of a synthetic hexapeptide, four preparative runs were sufficient to obtain enough peptide (140 pmol) for sequence and amino acid analyses. Analytical and micropreparative CE of salmon calcitonin and elcatonin was performed by Camilleri et al. (122) in H_2O and D_2O. Other examples of micropreparative fraction collection by CE are reported by the following authors: Grimm and Herold (123), collection of peptides for protein sequencing; Herold and Wu (124), automated peptide fraction collection device; Prusik et al. (125), continuous free-flow zone electrophoresis and CE used to prepare fractions of synthetic growth hormone releasing peptide; Kasicka and Prusik (126), analysis and preparation of other synthetic biopeptides; Lee et al. (127), preparative CE of native endorphins and enkephalins; and Moore and Jorgenson (128,129), two- and three-dimensional separation of fluorescein-5-isothiocyanate (FITC)-labeled peptides using HPLC and optically gated CE.

Some conclusions related to the *complementary use of CE and HPLC* for peptides are as follows:

- Selectivity differences help to minimize coelution.
- Selectivity differences, minimal sample requirements, and high efficiency and resolution make CE ideal for checking the purity of HPLC fractions prior to sequencing.

19.12. ENHANCED DETECTION METHODS

19.12.1. Capillary Electrophoresis and Mass Spectrometry

CE/MS is an increasingly popular method of characterizing peptide and protein samples. Some published examples are the following: Vinther et al. (130), CE, HPLC, and MS with bovine aprotinin peptides; Wahl et al. (131), Fourier transform–ion cyclotron resonance–MS (FT–ICR–MS) to characterize somatostatin; Craig et al. (132), CE–MS of phosphopeptide–peptide mixtures); Fang et al. (133), on-line CE–TOF (time-of-flight/MS with ubiquitin, leucine enkephalin, and other peptides; Weinmann et al. (134), CE and off-line MS with angiotensin, bradykinin, oxytocin, and sleep-inducing peptide; Hsieh et al. (135), low-level detection of Lys-bradykinin using CE–ion spray MS; Perkins and Tomer (136), CE–MS for peptide mixtures and snake venom; and Rosnack et al. (137), CE–electrospray ionization MS for heterogeneity of synthetic HPLC-purified peptides. Arnott et al. (138) presented an overview of CE–matrix-assisted laser desorption ionization MS and electrospray ionization MS for protein sequencing. (CE–MS is covered by Tomer in Chapter 12 of this monograph.)

19.12.2. Capillary Electrophoresis and Laser-Induced Fluorescence Detection of Peptides

The resolving power of CE in concert with the sensitivity of laser-induced fluorescence (LIF) provide a powerful combination for analyzing minute quantities of peptides. Recently published papers include: Shippy et al. (139), 2.3 nM detection limit of labeled neuropeptides using CE–LIF; Lim et al. (140), CE–LIF of labeled insulin tryptic digests; Fadden and Haystead (141), quantitative selective labeling at the attomole level of phosphoserine on peptides and proteins; Pinto et al. (142), picomole detection of labeled insulin peptides using a solid-phase reactor; Sweedler et al. (143), LIF detection of Texas Red–derivatized peptides at trace levels; and Timperman et al. (144), native fluorescence LIF detection and spectral differentiation of containing peptides.

19.12.3. Capillary Electrophoresis of Peptides Combined with Other Sensitive Types of Detection

Becklin and Desiderio (145) found that the amount of UV absorbance at 200 nm for 11 nonaromatic synthetic peptides was directly proportional to the number of peptide bonds. Olefirowicz and Ewing (146) used indirect amperometric detection for the analysis of dipeptides. Ye and Baldwin (18) combined CE with electrochemical detection (CE–EC) for detection of di- to hexaglycines and aspartame. Zhou et al. (147) and Stamler and Loscalzo (148) used CE–EC to detect glutathione/glutathione disulfide and related peptides. Attomole detection of tryptic peptides was achieved by CE with chemiluminescence detection of labeled peptides by Ruberto and Grayeski (149). Liao et al. (150) used Cu(II)-catalyzed luminol indirect chemiluminescence for low-level detection of peptides. Low-energy beta-emitting radiolabeled peptides were monitored by Tracht et al. (151) using radioactivity detection with a solid scintillator.

19.13. PHYSICOCHEMICAL MEASUREMENTS

19.13.1. Electrophoretic Mobility Determination and Prediction

Grossman's group (152,153) developed a semiempirical model for the effect of size and charge on the electrophoretic mobilities of peptides. In their study, 40 synthetic peptides (ranging in size from 3 to 39 amino acids and in charge from 0.33 to more than 14 charge units) were analyzed using citric acid buffer at pH 2.5. A plot of electrophoretic mobility vs. the combined size and charge parameter for the 40 peptides is shown in Figure 19.13A. The correlation coefficient for a linear fit was 0.989. This relationship should be useful for the prediction of peptide mobility and the detection of charge affecting structural modifications.

Deyl et al. (154) determined that relative retention times of peptides and proteins in CZE are related to pI, molecular weight, and the Offord mass/charge parameter. They found that a graph of relative retention time vs. pI was almost rectilinear, and was valid over a broad pH range (6.86 to 10.5). Zhu et al. (155) separated substance P and seven fragments using a phosphate buffer at pH 2.5. Figure 19.13B shows a plot of $m^{2/3}/z$ (m = mass and z = charge) vs. migration time for their data. The migration times varied linearly with $m^{2/3}/z$ (correlation coefficient, $r^2 = 0.99$). Zhu et al. used the substance P data to predict migration behavior of nine other bioactive peptides. A combined data plot showed slightly poorer correlation ($r^2 = 0.96$).

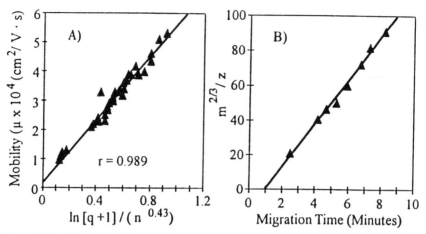

Figure 19.13. (A) Plot of measured electrophoretic mobility vs. $\ln(q+1)/n^{0.43}$ for 40 synthetic peptides. (B) Plot of $m^{2/3}/z$ vs. measured migration time for eight substance P fragments. [Redrawn with permission from Grossman et al. (152,153) and Zhu et al. (155).]

Issaq et al. (10) demonstrated that the migration velocity was directly proportional to the charge-to-mass ratio for six dipeptides. Rickard et al. (156) further developed the correlation of electrophoretic mobilities with physicochemical properties of proteins and peptides. Adamson et al. (157) analyzed multiple phosphoseryl-containing casein peptides and found a linear relationship between electrophoretic mobility and $q/m^{2/3}$. They concluded that CE is an efficient technique for monitoring peptide phosphorylation, dephosphorylation, deamidation, and truncation. Basak and Ladisch (158) determined that molecular size, charge, and shape were the main properties determining electrophoretic mobility. Chen et al. (159) developed a semiempirical equation relating migration times and the square root of the molecular weight divided by the number of charged groups. CE with double UV detection was applied to determination of effective peptide mobilities by Kasicka et al. (160).

19.13.2. Reaction Monitoring, Kinetics, Binding Constants, and Stoichiometries

Interest in using CE to study other physiochemical properties of proteins and peptides has been growing at a rapid rate. CE can achieve high-speed, high-resolution separation using an aqueous buffer that will not adversely affect the physiochemical parameter being studied. On-line monitoring of reactions is one area of interest. Landers et al. (161) used CE to monitor

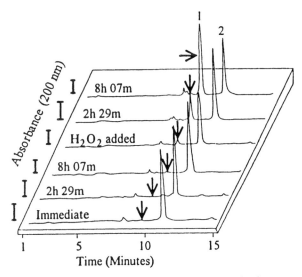

Figure 19.14. Use of CE to monitor time course of Ctc peptide dimerization process: blank period measured and hydrogen peroxide added where indicated. Conditions—buffer: 20 mM citrate, pH 2.5; capillary, 50 μm (i.d.) × 57 cm (L_t) FS; voltage: 25 kV; temperature: 28 °C; detection: UV at 200 nm; injection: pressure, 3 s at 0.5 psi. Peaks: (1) Ctc-Ctc; (2) Ctc (single headed arrow indicates dimer location; ↕ = 0.005 AU). [Redrawn with permission from Landers et al. (161).]

peptide homo- and heterodimer formation. Two synthetic peptides with N-terminal and C-terminal cysteine (Ctc) residues were oxidized separately under mild (air) and stronger (H_2O_2) conditions to give the disulfide dimers. The use of CE to monitor the time course of Ctc peptide dimerization process (161) is presented in Figure 19.14. After addition of H_2O_2, the monomer was converted to the Ctc-Ctc dimer. The reaction kinetics were determined. McLaughlin et al. (8) demonstrated the use of CE to monitor fast reactions. Using the short (7.7 cm) end of a capillary allowed fast separation (less than 1 min), which permitted sampling the reaction mixture every 2 min during the course of Met-Leu-Phe acetylation.

Other physicochemical studies include: Meyer et al. (162), separation of cis/trans isomers and determination of the rate constants for cis/trans interconversion; Moore and Jorgenson (163), kinetics of isomerization at room temperature for cis/trans forms of proline containing dipeptides and oligopeptides; Lim and Sim (164), metabolism of exogenous angiotensin peptides in lungs and plasma; Perron and Page (165), enzymatic specificity of carboxypeptidase A on a coated capillary; Heegaard and Robey (166), analysis of DNA binding to a peptide and estimated binding constants; Heegaard and

Robey (167), bimolecular noncovalent interactions; Chu et al. (168), affinity CE to determine binding stoichiometries of proteins and peptide–ligand interactions; and Pritchett et al. (169), LIF peptide immunoassay.

On-line microreactions have been carried out by CE. An immobilized enzyme fused-silica capillary microreactor was used for on-line peptide mapping of protein hydrolysates by Licklider and Kuhr (170). Enzyme (pepsin A, trypsin, or carboxypeptidase Y) was immobilized on the microreactor by biotin–avidin linkage and applied to the digestion of proteins. Chang and Yeung (171) digested proteins with pepsin on-column and monitored the peptide fragments using native fluorescence with LIF detection. Guzman and coworkers used CE combined with localized brain perfusion sampling techniques to analyze *in vivo* release of neuropeptides (172).

19.14. CONCLUDING REMARKS AND FUTURE TRENDS

The role of CE in peptide chemistry continues to expand, particularly in the areas of peptide isolation, characterization, purity determination, and mapping of protein digests. Among the important trends are the following: development of hyphenated techniques, especially CE–MS for high-resolution separation and analyte identification; sensitivity and detection enhancement strategies for direct analysis of biological fluids; characterization of peptide-binding reactions; immunological studies (e.g., affinity CE); micropreparative isolation; analysis of hydrophobic peptides using MEKC; and chiral discrimination of peptides. CE is rapidly becoming the method of choice for many types of peptide analysis.

ADDENDA

The numbers of the following sections correspond to those above in the body of this chapter.

19.1. Introduction. A number of recent papers focus on the practical application of CE to peptide analysis. The principles and practice of CZE were discussed by Wheat (173) with respect to peptide separation problems relevant to biotechnology. The use of CE for the investigation of peptide uptake by dairy starter bacteria was reported by Moore et al. (174). Recio et al. (175) used CE to study the formation of caseinomacropeptides in treated milk. Jansen and van der Hoeven (176) used CE to study the breakdown of serum proteins by oral bacterial species. Quantitation of collagen types I, III, and V in tissue slices by CE after cyanogen bromide solubilization was reported by

Deyl et al. (177). Mikšík et al. (178) applied CE to study large cyanogen bromide peptide fragments of fiber-forming collagens. The analysis by CE of imidodipeptides in the urine of prolidase-deficient patients was reported by Zanaboni et al. (179).

19.2. Effects of capillary dimensions, applied voltage, and temperature. Recent papers describe the combination of the electric field with other forces to drive the separation of peptides in a capillary. Kašička et al. (180) combined CZE with EOF control (by an externally applied electric field) to separate synthetic peptides. Wu et al. (181) utilized open-tube capillary electrochromatography (CEC) or CE with simultaneous application of hydrodynamic flow coupled with MS to produce rapid (< 6 min) separations of bioactive peptide mixtures and a tryptic map of myoglobin. Bergmann et al. (182) evaluated the use of on-line ITP–CE with hydrodynamic counterflow in the analysis of various basic peptides, basic proteins, and recombinant human interleukin-3. Two-dimensional separations of a tryptic digest of the contents of a single B2 neuron (from the marine mollusk *Aplysia californica*) were achieved by Liu and Sweedler (183) using CE coupled to channel gel electrophoresis.

19.4. Buffer additives. Newly reported studies have added more choices to the rapidly expanding number and type of buffer additives useful for the separation of peptides by CE. Selectivity changes in the separation of proteins and peptides by CE using high-molecular-mass polyethyleneimine were observed by Cifuentes et al. (184). Separation of toxic peptides (microcystins) using CE was achieved with the aid of organic mobile phase modifiers by Onyewuenyi and Hawkins (185). Castagnola et al. (186) observed that the structural properties of 2,2,2-trifluoroethanol (TFE) are useful for enhancing the separation of large apolar peptides by CE. Shihabi (187) described the use of acetonitrile–salt mixtures and sample stacking to achieve better resolution than obtainable with 100% aqueous, low-ionic-strength solutions.

19.5. Use of ion-pairing reagents. Improved mapping of peptide fragments with a net positive charge resulted when phytic acid was used as an ion-pairing buffer additive in CE, according to Kornfelt et al. (188). The use of ion-pairing reagents in CE to improve the separation of hydrophobic peptides was described by Martin (189).

19.6. Use of micellar electrokinetic chromatography for hydrophobic and neutral species. Dual usage of CZE and MEKC for selectivity differences and MEKC for the separation of hydrophobic peptides has been appearing in the literature. Szoko (190) recently compared CZE and MEKC for the analysis of proteins and peptides. Idei et al. (191,192) published two comparative studies that applied CZE and MEKC to the analysis of a somatostatin analogue and branched-chain polypeptides. Petersson et al. (193) showed the separation of bioactive cyclic peptides using MEKC. Strickland et al. (194) compared the use of CZE and MEKC for the analysis of peptides, proteins, and other constituents of cheddar cheese. Panak et al. (195) described the simultaneous direct determination of glutathione, lipoamide, and lipoic acid in human blood by MEKC.

19.7. Chiral selectors. Enantioseparation of amino acids, dipeptides, and tripeptides using vancomycin as a chiral selector in CE was described by Wan and Blomberg (196–198). Racemization of the eight stereoisomers of a tripeptide during solid-phase peptide synthesis was investigated using CE, as reported by Riester et al. (199). Zhang et al. (200) used CE to study the behavior of L- and D-alanine-scanning analogues of a yeast tridecapeptide pheromone.

19.8. Sensitivity-enhancement strategies. New ways of increasing sensitivity during the separation of peptides by CE have become available. The use of wide-bore capillaries for "off-line" preparative CE and collection of tryptic digest fractions prior to identification with MALDI–TOF–MS (matrix-assisted laser desorption ionization time-of-flight mass spectrometry) was described by Yin et al. (201). Smith (202) described CE methods for analytical and micropreparative collection of peptides. Tomlinson et al. (203) developed a unique method to load, clean up, and separate synthetic peptides and major histocompatibility complex (MHC) class I peptides using moving boundary transient ITP conditions for membrane preconcentration in small-diameter capillaries. Strategies for isolating and sequencing MHC class I peptides using mPC (membrane preconcentration)–CE–tandem MS were presented by Tomlinson et al. (204). Tomlinson et al. (205) used modified moving boundary transient ITP conditions and membrane preconcentration–CE for rapid loading of large sample volumes of synthetic and biological peptides into a mass spectrometer for detection and identification. A strategy for isolating and sequencing biologically derived MHC class I peptides using mPC–CE tandem MS for peptide sequencing was described by Tomlinson et al. (206). Naylor and Tomlison (207a) applied mPC–CE–tandem–MS in the analysis of peptides and other biologically derived metabolites and biopolymers.

19.9. Use of coated capillaries and wall coatings. The advent of commercially available coated capillaries and new chemistries has made it possible to achieve better resolution and reproducibility in peptide separations by CE. Righetti and Nembri (207b) achieved fast, highly efficient separations of tryptic digests of bovine β-casein using an isoelectric buffer combined with hydroxyethylcellulose to eliminate wall adsorption effects. Li et al. (208) described the use of a polybrene capillary coating in resolving the large number of digest products formed upon tryptic digestion of hemoglobin variants; prior separation of the hemoglobin α and β chains was not required. Bayer and Engelhardt (209) used organic-polymer-coated capillaries that reversed the EOF at low pH to produce high-efficiency separations of proteins and peptides. Huang et al. (210) described the effects of neutral and charged surface coatings of CE separations of peptides and other pharmaceutically active compounds. The separation of proteins and peptides by CEC in diol- and octadecyl-modified etched capillaries was shown by Pesek et al. (211).

19.11. Use of capillary electrophoresis and high-performance liquid chromatography as complementary techniques. The difference in selectivity between CE and HPLC has received considerable deserved attention in the literature. Miller and Rivier (212) published a comprehensive review of the uses of CE and HPLC in peptide chemistry. The separation of insulin nonapeptides and other biologically active peptides by CE and HPLC was compared by Pacakova et al. (213). Janaky et al. (214) applied CE and HPLC as separation methods in the analysis of a new peptide family: the galanins. Hynek et al. (215,216) also applied HPLC and CZE for peptide mapping of pepsin isoenzymes and for the fast detection of phosphorylation of human pepsinogen A, human pepsinogen C, and swine pepsinogen. The carbohydrate-dependent microheterogeneity of human recombinant factor VIIa was studied by Klausen and Kornfelt (217) using CE and HPLC. Jiskra et al. (218) monitored glycosylation of the peptides dalargin and desmopressin using CE and HPLC.

19.12. Enhanced detection methods. A number of recent publications have presented new methods of enhanced detection in CE. The acetate salt content of a basic antifungal lipopeptide was determined using CE with indirect UV detection by Zhou and Doveletoglou (219). Solid phase labeling of peptide toxins with fluorescent probes and their ultrasensitive determination based on CZE were discribed by Pinto et al. (220). Timperman and Sweedler (221) used CE with wavelength-resolved fluorescence detection to identify the tyrosine and tryptophan content of peptides. CE with pulsed amperometric detection of carbohydrates and glycopeptides was studied by Weber and Lunte (222). Olson et al. (223) described use of on-line nuclear magnetic resonance (NMR) detection for peptide analysis by CZE. Kelly et al. (224) demonstrated methods of optimizing detection sensitivity of bioactive peptides and CNBr glycopeptides from ovalbumin using CE–ESI (electrospray ionization)–MS at submicroliter flow rates combined with coated capillaries. Separation and detection of closely related peptides by MEKC coupled with ESI–MS using the partial filling technique was described by Koezuka et al. (225). Kelly et al. (226) studied the primary structure of bovine adenosine deaminase using HPLC, CE, and ESI–MS analysis of the intact protein and chemical and proteolytic digests, and sequencing of the resulting peptides by tandem MS. Romi-Lebrun et al. (227) used CE, HPLC, and MS to charcterize four toxins from scorpion (*Buthus martensi*) venom.

19.13. Physicochemical measurements. The ability of CE to produce fast, quantitative, stoichiometric answers has lead to its widespread use in physicochemical measurements. CE was used to measure inhibition of dipeptidyl peptidase IV (DP IV) by anti–DP IV antibodies and nonsubstrate X-X-Pro-oligopeptides by Hoffmann et al. (228). Zhang et al. (229) observed departures from linearity in the plot of log molecular weight vs. migration time for the smaller peptides (intrinsic net charge) during the separation of myoglobin molecular mass markers using non-gel-sieving CE. Novotná et al. (230) showed that CZE of collagen type I CNBr peptides using acid buffers produced separations strictly according to increasing molecular weight and increasing hydrophobicity. Kajiwara et al. (231) used CE to show evidence of metal-binding activities of a pentadecapeptide from *Panax* ginseng. Castagnola et al. (232) determined peptide dissociation constants and Stokes radii at different protonation stages by CE. Thunecke et al. (233) used CE to study the kinetics of the cis–trans isomerization of peptidyl–proline dipeptides by CZE, HPLC, and NMR. Rush et al. (234) used HPCE to determine that the sample matrix affects electrophoretic migration time, peak shape, and resolution, as well as the physical stability of the product glycopeptides. Liang et al. (235) proposed a model for peptide migration in CZE.

LIST OF ACRONYMS, ABBREVIATIONS, AND SYMBOLS

Acronym, Abbreviation, or Symbol	Definition
ACTH	adrenocorticotropic hormone
BSA	bovine serum albumin
CCE	chiral capillary electrophoresis
CD	cyclodextrin
CE	capillary electrophoresis
CEC	capillary electrochromatography
CHES	2-(cyclohexylamino) ethanesulfonic acid
Ctc	C-terminal cysteine (residue)
CZE	capillary zone electrophoresis
DPIV	dipeptidyl peptidase IV

EC	electrochemical (determination)
EDTA	ethylenediaminetetraacetic acid
EOF	electroosmotic flow
ESI	electrospray ionization
FITC	fluorescein-5-isothiocyanate
FS	fused-silica (capillary)
HEPES	4-(2-hydroxyethyl)-1-piperazineethanesulfonic acid
HPLC	high-performance liquid chromatography
HSA	hexanesulfonic acid
IPR	ion-pairing reagent
ITP	isotachophoresis
LC	liquid chromatography
L_d	length of capillary to the detector
LIF	laser-induced fluorescence
L_t	total length of the capillary
MALDI–TOF–MS	matrix-assisted laser desorption ionization time-of-flight mass spectrometry
MEKC	micellar electrokinetic chromatography
MES	2-morpholinoethanesulfonic acid
MHC	major histocompatibility complex
mPC	membrane preconcentration
MS	mass spectrometry
NACE	nonaqueous capillary electrophoresis
NMR	nuclear magnetic resonance
PEG	poly(ethylene glycol)
PSA	pentanesulfonic acid
SDS	sodium dodecyl sulfate
S/N	signal-to-noise ratio
TBAOH	tetrabutylammonium hydroxide
TFA	trifluoroacetic acid
TFE	2,2,2-trifluoroethanol
TOF	time-of-flight (mass spectrometry)
UV	ultraviolet

REFERENCES

1. G. M. McLaughlin, J. A. Nolan, J. L. Lindahl, R. H. Palmieri, K. W. Anderson, S. C. Morris, J. A. Morrison, and T. J. Bronzert, *J. Liq. Chromatogr.* **16**, 961–1021 (1992).
2. A. S. Cohen, A. Paulus, and B. L. Karger, *Chromatographia* **24**, 15–24 (1987).
3. H. G. Lee and D. M. Desiderio, *J. Chromatogr. B. Biomed. Appl.* **662**, 35–45 (1994).

4. J. H. Wahl, D. R. Goodlett, H. R. Udseth, and R. D. Smith, *Electrophoresis* **14**, 448-457 (1993).
5. D. A. Wolze, G. M. McLaughlin, D. K Hauffe, M. S. Bello, J. E. Gratteau, R. J. Gomez, T. D. Pham., R. E. Moore, and L. Thompson, *Symp. High Perform. Capillary Electrophor. 8th*, 1966, Orlando, FL, Abstr. P322 (1996).
6. G. M. McLaughlin, R. H. Palmieri, and K. W. Anderson, in *Techniques in Protein Chemistry* (J. Villafranca. ed.), p. 3-19 Academic Press, San Diego, CA, 1990.
7. N. Chen, L. Wang, and Y. K. Zhang, *J. Liq. Chromatogr.* **16**, 3609-3622 (1993).
8. G. M. McLaughlin, K. Srinivasan, J. Horvath, and M. Bello, *6th Annu. Frederick Conf. Capillary Electrophor.* Frederick, MD, Abstr. 25 (1995).
9. Y. K. Zhang, N. Chen, and L. Wang, *J. Liq. Chromatogr.* **16**, 3689-3697 (1993).
10. H. J. Issaq, G. M. Janini, I. Z. Atamna, G. M. Muschik, and J. Lukszo, *J. Liq. Chromatogr.* **15**, 1129-1142 (1992).
11. S. Ma, F. Kalman, A. Kalman, F. Thunecke, and C. Horvath, *J. Chromatogr. A* **716**, 167-182 (1995).
12. M. J. Field, R. H. Palmieri, R. Keck, J. I. Ohms, J. V. O'Connor, *Symp. High Perform. Capillary Electrophor. 2nd*, 1990, San Francisco, Abstr. P214 (1990).
13. R. M. McCormick, *Anal. Chem.* **60**, 2322-2328 (1988).
14. M. Strickland and N. Strickland, *Am. Lab. November* **22** (7), 60-65 (1990).
15. H. G. Lee and D. M. Desiderio, *J. Chromatogr. A* **667**, 271-283 (1994).
16. P. D. Grossman, K. J. Wilson, G. Petrie, and H. H. Lauer, *Anal. Biochem.* **173**, 265-270 (1988).
17. G. L. Hortin, T. Griest, and B. M. Benutto, *Bio Chromatography* **5**, 118-120 (1990).
18. J. Ye and R. P. Baldwin, *Anal. Chem.* **66**, 2669-2674 (1994).
19. M. H. J. M. Langenhuizen and P. S. L. Janssen, *J. Chromatogr.* **638**, 311-318 (1993).
20. C. Miller, J. F. Hernandez, A. G. Craig, J. Dykert, and J. Rivier, *J. Anal. Chim. Acta* **249**, 215-225 (1990).
21. J. E. Rivier, C. L. Miller, G. Tuchscherer, A. Craig, J. F. Hernandez, J. Dykert, F. Raschdorl, and M. Mutter, *Peptides (N.Y.)* **2**, 80-86 (1990).
22. E. Fukuoka, K. Suzuki, S. Yoshinari, H. Okunishi, and M. Miyazaki, *Bunseki Kagaku* **43**, 131-137 (1994).
23. A. Cifuentes and H. Poppe, *Electrophoresis* **16**, 516-524 (1995).
24. A. Pessi, A. E. Bianchi, L. Chiappinelli, A. Niardi, and S. J. Fanali, *J. Chromatogr.* **557**, 307-313 (1991).
25. T. M. Chen, R. C. George, and M. H. Payne, *J. High Resolut. Chromatogr.* **13**, 782-784 (1990).
26. D. Yannoukakos, H. E. Meyer, H. E. C. Vasseur, C. Draincourt, H. Wajcman, and E. Bursaux, *Biochem. Biophys. Acta* **1066**, 70-76 (1991).
27. P. Ferranti, A. Malomi, P. Pucci, S. J. Fanali, A. Niardi, and L. Ossicini, *Anal. Biochem.* **194**, 1-8 (1991).

28. J. R. Florance, Z. D. Konteatis, M. J. Macielag, R. A. Lessor, and A. Gaides, *J. Chromatogr.* **559**, 391–399 (1991).
29. G. M. Mclaughlin, W. A. Ausserer, K. Srinivasan, K. W. Anderson, J. Horvath, D. C. Siu, and R. M. McCormick, *Symp. High Perform. Capillary Electrophor. 7th.* Würzburg, Abstr. P-127 (1995). *J. Chromatogr. A* 1997.
30. R. G. Nielsen and E. C. Rickard, *J. Chromatogr.* **516**, 99–114 (1990).
31. R. G. Nielsen and E. C. Rickard, *ACS Symp. Ser.* **34**, 36–49 (1990).
32. N. Chen, L. Wang, and Y. K. Zhang, *J. Microcolumn Sep.* **7**, 193–198 (1995).
33. H. G. Lee and D. M. Desiderion, *J. Chromatogr. B: Biomed. Appl.* **655**, 9–19 (1994).
34. F. A. Chen, L. Kelly, R. Palmieri, R. Biehler, and H. Schwartz, *J. Liq. Chromatogr.* **15**, 1143–1161 (1992).
35. M. Idei, I. Mezo, Z. Vadasz, A. Horvath, I. Teplan, and G. Keri, *J. Liq. Chromatogr.* **15**, 3181–3192 (1992).
36. J. Florance, *Am. Lab.* **23**, 321, 32N, 320 (1991).
37. J. Florance, *Int. Lab.* June, 38, 40 (1991).
38. R. S. Sahota and M. G. Khaledi, *Anal. Chem.* **66** (7), 1141–1146 (1994).
39. J. A. Bullock, *J. Microcolumn Sep.* **3**, 241–248 (1991).
40. R. P. Oda, B. J. Madden, J. C. Morris, T. C. Spelsberg, and J. P. Landers, *J. Chromatogr. A* **680**, 341–351 (1994).
41. J. Varghese and R. B. Cole, *J. Chromatogr. A* **652**, 369–376 (1993).
42. J. Liu, K. A. Cobb, and M. Novotny, *J. Chromatogr.* **519**, 189–197 (1990).
43. Y. J. Yao and S. F. Y. Li, *J. Chromatogr. A* **663**, 97–104 (1994).
44. M. Thorsteinsdottir, I. Beijersten, and D. Westerlund, *Electrophoresis* **16**, 564–573 (1995).
45. S. Ma, F. Kalman, A. Kalman, F. Thunecke, and C. Horvath, *J. Chromatogr. A* **716**, 167–182 (1995).
46. R. J. Krueger, T. R. Hobbs, K. A. Mihal, J. Tehrani, and M. G. Zeece, *J. Chromatogr.* **543**, 451–461 (1991).
47. H. J. Issaq, G. M. Janini, I. Z. Atamna, G. M. Muschik, and J. Lukszo, *J. Liq. Chromatogr.* **15**, 1129–1142 (1992).
48. H. Kajiwara, *J. Chromatogr.* **559**, 345–356 (1991).
49. M. Deacon, T. J. O' Shea, S. M. Lunte, and M. R. Smyth, *J. Chromatogr. A*, **652**, 377–383 (1993).
50. M. P. Richards and J. H. Beattie, *J. Capillary Electrophor.* **1**, 196–207 (1994).
51. C. Dette and H. Waetzig, *J. Chromatogr. A*, **700**, 89–94 (1995).
52. G. N. Okafo, H. C. Birrell, M. Greenaway, M. Haran, and P. Camilleri, *Anal. Biochem.* **219**, 201–206 (1994).
53. G. N. Okafo, D. Perrett, and P. Camilleri, *Biomed. Chromatogr.* **8**, 202–204 (1994).
54. R. S. Rush, P. L. Derby, T. W. Strickland, and M. F. Rohde, *Anal. Chem.* **65**, 1834–1842 (1993).

55. P. A. Hagono, W. R. G. Baeyens, and G. Van der Weken, *Biomed. Chromatogr.* **9**, 291 (1995).
56. S. Terabe, H. Utsumi, K. Otsuka, and T. Ando, *HRC & CC, J. High Resolut Chromatogr. Chromatogr. Commun.* **9**, 666–670 (1986).
57. K. Otsuka and S. Terabe, *J. Microcolumn Sep.* **1**, 150–154 (1989).
58. S. Terabe, N. Chen, and K. Otsuka, *Adv. Electrophoresis* **7**, 87–153 (1994).
59. N. Matsubara, K. Koezuka, and S. Terabe, *Electrophoresis* **16**, 580–583 (1995).
60. B. Ye, M. Hadjmohammadi, and M. G. Khaledi, *J. Chromatogr. A*, **692**, 291–300 (1995).
61. T. Yashima, A. Tsuchiya, and O. Morita, *Anal. Chem.* **64**, 2981–2984 (1992).
62. H. J. Gaus, A. G. Beck Sickinger, and E. Bayer, *Anal. Chem.* **65**, 1399–1405 (1993).
63. M. R. Karim, S. Shinagawa, and T. Takagi, *Electrophoresis* **15**, 1141–1146 (1994).
64. K. F. Greve, W. Nashabeh, and B. L. Karger, *J. Chromatogr. A* **680**, 15–24 (1994).
65. H. J. Issaq, K. C. Chan, G. M. Muschik, and G. M. Janini, *J. Liq. Chromatogr.* **18**, 1273–1288 (1995).
66. T. A. A. Van de Goor, P. S. L. Janssen, J. W. Van Nispen, M. J. M. Van Meeland, and F. M. J. Everaerts, *J. Chromatogr.* **545**, 379–389 (1991).
67. S. A. Swedberg, *J. Chromatogr.* **503**, 449–452 (1990).
68. N. Matsubara, K. Koezuka, and S. Terabe, *Electrophoresis* **16**, 580–583 (1995).
69. N. Matsubara and S. Terabe, *Chromatographia* **34**, 493–496 (1992).
70. I. Beijersten and D. Westerlund, *Anal. Chem.* **65**, 3484–3488 (1993).
71. A. Wainright, *J. Microcolumn. Sep.* **2**, 166–175 (1990).
72. K. D. Altria, D. M. Goodall, and M. M. Rogan, *Chromatographia* **34**, 19–24 (1992).
73. K. D. Altria, *J. Chromatogr.* **646**, 245–257 (1993).
74. S. Terabe, Y. Miyashita, O. Shibata, E. R. Barnhart, L. R. Alexander, D. G. Patterson, B. L. Karger, K. Hosoya, and N. Tanaka, *J. Chromatogr.* **516**, 23–31 (1990).
75. R. Kuhn, D. Riester, B. Fleckenstein, and K. H. Wiesmuller, *J. Chromatogr. A* **716**, 371–379 (1995).
76. R. Kuhn, F. Erni, T. L. Bereuter, and J. Hausler, *Anal. Chem.* **64**, 2815–2820 (1992).
77. M. A. Survay, D. M. Goodall, S. A. C. Wren, and R. C. Rowe, *J. Chromatogr.* **636**, 81–86 (1993).
78. M. A. Survay, D. M. Goodall, S. A. C. Wren, and R. C. Rowe, *Anal. Proc.* **30** (12), 477–479 (1993).
79. C. Y. Quang and M. G. Khaledi, *J. Chromatogr. A* **692**, 253–265 (1994).
80. C. Y. Quang, A. Malek, and M. G. Khaledi, *5th Annu. Frederick Conf. Capillary Electrophor.*, Frederick, MD, Abstr. 109 (1994).
81. R. Aebersold and H. D. Morrison, *J. Chromatogr.* **516**, 79–88 (1990).

82. C. Schwer and F. Lottspeich, *J. Chromatogr.* **623**, 345–355 (1992).
83. C. Schwer, B. Gas, F. Lottspeich, and E. Kenndler, *Anal. Chem.* **65**, 2108–2115 (1993).
84. F. Foret, E. Szoko, and B. L. Karger, *J. Chromatogr.* **608**, 3–12 (1992).
85. J. L. Liao, R. Zhang, and S. Hjertén, *J. Chromatogr. A* **676**, 421–430 (1994).
86. V. Kasicka, Z. Prusik, O. Smekal, J. Hlavacek, T. Barth, G. Weber, and H. Wagner, *J. Chromatogr. B: Biomed. Appl.* **656**, 99–106 (1994).
87. D. T. Witte, S. Nagard, and M. Larsson, *J. Chromatogr. A* **687**, 155–166 (1994).
88. T. Satow, A. Machida, K. Funakushi, and R. L. Palmieri, *J. High Resolut. Chromatogr.* **14**, 276–278 (1991).
89. A. M. Hoyt, S. C. Beale, J. P. Larmann, and J. W. Jorgenson, *J. Microcolumn Sep.* **5**, 325–330 (1993).
90. M. A. Strausbauch, B. J. Madden, P. J. Wettstein, and J. P. Landers, *Electrophoresis* **16**, 541–548 (1995).
91. M. A. Strausbauch, J. P. Landers, and P. Wettstein, *Anal. Chem.* **68**, 306–314 (1996).
92. A. J. Tomlinson, L. M. Benson, W. D. Braddock, R. P. Oda, and S. Naylor, *J. High Resolut. Chromatogr.* **18**, 381–383 (1995).
93. A. J. Tomlinson, L. M. Benson, W. D. Braddok, R. P. Oda, and S. Naylor, *J. High Resolut Chromatogr.* **18**, 384–386 (1995).
94. M. Thorsteinsdottir, R. Isaksson, and D. Westerlund, *Electrophoresis* **16**, 557–563 (1995).
95. M. Huang, J. Plocek, and M. V. Novotny, *Electrophoresis* **16**, 396–401 (1995).
96. T. M. Phillips and P. L. Kimmel, *J. Chromatogr. B: Biomed. Appl.* **656**, 259–266 (1994).
97. W. Nashabeh and Z. El Rassi, *J. Chromatogr.* **536**, 31–42 (1991).
98. J. T. Smith and Z. El Rassi, *Electrophoresis* **14**, 396–406 (1993).
99. M. Huang, G. Yi, J. S. Bradshaw, and M. L. Lee, *J. Microcolumn Sep.* **5**, 199–205 (1993).
100. G. Piccoli, M. Fiorani, B. Biagiarelli, F. Palma, L. Vallorani, R. De Bellis, and V. Stocchi, *Electrophoresis* **16**, 625–629 (1995).
101. M. Castagnola, L. Cassiano, R. Rablno, D. Rossetti, and F. A. Bassi, *J. Chromatogr.* **572**, 51–58 (1991).
102. G. M. Mclaughlin, K. Srinivasan, W. A. Ausserer, K. W. Anderson, D. C. Siu, J. Horvath, and R. M. McCormick, *Symp. High Perform. Capillary Electrophor, 7th*, Würzburg, Abstr. P-128 (1995), *J. Chromatogr. A* 1997.
103. P. D. Grossman, J. C. Coburn, H. H. Lauer, R. G. Nielsen, R. M. Riggin, G. S. Sittampalam, and E. C. Rickard, *Anal. Chem.* **6**, 1186–1194 (1989).
104. B. C. Guarino and D. Phillips, *Am. Lab.* **23**, 68–69 (1991).
105. G. M. Janini, H. J. Issaq, and J. Lukszo, *J. High Resolut. Chromatogr.* **17**, 102–103 (1994).

106. M. Strickland, *Am. Lab.* **23**, 70–75 (1991).
107. S. E. Rudnick, V. J. Hilser, Jr., and G. D. Worosila, *J. Chromatogr.* **A672**, 219–229 (1994).
108. M. Stromqvist, *J. Chromatogr. A* **667**, 304–310 (1994).
109. M. Idei, I. Mezo, Z. Vadasz, A. Horvath, I. Teplan, and G. Keri, *J. Chromatogr.* **648**, 251–256 (1993).
110. E. Yildiz, G. Gruebler, S. Hoerger, H. Zimmermann, H. Echner, S. Stoeva, and W. Voelter, *Electrophoresis* **13**, 683–686 (1992).
111. N. Sutcliffe and P. H. Corran, *J. Chromatogr.* **636**, 95–103 (1993).
112. Y. K. Zhang, N. Chen, and L. Wang, *Biomed. Chromatogr.* **7**, 75–77 (1993).
113. P. M. Young, N. E. Astephen, and T. E. Wheat, *LC–GC* **10**, 26–28, 30–32 (1992).
114. S. E. Rudnick, V. J. Hilser, and G. D. Worosila, *J. Chromatogr. A* **672**, 219–229 (1994).
115. R. J. Kruegar, T. R. Hobbs, K. A. Mihal, J. Tehrani, and M. G. Zeece, *J. Chromatogr.* **543**, 451–461 (1991).
116. H. Cui, J. Leon, E. Reusaet, and A. Bult, *J. Chromatogr. A* **704**, 27–36 (1995).
117. S. Buoen, J. A. Eriksen, H. Revhelm, and J. S. Schanche, *Peptides (N. Y.)* **283** (1), 331–333 (1990).
118. J. P. Larmann, A. V. Lemmo, A. W. Moore, and J. W. Jorgenson, *Electrophoresis* **14**, 439–447 (1993).
119. L. D. Ward, G. E. Reid, R. L. Moritz, and R. J. Simpson, *J. Chromatogr.* **519**, 199–216 (1990).
120. H. G. Lee and D. M. Desiderio, *J. Chromatogr. A* **686**, 309–317 (1994).
121. T. Bergman, B. Agerberth, and H. Jornvall, *FEBS Lett.* **283**, 100–103 (1991).
122. P. Camilleri, G. N. Okafo, C. Southan, and R. Brown, *Anal. Biochem.* **198**, 36–42 (1991).
123. R. Grimm and M. Herold, *J. Capillary Electrophor.* **1**, 79–82 (1994).
124. M. Herold and S. Wu, *LC–GC* **12**, 531–533 (1994).
125. Z. Pruslk, V. Kasicka, P. Mudra, J. Stepanek, O. Smekal, and J. Hlavacek, *J. Electrophor.* **11**, 932–936 (1990).
126. V. Kasicka and Z. Prusik, *Am. Lab.*, 22–28 (1994).
127. H. G. Lee, J. L. Tseng, R. R. Becklin, and D. M. Desiderio, *Anal. Biochem.* **229**, 188–197 (1995).
128. A. W. Moore, Jr. and J. W. Jorgenson, *Anal. Chem.* **67**, 3456–3463 (1995).
129. A. W. Moore, Jr. and J. W. Jorgenson, *Anal. Chem.* **67**, 3448–3455 (1995).
130. A. Vinther, S. E. Bjorn, H. H. Sorenson, and H. J. Soeeberg, *J. Chromatogr.* **516**, 175–184 (1990).
131. J. H. Wahl, S. A. Hofstadler, and R. D. Smith, *Anal. Chem.* **67**, 462–465 (1995).
132. A. G. Craig, C. A. Hoeger, C. L. Miller, T. Goedken, J. E. Rivier, and W. H. Fischer, *Biol. Mass Spectrom.* **23**, 519–528 (1994).

133. L. Fang, R. Zhang, E. R. Williams, and R. N. Zare, *Anal. Chem.* **66**, 3696–3701 (1994).
134. W. Weinmann, C. E. Parker, L. J. Deterding, D. I. Papac, J. Hoyes, M. Przybylski, and K. B. Tomer, *J. Chromatogr. A* **680**, 353–361 (1994).
135. F. Y. L. Hsieh, J. Cai, and J. Henion, *J. Chromatogr. A* **69**, 206–211 (1994).
136. J. R. Perkins and K. B. Tomer, *Anal. Chem.* **66**, 2835–2840.
137. K. J. Rosnack, J. G. Stroh, D. H. Singleton, B. C. Guarino, and G. C. Andrews, *J. Chromatogr. A* **675**, 219–225 (1994).
138. D. Arnott, J. Shabanowitz, and D. F. Hunt, Clin. Chem. (*Winston-Salem, N. C.*) **39**, 2005–2010 (1993).
139. S. A. Shippy, J. A. Jankowski, and J. V. Sweedler, *Anal. Chim. Acta.* **307**, 163–171 (1995).
140. H. B. Lim, J. Lee, and K. J. Lee, *Electrophoresis* **16**, 674–678 (1995).
141. P. Fadden and T. A. J. Haystead, *Anal. Biochem.* **225**, 81–88 (1995).
142. D. M. Pinto, E. A. Arriaga, S. Sia, Z. Li, and N. J. Dovichi, *Electrophoresis* **16**, 534–540 (1995).
143. J. V. Sweedler, R. Fuller, S. Tracht, A. T. Timperman, V. Toma, and K. Khatib, *J. Microcolumn Sep.* **5**, 403–412 (1993).
144. A. T. Timperman, K. E. Oldenburg, and J. V. Sweedler, *Anal. Chem.* **67**, 3421–3426 (1995).
145. R. R. Becklin and D. M. Desiderio, *Anal. Lett.* **28**, 2175–2190 (1995).
146. T. M. Olefirowicz and A. G. Ewing, *J. Chromatogr.* **499**, 713–719 (1990).
147. J. Zhou, T. J. O'Shea, and S. M. Lunte, *J. Chromatogr. A* **680**, 271–277 (1994).
148. J. S. Stamler and J. Loscalzo, *Anal. Chem.* **64**, 779–785 (1992).
149. M. A. Ruberto and M. L. Grayeski, *J. Microcolumn Sep.* **6**, 545–550 (1994).
150. S. Y. Liao, Y. C. Chao, and C. W. Whang, *J. High Resolut. Chromatogr.* **18**, 667–669 (1995).
151. S. Tracht, V. Toma, and J. V. Sweedler, *Anal. Chem.* **66**, 2382–2389 (1994).
152. P. D. Grossman, J. C. Colburn, and H. H. Lauer, *Anal. Biochem.* **179**, 28–33 (1989).
153. P. D. Grossman, in *Capillary Electrophoresis: Theory and Practice* (P. D. Grossman and J. C. Colburn. eds.) pp. 125–126. Academic Press, San Diego, CA, 1992.
154. Z. Deyl, V. Rohlicek, and R. Struzinsky, *J. Liq. Chromatogr.* **12**, 2515–2526 (1989).
155. M. Zhu, D. Hansen, S. Burd, V. Huebner, P. Balasubramanian, and A. J. C. Chen, *Bio-Rad Lab. Bull.* **1482** (1989).
156. E. C. Rickard, M. M. Strohl, and R. G. Nielsen, *Anal. Biochem.* **197**, 197–207 (1991).
157. N. Adamson, P. F. Riley, and E. C. Reynolds, *J. Chromatogr.* **646**, 391–396 (1993).
158. S. B. Basak and M. R. Ladisch, *Anal. Biochem.* **226**, 51–58 (1995).
159. N. Chen, L. Wang, and Y. K. Zhang, *Chromatographia* **37**, 429–432 (1993).

160. V. Kasicka, Z. Prusik, P. Mudra, and J. Stepanek, *J. Chromatogr. A* **709**, 31–38 (1995).

161. J. P. Landers, R. P. Oda, J. A. Liebenow, and T. C. Spelsberg, *J. Chromatogr. A* **652**, 109–117 (1993).

162. S. Meyer, A. Jabs, M. Schutkowski, and G. Fischer, *Electrophoresis* **15**, 1151–1157 (1994).

163. A. W. Moore, Jr. and J. W. Jorgenson, *Anal. Chem.* **67**, 3464–3475 (1995).

164. B. C. Lim and M. K. Sim, *J. Chromatogr. B: Biomed. Appl.* **655**, 127–131 (1994).

165. M. J. Perron and M. Page, *J. Chromatogr. A* **662**, 383–388 (1994).

166. N. H. Heegaard and F. A. Robey, *J. Liq. Chromatogr.* **16**, 1923–1939 (1993).

167. N. H. Heegaard and F. A. Robey, *Am. Lab.* **26**(9), 28T–28X (1994).

168. Y.-H. Chu, W. J. Lees, A. Stassinopoulos, and C. T. Walsh, *Biochemistry* **33**, 10616–10621 (1994).

169. T. J Pritchett, R. A. Evangelista, and F. T. A. Chen, *J. Capillary Electrophor.* **2**, 145–149 (1995).

170. L. Licklider and W. G. Kuhr, *Anal. Chem.* **66**(24), 4400–4407 (1994).

171. H. T. Chunng and E. S. Yeung, *Anal. Chem.* **65**, 2947–2951 (1993).

172. N. A. Guzman, L. Hernandez, and J. P. Advis, *Curr. Res. Protein Chem.* **3**, 203–216 (1989).

173. T. E. Wheat, *Mol. Biotechnol.* **5**, 263–273 (1996).

174. I. L. Moore, G. G. Pritchard, and D. E. Otter, *J. Chromatogr. A* **718**(1), 211–215 (1996).

175. I. Recio, R. Lopez-Fandino, A. Olano, C. Olieman, and M. Ramos, *J. Agr. Food Chem.* **44**(12), 3845–3848 (1996).

176. H. J. Jansen and J. S. van der Hoeven, *J. Clin. Periodontol.* **24**(5), 346–353 (1997).

177. Z. Deyl, J. Novotna, I. Mikšik, D. Jelínková, M. Uhrová, and M. Suchanek, *J. Chromatogr. B: Biomed. Appl.* **689**(1), 181–194 (1997).

178. I. Mikšík, J. Novotná, M. Uhrová, D. Jelínková, and Z. Deyl, *J. Chromatogr. A* **772**, 213–220 (1997).

179. G. Zanaboni, R. Grimm, K. M. Dyne, A. Rossi, G. Cetta, and P. Ladarola, *J. Chromatogr. B* **683**, 97–107 (1996).

180. V. Kašička, Z. Prusík, P. Sázelová, T. Barth, E. Brynda, and L. Machová, *J. Chromatogr. A* **772**, 221–230 (1997).

181. J. T. Wu, P. Huang, M. X. Li, M. G. Qian, and D. M. Lubman, *Anal. Chem.* **69**, 320–326 (1997).

182. J. Bergmann, U. Jaehde, M. Mazereeuw, U. R. Tjaden, and W. Schunack, *J. Chromatogr. A* **734** (2), 381–38 (1996).

183. M. Liu and J. V. Sweedler, *Anal. Chem.* **68**(22), 3928–3933 (1996).

184. A. Cifuentes, H. Poppe, J. C. Kraak, and F. Bedia-Erim, *J. Chromatogr. B: Biomed. Appl.* **681**(1), 21–27 (1996).

185. N. Onyewuenyi and P. Hawkins, *J. Chromatogr. A* **749**(1–2), 271–277 (1996).

186. M. Castagnola, L. Cassiano, I. Messana, M. Paci, D. V. Rossetti, and B. Giardina, *J. Chromatogr. A* **735**(1–2), 271–281 (1996).
187. Z. K. Shihabi, *J. Chromatogr. A* **744**(1–2), 231–240 (1996).
188. T. Kornfelt, A. Vinther, G. N. Okafo, and P. Camilleri, *J. Chromatogr. A* **726**(1–2), 223–228 (1996).
189. L. M. Martin in P. T. P. Kaumaya and R. S. Hodges, (Editors), *Pept. Chem. Struc. Biol. Proc. Am. Pept. Symp.*, *14th* 1995, Mayflower Scientific, Kingswinford, 1996, pp. 144–145.
190. E. Szoko, *Electrophoresis* **18**(1), 74–81 (1997).
191. M. Idei, G. Mezo, Z. Vadasz, A. Horvath, J. Seprodi, J. Erchegyi, I. Teplan, and G. Keri, *Electrophoresis* **17**(4), 758–761 (1996).
192. M. Idei, G. Dibo, K. Bogdan, G. Mezo, A. Horvath, J. Erchegyi, G. Meszalaros, G. Teplan, G. Keri, and F. Hudecz, *Electrophoresis* **17**(8), 1357–1360 (1996).
193. M. Petersson, K. Walhagen, A. Nilsson, K.-G. Wahlund, and S. Nilsson, *J. Chromatogr. A* **769**, 301–306 (1997).
194. M. Strickland, B. C. Weimer, and J. R. Broadbent, *J. Chromatogr. A* **731**(1–2), 305–313 (1996).
195. K. C. Panak, O. A. Ruiz, S. A. Giorgieri, and L. E. Diaz, *Electrophoresis Weinheim* **17**, 1613–1616 (1996).
196. H. Wan and L. G. Blomberg, *Electrophoresis* **17**(12), 1938–1944 (1996).
197. H. Wan and L. G. Blomberg, *J. Microcolumn Sep. 8*, 339–344 (1996).
198. H. Wan and L. G. Blomberg, *J. Chromatogr. A* **758**, 303–311 (1997).
199. K. H. Riester, D. Wiesmuller, D. Stoll, and R. Kuhn, *Anal. Chem.* **68**(14), 2361–2365 (1996).
200. Y. L. Zhang, J. M. Becker, and F. R. Naider, *Anal. Biochem.* **241**(2), 220–227 (1996).
201. H. Yin, C. Keeley-Templin, and D. M. McManigill, *J. Chromatogr. A* **744**(1–2), 45–54 (1996).
202. J. Smith, *Methods Mol. Biol.* (Totowa), **64**, (Protein Sequencing Protocols), 91–99 (1997).
203. A. J. Tomlinson, L. M. Benson, S. Jameson, and S. Naylor, *Electrophoresis* **17**(12), 1801–1807 (1996).
204. J. Tomlinson, S. Jameson, and S. Naylor, *J. Chromatogr. A* **744**(1–2), 273–278 (1996).
205. J. Tomlinson, L. M. Benson, S. Jameson, and S. Naylor, *Electrophoresis* **17**(12), 1801–1807 (1996).
206. A. J. Tomlinson, S. Jameson, and S. Naylor, *J. Chromatogr. A.* **744**(1–2), 73–278 (1996).
207a. S. Naylor and A. J. Tomlinson, *Biomed. Chromatogr.* **10**(6), 325–330 (1996).
207b. P. Righetti and F. Nembri, *J. Chromatogr. A* **772**, 203–211 (1997).
208. M. X. Li, L. J. Liu, J. T. Wu, and D. M. Lubman, *Anal. Chem.* **69**(13), 2451–2456 (1997).

209. H. Bayer and H. Engelhardt, *J. Microcolumn Sep.* **8**(7), 479–484 (1996).
210. M. Huang, M. Bigelow, and M. Byers, *LC-GC-Int.* **9**(10), 658–664 (1996).
211. J. J. Pesek, M. T. Matyska, and L. Mauskar, *J. Chromatogr. A* **763**(1), 307–314 (1997).
212. C. Miller and J. Rivier, *Biopolymers* **40**, 265–317 (1996).
213. V. Pacakova, J. Suchankova, and K. Stulik, *J. Chromatogr. B: Biomed. Appl.* **681**(1), 69–76 (1996).
214. T. Janaky, E. Szabo, L. Balaspiri, B. Adi, and B. Penke, *J. Chromatogr. B: Biomed. Appl.* **676**(1), 7–12 (1996).
215. R. Hynek, V. Kašička, Z. Kucerova, and J. Kas, *J. Chromatogr. B: Biomed. Appl.* **681**(1), 37–45 (1996).
216. R. Hynek, V. Kašička, Z., Kucerova, and J. Kas, *J. Chromatogr.* **688**, 213–220 (1997).
217. N. K. Klausen and T. Kornfelt, *J. Chromatogr. A* **718**(1), 195–202 (1995).
218. J. Jiskra, V. Pacáková, M. Tichá, K. Štulík, and T. Barth, *J. Chromatogr. A* **761**(1–2), 285–296 (1997).
219. J. Zhou and A. Doveletoglou, *J. Chromatogr. A* **763**(1–2), 279–284 (1997).
220. D. Pinto, E. Arriaga, S. Sia, Z. Li, and N. Dovichi, *Ing. Cienc. Quim.* **16**, 28–29 (1996).
221. A. T. Timperman and J. V. Sweedler, *Analyst* **121**(5), 45R–52R (1996).
222. P. L. Weber and S. M. Lunte, *Electrophoresis* **17**(2), 302–309 (1996).
223. D. L. Olson, T. L. Peck, A. G. Webb and J. V. Sweedler in P. T. P. Kaumaya and R. S. Hodges, (Editors), *Pept.: Chem., Struc. Biol., Proc. Am. Pept. Symp., 14th 1995*, Mayflower Scientific, Kingswinford, 1996, pp. 730–731.
224. J. Kelly, L. Ramaley, and P. Thibault, *Anal. Chem.* **69**, 51–60 (1997).
225. K. Koezuka, H., Ozaki, N., Matsubara, and S. Terabe, *J. Chromatogr. B* **689**, 3–11 (1997).
226. M. A. Kelly, M. M. Vestling, C. M. Murphy, S. Hua, T. Sumpter, and C. Frenselau, *J. Pharm. Biomed. Anal.* **14**(11), 1513–1519 (1996).
227. R. Romi-Lebrun, M. F. Martin-Eauclaire, P. Escoubas, F. Q. Wu, B. Lebrun, M. Hisada, and T. Nakajima, *Eur. J. Biochem.* **245**(2), 457–464 (1997).
228. T. Hoffmann, D. Reinhold, T. Kaehne, J. Faust, K. Neubert, R. Frank, and S. Ansorge, *J. Chromatogr. A* **716** (1–2), 355–362 (1995).
229. Y. Zhang, H. K. Lee, and S. F. Y. Li, *J. Chromatogr. A* **744**(1–2) 249–257 (1996).
230. J. Novotná, Z. Deyl, and I. Mikšík, *J. Chromatogr. B: Biomed. Appl.* **681**(1), 77–82 (1996).
231. H. Kajiwara, A. M. Hemmings, and H. Hirano, *J. Chromatogr.* **687**, 443–448 (1996).
232. M. Castagnola, D. V. Rossetti, L. Casslano, F. Misiti, L. Pennacchietti, B. Giardina, and I. Messana, *Electrophoresis (Weinheim)* **17**, 1925–1930 (1996).

233. F. Thunecke, F. Kálmán, S. Ma, A. S. Rathore, and C. Horváth, *J. Chromatogr. A* **744**(1–2), 259–272 (1996).
234. R. S. Rush, H. J. Boss, V. Katta, M. F. Rohde, *Electrophoresis* **18**(5), 751–756 (1997)
235. H. Liang, Z. Wang, J. Gu, R. Fu, and B. Lin,. *J. Bejing Inst. Technol. Eng. Ed.* **5**, 35–39 (1996).

CHAPTER

20

CAPILLARY ELECTROPHORESIS OF PROTEINS

FRED E. REGNIER and SHEN LIN

Department of Chemistry, Purdue University, West Lafayette, Indiana 47907

20.1.	Conventional Electrophoresis of Proteins	684
20.2.	Electrophoresis of Proteins in Fused-Silica Capillaries	685
20.3.	Surface Modification of Capillaries	686
	20.3.1. Dynamic Modification	687
	20.3.2. Adsorbed Coatings	694
	20.3.3. Covalent Modification	697
	20.3.4. Conclusions on Coatings	702
20.4.	Zone Electrophoresis in Open-Tubular Capillaries	703
	20.4.1. Serum Proteins	703
	20.4.2. Glycoproteins	704
	20.4.3. Lipoproteins	705
	20.4.4. Antibodies/Immunology	706
	20.4.5. Hemoglobins	707
	20.4.6. rec-Proteins	708
	20.4.7. Food Proteins	710
	20.4.8. Structure Analysis/Folding	711
	20.4.9. Miscellaneous Topics	711
20.5.	Capillary Isoelectric Focusing	712
20.6.	Capillary Gel Electrophoresis	713
	20.6.1. Sodium Dodecyl Sulfate–Capillary Gel Electrophoresis	713
	20.6.2. Native Capillary Gel Electrophoresis	715
20.7.	Micellar Electrokinetic Chromatography	715
20.8.	Selectivity	715
20.9.	Mass Spectrometry	716
20.10.	The Future	719
List of Acronyms and Abbreviations		720
References		722

High Performance Capillary Electrophoresis, edited by Morteza G. Khaledi. Chemical Analysis Series, Vol. 146.
ISBN 0-471-14851-2 © 1998 John Wiley & Sons, Inc.

20.1. CONVENTIONAL ELECTROPHORESIS OF PROTEINS

Protein chemists have relied on electrophoresis for more than half a century as one of their most valued tools for purification, characterization, and analysis of proteins. Although the term "protein electrophoresis" implies a single technique, it is not. A variety of electrophoretic techniques have been used in protein separations, ranging from (i) sodium dodecyl sulfate–polyacrylamide gel electrophoresis (SDS–PAGE) and (ii) gel electrophoresis at near-physiological conditions to retain native three-dimensional (3D) structure to (iii) isoelectric focusing (IEF) and (iv) two-dimensional (2D) gel electrophoresis. Zonal electrophoresis of proteins was also described in early studies of electrophoresis but did not become a significant tool until the advent of capillary zone electrophoresis (CZE).

The success of conventional electrophoretic systems is based on a series of inherent features that life scientists find very desirable or perhaps even essential. One is the ease of preparative separations in slab and tube gel systems. Although analytical methods for proteins are important, most life scientists are separating proteins to recover them for further studies. Many researchers now have the skill and techniques to locate proteins and either excise them from gels or use them directly. New techniques that do not adapt readily for this purpose have a low probability of being accepted by the life science community.

A second major advantage of gel systems is that it is relatively easy to maintain the integrity of a separation while components are being subjected to further manipulation. Dyes may be infused into gels where the separated proteins are denatured and stained, substrates may infused into gels to perform enzyme assays, or proteins may be electroblotted onto adsorptive membranes for immunological assays. All of these operations are achieved with minimal alteration of zone integrity. Once a protein zone has been located it is even possible to bring the protein back into solution for further analysis.

The third major advantage of conventional slab and tube gel electrophoresis systems is the open nature of the format. The gel acts as both a fraction collector and a transport medium, which, after a separation in one dimension, may be used to carry sample components into another separation dimension without disturbing the separation in the first dimension. Not only does this allow 2D separations, but further fractionation of all components from the first dimension occurs simultaneously in the second dimension.

A fourth advantage of conventional electrophoretic systems is that they allow samples to be profiled, i.e., they give one a picture of everything that is present. This is very desirable in a system that is used to examine complex biological extracts.

20.2. ELECTROPHORESIS OF PROTEINS IN FUSED-SILICA CAPILLARIES

It has been noted in preceding chapters that the internal surface of fused-silica capillaries is rich in silanol groups that ionize over a broad titration curve from pH 3.5 to 9. At high pH these weakly acidic groups may be present at concentrations ranging up to $8\,\mu M/m^2$ (1,2). This causes the zeta potential of capillary walls and concomitant electroosmotic flow (EOF) to grow with increasing pH in direct proportion to the ionization of surface silanols (3). Thus, voltage-driven transport in fused-silica capillaries may occur in three ways: (i) by electrophoresis of analytes directly or in association with an ionic species; (ii) by EOF; or (iii) by a combination of the two.

In the case of proteins, surface charges make additional contributions to the separation. First, some proteins are adsorbed at silanol-rich surfaces, denature, and are rapidly release back into solution (4). The resulting change in structural conformation generally impacts electrophoretic mobility, i.e., the denatured protein elutes at a position different than that of the native protein. When there is only partial denaturation, multiple conformers of the same protein will be present. This is undesirable in that (i) some of the peaks will be artifacts of the separation process, (ii) artifact generation will probably be related to the sample matrix and history of the column, and (iii) analyte artifacts may interfere with the determination of other analytes. A second problem is that silanol groups cause fused-silica columns to act as a weak cation exchanger. Severe band spreading and diminished recovery occur when proteins with an isoelectric point higher than the buffer pH interact with the capillary wall. The extent of this problem is seen in Figure 20.1, showing a case where uncoated capillaries were used at pH 7 (5). The pI's of chymotrypsinogen, RNase A, cytochrome c, and lysozyme are all higher than 7, whereas that of ovalbumin is below 7. At pH 7.0, approximately one-third of all proteins are positively charged and therefore can potentially interact with the wall of a fused-silica capillary. Still another problem is that adsorbed proteins change the zeta potential at the capillary wall, impacting both EOF and column efficiency. This is important because alterations in electroosmotic velocity change the elution time of all substances in the electropherogram and compromise reproducibility. Finally, it has been shown that an axially heterogeneous distribution of zeta potential can diminish the separation efficiency of a column by triggering nonuniform flow along the length of the capillary (6). This

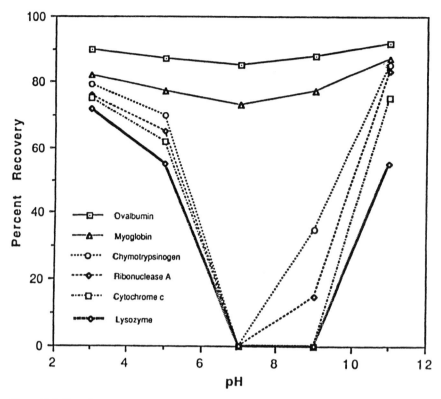

Figure 20.1. Protein recovery vs. pH for six proteins ranging in pI values from 3.1 to 11.1 on an uncoated 75 μm i.d. × 100 cm capillary with two detectors 50 cm apart. Conditions: pH 3 and 5; 0.01 M acetate; pH 7; 0.01 M phosphate buffer; pH 9 and 11; 0.01 M diazomethane; detection, UV at 214 nm; 300 V/cm, 30 μA. [Reprinted from J. K. Towns and F. E. Regnier, *Anal. Chem.* **64**, 2473 (1992), with permission.]

axially heterogeneous distribution of zeta potential results from the inlet of a capillary being fouled with sample protein.

20.3. SURFACE MODIFICATION OF CAPILLARIES

The problems just noted point to the need for surface modification in capillary electrophoresis of proteins, particularly in fused-silica capillaries. Techniques that have been successfully applied are (i) dynamic modification procedures that use extremes in pH or competing ions, (ii) adsorbed oligomer and polymer coatings, and (iii) covalent derivatization.

As important as the need to modify capillaries to increase efficiency and analyte recovery is the need to monitor that this goal has, in fact, been met. Standard practice is to access the "goodness" of a column by peak shape and number of theoretical plates. Recent reports indicate that these are inadequate measures of column efficacy (5,7). Recovery of a protein can be poor and the peaks still be sharp and symmetrical. In the entire scientific literature examined for the preparation of this chapter, protein recovery was reported in only two cases. Use of plate counts for column evaluation suffers from the problem that variables other than surface silanols determine efficiency. The dectector, data aquisition system, sample volume, sample mass, sample concentration, nature of the sample buffer, nature of the operating buffer, column dimensions, operating voltage, and mode of heat dissipation are some of the variables that contribute to column efficiency.

There is an urgent need to develop more scientifically sound methods for the evaluation of column efficacy!

20.3.1. Dynamic Modification

One of the simpler methods for modifying the surface of capillaries is through the use of mobile phase (buffer) additives. The concept in this approach is that one can in some way negate the predominantly electrostatic interactions of proteins with the walls of capillaries. This may be done in three ways: (i) by modification of the charge at the capillary wall; (ii) by modification of the charge on the protein; or (iii) by a combination of these two effects.

20.3.1.1. Extremes in pH

When the operating pH of fused-silica capillaries is reduced to less than 3.0, silanol ionization drops to almost zero and EOF is drastically reduced. Both peptides and proteins have been separated under these conditions. There are three important features of protein separations at extremes in pH that should be noted. First, most proteins are at least partially denatured at extremes in pH. It is possible that this may compromise the separation by dissociation of multimeric proteins into subunits or multiple conformers may be formed. Second, only the basic or acidic amino acids are ionized below pH 3 and above pH 10, respectively. This reduction in the charge diversity of proteins observed at near physiological pH will diminish selectivity of the electrophoretic separation system. Third, polypeptides undergo ion pairing with the acid used to control the pH under acidic conditions. The relative hydrophilicity or hydrophobicity of this acid will have a strong influence on selectivity.

When the operating pH of the column is 9.0 or greater, ionization of amino groups in the protein is repressed, carboxyl groups are fully ionized, and surface silanols are fully ionized. Under these conditions most proteins are electrostatically repelled from the capillary wall.

20.3.1.2. Competing Ions

A second approach to overcoming the negative effects of silanols is with competing ions such as salts, amines, or zwitterions. It has been shown that the tailing peaks so often observed with fused-silica capillaries can be sharpened by increasing mobile phase salt concentration to 100 mM or greater. The problem with this approach is that Joule heating increases in direct proportion to ionic strength.

Ideally a single coating procedure should work with all proteins and it should function in the pH range from 5 to 9, i.e., within 2 pH units of physiological pH. One of the problems with the current literature is that the effectiveness of dynamic coatings is accessed with a small number of proteins, and in some cases with the variants of a single protein. It is not possible to know in these cases how well the coating will do with a broad spectrum of proteins that (i) range in pI from 3–11, (ii) have molecular weights from 10–1000 kDa, and (iii) vary in degree and type of posttranslational modification. If a particular dynamic coating works with only a small number of proteins, it is doubtful that this approach will ever be widely used by the life science community.

Dynamic modification with amines (Table 20.1) provides an effective means of minimizing silanol effects (Figure 20.2) (8). Amines appear to function by ion pairing with anionic silanol groups at the silica surface. Ion pair formation increases both efficiency and resolution by (i) decreasing EOF and (ii) competing with proteins for anionic groups at the surface. In general, decreases in EOF are inversely proportional to the concentration of amine additives. It has also been observed that the higher the binding constant of the amine for the surface, the lower the concentration required to decrease EOF. Because diamines and polyamines bind with higher affinity than monoamines, they are more effective in reducing EOF. For example, 0.07 mM spermine $[H_3N^+(CH_2)_3NH(CH_2)_4NH(CH_2)_3N^+H_3]$ and 200 mM ethylamine $[CH_3CH_2N^+H_3]$ reduce EOF by approximately the same extent (9). It has also been shown that EOF is reduced in direct proportion to (i) the CH_2/NH ratio in the amine additive, (ii) the total number of $CH_2(CH_3)$ groups, and (iii) the molecular weight of the amine. The mechanism by which the CH_2/NH ratio and $CH_2(CH_3)$ group content contributes to reductions in EOF is not readily apparent. With higher concentrations of polyamines it is even possible to reverse the direction of EOF, particularly below pH 6.0. This is the result of

Table 20.1. Amines Used to Dynamically Modify Fused-Silica Capillaries for Protein Separations

Compound	Structure	Reference
Hydroxylamine	H_3N^+OH	9
Ethylamine	$CH_3CH_2N^+H_3$	9
Triethylamine	$(CH_3CH_2)_3NH^+$	8,47
Triethanolamine	$(HOCH_2CH_2)_3N^+$	47
Morpholine	$O(CH_2CH_2)_2N^+H_2$	8,40
Glucosamine	$C_6H_{14}O_5N^+$	47
Galactosamine	$C_6H_{14}O_5N^+$	47
Chitosan	$(C_6H_{14}O_5N^+)_n$	14
Ethylenediamine	$H_3N^+(CH_2)_2N^+H_3$	12
Putrescine (1,4-diaminobutane)	$H_3N^+(CH_2)_4N^+H_3$	9,16,48
N,N,N',N',-Tetramethyl-1,3-butanediamine	$(CH_3)_2N^+(CH_2)_2CH(CH_3)N^+(CH_3)_2$	13
Cadaverine (1,5-diaminopentane)	$H_3N^+(CH_2)_5N^+H_3$	9
FC-134$^-$	$C_8F_{17}O_2NH(CH_2)_3N^+(CH_3)_3I^-$	49,50
Hexamethonium bromide	$(CH_3)_3N^+Br^-(CH_2)_6N^+Br^-(CH_3)_3$	15
Hexamethonium chloride	$(CH_3)_3N^+Cl^-(CH_2)_6N^+Cl^-(CH_3)_3$	15
Decamethonium bromide	$(CH_3)_3N^+Cl^-(CH_2)_{10}N^+Br^-(CH_3)_3$	15
Hexadecyltrimethyl-ammonium bromide	$CH_3(CH_2)_{15}N(CH_3)_3Br$	40,51,52
Decyltrimethyl-ammonium bromide	$CH_3(CH_2)_9N(CH_3)_3Br$	53
Agmatine	$H_3N^+(CH_2)_4NHCH(=NH)N^+H_3$	9
Spermidine	$H_3N^+(CH_2)_3NH(CH_2)_4N^+H_3$	9
Spermine	$H_3N^+(CH_2)_3NH(CH_2)_4NH(CH_2)_3N^+H_3$	9
1,4,7,10-Tetraazacyclo-dodecane (Cyclen)	$(CH_2CH_2NH)_4$—	8
1,4,8,11-Tetraazacyclo-tetradecane (Cyclam)	$C_{10}H_{24}N_4$	8
1,4,8,12-Tetraazacyclo-pentadecane ([15]aneN$_4$)	$C_{11}H_{26}N_4$	8

silanol ionization being low and the $N^+/$—SiO^- ratio at the surface being greater than 1. Single—SiO^- groups bind polyamines containing multiple N^+ groups, giving the surface a net positive charge. It is worth noting that polyamines and ionic surfactants are also useful as charge-masking agents in covalently coated capillary columns. Spermine (10), the tetraazamacrocycle

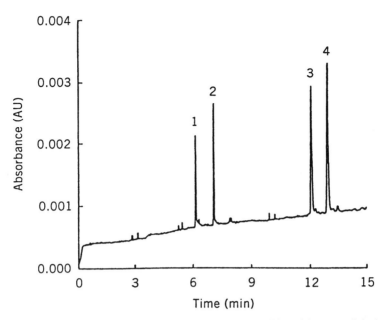

Figure 20.2. Separation of basic proteins using a polyamine additive with a cross-linked polyacrylamide column. Buffer: 20 mM phosphate/60 mM Cyclen, pH 5.5. Peaks: (1) lysozyme; (2) cytochrome c; (3) ribonuclease; and α-chymotrypsinogen. [Reprinted from A. Cifuentes, J. M. Santos, M. deFrutos, and J. C. Diez-Masa, *J. Chromatogr. A.* **652**, 161 (1993), with permission.]

Cyclen (1,4,7,10-tetraazacyclododecane) (8), and cetyltrimethylammonium chloride (11) have been used to mask residual charge in cross-linked polyacrylamide coatings bonded to fused-silica capillaries.

As noted above, adsorption of basic proteins to the walls of fused-silica capillaries is a major problem. A series of papers have addressed this issue with di- and polyamine additives (Table 20.1). For example, it has been shown that 80 mM ethylenediamine as a buffer additive improves the separation of the basic proteins lysozyme, cytochrome c, trypsinogen, ribonuclease, and α-chymotrypsinogen A in uncoated capillaries over the pH range from 6.5 to 9.5 (12). Efficiencies were on the order of 100,000 plates, and maximum resolution was achieved in a 70 min separation at 120 V/cm. Between runs, the capillary was first flushed with one column volume of 1 M KOH and then with carrier electrolyte for 5 min. Separations of acidic proteins were not reported, so it is not possible to assess the general utility of this procedure.

It has been reported that N,N,N',N'-tetramethyl-1,3-butanediamine is also an effective additive for efficient electrophoretic separations of basic proteins between pH 4 and 7 in bare fused-silica capillaries (13). Well-resolved, symmetric peaks were obtained at pH 6.5 in 40–100 min with additive concentrations of 20–120 mM. Higher peak capacities were obtained at higher additive concentrations and longer running times. Efficiency increased with additive concentration, approaching 200,000–400,000 plates at 120 mM. Depending on electrolyte pH and additive concentration, EOF could either be in an anodic or cathodic direction. Anodic migration was favored by lower pH and higher additive concentration.

A still larger polyamine can be generated by deacylation of the natural polymer chitin to form chitosan [$(1 \rightarrow 4)$-2-amino-2-deoxy-β-D-glucan] (14). Adsorption of this cationic polymer at a concentration of 0.1% to the capillary reversed the charge at the wall and the direction of EOF. The fact that reversed flow persists even after removal of chitosan from the buffers indicates that the polymer is strongly adsorbed to the capillary. Because the pK_a of amino groups in chitosans is approximately 6.3 and the polymer flocculates above this pH, the operating range with this additive is below pH 6. Although efficiencies on the order of 400,000 plates are achieved with most proteins below pH 5, plate counts decrease dramatically with lysozyme above pH 3.5. Separation times for a mixture of nine basic proteins were roughly 15 min.

There is great interest today in the separation of protein glycoforms. Using ovalbumin as a model, it has been shown that 100–300 µM concentrations of α, ω-bis-quaternary ammonium alkane additives, such as hexamethonium bromide, hexamethonium chloride, and decamethonium bromide (Table 20.1), dramatically impact the separation of glycosylation variants of this protein in the presence of 100 mM borate buffer at pH 8.4 (15). Glycoforms of human chorionic gonadotropin have also been resolved by this method. We should note the significant difference between hexamethonium chloride and hexamethonium bromide as additives. One-third the concentration of the chloride additive gave a separation in 20 min equivalent to a 33 min separation with the bromide. When alkyl chain lengths were compared, decamethonium bromide was more effective than the hexamethonium bromide analogue, giving equivalent separations in 18 and 33 min, respectively. Surprisingly, these two additives had roughly the same effect on EOF. This suggests that alteration of EOF is not the only parameter involved in additive-induced enhancement of resolution.

In a further examination of the resolution of ovalbumin glycoforms with di- and polyamine additives in the presence of borate buffer, work has focused on the use of putrescine, cadaverine, spermidine, and spermine (Table 20.1) (9). Surprisingly, it was found that although both spermidine [$H_3N^+(CH_2)_3 \cdot NH(CH_2)_4N^+H_3$] and spermine [$H_3N^+(CH_2)_3NH(CH_2)_4NH(CH_2)_3N^+H_3$] reduced EOF almost equally well, spermine produced much larger differences in selectivity. When the structure of all the di- and polyamines was examined together, selectivity was found to be related in a nonlinear fashion to (i) the CH_2/NH ratio in the amine additive, (ii) the total number of $CH_2(CH_3)$ groups in the additive, and (iii) the molecular weight of the amine. Again it is seen that factors other than reducing EOF play a role in additive mediated enhancement of resolution. There are several possible explanations for this phenomenon. One would be ion pair formation with the protein analytes and concomitant alteration of their electrophoretic mobility (16). Another would be a two-step process involving (i) the formation of borate esters with vicinal diols in the oligosaccharide portion of these proteins and (ii) subsequent ion pair formation with the di- or polyamine additive to alter electrophoretic mobility. In either case, ion pair formation plays a role.

Although most separations of protein glycoforms have been achieved with amine additives in borate buffer, it has been observed with human recombinant factor VIIa (rFVIIa) that replacement of borate with phosphate buffer resulted in shorter migration times and resolution of variants was improved. If borate ester formation occurs, it does not improve the separation of rFVIIa (16). Similar findings have been reported with human recombinant tissue plasminogen activator (rtPA). N-Acetylneuraminic acid in the oligosaccharide portion of the molecule appears to play a role in the separation of rFVIIa. Treating the sample with neuraminidase reduced the number of peaks from 10 to 2. This supports the idea that the various peaks of rFVIIa are in fact glycoforms.

Dynamic modification by polyanions has also been reported (17). In the case of polyanions, ion pairing appears to occur with amino groups on analytes in solution—in contrast to polycations, in which ion pairing occurs at the capillary wall. When 15 mM phytic acid [myo-inositol hexakis(dihydrogen phosphate)] is added to 150 mM sodium borate buffer (pH 9.5), it is possible to separate a mixture of proteins ranging from lysozyme to β-lactoglobulin B in an uncoated fused-silica capillary. It may be calculated that the six phosphate groups of phytate, which vary in pK_a from 1.9 to 9.5, present 9–11 negative charges at pH 9. Despite the fact that lysozyme has a net positive charge at pH 9.5, it elutes after the neutral marker in the presence of 15 mM phytate. It is clear from this behavior that the net charge of lysozyme has been altered by ion pairing with the polyanionic phytate. The net effect of this phenomenon is a compression of the electropherogram. By making basic proteins more acidic through ion pairing with acidic additives, their electrophoretic mobility will be nearer to that of an acidic protein.

20.3.1.3. Surfactants

Dynamic surface modification with surfactants has been used in several ways. In all the cases examined above, the silanol problem has been approached by attempting to titrate the surface charge with counterions. Another strategy is to covalently sequester many of the surface silanols with octadecyl- or dimethylsilane and then dynamically coat the hydrophobic surface of the column with nonionic surfactants during use (3,18). It has been shown with reversed-phase chromatography columns that nonionic surfactants stereospecifically adsorb to hydrophobic surfaces and create a hydrophilic surface layer (19). Alkyl groups of the surfactant are directed into the alkylsilane layer, while the hydrophilic portion of the surfactant projects into the aqueous phase to shield surfaces from the approach of proteins. This coating functions in the same manner in electrophoretic systems, with the exception that the charge from residual surface silanols projects through the hydrophilic coating and is used to drive EOF, albeit at a velocity substantially less than that with the uncoated capillaries. EOF is relatively constant across the pH range from 4 to 10, in contrast to bare-silica capillaries. Although surfactants are strongly adsorbed to octadecyl-derivatized capillaries from a 0.01% solution, both EOF and the adsorptive properties of columns change substantially within a few hours when the surfactant is deleted from the mobile phase. Greater than 95% recovery of a broad spectrum of proteins was achieved in the CZE mode from Brij 35–coated capillaries. Relative EOF at basic pH was manipulated from 0.033 to 0.33 times that of uncoated capillaries by increasing surfactant size (20), using methylcellulose (20), or using one of the polymeric pluronic surfactants (18,20). In the case of pluronic surfactants, much higher separation efficiency was achieved by using dimethylsilane-derivatized capillaries (18). This ability to select EOF has proven to be particularly effective in isoelectric focusing.

Surfactants have also been used with uncoated capillaries. It has been shown that electrophoretic mobility increases with the anionic detergents sodium deoxycholate (SDC) and sodium dodecyl sulfate (SDS), and decreases in the presence of the nonionic detergent Triton X-100 (21,22). Clearly the surfactant is adsorbing to the silica surface and increasing surface charge. Depending on structure and concentration, surfactants can play an additional role in capillary electrophoresis (CE) by binding to the protein. At low concentrations, surfactants bind to hydrophobic sights on the surface of proteins without denaturation, increasing both the Stokes radius and charge in the case of ionic detergents. Fewer than 10 surfactant molecules are probably bound. In the case of SDC and SDS, the number of surfactants bound is structure specific and would increase the anionic character of the protein. These phenomena have been clearly demonstrated in the CZE of plasma apolipoproteins (21). The electrophoretic mobility of apolipoprotein A (apoA) is very different in SDS and SDC because apoA binds four times as much SDS as SDC. Due to the fact that the binding stoichiometry of the various lipoproteins is different for SDS and SDC, it is apparent that selectivity in CZE mirrors these differences. Selectivity changes with nonionic detergents resulted from relative differences in Stokes radius based on binding stoichiometry.

20.3.1.4. Neutral Polymers

It is probable that most of the additives spend some finite amount of time bound to the silica surface. For this reason, there is no clear line of demarcation between dynamic coatings and adsorbed coatings. Polymers such as poly(vinyl alcohol) (PVA), dextrans, hydroxyethyl cellulose, hydroxypropyl methylcellulose (23), and poly(ethylene oxide) (PEO) (24), fall in this category. Although they may be used in the dynamic mode, more effective coatings are produced in the permanent coating mode.

20.3.2. Adsorbed Coatings

Most of the dynamic coating methods described above involve adsorption at a surface to create some sort of a coating. The difference between these "dynamic coatings" and "adsorbed coatings" is their degree of permanence. In the case of the adsorbed coatings described in the following subsections, they are essentially permanently bound. Permanence is achieved either by the binding affinity of the coating agent or by cross-linking adjacent adsorbed species into a continuous, permanent film.

20.3.2.1. Neutral Polymers

20.3.2.1.1. Poly(vinyl alcohol). PVA is one of a small group of polymers that becomes pseudocrystalline and precipitates from aqueous solutions when heated. This phenomenon is the result of eliminating water from the structure with increasing temperature and is accompanied by the formation of intermolecular hydrogen bridges. The requisite hydrogen bond formation necessary for PVA immobilization in fused-silica capillaries (23) will not occur if the polymer contains residual acetate groups from the hydrolysis of the poly(vinyl acetate) precursor used in the synthesis of PVA (25). Deposition of the coating is achieved by flushing the capillary with a 10% aqueous solution of 50 kDa PVA (99 + % hydrolyzed), pushing the coating solution out of the column with a stream of nitrogen, and then heating the capillary at 140 °C under a stream of nitrogen (23). This process produced capillaries with efficiencies of greater than 10^6 plates, essentially no EOF, stability up to pH 10, and the ability to separate either basic or acid proteins (Figure 20.3) (23).

20.3.2.1.2. Cellulose Acetates. Cellulose diacetate. (CDA) and cellulose triacetate (CTA) have been used as film-forming agents for more than half

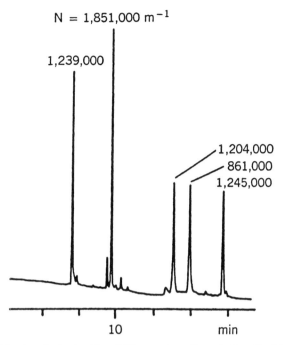

Figure 20.3. Efficiencies obtained with a PVA-coated capillary. Buffer: 50 mM sodium phosphate, pH 3.0. Conditions: 30 kV (357 V/cm), 69 µA; 20 °C. [From M. Gilges, M. H. Kleemiss, and G. Schomburg, *Anal. Chem.* **66**, 2038 (1994). (23).]

a century. One mode of film formation with CDA is by evaporation of the organic solvent used to dissolve the polymer. Using a procedure very similar to that used in the preparation of coated capillaries for gas chromatography, it has been possible to prepare CDA-coated fused-silica and polypropylene capillaries by flushing columns with a 1% (w/v) solution of CDA in acetone, pushing the coating solution out with a stream of helium, and drying for 30 min with helium (26). Coating thickness depends on the viscosity of the coating solution and the speed with which the solution is expelled from the capillary. Although CDA forms porous films, the pores are apparently too small to allow proteins to gain access to the underlying silica surface. Excellent separations of basic proteins with efficiencies up to 10^6 plates/m were achieved at pH 4.0–4.3. Unfortunately, the coating is destroyed above pH 7.5 and cannot be used for CZE at alkaline pH or for isoelectric focusing.

20.3.2.1.3. Poly(ethylene oxide). PEO has been widely used in liquid chromatography to suppress adsorption of proteins to high-surface-area, porous silica particles. This ether-rich polymer apparently forms hydrogen bonds with silanol groups, with concomitant adsorption of a PEO layer at silica surfaces. In the case of CZE, the fused-silica capillary was pretreated with 0.1 M NaOH and 0.1 M SDS at the beginning of each day. The typical coating protocol was to flush the capillary with 1.0 M HCl, then follow with a solution of 0.2% PEO, and finally wash with the electrophoretic buffer. The entire coating process was repeated before each CZE run, requiring approximately 5–15 min for each step. EOF was reduced roughly 60–70%. Columns thus treated functioned well in the separation of basic proteins (24).

20.3.2.2. Polyamines

It is well known that polyamines such as polyethyleneimine (PEI) adsorb tenaciously to silica surfaces. Binding affinity generally increases with molecular weight. High-molecular-weight PEI ($M_r = 0.6–1 \times 10^6$) can be coated on silica by flushing the capillary with an aqueous solution of the polymer. An electrostatically adsorbed layer of PEI is obtained that appears to be very stable and can be used in the pH range from 3–11. Good reproducibility within and between columns was achieved. Column efficiencies of 300,000–500,000 plates were achieved with basic proteins. Several important properties of PEI-coated capillaries should be noted. One is that EOF is toward the anode. This means that proteins will elute in the order of increasing pI. The second is that EOF will not be constant in PEI-coated capillaries. PEI has a broad titration curve ranging from 5 to 10. EOF will be relatively large at pH 3–5 and approach zero at pH 10. When PEI adsorbed to the inner wall of a fused-silica capillary was cross-linked with an excess of diepoxide, a stable quaternary amine layer was formed that gave constant EOF toward the anode across a pH range from 3 to 11 (27). The separation characteristics of these columns are similar to the dynamically coated polyamines that show anodally directed EOF, i.e., proteins elute in the order of increasing pI. The limitation of these columns is that negatively charged proteins will adsorb. This means that each run must be accompanied by a protein-desorption step. This column differs from the adsorbed PEI columns in that the PEI layer was quaternized during cross-linking. This causes EOF of the quaternized column to be almost constant for pH ranging from 3 to 11.

20.3.2.3. Hydrocarbons

Hydrocarbon coatings have been applied to fused-silica capillaries by both deposition of performed polymer, as in the case of polyethylene (28), and *de novo* synthesis (29). Polyethylene is difficult to dissolve in anything except hot halogenated solvents. When a capillary is filled with a hot saturated solution of polyethylene and then cooled, a thick layer of hydrocarbon is deposited at the surface (28). In the case of poly (styrene–divinylbenzene), monomers are deposited on the surface of either the bare silica or a column that has been derivatized with methacrylate and polymerization initiated (29). The layer of polymer produced by either of these procedures may be thought of as a "tube-in-a-tube." The extremely hydrophobic character of these hydrocarbons preclude their direct use in protein separations. Following either dynamic modification with a nonionic surfactant or some secondary coating step, it is anticipated that they will be effective coatings for protein separations.

20.3.3. Covalent Modification

20.3.3.1. Simple Silanes

Although derivatization with organosilane monomers has been used extensively to passivate the surface of fused-silica capillaries, these columns suffer from the problem that the siloxane bond hydrolyzes slowly at pH values of > 7. This not only allows the coating to leach from the surface, but negative charge is generated in the process. It is not possible to prepare columns of good long-term stability with simple silanes. For this reason, few consider organsilane monomer derivatization to be adequate alone in coating capillaries. More frequently, silanes are used as coupling agents for polymers. Hydrolytic stability is not as big an issue in this case because the silane is part of a coating that holds it in place following hydrolysis until it rebinds.

20.3.3.2. Polymers

Polymers provide an excellent solution to the surface deactivation problem with silica because they form a continuous skin across the surface that can be sufficiently thick to preclude proteins reaching the surface.

20.3.3.2.1. Cellulose. Hydrolytically stable cellulose-derived coatings for CZE of proteins have been prepared in a number of ways. Methyl cellulose (30) or hydroxypropyl cellulose (28,31,32) is generally derivatized with a vinyl, allyl, or methacryl group and then grafted to a fused-silica capillary through an organosilane coupling agent carrying a vinyl or methacryl group. The most direct approach has been to coat the surface of a γ-glycidoxypropylsilane-derivatized column with hydroxypropyl cellulose and couple the hydroxyl groups on the polysaccharide directly to the column using boron trifluoride (BF_3) as a catalyst (28). Although plate counts of greater than 300,000 were achieved with basic proteins, efficiency was inferior to that of columns prepared with poly(ethylene–propylene glycol).

In a more elaborate procedure, allylglycidyl ether was coupled to methylcellulose in base, and the resulting allylmethylcellulose (AMC) with 10% ammonium persulfate and N,N,N',N'-tetramethylethylenediamine (TEMED) forced into γ-methacrylpropyl silane-derivatized columns (30). Polymerization of AMC was allowed to proceed for 17 h, after which the highly viscous solution of unbonded AMC polymer was forced out of the column with an SDS solution at pressures of 100–150 bar. Columns of immobilized dextran were prepared in essentially the same way. EOF was reported to be eliminated in capillaries thus coated. Columns appeared to be stable under very harsh operating conditions, the most extreme being 0.05 M NaOH with 5% SDS or 1 M HCl.

Cellulose-derived capillaries have also been prepared by coupling 2-hydroxyethyl methacrylate (HEMA) derivatized hyroxypropyl cellulose (HPC) to columns derivatized with 7-oct-1-enyltrimethoxysilane (OES)(32). In this process, HPC is both derivatized with HEMA and the product grafted to the organosilane-derivatized OES surface simultaneously in a reaction initiated by 2,2'-azobisisobutyronitrile (AIBN) (33). Multiple simultaneous AIBN-initiated reactions involving (i) derivatization of cellulose with HEMA, (ii) grafting this product directly to OES, (iii) HEMA polymerization from the immobilized cellulose containing excess HEMA, and (iv) HEMA polymerization from the OES surface alone produce a hybrid coating that probably has the characteristics of both a HEMA polymer and a cellulosic. The exact composition of the coating will be a function of the relative rates of reaction of OES and HEMA, the relative concentration of HEMA and HPC, and the AIBN concentration relative to the reactants. The coating was stable for several weeks in the pH range from 2 to 10, diminished EOF, and was used primarily in the separation of peptides, glycoproteins, and derivatized oligosaccharides.

We should note that although widely different procedures were used to prepare the columns described above, they do not appear to vary markedly in separation properties. The message to be derived from this is that there does not appear to be a single "magic" coating.

20.3.3.2.2. Acrylates. Acrylamines are widely used to create polymeric coatings on fused-silica capillaries in a two-to-three-step process. In the first step, an organic coupling agent is attached to the silica surface that will subsequently anchor the acrylate polymer to the capillary wall. Next, acrylamide monomers are polymerized at the derivatized silica surface to create the polyacrylamide layer. As a final step, the acrylamide polymer is sometimes cross-linked. In other cases, both polymerization and cross-linking are achieved simultaneously.

Step One. Coupling agents that are attached to the silica surface in the first step of the coating process are generally vinyl, acrylate, or methacrylate moieties. Three different strategies have been used in the actual derivatization chemistry. The simplest is to use a commercially available organosilane of the general formula X_3SiY, in which X is a halide or alkoxide and Y is either a $-(CH_2)_3OCOC(CH_3)=CH_2$ (34) or a $-CH=CH_2$ group. It has also been reported that OES may be used as a reactive surface-coupling agent, although the reactivity of this monomer is substantially lower than that of vinyl, acryloyl, and methacryloyl groups (28). The objection to this approach is that siloxane bonds slowly hydrolyze and as a consequence coupling agents and polymers attached to them will leach from the surface during use unless the polymer is cross-linked.

One alternative is to use direct Si—C linkages to the silica surface. It has been shown that linear polyacrylamide coatings anchored to the surface through Si—C bonding are substantially more stable at pH 10 than those anchored by Si—O—Si bonding (35). Si—C bonding formation can be achieved by either of two strategies. In one, silanol groups are halogenated with thionyl chloride, and the halogenated silica is reacted with the Grignard reagent vinylmagnesium bromide to attach vinyl groups directly to the surface (36,37). Another scheme is to first modify the silica surface with triethoxysilane and then attach the vinyl group in a Pt-catalyzed hydrosilylation of allyl methacrylate (38,39).

Another tactic is to incorporate the vinyl coupling agent into a polymer, vinyl siloxanediol

$$HO-SiR_1R_2-O-(SiR_1R_2-O)_x-SiR_1R_2-OH$$

(in which R_1 is a methyl group and R_2 is a vinyl group), and bond the polysiloxane polymer to the silica (10). The inner wall of the capillary was coated with a mixture of the polymer and a silane cross-linker. The cross-linker reacted with the Si—OH groups of both the polysiloxanediol and the silica surface to form Si—O—Si bonds. The resulting vinyl-containing polymer layer is estimated to be 0.2 μm thick.

Step Two. Linear polyacrylamide may be grafted to any of the derivatized surfaces described above by incorporation of the surface-bound monomer (either a vinyl or methacryloyl group) into the acrylate polymer during polymerization. This is generally achieved by filling the capillary with an aqueous solution containing 3–6% monomer, a free radical initiator such as ammonium persulfate, and TEMED. Polymerization is complete in 4–12 h, after which nongrafted polymer formed in the capillary lumen is pushed out of the column. One of the problems with polyacrylamide is that the acrylamide bond is not very stable. During the course of a few weeks of operation, sufficient hydrolysis occurs that EOF increases sharply and efficiency diminishes (11,37).

There are several solutions to the problem of acrylamide stability. One is to cross-link polyacrylamide with paraformaldehyde at pH 10 in a third step (10). This cross-linking through the amide nitrogen increases the hydrolytic stability of the coating in addition to producing a highly viscous, hydrophilic surface layer. A second strategy is to use a monomer that is inherently more hydrolytically stable. Through the use of acryloylaminoethoxyethanol monomer, it has been demonstrated that a polyacrylamide coating can be produced with two times the stability of conventional acrylamide coatings (39). The other attractive feature of this monomer is that it yields a more hydrophilic polymer.

Cross-linking polyacrylamide at the capillary wall has the attractive feature of producing a more impermeable, physically stable coating. Cross-linked polyacrylamide is easily formed by using a mixture of acrylamide and bisacrylamide. The problem is to do this on the capillary wall and not throughout the lumen of the capillary. Bisacrylamide cross-linked polyacrylamide coatings have been produced by pumping bisacrylamide/acrylamide (4.2%: 10%) solution with free radical initiator and catalyst into a methacryloyl-derivated capillary and then flushing the column after 1 min of residence to preclude formation of a solid gel with the associated plugging. This process was repeated three times to achieve a homogeneous coating of sufficient thickness on the capillary wall (40).

Although the objective in the studies described above was to provide neutral, hydrophilic capillaries, mobile phase additives such as SDS (11,22), ethylaminocaproic acid, spermine (10), and morpholine (40), were used in a large number of cases. In many cases, the effect of the additive is concentration dependent and has an impact on selectivity, as was not the case with uncoated capillaries (11). Again, these additives play a role in the separation by either dynamically modifying the capillary wall or altering the charge of proteins through some type of pairing, or by some combination of the two Another approach to obtaining high-EOF, anionic acrylate–derivatized columns was by substituting 2-acrylamido-2-methyl propanesulfonate for acrylamide in the polymerization step of coating (41). EOF is pH independent with this coating.

20.3.3.2.3. Epoxy Coatings. Epoxy polymers are easily attached to inorganic materials through γ-glycidyltrimethoxypropylsilane. Fused-silica capillaries derivatized with this oxirane react readily with diglycidyl ethylene glycol $CH_2OCHCH_2O-CH_2CH_2-OCH_2CHOCH_2$ under either anionic and cationic catalysis to produce a polymeric coating of the following general formula:

$$\equiv \text{Si}(CH_2)_3O-(CH_2CHOHCH_2O)_m-(CH_2\overset{\overset{\displaystyle O^-}{|}}{C}HCH_2O)_n$$
$$-(CH_2CH_2O)_o-(CH_2CHOHCH_2OH)_p$$

This coating eliminates protein adsorption and reduces EOF approximately 80%, allowing the separation of proteins in a pH range from 2 to 10 (42). An alternative approach has been reported in which a thin film of diglycidyl bisphenol A was deposited on a glycidylsilane-derivatized capillary and cross-linked with ethylenediamine (43). The coating was stable for more than 2 months.

Yet another variation is to directly couple dextrans to a glycidoxy-derivatized capillary wall by flushing the column sequentially with a dioxane solution of 150 kDa dextran containing boron trifluoride (BF_3) and then a solution containing PEG [poly(ethylene glycol)] diglycidyl ether and BF_3 (44). In this process the dextran is first coupled directly to the silica surface and then cross-linked with the diepoxide. The columns exhibited reduced EOF and were useful in the separation of both acidic and basic protein.

20.3.3.2.4. Polyvinylimidazole. The cationic polymer polyvinylimidazole (PVI) has been used as an efficient coating agent for high-performance chromatography supports. This polymer has also been used in CE coatings, where it plays the role of repelling cationic proteins from the capillary wall and sterically limiting contact with the silica surface of the column (45). PVI was attached to the column wall by polymerization in capillaries derivatized with γ-methacryloxypropyltrimethoxysilane. When 1-vinylimidazole was copolymerized with 1-vinyl-2-pyrrolidone, the degree of charge at the wall could be controlled. In some instances the coating was also cross-linked with epichlorohydrin. Because the pK_a of imidazole is roughly 6, ionization of PVI and inhibition of EOF will increase substantially below pH 7. EOF was reduced approximately 80% with polyvinylpyrrolidone (PVP) and nearly 90% with PVI. PVI-containing coatings were effective in the separation of basic proteins but adsorbed acidic proteins. In contrast, PVP-coated capillaries could be used in the separation of both acidic and basic proteins at pH5. Above pH 7 these capillaries are not stable.

20.3.3.2.5. Polyamines. We noted above that polyamines may be adsorbed to the surface of silica and cross-linked with multifunctional epoxides into permanent films. Similar coatings have been prepared by applying oxiranes to the silica surface in the form of γ-glycidoxypropylsilane and covalently grafting either hydroxyethylated or alkylated PEI derivatives through these surface oxiranes in a nucleophilic ring opening (46). The dilemma with all polyamine-coated columns is simply the reverse of the uncoated fused-silica problem: polyamines present a cationic surface, whereas that of silica is anionic. Efforts to overcome the adsorption problem in PEI-based columns have been directed at grafting a nonionic, hydrophilic polymer layer of diepoxy–PEG 600 to the surface of PEI, which will shield it from contact with negatively charged proteins. It has been shown that negatively charged proteins that migrate toward the anode in the direction of the EOF are effectively separated in less than 25 min, with plate counts of approximately 100,000 (46).

20.3.4. Conclusions on Coatings

Although coatings are used in capillary gel electrophoresis, their major application is in CZE and isoelectric focusing. All the coatings discussed above were used in CZE with only a few exceptions. The numerous coatings described in the scientific literature for CZE of proteins enable us to draw a series of conclusions. Based on the fact that all these coatings were used to separate proteins with high efficiency, we may conclude that (i) the protein adsorption problem has been solved; (ii) it is possible to prepare capillaries with EOF approaching zero; (iii) there are many ways to solve the adsorption problem, i.e., almost an infinite array of coatings are possible; and (iv) there is now a large database on ways to prepare coatings for CZE of proteins (47–53).

Unfortunately, this does not necessarily mean that CZE of proteins will now undergo rapid acceptance by life scientists. CZE does not currently have the broad utility life scientists expect from their experience with conventional electrophoretic techniques. The majority of the coatings described above were used for the separation of pure reference proteins that are referred to as either "basic" or "acidic" proteins, i.e., a group of proteins are being separated that have high electrophoretic mobility at the operating pH of the CZE system. Although selectivity and resolution in this window are superb, this restricts the technique to the analysis of some fraction of the total number of components in a sample. The problem with capillaries of near-zero EOF is that proteins of intermediate isoelectric points, which are neither "basic" or "acidic," will have extremely long elution times when the CZE system is operated at near-physiological pH. This means that CZE with neutral capillaries will, by necessity, be of greater utility in the targeted component mode of analysis more familiar to analytical chemists than the profiling mode commnoly used by life scientists.

Capillaries with charged coatings have problems that are equally as large. They adsorb all proteins of opposite charge. Again, their primary mode of use must be in targeted component analysis.

20.4. ZONE ELECTROPHORESIS IN OPEN-TUBULAR CAPILLARIES

Although zone electrophoresis is perhaps the simplest of the electrophoretic separation modes, it is the only one that has never been widely used for protein separations. This is because thermally induced convection in 0.1–1 cm i.d. columns presented an insurmountable problem before the advent of capillary columns. The ability to eliminate thermal gradients at high voltage in CZE systems gives protein chemists a potential new high-resolution, high-speed tool for defining the composition of complex biological samples and characterizing protein structure.

The general nature of CZE has been described in other chapters of this book, and the impact of surface coatings on protein separations by CZE has been examined above. The discussion in the following subsections will focus on applications of CZE in the study of specific classes of proteins.

20.4.1. Serum Proteins

Analysis of serum proteins by electrophoresis has been practiced in clinical laboratories for almost half a century. The classic five-component paper electropherogram showing the serum protein fractions designated albumin α_1, α_2, β, and γ has now been produced by CZE (54–57). Although many of these separations were achieved in borate buffer, resolution can be increased to 10 protein zones by switching to an operational electrolyte containing 0.1 M methylglucamine and 0.1 M ϵ-aminocaproic acid or 0.1 M methylglucamine and 0.1 M γ-aminobutyric acid (58).

Although the composition of cerebrospinal fluid (CSF) is related to that of serum, the analysis of CSF has long been used as an aid in the diagnosis of central nervous system disorders. Protein adsorption to the capillary walls has been controlled in uncoated capillaries by the Beckman electrophoresis buffer (54) and in a commercial column with a hydrophilic coating along with the incorporation of methylcelllulose into the operating buffer (59). Using borate buffer at pH 10 in the latter case, Cowdry et al. observed substantial EOF toward the cathode. The elution order of proteins and peptides is therefore cations, neutrals, and finally anions. A surprisingly large number of CSF specific acidic species were noted relative to serum. With a total run time of approximately 30 min, 20–25 peaks were observed.

Human serum albumin (HSA) is the major component of human serum. It has been shown that as the operating pH of the column approaches the pI of HSA (5.2), the resolution of clinical grade HSA increases (60). Eight peaks were separated by using a neutral coated capillary. Mass spectral analysis of these fractions indicated the major components to be HSA with a free thiol at cysteine 34, a variant having cysteine 34 blocked with cysteine, amino–

terminally degraded HSA, and probably combinations of these variations. Surprisingly, glycosylated forms of HSA were not resolved from HSA.

20.4.2. Glycoproteins

Analysis of microheterogeneity in glycoproteins is a challenging problem. A single polypeptide can exist in several to a hundred different glycoforms, which may vary in position of glycosylation along the peptide chain, size of oligosaccharides attached to the protein, sequence of a specific oligosaccharide, degree of branching, and end group substitution. The biological significance of glycoforms is not even known in all cases.

Glycoforms vary sufficiently in structure of charge and are easily separated by CZE in many cases. Glycosylation variants of RNases A and B, horseradish peroxidase, and ovalbumin have been separated in cellulose-coated capillaries at acidic pH with a small amount of 1-propanol added to the buffer (32). It is claimed that neither high pH, higher ionic strength, nor Tris–borate buffers improved the separation of these proteins, as was claimed in the studies outlined below. Because CZE is a fairly new analytical technique discrepancies such as this still need to be resolved.

We noted earlier that mobile phase additives can amplify the separation of glycoforms by pairing with monosaccharide residues in the protein. For example, di- and polyamine additives such as decamethonium bromide and spermine substantially enhanced the separation of ovalbumin variants in the presence of 100 mM borate buffer (9,15,16,61,62). Removal of N-acetylneuraminic acid from the oligosaccharide portion of human recombinant factor VIIa (rFVIIa) reduced the number of peaks, confirming the importance of ion pair formation with the amine additive (16). It is possible in the case of ovalbumin that pairing by both amines and borate is important, with the role of borate being a structure-dependent formation of esters with vicinal diols. In contrast, it has been observed with rFVIIa and human recombinant tissue plasminogen activator (rtPA) that superior resolution is achieved with phosphate buffer.

Human chorionic gonadotropin (hCG) is a dimer of approximately 38 kDa composed of an α and β subunit. Both subunits have two N-glycosylated asparagines, whereas the β subunit has four additional O-glycosylated serine residues. More than 30% of the mass of this hormone results from carbohydrate. When 5 mM diaminopropane was used in 25 mM borate at pH 8.8, hCG was resolved into eight distinct peaks. When analyzed individually, the α subunit was resolved into four peaks and the β subunit into seven peaks.

Determining the position and number of glycosylation sites is an important part of characterizing a glycoprotein. In the case of O-linked oligosaccharides of bovine submaxillary mucin, Weber et al. reported that β-elimination occurs at O-linked serine and threonine residues in alkaline sulfite, with the formation of cysteic and α-amino-β-sulfonylbutyric acid residues in the polypeptide (63). Following hydrolysis and derivatization with naphthalene-2,3-dicarboxyaldehyde (NDA), stoichiometry of the sulfonated NDA amino acids was determined by CZE. Selective cleavages in the glycoprotein with glycosidic enzymes accompanied by electrophoretic analyses of the products provide an even more effective way to study the structural and immunogenic properties of glycoproteins (64). β-N-Acetyl glucosaminidase, neuraminidase, and the endoglycosidases D, F, and H all cleave at different sites in the oligosaccharide structure. Enzymes achieve a selectivity of cleavage that is currently impossible to duplicate by purely chemical means.

The structure of glycoproteins is frequently too complex for direct resolution and analysis of glycoforms. In such cases, oligosaccharides are cleaved from the protein and analyzed individually. N-glycans which had been liberated from distinct glycoproteins by either (polypeptide N-glycosidase) (PNGase) F treatment or hydrozinolysis and isolated by anion exchange chromatography were subjected to CZE using 80 mM ammonium sulfate and 20 mM phosphate buffer (pH 7.0) with 2 mM 1,5-diaminopentane additive (65). N-linked oligosaccharide mixtures from recombinant human urinary erythropoietin, α_1-acid glycoprotein, and bovine fetuin were resolved into 15–20 components and compared against an N-glycan-mapping database. Presumably these separations are based on sialic acid content of the oligosaccharides.

20.4.3. Lipoproteins

There are three principal classes of lipoproteins: the high-density lipoproteins (HDL), the very-low-density lipoproteins (VLDL), and the low-density lipoproteins (LDL). They are of great interest because of their relationship to coronary heart disease (CHD). Preliminary separations of lipoproteins are generally achieved with ultracentrifugation. Following centrifugation, LDL and HDL have been separated by CZE in < 20 min using uncoated capillaries with 50 mM borate buffer (pH 10) containing 3.5 mM SDS and 20% acetonitrile (66). It should be noted here that SDS is being used at subcritical micelle concentration. Among the lipoproteins, lipoprotein a [Lp(a)] has been recognized as a significant marker for predicting the risk of CHD. In an effort to find a more efficient analytical method, CZE has been applied to the detection of Lp(a) under the same conditions as are used for HDL and LDL. Both Lp(a) and its reduction products Lp(a$^-$) and apolipoprotein a [apo(a)] are separated.

It is generally not possible to determine HDL directly in serum. Apolipoprotein A-I (apoA-I) is the major protein constituent of HDL and is easier to monitor. Decreased levels of apoA-I in human serum are also indicators of arteriosclerotic risk. It was found that the BioRad LLV buffer was optimum to determine apoA-I in serum directly. A major and minor apoA-I constituent were found (57,67,68).

The use of detergents and organic solvents to solubilize lipoproteins and assist in their analysis has been noted above (66). Detergents are a problem when the protein will be analyzed by mass spectrometry, thus favoring the use of organic solvents. The problem with organic solvents such as 2-propanol is that they prolong analysis time by decreasing EOF. This issue has been resolved by using a small amount of pressure to induce flow. When this approach was employed, the human lung surfactant protein SP-C was solubilized with 70% 2-propanol and analyzed by CZE–electrospray–mass spectrometry (69).

20.4.4. Antibodies/Immunology

Analytical immunology in the form of ELISA (enzyme-linked immunosorbent assay), RIA (radioimmunoassay), immunoaffinity chromatography, or immunoelectrophoresis is one of the major analytical tools of life scientists. Developing an immunological assay has many components, including antibody purification, reagent synthesis, and constituting the reagents in a manner that will report the presence of antigen. All of these activities require careful analysis and quantitation. CZE can be a strong analytical tool in immunological assay development (70–72). More recently, CZE is being used directly in immunological assays.

All immunological assays are based on antigen(Ag)–antibody(Ab) complex formation. Frequently the monovalent Fab fragment of IgG (immunoglobulin G) is substituted for the bivalent whole antibody. The problem is to detect and quantify Ag–Ab complex formation. This is frequently done by labeling either Ab* or Ag* with a unique spectral or radioactive tag. Subsequent to complex formation, the assay medium in the specific case of a labeled Ab* will contain Ag–Ab* and Ab*. Following a separation step, either Ag–Ab* or Ab* is quantified. CE may be used in this process to separate Ag–Ab* from Ab* and provide a means for quantitation.

The algal protein B-phycoerythrin is frequently used as a fluorescence tag in immunological assays. After immunological complex formation outside the instrument, CZE has been used in a competitive immunological assay to separate digoxigenin-labeled B-phycoerythrin from the immunological complex containing the labeled digoxigenin antigen and the Fab antibody (73). Analysis in serum at clinically useful levels of 10^{-9} to 10^{-10} M were achieved using laser-induced fluorescence (LIF). The detection limit was in the range of 10^{-11} M. A similar competitive binding assay using fluorescence-tagged antigen and LIF has been used in the determination of insulin (74), morphine (75), and cortisol (76). CZE seems to work well in the competitive binding assay format for the determination of small molecules because there is a substantial difference between Ag* and Ab–Ag*. Unfortunately this is not always true in the case of protein antigens. CZE frequently can not separate Ag* and Ab–Ag*. Another serious problem is that isoforms of the antibody will cause the Ab–Ag* or Ab*–Ag complex to appear as multiple peaks (77).

This problem has been approached in several ways. One is to use isoelectric focusing (IEF). Human growth hormone has been assayed directly with a detection limit of 0.1 ng/mL using fluorescent-labeled monoclonal antibody and capillary IEF with LIF detection (78). Another strategy is to alter the charge of the antibody. Through both fluorescent labeling and succinylation to produce Ab^{*n-}, it has been possible to increase the separation between the complex of immunoglobulin $A-Ab^{*n-}$ and Ab^{*n-} (79). A similar charge-shifting approach is to use a competitive binding assay in which the antigen is charge modified, i.e., Ag^{n-}, and the antibody is fluorescent labeled. In this case Ab^*-Ag^{n-} must be separated from Ab* (54). Still another charge-shifting scheme is to use two labeled antibodies. One antibody, Ab^{n-}, plays the role of shifting the electrophoretic mobility of the complex. The second, Ab*, serves as a fluorescent reporter. The complex $Ab^{n-}-Ag-Ab^*$ will have a very different mobility in CZE than Ab*; Ab^{n-} is prepared by conjugating a polyanionic species to the antibody directed against the antigen. Although monoclonal Fab antibody is best in most cases, it is frequently possible to use highly purified polyclonal antibodies. We should again note that in all cases cited above immunological complex formation was achieved outside the capillary and the role of the electrophoresis system was separation of "bound from free" species and quantitation.

The mobility shifting techniques described above can also be used with binding proteins other than antibodies, such as protein A and G. For example, the complex of IgG with fluorescent-tagged recombinant protein A has a different electrophoretic mobility than the labeled protein A alone (80). CZE with LIF detection was used to quantify IgG in crude cell culture medium by this scheme.

20.4.5. Hemoglobins

There is great interest in hemoglobins (Hb) for a number of reasons. One is that hemoglobins are the dominant oxygen-carrying proteins in mammals. Another is that mutations in humans induce variations in the structure of several hemoglobins. At certain levels of oxygenation and blood pH, the Hb S and Hb C variants undergo a conformational transition that manifests itself as a change in the shape of the red blood cell, i.e., the classic sickle shape in sickle-cell anemia. The various anemias that result from variant hemoglobins can be lethal. A sufficiently large segment of the human population carries mutations for Hb that infant populations are monitored in many countries to aid in identification, treatment, and health management programs.

Hb variants were first recognized by paper and gel electrophoresis. These electrophoretic methods were so slow that they have now been displaced by liquid chromatography. The interest in CZE is that it will provide an even faster, higher-resolution method for mass screening. The major Hb species of interest in sickle-cell anemia are Hb A_1 ($pI = 7.10$), Hb A_2 ($pI = 7.40$), Hb F ($pI = 7.15$), and Hb S ($pI = 7.25$). CZE separations have been achieved in uncoated capillaries between pH 8 and 9 by using 1.0 M Tris, 50 mM vernol, or 20 mM borate buffer with detection by absorbance at 415 nm (81). The best separation was obtained in borate buffer at pH 8.5. Problems such as incomplete resolution of sample components, poor reproducibility of retention time, and poor recovery are yet to be solved.

Thalassemias present a slightly different analytical problem in that differing ratios of globin chains are produced. This class of anemias is associated with reduced α-globin chain production and is characterized by the appearance of abnormal Hb species in varying amounts depending on the nature and severity of the genetic defect. Still other anemias from primary structure alteration in the β-chain (Hb E) and α-chain (Hb C) are found in Asian populations. Hb E, Hb C, and Hb G Philadelphia were distinguished by altered mobility of the individual globin chains in CZE under denaturing conditions using linear polyacrylamide-coated capillaries (82,83). Hemoglobins were denatured and separated in 10 mM phosphate buffer (pH 2.5) using 7 M urea and 0.1% Triton X-100.

20.4.6. rec-Proteins

Proteins produced by recombinant DNA technology are frequently accompanied by structural variants resulting from expression errors, *in vivo* modifications, posttranslational errors, improper folding, aggregation, and chemical modifications that occur during purification. The presence of variant forms of a protein in a human therapeutic product is undesirable because they may be of diminished biological activity or immunogenic. Federal regulatory agencies generally require demonstration that species of very similar structure are not present. Proving that this is true is a challenging analytical problem.

Human growth hormone (hGH) was one of the first therapeutic proteins to be produced biosynthetically by recombinant DNA technology. CZE has been applied to the problem of determining the absence of hGH variants in preparations. Isoforms resulting from primary structure cleavage, mono-deamidation at two different positions, amino terminal succinylation, and a His to Gln replacement at position 18 have been found in hGH produced by *Escherichia coli* (84,85). Separations were achieved in uncoated capillaries with 100 mM phosphate buffer (pH 6.0). A detection limit of 0.03% was calculated for impurities.

Recombinant human interleukin-4 (rhIL-4) is a cytokine that has been investigated for cancer therapy; rhIL-4 is a monomeric protein (M_r 15,400 Da) of $pI = 9.2$ with three intrachain disulfide bonds. CZE of rhIL-4 mixtures prepared by *in vitro* degradation was carried out with an uncoated capillary using 50 mM 1,3-diaminopropane and phosphate buffers with pH ranging from 4.5 to 8.0 (86). Resolution of degradation products by CZE appeared to be superior to that HPLC.

TNF-β, a tumor necrosis factor, is produced by stimulated lymphoid cells and has been recognized as able to kill tumor tissue. This cytokine has a relatively high pI (~ 9) and has been produced by recombinant DNA technology in both *E. coli* and mammalian cells. Recombinant human tumor necrosis factor (rhTNF-β) was analyzed by CZE with polybrene-coated capillaries at pH 4.5 with 50 mM acetate buffer (87). Polybrene is a relatively hydrophobic, cationic polymer frequently used in sequencers to adsorb peptides. It is surprising that it does not adsorb rhTNF-β in this case. Columns were flushed with polybrene after each run. The particular sample examined had at least three minor impurities, which were partially resolved by CZE. None of the impurities were identified, and so their structural relationship with rhTNF-β remains unknown.

Human antithrombin III (hAT III) is a single-chain glycoprotein that inhibits serine proteins of the blood-clotting cascade and is thus of therapeutic importance. After removal of immunoglobulins from the culture supernatant by protein G chromatography, hAT III was determined by CZE in an uncoated capillary at pH 2.0 with 50 mM phosphate buffer using 0.1% hydroxypropyl methylcellulose (88). CZE was used to monitor both the biosynthesis and purification of hAT III. Heparin-stimulated intermolecular complex formation between hAT III and thrombin was monitored with capillary gel electrophoresis.

Hirudin is a natural 65-amino-acid polypeptide ($pI = 4.4$) of approximately 7 kDa that is derived from the leech (*Hirudo medicinalis*). It is a potent thrombin inhibitor and as a consequence is used as an anticoagulant. During the production of the recombinant desulfatohirudin variant (r-Hr) for anticoagulation therapy in humans by recombinant DNA technology, seven variants may result as either by-products or degraded forms of r-Hr. Based on the electrophoretic mobility of synthetic analogues, all the major variants were separated in an uncoated capillary with 60 mM acetate buffer (pH 4.4) containing 0.3% PEG 20,000 and 0.1 mM Zn^{2+} (89).

20.4.7. Food Proteins

CZE has become a valuable technique for the analysis of milk proteins (90–93). Bovine milk contains 3–3.5% protein, of which 80% consists of caseins. Serum proteins make up the remaining 20%. The casein fraction can be divided in α_1-, α_2-, β-, and κ-casein components. The major serum proteins are β-lactoglobulin (A, B, and C) and α-lactoglobulin. Using uncoated capillaries and either 10 mM phosphate (pH 7.4) or 150 mM borate (pH 8.5), with 0.05% Tween 20, it is possible to at least partially resolve whey into five major peaks: α-lactoglobulin, β-lactoglobulin A, β-lactoglobulin B, bovine serum albumin (BSA), and IgG (94). Although the β-lactoglobulin variants and BSA are only partially resolved in this case, it is possible to separate all the variants of β-lactoglobulin in the absence of BSA by changing the separation conditions slightly. All three β-lactoglobulin variants were separated from each other and from other whey proteins in an uncoated capilllary using a 50 mM 2-morpholinoethanesulfonic acid (MES) buffer (pH 8.0) with 0.1% Tween 20 (95). This is remarkable in that β-lactoglobulin A and B vary by an Asp to Gly substitution at position 64 and a Val to Ala substitution at position 118, respectively. β-Lactoglobulin C has the same substitution as the B variant with an additional Gln to His substitution at position 59. The C variant is an important indicator of the functional properties of milk even though present in small quantities.

Analysis of fish muscle sarcoplasmic (FMS) proteins is of major interest for species identification as used in regulation, quality issues, and pricing. FMS proteins were profiled in < 9 min using uncoated capillaries with 50 mM phosphate buffer between pH 6.6 and 7.5 (96). Between runs, capillaries were sequentially washed under pressure with 1.0 N NaOH, water, and the operating buffer. This method allowed frequently encountered species to be differentiated and the impact of storage conditions to be monitored.

Cereal proteins are also widely monitored as indicators of agronomic performance, functional properties of the derived proteins, and food quality. The strong relationship between plant variety and food quality makes varietal identification extremely important. Protein profiling is one of the most dependable approaches to varietal identification. Wheat cultivar differentiation has now been examined with CZE by a number of laboratories (97–100). A typical CZE method uses an uncoated capillary and 100 mM phosphate buffer (pH 2.5) with 30% ethanol and an unspecified polymer additive from BioRad (100). It was found with gliadins from hard red winter wheat, hard red spring wheat, and soft wheat classes that cultivars from closely related sister lines or intercrossings were readily differentiated whereas cultivars that were not genetically close exhibited very different electropherograms in 10 min analyses. After each run the capillary was rinsed under pressure for

4 min each with 1 M nitric acid, 0.1 N NaOH, distilled water, and running buffer.

Rapid characterization of soy protein and soy hydrolysates by CZE with borate buffer at pH 8.5 have shown that CE can be used effectively for monitoring protein hydrolysis during processing and for fingerprinting various types of protein products (101).

20.4.8. Structure Analysis/Folding

Protein folding *in vitro* is not well understood. *In vitro* folding generally occurs immediately after the polypeptide leaves the ribosome during biosynthesis or is excreted from the cell. One reason for the great interest in protein folding today is that polypeptides are frequently stored as inactive structural conformers in genetically engineered host cells. They must be refolded *in vitro* to reach their native 3D structure and become biologically active. The 3D structure may also be altered during purification, formulation, or storage.

One of the problems associated with studying protein folding with separation systems is that such folding may be occurring in a time frame shorter than the time needed for analysis. This frequently precludes analyses with conventional gel electrophoresis. CZE is attractive for folding studies because it is much faster. Trypsinogen provides a model system in which it has been shown that fully reduced protein at neutral pH will eventually reform native 3D structure with the accompanying disulfide bonds. CZE has been found to differentiate native trypsinogen from both the reduced and partially refolded intermediates (102). This fact has been exploited in assessing the conformational stability of iron-free transferrin in the presence of increasing concentrations of urea (103). Capillary IEF (CIEF) has also been used to look at the problem of transferrin structure (104). One of the problems with folding studies of glycoproteins that have multiple glycoforms, such as transferrin, is that it is difficult to differentiate between the structural transitions of the variants.

20.4.9. Miscellaneous Topics

Metabolites can be generated as products of intermediary metabolism that interact with proteins in hair. As a consequence, these hair keratin conjugates can provide a history of substances consumed and their impact on metabolism. When CZE was used to analyze the alkali-extractable proteins, noticeable differences were observed between controls and alcohol-consuming rats (105). Analyses carried out on the "high-sulfur" proteins at pH 3.5 with 50 mM phosphate buffer in uncoated capillaries revealed the presence of two sharp peaks absent in the controls. Similar CZE separations on the "low-sulfur" proteins at pH 9.2 with 50 mM borate buffer showed a single new protein. It was suggested that CZE could provide both a new tool for the study of keratin protein modification and a method to monitor alcohol abuse.

20.5. CAPILLARY ISOELECTRIC FOCUSING

The unique, structure-specific isoelectric point (pI) of proteins is due to the following: (i) they are composed of both cationic and anionic amino acids; (ii) the ratio of cationic to anionic amino acids varies between proteins; (iii) the ionization of amino acids varies with the environment in which they are located in the 3D structure of the protein; and (iv) the environment and electrostatic shielding around amino acids varies with structural conformation. Some other unique features of proteins are as follows: (v) above their pI they are negatively charged; (vi) below their pI they are positively charged; and (vii) the degree to which they accumulate charge and electrophoretic mobility (μ) as a function of pH [$d\mu/d$(pH)] is related to their amino acid composition and structure. These unique properties of proteins make them ideal candidates for a separation mechanism based on the isoelectric point (106–124). An example of the CIEF separation of monoclonal antibodies is seen in Figure 20.4 (123).

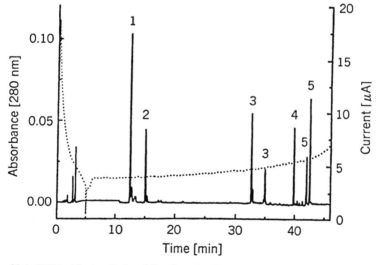

Figure 20.4. CIEF with chemical mobilization. Capillary: μ-SIL DB-1, 27 cm length. Carrier ampholyte solution: 4% Pharmalyte 3–10, 1% TEMED in 0.8% methylcellulose (MC). Analyte: 20 mM H$_3$PO. Catholyte: 20 mM NaOH (focusing)/20 mM NaOH + 30 mM NaCl (mobilization). Focusing for 5 min at 20 kV; cathodic mobilization at 20 kV. Peaks: (1) cytochrome c; (2) RNase A; (3) myoglobin; (4) carbonic anhydrase; (5) β-lactoglobulins A and B. [From C. Schwer, *Electrophoresis* **16**, 2121 (1995), (132).]

Numerous applications of CIEF have now been examined, particularly with regard to isoforms of proteins. It has been reported that CIEF can resolve at least seven glycoforms of recombinant human tissue plasminogen activator (125). Presumably these species differ in sialylation. Isoenzymes of arginase have also been resolved by CIEF (126). In another case, the absence of structural variants of recombinant granulocyte macrophage colony-stimulating factor was confirmed by CIEF (127). Based on the historic use of gel-based IEF to examine hemoglobin variants, CIEF has been investigated as a tool for monitoring hemoglobinopathies and glaciated hemoglobins associated with diabetes (82,112,114,128,129). High-resolution separations comparable to conventional gel-based systems were obtained in all cases. CIEF has also been used in histone fractionation (130).

20.6. CAPILLARY GEL ELECTROPHORESIS

20.6.1. Sodium Dodecyl Sulfate–Capillary Gel Electrophoresis

As discussed below in section 20.8, the function of filling the electrophoresis channel with a cross-linked gel or soluble polymers is to increase the size selectivity of the system. Protein chemists of the 1960s needed a simple, rapid technique to estimate the molecular weight of proteins. Because native proteins vary in charge, shape, and size, it is difficult to estimate their molecular weight by gel electrophoresis. The solution to this "problem" was to denature them with 0.1% SDS. Denaturation under reducing conditions with SDS generally causes all proteins to assume the same globular shape and have a constant mass-to-charge ratio. The only remaining difference is their size, which is related to molecular weight (M_r). Linear plots of electrophoretic mobility vs. log M_r are generally obtained across one log of M_r (Figure 20.5) (131).

SDS–CGE can vary from conventional SDS–PAGE in several important respects (132,133). One is that CGE channels do not have to be filled with cross-linked gel as in the case SDS–PAGE. Sieving effects may be achieved equally well with non-cross-linked polymers (134–136). The second is that a broad variety of polymers are available in CGE. PEO (137,138), dextrans (139,140) pullulan (141), and linear polyacrylamide (142) are some of the polymers that have been used effectively (143). In general, proteins which vary some 5–20% in molecular weight may be resolved with SDS–CGE (144). Resolution is poorer at the sieving limits of the system. Glycoproteins present a problem because they do not adsorb SDS to the same extent as do non-glycosylated proteins. SDS–CGE provides some major advantages over SDS–PAGE. First, it is much faster. CGE separations are accomplished in

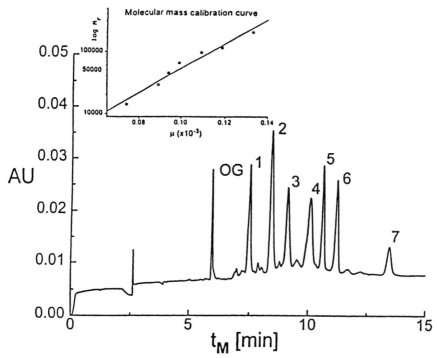

Figure 20.5. SDS–CGE pattern of a test mixture of seven proteins on an eCAP SDS 14–200 gel. Peaks: (1) α-lactalbumin; (2) carbonic anhydrase; (3) ovalbumin; (4) BSA; (5) phosphorylase b; (6) β-galactosidase; (7) myosin; [Reprinted from A. Guttman, P. Shieh, J. Lindahl, and N. Cooke, *J. Chromatogr. A* **676**, 227 (1994), with permission.]

10–20 min. Second, multiple separations may be executed on a single column. And third, the gel may be replaced in soluble polymer CGE systems. This is a vary important attribute. Cross-linked capillary PAGE columns often have "air bubble" formation problems and rapidly deteriorate after a few days of use. This is not a problem with replaceable gels.

Although not of particularly high resolution, SDS–PAGE has been the dominant form of analytical electrophoresis used by life scientists to determine protein purity. Matrix-assisted laser desorption ionization–mass spectrometry (MALDI–MS) is a far superior technique for determining protein molecular weight and will no doubt displace SDS–CGE and and SDS–PAGE for this purpose in the future. The utility of MALDI–MS in determining protein purity remains to be determined.

20.6.2. Native Capillary Gel Electrophoresis

PAGE or agarose gel electrophoresis under native conditions has been a widely used variant of gel electrophoresis. Because SDS generally dissociates multiple-subunit proteins, it is impossible to access the native shape and structure of a protein electrophoretically. The problem in estimating the size of native proteins by gel electrophoresis is that they vary widely in net charge and charge-to-size ratio. Thus, there is no relationship between electrophoretic mobility and size, as is true in the case of SDS–CGE. This problem is overcome by examining the electrophoretic mobility of the target protein and several standards in a series of gels that vary in gel density, i.e., sieving properties. When electrophoretic mobility is plotted against gel density in what is known as a "Ferguson plot," the slope of the curve is related to molecular size. Although tedious, this simple electrophoretic approach is much less expensive and simpler than analytical ultracentrifugation or light-scattering methods.

Theoretically, CGE of native proteins should be of higher resolution than CZE because the former probes all the important structural features of the protein: charge, shape, and size. Experimentally, this seems to be true (144).

20.7. MICELLAR ELECTROKINETIC CHROMATOGRAPHY

As has been discussed in other chapters of this book, MEKC involves the partitioning of analytes into mobile phase micelles according to their hydrophobicity and ability to form hydrogen bonds with the surfactant composing the micelle. It is generally assumed that substances of greater than 5000 Da do not partition between micelles and the solution. For example, SDS essentially coats proteins in a 1.4/1.0 (w/w) ratio. It has been reported that this is not the case with metallothionein (MT), the M_r of which is approximately 6500 Da (145). Uncoated capillaries resolved four isoforms of sheep MT at pH 8.5 with 20–300 mM borate buffer and 75 mM SDS. Optimum separation conditions were slightly different for rabbit MT isoforms.

20.8. SELECTIVITY

Selectivity in CE depends on both protein structure and the separation method. The equation $\mu = q/6\pi\eta a$ indicates that electrophoretic mobility (μ) is determined by the net charge (q) of the analyte, viscosity of the medium (η), and a size term (a). Anything that causes either q or a to vary within a mixture of proteins will obviously alter selectivity. Because gel and zone electrophoretic separations are based on electrophoretic mobility, selectivity may be manipulated by charge and size alterations of proteins. Proteins differ widely in q and relative charge as a function of pH. This means that very large selectivity differences may be achieved in both CZE and CGE by pH alterations. Using insulin and its acylated derivatives as a model, Gao et al. have validated the concept that addition of single charges in a protein impact electrophoretic mobility (146).

It is also possible to alter the relative charge of a protein with pairing agents as noted above in Section 20.4 on CZE. Polyamines pair with carboxyl groups and increase the cationic character of a protein. In similar fashion, polyanions pair with proteins and increase their anionic character. The degree to which charge is altered is a function of the number of pairing agents associated with the protein. Phytate and borate complexation are both examples in which the anionic character of proteins is increased. Hydrophobic pairing with either cationic or anionic surfactants can also alter selectivity as noted above.

Size- and shape-induced changes in proteins are achieved in a number of ways. One is by a change in structural conformation. Although generally large, these changes may also be small and subtle. In the case of staphylococcal nuclease A, single amino acid substitutions caused slight differences in hydrodynamic radius of the protein that are detectable by CZE (147). It has been shown that for the native form of this enzyme, the impact on electrophoretic mobility of a 2 Å change in the molecular radius is equivalent to that of a unit change in charge (148). Pairing agents and counterions frequently alter surface solvation and thus the size of a protein. Finally, there is the case where such large amounts of pairing agent are added that there is a global change in protein strucutre. SDS denaturation is such a case.

Selectivity is much harder to alter in CIEF. Only by complexation with ampholyte or urea-induced conformation changes can the structure and pI of a protein be changed.

20.9. MASS SPECTROMETRY

Interfacing CE systems to mass spectrometers has been discussed elsewhere in this monograph (see Chapter 12) and will not be discussed here. The focus here will be with the types of information MS supplies relative to protein structure. MS of proteins today centers primarily on two types of ionization techniques: electrospray ionization (ESI) and MALDI (149). Californium-252 (^{252}Cf) plasma desorption MS (150) and fast atom bombardment (151) are also used with CE but will probably decline in popularity because they are not well supported commercially.

Although protein molecular weight is the primary information obtained from CE–MS systems today, it is very valuable in the case of proteins produced by recombinant DNA technology. Based on the DNA sequence, it is possible to predict the amino acid sequence of proteins. The objective of the MS analysis is generally to confirm, not to elucidate, the structure. When coupled to a high-resolution separation system, mass spectrometers can easily determine the molecular weight of all components, detect proteolysis at either end of the protein, identify deamidation, find glycosylation variants, and with peptide mapping confirm primary structure, disulfide bond position, and site-specific glycosylation (149).

MALDI is based on the use of laser energy to ionize proteins that have been deposited in a light-absorbing matrix on the surface of an electrode in the ion source of a mass spectrometer, generally a time-of-flight (TOF) type. When a laser pulse is focused on the electrode plate, the matrix absorbs energy from the beam and causes a cloud of ions, including analyte, to be expelled from the electrode (152). These ions are accelerated away from the electrode and mass analyzed in the mass spectrometer. MALDI is a "soft" ionization technique that produces predominantly singly charged ions of m/z ratio ranging up to 100,000 or higher (153). This has the advantage that individual ions from mixtures may be relatively easily discriminated, as in the case of direct mass spectral analysis of tryptic digests (154). Identification of glycosylation variants on a single tryptic peptide is another example of the value of being able to analyze multiple species in a sample by MALDI–MS. The problem in the analysis of mixtures by MALDI is that all analytes are not ionized with equal efficiency or they quench each other. At the present time it has not been possible to build a continuous-flow MALDI interface that allows high sensitivity. Thus, samples must be collected from the CE and introduced into MALDI–TOF mass spectrometers (Figure 20.6) (154–156). Special fraction collectors for CZE (157) and CIEF systems (158) have even been built.

ESI, in contrast to MALDI, is inherently a continuous-flow interface. Effluent is sprayed under potential from the end of a capillary where the droplets aquire charge (159–161). Charge repulsion within the droplet causes it to undergo multiple disintegrations during solvent evaporation, concentrating the charge. Finally, when all solvent has been removed, the remaining macromolecular ions are multiply charged. The advantage of ESI is that multiply charged macromolecules may be mass analyzed at relatively low mass. This means that ESI can be used for the analysis of high-molecular-weight analytes with MS instruments of low mass range [see Figure 20.7 (162); also Cole et al. (163)]. The molecular weight of the parent ion is computed from the mass of these multiply charged species. Sophisticated algorithms even allow the analysis of mixtures of analytes, although the analysis of mixtures is more difficult.

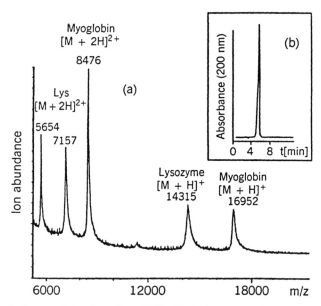

Figure 20.6. Aminopropyltrimethoxysilane (APS)–CE–MALDI–MS of hen egg white lysozyme: (a) MALDI–MS with 1 pmol of myoglobin spiked onto the target as the internal standard; (b) APS–CE separation. [Repinted from W. Weinmann, C. E. Parker, L. J. Deterding, D. I. Papac, J. Hoyes, M. Przybylski, and K. B. Tomer, *J. Chromatogr. A* **680**, 353 (1994), with permission.]

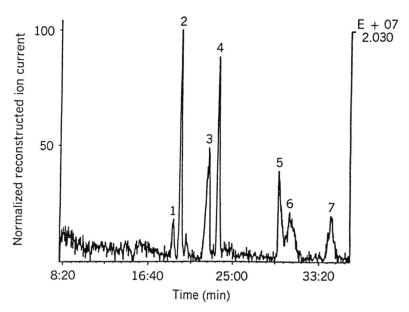

Figure 20.7. CZE–MS full scan (m/z 600–2000) reconstructed ion electropherogram of a 150 nL injection of 12 μm of each the following– (1) lysozyme; (2) cytochrome *c*; (3) RNase; (4) myoglobin; (5) β-lactoglobulin A ; (6) β-lactoglobulin B; (7) carbonic anhydrase. [Reprinted from T. J. Tnompson, F. Foret, P. Vouros, and B. L. Karger, *Anad. Chem.* **65**, 900 (1993), with permission.]

In addition to ESI interface design issues, attention must also be given to additives used to suppress column effects, background electrolytes (BGE) in the CE buffer, and liquid sheath ionic (LSI) species in the sheathing liquid added in the interface. Migration of counterions from the sheath liquid into the column exit can form ionic boundaries that move back against the oncoming protein zones. These ionic boundaries can either delay or invert elution order and contribute to loss of resolution (164). This problem is most severe with columns of low EOF. Strategies to deal with this problem are the use of the following: (i) a counterion common that both the BGE and LSI; (ii) an LSI counterion with a pK_a and electrophoretic mobility similar to that in the BGE; (iii) hydrodynamic counter flow that moves faster than the boundary; or (iv) a capillary of higher EOF.

Molecular weight by itself carries relatively little structural information. Ladders of ions derived from cleavages at peptide bonds enhance sequencing efforts by MS. Fragmentation of specific ions within the mass spectrometer by collision-induced dissociation (CID) has been widely used to sequence polypeptides. When proteolytic digests are separated on a 20 μm i.d. column and the individual peptides analyzed by ESI–MS/MS with CID, it is possible to derive sequence information at the subfemtomole level (165). A new approach is to couple CE through an electrospay interface to a Fourier transform ion cyclotron resonance (FTICER) mass spectrometry with sustained off-resonance irradiation (166). This allows multiple irradiation frequencies to be broadcast simultaneously, which yields fragmentation of species at different m/z ratio values from the same waveform. Initial studies on gramicidin S and mellifin produced structural information. Further work with this CE–ESI–FTICR–MS instrument has shown that mass spectra could be obtained from both the α-and β-globin chains of hemoglobin obtained from 5 to 10 human erythrocytes (167).

20.10. THE FUTURE

A probable area where major advances will be made in the future involves the use of separation systems, enzymology, immunology, and MS together to study protein structure. Although this has already begun in the case of primary structure analysis (168,169), immobilized enzymes either directly on the MALDI plate or in tandem with an CE–MS or LC–MS system will revolutionize glycoprotein analysis. Because so many monosaccharides are of identical mass, it is only when coupled with the hydrolytic specificity of combinations of enzymes that the mass spectrometer can identify oligosaccharide structure. Incorporating immobilized antibodies into CE–MS and LC–MS systems will be equally efficacious. As in the case of enzymes, the immunosorbent recognizes structures that are then analyzed by the mass spectrometer (170). It is also likely that future separations will be carried out with microfabricated systems in which entire separations are achieved in a few minutes or less, i.e., peak widths will be 1–2s in width. Scan times of 0.1s or less will be necessary in future generations of CE–MS instruments to aquire sufficient scans across a peak for single-ion monitoring (171).

LIST OF ACRONYMS AND ABBREVIATIONS

Acronym or Abbreviation	Definition
Ab	antibody
Ag	antigen
AIBN	2,2′-azobisobutyronitrile
AMC	allylmethylcellulose
apo(a)	apolipoprotein a
apo A	apolipoprotein A
apoA-I	apolipoprotein A-I
APS	aminopropyltrimethoxy-silane
AU	absorbance units
BGE	background electrolytes
BSA	bovine serum albumin
CDA	cellulose diacetate
CE	capillary electrophoresis
CGE	capillary gel electrophoresis
CHD	coronary heart disease
CID	collision-induced dissociation
CIEF	capillary isoelectric focusing
CSF	cerebrospinal fluid
CTA	cellulose triacetate
CZE	capillary zone electrophoresis
2D,3D	two-, three-dimensional
ELISA	enzyme-linked immunosorbent assay
EOF	electroosmotic flow
ESI	electrospray ionization
FMS	fish muscle sarcoplasmic (proteins)
FS	fused-silica (capillary)
FTICR	Fourier transform ion cyclotron resonance
G-CSF	granulocyte colony-stimulating factor
hAT III	human antithrombin III
Hb	hemoglobin

hCG	human chorionic gonadotropin
HDL	high-density lipoproteins
HEMA	2-hydroxyethyl methacrylate
hGH	human growth hormone
HPC	hydroxypropyl cellulose
HSA	human serum albumin
IEF	isoelectric focusing
IgG	immunoglobulin G
LC	liquid chromatography
LDL	low-density lipoproteins
LIF	laser-induced fluorescence
Lp(a)	lipoprotein a
Lp(a$^-$)	reduced lipoprotein a
LSI	liquid sheath ionic (species)
MALDI	matrix-assisted laser desorption
MC	methylcellubse ionization
MES	2-morpholinoethanesulfonic acid
MS	mass spectrometry
MT	metallothionein
NDA	naphthylene-2,3-dicarboxyaldehyde
OES	7-oct-1-enyltrimethoxysilane
OG	orange-G
PAGE	polyacrylamide gel electrophoresis
PEG	poly(ethylene glycol)
PEI	polyethyleneimine
PEO	poly(ethylene oxide)
PNGase	polypeptide N-glucosidase
PVA	poly(vinyl alcohol)
PVI	polyvinylimidazole
PVP	polyvinylpyrrolidone
rFVIIa	(human) recombinant factor VIIa
rHIL-4	recombinant human interleukin-4
RIA	radioimmunoassay
r-Hr	recombinant desulfatohirudin variant
rhTNF-β	recombinant human tumor necrosis factor β
rtPA	(human) recombinant tissue plasminogen activator radioimmunoassay
SDC	sodium deoxycholate
SDS	sodium dodecyl sulfate
TEMED	N,N,N',N'-tetramethylethylenediamine
TOF	time-of-flight (mass spectrometer)
UV	ultraviolet
VLDL	very-low-density lipoproteins

REFERENCES

1. K. K. Unger, N. Becker, and P. Roumeliotis, *J. Chromatogr.* **125**, 115 (1976).
2. K. K. Unger, Packings and Stationary Phases in Chromatographic Techniques. Dekker, New York, 1979.
3. J. K. Town and F. E. Regnier, *Anal. Chem.* **63**, 1126 (1991).
4. John M. Heimlich, Ph. D. thesis, Department of Pharmacy, Purdue University, West Lafayette, IN (1997).
5. J. K. Towns and F. E. Regnier, *Anal. Chem.* **64**, 2473 (1992).
6. J. W. Jorgenson and K. D. Lukacs, *Science* **222**, 266 (1983).
7. H. Engelhardt and M. A. Cunat-Walter, *J. Chromatogr. A* **717**, 15 (1995).
8. A. Cifuentes, J. M. Santos, M. deFrutos, and J. C. Diez-Masa, *J. Chromatogr. A.* **652**, 161 (1993).
9. M. E. Legaz and M. M. Pedrosa, *J. Chromatogr. A* **719**, 159 (1996).
10. D. Schmalzing, C. A. Piggee, F. Foret, E. Carrilho, and B. L. Karger, *J. Chromatogr. A* **652**, 149 (1993).
11. M. A. Strege and A. L. Lagu, *J. Chromatogr.* **630**, 337 (1993).
12. L. Song, Q. Ou, and W. Yu, *J. Chromatogr.* **657**, 175 (1993).
13. D. Corradini and G. Cannarsa, *Electrophoresis* **16**, 630 (1995).
14. Y. J. Yao and S. F. Y. Li, *J. Chromatogr. A* **663**, 97 (1994).
15. R. P. Oda, B. J. Madden, T. C. Spelsberg, and J. P. Landers, *J. Chromatogr. A* **680**, 85 (1994).
16. N. K. Klausen and T. Kornfelt, *J. Chromatogr. A* **718**, 195 (1995).
17. G. N. Okafo, H. C. Birrell, M. Greenaway, M. Garan, and P. Camilleri, *Anal. Biochem.* **219**, 201 (1994).
18. C. L. Ng, H. K. Lee, and S. F. Y, Li, *J. Chromatogr. A* **659**, 427 (1994).
19. C. P. Desilets, M. A. Rounds, and F. E. Regnier, *J. Chromatogr.* **544**, 25 (1991).
20. X. W. Yao, D. Wu, and F. E. Regnier, *J. Chromatogr.* **636**, 21 (1993).
21. T. Tadey and W. C. Purdy, *J. Chromatogr.* **652**, 131 (1993).
22. M. R. Karim, S. Shinagawa, and T. Takagi, *Electrophoresis* **15**, 1141 (1994).
23. M. Gilges, M. H. Kleemiss, and G. Schomburg, *Anal. Chem.* **66**, 2038 (1994).
24. M. Iki and E. S. Yeung, *J. Chromatogr. A* **731**, 273 (1996).
25. J. F. Kenny and G. W. Willcockson, *J. Polym. Sci., Part A-1* **4**, 679 (1966).
26. M. H. A. Busch, J. A. Kraak, and H. Poppe, *J. Chromatogr. A* **695**, 287 (1995).
27. J. K. Towns and F. E. Regnier, *J. Chromatogr.* **516**, 69 (1990).
28. Z. Zhao, A. Malki, and M. L. Lee, *Anal. Chem.* **65**, 2747 (1993).
29. L. A. Colon, Y. Guo, and A. Fermier, *Anal. Chem.* **69**, 461A (1997).
30. S. Hjertén and K. Kubo, *Electrophoresis* **14**, 390 (1993).
31. M. Huang and M. L. Lee, *J. Microcolumn Sep.* **4**, 491 (1992).
32. M. Huang, J. Plocek, and M. Novotony, *Electrophoresis* **16**, 396 (1995).

33. N. Nishioka, K. Minami, and K. Losai, *Polym. J.* **13**, 591 (1983).
34. S. Hjerten, *J. Chromatogr.* **347**, 191 (1985).
35. M. Nakatani, A. Shibukawa, and T. Nakagawa, *Electrophoresis* **16**, 1451 (1995).
36. K. A. Cobb, V. Dolnik and M. Novotny, *Anal. Chem.* **62**, 2478 (1990).
37. H. Engelhardt and M. A. Cunat-Walter, *J. Chromatogr. A* **716**, 27 (1995).
38. M. C. Montes, C. VanAmen, J. J. Pesek, and J. E. Sandoval, *J. Chromatogr. A* **688**, 31 (1994).
39. M. Chiari, M. Nesi, J. E. Sandoval, and J. J. Pesek, *J. Chromatogr. A* **717**, 1 (1995).
40. A. Cifuentes, M. deFrutos, J. M. Santos, and J. C. Diez-Masa, *J. Chromatogr. A* **655**, 63 (1993).
41. P. Sun, A. Landman, G. E. Barker, and R. A. Hartwick, *J. Chromatogr. A* **685**, 303 (1994).
42. J. K. Towns, J. M Bao, and F. E. Regnier, *J. Chromatogr.* **599**, 227 (1992).
43. Y. Liu, R. Fu, and J. Gu, *J. Chromatogr. A* **698**, 498 (1995).
44. Y. Mechref and Z. El Rassi, *Electrophoresis* **16**, 617 (1995).
45. R. J. Xu, C. Vidal-Madjar, B. Sebille, and J. C. Diez-Masa, *J. Chromatogr. A* **730**, 289 (1996).
46. J. T. Smith and Z. El Rassi, *Electrophoresis* **14**, 396 (1993).
47. D. Corradini, A. Rhomberg, and C. Corradini, *J. Chromatogr. A* **661**, 305 (1994).
48. J. P. Landers, R. P. Oda, B. J. Madden, and T. C. Spelsberg, *Anal. Biochem.* **205**, 115 (1992).
49. A. Emmer, M. Jasson, and J. Roeraade, *J. High Resolut. Chromatogr.* **14**, 738 (1991).
50. A. Emmer, M. Jansson, and J. Roeraade, *J. Chromatogr. A* **672**, 231 (1994).
51. K. Altria and C. Simpson, *Anal. Proc.* **23**, 453 (1986).
52. T. Tsuda, HRC & CC, *J. High Resolut. Chromatogr. Chromatogr. Commun.* **10**, 622 (1987).
53. X. Huang, J. A. Luckey, M. J. Gordon and R. N. Zare, *Anal. Chem.* **61**, 766 (1989).
54. M. J. Gordon, K. J. Lee, A. A. Arias, and R. N. Zara, *Anal. Chem.* **63**, 69 (1991).
55. F. T. A. Chen, C. M. Liu, Y. Z. Hsieh, and J. C. Sternberg, *Clin. Chem. (Winston-Salem. N. C.)* **37**, 14 (1991).
56. F. T. A. Chen and J. C. Sternberg, *Electrophoresis* **15**, 13 (1994).
57. R. Lehmann, H. Liebich, G. Grubler, and W. Voelter, *Electrophoresis* **16**, 998 (1995).
58. V. Dolnik, *J. Chromatogr. A* **709**, 99 (1995).
59. G. Cowdry, M. Firth and G. Firth, *Electrophoresis* **16**, 1922 (1995).
60. K. A. Denton and R. Harris, *J. Chromatogr. A* **705**, 335 (1995).
61. D. E. Morbeck, B. J. Madden, and D. J. McCormick, *J. Chromatogr. A* **680**, 217 (1994).
62. H. B. Hines and E. E. Brueggemann, *J. Chromatogr. A* **670**, 199 (1994).

63. P. L. Weber, C. J. Bramich, and S. M. Lunte, *J. Chromatogr. A* **680**, 225 (1994).
64. M. M. Schmerr and K. R. Goodwin, *J. Chromatogr. A* **652**, 199 (1993).
65. P. Hermentin, R. Doenges, R. Witzel, C. H. Hokke, J. F. G. Vliegenthart, J. P. Kamerling, H. S. Conradt, M. Nimtz, and D. Brazel, *Anal. Biochem* **221**, 29 (1994).
66. A. Z. Hu, I. D. Cruzado, J. W. Hill, C. J. McNeal, and R. D. Macfarlane, *J. Chromatogr. A* **717**, 33 (1995).
67. A. Goux, A. Athias, L. Persegol, L. Lagrost, P. Gambert, and C. Lallemant, *Anal. Biochem.* **218**, 320 (1994).
68. H. M. Liebich, R. Lehmann, A. E. Weiler, G. Grubler, and W. Voelter, *J. Chromatogr. A* **717**, 25 (1995).
69. W. Weinmann, C. Maier, K. Baumeister, M. Przybylski, C. E. Parker, and K. B. Tomer, *J. Chromatogr. A* **664**, 271 (1994).
70. A. M. Arentoft, H. Frokiaer, S. Michaelsen, H. Sorensen, and S. Sorensen, *J. Chromatogr. A* **652**, 189 (1993).
71. P. R. Banks and D. M. Paquette, *J. Chromatogr. A* **693**, 145 (1995).
72. H. Frokiaer, P. Moller, H. Sorensen, and S. Sorensen, *J. Chromatogr. A* **680**, 437 (1994).
73. F.-T. A. Chen and S. L. Pentoney, *J. Chromatogr. A* **680**, 425 (1994).
74. N. M. Schultz and R. T. Kennedy, *Anal. Chem.* **65**, 3161 (1993).
75. R. A. Evangelista and F.-T. A. Chen, *J. Chromatogr. A* **680**, 587 (1994).
76. M. Fuchs, *Anal. Chem.* (in press).
77. R. Vincentelli and N. Bihoreau, *J. Chromatogr.* **641**, 383 (1993).
78. K. Shimura and B. Karger *Anal. Chem.* **66**, 13 (1994).
79. F.-T. A. Chen, *J. Chromatogr. A* **680**, 419 (1994).
80. R. Lausch, O.-W. Reif, P. Riechel, and T. Scheper, *Electrophoresis* **16**, 636 (1995).
81. A. Sahin, Y. R. Laleli, and R. Ortancil, *J. Chromatogr. A* **709**, 121 (1995).
82. M. Zhu, T. Wehr, V. Levi, R. Rodriguez K. Shiffer, and Z. A. Cao, *J. Chromatogr. A* **652**, 119 (1993).
83. M. Castagnola, I. Messana, L. Cassiano, R. Rabino, D. V. Rossetti, and B. Giardina, *Electrophoresis* **16**, 1492 (1995).
84. J. Frenz, S.-L. Wu, and W. Hancock, *J. Chromatogr.* **480**, 379 (1989).
85. P. Dupin, F. Galinou, and A. Bayol, *J. Chromatogr. A* **707**, 396 (1995).
86. J. Bullock, *J. Chromatogr.* **633**, 235 (1993).
87. Y. J. Yao, K. C. Loh, M. C. M. Chung, and S. F. Y. Li, *Electrophoresis* **16**, 647 (1995).
88. O.-W. Reif and R. Freitag, *J. Chromatogr, A* **680**, 383 (1994).
89. C. Dette and H. Watzig, *J. Chromatogr. A* **700**, 89 (1995).
90. F. T. A. Chen and J. H. Zang, *J. Assoc. Off. Anad. Chem.* **75**, 905 (1992).
91. A. Cifuentes, M. deFrutos, and J. C. Diez-Masa, *J. Dairy Sci.* **76**, 1870 (1993).
92. N. DeJong, S. Visser, and C. Olieman, *J. Chromatogr. A* **652**, 207 (1993).
93. F.-T. A. Chen and A. Tusak, *J. Chromatogr. A* **685** 331 (1994).

94. N. M. Kinghorn, C. S. Norris, G. R. Paterson, and D. E. Otter, *J. Chromatogr. A* **700**, 111 (1995).
95. G. R. Paterson, J. P. Hill, and D. E. Otter, *J. Chromatogr. A* **700**, 105 (1995).
96. E. L. LeBlanc, S. Singh, and R. J. LeBlanc, *J. Food Sci.* **59**, 1267 (1994).
97. J. A. Bietz and E. Schmalzried, *Cereal Foods World* **37**, 555 (1992).
98. J. A. Bietz and E. Schmalzried, *Cereal Foods World* **38**, 615 (1993).
99. W. E. Werner, J. E. Wictorowicz, and D. D. Kasarda, *Cereal Chem.* **71**, 397 (1994).
100. G. Lookhart and S. Bean, *Cereal Chem.* **72**, 42 (1995).
101. T. M. Wong, C. M. Carey, and S. H. C. Lin, *J. Chromatogr. A* **680**, 413 (1994).
102. M. A. Strege and A. L. Lagu, *J. Chromatogr. A* **652**, 179 (1993).
103. F. Kilar and S. Hjertén, *J. Chromatogr.* **638**, 269 (1993).
104. J. Wu and J. Pawliszyn, *J. Chromatogr. A* **652**, 295 (1993).
105. D. Jelinkova, Z. Deyl, I. Miksik, and F. Tagliaro, *J. Chromatogr. A* **709**, 111 (1995).
106. H. Svensson, *Acta Chem. Scand.* **15**, 325 (1961).
107. J. Wu and J. Pawliszyn, *J. Anal. Chem.* **64**, 224 (1992).
108. J. Wu and J. Pawliszyn, *Anal. Chem.* **66**, 867 (1994).
109. J. Wu and J. Pawliszyn, *J. Liq. Chromatogr.* **16**, 1891 (1993).
110. J. Wu and J. Pawliszyn, *Anal. Chem.* **67**, 2010 (1995).
111. J. R. Mazzeo, J. A. Martineau, and I. S. Drull, *Anal. Biochem.* **208**, 323 (1993).
112. X.-W. Yao and F. E. Regnier, *J. Chromatogr.* **632**, 185 (1993).
113. S. Molteni and W. Thormann, *J. Chromatogr.* **638**, 187 (1993).
114. S. Molteni, H. Frischknecht, and W. Thormann, *Electrophoresis* **15**, 22 (1994).
115. K. G. Moorhouse, C. A. Eusebio, G. Hunt, and A. B. Chen, *J. Chromatogr. A* **727**, 61 (1995).
116. B. Potocek, B. Gas, E. Kenndler, and M. Stedry, *J. Chromatogr. A* **709**, 51 (1995).
117. S.-M. Chen and J. E. Wiktorowicz, *Anal. Biochem.* **206**, 84 (1992).
118. J. Wu and J. Pawliszyn, *Electrophoresis* **24**, 469 (1993).
119. T.-L. Huang, P. C. H. Shieh, and N. Cooke, *Chromatographia* **39**, 543 (1994).
120. S. Hjertén and M.-D. Zhu, *J. Chromatogr.* **346**, 265 (1985).
121. F. Kilar and S. Hjertén, *J. Chromatogr.* **480**, 351 (1989).
122. M.-D. Zhu, R. Rodriguez, and T. Wehr, *J. Chromatogr.* **559**, 497 (1991).
123. C. Schwer, *Electrophoresis* **16**, 2121 (1995).
124. K. Slais and Z. Friedl, *J. Chromatogr. A* **695**, 113 (1995).
125. K. G. Moorhouse, C. A. Eusebio, G. Hunt, and A. B. Chen *J. Chromatogr. A* **717**, 61 (1995).
126. M. M. Pedrosa and M. E. Legaz, *Electrophoresis* **16**, 659 (1995).
127. G. G. Yowell, S. D. Fazio, and R. V. Vivilecchia, *J. Chromatogr. A* **652**, 215 (1993).
128. J. Wu and J. Pawliszyn, *Electrophoresis* **16**, 670 (1995).
129. M. Conti, C. Gelfi, and P. G. Righetti, *Electrophoresis* **16**, 1485 (1995).

130. A. Bossi, C. Gelfi, A. Orsi, and P. G. Righetti, *J. Chromatogr. A* **686**, 121 (1994).
131. A. Guttman, P. Shieh, J. Lindahl, and N. Cooke, *J. Chromatogr. A* **676**, 227 (1994).
132. A. S. Cohen and B. L. Karger, *J. Chromatogr.* **387**, 409 (1987).
133. A. Guttman, J. Horvath, and M. Cooke, *Anal. Chem.* **65**, 199 (1993).
134. K. Tsuji, *J. Chromatogr.* **550**, 823 (1991).
135. H.-J. Bode, *Anal. Biochem.* **83**, 364 (1977).
136. A. Widhalm, C. Schwer, D. Blass, and E. Kenndler, *J. Chromatogr.* **549**, 446 (1991).
137. M. Zhu, D. Hansen, S. Burd, and F. Gannon, *J. Chromatogr.* **480**, 331 (1989).
138. K. Benedek and S. Thiede, *J. Chromatogr. A* **676**, 209 (1994).
139. K. Ganzler, K. S. Greve, A. S. Cohen, and B. L. Karger, *Anal. Chem.* **64**, 2665 (1992).
140. T. Takagi and M. R. Karim, *Electrophoresis* **16**, 1463 (1995).
141. M. Nakatani, A. Shibukawa, and T. Nakagawa, *J. Chromatogr. A* **672**, 213 (1994).
142. H. E. Schwartz, K. J. Ulfelder, F. J. Sunzeri, M. P. Busch, and R. G. Brownlee, *J. Chromatogr.* **559**, 267 (1992).
143. Y. C. Bae and D. Soane, *J. Chromatogr. A* **652**, 17 (1993).
144. D. Wu and F. E. Regnier, *J. Chromatogr. A* **608**, 349 (1992).
145. J. H. Beattie and M. P. Richards, *J. Chromatogr. A* **700**, 95 (1995).
146. J.-M. Gao, M. Mrksich, F. A. Gomez, and G. M. Whitesides, *Anal. Chem.* **67**, 3093 (1995).
147. F. Kalman, S. Ma, A. Hodel, R. O. Fox, and C. Horvath, *Electrophoresis* **16**, 595 (1995).
148. F. Kalman, S. Ma, A. Hodel, R. O. Fox, and C. Horvath, *J. Chromatogr. A* **705**, 135 (1995).
149. J. S. Andersen, B. Svensson, and P. Roepstorff, *Nat. Biotechnol.* **14**, 449 (1996).
150. W. Weinmann, C. E. Parker, K. Baumeister, C. Maier, K. V. Tomer, and M. Przybylski, *Electrophoresis* **15**, 228 (1994).
151. R. M. Caproli and K. B. Tomer, in *Continuous Flow Fast Atom Bombardment Mass Spectrometry* (R. M. Caproli, ed.), pp. 93–120. Wiley, New York, 1990.
152. J. A. Castoro, R. W. Chiu, C. A. Monnig, and C. L. Wilkins, *J. Am. Chem. Soc.* **114**, 109 (1992).
153. P. A. van Veelen, U. R. Tjaden, J. van der Greef, A. Ingendoh, and F. Hillenkamp, *J. Chromatogr.* **647**, 367 (1993).
154. A. Apffel, J. Chakel, S. Udiavar, W. S. Hancock, C. Souders, and E. Pungor, *J. Chromatogr. A* **717**, 41 (1995).
155. W. Weinmann, C. E. Parker, L. J. Deterding, D. I. Papac, J. Hoyes, M. Przybylski, and K. B. Tomer, *J. Chromatogr. A* **680**, 353 (1994).
156. S. D. Patterson, *Anal. Biochem.* **221**, 1 (1994).

157. K. L. Walker, R. W. Chiu, C. A. Monnig, and C. L. Wilkins, *Anal. Chem.* **67**, 4197 (1995).
158. F. Foret, O. Muller, J. Thorne, W. Gotzinger, and B. L. Karger, *J. Chromatogr. A* **716**, 157 (1995).
159. R. D. Smith, J. H. Wahl, D. R. Goodlett, and S. A. Hofstadler, *Anal. Chem.* **65**, 574A (1993).
160. J. H. Wahl, D. C. Gale, and R. D. Smith, *J. Chromatogr. A* **659**, 217 (1994).
161. J. Cai and J. Henion, *J. Chromatogr. A* **703**, 667 (1995).
162. T. J. Thompson, F. Foret, P. Vouros, and B. L. Karger, *Anal. Chem.* **65**, 900 (1993).
163. R. B. Cole, J. Varghese, R. M. McCormick, and D. Kadlecek, *J. Chromatogr.* **680**, 363 (1994).
164. B. Foret, T. J. Thompson, P. Vouros, B. L. Karger, P. Gebauer, and P. Bocek, *Anal. Chem.* **66**, 4450 (1994).
165. D. Figeys, I. VanOostveen, A. Ducret, and R. Aebersold, *Anal. Chem.* **68**, 1822 (1996).
166. S. A. Hofstadler, J. H. Wahl, R. Bakhtiar, G. A. Anderson, J. E. Bruce, and R. D. Smith, *J. Am. Soc. Mass: Spectrom.* **5**, 894 (1994).
167. S. A. Hofstadler, F. D. Swanek, D. C. Gale, A. G. Ewing, and R. D. Smith, *Anal. Chem.* **67**, 1477 (1995).
168. L. Licklider, W. G. Kuhr, M. P. Lacey, T. Keough, M. P. Purdon, and R. Takigiku, *Anal. Chem.* **67**, 4170 (1995).
169. H.-T. Chang and E. S. Yeung, *Anal. Chem.* **65**, 2947 (1993).
170. Y. L. F. Hsieh, H. Q. Wang, C. Elicone, J. Mark, S. Martin, and F. E. Regnier, *Anal. Chem.* **68**, 455 (1996).
171. J. F. Banks and T. Dresch, *Anal. Chem.* **68**, 1480 (1996).

CHAPTER

21

CAPILLARY ELECTROPHORESIS OF CARBOHYDRATES

MILOS V. NOVOTNY

Department of Chemistry, Indiana University, Bloomington, Indiana 47405

21.1.	Introduction	729
21.2.	The Goals of Analytical Glycobiology	730
21.3.	Instrumental Aspects	732
21.4.	Sample Derivatization	736
21.5.	Electromigration Mechanisms	744
21.6.	Selected Applications	753
	21.6.1. Intact Glycoproteins and Glycopeptides	753
	21.6.2. Glycoprotein Compositional Analysis and Oligosaccharide Mapping	755
	21.6.3. Analysis of Polysaccharides	757
List of Acronyms and Abbreviations		761
References		761

21.1. INTRODUCTION

The development of new analytical methodologies is often a prelude to major advances in biochemical research. This has certainly been true for the modern studies of proteins and nucleic acids, the areas that achieved significant progress due to various techniques of chromatography, electrophoresis, mass spectrometry, and nuclear magnetic resonance spectrometry, among others. More recently, the field of glycobiology, a scientific area that deals with various aspects of carbohydrates, appears to follow a similar trend. Since the early 1970s, when word started to spread of the paramount biological importance of glycoconjugates as Nature's "recognition molecules" (1), a search for appropriate methodologies has greatly intensified. It has become increasingly

---

High Performance Capillary Electrophoresis, edited by Morteza G. Khaledi. Chemical Analysis Series, Vol. 146.
ISBN 0-471-14851-2 © 1998 John Wiley & Sons, Inc.

clear that significant progress in this "last great frontier of biochemistry" (2) is likely to be dependent on a series of powerful, complementary analytical methods. Due to the extreme complexity of many glycoconjugate mixtures, the modern separation methods such as high-performance liquid chromatography (HPLC) and high-performance capillary electrophoresis (HPCE) will undoubtedly be central to future methodological developments in this area.

The early pioneering studies in HPCE provided the basis for extensive development of commercial instrumentation, column studies, new detection capabilities, and numerous applications across an amazing range of molecular sizes, from small ions to large biopolymers. During the last several years, HPCE has also been applied to carbohydrates. While HPCE of glycoconjugates has not yet reached its methodological maturity, the initial studies in the area permit a highly optimistic view of future development. Due to its unprecedented resolving power, HPCE can deal effectively with complex mixtures of glycoconjugates that were previously hard to separate chromatographically. Empowered with ultrasensitive means of detection such as laser-induced fluorescence or electrospray mass spectrometry, HPCE becomes capable of addressing issues such as the role of glycoproteins in biomolecular recognition, receptor biochemistry, blood coagulation, and biochemical repair mechanisms, among others. Due to its quantitative capabilities, speed of analysis, and easy automation, the method also has the potential to be efficacious in clinical research and the diagnosis of carbohydrate-related metabolic disorders.

This chapter will review the recent methodological development in the HPCE of carbohydrate molecules. First, the main aspects of analytical and structural glycobiology will briefly be outlined with the emphasis on the unique role of HPCE and its related detection techniques in the separation and detection of glycoprotein glycoforms, the products of site-specific chemical degradations (glycans, in particular), and polysaccharides. Since fluorescence-tagging procedures are becoming central to the effective use of HPCE in glycoconjugate research, the current approaches in this area will be reviewed and their applications briefly demonstrated .

21.2. THE GOALS OF ANALYTICAL GLYCOBIOLOGY

At this time, glycoconjugates are known to be implicated in virtually all important events in cellular biology. The ubiquity and numerous ways in which Nature utilizes saccharides have their parallel in the structural complexity of glycoconjugate molecules. Unlike the proteins and polynucleotides (both linear polymers), different oligosaccharides exhibit a tendency toward branching. Different linkage forms, anomericity, and optical activity can further add structural uniqueness to the molecules of interest. Through a variety of techniques, it is now known that some saccharide mixtures can be exceedingly complex. This is a considerable challenge to even the best analytical separation techniques. The second major problem of glycoconjugate analysis is compounded by the limited capacity of an unmodified saccharide molecule to generate an easily recognizable spectroscopic signal for detection.

Capillary electrophoresis (CE), with its great resolving power, can clearly be an attractive tool of carbohydrate analysis. However, it is essential that the analyzed carbohydrates possess an electric charge for migration. While some glycoconjugates, such as the oligosaccharides containing sialic acids, or glycosaminoglycans with their various sulfate or carboxylic groups, naturally fulfill this requirement, the neutral oligosaccharides can be charged through boration of their vicinal hydroxy groups for the purpose (3,4). In order to detect carbohydrates at the low amounts compatible with good practice of HPCE, it is essential to convert them to the derivatives exhibiting fluorescence, or at least ultraviolet (UV) absorbance.

The demands for component resolution and measurement sensitivity may vary with the nature of a glycoconjugate under investigation. Certain investigations, such as those on the membrane glycoproteins, require the highest sensitivity that an analytical technique can offer. Processes as important as antigen–antibody interaction, cell-to-cell recognition, protein targeting, and functions of the receptor proteins have thus far been poorly understood because of the limited availability of the material collected for structural analysis. Since protein glycosylation is largely a posttranslational event, biological amplification is not feasible. In contrast, somewhat less stringent sensitivity requirements may exist in certain cases of biotechnologically produced glycoprotein pharmaceuticals. Here, the resolving power of HPCE is likely to be the major advantage of this method in separating various glycoforms of native glycoproteins or in mapping their oligosaccharides.

In order to elucidate the relationship between bioactivity and the structure of a glycoprotein or its glycoforms, it is essential to determine the sites of attachment of its oligosaccharide chains to the polypeptide backbone and to characterize the oligosaccharide class (N-linked vs. O-linked; high mannose; complex; hybrid; biantennary; triantennary; etc.) at a specific site. To begin with, glycoproteins must first be isolated from biological materials by means of dialysis, preparative chromatography, gel electrophoresis, isoelectric focusing, lectin binding, etc., or, in the most typical case, through an effective combination of these tools. Due to the great variety of glycoprotein structures and analytically different situations, it would be illusory to provide a general scheme or a general isolation strategy. It should, however, be noted that the isolation step, if improperly chosen, could become a bottleneck to the overall high-sensitivity carbohydrate analysis. This chapter will primarily emphasize the use of HPCE as a *powerful end method* in analytical and structural glycobiology. However, one must stress the importance also of appropriate isolation techniques for minimizing the losses that occur through sample degradation and adsorption. Sample treatment procedures that involve a minimum number of transfers, together with the use of microcolumn separation techniques and microreactors, are likely to become widely applicable prior to HPCE.

Once a glycoprotein has been isolated, it becomes feasible to use capillary isoelectric focusing or zone electrophoresis in assessing the complexity of its various glycoforms. For a further structural determination, specific or general degradation steps may be applied: (i) a site-specific proteolysis (e.g., with trypsin), yielding a mixture of glycopeptides for further characterization; (ii) removal of oligosaccharides from the polypeptide through an enzymatic hydrolysis or hydrazine treatment (5), followed by the development of an oligosaccharide map; and (iii) chemical hydrolysis yielding a monosaccharide mixture for further separation (compositional analysis). Alternatively, various of these approaches can be combined in micropreparative separations for a subsequent structural analysis, sequencing, selective sample derivatization, etc. In addition, sugar-specific or bond-specific enzymes can be applied in conjunction with highly sensitive analytical tools, such as HPCE–LIF (laser-induced fluorescence), to yield the structural details on a given glycan.

HPCE, with its increasing range of detection and ancillary techniques, is likely to become both complementary and competitive with the more established or previously reported methods of glycoconjugate analysis, such as gas chromatography–mass spectrometry, HPLC using aminopyridyl precolumn labeling (6,7), or pulsed amperometric detection (8,9), fast atom bombardment and electrospray mass spectrometry, and high-field nuclear magnetic resonance.

21.3. INSTRUMENTAL ASPECTS

CE is a relatively simple analytical technique. Its essential part is the separation capillary, whose ends are connected to an electrode and to a voltage supply. The samples of interest are introduced either pneumatically or electrokinetically. An important part of the instrument is a detector situated at the capillary's end. Because of the method's instrumental simplicity, home-built setups are quite popular in research laboratories. They have been largely modeled according to the system described originally by Jorgenson and Lukacs (10) in 1981. During the recent years, commercial HPCE instruments have become increasingly sophisticated. Their distinct advantages include the following: (i) autosampling arrangements for routine analyses; (ii) thermal control of the separation capillary; and (iii), with some instruments, the capability of recovering the separated fractions for further investigations. As demonstrated first by Jorgenson (11), capillary zone electrophoresis can attain its separation efficiency only if the input sample mass and volume are kept at very low levels. In the common analytical practice of HPCE, this roughly translates into measuring nanogram-to-picogram quantities per component as the maximum amounts in 50–80 µm i.d. separation capillaries. Miniaturized UV absorbance detectors, being the most commonly used concentration-sensitive devices of adequate mass sensitivity in microcolumn liquid chromatography, also work sufficiently well in HPCE. Thus, detection of glycoproteins and their glycopeptide fragments through UV absorbance at 205–220 nm is just as feasible as with other peptides and proteins. UV detection of underivatized carbohydrates at 195 nm in HPCE is likely to be confined to the simplest applications, although, as shown by Hoffstetter-Kuhn et al. (12), borate complexation somewhat enhances the absorbance signal. Sugar derivatization for increased UV absorption is also feasible, as demonstrated with the use of 2-aminopyridine (13), 6-aminoquinoline (14), 3-methyl-1-phenyl-2-pyrazolin-5-one (15), and aminobenzoic acid derivatives (16). Considerable attention is still being paid to the development of new detection techniques for HPCE, and it is likely that future improvements in the area will aid glycoconjugate analysis.

A key component of the HPCE system is the separation column itself. For most analytical work with glycoconjugates, the inner column diameters of fused silica capillaries are around 50–80 µm, although departures from this range may be increasingly more common in future studies. Very small column diameters are of interest in connection with manipulating extremely small samples, such as in the analysis of single biological cells. Capillaries with greater inner diameters (100–150 µm) are desirable for micropreparative isolations. For example, recovered glycoconjugates can be subjected to laser desorption mass spectrometry for their molecular-weight determination or other structural studies. In addition, large-bore capillaries are known to permit effective recoveries of picomole amounts of peptides for sequencing purposes.

It should be emphasized that departures from the usual column diameters are not without certain penalties. For both open tubes and gel-filled capillaries in HPCE, the column diameter has an appreciable effect on column performance (17). For larger columns, slow dissipation of the Joule heat generated during electrophoresis can have adverse effects on both the component resolution and integrity of biological samples.

Among the detection techniques for HPCE of carbohydrates, three measurement principles appear particularly attractive in terms of high sensitivity: (i) LIF; (ii) electrochemical (amperometric) detection; and (iii) mass spectrometry. Pulsed amperometric detection of sugars is a common and effective method used in HPLC. Some recent communications (18,19) indicate that electrochemical detection may become also common in HPCE. Naturally, the use of mass spectrometry extends well beyond detection to the actual structural information on glycoconjugates.

Laser-based detectors are of great interest to the users of miniaturized separation systems, including HPCE, as the highly collimated laser radiation can be easily focused into a very small area. When there is a good match between a laser's radiation and the spectral properties of the measured solutes, LIF provides a significant improvement in detection sensitivity compared to conventional fluorescence. Moreover, lasers are rapidly becoming convenient, reliable, and relatively inexpensive light sources. The commonly held view of lasers as highly sophisticated, temperamental, and costly devices is no longer justifiable in biochemical analysis. In fact, LIF detectors for HPCE have now become available commercially.

A typical instrumental setup for HPCE with LIF detection used in carbohydrate analysis in our laboratory has evolved as a combination of the instruments reported by Diebold and Zare (20) at Stanford as well as by Sepaniak and Yeung (21) at Iowa State University. The helium–cadmium laser has been used here as a light source either in its UV mode (325 nm) or the blue mode (442 nm). Through the appropriate optical components, the laser beam is focused onto the end section of a fused-silica capillary. Fluorescence emission is typically collected through an optical fiber situated at a right angle to the incident beam. A similar, or even a simpler, instrumental arrangement can be employed for an argon-ion laser, which is operated at the wavelengths different from those of the helium–cadmium laser.

Using chemical species that feature excitation maxima at or near the outputs of the most readily available lasers confers a strong advantage on the researcher. From a practical point of view, the modern versions of the helium–cadmium and the air-cooled argon-ion laser are the most desirable match for work with HPCE at excitation wavelengths below 500 nm. The intensity of the fluorescence signal produced is directly proportional to that of the laser light illuminating the cell. However, it is not advisable to use powers of more than a few milliwatts in LIF measurements; most fluorophores tend to photobleach at high intensities, and furthermore the scattered light may cause some noise enhancement.

With the most elaborate laser-detector designs, it has become feasible to achieve sensitivities at the level of single molecules (22,23) confined to very small sample volumes. An ordinary detection system represents a compromise, emphasizing simplicity and low cost, rather than the maximum achievable sensitivity. However, depending on the fluorophore utilized, detection limits in the subfemtomole-to-subattomole range are quite easy to achieve. Sugars may also be measured in their native forms by means of indirect fluorometry (24). Indirect detection methods, though, provide considerably lower sensitivity, and they also suffer from a limited linear dynamic range in quantitative measurements.

As laser technologies advance over time, they will undoubtedly provide further incentive for the development of additional tagging reagents. Diode lasers featuring higher-wavelength outputs are now frequently mentioned in the context of HPCE. Additionally, the development of reliable lasers in the UV region could be of interest in relation to detection of the native fluorescence of glycoproteins.

Mass spectrometry (MS), with its various ionization methods, has traditionally been among the key techniques used for the structural elucidation of proteins and carbohydrates. Various types of mass spectrometers, including quadrupole mass analyzers, ion traps, time-of-flight mass analyzers, and sector instruments, have now been shown to be technically feasible for HPCE–MS coupling. Besides the accurate molecular-weight information supplied by MS, the tandem (MS–MS) instruments can often solve the primary structures of both peptides and oligosaccharides.

The instrumental aspects of coupling HPCE with MS have received considerable attention lately. The two (on-line) ionization methods of interest are fast atom bombardment (FAB) ionization (25,26) and electrospray ionization, each of which has unique interfacing problems with different instruments. The HPCE–FAB–MS combination appears to be sensitive at picomole-to-femtomole levels. An array detector based on the ion-counting principle has been shown to increase the sensitivity of this combination considerably further (27).

The electrospray ionization principle has truly revolutionized modern MS of biological molecules through its inherent sensitivity and ability to record large molecular entities within a relatively small mass scale. Correspondingly, researchers in the area have reoriented much of their attention to the coupling of HPCE to electrospray mass spectrometry. The potential of electrospray mass spectrometry in glycobiology was exemplified by Duffin et al. (28), who dealt with both the intact glycoprotein (ovalbumin) and its N-glycanase digests. The oligosaccharides containing sialic acids were primarily studied in the negative-ion mode of detection, whereas the remaining glycans were cationized by adding sodium acetate or ammonium acetate to the sample for the positive-ion detection. Cleavages of the individual glycosidic bonds were observable in the tandem (MS–MS) instrumental arrangement, yielding much information on the sequence of sugar units. At this stage, the methodology does not differentiate the isomeric sugars and linkage positions.

A successful use of directly coupled micro-LC to electrospray MS in the analyses of glycopeptides (29–31), generated from glycoproteins by the action of proteases, is likely to be a prelude to investigations of similar mixtures by CE–MS. To some extent, a CE buffer compatibility with the electrospray process has been a significant issue. Most importantly, the CE–MS of peptides using miniaturized forms of electrospray (32,33) has recently been demonstrated at subfemtomole levels. Some interesting opportunities have also arisen (33,34) from the recent employment of powerful mass analyzers such as the ion cyclotron resonance (Fourier-transform MS) and the ion trap mass analyzers.

During a relatively short period of its existence, matrix-assisted laser desorption ionization (MALDI)–MS has increasingly shown its relevance to glycoconjugate analysis (35–41)—in molecular-weight determinations, branching analysis, and sequencing. Instrumentally, such investigations have been shown feasible using a magnetic sector instrument with an array detector (40), or more conveniently with a reflector/time-of-flight mass spectrometer [through post-source-decay fragmentation processes (41)]. Most importantly, MALDI–MS has been widely demonstrated at subpicomole sensitivities, conveniently matching the usual working range of HPCE. Trapping CE fractions for subsequent MALDI–MS analysis has received considerable attention lately (42–44), although a direct coupling of these methods remains somewhat remote. Improvements in the molecular design of MALDI matrices for carbohydrate analysis (45) are also desirable to reach even more sensitive determinations.

During the last few years, MS has assumed a considerable role in the structural analysis of glycoconjugates. Additional techniques for CE–MS must now be developed to diffuse the current frustration with perfectly separated but structurally ill-defined complex mixtures.

21.4. SAMPLE DERIVATIZATION

Native carbohydrates are difficult to detect in HPCE due to the absence of chromophoric groups in their molecules. Usually, they are very weak UV absorbers at very low wavelengths (around 200 nm), with the exception of their enhanced absorbances due to borate complexation (12) and the unsaturation in uronic acid residues (at 230 nm) in certain disaccharides obtained from glycosaminoglycan samples. Introducing a spectroscopically detectable moiety into sugar molecules through derivatization is now a common remedy for their lack of chromophores. While appealing in principle, the sample derivatization approach is not without certain penalties: (i) procedural complications; (ii) the possibility of analyte decomposition; (iii) quantitative discrepancies; and (iv) structural complications (if further investigations of the derivatized carbohydrates are required). Nevertheless, the advantages of enhanced sensitivity are crucial to the success of HPCE-based procedures, producing considerable interest in the development of optimum tagging procedures. For highly sensitive detection procedures, fluorescent tagging is obviously more attractive than UV tags. Fluorescent labeling finds its optimum utilization when the compound's excitation maximum coincides with the wavelength of a laser light source. A considerable effort is thus being extended to "engineer" the labeling chemistries toward that goal.

Different glycoconjugates possess a variety of hydroxy groups, which at first glance would seem to present natural sites for derivatization. In practice, however, this does not appear to work due to the different reactivities of different groups, the incidence of multiple (and incomplete) tagging, and potential complications arising from the advanced structure of complex carbohydrates. Amino sugars, though, present easily taggable moieties for a variety of reagents that are applicable to high-sensitivity measurements of primary amines. Carbohydrates with a reducing end can be converted to amino derivatives through reductive amination at the analytical scale (46,47). The reaction that follows with an amine-selective reagent then leads to a suitable derivative. The end groups of certain saccharide structures can also be derivatized directly: the carbonyl groups can be attached to a chromophore through the Schiff-base mechanism, and the open structural form of sialic acids reacts with an aromatic diamine to yield a fluorescent aromatic system (48).

In one of the first applications of HPCE to carbohydrates, Honda et al. (13) used the reaction between the reducing sugars and 2-aminopyridine to form derivatives that absorb strongly at 240 nm. Since the aminopyridyl derivatives also fluoresce, it is feasible to increase detection sensitivity through fluorometric detection (49). An analogous reaction also proceeds with 6-aminoquinoline, as shown by Nashabeh and El Rassi (50), with some improvement in sensitivity. With N-acetylglucosamine used as a model sugar (50), both derivatization routes are shown in the accompanying scheme (I).

3-Methyl-1-phenyl-2-pyrazolin-5-one (PMP) was demonstrated as an additional UV-tagging reagent for HPCE of carbohydrates by Honda et al. (15,51), using different complexation agents to provide the necessary charge for the sugar analytes. An example of a simple sugar analysis is shown in Figure 21.1 (15) with PMP–aldohexoses as a model mixture. Improvements with this type of derivatization were further realized by using 1-(p-methoxy)phenyl-3-methyl-5-pyrazolone (52), due to this reagent's greater reaction yields and enhanced UV absorbance. The pyrazolone-based reagents appear particularly suitable for tagging sialylated glycans. 4-Aminobenzoic acid, ethyl-p-aminobenzoic acid, and 4-aminobenzonitrile (16) represent additional reagents for the determination of a variety of sugar structures, including

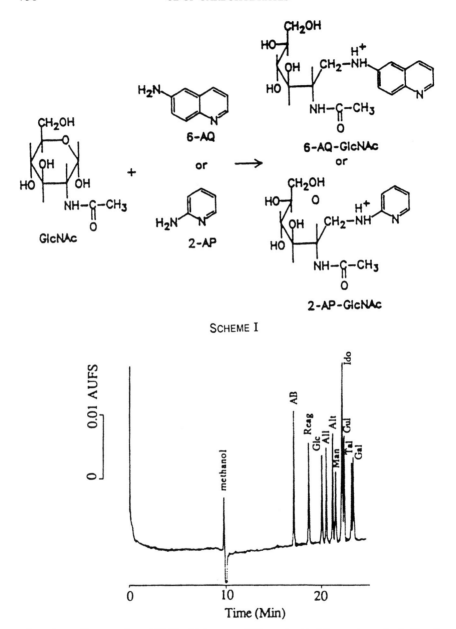

Figure 21.1. CE separation of PMP–aldohexoses: (AB) amobarbital (internal standard); (Reag) PMP reagent peak; (Glc) glucose; (All) allose; (Alt) altrose; (Man) mannose; (Ido) idose; (Gul) gulose; (Tal) talose; (Gal) galactose. Capillary: 50 μm i.d. × 78 cm. Buffer: 0.2 M borate, pH 9.5. Detection: UV at 245 nm. [Reproduced from S. Honda, S. Suzuki, A. Nose, K. Yamamoto, and K. Kakehi, *Carbohydr. Res.* **215**, 193 (1991), with permission of Elsevier Science Publishers.]

ketoses. The best UV-absorbing tags appear to result in detection limits of low picomoles.

In the search for ultrasensitive methodologies for glycoanalysis, our laboratory has introduced 3-(4-carboxybenzoyl)-2-quinolinecarboxaldehyde (CBQCA) as a fluorogenic reagent for determination of amino sugars (53) and reductively aminated, neutral carbohydrates (46,47). The derivatization scheme (II) for reducing sugars first produces 1-amino-1-deoxyalditols through reductive amination. In a subsequent step, a catalyzed ring formation with CBQCA, a reactive ketoaldehyde, occurs (47). The isoindoles similar to

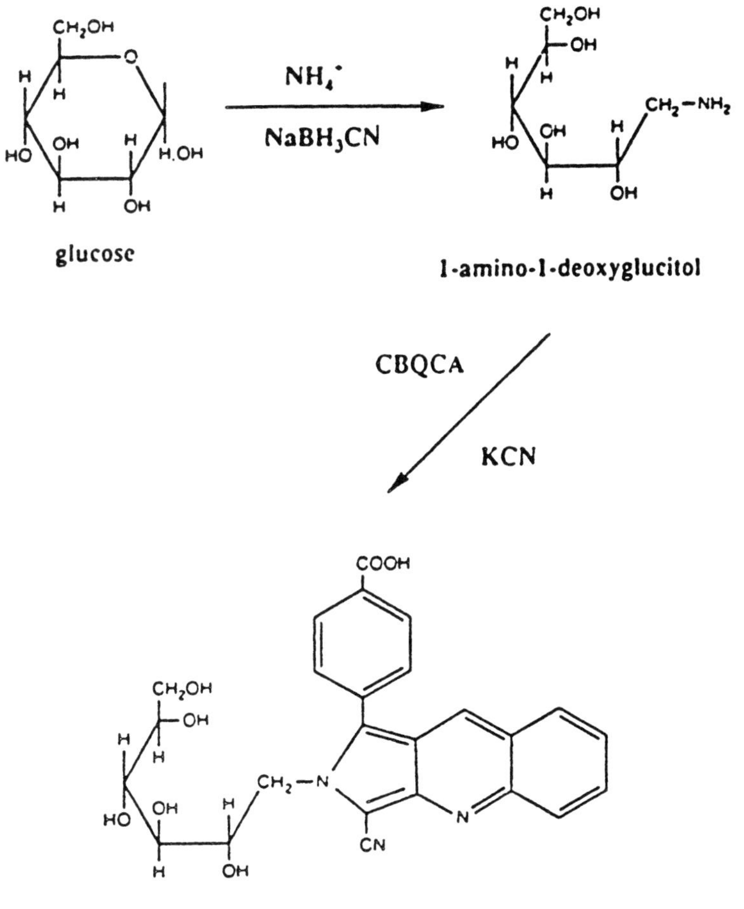

SCHEME II

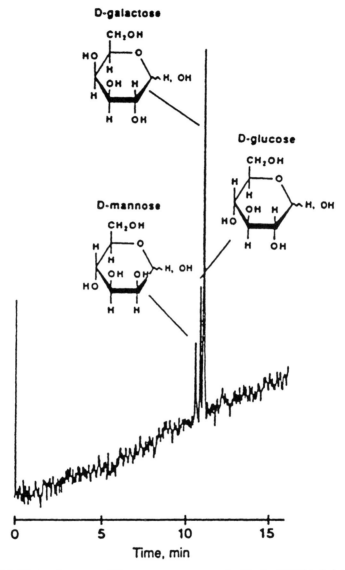

Figure 21.2. Electropherogram of three common monosaccharides (as CBQCA derivatives), at very high sensitivity (subattomole detection limits). Capillary: 50 μm i.d. × 90 cm (60 cm effective length). Buffer: 10 mM phosphate/10 mM borate, pH 9.40. [Reproduced from J. Liu, O. Shirota, D. Wiesler, and M. Novotny, *Proc. Natl. Acad. Sci. U.S.A.* **88**, 2302 (1991), with permission.]

the aforementioned structure are highly fluorescent. In addition, the closeness of these compounds' excitation maximum to the output wavelength of the helium–cadmium laser (442 nm) and one of the readily available wavelengths (457 nm) of the argon-ion laser makes this type of derivative eminently suitable for high-sensitivity work. A wide separation between the fluorescence excitation and emission maxima and the fact that CBQCA is a fluorogenic rather than just a tagging reagent are further advantages of this derivatization agent. The need to employ reductive amination prior to the isoindole formation is, however, a disadvantage. A demonstration of the high detection (subattomole) sensitivity is given in Figure 21.2 for three model monosaccharides (47). Some applications of CBQCA will be discussed below.

Sulfonated aminonaphthalenes are valuable reagents that have received considerable attention lately. In particular, a highly charged fluorescent tag, 8-aminonaphthalene-1,3,6-trisulfonic acid (ANTS) has been popular in tagging a variety of oligosaccharide structures in both glycoprotein analyses and oligosaccharide profiling. As shown in the accompanying scheme (III) (54), ANTS first forms a Schiff base with the sugar of interest, with subsequent stabilization of the fluorescent product through the action of sodium cyanoborohydride. The advantages of ANTS derivatization are its relative simplicity and the high efficiency of separation (55–57) with chemically modified capillaries. Its fluorescence excitation maximum is close to one of the output wavelengths (325 nm) of the helium–cadmium laser. However, a distinct disadvantage of this laser source is its relatively high operational cost associated with the typical lifetime of its laser tube. For this reason, a recently reported reagent, 9-aminopyrene-1,4,6-trisulfonate (58,59), appears promising because the excitation maximum of its product is in the blue spectral region while its other properties are highly reminiscent of ANTS. There is some interest in developing *specific* tags for certain sugars, such as amino sugars and acidic glycoconjugates. In a report by Mechref and El Rassi (60), acidic sugars were first treated with a carbodiimide, forming an intermediate that was further reacted with either a sulfanilic acid (aUV tag) or 7-aminonaphthalene-1,3-disulfonic acid (ANDSA), a fluorescent tag. An example of separating several acidic monosaccharides (detected fluorometrically, with excitation at 315 nm) is shown in Figure 21.3 (60).

Due to their general importance in the molecular recognition processes, sialic acids deserve special attention. To quantify them effectively, it may be advisable to cleave them first from the intact glycans and determine separately. In our laboratory, we have recently synthesized and evaluated substituted aromatic diamines as fluorogenic reagents for determination of sialic acids [see Shirota (61)]. The method utilizes a well-known capability of α-keto acids to form a quinoxaline ring structure upon a coupling reaction with *ortho-*

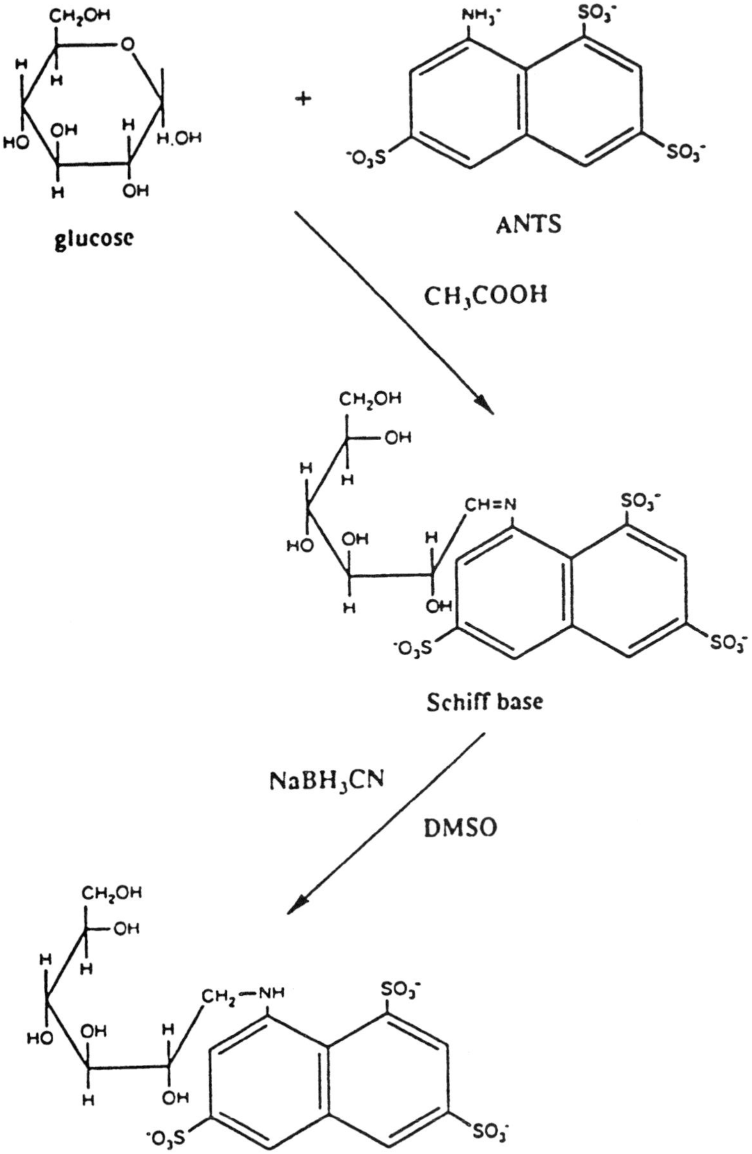

SCHEME III

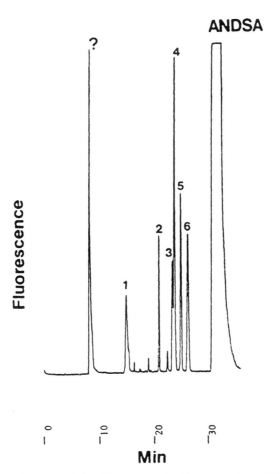

Figure 21.3. Electropherogram of six acidic monosaccharides prepared as ANDSA derivatives. Peaks: (1) N-acetylneuraminic acid; (2) gluconic acid; (3) galacturonic acid; (4) galactonic acid; (5) glyceric acid; (6) glucuronic acid. Capillary: 50 μm i.d. × 80 cm. Buffer: 100 mM borate, pH 10.0. Detection: fluorescence; excitation, 315 nm; emission, above 400 nm. [Reproduced from Y. Mechref and Z. El Rassi, *Electrophoresis* **15**, 627 (1994), with permission of VCH Publishers.]

diamines. As shown in the accompanying scheme (IV), an *in situ* reduction of a dinitro compound to a diamine (the actual reagent) is accomplished prior to derivatization because of the limited stability of the diamines during storage. The reaction with N-acetylneuraminic acid (NANA) yields a fluorescent derivative that is suitable for detection with the 325 nm helium–cadmium laser (62).

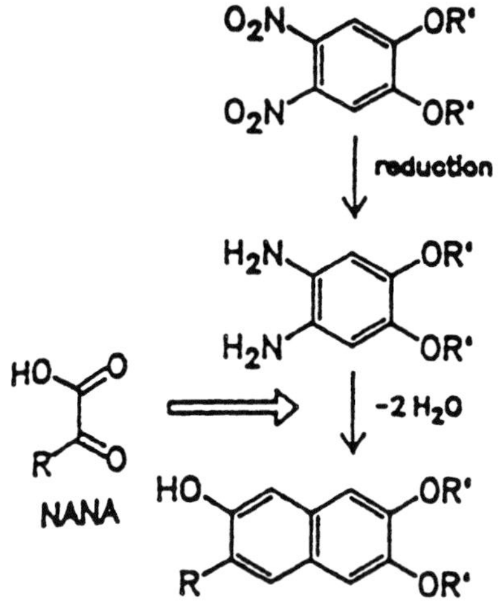

SCHEME IV

A search for the ideal reagents in sugar derivatization has barely begun. At present, all reported derivatization procedures have some distinct advantages and disadvantages. A parallel use of some reagents is likely to be found complementary. While the UV tags may prove adequate for some determinations, the challenges of ultrahigh sensitivity will undoubtedly need the best of laser fluorescence technology. The reagents with charged functional groups appear highly beneficial from the point of view of electrophoretic migration. Most importantly, the effectiveness of derivatization with minute quantities of glycans still remains to be addressed in more detail.

21.5. ELECTROMIGRATION MECHANISMS

In the current practice of HPCE, three types of media are particularly important to biomolecular separations: (i) free buffer solutions; (ii) gel-filled capillaries; and (iii) entangled polymer matrices. Each type can be crucial to success in a particular application involving glycoconjugates. Some simple sugar mixtures, such as those encountered in a typical mono- or disaccharide analysis, can be easily separated due to their charge-to-mass ratio (z/m), but wherever some degree of selectivity is required, micellar electrokinetic chromatography or, potentially, electrochromatography in packed capillaries can become the most viable options. In a number of recent investigations, the role of unique buffer additives in optimizing the carbohydrate separations has been stressed.

Glycoconjugates vary considerably in their polarity, necessitating different analytical approaches. On the one hand, we may deal with highly polar species, such as some polysaccharides, charged uniformly with the sulfate or carboxylic groups (heparins, heparan sulfates, polygalacturonates, hyaluronates, etc.), for which electrophoresis is a natural technique. On the other hand (at the other extreme), the relatively lipophilic molecules such as glycolipids or hydrophobic polysaccharides must be approached quite differently. The role of buffer additives becomes important in adjusting the solute's charge and electrophoretic mobility to be in tune with the analytical requirements for the speed of analysis and resolution. It has been shown (63) that suitable buffer additives can either induce or reduce electrophoretic mobility of certain polysaccharides. Since numerous glycoconjugates behave as uncharged, neutral molecules with pK values around 11.0, borate complexation is typically utilized to impart, for the sake of successful electrophoresis, a negative charge on the solutes of interest. Serendipitously, the negatively charged saccharides, large or small, are generally repelled by the negatively charged surfaces, and consequently the surface adsorption problems in HPCE are much less severe in carbohydrate analysis than in, say, analysis of proteins. The borate complexation has also been utilized (64) in improving resolution of various glycoforms in certain glycoproteins.

Glycolipids are relatively small molecules that represent amphiphilicity of natural glycoconjugates at the level of biological membranes. Certain capillary electromigration techniques based on electrokinetic phenomena can increasingly deal with such solutes. For example, gangliosides possessing different charged sugars as the polar heads on one side and a lipophilic tail on the other can be electrophoretically separated in their micellar forms (65). Addition of an appropriate cyclodextrin derivative into a buffer medium can result in a differential complexation with various gangliosides and, subsequently, their effective resolution (66). El Rassi and Nashabeh (67) have further improved the analytical scope of HPCE for gangliosides by placing a spectroscopic tag at the sialic acid end of the solute molecules.

In dealing with neutral glycoconjugates, the value of borate complexation has now been shown in numerous studies, ranging from the electromigration of native structures and spectroscopically modified oligosaccharides to some polysaccharide molecules. The vicinal hydroxy groups on the individual sugar links can complex with a borate according to the accompanying scheme (V). Alternatively, other anionic species can cause comparable complexation effects, although this possibility has not been systematically explored.

It has been shown (52) with a model mixture of CBQCA-labeled glucosamine and galactosamine that the borate complexation can enhance

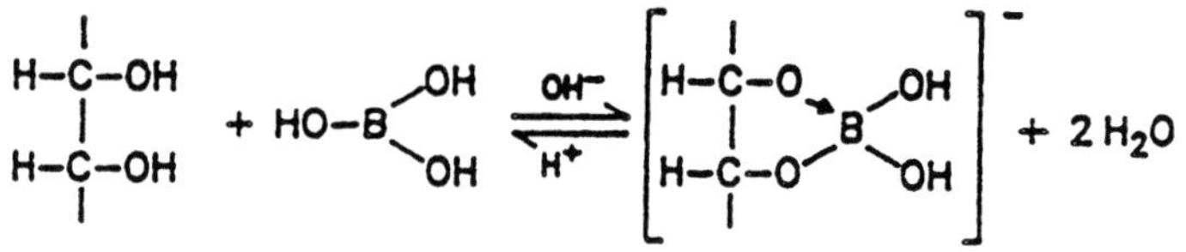

SCHEME V

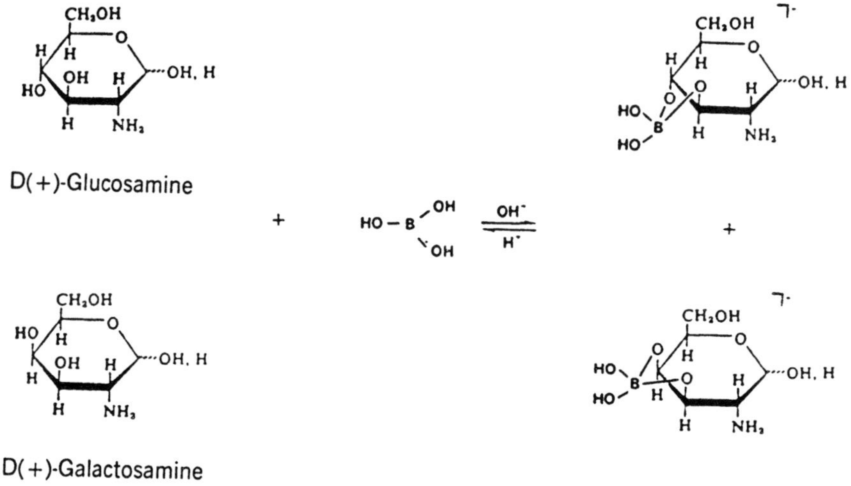

D(+)-Glucosamine

D(+)-Galactosamine

SCHEME VI

steric differences in the solutes, leading to an effective electrophoretic resolution of the solutes with the otherwise identical charge-to-mass ratio (see Scheme (VI). The result of borate complexation is shown in Figure 21.4. A similar effect on enhancing the separation selectivity through borate complexation was also demonstrated with spectroscopically unmodified sugars (12).

While borate complexation has been important as a way to achieve effective separations of neutral oligosaccharides, it may not be widely appreciated that the overall charge on a spectroscopic tag (seemingly, a small part of an end-labeled oligosaccharide) still has a very profound effect on the migration behavior. As demonstrated in Figure 21.5 (55), a very different electrophoretic profile is seen with (a) 2-aminopyridine, (b) 5-aminonaphthalene-2-sulfonate, and (c) 8-aminonaphthalene-1,3,6-trisulfonate tags, with the same Tris–borate buffer. A detailed study (55) of the effect of borate complexation with model

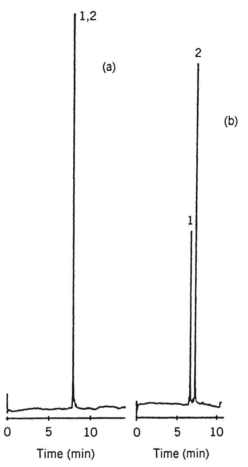

Figure 21.4. CE separation of CBQCA-derivatized amino sugars of closely related structures. Peaks: (1) D(+)-glucosamine; (2) D(+)-galactosamine. (a) Buffer: 0.04 M Na_2HPO_4/TrisHCl, pH 9.10. Capillary: 25 μm i.d. × 70 cm (43 cm effective length). Sample concentration: 5.88×10^{-4} M. Hydrodynamic injection: 10 s. Applied voltage: 20 kV (8 μA). (b) Buffer: 0.2 M $Na_2B_4O_7 \cdot 10H_2O$/0.02 M Na_2HPO_4, pH = 9.12. Applied voltage: 24 kV (9 μA). Other conditions are the same as in part a. [Modified from J. Liu, O. Shirota, and M. Novotny, *Anal. Chem.* **63**, 413 (1991), with permission of the American Chemical Society.]

oligosaccharides of different sugar linkages (dextrans vs. dextrins) revealed some differences (Figure 21.6), the knowledge of which is essential to the optimization efforts for different oligosaccharides. The extent of borate complexation will depend on the proximity of participating hydroxy groups in a sugar unit as well as certain aspects of advanced structures (e.g., helicity).

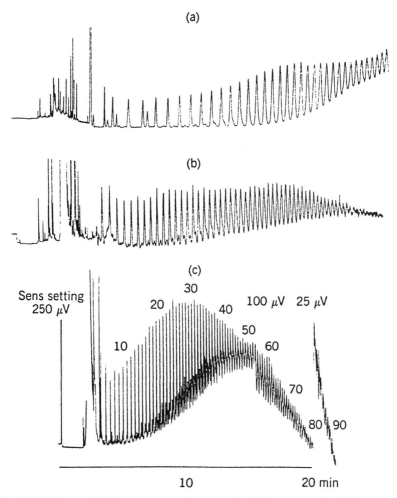

Figure 21.5. Influence of a fluorescent tag on the separation of a dextran standard with an average molecular weight of 18,300 Da. The reagents used were (a) 2-aminopyridine, (b) 5-aminonaphthalene-2-sulfonate, and (c) 8-aminonaphthalene-1,3,6-trisulfonate. Conditions: -500 V/cm (10 µA) with 0.1 M borate–tris at pH 8.65 as the electrolyte. The effective length of the separation capillary was 35 cm. [Reproduced from M. Stefansson and M. Novotny, *Anal. Chem.* **66**, 1134 (1994), with permission of the American Chemical Society.]

Also, the borate complexation will not be uniform throughout the length of a heterogeneous oligosaccharide.

Stefansson and Novotny (56) have recently described electrophoretic separations of monosaccharide enantiomers in borate–oligosaccharide complexation media. Fast separations of certain sugars are shown in

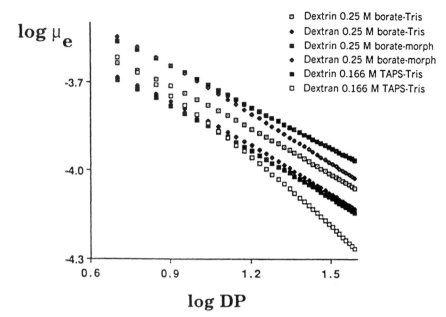

Figure 21.6. Influence of buffer composition on the electrophoretic mobility (μ_e) of ANTS-derivatized dextran and dextrin oligomers (DP = degree of polymerization; TAPS = 3-{[tris(hydroxymethyl)methyl]amino}-1-propanesulfonic acid.) [Reproduced from M. Stefansson and M. Novotny, *Anal. Chem.* **66**, 1134 (1994), with permission of the American Chemical Society.]

Figure 21.7. The linear and cyclic dextrins of the general poly[gly-(1–4)-α-D-Glu] structure were both found effective. The separation selectivity was influenced by the type and concentration of the chiral additive, the concentration of boric acid, and the chemical nature of a fluorescence-tagging reagent. A mechanism for the chiral recognition was proposed for β-cyclodextrin in which a hydrophobic naphthalene derivative tag is inserted into the cyclic saccharide's cavity while the oligosaccharide's hydroxy groups interact more or less selectivily with a monosaccharide structure. With the linear dextrins used as chiral selectors, the derivatized monosaccharides are believed to be inserted into a relatively flexible (56,68) borate-induced dextrin helix, acting as a host molecule. The enantioselectivity was strongly influenced by the number of glucose units in the host oligosaccharide. This selectivity mechanism was tentatively verified later (68) with certain chiral pharmaceuticals.

With a variety of complex oligosaccharide mixtures, including dextrans, amylose, laminarin, and amylopectin fragments (55,57), the selectivity and

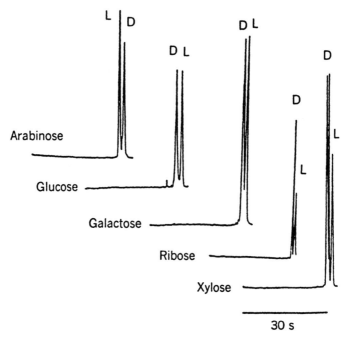

Figure 21.7. Fast enantiomeric separations of derivatized monosaccharides. Electrolyte: 5 mM β-cyclodextrin in 20 mM borate. The extra peak in the xylose separation is the reagent peak. [Reproduced from M. Stefansson and M. Novotny, *J. Am. Chem. Soc.* **115**, 11573 (1993), with permission of the American Chemical Society.]

time of analysis in HPCE appear to be primarily dependent on the number of charges on a fluorescent tag and the concentration of boric acid. However, the notable exceptions causing problems in electrophoresis are two classes of very important polysaccharides: chemically modified celluloses and uniformly charged saccharides (e.g., heparins). Due to the lack of vicinal hydroxy groups in their molecules, chemically modified (water-soluble) celluloses do not complex readily with borate. Heparins, on the other hand, are totally ionized throughout the entire pH range due to the abundance of sulfate groups. Consequently, their electrophoretic mobility is high, resulting in comigration. It has been shown (63) that these hard-to-analyze classes of polysaccharides can be successfully dealt with in HPCE. The relative hydrophobicity of modified celluloses was utilized in binding conventional ionic detergents to provide the needed charge. Similarly, highly charged polysaccharides were modified by polycationic additives (69). For both systems, simple migration models based on secondary thermodynamic equilibria (ion-pairing and adsorption interactions) were presented. The choice of a suitable polycationic additive can be crucial to success of high-efficiency analysis, as shown with very complex polysaccharide mixtures (69).

The current dilemma as to whether to use free-solution media or gel-like structures in complex carbohydrate analyses is gradually being addressed. The early results in HPCE of carbohydrates were obtained with unmodified separation capillaries. These may still be adequate for relatively simple applications. A strong electroosmosis usually complicates resolution of even moderately sized oligosaccharides (46,54), while borate complexation additionally results in charge-to-mass ratios unfavorable to separation. The use of surface-modified capillaries was shown to improve the resolution of lower oligosaccharides separated as aminopyridyl derivatives (49). Earlier we advocated in this laboratory (17,46) that the gel-filled capillaries be used to overcome the unfavorable charge-to-mass ratios for higher oligosaccharides. However, the advances made with buffer optimization in the open-tubular format (with coated capillaries) (55,57) and the use of entangled polymer matrices (69) are likely to make the permanent gels obsolete.

Very large glycoconjugate molecules, such as various polysaccharides and glycosaminoglycans, cannot be passed through the tightly arranged and crosslinked polyacrylamide matrices, the media that are common in the HPCE separations of large oligonucleotides. In highly viscous solutions of various polymers, above the threshold entanglement concentration (70), a "dynamic sieving effect" is encountered, which allows relatively unhindered migration of the charged solutes through uncharged, relatively hydrophilic and flexible polymer networks. Such media have recently shown considerable promise (69) in separating polysaccharides. As demonstrated in Figure 21.8 with a sample of hyaluronic acid, there is a superb resolution of the sample components using this principle. The principal advantage of using entangled polymer solutions as HPCE separation media is that, unlike the permanently fixed gels, they can be replaced following an analysis or a series of analytical runs.

It now appears that various separation media used in HPCE of glycoconjugates may be complementary to each other in solving different analytical tasks. Because of their simplicity, separations in the open-tubular format should be attempted first, and it may well be that, for a majority of glycoprotein-related analyses, these will provide adequate solutions. Most glycans released from glycoproteins do not exceed 20–30 sugar units in size. For larger molecules, such as various natural polysaccharides, the entangled polymer solutions used as sieving media appear to be effective. The use of secondary equilibria (through buffer additives) can also be helpful in enhancing the separation selectivity. Further explorations into new polymeric media and selective interactions are still desirable.

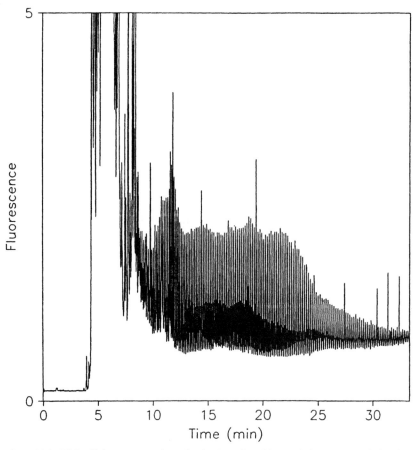

Figure 21.8. High-efficiency separation of a hyaluronic acid sample in an entangled polymer matrix. Conditions 45 cm capillary coated with polyacrylamide, 50 μm i.d.; 25 kV (416 V/cm). Buffer: 25 mM phosphate with 5% linear polyacrylamide, pH 4.6. Hyaluronate oligomers derivatized with aminopyrene-trisulfonic acid. Detection: LIF, 457–514 nm.

An interesting direction for oligomeric separations is the use of so-called *end-label, free-solution electrophoresis* ((ELFSE), as originally proposed for DNA sequencing (71). In the process of electrophoresis, the electrophoretic mobility of a solute is proportional to its charge-to-friction ratio, and this ratio is known to be constant and independent of the solute size for uniformly charged polyelectrolytes. While the use of sieving media has been the common remedy in separating such polyelectrolytes, yet another strategy is to impart additional charge or friction on the solute. This has been utilized recently by

Sudor and Novotny (72) in resolving model carrageenan oligomers. Using 6-aminoquinoline as a fluorescence tag, they were able to reverse electromigration of the carrageenan oligomers, as compared to their migration order with the negatively charged ANTS tag. A superior resolution of the lower oligomers was achieved following the ELFSE approach.

21.6. SELECTED APPLICATIONS

21.6.1. Intact Glycoproteins and Glycopeptides

The possibility of multiple substitution by oligosaccharide chains on the same polypeptide entity often leads to a number of glycoforms. Such microheterogeneity thus presents a formidable analytical challenge, and HPCE is currently one of several available techniques to address the resolution of glycoforms. In particular, HPCE has the potential of a rapid fingerprinting technique to assess many protein variants or the suitability of a biotechnologically produced glycoprotein pharmaceutical.

Kilar and Hjertén (73) were probably the first to demonstrate the merits of HPCE in glycoform separations. Using human transferrin as a model glycoprotein, they used both the isoelectric focusing and zonal modes of HPCE. To provide suitably charged molecules for such experiments, the negatively charged sialic acid residues were first removed enzymatically. Comparing different HPCE techniques was also reported by Yim (74), who studied the human recombinant plasminogen activator with its many possible glycoforms.

Depending on the glycoprotein type, separation of the individual glycoforms can be a task of varying difficulty. For instance, the case of RNase B (75,76) presents a relatively small number of glycoforms, whereas ovalbumin appears as an exceedingly complex example (64). Other reports in the literature show the glycoforms (through UV detection) from pepsin (64), a recombinant granulocyte-colony-stimulating factor (77), human erythropoietin (78,79), and immunoglobulin G (IgG) (80). Although HPCE alone usually falls short of separating very complex glycoform patterns, this does not diminish its value as a rapid fingerprinting technique.

The choice of appropriate buffer systems and column types is important in the optimization of glycoprotein profile runs. This has been demonstrated by the examples of RNase B (75) and ovalbumin and pepsin glycoforms (64). In both applications, borate complexation was found to play a beneficial role.

In the report by Rudd and coworkers on RNase B, a useful application was demonstrated in which HPCE was effectively utilized to monitor the course of enzymatic cleavage of the mannose residues. This is shown in Figure 21.9 (75):

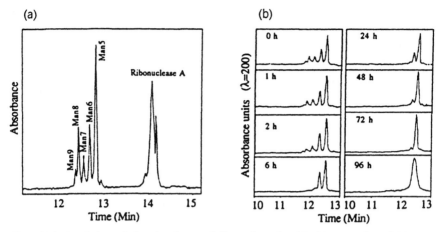

Figure 21.9. Resolution of the glycoforms of ribonuclease by CE: (a) separation of RNase A (non-glycosylated) from the glycoforms of RNase B; (b) the course of RNase B enzymatic digestion. Capillary: 75 μm i.d. × 72 cm. Detection: UV at 200 nm. Buffer: 20 mM phosphate with 50 mM sodium dodecyl sulfate and 50 mM borate, pH 7.2. [Reproduced from P. M. Rudd, I. G. Scragg, E. Coghill, and R. A. Dwek, *Glycoconjugate J.* **9**, 86 (1992), with the permission of the *Glycoconjugate Journal*.]

in part a, a mixture of RNase A (the non-glycosylated variety) and RNase B, with its glycoforms, is displayed; in part b, a series of runs are shown, comparing the reaction products of RNase B and an α(1–2)-mannosidase analyzed at different times.

HPCE has become recently recognized as a powerful tool for peptide mapping. Recognizing the complementarity of HPCE with the modern techniques of liquid chromatography (81,82), and reversed-phase LC, in particular, Jorgenson and coworkers (83,84) put a major effort into coupling these techniques into an integrated system with enormous resolving power. Additionally, peptide-mapping techniques have been scaled down to work with very small quantities of the initial protein materials (85–88), and further sensitivity increases through fluorescent tagging and LIF (87) are likely to benefit the field of glycoprotein analysis.

A more detailed description of techniques and systems in peptide analysis by HPCE would be beyond the scope of this chapter. Due to their different electromigration behavior, glycosylated peptides can often be easily traced in complex electropherograms comparing differently glycosylated proteins. Peptide-mapping procedures thus add significantly to our capabilities of studying protein microheterogeneities.

In an application to human α_1-acid glycoprotein, Nashabeh and El Rassi (89) combined the peptide-mapping HPCE procedure with the time-honored

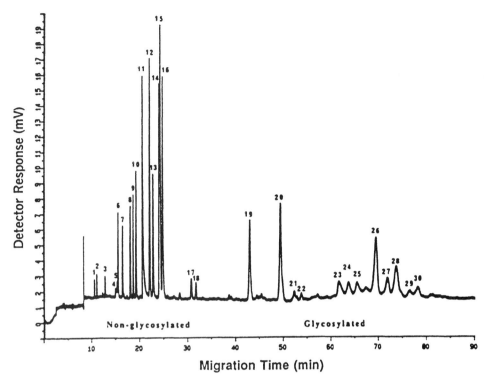

Figure 21.10. A tryptic map of recombinant human erythropoietin recorded under ion pairing with heptanesulfonic acid. Capillary: 50 μm i.d. × 75 cm (50 cm effective length). Buffer: 40 mM sodium phosphate, pH 2.5, plus 100 mM heptanesulfonic acid as the ion-pairing agent. [Reproduced from R. S. Rush, P. L. Cerby, T. W. Stickland, and M. F. Rhode, *Anal. Chem.* **65**, 1834 (1993), with permission of the American Chemical Society.]

lectin chromatography approach to distinguish glycopeptides from other cleavage fragments. Another application (90) uniquely devoted to the analysis of glycopeptides has utilized an ion-pairing technique to permit separation of the tryptic fragments obtained from a recombinant human erythropoietin into the glycosylated and non-glycosylated fractions (Figure 21.10). These two applications are good examples of combining selective principles with the efficiency of HPCE.

21.6.2. Glycoprotein Compositional Analysis and Oligosaccharide Mapping

The most essential pieces of information concerning a purified glycoprotein pertain to the knowledge of its monosaccharide composition, variation of the

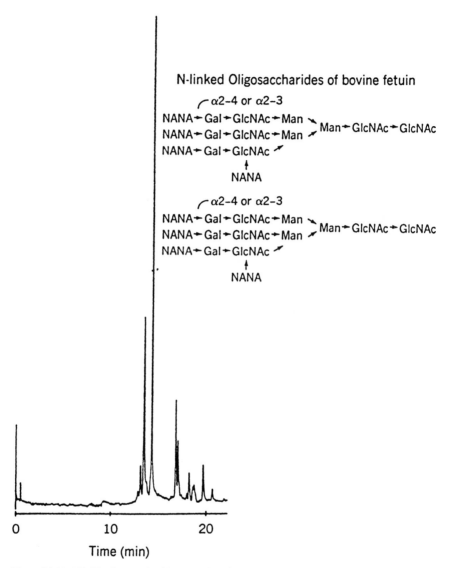

Figure 21.11. HPCE oligosaccharide map of N-linked oligosaccharides released from bovine fetuin by hydrazinolysis. (NANA = N-acetylneuraminic acid.) Capillary: 50 μm i.d. × 90 cm (60 cm effective length). Buffer: 20 mM phosphate/20 mM borate, pH 9.50. [Reproduced from J. Liu, O. Shirota, D. Wiesler, and M. Novotny, *Proc. Natl. Acad. Sci. U.S.A.* **88**, 2302 (1991), with permission.]

individual glycan moieties, and their incorporation within the polypeptide backbone. HPCE appears capable of delivering such data at very high sensitivity. However, just as with other analytical methodologies, the complete analysis of glycoproteins may involve a number of sample manipulation steps (hydrolysis, chromatographic preconcentration, treatment with enzymes, etc.) that will all influence the overall sensitivity.

Several examples of the potential of HPCE in glycoprotein glycan analysis have recently been presented in the literature. Using a borate-complexing medium, Honda et al. (49) analyzed monosaccharides from ovalbumin as aminopyridyl derivatives. Liu et al. (47) employed CBQCA reagent and LIF detection to analyze various types of hydrolyzates of bovine fetuin.

The enzymatically released oligosaccharides were also analyzed in ovalbumin by Honda et al. (49), and in α_1-acid human glycoprotein by Nashabeh and El Rassi (89). Liu et al. (47) recorded N-linked oligosaccharides, prepared by hydrazinolysis, as CBQCA derivatives. The oligosaccharide map (N-linked species) from bovine fetuin is displayed in Figure 21.11 as an example.

Following the implementation of suitable sample treatment procedures, it is expected that HPCE with LIF will become a very useful, highly sensitive tool in the analysis of minute quantities of glycoproteins. This important area has thus far been hampered by (i) slow progress in devising highly effective derivatization techniques at very low analyte levels and (ii) limited availability of appropriate oligosaccharide standards for profile comparisons. An example of comparative runs is shown in Figure 21.12 (91) with ANTS-derivatized oligosaccharides available as the human IgG "library" and a neutral biantennary oligosaccharide isolated from human fibrinogen. As such deficiencies with standards are gradually overcome, more "real-world applications" will undoubtedly appear in the literature. Thus far, the most common cases of HPCE oligosaccharide analyses come from the model glycoproteins whose structural details were previously established by techniques other than HPCE.

21.6.3. Analysis of Polysaccharides

The importance of polysaccharides to biochemistry, medicine, the food industry, nutrition, etc., has been steadily increasing. Unfortunately, the analytical methodologies in this area have not been adequately developed to permit the needed structure–property investigations. The most pressing problems have been the limited separation capabilities and sensitivity of detection. The combination of HPCE with LIF offers unique possibilities in this long-neglected area.

Important lessons for the HPCE separations of polysaccharides and large proteoglycan molecules can be derived from a fairly extensive literature on their degradation products in slab gels. Intact polysaccharides and proteo-

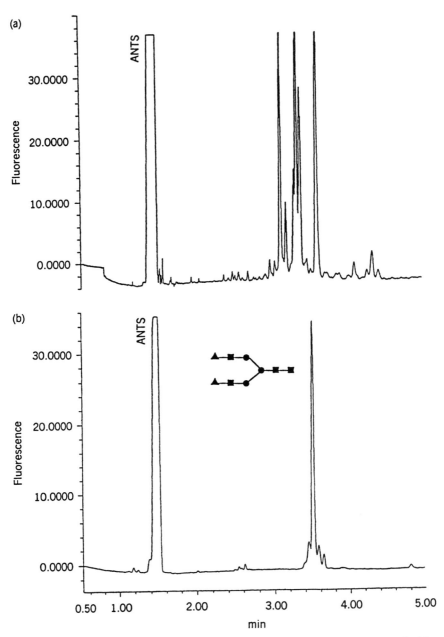

Figure 21.12. CE separation of ANTS-derivatized complex carbohydrates: (a) oligosaccharide "library" from human IgG; (b) neutral biantennary oligosaccharide isolated from human fibrinogen. Key: (■) N-acetylglucosamine; (▲) galactose; (●) mannose. Conditions: 50 mM phosphate buffer, pH 2.5. Capillary: 50 μm i.d. × 27 cm (20 cm effective length). [Reproduced from A. Klockow, H. M. Widmer, R. Amado, and A. Paulus, *Fresenius J. Anal. Chem.* **350**, 415 (1994), with permission of Springer-Verlag.]

glycans are biomolecules of enormous complexity, existing moreover in various aggregated forms. It seems that comprehensive studies need both the capabilities of separating various intact molecules and the availability of reproducible techniques for their controlled degradation and further separation. Potentially, HPCE can play a role in both directions. However, at this stage, studies of polymer degradation products appear methodologically simpler than investigations of large polysaccharide polyelectrolytes that often exhibit peculiar physical and chemical properties.

Sulfated residues of chondroitins, dermatans, keratans, heparans, and heparins, as well as hyaluronic acid and its derivatives, provide the natural charged moieties for electrophoretic separations. Various glycosaminoglycans can be depolymerized to a convenient molecular size through the use of several polysaccharide lyases. The hydrolysis products featuring unsaturation can be detected by UV absorption at 232 nm, as demonstrated by Linhardt's group [see Al-Hakim and Linhardt (92); Ampofo et al. (93)] and Carney and Osborne (94). This general approach has been exemplified with an HPCE separation of chondroitin sulfate disaccharides hydrolyzed from a beagle cartilage sample (94). In the low-molecular-weight range, separations of this kind can be accomplished in open tubes.

Examples of oligosaccharide separations performed in a free buffer medium include those of maltodextrin (54,68), branched oligosaccharides of the xyloglucan type (50), and others. As the number of sugar units in oligosaccharide molecules increases, separations become more difficult. Yet, fairly complex mixtures of various oligosaccharides have recently been shown at high resolution (55,57). The resolving power of a typical HPCE system is exemplified by an electropherogram of corn amylose oligomers (Figure 21.13) (57) showing minor peaks between the major oligomers detected as ANTS derivatives. The mixtures of various other natural materials (amylopectin, lichenan, pullulan, etc.) often show satellite peaks in addition to the major oligosaccharide patterns. Based on information from other measurement techniques, branching of glycosaccharide chains has been suggested in such materials, with little experimental proof. It now appears that HPCE (57), in conjunction with the analytical use of debranching enzymes, can address this significant issue.

To increase the scope of HPCE applications to larger polysaccharides, it is essential to resolve effectively the molecules of interest according to their molecular size because of the possibilities of sample polydispersity, while other structural types due to branching, modification of some sugar units, etc., can further complicate the overall resolution. Fluorescent tagging provides the best opportunity for detection. However, the fact that there may be only a limited number of derivatization sites in a polymer molecule will challenge even the best detection technologies.

Separation of large oligosaccharides in entangled polymer solutions has already been exemplified with a run of a hyaluronic acid sample (shown in

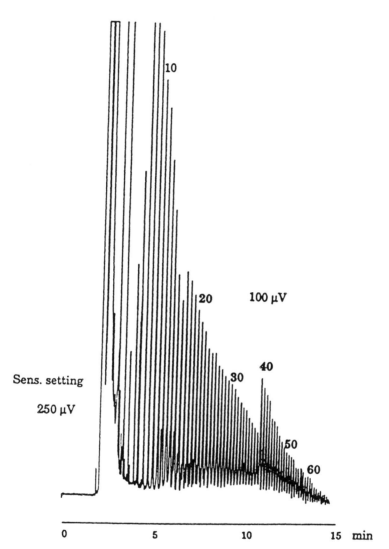

Figure 21.13. Separation of the corn amylose oligomers. Conditions: 0.2 M borate–Tris, pH 8.65; − 500 V/cm (18 μA). Capillary: 50 μm i.d. and (35 cm effective length). The numbers refer to the degree of polymerization. [Reproduced from M. Stefansson and M. Novotny, *Carbohydrate Res.* **258**, 1 (1994), with permission of Elsevier Science Publishers.]

Figure 21.8). The work in this area remains preliminary. However, it can be safely concluded that HPCE of fluorescently labeled polysaccharides has significant potential. Separations of very large molecules can also be approached by HPCE utilizing a pulsed-field mode of operation (95,96). Polysaccharide analysis probably represents the most challenging area for HPCE techniques.

LIST OF ACRONYMS AND ABBREVIATIONS

Acronym or Abbreviation	Definition
ANDSA	7-aminonaphthalene-1,3-disulfonic acid
ANTS	8-aminonaphthalene-1,3,6-trisulfonic acid
CBQCA	3-(4-carboxybenzoyl)-2-quinolinecarboxaldehyde
CE	capillary electrophoresis
ELFSE	end-label, free-solution electrophoresis
FAB	fast atom bombardment
HPCE	high-performance capillary electrophoresis
HPLC	high-performance liquid chromatography
IgG	immunoglobulin G
LC	liquid chromatography
LIF	laser-induced fluorescence
MALDI	matrix-assisted laser desorption ionization
MS	mass spectrometry
NANA	N-acetylneuraminic acid
PMP	3-methyl-1-phenyl-2-pyrazolin-5-one (also called 1-phenyl-3-methyl-5-pyrazolone, whence the acronym)
TAPS	3-{[tris(hydroxymethyl)methyl]amino}-1-propanesulfonic acid
Tris	tris(hydroxymethyl)aminomethane
UV	ultraviolet

ACKNOWLEDGMENTS

Over the last several years, our research on glycoconjugates has been supported by grants #GM24349 from the Institute of General Medical Sciences, the U.S. Department of Health and Human Services, and #CHE-9321431 from the National Science Foundation.

REFERENCES

1. N. Sharon and H. Lis, *Sci. Am.*, January, 82–89 (1993).
2. G. W. Hart, *Curr. Opin. Cell Biol.* **4**, 1017 (1992).

3. A. B. Foster, *Methods Carbohydr. Chem.* **12**, 81 (1957).
4. J. G. Dawber and G. E. Hardy, *J. Chem. Soc. Faraday Trans.* **80**, 2467 (1984).
5. T. Patel, J. Bruce, A. Merry, C. Bigge, M. Wormald, A. Jaques, and R. Parekh, *Biochemistry* **32**, 679 (1993).
6. S. Hase, H. Oku and T. Ikenaka, *Anal. Biochem.* **167**, 321 (1987).
7. S. Hase and T. Ikenaka, *Anal. Biochem.* **184**, 135 (1990).
8. D. C. Johnson, *Nature (London)* **321**, 451 (1986).
9. M. R. Hardy and R. R. Townsend, *Proc. Natl. Acad. Sci. U.S.A.* **85**, 3289 (1988).
10. J. W. Jorgenson and K. D. Lukacs, *Anal. Chem.* **53**, 1298 (1981).
11. J. W. Jorgenson, in *New Directions in Electrophoretic Methods* (J. W. Jorgenson and M. Phillips, eds.), pp. 182–198. American Chemical Society, Washington, DC 1987.
12. S. Hoffstetter-Kuhn, A. Paulus, E. Gassman, and H. M. Widmer, *Anal. Chem.* **63**, 1541 (1991).
13. S. Honda, S. Iwase, A. Makino, and S. Fujiwara, *Anal. Biochem.* **176**, 72 (1989).
14. W. Nashabeh and Z. El Rassi, *J. Chromatogr.* **514**, 57 (1990).
15. S. Honda, S. Suzuki, A. Nose, K. Yamamoto, and K. Kakehi, *Carbohydr. Res.* **215**, 193 (1991).
16. P. Oefner and H. Scherz, in *Adv. Electrophor.* 157–224 (1994).
17. J. P. Liu, V. Dolnik, Y.-Z. Hsieh, and M. Novotny, *Anal. Chem.* **64**, 1328 (1992).
18. L. A. Colon, R. Dadoo, R. N. Zare, *Anal. Chem.* **65**, 476 (1993).
19. R. P. Baldwin and J. Hong, *Proc. Int. Symp. on Capillary Chromatogr. Electrophor. 17th*, Wintergreen, VA, *1995*, p. 18 (1995).
20. G. Diebold and R. N. Zare, *Science* **196**, 1439 (1977).
21. M. Sepaniak and E. Yeung, *J. Chromatogr.* **190**, 377 (1980).
22. Y. F. Cheng and N. J. Dovichi, *Science* **242**, 562 (1988).
23. N. W. C. Chan, K. Stangier, R. Sherburne, D. E. Taylor, Y. Zhang, N. J.Dovichi, and M. M. Palcic, *Glycobiology* **5**, 683 (1995).
24. T. W. Garner and E. S. Yeung, *J. Chromatogr.* **515**, 639 (1990).
25. M. A. Moseley, L. J. Deterding, J. S. M. de Wit, K. Tomer, R. T. Kennedy, N. Bragg, and J. W. Jorgenson, *Anal. Chem.* **61**, 1577 (1989).
26. R. M. Caprioli, W. T. Moore, M. Martin, B. B. DaGue, K. Wilson, and S. Moring, *J. Chromatogr.* **480**, 247 (1989).
27. N. J. Reinhoud, E. Schroder, V. R. Tjaden, V. M. A. Niessen, M. D. Ten Noever de Brauw, and J. van der Greef, *J. Chromatogr.* **516**, 147 (1990).
28. K. L. Duffin, J. K. Welply, E. Huang, and J. D. Henion, *Anal. Chem.* **64**, 1440 (1992).
29. V. Ling, A. W. Guzzetta, E. Canova-Davis, J. T. Stults, W. S. Hancock, T. Covey, and B. Shushan, *Anal. Chem.* **63**, 2909 (1991).

30. M. J. Huddleston, M. F. Bean, and S. A. Carr, *Anal. Chem.* **65**, 877 (1993).
31. S. A. Carr, M. J. Huddleston, and M. F. Bean, *Protein Sci.* **2**, 183 (1993).
32. M. R. Emmett and R. M. Caprioli, *J. Am. Soc. Mass Spectrom.* **5**, 605 (1994).
33. M. S. Kriger, K. D. Cook, and R. S. Ramsey, *Anal. Chem.* **67**, 385 (1995).
34. S. A. Hofstadler, F. D. Swanek, D. C. Gale, A. G. Ewing, and R. D. Smith, *Anal. Chem.* **67**, 1477 (1995).
35. K. K. Mock, M. Davey, and J. S. Contrell, *Biochem. Biophys. Res. Commun.* **177**, 644 (1991).
36. B. Spengler, D. Kirsch, R. J. Kaufmann, and J. Lemoine, *J. Mass Spectrom.* **30**, 782 (1995).
37. C. W. Sutton, J. A. O'Neill, and J. S. Cottrell, *Anal. Biochem.* **218**, 34 (1994).
38. M. C. Huberty, J. E. Vath, W. Yu, and S. A. Martin, *Anal. Chem.* **65**, 2791 (1993).
39. B. Stahl, M. Steup, M. Karas, and F. Hillenkamp, *Anal. Chem.* **63**, 1463 (1991).
40. R. S. Bordoli, K. Howes, R. G. Vickers, R. H. Bateman, and D. J. Harvey, *Rapid Commun. Mass Spectrom.* **8**, 585 (1994).
41. G. Talbo and M. Mann, *Rapid Commun. Mass Spectrom.* **10**, 100 (1996).
42. J. A. Castoro, R. W. Chiu, C. A. Monnig, and C. L. Wilkins, *J. Am. Chem. Soc.* **114**, 7571 (1992).
43. R. W. Chiu, K. L. Walker, J. Hagen, C. A. Monnig, and C. L. Wilkins, *Anal. Chem.* **67**, 4190 (1995).
44. K. L. Walker, R. W. Chiu, C. A. Monnig, and C. L. Wilkins, *Anal. Chem.* **67**, 4197 (1995).
45. P. Chen, A. G. Baker, and M. V. Novotny, *Anal. Biochem* **244**, 144 (1997).
46. J. Liu, O. Shirota, and M. Novotny, *J. Chromatogr.* **559**, 223 (1991).
47. J. Liu, O. Shirota, D. Wiesler, and M. Novotny, *Proc. Natl. Acad. Sci. U.S.A.* **88**, 2302 (1991).
48. O. Shirota, D. Rice, and M. Novotny, *Anal. Biochem.* **205**, 189 (1992).
49. S. Honda, A. Makino, S. Suzuki, and K. Kakehi, *Anal. Biochem.* **191**, 228 (1990).
50. W. Nashabeh and Z. El Rassi, *J. Chromatogr.* **600**, 279 (1992).
51. S. Honda, K. Yamamoto, S. Suzuki, M. Ueda, and K. Kakehi, *J. Chromatogr.* **558**(3), 27 (1991).
52. K. Kakehi, S. Suzuki, S. Honda, and Y. C. Lee, *Anal. Biochem.* **199**, 256 (1991).
53. J. Liu, O. Shirota, and M. Novotny, *Anal. Chem.* **63**, 413 (1991).
54. C. Chiesa and Cs. Horvath, *J. Chromatogr.* **645**, 337 (1993).
55. M. Stefansson and M. Novotny, *Anal. Chem.* **66**, 1134 (1994).
56. M. Stefansson and M. Novotny, *J. Am. Chem. Soc.* **115**, 11573 (1993).
57. M. Stefansson and M. Novotny, *Carbohydr. Res.* **258**, 1 (1994).
58. R. A. Evangelista, M.-S. Liu, and F.-T. A. Chen, *Anal. Chem.* **67**, 2239 (1995).
59. F.-T. A. Chen, and R. Evangelista, *Anal. Biochem.* **230**, 273 (1995).
60. Y. Mechref and Z. El Rassi, *Electrophoresis* **15**, 627 (1994).

61. O. Shirota, Ph.D. Thesis, Indiana University, Bloomington, 1992.
62. M. V. Novotny and J. Sudor, *Electrophoresis* **14**, 373 (1993).
63. M. Stefansson and M. Novotny, *Anal. Chem.* **66**, 1134 (1994).
64. J. P. Landers, B. J. Madden, R. P. Oda, and T. C. Spelsberg, *Anal. Biochem.* **205**, 115 (1992).
65. Y. Liu and K. F. Chan, *Electrophoresis* **12**, 402 (1991).
66. Y. W. Yoo, Y.-S. Kim, G. J. Jhon, and J. Park, *J. Chromatogr.* **652**, 431 (1993).
67. Z. El Rassi and W. Nashabeh, in *Carbohydrate Analysis* (Z. El Rassi, ed.), J. Chromatogr. Lib., Vol. 58, p. 351. Elsevier, Amsterdam, 1995.
68. H. Soini, M. Stefansson, M.-L. Riekkola, and M. V. Novotny, *Anal. Chem.* **66**, 3477 (1994).
69. M. Stefansson, J. Sudor, J. Chmelikova, M. Hong, J. Chmelik, and M. V. Novotny, *Anal. Chem.* **69**, 3846 (1997).
70. P. D. Grossman and D. S. Soane, *Biopolymers* **31**, 1221 (1991).
71. P. Meyer, G. W. Slater, and G. Drouin, *Anal. Chem.* **66**, 1777 (1994).
72. J. Sudor and M. V. Novotny, *Anal. Chem.* **67**, 4205 (1995).
73. F. Kilar and S. Hjertén, *J. Chromatogr.* **480**, 351 (1989).
74. K. W. Yim, *J. Chromatogr.* **559**, 401 (1991).
75. P. M. Rudd, I. G. Scragg, E. Coghill, and R. A. Dwek, *Glycoconjugate J.* **9**, 86 (1992).
76. M. Huang, J. Plocek, and M. V. Novotny, *Electrophoresis* **16**, 396 (1995).
77. E. Watson and F. Yao, *J. Chromatogr.* **630**, 442 (1993).
78. A. D. Tran, S. Park, P. J. Lisi, O. T. Huynh, R. R. Ryall, and P. A. Lane, *J. Chromatogr.* **542**, 459 (1991).
79. E. Watson and F. Yao, *Anal. Biochem.* **210**, 389 (1993).
80. R. Vincentelli and N. Bihoreau, *J. Chromatogr.* **641**, 383 (1993).
81. J. Frenz, S.-L. Wu, and W. S. Hancock, *J. Chromatogr.* **480**, 379 (1989).
82. R. G. Nielsen, R. M. Riggin, and E. C. Rickard, *J. Chromatogr.* **480**, 393 (1989).
83. J. P. Larmann, Jr., A. V. Lemmo, A. W. Moore, Jr., and J. W. Jorgenson, *Electrophoresis* **14**, 439 (1993).
84. A. J. Moore, Jr. and J. W. Jorgenson, *Anal. Chem.* **67**, 3448 (1995).
85. K. Cobb and M. Novotny, *Anal. Chem.* **61**, 2226 (1989).
86. K. Cobb and M. Novotny, *Anal. Chem.* **64**, 879 (1992).
87. K. Cobb and M. Novotny, *Anal. Biochem.* **200**, 149 (1992).
88. L. Licklider, W. G. Kuhr, M. P. Lacey, T. Keough, M. Purdon, and R. Takigiku, *Anal. Chem.* **67**, 4170 (1995).
89. W. Nashabeh and Z. El Rassi, *J. Chromatogr.* **573**, 31 (1991).
90. R. S. Rush, P. L. Cerby, T. W. Stickland, and M. F. Rhode, *Anal. Chem.* **65**, 1834 (1993).
91. A. Klockow, H. M. Widmer, R. Amado, and A. Paulus, *Fresenius J. Anal. Chem.* **350**, 415 (1994).

92. A. Al-Hakim and R. J. Linhardt, *Anal. Biochem.* **195**, 68 (1991).
93. S. A. Ampofo, H. M. Wang, and R. J. Linhardt, *Anal. Biochem.* **199**, 249 (1991).
94. S. L. Carney and D. J. Osborne, *Anal. Biochem.* **195**, 132 (1991).
95. J. Sudor and M. Novotny, *Proc. Natl. Acad. Sci. U.S.A.* **90**, 9451 (1993).
96. J. Sudor and M. Novotny, *Nucleic Acids Res.* **23**, 2538 (1995).

CHAPTER

22

DNA SEQUENCING BY MULTIPLEXED CAPILLARY ELECTROPHORESIS

EDWARD S. YEUNG and QINGBO LI

Ames Laboratory–USDOE and Department of Chemistry, Iowa State University, Ames, Iowa 50011

22.1.	**Highly Multiplexed DNA Sequencing**	768
	22.1.1. Optics	770
	22.1.2. The Imaging Detector	771
22.2.	**Acceleration of Electrophoretic Runs**	774
	22.2.1. A Novel Sieving Medium	775
	22.2.2. The Column Treatment Protocol	779
22.3.	**Base Calling and Data Handling**	780
22.4.	**System Integration**	783
22.5.	**Future Prospects**	785
List of Acronyms and Symbols		786
References		787

The Human Genome Project is an initiative to sequence the entire human genome, which consists of 3×10^9 bp of nucleic acids. The ability to decipher our entire genetic code promises to revolutionize biology by bringing our understanding to the finest molecular details. From a practical standpoint, the diagnosis of diseases and the development of treatment should both benefit from these advances.

The magnitude of the problem is immense. Just to read the A, C, T, and G alphabets out loud at four per second will take 24 years. To print one human genome at 60 characters per line and 50 lines per page will result in an encyclopedia occupying 142 ft of shelf space. At the present time, a good molecular biology laboratory can sequence about 3×10^6 bp of genomic DNA per year. This means that it will take 1000 years to sequence the genome once.

High Performance Capillary Electrophoresis, edited by Morteza G. Khaledi. Chemical Analysis Series, Vol. 146.
ISBN 0-471-14851-2 © 1998 John Wiley & Sons, Inc.

At that rate, the practical utility is debatable. The cost of sequencing at $1–5/bp is also prohibitive. Recent developments have shown that major centers can be set up to scale up current technology substantially in terms of speed and throughput (1). Still, one needs a major departure from current technology to go the next step in sequencing the human genome economically and in a timely manner. This will become even more critical as we start to use the information clinically, when the process will have to be repeated for every patient.

It is important to note that an integrated approach is critical to success. What works best when single capillaries are used may not be transferable to a large-scale multiplexed capillary array. It may therefore be necessary to sacrifice some of the performance level of each of the critical technologies to achieve a workable compromise for large-scale applications. One does not necessarily strive for the best detection limits, the most efficient separations, or the most sophisticated hardware and software. We need, however, a set of technologies that can function in concert to achieve the final goal of high-speed, high-throughput DNA sequencing.

22.1. HIGHLY MULTIPLEXED DNA SEQUENCING

It is obvious that irrespective of whichever basic technology is eventually selected to sequence the entire human genome, there are substantial gains to be made if a high degree of multiplexing of parallel runs can be implemented. Such multiplexing should not involve expensive instrumentation and should not require additional personnel, or else the main objective of cost reduction will not be satisfied even though the total time for sequencing is reduced. Unfortunately, slab gels are not readily amenable to a high degree of multiplexing and automation. Difficulties include the following: uniform gel preparation over a large area; reproducibility over different gels; reusability; loading of sample wells; the large physical size of the medium; uniform cooling; the large amounts of media, buffer, and samples needed; and long run times for extended reading of bases. The introduction of ultrathin slab gels (2,2a) does address some of these difficulties. Still, it is hard to imagine how 1000 lanes can be run simultaneously in one instrument or how these gels can be used repeatedly.

Several research groups have shown that capillary gel electrophoresis (CGE) is an attractive alternative to slab gel electrophoresis (SGE) for DNA sequencing (3–12). The medium used (cross-linked polyacrylamide), buffer composition, separation mechanism, sequencing chemistry, and tagging chemistry for CGE are all derived from proven SGE schemes. A 25-fold increase in the sequencing rate per capillary has already been demonstrated. This is a direct consequence of the small internal diameter of the capillary tubes, typically around 50–75 μm, greatly reducing Joule heating associated with the electric current. Gel distortions and temperature gradients that can affect resolution of the bands are thus virtually absent. More importantly, much higher electric fields can be applied to speed up the separation. For comparison, conventional SGE are limited to field strengths below 50 V/cm, whereas CGE have been successfully used for sequencing up to 500 V/cm (5). The unique aspect ratio of capillaries (25–50 cm long) provides uniform field strengths, and the large surface-to-volume ratio favors efficient heat removal. These combine to produce much sharper bands than are possible in slab gels.

Part of the improvement in sequencing speed in CGE is counteracted by the inherent ability of slab gels for accommodating multiple lanes in a single run. Fortunately, the capillary format is in fact well suited for multiplexing. Already mentioned is the substantial reduction of Joule heating per lane, even given the high applied fields, so that the overall cooling requirement and electrical requirement remain manageable. The cost of materials per lane is much reduced because everything, including sample sizes, is smaller. The reduced band dimensions are ideal for excitation by laser beams and for imaging onto solid-state array detectors. The use of electromigration injection provides reproducible sample introduction with little band spreading and with little effort. The small band sizes, however, put stringent requirements on detection. It is not possible to use large amounts of DNA fragments because of overloading and plugging of the gel pores (4).

Parallel sequencing runs in a set of 24 capillaries have been demonstrated recently (13–15). To provide sensitive laser-excited fluorometric detection, a confocal illumination geometry couples a single laser beam to a single photomultiplier tube. Observation is done one capillary at a time, and the capillary bundle is translated across the excitation/detection region at 20 mm/s by a mechanical stage. This provides adequate observation time for each capillary (to achieve a reasonable signal-to-noise ratio, S/N, for base calling) but fast enough to repeat the scan every second (to achieve reasonable temporal resolution). Those studies clearly demonstrate that capillaries can be run in parallel for DNA sequencing.

There are some subtle features inherent to the confocal excitation scheme (13–15) that may limit its use for very large numbers (hundreds) of capillaries. Since data acquisition is sequential and not truly parallel, the ultimate sequencing speed will be determined by the observation time needed per DNA band for an adequate S/N. Having more capillaries in the array or being able to translate the array across the detection region faster will not increase the overall sequencing speed. The use of a translational stage may become problematic for a large capillary array. The confocal geometry also requires critical alignment and positioning. The translational movement has to maintain flatness to the order of the confocal parameter (around 25 µm) or else the cylindrical capillary walls will distort the spatially selected image.

In another setup (16,17), multiple sheath-flow and four-color detection has been developed. Twenty gel-filled capillaries were lined up and were placed into a flow cell to couple with 20 large-i.d., gel-free capillaries. Two laser beams are combined into one to cross the flow streams in a line for excitation, and a charge-coupled device (CCD) was used for simultaneous detection perpendicular to the excitation beam. The main advantage is superior stray-light rejection in the sheath flow (18,19). The main challenges in scaling up to thousands of capillaries are alignment of the individual sheath flows, possible turbulence in the flow paths, and improper matching of the laser beam waist over a long distance with the core diameters containing the eluted fragments. A more compact version of the sheath-flow multiple capillary scheme has been reported (20). As shown below, a workable multiplexed capillary electrophoresis (CE) system should have provisions for flushing out and replacing the sieving matrix at regular intervals to maintain good separation efficiency, perhaps even after every run. While technically feasible, the sheath-flow cell will probably have to be dismantled during the flush cycle to avoid cross-contamination. Apparently, the separation medium has to be immobilized on the column (via polymerization *in situ*) to isolate it from the sheath flow (20). There is therefore good incentive to perform detection directly on the capillary columns: one is thus trading sensitivity for convenience. We recognize that the Sanger reaction products from standard front-end protocols contain pretty much fixed *concentrations* of DNA fragments. That is, if the *concentrations*, of the template, primer, polymerase, or terminators were altered substantially, one would not obtain the nice even distribution of fragments in the ladder. So, what is needed is adequate concentration sensitivity for today's Sanger mixtures and not necessarily the ultimate absolute sensitivity.

Other multiplexed CE schemes have been suggested recently. An interesting format is the use of machined channels (21,22). We note that it will still be necessary to have 30–50 cm long channels to provide adequate resolution of the fragments. The glass or silicon surfaces of these channels will need new surface treatment procedures to avoid electroosmotic flow or other interactions with the DNA. Sample injection and sieving-matrix replacement will need to be implemented. With today's technology, multiple capillaries still provide the most mature format for multiplexed CE runs.

22.1.1. Optics

Initial work in our laboratory (23,24) centered around excitation via optical fibers inserted into the ends of the separation capillaries (axial-beam mode) and detection perpendicular to the capillary array with a CCD camera.

Advantages include easy and efficient coupling of the laser beam to each capillary, the complete lack of moving parts, and a mechanical alignment system that is rugged and reproducible. The main disadvantage is the intrusion of the optical fiber into the separation capillary, affecting electroosmotic flow and increasing the likelihood of contamination and column blockage.

We recently evaluated the various design parameters that are important for parallel laser-excited fluorescence detection in large CE arrays. A functional system with an array of 100 capillaries was then successfully assembled and tested (25). A schematic diagram of our fluorescence detection system for multicapillary electrophoresis is shown in Figure 22.1.

The optical design is based on several important principles. First, the laser is a broadband continuous-wave Ar-ion laser identical to those used in the commercial fluorescence-based DNA sequencer. Eventual adaptation to DNA sequencing based on proven dye technology is straightforward. Second, a beam expander is used to lengthen the focused line to cut across a large number of capillaries. Third, a camera lens is used for light collection and imaging. This offers a larger observation field compared to a microscope objective to image the larger number of capillaries. A wide-angle lens provides a lower magnification ratio to match the 75 µm capillaries to the 25 µm pixels in the CCD element. An extension tube is used to allow close focusing and thus a better effective f-number without additional image distortion. Actually, even at the 4000 capillary level, the image resolution required is substantially below that found in high-quality photographic applications.

Although separation is performed in 102 discrete capillaries, the optical signals can interfere with each other. Spread of the excitation beam over to adjacent capillaries is not a problem, since each capillary is excited along the same focused line by design. The reverse of this is of concern, however, i.e., fluorescence from adjacent capillaries being refracted by the cylindrical walls to reach the CCD camera. Immersing the capillaries in water, which roughly matches the refractive index, is a solution. Fairly even signals from all the capillaries are obtained (26). Crosstalk between the separation channels is another important issue, especially for a multiplexed detection system. For example, if one capillary (150 µm o.d.) is focused down to the size of only one pixel of the CCD, crosstalk cannot be avoided. However, if the image size of one capillary covers two pixels so that the liquid core exactly matches one pixel and the adjacent capillary walls cover the next pixel, virtually no crosstalk ($< 1\%$) can be observed (26).

22.1.2. The Imaging Detector

We investigated the possibility of using another type of charge-transfer device—a charge-injection device (CID)—as an array detector for DNA

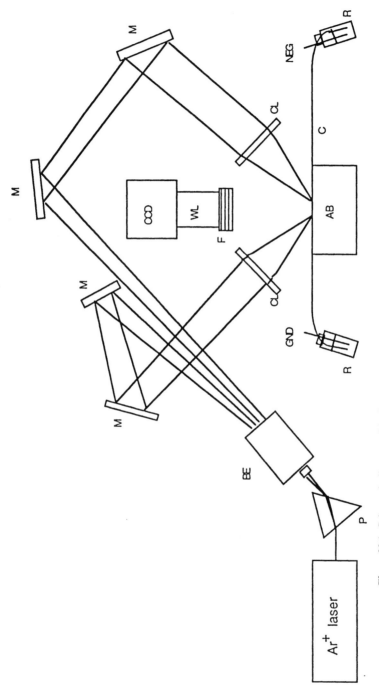

Figure 22.1. Schematic diagram of the fluorescence detection system for multicapillary electrophoresis. Key: CCD, charge-coupled device camera; F, cutoff and long-pass filters; WL, wide-angle lens with extension tube; BE, beam expander; C, capillary; M, mirror; CL, planoconvex cyiindrical lens; AB, aluminum block; P, prism; R, buffer reservoir; GND, ground potential; NEG, negative high voltage. [Reproduced from K. Ueno and E. S. Yeung, *Anal. Chem.* **66**, 1424 (1994), with permission.]

sequencing (27). CID is a solid-state imaging device similar to a CCD but has unique characteristics (28,29). The most important features of the CID include random pixel addressing, flexibility of user-programmable architecture, large dynamic range, low dark current, antiblooming imaging, high tolerance to irradiation, high quantum yield over a wide wavelength range, and nondestructive readout. When only a single capillary or a small number of capillaries are involved, there is no obvious reason for using a CID while various phototubes, avalanche photodiodes, or CCD cameras are available. However, when a large number of capillaries need to be monitored simultaneously in an array format, the unique features of a CID camera can make a significant difference.

For CID cameras, the limit of detection (LOD) is determined by background fluctuations, integration time, duty cycle, readnoise, dark current, and quantum yield at the specific wavelength range. For the case of on-column detection in a polymer-matrix-filled capillary, we found that the dominant source of noise was background fluorescence from the polymer matrix (27). The larger read noise of the CID is therefore no longer the primary concern. In fact, the dark current of our CID camera at ambient temperature with 450 ms exposure time is also small compared to the background fluorescence. So, at the level of sensitivity needed for DNA sequencing (see our results below), cooling is not even needed! This makes the CID simpler and more compact to incorporate into an automated DNA sequencing instrument.

The flexible reading mode of CID has additional advantages. If the light intensity is higher at one side of the capillary array than at the other side due to absorption or light scattering (26), the corresponding subarray can be scanned from the side where the light intensity is higher so that S/N is more even across the capillary array. If the capillary array is illuminated by a Gaussian-shaped light profile (25), the fluorescence from the center of the capillary array may saturate the middle pixels before the pixels on the sides have accumulated sufficient charge. What one can do is to read the middle pixels with shorter integration times and read the pixels on the sides after longer integration times. This can be accomplished easily by software programming (30).

In DNA separation by CE, different sizes of fragments migrate at different velocities. The larger fragments elute later and stay at the detection window longer. In addition, peaks of the larger fragments become broader because of a loss in separation efficiency. If the camera runs at a slower frame rate to monitor a larger fragment, the total fluorescence from that fragment can be concentrated into fewer sampling points without loss of resolution. The S/N will thus be improved. This is very useful because the longer fragments in a DNA sequencing sample are typically at lower concentrations than the shorter fragments due to the nature of the polymerase reaction. To demonstrate this, we monitored the separation of DNA sequencing fragments

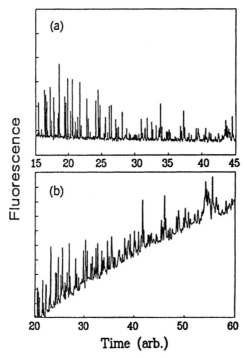

Figure 22.2. Detection of DNA fragments after Sanger reaction (a) without and (b) with exposure-time gradient. A linear gradient from 0.5 to 1.5 s exposure time per frame over the course of the separation was employed. [Reproduced from Q. Li and E. S. Yeung, *Appl. Spectrosc.* **49**, 825 (1995), with permission.]

through a band-pass filter to deliberately reduce the number of peaks and to reduce the signal levels. From the electropherograms in Figure 22.2, we can see that the S/N is improved for the large fragments after implementing an exposure-time gradient. There is a sloping baseline in Figure 22.2b, since the background is also integrated for progressively longer periods. However, this is corrected by software after the run.

22.2. ACCELERATION OF ELECTROPHORETIC RUNS

It is already an amazing feat that the conventional gel medium works over such a large range of molecular sizes with single-base resolution. It will be even more challenging if one wants to extend the base numbers that can be read per run to reduce front-end manipulations. What is needed are entirely new sieving matrices or the equivalent of gradient gels (31–33) used for protein separations.

Early on, cross-linked polymers such as polyacrylamide PA (3–12) were used as matrices in CGE because of their known utility in slab gels for the separation of proteins and DNA. However (34), due to their instability over time, irreproducibility in the polymerization processes, and the fragile nature of the medium, cross-linked PA matrices in CE have not been reported to last for more than a few runs and are therefore not suitable for large-scale DNA sequencing, especially in multiplexed operation. The longevity issue also applies to ultrathin gels (2,2a). The preparation of linear PA polymer solutions is also difficult to control and to reproduce. The polymerization process depends critically on oxygen content (35), temperature (36), the time needed for a complete reaction, reagent purity, and contamination. So, despite the distinct advantage of higher resolution for the larger fragments in cross-linked PA, alternative sieving matrices are urgently needed.

Low- to moderate-viscosity entangled polymers have been used to overcome some of the above problems (37–46). Unlike cross-linked gels, they are replaceable and more stable for use at higher temperatures and electric field strengths. Linear PA (0% C) has been used for the size separation of DNA or proteins by sieving (36–38,47). In addition, methylcellulose (39,40), hydroxyalkyl cellulose (41–43), polyhydroxy methacrylate and poly[(ethylene glycol)–methacrylate] (44), and poly(vinyl alcohol) (45,46) have been employed for DNA separations.

A separate problem is the internal coating of the capillary tubes. Without exception, the fused-silica capillaries used in DNA sequencing by CE have all been pretreated with a bonded coating. These are mostly variations of a bonded polyacrylamide layer (48). The reason is that electroosmotic flow (EOF) exists in bare fused-silica capillaries. Even when EOF is slow, the fact that it is opposite to the migration direction of DNA fragments means long separation times. Further, since the net motion is dictated by $(\mu_{DNA} - \mu_{EOF})$, the corresponding difference in mobilities, the large fragments are affected much more severely than the small fragments. Unfortunately, the coating degrades with usage (36). This is not surprising since PA, when used as the sieving medium, also breaks down with time on interaction with the typical buffers used for DNA sequencing (34,49). There is definitely a need for better surface-treatment procedures so that the capillary columns will retain their integrity over many runs.

22.2.1. A Novel Sieving Medium

Poly(ethylene oxide) (PEO), with a large range of molecular masses, is available commercially. In addition, it is easy to prepare homogeneous solutions of PEO in typical electrophoretic buffers to provide highly reproducible separation matrices (50). Initially, to evaluate the applicability of this alternative polymer matrix for DNA sequencing, we studied the separation performance by using the double-stranded (ds) fragments of pBR 322 DNA/HaeIII digest (intercalated with ethidium bromide and detected by fluorescence excited at 543 nm).

We found that the resolution with PEO matrices is much better than that with cellulose-type matrices (39–43). Also, separation performed in PEO matrices provides highly reproducible results for at least 10 runs without replacement. From our experience, the capillary can be used for over 2 weeks and more than 50 runs without any degradation. The reproducibility among different capillaries and different batches of polymers is also excellent. Compared to homemade non-cross-linked PA, PEO is likely to be more stable since there is no further polymerization of these commercial preparations. There are non-cross-linked PA preparations available commercially. However, they are not yet available with a wide range of M_n, which turns out to be important for DNA separations (*vide infra*).

We found that the smaller fragments can be well separated in matrices prepared from individual polymers with low M_n. For larger DNA fragments, better resolution is achieved in matrices prepared from polymers with higher M_n. We therefore tested mixtures of polymer sizes for the same separation. Figure 22.3 shows that a mixed-polymer solution can indeed provide high separation efficiencies over a large range of DNA sizes. In the mixed-polymer matrices, a polymer network with random pore sizes is formed. This provides optimum pore sizes for a large range of DNA fragments.

Figure 22.3 highlights the excellent separation performance among the fragments with 434, 453, and 458 bp. The results also show that this matrix can be used to separate certain normal DNA samples from mutated samples, since fragments with 234 bp from two different samples (peaks 14 and 23) are well separated. Independent of DNA sequencing applications, the highly reproducible polymer matrices developed here for high-resolution separation of restriction fragment digests or polymerase chain reaction (PCR) products should be of value for DNA typing (51).

We have successfully used these mixed polymers (52) for actual DNA sequencing by CE. For comparison, commercial non-cross-linked PA (10% T solution, 700,000–1,000,000 M_n) was diluted to form a 6% T, 1 × TBE (tris-borate-EDTA), 3.5 M urea matrix. Excitation was provided by a He–Ne laser at 543 nm and 2 RG610 long-pass filters were used to select primarily the C and T fragments. The capillary length was 45 cm, with a 35 cm effective length (to the detector). Lyophilized DNA samples were denatured, electrokinetic injection was performed at 6 kV for 12 s, and the separation was run at 13 kV. The results are shown in Figures 22.4 and 22.5 for the

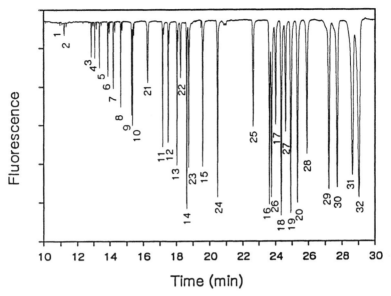

Figure 22.3. Electrophoretic separation of the mixture of pBR 322 DNA/*Hae*III, pBR 328 DNA/*Bgl*I, and pBR 328 DNA/*Hin*fI digests. The matrix is 0.7% each M_n 300,000; 600,000; 1,000,000; 2,000,000; 5,000,000; and 8,000,000. Buffer: 1 × TBE, pH = 8.2. Capillary: 75 μm i.d., total length 50 cm, effective length 42 cm. Applied potential: − 10 kV. Peak assignments (base pairs): (1) 18; (2) 28; (3) 51; (4) 57; (5) 64; (6) 80; (7) 89; (8) 104; (9) 123; (10) 124; (11) 184; (12) 192; (13) 213; (14) 234; (15) 267; (16) 434; (17) 458; (18) 504; (19) 540; (20) 587; (21) 154; (22) 220; (23) 234 (from pBR 328); (24) 298; (25) 394; (26) 453; (27) 517; (28) 653; (29) 1033; (30) 1230; (31) 1766; (32) 2176. [Reproduced from H.-T. Chang and E. S. Yeung, *J. Chromatogr.* **669**, 113 (1995), with permission.]

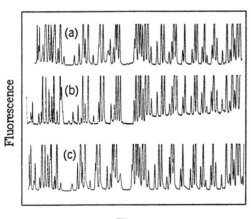

Figure 22.4. Separation of DNA fragments from Sanger reaction by CE from base pairs 28 to 108. The span of the abscissa is different in each case and is specified here: (a) PEO in coated capillary, 14–19 min; (b) PA in coated capillary, 37–55 min; and (c) PEO in HCl-reconditioned bare capillary, sixth run, 9.5–13 min.

778   DNA SEQUENCING BY MULTIPLEXED CE

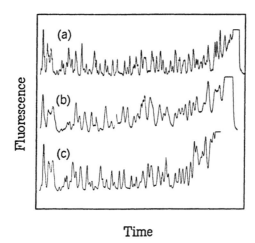

Figure 22.5. Separation of DNA fragments from Sanger reaction by CE from base pair 420 upward. The span of the abscissa is different in each case and is specified here: (a) PEO in coated capillary, 39–52 min; (b) PA in coated capillary, 111–129 min; and (c) PEO in HCl-reconditioned bare capillary, sixth run, 26–33 min.

regions of base pairs 28–108 and >420, respectively. In constructing Figures 22.4 and 22.5, the ordinate of each electropherogram was adjusted so that they roughly matched each other and so as to emphasize the small peaks, as those would cause the most problems in base calling due to overlap and inadequate S/N. The abscissa of each electropherogram was also adjusted to plot the same base-pair region in each case. However, the actual timescale for each is different and all are listed in the captions.

Figure 22.4a,b shows that for the small fragments, PEO provides a resolving power quite close to that of PA. PA is able to resolve certain partially overlapped peaks as well as shoulders next to large peaks in the first half of the plots. It should be noted that the run in Figure 22.4a does provide adequate resolution for base calling. The major difference is separation speed. The PEO plot (Figure 22.4a) was from 14 to 19 min, whereas the PA plot (Figure 22.4b) was from 37 to 55 min. This is expected from the higher viscosity of the PA matrix. Very striking is the difference in separation for the large fragments (Figure 22.5a,b). The PEO matrix clearly provides better resolution and may even be extending the convergence limit to higher base numbers. In the middle range (108–420 bp, data not shown), there is a one-to-one correspondence between the resolution of DNA fragments in PEO compared to that in PA, although some degradation is already evident in PA past 320 bp.

Apparently, the maximum length of our PA polymer is not sufficient to form dynamic pore sizes large enough for the large DNA fragments. In fact, a PEO molecule of the same M_n should be longer than PA because of the specific atomic arrangement along the backbone. The same is true when PEO is compared with any other polymer that has been used for CE sieving. Incidentally, any of the PEO polymers costs only around \$1/g, and we are using 2–3% solutions in the 2 µL range per capillary.

Other workers (36,47) have demonstrated base calling in CE with 6% TPA out to > 400 bp. Presumably, the home-prepared PA matrices have higher M_n components and thus include larger pore sizes in solution. In fact, it may well be true that a wide range of polymer lengths are (fortuitously) produced to allow non-cross-linked PA to separate a large range DNA sizes with good efficiency.

22.2.2. The Column Treatment Protocol

Even though in principle one can replace the sieving medium after every run to allow repeated usage of the capillaries, the coating (48), gradually degrades (34,36,49). We are not aware of any report of instances where a capillary column has been used for tens of runs in DNA sequencing. Since the main purpose of coating the capillary column is to eliminate EOF, we proceeded to evaluate alternative approaches toward this goal. In the zone electrophoretic separation of small molecules and even proteins, it is common to use a dynamic coating of a linear polymer at the capillary wall. This suggests that DNA sequencing can be performed in a bare fused-silica column. Even if there is some residual surface charge, the high viscosity of the polymer matrix will lead to further reduction of this undesirable counterflow during DNA separation (53). We therefore started with commercial capillary columns and washed out the inside of each with methanol. Immediately afterward, the polymer matrix was introduced into the capillaries and a DNA sequencing run was started.

There is practically no difference in the electropherograms with or without a coating on the capillary wall. Unfortunately, while we can reproduce these results every time for a new capillary column, the resolution invariably starts to degrade immediately after one or two runs. The migration times become progressively longer and the largest fragment peaks become unrecognizable. Naturally, having to replace the capillary array after every run is not a viable option for high-throughput DNA sequencing.

Finally, we realized that the symptoms associated with column degradation are those of increased EOF in subsequent runs. For a dry, fresh column (fused-silica) surface, the silanol groups should be fully protonated. A methanol wash further minimizes any adsorbed moisture. So, initially the EOF should be negligible. However, the separation matrix will eventually cause ionization of the silanol groups to increase the EOF. The remedy is to flush the column in between runs with 0.1 N HCl so as to retitrate the surface silanol groups back to their original protonated state (52).

The performance of a bare fused-silica CE column in separating DNA fragments after six cycles of PEO fill, DNA electrophoresis, pressure removal of PEO, and 0.1 N HCl conditioning is shown in Figure 22.4c and 22.5c. There is no obvious difference in resolution between this electropherogram and those for coated column/PEO (Figures 22.4a and 22.5a) or for a fresh bare column/PEO. So far we have observed reproducible performance in over 30 runs in a 3 week period. There are slight random variations in migration times, but no systematic change over time. This is to be expected since we relied on a manual flush/fill operation and have not yet defined the time and concentrations required to completely rejuvenate the column surface. Since no bonded coating is used, there is nothing to degrade and the column should in principle last indefinitely.

A surprising result is that the migration times observed in the HCl-treated capillaries are much shorter than those found for any polymer matrix/surface preparation reported either here or in the literature. One possible explanation is that the coating (bonded) on the capillaries and the fresh fused-silica surface both have residual EOF to oppose DNA migration. The HCl treatment was able to eliminate EOF entirely. Regardless of the mechanism, the bottom line is that base 28 through base 420 eluted within a time span of only 16 min for an average rate of 25 bp/min. This is faster by a factor of 3–5 compared to reported results using non-cross-linked PA in coated capillaries. Even more recent results show that 30 bp/min is possible.

22.3. BASE CALLING AND DATA HANDLING

Since the original demonstration of base calling (54) by means of fluorescence-based labeling of the fragments from the Sanger reaction (55), many variations have been developed. It is unlikely that CE will provide migration times that are reproducible enough among a group of capillaries to allow running the four-base set of fragments in separate capillaries (13,53). So, one needs a color scheme to sort out the four bases run in one column. Because of difficulties in controlling the polymerase and in detection S/N, we consider the one-color four-intensity scheme (56,57) to be least desirable at the present time. The two-color two-intensity scheme is a hybrid between the above two methods (14,36,56). It retains the advantage of a simpler optical arrangement, better light collection, and straightforward algorithm. However, uniform incorporation by the polymerase must also be assumed. So, the most mature technology is still the original four-label scheme (54).

We should note that the four standard dyes are by no means spectrally distinct, either in excitation or in emission (54). A complicated set of ratios have to be employed for base calling. A double-rotating-filter system analogous to the commercial hardware has been demonstrated (5). Alternatively, the two laser beams can be made to be collinear, and wedged prisms can be used to obtain four distinct images on an array detector (17). A single laser can also be used, with minimal compromise in sensitivity (2a,57). Others have dispersed the fluorescence with a monochromator prior to array detection with one (58) or two (47) excitation wavelengths. Monochromator-based spectral identification of the labels in principle offers the best selectivity. However, one needs to disperse the total fluorescence over many pixels to obtain a spectrum. This adds to the amount of raw data acquired and increases the acquisition time and the data workup effort. Monochromators also do not have the favorable f-numbers for light collection that simple filters do.

We note that the so-called two-color sequencing scheme (59) is actually a four-label method. The optics is simplified and the ratio-based base-calling algorithm is fairly straightforward (5). One reason why this is not more widely used is that the four labels have emission bands that are very closely spaced. Even though the intensity ratios (used for base calling) are relatively independent of the incorporation rate of the polymerase reaction, spectral interference and a low S/N (low transmission of the band-pass filters) can lead to ambiguities.

Since the commercial instruments use both the 488 and 515 nm light from the Ar-ion laser for excitation, we have adapted this strategy in an arrangement with separation in space (25,27), rather than separation in time, to maximize the duty cycle. Therefore, two sets of electropherograms were obtained in the capillary array. These can be readily normalized to correct for the migration time difference for a particular fragment crossing the two windows (27,47). Instead of using narrowband filters, we used two cutoff filters to maximize light throughput—one a sharp-cut Raman-edge filter on top of the 488 nm laser location, and one at 610 nm on top of the 514 nm laser location. The ratios of peak heights for a given capillary cluster around distinct values to allow accurate base calling of around 97% for 300 bases (60). Basically, the corresponding CID pixels generate distinct electropherograms over the run. The data were loaded into standard chromatography software to correct for the baseline and to detect individual peaks. The peak height for each detected feature was also determined by the software. The peak positions and peak heights were imported to a spreadsheet program and matched for the two excitation channels based on the relative migration distances to the two windows (27). The ratio of the matched peak heights was calculated. The bases were called based on this ratio.

We then realized that there are two aspects of this base-calling scheme that need improvement. Matching of the migration times by normalization to the relative distance traveled depends on having uniform velocities (temperature, matrix homogeneity) along the entire capillary. The two laser beams also

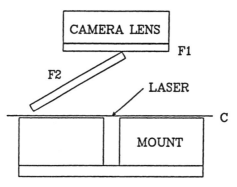

Figure 22.6. Optical arrangement that splits a line of excited fluorescence in the capillaries (C) into two emission channels—one through a Raman-edge long-pass filter (F1), and the other through that plus a 610 nm long-pass filter (F2). [Reproduced from Q. Li and E. S. Yeung, *Appl. Spectrosc.* **49**, 1528 (1995).]

produce scattered light that can interfere with each other (a 514 nm laser line transmitted by the Raman-edge cutoff filter) and decrease the S/N. So, we tried to use the 488 nm laser line by itself. All four standard dye labels absorb at 488 nm, although not all four do so equally efficiently (2a,57,58).

Our ultimate solution is to use one laser and one window in excitation (60). In Figure 22.6, we show an elegant solution that splits the image into two emission channels with maximum light throughout. A 488 nm cutoff filter is placed at the camera to eliminate all stray laser light. A 610 nm long-pass filter is tilted 30° and covers roughly half the camera lens. The subtle feature is that a tilted optical element will deflect (shift) the image to the side depending on the tilt angle and its thickness. We only need to shift the image line a few pixels away from the direct image. The shifted image automatically has the correct wavelength selection, as before. Base-calling performance is similar to the two schemes described above, but there is no longer a time difference between the electropherograms from the two channels. Residual errors were still due to the inability of our simple chromatography software to identify peaks with resolution $1.0 < R$. The extra effort and potential errors in matching the peaks from the two channels are eliminated entirely, however. The base-calling accuracy improved to 99.5% (to 250 bp), 99.3% (to 330 bp), and 97.1% (to 350 bp).

Very recently, we devised another novel approach to base calling based on the one-excitation laser/two-emission wavelength data described above (60). We generated a "ratiogram," which is simply the ratio of signals from the two channels calculated point by point at each data interval. Such ratiograms have been used in liquid chromatography to determine peak purity when diode array detectors or rapid-scan multiwavelength detectors are utilized. The idea is that the ratio (intensities at two independent wavelengths) is independent of concentration (which varies across the peak) and can be used to sort out the unresolved components in the merged peaks. As long as the overall S/N is good, even peaks with resolution $R < 0.5$ can in many cases be identified by noting the ratios at the leading edge and at the trailing edge of the merged peak. We still need the usual electropherogram to determine where the ratios are meaningful and where the signals are at the noise level and are therefore meaningless. However, peaks need not be resolved by the chromatography software, and errors in determining peak heights (when unresolved) are avoided.

An actual example of such a ratiogram is plotted in Figure 22.7 on top of the electropherogram obtained through the 488 nm cutoff filter, which should record all peaks regardless of the label. The called bases are typed below. The accuracy is 99% through 340 bases. Note the feature at marker 800. It is clearly broader than the surrounding features, indicating an overlapping set of fragments. The ratiogram clearly shows that the leading edge is "A" and the trailing edge is "G" in character. So, even though this feature led to base-calling errors in the peak-height scheme using chromatography software, it was correctly called in this novel scheme. Other noteworthy portions are the regions around markers 600 and 1180, where partially resolved features are correctly called in a similar fashion.

We found that defining the baseline for the peaks is critical to base-calling accuracy. All three errors (< 340 bp) occured in stretch around 260 bp, where our simple baseline-selection algorithm resulted in negative values for some of the fluorescene intensities in the 488 nm cutoff channel. Similarly, there was a series of errors around 350 bp, where there was insufficient background subtraction. Refinement in the software should allow accurate base calling to > 400 bp, since the peak resolution there is still better than 0.5.

22.4. SYSTEM INTEGRATION

In anticipation of the need to flush out, recondition, and refill the capillary columns after each of many runs and the need to inject multiple samples separately into the array, we have developed a pressure cell suitable for these operations (Figure 22.8). Note the pressure needed for this relatively low-viscosity sieving matrix is only 100–400 psi, depending on the time allowed for each operation. This cell has been used for single-capillary reconditioning operation in most of the experiments described above. For 100 capillaries in a bundle the exit end can be gathered together to form a close-packed group of

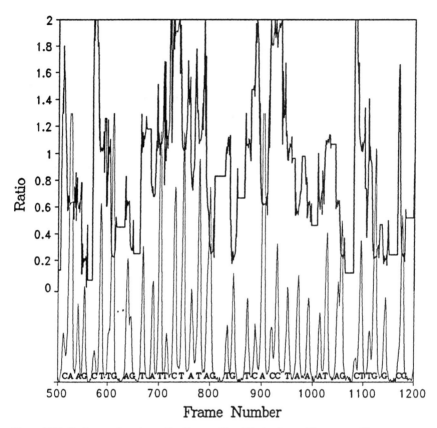

Figure 22.7. Ratiogram (top, heavy line) base calling. The horizontal lines are artifacts to prevent division errors when S/N is too low to determine a meaningful ratio. The raw signal from the 488 nm excitation/Raman-edge filter is plotted below (light line). [Reproduced from Q. Li and E. S. Yeung, *Appl. Spectrosc.* **49**, 1528 (1995), with permission.]

only 2 mm in diameter. This can go through the same fitting as shown in Figure 22.8 to implement the same recycling process. Recently, we have constructed a larger pressure cell to fit a standard 96-well microtiter plate for simultaneous operation.

It is worth noting that we have actually tried sequencing runs derived from pressure injection of the DNA sample in the pressure cell (100 psi, 40 s). The high viscosity of the polymer matrix means that only a very narrow plug of sample was injected in a controlled manner. The electropherograms showed the same resolution as the electrokinetically injected runs in the traces in

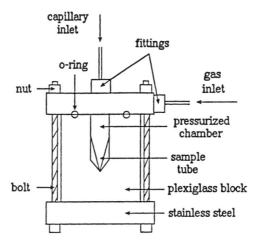

Figure 22.8. Pressure injection/flush cell for one capillary. About 100–500 psi can be used. Fittings are standard chromatographic plumbing. [Reproduced form E. N. Fung and E. S. Yeung, *Anal. Chem.* **67**, 1913 (1995), with permission.]

Figures 22.4 and 22.5. Actually one would expect that pressure injection will avoid any electroinjection bias (61) that discriminates against the larger DNA fragments. Finally, pressure injection avoids cross-contamination of the sample from the electrodes and avoids the need to make electrical contacts with each sample vial in a large array entirely.

22.5. FUTURE PROSPECTS

Clearly, CE has advanced to the stage where high-speed high-throughout DNA sequencing at a reasonable cost appears to be within reach. These various technologies need to be packaged in a user-friendly instrument. A 100 capillary instrument at 30 bp/min per channel is already 150 times faster than the ABD 373 instrument (50 bp/h and 24 lanes) or 40 times faster than the ABD 377 instrument (200 bp/h and 24 lanes).

The technology can readily be scaled up to 1024 capillaries, which is a convenient number for available array detectors. The total current used will be in the 10–50 mA range and is well within the capabilities of a rack-mount high-voltage (HV) power supply. The array width will only be 15 cm, which is easily spanned by the same wide-angle lens. The number of data points per frame is around 5000, which takes less than 10 ms for transfer in the new CCD systems. At 0.2 ms exposure time for each frame, the total data collected over a 10 min run is only 15 Mbytes, which is easily accommodated by any personal computer in RAM (random-access memory). To achieve a 30 bp/min sequencing rate continuously, one can imagine having parallel stations where the capillary array is reconditioned, samples are injected, and electrophoresis prerun to the primer peak before being brought to the optical system. Then, raw sequencing of 3×10^9 bp should take only 70 days.

For sequencing the human genome, CE has eliminated the bottleneck involved in separating the Sanger fragments and reading the fluorescent dye labels. Naturally, both the front end and the back end must be speeded up to the same level before the full potential of multiplexed CE can be realized. A thousand samples must be delivered every 10 min to the capillaries. Sequence assembly from the vast number of 500-base pieces must be implemented at high speeds. Much progress has been made recently in robotics and in informatics. Sequencing the entire genome in another 10 years thus appears to be feasible.

LIST OF ACRONYMS AND SYMBOLS

Acronym or Symbol	Definition
bp	base pairs
CCD	charge-coupled device
CE	capillary electrophoresis
CGE	capillary gel electrophoresis
CID	charge-injection device
DNA	deoxyribonucleic acid
ds	double-stranded
EOF	electroosmotic flow
HV	high-voltage
LOD	limit of detection
PA	polyacrylamide
PCR	polymerase chain reaction
PEO	poly(ethylene oxide)
RAM	random-access memory
SGE	slab gel electrophoresis
S/N	signal-to-noise ratio
TBE	tris-borate-EDTA
USDOE	U.S. Department of Energy

ACKNOWLEDGMENT

The Ames Laboratory is operated for the U. S. Department of Energy (USDOE) by Iowa State University under Contract No. W-7405-Eng-82. This work was supported by the Director of Energy Research, Office of Health and Environmental Research, USDOE.

REFERENCES

1. R. D. Fleischmann et al., *Science* **269**, 496 (1995).
2. J. Stegemann, C. Schwager, H. Erfle, N. Hewitt, H. Voss, J. Zimmermann, and W. Ansorge, *Nucleic Acids Res.* **19**, 675 (1991); R. L. Brumley and L. M. Smith, *ibid.*, 4121; D. E. Kolner, D. A. Mead, and L. M. Smith, *BioTechniques* **13**, 338 (1992). A. J. Kostichka, M. L. Marchbanks, R. L. Brumely, Jr., H. Drossman, and L. M. Smith, *Bio/Technology* **10**, 78 (1992).
3. A. S. Cohen, D. R. Najarian, A. Paulus, A. Guttman, J. A. Smith, and B. L. Karger, *Proc. Natl. Acad. Sci. U.S.A.* **85**, 9660 (1988).
4. H. Drossman, J. A. Luckey, A. J. Kostichka, J. D'Cunha, and L. M. Smith, *Anal. Chem.* **62**, 900 (1990).
5. H. Swerdlow, J. Z. Zhang, D. Y. Chen, H. R. Harke, R. Grey, S. Wu, N. J. Dovichi, and C. Fuller, *Anal. Chem.* **63**, 2835 (1991).
6. D. Y. Chen, H. P. Swerdlow, H. R. Harke, J. Z. Zhang, and N. J. Dovichi, *J. Chromatogr.* **559**, 237 (1991).
7. Y. Baba and M. Tsuhako, *Trends Anal. Chem.* **11**, 280 (1992).
8. J. A. Luckey, H. Drossman, A. J. Kostichka, D. A. Mead, J. D'Cunha, T. B. Norris, and L. M. Smith, *Nucleic Acids Res.* **18**, 4417 (1990).
9. J. A. Luckey and L. M. Smith, *Anal. Chem.* **65**, 2841 (1993).
10. Y. Baba, T. Matsuura, K. Wakamoto, Y. Morita, and Y. Nishitsu, *Anal. Chem.* **64**, 1221 (1992).
11. H. Lu, E. Arriaga, D. Y. Chen, N. J. Dovichi, *J. Chromatogr. A* **680**, 497 (1994).
12. H. Lu, E. Arriaga, D. Y. Chen, D. Figeys, and N. J. Dovichi, *J. Chromatogr. A* **680**, 503 (1994).
13. X. C. Huang, M. A. Quesada, and R. A. Mathies, *Anal. Chem.* **64**, 967 (1992).
14. R. A. Mathies, X. C. Huang, and M. A. Quesada, *Anal. Chem.* **64**, 2149 (1992).
15. J. Bashkin, D. Barker, D. Roach, M. Bartosiewicz, J. Leong, T. Zarella, and R. Johnston, *DOE Hum. Genome Workshop IV*, Santa Fe, NM, 1994, Abstr. No. 130 (1994).
16. H. Kambara and S. Takahashi, *Nature (London)* **361**, 565 (1993).
17. S. Takahashi, K. Murakami, T. Anazawa, and H. Kambara, *Anal. Chem.* **66**, 1021 (1994).

18. N. J. Dovichi, J. C. Martin, J. H. Jett, M. Trkula, and R. A. Keller, *Anal. Chem.* **56**, 348 (1984).
19. N. J. Dovichi and F. Zarrin, *Anal. Chem.* **57**, 2690 (1985).
20. N. J. Dovichi, J. Zhang, J. Zhao, J. Rong, R. Liu, J. Elliott, S. Bay, P. Roos, and L. Coulson, *DOE Hum. Genome Workshop IV*, Santa Fe, NM, *1994*, Abstr. No. 131 (1994).
21. R. A. Mathies, J. Ju, H. Zhu, S. M. Clark, A. T. Woolley, Y. Wang, S. C. Benson, and A. N. Glazer, *DOE Hum. Genome Workshop IV*, Santa Fe, NM, Abstr. *1994*, No. 133 (1994).
22. J. Balch, C. Davidson, J. Gingrich, M. Sharaf, L. Brewer, J. Koo, D. Smith, M. Albin, and A. Carrano, *DOE Hum. Genome Workshop IV*, Santa Fe, NM, *1994*, Abstr. No. 134 (1994).
23. J. A. Taylor and E. S. Yeung, *Anal. Chem.* **64**, 1741 (1992).
24. J. A. Taylor and E. S. Yeung, *Anal. Chem.* **65**, 956 (1993).
25. K. Ueno and E. S. Yeung, *Anal. Chem.* **66**, 1424 (1994).
26. X. Lu and E. S. Yeung, *Appl. Spectrosc.* **49**, 605 (1995).
27. Q. Li and E. S. Yeung, *Appl. Spectrosc.* **49**, 825 (1995).
28. J. V. Sweedler, R. B. Bilhorn, P. M. Epperson, G. R. Sims, and M. B. Denton, *Anal. Chem.* **60**, 282A (1988).
29. P. M. Epperson, J. V. Sweedler, R. B. Bilhorn, G. R. Sims, and M. B. Denton, *Anal. Chem.* **60**, 327A (1988).
30. L. B. Koutny and E. S. Yeung, *Anal. Chem.* **65**, 148 (1993).
31. K. Altland and A. Altland, *Electrophoresis* **5**, 143 (1984).
32. M. J. Dunn and A. H. M. Burghes, *Electrophoresis* **4**, 97 (1983).
33. A. H. M. Burghes, M. J. Dunn, and V. Dubowitz, *Electrophoresis* **3**, 354 (1982).
34. H. Swerdlow, K. E. Dew-Jager, K. Brady, R. Gray, N. J. Dovichi, and R. Gesteland, *Electrophoresis* **13**, 475 (1992).
35. A. Chrambach and D. Rodbard, *Science* **172**, 440 (1971).
36. M. C. Ruiz-Martinez, J. Berka, A. Belenkii, F. Foret, A. W. Miller, and B. L. Karger, *Anal. Chem.* **65**, 2851 (1993).
37. D. N. Heiger, A. S. Cohen, and B. L. Karger, *J. Chromatogr.* **516**, 33 (1990); A. S. Cohen, D. R. Najarian, and B. L. Karger, *ibid*; 49.
38. K. Ganzler, K. S. Greve, A. S. Cohen, B. L. Karger, A. Guttman, and N. C. Cooke, *Anal. Chem.* **64**, 2665 (1992).
39. W. A. M. Crehan, H. T. Rasmussen, and D. M. Northrop, *J. Liq. Chromatogr.* **15**, 1063 (1992).
40. M. Zhu, D. L. Hansen, S. Burd, and F. Gannon, *J. Chromatogr.* **480**, 311 (1989).
41. S. Nathakarnkitkool, P. J. Oefner, G. Bartsch, M. A. Chin, and G. K. Bonn, *Electrophoresis* **13**, 18 (1992).
42. K. J. Ulfelder, H. E. Schwartz, J. M. Hall, and F. Sunzeri, *Anal. Biochem.* **200**, 260 (1992).

43. P. D. Grossman and D. S. Soane, *J. Chromatogr.* **559**, 257 (1991).
44. T. Zewert and M. Harrington, *Electrophoresis* **13**, 817 (1993).
45. M. H. Kleemiss, M. Gilges, and G. Schomburg, *Electrophoresis* **14**, 515 (1993).
46. A. Chrambach and A. Aldroubi, *Electrophoresis* **14**, 18 (1993).
47. S. Carson, A. S. Cohen, A. Belenkii, M. C. Ruiz-Martinez, J. Berka, and B. L. Karger, *Anal. Chem.* **65**, 3219 (1993).
48. S. Hjertén, *J. Chromatogr.* **347**, 191 (1985).
49. M. Starita-Geribaldi, A. Houri, and P. Sudaka, *Electrophoresis* **14**, 773 (1993).
50. H.-T. Chang and E. S. Yeung, *J. Chromatogr.* **669**, 113 (1995).
51. P. E. Williams, M. A. Marino, S. A. Del Rio, L. A. Turni, and J. M. Devaney, *J. Chromatogr. A* **680**, 525 (1994).
52. E. N. Fung and E. S. Yeung, *Anal. Chem.* **67**, 1913 (1995).
53. T. T. Lee and E. S. Yeung, *Anal. Chem.* **63**, 2842 (1991).
54. L. M. Smith, J. Z. Sanders, R. J. Kaiser, P. Hughes, C. Dodd, C. R. Connell, C. Heiner, S. B. Kent, and L. E. Hood, *Nature (London)* **321**, 674 (1986).
55. F. Sanger, S. Nicklen, and A. R. Coulson, *Proc. Natl. Acad. Sci. U.S.A.* **74**, 5463 (1977).
56. D. Chen, H. R. Harke, and N. J. Dovichi, *Nucleic Acids Res.* **20**, 4873 (1992).
57. R. Tomisaki, Y. Baba, M. Tsuhako, S. Takahashni, K. Murakami, T. Anazawa, and H. Kambara, *Anal. Sci.* **10**, 817 (1994).
58. A. E. Karger, J. M. Harris and R. F. Gesteland, *Nucleic Acids Res.* **19**, 4955 (1991).
59. J. M. Prober, G. L. Trainor, R. J. Dam, F. W. Hobbs, C. W. Robertson, R. J. Zagursky, A. J. Cocuzza, M. A. Jensen and K. Baumeister, *Science* **238**, 336 (1987).
60. Q. Li and E. S. Yeung, *Appl. Spectrosc.* **49**, 1528 (1995).
61. K. Kleparnik, M. Garner, and P. Boček, *J. Chromatogr. A* **698**, 375 (1995).

CHAPTER

23

CHIRAL SEPARATIONS BY CAPILLARY ELECTROPHORESIS

FANG WANG and MORTEZA G. KHALEDI

Department of Chemistry, North Carolina State University, Raleigh, North Carolina 27695-8204

23.1.	Introduction	791
23.2.	Chiral Resolution	793
23.3.	Types of Chiral Selector	795
	23.3.1. Cyclodextrins and Their Derivatives	796
	23.3.2. Polysaccharides	799
	23.3.3. Chiral Crown Ethers	801
	23.3.4. Antibiotics	802
	23.3.5. Proteins	804
	23.3.6. Chiral Surfactants	804
23.4.	Effect of the Chiral Selector Concentration	806
23.5.	pH and Ionic Strength	807
23.6.	Organic Solvents	808
23.7.	The Counter–Electroosmotic Flow Scheme	811
23.8.	Capillary Electrochromatography	812
23.9.	Other Methods	813
23.10.	Chiral Separation Efficiency	814
23.11.	Conclusions and Future Trends	816
Addendum		817
List of Acronyms		818
References		819

23.1. INTRODUCTION

Direct chiral separation of enantiomeric isomers in a racemic mixture has been a challenging task for separation scientists for the past several decades.

High Performance Capillary Electrophoresis, edited by Morteza G. Khaledi. Chemical Analysis Series, Vol. 146.
ISBN 0-471-14851-2 © 1998 John Wiley & Sons, Inc.

Developments in the area of chiral separation have made it possible to obtain knowledge about the efficacy and toxicity of optical isomers in biological systems. Since the first direct separation of some volatile derivatives of amino acid enantiomers by capillary gas chromatography (GC) by Gil-Av et al. (1), chromatographic methods, especially high-performance liquid chromatography (HPLC), have been predominantly used in both analytical and preparative scale chiral separation for some 30 years.

Chiral separation by capillary electrophoresis (CE) has grown rapidly since it was first reported by Gassman et al. in 1985 (2). Enantiomers have identical charge/mass ratio and can not be separated without the presence of a chiral selector in the buffer solution of CE. Chiral separation primarily occurs as two optical isomers bind to a chiral selector to different extents. Various CE modes such as capillary zone electrophoresis (CZE), capillary gel electrophoresis (CGE), micellar electrokinetic chromatography (MEKC), isotachophoresis (ITP), and more recently packed capillary electrochromatography (CEC) have been used for chiral separation. A chiral selector can be added into a running buffer (as in CZE and MEKC), incorporated into a gel matrix, or immobilized as a stationary phase (as in CEC). The most convenient way to perform chiral separation is the addition of a chiral selector into a supporting electrolyte. As compared to HPLC, chiral separation by CE has the advantages of higher separation efficiency, speed of analysis, and flexibility of rapid incorporation of various chiral selectors. In addition, due to the inherent microsize nature of the technique, expensive and exotic chiral reagents can be incorporated into the separation scheme. In general, method development is much faster in CE than in HPLC. The chiral recognition mechanisms are not well understood for various selectors and classes of enantiomers. As a result, the separation conditions are often chosen on the basis of trial and error. The major limitation of CE is that preparative chiral separation is not feasible. Admittedly, the transfer of existing knowledge from HPLC chiral separation has been instrumental in the rapid development occurring in CE investigations. In principle, the separation process in both techniques is based upon differential enantiomer–selector interactions. Consequently, the knowledge that is achieved on methods developed in one technique should be adaptable in the other. For example, binding constants between enantiomers and selectors as well as enantioselectivity can be readily determined in CE. Such information can be quite useful in selecting or even designing new chiral stationary phases in HPLC.

This chapter provides an overview of the major developments in CE chiral separation over the past 10 years. It is not intended to be an exhaustive review, as this would not be possible within the limited space available here; however, key references are covered. For the details of applications of CE chiral separation, readers can refer to some recent review papers (3–5).

23.2. CHIRAL RESOLUTION

The principle of chiral separation can be explained by the following two thermodynamic chemical equilibria

$$E_1 + CS \xrightleftharpoons{K_1} E_1CS \qquad (23.1)$$

$$E_2 + CS \xrightleftharpoons{K_2} E_2CS \qquad (23.2)$$

where E_1 and E_2 are two enantiomers of a racemic mixture, respectively; CS is a chiral selector; and K_1 and K_2 are the binding constants between the chiral selector and the enantiomers, respectively. The electrophoretic mobility (μ) of an enantiomer at a given concentration of the chiral selector is expressed as (6)

$$\mu = \frac{\mu^f + \mu^c K[CS]}{1 + K[CS]} \qquad (23.3)$$

where μ^f and μ^c are the electrophoretic mobilities at the concentrations of the chiral selector at 0 and ∞, respectively; [CS] is the equilibrium concentration of the chiral selector.

The relationship between the mobility difference ($\Delta\mu$), or separation selectivity, and concentration of the chiral selector can be expressed by (7,8)

$$\Delta\mu = \frac{(\mu^f - \mu^c)\Delta K[CS]}{(1 + K_1[CS])(1 + K_2[CS])} \qquad (23.4)$$

As can be seen from Eq. (23.4), separation selectivity, $\Delta\mu$, is proportional to the mobility difference of the racemate in the free (μ^f) and totally complexed (μ^c) forms, and their binding constant difference (ΔK). It demonstrates that no chiral separation can be achieved if there is no complexation between enantiomers and the chiral selector. In addition, the two enantiomers should bind to the chiral selector to different extents in order to be separated. Therefore, the choice of a chiral selector is crucial for chiral separation, as it controls three terms, K, ΔK, and ($\mu^f - \mu^c$). The other experimental factor that influences $\Delta\mu$ is the selector concentration, [CS], which should be optimized in order to achieve better separations.

The resolution equation in CZE is also valid in chiral separation as (9,10)

$$R_s = \frac{\sqrt{N}}{4}\left(\frac{\Delta\mu}{\mu_{avg} + \mu_{eo}}\right) \qquad (23.5)$$

where N is the number of theoretical plates; μ_{avg} is the average electrophoretic mobility of two enantiomers; and μ_{eo} is the mobility of electroosmotic flow (EOF). Chiral resolution can be improved by enhancing separation efficiency ($\sqrt{N}$), maximizing separation selectivity ($\Delta\mu$), optimizing retention (μ), and controlling EOF. In order to maximize $\Delta\mu$, several parameters such as the type and concentration of chiral selector as well as pH (for ionizable solutes) would have to be optimized. Other experimental conditions such as the buffer ionic strength and temperature can also play a role through their effects on retention (μ_{avg}), selectivity ($\Delta\mu$), and/or EOF (μ_{eo}). EOF plays an important role in chiral separation, as suggested in Eq. (23.5). Two general migration schemes are recognized in CE: one is co-electroosmotic flow (co-EOF), where the ions and EOF migrate in the same direction as the EOF, and the other is counter-electroosmotic flow (counter-EOF), where the ions migrate in the opposite direction of EOF. Figure 23.1 shows diagrams of these two schemes for positively charged enantiomers. In the case of co–EOF, for example, basic racemates are positively charged at a lower pH range and there is a weak EOF from the anode to the cathode. This co-EOF setup has been the most commonly used scheme for separations of basic racemates with different types

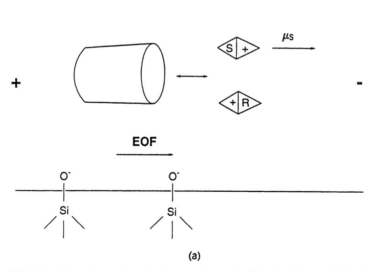

Figure 23.1. Migration schemes for cationic enantiomers in CE using CDs as chiral selectors: (a) co-EOF; (b) counter-EOF. μ_S = electrophoretic mobility of free form of enantiomer; μ_{CD} = electrophoretic mobility of an anionic CD.

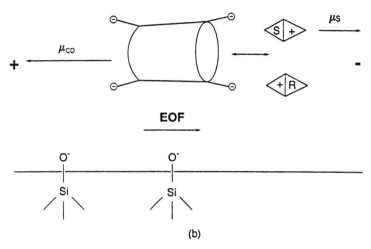

Figure 23.1. (*Continued*)

of chiral selectors. In the counter–EOF case, however, higher resolution can be achieved as the term $(\mu_{avg} + \mu_{eo})$ in the denominator of the resolution equation (23.5) becomes smaller. According to Eq. (23.5), higher resolution results if the analytes migrate in the direction opposite to the EOF. This is achieved at the expense of longer analysis times. In certain situations, chiral separation might be achieved through controlling EOF even when other parameters such as selector concentration or pH are not at optimum conditions.

Among the various sources of band broadening in CE (see Chapter 2 by Kenndler in this monograph), electrodispersion caused by mobility mismatch can be particularly pronounced in CE chiral separation, especially when charged selectors are used. The influence of the kinetics of enantiomer–selector interactions is not known.

23.3. TYPES OF CHIRAL SELECTOR

The mechanism for chiral recognition, the nature of enantiomer–selector interactions, and consequently, enantioselectivity, varies greatly among different types of chiral selector. Although selection of the chiral selector in CE is still based on trial and error, much of the existing knowledge in chiral separation by HPLC has been successfully utilized in CE separations. In contrast to HPLC, the solubility of a chiral selector in the separation media in CE should be large so that a variety of optical isomers can be separated over a wide concentration range of the chiral selector. There exists an optimum chiral selector concentration for a pair of chiral isomers that provides the best resolution. This optimum value is inversely proportional to the average binding constants between enantiomers and the chiral selector. Several types of chiral selector have been used in CE chiral separation, and these are discussed in the following subsections.

23.3.1. Cyclodextrins and Their Derivatives

Cyclodextrins (CDs) have been widely used in both normal-phase and reversed-phase HPLC chiral separations. They have also been the most successful chiral selectors in CE since their first introduction (11,12). CDs are cyclic oligosaccharides composed of different numbers of D-glucose units, which connect a ring structure through $\alpha(1,4')$-glucosidic bonds. Native CDs with six, seven, and eight glucose units, correspoding to α-, β-, and γ-CD, respectively, are commonly used in CE. Their major physical and chemical properties are listed in Table 23.1 (12a,12b). Figure 23.2 shows the geometric structure of β-CD. It looks like a truncated cone. The wider rim consists of the secondary hydroxyl groups (on C_2 and C_3), while the narrower rim consists of primary

Table 23.1. Properties of Native CDs

Characteristics	α-CD	β-CD	γ-CD
Number of glucose units	6	7	8
Molecular weight (Da)	972	1135	1297
External diameter (nm)	1.37	1.53	1.69
Internal diameter (nm)	0.57	0.78	0.95
$[\alpha]_D^{25}$	+150.5	+162.0	+177.4
pK_a hydroxyl groups	12.1–12.6	12.1–12.6	12.1–12.6
Solubility in water (%, w/v)	14.5	1.8	23.2
Solubility in ethanol (%, w/v)	Very low	0.1	Very low
Solubility in DMF (%, w/v)	Excellent	Excellent	Excellent
Encapsulated molecules	Benzene	Naphthalene	Anthracene

Source: Data taken from Saenger (12a) and Schurig and Nowotny (12b).

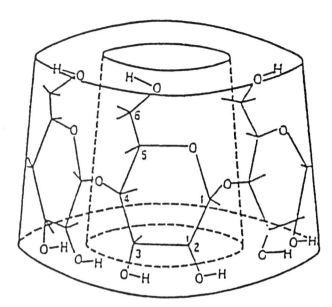

Figure 23.2. Structure of β-CD. [From H. Nishi and S. Terabe, *J. Chromatogr. A* **694**, 245 (1995), with permission.]

hydroxyl groups (on C_6). The cavity consists of the C—H groups (on C_3 and C_5) and glucoside ether linkages. Therefore, CDs can include hydrophobic groups into their cavities while the hydroxyl groups on the rims provide sites for hydrophilic interactions with the polar groups of the guest molecules. The cavity sizes of α-, β-, and γ-CD are such that they can accomodate guest molecules with one, two, and three aromatic rings, respectively. β-CD is the most readily available and widely used type. Unfortunately, it has low aqueous solubility—around 16 mM. β-CD can be chemically derivatized to improve aqueous solubility and increase the diversity in enantioselectivity for chiral separation. The most frequently used neutral β-CD derivatives include monomethyl-, dimethyl-, trimethyl-, and hydroxypropyl-β-CD.

The first application of CD for CE chiral separation was reported by Snopek et al. (11). Chiral separations of pseudoephedrine, norpseudoephedrine, *o*-acetylpseudoephedrine, and *p*-hydroxynorpseudoephedrine were achieved by capillary isotachophoresis (ITP) with β-CD and heptakis(2,6-di-*O*-methyl)-β-CD as chiral additives to the leading electrolyte. Fanali first reported the use of CDs as chiral selectors in CZE (12). Chiral separations of some sympathomimetic drugs, such as ephedrine, norephedrine, epinephrine, norepinephrine, and isoproterenol, were achieved by di-O-Me-β-CD with a short capillary coated with linear polyacrylamide. CDs have become the most extensively used selectors in CZE for chiral separations of a large number of drugs with different pharmaceutical functions. In addition, CD polymers have been used as chiral selectors for CE chiral separations (13).

The derivatization of native CDs with charged substituent groups, such as sulfate, sulfobutyl, carboxyl, amino, and aminocarboxyl groups, makes it possible to separate neutral enantiomers. In addition to inclusion interactions, hydrogen bonding, and dipolar interactions, the presence of charged CDs also introduces electrostatic effects into the host–guest complexation that subsequently changes enantioselectivity. Incorporation of coulombic interaction is needed for those charged analytes that have otherwise weak interactions with uncharged CDs.

Terabe reported the first application of 2-O-carboxymethyl-β-CD for the achiral separation of xylidine isomers in phosphate buffer (pH 7.0) (14). Chiral separations of six dansyl–amino acids were achieved with mono(6-β-aminoethylamino-6-deoxy)-β-CD at pH 3.0 in the presence of 0.1% hydroxypropyl cellulose to suppress EOF (15). The application of sulfobutyl ether β-CD in CE was reported by Tait et al. (16). The change in the type of CD (especially charged CD) may result in a change in the $(\mu^f - \mu^c)$ term. For chiral separation of cationic enantiomers with neutral CD, for example, the complexed enantiomers elute between t_f and t_c, which are the elution times for the enantiomers at CD concentrations equal to zero and infinity, respectively. As in MEKC, the time between t_f and t_c can be defined as the "separation window." The value of t_c should be smaller than t_{eo} since the net charge of the cationic analyte–uncharged CD complexes is still positive. Compared with neutral CDs, the polyanionic CD could be used to separate cationic solutes with a larger separation window, i.e., there is a possibility that the diastereomeric complexes between the CD and analytes could elute after t_{eo}, which is impossible for a basic solute and a neutral CD system. The change of the sign of μ^c from positive (for the cationic enantiomer–neutral CD complexes) to negative (for the cationic enantiomer–multiple anionic CD complexes) can increase the $(\mu^f - \mu^c)$ term in the selectivity equation (23.4). This could result in an increase in separation selectivity and resolution. Schmitt and Engelhardt (17) used carboxymethyl-, carboxyethyl-, and succinyl-β-CDs for chiral separations of neutral and cationic racemates. At pH < 4, all carboxylic groups on the CDs were protonated and the CDs were used under the neutral charge condition for chiral separation of cationic analytes. At pH > 5, the carboxylic groups were ionized and the CDs could be used to separate both cationic and neutral analytes. The change in the CD charges had an effect on the migration order of some cationic enantiomers. The application of sulfated β-CD for chiral separation was reported by Stalcup and Gahm (18). A large number of neutral and basic chiral solutes were separated at one sulfated CD concentration under the counter-EOF condition. Recently, chiral separation of basic solutes in aqueous CE based on the charges of the basic solutes with a sulfated β-CD was achieved by our group (19). At pH > 2.0, the charges of the sulfate groups on the CD do not change; thus the charges of test solutes could be controlled by varying the buffer pH. At pH 2.5, basic compounds were positively charged and interacted more strongly with the negatively charged CD than with the neutral CDs due to the extra electrostatic interactions. Chiral separations of acebutolol, metanephrine, normetanephrine, nafronyl, labetalol, and nadolol were achieved. However, for compounds such as trimipramine, nefopam, and thioridazine, severe peak tailing was observed even at the micromolar concentration of the CD. This is due to strong complexation between the CD and the enantiomers, as well as electrodispersion caused by the mobility mismatch between the solutes and the running buffer. At pH 11.6, most of the basic test compounds were neutral and had much weaker interactions with the negatively charged CD. Chiral separations of the compounds with large binding constants were achieved at the basic pH and 20 mM sulfated β-CD. Under these conditions, the extent of binding between the charged CD and hydrophobic enantiomers is moderate; thus selection of optimum CD concentration is not as crucial as for strongly bound enantiomers (*vide infra*). In addition, the elimination of electrostatic interactions between the CD and the amines improved the separation efficiency, with a theoretical plate number of $\sim 250{,}000$.

23.3.2. Polysaccharides

The application of neutral polysaccharides in CE was first reported in 1992 by D'Hulst and Verbeke for the chiral separation of nonsteroidal anti-inflammatory and coumarinic anticoagulant, and diastereomeric cefalosporin antibiotics (20,21). The separation was carried out in a neutral phosphate buffer with maltodextrins or maltooligosaccharides. Novotny's group reported chiral separations of fluoxetine, ibuprofen, ketoprofen, norverapamil, simendan, verapamil, and warfarin with maltodextrin oligosaccharides [see Soini et al. (22)]. Dextrins and dextrans were also used to separate nonsteroidal anti-inflammatory drugs and dansyl–amino acids by Quang and Khaledi (23).

A wide range of racemates can be separated by anionic polysaccharides. The additional electrostatic interactions strengthen analyte–polysaccharide complexation and enantioselectivity. Usually, the strength of comlexation between acyclic dextrins and analytes is 2–3 orders of magnitude smaller than those with CDs (24). To date, three types of anionic polysaccharides—heparin (25–28), sulfated dextran (26,29), and sulfated chondroitin (26)—have been used in CE chiral separation. Figure 23.3 shows the structure of these charged

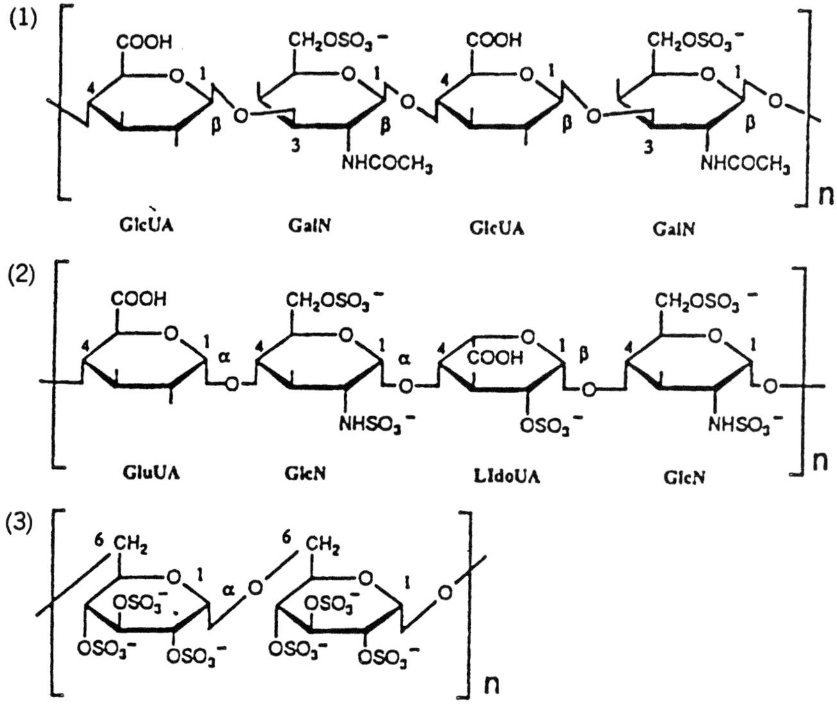

Figure 23.3. Structures of anionic polysaccharides: (1) chondroitin sulfate C; (2) heparin; (3) dextran sulfate. [From H. Nishi, K. Nakamura, H. Nakai, and T. Sato, *Anal. Chem.* **67**, 2334 (1995), with permission.]

chiral selectors. Sulfated chondroitins can be used at both acidic and basic pH, whereas heparin and sulfated dextran can only be used at pH > 5.0 for chiral separation of basic solutes under typical experimental conditions in CE (i.e. with an uncoated fused-silica capillary and the EOF direction from anode to cathode). Otherwise, the polarity of the electrodes should be reversed to detect the peaks. This is because the number of sulfate groups per D-glucose unit is 1, 1–2, and 3 for sulfated chondroitin C, heparin, and sulfated dextran, respectively. At pH < 5.0 and with an uncoated fused-silica capillary, the EOF is not large enough to carry the negatively charged enantiomer–sulfated polysaccharide complexes toward the cathode (i.e., the detection zone) when positive voltage was applied. This is a situation similar to that of chiral separation of basic solutes with polyanionic CDs. A typical chiral separation using charged polysaccharides is shown in Figure 23.4.

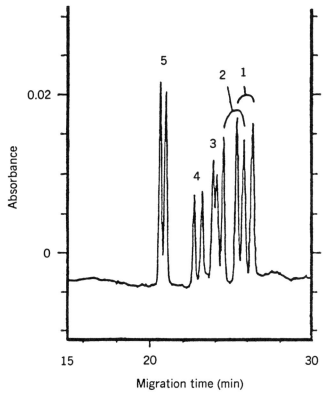

Figure 23.4. Chiral separation with anionic polysaccharides. Buffer: 3% chondroitin sulfate C in 20 mM phosphate–borate buffer, pH 2.4. Solutes: (1) trimetoquinol; (2) trimetoquinol isomer; (3) norlaudanosoline; (4) laudanosoline; (5) laudanosine. [From H. Nishi, K. Nakamura, H. Nakai, and T. Sato, *Anal. Chem.* **67**, 2334 (1995), with permission.]

23.3.3. Chiral Crown Ethers

Crown ethers are macrocyclic polyethers that can form stable host–guest complexes with small inorganic cations and alkylamine cations. The only crown ether presently used in CE chiral separation is 18-crown-6-tetracarboxylic acid. The structure of this chiral selector is shown in Figure 23.5. The six oxygen atoms inside the ring can interact with ammonium or primary amine functional groups through hydrogen bonding and dipolar interactions. Since the inclusion does not involve interactions among chiral centers of the crown ether and analytes, it is believed that the interactions between the carboxylic acid groups on the rim of the crown ether and the chiral centers, which are directly attached to the amine group of the guest molecule, play

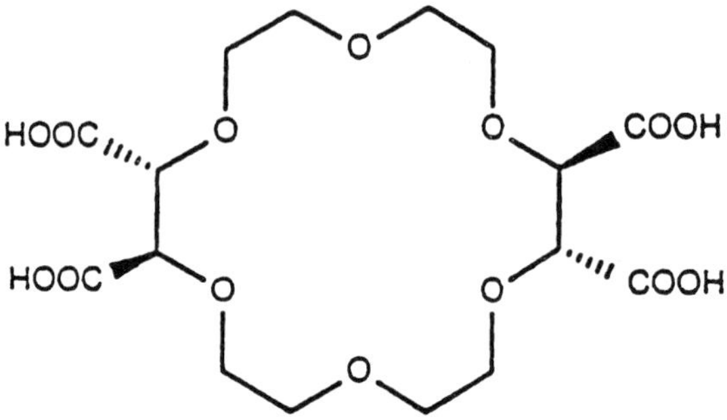

Figure 23.5. Structure of a chiral crown ether, 18-crown-6-ether tetracarboxylic acid. [From R. Kuhn, F. Erni, T. Bereuter, and J. Hausler, *Anal. Chem.* **64**, 2815 (1992), with permission.]

a major role in chiral recognition. Separation is performed at low pH (2–3): the amine groups on isomers are protonated and are included in the cavity of the crown ether. At low pH, however, the carboxylic groups of the crown ether are also protonated. Chiral separations of amino acids (30,31), amines (32), and di- (33) and tripeptides (34) have been reported at low pH values.

23.3.4. Antibiotics

Recently, the use of macrocyclic antibiotics as chiral selectors in both CE and HPLC was reported by Armstrong's group (35–41). There are two types of antibiotics that have been used in CE: the ansamycins and the oligophenolic glycopeptides. The ansamycins used are rifamycins (36,42). The glycopeptides include vancomycin (35,37,40,41,43), thiostrepton (35), ristocetin (39,41), and teicoplanin (41). The properties of the antibiotics used in CE chiral separations are listed in Table 23.2. The structure of vancomycin is shown in Figure 23.6. The characteristics of antibiotics can be summarized as follows: (i) with the exception of rifamycins, they have relatively low UV–Vis molar absorptivity for direct detection at $\lambda > 250$ nm; (ii) they have multiple functional groups, namely, acidic groups, basic groups, sugar moieties, aromatic rings and tail, and stereogenic centers, which are important for chiral separations; and (iii) they are stable in aqueous buffer for CE experiments.

Rifamycin B is negatively charged at pH 7.0 and has been used to separate cationic compounds. Indirect detection has to be used due to its own strong UV–Vis background (36,42). Glycopeptide antibiotics are more useful for chiral separations of anionic and neutral racemates. More than 300 solutes (mostly acidic chiral compounds) have been separated by this type of antibiotic (41). Molecular "space-filling" modeling shows that they have a semi-rigid structure, like a C-shaped "basket" in profile, which consists of fused

Table 23.2. Properties of Some Antibiotics

Characteristics	Vancomycin	Ristocetin A	Teicoplanin
Molecular weight (Da)	1449	2066	1877
Number of stereogenic centers	18	38	23
Number of macrocycles	3	4	4
Number of monomer sugars	2	6	3
Number of hydroxyl groups	9(3)[a]	21(4)[a]	15(4)[a]
Number of amine groups	2	2	1
Number of carboxylic acids	1	0	1
Number of amido groups	7	6	7
Number of aromatic groups	5(2)[b]	7	7(2)[b]
pI	7.2	7.5	4.2, 6.5

Source: Data taken from Gasper et al. (41).

[a] The numbers in parentheses correspond to phenolic moieties.
[b] The number in parentheses correspond to the number of chlorinated substituents attached to the aglycon baskets.

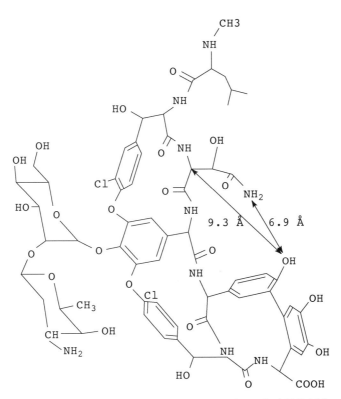

Figure 23.6. Structure of vancomycin. [From M. P. Gasper, N. A. Berthod, U. B. Nair, and D. W. Armstrong, *Anal. Chem.* **68**, 2501 (1996), with permission.]

macrocyclic rings and pendant sugar groups (41). Armstrong's group has also shown that two important characteristics in morphology determine the enantioselectivity of these macrocyclic antibiotics: (i) the openness of the C-shaped basket and (ii) the degree of helical twist.

The same group has examined the effects of various experimental parameters such as the concentration of antibiotics, pH, ionic strength of buffer, organic modifiers, and micelles on chiral separation. The buffer pH determines the charges and the migration behaviors of antibiotics and some of the analytes. The effect of organic modifiers on chiral separation is different from one antibiotic to another. Teicoplanin requires a small amount of acetonitrile for most racemates to achieve chiral separation; however, it precipitates from the solution when alcoholic modifiers are added. Enantioselectivity can rarely be enhanced in a vancomycin system with the addition of small amounts of miscible organic modifiers. The reason that a small amount of acetonitrile changes enantioselectivity in the teicoplanin system is that organic modifiers prevent the aggregation of teicoplanin [the critical micelle concentration (CMC) is ~ 0.18 mM], whereas there is no aggregation of vancomycin. The addition of sodium dodecyl sulfate (SDS) micelle might form comicelles with some of the antibiotics, which can change enantioselectivity in either direction.

23.3.5. Proteins

Proteins have been one of the important chiral stationary phases in HPLC. In CE chiral separation, proteins have been mainly used as running buffer additives; however, they can also be trapped in a gel matrix or immobilized on a capillary wall or on a solid support (used as stationary phases in CEC). Various proteins such as α_1-acid glycoprotein (AGP) (44), avidin (45), bovine serum albumin (BSA) (44,46–49), cellobiohydrolase I (50), conalbumin (44), fungal cellulase (47), human albumin (HA) (51,52), orosomucoid (47), and ovomucoid (OVM) (44,53) have been used for chiral separations of amino acids, and acidic and basic pharmaceutical compounds in aqueous media. In some cases, organic solvents have been added to improve chiral separation. Usually, the concentration of proteins used for chiral separation is in the micromolar range due to the strong binding between pharmaceutical compounds and proteins. However, there are some problems with proteins as chiral selectors in CE: (i) their low separation efficiency; (ii) a higher UV–Vis background at low wavelengths; and (iii) adsorption of proteins onto the capillary wall.

23.3.6. Chiral Surfactants

Chiral surfactants represent a different type of charged selectors. They are used above their CMC in the CE buffer solution. Two types of chiral surfactants, natural surfactants and synthetic surfactants, have been used in chiral MEKC as individual and mixed micellar systems.

Natural surfactants include bile salts and saponins. Bile salts are biological surfactants produced by the liver. They consist of a hydroxyl-substituted steroidal frame and an ionizable side chain. The first application of bile salts for chiral separation in CE was reported by Terabe and coworkers (54). The separation of dansyl–amino acid racemates was carried out at pH 3.0 phosphate buffer. It took somewhat more than an hour for chiral separation of six amino acids. The separation efficiency was very low. However, it demonstrated the potentials of chiral separation using chiral surfactants. Among four bile salts, sodium taurodeoxycholate has been the most effective for chiral separations at pH 7.0. Because cholate and its derivatives can be negatively charged at certain pH's, they have been used for chiral separation of some fused-ring basic compounds such as the diltiazem series compounds (55) and trimethoquinol series compounds (55,56). They have also been used to separate chiral binaphthyl compounds (55,57,58). Saponins such as digitonin, glycyrrhizic acid, and β-escin have been used for chiral separation (59,60). Enantiomers of dansyl–amino acids have been separated using phospholipids (61).

Chiral synthetic surfactants include a chiral head group that is either an amino acid derivative or a sugar moiety attached to long-alkyl-chain tail groups and their polymerization products (62). These surfactants are used in the micellar form at concentrations above their CMC. Chiral surfactants with head groups such as alaninate (63), glutamate (64), serine (65), tartarate (66), threoninate (67), and valinate (59,68–70) attached to a hydrophobic dodecanoyl group have been used for chiral separations of amino acids, benzoin, and warfarin. The problem with this type of straight-chain surfactant is that enantioselectivity and resolution are low. Therefore, some research groups have tried to modify this type of surfactant to improve separation selectivity. Another chiral surfactant, namely N-dodecoxycarbonylvaline (S or R form) (Figure 23.7) was synthesized by Mazzeo and coworkers (71,72). It has been used to separate adrenergic drugs and β-blockers. Dodecyl β-D-

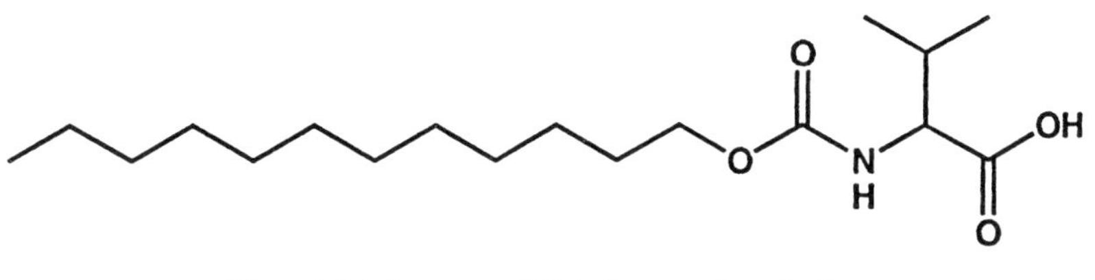

Figure 23.7. Structure of N-dodecoxycarbonylvaline.

glucopyranoside monophosphate and monosulfate anionic surfactants were synthesized by Tickle et al. (73). The advantage of these surfactants is that their CMC is about 0.5 and 1.0 mM, respectively. They were used to separate dansyl–amino acids, cromakalim, Troger's base, mephenytoin, and binaphthyl compounds. Another approach is the use of polymerized micelles with chiral head groups. Wang and Warner reported that the polymer micelles are more stable, more rigid, and easier to control as to their size than normal micelles because there is no dynamic equilibrium between the surfactant monomers and the micelles (62). The CMC for the polymerized micelles is zero. A high percentage of organic solvents can be incorporated into the running buffer without breaking down the micelles. Chiral separations of binaphthyl compounds and laudanosine were reported.

The combination of CD with a micellar system in CE, namely, CD–MEKC, has also been used in chiral separation (74–83). As mentioned in Chapter 3 of this monograph, CD–MEKC is a useful method for the separation of highly hydrophobic compounds, especially polycyclic aromatic hydrocarbons (PAHs). The most effective system for chiral separations has been the γ-CD–SDS system. Apparently, the large cavity of γ-CD can accomodate a variety of chiral molecules as well as the SDS monomers at the same time. Other mixed systems such as bile salts with CD (84) and polymerized chiral micelle with γ-CD (85) have also been used for chiral separation.

23.4. EFFECT OF THE CHIRAL SELECTOR CONCENTRATION

According to the theory developed by Wren and Rowe (7,8), there exists an optimum concentration of chiral selector at which maximum separation selectivity, $\Delta\mu$, and consequently best resolution are achieved [Eq. (23.4)]. This is shown in Figure 23.8 for three pairs of hypothetical optical isomers with different binding affinities toward the selector, while relative differences in binding constants are kept constant at 10% and the $(\mu^f - \mu^c)$ value is the same. The highest separation selectivity can be achieved at $[CS]_{opt}$ as (86)

$$[CS]_{opt} = \frac{1}{\sqrt{K_1 K_2}} \qquad (23.6)$$

The larger the binding constants, the smaller is the optimum concentration of the chiral selector. For example, the binding constants of certain hydrophobic compounds to CDs (e.g., tricyclic antidepressants) can be as large as 10^4–10^5 in aqueous solution. Thus, the optimum concentration of the chiral selector for these chiral compounds would be in the micromolar range. In

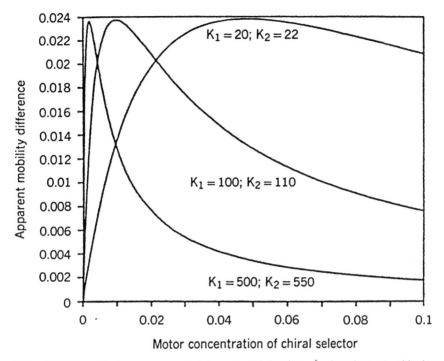

Figure 23.8. Theoretical curves generated from Eq. (23.4) using $\mu^f = 2$ and $\mu^c = 1$, with the binding constants as shown. [From S. A. C. Wren and R. C. Rowe, *J. Chromatogr.* **603**, 235 (1992), with permission.]

addition, for compounds with larger binding constants, $\Delta\mu$ varies rapidly. Therefore, small changes in CD concentration can lead to great variation in separation selectivity ($\Delta\mu$). The small value of optimum concentration and the rapid changes in $\Delta\mu$ would also make it difficult to reproduce results. On the other hand, the optimum concentrations for compounds with smaller binding constants occur at higher values. This can be troublesome if the optimum concentration lies outside the solubility range of the chiral selector. In such cases, a change in the type of selector is needed.

23.5. pH AND IONIC STRENGTH

The pH of the buffer solution determines the charges on the analytes; thus it can have a pronounced effect on enantioselectivity and resolution. Vigh's group has derived fundamental equations that quantitatively describe the dependence of mobility, enantioselectivity, and resolution on pH and chiral selector concentration simultaneously [see Rawjee et al. (87–89); also Rawjee and Vigh (90)]. In order to utilize these equations, a number of initial experiments must be performed to determine solute pK_a and the binding constants between enantiomers in the charged and uncharged states and the chiral selector.

The effect of ionic strength on chiral separation can be examined through its influences on (i) the electrophoretic mobility of enantiomers and EOF, and (ii) separation efficiency. Mobility is directly related to the ζ potential of a charged ion, and the ζ potential is inversely proportional to the ionic strength of the buffer (91). Therefore, both electrophoretic mobility and EOF decrease with an increase in ionic strength. Smaller mobility and EOF lead to better chiral resolution. In addition, ionic strength results in a reduction of electrodispersion and consequently higher separation efficiency (up to a limit where Joule heating becomes significant). The other advantage is that the higher the ionic strength, the greater the stacking of the sample zone and the higher the sample capacity.

23.6. ORGANIC SOLVENTS

In the past decade, most chiral separations have been carried out in pure aqueous buffers. Organic cosolvents, used as buffer additives, can change the binding constants between a chiral selector and enantiomers as well as the EOF. Wren and Rowe have studied the effect of organic additives on chiral separation. They assumed that the increase in the percentage of the organic additives could reduce the binding constants between neutral CD and chiral compounds (8).

Although the application of organic solvents in HPLC chiral separation has been a common tool to manipulate enantioselectivity and/or resolution, chiral separation in pure organic solvents in CE has been investigated only recently (92–98). The first chiral separation by nonaqueous capillary electrophoresis (NACE) was reported by Ye and Khaledi for chiral separation of enantiomers of trimipramine using β-CD in pure N-methylformamide (NMF) (92). Wang and Khaledi reported chiral separations of a group of basic pharmaceutical racemates with β-CD in formamide (FA) media (93,94). The effect of separation media on chiral separation was studied by measuring the binding constants of these basic enantiomers with β-CD (94). The binding constants of mianserin, thioridazine, and trimipramine with β-CD in three organic solvents, FA, NMF, and N,N-dimethylformamide (DMF), were compared to those in pure water and in 6 M urea in water. They decrease systematically from around 10^4 in water to 10 in FA and about 10^{-2} in DMF.

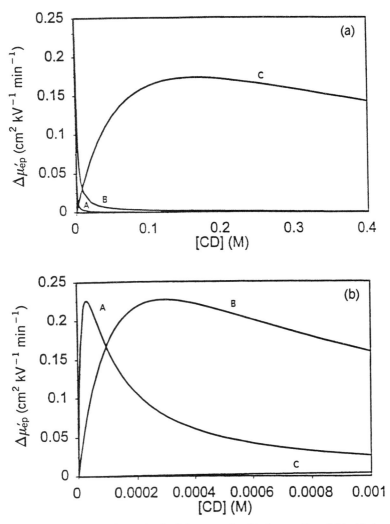

Figure 23.9. (a,b) Theoretical enantioselectivity curves for chiral separation of thioridazine with β-CD in three different solvents: (A) Tris–citrate in pure water, pH 3.02; (B) Tris–citrate in 6 M urea water, pH 3.90; (C) Tris–citrate in FA, apparent pH = 5.4. [From F. Wang and M. G. Khaledi, *Anal. Chem.* **68**, 3460 (1996), with permission.]

The effect of CD concentration on separation selectivities in aqueous and nonaqueous media is shown in Figure 23.9 for thioridazine. The behavior follows the theory developed by Wren and Rowe [see Section 23.2, above, and Eq. (23.4)] (7,8). There is an optimum CD concentration at which maximum selectivity is observed in pure water, 6 M urea, and FA (94). In the aqueous buffer, the variation in separation selectivity is dramatic and optimum concentration occurs at very low concentration (about 30 μM for thrioridazine) (Figure 23.9b). The optimum can be easily missed in a procedure for method development that is based upon trial and error. The optimum concentration shifts to higher values as the binding constants of the solute decrease during the change of the media to 6 M urea in water and FA. In FA, however, the enantioselectivity has an asymptotic behavior as a function of CD concentration, and highest enantioselectivity occurs over a wide range of CD concentration; thus, selection of the optimum concentration is not as crucial as it is with the aqueous media (Figure 23.9a).

Another important issue is the difference in mechanisms in chiral separations in the aqueous and nonaqueous media (94). According to the three-point interaction theory for chiral separation, in aqueous buffer the inclusion complexation of enantiomers into the CD cavity, driven by hydrophobic interaction, plays an important role in the separation process. The hydrogen-bonding interactions with the hydroxyl groups on the CD rims would provide additional points of interaction. It is not clear whether the same mechanism is applicable for the chiral separation in the amide solvents. Obviously, the hydrophobic interaction between the solutes and the CD does not exist in the NACE systems. It is thus quite logical to conclude that the inclusion complexation occurs only to a small extent, if at all. It seems that the NACE systems are mainly effective for the tricyclic compounds. The best chiral separations were achieved using β-CD and its derivatives, whereas partial or no separation was achieved using γ-CD. This is quite interesting, since tricyclic compounds are typically too large for β-CD and fit γ-CD better. This observation might suggest that inclusion complexation is not the primary mechanism. Instead, solutes might "lay flat on the mouth" of the CD and interact with the hydroxyl groups on the rims through polar–polar interaction.

This mechanism was first suggested by Armstrong's group for the HPLC separations of chiral amines on CD columns using nonaqueous eluents consisting of acetonitrile (95%) and methanol (5%) (99–101). They proposed that the inclusion complexation would not occur under these conditions. However, acetonitrile is an aprotic solvent. This results in strong hydrogen bonding between the enantiomers and the hydroxyl groups of CD in acetonitrile. In fact, the addition of 5% methanol was to reduce the strong solute–CD stationary phase interactions due to the hydrogen bonding in acetonitrile. This is not the case, however, for basic solvents such as FA and NMF that were used in this study. These hydrogen-bonding acceptors should compete with the enantiomeric amines for the hydroxyl groups on the CD. More studies are needed to investigate the interaction mechanism in NACE.

Recently, charged chiral selectors have been applied in NACE chiral separation. Quinine was used as a chiral selector in methanol media for chiral separations of some derivatized amino acids, 1,1'-binaphthyl-2,2'-diyl hydrogen phosphate, and N-[1-(1-naphthyl)ethyl]phthalamic acid by Stalcup and Gahm (95). A polyanionic β-CD, sulfated β-CD, has also been used in FA for chiral separation of basic pharmaceutical amines by Wang and Khaledi (94,98). Chiral separation of labetalol by sulfated CD was carried out in both aqueous capillary electrophoresis (ACE) and NACE. In ACE, only three of four peaks of labetalol were resolved in either co–EOF or counter–EOF setups. Chiral resolution of all four isomers of labetalol was achieved in FA, apparently due to enhanced enantioselectivity for the two pairs of the enantiomers.

23.7. THE COUNTER–ELECTROOSMOTIC FLOW SCHEME

According to the resolution equation (23.5), EOF plays an important role in chiral separation. For basic analytes, most chiral separations of basic racemates have been carried out in aqueous buffers at pH 2.5 or 3.0 when analytes are positively charged and EOF is small. Buffer additives such as methylhydroxyethyl cellulose, hydroxyethyl cellulose (102), cetyltrimethylammonium bromide (CTAB), cetylpyridinium chloride (103), and poly(vinyl alcohols) (104) have been used in the CD systems to suppress EOF in order to increase chiral resolution. In the case of counter-EOF, for uncoated capillary, it is important to adjust the buffer components to create a situation that the movement of the enantiomer–chiral selector complexes is opposite to that of the EOF (18,105–107). Quang and Khaledi (108,109) used organic cationic reagents such as tetramethylammonium (TMA^+) and tetrabutylammonium (TBA^+) in order to reverse the EOF to achieve chiral separation of a large number of basic solutes at one concentration of neutral CDs. In contrast to long-chain surfactants like CTAB, short-chain organic cations can reverse EOF and would not interfere with the chiral recognition process due to their smaller affinity to interact with CDs. Figure 23.10 shows the chiral separations of 12 chiral amines at one hydroxypropyl-β-CD (HP-β-CD) concentration in a TMA^+–phosphate buffer. The acebutolol isomers could not be separated due to lack of adequate enantioselectivity and/or small binding with the chiral selectors. Only two of the four isomers of labetalol (which has two chiral centers) were resolved due to inadequate separation selectivity.

Another method for creating a counter-EOF scheme is to change the sign of the average mobility of the enantiomers by using CDs with opposite charges to that of enantiomers. For example, polyanionic CDs can be used to complex with cationic analytes effectively to reverse the sign of μ_{avg} from positive to

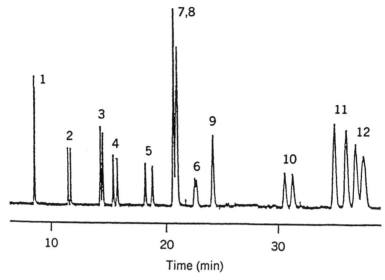

Figure 23.10. Chiral separation of 12 amines in a TMA$^+$ buffer. Buffer: 20 mM HP-β-CD in 50 mM TMA$^+$–phosphate at pH 2.50. Peaks: (1) doxylamine; (2) chlorpheniramine; (3) norepinephrine; (4) epinephrine; (5) isoproterenol; (6) atenolol; (7) pindolol; (8) oxprenolol; (9) acebutolol; (10) alprenolol; (11) propanolol; (12) labetalol. [From C. Quang and M. G. Khaledi, *J. Chromatogr. A* **692**, 253 (1995), with permission.]

negative to create a counter-EOF setup (18,105–107). A large number of basic optical isomers were resolved even though the concentration of the CD was not optimized for individual compounds. The same strategy might be used for chiral separation of acidic compounds with cationic CDs.

23.8. CAPILLARY ELECTROCHROMATOGRAPHY

CEC can be carried either by immobilization of the chiral selector onto the capillary wall or by packing the capillary with chiral stationary phases. The immobilization of different types of CD on the capillary wall for chiral separation of acidic racemates was reported by Schurig's group [see Mayer and Schurig (110); also Mayer et al. (111)]. They found that the thickness of the coating layer affects the separation efficiency due to the mass transfer. Armstrong and coworkers used the immobilized CD capillary for chiral separation in GC, supercritical fluid chromatography (SFC), and CE (112).

Lloyd et al. used both proteins and β-CD HPLC chiral stationary phases in CEC for chiral separations (113–115). An α_1-acid glycoprotein stationary phase was used to separate some neutral and cationic racemates in a normal phosphate buffer in the pH range from 4.45 to 7.5 (113). A human serum albumin chiral stationary phase was used to separate three cationic phenothiazine derivatives (115). The effects of pH, ionic strength, and organic additives on EOF and enantioselectivity were studied. It was shown that EOF plays an important role in CEC chiral separation. Anionic racemates could not be separated in the normal phosphate buffer. Some neutral and anionic racemates were separated by a β-CD chiral stationary phase in a triethylammonium acetate (TEAA) phosphate buffer. The presence of TEAA causes the reversal of the EOF direction (114). The separation efficiency for both protein chiral stationary phases was very low. It ranged from 25,000 to 75,000 plates for the CD chiral stationary phase. Recently, CEC chiral separations of chlorthalidone and mianserin were reported by Zare's group [see LePievre et al. (116)]. HP-β-CD was used either as a mobile phase additive with an achiral stationary phase or as a chiral stationary phase for the chiral separation of chlorthalidone. Chiral separation of mainserin was achieved with the chiral stationary phase. The effect of acetonitrile concentration on chiral separation of mainserin was also studied. In addition to the difficulties in column preparation, the key problem with chiral separation by CEC is the need for various chiral stationary phases. The main advantage of CE is the flexibility and versatility of incorporating a wide range of chiral recognition agents in the buffer solutions.

23.9. OTHER METHODS

Indirect chiral separation methods involve chemical derivatization of a racemic mixture prior to CE separation. As in HPLC and GC chiral separations, if a chiral selector for a direct chiral separation of a racemate is not available, an indirect method is needed. Compared with the direct method, the indirect separation method is tedious, and requires optically pure derivatizing reagents, and avoids stereo-transformation of reactants and enantiomers involved in the derivatization reaction.

The indirect chiral separation in CE has been used for chiral separation of a number of amino acids with derivatization reagents such as 2,3,4,6-tetra-O-acetyl-β-D-glucopyranosyl isothiocyanate (GITC) (117), 1-fluoro-2,4-dinitrophenyl-5-L-alanine, or 1-fluoro-2,4-dinitro-5-D-alanine (118), and (+)-O,O'-dibenzoyl-L-tartaric anhydride (119,120). Chiral separations of some phenethylamines were achieved by derivatization with GITC in the presence of 100 mM SDS and 20% methanol. The indirect method was also used for chiral separations of carnitine (121), some amphetamines (122), and some aldose racemates (123).

23.10. CHIRAL SEPARATION EFFICIENCY

In addition to the longitudinal diffusion in CE chiral separation, the other major factors contributing to band broadening are electrodisperson, solute–wall interactions, and Joule heating. Electrodispersion, caused by the mobility mismatch between the analytes and supporting electrolytes, has been investigated in CE chiral separations with neutral CD (124,125) and in CE chiral separations with polyanionic CD (126). Ordinarily, the mismatch could be avoided by (i) increasing the ionic strength of the background electrolyte or (ii) using a co-ion with the same electrophoretic mobility as that of the analytes (125). Vigh's group used a multiequilibrium method to calculate the electrophoretic mobility of a negatively charged form of a zwitterionic buffer [see Rawjee et al. (124); Williams and Vigh (125)]. The theoretical equation deduced by Vigh's group shows that the mobility of the negatively charged buffer species is related to the analytical concentration of CD (C_{CD}), pH, and the bulk buffer concentration (C_z). When pH < pH_{opt} and/or $C_z < C_{z/opt}$ the peaks are fronting, whereas at pH > pH_{opt} and/or $C_z > C_{z/opt}$ the peaks begin to tail. Here, the pH_{opt} and $C_{z/opt}$ are defined as the pH and concentration of buffer giving the best separation efficiency; they vary according to the properties of analytes and buffer types. Both calculation and experiments have shown that the matching of electrophoretic mobilities can be achieved by changing the pH and/or the buffer concentration, which can improve the peak shapes. This is called the dynamic mobility-matching method. Figure 23.11 shows the effect of buffer concentration on peak shapes at optimized buffer pH. Fronting peaks were observed when the concentration of buffer was lower than the optimum (Figure 23.11a), whereas tailing peaks were obtained at higher-than-optimum buffer concentration (Figure 23.11c, p. 816). Symmetrical peaks could be found only at the right buffer concentration, at which the mobility mismatching was eliminated or at least minimized (Figure 23.11b). However, this method is time consuming and complicated. In another report, the same group used weak basic analytes as an example to show the effect of independently varying the buffer pH (controlling selectivity) and co-ion in buffer (elimination of mobility mismatching) (125). According to their method, for a weak basic analyte the buffer electrolyte can be a weak acid titrated with a strong base to control the pH of the buffer. The effect of cationic co-ions on the peak shape in a phosphate buffer at pH 2.2 was shown for five cationic racemate analytes.

Lurie et al. (126) observed significant peak tailing with polyanionic CD in CE for chiral separations of some basic solutes. The mobility mismatch was considered to cause the tailing. Another possibility that could cause band broadening is multiple complexation between the analytes and the multicharged CD. The lower separation efficiency problem also exists in a counter-

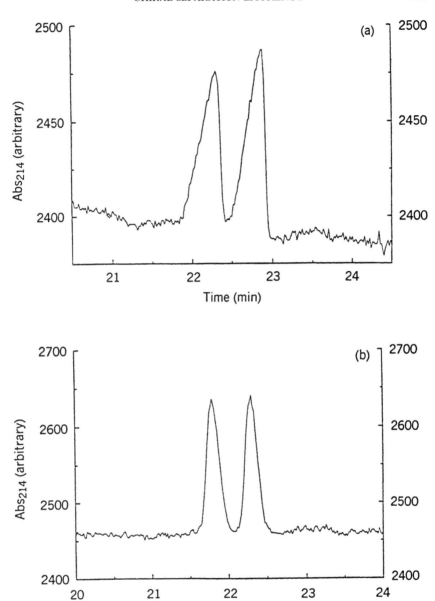

Figure 23.11. Peak shape in chiral separation of fenoprofen. (a) Conditions: 15 mM β-CD with 0.2% HEC in 50 mM MES aqueous buffer, pH 4.65. Field strength: 600 V/cm. (b) The same as in part (a) except $C_{MES} = 113$ mM: (c) The same as in part (a) except $C_{MES} = 600$ mM. [From Y. Y. Rawjee, R. L. Williams, and G. Vigh, *Anal. Chem.* **66**, 3777 (1994), with permission.]

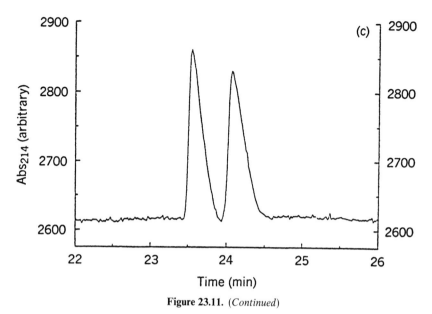

Figure 23.11. (*Continued*)

EOF setup. In general, the counter-EOF setup achieves high resolution at the cost of longer analysis time. Also, separation efficiency deteriorates with the increase in separation time due to the longitudinal diffusion. The problem with reversing EOF by short-chain surfactants is that the increase in current can lead to lower separation efficiency due to Joule heating. The same current problem occurs when the average electrophoretic mobility of cationic enantiomers is reversed by polyanionic chiral selectors. When a polyanionic CD is used for the counter-EOF setup, higher CD concentration has to be used to form negatively charged complexes. The increase in the concentration of charged CD will cause a dramatic increase in current because the degree of substitution of anionic CDs is usually larger than 1. Band broadening is also more pronounced when the charges on the chiral selector are opposite to that of the chiral molecules. Such problems can be solved by using a coated capillary to control or reverse EOF without an increase in current.

23.11. CONCLUSIONS AND FUTURE TRENDS

CE chiral separation is a highly efficient, simple, fast, and convenient analytical method. The available knowledge on chiral separation by HPCL has greatly facilitated the process of method development in CE. The inherent feasibility and flexibility of adjusting the compositions of buffer solutions in CE are significant advantages that allow rapid screening of a large number of experimental parameters. A wide variety of chiral selectors in aqueous and nonaqueous solvents have been investigated over the past few years. This trend will likely continue in the future so as to further extend the applicability of the technique to new chiral molecules. The microsize nature of CE should allow investigation of exotic and expensive chiral selector reagents. In spite of much progress in CE and HPLC chiral separations, selection of the right type of chiral selector and other important experimental parameters is still based on trial and error. Obviously, an in-depth understanding of chiral recognition mechanisms for various chiral selectors is still a long-term goal whose attainment will greatly facilitate the optimization process and will lead to better separations. Currently, the need for investigating experimental designs for rationale method development and systematic optimization of the large number of experimental parameters in chiral separation deserves more attention.

ADDENDUM

The interest in chiral separations by CE continues to grow. An entire issue of *Electrophoresis* has been devoted to this subject (127). Vespalec and Bocek have reviewed the progress in CE chiral separation since 1993 (128). The review by Shamsi and Warner emphasized monomeric and polymeric chiral surfactants for CE chiral separations (129). However, CDs remain the most widely used chiral selectors (128,130–151). Several of the most recent publications focused on the applications of charged CDs to separate oppositely charged enantiomers (130–133,135,136,139,140,142–146, 148–150).

In addition, a new type of cationic CDs, called histamine-modified β-CD, was used to separate dansyl-amino acids (146). A zwitterionic CD was also used to separate the enantiomers of chlorthalidone, clenbuterol, carprofen, flurbiprofen, benzoin and its derivatives (142). The charges on the CD were controlled by pH of the buffer electrolyte. At pH 2.3, this CD is partially protonated and showed better separations than those of mono-(6-amino-6-deoxy)-β-CD. At pH 11.2, the CD is negatively charged, and separation of hydrobenzoin enantiomers was obtained. The elution order of enantiomers at pH 11.2 was the opposite of that at pH 2.3. Interestingly, Fanali and Camera used cationic CDs to separate basic enantiomers (135). For an untreated capillary, EOF was negative in the presence of the cationic CDs. Thus, a counter-EOF setup was established. However, due to less enantioselectivity between enantiomers and CDs of the same charges, addition of 100 mM

TMA$^+$ was necessary to improve the separation through a stronger counter-EOF (135).

Ingelse et al. (152) studied the effect of buffer anions on the binding constants between anionic sulfonamides and neutral CDs and a polymeric CD. The results showed that different co-ions (anions in this case) in the running buffers can significantly affect the magnitude of the binding constants. It was also demonstrated that this effect differs among the CDs.

As different numbers of charged substituent groups can be attached to CDs in the synthetic process, most of the charged CDs show a distribution in the degree of substitution. The purity of charged CDs has been measured by direct injection of the CDs into capillary with an indirect detection mode (153–155). The results showed that charged CDs have a wide distribution with different numbers of charged substituent groups (133,142,143).

Other types of chiral selectors have also been investigated. Anionic polysaccharides were used to separate neutral and basic racemic mixtures (156,157). Chiral crown ether has been uniquely effective for the separation of primary amine enantiomers (158,159). Antibiotics were used to separate peptides (160) and arylpropionic acids (161). For the arylpropionic acids, vancomycin proved to be better than trimethyl-β-CD and heptamethylamino-β-CD. Among the protein chiral selectors, α_1-acid glycoprotein (162,163), BSA (164) and pepsin (165) were successfully applied in CE by a partial filling technique for chiral separations of some pharmaceutical enantiomers.

LIST OF ACRONYMS

Acronym	Definition
ACE	aqueous capillary electrophoresis
AGP	α_1-acid glycoprotein
BSA	bovine serum albumin
CD	cyclodextrin
CE	capillary electrophoresis
CEC	capillary electrochromatography
CGE	capillary gel electrophoresis
CMC	critical micelle concentration
co–EOF	co–electroosmotic flow
counter–EOF	counter–electroosmotic flow
CS	chiral selector
CTAB	cetyltrimethylammonium bromide
CZE	capillary zone electrophoresis
DMF	N,N-dimethylformamide
EOF	electroosmotic flow
FA	formamide

GC	gas chromatography
GITC	2,3,4,6-tetra-*O*-acetyl-β-D-glucopyranosyl isothiocyanate
HA	human albumin
HP-β-CD	hydroxypropyl-β-cyclodextrin
HPLC	high-performance liquid chromatography
ITP	isotachophoresis
MEKC	micellar electrokinetic chromatography
NACE	nonaqueous capillary electrophoresis
NMF	*N*-methylformamide
OVM	ovomucoid
PAHs	polycyclic aromatic hydrocarbons
SDS	sodium dodecyl sulfate
SFC	supercritical fluid chromatography
TBA$^+$	tetrabutylammonium
TEAA	triethylammonium acetate
TMA$^+$	tetramethylammonium
UV–Vis	ultraviolet–visible (range)

REFERENCES

1. E. Gil-Av, B. Feibush, and R. Charles-Sigler, *Tetrahedron Lett.* **10**, 1009 (1966).
2. E. Gassman, J. E. Kuo, and R. N. Zare, *Science* **230**, 813 (1985).
3. H. Nishi, *J. Chromatogr. A* **735**, 57 (1996).
4. S. Fanali, *J. Chromatogr. A* **735**, 77 (1996).
5. H. Nishi and S. Terabe, *J. Chromatogr. A* **694**, 245 (1995).
6. A. Guttman, A. Paulus, A. S. Cohen, N. Grinberg, and B. L. Karger, *J. Chromatogr.* **448**, 41 (1988).
7. S. A. C. Wren and R. C. Rowe, *J. Chromatogr.* **603**, 235 (1992).
8. S. A. C. Wren and R. C. Rowe, *J. Chromatogr.* **609**, 363 (1992).
9. J. C. Giddings, *Sep. Sci.* **4**, 181 (1969).
10. J. W. Jorgenson and K. D. Lukacs, *Anal. Chem.* **53**, 1298 (1981).
11. J. Snopek, I. Jelinek, and E. Smolkova-Keulemansova, *J. Chromatogr.* **438**, 211 (1988).
12. S. Fanali, *J. Chromatogr.* **474**, 441 (1989).
12a. W. Saenger, *Angew. Chem. Int. Ed. Engl.* **19**, 344 (1980).
12b. V. Schurig and H.-P. Nowotny, *Angew. Chem. Int. Ed. Engl.* **29**, 939 (1990).
13. H. Nishi, K. Nakamura, H. Nakai, and T. Sato, *J. Chromatogr. A* **678**, 333 (1994).
14. S. Terabe, H. Ozaki, K. Otsuka, and T. Ando, *J. Chromatogr.* **332**, 211 (1985).
15. S. Terabe, *Trends Anal. Chem.* **8**, 129 (1989).

16. R. J. Tait, D. O. Thompson, V. J. Stella, and J. F. Stobaugh, *Anal. Chem.* **66**, 4013 (1994).
17. T. Schmitt and H. Engelhardt, *Chromatographia* **37**, 475 (1993).
18. A. M. Stalcup and K. H. Gahm, *Anal. Chem.* **68**, 1360 (1996).
19. F. Wang and M. G. Khaledi, *J. Microcolumn Sep.* Submitted.
20. A. D'Hulst and N. Verbeke, *J. Chromatogr.* **608**, 275 (1992).
21. A. D'Hulst and N. Verbeke, *Chirality* **6**, 255 (1994).
22. H. Soini, M. Stefansson, M.-L. Riekkola, and M. V. Novotny, *Anal. Chem.* **66**, 3477 (1994).
23. C. Quang and M. G. Khaledi, *J. High Resolut. Chromatogr.* **17**, 609 (1994).
24. R. B. Friedman, *Amylodextrin Oligosaccharides*, ACS Symp. Ser. No. 458. American Chemical Society, Washington, DC, 1991.
25. A. M. Stalcup and N. M. Agyei, *Anal. Chem.* **66**, 3054 (1994).
26. H. Nishi, K. Nakamura, H. Nakai, and T. Sato, *Anal. Chem.* **67**, 2334 (1995).
27. N. M. Agyei, K. H. Gahm, and A. M. Stalcup, *Anal. Chim. Acta* **307**, 185 (1995).
28. A. M. Abushoffa and B. J. Clark, *J. Chromatogr. A* **700**, 51 (1995).
29. H. Nishi, K. Nakamura, H. Nakai, T. Sato, and S. Terabe, *Electrophoresis* **15**, 1335 (1994).
30. R. Kuhn and S. Hoffstetter-Kuhn, *Chromatographia* **34**, 505 (1992).
31. R. Kuhn, F. Erni, T. Bereuter, and J. Hausler, *Anal. Chem.* **64**, 2815 (1992).
32. P. Castelnovo and C. Albanesi, *J. Chromatogr. A* **715**, 143 (1995).
33. M. G. Schmid and G. Gubitz, *J. Chromatogr. A* **709**, 81 (1995).
34. R. Kuhn, D. Riester, B. Fleckenstein, and K.-H. Wiesmuller, *J. Chromatogr. A* **716**, 371 (1995).
35. D. W. Armstrong, Y. Tang, S. Chen, Y. Zhou, C. Bagwill, and J.-R. Chen, *Anal. Chem.* **66**, 1473 (1994).
36. D. W. Armstrong, K. L. Rundlett, and G. L. Reid, III, *Anal. Chem.* **66**, 1690 (1994).
37. D. W. Armstrong and Y. Zhou, *J. Liq. Chromatogr.* **17**, 1695 (1994).
38. D. W. Armstrong, K. L. Rundlett, and J.-R. Chen, *Chirality* **6**, 496 (1994).
39. D. W. Armstrong, M. P. Gasper, and K. L. Rundlett, *J. Chromatogr. A* **689**, 285 (1995).
40. K. L. Rundlett and D. W. Armstrong, *Anal. Chem.* **67**, 2088 (1995).
41. M. P. Gasper, N. A. Berthod, U. B. Nair, and D. W. Armstrong, *Anal. Chem.* **68**, 2501 (1996).
42. T. J. Ward, C. Dann, III, and K. A. Blaylock, *J. Chromatogr. A* **715**, 337 (1995).
43. R. Vespalec, H. Corstjens, H. A. H. Billiet, J. Frank, and K. C. A. M. Luyben, *Anal. Chem.* **67**, 3223 (1995).
44. Y. Tanaka and S. Terabe, *J. Chromatogr. A* **694**, 277 (1995).
45. Y. Tanaka, N. Matsubara, and S. Terabe, *Electrophoresis* **15**, 848 (1994).
46. G. E. Barker, P. Russo, and R. A. Hartwick, *Anal. Chem.* **64**, 3024 (1992).

47. S. Busch, J. C. Kraak, and H. Poppe, *J. Chromatogr.* **635**, 119 (1993).
48. P. Sun, N. Wu, G. Barker, and R. A. Hartwick, *J. Chromatogr. A* **648**, 475 (1993).
49. P. Sun, G. Barker, R. A. Hartwick, N. Grinberg, and R. Kaliszan, *J. Chromatogr. A* **652**, 247 (1993).
50. L. Valtcheva, J. Mohammed, G. Pettersson, and S. Hjertén, *J. Chromatogr.* **638**, 263 (1993).
51. R. Vespalec, V. Sustacek, and P. Bocek, *J. Chromatogr.* **638**, 255 (1993).
52. D. K. Lloyd, S. Li, and P. Ryan, *Chirality* **6**, 230 (1994).
53. Y. Ishihaama, Y. Oda, N. Asakawa, Y. Yoshida, and T. Sato, *J. Chromatogr. A* **666**, 193 (1994).
54. S. Terabe, M. Shibata, and Y. Miyashita, *J. Chromatogr.* **480**, 403 (1989).
55. H. Nishi, T. Fukuyama, M. Matsuo, and S. Terabe, *J. Chromatogr.* **515**, 233 (1990).
56. H. Nishi, T. Fukuyama, M. Matsuo, and S. Terabe, *Anal. Chim Acta* **236**, 281 (1990).
57. H. Nishi, T. Fukuyama, M. Matsuo, and S. Terabe, *J. Microcolumn Sep.* **1**, 234 (1989).
58. R. O. Cole, M. J. Sepaniak, and W. L. Hinze, *J. High Resolut. Chromatogr.* **13**, 579 (1990).
59. K. Otsuka and S. Terabe, *J. Chromatogr.* **515**, 221 (1990).
60. Y. Ishihama and S. Terabe, *J. Liq. Chromatogr.* **16**, 933 (1993).
61. N. Nimura, H. Itoh, C. Mitsuno, and T. Kinoshita, *Proc. Sep. Sci.*, Tokyo, *1994*, p. 283 (1994).
62. J. Wang and I. M. Warner, *Anal. Chem.* **66**, 3773 (1994).
63. A. Dobashi, T. Ono, S. Hara, and J. Yamaguchi, *J. Chromatogr.* **480**, 413 (1989).
64. K. Otsuka, M. Kashihara, Y. Kawaguchi, R. Koike, T. Hisamitsu, and S. Terabe, *J. Chromatogr. A* **652**, 253 (1993).
65. K. Otsuka, M. Karuhara, M. Higashimori, and S. Terabe, *J. Chromatogr. A* **680**, 317 (1994).
66. D. D. Dalton, D. R. Taylor, and D. G. Waters, *J. Chromatogr. A* **712**, 365 (1994).
67. A. Dobashi, M. Masaki, and Y. Dobashi, *Anal. Chem.* **67**, 3011 (1995).
68. A. Dobashi, T. Ono, S. Hara, and J. Yamaguchi, *Anal. Chem.* **61**, 1984 (1989).
69. K. Ostuka and S. Terabe, *J. Chromatogr.* **559**, 209 (1991).
70. K. Ostuka and S. Terabe, *Electrophoresis* **11**, 982 (1990).
71. J. R. Mazzeo, E. R. Grover, M. E. Swartz, and J. S. Petersen, *J. Chromatogr. A* **680**, 125 (1994).
72. J. R. Mazzeo, M. E. Swartz, and E. R. Grover, *Anal. Chem.* **67**, 2966 (1995).
73. D. C. Tickle, G. N. Okafo, P. Camilleri, R. F. D. Jones, and A. J. Kirby, *Anal. Chem.* **66**, 4121 (1994).

74. S. Terabe, Y. Miyashita, O. Shibata, E. R. Banhart, L. R. Alexander, D. G. Patterson, B. L. Karger, K. Hosoya, and N. Tanaka, *J. Chromatogr.* **516**, 23 (1990).
75. T. Ueda, F. Kitamura, R. Mitchell, T. Metcalf, T. Kuwana, and A. Nakamoto, *Anal. Chem.* **63**, 2979 (1991).
76. S. Terabe, Y. Miyashita, Y. Ishihama, and O. Shibata, *J. Chroamtogr.* **636**, 47 (1993).
77. T. Ueda, F. Kitamura, R. Mitchell, T. Metcalf, T. Kuwana, and A. Nakamoto, *J. Chromatogr.* **593**, 265 (1992).
78. H. Wan, P. E. Andersson, A. Engström, and L. G. Blomberg, *J. Chromatogr. A* **704**, 179 (1995).
79. H. Siren, J. H. Jumppanen, K. Manninen, and M.-L. Riekkola, *Electrophoresis* **15**, 779 (1994).
80. J. Prunonosa, R. Obach, A. Diez-Cascon, and L. Gouesclou, *J. Chromatogr.* **574**, 127 (1992).
81. H. Nishi, T. Fukuyama, and S. Terabe, *J. Chromatogr.* **553**, 503 (1991).
82. R. Furuta and T. Doi, *Electrophoresis* **15**, 1322 (1994).
83. R. Furuta and T. Doi, *J. Chromatogr. A* **676**, 433 (1994).
84. A. Aumatell and R. J. Wells, *J. Chromatogr. A* **688**, 329 (1994).
85. J. Wan and I. M. Warner, *J. Chromatogr. A* **711**, 297 (1995).
86. S. G. Penn, E. T. Bergstrom, and D. Goodall, *Anal. Chem.* **66**, 2866 (1994).
87. Y. Y. Rawjee, D. U. Staerk, and G. Vigh, *J. Chromatogr.* **635**, 291 (1993).
88. Y. Y. Rawjee, R. L. Williams, and G. Vigh, *J. Chromatogr. A* **652**, 233 (1993).
89. Y. Y. Rawjee, R. L. Williams, L. A. Buckingham, and G. Vigh, *J. Chromatogr. A* **688**, 273 (1994).
90. Y. Y. Rawjee and G. Vigh, *Anal. Chem.* **66**, 619 (1994).
91. G. M. Janini and H. J. Issaq, in *Capillary Electrophoresis Technology* (N. A. Guzman, ed.), Chromatogr. Sci. Ser., Vol. 64, p. 119 (1993).
92. B. Ye and M. G. Khaledi, *Symp. High Perform. Capillary Electrophor.*, *6th*, San Diego, CA, 1994, Abstr. No. 133 (1994).
93. M. G. Khaledi, F. Wang, and C. Quang, *Pittsburgh Conf.*, New Orleans, LA, *1995*, Abstr. No. 51 (1995).
94. F. Wang and M. G. Khaledi, *Anal. Chem.* **68**, 3460 (1996).
95. A. M. Stalcup and K. H. Gahm, *J. Microcolumn Sep.* **8**, 145 (1996).
96. I. E. Valko, H. Siren, and M.-L. Riekkola, *J. Chromatogr. A* **737**, 263 (1996).
97. I. Bjoernsdottir, S. Terabe, and S. H. Hansen, *Symp. High Perform. Capillary Electrophor.*, *8th*, Orlando, FL, *1996*, Abstr. No. 367 (1996).
98. F. Wang and M. G. Khaledi, *J. Chromatogr.* Submitted.
99. D. W. Armstrong, S. Chen, C. Chang, and S. Chang, *J. Liq. Chromatogr.* **15**, 545 (1992).
100. J. Zukowski, M. Pawlowska, M. Nagatkina, and D. W. Armstrong, *J. Chromatogr.* **629**, 169 (1993).

101. S. C. Chang, G. L. Reid, III, S. Chen, C. D. Chang, and D. W. Armstrong, *Trends Anal. Chem.* **12**, 144 (1993).
102. J. Snopek, H. Soini, M. Novotony, E. Smolkova-Keulemasova, and I. Jelinek, *J. Chromatogr.* **559**, 215 (1991).
103. H. Soini, Marja-Liisa Reikkola, and M. V. Novotny, *J. Chromatogr.* **608**, 265 (1992).
104. D. Belder and G. Schomburg, *J. High Resolut. Chromatogr.* **15**, 686 (1992).
105. R. J. Tait, D. O. Thompson, and J. F. Stobaugh, *Anal. Chem.* **66**, 4013 (1994).
106. C. Dette and S. Terabe, *Electrophoresis* **15**, 799 (1994).
107. T. Schmitt and H. Engelhardt, *Chromatographia* **37**, 343 (1995).
108. C. Quang and M. G. Khaledi, *Anal. Chem.* **65**, 3354 (1993).
109. C. Quang and M. G. Khaledi, *J. Chromatogr. A* **692**, 253 (1995).
110. S. Mayer and V. Schurig, *J. Liq. Chromatogr.* **16**, 915 (1993).
111. S. Mayer, M. Schleimer, and V. Schurig, *J. Microcolumn Sep.* **6**, 43 (1994).
112. D. W. Armstrong, Y. Tang, T. Ward, and M. Nichols, *Anal. Chem.* **65**, 1114 (1993).
113. S. Li and D. K. Lloyd, *Anal. Chem.* **65**, 3684 (1993).
114. S. Li and D. K. Lloyd, *J. Chromatogr. A* **666**, 321 (1994).
115. D. K. Lloyd, S. Li, and P. Ryan, *J. Chromatogr. A* **694**, 285 (1995).
116. F. Lelievre, C. Yan, R. N. Zare, and P. Gareil, *J. Chromatogr. A* **723**, 145 (1996).
117. H. Nishi, T. Fukuyama, and M. Matsuo, *J. Microcolumn Sep.* **2**, 234 (1990).
118. A. D. Tran, T. Blane, and E. J. Leopold, *J. Chromatogr.* **516**, 241 (1990).
119. W. Schutzner, S. Fanali, A. Rizzi, and E. Kenndler, *J. Chromatogr.* **639**, 375 (1993).
120. W. Schutzner, S. Fanali, A. Rizzi, and E. Kenndler, *Electrophoresis* **15**, 769 (1994).
121. C. Vogt, A. Georgi, and G. Werner, *Chromatographia* **40**, 287 (1995).
122. S. Cladrowa-Runge, R. Hirz, E. Kenndler, and A. Rizzi, *J. Chromatogr. A* **710**, 339 (1995).
123. C. R. Noe and J. Freissmuth, *J. Chromatogr. A* **704**, 503 (1995).
124. Y. Y. Rawjee, R. L. Williams, and G. Vigh, *Anal. Chem.* **66**, 3777 (1994).
125. R. L. Williams and G. Vigh, *J. Chromatogr. A* **730**, 273 (1996).
126. I. S. Lurie, R. F. X. Klein, T. A. D. Cason, M. J. LeBelle, R. Brenneisen, and R. E. Weinberger, *Anal. Chem.* **66**, 4019 (1994).
127. S. Fanali (Ed.), *Electrophoresis* (*Spec. Issue Chiral Sep. CE*) **18**, 843–1043 (1997).
128. R. Vespalec and P. Bocek, *Electrophoresis* **18**, 843 (1997).
129. S. A. Shamsi and I. M. Warner, *Electrophoresis* **18**, 853 (1997).
130. S. Fanali and E. Camera, *Chromatographia* **43**, 247 (1996).
131. S. Cladrowa-Runge and A. Rizzi, *J. Chromatogr. A* **759**, 157 (1997).
132. K. Gahm, L. W. Chang, and D. W. Armstrong, *J. Chromatogr. A* **759**, 149 (1997).
133. G. Schulte, B. Chankvetadze, and G. Blaschke, *J. Chromatogr. A* **771**, 259 (1997).
134. C. Vogt and S. Kiessig, *J. Chromatogr. A* **745**, 53 (1996).

135. S. Fanali and E. Camera, *J. Chromatogr. A* **745**, 17 (1996).
136. M. Jung and E. Francotte, *J. Chromatogr. A* **755**, 81 (1996).
137. J. Liu, H. Coffey, D. J. Detlefsen, Y. Li, and M. Lee, *J. Chromatogr. A* **763**, 261 (1997).
138. L.Liu, L. M. Osborne, and M. A. Nussbaum, *J. Chromatogr. A* **745**, 45 (1996).
139. S. Surapaneni, K. Ruterbories, and T. Lindstrom, *J. Chromatogr. A* **761**, 249 (1997).
140. C. Desiderio, C. M. Polcaro, and S. Fanali, *Electrophoresis* **18**, 227 (1997).
141. S, Ma and C. Horvath, *Electrophoresis* **18**, 873 (1997).
142. F. Lelievre, C. Gueit, P. Gareil, Y. Bahaddi, and H. Galons, *Electrophoresis* **18**, 891 (1997).
143. K. Ishibuchi, S. Izumoto, H. Nishi, and T. Sato, *Electrophoresis* **18**, 1007 (1997).
144. S. Nilsson, L. Schweitz, and M. Petersson, *Electrophoresis* **18**, 884 (1997).
145. H. Jakubetz, M. Juza, and V. Schurig, *Electrophoresis* **18**, 897 (1997).
146. G. Galacerna, R. Corradini, A. Dossena, R. Marchelli, and G. Vecchio, *Electrophoresis* **18**, 905 (1997).
147. T. Blitzke, H. Wilde, and C. Vogt, *Electrophoresis* **18**, 978 (1997).
148. D. J. Skanchy, R. Wilson, T. Poh, G. Xie, C. W. Demarest, and J. F. Stobaugh, *Electrophoresis* **18**, 985 (1997).
149. Z. Juvancz, K. E. Markides, and L. Jicsinszky, *Electrophoresis* **18**, 1002 (1997).
150. M. Fillet, P. Hubert, and J. Crommen, *Electrophoresis* **18**, 1013 (1997).
151. M. Frost, H. Kohler, and G. Blaschke, *Electrophoresis* **18**, 1026 (1997).
152. B. A. Ingelse, H. C. Claessens, S. van der Wal, A. L. L. Duchateau, and F. M. Everaerts, *J. Chromatogr. A* **745**, 45 (1996).
153. R. J. Tait, D. J. Skanchy, D. O. Thompson, N. C. Chetwyn, D. A. Dunshee, R. A. Rajewskii, V. J. Stella, and J. F. Stobaugh, *J. Pharm. Biomed. Anal.* **10**, 615 (1992).
154. B. Chankvetadze, G. Endresz, and G. Blaschke, *J. Chromatogr. A* **704**, 234 (1995).
155. G. Weseloh, H. Bartsch, and A. Konig, *J. Microcolumn Sep.* **7**, 355 (1995).
156. H. Nishi, *J. Chromatogr. A* **735**, 345 (1996).
157. M. Jung, K. O. Bornsen, and E. Francotte, *Electrophoresis* **17**, 130 (1996).
158. Y. Mori, K. Ueno, and T. Umeda, *J. Chromatogr. A* **757**, 328 (1997).
159. H. Nishi, K. Nakamura, H. Nakai, and T. Sato, *J. Chromatogr. A* **757**, 328 (1997).
160. H. Wan and L. G. Blomberg, *Electrophoresis* **18**, 943 (1997); K. Gahm, L. W. Chang, and D. W. Armstrong, *J. Chromatogr. A* **759**, 149 (1997).
161. S. Fanali, C. Desiderio, and Z. Aturki, *J. Chromatogr. A* **772**, 185 (1997).
162. A. Amini, C. Pettersson, and D. Westerlund, *Electrophoresis* **18**, 950 (1997).
163. Y. Tanaka and S. Terabe, *Chromatographia* **44**, 119 (1997).
164. D. Eberle, R. P. Hummel, and R. Kuhn, *J. Chromatogr. A* **759**, 185 (1997).
165. S. Fanali, G. Caponecchi, and Z. Aturki, *J. Microcolumn Sep.* **9**, 9 (1997).

CHAPTER

24

CAPILLARY ELECTROPHORESIS OF INORGANIC IONS

JEFFREY R. MAZZEO

GelTex Pharmaceuticals, Waltham, Massachusetts 02154

24.1.	Introduction	825
24.2.	Indirect-Ultraviolet Detection: General Principles	826
	24.2.1 Indirect-Ultraviolet Detection: Anion Determinations	831
	24.2.2. Indirect-Ultraviolet Detection: Cation Determinations	837
24.3.	Direct-Ultraviolet Detection: Anion Determinations	841
24.4.	Direct-Ultraviolet Detection: Cation Determinations	843
24.5.	Other Detection Modes	844
24.6.	Concluding Remarks	847
List of Acronyms and Symbols		849
References		849

24.1. INTRODUCTION

The determination of inorganic ions such as chloride, sulfate, sodium, and calcium has traditionally been performed by a variety of analytical techniques. These techniques include ion-exchange chromatography with conductivity or indirect-UV detection, atomic absorption, inductively coupled plasma emission, and ion-selective electrodes. However, since the early 1990s capillary electrophoresis (CE) has emerged as a viable alternative to these traditional methods for the determination of inorganic ions. The general advantages of high-resolution, fast analysis times and a simple experimental setup inherent to CE make it an attractive technique for this application.

The challenge for those employing CE in general has been the ability to detect analytes at very low levels (sub-mg/L) using commercial systems equipped with UV–Vis absorption detection. In the case of inorganic ions,

High Performance Capillary Electrophoresis, edited by Morteza G. Khaledi. Chemical Analysis Series, Vol. 146.
ISBN 0-471-14851-2 © 1998 John Wiley & Sons, Inc.

this challenge is made more difficult by the fact that many of these analytes have very little absorbance in the UV–Vis region. Therefore, methods for determination of these ions have relied upon indirect-UV detection (1). More recently, the use of conductivity detection has been investigated for inorganic ion determinations by CE, and a commercial system is now available (2–4).

24.2. INDIRECT-ULTRAVIOLET DETECTION: GENERAL PRINCIPLES

Jandik et al. outlined the important electrolyte conditions for ion analysis by CE with indirect-UV detection (1). A classic example that illustrates the various conditions is the use of chromate in combination with tetradecyltrimethylammonium bromide (TTAB) for the separation and detection of inorganic anions (5) (Figure 24.1). In this electrolyte, TTAB reverse the direction of the electroosmotic flow (EOF) to decrease the analysis time. Chromate,

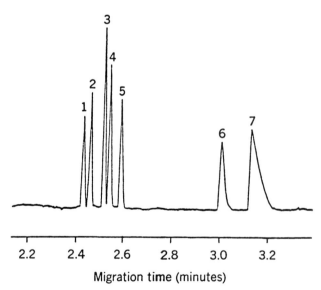

Figure 24.1. Separation of inorganic anions with chromate/TTAB electrolyte. Conditions: 5 mM sodium chromate/0.5 mM TTAB, pH 8.0; 75 μm i.d. × 60 cm capillary; 30 s injection; −20 kV; detection—UV, 254 nm. Peaks: (1) bromide (4 μg/mL); (2) chloride (2 μg/mL); (3) sulfate (4 μg/mL); (4) nitrite (4 μg/mL); (5) nitrate (4 μg/mL); (6) fluoride (1 μg/mL); (7) phosphate (6 μg/mL). [From J. Romano, P. Jandik, W. R. Jones, and P. E. Jackson, *J. Chromatogr.* **546**, 411 (1991), with permission.]

the electrolyte co-ion (i.e., the electrolyte ion with the same charge as that of the analyte ions), is the visualization reagent. Furthermore, it provides a good mobility match with the ions of interest, thereby resulting in symmetrical peaks.

Again in the nomenclature of Jandik et al. (1), the co-electroosmotic flow condition requires the direction of EOF to be in the same direction as the ions' electrophoretic mobilities. In the case of cation analysis, this condition already exists, since the EOF is generated by the movement of solvated cations toward the cathode. The EOF must be reversed for anion analysis, and this reversal has been achieved by adding to the electrolyte a cationic compound that binds to the silica wall. This binding leads to a net positive charge on the wall and EOF toward the anode. Cationic additives that have been used include quaternary amine surfactants such as TTAB and cetyltrimethylammonium bromide (CTAB), and diquaternary amine compounds such as hexamethonium bromide (5–7).

Having the EOF and electrophoretic mobilities of the analytes in the same direction leads to several advantages. Consider an anion with an electrophoretic mobility of -5×10^{-4} cm²/Vs. A typical value for a robust EOF in an uncoated fused-silica capillary is $+6 \times 10^{-4}$ cm²/Vs. Since the apparent mobility of an analyte ion (μ_{app}) is equal to the vector sum of its own electrophoretic mobility (μ_{ep}) plus the electroosmotic mobility (μ_{eo}),

$$\mu_{app} = \mu_{ep} + \mu_{eo} \tag{24.1}$$

the apparent mobility of the anion would be $+1 \times 10^{-4}$ cm²/Vs. If the direction of EOF is exactly reversed but has the same magnitude, the anion would have an apparent mobility of -11×10^{-4} cm²/Vs. Therefore, by simply reversing the direction of EOF, the analysis time is reduced by an order of magnitude. Another potential advantage is improved migration time reproducibility. In most cases, migration time variability is due to changes in the EOF. Taking the two cases described above for an anion with normal and reversed EOF, we can assume a 10% decrease in EOF. Under normal EOF conditions, the apparent mobility of the anion would decrease by 60% ($+0.4 \times 10^{-4}$ cm²/Vs vs. $+1 \times 10^{-4}$ cm²/Vs). Under reversed EOF conditions, the apparent mobility of the anion would only decrease by 5% (-10.4×10^{-4} cm²/Vs vs. -11×10^{-4} cm²/Vs).

Plate counts (N) are directly proportional to apparent mobility (8):

$$N = \frac{\mu_{app} V l}{2DL} \tag{24.2}$$

where V is the applied voltage; l is the capillary length from injection to detection; and L is the total capillary length. Therefore, plate counts are higher when the EOF and electrophoretic mobilities are in the same direction. Higher plates will lead to better detectability. However, one disadvantage of having the EOF in the same direction as the analytes' electrophoretic mobilities is that resolution (R_s) is compromised. This point is emphasized upon examination of the resolution equation for capillary electrophoresis (8):

$$R_s = \left(\frac{N^{1/2}}{4}\right) \left|\frac{\mu_{ep,1} - \mu_{ep,2}}{\mu_{ep,avg} + \mu_{os}}\right| \qquad (24.3)$$

where μ_{ep} is electrophoretic mobility and μ_{os} is electroosmotic mobility. This equation assumes that diffusion is the sole source of band broadening. To maximize resolution, it is desirable that the EOF be opposite in direction to the analyte's electrophoretic mobilities. In practice, this disadvantage does not pose a significant problem, since the differences in electrophoretic mobility of inorganic ions are sufficient to allow separation even under co-electroosmotic conditions. An exception is the separation of certain inorganic cations, which require the addition of complexing agents to achieve separation.

The first co-ion condition requires the electrolyte co-ion to have an electrophoretic mobility similar to the mobilities of the ions of interest. This situation ensures symmetrical peaks. If the electrolyte co-ion mobility is greater than a given analyte mobility, the analyte peak will exhibit tailing. If the electrolyte co-ion mobility is smaller than the analyte mobility, the peak will exhibit fronting. An example is shown in Figure 24.2. In this case, the analytes are five inorganic cations and ammonium ion, and the electrolyte co-ion is imidazole. 18-Crown-6 was included in the electrolyte to permit separation of ammonium and potassium. Note that (i) ammonium and potassium exhibit fronting; (ii) sodium, calcium, and magnesium exhibit no asymmetry; and (iii) lithium exhibits tailing. These results indicate that imidazole has a mobility close to that of sodium, calcium, and magnesium.

The extent of peak asymmetry is also related to the ratio of conductivities of the sample zone and the surrounding electrolyte zone. The absolute difference in conductivity is related both to the difference in mobility between the analyte and the electrolyte co-ion and to the total concentration of analyte by the following equation (9):

$$\Delta\kappa = \frac{c_{analyte}}{\mu_{analyte}}\left(\mu_{co-ion} - \mu_{analyte}\right)\left(\mu_{counterion} - \mu_{analyte}\right) \qquad (24.4)$$

where $\Delta\kappa$ is the conductivity in the surrounding electrolyte minus the conductivity in the sample zone; c is the concentration of the analyte ion; $\mu_{analyte}$ is the

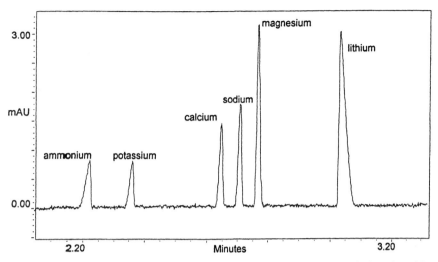

Figure 24.2. Separation of inorganic cations with imidazole/HIBA (hydroxyisobutyric acid) electrolyte. Conditions: 5 mM imidazole/6.5 mM HIBA/2 mM 18-crown-6, pH 4.5; 75 μm i.d. × 60 cm capillary; 30 s injection; 20 kV; detection—UV, 214 nm. Sample: 1 μg/mL each cation.

mobility of the analyte; $\mu_{co\text{-ion}}$ is the mobility of the co-ion; and $\mu_{counterion}$ is the mobility of the counterion (i.e., the electrolyte ion of charge opposite to that of the analyte). Note that the mobility of an anion has a negative sign whereas the mobility of a cation has a positive sign.

This equation gives the difference in conductivity between the sample zone and the bulk electrolyte for a given analyte concentration and mobility difference. However, the total concentration of electrolyte co-ion must also be considered, since it is the ratio of conductivities that will determine the extent of asymmetry. Therefore, increasing the electrolyte co-ion concentration, all else remaining constant, will decrease the amount of asymmetry. In practice, there is a maximum concentration of co-ion that can be successfully employed. This concentration is limited by Joule heat-induced band broadening, which leads to decreased resolution. Furthermore, with indirect-UV detection, if the co-ion concentration is too high, the linear range of the UV detector may be exceeded, leading to decreased detectability (see below). Noise levels will also increase at higher co-ion concentrations.

The second co-ion condition requires that the co-ion absorb at a wavelength where the analyte ions have minimal absorbance. If this condition is not met, detectability will be compromised. For most inorganic cations, which have no UV absorbance at any wavelength, co-ions can be chosen that have UV absorbance at wavelengths as low as 185 nm. For inorganic anions, many of which absorb at wavelengths less than 220 nm, co-ions with absorbance at 254 nm have been used.

The limit of detection with indirect-UV detection is generally expressed as (10)

$$C_{\lim} = \frac{\text{noise}}{aL(\text{TR})} \tag{24.5}$$

where a is the molar absorbtivity of the electrolyte co-ion and L is the path length; TR, the transfer ratio, describes the number of co-ions that are displaced per analyte ion. TR is given by (11)

$$\text{TR} = \frac{z_a}{z_b} \cdot \frac{\mu_b(\mu_a + \mu_c)}{\mu_a(\mu_b + \mu_c)} \tag{24.6}$$

where indices a, b, and c refer to the analyte a, electrolyte co-ion b, and electrolyte counterion c, respectively, and z is the valence.

Given Eqs. (24.5) and (24.6), several points about the electrolyte co-ion can be made. First, it should have high molar absorbtivity; however, it cannot be too high, since absorbance detectors for CE have a limited linear range. For instance, consider a detector and capillary configuration that shows linear response over 0.200 AU. The maximum concentration of co-ion that can be used is that concentration which will give an absorbance of 0.200 AU. For co-ion A, assume this corresponds to 5 mM concentration. A co-ion with 10 times the molar absorbtivity of co-ion A can only be used at 0.5 mM. At such a low co-ion concentration, severe peak asymmetry will result for all analytes that do not exactly match the co-ion mobility. Such peak asymmetry will decrease sensitivity and resolution. Second, the charge of the co-ion should ideally be 1 to ensure an optimum TR. Third, for an optimum TR, the co-ion mobility should be larger than the analyte mobility. However, this condition would lead to substantial peak tailing, compromising resolution and, potentially, sensitivity. In general, most published work has opted for a co-ion that matches the analyte ions' mobilities, trading off any potential gain in sensitivity by choosing a high-mobility co-ion for good resolution and peak symmetry.

Note that the limit of detection is independent of co-ion concentration. However, linearity is strongly dependent on co-ion concentration. Higher co-ion concentration will lead to greater linearity at the high analyte concentration end. Again, this extended linearity must be balanced against increased Joule heating and noise at higher co-ion concentration. Most electrolytes use from 1 to 10 mM co-ion. In general, this translates to 3 orders of magnitude in linearity. Note that these concentrations are roughly 10 times less than those typically used in capillary zone electrophoresis (CZE) with direct-UV detection. Therefore, peak asymmetry due to mobility mismatch is more pronounced with indirect-UV detection.

Finally, injection technique plays a key role in determining the limit of detection. By means of hydrostatic injection, 1–2% of the capillary volume is filled with sample. In many cases, the conductivity of the sample is less than the conductivity of the electrolyte and sample stacking is achieved. This stacking both increases plate counts and decreases the detection limit. Detection limits with hydrostatic injection are in the mid-μg/L range. Detection limits can be decreased to the low-μg/L range by ITP stacking. An ion of lower mobility than the sample ions is added to the sample, and electrokinetic injection is performed (12).

24.2.1. Indirect-Ultraviolet Detection: Anion Determinations

The most widely used electrolyte for inorganic anions employs chromate as the electrolyte co-ion and TTAB to reverse the EOF (5,12). Chromate is a high-mobility anion and exhibits strong UV absorbance in the 200–400 nm range. Detection is predominantly performed at 254 nm, where the common inorganic ions such as chloride, bromide, sulfate, nitrate, nitrite, fluoride, and phosphate exhibit no absorbance. Furthermore, most commercial CE systems exhibit a wide linear range at 254 nm and low noise. At concentrations $\geq$ 0.3 mM, TTAB provides a robust EOF toward the anode, permitting fast separations.

The first formulation for the chromate/TTAB electrolyte consisted of 5 mM sodium chromate/0.5 mM TTAB, pH 8.0. Although quite useful for many samples, this formulation suffers from several disadvantages. First, the quaternary surfactant is added in the bromide form. Samples with negligible bromide levels will exhibit a negative peak where bromide normally migrates in the electrolyte. This negative peak decreases the sensitivity for bromide and can also cause problems with integration of chloride. Second, the migration order is not favorable, with thiosulfate migrating first, bromide second, and chloride third. It would be ideal if chloride migrated first so that it could be used as a reference peak. Most samples will contain some chloride, while a few contain thiosulfate and/or bromide. The data system can be programmed to call the first peak chloride for all samples, and the rest of the ions can be identified based on their migration time ratio with chloride. This referencing can minimize problems with migration time drifts due to changes in EOF. However, it should be realized that a given change in EOF will affect ions of different mobility to different extents. Therefore, using a reference peak for migration time ratioing will not allow successful identification of all peaks over a wide migration time window. Finally, this formulation has no buffering capacity, which in CE is important since electrolysis of water is occuring at the inlet and outlet vials.

To address these limitations, work has been undertaken to reformulate the electrolyte (13). First, tetradecyltrimethylammonium hydroxide (TTAOH) has been used in place of TTAB. This form of the surfactant can be prepared by passing the bromide form through an anion exchange resin in the hydroxide form. However, when TTAOH is added to the electrolyte, the pH increases to about 12. To reduce the pH, various inorganic acids can be added, such as acetic or boric acid. Unfortunately, a negative peak appears in the electrolyte where the acid counterion would migrate. Ideally, an acid should be chosen that has a very low mobility such that the negative peak is well removed from the area where the ions of interest migrate. The zwitterionic buffer 2-(N-cyclohexylamino)ethanesulfonic acid (CHES) has been used for pH reduction. CHES contains an amine group with a pK_a (protonated amine) of 9.3, and a strong sulfonic acid group. When CHES is added to the high pH electrolyte, it reduces the pH down into the region of its pK_a where its net charge is low. For instance, at pH 9.3, CHES has a net charge of -0.5. Because of this low net charge, a negative peak is obtained in the electrolyte well removed from the ions of interest. Furthermore, at pH's near its pK_a, CHES is an excellent buffer.

Efforts have also focused on optimizing the concentration of TTAOH in the electrolyte. Figure 24.3 plots the migration time of the seven common inorganic anions referenced to chloride vs. TTAOH concentration in the range 0.5–0.6 mM. As can be readily seen, large changes in selectivity are noted when the concentration of TTAOH is varied over this range. Bromide, sulfate, and nitrate exhibit the strongest interactions with the surfactant as evidenced by their positive deviation from linearity. Others have also discussed the interaction of quaternary surfactants with inorganic anions (6,12).

Based on these studies, a formulation of 5 mM sodium chromate/3.5 mM TTAOH/10 mM CHES, pH 9.0, was chosen. A separation of the seven common inorganic anions with this formulation is shown in Figure 24.4. Note that chloride migrates first. Thiosulfate migrates just after nitrate. Over 45 injections, the reproducibility of migration time for the seven anions was less than 0.6% RSD. The critical pair resolution with this electrolyte (chloride/bromide, $R_s = 2.9$) is significantly better than that afforded by the old formulation (sulfate/nitrite, $R_s = 1.3$). The negative peak for CHES migrates well after propionate, an organic acid. Detection limits with hydrostatic injection are about 100 μg/L for each anion. Finally, this electrolyte has excellent buffering capacity.

Figure 24.3. Migration time (MT) normalized to chloride vs. concentration of TTAOH. Conditions: 5 mM sodium chromate/0.5–6 mM TTAOH/10–100 mM CHES, pH 9.0; 75 μm i.d. × 60 cm capillary; 30 s injection; −20 kV; detection—UV, 254 nm.

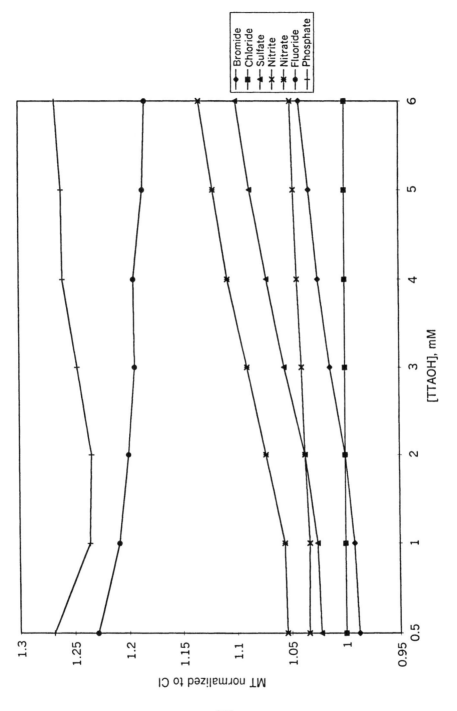

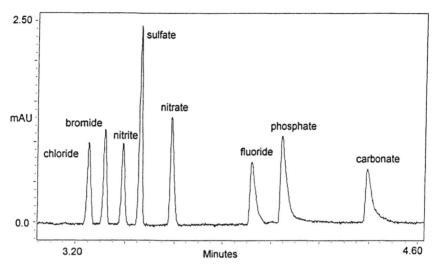

Figure 24.4. Separation of inorganic anions with chromate/TTAOH electrolyte. Conditions: 5 mM sodium chromate/3.5 mM TTAOH/10 mM CHES, pH 9.0; 75 μm i.d. × 60 cm capillary; 30 s injection; −20 kV; detection—UV, 254 nm. Analyte concentrations same as in Figure 24.1.

An alternative electrolyte to chromate with TTAOH is based on pyromellitic acid (PMA; or 1,2,4,5-benzenetetracarboxylic acid) as the electrolyte co-ion (7). This electrolyte consists of 2.25 mM PMA/6.5 mM sodium hydroxide/0.75 mM hexamethonium hydroxide/1.6 mM triethanolamine, pH 7.7. PMA is the electrolyte co-ion; hexamethonium hydroxide reverses the EOF; sodium hydroxide adjusts the electrolyte pH; and triethanolamine is a buffering agent. A separation of the seven common inorganic anions with this electrolyte is shown in Figure 24.5. According to Haddad's group, PMA offers better sensitivity than chromate does but a poorer mobility match for the common inorganic anions [see Cousins et al. (14)].

A comparison was made between the PMA/hexamethonium hydroxide electrolyte and chromate/TTAOH on the same commercial CE system under the same conditions (13). Table 24.1 gives the migration times, resolutions, and asymmetries for each electrolyte. Peaks are more fronted with the PMA electrolyte (asymmetry values less than 1), indicating a poor mobility match. Migration times are about 2.2 times longer with PMA electrolyte, since hexamethonium hydroxide gives a lower EOF than does TTAB. The analyte pair with the lowest resolution with chromate is chloride/bromide, with a resolution of 2.9. The analyte pair with the lowest resolution with PMA is

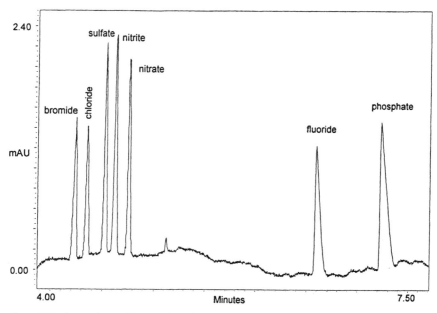

Figure 24.5. Separation of inorganic anions with PMA electrolyte. Conditions: 2.25 mM PMA/6.5 mM sodium hydroxide/0.75 mM hexamethonium hydroxide/1.6 mM triethanolamine, pH 7.7. Other conditions same as in Figure 24.4.

sulfate/nitrite, with a resolution of 1.9. The signal-to-noise ratio (S/N) was compared for chloride and phosphate with both electrolytes (S is defined as peak height in AU; N, as peak-to-peak in AU): S/N was 30 for both chloride and phosphate with chromate, and was 24 and 28 for chloride and phosphate, respectively, with PMA.

Other UV-active co-ions used for inorganic anions include naphthalene-trisulfonate, -disulfonate, and -sulfonate (15), trimellitate, phthalate, and benzoate (14), ribonucleotides for polyphosphonates and polyphosphates (16), and phthalate for condensed phosphates (17).

Many researchers have explored other ways of reversing or reducing EOF. Stathakis and Cassidy investigated several cationic polyelectrolytes for EOF reversal (18,19). Not only did these polymers reverse the EOF, but they also provided selectivity changes as a function of concentration, as seen with cationic surfactants.

Typically, a concentration of at least 0.25 mM TTAB is required for a robust reversed EOF. Benz and Fritz found that a concentration of only 0.03 mM, in combination with 3–5% 1-butanol, could achieve the same

Table 24.1. Comparison of Chromate Electrolyte[a] to PMA Electrolyte[b] for Seven Inorganic Anions[c]

Analyte	Chromate		
	Migration Time	Asymmetry	Resolution
Chloride	2.484	0.7	—
Bromide	2.542	0.8	2.9
Nitrite	2.598	1.2	3.0
Sulfate	2.665	0.7	3.8
Nitrate	2.767	1.8	5.2
Fluoride	3.009	3.4	8.9
Phosphate	3.128	4.7	3.4

Analyte	PMA		
	Migration Time	Asymmetry	Resolution
Bromide	4.298	0.3	—
Chloride	4.418	0.4	2.3
Sulfate	4.608	0.3	3.5
Nitrite	4.706	0.5	1.9
Nitrate	4.837	0.6	2.8
Fluoride	6.734	3.4	25.7
Phosphate	7.405	5.8	5.5

[a] 5.0 mM sodium chromate/3.5 mM TTAOH/10 mM CHES, pH 9.0.
[b] 2.25 mM PMA/6.5 mM sodium hydroxide/0.75 mM hexamethonium hydroxide/1.6 mM triethanolamine, pH 7.7.
[c] Conditions: Waters Quanta 4000E; 75 μm i.d. × 60 cm capillary; 30 s hydrostatic injection; detection—UV, 254 nm; 25 °C; −20 kV; 1–5 ppm each anion in water.

reversed EOF rate (20). Haddad's group found that selectivity changes could be obtained by using mixtures of TTAB and dodecyltrimethylammonium bromide, while maintaining the total surfactant concentration constant [see Harakuwe et al. (21)]. Instead of including the EOF reversal agent in the electrolyte, Oefner employed a rinsing step with 0.0003% hexadimethrine bromide to reverse the EOF (22). Tindall and Perry investigated the use of coated capillaries with reduced EOF for inorganic anion analysis (23).

Applications of indirect detection of anions are numerous. Romano et al. used 5 mM sodium chromate/0.5 mM TTAB, pH 10.0, for the determination of thiosulfate, chloride, sulfate, and several organic acids in Kraft black liquor, an important sample in the pulp and paper industry (24). Romano and Krol used 4 mM sodium chromate/0.3 mM TTAB, pH 8.1, to determine anions in drinking water and wastewater (25). Swartz used 5 mM chromate/0.4 mM TTAB, pH 8.0, to determine anions in prenatal vitamin tablets (26). Other application areas include anions in urine (27), the pulp and paper industry (28), atmospheric aerosols (29), the eye-care pharmaceutical industry (30), and detergent products (31).

Bondoux and Jones compared CE and ion chromatography (IC) for the determination of anions at the ppb level in nuclear power plant water (32). For inorganic anions at the low-ppb level, 10 mM chromate/0.5 mM TTAOH was used as the electrolyte and electrokinetic injection was performed, with sodium octane sulfonate added to the sample for isotachophoretic (ITP) stacking. IC gave approximately 10 times lower detection limits, but CE offered greater flexibility, lower cost of operation, and sufficient sensitivity for the samples analyzed.

An application demonstrating the high resolving power of CE was reported by Lucy and McDonald (33). The two isotopes of chloride were baseline resolved using 5 mM chromate/3 mM borate, pH 9.7. In this electrolyte, the EOF was not reversed to obtain the highest resolution.

24.2.2. Indirect-Ultraviolet Detection: Cation Determinations

As stated previously, for cation analysis there is no need to reverse the EOF for fast analysis. However, in the co-electroosmotic mode, many metal cations have insufficient differences in electrophoretic mobility to permit separation. Therefore, neutral or anionic complexing agents are added to the buffer that afford selective decreases of the apparent mobilities of the metal cations being analyzed. Separation is based both on differences in electrophoretic mobility and on differences in complex formation constants. Complexing agents that have been used include α-hydroxyisobutyric acid, glycolic acid, lactic acid, citric acid, tartaric acid, EDTA (edetic acid) and tropolone. For separation of potassium from ammonium, 18-crown-6 is added to the electrolyte. This crown ether selectively complexes with potassium, decreasing its apparent mobility (34).

In the analysis of alkali and alkaline earth metals, a high-mobility co-ion must be chosen. Beck and Engelhardt first described the use of imidazole as the electrolyte co-ion (35). A rather simple electrolyte containing 5 mM imidazole adjusted to pH 4.5 with sulfuric acid permits separation of potassium, sodium, barium, calcium, magnesium, and lithium in three minutes (Figure 24.6). Apparently, some complexation of the divalent metals with sulfate occurs, since migration order does not correlate with ionic equivalent conductance. With hydrostatic injection and indirect detection at 214 nm, detection limits were 100 µg/L for the cations.

Weston et al. used 5 mM 4-methylbenzylamine/6.5 mM HIBA, pH 4.4, for the simultaneous separation of alkali, alkaline earth, and transition metals and lanthanides (36,37). Detection limits were ~ 100 µg/L for the 28 metals tested.

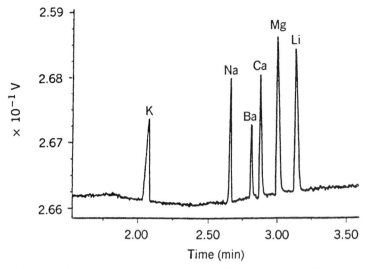

Figure 24.6. Separation of inorganic cations with imidazole electrolyte. Conditions: 5 mM imidazole, pH 4.5, with sulfuric acid; 75 μm i.d. × 63 cm capillary; 30 s injection; +25 kV; detection—UV; 214 nm. Sample: 1 μg/mL each cation. [From W. Beck and H. Engelhardt, *Chromatographia* **33**, 313 (1992), with permission.]

Shi and Fritz used 4-methylbenzylamine as the electrolyte co-ion and investigated the use of tartrate, lactate, and HIBA as complexing agents for several cations (38). An electrolyte containing 15 mM lactic acid/8 mM 4-methylbenzylamine/5% methanol, pH 4.25, permitted the separation of 27 alkali, alkaline earth, transition and rare earth metals (Figure 24.7). Chen and Cassidy used 6 mM N,N-dimethylbenzylamine/4.2 mM HIBA/0.2 mM Triton X-100, pH 5.0, for the separation of 26 metal ions (39). Lin et al. used 5 mM imidazole in combination with HIBA, lactate, acetate, oxalate, malonate, succinate, or citrate for the separation of potassium, barium, calcium, sodium, magnesium, and lithium (40). As expected, the monovalent acids caused the weakest complexation with the divalent cations whereas triprotic citrate caused the strongest.

When acidic complexing agents are used, the degree of complexation is dependent on both the concentration of ligand and the ligand's charge state. The charge state is dependent on pH, which is dependent on ligand concentration. Therefore, optimization of complexing agent concentration is difficult when one is using a simple univariate approach. To overcome this problem,

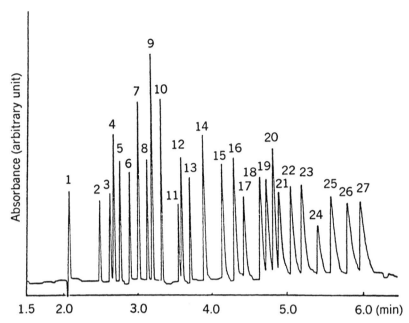

Figure 24.7. Separation of 27 inorganic cations with 4-methylbenzylamine/lactate electrolyte. Conditions: 8 mM 4-methylbenzylamine/15 mM lactic acid/5% methanol, pH 4.25; 75 μm i.d. × 60 cm capillary; 30 s injection; +30 kV; detection—UV, 214 nm. Peaks: (1) potassium; (2) barium; (3) strontium; (4) sodium; (5) calcium; (6) magnesium; (7) manganese(II); (8) cadmium(II); (9) lithium; (10) cobalt(II); (11) lead(II); (12) nickel(II); (13) zinc(II); (14) lanthanum(III); (15) cerium(III); (16) praseodymium(III); (17) neodymium(III); (18) samarium(III); (19) gadolinium(III); (20) copper(II); (21) terbium(III); (22) dysprosium(III); (23) holmium(III); (24) erbium(III); (25) thulium(III); (26) ytterbium(III); (27) lutetium(III). Sample: 1–5 μg/mL each cation. [From Y. Shi and J. S. Fritz, *J. Chromatogr.* **640**, 473 (1993), with permission.]

Quang and Khaledi developed a migration model and used computer simulation to predict the migration time of 14 metal cations as a function of pH and HIBA concentration (41). Imidazole was used as electrolyte co-ion. Optimal conditions were 6 mM imidazole/12 mM HIBA, pH 3.95. These conditions permitted separation of the 14 cations in less than 4 min.

Lee and Lin investigated imidazole, benzylamine, ephedrine, and pyridine as co-ions and glycolic acid, HIBA, and succinic acid as complexants for the separation of 19 metal cations (42). Using electrokinetic injection, these authors noted that detection limits were in the low-μg/L range. Imidazole or pyridine with glycolic acid as complexant was found to give the optimum separations. Massart's group used 5 mM imidazole/6.5 mM HIBA/0.53 mM 18-crown-6/20% methanol, pH 4.5, for the separation of 12 metal ions [see

Yang et al. (43)]. Wang and Li employed 10 mM pyridine as co-ion and concentrations of EDTA ranging from 0.6 to 1.0 mM for separation of seven alkali and alkaline earth metals (44). Kaniansky's group used benzimidazole as the co-ion and tartrate/18-crown-6 as complexants for the separation of alkali and alkaline earth metals and ammonium [see Simunikova et al. (45)].

Carpio et al. employed guanidine–HCl as co-ion and tropolone as the complexant for the determination of alkali and alkaline earth metals in semiconductor-grade hydrogen peroxide (46). In comparison to an electrolyte containing 4-methylbenzylamine and HIBA, the guanidine/tropolone electrolyte was found to be more tolerant to samples that contained more than 3% hydrogen peroxide. This advantage was attributed to the fact that tropolone is a neutral complexant unaffected by pH, in contrast to HIBA, whose complexation is strongly dependent on pH. Electrokinetic injection was performed, resulting in detection limits of about 1 ppb for each cation in the peroxide solution.

Riviello and Harrold used cupric ion as the co-ion for metal analysis (47). The electrolyte was 4 mM copper(II) sulfate/4 mM 18-crown-6/4 mM formic acid, pH 3.0. Subsequently, Wojtusik and Harrold used 5 mM dimethyldiphenylphosphonium hydroxide/4 mM 18-crown-6/5 mM 21(N-morpholino)-ethanesulfonic acid (MES), pH 6.0, and found this electrolyte to provide two to five times lower detection limits than those of the cupric ion electrolyte (48).

Imidazole appears to offer the best mobility match for alkali and alkaline earth metals, based on its recent popularity in studies on the effect of various complexing agents and recent applications. In many applications, the sample contains an excess of sodium compared to the other metals. Therefore, it is preferable for the co-ion mobility to closely match sodium so that a symmetrical peak is obtained. Massart's group compared imidazole/sulfuric acid to 4-methylbenzylamine/HIBA and found that sodium gave a more symmetrical peak with the imidazole electrolyte [see Yang et al. (49)]. With 4-methylbenzylamine, sodium exhibited fronting. A method based on imidazole/sulfuric acid was validated and applied to the analysis of beverages and multiple electrolyte solutions for parenteral use. CE was compared to flame atomic spectroscopy, and the results agreed well for several samples.

Altria et al. used imidazole/sulfuric acid for determination of the cationic counterions of acidic drugs (50). Chromate/TTAB was used for determinations of anionic counterions of basic drugs. Huie's group used imidazole for the determination of fluoro- and oxalatoaluminum complexes [see Wu et al. (51)]. Ma and colleagues used 100 mM imidazole adjusted to pH 7.4 with sulfuric acid for the determination of ionized and total calcium in human serum [see Zhang et al. (52)]. Untreated serum was analyzed first for ionized calcium. After treatment with trichloroacetic acid, the serum was analyzed for total calcium. Transfiguracion et al. measured potassium, sodium, magnesium, and calcium ions in rat airway surface fluid with 10 mM imidazole, pH 3.5, and 8% 2-propanol (53).

Gupta and Tan used 4-methylbenzylamine/HIBA and electrokinetic injection for low-ppb determinations of cations in high-purity water used in the microelectronics industry (54). Results obtained by CE, inductively coupled plasma mass spectrometry (ICP–MS) and graphite furnace atomic absorption (AA) agreed well.

24.3. DIRECT-ULTRAVIOLET DETECTION: ANION DETERMINATIONS

Most of the principles that apply to indirect-UV detection also apply to direct UV detection. The exception is that the electrolyte co-ion should be UV transparent. In the low UV, many of the common inorganic anions will absorb. For instance, consider the two electropherograms in Figure 24.8 a,b. The electrolyte for these separations was 10 mM sodium sulfate/3.5 mM TTAOH/10 mM CHES, pH 9. In Figure 24.8a, direct detection was performed at 185 nm. All four anions were detected. In Figure 24.8b, detection was at 214 nm for the same sample. Chloride was no longer detected and the response for bromide was reduced.

The high specificity of direct UV detection for anions has been exploited for the determination of nitrate and nitrite in human blood plasma (55). Detection was performed at 214 nm. The electrolyte contained 750 mM sodium chloride/1 mM TTAB. The very high level of sodium chloride was required to obtain stacking with hydrostatic injection, since plasma contains about 3500 ppm sodium. Detection limits of 0.1 ppm were obtained for each anion. The method was successfully applied to plasma samples whose nitrite and nitrate levels averaged 0.15 and 3.2 mg/L, respectively. Acceptable precision was obtained with the method. Similarly, Guan et al. developed a method for determination of nitrite and nitrate using an electrolyte containing 20 mM tetraborate/1.1 mM cetyltrimethylammonium chloride, pH 8.94 (56). With electrokinetic injection and detection at 200 nm, detection limits were 1 µg/L for both anions.

Soga et al. used DB-WAX-coated capillaries (J & W Scientific) with suppressed EOF for analysis of 11 inorganic anions with detection at 200 nm (57). The electrolyte contained 20 mM phosphate, pH 8.0. Detection limits ranged from 14 to 260 µg/L with hydrostatic injection for bromide, iodide, chromate, nitrate, thiocyanate, molybdate, tungstate, bromate, chlorite, arsenate, and iodate. Martinez and Aguilar developed a method for the determination of chromate ion in chromium-plating baths (58). The electrolyte

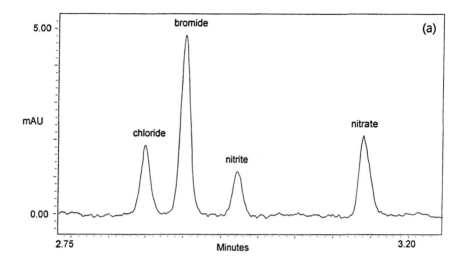

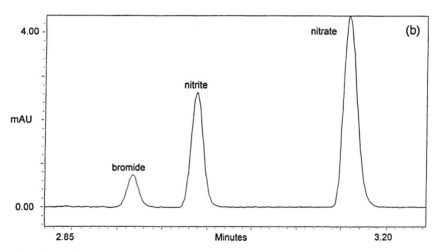

Figure 24.8. Direct-UV detection of inorganic anions at 185 nm (a) and 214 nm (b). Conditions: 10 mM sodium sulfate/3.5 mM TTAOH/10 mM CHES, pH 9.0; 75 μm i.d. × 60 cm capillary; 30 s injection; −20 kV Analyte concentrations same as in Figure 24.1.

contained 10 mM carbonate/5 mM TTAB, pH 10.0, with detection at 273 nm. The detection limit was 1.2 mg/L. Subsequently, the detection limit was reduced to 11 μg/L by increasing the concentration of carbonate buffer to 100 mM to increase the amount of stacking (59).

24.4. DIRECT-ULTRAVIOLET DETECTION: CATION DETERMINATIONS

A paper by Timberbaev reviewed the determination of metal ions by CE as complexes formed prior to separation (60). These performed complexes can be directly detected with UV detection. The resulting complexes are usually anionic, and their mobilities depend on the ligand employed. Several migration modes are possible, including the addition of surfactants for micellar electrokinetic chromatography (MEKC) separations. An example of a complexing agent that forms negatively charged complexes with metals is 8-hydroxyquinoline-5-sulfonic acid (HQS). This reagent was first used by Swaile and Sepaniak for metal determinations, and the resulting complexes were detected by laser-induced fluorescence (61). Timberbaev et al. determined the same complexes by direct-UV detection at 254 nm (62). Nickel, cobalt, zinc, cadmium, iron, and copper were separated as the HQS complexes in an electrolyte containing 10 mM borate/0.1 mM HQS, pH 9.2. To reduce EOF and permit detection at the anode, the capillary was periodically rinsed with a solution of TTAB. Detection limits were in the low-µg/L range. The same group compared three modes of migration for the HQS complexes: reduced EOF with detection at the anode; fast EOF with detection at the cathode; and MEKC with sodium dodecyl sulfate (SDS) and detection at the cathode (63). As would be expected, fast EOF, in opposition to the complexes' electrophoretic mobilities, provided the best resolution. As the authors pointed out, it is questionable as to whether or not the anionic complexes would partition into negatively charged SDS micelles.

Another example is metal–cyanide complexes (64–66). Aguilar et al. determined gold(I) and silver(I) cyanide in ores with direct detection at 214 nm (65). CE and AA showed excellent agreement in the analysis of two ore samples. Buchberger and Haddad optimized conditions for the separation of the cyanide complexes of gold, platinum, iron(II), iron(III), palladium, copper(I), cobalt, silver, chromium and nickel (66). Optimum conditions were 5 mM potassium hydrogen phosphate/5 mM triethanolamine/0.8 mM hexamethonium bromide, pH 8.5 (Figure 24.9).

Other complexing agents that have been studied include EDTA (67), several aminopolycarboxylic reagents (68), 2,6-diacetylpyridine bis(N-methylene)-pyridiniohydrazone (69), β-diketone (70), and 4-(2-pyridylazo)resorcinol (71).

There are several disadvantages to this mode of operation. First, the ligand must be added to the sample to form the complexes, and also included in the electrolyte for those metals that form unstable complexes. Separation of several metals at once is difficult, as alkali metal ions do not form stable complexes with the ligands reported to date. Peak shapes are often poor, indicative of poor solubility of the complexes or slow formation kinetics.

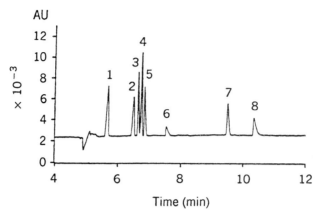

Figure 24.9. Separation of metal–cyanide complexes. Conditions: 5 mM potassium hydrogen phosphate/5 mM triethanolamine/0.8 mM hexamethonium bromide, pH 8.5, with phosphoric acid; 75 μm i.d. × 60 cm capillary; 20 s injection; −20 kV; detection—UV, 214 nm. Peaks: (1) Fe(II); (2) Pd; (3) Co; (4) Pt; (5) Fe(III); (6) Cr; (7) Au; (8) Ag. [From W. Buchberger and P. R. Haddad, *J. Chromatogr.* **687**, 343 (1994), with permission.]

However, the main advantage is the applicability toward metals difficult to determine with indirect-UV methods, such as gold and silver, which possess a slow rate of complexation with typical complexing agents.

24.5. OTHER DETECTION MODES

Several other detection modes have also been reported for the determination of inorganic ions by CE. Given the success of indirect-UV methods, it is not surprising that indirect-fluorescence detection has also been reported (72,73). Bachmann et al. demonstrated the use of Ce(III) as an electrolyte co-ion for alkali and alkaline earth metal determinations by CE with indirect-fluorescence detection (73). The electrolyte was 0.5 mM cerium(III) sulfate. At such a low co-ion concentration, only those metals with mobilities similar to cerium(III) gave symmetrical peaks. Detection limits were in the 0.1–0.3 μM range with electrokinetic injection. Bruno et al. reported similar detection limits for metals using a hologram-based refractive index detector (74).

Three recent reports described the coupling of CE to ICP–MS for metal determinations (75–77). The high elemental specificity of ICP–MS places minimal reliance on the CE separation of different metals. Instead, the purpose of the CE separation is to separate the different charge states of a given metal atom, i.e., metal speciation. Detection limits reported by two groups were in the high-ng/L to low-μg/L range (75,76). Another group used the combination to study the metal content of proteins (77).

Many metals and inorganic anions can be detected by amperometry. Cassidy's group used mercury film ultramicroelectrodes for the determination of several metals by reductive amperometry [see Lu and Cassidy (78)]; Lu et al. detected nitrite oxidatively at a carbon fiber electrode (79). Of broader applicability is potentiometric detection, which has been investigated by Nann and colleagues (80–82). The problem is that the response is nonlinear. To linearize the response, the emf value of the response must be delogarithmized and the resulting curve integrated.

Given its popularity in IC, it is not surprising that conductivity detection has been described for CE. With conductivity detection, sensitivity is maximized when the difference in conductivity between the analyte and electrolyte co-ion is maximized. However, as we have already discussed, matching the co-ion mobility (which is directly related to conductivity) to the analyte mobility is necessary for good peak symmetry. Therefore, a contradiction exists. High sensitivity is obtained at the expense of peak shape and resolution with conductivity detection. Jones et al. have discussed this trade-off and described approaches to minimize it (83). High concentrations of low-conductivity electrolyte co-ions are employed. Consider the separation in Figure 24.10 of 37 anions with direct conductivity detection. The electrolyte co-ion was CHES, which (as discussed above in Section 24.2.1) has very low conductivity near its pK_a; 50 mM of this co-ion was used to minimize peak asymmetry. Note that the concentration of ion injected in Figure 24.10 increases from low ppm to higher ppm, while the peak heights decrease. This progression is because early peaks have the largest difference in conductivity with the co-ion—and therefore the largest sensitivity.

The advantage of conductivity detection is high sensitivity. Jones et al. reported low-ng/L detection limits for anions with ITP stacking (83). A plug of hydroxide ion is injected before the sample to act as the ITP leading ion. Since peak asymmetry is also related to analyte concentration [Eq. (24.4)], this high sensitivity also helps to minimize peak asymmetry. In many cases, samples of mid-mg/L concentration can be diluted for improved peak symmetry and resolution. Cations were detected with histidine as the electrolyte co-ion (83).

Suppressed conductivity detection for CE has been described by two groups (84,85). In this case, the electrolyte co-ion is a weak acid for anions or a weak base for cations. The conductivity of the electrolyte is suppressed after the separation by addition of the acid (base), forming a neutral species with no conductivity. As a result, a low detector background is obtained. Without ITP stacking, such a system gave detection limits of 1–10 μg/L (85). The main disadvantage of suppressed conductivity is that the EOF cannot be reversed for anion determinations, and it must be reversed for cation determinations, or

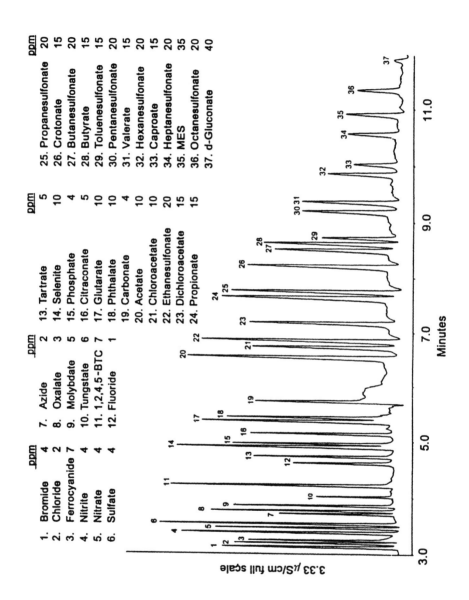

else the suppressor column would not function properly (85). Therefore, the analyte ions must migrate against the flow. As discussed previously, this is advantageous for resolution, but leads to long run times. To detect high-mobility ions, electrolytes with very low ionic strength must be used to generate a fast enough EOF so that the ions can be detected. Figure 24.11 demonstrates the separation of several inorganic anions at the 10 µM level with suppressed conductivity detection. The electrolyte was 2 mM sodium tetraborate. Note that the higher-mobility anions migrate last and exhibit tailing. These high-mobility anions tail because their apparent mobility is lower than that of borate under these conditions.

24.6. CONCLUDING REMARKS

The determination of inorganic ions by CE clearly offers many advantages vs. competing techniques. These advantages include fast analysis, a simple experimental setup, minimal solvent consumption, high resolution, and adequate sensitivity for most samples. Methods based on indirect-UV detection currently enjoy the most popularity. Advantages of this approach include universality, simplicity, and the availability of many different chromophoric ions covering a wide range of mobilities. In general, sensitivity is adequate for most samples, and it can be increased with ITP stacking. The main disadvantage is that because only a small amount of a chromophore can be included in the electrolyte, only a narrow range of analyte mobilities can be successfully separated with high efficiency in one run. Advances are expected to be made in the development of new visualization reagents, coated capillaries for reversed EOF with anions, and different complexing agents for cations. A recent report described the ability to perform CE of inorganic anions in nonaqueous media (86). Such media offer different selectivity from aqueous electrolytes, and more work can be expected in this area. Direct-UV detection will continue to be popular for the determination of specific ions in difficult matrices. Conductivity detection seems to offer the advantage of improved sensitivity vs. indirect-UV detection, but advances must be made to overcome the trade-off that exists between detectability and separation. Furthermore, the added complexity of such systems raises issues about ease of use, ruggedness, and long-term stability. However, because of the sub-µg/L detection limits possible, work will

---

Figure 24.10. Conductivity detection of 37 anions. Conditions: 50 mM CHES/20 mM lithium hydroxide/0.03% Triton X-100; 1 min rinse between injections with 1 mM CTAB solution; 50 µm i.d. × 50 cm ConCap capillary; 25 mbar injection for 12 s. (BTC) = benzenetetracarboxylic acid. [From W. R. Jones, J. Soglia, M. McGlynn, C. Haber, J. Reineck, and C. Krstanovic, *Am. Labo.* 25, (1996), with permission.]

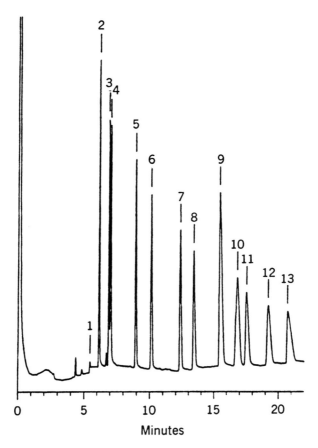

Figure 24.11. Suppressed conductivity detection of inorganic anions. Conditions: 2 mM borax buffer; 75 μm i.d. × 60 cm capillary; 1 psi injection for 2 s; Peaks: (1) bicarbonate; (2) chlorite (0.67 μg/mL); (3) fluoride (0.19 μg/mL); (4) phosphate (0.95 μg/mL); (5) chlorate (0.83 μg/mL); (6) perchlorate (0.99 μg/mL); (7) nitrate (0.62 μg/mL); (8) nitrite (0.46 μg/mL); (9) sulfate (0.96 μg/mL); (10) chloride (0.35 μg/mL); (11) iodide (1.27 μg/mL); (12) bromide (0.80 μg/mL); (13) chromate (1.16 μg/mL). [From N. Avdalovic, C. A. Pohl, R. D. Rocklin, and J. R. Stillian, *Anal. Chem.* **65**, 1470 (1993), with permission.]

continue in this area with the goal of maximizing separation and detection performance.

The true measure of success of any new analytical technique is how often it is applied to "real" samples. Although several examples exist in the literature attesting to the utility of CE for the determination of inorganic ions, it remains to be seen how CE will compete with other techniques. Clearly, the potential is there for widespread use and the eventual placement of CE at the forefront of inorganic ion determinations.

LIST OF ACRONYMS AND SYMBOLS

Acronym or Symbol	Definitions
AA	atomic absorption
AU	absorbance units
BTC	benzenetetracarboxylic acid
CE	capillary electrophoresis
CHES	2-(N-cyclohexylamino)ethanesulfonic acid
CTAB	cetyltrimethylammonium bromide
CZE	capillary zone electrophoresis
EDTA	edetic acid (or ethylenediaminetetraacetic acid)
emf	electromotive force
EOF	electroosmotic flow
HIBA	hydroxyisobutyric acid
HQS	8-hydroxyquinoline-5-sulfonic acid
IC	ion chromatography
ICP–MS	inductively coupled plasma mass spectrometry
ITP	isotachophoresis
MEKC	micellar electrokinetic chromatography
MES	1-(N-morpholino)ethanesulfonic acid
PMA	pyromellitic acid (or 1,2,4,5-benzenetetracarboxylic acid)
RSD	relative standard deviation
SDS	sodium dodecyl sulfate
S/N	signal-to-noise ratio
TR	transfer ratio
TTAB	tetradecyltrimethylammonium bromide
TTAOH	tetradecyltrimethylammonium hydroxide
UV–Vis	ultraviolet–visible (region)

REFERENCES

1. P. Jandik, W. R. Jones, A. Weston, and P. R. Brown, *LC/GC* **9**, 634 (1991).
2. P. K. Dasgupta and L. Bao, *Anal. Chem.* **65**, 1003 (1993).
3. N. Avdalovic, C. A. Pohl, R. D. Rocklin, and J. R. Stillian, *Anal. Chem.* **65**, 1470 (1993).
4. W. R. Jones, J. Soglia, M. McGlynn, C. Haber, J. Reineck, and C. Krstanovic, *Am. Labo.* 25 (1996).
5. W. R. Jones and P. Jandik, *J. Chromatogr.* **546**, 445 (1991).
6. T. Kaneta, S. Tanaka, and M. Taga, *J. Chromatogr.* **653**, 313 (1993).
7. M. P. Harrold, M. J. Wojtusik, J. Rivello, and P. Henson, *J. Chromatogr.* **640**, 463 (1993).

8. J. W. Jorgenson and K. D. Lukacs, *Anal. Chem.* **53**, 1298 (1981).
9. S. Hjerten, *Electrophoresis* **11**, 665 (1990).
10. P. Jandik and G. Bonn, *Capillary Electrophoresis of Small Molecules and Ions*. VCH Publishers, New York, 1993.
11. M. W. F. Nielen, *J. Chromatogr.* **588**, 321 (1991).
12. G. Bondoux, P. Jandik, and W. R. Jones, *J. Chromatogr.* **602**, 79 (1992).
13. J. R. Mazzeo, J. Krol, E. Grover, and M. Benvenuti, 1995–1996.
14. S. M. Cousins, P. R. Haddad, and W. Buchberger, *J. Chromatogr.* **671**, 397 (1994).
15. S. A. Shamsi and N. D. Danielson, *Anal. Chem.* **66**, 3757 (1994).
16. S. A. Shamsi and N. D. Danielson, Anal. Chem. **67**, 1845 (1995).
17. F. S. Stover and S. S. Keffer, *J. Chromatogr.* **657**, 450 (1993).
18. C. Stathakis and R. M. Cassidy, *Anal. Chem.* **66**, 2110 (1994).
19. C. Stathakis and R. M. Cassidy, *J. Chromatogr.* **699**, 353 (1995).
20. N. J. Benz and J. S. Fritz, *J. Chromatogr.* **671**, 437 (1994).
21. A. H. Harakuwe, P. R. Haddad, and W. Buchberger, *J. Chromatogr.* **685**, 161 (1994).
22. P.J. Oefner, *Electrophoresis* **16**, 46 (1995).
23. G. W. Tindall and R. L. Perry, *J. Chromatogr.* **696**, 349 (1995).
24. J. Romano, P. Jandik, W. R. Jones, and P. E. Jackson, *J. Chromatogr.* **546**, 411 (1991).
25. J. P. Romano and J. Krol, *J. Chromatogr.* **640**, 403 (1993).
26. M. E. Swartz, *J. Chromatogr.* **640**, 441 (1993).
27. B. J. Wildman, P. Jackson, W. R. Jones, and P. G. Alden, *J. Chromatogr.* **546**, 459 (1991).
28. D. R. Salomon and J. Romano, *J. Chromatogr.* **602**, 219 (1992).
29. E. Dabek-Zlotorynska and J. F. Dlouhy, *J. Chromatogr.* **671**, 389 (1993).
30. R. R. Chadwick, J. C. Hsieh, K. S. Resham, and R. B. Nelson, *J. Chromatogr.* **671**, 403 (1993).
31. J. M. Jordan, R. L. Moese, R. Johnson-Watts, and D. E. Burton, *J. Chromatogr.* **671**, 445 (1993).
32. G. Bondoux and T. Jones, *LC–GC Int.* **7**, 29 (1994).
33. C. A. Lucy and T. L. McDonald, *Anal. Chem.* **67**, 1074 (1995).
34. K. Bachmann, J. Boden, and I. Haumann, *J. Chromatogr.* **629**, 259 (1992).
35. W. Beck and H. Engelhardt, *Chromatographia* **33**, 313 (1992).
36. A. Weston, P. R. Brown, P. Jandik, W. R. Jones, and A. L. Heckenberg, *J. Chromatogr.* **593**, 289 (1992).
37. A. Weston, P. R. Brown, A. L. Heckenberg, P. Jandik, and W. R. Jones, *J. Chromatogr.* **602**, 249 (1992).
38. Y. Shi, and J. S. Fritz, *J. Chromatogr.* **640**, 473 (1993).
39. M. Chen and R. M. Cassidy, *J. Chromatogr.* **640**, 425 (1993).

40. T.-I. Lin, Y.-H. Lee, Y.-C. Chen, *J. Chromatogr.* **654**, 167 (1993).
41. C. Quang and M. G. Khaledi, *J. Chromatogr.* **659**, 459 (1994).
42. Y.-H. Lee and T.-I. Lin, *J. Chromatogr.* **675**, 227 (1994).
43. Q. Yang, J. Smeyers-Verbeke, W. Wu, M. S. Khots, and D. L. Massart, *J. Chromatogr.* **688**, 339 (1994).
44. T. Wang and S. F. Y. Li, *J. Chromatogr.* **707**, 343 (1995).
45. E. Simunikova, D. Kaniansky, and K. Loksikova, *J. Chromatogr.* **665**, 203 (1994).
46. R. A. Carpio, P. Jandik, and E. Fallon, *J. Chromatogr.* **657**, 185 (1993).
47. J. M. Riviello and M. P. Harrold, *J. Chromatogr.* **652**, 385 (1993).
48. M. J. Wojtusik and M. P. Harrold, *J. Chromatogr.* **671**, 411 (1994).
49. Q. Yang, M. Jimidar, T. P. Hamoir, J. Smeyers-Verbeke, and D. L. Massart, *J. Chromatogr.* **673**, 275 (1994).
50. K. D. Altria, D. M. Goodall, and M. M. Rogan, *Chromatographia* **38**, 637 (1994).
51. N. Wu, W. J. Horvath, P. Sun, and C. W. Huie, *J. Chromatogr.* **635**, 307 (1993).
52. R. Zhang, H. Shi, and Y. Ma, *J. Microcolumn Sep.* **6**, 217 (1994).
53. J. C. Transfiguracion, C. Dolman, D. H. Eidelman, and D. K. Llyod, *Anal. Chem.* **67**, 2937 (1995).
54. P. Gupta and S. H. Tan, *Ultrapure Water*, May/June, 54 (1994).
55. T. Ueda, T. Maekawa, D. Sadamitsu, S. Oshita, K. Ogino, and K. Nakamura, *Electrophoresis* **16**, 1002 (1995).
56. F. Guan, H. Wu, and Y. Luo, *J. Chromatogr.* **719**, 427 (1996).
57. T. Soga, Y. Inoue, and G. A. Ross, *J. Chromatogr.* **718**, 421 (1995).
58. M. Martinez and M. Aguilar, *J. Chromatogr.* **676**, 443 (1994).
59. B. Baraj, M. Martinez, A. Sastre, and M. Aguilar, *J. High Resolut. Chromatogr.* **18**, 675 (1995).
60. A. R. Timerbaev, *J. Capillary Electrophor.* **2**, 14 (1995).
61. D. F. Swaile and M. J. Sepaniak, *Anal. Chem.* **63**, 179 (1991).
62. A. R. Timerbaev, W. Buchberger, O. P. Semenova, and G. K. Bonn, *J. Chromatogr.* **630**, 379 (1993).
63. A. Timerbaev, O. Semenova, and G. Bonn, *Chromatographia* **37**, 497 (1993).
64. W. Buchberger, O. P. Semenova, and A. R. Timerbaev, *J. High Resolut. Chromatogr.* **16**, 153 (1993).
65. M. Aguilar, A. Farran, and M. Martinez, *J. Chromatogr.* **635**, 127 (1993).
66. W. Buchberger and P. R. Haddad, *J. Chromatogr.* **687**, 343 (1994).
67. B. Baraj, M. Martinez, A. Sastre, and M. Aguilar, *J. Chromatogr.* **695**, 103 (1995).
68. A. R. Timerbaev and O. P. Semenova, *J. Chromatogr.* **690**, 141 (1995).
69. A. R. Timerbaev, O. P Semenova, G. K. Bonn, and J. S. Fritz, *Anal. Chim. Acta.* **296**, 119 (1994).
70. K. Saitoh, C. Kiyohara, and N. Suzuki, *J. High Resolut. Chromatogr.* **14**, 245 (1991).
71. T. Saitoh, H. Hoshino, and T. Yosuyanagi, *J. Chromatogr.* **469**, 175 (1989).

72. L. Gross and E. S. Yeung, *Anal. Chem.* **62**, 427 (1990).
73. K. Bachmann, J. Boden, and I. Haumann, *J. Chromatogr.* **626**, 259 (1992).
74. B. Krattiger, G. J. M. Bruin, and A. E. Bruno, *Anal. Chem.* **66**, 1 (1994).
75. J. W. Olesik, J. A. Kinzer, and S. V. Olesik, *Anal. Chem.* **67**, 1 (1995).
76. Y. Liu, V. Lopez-Avila, J. J. Zhu, D. R. Wiederin, and W. F. Beckert, *Anal. Chem.* **67**, 2020 (1995).
77. Q. Lu, S. M. Bird, and R. M. Barnes, *Anal. Chem.* **67**, 2949 (1995).
78. W. Lu and R. M. Cassidy, *Anal. Chem.* **65**, 1649 (1993).
79. W. Lu, R. M. Cassidy, and A. S. Baranski, *J. Chromatogr.* **640**, 433 (1993).
80. A. Nann and W. Simon, *J. Chromatogr.* **633**, 207 (1993).
81. A. Nann, I. Silvestri, and W. Simon, *Anal. Chem.* **65**, 1662 (1993).
82. A. Nann and E. Pretsch, *J. Chromatogr.* **676**, 437–442 (1994).
83. W. R. Jones, J. Soglia, M. McGlynn, C. Haber, J. Reineck, and C. Krstanovic, *Am. Labo.*, 25 (1996).
84. P. K. Dasgupta and L. Bao, *Anal. Chem.* **65**, 1003 (1993).
85. N. Avdalovic, C. A. Pohl, R. D. Rocklin, and J. R. Stillian, *Anal. Chem.* **65**, 1470 (1993).
86. H. Salimi-Moosavi and R. M. Cassidy, *Anal. Chem.* **67**, 1067 (1995).

CHAPTER

25

THE ANALYSIS OF PHARMACEUTICALS BY CAPILLARY ELECTROPHORESIS

K. D. ALTRIA

Analytical Sciences, GlaxoWellcome R & D, Ware, Herts SG12 ODP, United Kingdom

25.1.	Introduction	853
25.2.	Application Areas of Capillary Electrophoresis in Pharmaceutical Analysis	854
	25.2.1. Determination of Drug-Related Impurities	855
	25.2.2. Assay of the Main Component	858
	25.2.3. Chiral Analysis	862
	25.2.4. Determination of Inorganic Ions	865
	25.2.5. Dissolution Analysis Testing	868
	25.2.6. Analysis of Excipients	868
25.3.	Future Directions	873
25.4.	Conclusions	875
List of Acronyms		875
References		875

25.1. INTRODUCTION

The analysis of pharmaceuticals is currently predominantly performed by high-performance liquid chromatography (HPLC). This situation exists following the widespread advent of commercial HPLC systems that offer the possibility of routine, reliable operation to meet the requirements of a modern pharmaceutical analysis laboratory where the sample throughput is forever escalating. The alternative separative techniques available prior to HPLC were gas chromatography (GC) for volatile compounds or thin-layer chromatography (TLC) where high-resolution or low-level quantitation were not overriding requirements. However, there is now an increasing presence of

---

High Performance Capillary Electrophoresis, edited by Morteza G. Khaledi. Chemical Analysis Series, Vol. 146.
ISBN 0-471-14851-2 © 1998 John Wiley & Sons, Inc.

capillary electrophoresis (CE) instrumentation in many pharmaceutical analysis laboratories, as this offers a real and attractive alternative to HPLC. CE can, in many instances, have distinct advantages over HPLC in terms of rapid method development, reduced operating costs, and increased simplicity. In addition, a single set of CE operating conditions may be appropriate for a wide range of pharmaceuticals, leading to very significant efficiency gains. However, the major strength of CE is that the basis of the separation principles is different from that of the chromatographic separations obtained in HPLC, TLC, and GC. Therefore, the use of CE and HPLC together is a powerful combination.

The scope of CE application in the area of pharmaceutical analysis is identical to that of HPLC, and therefore often a choice between the two techniques is required. The application range includes determination of related impurities, enantiopurity determinations, main-component assay, determination of inorganic ions, trace level quantitations, analysis of dissolution test sample solutions, and raw-material testing. The performance of an optimized CE method is similar to that expected for HPLC methods. Validation of CE methods is possible, and CE methods have been accepted by regulatory authorities. A draft USP (United States Pharmacopeia) chapter covering CE has recently been published (1).

To date there are more than 300 papers concerned with the CE analysis of pharmaceuticals. Smith and Evans (2) and Altria and Rogan (3) have written recent review articles. Many early papers were concerned with demonstrating that the analysis by CE was possible. Lately the emphasis has been toward highlighting the effective use of CE for routine applications. The present chapter summarizes the utility of CE in the various pharmaceutical analysis application areas and indicates the likely performance levels. In addition, a brief section (25.3) on possible future directions is included that concentrates largely on the significant likely impact of capillary electrochromatography (CEC), which is a hybrid technique between CE and HPLC and will undoubtedly emerge as a useful alternative to either CE or HPLC.

25.2. APPLICATION AREAS OF CAPILLARY ELECTROPHORESIS IN PHARMACEUTICAL ANALYSIS

A survey concerning the usage of CE in major British and U. S. pharmaceutical companies was conducted in 1993. This covered a variety of aspects and graded the various application areas of CE within the pharmaceutical companies surveyed (4). Figure 25.1 shows results compiled from the responses. Examples from each of these applications will be covered, together with application areas that have developed subsequent to this survey.

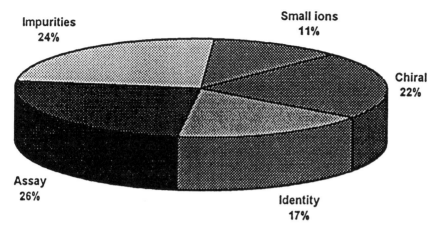

Figure 25.1. Current application areas of CE within the pharmaceutical industry. [Reproduced from K. D. Altria and M. Kersey, *LG–CG Int.* **8**, 201 (1995), with permission.]

25.2.1. Determination of Drug-Related Impurities

As noted earlier, the combined use of CE and HPLC is a powerful combination, and this is most often demonstrated in the area of impurities determination. For example, if independent HPLC and CE methods confirm the number and levels of impurities in a test sample, then the results can be viewed with considerable confidence. Data such as these can form a pivotal portion of the validation data of either the CE or the HPLC method. Several papers have cross-correlated (5–9) the impurity levels present in drug substance and pharmaceutical formulation samples. Ten batches of salbutamol sulfate were analyzed (Table 25.1) by CE and HPLC for levels of specific dimeric impurities at levels ranging from 0.05 to 0.8% w/w. Statistical analysis of the results (5) showed them to be equivalent.

The overriding requirement in impurities determination methods is the ability to quantify impurities present at the 0.1% level and lower. This is possible using commercial CE instrumentation with standard capillaries. Sensitivity is enhanced by use of low-UV wavelength detection (5) and/or modified large-bore capillaries (10). Higher sample concentrations can be employed; however, this can lead to problems relating to capillary overload and detector signal saturation. For example impurities at the 0.01% level were determined in codeine samples using 10 mg/mL sample concentrations (11).

The analysis times for selective impurity determinations are similar to those for HPLC and are on the order of 10–40 min. Figure 25.2 shows the separation

Table 25.1. Levels of Salbutamol Sulfate Impurities in Drug Substance Determined by CE and HPLC

Batch	Bis(ether) (%w%w)		Dimer (%w%w)	
	CE	HPLC	CE	HPLC
1	0.14	0.16	0.08	0.08
	0.14	0.16	0.08	0.08
2	0.10	0.11	0.06	0.07
	0.10	0.11	0.07	0.06
3	0.20	0.19	0.13	0.11
	0.20	0.19	0.14	0.10
4	0.12	0.13	0.07	0.06
	0.15	0.14	0.08	0.05
5	0.13	0.14	0.08	0.06
	0.12	0.13	0.07	0.05
6	0.31	0.28	0.18	0.17
	0.31	0.26	0.19	0.15
7	0.07	0.09	0.05	0.04
	0.08	0.10	0.06	0.03
8	0.38	0.38	0.20	0.18
	0.44	0.38	0.22	0.19
9	0.37	0.38	0.19	0.18
	0.37	0.35	0.19	0.19
10	0.75	0.66	0.39	0.33
	0.77	0.67	0.40	0.35

(12) of a basic drug from a range of its impurities using a triethanolamine/phosphoric acid electrolyte, pH 2.5, with UV detection of 230 nm, the trace level impurities are present at 0.01% of the main component peak. This separation gives markedly different selectivity to a HPLC counterpart method, as the CE method relies on the change and size of the solute whereas the HPLC selectivity is based on partitioning differences. This can allow separation of impurities previously unresolved by chromatographic techniques (2,6).

Accordingly the impurity separation profile and peak order are somewhat similar when obtained by a micellar electrokinetic chromatography (MEKC) method and a reversed-phase HPLC method. MEKC methods have been successfully employed to determine levels of impurities present in cephalosporin drug substance samples (8) and in Injection solutions (9). The results obtained showed no significant difference between MEKC and HPLC data generated on identical samples (9). A MEKC method has also been used to

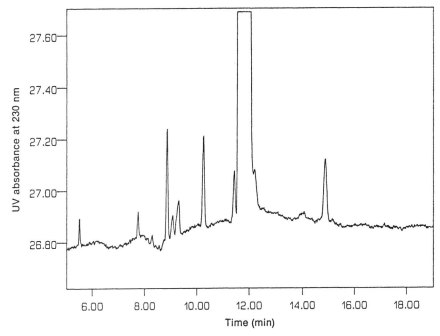

Figure 25.2. Separation of a basic drug from a range of its impurities using a triethanolamine/phosphoric acid electrolyte, pH 2.5. Conditions: 50 mM phosphoric acid adjusted to pH 2.5 with triethanolamine: sample concentration, 0.5 mg/mL in water; UV detection, 230 nm; +10 kV; 30 °C; capillary, 50 μm i.d. × 37 cm. [Reproduced from K. D. Altria, *J. Chromatogr.* **735**, 43 (1996), with permission.]

quantify levels of a synthetic impurity (*p*-toluene sulfonic acid) in Bristol-Myers-Squibb drug substance batches (13).

Optimization of the ionic strength can beneficially alter the shape of the main peak and allow quantitation of closely resolved trace impurities. The use of combinations of ion-pair reagents was shown to optimize the peak shape for remoxipride and related compounds and enabled detection of a range of impurities below the 0.1% level (14).

The impurity profile of a specific batch of a material can be used as an identifier. Examples of this include the analysis of drug seizure samples (15) and antibiotic samples obtained from various pharmaceutical companies (16). The seized drugs country of origin can often be determined by the CE impurity profile of each sample. Confirmation of the identity of specific impurities in separations can be achieved by various means. Test solutions can be manually spiked with an authentic impurity standard. Alternatively spiking of various solutions can be performed automatically by the autosampler (17). Capillary electrophoresis–mass spectrometry (CE–MS) can also be used to confirm the identity and structure of trace impurities, although sensitivity issues can be demanding. The majority of commercial CE systems incorporate diode array detectors (DAD), which can be used to obtain spectral verification of impurity identifications (12).

25.2.2. Assay of the Main Component

Figure 25.1 shows that main-component assay was the most frequent application within pharmaceutical companies. Acceptable precision and accuracy can be obtained using optimized and well-controlled methods. The use of internal standards is recommended, as the volumes injected in CE are strongly related to the viscocity of the sample and calibration solutions. Errors are also minimized by the use of relatively concentrated samples to generate large, easily integrated peaks. Peak areas are generally used in quantitative work, as they offer an extended linear dynamic range when compared to peak heights. Adopting these approaches can enable sub-1% RSD values to be obtained, which is necessary if CE is to compare favorably with HPLC (where sub-1% RSD values are routinely obtained).

A large number of reports (3,9,18–21) have demonstrated accuracy by showing equivalence between CE results and either HPLC or sample label claim data or both. For example, equivalent results were obtained by CE and HPLC for a range of bronchodilator formulations that matched the label claims (18). Good CE precision and linearity data were also achieved in this study for both CE and HPLC.

Often a single set of CE operating conditions can be employed to analyze a range of different drugs. For example, use of a phosphate buffer, pH 2.5, is suitable for the analysis of an extremely wide range of basic drugs (19). These drugs can be accurately quantified using internal standards such as imidazole, aminobenzoate, or aspartame. Validation of a general method for quantitation of range basic drugs included the following: assessments of precision (typically 1% RSD using internal standards); linearity (correlation coefficients of greater than 0.999); method robustness (factors evaluated using experimental designs); sensitivity (sufficient to analyze placebo formulations); accuracy (cross-validation with HPLC can label claim); sample and reagent stability (for 3 months of storage for electrolytes); and separation repeatability on different capillaries on different days with different analysts and reagents (19). Figure 25.3 shows separation of a wide range of compounds by this method. A similar phosphate buffer, pH 2.5, has been used to determine levels of sumatriptan in injection solutions (20). The data obtained cross-correlated well with HPLC data (Table 25.2). Good CE injection precision was obtained using an internal standard.

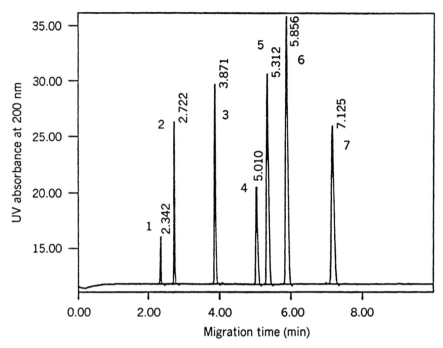

Figure 25.3. Separation of wide range of basic drugs using a low-pH phosphate buffer. Conditions: 50 mM NaH_2PO_4, pH 2.5: UV detection, 200 nm; +25 kV; 30 °C; capillary, 50 μm i.d. × 57 cm. [Reproduced from K. D. Altria, P. Frake, I. Gill, T. Hadgett, M. A. Kelly, and D. R. Rudd, *J. Pharm. Biomed. Anal.* **13**, 951 (1995), with permission.]

Table 25.2. Cross-Correlation of CE and HPLC Results for Sumatriptan Content in Injection Solutions[a]

	Sumatriptan Content (mg/mL)	
Sample	CE	HPLC
Batch 2, Sample 1	11.5, 11.6	11.6, 11.6
Batch 2, Sample 2	11.6, 11.6	11.7, 11.7
Batch 3, Sample 1	11.7, 11.8	11.8, 11.8
Batch 3, Sample 2	11.6, 11.6	11.7, 11.7
Batch 4, Sample 1	11.7, 11.8	11.8, 11.8
Batch 4, Sample 2	11.7, 11.6	11.7, 11.7

[a] The two results given in each column are duplicate results.

Source: Reproduced with permission from Altria and Filbey (20).

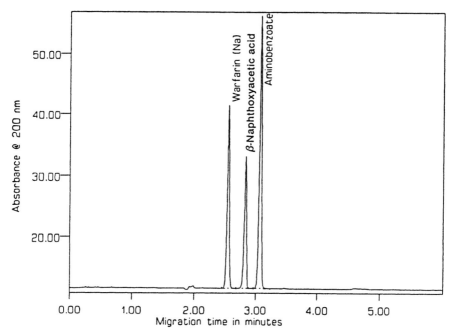

Figure 25.4. Separation of warfarin and two internal standards using a borate buffer. Conditions: 10 mM borate; sample concentration, 0.1 mg mL$^-$ in water; UV detection, 200 nm; +6.5 kV; 30 °C; capillary, 75 μm i.d. × 27 cm (total length). [Reproduced from K. D. Altria, S. M. Bryant, and T. A. Hadgett, *J. Pharm. Biomed. Anal.* **15**, 1091 (1997), with permission.

Similar to the use of the pH 2.5 phosphate buffer for the separation of basic drugs, there is a general method for the separation of a range of acidic drugs using a borate buffer that has a natural pH of 9.5 (21). Appropriate internal standards may be aminobenzoic acid or β-naphthoxyacetic acid. Figure 25.4 shows separation of warfarin and these two internal standards by using a borate buffer, with UV detection at 200 nm (21). This general buffer has been used by other workers to analyze a wide variety of drugs including antibiotics (22), vitamins (23), and diols (24). Alternative high-pH electrolytes include Tris or triethanolamine (25).

MEKC systems have been widely employed, as they are able to simultaneously separate both neutral and charged solutes. This is a useful feature, especially when combination products are being analyzed that contain several components of variable ionic nature. For example, MEKC has been used to simultaneously quantify seven components in analgesic formulations (26). Figure 25.5 shows the separation achieved for the multicomponent mixture.

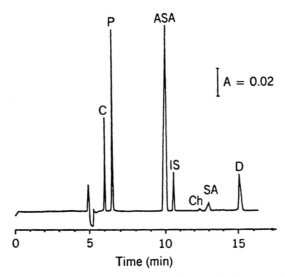

Figure 25.5. MEKC separation of seven components in an analgesic formulation. C = caffeine; P = paracetamol; ASA = acetylsalicylic acid; IS = internal standard (propylhydroxybenzoate); Ch = chlorpheniramine; SA = salicyclic acid; D = dextropropoxyphene. Conditions: 25 mM sodium cholate in 20 mM borate. [Reproduced from S. Boonkerd, M. Lauwers, M. R Detaevernier, and Y. Michotte, *J. Chromatogr.* **695**, 97 (1995), with permission.]

MEKC is also an appropriate choice for the analysis of water-insoluble compounds such as steroids (27) and fat-soluble vitamins (28).

A MEKC method has been validated to full USP validation specifications for the determination of hydrochlorothiazide and chlorothiazide in tablet formulations (29). Sub-1% RSD values were obtained for injection precision without using an internal standard. Good linearities were obtained and selectivity for all likely impurities was demonstrated. The analysis was performed on different occasions by different analysts. Recovery data was demonstrated by analyzing spiked placebo samples.

An intercompany method-transfer exercise has been conducted using a MEKC method for the detemination of paracetamol (acetaminophen) in capsules (30). The method employed an sodium dodecyl sulfate (SDS)–based electrolyte and used acetophenone as an internal standard. All of the seven independent pharmaceutical companies repeated the separation and obtained assay data (305 mg/capsule) in good agreement with the paracetamol label claim for the capsules (300 mg/capsule) and HPLC data (304 mg/capsule).

It is most desirable to include an internal standard in the sample solution to improve quantitation, as this significantly reduces injection-related impreci-

Table 25.3. Analysis of Levothyroxine

Factor	Peak Areas	Peak Area Ratios
Injection precision ($n = 10$)	1.2% RSD	0.6% RSD
Average recovery	91.3%	103.0%
Linearity–correlation	0.99856	0.99991
Linearity–intercept	+4.6%	−0.8%
Factor	Migration Time (MT) Levothyroxine	Relative Migration Time (RMT) Levothyroxine
Calibration with no placebos	4.1 min	1.21
Calibration with placebos	4.3	1.21
MT ($n = 10$)	1.3% RSD	0.3% RSD

sion (31,32). Table 25.3 shows the improvements obtained (33) in both injection precision and linearity when an internal standard was used for the analysis of levothyroxine. The injection volume in CE is highly related to the sample solution viscosity; therefore, if the sample solution is more viscous than the calibration solution, incorrect data will apparently be achieved. However, Table 25.3 shows that this will be corrected by use of an internal standard in the sample and standard solutions.

25.2.3. Chiral Analysis

Chiral analysis is an attractive application, as CE can have distinct advantages over HPLC in terms of reduced method development time, operating costs, and method ruggedness. (For further details refer to Chapter 23 by Wang and Khaledi in this monograph.). Detection levels of <0.1% of the undesired enantiomer are attainable, which is performance equivalent to that of HPLC. Methods for determining chiral purity have been sucessfully validated within several pharmaceutical companies (34–36).

A number of selectivity options are available for chiral separations including the use of cyclodextrins, crown ethers (37), antibiotics (38), and proteins (39). The vast majority of quantitative applications to pharmaceuticals has involved the addition of cyclodextrin to the separation buffer.

Figure 25.6a shows the use of a low-pH electrolyte containing α-cyclodextrin to determine 0.1% L-tryptophan in the presence of D-tryptophan. This method was successfully transferred (40) to a different instrument type (Figure 25.6b, p. 864), and the use of aspartame as an internal standard allowed quantitation of D-tryptophan contents with RSD values below 1%.

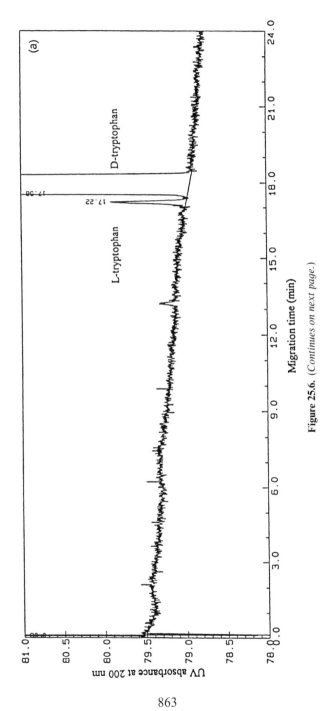

Figure 25.6. (*Continues on next page.*)

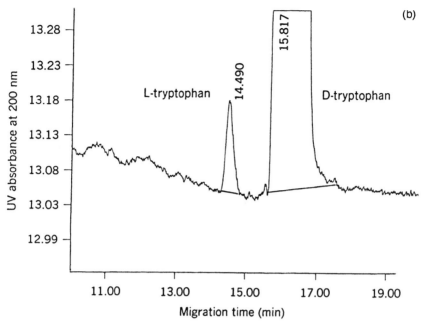

Figure 25.6. Separations of 0.1% L-tryptophan present in D-tryptophan. Conditions: 75 mM α-cyclodextrin dissolved in 25 mM triethanolaminephosphoric acid, pH 2.5; sample concentration; 0.1 mg mL^{-1} in water; +20 kV; UV detection, 200 nm; 30 °C; capillary 50 μm i.d. × 37 cm (total length). [Reproduced from K. D. Altria, P. Harkin, and M. Hindson, *J. Chromatogr. B* **686**, 103 (1996), with permission.]

Additional selectivity options to the use of low-pH cyclodextrin-containing electrolytes are available for the chiral resolution of neutral or acidic compounds. For example, a range of ionizable synthetic cyclodextrins (41) is commercially available, and these can have beneficial effects as compared to native cyclodextrins. Combinations of SDS and cyclodextrins have been used to good effect in chiral MEKC separations. For instance, trace levels of BMS-180431-09 inactive enantiomer were determined using a chiral MEKC method (36).

Analysis times for chiral CE separations can be rapid if sufficient selectivity is obtained. A combination of a high voltage and short capillary length enabled separation of dansyl-L-phenyl enantiomers within 70 s (42). Alternatively, the use of only a 7 cm detection length on a 27 cm capillary enabled the rapid chiral separation of picumeterol and clenbuterol to be obtained by using a low-pH electrolyte containing hydroxypropyl-β-cyclodextrin (Figure 25.7) (43).

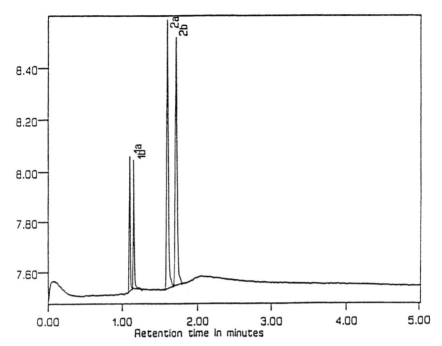

Figure 25.7. Rapid chiral separation of picumeterol and clenbuterol using a short-end injection procedure. Peaks: (1a) and (1b) picumeterol enantiomers; (2a) and (2b) clenbuterol enantiomers. Conditions: 50 mM NaH_2PO_4, pH 2.5, containing 30 mM dimethyl-β-cyclodextrin; UV detection 214 nm; -20 kV; 50 μm i.d. × 34 cm injection, -50 mbar for 2 s. [Reproduced from K. D. Altria, B. Clark, and, M. A. Kelly, *Chromatotgraphia* **43**, 153 (1996), with permission.]

A chirally selective method for the resolution of clenbuterol was successfully transferred between independent pharmaceutical companies in a collaborative excercise (44). All participating companies obtained similar selectivity and achieved acceptable measurements of migration time precision (1–2% RSD), relative migration time (less than 0.5%), peak area ratio precision (less than 1% RSD), and detector linearity.

25.2.4. Determination of Inorganic Ions

The use of indirect-UV detection in CE allows the relatively rapid and simple determination of inorganic ions. (For further details see Chapter 24 by Mazzeo in this monograph.). This ability is used to quantify levels of anions such as chloride, sulfate, and nitrate or metal ions such as sodium, potassium, or calcium present in drug substance samples. This is an important application, as the majority of drugs are ionic salts and need to be characterized fully. For instance, basic drugs are often manufactured as the chloride, hydrochloride, sulfate, or maleate salt. Similarly, acidic drugs are often manufactured as the sodium, potassium, or magnesium salt. CE has been used to quantitatively determine the percent w/w content of the drug counterion. In addition, CE has been used to monitor levels of inorganic contaminants present in drug substance samples.

Levels of chloride and sulfate were determined in drug substance samples by using an electrolyte containing chromate to provide the background UV signal for indirect detection (45). A cationic surfactant, tetradecyltrimethylammonium bromide (TTAB), was added to reverse the electroosmotic flow (EOF) direction. Calibration solutions were prepared using AnalaR grade reagents such as NaCl. Acceptable levels of precision (1–2% RSD), sensitivity (1 mg/L), and cross-validation with HPLC and the theoretical chloride or sulfate content were reported (45). Subsequent development of the method (46) involved addition of 1 mM borate to the chromate–TTAB system to buffer the system and reduce buffer-depletion effects, and use of another anion such as nitrate to act as an internal standard also improved both precision and linearity (Table 25.4). Figure 25.8 shows the separation of a 50 ppm test mixture of chloride and the internal standard, nitrate. Similar performance levels have been reported for the determination of sulfate content in washing powders (47).

An electrolyte containing imidazole and sulfuric acid has been used to quantify sodium or potassium counterions in acidic drugs (45). The imidazole

Table 25.4. Performance Data Obtained for Chloride Assay Using Nitrate as an Internal Standard

Factor	Result
10 replicate injections of 50 ppm test mixture	
Migration time (min)	0.60% RSD
Relative migration time (using nitrate as the internal standard)	0.11% RSD
Peak area (chloride)	4.19% RSD
Peak area ratio (using nitrate as the internal standard)	0.52% RSD
Linearity (25–75 ppm Cl^-) calculated using peak area ratios	
Correlation Coefficient	0.9998
Calibration preparation (10 separate preparations)	
Response factor precision	1.31% RSD
Accuracy (assay of KCl, theoretical Cl^- content = 47.7% w/w)	
Calculated percent w/w in sample	47.9% w/w
Precision of measurement ($n = 4$)	1.10% RSD

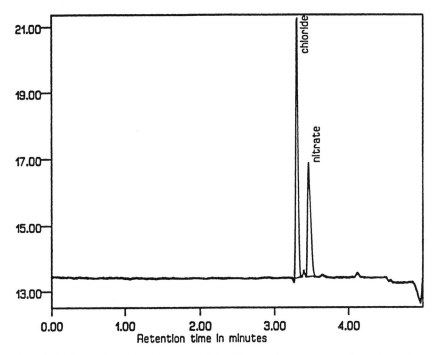

Figure 25.8. Separation of a solution containing 50 ppm of chloride and nitrate. Conditions: 1 mM borate/5 mM chromate/0.5 mM TTAB; indirect-UV detection at 254 nm; −10 kV; 30 °C; capillary 75 μm i.d. × 37 cm (total length). [Reproduced from K. D. Altria, J. Elgey, P. Lockwood, and D. Moore, *Chromatographia.* **42**, 332 (1996), with permission.]

provided the background signal for indirect-UV detection at 214 nm. Levels of metal ions were quantified in a number of drug substance samples. Analytical performance criteira such as precision (1–2% RSD), sensitivity (1 mg/L), linearity (correlation coefficients of 0.999), and accuracy ion-exchange chromatography were shown to be comparable with data obtained in IEC. A modified form of this method using a formic acid/imidazole electrolyte was validated for the quantitation of the potassium content in a drug substance, with sodium used as an internal standard (48). Precision data of 1% RSD were obtained for replicate injections and sample analysis repeatability (48). The robustness of the method was assessed in an extensive study employing experimental designs (49).

An intercompany collaboration excercise was conducted (50) on the use of the formic acid/imidazole electrolyte for the quantitation of the sodium in the cephalosporin-type antibiotic sodium cepthalothin. Each of the six independent companies repeated the separations and obtained accurate results confirming the theoretical percent w/w sodium content.

The presence of trace levels of inorganic contaminants can alter the physical form of the drug substance material and is therfore of concern. CE has been used to monitor trace levels of inorganic anions such as chloride and nitrate, and simple organic acids such as maleate or acetate (51). Detection limits of single-figure ppm (mg/L) levels were shown using a chromate electrolyte. The quality of the input water used in chemical processing is critical and can be monitored by CE (46). For example, Figure 25.9a,b shows the separations obtained for water used in two of our laboratories. The laboratory having the contaminated water had experienced considerable processing difficulties, which were resolved when the water purification unit was replaced.

25.2.5. Dissolution Analysis Testing

The rate at which a solid pharmaceutical dosage form releases the drug into solution is an important property, and this dissolution testing is widely performed (52). This is an ideal application area for CE, as a large number of samples are generated that require a rapid and cost-effective means of analysis. Generally analysis is performed by simple UV measurements. However, simple UV measurements are not selective, and a separative technique is generally required in circumstances where UV-active excipients or drug combinations are involved. CE has been used to analyze dissolution test sample solutions, and the results have been compared to UV absorbance measurements (Figure 25.10a,b, pp. 871 and 872) for the rate of drug release from tablets (53). Short-length separation capillaries (27 cm) and low-UV detection wavelengths (200 nm) enabled acceptable analytical performance to be obtained.

Dissolution test solution analysis for low-strength tablets has also been performed (54). The dissolution rates of 20 mg clenbuterol and 100 mg levothyroxine tablets were monitored. Sample solutions were preconcentrated using solid-phase extraction disks prior to CE analysis. The CE results were successfully compared to existing HPLC methods (54).

25.2.6. Analysis of Excipients

The use of CE to profile and quantify pharmaceutical excipients and raw materials is possible with standard commercial equipment (46). The use of indirect-UV detection can allow quantitation of inorganic ion contents of a range of materials such as buffers, sweetners, and electrolytes in liquid formations. Levels of other excipient materials such as sugars, lactose cyclodextrins, and lecithins can be determined by CE (46). Figure 25.11 (p. 873)

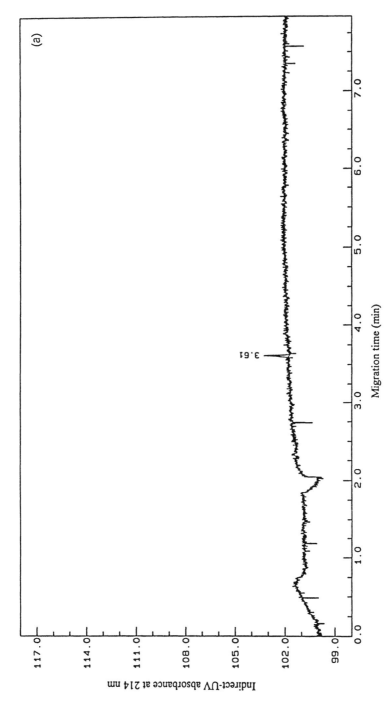

Figure 25.9. (*Continues on next page.*)

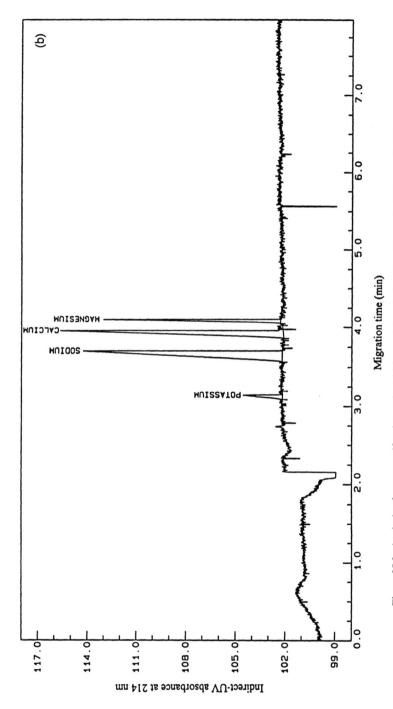

Figure 25.9. Analysis of water purification units within our laboratories. Conditions: 1 mM borate/5 mM chromate/0.5 mM from too water TTAB; indirect-UV detection at 254 nm; −10 kV; 30 °C; capillary 75 μm i.d. × 37 cm (total length). [Reproduced from K. D. Altria, J. Elgey, P. Lockwood, and D. Moore, *Chromatographia.* **42**, 332 (1996), with permission.]

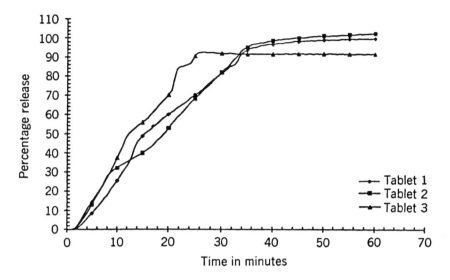

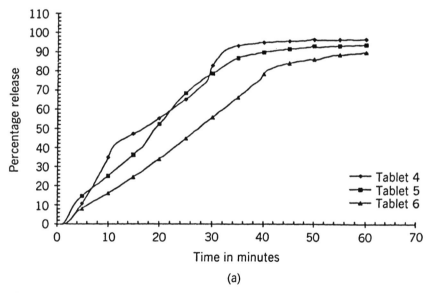

(a)

Figure 25.10. Dissolution profiles obtained by CE and UV measurements: (a) dissolution profiles for six ranitidine tablets from UV data; (b) dissolution profiles for six ranitidine tablets from CE data. [Reproduced from K. D. Altria, E. Traylen, and N. Turner, *Chromatographia* **41**, 393 (1995), with permission.]

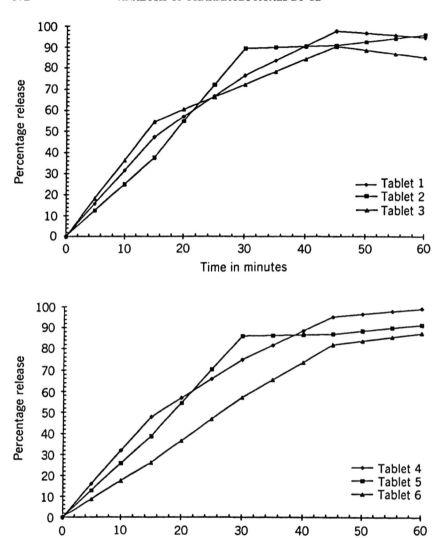

Figure 25.10. (*Continued*)

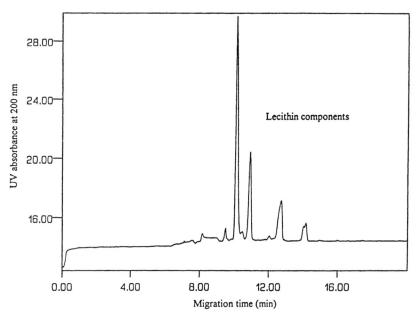

Figure 25.11. Separation of the components of U.S. National Formulary (USNF)–grade lecithin by CE. Conditions: (75 mM sodium cholate/10 mM Na_2HPO_4/6 mM borate) : isopropanol (70:30 v/v) adjusted to pH 8.5 with 1 M HCl; sample concentration, 1 mg mL^{-1} lecithin in methanol; indirect-UV detection at 200 nm; +20 kV; 50 °C; capillary 50 µm i.d. × 37 cm (total length). [Reproduced from K. D. Altria, J. Elgey, P. Lockwood, and D. Moore, *Chromatographia*. **42**, 332 (1996), with permission.]

shows profiling of lecithin with indirect-UV detection at 200 nm: the separation of these water-insoluble species is achieved using a MEKC electrolyte containing sodium cholate and methanol.

Table 25.5 shows assay results for the analysis of the metal ion content in a range of common excipients and raw materials. The data obtained compare well with the theoretical cation content, and the analytical method is simple and rapid.

25.3. FUTURE DIRECTIONS

The application of CE to the analysis of pharmaceuticals will in all likelihood continue at the current rapid rate. The development of novel CE-specific additives such as ionizable cyclodextrins (41) and chirally selective surfactants (55) will no doubt assist in terms of increased selectivity options. The increased

Table 25.5. Analysis of Metal Ion Content in Ionic Raw Materials and Excipients

Sample	Theoretical Percent w/w	Calculated Percent w/w
Na_2CO_3	43.4% w/w Na	43.9% w/w Na
Na_2HPO_4	32.4% w/w Na	32.0% w/w Na
$NaH_2PO_4, 2H_2O$	14.7% w/w Na	14.9% w/w Na
KI	24.5% w/w K	24.5% w/w K

advent of nonaqueous CE should lead to the more efficient analysis of water-insoluble compounds, which is currently an area of potential weakness (56). Perhaps the most significant advances in the next few years are expected to result from the development of capillary electrochromatography (CEC), which takes advantage of both the miniaturization benefits of CE and the selectivity array of HPLC columns (57). In CEC the capillaries used in CE are packed with HPLC reversed-phase material. Application of a voltage across the capillary results in genertion of a strong EOF, which sweeps solutes along the capillary while they differentially partition with the stationary phase, resulting in highly efficient separations (58). Figure 25.12 shows the efficient CEC separation (59) of a prostaglandin and a range of closely related impurities.

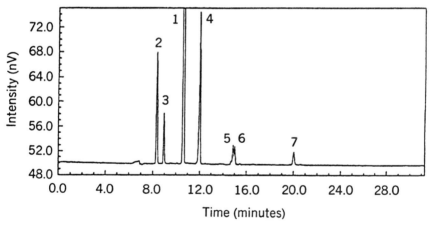

Figure 25.12. Separation of a prostaglandin from six related impurities by electrochromatography. Conditions: 70% ACN/30% 10 mM Na_2HPO_4, pH 9.9; UV detection at 270 nm; 30 kV; 50 μm i.d. × 40 cm capillary filled with 1.8 mm Zorbax SBC8. Peaks: (1) = prostaglandin; (2–7) related impurities. [Reproduced from N. W. Smith and M. B. Evans, *Chromatographia* **38**, 649 (1994), with permission.]

25.4. CONCLUSIONS

The range of applications of CE in the area of pharmaceutical analysis is at least as extensive as that of HPLC. One gains a number of distinct advantages when adopting CE testing, among which are speed of analysis and method development, reduction in consumable expenses, and simplicity of operation. All of these factors are especially important in the area of chiral separations, where CE is often the technique of choice. The disadvantages of CE should not be forgotten, and they are generally accepted as being poorer injection precision (hence the need to incorporate internal standards) and the limited amount of staff training and experience as compared to HPLC. But the highly complementary nature of the two techniques means that CE and HPLC should be viewed as two kindred approaches rather than as competitors. The advent of CEC in the next few years will undoubtedly further increase our choices.

LIST OF ACRONYMS

Acronym	Definition
CE	capillary electrophoresis
CEC	capillary electrochromatography
DAD	diode array detectors
EOF	electroosmotic flow
GC	gas chromatography
HPLC	high-performance liquid chromatography
IEC	ion-exchange chromatography
MEKC	micellar electrokinetic chromatography
MS	mass spectrometry
MT	migration time
RMT	relative migration time
SDS	sodium dodecyl sulfate
TLC	thin-layer chromatography
TTAB	tetradecyltrimethylammonium bromide
USNF	U. S. National Formulary
USP	*United states Pharmacopeia*
UV	ultraviolet

REFERENCES

1. Anonymous, *Pharm. Forum* **22**, 1727 (1995).
2. N. W. Smith and M. B. Evans, *J. Pharm. Biomed. Anal.* **12**, 579 (1994).

3. K. D. Altria and M. M. Rogan, *Introduction to Capillary Electrophoresis of Pharmaceuticals*, p. 1–55. Beckman Instruments, Fullerton, CA, 1994.
4. K. D. Altria and M. Kersey, *LG–CG Int.*, **8**, 201 (1995).
5. K. D. Altria, *J. Chromatogr.* **634**, 323 (1993).
6. A. Pluym, W. Van Ael, and M. De Smet, *Trends Anal. Chem.* **11**, 27 (1992).
7. R. C. Willaims, J. G. Edwards, and C. R. Ainsworth, *Chromatographia* **38**, 441 (1994).
8. G. C. Penalvo E. Julien, and H. Fabre, *Chromatographia* **42**, 159 (1996).
9. P. Emaldi, S. Fapanni, and A. Baldini, *J. Chromatogr. A* **711**, 339 (1995).
10. K. D. Altria, *LC–GC Int.* **6**, 164 (1993).
11. I. Bjørnstdottir and S. H. Hansen, *J. Pharm. Biomed. Anal.* **13**, 687 (1995).
12. K. D. Altria, *J. Chromatogr.* **735**, 43 (1996).
13. P. A. Shah and L. Quinones, *J. Liq. Chromatogr.* **18**, 1349 (1995).
14. O. Stalberg, D. Westerlund, U.-B. Rodby, and S. Schidmt, *Chromatographia* **41**, 287 (1995).
15. V. G. Trennery, R. J. Wells, and J. Robertson, *J. Chromatogr. Sci.* **1**, 32 (1994).
16. C. L. Flurer and K. A. Wolnik, *J. Chromatogr.* **663**, 259 (1994).
17. K. D. Altria and D. C. M. Luscombe, *J. Pharm. Biomed. Anal.* **11**, 415 (1993).
18. M. T. Ackermans, J. L. Beckers, F. M. Everaerts, and I.G.J.A. Seelen, *J. Chromatogr.* **590**, 341 (1992).
19. K. D. Altria, P. Frake, I. Gill, T. Hadgett, M. A. Kelly, and D. R. Rudd, *J. Pharm. Biomed. Anal.* **13**, 951 (1995).
20. K. D. Altria and S. D. Filbey, *J. Liq. Chromatogr.* **16**, 2281 (1993).
21. K. D. Altria, S. M. Bryant, and T. A. Hadgett, *J. Pharm. Biomed. Anal.* **15**, 1091 (1997).
22. C. L. Flurer, *J. Pharm. Biomed. Anal.* **13**, 809 (1995).
23. S. Boonkerd, M. R. Detaevernier, and Y. Michotte, *J. Chromatogr.* **670**, 209 (1994).
24. J. P. Landers, R. P. Oda, and M. D. Schuchard, *Anal. Chem.* **64**, 2846 (1992).
25. I. Bechet, M. Fillet, P. Hubert, and J. Crommen, *J. Pharm. Biomed. Anal.* **13**, 497 (1995).
26. S. Boonkerd, M. Lauwers, M. R. Detaevernier, and Y. Michotte, *J. Chromatogr.* **695**, 97 (1995).
27. H. Nishi, T. Fukuyama, M. Matsuo, and S. Terabe, *J. Chromatogr.* **513**, 279 (1990).
28. C. P. Ong, C. L. Ng, H. K. Lee, and S. F. Y. Li, *J. Chromatogr.* **547**, 345 (1991).
29. B. R. Thomas, X. G. Fang, X. Chen, R. J. Tyrell, and S. Ghobdane, *J. Chromatogr.* **657**, 383 (1994).
30. K. D. Altria, N. G. Clayton, R. C. Harden, M. Hart, J. Hevizi, J. Makwana, and M. J. Portsmouth, *Chromatographia* **39**, 180 (1994).
31. K. D. Altria and H. Fabre, *Chromatographia* **40**, 313 (1995).
32. E. V. Dose and G. A. Guiochon. *Anal. Chem.* **63**, 1154 (1991).

33. K. D. Altria and J. Bestford, *J. Capillary Electrophor.* **3**, 13 (1996).
34. R. C. Rickard and R. J. Bopp. *J. Chromatogr.* **680**, 609 (1994).
35. K. D. Altria, A. R. Walsh, and N. W. Smith, *J. Chromatogr.* **645**, 193 (1993).
36. J. E. Noroski, D. J. Mayo, and M. Moran, *J. Pharm. Biomed. Anal.* **13**, 54 (1995).
37. E. Hohne, G. J. Krauss, and G. Gubitz, *J. High Resolut. Chromatogr.* **15**, 698 (1992).
38. D. W. Armstrong, M. P. Gasper, and K. L. Rundlett, *J. Chromatogr.* **689**, 285 (1995).
39. S. Sun, G. E. Barker, R. A. Hartwick, N. Grinberg, and R. Kaliszan, *J. Chromatogr.* **652**, 247 (1993).
40. M. D. Altria, P. Harkin, and K. Hindson, *J. Chromatogr. B* **684**, 103 (1996).
41. T. Schmitt and H. Engelhardt, *J. Chromatogr.* **697**, 561 (1995).
42. M. J. Sepania, R. O. Cole, and B. K. Clar, *J. Liq. Chromatogr.* **15**, 1023 (1992).
43. K. D. Altria, B. Clark, and M. A. Kelly, *Chromatographia.* **43**, 153 (1996).
44. K. D. Altria, R. C. Harden, M. Hart, J. Hevizi, P. A. Hailey, J. Makwana, and M. J. Portsmouth, *J. Chromatogr.* **641**, 147 (1993).
45. K. D. Altria, D. M. Goodall, and M. M. Rogan, *Chromatographia.* **38**, 637 (1994).
46. K. D. Altria, J. Elgey, P. Lockwood, and D. Moore, *Chromatographia.* **42**, 332 (1996).
47. J. M. Jordan, R. L. Moese, R. Johnson-Watts, and D. E. Burton, *J. Chromatogr.* **671**, 445 (1994).
48. K. D. Altria, T. Wood, R. Kitscha, and A. Roberts-McIntosh, *J. Pharm. Biomed. Anal.* **13**, 33 (1995).
49. S. D. Filbey and K. D. Altria, *J. Capillary Electrophor.* **1**, 190 (1994).
50. K. D. Altria, N. G. Clayton, R. C. Harden, M. Hart, J. Hevizi, J. Makwana, and M. J. Portsmouth, *Chromatographia.* **40**, 47 (1995).
51. J. B. Nair, and C. G. Izzo, *J. Chromatogr.* **640**, 445 (1993).
52. A. C. Mehta, *Anal. Proc.* **31**, 245 (1994).
53. K. D. Altria, E. Traylen, and N. Turner, *Chromatographia.* **41**, 393 (1995).
54. C. N. Carducci, S. E. Lucangioloi, V. G. Rodriguez, and F. Otero, *J. Chromatogr. A* **730**, 313 (1996).
55. J. R. Mazzeo, E. R. Grover, M. E. Swartz, and J. S. Petersen, *J. Chromatogr.* **680**, 125 (1994).
56. R. Sahota and M. Khaledi, *Anal. Chem.* **66**, 1141 (1994).
57. J. H. Knox and I. H. Grant, *Chromatographia.* **24**, 135 (1987).
58. N. W. Smith and M. B. Evans, *Chromatographia.* **38**, 649 (1994).
59. N. W. Smith and M. B. Evans, *Chromatographia.* **41**, 197 (1995).

CHAPTER
26

ON-LINE IMMUNOAFFINITY CAPILLARY ELECTROPHORESIS FOR THE DETERMINATION OF ANALYTES DERIVED FROM BIOLOGICAL FLUIDS

NORBERTO A. GUZMAN

The R. W. Johnson Pharmaceutical Research Institute, Raritan, New Jersey 08869

ANDY J. TOMLINSON and STEPHEN NAYLOR

Biomedical Mass Spectrometry Facility, Department of Biochemistry and Molecular Biology, Mayo Clinic, Rochester, Minnesota 55905

26.1.	Introduction	879
26.2.	Nonspecific On-Line Preconcentration Capillary Electrophoresis	881
26.3.	Specific On-Line Preconcentration Capillary Electrophoresis	883
	26.3.1. History of Affinity Chromatography	883
	26.3.2. Construction and Use of the Immunoaffinity Analyte Concentrator	884
	26.3.3. Specific Examples of Immunoaffinity Capillary Electrophoresis	888
26.4.	Conclusions	894
List of Acronyms		894
References		895

26.1. INTRODUCTION

An understanding of the biochemical constituents present in biological fluids, cells, and tissues is important in assessing the state of individual health. In particular, a significant perturbation in the homeostasis of such compounds often triggers a variety of responses that can ultimately produce a disease condition (1). Hence, although the molecular basis of many disease states is not understood, it is recognized that the analysis of biochemical constituents of body fluids, cells, and tissues is important in making a medical diagnosis (1). In

---

High Performance Capillary Electrophoresis, edited by Morteza G. Khaledi. Chemical Analysis Series, Vol. 146.
ISBN 0-471-14851-2 © 1998 John Wiley & Sons, Inc.

addition, the *in vivo* screening of both therapeutic and narcotic drugs is very necessary since it allows the determination of concentration levels, efficacy, and pharmacological and toxicological effects of such compounds, as well as their metabolites.

In many instances the limited availability of biological fluids significantly hinders the analysis of the relevant biochemical compounds. Furthermore, since inflammation, biochemical imbalance, and disease are often localized, only subfemtomole quantities of biologically relevant compounds are normally available. This necessitates that analytical methods capable of handling low-nanoliter sample volumes must be available. It is also important that detection limits at the attomole (10^{-18} mol), zeptomole (10^{-21} mol), or even yoctomole (10^{-24} mol) level can be attained.

The inherent characteristics of capillary electrophoresis (CE), such as separation selectivity, small sample size, high speed of analysis, excellent mass sensitivity, low reagent consumption, high resolution, and minimal sample loss, clearly indicate the potential of this technique to overcome the limitations of conventional analytical approaches. This is particularly true when CE is coupled to suitable sensitive detection devices such as a mass spectrometer (CE–MS) (2), laser-induced fluorescence (CE–LIF) (3,4), electrochemical detector (5), and multiwavelength ultraviolet (UV) array detectors (6,7).

Despite the unique features of CE and the numerous reports of its use in the analysis of biological fluids, it is perceived to have one limited drawback. It is often reported that optimal CE performance is obtained when small volumes are analyzed, typically <2% of the total CE capillary volume. Hence, capillary sample introduction volumes of ∼1–100 nL are typical. Ultimately, this leads to poor concentration limits of detection (CLOD) and makes it difficult to provide suitable sensitivity levels of detection for many analytes (8). Furthermore, apart from CE–MS, sensitive detectors such as electrochemical and CE–LIF devices are nonspecific and afford very limited structural information for the detected analyte. However, when biologically derived compounds of known structure and chemical behavior are analyzed, method specificity coupled with high analyte detection sensitivity are of prime importance. These criteria can be satisfied by use of specific immunoaffinity extraction protocols in conjunction with the sensitive detection devices listed above.

In this chapter, we discuss the use of immunoaffinity capillary electrophoresis (IACE) in the on-line concentration mode for the analysis of constituents of biological fluids. In particular, we describe the use of on-line IACE to overcome poor CLOD associated with conventional CE, as well as to enhance the specificity of detection of analytes derived from biological sources.

26.2. NONSPECIFIC ON-LINE PRECONCENTRATION CAPILLARY ELECTROPHORESIS

In attempts to overcome the poor CLOD of CE, a number of workers have developed enhanced capillary sample introduction techniques that include analyte stacking, field amplification, and transient isotachophoresis (tITP) (9–11). All of these techniques are implemented as the CE voltage is applied across the capillary. This results in analyte zones being stacked or focused due to variation of ion mobilities in various field strengths or chemical microenvironments. However, since these techniques are carried out within a conventional capillary, the maximum sample volume that can be analyzed is predetermined by the total capillary volume. Hence, even in most favorable cases such optimized sample introduction techniques can normally only introduce ~1–2 µL of sample without undue loss of CE performance (8).

Another approach to circumvent poor CLOD is to undertake off-line sample pretreatment and analyte concentration. However, if possible, this should be avoided for dilute biologically derived solutions since losses to exposed surfaces (e.g., walls of microcentrifuge tubes, pipet tips, or solid extraction phases) can be substantial. Furthermore, excessive handling of a concentrated solution of biopolymers can lead to denaturation, aggregation, precipitation, and ultimately poor analyte recovery. Therefore, minimal sample handling is advisable. This can be achieved using an analyte concentrator (8,12–23) or membrane preconcentration cartridge (24–29) on-line with the CE capillary. These devices usually consist of an adsorptive phase at the inlet of the CE capillary and serve to enrich trace levels of analytes prior to component separation by CE. Introduction of sample volumes in excess of 100 µL into a capillary with <300 nL total volume has been reported (29). Recently we reviewed all general aspects of nonspecific on-line preconcentration (8). In particular, both the analyte concentrator and preconcentration CE cartridge are multipurpose devices that possess many unique features and includes the following: (i) minimizing sample handling, (ii) concentrating samples present in dilute solutions or present in low concentrations in biological fluids; (iii) serving as a desalting system and as a cleanup procedure; (iv) enabling the derivatization of samples to enhance detection sensitivity; (v) serving as an affinity or chromatographic column; (vi) serving as a microreactor chamber to perform a variety of macromolecule interactions including peptide mapping; (vii) presenting a practical approach to increasing the concentration limits of detection using UV detection; and (viii) obtaining additional structural information of the analyte when connected on-line with CE–MS or other detectors such as nuclear magnetic resonance (CE–NMR) or circular dichroism (CE–CD).

Some modifications to the original designs of analyte preconcentration devices have been made that decrease or remove potential problems asso-

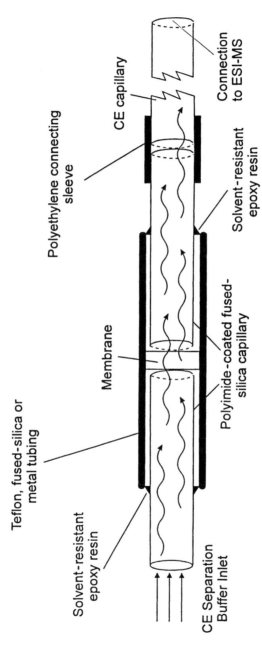

Figure 26.1. Schematic of the membrane preconcentration cartridge used on-line with CE, termed mPC–CE. A suitably impregnated membrane is enclosed in a Teflon cartridge and sandwiched between the inlet leg of the capillary and the connecting capillary to the conventional CE capillary.

ciated with primary designs. One new concept is based on the development of an adsorptive phase made of an appropriate coated/impregnated membrane, installed for convenience in a Teflon cartridge system, and thus replacing the beads and the frits of the original cartridge (see Figure 26.1). A major advantage of this new concept for an on-line analyte concentrator is the reduction in the volume of elution solvent required for the efficient removal of analytes from the membrane (27–29). The technology that utilizes a coated/impregnated membrane as a preconcentration capillary electrophoresis cartridge has been termed mPC–CE, and its compatibility with MS detection (mPC–CE–MS) has been demonstrated (27,28). This approach has proven invaluable for dramatically enhancing CLOD in CE, as well as for preconcentrating structurally unknown compounds prior to analysis by CE and CE–MS (30). However, once an analyte has been identified and targeted for analysis and is present in a complex biological matrix, a more specific on-line concentration method is appropriate.

26.3. SPECIFIC ON-LINE PRECONCENTRATION CAPILLARY ELECTROPHORESIS

26.3.1. History of Affinity Chromatography

Affinity chromatography utilizing a solid support containing an immobilized liquid that specifically interacts with a corresponding counterpart molecule has had a varied and widespread history (31). Such interactions are best represented by antibody–antigen and lectin–carbohydrate pairings where high-affinity binding occurs. Indeed, immobilized antibodies have long been used as affinity reagents. However, original approaches suffered from low binding capacity and specificity limited by the quality of the antibodies. Harsh elutions were also often necessary, causing irreversible conformational changes in the antibody. As a result, antibody immunoadsorbents for some time did not achieve the widespread use originally expected. The introduction of monoclonal antibodies, however, began a new era, allowing columns with high specificity and high binding capacity that required only mild elution conditions. The nearly infinite supply of reagents made commercial applications feasible. Furthermore, many of the difficulties associated with generating monoclonal antibodies by B-cell immortalization can be overcome by engineering and expressing antibody fragments in *Escherichia coli*, using phage display (32). Also, recombinant Protein A is now commercially available. This protein has been modified to favor an optimal orientation of the ligand when coupled to a solid support and an increased capacity for binding to monoclonal antibodies (33). Today, affinity columns have found their greatest use in purifying (and characterizing) high-priced biologicals isolated from tissue and recombinant cell culture.

26.3.2. Construction and Use of the Immunoaffinity Analyte Concentrator

The concept of analyte concentrators that contain covalently bound antibodies was first demonstrated by Kasicka and Prusik (13) in conjunction with capillary isotachophoresis (CITP) and subsequently described for capillary zone electrophoresis (CZE) by Guzman et al. (14,15) and Cole and Kennedy (20). In these studies, specific antibodies were covalently bound to either a solid phase (13), glass beads (15), multiple capillary bundles, or, more recently, a piece of solid glass predrilled with a laser beam (23). A summary of the various studies using affinity capillary electrophoresis (ACE) is presented in Table 26.1 (34–56). Guzman and colleagues originally used controlled-pore silylated glass beads reacted with 1,4-phenylene diisothiocyanate (DITC) and conjugated with the appropriate antibody (14). The glass beads were held in place by sintered glass frits. Unfortunately, the use of such frits often leads to restricted flow through the analyte concentrator chamber and clogging of the system by sample-derived particulates. This prompted the development and design of new analyte concentrators (15,23). One design consisted of a bundle of microcapillaries (see Figure 26.2a). Between 5 and 14 polyimide-coated capillaries (25 µm i.d. × 150 µm o.d.) were tightly handfitted into a rigid plastic tubing of ~5 cm in length and 400–800 µm i.d. Approximately 5 mm portions of this assembly were cut to produce individual analyte concentrators. Monoclonal antibodies can then be covalently bound to the surfaces of each microcapillary by using the DITC method just described. The second modified design consisted of a series of through-holes (~25µm i.d.) fabricated from a solid-glass rod using a laser beam (Figure 26.2b). Antibodies can then be covalently bound to the surface of each microcapillary by employing the same DITC coupling chemistry (Figure 26.2c).

This latter design has several advantages over the other methods of construction and includes the following: (i) easier fabrication of the cartridge; (ii) better consistency of the electroosmotic flow enabling a high reproducibility of the peak area and migration time; (iii) reduced possibility of clogging the system; (iv) increased number of uses; (v) greater stability of the chemistries since no heat was involved in the production of the cartridge; and (vi) increased length (up to 5 mm) without changing significantly the electroosmotic flow. The fabrication of a cartridge (containing microbeads) greater than 3 mm in length can yield a product having consistently low reproducibility of the electroosmotic flow after several uses, an increased tendency for blocking the system, and a short life span. This is due primarily to (i) use of irregularly made

Table 26.1. Determination of Metabolic Intermediates and Drugs in Biological Fluids by ACE[a]

Specimen	Metabolic Intermediate and/or Drug	Reference
Urine	Methamphetamine	14
Standards	Peptide mapping	22,34–38
Serum	Amyloid P component	39
Standards	Nucleic acid mapping	40
Standards	Lactobionic acid	41
Standards	Glucose-6-phosphate dehydrogenase	42
Standards	Carbonic anhydrase	43,44
Standards	Procainamide nitrosoprocainamide	45
Standards	Proteins	12
Standards	Porphyrins	46
Standards	C-reactive protein	47
Standards	Growth hormone	48
Standards	Insulin	20,49
Standards	Monosaccharides	50
Serum	Cortisol	51,52
Serum	IgE	23
Standards	Nitrophenyl-β-galactoside	53
Standards	Leucine amino peptidase	54
Standards	Atrazine	55
Urine	Glycopeptides	56

[a] Affinity capillary electrophoresis (or affinity microchip electrophoresis) utilizes the interaction and separation by CE of an antigen–antibody, sugar–lectin, substrate–enzyme, ligand–macromolecule, or any other ligand with high affinity for another biomolecule. The specific binding can be carried out outside the capillary (off-line), and the resulting components of the reaction can then be separated by CE. Alternatively, the specific binding can be carried out within the capillary (on-line), in which all the reagents are part of the separation buffer and the enzyme, for example, is added to the capillary to react with the substrate to form the corresponding product and then separated by CE. In addition, a more convenient way to do on-line ACE is to immobilize the antibody, enzyme, receptor, or the corresponding ligands in order to perform the reaction on the microreactor and then separate the bound and eluted substance or the resulting products by CE.

frits; (ii) compacting of the microbeads due to increasing pressure; and (iii) difficulty in reproducing the exact dimensions and behavior of the cartridge.

Maintenance of antibody activity is crucial to the successful construction of the immunoaffinity analyte concentrator. When covalently bonded to solid supports, polyclonal antibodies often lose antigen-binding capacity. Monoclonal antibodies lack the diversity of structures found in polyclonal antisera. Thus, the antibody with the best binding properties in solution may not

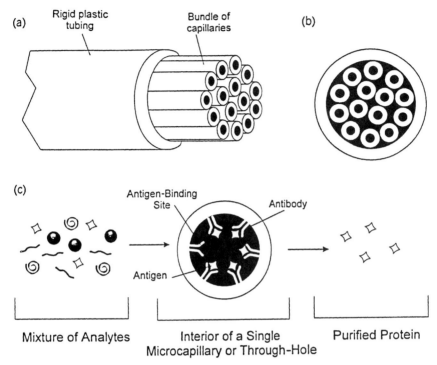

Figure 26.2. Various immunoaffinity analyte concentrator designs. (a) Schematic of an analyte concentrator consisting of a bundle of microcapillaries all contained in plastic tubing. (b) A solid piece of glass that has been drilled with a laser beam to produce a series of small-diameter passages of ~25 μm i.d. (c) After covalent binding of the appropriate antibody, high analyte selectivity and concentration can be achieved.

necessarily retain its affinity when coupled. It is well known that the efficiency of antigen binding to a solid support can vary widely when analysts are determining the amount of bound antigen per milligram of bound antibody. In addition, the chemistry used for the immobilization of the antigen or antibody (or any other ligand) will significantly affect the efficiency of the system. Some investigators have suggested that coupling an antibody via its Fc portion to immobilized Protein A (or Protein G) would affect binding capacity less than covalent linkage. In any case, the choice of a monoclonal for affinity should be based on the antibody's performance when linked to the solid support and not its efficiency in solution. At the present time, three approaches have been used for immobilizing proteins to the CE microreactor (see Figure 26.3): (a) direct chemical coupling of the antibody to the microbead's surface using DITC chemistry described previously; (b) coupling the antibody through a biotin–

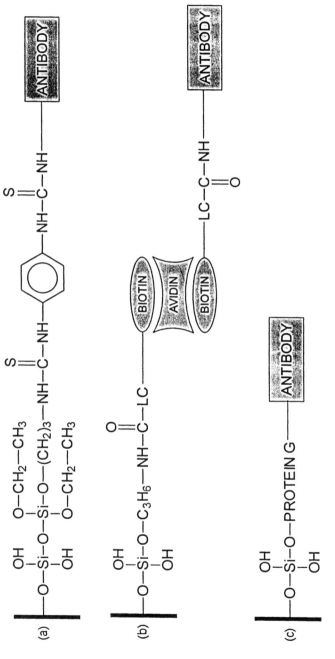

Figure 26.3. A schematic representation of the various chemical reactions used to covalently immobilize an antibody to the surface of a glass bead. (a) Glass beads were silylated with 3-aminopropyltriethoxysilane and then reacted with DITC before being conjugated to a monoclonal antibody. [For details, see Guzman et al. (14).] (b) The antibody was immobilized using the biotin–avidin complex. LC refers to a long-chain spacer chemical unit necessary to facilitate an optimal interaction between the immobilized antibody and the correspondent antigen. [For details, see Kuhr et al. (35,36).] (c) Binding of Protein G (or A) to the silyl group of the capillary and coupling the antibody to Protein G.

avidin complex to the capillary wall; or (c) the Protein G–antibody approach routinely used in various immunological purifications.

Finally, after manufacturing the immunoaffinity analyte concentrator, the biological fluid such as serum is pressure introduced onto the immunoaffinity–analyte concentrator, followed by a cleanup procedure consisting of a wash buffer to remove salts and other nonrelevant matrix components. Once the cartridge has been washed with separation buffer, it is equilibrated with 10 column volumes of the same buffer. Subsequently an aliquot of ~ 100 nL of an optimized buffer system is applied to elute the analyte. In the case of proteins, the elution buffer consists of 75 mM 4-(2-hydroxyethyl)-1-piperazine ethanesulfonic acid (HEPES)/NaOH, pH 7.2, containing 2 M $MgCl_2$ and 25% ethylene glycol (23). After elution of the protein from the immunoaffinity analyte concentrator chamber by sequential use of elution buffer, followed by separation buffer under hydrodynamic pressure, the CE voltage is applied across the capillary.

26.3.3. Specific Examples of Immunoaffinity Capillary Electrophoresis

For purposes of clarification, the terminology of ACE should be subdivided into three categories: (i) the interaction of an antigen with an antibody or a fraction of an antibody (or other ligands with high affinity for each other) in a test tube (off-line) and then separated by CE; (ii) the interaction of an antigen with an immobilized antibody (or other immobilized ligands of high affinity for each other) in a microreactor (on-line); and (iii) the interaction of antibody or ligand in solution throughout the length of the CE capillary (57). Here, we will address only the interaction of an antigen with an immobilized antibody (or other immobilized ligands) using the microreactor technique that we have termed on-line immunoaffinity capillary electrophoresis (IACE) (see Section 26.1).

The first published application of on-line IACE was reported in 1991 (14). The report described the use of an immobilized antibody for the determination of methamphetamine in urine samples. The antigen–antibody reaction was allowed to take place in a microreactor approximately 5 mm in length and located near the injection side of the capillary. The microreactor was fabricated using controlled porous glass to which the purified monoclonal antibody was immobilized (Figure 26.4a). Once the methamphetamine was specifically bound to the immobilized antibody, fresh buffer was applied to the column to eliminate the excess of salt and nonrelevant matrix components present in urine. The capillary column was then equilibrated with fresh buffer, and the bound methamphetamine was eluted from the antigen–antibody complex using nanoliter quantities of an acidic buffer. Improvements in the elution procedure have since been made, and a more effective buffer is currently being used (23): 100 nL of HEPES/NaOH buffer, pH 7.2, containing

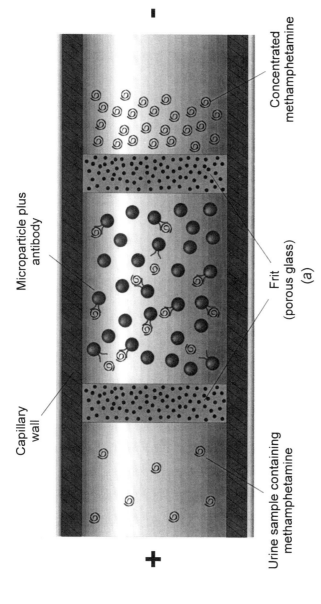

Figure 26.4. (a) Schematic of the immunoaffinity analyte concentrator cartridge made of controlled porous glass containing an immobilized antibody directed against methamphetamine. (b) A urine sample spiked with methamphetamine was applied to the capillary and then washed with separation buffer. The bound methamphetamine was eluted from the cartridge with 50 nL of HEPES/NaOH buffer, pH 7.2, containing 2 M MgCl$_2$ and 25% ethylene glycol. The separation was carried out in 50 mM sodium tetraborate buffer, pH 8.3.

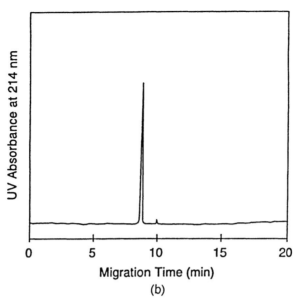

Figure 26.4. (*Continued*)

2 M $MgCl_2$ and 25% ethylene glycol, although a viscous solution, seems to be an optimized elution system. As shown in Figure 26.4b, a single peak of the eluted material was obtained under these experimental conditions.

Even though nonaffinity microreactors can adsorb many more substances than an affinity microreactor can, the preconcentration step is carried out with low selectivity. A big advantage of an affinity microreactor is the high selectivity of preconcentration. In principle, one substance or one group of closely related substances should be captured by the immobilized antibody, metal, lectin, or any appropriate affinity ligand. If the affinity constant and the number of affinity ligands per surface area of an affinity cartridge are larger than those of a nonaffinity cartridge, then there is a greater increase in concentration of the analyte. Efficiency of binding of ligands per surface area is more important than the length of the cartridge.

The main attractive features of the affinity microreactor conducted on a solid support on-line with CE are as follows: (i) the accomplishment of an affinity reaction in a short period of time; (ii) the consumption of smaller amount of reagent; (iii) the separation of the main compound from non-relevant matrix components; and (iv) the reusability of the cartridge. These features have prompted several investigators to explore other principles of affinity.

A second approach came from Cai and El Rassi (58), who developed an open-tubular precapillary that was treated with iminodiacetic acid metal with chelating properties for the selective on-line preconcentration of dilute protein samples. Only proteins with affinity for the chelated metal were concentrated. In this particular study, carbonic anhydrase was used as a model protein to illustrate the principle of specific capture and concentration using Zn(II) as the metallic ligand. As a desorbing agent, to remove the bound protein from the complex, EDTA (ethylenediaminetetraacetic acid) was used since it forms stronger complexes with metals and thus competes with the binding. Unfortunately, in this example, instead of using an analyte concentrator containing a membrane or microbeads similar to the one described in Figures 26.1 or 26.4a, a piece of a capillary (50 μm i.d. and 20 cm in length) was used as the affinity capture microreactor in tandem with a 60 cm separation capillary. The exposure of analytes to the modified capillary was low since only a 25-fold greater concentration of proteins was obtained.

Cole and Kennedy (20) have also demonstrated the efficacy of IACE. They applied a selective preconcentration step for CE using Protein G immunoaffinity electrophoresis chromatography. However, in this case the immunoaffinity was carried out off-line in a larger capillary. In addition, a required collection of desorbed fractions was necessary before injection into the CE capillary.

More recently, Guzman reported the use of a cartridge made of multiple capillary bundles or a piece of solid glass predrilled with a laser beam to form multiple through-holes (Figure 26.2a,b) (23). There are no beads or frits in this model to prevent clogging, allowing a more uniform electroosmotic flow. Specific antibodies directed against immunoglobulin E (IgE) were covalently bound to the surface of every microcapillary. Serum was applied directly into the capillary. After binding, the capillary was rinsed with buffer and IgE was eluted with the buffer containing 2 M $MgCl_2$, as described above: the electropherogram is shown in Figure 26.5. The fraction was collected and subjected to ELISA (enzyme-linked immunosorbent assay), and IgA, IgG, and IgM could not be detected, demonstrating the high specificity of this approach.

Another approach using the concept of on-line ACE is the use of immobilized lectins and enzymes on the surface of microbeads or directly on the inner surface of the capillary. Lectins are carbohydrate-binding proteins of a nonimmunoglobulin nature. The use of lectins of plant and animal origin for studying animal cell glycoconjugates has a long and productive history (59,60). By definition, most lectins are multivalent proteins having multiple subunits, and the interaction of a conjugate with a lectin is governed by the binding specificity and affinity of each subunit for a glycoconjugate. Today, many different lectins are commercially available and used for purification of glycoproteins, glycopeptides, and glycolipids by lectin chromatography.

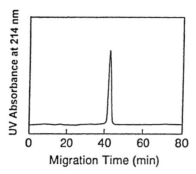

Figure 26.5. Immunoaffinity microreactor electropherogram of serum IgE. Approximately 20 µL of a serum sample containing high titers of IgE were applied to the capillary (containing the cartridge with immobilized antibody against IgE), followed by a cleanup procedure consisting of separation buffer to remove salts and other serum constituents. The bound IgE was eluted from the cartridge with 100 nL of 50 mM HEPES/NaOH buffer, pH 7.2, containing 2 M $MgCl_2$ and 25% ethylene glycol. The separation was carried out in 50 mM sodium tetraborate buffer, pH 8.3. A typical electropherogram was obtained using the cartridge fabricated from a solid glass rod that had a plurality of small-diameter passages or through-holes. [After N. A. Guzman, *J. Liq. Chromatogr.* **18**, 3751 (1995), with permission]

Many lectins with well-characterized sugar-binding activities can be immobilized in any of the various models of the analyte concentrator–microreactor cartridge and employed for affinity interaction. In fact, we have immobilized concanavalin A into controlled porous glass beads and fabricated an analyte concentrator cartridge [see Figure 26.6 and Guzman (56)]. In order to differentiate this affinity method from on-line IACE, we have termed it on-line lectin affinity capillary electrophoresis (LACE). As shown in Figure 26.7, we have demonstrated the usefulness and potential diagnostic value of on-line LACE. We analyzed urine from a pooled sample of six urines collected from healthy individuals by using the concanavalin affinity cartridge. The glycopeptides were eluted from the analyte concentrator with 100 nL of 50 mM Tris–HCl (pH 7.4) containing 0.3 M α-D-mannopyranoside and 25% ethylene glycol. Even after concentration of ∼ 50 µL of normal urine, followed by elution, only minor responses in the electropherogram were observed (Figure 26.7a). However, when the same volume of urine collected from a patient suffering from pancreatic cancer and complications to the liver and other organs was analyzed by LACE, a significantly different electropherogram was detected (Figure 26.7b). The actual identity of these compounds is under investigation using LACE coupled with MS.

Very recently, Vaughan et al. (32) immobilized enzymes onto the surface of microbeads of the inner surface of the capillary to perform on-line enzyme

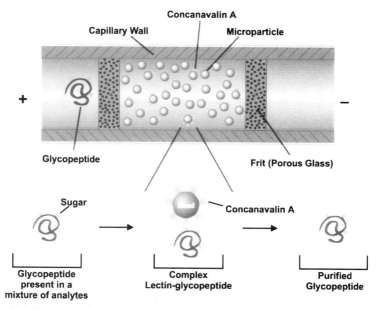

Figure 26.6. Schematic of a microreactor containing an immobilized lectin. Controlled porous glass beads were activated, conjugated with concanavalin A, and installed inside a portion of a fused-silica capillary. [For details, see Guzman (56).]

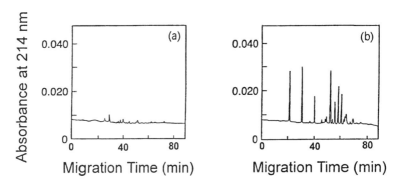

Figure 26.7. Immunoaffinity microreactor electropherograms of urine glycopeptides. Approximately 50 μL of a urine sample containing high levels of suspected glycopeptides were applied to the capillary (containing the cartridge with immobilized concanavalin A), followed by a cleanup procedure consisting of use of a separation buffer to remove salts and other urine constituents. The bound glycopeptides were eluted from the cartridge with 100 nL of 50 mM Tris–HCl buffer, pH 7.4, containing 0.3 M α-D-mannopyranoside and 25% ethylene glycol. The separation was carried out in 50 mM sodium tetraborate buffer, pH 8.3. A typical electropherogram was obtained from a freshly collected clean-catch urine specimen obtained from a pool of six normal individuals (panel a) or from a urine specimen obtained from a patient suffering from pancreatic cancer and complications to the liver and other organs (panel b). [After N. A. Guzman, *J. Liq Chromatogr.* in press, with permission.]

affinity capillary electrophoresis (EACE). In particular, they demonstrated that EACE can be used to quantitate the chemical conversion of substrate into product, as well as to evaluate the binding and kinetic constants of an enzyme reaction (53)

26.4. CONCLUSIONS

The use of IACE, LACE, and EACE is still in its infancy. However, it is clear that they show great potential and promise for the rapid processing of biological samples. Furthermore, given their high selectivity and specificity, these techniques offer the potential to detect analytes present in only very minute quantities in biological matrices and tissue. Also, the future holds the promise of further miniaturizing with such approaches as microchip technology, affording a very rapid and convenient screen of biological fluids for clinically relevant biomarkers.

LIST OF ACRONYMS

Acronym	Definition
ACE	affinity capillary electrophoresis
CD	circular dichroism
CE	capillary electrophoresis
CITP	capillary isotachophoresis
CLOD	concentration limits of detection
CZE	capillary zone electrophoresis
DITC	1,4-phenylene diisothiocyanate
EACE	enzyme affinity capillary electrophoresis
EDTA	ethylenediaminetetraacetic acid
ELISA	enzyme-linked immunosorbent assay
HEPES	4-(2-hydroxyethyl)-1-piperazineethanesulfonic acid
IACE	immunoaffinity capillary electrophoresis
IgE	immunoglobulin E
LACE	lectin affinity capillary electrophoresis
LIF	laser-induced fluorescence
mPC–CE	membrane preconcentration capillary electrophoresis
MS	mass spectrometry
NMR	nuclear magnetic resonance
tITP	transient isotachophoresis
UV	ultraviolet

ACKNOWLEDGMENTS

We would like to acknowledge funding from Mayo Foundation and Beckman instruments (to AJT and SN) for some of this work.

REFERENCES

1. C. A. Burtis and E. R. Ashwood eds. *Tietz Textbook of Clinical Chemistry,* 2nd ed. Saunders, Philadelphia, 1994.
2. J. Cai and J. D. Henion, *J. Chromatogr. A* **703**, 667 (1995).
3. M. C. Roach, P. Gozel, and R. N. Zare, *J. Chromatogr.* **426**, 129 (1988).
4. Y. Ma, Z. Wu, H. C. Furr, C. Lammi-Keefe, and N. E. Craft, *J. Chromatogr.* **616**, 31 (1993).
5. R. A. Wallingford and A. G. Ewing, *Anal. Chem.* **61**, 98 (1989).
6. D. N. Heiger, P. Kaltenbach, and H.-J. Sievert, *Electrophoresis* **15**, 1234 (1994).
7. P. Wernly and W. Thormann, *J. Chromatogr.* **608**, 251 (1992).
8. A. J. Tomlinson, N. A. Guzman, and S. Naylor, *J. Capillary Electrophor.* **2**, 247 (1995).
9. B. J. Wanders and F. M. Everaerts in *Handbook of Capillary Electrophoresis* (J. P. Landers, ed.) p.111. CRC Press, Boca Raton, FI, 1994.
10. R. L. Chien and D. S. Burgi, *Anal. Chem.* **64**, 489A (1992).
11. P. Gebauer, W. Thormann, and P. Bocek *J. Chromatogr.* **608**, 47 (1992).
12. J. Cai and Z. El Rassi, *J. Liq. Chromatogr.* **16**, 2007 (1993).
13. V. Kasicka and Z. Prusik, *J. Chromatogr.* **273**, 117 (1983).
14. N. A. Guzman, M. A. Trebilcock, and J. P. Advis, *J. Liq. Chromatogr.* **14**, 997 (1991).
15. N. A. Guzman, C. L. Gonzalez, M. A. Trebilcock, L. Hernandez, C. M. Berck, and J. P. Advis, in *Capillary Electrophoresis Technology* (N. A. Guzman, ed.) Chromatogr. Sci. Soc., Vol. 64, p.643. Dekker, New York, 1993.
16. M. E. Swartz and M. Merion, *J. Chromatogr.* **632**, 209 (1993).
17. A. M. Hoyt, Jr., S. C. Beale, J. P. Larmann, Jr., and J. W. Jorgenson, *J. Microcolumn Sep.* **5**, 325 (1993).
18. A. J. Tomlinson, L. M. Benson, W. D. Braddock, R. P. Oda, and S. Naylor, *J. High Resolut. Chromatogr.* **17**, 729 (1994).
19. A. J. Tomlinson, W. D. Braddock, L. M. Benson, R. P. Oda, and S. Naylor, *J. Chromatogr. B: Biomed. Appl.* **669**, 67 (1995).
20. L. J. Cole and R. T. Kennedy, *Electrophoresis,* **16**, 549 (1995).
21. J. H. Beattie, R. Self, and M. P. Richards, *Electrophoresis* **16**, 322 (1995).

22. N. A. Guzman, in *Capillary Electrophoresis: An Analytical Tool in Biotechnology* (P. G. Righetti, ed.) Chapter 45 pp. 101–121. CRC Press, Boca Raton, Fl, 1995.
23. N. A. Guzman, *J. Liq. Chromatogr.* **18**, 3751 (1995).
24. A. J. Tomlinson, L. M. Benson, R. P. Oda, W. D. Braddock, B. L. Riggs, J. A. Katzmann, and S. Naylor, *J. Capillary Electrophor.* **2**, 97 (1995).
25. A. J. Tomlinson, L. M. Benson, W. D. Braddock, R. P. Oda, and S. Naylor, *J. High Resolut. Chromatogr.* **18**, 381 (1995).
26. A. J. Tomlinson, and S. Naylor, *J. High Resolut. Chromatogr.* **18**, 381 (1995).
27. A. J. Tomlinson and S. Naylor, *J. Capillary Electrophor.* **2**, 225 (1995).
28. A. J. Tomlinson and S. Naylor, *J. Liq. Chromatogr.* **18**, 3591 (1995).
29. A. J. Tomlinson, L. M. Benson, S. Jameson, and S. Naylor, *Electrophoresis* **17**, 1801 (1996).
30. A. J. Tomlinson, S. Jameson, and S. Naylor, *J. Chromatogr.* **744**, 273 (1996).
31. C. Jones, A. Patel, S. Griffin, J. Martin, P. Young, K. O'Donnell, C. Silverman, T. Porter, and I. Chaiken, *J. Chromatogr. A* **707**, 3 (1995).
32. T. J. Vaughan, A. J. Williams, K. Pritchard, J. K. Osbourn, A. R. Pope, J. C. Earnashaw, J. McCafferty, R. A. Hodits, J. Wilton, and K. S. Johnson, *Nat. Biotechnol.* **14**, 309 (1996).
33. C. M. Johansson and H. J. Johansson, *IBC Conf. Monoclonal Antibody Purif.*, La Jolla, CA, *1996,* Abstr. M-112 (1996).
34. L. N. Amankwa and W. G. Kuhr, *Anal. Chem.* **64**, 1610 (1992).
35. W. G. Kuhr, L. Licklider, and L. Amankwa, *Anal. Chem.* **65**, 277 (1993).
36. L. N. Amankwa and W. G. Kuhr, *Anal. Chem.* **65**, 2693 (1993).
37. H.-T. Chang and E. S. Yeung, *Anal. Chem.* **65**, 2947 (1993).
38. L. Licklider and W. G. Kuhr, *Anal. Chem.* **66**, 4400 (1994).
39. N. H. H. Heegaard and F. A. Robey, *Anal. Chem.* **64**, 2479 (1992).
40. W. Nashabeh and Z. El Rassi, *J. Chromatogr.* **596**, 251 (1992).
41. S. Honda, A. Taga, S. Suzuki, and K. Kahehi, *J. Chromatogr.* **597**, 377 (1992).
42. J. Bao and F. E. Regnier, *J. Chromatogr.* **68**, 217 (1992).
43. Y.-H. Chu, L. Z. Avila, H. A. Biebuyck, and G. M. Whitesides, *J. Med. Chem.* **35**, 2915 (1992).
44. L. Z. Avila, Y.-H. Chu, E. C. Blossey, and G. M. Whitesides, *J. Med. Chem.* **36**, 126 (1993).
45. C. V. Thomas, A. C. Cater, and J. J. Wheeler, *J. Liq. Chromatogr.* **16**, 1903 (1993).
46. G. E. Barker, W. J. Horvath, C. W. Huie, and R. A. Hartwick, *J. Liq. Chromatogr.* **16**, 2089 (1993).
47. N. H. H. Heegaard and F. A. Robey, *J. Immunol. Methods* **166**, 103 (1993).
48. K. Shimura and B. L. Karger, *Anal. Chem.* **66**, 9 (1994)
49. N. M. Schultz and R. T. Kennedy, *Anal. Chem.* **65**, 3161 (1993)
50. K. Shimura and K. Kasai, *Anal. Biochem.* **227**, 186 (1995)

REFERENCES

51. D. Schmalzing, W. Nashabeh, X.-W. Yao, R. Mahatre, F. E. Regnier, N. B. Afeyan, and M. Fuchs, *Anal. Chem.* **67**, 606 (1995).
52. D. Schmalzing, W. Nashabeh, and M. Fuchs, *Clin. Chem. (Winston-Salem, N. C.)* **41**, 1403 (1995).
53. Y. Yoshimoto, A. Shibukawa, H. Susagawo, S. Nitta, and T. Nakagawa, *J. Pharm. Biomed. Anal.* **13**, 483 (1995).
54. S. H. Hansen, I. Bjørnsdottir, and J. Tjørnelund, *J. Pharm. Biomed. Anal.* **13**, 489 (1995).
55. K. Ensing and A. Paulus, *J. Pharm. Biomed. Anal.* **14**, 305 (1995).
56. N. A. Guzman, *J. Liq. Chromatogr.* (1998).
57. Y.-H. Chu, L. Z. Avila, H. A. Biebuyck, and G. M. Whitesides, *J. Org. Chem.* **58**, 648 (1993).
58. J. Cai and Z. El Rassi, *J. Liq. Chromatogr.* **15**, 1179 (1992).
59. N. A. Guzman, R. A. Berg, and D. J. Prockop, *Biochem. Biophys. Res. Commun.* **73**, 279 (1976).
60. F. Lampreave, M. A. Alava, and A. Piñeiro, *Trends Anal. Chem.* **15**, 122 (1996).

CHAPTER

27

MICROBIOANALYSIS USING ON-LINE MICROREACTORS–CAPILLARY ELECTROPHORESIS SYSTEMS

LARRY LICKLIDER and WERNER G. KUHR

Department of Chemistry, University of California, Riverside, Riverside, California 92521

27.1.	Introduction	899
27.2.	Sampling Single Biological Cells	900
	27.2.1. Techniques	900
	27.2.2. Sampling with On-Line Reactions	901
27.3.	On-Capillary Assays	904
	27.3.1. General Techniques	904
	27.3.2. Applications to Single-Cell Analysis	906
27.4.	On-Line Capillary Microreactors–Capillary Electrophoresis Systems	909
	27.4.1. Advantages Over the Single-Capillary System	909
	27.4.2. Applications to Biopolymer Characterization	912
	27.4.3. A Microchip Electrophoresis System with an On-Chip Reactor	917
27.5.	Conclusions	919
List of Acronyms and Symbols		920
References		920

27.1. INTRODUCTION

In addition to being one of the most powerful bioanalytical separation methods (1–4), capillary electrophoresis (CE) excels as a tool for sampling from microenvironments (5–13). These attributes have lent impetus to the integration of CE with miniaturized enzymatic and chemical reactors for bioanalytical applications with comprehensive demands for efficient sample utilization, high selectivity, sensitivity, and speed (14–19). Toward this goal,

---

High Performance Capillary Electrophoresis, edited by Morteza G. Khaledi. Chemical Analysis Series, Vol. 146.
ISBN 0-471-14851-2 © 1998 John Wiley & Sons, Inc.

the concomitant development of on-line sample-handling procedures has allowed microreactor optimization for volumes that are commensurate with the nanoliter (nL; 10^{-9} L) to picoliter (pL; 10^{-12} L) sampling capability of CE (13,17,20–22a). As a prerequisite for CE analysis, derivatization with a fluorophore (23,24) an enzyme-mediated reaction (14,25), or a reaction with an immunoaffinity reagent (26–30) can provide improved sensitivity and additional dimensions of selectivity for detection and identification, as well as the capability to assay or to characterize biological activities. Off-line preparation of the reaction mixture can become much less effective for ultramicro sample quantities when relatively large (10^{-3} to 10^{-6} L) reaction volumes are used. This situation produces low reaction rates and also raises detection limits, since only a minute fraction of the reaction mixture can be sampled at once. The on-line integration of ultralow-volume reactors with CE can provide essential sample-handling capabilities for nanomole (nmol; 10^{-9} mol) or smaller sample amounts, in order to limit dilution and to provide optimal conditions for generating, sampling, separating, and detecting the reaction products. Burgeoning bioresearch applications for on-line microreactors–CE systems can be found in the areas of chemical analysis of single biological cells (22,31,32), enzyme assays (33,34), measurement of receptor–ligand binding constants (29,35–37), and structural analysis of biopolymers (38–41). Recently, the possibilities for conducting a wide variety of extremely-low-volume bioanalyses in a readily automated format have been expanded by microfabrication of CE-based devices for generating, separating, and detecting specific reaction products (42). A summary of the literature in the area of on-line microreactors for bioanalysis is given in this review, highlighting applications in the analysis of single biological cells and structural analysis of biopolymers at the picomole and smaller scale.

27.2. SAMPLING SINGLE BIOLOGICAL CELLS

27.2.1. Techniques

Sample introduction techniques in CE are based on either electromigration or hydrodynamic flow (1,43–45). The basic method for increasing their efficiency, particularly in regard to relatively low sample concentrations, is described in detail by Chien in Chapter 13 of this monograph. Generally, sampling efficiency is optimized by limiting the length of the introduced sample plug (less than 1% of the capillary length) and applying stacking or field-amplified conditions for sample injection as detailed in Chapter 13 (46,47).

Extremely limited analysis volumes have highlighted the advantages of CE for direct sampling from microenvironments. Small capillary diameters (2 and 5 µm) and a tapered capillary tip were employed to penetrate single large invertebrate neurons (~ 200 µm diameter) and allow electromigration of as little as 270 fL (10^{-15} L) of cytoplasm sample (6,7). This allowed measurement of several hundred attomoles (amol; 10^{-18} mol) of the neurotransmitters dopamine and serotonin using on-capillary amperometric detection. Since most biological cells have sufficiently small diameters to allow the introduction of the entire cell, most CE-based procedures for chemical analysis of single biological cells have adopted a cell injection procedure for analysis of the cell contents (8). The procedure for whole-cell injection and for lysing the cell in the capillary is shown in Figure 27.1. Ewing et al. employed electromigration to inject single intact neurons of *Planorbis corneus* (a snail) for measurement of dopamine and serotonin in the low-amol range using on-capillary amperometric detection (9). As is evident from the variety of chemical analyses that have been performed with injected cells, adaptations of the whole-cell injection technique have been very effective in preventing dilution (31).

Recently, the feasibility of employing a CE capillary to deliver ultralow volumes has been demonstrated by positioning the outlet of the capillary over an intact biological cell (48). The presence of analytes that elute from the capillary following a separation was transduced in a qualitative fashion from the cell response. The response upon binding of an analyte to the cell membrane was detected either as changes in the intracellular concentration of a fluorescent calcium indicator or as changes in the transmembrane current. The applied voltage was decoupled near the capillary outlet by creating an electrical connection on the capillary to avoid disturbing the detection region at the outlet (5,49).

27.2.2. Sampling with On-Line Reactions

A microscope and micromanipulation of the capillary have been necessary for direct sampling of intact cells, yet compensation for the labor involved in sampling individual cells has included opportunities for performing a variety of chemical and enzyme-mediated reactions on-line with a CE analysis of the reaction products (11,22,31,50). *In vivo* derivatization of intercellular thiols to form more easily detected fluorescent products proved to be effective with CE–LIF (laser-induced fluorescence) analyses of individual cells. Either monobromobimane (MBB) (11) or 2,3-naphthalenedicarboxaldehyde (NDA) (51) were added to the extracellular medium to allow their diffusion across cell membranes and reaction with intracellular thiols which are present in millimolar concentrations. Although this effectively avoids diluting the cell and may allow less abundant cell components to be targeted with *in vivo* derivati-

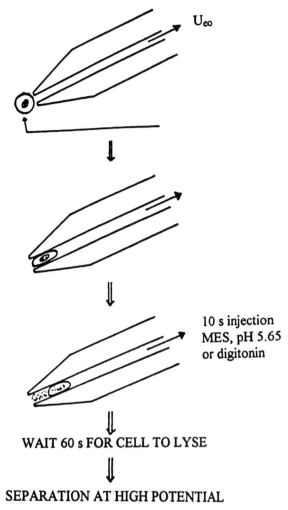

Figure 27.1. Procedure for whole-cell injection, lysing, and separation with CE; MES = 2-(N-Morpholino)ethanesulfonic acid. [From A. G. Ewing, T. G. Strein, and Y. Y. Lau, *Acc. Chem. Res.* **25**, 445 (1992), with permission.]

zation reagents, an obvious drawback is the limited flexibility in setting reaction conditions. However, several hundred amol of MBB–glutathione was determined in single human erythrocytes (< 100 fL) by *in vitro* derivatization, whole-cell sampling, and CE–LIF analysis (11). Figure 27.2 displays a result for the CE–LIF analysis of the MBB-derivatized contents of a single human

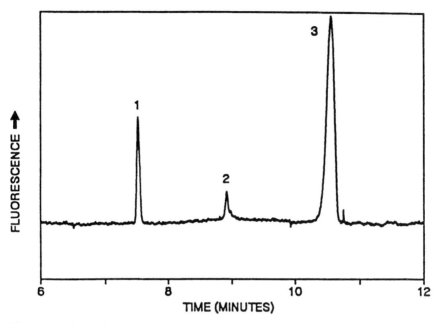

Figure 27.2. Electropherogram of MBB-derivatized contents of a single human erythrocyte using an Ar ion laser to excite fluorescence for detection. [From B. L. Hogan and E. S. Yeung, *Anal. Chem.* **64**, 2843 (1992), with permission.]

erythrocyte. NDA–glutathione and NDA–glutamylcysteine have also been determined in single-cell volumes of several pL under similar experimental conditions (51).

A procedure was reported to perform fluorescence derivatization of primary amines after electromigrating single PC12 cells (several pL volume) into the capillary. The cells reportedly adhered to the capillary wall near the inlet end, which ensured their contact with a plug injection of internal standards and reagents needed to lyse the cell or to form fluorescent derivatives (22). This procedure afforded considerable flexibility to set reaction conditions and to characterize the reaction products. CE–LIF analysis of the reaction mixture allowed determinations of dopamine (a neurotransmitter that is produced in large quantities in the PC12 cell line) at the femtomole (fmol; 10^{-15} mol) level and a number of intracellular amino acids in the amol range. Derivatization in the capillary was reported to reduce dilution of the reaction mixture more than an order of magnitude, in comparison with an off-line procedure that utilized a pneumatic microsyringe to transfer large invertebrate neurons (1 nL volume) and low-nL volumes of reagents and internal standards to 500 nL vials for fluorescence derivatization. CE–LIF and open-tubular liquid chromatography with LIF detection were used to characterize the microvial reaction mixtures (10,52). In light of the extremely limited material present in mammalian cell lysates (e.g., human erythrocytes are 4 orders of magnitude smaller by volume than invertebrate neurons), the sample portion remaining in the reaction vial following sampling represents a significant limitation to analysis of trace components in cell lysates (53). Accordingly, the technique of lysing a cell in the CE capillary inlet has been adopted for many types of single-cell analyses where the practical mass limit of detection depends on using the smallest possible volume for preparation of a reaction mixture.

27.3. ON-CAPILLARY ASSAYS

27.3.1. General Techniques

The basic mechanism operating in a CE separation, e.g., differences in migration rates that correspond to electrophoretic mobility differences, also provides a basis for rapid initiation of reactions following sample introduction (3,21,33,36,54). Pioneering work in on-capillary enzymatic reactions was reported by Bao and Regnier (33). This work is discussed by Harmon and Regnier in Chapter 28 of the present monograph along with many applications of electrophoretically mediated microanalysis (EMMA). The unique biological specificities of enzymes and of immunoaffinity techniques are well suited to small-volume determinations since minimal sample preparation is necessary. Additional selectivity has been obtained by performing an initial separation to resolve a sample mixture before electrophoretically mixing the reactants to initiate reactions (55,56).

Absorbance detection for on-capillary assays of complex biological matrices has the disadvantage of limited detector selectivity, which is compounded by substantial losses in sensitivity with decreasing capillary diameter (57). Consequently, on-capillary LIF excitation has been employed for EMMA with detection limits in the zeptomole (zmol; 10^{-21} mol) range and synthetic fluorogenic substrates have been employed for enzymes which do not form intrinsically fluorescent products (58). Fluorescence measurements in complex biological matrices can provide additional selectivity when implemented with time-resolved LIF detection methods (58). Figure 27.3 shows a comparison of electropherograms obtained for *Escherichia coli* supernatant without (top) and with (bottom) delayed integration of the fluorescence decay.

Single-enzyme molecule detection has been demonstrated after careful optimization of assay conditions for CE–LIF detection of the fluorescent product generated after a 1 h incubation period (59). In this application the

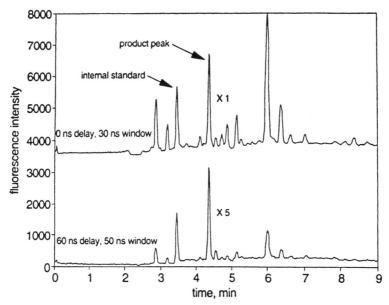

Figure 27.3. Dual electropherograms for *E. coli* supernatant assayed for leucine aminopeptidase. Incubation under zero potential for 10 min. Data was collected with no delay (upper) and 60 ns delay (lower). [From K. J. Miller, I., Leesong, J. Bao, F. E. Regnier, and F. E. Lytle, *Anal. Chem.* **65**, 3269 (1993), with permission.]

rate of product formation is proportional to the enzyme turnover number, which ranges from 1 to 10^3 per second for the majority of enzymes, 10^7 per second being the upper limit (60). The minute reaction volume in the analysis capillary is a distinct advantage for measuring single enzyme molecules in exceedingly dilute ($\sim 10^{-17}$ M) solutions because the measurement precision is essentially shot noise limited and therefore is proportional to the square root of the number of product molecules in the detection volume (61).

CE has also proven to be rapid, sensitive means of obtaining kinetic and thermodynamic binding constants for a wide variety of receptor–ligand systems (3,26,35–37, 62–64). On-capillary binding assays are facilitated by having sufficiently rapid formation and dissociation rates, which permit equilibrium to be established on the CE timescale (34,65). Extensive coverage of on-capillary binding assays is found in Chapter 29 by Gao et al. in this monograph [see also Chu et al. (66)]. Generally, migration times of the analytes are measured as the concentration of a buffer additive is varied. Alternatively, absorbance measurements of the free or bound analyte can be used to determine binding constants, in which cases the analyte concentration must be known (63). Affinity capillary electrophoresis (ACE) has also been used for estimates of binding rate constants by comparison of the relative peak widths obtained under real and simulated conditions (29,34,65).

An on-capillary immunoaffinity technique has been developed and utilized for analysis of cyclosporin in tears (67). Immobilization of a monoclonal antibody (MAB) to the capillary inner wall at the inlet end allowed convenient sample introduction and repeatable use of the same MAB preparation. After a 20 nL tear sample was introduced and during a short incubation, cyclosporin and several of its metabolites formed immunocomplexes with the MAB. Then the remaining sample components were rinsed from the capillary. Finally, the immunocomplexes were dissociated by introduction of low-pH buffer, and the antigens were separated in the same buffer. Recoveries based on standards added to tear samples were determined to be better than 80% at the low ng/mL levels monitored in actual tear samples. Immunoaffinity electropherograms shown in Figure 27.4 were obtained by UV absorbance detection and demonstrate that rapid binding and dissociation rates were possible for five antigens in tear fluid. The simplicity of this design for immunoaffinity capillary electrophoresis (IACE) is an advantage over a design reported earlier for antibody immobilization on a solid support packed into a short capillary segment for IACE analyses of uric acid and methamphetamine in urine (15). The use of a packed capillary segment in-line with the analysis capillary can induce substantial effects on the CE analysis (68).

CE conditions (pH, ionic strength, buffer additives) can provide considerable flexibility for accomplishing a wide range of bioanalyses (69–73). However, most bioanalyses must contend to some extent with sample components that adsorb readily to the capillary wall. This phenomenon produces shifts in migration times, peak broadening, and—in worst cases—loss of the sample. Many procedures have been reported for modification of the capillary wall to prevent interactions with solutes (4,74,75). Regnier's group performed EMMA with a polymer/detergent coating applied to the silica surface or in polyacrylamide gel-filled capillaries to block adsorption of proteins [see Wu and Regnier (55); also Towns et al. (76)]. Protein adsorption has also been minimized by performing CE in relatively high ionic strength buffers containing certain buffer additives (77–79). A common strategy to lessen protein–wall interactions is to work at a buffer pH that is above the pI of the protein or below the pK_a of the silica (80,81).

27.3.2. Applications to Single-Cell Analysis

Enzyme-mediated and immunospecific reactions initiated in the CE capillary have permitted selected components in single-cell lysates to be monitored by

(a)

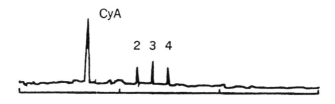

(b)

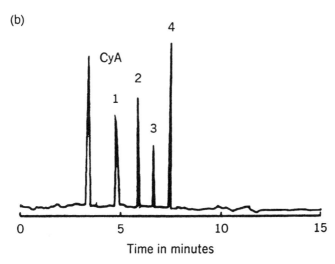

Figure 27.4. IACE electropherograms for tear fluid samples from (a) a patient with no clinical signs of cyclosporin A (CyA) toxicity and (b) a patient during an episode of systemic toxicity. Identified peaks are for CyA; and 1–4 are for CyA metabolites. [From T. M. Phillips and J. J. Chmielinska, *Biomed. Chromatogr.* **8**, 244 (1994), with permission.]

formation, separation, and detection of specific products. Mixing the reactants electrophoretically allows for a preseparation of classes of analytes such as isoenzymes that share a common reactivity (55). By application of an EMMA technique with on-capillary LIF detection it has been possible to distinguish zmol quantities of lactate dehydrogenase (LDH) isoenzymes in single human erythrocytes (56). An on-capillary immunoassay of glucose-6-phosphate dehydrogenase (G6PDH) in single human erythrocytes was carried out that

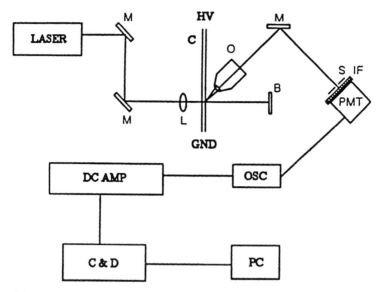

Figure 27.5. Instrument schematic for particle-counting immunoassay in a capillary. Major components are a CE system (C, 65 cm length, 20 μm i.d. capillary; GND, ground; HV, high-voltage power supply), a light-scattering detection system (Ar ion laser; M, mirror; B, beam stop; L, focusing lens; O, 10 × microscope objective; S, 1 mm slit; IF, 488 nm interference filter; PMT, photomultiplier tube), and a data-collection system (OSC, oscilloscope; DC AMP, coupled amplifier and converter; C&D, counter and discriminator; PC, personal computer). [From Z. Rosenzweig, and E. S. Yeung, *Anal. Chem.* **66**, 1772 (1994), with permission.]

utilized antibody-coated latex particles added to the CE buffer (82). The instrument schematic for that experiment is shown in Figure 27.5. The specificity for the antigen was demonstrated by injecting a sample containing G6PDH and LDH and then separating these before allowing an incubation period. Only the G6PDH zone produced agglutinated particle counts attributed to the immunoreaction. After sampling and lysing of a cell, electrophoresis was performed for several minutes in 20 μm i.d. capillary to allow G6PDH to mix with the particles suspended in the electrophoresis buffer. The reaction was then reportedly complete following a 30 min incubation, as was evident from observation of maximum agglutination of the particles upon electromigration to the detection end of the capillary. By careful calibration of a laser-based light-scattering detector for particle counting, a detection limit of ∼ 600 G6PDH molecules in single erythrocytes was determined.

DNA labeled with a fluorescent dye has been added to the CE buffer and utilized for a DNA–protein binding assay of amol quantities of a DNA-binding protein in single sea urchin eggs (50). Sampling of an egg was followed by an injection of the fluorescent DNA probe and buffer additives needed to lyse the cell membrane or to promote specific complex formation. The reaction mixture was incubated for a short time before separation and detection of the monovalent and divalent protein–DNA probe complexes. The mass sensitivity of the on-capillary assay was reported to be at least 2 orders of magnitude better than conventional gel electrophoresis mobility shift assays of DNA-binding proteins.

27.4. ON-LINE CAPILLARY MICROREACTORS–CAPILLARY ELECTROPHORESIS SYSTEMS

27.4.1. Advantages Over the Single-Capillary System

The application of electrophoresis for mixing reactants in the analysis capillary complements its nanoliter or smaller sampling capabilities and also can accomplish a separation of individual reaction zones for greatest efficiency in assaying mixtures (55,66). However, incompatibilities can be expected if the optimal reaction conditions (pH, ionic strength, and buffer additives) differ significantly from conditions that facilitate analysis of the reaction mixture. A report of an on-capillary enzymatic digestion of protein utilizing pepsin had a detection limit of 10 fmol by CE–LIF analysis of the underivatized reaction mixture (81). Although pepsin has its maximum activity near pH values that allow a separation of complex protein digests, it is a relatively nonspecific enzyme, which limits its utility in this application (83). Trypsin, which is widely used to cleave specific peptide bonds in proteins, cannot be useful in an on-capillary digestion. This is because trypsin has optimal activity at pH values between 8 and 9 whereas the optimal separation buffer pH for tryptic digests is found in a much lower range of pH values—between 2 and 5 (84).

Another critical consideration for performing a reaction in the analysis capillary must be the rate of the reaction. Long incubation times promote dilution of the reaction zone by diffusion and also idle the analysis capillary. Quite stringent temporal demands can be placed on the operation of the CE capillary. For example, the excellent temporal resolution (< 60 s) possible for monitoring *in vivo* biochemical events with an on-line microdialysis–CE system can be seen as the forte of CE in its application (12,85). Recently, an in-line injection tee was utilized to introduce *o*-phthaldialdehyde for rapid derivatization of primary amines on-line with an microdialysis–CE–LIF analysis (22a). To allow rapid sampling by a 25 μm i.d. CE separation capillary, the reaction mixture eluted from a short length (8 cm) of 75 μm i.d. fused-silica reaction capillary into a gap junction (75 μm width) after an optimized

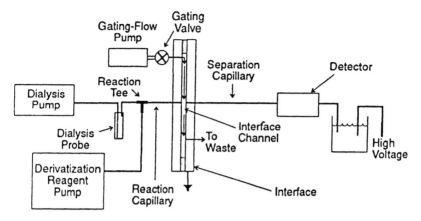

Figure 27.6. Block diagram of a microdialysis–CE–LIF system. [From M. W. Lada and R. T. Kennedy, *Anal. Chem.* **68**, 2790 (1996), with permission.]

reaction time (∼ 2 min). As shown in the instrument schematic in Figure 27.6, continuous cross-flow of electrophoresis buffer through the gap junction prevented the derivatization mixture from entering the CE capillary during a separation that required between 45 and 180 s. During sampling, the flow of buffer through the gap junction was stopped to allow electromigration of sample into the CE capillary. The extremely low flow rates for the microdialysis procedure (79 nL/min) allowed relative recoveries of nearly 100% for low-molecular-weight analytes present at micromolar levels in the extracellular fluid. Perfusion of the tissue sample at such low flow rates improves the absolute recovery (mass sensitivity) by minimizing dilution of the dialysate sample and also minimizes the impact of the microdialysis procedure on the tissue. The interface between derivatization and CE capillaries maintained independent solution conditions in each without sacrificing maximum temporal resolution (between 45 and 180 s), which was limited by the separation parameters.

The principle advantages of on-capillary reactions have been the powerful sample-handling capabilities for extremely limited sample volumes. To retain these advantages and address the need for reaction conditions and analysis conditions to be independently optimized, it has been advantageous to couple the CE capillary with a microreactor capillary of similar diameter. The feasibility of transferring samples efficiently at a capillary junction on the sampling end of the CE capillary has been demonstrated for electrokinetic (electromigration) (13,20,49,86) and hydrodynamic injections (17,39). Fluid gap junctions (either with or without cross-flow supplied to the junction) have been used to integrate CE analyses with a variety of capillary microreactors

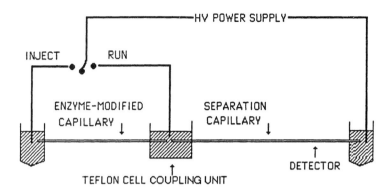

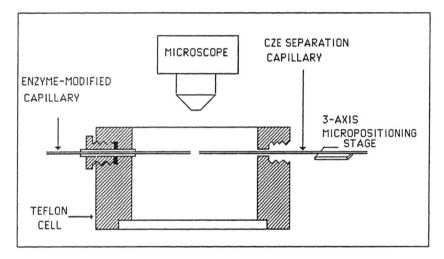

Figure 27.7. Schematic of the coupled-capillary CE instrument (upper). Electrical connections to the high-voltage power supply allow the electrophoresis voltage to be applied across both capillaries during sampling or across the CE capillary for a separation. Two 50 μm i.d. fused-silica capillaries are coupled in the Teflon cell (lower) via a fluid gap junction. The enzyme-modified capillary is secured to the cell, and the CZE (capillary zone electrophoresis) capillary is mounted on a micropositioning stage that allows submicrometer step resolution. The stereomicroscope is used to inspect the junction for alignment to within a predetermined width (50–100 μm). [From L. Licklider and W. G. Kuhr, *Anal. Chem.* **66**, 4403 (1994), with permission.]

(18,22a,41,87,88), with a microdialysis capillary (85), with an affinity preconcentration capillary (30), and with capillary liquid chromatography (89).

To illustrate the feasibility of sampling extremely limited reaction volumes, Figure 27.7 contains a schematic of an instrument coupling a microreactor capillary and a CE capillary via a fluid gap junction. To transfer a microreactor sample for CE in this design, it was necessary first to study electromigration in the gap junction using quantitative spectroscopic imaging of a dye sample (13). With precise alignment and a 50–100 µm gap width, the major parameter found to affect sampling efficiency was the electric field strength in the gap junction. Before sampling is done from the microreactor, the fluid in the coupling cell (Figure 27.7) is replaced with a fluid of lower ionic strength than the CE buffer. Application of a high voltage across both capillaries then results in a relatively high field strength in the gap junction. After electromigration of sample into the CE capillary, the gap fluid is replaced with the higher-ionic-strength CE buffer before the CE analysis. The fluid gap allows independent operation of each capillary for greatest efficiency, since it serves as the CE inlet reservoir with an electrical connection to the high-voltage power supply at the coupling cell. In contrast, hydrodynamic sampling from a microreactor requires that the capillaries tightly butt together in a tandem configuration that prevents independent optimization of conditions in each capillary. A capillary junction at the detection end of the CE capillary also requires each capillary to function in tandem in order to efficiently transfer the separation medium to a detection capillary. The post-CE junction always results in conflicts between maintenance of separation efficiency and the conditions needed in the detection capillary, e.g., an electrical connection to facilitate electrochemical detection (90), introduction of a matrix for mass spectrometry (91), introduction of derivatization reagents for fluorescence detection (92–95), or introduction of an enzyme to conduct substrate assays (96,97).

27.4.2. Applications to Biopolymer Characterization

One of the most important microbioanalysis applications of CE has been the separation of the complex fragment mixtures that result from enzymatic degradation of proteins (3,98) and DNA (99,100). Several proteolytic enzymes, trypsin in particular, selectively cleave peptide bonds to release specific fragments that furnish a unique fingerprint or peptide map following a CE analysis of the peptide digest. Tryptic peptide mapping using CE has allowed purity confirmation of protein sequences produced by recombinant DNA techniques (101,102) and has been utilized to distinguish structural variants among proteins (84,103–107). Microscale procedures for peptide mapping of proteins have become increasingly popular due to the use of polyacrylamide gel electrophoresis (108) and high-performance liquid chromatography (109) to isolate pmol levels of proteins.

There has been considerable success in developing procedures for digestion of pmol and smaller protein samples using trypsin immobilized by covalent bonds to a solid support (14,16,110). Immobilized enzymes are particularly well suited for digestion of minute samples because the peptide digests can be readily recovered from extremely low reaction volumes without contamination due to enzymes autodigestion. Immobilized enzymes generally retain their stability over relatively long periods of use, and very high reaction rates can be obtained with utilization of support materials having high surface areas (111,112). CE-based peptide mapping analyses of low-pmol protein samples have been accomplished with trypsin immobilized in a relatively large volume ($\sim$ 1 mL) of agarose packed into a 1 mm i.d. column (14). A protein sample was transported by gravity through the column, and the reaction products were collected at the outlet in a large volume that required an off-line volume reduction step before transfer to the CE analysis. The digestion procedure was optimized for any relatively complex protein by preceding the digestion with off-line preparation of the protein to irreversibly unfold its polypeptide chains (113). Preparation consisted of dissolving purified protein in several µL of solution containing either performic acid or tributylphosphine with 2-methylaziridine. Excess reagents were removed before introducing the denatured protein to the immobilized trypsin column.

To fully exploit the ultramicro sampling capability of CE, an immobilized enzyme microreactor must have—

1. A volume commensurate with the CE analysis
2. A large surface area for enzyme immobilization
3. Mild conditions for enzyme immobilization
4. Efficient mass transport to facilitate interaction with enzyme.

The first requirement demands that the microreactor be coupled on-line with the CE capillary in order to avoid dilution of nL or smaller sample volumes. All four requirements have been met with the development of immobilized enzyme fused-silica capillary microreactors (ECM) of 50 µm i.d. for protein digestion (38,87,88) and for modification of nucleic acids (17,40). The use of an open-tubular ECM design has simplified the mass transport properties in the ECM and also facilitates sampling by either electromigration or hydrodynamic flow (13,17). Capillaries of 50 µm i.d. provide a high ratio of surface area to volume for immobilization of the enzyme at the capillary interior wall. For enzymatic digestion of proteins, ECM designs have utilized avidin–biotin technology to attach a variety of biotinylated enzymes to the wall (16, 88, 114). Biotin is bound covalently to the wall before addition of avidin, which is a tetrameric protein with four identical biotin-binding sites. Finally, a biotinylated enzyme is rinsed through under exceptionally mild conditions and immobilized via free biotin-binding sites on the wall. Insertion of a long-chain spacer between the enzyme and its biotin tether can be helpful in retaining the affinity of substrates for the immobilized enzyme (115) and also facilitating the biotin–avidin interaction (114). Avidin has a cross-sectional area of ~ 2000 Å^2; therefore a monolayer surface coverage of avidin in an ECM of 20 nL/cm (length) would correspond to an avidin concentration of $\sim 6.5\,\mu$M in the ECM. This would allow binding of a biotinylated enzyme to produce an equivalent enzyme concentration.

The first reported on-line ECM–CE procedure utilized a variety of enzymes for analyses of low-pmol quantities of nucleic acids (17,40). The integration of the two capillaries relied on a tandem coupling configuration in which the ECM is tightly attached to the CE capillary. Enzyme immobilization was performed under relatively mild conditions by acylation of amino groups on the enzyme using aldehyde groups already incorporated into the ECM wall. To show the utility of employing enzymes with different substrate specificities for identification of a nucleotide or nucleotide sequence of interest, dinucleotide mixtures were injected by hydrodynamic flow. The reaction time and the corresponding product yield depended on the flow rate and ECM length. Once the reaction mixture entered the CE capillary, the CE capillary was detached and the inlet end was dipped in a buffer reservoir for application of the separation voltage (17). A linear relationship between the detected product and the substrate concentration injected was demonstrated over a small range. Such an enzymophoresis format could conceivably become extremely useful in biosensing applications (111,112).

The enzymophoresis format was also used for efficient generation and separation of a complex mixture of oligonucleotides generated by digestion of a transfer ribonucleic acid (tRNA^{Phe}). The electropherogram in Figure 27.8 shows a result obtained for the on-line digestion and mapping experiment with UV absorbance detection. It has been possible to co-immobilize two ribonuclease enzymes, one of which displayed a strong pH dependence for its activity, and thus vary substrate specificity as a function of pH (40). There are significant constraints on the choice of solution conditions in the ECM, since these must be virtually identical to the CE analysis conditions in a tandem capillary configuration.

Peptide-mapping experiments utilizing an on-line ECM–CE procedure have utilized electromigration to sample from the ECM via a fluid gap junction. This allows independent conditions (electric field, pH, buffer salts, additives) to be maintained in each capillary. The schematic in Figure 27.7 shows the instrumental design used in our laboratory. A series of electropherograms taken for an on-line trypsin ECM–CE–UV absorbance peptide-mapping experiment are shown in Figure 27.9 (88). A 50 cm length of ECM was loaded with ~ 50 pmol of bovine α-acid glycoprotein, a heavily glycosylated protein of ~ 40 kDa, which was incubated for 11 h. The four electropherograms correspond to digest aliquots electromigrated from the ECM with sampling times (form top) of 75, 45, 30, and 30 s. The bottom electropherogram

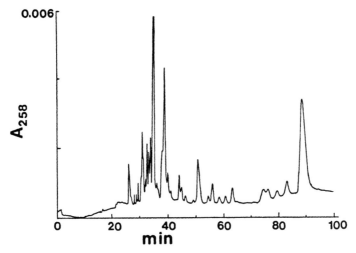

Figure 27.8. On-line digestion and mapping of tRNA^{Phe} by a tandem RNase T$_1$ capillary enzyme reactor, 20 cm long and 50 μm i.d., coupled with an 80 cm long and 50 μm i.d. polyether-modified separation capillary (50 cm effective length to the detector); tRNA^{Phe} was introduced at an unspecified concentration and allowed to flow hydrodynamically in the direction of the separation capillary for 22 min. Subsequently, the enzyme capillary was disconnected and the separation proceeded in the CE capillary. The buffer used was 0.1 M MES/0.1 M HIS/5 mM spermine, pH 7.0. The separation voltage was 15 kV. [From W. Nashabeh and Z. El Rassi, *J. Chromatogr.* **596**, 257 (1992), with permission.]

demonstrated that removal of the ECM from the coupling cell before we began the separation was unnecessary to prevent siphoning from the ECM during the analysis. It is apparent that siphoning of sample from the ECM does not occur during the separation, since fewer components are not found in the bottom trace, which corresponds to removal of the ECM after sampling. Each 30 s injection corresponds to ~10 nL of the total ECM sample, which corresponds to ~500 fmol of protein.

The digestion of complicated protein structures such as tightly folded (globular) proteins or glycosylated proteins required optimization of mass transfer processes in the ECM. This was accomplished by vibrating the ECM at low intensity during incubation of a protein sample. The rate of digestion of a globular protein, cytochrome *c*, has been investigated by varying the frequency of vibration using a piezoelectric transducer. The digestion rate was found to increase dramatically with frequency. This allowed the digestion to proceed on the same timescale as the separation of the tryptic peptides (116).

On-line detection with electrospray ionization mass spectrometry (ESI–MS) can identify the separated peptides by their mass and has shown

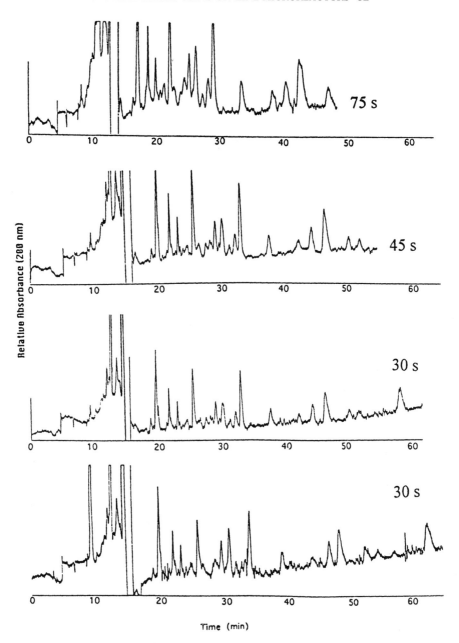

considerable promise in rapid mass fingerprinting procedures to identify a protein from its tryptic peptides (117,118). The capability to set CE conditions entirely independent of the ECM is essential for CE–MS detection, since the best sensitivity (10^{-6} M) is obtained with volatile buffers of low pH and ionic strength (119,120). The overall efficiency of on-line trypsin ECM–CE–ESI–MS peptide mapping was investigated with a bovine insulin B-chain sample. This combination allowed generation, separation, and detection of the tryptic peptides in < 1 h (38). Figure 27.10 contains a representative total ion electropherogram and the accompanying selected mass spectra showing the mass/charge ratio (m/z) of each tryptic peptide detected in a 10 s on-line injection and separation of digest. Several nL of digest was consumed for a single CE–MS analysis corresponding to ∼ 1 pmol of the digest sample.

27.4.3. A Microchip Electrophoresis System with an On-Chip Reactor

Microfabricated CE-based system that are configured as intersecting micrometer sized channels in glass or fused quartz utilize electromigration to sample from an external reservoir, to mix the sample with reagents in a microreactor channel, and to transfer an aliquot of the reaction mixture for CE with on-chip LIF detection (18,41). Thus, the entire analysis can be carried out with a sample volume of several hundred pL by precisely controlling the voltages applied at buffer reservoirs (20,86,121). The exciting developments in this area are reviewed by Ramsey in Chapter 18 of this monograph, including procedures necessary to maintain a high degree of control over reaction conditions on-chip. A reaction channel of several mm length gave a reaction volume of 700 pL for enzymatic digestion of 30 amol of a plasmid DNA sample in an on-chip DNA restriction fragment analysis (41). The low reaction volume and the effectiveness of electrophoretic mixing served to minimize diffusional limitations on the reaction rate and allowed reaction times as short as 9 s. Reaction times were decided by the electrophoretic mobilities of the reactants and the field strength in the reaction channel.

Figure 27.9. Electropherograms showing an on-line peptide-mapping experiment to investigate (from the top) the duration of injection (75,45,30 s) and (at bottom) the effect of removing the microreactor (50 cm length, 360 μm o.d., 50 μm i.d.) from the coupling well after a 30 s injection. Tryptic digest of bovine α_1-acid glycoprotein, 2 mg/mL in 0.1 M Tris HCl, pH 8.1, and 11 h incubation. Before each run, −15 kV was applied over the coupled system to inject a digest sample into a 62 cm long, 360 μm o.d., 50 μm i.d. polybrene-modified capillary for CE. The coupling cell was filled with 20 mM ammonium citrate, pH 3.2, for injection, and with 50 mM citric acid/250 mM betaine, pH 3.2, during each separation. The gap width was 50 μm for the 75 s injection and 100 μm for the other injections. [From L. Licklider and W. G. Kuhr, *Anal. Chem.* **66**, 4406 (1994), with permission.]

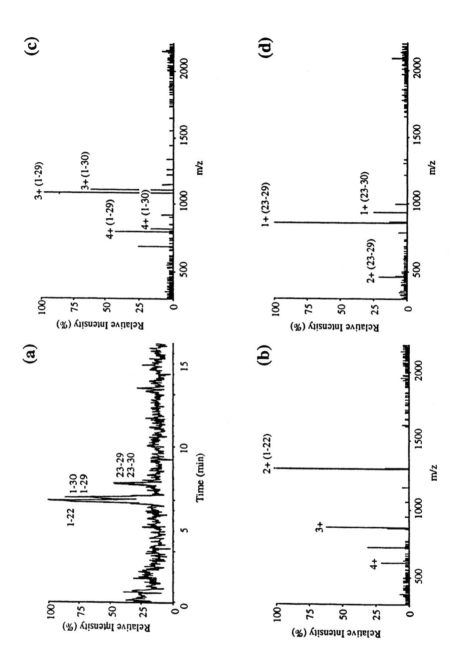

CE analyses that occur on the millisecond-to-second timescale and their concomitant requirements for high separation efficiency have placed great demands on the sampling efficiency provided by microchip CE designs (20,86). The channel intersection utilized in microchip designs permits precise sample injections and also allows reagents and additives to be dispensed from multiple reservoirs. By the influx of charged species and/or bulk solution from reservoirs, solution conditions can be dynamically adjusted in the channels. The stringent requirements for the separation medium used for the separation and detection of restriction enzyme fragments were met by the influx of additives from three different buffer reservoirs (41). On-chip sample preparation will likely benefit from the flexibility that microchip designs allow for control over solution conditions.

27.5. CONCLUSIONS

The material covered here included a wide variety of bioanalyses based on the integration of miniaturized enzymatic and chemical reactors with CE analyses. In analyses of single biological cells, on-line sampling procedures have enabled on-capillary reaction-based analyses that due to sheer minuscule sample size cannot be achieved by combining off-line reactions and CE. On-capillary reactions and capillary microreactors integrated with CE can contribute unprecedented efficiencies in generating, separating, and detecting specific reaction products. The absence of sample handling and the minimal volume of on-capillary bioanalyses are particularly important advantages with ultramicroscale sample amounts that now extend to single enzyme molecules.

Independent optimization of the reaction and analysis conditions has been especially important in applications with comprehensive demands for efficient sample utilization, high selectivity, sensitivity, and speed. Multiple capillary approaches have been highly successful in combining relevant reaction chemistries with integrated sampling and analysis of the reaction products. The success of microchip electrophoresis devices supports an expectation that a virtually unlimited variety of bioanalyses may be accomplished with speed and precision in an automated format.

Figure 27.10. On-line trypsin microreactor–CE–MS peptide-mapping experiment. (a) Total-ion electropherogram (TIE) showing separation of the tryptic fragments of bovine insulin. The numbers correspond to the sequence for each fragment numbered from the N terminus of the B-chain. The ion spray mass spectra correspond to peaks in the TIE, (b) the early-eluting component, and (c,d) unresolved later-eluting components. [From L. Licklider, W. G. Kuhr, M. P. Lacey, T., Keough, M. P. Purdon, R. Takigiku, *Anal. Chem.* **67**, 4176 (1995), with permission.]

LIST OF ACRONYMS AND SYMBOLS

Acronyms or Symbol	Definition
ACE	affinity capillary electrophoresis
CE	capillary electrophoresis
CyA	cyclosporin A
CZE	capillary zone electrophoresis
ECM	enzyme (fused-silica) capillary microreactors
EMMA	electrophoretically mediated microanalysis
ESI–MS	electrospray ionization mass spectrometry
G6PHD	glucose-6-phosphate dehydrogenase
HV	high-voltage (power supply)
LDH	lactate dehydrogenase
LIF	laser-induced fluorescence
MAB	monoclonal antibody
MBB	monobromobimane
MES	2-(N-morpholino)ethanesulfonic acid
m/z	mass/charge ratio
NDA	2,3-naphthalenedicarboxaldehyde
TIE	total-ion electropherogram
UV	ultraviolet

REFERENCES

1. J. W. Jorgenson and K. D. Lukacs, *Anal. Chem.* **53**, 1298 (1981).
2. J. W. Jorgenson and K. D. Lukacs, *Science* **222**, 266 (1983).
3. P. D. Grossman, J. C. Colburn, H. H. Lauer, R. G. Nielsen, R. M. Riggin, G. S. Sittampalam, and E. C. Rickard, *Anal. Chem.* **61**, 1186 (1989).
4. C. A. Monnig and R. T. Kennedy, *Anal. Chem.* **66**, 280R (1994).
5. R. A. Wallingford and A. G. Ewing, *Anal. Chem.* **59**, 678 (1987).
6. T. M. Olefirowicz and A. G. Ewing, *Anal. Chem.* **62**, 1872 (1990).
7. J. B. Chien, R. A. Wallingford, and A. G. Ewing, *J. Neurochem.* **54**, 633 (1990).
8. T. M. Olefirowicz and A. G. Ewing, *Chimia* **45**, 106 (1991).
9. A. G. Ewing, T. G. Strein, and Y. Y. Lau, *Acc. Chem. Res.* **25**, 440 (1992).
10. R. T. Kennedy, M. D. Oates, B. R. Cooper, B. Nickerson, and J. W. Jorgenson, *Science* **246**, 57 (1989).
11. B. L. Hogan and E. S. Yeung, *Anal. Chem.* **64**, 2841 (1992).
12. B. L. Hogan, S. M. Lunte, J. F. Stobaugh, and C. E. Lunte, *Anal. Chem.* **66**, 596 (1994).
13. W. G. Kuhr, L. J. Licklider, and L. N. Amankwa, *Anal. Chem.* **65**, 275 (1993).

14. K. A. Cobb and M. Novotny, *Anal. Chem.* **61**, 2226 (1989).
15. N. A. Guzman, M. A. Trebilcock, and J. P. Advis, *J. Liq. Chromatogr.* **14**, 997 (1991).
16. L. N. Amankwa and W. G. Kuhr, *Anal. Chem.* **64**, 1610 (1992).
17. W. Nashabeh and Z. El Rassi, *J. Chromatogr.* **596**, 241 (1992).
18. S. C. Jacobson, R. Hergenroder, A. W. Moore, and J. M. Ramsey, *Anal. Chem.* **66**, 4127 (1994).
19. D. J. Harrison, K. Fluri, K. Seiler, Z. H. Fan, C. S. Effenhauser, and A. Manz, *Science* **261**, 895 (1993).
20. D. J. Harrison, A. Manz, Z. Fan, H. Ludi, and H. M. Widmer, *Anal. Chem.* **64**, 1926 (1992).
21. B. J. Harmon, D. H. Patterson, and F. E. Regnier, *Anal. Chem.* **65**, 2655 (1993).
22. S. D. Gilman and A. G. Ewing, *Anal. Chem.* **67**, 58 (1995).
22a. M. W. Lada and R. T. Kennedy, *Anal. Chem.* **68**, 2790 (1996).
23. B. Nickerson and J. W. Jorgenson, *HRC & CC, J., High Resolut. Chromatogr., Chromatogr. Commun.* **11**, 878 (1988).
24. L. N. Amankwa, M. Albin, and W. G. Kuhr, *Trends Anal. Chem.* **11**, 114 (1992).
25. H. E. Schwartz, K. Ulfelder, F. J. Sunzeri, M. P. Busch, and R. G. Brownlee, *J. Chromatogr.* **559**, 267 (1991).
26. N. H. H. Heegaard and F. A. Robey, *Anal. Chem.* **64**, 2479 (1992).
27. N. M. Schultz, *Anal. Chem.* **65**, 3161 (1993).
28. N. H. H. Heegaard and F. A. Robey, *J. Liq. Chromatogr.* **16**, 1923 (1993).
29. N. H. Heegaard, *J. Chromatogr.* **680**, 405 (1994).
30. L. J. Cole and R. T. Kennedy, *Electrophoresis* **16**, 549 (1995).
31. E. S. Yeung, *Acc. Chem. Res.* **27**, 409 (1994).
32. O. Orwar, H. A. Fishman, N. E. Ziv, R. H. Scheller, and R. N. Zare, *Anal. Chem.* **67**, 4261 (1995).
33. J. Bao and F. E. Regnier, *J. Chromatogr.* **608**, 217 (1992).
34. L. Z. Avila, Y. H. Chu, E. C. Blossey, and G. M. Whitesides, *J. Med. Chem.* **36**, 126 (1993).
35. Y.-H. Chu and G. M. Whitesides, *J. Org. Chem.* **57**, 3524 (1992).
36. Y.-H. Chu, L. Z. Avila, H. A. Biebuyck, and G. M. Whitesides, *J. Med. Chem.* **35**, 2915 (1992).
37. Y. H. Chu, L. Z. Avila, H. A. Biebuyck, and G. M. Whitesides, J. Org. Chem. **58**, 648 (1993).
38. L. J. Licklider, W. G. Kuhr, M. P. Lacey, T. Keough, M. P. Purdon, and R. Takigiku, *Anal. Chem.* **67**, 4170 (1995).
39. L. N. Amankwa, K. Harder, F. Jirik, and R. Aebersold, *Protein Sci.* **4**, 113 (1995).
40. E. Mechref and Z. El Rassi, *Electrophoresis* **16**, 2164 (1995).
41. S. C. Jacobson and J. M. Ramsey. *Anal. Chem.* **68**, 720 (1996).

42. J. M. Ramsey, S. C. Jacobson, and M. R. Knapp, *Nat. Med.* **1**, 1093 (1995).
43. D. J. Rose and J. W. Jorgenson, *Anal. Chem.* **60**, 642 (1988).
44. X. Huang, W. F. Coleman, and R. N. Zare, *J. Chromatogr.* **480**, 95 (1989).
45. D. S. Burgi and R. L. Chien, *Anal. Chem.* **63**, 2042 (1991).
46. R. L. Chien and D. S. Burgi, *J. Chromatogr.* **559**, 141 (1991).
47. D. Burgi and R.-L. Chien, *Anal. Chem.* **63**, 2042 (1991).
48. J. B. Shear, H. A. Fishman, N. L. Allbritton, D. Garigan, R. N. Zare, and R. H. Scheller, *Science* **267**, 74 (1995).
49. M. C. Linhares and P. T. Kissinger, *Anal. Chem.* **63**, 2076 (1991).
50. J. Xian, M. G. Harrington, E. H. Davidson, *Proc. Natl. Acad. Sci. U.S.A.* **93**, 86 (1996).
51. O. Owar, H. A. Fishman, N. E. Ziv, R. H. Scheller, and R. N. Zare, *Anal. Chem.* **67**, 4261 (1995).
52. R. T. Kennedy and J. W. Jorgenson, *Anal. Chem.* **60**, 1521 (1988).
53. B. R. Cooper, J. A. Jankowski, D. J. Leszczyszyn, R. M. Wightman, and J. W. Jorgenson, *Anal. Chem.* **64**, 691 (1992).
54. L. Z. Avila and G. M. Whitesides, *J. Org. Chem.* **58**, 5508 (1993).
55. D. Wu and F. E. Regnier, *Anal. Chem.* **65**, 2029 (1993).
56. Q. Xue and E. S. Yeung, *Anal. Chem.* **66**, 1175 (1994).
57. M. Albin, P. D. Grossman, and S. E. Moring, *Anal. Chem.* **65**, 489 (1993).
58. K. J. Miller, I. Leesong, J. Bao, F. E. Regnier, and F. E. Lytle, *Anal. Chem.* **65**, 3267 (1993).
59. Q. Xue and E. S. Yeung, *Nature (London)* **373**, 681 (1995).
60. A. Fersht, *Enzyme Structure and Mechanism*, p. 121. Freeman, New York, 1985.
61. D. B. Craig, J. C. Y. Wong, and N. J. Dovichi, *Anal. Chem.* **68**, 697 (1996).
62. S. Honda, A. Taga, K. Suzuki, S. Suzuki, and K. Kakehi, *J. Chromatogr.* **597**, 377 (1992).
63. J. C. Kraak, S. Busch, and H. Poppe, *J. Chromatogr.* **608**, 257 (1992).
64. D. J. Rose, *Anal. Chem.* **65**, 3545 (1993).
65. V. Matousek and V. Horejsi, *J. Chromatogr.* **245**, 271 (1982).
66. Y.-H. Chu, D. P. Kirby, and B. L. Karger, *J. Am. Chem. Soc.* **117**, 5419 (1995).
67. T. M. Phillips and J. J. Chmielinska, *Biomed. Chromatogr.* **8**, 242 (1994).
68. M. A. Strausbauch, J. P. Landers, and P. J. Wettstein, *Anal. Chem.* **68**, 306 (1996).
69. P. Grossman, D. J. C. Colburn, and H. H. Lauer, *Anal. Biochem.* **179**, 28 (1989).
70. R. G. Nielsen, E. C. Rickard, P. F. Santa, D. A. Sharknas, and G. S. Sittampalam, *J. Chromatogr.* **539**, 177 (1991).
71. E. C. Rickard, M. M. Strohl, and R. G. Nielsen, *Anal. Biochem.* **197**, 197 (1991).
72. J. P. Landers, R. P. Oda, T. C. Spelsberg, J. A. Nolan, and K. J. Ulfelder, *Biotechniques* **14**, 98 (1993).
73. S. K. Basak and M. R. Ladisch, *Anal. Biochem.* **226**, 51 (1995).

74. W. G. Kuhr and C. A. Monnig, *Anal. Chem.* **64**, 389R (1992).
75. X. W. Yao, D. Wu, and F. E. Regnier, *J. Chromatogr.* **636**, 21 (1993).
76. J. K. Towns, J. Bao, and F. E. Regnier, *J. Chromatogr.* **559**, 227 (1992).
77. J. S. Green and J. W. Jorgenson, *J. Chromatogr.* **478**, 63 (1989).
78. J. A. Bullock and L.-C. Yuan, *J. Micorcolumn Sep.* **3**, 241 (1991).
79. G. N. Okafo, H. C. Birrell, M. Greenaway, M. Haran, and P. Camilleri, *Anal. Biochem.* **219**, 201 (1994).
80. P. D. Grossman, K. J. Wilson, G. Petrie, and H. H. Lauer, *Anal. Biochem.* **173**, 265 (1988).
81. H.-T. Chang and E. S. Yeung, *Anal. Chem.* **65**, 2947 (1993).
82. Z. Rosenzweig and E. S. Yeung, *Anal. Chem.* **66**, 1771 (1994).
83. J. C. Powers, A. D. Harley, and D. V. Myers, in *Acid Proteases; Structure, Function, and Biology* (J. Tang, ed.)., Vol. 95, p. 141. Plenum, New York, 1977.
84. T. E. Wheat, P. M. Young, and N. E. Astephen, *J. Liq. Chromatogr.* **14**, 987 (1991).
85. M. W. Lada and R. T. Kennedy, *J. Neurosci. Methods* **63**, 147 (1995).
86. S. C. Jacobson, R. Hergenroder, L. B. Koutny, R. J. Warmack, and J. M. Ramsey, *Anal. Chem.* **66**, 1107 (1994).
87. L. A. Amankwa and W. G. Kuhr, *Anal. Chem.* **65**, 2693 (1993).
88. L. J. Licklider and W. G. Kuhr, *Anal. Chem.* **66**, 4400 (1994).
89. A. V. Lemmo and J. W. Jorgenson, *Anal. Chem.* **65**, 1576 (1993).
90. R. A. Wallingford and A. G. Ewing, *Anal. Chem.* **59**, 1762 (1987).
91. M. A. Moseley, L. J. Deterding, K. B. Tomer, and J. W. Jorgenson, *J. Chromatogr.* **516**, 167 (1990).
92. S. L. Pentoney, X. Huang, D. S. Burgi, and R. N. Zare, *Anal. Chem.* **60**, 2625 (1988).
93. B. Nickerson and J. W. Jorgenson, *J. Chromatogr.* **480**, 157 (1989).
94. M. Albin, R. Weinberger, E. Sapp, and S. Moring, *Anal. Chem.* **63**, 417 (1991).
95. S. D. Gilman, J. J. Pietron, and A. G. Ewing, *J. Microcolumn Sep.* **6**, 373 (1994).
96. A. Emmer and J. Roeraade, *J. Chromatogr. A* **662**, 375 (1994).
97. A. Emmer and J. Roeraade, *Chromatographia* **39**, 271 (1994).
98. E. C. Rickard and J. K. Towns, in *New Methods in Peptide Mapping for the Characterization of Proteins* (W. S. Hancock, ed.), p. 97. CRC Press, Boca Raton, FL, 1996.
99. A. S. Cohen, D. Najarian, J. A. Smith, and B. L. Karger, *J. Chromatogr.* **458**, 323 (1988).
100. A. S. Cohen, D. R. Najarian, A. Paulus, A. Guttman, J. A. Smith, and B. L. Karger, *Proc. Natl. Acad. Sci. U.S.A.* **85**, 9660 (1988).
101. J. Frenz, S.-L. Wu, and W. S. Hancock, *J. Chromatogr.* **480**, 379 (1989).
102. R. G. Nielsen, R. M. Riggin, and E. C. Rickard, *J. Chromatogr.* **480**, 393 (1989).
103. P. Ferranti, A. Malorni, P. Pucci, S. Fanali, A. Nardi, and L. Ossicini, *Anal. Biochem.* **194**, 1 (1991).

104. C. Grenot, A. De Montard, T. Blanchere, M. R. DeRavel, E. Mappus, and C. Y. Cuilleron, *Biochemistry* **31**, 7609 (1992).
105. H. Tu, Q. Xia, and W. Li, *Shengwu Huaxue Zazhi* **8**, 570 (1992).
106. G. A. Ross, P. Lorkin, and D. Perrett, *J. Chromatogr.* **636**, 69 (1993).
107. I. J. Chang, H. B. Gray, and M. Albin, *Anal. Biochem.* **212**, 24 (1993).
108. M. W. Hunkapiller, E. Lujan, F. Ostrander, and L. E. Hood, *Methods Enzymol.* **91**, 227 (1983).
109. F. E. Regnier, *Methods Enzymol.* **91**, 137 (1983).
110. C. B. Quern, *J. NIH Res.* **7**, 59 (1995).
111. L. D. Bowers, *Anal. Chem.* **58**, 513A (1986).
112. M. N. Gupta and B. Mattiasson, *Methods Biochem. Anal.* **36**, 1 (1992).
113. K. A. Cobb and M. V. Novotny, *Anal. Chem.* **64**, 879 (1992).
114. N. M. Green, *Adv Protein Chem.* **29**, 121 (1975).
115. P. Zhuang and D. A. Butterfield, *Biotechnol. Prog.* **8**, 204 (1992).
116. L. Licklider and W. G. Kuhr, in preparation.
117. C. G. Edmonds, J. A. Loo, R. R. O. Loo, H. R. Udseth, C. J. Baringa, and R. D. Smith, *Biochem. Soc. Trans.* **19**, 943 (1991).
118. J. R. D. Yates, S. Speicher, P. R. Griffin, and T. Hunkapiller, *Anal. Biochem.* **214**, 397 (1993).
119. P. Thibault, C. Paris, and S. Pleasance, *Rapid Commun. Mass Spectrom.* **5**, 484 (1991).
120. M. A. Moseley, J. W. Jorgenson, J. Shabanowitz, and D. F. Hunt, *J. Am. Soc. Mass Spectrom.* **3**, 289 (1992).
121. S. C. Jacobson, R. Hergenroder, L. B. Koutny, and J. M. Ramsey, *Anal. Chem.* **66**, 1114 (1994).

CHAPTER

28

ELECTROPHORETICALLY MEDIATED MICROANALYSIS

BRYAN J. HARMON

Biotechnology Process Engineering Center, Massachusetts Institute of Technology, Cambridge, Massachusetts 02139

FRED E. REGNIER

Department of Chemistry, Purdue University, West Lafayette, Indiana 47907

28.1.	Coupling On-Line Chemical Reactions to Capillary Electrophoresis	926
28.2.	Determinations of Enzymes by Electrophoretically Mediated Microanalysis	928
	28.2.1. Constant-Potential Mode	928
	28.2.2. Zero-Potential Mode	931
	28.2.3. Simultaneous Determinations of Multiple Enzymes by Electrophoretically Mediated Microanalysis	935
	28.2.4. Alternative Detection Methods	936
28.3.	Enzymatic Determinations of Substrates by Electrophoretically Mediated Microanalysis	938
28.4.	Complexometric Determinations of Inorganic Ions by Electrophoretically Mediated Microanalysis	941
28.5.	Determinations of Single Cells by Electrophoretically Mediated Microanalysis	941
List of Acronyms		942
References		942

The trend today in biotechnology, biochemistry, cell biology, and molecular biology is toward a decrease in both sample volume and analyte concentration in analytical determinations. This trend is driven by efforts to (i) measure chemical events in single cells and cellular organelles and (ii) study substances that are biologically active *in vivo* at 10^{-12} M or less. Consequently,

---

High Performance Capillary Electrophoresis, edited by Morteza G. Khaledi. Chemical Analysis Series, Vol. 146.
ISBN 0-471-14851-2 © 1998 John Wiley & Sons, Inc.

analyte dilution during analysis, application of ultrahigh-sensitivity detection systems, and integration of muiltiple steps required in such determinations will be major issues in the future. This chapter will examine how electrophoretically mediated microanalysis (EMMA) paired with high-sensitivity laser-induced fluorescence, electrochemical, and chemiluminescence detection systems may provide a solution to these problems. It will be shown that obstacles often associated with high-sensitivity detection systems, including absence of detectable analyte functionality and interference from matrix components, can be overcome in the EMMA format, particularly through the use of biospecific reagents widely applied in biochemistry and clinical chemistry.

Clinical chemistry teaches us that a large number of analytes, some of which have no directly detectable functionality, may be determined in serum with a relatively small number of similar methods. This remarkable feat is achieved in one of the following ways: (i) by selecting enzymes, cofactors, or substrates that are specific for the analyte and that yield detectable species; (ii) by the use of artificial substrates that are converted to detectable species; or (iii) through coupled enzyme reactions that produce a detectable species. It will be shown below that the very large, rich source of chemical-reaction-based analytical methods developed in biochemistry and clinical chemistry laboratories over the past four decades may be adapted to microanalysis through electrophoretic mixing in capillaries. Electrophoretic mixing is a process in which a zone of higher electrophoretic mobility is caused to overtake and merge with a zone of lower electrophoretic mobility. When electrophoretic mixing initiates a chemical reaction in a microchannel and the product of that reaction is subsequently transported to a detector, the process has been termed *electrophoretically mediated microanalysis*.

28.1. COUPLING ON-LINE CHEMICAL REACTIONS TO CAPILLARY ELECTROPHORESIS

Although capillary electrophoresis has emerged as a powerful separation technique due to its minimal sample requirements, extremely high efficiencies, and rapid analysis times, greater information is often sought than can be provided by a simple electrophoretic separation. More specific information (e.g., determinations of enzymatic activities, enzymatic assays of substrates, and complexometric determinations of inorganic ions) and/or the formation of species that can be detected more sensitively (e.g., fluorophore labeling) can be obtained by coupling chemical reactions to the small dimensions of capillary electrophoretic systems in order to perform microscale-reaction-based chemical analyses. Initial attempts to combine on-line chemical reactions with capillary electrophoretic separations have generally utilized a reactor capillary coupled to the separation capillary. One approach has been the use of immobilized enzymes in precolumn reactors, most commonly the immobilization of proteolytic enzymes to produce protein hydrolysates for on-line peptide mapping of minute quantities of protein (1–3). However, the use of heterogeneous reactions is generally limited to substrate analysis and consequently lacks the flexibility of homogeneous methodologies for the performance of other classes of chemical reactions.

The most prevalent approach to the implementation of homogeneous reactions in capillary electrophoresis has been the use of postcolumn reactors in which the effluent from the separation capillary is merged with a stream of analytical reagents prior to detection. Although this methodology has been most commonly applied to on-line fluorophore labeling of analytes (4–9). Emmer and Roeraade (10,11) have reported the determination of enzymatic activities of glucose-6-phosphate dehydrogenase (G6PDH) and 6-phosphogluconate dehydrogenase by coupling the separation capillary to a flow of substrate solution. However, off-column reactors present several potential problems. The additive mixing of reagent solutions dilutes the analyte of interest, thereby decreasing the sensitivity of such techniques. Turbulent mixing of the effluent and analytical reagent streams and transport by laminar flow of reaction products to the detector also result in band broadening and loss of efficiency and resolution. This effect is most profound for analyses involving slow reaction kinetics (e.g., determinations of dilute enzyme solutions), as long incubation times in the postcolumn reactor are necessary to accumulate sufficient product for detection. Furthermore, this methodology requires the fabrication of a microjunction connecting the separation capillary and analytical reagent stream.

Research by Whitesides' group [see Chu et al. (12); also Avila et al. (13)] in affinity capillary electrophoresis has suggested that differences in electrophoretic mobility can be exploited to mix reagents on-column under the influence of an applied potential without experiencing the concurrent dilution or turbulence encountered in off-column reactors. Since Bao and Regnier (14) first described the methodology of EMMA for the determination of G6PDH, several reports (15–32) have exploited this phenomenon of electrophoretic mixing for the performance of on-column, homogeneous reaction-based chemical analysis using conventional capillary electrophoretic systems. In a typical EMMA experiment, electrophoretic mixing is utilized to merge zones containing the analyte and its analytical reagents; the reaction is then allowed to proceed either in the presence or absence of an applied potential; and, finally, the reaction product is transported under the influence of an applied electric field to the detector. EMMA-type methods have been reported for the determination of enzymatic activities (14–27), enzymatic assays of substrates (24–29), and complexatory determinations of inorganic ions (30–32). Since EMMA determinations of enzymes have been of greatest interest, this chapter emphasizes such analyses with brief sections describing other applications.

28.2. DETERMINATIONS OF ENZYMES BY ELECTROPHORETICALLY MEDIATED MICROANALYSIS

The following discussion describes the use of the EMMA methodology for kinetic determinations of enzymatic activities. More detailed descriptions of EMMA have been previously published utilizing both static mathematical models (24–26) and a computer-implemented dynamic model (27). EMMA determinations of enzymatic activities have been performed in two basic formats: (i) under the influence of a constant potential, or (ii) with an intermittent period in which the product is allowed to accumulate in the absence of an electric field.

28.2.1. Constant-Potential Mode

Since EMMA employs an electrophoretic capillary as a microreactor in which reagent zones are merged on the basis of differential electrophoretic mobility, the enzymatic analyte and appropriate substrate(s) must be introduced into the capillary as distinct zones in appropriate positions so that the various reagents approach and engage each other under the influence of an applied electric field. As initially described by Bao and Regnier (14), the most common mode of electrophoretic mixing has been the *continuous-engagement* or *reagent-filled capillary* approach. In this methodology, the capillary and/or appropriate buffer reservoirs are filled with the analytical reagents (i.e., substrates), and the analyte (i.e., enzyme) is introduced as a plug by means of traditional capillary electrophoretic sample introduction techniques such as hydrodynamic or electrokinetic injection. For the determination of alcohol dehydrogenase (ADH) (25) depicted in Figure 28.1, the enzyme demonstrated a lower net mobility than either of the required substrates—ethanol and NAD^+. Thus, the substrates were maintained in the anodic buffer reservoir, and ADH was injected at the anodic inlet so that the reagent zones electrophoretically merged under potential. Substrates that exhibit lower net mobilities than the enzyme must be initially placed in the capillary and cathodic buffer reservoir. The continuous-engagement mode offers maximal incubation time as electrophoretic mixing continues throughout the analyte's traversal of the

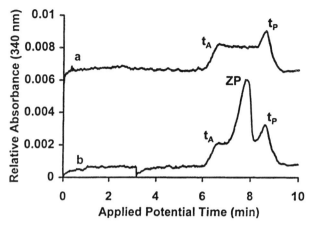

Figure 28.1. (a) Constant-potential and (b) zero-potential EMMA determinations of ADH: t_A and t_P indicate the temporal limits of the product detection window defined by the net mobilities of the analyte and product, respectively, while ZP signifies the product formed during a 3.8 min zero-potential incubation performed following the application of potential for 3 min; t_P also indicates the product formed at the enzyme and substrate interface prior to the initial application of an electric field. [Adapted from B. J. Harmon, Ph.D. Thesis, Purdue University, West Lafayette, IN (1994).]

capillary. Alternatively, Avila and Whitesides (18) have described "transient-engagement" or "plug–plug" electrophoretic mixing, in which an analytical reagent zone of finite width is utilized. As a result, the incubation proceeds only during that fleeting time period for which the finite analyte and analytical reagent regions are merged.

Since solutes electrophorese essentially independently of the bulk solution, electrophoretic mixing allows an analyte zone to encounter several times its own volume in the analytical reagent without experiencing the concurrent change in volume and resulting dilution experienced in additive mixing methods. Since the enzyme is not depleted in the reaction and the substrate is replenished by electrophoretic mixing, a relatively constant rate of product formation is observed as the analyte zone traverses the analytical reagent region. However, the enzyme and product generally also differ in electrophoretic mobility, and thus the product does not accumulate in the vicinity of the enzyme zone under potential. As a result, the observed product profile for an EMMA determination of an enzyme performed under constant potential is typically a plateau, as seen in Figure 28.1a, rather than a peak.

Although the reaction continues throughout the enzyme zone's traversal of the capillary in the continuous-engagement mode, only product that is formed between the point of injection and detector can be observed (assuming that the product migrates in the same direction as the enzyme). Consequently, the product plateau is observed only during an exclusive "detection time window" whose temporal limits, t_P and t_A in Figure 28.1, are defined by the net mobilities of the product and analyte, respectively. If the reaction occurs at the point of injection, the product traverses essentially the entire length of the capillary and is observed at a detection time corresponding to its net migration velocity (v_P):

$$t_P = \frac{L_d}{v_P} = \frac{L_d}{(\mu_{ep,P} + \mu_{eo})E} \qquad (28.1)$$

where L_d is the effective length of the capillary (i.e., distance from the injection inlet to the detection position); $\mu_{ep,P}$ is the electrophoretic mobility of the product; μ_{eo} is the electroosmotic mobility; and E is the electric field strength. For analyses in which the enzyme and substrate zones are initially positioned adjacently, t_P is typically signified by an additional accumulation of product (indicated by the peak labeled t_P in Figure 28.1a), which forms due to diffusional interpenetration of the reagent zones and concurrent reaction prior to the initial application of the electric field (14). At the opposite extreme, if the reaction occurs just as the analyte reaches the detection position, the analyte itself traverses essentially the entire effective length of the capillary and the product is detected at a time reflective of the net migration velocity of the analyte (v_A):

$$t_A = \frac{L_d}{v_A} = \frac{L_d}{(\mu_{ep,A} + \mu_{eo})E} \qquad (28.2)$$

where $\mu_{ep,A}$ is the electrophoretic mobility of the analyte. Any product that is formed between the point of injection and the detection position must be observed within the detection time window defined by t_P and t_A. In the determination of ADH, the detected product (NADH) possessed a lower net mobility than the enzyme, and thus t_A and t_P corresponded to the initial and final observed product, respectively, as indicated in Figure 28.1a.

In the continuous-engagement mode, both the height (H_P) and area (A_P) of the plateau product profile obtained under constant potential are directly proportional to the activity of the enzyme (assuming that Michaelis–Menten kinetics and enzyme-saturating concentrations of substrates are maintained), and thus either parameter can be utilized for quantitation (25):

$$H_P = \frac{\varepsilon_P b}{(\mu_{ep,P} - \mu_{ep,A})E\pi r^2} k_{cat}[E]V_{inj} \qquad (28.3)$$

$$A_P = \frac{\varepsilon_P b L_d}{(\mu_{ep,A} + \mu_{eo})(\mu_{ep,P} + \mu_{eo})E^2\pi r^2} k_{cat}[E]V_{inj} \qquad (28.4)$$

where ε_P is the molar absorptivity of the product; b is the cell pathlength (i.e., typically the capillary diameter for on-column absorbance detection); r is the capillary radius; k_{cat} is the turnover number of the enzyme for the substrate; [E] is the total concentration of enzyme; and V_{inj} is the injection volume. As suggested by Eq. (28.3) and (28.4), the sensitivity of this method is directly proportional to the turnover number of the enzyme and the injection volume, with the height of the plateau is inversely proportional to the rate of electrophoretic separation of the analyte and product.

Although EMMA techniques have generally employed zonal analyte injections, moving-boundary EMMA (23) has been described for the determination of leucine aminopeptidase (LAP). In this technique, the capillary was initially filled with the analyte solution while the faster-migrating substrate (L-leucine-p-nitroanilide) was maintained in the inlet reservoir. Upon application of an electric field, electrophoretic merging of the reagents proceeded and the detected product (p-nitroaniline) was transported to the detector. In contrast to the plateau product profile obtained for constant-potential EMMA enzyme determinations utilizing zonal injections, the constant-potential moving-boundary EMMA method yielded a triangular product profile whose area, maximum height, inclining slope, and declining slope were each directly proportional to the enzymatic activity. The moving-boundary technique offered more than an order of magnitude greater concentration sensitivity than a zonal injection method and thus could be exploited for rapid enzymatic determination, as the use of elevated electric field strengths and short capillaries allowed for a 24 s kinetic determination of LAP (23).

28.2.2. Zero-Potential Mode

In order to prevent electrophoretic broadening of the product zone and the concurrent loss in sensitivity indicated by Eq. (28.3), the potential can be removed at any time following electrophoretic mixing. In the absence of an electric field, there is no electrophoretic separation of the enzyme and product, and thus product accumulates in the vicinity of the enzyme zone. This EMMA methodology is termed the *zero-potential* mode (14). Depending upon the turnover number of the enzyme for the particular substrates(s), an enzyme can typically generate thousands of times its own concentration of product in a few minutes of zero-potential incubation. Detection is achieved by reapplying a potential and electrophoretically transporting the product to the detector. As seen in Figure 28.1b, which depicts the zero-potential EMMA determination of ADH (25), the zero-potential product accumulation appears as a peak superimposed upon the plateau product profile resulting from the applied potential time periods required for electrophoretic mixing and detection. The area of the zero-potential peak is directly proportional to the activity of the enzyme if we assume that Michaelis–Menten kinetics and enzyme-saturating concentrations of substrates are maintained (25):

$$A_P = \frac{\varepsilon_p b}{(\mu_{ep,P} + \mu_{eo})E\pi r^2} k_{cat}[E]V_{inj}t_{ZP} \tag{28.5}$$

where t_{ZP} is the duration of the zero-potential incubation. Equation (28.5) indicates that the sensitivity of the zero-potential method is directly proportional to the turnover number of the enzyme, the injection volume, and the incubation time. Zero-potential EMMA methods employing UV–Vis absorbance detection of reaction products typically yield lower limits of detection on the order of 10^{-17} to 10^{-19} mol of enzyme (14,15).

Although Eq. (28.5) suggests that the sensitivity of a zero-potential EMMA determination can be increased by proportionately lengthening the incubation time, the sensitivity of zero-potential EMMAs is also limited by product diffusion during prolonged incubations. In order to minimize this restriction, Wu and Regnier (15) have utilized capillary gel electrophoretic systems for EMMA determinations of alkaline phosphatase (ALP) (shown in Figure 28.2) and β-galactosidase. The high viscosity of polyacrylamide gel matrices restricted diffusion and consequent losses of efficiency during zero-potential incubations. As a result, longer incubation periods were allowed and concurrent increases in sensitivity were observed relative to EMMA determinations utilizing open-tubular capillaries. A 2 h zero-potential mode offered a lower limit of detection of approximately 5×10^{-20} mol of ALP by absorbance detection of p-nitrophenol.

Although differential electrophoretic mobility between an analyte and its product results in broadening of the product zone under the influence of an electric field, it also offers EMMA a unique capability to selectively control the detection time of a zero-potential product (26). Upon the application of an electric field, the analyte traverses a portion of the capillary from the injection point to the position at which reaction occurs (d_{rxn}):

$$d_{rxn} = (\mu_{ep,A} + \mu_{eo})Et_{rxn} \tag{28.6}$$

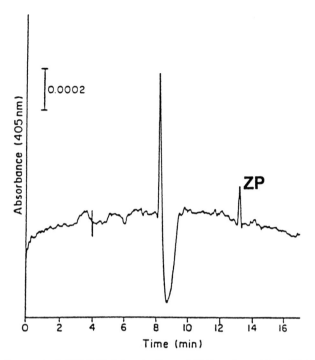

Figure 28.2. Zero-potential EMMA determinations of ALP performed in a 4% linear polyacrylamide gel-filled capillary: ZP indicates the product formed during a 2 h zero-potential incubation. [Adapted from D. Wu and F. E. Regnier, *Anal. Chem.* **65**, 2029 (1993).]

where t_{rxn} is the time of applied potential after which the reaction occurs. Following the chemical reaction, either in the presence or absence of potential, the resulting product migrates the remaining distance to the detection position under the influence of an applied electric field. Therefore, the product is detected in a total applied potential time period (t_{tot}) that is equal to the sum of the times during which it was effectively transported with the velocity of the analyte (i.e., t_{rxn}) and during which it migrated with the velocity of the product to the detector (t_{det}):

$$t_{tot} = t_{rxn} + t_{det} = \frac{d_{rxn}}{v_A} + \frac{L_d - d_{rxn}}{v_P} = \frac{L_d + (\mu_{ep,P} - \mu_{ep,A})Et_{rxn}}{(\mu_{ep,D} + \mu_{eo})E} \quad (28.7)$$

or, in terms of the temporal limits of the detection window,

$$t_{tot} = t_P + \frac{t_A - t_P}{t_A} t_{rxn} \quad (28.8)$$

Equations (28.7) and (28.8) indicate that, if an analyte and its product differ in electrophoretic mobility, the detection time of the reaction product can be selectively maneuvered within the detection time window defined by t_A amd t_P by controlling when the reaction is allowed to occur (i.e., by selecting the value of t_{rxn} at which to perform the zero-potential incubation). In contrast, a nonreacting interferant must traverse the entire capillary with its single net migration velocity regardless of when the analytical reaction occurs. This selective control of product detection time can be utilized to resolve EMMA product profiles from nonreacting matrix interferants, as shown in Figure 28.3, which depicts the control of product detection time by use of the zero-potential mode for the determination of LAP (26). The constant-potential EMMA determination of LAP is shown in Figure 28.3a. Although, due to the low activity of LAP in the sample, the product plateau was not clearly apparent, it extended from 7.0 (t_P) to 15.0 (t_A) min. In zero-potential mode determinations depicted in Figure 28.3b–d, the time at which potential was removed (i.e., t_{rxn}) was altered to selectively maneuver the detection time of the zero-potential product (the peak indicated by ZP) within this detection time window. However, since the nonreacting matrix components migrated with the same velocity prior to and following the zero-potential incubation, their observation times were not affected by the chosen values of t_{rxn}.

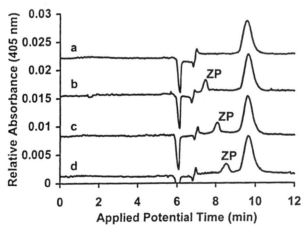

Figure 28.3. (a) Constant-potential and (b–d) zero-potential EMMA determinations of LAP: ZP indicates the product formed during 10 min zero-potential incubations performed following the application of potential for (b) 1, (c) 2, and (d) 3 min. [Adapted from B. J. Harmon, I. Leesong, and F. E. Regnier, *Anal. Chem.* **66**, 3797 (1994).]

28.2.3. Simultaneous Determinations of Multiple Enzymes by Electrophoretically Mediated Microanalysis

Since it is frequently the case that multiple enzymes in a sample must be assayed, EMMA can exploit the inherent separative capability of capillary electrophoresis to achieve simultaneous multiple enzyme analyses (14,19,22,26) by performing either pre- or postreaction separations. If two or more enzymes that yield the same reaction product differ in electrophoretic mobility, they can be electrophoretically separated in the presence of their necessary substrate(s) prior to undergoing concurrent zero-potential incubations. The reaction products then accumulate at the unique spatial positions of their respective enzymes and thus reach the detection position at different times upon the reapplication of an electric field. Alternatively, if the enzymes cannot be electrophoretically separated but yield products differing in electrophoretic mobility, zero-potential product accumulations that form at the same time and position within the capillary can be separated during their postreaction transport to the detector.

Numerous assays of proteases, peptidases, phosphatases, sulfatases, and oligosaccharide-cleaving enzymes have been developed that are based upon the production of nitrophenol. Consequently, simultaneous EMMA assays of such enzymes can be performed by separating the enzymes prior to undergoing concurrent zero-potential incubations. Figure 28.4 depicts the simulta-

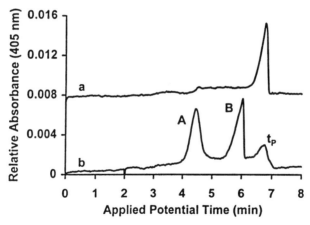

Figure 28.4. Simultaneous zero-potential EMMA determinations of ALP and β-galactosidase performed (a) without and (b) with an initial application of potential for 2 min to separate the analytes prior to a 10 min zero-potential incubation: A and B indicate the zero-potential nitrophenol products corresponding to alkaline phosphatase and β-galactosidase, respectively, while t_p signifies the product formed at the enzyme and substrate interfaces prior to the initial application of an electric field. [Adapted from B. J. Harmon, I. Leesong, and F. E. Regnier, *Anal. Chem.* **66**, 3797 (1994).]

neous zero-potential EMMA determinations of ALP and β-galactosidase (26) utilizing nitrophenol-releasing substrates (p-nitrophenyl phosphate and o-nitrophenyl-β-galactopyranoside, respectively) to yield products of similar electrophoretic mobility and detection characteristics (p-nitrophenol and o-nitrophenol, respectively). In Figure 28.4a, the zero-potential incubation was performed at the point of injection so that the reaction products comigrated essentially the entire length of the capillary. However, in Figure 28.4b, ALP and β-galactosidase were electrophoretically separated by the application of potential prior to initiating the zero-potential time period, and as a result their respective nitrophenol accumulations (indicated by peaks A and B, respectively) were resolved during their subsequent transport to the detector. In a similar manner, multiple NAD^+- or $NADP^+$-linked oxidoreductases (22) and the isoenzymes of a single oxidoreductase (19) which could be separated prior to simultaneous zero-potential incubations have been assayed by EMMA.

28.2.4. Alternative Detection Methods

The small dimensions of capillary electrophoretic systems and the amplifying effect of enzymatic reactions make EMMA an inherently sensitive technique for the kinetic determination of enzymatic activities. Furthermore, the use of specific chemical reactions and the ability to maneuver EMMA peaks independently of nonreacting matrix interferents makes EMMA a highly selective technique. However, the sensitivity and selectivity of EMMA can be further enhanced by the choice of detection method.

Three orders of magnitude lower detection limits relative to UV–Vis absorbance detection can be typically obtained by utilizing fluorescence detection. Yeung's group reported detection limits of 10^{-21} and 10^{-22} mol for determinations of lactate dehydrogenase (LDH) [see Xue and Yeung (19)] and G6PDH [see Tan and Yeung (20)], respectively, by detection of the laser-induced native fluorescence of zero-potential reaction products (NADH and NADPH, respectively). Whereas natural substrates frequently do not yield fluorescent products and thus are not appropriate for fluorescence-based EMMA methods, synthetic substrates yielding fluorescent products are currently available for several enzymes. Miller et al. (16) have reported a detection limit of 10^{-21} mol of LAP using laser-induced fluorescence detection of the product (4-methoxy-β-naphthylamine) of a synthetic substrate (L-leucine-4-methoxy-β-naphthylamine) accumulated during a 45-min zero-potential incubation. This study also exploited the detection method to enhance selectivity. Time resolution of the laser-induced fluorescence (i.e., a 60 ns delay in signal integration) allowed for preferential monitoring of the reaction product (26 ns lifetime) relative to interfering matrix substances exhibiting shorter fluorescent lifetimes.

In a study by Xue and Yeung (33), dilute solutions of LDH were mixed off-column with substrates, and the capillary was filled with the mixture ($\sim 10^{-17}$ M LDH). Following 1 h in the absence of potential in which product was allowed to accumulate, the resultant NADH accumulations were electrophoretically transported to the laser-induced fluorescence detector. Product peaks (typically 2×10^7 molecules of NADH) corresponding to individual enzyme molecules were detected, and it was observed that the activities of individual enzyme molecules varied by a factor of 4. Although reagent mixing was not performed electrophoretically and thus this analysis was not an EMMA methodology, it illustrated the sensitivity obtainable by combining the microdimensions of capillary electrophoretic systems, the amplifying effect of enzymatic reactions, and laser-induced fluorescence detection.

Electrochemical detection in capillary electrophoresis has also been shown to be orders of magnitude more sensitive than absorbance methods for electrochemically active species. Wu et al. (17) observed that the lower limit of detection by electrochemical detection for *p*-aminophenol, an electrochemically active product of ALP, was 100-fold lower than for *p*-nitrophenol with absorbance detection. However, only a 10-fold decrease in the detection limit was obtained for the EMMA determination of ALP using electrochemical detection compared to the absorbance detection method utilizing similar incubation times. This loss of expected sensitivity was due to the turnover number of ALP for the substrate producing the electrochemically active product (*p*-aminophenyl phosphate) being an order of magnitude lower than for the *p*-nitrophenol-yielding substrate (*p*-nitrophenyl phosphate). Thus, as indicated by Eq. (28.5), detection sensitivity is a function of the turnover number of the enzyme for the chosen substrate as well as the relative molar response of the resultant product in the chosen detection system.

A limitation of the EMMA methodology is that it generally requires that electrophoretic mixing and separation, chemical reaction, and product detection be performed at relatively similar capillary electrophoretic conditions (e.g., pH, ionic strength, or buffer additives), whereas the use of a postcolumn reactor offers a greater capability to dramatically alter the reaction and detection conditions from those of the electrophoretic separation. For this reason, Regehr and Regnier (21) have coupled the use of EMMA and postcolumn reactors for the performance of enzymatic assays with chemiluminescence detection. Using a zero-potential EMMA format, these authors reported that H_2O_2 was either produced (e.g., determinations of glucose oxidase and galactose oxidase) or consumed (e.g., determinations of catalase) by the analytical enzymatic reaction.

However, since the optimal pH for the subsequent chemiluminescence reaction differed considerably from that of the enzymatic reactions, chemiluminescence was produced by postcolumn mixing of the H_2O_2 with alkaline luminol solution. A detection limit of 10^{-20} mol was reported for the determination of catalase using this methodology. Chemiluminescence detection is appropriate for the determination of numerous enzymes, including those that either produce or consume H_2O_2, ATP, or NADH or which can be coupled to such a reaction.

28.3. ENZYMATIC DETERMINATIONS OF SUBSTRATES BY ELECTROPHORETICALLY MEDIATED MICROANALYSIS

In addition to kinetic determinations of enzymatic activities, the EMMA methodology is appropriate for other reaction-based chemical analyses of interest, including end-point enzymatic determinations of substrates. In these assays, the capillary and/or appropriate buffer reservoirs are filled with a solution containing enzyme and any necessary cofactors, the substrate is injected, electrophoretic mixing is initiated by the application of an electric field, and the resulting product is electrophoretically transported to the detector. Figure 28.5 depicts continuous-engagement, constant-potential EMMA assays of ethanol (28) determined by its oxidation to acetaldehyde catalyzed by ADH with the concurrent reduction of NAD^+ to NADH monitored at 340 nm. In contrast to the constant reaction rates and resulting plateau product profiles observed for constant-potential determinations of enzymes employing zonal injections, EMMA determinations of substrates demonstrate skewed product profiles, as seen in Figure 28.5. This profile results from the consumption of the analyte during its traversal of the enzyme region and the resultant decrease in rate of product formation. Detailed models of EMMA enzymatic determinations of substrates have been previously presented (24–27). Due to the nonamplifying nature of enzymatic reactions with respect to substrate concentration, the sensitivity of EMMA determinations of substrates does not approach that obtained for determinations of enzymes [i.e., a detection limit of 3×10^{-13} mol of ethanol was reported by absorbance detection of NADH (28)].

EMMA enzymatic determinations of substrates also offer the capability to selectively alter the detection time of the reaction product as well as to perform simultaneous assays of multiple analytes. Figure 28.6 demonstrates the control of product detection time for the constant-potential EMMA determination of ethanol (26). In Figure 28.6a, the substrate and enzyme zones were initially positioned adjacently so that electrophoretic mixing and reaction

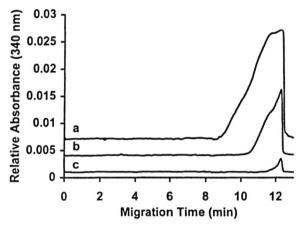

Figure 28.5. Constant-potential EMMA determinations of (a) 3, (b) 1, and (c) 0.1 mgmL^{-1} solutions of ethanol. [Adapted from B. J. Harmon, D. H. Patterson, and F. E. Regnier, *J. Chromatogr. A* **657**, 429 (1993).]

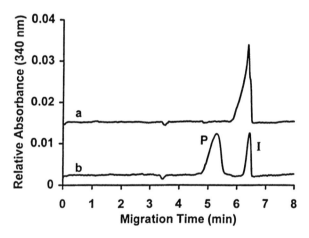

Figure 28.6. Constant-potential EMMA determinations of an ethanol solution containing interferant NADH performed (a) without and (b) with delaying of the electrophoretic engagement of the ethanol and alcohol dehydrogenase zones: P and I indicate product and interferant NADH, respectively. [Adapted from B. J. Harmon, I. Leesong, and F. E. Regnier, *Anal. Chem.* **66**, 3797 (1994).]

proceeded immediately. As a result, the product NADH comigrated with interferant NADH contained in the sample. However, since the analyte possessed a greater net mobility than NADH, the product was selectively maneuvered to an earlier detection time by placing a buffer solution devoid of enzyme between the ethanol and ADH zones prior to the initial application of

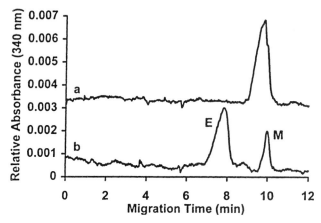

Figure 28.7. Simultaneous constant-potential EMMA determinations of ethanol and malate performed (a) without and (b) with delaying of the electrophoretic engagement of the ethanol and alcohol dehydrogenase zones: E and M indicate product NADH corresponding to ethanol and malate, respectively. [Adapted from B. J. Harmon, I. Leesong, and F. E. Regnier, *Anal. Chem.* **66**, 3797 (1994).]

the electric field in order to delay the onset of the reaction. As shown in Figure 28.6b, this methodology allowed the product and interferant NADH accumulations (the peaks indicated by P and I, respectively) to be resolved. Figure 28.7 depicts simultaneous determinations of ethanol and malate (26) catalyzed by ADH and malic dehydrogenase, respectively, to yield a common detected product (NADH). In Figure 28.7a, the reactions were each allowed to proceed at the point of injection so that their respective products comigrated. However, in Figure 28.7b, the NADH accumulations corresponding to ethanol and malate (the peaks indicated by E and M, respectively) were resolved by delaying the onset of the ethanol–ADH reaction, as previously described.

As an alternative to immobilized-enzyme precolumn reactors, Chang and Yeung (29) have utilized an EMMA-type methodology for the performance of on-column protein digestion and peptide mapping. Electrophoretic mixing was utilized to merge zones containing β-lactoglobulin and pepsin; then, following a zero-potential incubation, the resulting peptic fragments were electrophoretically separated and transported to the detector by the reapplication of an electric field. Laser-induced fluorescence detection allowed peptide mapping of as little as 10^{-14} mol of protein.

28.4. COMPLEXOMETRIC DETERMINATIONS OF INORGANIC IONS BY ELECTROPHORETICALLY MEDIATED MICROANALYSIS

The EMMA methodology has also been applied to the chemical analysis of inorganic ions via colorimetric complexatory reactions by filling the capillary and/or appropriate buffer reservoirs with a solution containing the complexing agent, injecting the analyte, and applying an electric field to initiate electrophoretic mixing and transport the resulting product to the detector. Although complexation reactions are often nearly instantaneous, continuous-engagement methods must be utilized due to the reversibility of such reactions. Liu and Dasgupta described an EMMA-type flow injection analysis method for the assay of Fe^{2+} with 1,10-phenanthroline utilizing both hydrodynamic (30) and valve-based (31) injections. Simultaneous determinations of multiple cations (e.g., Fe^{2+} and Cu^{2+}) were described either by electrophoretically separating their 1,10-phenanthroline complexes or by employing multiple complexing agents (e.g., 1,10-phenanthroline and ethylenediamine), which each displayed a preference for one of the analytes and yielded complexes of differing electrophoretic mobility. In a similar study, Patterson et al. (32) reported the continuous-engagement, constant-potential EMMA determinations of Ca^{2+} by its reaction with o-cresolphthalein complexone.

28.5. DETERMINATIONS OF SINGLE CELLS BY ELECTROPHORETICALLY MEDIATED MICROANALYSIS

Due to its minimal sample requirements and the sensitivity of laser-induced fluorescence and electrochemical detection, capillary electrophoresis has demonstrated the capability to provide analytical measurements of biological microenvironments at the single-cell level (34–36). Since chemical species of interest are typically present in single cells at 10^{-15} to 10^{-21} mol levels, such analyses challenge the detection limits of the most sensitive detection methods. Furthermore, due to the small volumes of cells (e.g., typically 10^{-9} to 10^{-13} L for large invertebrate neurons and erythrocytes, respectively) and the difficulties of manipulating minute volumes of solutions, off-column addition of reagents for the performance of reaction-based chemical analysis or derivatization reactions to improve the detection characteristics of analytes typically results in the dilution of a single cell's contents by several orders of magnitude. Thus, the capability to electrophoretically mix analytes and reagents without concurrent dilution makes EMMA well suited to such analyses. Xue and Yeung (19) have exploited the sensitivity of laser-induced fluorescence detection of NADH and the separative capability of EMMA to profile the activities of isoenzymes of LDH in single human erythrocytes. Following on-column lysis of a single cell, LDH isoenzymes were electrophoretically separated and mixed with the required substrates prior to undergoing simultaneous zero potential incubations. Thus, upon the reapplication of potential, NADH accumulations corresponding to the activities of the individual isoenzymes were resolved. Tan and Yeung (20) employed a similar methodology to assay the activity of G6PDH in single erythrocytes. Although EMMA substrate and complexometric analyses have not been reported at the single-cell level, Gilman and Ewing (37) have described the use of electrophoretic mixing and on-column cell lysis followed by a zero-potential incubation for on-column fluorophore-labeling of the contents of single rat pheochromocytoma (PC12) cells, thereby allowing dopamine and five amino acids to be quantitated at 10^{-15} mol/cell levels.

LIST OF ACRONYMS

Acronym	Definition
ADH	alcohol dehydrogenase
ALP	alkaline phosphatase
EMMA	electrophoretically mediated microanalysis
G6PDH	glucose-6-phoshate dehydrogenase
LAP	leucine aminopeptidase
LDH	lactate dehydrogenase
NAD^+	nicotinamide adenine dinucleotide
NADH	reduced nicotinamide adenine dinucleotide
$NADP^+$	nictoinamide adenine dinucleotide phosphate
UV–Vis	ultraviolet–visible (region)
ZP	zero-potential (product)

REFERENCES

1. K. A. Cobb and M. Novotny, *Anal. Chem.* **61**, 2226 (1989).
2. L. A. Amankwa and W. G. Kuhr, *Anal. Chem.* **64**, 1610 (1992).
3. L. Licklider and W. G. Kuhr, *Anal. Chem.* **66**, 4400 (1994).
4. S. L. Pentoney, Jr., X. Huang, D. S. Burgi, and R. N. Zare, *Anal. Chem.* **60**, 2625 (1988).
5. D. J. Rose, Jr. and J. W. Jorgenson, *J. Chromatogr.* **447**, 117 (1988).
6. T. Tsuda, Y. Kobayashi, A. Hori, T. Matsumoto, and O. Suzuki, *J. Chromatogr.* **456**, 375 (1988).
7. B. Nickerson and J. W. Jorgenson, *J. Chromatogr.* **480**, 157 (1989).

8. S. D. Gilman, J. J. Pietron, and A. G. Ewing, *J. Microcolumn Sep.* **6**, 373 (1994).
9. R. Zhu, and W. Th. Kok, *J. Chromatogr. A* **716**, 123 (1995).
10. Å. Emmer and J. Roeraade, *J. Chromatogr. A* **662**, 375 (1994).
11. Å. Emmer and J. Roeraade, *Chromatographia* **39**, 27 (1994).
12. Y. Chu, L. Avila, H. Biebuyck, and G. Whitesides, *J. Med. Chem.* **35**, 2915 (1992).
13. L. Avila, Y. Chu, E. Blossey, and G. Whitesides, *J. Med. Chem.* **36**, 126 (1993).
14. J. Bao and F. E. Regnier, *J. Chromatogr.* **608**, 217 (1992).
15. D. Wu and F. E. Regnier, *Anal. Chem.* **65**, 2029 (1993).
16. K. J. Miller, I. Leesong, J. Bao, F. E. Regnier, and F. E. Lytle, *Anal. Chem.* **65**, 3267 (1993).
17. D. Wu, F. E. Regnier, and M. C. Linhares, *J. Chromatogr. B: Biomed. Appl.* **657**, 357 (1994).
18. L. Z. Avila and G. M. Whitesides, *J. Org. Chem.* **58**, 5508 (1993).
19. Q. Xue and E. S. Yeung, *Anal. Chem.* **66**, 1175 (1994).
20. W. Tan and E. S. Yeung, *Anal. Biochem.* **226**, 74 (1995).
21. M. F. Regehr and F. E. Regnier, *J. Capillary Electrophor.* (in press).
22. D. H. Patterson, Ph.D. Thesis, p. 70. Purdue University, West Lafayette, IN (1994).
23. B. J. Harmon, I. Leesong, and F. E. Regnier, *J. Chromatogr. A* **726**, 193 (1996).
24. B. J. Harmon, D. H. Patterson, and F. E. Regnier, *Anal. Chem.* **65**, 2655 (1993).
25. B. J. Harmon, Ph.D. Thesis, Purdue University, West Lafayette, IN (1994).
26. B. J. Harmon, I. Leesong, and F. E. Regnier, *Anal. Chem.* **66**, 3797 (1994).
27. D. H. Patterson, B. J. Harmon, and F. E. Regnier, *J. Chromatogr. A* **732**, 119 (1996).
28. B. J. Harmon, D. H. Patterson, and F. E. Regnier, *J. Chromatogr. A* **657**, 429 (1993).
29. H.-T. Chang and E. S. Yeung, *Anal. Chem.* **65**, 2947 (1993).
30. S. Liu and P. K. Dasgupta, *Anal. Chim. Acta.* **268**, 1 (1992).
31. S. Liu and P. K. Dasgupta, *Anal. Chim. Acta.* **283**, 739 (1993).
32. D. H. Patterson, B. J. Harmon, and F. E. Regnier, *J. Chromatogr. A* **662**, 389 (1994).
33. Q. Xue and E. S. Yeung, *Nature (London)* **373**, 681 (1995).
34. R. T. Kennedy, M. D. Oates, B. R. Cooper, B. Nickerson, and J. W. Jorgenson, *Science* **246**, 57 (1989).
35. T. M. Olefirowicz and A. Ewing, *Anal. Chem.* **62**, 1872 (1990).
36. E. S. Yeung, *Acc. Chem. Res.* **27**, 409 (1994).
37. S. D. Gilman and A. G. Ewing, *Anal. Chem.* **67**, 58 (1995).

PART V

PHYSICOCHEMICAL STUDIES

CHAPTER

29

AFFINITY CAPILLARY ELECTROPHORESIS: USING CAPILLARY ELECTROPHORESIS TO STUDY THE INTERACTIONS OF PROTEINS WITH LIGANDS

JINMING GAO, MILAN MRKSICH, MATHAI MAMMEN, and GEORGE M. WHITESIDES

Department of Chemistry, Harvard University, Cambridge, Massachusetts 02138

29.1.	Background	948
29.2.	Principles of Affinity Capillary Electrophoresis	950
29.3.	Technical Issues	952
	29.3.1. Correction for Changes in Electroosmotic Flow	952
	29.3.2. Choice of Buffers	952
29.4.	Carbonic Anhydrase as a Model Protein	953
29.5.	Determination of Binding Affinity	953
	29.5.1. Using Affinity Capillary Electrophoresis to Determine the Binding Affinity of Benzenesulfonamide for Carbonic Anhydrase	953
	29.5.2. Binding of Families of Derivatives of Carbonic Anhydrase to a Positively Charged Ligand	955
	29.5.3. Using Competitive Affinity Capillary Electrophoresis to Study Binding of Small Neutral Ligands to a Receptor	955
	29.5.4. Binding of a Protein to Two Ligands: Immunoglobulin G_{2b}–N-Dinitrophenol Complexes	957
	29.5.5. Simultaneous Determination of Relative Binding Affinities of Several Isozymes of Carbonic Anhydrase to a Charged Ligand	959
	29.5.6. Simultaneous Determination of Relative Binding Affinities of Several Ligands to One Receptor	959
	29.5.7. Protein–Protein Interactions: Dimerization of Insulin	961
29.6.	Determination of Kinetic Parameters for Binding	963
29.7.	Using Affinity Capillary Electrophoresis to Measure the Effective Charge of a Protein	963
29.8.	Determination of Stoichiometry of Binding	964

High Performance Capillary Electrophoresis, edited by Morteza G. Khaledi. Chemical Analysis Series, Vol. 146.
ISBN 0-471-14851-2 © 1998 John Wiley & Sons, Inc.

29.9. Prospects and Limitations of Affinity Capillary Electrophoresis	966
Addendum	968
List of Acronyms or Abbreviations	969
References	970

29.1. BACKGROUND

Affinity capillary electrophoresis (ACE) is a procedure that uses shifts in electrophoretic mobilities of a receptor on association with a ligand to measure binding constants (1). ACE is especially useful in studying protein–ligand interactions because it requires only small quantities of proteins and provides thermodynamic and kinetic information about the complexes under physiological conditions (Table 29.1).

Affinity gel electrophoresis (AGE) (2–3) provides a historical and conceptual background for ACE. AGE uses gels that have ligands covalently attached; these ligands interact with proteins migrating in the gels. Because interaction of protein with immobilized ligand decreases the electrophoretic migration of the protein, analysis of the change in mobility of the protein as a function of the density of the immobilized ligand provides an estimate of the binding affinity. AGE can also be used to estimate the binding affinity of a soluble ligand by allowing it to compete for the protein with the immobilized ligand during electrophoresis. AGE uses small amounts of proteins, does not require radioactive or colorimetric ligands, and is applicable to a wide range of proteins. There are, however, several characteristics of the technique that have limited its use in studies of protein–ligand complexes: designing and preparing the affinity gels can be difficult; estimating the thermodynamic activity (the "effective concentration") of the ligands immobilized in the gels is not straightforward; understanding interactions between proteins and both gels and gel-immobilized ligands is complicated and difficult to relate to interactions between soluble ligands and proteins in solution; gel electrophoresis of proteins in nondenaturing conditions is also slow.

ACE (1) circumvents some of the limitations encountered in AGE (although ACE, of course, has its own limitations). ACE measures changes in the electrophoretic mobility of a soluble protein as a function of the concentration of a soluble, charged ligand in the buffer. Since ACE determines the affinity constants describing protein–ligand interactions in homogeneous solution, it is conceptually simpler than AGE and is compatible with a wider range of buffer conditions than is AGE. Further, the well-defined mechanism of separation on which CE is based makes it possible to design experiments rationally.

Table 29.1. Summary of Advantages and Disadvantages of CE and ACE in Analyzing Proteins and Protein–Ligand Interactions

Advantages	Disadvantages
• High resolution	• Narrow pathlength ($\sim 50\,\mu m$); requires (typically) $>\mu M$ concentrations of analytes
• Easily automated	• Adsorption of proteins with high values of pI or high MW to the walls of the uncoated capillaries limits applicability
• *Uncoated* capillaries are inexpensive	• Identification of analytes can be difficult
• Use of internal standards to measure electroosmotic flow (EOF) allows very reproducible measurement of mobilities	• Analytical, not preparative scale
• Small quantities of proteins and ligands required	• Background UV adsorption of analytes may be problematic in ACE
• Mobilities can be measured with impure samples	
• A wide range of types of buffer can be used	
• Buffers can simulate physiologically relevant conditions	
• Applicable to screening libraries of ligands	
• Separation based on charge and mass; rational design of experiment is possible	
• Applicable to certain mixtures of isozymes	
• When used with charge ladders, provides detailed information about electrostatics of interactions	
• Can provide information about *rates* of association and dissociation	

29.2. PRINCIPLES OF AFFINITY CAPILLARY ELECTROPHORESIS

The velocity with which a protein migrates in an electric field at unit field strength is defined as its electrophoretic mobility (μ, in cm^2 kV^{-1} s^{-1}). The value of μ for a protein is related linearly to the force on it in the electric field, and inversely to the hydrodynamic drag on it (4,5):

$$\mu \approx C_p \frac{Z}{M^\alpha} \tag{29.1}$$

The electrical force is proportional to the charge, Z, of the protein; this charge is the summation of charges carried by its electrostatic components, including the side chains of the protein, associated ligands, covalently attached saccharides, cofactors, and metal ions. The hydrodynamic drag is related to the mass (M) (or the molecular volume and shape) of a protein in solution (6). C_p is an electrophoresis constant and its value is dependent on protein shape and structure and conditions of the experiments; α is defined as the power constant of the molecular weight to which the electrophoretic mobility is inversely proportional. Recently, we have determined the values of C_p and α to be equal to 6.3 cm^2 min^{-1} kV^{-1} charge^{-1} kDa$^{0.48}$ and 0.48, respectively, under the condition of 25 mM Tris/192 mM Gly (pH = 8.3; T = 37 °C) (7).

The ACE experiment requires that the electrophoretic mobility of the protein–ligand complex differ from that of the protein (Figure 29.1). When rapid equilibration occurs between unbound and bound forms of the protein, the measured electrophoretic mobility is the *average* of the electrophoretic mobilities (appropriately weighted by the mole fractions) of the bound and unbound protein:

$$\mu = \theta \mu_{P,L} + (1 - \theta)\mu_P \tag{29.2}$$

Equation (29.2) describes this relationship for a monovalent binding system, where θ is the mole fraction of the protein bound with ligand, and $\mu_{P,L}$ and μ_P are the electrophoretic mobilities of the protein–ligand complex and the uncomplexed protein, respectively. Scatchard analysis of the change in electrophoretic mobility ($\Delta\mu_{P,L} = \mu - \mu_P$) as a function of the concentration of the ligand ([L]) in solution yields the binding constant (K_b) [Eq. (29.3)]:

$$\Delta\mu_{P,L}/[L] = K_b \Delta\mu_{P,L}^{max} - K_b \Delta\mu_{P,L} \tag{29.3}$$

The analysis based on Eq. (29.3) makes several assumptions: equilibrium is established between bound and unbound species; the rate constant for disso-

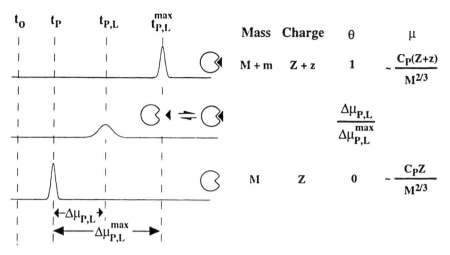

Figure 29.1. The electrophoretic mobility of a protein P (mass M and charge Z) varies with the occupancy of the binding site by the ligand L (▶, mass m and charge z). The molar fraction of the bound form of enzyme (θ) is calculated from the ratio of $\Delta\mu_{P,L}$ to $\Delta\mu_{P,L}^{max}$. Peaks broadening is observed when the value of the half-life of the protein–ligand complex is close to the migration time. The expressions for mobility assume that $(M+m)^{2/3} \approx M^{2/3}$ (that is, $M \gg m$) and that the values of C_P and α are the same for the protein and the protein–ligand complex.

ciation of the protein–ligand complex is fast relative to the time for the experiment; the concentration of the ligand in the CE buffer is sufficiently high relative to protein in the sample that the total concentration changes negligibly upon binding to protein; the interaction of the ligand and the receptor with the wall of the capillary does not significantly alter the binding of the ligand to the receptor; and the electric field does not affect the binding. In experiments where changes in the electroosmotic flow with higher values of [L] are much smaller than the changes in $\mu_{P,L}$, Eq. (29.4) can be used to determine the binding affinity of the protein–ligand pair.

$$\Delta t_{P,L}/[L] = K_b \Delta t_{P,L}^{max} - K_b \Delta t_{P,L} \qquad (29.4)$$

Association of a ligand can change the electrophoretic mobility of a protein in three ways: (i) by changing the hydrodynamic drag, while leaving the charge unchanged (binding of large, neutral ligands); (ii) by changing the charge, while leaving the hydrodynamic drag unchanged (binding of small, charged ligands); and (iii) by changing both the charge and hydrodynamic drag (binding of large, charged ligands, i.e., protein–protein interactions). Any of these three strategies can be used in rationally designing ACE experiments; we and others have used the last two strategies.

29.3. TECHNICAL ISSUES

29.3.1. Correction for Changes in Electroosmotic Flow

In the absence of changes in electroosmotic flow in the ACE experiment, it has been possible to use migration times directly to determine association constants [Eq. (29.4)]. ACE experiments, however, are often complicated by changes in the electroosmotic flow (indicated by changes in the migration time of a neutral marker) with increasing concentrations of the ligand in the buffer. This behaviour may be due to adsorption of the ligand on the wall of the capillary, changes in the dielectric constant of the buffer, or changes in the viscosity of the buffer. This effect can be corrected by adding to the sample a species whose electrophoretic mobility does not change with increasing concentrations of ligands—this species can be either neutral molecule or a protein—and calculating electrophoretic mobilities from the times of migration of the noninteracting marker (t_{marker}) and the receptor ($t_{receptor}$) using the following equation:

$$\mu = \frac{L_t L_d}{V}\left(\frac{1}{t_{marker}} - \frac{1}{t_{receptor}}\right) \qquad (29.5)$$

where L_t is the total length of the capillary; L_d is the length of the capillary from the injection end to the detector; and V is the voltage that is applied across the capillary. We have used this strategy employing mesityl oxide as neutral noninteracting marker to measure association constants for the binding of CA to benzenesulfonamide ligands (8), SH_3 to peptides (9), and IgG_{2b} to DNP (10).

29.3.2. Choice of Buffers

Proteins with high values of pI (p$I > 8$) or with high molecular weight ($>50\,kDa$) tend to adsorb to the walls of the uncoated fused-silica capillaries when the pH of the buffer is below ~ 7. Several strategies help to overcome this problem. The simplest method is to increase the pH of the electrophoresis buffer (while remaining in the range in which the protein is still in its native conformation). For values of pH higher than the pI of the protein, the protein will carry a net negative charge and will have a smaller tendency to adsorb electrostatically to the negatively charged wall. Buffer additives, such as the organic zwitterions in combination with inorganic salts developed by Bushey and Jorgenson (11), decrease the adsorption of proteins. Some of these zwitterions (e.g., betaine) are commercially available and inexpensive. The requirement that these zwitterions be used at high concentrations ($>0.5\,M$)

makes it necessary to examine the possibility that they might influence ligand–receptor interactions. We note, however, that a high concentration of zwitterionic materials and organic materials may be more representative of intracellular conditions than are the simple buffers often used in binding assays (10). For proteins with high values of pI, noncovalent adsorption of polycations on the wall of the capillary can reduce adsorption (12). Although covalently modified capillaries can markedly reduce adsorption in some instances, they are expensive and are often not stable over the course of many experiments.

29.4. CARBONIC ANHYDRASE AS A MODEL PROTEIN

We have used carbonic anhydrase (CA; EC 4.2.1.1, from human and bovine erthrocytes) (13) as a model protein for developing and testing new techniques in ACE. Several crystal structures of CA and CA–ligand complexes are known (14). CA has a molecular weight of 30 kDa, and a number of isozymes that differ in pI are commercially available. CA is inhibited by para-substituted arylsulfonamides with values of K_b ranging between 10^5 and $10^9 \, M^{-1}$; charged derivatives of these ligands are easily synthesized (Scheme I). Arylsulfonamide ligands bind CA in a 15 Å deep, conically shaped cleft; the protein does not undergo significant conformational changes upon binding the ligands (14).

29.5. DETERMINATION OF BINDING AFFINITY

29.5.1. Using Affinity Capillary Elecrtrophoresis to Determine the Binding Affinity of Benzenesulfonamide for Carobonic Anhydrase

We used ACE to determine the association constants of several benzenesulfonamide ligands [**1–7** (see Scheme I)] to CA. Binding of the charged ligands **1–3** to CA results in changes in electrophoretic mobility that correlate with the charge of the ligand. Ligands **1** and **2** carry opposite charges; upon binding of these ligands to CA, the electrophoretic mobility shifts by equal amounts but in opposite directions. Ligand **3** has a charge of -2 and results in a shift in the electrophoretic mobility of CA that is twice the magnitude of that observed upon binding of ligand **2**. Scatchard analysis of these changes in electrophoretic mobility as a function of the concentration of ligand added to the buffer yields association constants for the complexes [Eq. (29.3)].

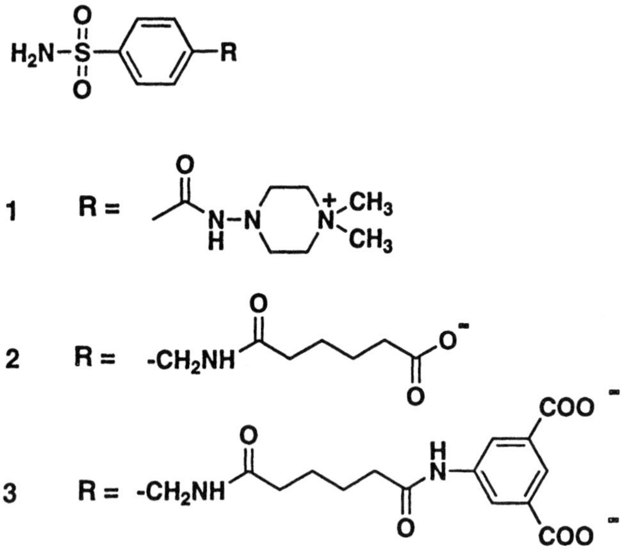

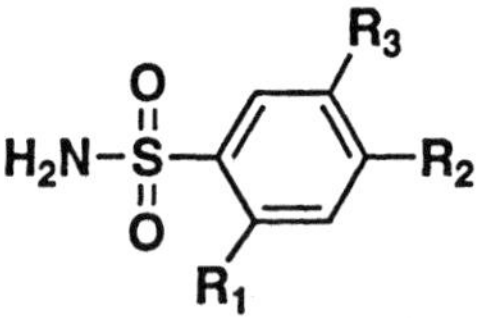

	R_1	R_2	R_3
4	H	H	H
5	H	CH_3	H
6	Cl	Cl	Cl
7	H	NO_2	H

Scheme I. Structures of arylsulfonamide ligands having different net charges that bind to CAII.

29.5.2. Binding of Families of Derivatives of Carbonic Anhydrase to a Positively Charged Ligand

Random acetylation of the ε-amino groups of lysine side chains of CA generates a mixture of derivatives of the protein. Conversion of each charged ε-ammonium group (—$CH_2NH_3^+$, $pK_a \sim 11$; a total of 18) to a neutral N-acetyl group (—$CH_2NHCOCH_3$) changes the charge of CA by approximately 1 unit (at pH 8.3). Making the assumption (well supported by experiment) that the hydrodynamic drag on the protein changes relatively little upon acetylation, we find that each peak in the electropherogram corresponds to a heterogeneous family of derivatives having the same *number* of N-acetylated lysine groups. The set of acylated proteins appear in the electropherogram as a series of peaks differing in mobility (and charge) by regular intervals; we call the electropherogram of this set of proteins a *charge ladder* (Figure 29.2). Upon addition of a sulfonamide ligand to the electrophoresis buffer, the electrophoretic mobilities of all the derivatives of CA change. Scatchard analysis of the changes in electrophoretic mobility as a function of the concentration of the sulfonamide ligand gives the binding affinity of the ligand to each family of derivatives simultaneously. This example demonstrates two strengths of CE: derivatives of proteins that differ only by 1 unit of charge can be prepared and resolved; and binding affinites of a single ligand to multiple proteins can be measured simultaneously. The combination of protein charge ladders and ACE is also usefull in quantitating the influence of charge on protein–ligand interactions (15).

29.5.3. Using Competitive Affinity Capillary Electrophoresis to Study Binding of Small Neutral Ligands to a Receptor

Since a small, electrically neutral ligand will not change the electrophoretic mobility of a protein upon binding [Eq. (29.1)], CE cannot differentiate between bound and unbound forms of the protein and therefore cannot measure directly the binding constants of these type of ligands. This limitation can, however, be overcome by using a charged ligand with a known binding affinity to compete with the neutral ligand for the protein. Typically, a sufficient concentration of the charged ligand is added to the sample and electrophoresis buffer to saturate the binding site of the protein; electropherograms are obtained with increasing concentrations of the neutral ligand in the electrophoresis buffer to compete with the charged ligand. Analysis of the average mobility of the protein as a function of the concentration of the neutral ligand yields its binding affinity relative to the charged ligand (16). Table 29.2

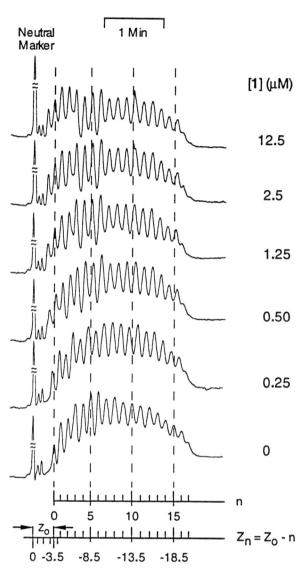

Figure 29.2. Electropherograms of binding a positively charged ligand (**1**) to the families of derivatives of CAII. The derivatives were obtained by nonspecific modification of ε-amino groups of Lys on CAII using acetic anhydride. Increasing concentrations of **1** (with charge $+1$) was added to the electrophoresis buffer (25 mM tris/192 mM Gly, pH = 8.3). The neutral marker was p-methoxybenzyl alcohol. The number of modified ε-amino groups (n) and net charge of the modified proteins (Z_n) are indicated below the electropherograms. The timescale shown on top of the figure applies to all electropherograms.

Table 29.2. Comparison of Binding Constants of Ligands 4–7 to CA Obtained by ACE and Other Methods

Ligand	ACE^a (10^6 M^{-1})	Values from the Literature^b (10^6 M^{-1})	Values form the Literature^c (10^6 M^{-1})
4	1.1	0.3	0.7
5	1.9	0.7	2.0
6	4.5	—	6.8
7	7.0	2.6	16.0

^{a} The conditions for ACE were 192 mM Gly and 25 mM Tris at 25 °C (pH 8.3).
^{b} Literature values are for bovine CA II (BCA II) measured at 25 °C in 100 mM Tris–HCl Buffer, pH 7.2 (17).
^{c} Literature values are for human CA II (HCA II) measured by fluorescence at 25 °C in 20 mM phosphate buffer, pH 7.5 (18).

compares values of K_b of arylsulfonamides to CA measured by ACE and other techniques (16). The values obtained from ACE agree well with those obtained from other measurements.

29.5.4. Binding of a Protein to Two Ligands: Immunoglobulin G$_{2b}$–N-Dinitrophenol Complexes

We used ACE to quantify the interaction between a bivalent anti-DNP rat monoclonal IgG$_{2b}$ antibody and charged, monovalent ligands containing an N-dinitrophenyl (DNP) group (Figure 29.3.) (10). Since the antibody has two binding sites for the DNP ligands, three forms of protein are possible: IgG not bound to ligand, IgG bound to one ligand, and IgG bound to two ligands. Analysis of these equilibria using ACE presented three challenges: (i) the change in mobility of the IgG on complexing a ligand with a single charge was small because of the high molecular weight of the IgG [Eq. (29.1)]; (ii) the IgG tended to adsorb slowly to the walls of the uncoated quartz capillary during the course of the experiment; and (iii) the independence of the two sites on IgG could not be assumed a priori. We overcame these problems using three strategies: (i) we synthesized ligands that were multiply charged (up to -9) at pH 8.3 (the pH of the experiment); (ii) we used buffers containing zwitterionic additives in combination with inorganic salts to reduce adsorption of proteins [a strategy first used by Bushey and Jorgenson (11)]; and (iii) we devised an analysis that could yield both association constants in the bivalent system simultaneously (without assuming any cooperativity between the two sites).

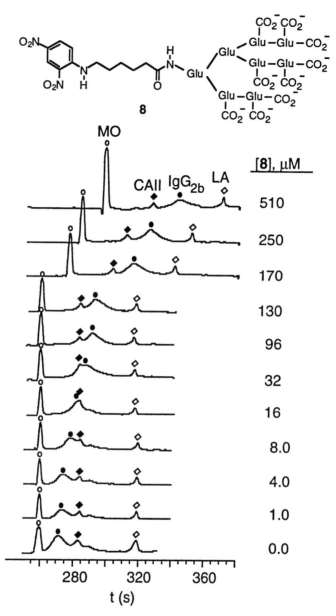

Figure 29.3. Electropherograms of IgG$_{2b}$ as a function of the concentration of ligand **8** in 25 mM Tris/192 mM Gly/10 mM K$_2$SO$_4$/0.5 M 3-quinuclidinopropanesulfonate, pH = 8.3. Two charged ligands containing DNP groups (**8**) bind to IgG$_{2b}$, thus changing its electrophoretic mobility. The peak of IgG$_{2b}$ is broad, probably due to different patterns of glycosylation (microheterogeneity). Ligand **8** has a total charge of -9 at pH 8.3. Mesityl oxide (MO), CAII, and α-lactalbumin (LA) were used as internal standards.

With increasing concentration of ligand in the buffer, the average electrophoretic mobility of the IgG changed sigmoidally. Because only a single peak was observed for all concentrations of ligand, equilibration was fast on the timescale of the experiment (10). We were able to determine association constants for a DNP ligand binding to each site of the IgG and showed that these binding events were noncooperative.

We believe that ACE can be used to quantitate many antibody–ligand interactions rapidly and conveniently. The use of ligands having high net charges will be useful in analyzing shifts in electrophoretic mobility for large proteins (>100 kDa).

29.5.5. Simultaneous Determination of Relative Binding Affinities of Several Isozymes of Carbonic Anhydrase to a Charged Ligand

Figure 29.4 shows the electropherograms of a mixture of four isozymes of CA (from human and bovine erythrocytes) obtained using an electrophoresis buffer containing a sulfonamide inhibitor having a charge of -1. The electrophoretic mobilities of the isozymes of CA increased with increasing concentrations of the charged ligand in the electrophoresis buffer, while the migration times of MO, as well as horse heart myoglobin and soybean trypsin inhibitor (used as protein markers), remained constant (up to 140 µM). Scatchard analysis of the changes in electrophoretic mobilities of the isozymes gave binding affinities of the ligand for each isozyme. This procedure may be useful to rapidly identify ligands that bind a single isozyme from a pool of several.

29.5.6. Simultaneous Determination of Relative Binding Affinities of Several Ligands to One Receptor

By analyzing the electrophoretic mobilities of several ligands in a sample using an electrophoresis buffer containing a single protein, ACE can provide the binding affinities of the receptor for each ligand in a single experiment. We used vancomycin and four peptidyl ligands as a model system (19). We measured changes in the electrophoretic mobility of the four ligands with increasing concentrations of vancomycin in the electrophoresis buffer to obtain four independent binding constants (Figure 29.5). Karger's group applied this approach to the screening of tight-binding ligands for vancomycin from a library comprising 100 peptides [see Chu et al. (20)]. The electrophoretic mobilities of the tighter-binding peptidyl ligands were retarded in the capillary, and their structures were identified by on-line mass spectrometry (MS) (21,22). This work demonstrates the usefulness of ACE–MS in screening and identifying tight-binding ligands (leads) from a library of compounds, although MS is still not a routine method of detection in CE.

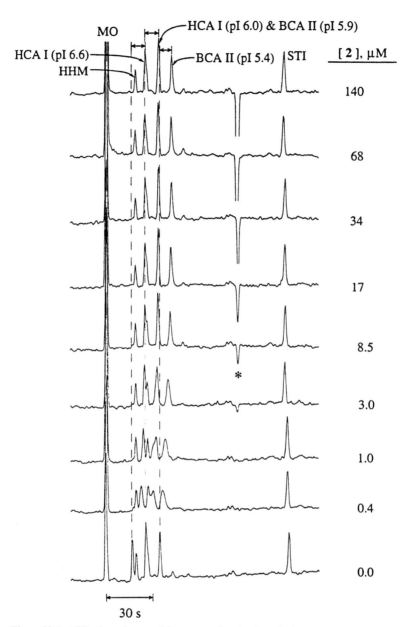

Figure 29.4. ACE of a mixture of isozymes of carbonic anhydrase: HCA I (human carbonic anhydrase, p*I* 6.6), HCA I (p*I* 6.0), BCA II (bovine carbonic anhydrase, p*I* 5.9), and BCA II (p*I* 5.4). The electrophoresis buffer used was Tris (25 mM)/Gly (192 mM) at pH 8.3. MO was added to measure EOF; HHM (horse heart myoglobin) and STI (soybean trypsin inhibitor) are protein markers that do not interact with affinity ligand **2**. The inverted peak migrates at the migration time of free **2** and is due to a lower free concentration of **2** in the migrating plug relative to the electrophoresis buffer because of binding of the ligand to CA.

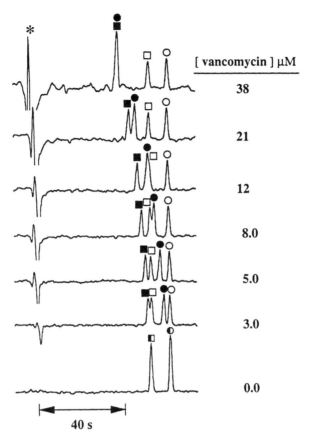

Figure 29.5. ACE of N-Fmoc-Gly-D-Ala-D-Ala (●), N-Fmoc-Gly-D-Ala-D-Ala (■), N-Fmoc-Gly-L-Ala-L-Ala (○), N-Fmoc-Gly-L-Ala-L-Ala (□) in 20 mM sodium phosphate buffer, pH 7.5, containing various concentrations of vancomycin. The asterisk (∗) indicates the position of the peak for unidentified neutral species carried through the capillary by EOF.

29.5.7. Protein–Protein Interactions: Dimerization of Insulin

Insulin is a peptide of MW = 5.7 kDa that dimerizes in aqueous solution (23). Based on Eq. (29.1) and a value of $\alpha = 2/3$, the electrophoretic mobility of monomeric insulin is a factor of 1.25 smaller than that of the dimer (we assume negligible changes in the values of pK_a of the charged side chains upon dimerization). We measured the electrophoretic mobility of insulin in Tris/Gly buffer at pH 8.3 as a function of its concentration over the range from 2 to 400 μM. The electrophoretic mobility of insulin increased with increasing

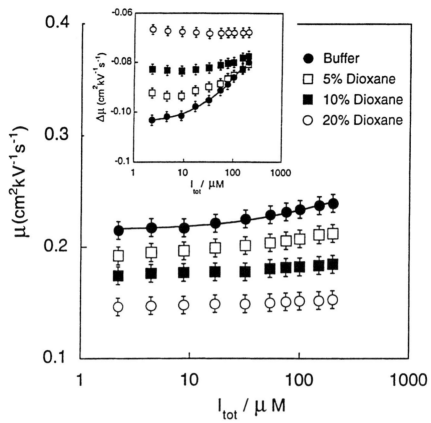

Figure 29.6. Measurement of the dimerization constants of bovine insulin. Plots of the electrophoretic mobility of insulin as a function of the concentration of insulin in several different buffers: (●) 25 mM Tris/192 mM Gly, pH = 8.4 (buffer A); (□) 5% dioxane in buffer A; (■) 10% dioxane in buffer A; (○) 20% dioxane in buffer A. Soybean trypsin inhibitor (STI) was used as protein marker to correct for changes in EOF; the inset shows plots of corrected mobilities after subtracting the mobility of STI.

concentration of insulin in the sample (Figure 29.6), while the mobility of the protein marker, soybean trypsin inhibitor, remained constant.

The determination of a dimerization constant is based on the assumption that the measured electrophoretic mobility of insulin is a weighted average of the electrophoretic mobilities of the monomer (μ_M) and dimer (μ_D) [$\mu = \mu_M(1 - \theta_D) + \mu_D\theta_D$], where θ_D is the mole fraction of dimer. Analysis of the electrophoretic mobility of insulin as a function of its concentration yields a dimerization constant of 6000 M^{-1} at pH 8.3, in good agreement with previously reported values (23). In agreement with work by Frederiq, the dimerization of insulin (as inferred by changes in μ) was completely inhibited in an electrophoresis buffer containing 20% dioxame (23).

29.6. DETERMINATION OF KINETIC PARAMETERS FOR BINDING

The shapes of peaks in the electropherogram depend, in part, on the kinetic parameters for complexation (k_{on} and k_{off}) and the migration time of the protein–ligand complex. We used CA and a charged sulfonamide ligand as a model system to illustrate how ACE can be used to obtain kinetic constants for association of ligands with proteins (16). We simulated the shapes of peaks using k_{on} and k_{off}, the concentration of the ligand in the buffer, and the relative mobilites of the protein and its complex as variables. By comparing qualitatively the shapes of the experimental peaks to the shapes of simulated peaks, we estimated the values of k_{on} and k_{off} for the binding. These estimated values were consistent with those obtained from studies using fluorescence spectroscopy (16). This method is useful only for systems where the rate constant for dissociation of the complex is comparable to the time of the electrophoresis experiment. The method will also be useful for analyzing the rate constants for association in systems where binding is slow but effectively irreversible during the course of the experiment, that is, when peaks for both the protein and protein–ligand complexes are resolved and sharp.

29.7. USING AFFINITY CAPILLARY ELECTROPHORESIS TO MEASURE THE EFFECTIVE CHARGE OF A PROTEIN

Protein charge ladders can be used to estimate the charge on a protein. In Section 29.5.2, we described acetylation as one method of generating charge ladders; another method relies on association of several small, differently charged ligands to a protein. In this latter method, a sufficient concentration of ligand (~ 100 times the K_d) is added to the buffer to saturate the binding site of the protein. Multiple experiments using differently charged ligands generate each peak of the charge ladder independently.

In both methods, the analysis assumes that the values of C_p, α, and M are constants that do not change upon modification of the protein. This assumption is reasonable for acylating agents or charged ligands that are small relative to the protein. The charge of a protein changes from a value of Z_0 to Z_n on modification or association with ligands. If the change in hydrodynamic drag on modification is small compared to the changes in mobility due to charge, the ratio of the mobilities of the modified and unmodified proteins eliminates the variables C_P, α, and M [Eq. (29.6)]:

$$\frac{\mu_0}{\mu_n} = \frac{Z_0}{Z_n} \qquad (29.6)$$

$$\Delta Z = Z_n - Z_0 = Z_0 \frac{\mu_n - \mu_0}{\mu_0} = Z_0 \left[\frac{(\Delta t/t)_n}{(\Delta t/t)_0} - 1 \right] \qquad (29.7)$$

Equation (29.7) relates the increments of charge to the electrophoretic mobilities (and hence migration times) of the derivatives of the protein. Analysis using Eq. (29.7) yields the total charge of the protein in the electrophoresis buffer (Figure 29.7) (24). This type of analysis on a charge ladder of CA indicates the total charge of CA to be -3.5 at pH 8.3 (24). Similar experiments found the charge of IgG_{2b} to be -8.0 at pH 8.3 (10) and of insulin to be -4.2 at pH 8.3 (23).

The alternate method of generating charge ladders using noncovalent modification has the disadvantages that it requires knowledge of the binding stoichiometry of the ligands and protein, and the syntheses of ligands having different charges. This method is useful, however, when covalent modification of the protein caused changes in structure or binding properties or the derivatized proteins cannot be resolved by CE (e.g., for proteins having high molecular weights).

29.8. DETERMINATION OF STOICHIOMETRY OF BINDING

ACE can be used to determine the stoichiometry in receptor–ligand complexes. One method analyzes the changes in concentration of free ligand in the sample containing protein caused by association of the ligand with protein. For example, when the same *total* concentration of ligand **2** was maintained in both the electrophoresis buffer and sample containing CA (4.2 μM), a negative peak appeared at the migration time of the free ligand. This negative peak was due to a lower *free* concentration of **2** in the migrating plug relative to the electrophoresis buffer due to binding of the ligand to CA. As the concentration of **2** was increased in the sample, the amplitude of the negative peak decreased until a positive peak (due to excess ligand) appeared. The increase in concentration of ligand **2** in the sample required to compensate the negative peak is equal to the amount of **2** that bound CA. The ratio between the increase in concentration of **2** in the sample and the concentration of CA gave a binding stoichiometry of 1:1 for CA binding **2** (25). The binding stoichiometries of several protein–ligand systems have been detemined by this method, including mouse monoclonal IgG–human serum albumin and streptavidin–biotin (25).

DETERMINATION OF STOICHIOMETRY OF BINDING 965

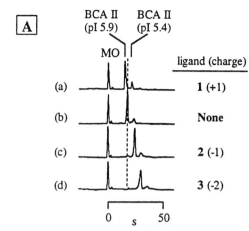

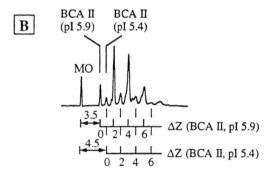

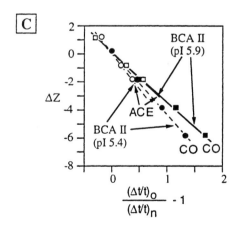

Figure 29.7. Measurement of the effective charge of bovine carbonic anhydrases (BCA II) by measuring mobilities of aggregates with ligands having different values of charge and using a charge ladder. The BCA II sample (from Sigma) contains two isozymes: BCA II (pI 5.9) and BCA II (pI 5.4) in about a 10:1 ratio; these isozymes differ by 1 unit of charge. (A) The charge ladder was generated by complexation of BCA II with charged ligands: (a) [**1**] = 1.0 mM ($Z = +1$); (b) buffer; (c) [**2**] = 1.0 mM ($Z = -1$); and (d) [**3**] = 0.5 mM ($Z = -2$). (B) Covalent modification of lysine ε-amino groups on BCA II by 4-sulfophenyl isothiocyanate generated two independent charge ladders of BCA II (pI 5.9) and BCA II (pI 5.4) [buffer = Tris (25 mM)/Gly (192 mM), pH = 8.3, $t_{EO} \approx 170$ s]. The charge increment for covalent modifications is -2 for this acylating agent. The migration times of the unmodified proteins are indicated and provide a scale in ΔZ. The timescale applies to electropherograms in both parts A and B. (C) Determination of effective charge from the analysis of charge ladders. The data shown for BCA II (pI5.9) and BCA II (pI 5.4) were estimated using covalent and noncovalent charge ladders generated by ACE and covalent (CO) methods, respectively.

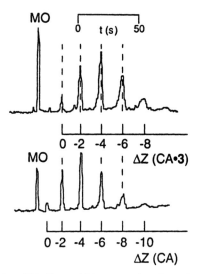

Figure 29.8. Determination of binding stoichiometry using charge ladders. The charge ladder of CA II (from human erythrocyte) was formed by acylation of its ε-amino groups of Lys using succinic anhydride ($\Delta Z = -2$). Treatment of this charge ladder with ligand **3** (charge of -2) at a concentration that saturates the active site of CA II shifted the mobility of each protein by an amount corresponding to a change in charge of -2. Comparison of the charge increments obtained by covalent modification and upon binding of ligand **3** indicated a 1:1 stoichiometry of binding of **3** to CA II.

More recently, we have used charge ladders of proteins to determine stoichiometries of binding (26). The charge ladder provides an internal scale in integral units of charge. Saturation of the derivatized proteins with a charged ligand results in a shift of the peaks in the ladder by an amount that is related to the net change in effective charge upon binding. The ratio between the total change in charge of members of the ladder and the charge of a single ligand provides the number of ligands that bind each protein (Figure 29.8). This method has the advantage that it does not require knowledge of the molecular weight of the protein; hence, proteins from impure samples can be analyzed without purification, provided that the peaks are resolved in the electropherogram.

29.9. PROSPECTS AND LIMITATIONS OF AFFINITY CAPILLARY ELECTROPHORESIS

ACE is a useful analytical method in biochemistry (Table 29.3) (23,25,27–56). It is a simple technique experimentally: it does not involve radioactive

Table 29.3. Studies on Binding of Ligands to Receptors Using ACE

Example	Reference
A. Protein–protein interactions	
• Human serum albumin (HSA) and anti-HSA	25
• Human growth hormone (hGH) and anti-hGH (or its fragment)	27
• IgG with protien A	28
• Insulin dimerization	23
B. Protein–DNA interactions	
• EcoR1 and oligonucleotide; peptide and oligonucleotide	29
• Transcription factor–oligonucleotide	30
C. Protien–peptide interactions	
• SH$_3$ domain and proline-rich peptides	31
• Antibody–antigen interaction	32,33
D. Protein–drug interactions	
• Bovine carbonic anhydrase (CA) and arylsulfonamides; glucose-6-phosphate dehydrogenase (G6PDH) and NADP$^+$; G6PDH and NADPH; IgG$_{2b}$ and 2,4-DNP	8,10,16,34
• Bovine serum albumin (BSA); bacteiral cellulase and tryptophan benzoin; pindolol; promethazine; warfarin	35
• HSA and kynurenine; tryptophan; 3-indolactic acid; 2,3-dibenzoyl tartaric acid; 2,4-dinitrophenyl glutamate	36
• Cellulase and β-blockers	37
• BSA and warfarin; leucovorin	38,39
• Albumin–ofloxacin	40
E. Protein–metal ion interactions	
• Calmodulin, parvalbumin, thermolysin, and Ca (II); carbonic anhydrase, thermolysin, and Zn(II)	41,42
• C-reactive protein and Ca(II)	43
F. Protein–carbohydrate interactions	
• Concanavalin A–monosaccharides	44
• Lectin–carbohydrate	45–47
G. Peptide–peptide interactions	
• Vancomycin and peptides	19,20,48,49
H. Peptide–carbohydrate interactions	
• Synthetic peptide and anionic carbohydrates	50
I. Peptide–dye interactions	
• Synthetic peptide and Congo Red	51
J. Carbohydrate–drug interactions	
• Methyl-β-cyclodextrin and propranolol	52
• β-Cyclodextrin and salbutanol	53
K. Oligonucleotide–oligonucleotide interactions	
• d(A)$_n$/d(T)$_n$	54–56

materials or chemically immobilized ligands, and the analyses are fast, reproducible, and capable of high resolution. Because ACE permits a wide range of buffer conditions, proteins can be studied in their native conformations under physiological conditions. Even complex and impure samples can be analyzed if the analyte of interest is well resolved in the electropherogram. The quantitative relationships between the electrophoretic mobility of a protein and the charge and mass of the protein make this technique useful for both fundamental studies of biomolecular recognition and applications in bioanalytical chemistry.

ACE currently has certain limitations (Table 29.1); the primary limitation is that many proteins do not elute through the capillary to give sharp peaks but interact strongly with the walls of the capillary to give broadened peaks. Uncoated capillaries (with or without zwitterionic and inorganic additives) and coated capillaries (with either covalent or noncovalent coatings) handle only a limited subset of proteins and have their own set of drawbacks. The mechanisms of adsorption of protein to the walls of the capillary are not yet well established but are under active investigation. Other technical problems with ACE are the high-background UV absorbance present when protein–protein or protein–DNA complexes are being analyzed and difficulties in identifying the analytes that emerge from the capillary when complex samples are used.

Many of the applications of ACE described in this chapter can be applied to problems in bioanalytical and medicinal chemistry. Although the theoretical basis for ACE is general and applicable to most receptor–ligand pairs, the aforementioned limitations still prevent wider application of the technique. Despite these limitations, CE is well suited for fundamental studies of the roles of electrostatic interactions in protein–ligand complexes. We believe this technique will permit detailed studies of electrostatic influences in biomolecular recognition using the CA–sulfonamide and other model systems.

ADDENDUM

After submission of this chapter, several advances have been reported in ACE that have broadened its application in studying the influence of electrostatics on molecular recognition. Mammen et al. have applied the principles of ACE to examine shielding of the charged capillary surface by small ions in solution (57). CE was used to measure the rate of EOF as a function of the concentration of various monovalent and divalent cations in aqueous solution. A model was described that treats the interactions between cations in solution and negatively charged siloxide groups on a surface in terms of discrete dissociation equilibria with characteristic dissociation constants, then interprets these dissociation constants in terms of well-recognized physical characteristics of the ions. In another example, Rao and coworkers used vancomycin and D-Ala-D-Ala (DADA)–containing ligands as a model system to examine the influence of electrostatics on the net charge of vancomycin and on the pK_a values of ionizable residues on vancomycin (58). They confirmed that the electrostatic interaction between the $-NH_2^+CH_3$ group of vancomycin and the $-CO_2^-$ group of DADA contributes ~ 1.4 kcal/mol to the free energy of binding. In addition, the interaction increases the value of pK_a of the N-terminal ammonium group from 7.1 to 8.8 and thus influences the net charge on vancomycin. These examples demonstrate that ACE/CE provides a useful tool with which to investigate electrostatic effects in physiologically relevent media.

In another application, ACE has been used to separate racemic mixtures of drugs via interactions with a chiral selector. For example, Liu and coworkers have used β-cyclodextrin and its derivatives as chiral selectors to distinguish water-soluble drug enantiomers (59). The separation was based on different affinities of the drug enantiomers for the cavity of the β-cyclodextrin. Tanaka et al. have used α_1-acid glycoprotein to separate a variety of racemic basic drugs (60). These applications demonstrate the power of ACE to analyze racemic drugs and may also be useful in examining the purity of drugs.

LIST OF ACRONYMS OR ABBREVIATIONS

Acronym or Abbreviation	Definition
ACE	affinity capillary electrophoresis
AGE	affinity gel electrophoresis
BCA	bovine carbonic anhydrase
CA	carbonic anhydrase
CE	capillary electrophoresis
DADA	D-Ala-D-Ala
DNP	N-dinitrophenol
EOF	electroosmotic flow
HCA	human carbonic anhydrase
HHM	horse heart myoglobin
IgG_{2b}	immunoglobulin G_{2b}
MO	mesityl oxide
MS	mass spectrometry
STI	soybean trypsin inhibitor
Tris	tris(hydroxymethyl)aminomethane
UV	ultraviolet

ACKNOWLEDGMENTS

This work was supported by the National Institutes of Health (GM 51559). Milan Mrksich was supported by an American Cancer Society postdoctoral fellowship. Mathai Mammen was supported by an Eli Lilly predoctoral fellowship.

REFERENCES

1. Y.-H. Chu, L. Z. Avila, J. Gao, and G. M. Whitesides, *Acc. Chem. Res.* **28**, 461 (1995).
2. K. Takeo, in *Advances in Electrophoresis*, (A. Chrambach, M. J. Dunn, and B. J. Radola, eds.), p. 229. VCH Publishers, New York, 1987.
3. K. Shimura, *J. Chromatogr.* **510**, 251 (1990).
4. P. D. Grossman, in *Capillary Electrophoresis: Theory and Practice* (P. D. Grossman and J. C. Colburn, eds.), pp. 111–118. Academic Press, San Diego, CA, 1992.
5. H. A. Abramson, L. S. Moyer, and M. H. Gorin, in *Electrophoresis of Proteins and the Chemistry of Cell Surfaces* (H. A. Abramson, L. S. Moyer, and M. H. Gorin, eds.), pp. 105–172. Reinhold, New York, 1942.
6. S. K. Basak, and M. R. Ladisch, *Anal Biochem.* **226**, 51 (1995).
7. J. Gao, and G. M. Whitesides, *Anal. Chem.* **69**, 575 (1997).
8. F. A. Gomez, L. Z. Avila, Y. Chu, and G. M. Whitesides *Anal. Chem.* **66**, 1785 (1994).
9. F. A. Gomez, J. K. Chen, A. Tanaka, S. L. Schreiber, and G. M. Whitesides *J. Org. Chem* **59**, 2885 (1994).
10. M. Mammen, F. A. Gomez, and G. M. Whitesides, *Anal. Chem.* **67**, 3526 (1995).
11. M. M. Bushey, and J. W. Jorgenson, *J. Chromatogr.* **480**, 301 (1989).
12. Y. J. Yao, K. S. Khoo, M. C. M. Chung, and S. F. Y. Li, *J. Chromatogr.* **680**, 431 (1994).
13. S. J. Dodgson, R. E. Tashian, G. Gros, and N. D. Carter, *The Carbonic Anhydrase: Cellular Physiology and Molecular Genetics*. Plenum, New York and London, 1991.
14. A. M. Cappalonga, R. S. Alexander, and D. W. Christianson, *J. Am. Chem. Soc.* **116**, 5063 (1994).
15. J. Gao, M. Mammen, and G. M. Whitesides, *Science* **272**, 535 (1996).
16. L. Z. Avila, Y. Chu, E. C. Blossey, and G. M. Whitesides, *J. Med. Chem.* **36**, 126 (1993).
17. A. Carotti, C. Raguseo, F. Campagna, R. Langridge, and T. E. Klein *Quant. Struct. Act. Relat.* **8**, 1 (1989).
18. P. W. Talor, R. W. King, and A. S. V. Burgen, *Biochemistry* **9**, 2638 (1970).
19. Y.-H. Chu, and G. M. Whitesides, *J. Org. Chem.* **57**, 3524 (1992).

20. Y. H. Chu, Y. M. Dunayevskiy, D. P. Kirby, P. Vouros, and B. L. Karger, *J. Am. Chem. Soc.* **118**, 7827 (1995).
21. T. J. Thompson, F. Foret, P. Vouros, and B. L. Karger, *Anal. Chem.* **65**, 900 (1993).
22. W. Nichols, J. Zweigenbaum, F. Garcia M. Johansson, and J. Henion, *LC–GC* **10**, 676 (1992).
23. J. Gao, M. Mrksich, F. A. Gomez, and G. M. Whitesides, *Anal. Chem.* **67**, 3093 (1995).
24. J. Gao, F. Gomez, R. Haerter, and G. M. Whitesides, *Proc. Natl. Acad. Sci. U.S.A.* **91**, 12027 (1994).
25. Y.-H. Chu, W. J. Lees, A. Stassinopoulos, and C. T. Walsh, *Biochemistry* **33**, 10616 (1994).
26. J. Gao, and G. M. Whitesides, Unpublished results.
27. K. Shimura, and B. L. Karger, *Anal. Chem.* **66**, 9 (1994).
28. R. Lausch, O. W. Reif, P. Riechel, and T. Scheper, *Electrophoresis* **16**, 636 (1995).
29. N. H. H. Heegaard, and F. A. Robey, *Am. Lab.*, 28 (1994).
30. J. Xian, M. G. Harrington, and E. H. Davidson, *Proc. Natl. Acad. Sci. U.S.A.* **93**, 86 (1996).
31. F. A. Gomez, J. K. Chen, A. Tanaka, S. L. Schreiber, and G. M. Whitesides, *J. Org. Chem.* **59**, 2885 (1994).
32. N. H. Heegaard, *J. Chromatogr.* **680**, 405 (1994).
33. J. Chen, and T. A. Fu, *J. Chromatogr.* **680**, 419 (1994).
34. Y. Chu, L. Z. Avila, H. A. Biebuyck, and G. M. Whitesides, *J. Med. Chem.* **35**, 2915 (1992).
35. S. Busch, J. C. Kraak, and H. Poppe, *J. Chromatogr.* **635**, 119 (1993).
36. L. Vaitcheva, J. Mohammad, G. Pettersson, and S. J. Hjertén, *J. Chromatogr.* **638**, 263 (1993).
37. R. Vespaleck, V. Sustacek, and P. Bocek, *J. Chromatogr.* **638**, 255 (1993).
38. J. C. Kraak, S. Busch, and H. Poppe, *J. Chromatogr.* **608**, 257 (1992).
39. G. E. Barker, P. Rosso, and R. A. Hartwick, *Anal. Chem.* **64**, 3024 (1992).
40. T. Arai, N. Nimura, and T. Kinoshita, *Biomed. Chromatogr.* **9**, 68 (1995).
41. H. Kajiwara, H. Hirano, and K. Oono, *J. Biochem. Biophys. Methods* **22**, 263 (1991).
42. H. Kajiwara, *J. Chromatogr.* **559**, 345 (1991).
43. N. H. H. Heegaard, and F. A Robey, *J. Immunol. Methods* **166**, 103 (1993).
44. K. Shimura, and K. I. Kasai, *Anal. Biochem.* **227**, 186 (1995).
45. R. J. Linhardt, X. J. Han, and J. R. Fromm, *Mol. Biotechnol.* **3**, 191 (1995).
46. R. Kuhn, R. Frei, and M. Christen, *Anal. Biochem.* **218**, 131 (1994).
47. S. Honda, A. Taga, K. Suzuki, S. Suzuki, and K. Kakehi, *J. Chromatogr.* **597**, 377 (1992).
48. J. Liu K. J. Volk, M. S. Lee, M. Pucci, and S. Handwerger, *Anal. Chem.* **66**, 2412 (1994).

49. J. L. Carpenter, P. Camilleri, D. Dhanak, and D. Goodall, *J. Chem. Soc., Chem. Commun.*, 804 (1992).
50. N. H. H. Heegaard, and F. A. Robey, *Anal. Chem.* **64**, 2479 (1992).
51. R. Biehler, and A. Jacobs, *Capillary Electrophoresis Technical Information.* A-1727. Beckman Instruments, Fullerton, CA, 1993.
52. S. A. C. Wren, and R. C. Rowe, *J. Chromatogr.* **603**, 235 (1992).
53. V. Lemesle-Lamache, M. Taverna, D. Wouessidjewe, D. Duchene, and D. Ferrier, *J. Chromatogr. A* **735**, 321 (1996).
54. Y. Baba, H. Inoue, M. Tsuhako, T. Sawa, A. Kishida, and M. Akashi, *Anal. Sci.* **10**, 967 (1994).
55. Y. Baba, M. Tsuhako, T. Sawa, and M. Akashi, *J. Chromatogr.* **632**, 137 (1993).
56. Y. Baba, M. Tsuhako, T. Sawa, M. Akashi, and E. Yashima *Anal. Chem.* **64**, 1920 (1992).
57. M. Mammen, J. Carbeck, E. E. Simanek, and G. M. Whitesides, *J. Am. Chem. Soc.* **119**, 3469 (1997).
58. J. Rao, I. Colton, and G. M. Whitesides, *J. Am. Chem. Soc.* (in press).
59. J. Liu, H. Coffey, J. D. Detlefsen, Y. Li, and M. S. Lee, *J. Chromatogr. A* **763**, 261 (1997).
60. Y. Tanaka and S. Terabe, *Chromatographia* **44**, 119 (1997).

CHAPTER

30

DETERMINATION OF PHYSICOCHEMICAL PARAMETERS BY CAPILLARY ELECTROPHORESIS

PIER GIORGIO RIGHETTI

Department of Agricultural and Industrial Biotechnologies, University of Verona, Strada Le Grazie, 37134 Verona, Italy

30.1.	Introduction	973
30.2.	Determination of pK Values of Weak Electrolytes	975
30.3.	Determination of pK Values of Amphoteric Compounds	980
30.4.	Assessment of pK Values of Silanols	981
30.5.	Determination of pI Values of Proteins	982
30.6.	Determination of Absolute Mobility and Its Relation to the Charge/Mass Ratio in Peptides (and Proteins)	984
30.7.	Determination of Binding Constants	988
30.8.	Determination of Diffusion Constants	991
30.9.	Viscosity Measurements	992
30.10.	Determination of T_m Values of Nucleic Acids	992
30.11.	Conclusions	995
List of Acronyms and Symbols		995
References		996

30.1. INTRODUCTION

The extraordinary usefulness of electrophoretic techniques for the analysis of macromolecules was first demonstrated by Tiselius (1) with his elegant moving-boundary apparatus, which resolved human serum proteins into albumin and the four globulin fractions, α_1, α_2, β, and γ. This original discovery gave rise to a steady development of new instruments and techniques with ever-increasing resolution. Currently, two-dimensional electrophoresis, combined with sophisticated computer-image analysis, can

---

High Performance Capillary Electrophoresis, edited by Morteza G. Khaledi. Chemical Analysis Series, Vol. 146.
ISBN 0-471-14851-2 © 1998 John Wiley & Sons, Inc.

resolve several thousand proteins among the products of a given cell type (2,3).

Most materials in aqueous solution acquire an electric charge due to ionization and therefore move in response to an external electric field. The charged entities may be simple ions, complex macromolecules, colloids, or living cells. The rate of migration depends on the amount of charge, the size and shape of the particle, and the properties of the solvent. Resolution is increased by the imposition of other constraints, such as molecular sieving or pH gradients. A systematic, logical classification of the rather bewildering choice of analytical and preparative techniques is difficult, but four classical modes of electrophoresis are generally recognized: moving-boundary electrophoresis (MBE), zone electrophoresis (ZE), isotachophoresis (ITP) and isoelectric focusing (IEF) (4).

Due to their nature, it is to be expected that electrokinetic transport phenomena should allow measurement of physicochemical parameters of the ions under analysis. For instance, ZE and ITP, in free solution, under carefully controlled conditions (temperature, pH, ionic strength, and dielectric constant of solvent) allow precise assessments of free mobilities (μ), which in turn are related to the ratio of net surface charge to mass of a given analyte (5,6). In sieving matrices, a number of other properties can be measured; thus, as early as the 1970s it was understood that a protein particle subjected to the frictional drag of a polyacrylamide matrix would move according to the charge/mass ratio (Z/r). While the two contributions could not be easily evaluated, Ferguson plots permitted independent evaluation of K_R (coefficient of retardation, related to mass) and Y_O (the y-intercept, correlated to mobility in free solution) (7). As methods evolved, more accurate measurements could be made: thus, electrophoresis of sodium dodecyl sulfate (SDS)–laden polypeptides, in polyacrylamide gels (PAG), permitted correct evaluation of their molecular mass (M_r) value (8). Parallel to SDS–PAGE electrophoresis (a method discriminating purely on a size basis), other methods purely based on surface charge fractionation (such as IEF) were developed that facilitated assessment of proteins' (and amphoteric compounds) net surface charge (at a precise point of the titration curve called the isoelectric point, pI) (2,3).

It is only natural, then, that with the advent of capillary zone electrophoresis (CZE), scientists have tried to transfer these physicochemical measurements to this novel system. This should not be taken as a simple exercise at "reinventing the wheel": there are distinct advantages in working with CZE, such as minute sample requirements, on-line sample detection and quantitation, and full control of environmental parameters. The following is a nonexhaustive account (through some selected examples) of what has been achieved so far in this field.

30.2. DETERMINATION OF pK VALUES OF WEAK ELECTROLYTES

Basic equations can be derived linking the electrophoretic migration of simple, singly charged cations or anions and of mono-monovalent ampholytes with the pK's of their ionizable groups. This was already attempted in the early 1980s, when Righetti et al. (9) measured the protolytic equilibria of doxorubicin via "electrophoretic titration curves" (pH/mobility curves). A theoretical model was then developed allowing precise pK determinations via the generation of pH/mobility curves, obtained by a two-dimensional (2D) technique coupling IEF to orthogonal sample electrophoresis (10). This model is fully valid in CZE as well, provided that the observed analyte mobility (μ_{ob}) is corrected for electrosmotic transport (μ_{eo}) (11).

The electrophoretic mobility can be simply derived as a function of pK and of the prevailing pH of the electrophoresis buffer. Let us consider the case of a neutral base B that binds a proton H^+ and becomes the cation BH^+:

$$[B] + [H^+] \rightleftharpoons [BH^+]$$

The following equations can be written:

$$\mu_{eff(c)} = \mu_{act}[BH^+]/[T] \tag{30.1}$$

$$K_c = \{[B][H^+]\}/[BH^+] \tag{30.2}$$

$$[T] = [B] + [BH^+] \tag{30.3}$$

where $\mu_{eff(c)}$ is the effective cation mobility; μ_{act} is a proportionality coefficient (it is the actual mobility per unit of charge and unit of concentration and for a given time); and [T] and [BH^+] are the total and cationic concentrations, respectively. By solving the system of Eq. (30.1)–(30.3), we obtain

$$\mu_{eff(c)} = \mu_{act}[H^+]/\{[H^+] + K\} \tag{30.4}$$

By using the well-known relationships $pH = \log 1/[H^+]$ and $pK = \log 1/K$, Eq. (30.4) can be rewritten as follows:

$$\mu_{eff(c)} = \mu_{act}[10^{-pH}/(10^{-pH} + 10^{-pK_c})]$$

$$\mu_{eff(c)} = \mu_{act}/(1 + 10^{pH - pK_c}) \tag{30.5}$$

By the same reasoning, we can derive for the anionic mobility [$\mu_{eff(a)}$]:

$$\mu_{eff(a)} = -\mu_{act}K_a/([H^+] + K_a) \tag{30.6}$$

which becomes

$$\mu_{\text{eff}(a)} = -\mu_{\text{act}}[10^{-pK_a}/(10^{-pH} + 10^{-pK_a})]$$

$$\mu_{\text{eff}(a)} = -\mu_{\text{act}}/(1 + 10^{pK_a - pH}) \tag{30.7}$$

As seen from Eq. (30.5) and (30.7), the pK's of simple cationic (pK_c) or anionic (pK_a) species can be derived in a pH/mobility plot by simply measuring the pH of 1/2 mobility in the cationic or anionic directions, respectively. These two equations are described by the upper and lower curves of Figure 30.1, which do not cross the pH axis. A simple verification of Eqs. (30.5) and (30.7) comes from the fact that they can be independently derived on the basis of the classical description of simple ionization equilibria. For instance, if Eq. (30.5) is multiplied and divided by 10^{pK_c}, we obtain

$$\mu_{\text{eff}(c)} = \mu_{\text{act}}[10^{pK_c - pH}/(1 + 10^{pK_c - pH})] \tag{30.8}$$

which is analogous to Eq. 1.68 given by Poland (12) in the derivation of the association constant for the binding of a single proton to a molecule. The two independent treatments are fully equivalent. While the pH/mobility curve is automatically generated by the 2D technique described above, it has to be constructed point by point in CZE by running the analyte at a series of pH values of background electrolyte encompassing the explored pK value. A note of caution: these experiments should be run at iso-ionic strength for the different buffers [in this regard see Figure 1 and Table 1 of Cai et al. (11)]. These authors have reported, in CZE, a pK value of 4.43–4.50 for aniline (literature data: 4.60–4.82) and a pK of 4.98–5.03 for p-anisidine (literature data: 5.01–5.36). The discrepancies might be due to different ionic strengths: 0.2 equiv L^{-1} in CZE vs. 0.1 up to 1 M for the literature data. Similar data have been obtained by Cross and Ricci (13) in the CZE analysis of sulfonamides: for sulfathiazole, pK's 7.2 vs. 7.5; for sulfamethazine, pK's 7.4 vs. 7.8; and for sulfapyridine, pK's 8.4 vs. 8.2 (the first value being the observed, the second the calculated pK value). Cleveland et al. (14,15) have additionally correlated the pH/mobility curve method for assessing ionization constants in CZE with UV–Vis spectrophotometric titration, which relies on the neutral and ionic states of a molecule having different absorption coefficients (where applicable!). In the case of p-nitrophenol, the two values were identical (pK 7.15 in both cases). When pK values were correlated to chromophores in a molecule, it was additionally shown (15) that they could be related to peak area (absorbance): the value thus derived for p-nitrophenol was 7.19, in good agreement with the above data.

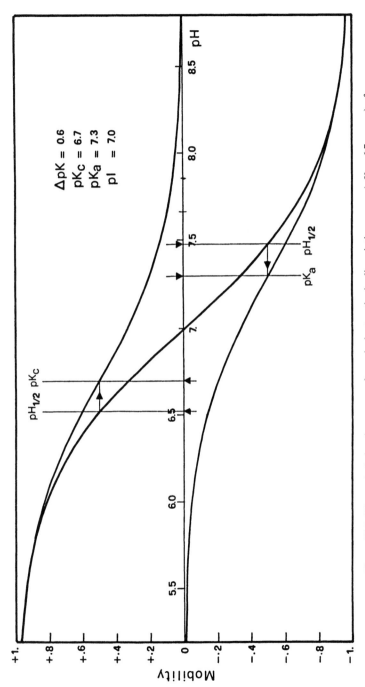

Figure 30.1. Theoretical titration curve of a univalent cationic dissociating group ($pK_c = 6.7$; upper) of a univalent anion ($pK_a = 7.3$; lower) and of an amphoteric molecule with the same ionizable functions, $pI = 7$, ΔpK 0.6 (intermediate tracing). Notice that the ampholyte behavior results from the addition of the anionic and cationic segments of the titration curve. Moreover, for the simple univalent ions pK equals $pH_{1/2}$; whereas for the amphoteric species the influence of the opposite charge dissociation should be accounted for by adding (for pK_c) or subtracting (for pK_a) a correction factor (arrows) to $pH_{1/2}$. In contrast the pI simply corresponds to the pH of zero mobility (crossover point). [From L. Valentini, E. Gianazza, and P. G. Righetti, *J. Biochem. Biophys. Methods* **3**, 323 (1980), with permission.]

Instead of constructing the entire pH/mobility curve for pK determination, Beckers et al. (16) proposed a simplified method consisting of measuring the effective electrophoretic mobility of a component for only two different electrolyte systems at different pH's at which such a component would show substantially different degrees of dissociation. If the ionic strengths and equivalent concentrations of the two electrolyte systems are known, all activity coefficients can be calculated. If Faraday's constant is used, measured mobilities can be replaced by absolute ionic mobilities, which in turn allow calculation of pK values according to the following equation (referring to the dissociation of a weak acid, HA):

$$pK = -\log \gamma_A - \log \gamma_H + pH - \log [A^-]/[HA] \qquad (30.9)$$

where the γ's are the activity coefficients of the A and H ions, respectively. In turn, the ratio of ionized vs. nonionized species can be calculated from values of effective (μ_{eff}) and absolute (μ_c) ionic mobilities at specific equivalent concentrations, according to

$$[A^-]/[HA] = \mu_{eff}/(\mu_c - \mu_{eff}) \qquad (30.10)$$

I should like to emphasize that these measurements in CZE are to be preferred to ITP determinations, as done in the past (17–21): calculations of mobilities and pK's are very laborious in ITP, since all zones have different pH, concentration, and temperature parameters, through which the data have to be recalculated by iterative processes. Conversely, in CZE, all parameters in the background electrolyte, such as ionic strength, pH, temperature, and electric field strength, can be considered as nearly constant. The two-pH-values method of Beckers et al. (16) has been adopted by Cao and Cross (22) for the determination of pK_1 values of eight dihydrofolate reductase inhibitors. These compounds were found to have rather close pK values, ranging from 1.065 to 1.341. From the determination of pK values, it was possible to calculate the effective charge Z of the various ions at the operative pH. The hydrodynamic radius r was also assessed, and an attempt at correlating the electrophoretic mobility μ_{ep} with the charge/mass ratio (Z/r) gave the plot of Figure 30.2, in very good agreement with the classic equation:

$$\mu_{ep} = Ze/(6\pi\eta r) \qquad (30.11)$$

where η is the solvent viscosity and e is the electronic charge.

The problem can also be turned around: having a family of compounds of known but very close pK values, what would be the optimum operative pH value (pH_{opt}) allowing for maximum resolution? Terabe et al. (23) addressed

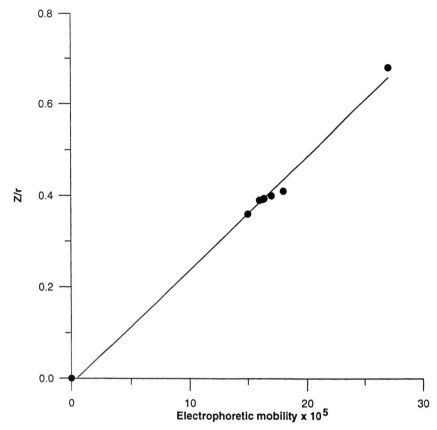

Figure 30.2. Correlation between the electrophoretic mobility (μ_{ep}) and the charge/mass ratio (Z/r) of eight dihydrofolate reductase inhibitors run at pH 2.1 in 250 mM phosphate buffer. [From J. Cao and R. F. Cross, *J. Chromatogr. A* **695**, 297 (1995), with permission.]

this problem when attempting separation of three oxygen isotopic benzoic acids (BA–$^{16}O_2$, with pK 4.19, from BA–$^{16}O^{18}O$ and BA–$^{18}O_2$). The solution:

$$pH_{opt} = pK - \log 2 \qquad (30.12)$$

According to Eq. (30.12), at an operative pH of 3.89 (precisely 0.30 pH units smaller than the pK value), excellent resolution was obtained. This is in contrast with reversed-phase high-performance liquid chromatography (RP-HPLC), where a maximum separation was seen in a pH range from 0.5 to 1.0 pH units higher than the pK of benzoic acid. Similar findings have been reported by Nielen (24) in the analysis of positional isomers of aminobenzoic acid (the para isomer, $pK_2 = 4.85$; the meta isomer, $pK_2 = 4.74$): optimum resolution was at pH values in proximity to and just below these pK values.

30.3. DETERMINATION OF pK VALUES OF AMPHOTERIC COMPOUNDS

The problem here is more complex. It can be easily solved for mono-monovalent amphoteres, to which this treatise is limited. Two cases will have to be distinguished:

(A) Case of $pK_c < pK_a$ (i.e., $K_c > K_a$). Assuming that the total effective mobility $[\mu_{eff(t)}]$ is the sum of $\mu_{eff(c)}$ and $\mu_{eff(a)}$ (where the subfixes c and a denote cations and anions, respectively), the final expression is

$$\mu_{eff(t)} = \mu_{act}[(10^{pK_c - pH} - 10^{pH - pK_a})/(1 + 10^{pK_c - pH} + 10^{pH - pK_a})] \quad (30.13)$$

(B) Case of $pK_c > pK_a$ (i.e., $K_c < K_a$). The final expression will be

$$\mu_{eff(t)} = \mu_{act}[(10^{pK_a - pH} - 10^{pH - pK_c})/(1 + 10^{pK_a - pH} + 10^{pH - pK_c})] \quad (30.14)$$

Equations (30.13) and (30.14) can be verified by considering that at the isoelectric point (pH = pI) the electrophoretic mobility is zero ($\mu_{eff(t)} = 0$) and therefore they both become

$$pI = (pK_c + pK_a)/2 \quad (30.15)$$

which is the well-known equation giving the isoelectric point as the arithmetical mean of the two pK's of a mono-monovalent ampholyte. It can also be seen that, at the two extremes of the pH scale, the total mobility $\mu_{eff(t)}$ tends to a limiting value of μ_{act}, in agreement with the definition given above of μ_{act} being a proportionality coefficient. In fact, when pH = 0, then $\mu_{eff(t)} \approx \mu_{act}$, and when pH = 14, then $\mu_{eff(t)} \approx -\mu_{act}$ [see Eqs. (30.13) and (30.14)]. How then can one calculate pK's from Eqs. (30.13) and (30.14)? One should construct a pH/mobility curve over an ample pH interval, encompassing the two pK values of the mono-monovalent ampholyte (see Figure 30.1, middle curve crossing the pH axis). Then one should measure the pH (pH$_{1/2}$) corresponding to 1/2 mobility in the cathodic or anodic directions, respectively, and insert it into Eqs. (30.13) and (30.14). For instance, for the measurement of a correct value of pK_c by substituting in Eqs. (30.13) and (30.14) the values $M_t = h/2$, pH = $pH_{1/2}$, and considering Eq. (30.15), one obtains

$$pK_c = pH_{1/2} - \log[1 - 3 \times 10^{-2(pI - pH_{1/2})}] \tag{30.16}$$

for $pK_c < pK_a$, and

$$pK_a = pH_{1/2} - \log[1 - 3 \times 10^{-2(pH_{1/2} - pI)}] \tag{30.17}$$

for $pK_c > pK_a$. A similar equation has been derived by Cai et al. (11) for CZE of amphoteres. Interestingly, in the case of p-aminobenzoic acid, they obtained a pK_1 of 2.61 vs. 2.41 and for pK_2, 4.87 vs. 4.87 (the second value referring to tabulated data). It must be emphasized, however, that their values are correct only because the ΔpK of their ampholyte is rather large (>2). For smaller ΔpK's (<1.5 pH unit), a correction factor must be applied accounting for the influence of the ionization of the cation on the anion (and vice versa). For the smallest possible ΔpK (0.6; see Figure 30.1), (25) this correction factor is as high as 0.18 pH units (10).

30.4. ASSESSMENT OF pK VALUES OF SILANOLS

This is indeed a particular case of the above treatment, namely, finding the pK value of weak electrolytes. It is treated separately here just to highlight the importance of the capillary wall in CZE separations. The invention of the fused-silica capillary by Dandenau and Zerenner (26) in the late 1970s for gas chromatography paved the way to modern CZE as well. In their search for the purest possible material, these authors abandoned quartz, which contains 60 ppm of metal oxide, in favor of fused silica, found to contain < 1 ppm metal contaminants. This last material comprises a number of acidic surface silanols (isolated, vicinal, geminal), inert siloxane bridges, and highly acidic hydrogen bonding sites (27). At any pH above 2, silanol groups begin to ionize and produce a solvent flux, called electroosmotic flow (EOF). The electroosmotic force in a capillary column is generated by the electric field and transmitted by the drag of ions acting in a thin sheath of charged fluid adjacent to the silica wall column. The origin of charge in this sheath is an unbalance between positive and negative ions in the bulk solution, which have to balance the fixed negative charge on the silica wall. The total density of ionogenic silanols on this surface is given as 8.31×10^{-6} M/m^2, corresponding to ca. 5 silanols/nm^2. There has been some debate as to the assessment of the pK value of such silanols. While Schwer and Kenndler (28) reported a pK of 5.3, recent data by Huang et al. (29) and Bello et al. (30) suggest that the pK value is indeed 6.3, i.e., 1 pH unit higher. This latter value is consistent with the fact that the point of zero EOF is found at pH 2.3 and with the fact that EOF does not plateau before reaching a pH of 9.0 in the electrolyte solution bathing the wall. The data of Bello et al. (30) differ from other data in that the electroosmotic (μ_{eo}) is extrapolated to zero voltage, since μ_{eo} was found to depend strongly on applied voltage. However, due to the heterogeneity of ionizing groups on the silica wall, the pK of 6.3 reported by these authors could represent a mean pK value and does not per se exclude the existence of a spectrum of pK values. Bello et al. (30) have additionally hypothesized that the radial electric field coexisting with the axial voltage drop could indeed drive and embed cations into the silica wall. If so, paradoxically, this wall could exhibit a pI value (typical of only amphoteric compounds!) and reverse its zeta potential and EOF below certain pH values. While no data seem to be available on fused silica, this phenomenon has been demonstrated for plastic capillaries [polytetrafluoroethylene (PTFE)]. Everaerts' group [see Wanders et al. (31); Van De Goor et al. (32)] has indeed reported a pI of 3.5 for PTFE, with reversal of zeta potential; similar data (pI of 3.25 for PTFE) have also been given by Rohlicek et al. (33).

30.5. DETERMINATION OF pI VALUES OF PROTEINS

A protein's pI is an important physicochemical parameter, since it is a particular point of the protein's titration curve at which its net surface charge is equal to zero. The pI value is very sensitive to any chemical modification (e.g., it acidifies upon acetylation, phosphorylation, deamidation, glycation, glycosilation, and carbamylation, but in particular cases it could also become more alkaline, like upon trisulfide formation) (34). Additionally, the pI is sensitive to changes in conformation in the native state, e.g., as induced even by neutral-to-neutral amino acid substitutions (35). The pI value can be measured with extreme precision by isoelectric focusing in immobilized pH gradients (3), down to a minimum of ΔpI = 0.001 pH units. In CZE, a few approaches have been reported for pI assessment under IEF conditions. In the simplest approach, unknown pI values can be assessed by plotting the pI values of a set of markers, cofocused with the proteins under investigation, vs. their relative mobility upon elution. In the vacuum method proposed by Chen and Wiktorowicz (36), this plot is linear and thus a relatively high precision (ca. $\pm$ 0.1 pH unit) is obtained (see Figure 30.3). According to these authors, pI values as low as 2.9 (for RNase T_1–wild type) and 2.75 (for unsulfated cholecystikinin flanking peptide) were determined. In another, more intriguing approach, in the focusing of transferrin (37), Kilàr (38) has proposed a novel method for pI assessments: monitoring the current in the mobilization step. If one simultaneously monitors the peaks of the mobilized stack of proteins and the rising current due to passage of the salt wave in the capillary,

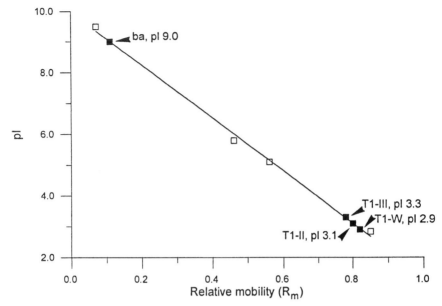

Figure 30.3. Calibration graph for pI determination using a set of marker proteins. The markers (open squares) are ribonuclease A (pI 9.45), carbonic anhydrase (pI 5.90), β-lactoglobulin (pI 5.1), and unsulfated cholecystikinin flanking peptide (pI 2.75). The four solid squares represent four unknown proteins whose pI's have been determined by linear interpolation in the calibration graph. [From S. M. Chen and J. E. Wiktorowicz, *Anal. Biochem.* **206**, 84 (1992), with permission.]

one can correlate a given pI value (which should be known from the literature a priori though) with a given current associated with the transit of a given peak at the detector port. The system can thus be standardized and used for constructing a calibration graph to be adopted in further work, without resorting to "internal standards." One such graph, correlating current with pI values, is shown in Figure 30.4: this appears to be a precise method, since the error is given as only about 0.03 pH unit. In another approach, Kleparnik et al. (39) proposed an absolute determination of pI values based on the application of an additional electroosmotic and/or hydrodynamic flow of a background electrolyte. The mobilities of a given amphoteric species are measured at various pH values so as to find the pH at which the substance moves through the capillary at zero mobility under the influence of only the additional flow. Since the number of experimental points needed to find the precise point of zero mobility could be quite high, it is kept to a minimum value with the help of an iterative procedure based on the *Regula Falsi* algorithm.

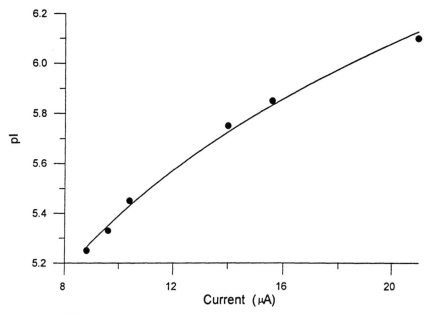

Figure 30.4. Calibration graphs for pI determination using the current during the mobilization step as a parameter in capillary IEF. The six experimental points represent six forms of transferrin containing different amounts of sialic acid and iron. [From F. Kilàr, *J. Chromatogr.* **545**, 403 (1991), with permission.]

30.6. DETERMINATION OF ABSOLUTE MOBILITY AND ITS RELATION TO THE CHARGE/MASS RATIO IN PEPTIDES (AND PROTEINS)

This field is quite controversial, and I will attempt to offer the different viewpoints of a variety of research groups, often in conflict. Already the problem of determining absolute mobilities is not an easy one, and to this purpose the technique of choice, as stated above, had been ITP (17–21). Since, however, ITP runs were quite cumbersome and difficult to interpret, CZE runs have now replaced ITP measurements. Some authors (40) went as far as to propose a double-detector system for measuring mobilities in CZE, thus obtaining data comparable to those in the literature, with a relative standard deviation (RSD) of ~1%. It should be remembered that, even when one is holding everything constant (ionic strength, pH, voltage gradient, solvent type, and dielectric constant), a most important parameter is the temperature: all ionic species show a similar dependence of their absolute mobilities on temperature (about 2% per °C).

Among analysts dealing with proteins and peptides, there has been much debate as to what relationship would give the best fit in correlating μ with Z/r. The earliest attempt is attributed to Offord (41), who has shown a linear relationship between paper electrophoretic mobilities of a series of charged peptides and their molecular mass to the power of $-2/3$. In principle, according to Eq. (30.11), there should be a linear relationship between μ_{ep} and Z/r. However, Grossman et al. (42) proposed a semiempirical model that related μ_{ep} of a range of positively charged peptides to their size, charge, and hydrophobicity. The effects of size and charge were determined independently and then combined to show the following relationship:

$$\mu_{ep} = [A \log(Z + 1)/n^{0.43}] + B \tag{30.18}$$

where n is the number of amino acids in the peptide, and A and B are constants related to the solvent system used. Compton and O'Grady (43,44) modified Eq. (30.11) by applying the Debye–Hückel–Henry theory to account for ionic effects:

$$\mu_{ep} = Ze\,\Phi(kr)/[6\pi\eta r(1 + k)] \tag{30.19}$$

where $\Phi(kr)$ is Henry's function (45) and k is the Debye screening parameter. By expressing the size dependence of μ_{ep} in terms of molecular mass (M) rather than radius (r), Compton and O'Grady obtained a general equation of the form

$$\mu_{ep} = K_1 Z/(K_2 M^{1/3} + K_3 M^{2/3}) \tag{30.20}$$

where K_1, K_2, and K_3 are three terms that include common physical constants, the solution ionic strength, and the frictional ratio (f/f_0). Further, they indicated that, after some simplifications and transformations of Eq. (30.19), a relationship similar to Eq. (30.18) is obtainable. On the other hand, if $K_2 \ll K_3$, Eq. (30.20) becomes similar to the correlation observed by Rickard et al. (46), where

$$\mu_{ep} = Z/M^{2/3} \tag{30.21}$$

Thus, it would appear that the different approaches in Grossman et al. (42) and Rickard et al. (46) provide similar results because both are particular cases deriving from a more general theory elaborated by Compton and O'Grady (43,44). There are at least two major discrepancies here that need to be resolved: (i) Is the relationship between μ_{ep} and Z/r linear or logarithmic? (ii) Which power function of molecular mass ($M^{1/3}$, $M^{2/3}$, or $M^{1/2}$) best fits this relationship?

The last question first: if we assume that the structure of the peptide in the capillary can be approximated by a sphere of constant density, then the radius of the sphere would be proportional to $M^{1/3}$. Hence, if frictional drag is governed by Stokes' law, then μ_{ep} would be related to the radius of the species, i.e., to $M^{1/3}$. However, if frictional drag is related to the surface area (or cross-sectional area) of the molecule, then μ_{ep} would be proportional to $M^{2/3}$. On the other hand, studies on synthetic polymers have shown that the radius of gyration is proportional to the square root of the number of monomers in the polymer multiplied by the length of a single residue (47). If frictional drag were proportional to the average radius of gyration, then μ_{ep} would be proportional to the square root of the number of residues, i.e., to ca. $M^{1/2}$. In an extensive analysis of multiple phosphoseryl-containing casein peptides, Adamson et al. (48) tried any possible fit and found the best correlation ($r^2 = 0.993$) to be between μ_{ep} and $Z/M^{2/3}$ (see Figure 30.5a). The other two possible correlations, $Z/M^{1/3}$ (Figure 30.5b, p. 988) and $Z/M^{1/2}$ (Figure 30.5c, p. 989) gave poor fits. This would appear to settle at least the second question, but evidence to the contrary has recently been provided by Kalman et al. (49), this time not even with peptides (which should be easier to handle) but with native proteins (*Staphylococcus* nuclease mutants). The best linear fit these authors could obtain was μ_{ep} vs. $Z/M^{1/3}$; thus the question of what power of M should be taken still remains open.

As for the first question (linear of log relationship), the story is quite interesting. According to Castagnola et al. (50,51), the log relationship proposed by Grossman et al. (42) is intrinsically erroneous because it is based on calculations of net chage assuming fixed pK values of amino acids (AA; those tabulated for free AA). Conversely, Castagnola et al. (51) performed microtitration of peptides, found experimentally accurate pK values, and used those for calculating Z. Under these last conditions, the correct relationship was found to be a linear one, as predicted by Eq. (30.11). These authors thus argued that the apparent good correlation between μ_{ep} and $\log(Z+1)$ was simply due to the fact that a log function is less sensitive to errors in pK values than is the correct linear relationship between μ_{ep} and Z, where a good fit is only obtained by using very accurate pK values. This observation could also explain the results of Hilser et al. (52), who also found a better correlation between μ_{ep} and $\log(Z+1)$ than between μ_{ep} and Z. In agreement with the arguments of Castagnola et al. (51), Adamson et al. (48), when replotting their mobility data according to the dependence $\log(Z+1)$, found a much worse fit than with the linear relationship (Figure 30.5d, p. 990). Yet, it cannot be denied that other subtle parameters might play a role in peptide separation. An interesting example has been offered by Meyer et al (53): when working with the thioxo peptide Ala-Phe-Ψ[CS-N]-Pro-Phe-4-nitroanilide, peak splitting was observed at 25 °C. Both peaks, however, merged as the temperature was in-

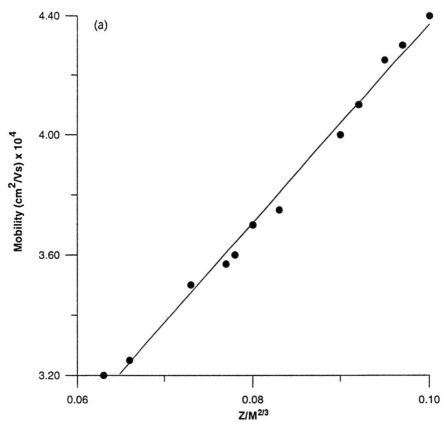

Figure 30.5. Study of the different possible relationships between absolute electrophoretic mobility and the Z/M ratio for 13 multiple phosphoseryl-containing peptides purified from casein tryptic and pancreatic digests: (a) plot of μ_{ap} vs. $Z/M^{2/3}$; (b) plot of μ_{cp} vs. $Z/M^{1/3}$; (c) plot of μ_{ep} vs. $Z/M^{1/2}$; (d) plot of μ_{ep} vs. $\ln(Z+1)/n^{0.43}$. Note the very good fit in panel a and the poor fit for all other relationships. [From N. Adamson, P. F. Riley and E. C. Reynolds, *J. Chromatogr.* **646**, 391 (1993), with permission.]

creased to 60 °C, but they reappeared when the heated sample was cooled down. This behavior was attributed to the electrophoretic separation of the cis/trans prolyl bond isomers of the thioxo peptide. These authors calculated that the mobility of the cis form is 1.027 times higher than that of the trans isomer. In fact, molecular modeling showed the cis conformer to be of more compact cylindrical shape than the trans isomer, behaving as a slightly more expanded nearly spherical object. Computations revealed a 1.015-fold elevation of the average volume of the trans isomer, hence the slightly reduced mobility as compared with the cis form.

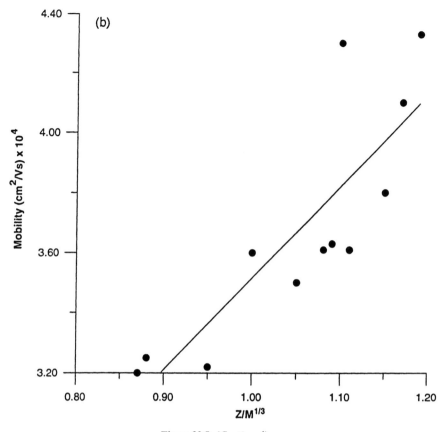

Figure 30.5. (*Continued*)

30.7. DETERMINATION OF BINDING CONSTANTS

Determination of association (K_A) or dissociation (K_D) constants constitutes an important part of electrophoretic techniques. In gel slab electrophoresis, a vast literature already exists (54,55). The basic equation is the one derived by Takeo (54):

$$1/r = 1/R_0(1 + C/K_D) \tag{30.22}$$

It is used either as such or in its extended form:

$$1/(R_0 - r) = 1/(R_0 - R_c)(1 + K_D/C) \tag{30.23}$$

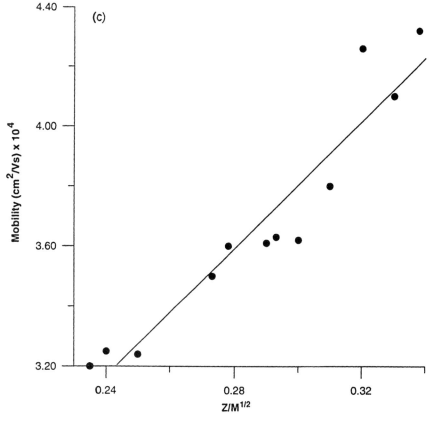

Figure 30.5. (*Continued*)

where r is the relative migration distance of the protein in the presence of affinity ligands in the electrophoresis gel; R_0 is the relative migration distance of the free protein in the absence of a ligand; R_c is the relative migration distance of the complex between the protein and ligand, or that obtained in the presence of an excess of the ligand when all protein molecules are bound to the ligand; and C is the total ligand concentration (molarity) in the gel. A series of runs is made, exploring an ample interval of ligand molarity in the gel and the relevant migration parameters measured. When Eq. (30.22) is plotted, taking $1/r$ as the ordinate and C as the abscissa, a straight line is obtained; its intercept on the abscissa gives the negative value of K_d. A variant of this method has been described called "affinity titration curves," by which,

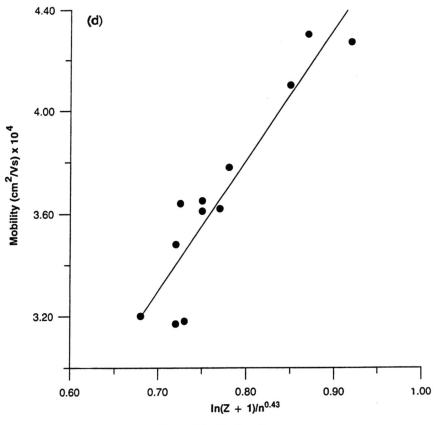

Figure 30.5. (*Continued*)

in a single gel, not only K_d is measured but also its dependence on the pH scale. A plot of K_d vs. pH will give a parabolic curve complementary to the classical bell-shaped curve giving the dependence of enzyme activity on pH (56,57).

The same philosophy has been applied to CZE experiments. Some examples: Okada (58) has measured complexation constants between polyoxyethylene and various cations and of benzo-12-crown-4 with different cations. Honda et al. (59) and Winzor (60), by affinity CZE, have derived the association constants among a few β-galactose-specific lectins and the sugar lactobionic acid.

30.8. DETERMINATION OF DIFFUSION CONSTANTS

Knowledge of the diffusion coefficient (D) of macromolecules is very important in biochemical and biophysical analysis because this quantity is related to the mass and shape (stokes radii) of these substances. In the early 1950s, Stokes (61) proposed a method based on a diffusion cell having a porous disk separating the pure solvent from the solution. The method gave only relative, not absolute, data. Another method much in vogue then was the use of the analytical centrifuge, by which, on the same time photograms used for measuring the sedimentation coefficient (by monitoring the movement of a boundary through Schlieren optics), one could also assess D values (62). More recent methods are based on dynamic laser light scattering (63).

In a classic paper in the 1950s, Taylor (64) expounded the concept of using the dispersion of a solute plug in a laminar Poiseuille flow for measuring the diffusion coefficient of solute molecules. Aris (65) extended Taylor's theory and proposed the following equation linking the dispersion coefficient (D^*) in a capillary to the true coefficient of molecular diffusion (D):

$$D^* = D + R^2 v^2/48D \tag{30.24}$$

where R is the tube radius and v is the mean velocity of the fluid in the capillary. Equation (30.24) provides the basis for obtaining the true D value of a solute molecule through an experimental measurement of its dispersion coefficient D^* in a fluid of known mean velocity flowing in a tube of known radius. Experimentally, precise assessements of v and D^* can be obtained through the following two equations:

$$L_d/v = m_1/2[3 - (1 + 4m_2/m_1^2)^{1/2}] \tag{30.25}$$

$$D^* = \tfrac{1}{2}v^2(m_1 - L_d/v) \tag{30.26}$$

where L_d is the length of the capillary to the detector, and m_1 and m_2 are the normalized first and second moments of the sample plug traveling inside the tube, respectively (the Taylor–Aris theory assumes that the distribution of the mean sample concentration over the capillary axis becomes Gaussian after a relatively short time). Substituting the experimental values of v and D^* (and the known value of R) into Eq. (30.24) allows calculation of the true coefficient of molecular diffusion D. Indeed five superimposed runs were obtained for estimating D of ovalbumin in a CZE unit: we obtained a $D = 0.759 \times 10^{-6}\,\text{cm}^2/\text{s}$ vs. a tabulated value of $0.776 \times 10^{-6}\,\text{cm}^2/\text{s}$ (error: 2%) (66). It should be noted that our method does not require use of applied voltage: the analyte plug is simply pumped through the capillary tube. Other methods, using "stopped migration" and "low electric fields" have been adopted for measuring D of small analytes, proteins, and oligonucleotides (67–70).

30.9. VISCOSITY MEASUREMENTS

The possibility of assessing the viscosity of a liquid by measuring its velocity in a tube is well known and follows from the classical Poiseuille equation (71,72). In order to implement this method, one needs tubings, a pump with connections, a marker, and a detector. Fortunately, a CZE unit contains all these components and hence can be used for viscosity measurements. We have derived an approximate equation for calculating the time taken by the boundary between two liquids to migrate from one end of the capillary to the UV window (73):

$$L_d \eta_2 / 2L_t + (1 - L_d/2L_t)\eta_1 = (\Delta t\, \Delta p d^2)/(32 L_t L_d) \qquad (30.27)$$

where d is the capillary diameter; Δp is the total pressure drop; Δt is the small time during which the boundary is migrating from the entrance of the capillary to the detector window; L_d is the length of the capillary to the detector port; L_t is the total capillary length; and η_1 and η_2 represent the known viscosity of a reference liquid and the unknown viscosity to be measured, respectively. By this method, viscosities of small analytes (e.g., sucrose solutions) and of macromolecular solutions (e.g., cellulose) can be assessed with a precision on the order of 3%. The boundary between the two liquids is usually detected by refractive index gradients, even in presence of non-UV-absorbing species. If the refractive index variation is minute, the boundary can still be detected by spiking one of the two solutions with traces of a strongly UV-absorbing compound (e.g., riboflavin).

30.10. DETERMINATION OF T_m VALUES OF NUCLEIC ACIDS

The melting temperature (T_m) of a DNA double helix is defined as the midpoint temperature at which each base pair is at 50/50 equilibrium between the helical and melted states. In the 1960s extensive measurements were performed (spectrophotometrically) to evaluate T_m's of nucleic acids, since this value is linked to the chemical composition of a DNA: a pure poly(GC) melting at 110 °C vs. a pure poly(AT) with a melting point of only 80 °C. Today's interest in T_m's stems from the possibility of detecting point mutations in DNA. In this last case, the separation technique of choice has been denaturing gradient gel electrophoresis (DGGE), first described by Fisher and Lerman (74). It is based on variation of electrophoretic mobility of a double-stranded (ds) DNA molecule through linearly increasing concentrations of denaturing agents (typically formamide and urea). As the DNA fragment proceeds through the gradient gel, it will reach a position where the concentration of the denaturing agent equals the T_m of its lowest melting domain, causing denaturation, partial unwinding, and consequent marked retardation due to frictional drag. The T_m and in fact the entire melting profile of DNA molecules of known sequence, together with calculations of the expected changes in electrophoretic mobility in gels under denaturing conditions, can be predicted with accuracy via the computational simulation program developed by Lerman and Silverstein (75). In the DGGE technique, nonisocratic conditions along the migration path are "tuned" to the slightly different T_m values of the duplexes present in the analyte, allowing separation of DNA fragments differing by as little as a single base substitution (76). In order to identify all mutants, the wild type and (presumptive) mutant chains are fully melted and reannealed while mixed in solution prior to analysis. This results in a spectrum of four ds-DNA fragments: two homo- and two heteroduplexes. A variant of DGGE employs temperature gradients along the migration path and is known as TGGE (77). In TGGE the temperature gradient is established in the space axis before electrophoresis, usually with the aid of two thermostatic baths set at the gel extremities, either parallel or perpendicular to the migration axis.

We have reported the possibility of performing CZE in a nonisocratic mode, i.e., under temperature-programmed conditions (called TGCE: thermal gradient capillary electrophoresis) (78). As a fundamental distinction, unlike TGGE, where the temperature gradient exists along the separation space and is controlled externally via circulating liquid and thermostats, the denaturing temperature gradient in the fused-silica capillaries is generated internally via ohmic heat produced by voltage ramps. The temperature increments (ΔT) produced inside the capillary by given voltage gradients (E, in V/cm) can be calculated according to

$$\Delta T = \kappa E^2 d^2 / 4\lambda \qquad (30.28)$$

where κ is the buffer specific electric conductivity at a reference temperature (25 °C); d is the capillary diameter; and λ the thermal conductivity of the buffer solution. All thermal theories assume the buffer conductivity to be linearly dependent on temperature:

$$\lambda = \kappa [1 + \alpha (T - T_0)] \qquad (30.29)$$

where α is the temperature coefficient of conductivity and T is the temperature

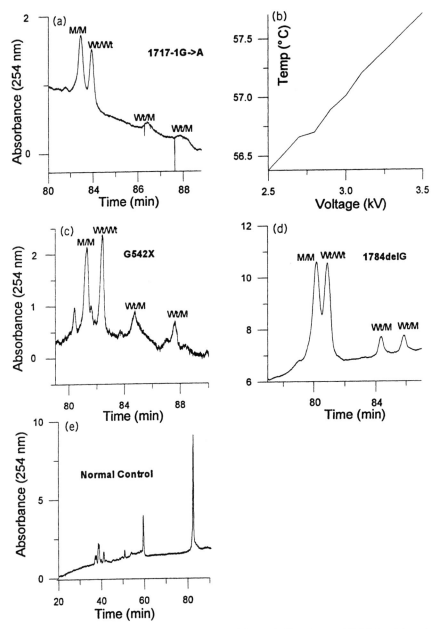

Figure 30.6. CZE analysis of 3 mutants in exon 11 of the cystic fibrosis gene: 1717-1G → A (panel a); G542X (c) and 1784delG (d); the normal control (NC) is in panel e. Unit: Bio Focus 2000 (Bio Rad, Hercules, California) equipped with a coated capillary, 60 cm long, 100 μm i.d. Background electrolyte: 8.9 mM Tris, 8.9 mM borate, 1 mM EDTA, 10 mM NaCl, containing 6 M urea and 6%T linear poly(AAEE: N-acryloyl aminoethoxyethanol) as the sieving entangled polymer. Sample injection: electrokinetic, 3 s at 4 kV, at a constant temperature plateau of 56.5 °C. (b) plot of the temperature profile over the applied voltage ramp. Abbreviations: M/M = mutated/mutated homoduplex; Wt/Wt = wild type/wild type homoduplex; Wt/M = wild type/mutated heteroduplex. [From C. Gelfi, P. G. Righetti, L. Cremonesi, and M. Ferrari, *Biotechniques* **21**, 926 (1996), with permission.]

inside the capillary (79). Thus, the experimental parameters needed for predicting the temperature increments linked to voltage ramps are the capillary diameter, its total length, the electric current values (μA) linked to a given applied voltage, and the buffer electric conductivity (κ) and its thermal coefficient of conductivity (α). Given these input parameters, a dedicated software allows precise determination of the capillary inner temperature (80). Graphs can then easily be constructed linking voltage ramps to temperature ramps. Figure 30.6 gives a unique example of such a separation: it can be appreciated that in all cases the two homo- and two heteroduplexes are fully resolved (81). Note that while in principle, by this technique, we could easily calculate the T_m from the temperature graph (see panel b) in reality we could not care less: the technique is much more important for detecting point mutations than for really assessing T_m values. From this point of view, the title of this section is misleading, but I hope the readers will forgive me if I used this trick to bring them into a unique realm of science!

30.11. CONCLUSIONS

Let me be telegraphic: after this excursus, exploring the variegated possibilities offered by CZE in measuring a variety of important parameters, is there any physicochemical parameter not accessible to CZE technology? I will let you find out!

LIST OF ACRONYMS AND SYMBOLS

Acronym or Symbol	Definition
AA	amino acids
CZE	capillary zone electrophoresis
2D	two-dimensional
DGGE	denaturing gradient gel electrophoresis
ds	double-stranded (DNA)
EDTA	ethylenediaminetetraacetic acid
EOF	electroosmotic flow
IEF	isoelectric focusing
ITP	isotachophoresis
MBE	moving-boundary electrophoresis
PAG	polyacrylamide gels
poly(AAEE)	poly(N-acryloyl aminoethoxyethanol)
PTFE	polytetrafluoroethylene (or Teflon)
RSD	relative standard deviation

RP–HPLC	reversed-phase high-performance liquid chromatography
SDS	sodium dodecyl sulfate
TGCE	thermal gradient capillary electrophoresis
ZE	zone electrophoresis
Z/r	charge/mass ratio

ACKNOWLEDGMENTS

I am indebted to various agencies [the Italian National Research Council (CNR), Target Project Biotechnology, the Italian Space Agency (ASI, grant ARS-96-214), and the Theleton (grant E. 555)] for supporting my group through several years of research, as covered in the present review.

REFERENCES

1. A. Tiselius, *Trans. Faraday Soc.* **33**, 524 (1937).
2. P. G. Righetti, *Isoelectric Focusing: Theory, Methodology and Applications.* Elsevier, Amsterdam, 1983.
3. P. G. Righetti, *Immobilized pH Gradients: Theory and Methodology.* Elsevier, Amsterdam, 1990.
4. M. Bier, O. A. Palusinski, R. A. Mosher, and D. A. Saville, *Science* **219**, 1281 (1983).
5. S. Hjertèn, *Chromatogr. Rev.* **9**, 122 (1967).
6. F. M. Everaerts, J. L. Beckers, and Th. P. E. M. Verheggen, *Isotachophoresis: Theory, Instrumentation and Practice,* J. Chromatogr. Libr., vol. 6. Elsevier, Amsterdam, 1976.
7. A. Chrambach and D. Rodbard, *Science* **172**, 440 (1971).
8. D. Rodbard, in *Methods of Protein Separation* (N. Catsimpoolas, ed.), Vol. 2, p. 145. Plenum, New York, 1976.
9. P. G. Righetti, M. Menozzi, E. Gianazza, and L. Valentini, *FEBS Lett.* **101**, 51 (1983).
10. L. Valentini, E. Gianazza, and P. G. Righetti, *J. Biochem. Biophys. Methods* **3**, 323 (1980).
11. J. Cai, J. T. Smith, and Z. El Rassi, *J. High Resolut. Chromatogr.* **15**, 30 (1992).
12. D. Poland, *Cooperative Equilibria in Physical Biochemistry*, p. 18. Clarendon Press, Oxford, 1978.
13. R. F. Cross and M. C. Ricci, *LC–GC Int.* **8**, 399 (1995).
14. J. A. Cleveland Jr., M. H. Benko, S. J. Gluck, and Y. M. Walbrohel, *J. Chromatogr. A* **652**, 301 (1993).
15. J. A. Cleveland, Jr., C. L. Martin, and S. J. Gluck, *J. Chromatogr. A* **679**, 167 (1994).

16. J. L. Beckers, F. M. Everaerts, and M. T. Ackermans, *J. Chromatogr.* **537**, 407 (1991).
17. T. Hirokawa and Y. Kiso, *J. Chromatogr.* **252**, 33 (1982).
18. T. Hirokawa, T. Tsuyoshi, and Y. Kiso, *J. Chromatogr.* **408**, 27 (1987).
19. J. Pospichal, M. Deml, and P. Bocek, *J. Chromatogr.* **390**, 17 (1987).
20. J. L. Beckers, *J. Chromatogr.* **320**, 147 (1985).
21. M. Polasek, B. Gas, T. Hirokawa, and J. Vacik, *J. Chromatogr.* **596**, 265 (1992).
22. J. Cao and R. F. Cross, *J. Chromatogr. A* **695**, 297 (1995).
23. S. Terabe, T. Yashima, N. Tanaka, and M. Araki, *Anal. Chem.* **60**, 1673 (1988).
24. M. W. F. Nielen, *J. Chromatogr.* **542**, 173 (1991).
25. H. Rilbe, *Acta Chem. Scand.* **25**, 2768 (1971).
26. R. D. Dandenau and E. H. Zerenner, *J. High Resolut. Chromatogr., Chromatogr. Commun.* **1**, 351 (1979).
27. W. H. Wilson, H. M. McNair, and K. J. Hyver, *J. Chromatogr.* **540**, 77 (1991)
28. C. Schwer and E. Kenndler, *Anal. Chem.* **63**, 1801 (1991).
29. T. L. Huang, P. Tsai, C. T. Wu, and C. S. Lee, *Anal. Chem.* **65**, 2887 (1993).
30. M. S. Bello, L. Capelli, and P. G. Righetti, *J. Chromatogr. A* **684**, 311 (1994).
31. B. J. Wanders, A. A. A. M. Van De Goor, and F. M. Everaerts, *J. Chromatogr.* **470**, 89 (1989).
32. A. A. A. M. Van De Goor, B. J. Wanders, and F. M. Everaerts, *J. Chromatogr.* **470** 95 (1989).
33. V. Rohlicek, Z. Deyl, and I. Miksik, *J. Chromatogr. A* **662**, 369 (1994).
34. J. Breton, N. Avanzi, B. Valsasina, L. Sgarella, A. La Flura, E. Breme, G. Orsini, E. Wenisch, and P. G. Righetti, *J. Chromatogr. A* **709**, 135 (1995).
35. G. Cossu, M. Manca, J. R. Strahler, S. M. Hanash, and P. G. Righetti, *J. Chromatogr.* **361**, 223 (1986).
36. S. M. Chen and J. E. Wiktorowicz, *Anal. Biochem.* **206**, 84 (1992).
37. F. Kilàr and S. Hjertèn, *Electrophoresis* **10**, 23 (1989).
38. F. Kilàr, *J. Chromatogr.* **545**, 403 (1991).
39. K. Kleparnik, K. Slais, and P. Bocek, *Electrophoresis* **14**, 475 (1993).
40. J. L. Beckers, Th. P. E. M. Verheggen, and F. M. Everaerts, *J. Chromatogr.* **452**, 591 (1988).
41. R. E. Offord, *Nature (London)* **211**, 591 (1966).
42. P. D. Grossman, J. C. Colburn, and H. K. Lauer, *Anal. Biochem.* **179**, 28 (1989).
43. B. J. Compton and E. A. O'Grady, *Anal. Chem.* **63**, 2587 (1991).
44. B. J. Compton, *J. Chromatogr.* **559**, 357 (1991).
45. D. C. Henry, *Proc. R. Soc. London Ser. A* **133**, 106 (1931).
46. E. C. Rickard, M. M. Strohl, and R. G. Nielsen, *Anal. Biochem.* **197**, 197 (1991).
47. C. H. Tanford, *Physical Chemistry of Macromolecules*, Chapter 3. Wiley, New York, 1961.

48. N. Adamson, P. F. Riley, and E. C. Reynolds, *J. Chromatogr.* **646**, 391 (1993).
49. F. Kalman, S. Ma, R. O. Fox, and C. Horvàth, *J. Chromatogr. A* **705**, 135 (1995).
50. M. Castagnola, L. Cassiano, R. Rabino, D. V. Rossetti, and F. Andreasi Bassi, *J. Chromatogr.* **572**, 51 (1991).
51. M. Castagnola, L. Cassiano, I. Messana, G. Nocca, R. Rabino, D. V. Rossetti, and B. Giardina, *J. Chromatogr. A* **656**, 87 (1994).
52. V. J. Hilser, Jr., G. D. Worosila, and S. E. Rudnick, *J. Chromatogr.* **630**, 329 (1993).
53. S. Meyer, A. Jabs, M. Schutkowski, and G. Fischer, *Electrophoresis* **15**, 1151 (1994).
54. K. Takeo, *Electrophoresis* **5**, 187 (1984).
55. K. Takeo, *Adv. Electrophoresis* **1**, 229 (1987).
56. K. Ek and P. G. Righetti, *Electrophoresis* **1**, 137 (1980).
57. K. Ek, E. Gianazza, and P. G. Righetti, *Biochim. Biophys. Acta* **626**, 356 (1980).
58. T. Okada, *J. Chromatogr. A* **695**, 309 (1995).
59. S. Honda, A. Taga, K. Susuki, S. Suzuki, and K. Kakehi, *J. Chromatogr.* **597**, 377 (1992).
60. D. J. Winzor, *J. Chromatogr. A* **696**, 160 (1995).
61. R. H. Stokes, *J. Am. Chem. Soc.* **72**, 763 (1950).
62. H. Neurath, *Chem. Rev.* **30**, 357 (1942).
63. B. J. Berne and R. Pecora, *Dynamic Light Scattering*. Wiley, New York, 1976.
64. G. Taylor, *Proc. R. Soc. London Ser. A* **219**, 186 (1953).
65. R. Aris, *Proc. R. Soc. London Ser. A* **235**, 67 (1956).
66. M. S. Bello, R. Rezzonico, and P. G. Righetti, *Science* **266**, 773 (1994).
67. Y. Walbroehl and J. W. Jorgenson, *J. Microcolumn Sep.* **1**, 41 (1989).
68. H. F. Yin, M. K. Kleemiss, J. A. Lux, and G. Schomburg, *J. Microcolumn Sep.* **3**, 331 (1991).
69. S. Terabe, O. Shibata, and T. Isemura, *J. High Resolut. Chromatogr.* **14**, 52 (1990).
70. Y. J. Yao and S. F. Y. Li, *J. Chromatogr. Sci.* **32**, 117 (1994).
71. R. B. Bird, W. E. Stewart, and E. N. Lightfoot, *Transport Phenomena*. Wiley, New York, 1960.
72. F. M. White, *Viscous Liquid Flow*. McGraw-Hill, San Francisco, 1974.
73. M. S. Bello, R. Rezzonico, and P. G. Righetti, *J. Chromatogr. A* **659**, 199 (1994).
74. S. G. Fisher and L. S. Lerman, *Proc. Natl. Acad. Sci. U.S.A.* **80**, 1579 (1983).
75. L. Lerman and K. Silverstein, *Methods Enzymol.* **155**, 482 (1987).
76. R. Fodde and M. Losekoot, *Hum. Mutat.* **3**, 83 (1994).
77. D. Riesner, K. Henco, and G. Steger, *Adv. Electrophor.* **4**, 169 (1991).
78. C. Gelfi, P. G. Righetti, L. Cremonesi, and M. Ferrari, *Electrophoresis* **15**, 1506 (1994).
79. M. S. Bello, M. Chiari, M. Nesi, and P. G. Righetti, *J. Chromatogr.* **652**, 323 (1992).
80. M. S. Bello, E. I. Levin, and P. G. Righetti, *J. Chromatogr.* **652**, 329 (1993).
81. C. Gelfi, P. G. Righetti, L. Cremonesi, and M. Ferrari, *Biotechniques* **21**, 926 (1996).

CHAPTER

31

APPLICATIONS OF MICELLAR ELECTROKINETIC CHROMATOGRAPHY IN QUANTITATIVE STRUCTURE–ACTIVITY RELATIONSHIP STUDIES: ESTIMATION OF LOG P_{ow} AND BIOACTIVITY

MORTEZA G. KHALEDI

Department of Chemistry, North Carolina State University, Raleigh, North Carolina 27695-8204

31.1.	Introduction	999
31.2.	Solute–Micelle Interactions and Hydrophobicity	1001
31.3.	Relationships Between Micellar Electrokinetic Chromatography Retention and log P_{ow}	1002
31.4.	Role of the Pseudostationary Phase	1005
31.5.	Prediction of Retention in Micellar Electrokinetic Chromatography from Solute Hydrophobicity	1011
31.6.	Quantitative Retention–Activity Relationships in Micellar Electrokinetic Chromtography	1011
31.7.	Conclusions	1012
	List of Acronyms	1013
	References	1014

31.1. INTRODUCTION

Over the past century, the field of quantitative structure–activity relationships (QSAR) (1) has grown rapidly. Overton (2) and Meyer (3) first reported the relationship between narcosis activity and the lipophilic nature of molecules. Hansch and Leo (4,5) revolutionized the modern discipline of QSAR by demonstrating the relationship between bioactivity and lipophilicity in the

---

High Performance Capillary Electrophoresis, edited by Morteza G. Khaledi. Chemical Analysis Series, Vol. 146.
ISBN 0-471-14851-2 © 1998 John Wiley & Sons, Inc.

form of:

$$\log(1/C) = m \log P_{ow} + s \tag{31.1}$$

where C is the concentration of the bioactive compound that induces a certain activity. The logarithm of partition coefficients between n-octanol and water, $\log P_{ow}$, has become the most widely accepted index for measuring lipophilicity of molecules (4).

Despite numerous reports by many workers using a variety of methods, measuring $\log P_{ow}$ is still a difficult task. The direct measurement of $\log P_{ow}$ using the conventional shake-flask (SF) method is time consuming, tedious, requires highly pure compounds in reasonable quantities, and has a limited dynamic range. Therefore, characterization of physicochemical properties of innumerable new compounds in QSAR studies can be severely restricted.

An alternative method is the indirect determination through linear relationships between $\log P_{ow}$ and the logarithm of the retention factor (k') in a chromatographic system, such as reversed-phase liquid chromatography (RPLC) and micellar electrokinetic chromatography (MEKC), where hydrophobic interaction is a predominant force contributing to solute retention as:

$$\log P_{ow} = m' \log k' + b' \tag{31.2}$$

This indirect method by RPLC has received much attention over the past 15 years due to the unique advantages of high-performance liquid chromatography (HPLC) such as small sample size requirement, speed, high sample throughput, better reproducibility, suitability for substances containing impurities and mixtures, wider dynamic range, and feasibility for automation (1,6). The use of RPLC for estimation of $\log P_{ow}$ suffers from several shortcomings: (i) the correlations are limited to congeneric compounds; (ii) determination of $\log P_{ow}$ values for ionizable compounds is difficult; (iii) the hydrophobic effect is not observed (or weakened) in hydro-organic eluents (especially for non-hydrogen-bonding mobile phases); and (iv) the retention factor is not a continuous scale and depends on mobile and stationary phases compositions. It is now widely believed that the most appropriate RPLC parameter for estimation of $\log P_{ow}$ is the retention factor with a purely aqueous mobile phase, k'_w. However, the direct measurement of k'_w is not trivial, if not impossible, for highly hydrophobic compounds with prohibitively long retention times. The estimation of k'_w values from the extrapolated relationship between $\log k'$ vs. Φ_{org} (volume percent of organic modifier) is not accurate due to the deviations from linearity. This problem is even more pronounced for ionizable compounds, especially for organic bases where a minimum is often observed in the $\log k'$ vs. Φ_{org} plots due to dual retention mechanisms (1,6).

In this chapter, the use of micellar electrokinetic chromatography (MEKC) for estimation of $\log P_{ow}$ is examined. In addition, the direct application of MEKC for prediction of biological activity in QSAR studies is discussed.

31.2. SOLUTE–MICELLE INTERACTIONS AND HYDROPHOBICITY

Micelles enhance the solubility of nonpolar organic compounds in aqueous media. The primary driving force for micellar solubilization is hydrophobic interaction, as the presence of nonpolar compounds in the bulk aqueous media disturbs the elaborate hydrogen bond network of water molecules. The hydrophobic solutes are then entropically repelled out of the bulk aqueous media and into the nonpolar environments of the micellar pseudophase. This unique mechanism of "interaction" is common among a number of processes in chemical, biological and environmental systems such as partitioning of solutes between biphasic solvent systems like octanol–water, distribution of analytes between the bulk aqueous mobile phase and the hydrocarbonaceous stationary phase in RPLC, transport of drugs across biological membranes, and adsorption of hydrophobic compounds onto soils and sediments. This common driving force is the physicochemical basis for the existence of linear free-energy relationships (LFER) among these systems. Examples that are relevant to this chapter are relationships between micelle–water and octanol–water partition coefficients of solutes as well as the correlations between micelle–water partition coefficients and bioavailability. These relationships are usually logarithmic since the free-energy change in distribution of a solute between two phases is directly related to the logarithm of the partition coefficient.

The first relationship between micelle–water partition coefficient (P_{mw}) and P_{ow} was reported by Collett and Koo more than 20 years ago for para-substituted benzoic acids in a nonionic micellar system (7). Similar studies were conducted by a number of other researchers for various types of ionic and nonionic micelles (8–12). A main obstacle to correlation analysis involving micellar solubilization for a wider range of compounds has been the difficulty of measuring P_{mw} using classical methods such as critical micelle concentration (CMC) determinations (10,11). The use of micelles as separation media in HPLC and in capillary electrophoresis (CE) has greatly facilitated studies involving of solute–micelles interactions.

The relationship between retention in micellar liquid chromatography (MLC) and $\log P_{ow}$ as well as $\log P_{mw}$ and $\log P_{ow}$ was investigated in our laboratory for various groups of aromatic compounds (13). In MLC, solute retention is a result of two competing partitioning equilibria—one between the bulk aqueous mobile phase and micelles, and the other one between the bulk aqueous mobile phase and the alkyl bonded stationary phase. The micelle–water partition coefficient can be determined from a linear relationship between the MLC retention factor and micelle concentration. In general, excellent correlations with $\log P_{ow}$ were found for MLC retetion and partitioning into cationic micelles of cetyltrimethylammonium bromide (CTAB). There results were consistent with those reported for $\log P_{mw}$ vs. $\log P_{ow}$ by other workers.

31.3. RELATIONSHIPS BETWEEN MICELLAR ELECTROKINETIC CHROMATOGRAPHY RETENTION AND LOG P_{ow}

The use of MEKC for estimation of partition coefficients to quantitate hydrophobicity of molecules provides new and exciting possibilities in QSAR research. In MEKC, solute retention, as measured by retention factor k', is directly related to micelle–water partition coefficient, P_{mw}, at low micelle concentrations (as was also discussed in Chapter 3 of this monograph):

$$k' \cong P_{mw} V_{mc}/V_{aq} \quad \text{or} \quad k' \simeq P_{mw} v([S] - \text{cmc}) \quad (31.3)$$

where V_{mc} is the micellar volume, V_{aq} is the aqueous phase volume, v is the surfactant molar volume, and [s] is the total molar cocentration of surfactant. Considering the earlier reports that indicated high linear correlations for the $\log P_{mw} - \log P_{ow}$ linear relationship between $\log k'$(MEKC) and $\log P_{ow}$ is not surprising.

Over the past few years, several papers have described the relationship between MEKC and $\log P_{ow}$. Takeda and coworkers determined partition coefficients (P_{mw}) of phthalate esters between bulk aqueous–SDS micelles by MEKC (14). They reported excellent linearity between $\log P_{mw}$ vs. $\log P_{ow}$ in purely aqueous and in 20% methanol-in-water solvents. Using the linear relationships, they predicted the migration times of the phthalates in MEKC from the known $\log P_{ow}$ values with less than 5% error for most of the analytes.

Chen and coworkers reported linear relationship between $\log k' - \log P_{ow}$ as well as between $\log P_{mw} - \log P_{ow}$ for a small set of aromatic solutes ($n < 10$) in MEKC with sodium dodecyl sulfate (SDS) as well as for magnesium dodecyl sulfate (MgDS) (15). The lines for SDS and MgDS were nearly parallel (i.e., same slopes), which indicated the same selectivity for the two micellar systems.

Ishihama and coworkers also showed linear $\log k'$ vs. $\log P_{ow}$ relationships in MEKC with a cationic (CTAB), an anionic (SDS), and a mixed nonionic–anionic (Brij 35–SDS) micellar system for a group of 18 aromatic solutes with different functional groups such as esters, ethers, or amides (16). Deviations from linearity for more hydrophobic solutes were attributed to uncertainties in the determination of t_{mc}, which would result in larger errors for longer-retained compounds. The correlation in the CTAB micelles was better than that observed for SDS. Addition of Brij 35 to SDS was effective in improving that correlation. This was attributed to the shielding of the SDS head groups by Brij 35 molecules in the mixed micellar system. They observed different migration orders for the three systems. Such variations in selectivity with the type of surfactant indicates the significance of selecting the appropriate type of micelles for log P_{ow} determination (*vide infra*).

Greenaway and coworkers also showed that log k' in MEKC with SDS micelles was statistically and significantly related to log P_{ow} (17). Adlard et al. reported high correlations between retention in MEKC with deoxycholic acid micelles and log P_{ow} for a group of 32 substituted aromatic compounds with different functional groups and various drug molecules. Using a calibration plot on the basis of the aromatic solutes, the latter group successfully predicted log P_{ow} of the drugs (18). They noted anomalous behavior for a number of compounds containing polar moieties such as nitro and keto substituents that had a longer retention than the more hydrophobic parent compound (i.e., certain compounds did not elute according to their hydrophobicity). This "anomalous" behavior is the result of specific interactions (mainly hydrogen bonding) between the solute and micelles (*vide infra*).

Herbert and Dorsey examined the correlation between MEKC retention using SDS micelles and log P_{ow} for more than 100 solutes with varying functional groups that also included ionic compounds (19). They reported an r^2 value of 0.84 for the whole set and higher correlation values for different subsets. In general, they reported different relationships for the non-hydrogen-bonding and hydrogen-bonding solutes. Such behavior is similar to that observed in RPLC, where different relationships are observed for congeneric sets of compounds. Such anomalies can be troublesome in determination of log P_{ow} values of new compounds, as the calibration plots would have to be constructed using standard solutes that belong to the same congeneric group and that exhibit similar retention behavior.

Smith and Vinjamoori also used MEKC with SDS micelles for rapid determination of log P_{ow}. Their test set contained highly hydrophobic compounds (log $P_{ow} \sim 5$–6); thus a small percentage of 2-propanol (10%) was added to the micellar solution to enhance the resolution for the more hydrophobic solutes (20). These authors reported an average error of 0.26 units in predicting log P_{ow} for 11 model compounds. They noted that while the shake-flask (SF) method requires milligram quantities of a pure sample, nanogram amounts of compounds even in a mixture would be sufficient using a microscale technique such as MEKC. In general, MEKC is more cost-effective as compared to the SF method and HPLC due to small solvent consumption. It was reported that once mobility of micelles was determined (for k' calculation) and the calibration plot was constructed, the throughput for determination of partition coefficients was 15 min per sample.

Jinno and Sawada investigated the relationship between retention in cyclodextrin-modified MEKC (CD–MEKC) and log P_{ow} for a group of polyaromatic hydrocarbons (21). Since cyclodextrins provide shape selectivity, elution order might not necessarily be according to solute hydrophobicity in CD–MEKC. This system would then have limited applications for estimation of hydrophobicity of compounds. Better correlations were observed with C_1- and C_{18}-modified capillaries as compared to untreated capillaries. However, the peaks with the C_1 and C_{18} capillaries were significantly tailed, which suggests strong wall interactions that might have actually contributed to the improved correlation coefficients.

Ishihama and coworkers introduced the use of microemulsion electrokinetic chromatography (MEEKC) for estimation of log P_{ow} (22,23). A microemulsion is a transparent solution with high solubilization capability. They used an oil-in-water microemulsion consisting of 1.44% (or 50 mM) SDS (as surfactant), 6.49 wt% butanol (as cosurfactant), and 0.82% heptane (as oil) in a borate–phosphate buffer at pH 7.0. They reported better correlations between retention and log P_{ow} for the microemulsion system as compared to that for the MEKC system with SDS micelles. For a group of 53 compounds that consisted of aromatic solutes as well as heteroaromatic compounds with substituents capable of forming hydrogen bonds, high correlation ($r = 0.996$) was observed between a MEEKC migration index and log P_{ow}. The thermodynamic parameters for solute partitioning in four different systems—microemulsion, SDS micelles, octanol–water, and dimyristoyl phosphocholine (DMPC) liposomes—were compared. Although the free-energy changes in the microemulsion, the micellar, and the octanol–water systems correlated with one another, little correlation was observed between the changes in enthalpies and entropies in the two chromatographic systems and those in octanol–water. On the contrary, the enthalpy and entropy changes in the microemulsion system have higher correlations with those in the gel-phase liposome system; however, the free-energy changes in these two systems had the lowest correlation. Based on the correlation analysis of the enthalpy and entropy changes, these authors concluded that "partitioning behavior" in the microemulsion system was similar to that in the gel-phase liposome system but different from that in octanol–water and the micellar systems. This conclusion, however, seems to contradict the principle of linear free-energy relationships, where linear relationships between free-energy changes in physicochemical and biological systems are interpreted as an indication of the existence of some commonality in their underlying forces or mechanisms. In another report, these authors extended the migration index vs. log P_{ow} relationship to anionic and cationic solutes [23,23a].

31.4. ROLE OF THE PSEUDOSTATIONARY PHASE

The influence of the surfactant type on the relationships between retention in MEKC and $\log P_{ow}$ has been investigated in our laboratory (24). Figure 31.1a–d illustrates the relationships between $\log k'$ and $\log P_{ow}$ for four different types of micelles—SDS, CTAB, sodium cholate (SC), and an anionic fluorocarbon surfactant, lithium perfluorooctane sulfonate (LiPFOS)—for a group of 60 aromatic solutes. It is evident that retention in the SC–MEKC system has the highest correlation with $\log P_{ow}$ ($r = 0.983$), whereas for the SDS and CTAB micelles different relationships were observed for three groups of compounds. The worst correlation was observed for the LiPFOS system ($r = 0.560$). The dependence of the retention vs. $\log P_{ow}$ relationship on the type of surfactant is not surprising since large variations in selectivity and elution patterns exist between different pseudostationary phases in MEKC (see

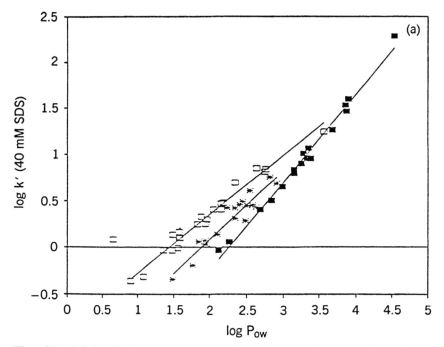

Figure 31.1. Relationships between $\log k'$ and $\log P_{ow}$ for a group of 60 compounds in different MEKC systems: (a) 40 mM SDS [$\log k' = (0.60 \pm 0.01) \log P_{ow} - 0.95$, $n = 60$, $r = 0.9301$, SE = 0.189]; (b) 10 mM CTAB [$\log k' = (0.60 \pm 0.01) \log P_{ow} - 1.31$, $n = 60$, $r = 0.8877$, SE = 0.248]; (c) 80 mM SC [$\log k' = (0.69 \pm 0.00) \log P_{ow} - 1.46$, $n = 60$, $r = 0.9829$, SE = 0.103]; (d) 40 mM LiPFOS [$\log k' = (0.30 \pm 0.02) \log P_{ow} - 0.86$, $n = 60$, $r = 0.5595$, SE = 0.352].

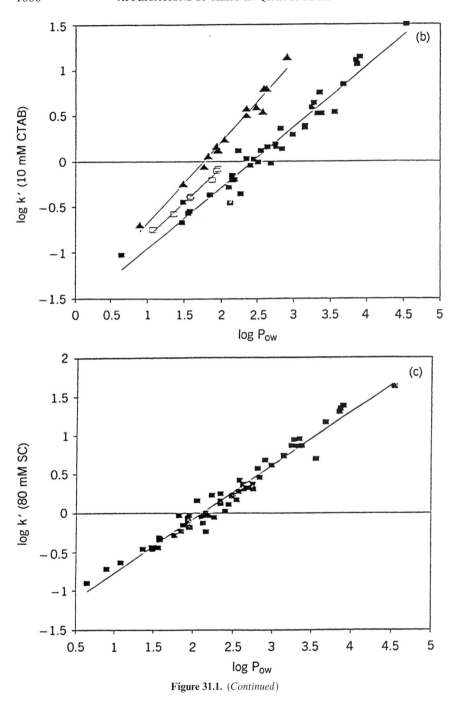

Figure 31.1. (*Continued*)

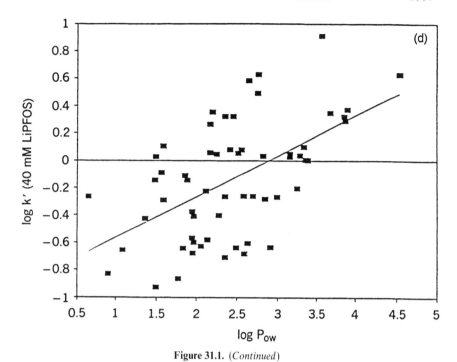

Figure 31.1. (*Continued*)

Chapter 3 of this monograph). Undoubtedly, hydrophobic interaction plays a major role; however, it is not the only factor that determines retention in MEKC. As discussed in Chapter 3, hydrogen-bonding interactions are also significant.

A comparison of the linear solvation-energy relationships (LSER) for various micellar and octanol–water systems provides an insight into the similarities and differences in retention behavior and the correlations between retention and $\log P_{ow}$ (25–27). The LSER models for retention factor k' in MEKC have the general form of

$$\log k' = (SP_0) + mV/100 + s\pi^* + b\beta + a\alpha \qquad (31.4)$$

where V is the solute molar volume (size); π^* is a measure of solute ability to engage in dipolar interactions; β is a measure of the solute hydrogen-bonding acceptor (HBA) strength; α is a measure of solute hydrogen-bonding donor (HBD) strength; and SP_0 is the regression constant, which is a function of the phase ratio. The four coefficients m, s, b, and a are measures of the cohesiveness, dipolarity/polarizability, HBD strength, and HBA strengths of the micellar solutions, respectively. Since retention factor, k' is directly proportional to the micelle–water partition coefficient, the LSER modeling of retention reflects solute–micelle interactions. The following five LSER models were observed for retention in the four micellar systems in MEKC and for octanol–water (O/W) partition coefficients:

40 mM LiPFOS:

$$\log k' = -1.51 + 2.44(V/100) - 0.25(\pi^*) + 0.16(\beta) - 0.98(\alpha) \quad (31.5)$$

40 mM SDS:

$$\log k' = -1.49 + 3.95(V/100) - 0.26(\pi^*) - 1.80(\beta) - 0.18(\alpha) \quad (31.6)$$

10 mM CTAB:

$$\log k' = -1.78 + 3.96(V/100) - 0.26(\pi^*) - 2.75(\beta) + 0.99(\alpha) \quad (31.7)$$

80 mM SC:

$$\log k' = -1.62 + 3.82(V/100) - 0.32(\pi^*) - 2.85(\beta) + 0.18(\alpha) \quad (31.8)$$

O/W:

$$\log P_{ow} = 0.17 + 5.62(V/100) - 0.66(\pi^*) - 3.90(\beta) + 0.14(\alpha) \quad (31.9)$$

The positive m coefficient is the largest in all LSER models. This suggests that solute size plays the predominant role in both micelle–water and octanol–water partitioning systems. The positive signs for the m coefficients indicate that bulkier molecules have stronger interactions with the organic phase (micelles or octanol). Note that the term $mV/100$ in Eq. (31.4) represents the energetically unfavorable process of separating solvent molecules in order to create a properly sized cavity for solutes. It is a direct measure of the difference in cohesive energies of the aqueous phase and the organic phase (micelles or octanol). Micelles have more a cohesive environment than does octanol, as indicated by the smaller m values. The three hydrocarbon-based micelles—SDS, CTAB, and SC—have nearly identical m coefficients. Thus, the cavity term does not contribute to the differences in the retention vs. $\log P_{ow}$ relationships for the three micellar systems.

The next important factor is solute basicity in all systems with the exception of LiPFOS, as the b coefficient is the second largest. The negative sign of the

b coefficient indicates that stronger HBA solutes have smaller affinity for either micelles or octanol. The LiPFOS micelles are the strongest HBD system (a larger b or less negative b coefficient), followed by SDS, CTAB, and SC. Among the five systems, octanol is the weakest HBD. The a coefficient is a measure of HBA strength and is significant for both LiPFOS and CTAB. The fluorocarbon micelle has the largest negative value and is considered the weakest HBA. The cationic micelles of CTAB have the largest positive a coefficient and therefore constitute the strongest HBA among the five systems. In addition to the m coefficient, both SC and CTAB have similar b coefficient values, which are closest to that for the O/W system. However, CTAB is a much stronger HBA than is octanol. On the basis of the LSER coefficients, one can conclude that SC micelles have closer interactive properties to octanol than do the other three micellar systems. It is not surprising, then, that the best correlation between retention and log P_{ow} was observed for the SC micelles. In contrast, the interactive properties of the LiPFOS micelles are very different from that of octanol, as is evident from the large differences in the LSER coefficients. The LiPFOS micelles are the most cohesive, the strongest HBDs, and the weakest HBAs among the five systems. On the other hand, octanol has the least cohesive character, is the weakest HBD, and is a stronger HBA than LiPFOS. Consequently, a poor correlation was observed between retention in LiPFOS–MEKC and log P_{ow}.

The congeneric behavior for the SDS and CTAB micelles, as shown by the existence of different lines for various groups of solutes, can also be interpreted in terms of the hydrogen-bonding characteristics of the two systems. SDS micelles are stronger HBDs than is 1-octanol; thus they exhibit more selective interactions towards HBA solutes. In fact, one can recognize a trend in the grouping of various solutes in the three log k' vs. log P_{ow} lines according to their HBA strengths (as measured by the Kamlet and Taft solvatochromic β values). For example, the first subgroup (Figure 31.1a, the line on the right with solid squares) consists of weak HBAs like hydrophobic substituted aromatic compounds such as alkylbenzenes, halogenated benzenes, and polycyclic aromatic hydrocarbons (PAHs) ($\beta \leq 0.20$). The second subgroup (the middle line with asterisks) contains HBA compounds with intermediate strengths such as aromatic ethers and some substituted nitrobenzenes ($0.20 \leq \beta \leq 0.35$). The third subgroup (the line on the left with open squares) comprises strong HBA compounds like alkyl aromatic ketones, benzonitrile, aromatic esters, aromatic alcohols, anilines, and some other nitrobenzenes ($\beta \geq 0.35$).

The congeneric behavior for CTAB (Figure 31.1b) can be interpreted in terms of the HBD strengths of the three subgroups. Since CTAB micelle is a stronger HBA than octanol, it selectively interacts with solutes with HBD functional groups. In Figure 31.1b the first subgroup (the right-hand line with

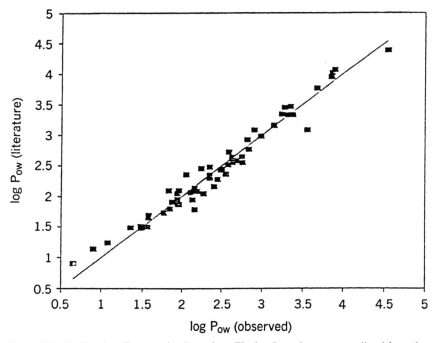

Figure 31.2. Predicted vs. literature $\log P_{ow}$ values. The $\log P_{ow}$ values were predicted from the relationship between retention in an MEKC system with the bile salt SC (80 mM) for a group of 60 compounds using the leave-one-out approach.

solid squares) consists of non-HBDs and weak HBDs ($\alpha \leq 0.17$) like anilines. The second subgroup (the middle line with open squares) contains solutes with intermediate HBD strength like benzyl alcohols ($0.33 \leq \alpha \leq 0.40$), while the left-hand line (solid triangles) includes phenols that are strong HBD with $\alpha \geq 0.54$.

One should also consider the differences in the micelle structures for alkylated surfactants like SDS and CTAB systems and bile salts such as SC. In general, micelles are dynamic aggregates with heterogeneous microenvironments. Solutes experience different microenvironment polarities as they are localized in or on micelles. Solute–micelle interactions occur through different mechanisms such as surface adsorption, partitioning into the micellar core, and comicellization. Therefore, the underlying interactive forces would be different between various groups of solutes depending on their locations. SDS and CTAB micelles are roughly spherical, with large aggregation numbers (between 60 and 80), whereas bile salt micelles such as SC are much smaller, with aggregation numbers around 2–10. There is a greater degree of heterogeneity in the SDS and CTAB micelle structures than in SC micelles. This leads to larger variations in the locations and microenvironment polarities in the alkyl chain micelles than in the bile salt.

In Figure 31.2 the $\log P_{ow}$ values that were predicted from the relationship with $\log k'$ in the SC–MEKC system are compared to the values in the literature. The high correlation for this plot indicates the suitability of the SC system for estimation of P_{ow} values for various types of compounds.

31.5. PREDICTION OF RETENTION IN MICELLAR ELECTROKINETIC CHROMATOGRAPHY FROM SOLUTE HYDROPHOBICITY

The relationships between $\log k'$ and $\log P_{ow}$ can also be used for the prediction of retention behavior in MEKC. There exists an extensive $\log P_{ow}$ database for thousands of compounds. For systems such as SDS and CTAB, where congeneric behavior is observed, one should first find the best $\log k'$ vs. $\log P_{ow}$ relationship for a given group of solutes. The situation is much easier for micellar systems like SC. By means of the leave-one-out approach, we have predicted retention factors of a group of aromatic solutes at 80 mM SC using the $\log k'$ vs. $\log P_{ow}$ relationship. As shown in Figure 31.3, a high correlation is observed between the predicted vs. observed retention factors for aromatic compounds in the MEKC–SC systems.

31.6. QUANTITATIVE RETENTION–ACTIVITY RELATIONSHIPS IN MICELLAR ELECTROKINETIC CHROMATOGRAPHY

In situations where hydrophobicity of molecules is the main driving force behind their biological response, one can also establish direct relationships between retention in MEKC and bioactivity. Yang et al. reported direct relationship between retention of a group of corticosteroids and two types of bioactivities: adsorption in the small intestine of rat and binding to human serum protein (24). Various micellar systems were investigated, and correlation as high as 0.96 was observed in the QRAR. The $\log P_{ow}$ and bioactivities of the corticosteroids were predicted by MEKC. Ishihama and coworkers also reported the relationships between retention in MEEKC and toxicity for a group of phenolic compounds (22). Since both hydrophobic and electronic effects influenced the biological response, a two-parameter equation had to be used with MEKC retention and pK_a as measures of hydrophobicity and electronic effects, respectively. Previously, Breyer and coworkers reported that bioactivity of this group of phenols could be described through a relationship with retention in MLC alone (28).

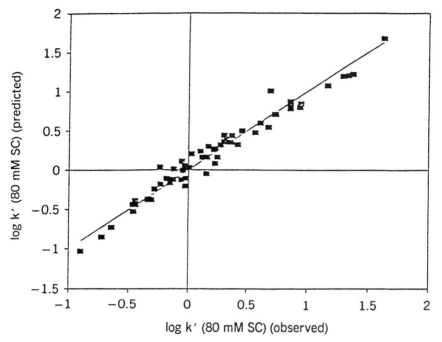

Figure 31.3. Predicted retention factors vs. observed retention factors for MEKC system with 80 mM SC using the leave-one-out technique.

31.7. CONCLUSIONS

MEKC has several potential advantages in structure–activity studies. First, MEKC offers the same features for physicochemical measurements as does HPLC (i.e., no sample purity requirement, small sample size, automation, etc.). Second, it is feasible with MEKC to adjust the composition of the micellar pseudostationary phase by simply changing the type of surfactant(s) in the system in order to provide better chemical models for the interactions in biological systems or for facilitating $\log P_{ow}$ determination. The composition of the pseudostationary phase in MEKC can be easily changed by rinsing the capillary with the new micellar solutions. Third, micelles are amphiphilic aggregates with an anisotropic microenvironment that provide both hydrophobic and electrostatic sites of interactions. In this respect they are more structurally similar to biomembranes than are n-octanol or RPLC stationary phases (the latter is anisotropic but not amphiphilic). As a result, micelles have long been known as simple chemical models for biomembranes. In MEKC, solute partitioning is between the bulk aqueous phase and the micelles, which resembles the hydrophobic partitioning into biomembranes. In RPLC, solutes partition from a mixed hydro-organic solvent into an alkyl bonded phase that is also enriched with the organic modifier from the mobile phase. Fourth, a great potential of MEKC is the possibility of standardization of retention. An advantage of $\log P_{ow}$ is that it is a single and continuous scale, at least in theory. In practice, there exist large variations in the $\log P_{ow}$ values (by as much as one unit), which reflects the difficulties of the measurements of estimations. In MEKC, the retention factor is directly related to the partition coefficients into micelles, which by definition is a single and continuous scale for a given surfactant system. We therefore expect that the number of applications of MEKC in QSAR-related studies will increase.

LIST OF ACRONYMS

Acronym	Definition
CD–MEKC	cyclodextrin-modified micellar electrokinetic chromatography
CE	capillary electrophoresis
CMC	critical micelle concentration
CTAB	cetyltrimethylammonium bromide
DMPC	dimyristoyl phosphocholine
HBA	hydrogen bond acceptor
HBD	hydrogen bond donor
HPLC	high-performance liquid chromatography
LFER	linear free-energy relationships
LiPFOS	lithium perfluorooctane sulfonate
LSER	linear solvation-energy relationships
MEEKC	microemulsion electrokinetic chromatography
MEKC	micellar electrokinetic chromatography
MgDS	magnesium dodecyl sulfate
MLC	micellar liquid chromatography
O/W	octanol–water (partitioning system)
PAHs	polycyclic aromatic hydrocarbons
QRAR	quantitative retention–activity relationship
QSAR	quantitative structure–activity relationships
RPLC	reversed-phase liquid chromatography
SC	sodium cholate
SDS	sodium dodecyl sulfate
SE	standard error
SF	shake-flask (method)

REFERENCES

1. R. Kaliszan, *Quantitative Structure–Chromatographic Retention Relationships* Wiley-Interscience, New York 1987.
2. E. Overton, *Phys. Chem.* **22**, 189 (1897).
3. H. Meyer, *Arch. Exp. Pathol. Pharmakol.* **42**, 109 (1899).
4. C. Hansch and A. Leo, *Substituent Constants for Correlation Analysis in Chemistry and Biology.* Wiley-Interscience, New York, 1979.
5. C. Hansch, in *Drug Design* (E. Aries, ed.), Vol. 1, p. 271. Academic Press, New York, 1971.
6. J. G. Dorsey and M. G. Khaledi, *J. Chromatogr.* **656**, 485 (1993).
7. J. L. Collet and L. Koo, *J. Pharm. Sci.* **64**, 1253 (1975).
8. E. Azaz and M. Donbrow, *J. Colloid Interface Sci.* **57**, 11 (1976).
9. H. Tomida, T. Totsuyanagi, and K. Ikeda, *Chem. Pharm. Bull* **26**, 2824 (1978).
10. C. Treiner, *Can. J. Chem.* **59**, 2518 (1981).
11. C. Treiner and A. K. Chattopadhyay, *J. Colloid Interface Sci.* **109**, 101 (1986).
12. K. T. Valsaraj and L. J. Thibodeaux, *Water Res.* **23**, 183 (1989).
13. M. G. Khaledi and E. D. Breyer, *Anal. Chem.* **61**, 1040 (1989).
14. S. Takeda, S. Wakida, M. Yamane, A. Kawahara, and K. Higashi, *Anal. Chem.* **65**, 2489 (1993).
15. N. Chen, Y. Zhang, S. Terabe, and T. Nakagawa, *J. Chromatogr.* **678**, 327 (1994).
16. Y. Ishihama, Y. Oda, K. Uchikawa, and N. Asakawa, *Chem. Pharm. Bull.* **42**, 1525 (1994).
17. M. Greenaway, G. Okafo, D. Manallack, and P. Camilleri, *Electrophoresis* **15**, 1284 (1994).
18. M. Adlard, G. Okafo, E. Meenan, and P. Camilleri, *J. Chem. Soc., Chem. Commun.*, 2241 (1995).
19. B. J. Herbert and J. G. Dorsey, *Anal. Chem.* **67**, 744 (1995).
20. J. T. Smith and D. V. Vinjamoori, *J. Chromatogr.* **669**, 59 (1995).
21. K. Jinno and Y. Sawada, *J. Liq. Chromatogr.* **18**, 3719 (1995).
22. Y. Ishihama, Y. Oda, K. K. Uchikawa, and N. Asakawa, *Anal. Chem.* **67**, 1588 (1995).
23. Y. Ishihama, Y. Oda, and N. Asakawa, *Anal. Chem.* **68**, 1028 (1996).
23a.Y. Ishihama, Y. Oda, and N. Asakawa, *Anal. Chem.* **68**, 4281 (1996).
24. S. Yang, J. G. Bumgarner, L. F. R. Kruk, and M. G. Khaledi, *J. Chromatogr.* **721**, 323 (1996).
25. S. Yang and M. G. Khaledi, *J. Chromatogr.* **692**, 301 (1995).
26. S. Yang and M. G. Khaledi, *Anal. Chem.* **67**, 499 (1995).
27. S. Yang, J. G. Bumgarner, and M. G. Khaledi, *J. Chromatogr.* **738**, 265 (1996).
28. E. D. Breyer, J. K. Strasters, and M. G. Khaledi, *Anal. Chem.* **63**, 828 (1991).

INDEX

Absolute mobility:
  charge/mass ratio, 985
  CZE measurements, 984
  determination, 984
  ITP measurements, 984
  Offord equation, 985
  peptides, 985
  precision, 984
Absorbed coatings:
  amines, 688, 690, 692, 689
  effect of EOF, 691
  hydrocarbons, 696–697
  neutral polymers, 694
  polyamines, 696
Absorption detection, indirect, 382
Acetate buffer, 55
Acetic acid, 527
Acetonitrile, 527, 529–531, 536, 540–541, 546, 550, 553, 705
Acetylation, of amino groups, 955
N-Acetylneuraminic acid, 743, 756
Acid–base equilibria, 29
Acidic peptides, 515, 644, 656
Acidic surface, 494
Acrylamide, 189
Acrylamide-bisacrylamide, 512
Acrylamide-coated capillaries, 229, 231
Acrylates, 698–700
Acryloylaminoethoxyethanol, 518
Activity coefficient, 33, 978
Adrenaline, 518
Adrenocorticotropic hormone, 649, 652
Adsorbed contaminants, 487
Adsorption, 486, 489
  basic compounds, 490
  of proteins, 952
  of SDS, 509
Adsorption and desorption kinetics, plate height, 151, 152

Adsorptive phase, 881, 883
Affinity capillary electrophoresis (ACE), 666, 884, 885, 888, 947
  advantages and disadvantages, 949
  choice of buffer, 952
  competitive, 955
  correction for changes in EOF, 952
  determination of binding constant, 950, 951
  determination of stoichiometry of binding, 964
  measuring effective charge of a protein, 963
  principles, 950
  prospects and limitations, 966
  separation of racemic mixtures, 969
  simultaneous determination of relative binding affinities, 957
Affinity capture microreactor, 891
Affinity cartridge, 891
Affinity CE–MS, 432, 957
  combinatorial library, 432, 437
  noncovalent complex, 432
Affinity chromatography, 883
Affinity column, 881
Affinity constant, 890
Affinity gel electrophoresis (AGE), limitation of, 948
Affinity ligand, 890
  in CGE, 205
Affinity macroligand, 198
Affinity microreactor, 890
Agarose, 189, 209, 229
Aggregation number, 83
Alcohol dehydrogenase (ADH), 928–930
  EMMA determination, 928, 932
Alkaline pH, 229
Alkaline phosphatase (ALP), EMMA determination, 932, 933
Alkaloids, nonaqueous CE, 543
Alkyl chain, 97

1016　　　　　INDEX

Alkylphenones–MEKC separation, 113
Alkylpolysiloxane, 492
Alkyltrimethylammonium bromide, 494
Amalgamated gold microelectrodes, 360
Amines for wall modification, 689
Amino acids, EC detection limits, 359, 360
Aminonaphthalene (ANTS), as fluorescence tag, 741, 742, 748, 753, 757, 759, 760
Aminoprop 20, 518
Aminopropyltrialkooxysilane, 514
Aminopropyltriethoxysilane, 514
Aminopropyltrimethoxysilane, 514
Aminopyrene-1,3,6-trisulfonate, 215
Aminopyrene-1,4,6-trisulfonate (ANDSA), as fluorescent tag, 741
Aminopyridine, as UV-Vis tag, 737, 748
Aminopyridyl, as UV-Vis tag, 737, 757
Aminosilylation, 518
Ammonium acetate buffer, 6
Ammonium persulfate, 698, 699
Ammonium sulfate, 705
Amperometric detection:
  carbohydrate detection, 733
  ion analysis, 845
Amperometry, 359
Ampholytes, 223, 225, 232, 234
Amphoteric compounds, 223
  determination of pK_a, 980
  isoelectric point, 980
Amylose oligomers, 759, 760
Analysis of single cell, 359
Analyte concentrator, 881, 884
Analyte preconcentration device, 881
Analyte stacking, 881
Analyte wall interaction, 483, 485, 486
  suppression, 496, 505, 517
Analytical separation performance, 507
ANDSA. *See* 9-Aminopyrene-1,4,6-trisulfonate (ANDSA)
Angiotensin analogues, 652, 662, 666
Anion analysis. *See* Inorganic ions
Anion determination, CE:
  addition of TTAB, 831
  addition of TTAOH, 832
  application of indirect detection, 835
  buffer formulation, 831, 832, 834
  chromate as co-ion, 831
  coated capillaries, 841
  common inorganic ions, 831, 832
  comparison with ion chromatography, 835
  direct UV detection, 841
  EOF reversal, 831
  negative peak, 831, 832
  nonaqueous separations, 541
  other co-ions, 835
  pyromellitic acid, 834
  reference peak, 831
  signal-to-noise ratio, 835
  zwitterionic buffer, 832
Anionic drugs, 505
Anionic polymers, 499
Anionic pseudostationary phase, 78
Anionic surfactants, 81, 97
Anodic EOF, 484
Antibiotics, chiral selectors:
  ansamycins, 802
  properties, 803
  vancomycin, 802, 803
Antibody, 706, 957
  activity, 885
  antigen, 883
  immunoadsorbents, 883
Anticoagulant peptide, 645
Anticonvective properties, 225
Antigen–antibody complex, 706
Antigen binding, 886
Antisense oligonucleotides, 625
Antisense therapeutic, 200
ANTS. *See* Aminonaphthalene (ANTS)
Apolipoprotein A, 693, 705, 706
Apparent mobility, 501, 827
Apparent pH, 538
Apparent p*I*, 244
Applications of MEKC, 128
Applications of permanent surface modifications, 507
Applied voltage, 35
  effects in peptide separation, 640–642, 657
APTS. *See* 8-Aminopyrene-1,3,6-trisulfonate
Aqueous media, characteristics, 526
Aqueous polymer, 514
Argon ion laser, 771, 781
Aromatic peptide pair, 653
Array detection, 769
Artificial substrates, 926
Aspartame, 663
Assessment of separations, 503
Association constant, K_A, 953
  determination, 988
Association rate constants, 963

Atomic absorption spectroscopy, 560
Atomic lamps, 311, 312
    Cd, 311
    deuterium, 311, 312
    Hg, 311, 323
    Hg–Xe, 323
    tungsten–halogen, 311
    Xe, 325
    Zn, 311
Attomole, 7, 880
Automation, in DNA sequencing, 201
Autoprotolysis constant, 533–535
Avalanche photodiodes, 773
Azetidinium-ion, 517
Azobisisobutyronitrile, 698

Background electrolyte (BGE), 252, 489, 719
    indirect detection, 378, 388
    parameters, 978
Background fluorescence, 773
Band broadening:
    contribution, 485
    microchip electrophoresis, 621
    proteins, 685
    retentive type, 486
Band broadening in CEC, 286
    extra column effects, 287
    mechanisms, 286
Band-broadening effects, validation, 561
Band broadening, MEKC, 141
Band dispersion. *See* Band broadening
Bare capillaries, 483
Base calling, 769, 780, 781
Baseline irregularity, 60, 61
Basic antifungal lipopeptides, 669
Basic compounds separation, 505
Basic peptides, 652, 667
Basic proteins, 494, 496
    separation, 496
Benzo-derivatives, as UV-Vis tag, 737
Benzoic acid isotopes, 979–980
Benzoic acids, separation of, 55, 56
Big CHAP, 105
Bile salts, 81, 99, 172
Binding:
    determination of kinetic parameters, 963
    ligand to receptor studies, 967
Binding affinity. *See* Binding constant
Binding assays, on-capillary, 905
Binding capacity, 886

Binding constant, 696, 905, 948
    determination, 950, 951, 953, 988–990
    in organic solvents, 808
    several isozymes to a ligand, 957
    several ligands to a receptor, 957
Binding of neutral ligands, 955
Binding site of protein, 955
Binding stoichiometry, determination, 964
Bioactivity, 215
    glycoproteins, 731
    relationship with log P_{ow}, 1000
Bioavailability, 1001
Biochemical reaction and analysis, microchip, 627
Biological activity, 215. *See also* Bioactivity
Biological fluids, 879, 880
    ACE determination of drugs in urine and serum, 885
Biological membrane, 1001
Biomolecular recognition, 968
Biopolymer characterization, 912
Biospecific reagents, 926
Biotin, 964
Biotin–avidin complex, 886, 887
Biotin–avidin interaction, 914
Biphasic solvents, 1001
BIS–acrylamide, 198
Bisphenol A, 514
Bond-specific enzymes, 732
Borate complexation, 104, 733, 736
    glycoproteins, 745–749
    oligosaccharides, 748
    polysaccharide separation, 745–749
Boron trifluoride, 697, 701
Borosilicate glass, 615
Boundaries, steady state, 254
Bound form, 955
Bovine adenosine deaminase, 669
Bovine aprotinin peptides, 662
Bovine β-casein, 668
Bovine erythrocytes, 953
Bovine fetuin, 705
Bovine serum, 593
Bovine serum albumin tryptic peptides, 639, 645, 652, 655–656
Bradykinin, 645
Brij 35 [polyoxyethylene (23) dodecanol], 81
Bubble capillaries, 174
Bubble cell, 313–316
Bubble formation
    in CEC, 282

1018　　　　INDEX

Bubble formation (*Contd.*)
  in CGE, 189, 198
Buffer:
  additives, 488
  aggressive, 516
  modifiers, 488
  selectors, 488
Buffer additives, 827, 832, 834
  concentration, 691
  cyclodextrins, 564, 566
  diamine, 704, 705
  effects in peptide separation, 647–654, 658
  polyamine, 690, 691, 696, 701–702
  polysaccharide separation, 745
  spermidine, 691
  surface binding, 684
  surfactant micelles, 566
Buffer choice, CE-MS, 414
Buffer co-ion, 157
Buffer depletion, validation, 561
Buffer viscosity, 500
Buffers, protein separation:
  acetate, 709
  BioRad LLV, 706
  borate, 691, 692, 703, 704, 710, 712
  phosphate, 692, 695, 704, 709, 710, 711
  Tris, 708
  vernol, 708
Bulk, property detection, 361

C18:
  polysiloxane-bonded columns, 518
  substituted silanization reagent, 512
Caffeine, 509
CAGE. *See* Capillary affinity gel electrophoresis (CAGE)
Calibration plot, p*I* determination, 983–984
Calmoduim, 649
Calmon calcitonin, 661
Camphor sulfonate, 552–553
Capacity factor, 84
Capillary affinity gel electrophoresis (CAGE), 198
  separation of dsDNA, 205
  separation of ssDNA, 198
Capillary array, 769, 773
Capillary bundles, 769, 886
Capillary coating:
  catalytic treatment, 509
  in cIEF, 228
  thermal treatment, 509

Capillary coiling, 38
Capillary dimensions, effect on glycoprotein separation, 733
Capillary dimensions in CGE, effects of length and i.d., 195
Capillary electrochromatography (CEC), 18, 277
  applications, 290
  band-broadening mechanisms, 286
  bubble formation, 282
  coupling with mass spectrometry, 296
  detection window, 280, 283
  extra column effects, 287
  general elution problem, 290
  gradient elution, 290
  Joule heating, 286
  mobile phase, 289
  molecular imprinted polymer, 297
  peak tailing, 280
  pharmaceuticals, 874
  polycyclic aromatic hydrocarbons, 290, 293
  polyimide coating, 281, 283
  pressurized reservoirs, 282
  solgel, 287
  solvent degassing, 282
  sonication, 281
  stationary phase, 287
  submicron particles, 296
  tetracycline separation, 290
Capillary electrophoresis, characteristics, 880
Capillary electrophoresis, microfabricated devices, 613
Capillary electrophoresis–mass spectrometry 16, 405–446. *See also* CE–MS
Capillary electrophoresis microreactor, 886
Capillary electrophoresis, separation:
  of carbohydrates, 730
  of glycoconjugates, 729
  of glycopeptides, 753–755
  of glycoproteins, 732, 753–755
  of polysaccharide analysis, 757
  of proteins, 683–727
Capillary gel electrophoresis (CGE), 185, 709, 713–715, 768
  application in EMMA, 932
  applications, 196
  carbohydrate separation, 751
  chemical and physical gels, 189
  effect of capillary dimensions, 195
  effect of constant current, 191, 192
  effect of constant power, 191

effect of temperature, 191
effect of voltage, 191
injection, 195
microfabricated devices, 614
operation parameters, 187
operation variables, 190
separation fundamentals, 187
size selectivity, 205
theory, 187
versus HPLC, 197
Capillary inner diameter, 573
effects in peptide separation, 640, 657
Capillary isoelectric focusing (cIEF), 10, 226
Capillary isotachophoresis, 251, 884
Capillary length, effects in peptide separation, 639–640, 657
Capillary rinsing procedure, 488
Capillary stability, 514
Capillary surface, 483
electroosmotic flow, 483
Capillary temperature, 157
effects in peptide separation, 642–643, 657
Capillary wall, 981–982
Capillary zone electrophoresis:
    absence of EOF, 27
    physicochemical parameters, 973–998
    in the presence of EOF, 58
    theory, 25–75
CAPS buffer, 144
Carbohydrates:
    amperometric detection, 733
    CGE analysis, 214
    CGE separation, 751
    coated capillary separation, 751
    derivatization, 736, 737
    detection limit, 359, 360
    electrochemical detection, 733
    fluorescent tags, 739
    HPCE separation, 730
    LIF, 7, 34
    LIF detection, 733, 734, 754, 757
    MALDI-MS detection, 736
    mass spectrometry, 735
    MS detection, 733, 735
    profiling, 215
    reaction schemes, 738, 739
    separation using end-label, free-solution electrophoresis (ELFSE), 752
    UV-Vis tags, 737
Carbon fiber electrode, 845

Carbonic anhydrase, 891, 953–955
Carbowax, 509
Carboxybenzoyl quinoline carboxaldehyde (CBQCA), as fluorescence tag, 739–741, 745, 747, 757
Carboxymethylated insulin B chain, 660, 662
Carboxypeptidase A, 665
Carrier ampholytes, 232
Carrier electrolyte, 690
Cartridge fabrication, 884
Cascade macromolecules, 83, 110
Casein peptides, 664, 666
Catechols, electrochemical detection, 360
Cathodic mobilization, 234
Cation determinations, CE, 827
    alkali and alkaline metals, 837, 840
    co-electroosmotic flow, 837
    complexing agents, 837, 843
    direct UV detection, 843
    α-hydroxyisobutyric acid, 837
    imidazole, 839, 840
    indirect detection, detection limit, 837
    laser-induced fluorescence, 843
    MEKC, 843
    metal–cyanide complexes, 843
    methylbenzylamine, 837
    nonaqueous separations, 540
Cationic additives, 827
Cationic analytes, 505
Cationic cyclodextrin, 817
Cationic double layer, 501
Cationic modifiers, 492
Cationic polybrene, 505
Cationic polyelectrolyte, 835
Cationic polymers, 490, 496
    stability, 505
Cationic pseudostationary phase, 78
Cationic surfaces, 507
Cationic surfactants, 81, 104, 490, 866
CCD camera, 770, 771, 773
    advantages and disadvantages, 771
CEC column preparation, 279
    drawn packed columns, 282
    sintering, 280
    slurry-packed columns, 280
    sonication, 281
CEC equipment, 278
    bubble formation, 279, 282–283
    frits, 279–280
    gradient elution, 293

CEC. *See* Capillary electrochromatography (CEC)
Cell component analysis, 10
Cell function, analysis using CGE, 214
Cell injection procedure, 901
Cell lysates, analysis of trace components, 904
Cells, 880
Cellulose, 775, 776
   chemical modification, 750
   electromigration, 750
   hydrophobicity of, 750
Cellulose acetate, 515, 694, 695
Cellulose derivatives, 189
CE–MS, 16, 405, 507, 881, 883
   interface, 507
   ion source, 507
CE–MS applications, 429, 440
   affinity CE, 432
   in vivo microdialysis, 437
   oligonucleotides, 432, 437
   peptides and proteins, 416, 420, 424, 430, 437
   pharmaceuticals, 430
CE–MS interface design, 408
   liquid junction interface, 409, 412, 413
   sheath flow interface, 409, 410, 411
   sheathless interfaces, 414, 422
CE–MS sample concentration:
   cIEF, 438
   isotachophoresis, 416
   precolumn and membranes, 416, 438
   sample stacking, 415
Centripetal force, 296
Cetyltrimethylammonium bromide, (CTAB), 492
CF–FAB. *See* Continuous flow–fast atom bombardment (CF–FAB)
CGE. *See* Capillary gel electrophoresis (CGE)
Channel manifolds, 615
CHAPS (3-[(3-cholamidopropyl)dimethylammonio]-1-propanesulfonate), 81
Charge-coupled device (CCD), 770
Charged cyclodextrins, 171
Charged solutes
   MEKC, 85
   nonaqueous CE, 540
Charged surface, 505
Charge-injection device, 771
Charge ladder, 955
Charge-masking agents, 689
Charge-to-mass ratio, 974

   in CGE, 198
Charge-transfer additives, 177
Charge-transfer complexation, 550
Chelated metals, 891
Chemical aggression stability, 512
Chemical composition, micellar, 95
Chemical etching, 359
Chemical gels, 189
Chemically cross-linked gels, 189
Chemically modified electrodes, 360
Chemical reaction and analysis, microchip, 627
Chemical reactions in capillary, on-line coupling, 926
Chemical selectivity, 89
Chemiluminescence, 331–333
   amino acids, 332, 333
   applications in EMMA, 937
   peptides, 332
Chemisorption, 489
   mechanism, 497
   modifiers, 495
Chiral analysis, pharmaceuticals, 862–865
Chiral crown ethers, 801, 802
Chiral micelle, 105
Chiral purity and impurity, validation, 560, 561
Chiral recognition mechanism, 795, 810
Chiral resolution, 793, 794
Chiral selector concentration:
   effect on mobility, equation, 793
   optimum, 806, 807
Chiral selectors in peptide separation, 653–654
   bile salt, 653
   crown ether, 653
   hydroxyethyl-β-CD, 653
Chiral separation by CE, 791–823
   advantages, 792
   antibiotics, 802
   capillary electrochromatography, 812
   chiral crown ethers, 801
   chiral surfactants, 804
   co-electroosmotic flow, 794
   counter-electroosmotic flow, 811
   cyclodextrins, 796
   effect of selector concentration, 806
   efficiency, 814
   enantiomer-selector binding constants, 793
   enantioselectivity, 793
   indirect method, 813
   limitation, 792
   migration, 794–795, 811

organic solvents, 808
pH and ionic strength, 807
polysaccharides, 799
proteins, 804
resolution, 793
selectivity, 793
separation selectivity, 793
types of chiral selector, 795
Chiral separation by nonaqueous CE, 550–553
Chiral surfactants, 804
  bile salts, 805
  saponins, 805
  synthetic, 805
Cholate, 81
Chromatography, microchip, 614, 615
Chymotrypsinogen, 685, 690
CID cameras:
  duty cycle, 773
  integration time, 773
  limit of detection, 773
  noise, 773
cIEF, 226
  calibration, 236
  capillary coating, 228
  coating stability, 229, 230
  control of EOF, 228
  detection, 232
  electroneutrality, 232
  gradient mobilization, 232
  isoform separation, 232
  mass spectrometry detection, 247
  mobilization schemes, 226, 227
  simultaneous focusing and mobilization, 232
  whole column detection, 238, 240
Circular dichroism, 881
CITP. See Capillary isotachophoresis
Cleanup procedure, 881, 888
Clinical analysis, 7
Clinical chemistry, 926
CMC. See Critical micelle concentration
Coated capillaries, 8, 481, 775, 841. See also Wall coating
  absorbed, 694
  Brij-35-coated, 693
  carbohydrate separation, 751
  in CGE, 190, 214
  cross-linked polyacrylamide, 690
  dynamic, 694
  hydrocarbon, 696
  hydrophilic coating, 693

peptide separation, 656–657, 660
  pH, 695
  versus uncoated in CGE, 205
Coated membrane, 883
Coating, 483
  cellulose-derivative, 518
  covalent, 516
  cross-linked, 516
  hydrolytically stable, 518
  methylcellulose, 516
  noncovalent, 518
  peak symmetries, 499
  resolution, 499
  static, 508
  universal CE separation, 508
Coating stability in cIEF, 229, 230
Coaxial sheath flow interfaces, 412, 416, 418
Co-electroosmotic flow condition or co-EOF, 490, 827
  chiral separation, 794
Cofactors, 926
Co-ion, 48
  absorbance, 829
  of BGE, 460
  concentration, 830
  condition, 828, 830
Collagen, 666
Collection, in 2D separation, 593
Column diameter and length, CGE, 195
Columns. See also CEC column preparation
  centripetal packed, 296
Column temperature, 4
  CGE, 192
Combinatorial library, 432, 437
Commercial coated capillaries, 516
  criteria, 508
Competitive affinity capillary electrophoresis, 955
Complexation, 550
  in CGE, 188
Complex carbohydrates, 214
Complexing agents, 837, 838, 840
  citric acid, 837
  EDTA, 837
  glycolic acid, 837, 839
  lactic acid, 837, 838
  succinic acid, 839
  tartaric acid, 837
Complexometry determination of inorganic ions, 926

Compositional analysis of glycoproteins, 755
Compound identification, 407
Comprehensive 2D systems, 588
Concanavalin A, 892
Concentration detection limit (CLOD), 356, 418, 880, 881
Concentration gradient, 39
  detection, 335
Concentration of negative charges, 505
Concentration of polymer in CGE, 205
Concentration overload, 48, 390
  MEKC, 145
Concentration sensitivity, 931
Conductivity, 361
  electrical, 45
  thermal, 45
Conductivity detection, ion analysis, 845
Conductivity difference, 460, 828, 829
Conductivity gradient, 48
Conductivity ratio, relationship to peak asymmetry, 829
Confocal excitation, 769
Confocal fluorescence detection, single molecule detection, 631
Congeneric behavior, 1001, 1009
Constant potential EMMA, 928–931
  plateau product profile relationships, 931
Contamination of the surface, 487
Continuous bed columns, 289
Continuous denaturing, CGE, 195
Continuous flow–fast atom bombardment (CF–FAB), 408
Continuous free-flow zone electrophoresis, 661
Continuous-wave Ar–ion laser, 771
Control of electroosmotic flow in cIEF, 228, 238
Convection current, 359
Convective mixing, 225
Conventional electrophoresis of proteins, 684
Conventional gel electrophoresis, 190
Cooling of the column, 4
Copolymerization, 198
  of surface, 512
Copper electrode, 359
Coronary heart disease, 705
Corticosteroids, MEKC separation, 110
Coulombic explosion, 408
Coulomb interaction, 229, 496, 505, 507
Coulometric efficiency, 359
Counter-electroosmotic flow (counter EOF):
  chiral separation, 811

Counterions, 693, 716, 719
  for ITP, 259
Coupled capillary CE, 911
Coupled column system, 587
Coupling techniques, 844
Covalent coating, 228
Covalently bonded:
  hydrophilic polymers, 514, 516
  hydroxylic polymers, 516
  polymeric layers, 512
  PVA, 514
Covalently bonded micelles, 83
Covalently bound antibodies, 884, 887
Covalent modification:
  polymers, 697–702
  proteins, 964
  simple silanes, 697
Covalent PEG, 514
Covalent Si-O-Si, 516
Creatinine–acetate buffer, 11
Criteria for commercial coated capillaries, 508
Criteria for 2D separation, 586
Critical micelle concentration, 81, 96
Cross correlation of CE data, validation, 559, 561
Cross-linked coating, 516
Cross-linked gels, 775
Cross-linked polyacrylamide gel, 205
Cross-linked polyethyleneimine, 496
Cross-linked polymers, 189, 775
Cross-linking, 508
Crown ethers, chiral, 801, 802
CTAB, 81, 492. See also Cetyltrimethylammonium bromide (CTAB)
Current:
  effect of column radius, 5
  effect of conductivity, 5
  effect of electric field strength, 5
Current density, 60, 254
Cyanogen bromide peptide, 666
Cyclodextrin-modified MEKC, 115, 118
Cyclodextrins, 864
  derivatives, 497
Cyclodextrins, chiral selectors:
  charged, 798
  molecular dimensions, 796
  neutral, 796, 798, 806, 811, 812
  non-aqueous CE, 529, 550–552, 808–811
  polyanionic, 798, 799, 811, 812, 814, 817
  properties, 796

structure of β-CD, 797
water solubility, 796
Cystic fibrosis, 195
  gene, 994
Cytochrome c, 685, 690
CZE. See also Capillary zone electrophoresis
  of proteins, 693, 696, 697, 702, 703, 704,
    705, 706, 707, 708
  theory, 25–75

Dalargin, 669
Dark current, 773
Debye–Huckel–Henry theory, 985
Decoupler, 358
Degree of dissociation or ionization, 29, 978
Denaturing condition in CGE, 190
Denaturing gradient gel electrophoresis
  (DGGE), 992–993
Dendrimers, 83, 174
Deoxycholate, 81
Derivatization:
  carbohydrates, 736
  glycobiology, 736
  glycoconjugates, 736–737
  on-column detection, 330, 331
  postcolumn detection, 331
  precolumn detection, 330, 331
  of samples, 881
Desalting, 881
  of sample, CGE, 196
Desorption, 496
Destacking, 251
  dynamics, 265
Detectability, ion analysis, 829
Detection, 456, 460, 538–539
  bubble cell, 456
  electrochemical detection, 539
  glycoproteins, 733
  laser-induced fluorescence, 456, 539
  limit, 456, 460, 474
  mass spectrometry, 474–476, 538–539
  sample loading, 451
  sensitivity, 451, 474, 477
  short optical path, 450
  single molecule, 7
  UV absorbance, 538
  validation, 561
  z-shaped cell, 456
Detection cell, 305, 322
  bubble cell, 313–316

capillary, 305, 314
  on-column, 322
  postcolumn, 323, 324
  sheath-flow cell, 306, 323–325
  z-cell, 313–315
Detection in cIEF, 232
Detection limit, 7, 880
  MS detection, 414
  EC detection, 358, 359
  inorganic ions, 830–831, 837, 840, 841
  optical detectors, 308
Detection methods, 844
  chemiluminescence, 331–333
  fluorescence, 321–331
  hyphenated, 341
  light scattering, 343
  living cells, 346
  multidimensional fluorescence, 327–329
  NMR, 342, 343
  radionuclide, 338–340
  Raman, 341, 342
  refractive index, 333–335
  thermooptical, 336–338
  UV-Vis absorbance, 304–320
Detection of peptide separation, 662–663
  chemiluminescence, 664
  electrochemical detection, 663
  indirect amperometric detection, 663
  laser-induced fluorescence, 662, 666
  mass spectrometry, 662, 668, 669
  NMR, 669, 670
  radioactivity detection, 664
Detection schemes, 356
  end column, 305, 306, 333, 346
  entire column, 306
  on-column, 305–308
  postcolumn, 305, 306, 343
Detection sensitivity, 926
Detection time window, 930
Detection window:
  CEC, 280, 283
  width, 144–145
Detection, in CEC:
  in-column, 284
  on-column, 283
Detection figures of merit, 307
Detector time constant, 144–145
Determination of binding constant, 950, 951.
    See also Binding constant

Determination of enzymatic activities, 928. *See also* EMMA
Determination of isoelectric point (p*I*), 238, 241
Determination of stoichiometry of binding, 964. *See also* Binding constant
Dextrans, 490, 695, 701, 713
   CGE, 190, 205, 211
   electropherogram, 748
   in oligosaccharide separation, 749
Dextrins, in oligosaccharide separation, 749
Dicumylperoxide, 508
Diepoxy-PEG, 514, 515
Diethylenetriamine, 494
Differential partitioning, 79
Diffraction grating, 313
Diffuse layer, 58, 482
   mechanism of, 63
Diffusion coefficients, 147, 178
   measurement, 178, 991–992
   of micelles, 178
Diffusion, plate height, 147
Diglycidyl, 514
Dilute polymer solutions, CGE, 190
Dilution tee, 607
Dimerization constant, 962
Dimerization of insulin, 961
Dimethylacetamide, 527, 531–532, 536, 539
Dimethylformamide, 527, 529, 531–532, 536, 539, 541
Dimethylsulfoxide, 527, 531–532, 536, 539
Dimyristoyl phosphacholine (DMPC), 1004
Dinitrophenol complexes, 957
Diode array detection, 562, 564
Diode lasers, 735
Dipeptides, 651, 653, 663, 664
Dipolar interactions, 95
Dipole–dipole interactions, 497
Direct chemical coupling, 886
Direct UV detection, 841
Dispersion, 503
Dispersion coefficient, determination, 991–992
Dissipation of power, 157
Dissociation constants, determination, 254, 988–990
Dissociation of silanols, 483
Dissociation of weak acid, 978
Dissociation rate constant, 963
DNA, 717
   ladder in CGE, 205, 206
   migration in polymer networks, 188
   restriction fragments, 189, 190
   restriction mapping in CGE, 204
   separation using capillary affinity gel electrophoresis (CAGE), 205
   separation using chemical gels, 205
   separation using entangled polymers, 205
   separation using physical gels, 205
   sequencing using CGE, 190, 201
   structure, 204
DNA adduct, 474
DNA analysis, 15
DNA diagnostics, CGE, 196–201
DNA double helix, T_m values, 992–995
DNA fragments, 776
DNA restriction fragment, effect of temperature, 191, 193, 194
DNA separation, 12, 15
DNA separations, microchip, 624
   double-stranded DNA, 624
   electropherogram, 630
   one-color detection, 625
   sequencing fragments, 625
   single-stranded DNA, 624
DNA sequencing, 15, 767
   automation using CGE, 201
   base calling, 780
   capillary coating, 779
   electrokinetic injection, 776
   EOF elimination, 779
   pressure injection, 784, 785
   ratiogram, 782, 784
   rinsing procedure, 779
   sieving matrices, 775
   using end-label, free-solution electrophoresis (ELFSE), 752
Dodecanophenone, 84
Dodecyl-D-mannoside, 512
Dodecyltriethoxysilane, 512
Dodecyltrimethylammonium bromide (DTAB), 495
Double layer, effect of coating, 489
Double-stranded DNA (dsDNA), CGE separation of, 204
"Double-tee" injector, 620, 621
Drawn packed columns, 282
Drug analysis, 853–877. *See also* Pharmaceutical analysis
Drug impurities:
   CE, 855
   HPLC, 855

MEKC, 856
  profile, 857
Drug transport, 1001
Drugs:
  determination by ACE, 885
  nonaqueous CE, 541
DTAB (dodecyltrimethylammonium bromide), 81, 495
Dyes, 684
Dynamic coating, 228, 483, 489
  method and application, 489
  proteins and peptides, 494
Dynamic formation of SDS layer, 509
Dynamic modification:
  by amines, 689
  development, 516
  hydroxylic polymers, 494
Dynamic pore structure in CGE, 191
Dynamic reserve, 380
  detection limits, 381
Dynamic wall modification:
  competing ions, 688
  extremes in pH, 687–688
  neutral polymers, 694
  surfactants, 692

EC detection methods for electrical isolation, 356. *See also* UV-Vis absorbance
Effective charge determination, 978
  of a protein, 963
Effective mobility, 254, 978
  relationship with diffusion coefficient, 41, 42
Efficiency, 40–42, 54, 685, 687, 690, 691, 694, 695, 696, 698
  of binding, 890
  in CGE, 195
  enhancement, 490, 505
  microchip electrophoresis, 620, 621
Efficiency in MEKC, 143
  bile salts, 172
  dendrimers, 174
  effect of counterions, 161
  effect of electrolyte concentration, 164
  effect of organic solvents, 165
  effect of surfactant concentration, 160, 175
  microemulsions, 171
  mixed media, 172
  mixed micelles, 169
  polyelectrolytes, 173
  polymerized micelles, 173

Einstein–Smoluchowski equation, 40
Einstein equation, 146
Elcatonin, 661
Electrical conductivity, 45
Electrical double layer:
  effect of electrolyte concentration, 286
  minimum particle diameter, 286
  overlap, 285
  thickness, 285
Electrically driven pumping, 278
Electric double layer, 58
Electric field:
  nonuniform distribution, 463
  vertical, 501
Electric field strength, 4–5, 27, 35
Electroblotting of proteins, 684
Electrochemical cell, decoupling the separation capillary, 356
Electrochemical detection, 355, 539
  application in EMMA, 937
  indirect, 383
  of carbohydrates, 733
  on-column and off-column schemes, 356, 357
Electrochromatography on microfabricated devices, open channel, 614, 626
Electrode construction, 359
Electrode isolation, EC detection, 358
Electrodispersion, MEKC, 157
Electrokinetically packed columns, 282
Electrokinetic chromatography, 77
Electrokinetic focusing, 631
Electrokinetic injection, 37, 451–456, 464
  injection bias, 454
  CGE, 190, 195
Electrolyte co-ion, 828, 830
  concentration, 830
  mobility, 830
Electrolytes in ITP:
  background, 262
  leading, 252
  strong, 252
  terminating, 252
Electromigration:
  biomolecules, 744
  cellulose, 750
  effect of spectral tags, 746–749
  heparins, 750, 759
  polysaccharides, 745
Electromigration dispersion, 48, 390
  MEKC, 157

Electromigration factor. *See* Reduced mobility
Electromotive force, 223
Electroneutrality in cIEF, 232
Electron transfer kinetics, 360
Electroosmosis, 37, 453
  microchip, 617
  velocity, 453, 464, 466
Electroosmotic flow, 6, 27, 58, 482, 527–533, 866
  in CGE, 189
  cIEF, 228
  DNA sequencing, 775, 779
  effect of nonaqueous solvents, 529, 531
  electrical modification, 505
  equation, 527
  Hagen–Poiseuille profile, 482
  influence of surface coating, 507
  magnitude and reproducibility, 488
  manipulation, 483, 484
  marker, 84
  mechanism, 482
  MEKC, 83
  modification, 501
  parabolic profile, 482
  pH effect, 482, 981–982
  plate count, 501
  protein separation, 685, 688, 691, 692, 693, 694, 696, 698, 700, 701, 706, 719
  pumping, 278
  reduction, 843
  reversal, 490, 826
  suppression, 492, 501, 508
  unstable, 496
  velocity, 6, 27, 47, 685
Electroosmotic flow, CEC, 278, 285
Electrophoresis:
  electric field strength, 453, 463–466
  free flow, 622
  microchip, 614, 615, 617
  mobility, 454, 457
  in nonaqueous media, 526
  velocity, 453, 460, 466
Electrophoretically mediated microanalysis, 926–942. *See also* EMMA
Electrophoretic dispersion, 151
Electrophoretic mixing, 926
Electrophoretic mobility, 4, 5, 692, 693, 715, 716
  chiral separation, 793
  correlation to charge/mass ratio, 975–980

Electrophoretic mobility in nonaqueous CE, 533, 535
  pH effect, 536–537
Electrophoretic techniques:
  2D gel electrophoresis, 684
  3D gel electrophoresis, 684
  isoelectric focusing, 684, 707, 711, 712–713
  SDS–PAGE, 684, 713, 714
Electrophoretic titration curves, 975
Electrophoretic velocity. *See* Velocity
Electrospray interfaces, microchip coupling, 631
Electrospray ionization, 735
Electrospray–mass spectrometry, in glycobiology, 735
Electrostatic interactions, 482, 494
  proteins, 968
ELFSE. *See* End-label, free-solution electrophoresis (ELFSE)
Elimination of EOF, 232
ELISA, 12. *See also* Enzyme-linked immunosorbent assay (ELISA)
Elution pattern, MEKC, 89
Elution window, 84
EMMA (electrophoretically mediated microanalysis), 904, 926–942
  complexometric determination of inorganic ions, 941
  continuous engagement, 928
  detection methods, 936
  determination of enzyme activity at constant potential mode, 928
  determination of enzyme activity at zero potential mode, 931
  determination of single cells, 941
  differential electrophoretic mobility, 932
  enzymatic determination of substrates, 938
  moving boundary, 931
  product detection time, 932
  selectivity of detection time, 932
  sensitivity and selectivity, 936
  simultaneous multiple enzymes analysis, 935
  use of polyacylamide gel, 932
Enalapril, 569
Enantiomer mobility, 793
Enantiomer-selector binding constants, 793, 806, 809
Enantiomeric separations, 496. *See also* Chiral separation
  monosaccharides, 749–750
  domperidone, 570–571

naproxen, 575
tryptophan, 566
Enantioselectivity, 496
End-column detection, 356, 357, 358
End-label, free-solution electrophoresis (ELFSE), 752
  carbohydrate separation, 752
  DNA sequencing, 752
Endorphins, 661
Enhanced detection sensitivity, CE–MS, 422
Enkephalin-related peptides, 649, 652, 656, 661
Entangled polymers, 14, 775
  carbohydrate separation, 751, 752
  CGE, 205, 211
  networks, 190
  pore structure, 205
Environmental samples, 474
Enzymatic activity, 926
Enzymatic assays of substrates, 926
Enzymatic degradation of proteins and DNA, 912
Enzymatic digestion, on-line, 909
Enzyme affinity capillary electrophoresis, 893, 894
Enzyme assays:
  chemiluminescence detection, 937–938
  inhibition constants, 631
  kinetics information, 631
Enzyme-linked immunosorbent assay (ELISA), 12, 891
Enzyme-mediated reactions, 901
Enzyme reactions, 926
Enzymes:
  bond-specific, 732
  sugar-specific, 732
  turnover number, 931
Enzyme-saturating substrate concentrations, 930, 932
Enzymophoresis, 914
EOF. *See* Electroosmotic flow
Epinephrine, 518
Epoxy, 516, 518, 700–701
Equine serum, 593
Erythrocytes, single, 942
Erythropoietin, 753
Escherichia coli, 883
ESI, 407, 408. *See also* Electrospray ionization
Etching, 487
Ethanol, 518
  EMMA assays, 938

Ethidium bromide, 205, 208
Ethylenediamine, 488, 514
Ethylenediaminetetraacetic acid (EDTA), 891
Ethyleneoxide, 512
Extracolumn dispersion, 36, 37, 151
Extraneous injection, 455

FAB. *See* Continuous flow–fast atom bombardment (CF–FAB)
Fast atom bombardment (FAB), 474, 735
Fast CZE, 601
Ferguson plot, 715, 974
Fetuin, 650, 662
Fiber-optic cable, 313
Fick's second law, 39
Field amplification, 144, 881
Field-amplified stacking, 457, 460–465
  field enhancement factor, 463, 467
  laminar flow, 463
  MEKC, 127
Field enhancement factor, 463
Field strength gradient, 191
Film-forming agents, 694–695
Fingerprinting in CE, glycoproteins, 753
Fish muscle sarcoplasmid, 710
Fixation of synthesized oligomers, 512
Fixed multimolecular layers, 498
Flow gating interface, 596
Flow injection analysis, microfabricated devices, 614
Flow profile:
  laminar flow, 278
  packed capillary, 279
  plug flow, 278
Fluid-impervious polymer tubing, 518
Fluid manipulation, microchips, 617
Fluorescamine, 591
Fluorescence, 321–331
  amino acids, 325
  background, 322
  detection limit, 321–323, 327, 331
  detector cell, 322
  excitation source, 323, 325
  indirect, 381
  optics, 326, 327
Fluorescence tags, 737
  aminonaphthalene (ANTS), 741, 742, 748, 753, 757, 759, 760
  9-aminopyrene-1,4,6-trisulfonate (ANDSA), 741

1028 INDEX

Fluorescence tags (*Contd.*)
　for carbohydrates, 739
　　3-(4-carboxybenzoyl)-2-quinoline carboxaldehyde (CBQCA), 739–741, 745, 747, 757
　　sulfonated aminonaphthalenes, 741
Fluorescent derivatization of primary amines, 903
Fluorocarbon surfactants, 102
Fluorophore labeling, 926
　on-line, 927
Fluparoxan, 562
Flush buffer, 589
Focusing in cIEF, 232
　procedure, 225, 226
Folding of proteins, 215
Formamide, 527–529, 539, 541
　as denaturant, 190
Four-color detection, 770
Free energy change, 1001
Free flow electrophoresis:
　microchip, 622, 624
　microfabricated devices, 613
Free solution electrophoresis, carbohydrates, 752
Frits, 278–280
Frontal electropherogram, 617
Fronting, 828
　indirect detection, 390
Fullerenes, 550
　MEKC separation, 112
Fundamental resolution equation, MEKC, 85
Fused-silica capillary:
　adsorption of proteins, 685
　EOF, 685, 691, 700, 701
　modification, covalent, 697–698
　modification, dynamic, 687, 688, 692, 693
　modification, surface, 686–687, 692, 696
　nonuniform flow, 685
　pH, 685, 687–688, 691, 692, 694, 696
　voltage-driven transport, 685
　zeta potential, 685
Fused-silica capillary microreactors, 913

Galactose oxidase, EMMA determination, 937
Galactosidase, EMMA determination, 932
Galanins, 669
Gangliosides, 745
Gastrin peptides, 652
Gaussian curve, 40, 43
Gel-based IEF, 225

Gel electrophoresis, protein separations, 684
Gel-filled capillaries, 189, 198
　carbohydrate separation, 751
Gel IEF, 235
Gel matrix, presence of complexing agent, 188
Gel polymerization, 198
Gel stability, CGE, 198
Gene expression, 11
General elution problem, in CEC, 290
Genomic DNA, 767
Glass beads, 884, 887
Glass substrates, 617
Globins, 645
Glucose, MEKC, 115
Glucose oxidase, EMMA determination, 937
Glutathione/glutathione disulfide, 663, 668
γ-Glycidoxypropyltrialkoxysilane, 514
Glycoconjugate analysis, 729, 891
　borate complexes, 745–746
　branching analysis, 736
　derivatization, 736
　electrospray–MS, 735
　goals, 730
　HPCE, 730
　HPLC, 730
　MALDI–MS, 736
　molecular weight determination, 736
　neutral, 745
　sample preparation, 731
　sequencing, 736
Glycoform:
　hydrolysis, 732, 759
　proteolysis, 732
　separation, 753, 754
　separation by ion pairing, 755
　separation using isoelectric focusing, 753
Glycolipids, 745
Glycopeptides, 651, 656
　electropherogram, 892
　HPCE separation, 753–755
Glycoproteins, 704–705, 731
　borate complexation, 745–749
　CGE, 214
　compositional analysis, 755
　compositional analysis by HPCE, 755
　detection, 733
　effect of capillary dimensions on separation, 733
　fingerprinting in CE, 753

glycoforms, 753
HPCE separation, 732, 753–755
HPCE separation instrumentation, 732
microheterogeneity, 753
MS detection, 735
native fluorescence, 735
oligosaccharide mapping, 755
proteases, 736
Glycosaminoglycan, 731, 759
  separation using polyacrylamide, 751
  separation using polymers, 751
Glycosidic bonds, cleavage, 735
Glycosylated derivatives, 496
Glycosylated proteins, 214, 731
Golay equation, 51
Gold–mercury amalgam, 360
Gold microdisk electrode, 360
"Good buffers," 5, 641
  CHES [2-(cyclohexylamino)ethanesulfonic acid], 641
  HEPES [4-(2-hydroxyethyl)-1-piperazineethanesulfonic acid], 641
  MES (2-morpholinoethanesulfinic acid), 641
Gradient elution, 508, 620
  in CEC, 290
Gradient mobilization, 232
Gradient modes in CGE, 191
Grating monochromator, 312
Grignard reaction, 512
Grounding capillary, 358

HACs, 508. See also Hydroxyalkyl cellulose (HAC)
HCl conditioning, 780
Head-first motion, 188
Heart-cutting technique, 588
Heat dissipation, 157, 226
  in CGE, 195
HEC (hydroxyethyl cellulose), 492, 505
Height and area relationships, 931
Helium–cadmium laser, 734, 741
Helium–neon laser, 597, 776
Hemoglobins, 707–708
  variants, 651, 669
Heparins, 750
  electromigration, 750, 759
Heptanesulfonic acid, 8
Heptapeptides, 652
Heteroconjugation, 536
Heterocyclic peptides, 655

Heterogeneous reactions, 927
Heterozygous, CGE separation, 209
Hexamethonium bromide, 494, 495, 691
Hexamethonium chloride, 691
Hexamethyldisilizane, 508
HIBA, 837, 838, 839. See also Hydroxybutyric acid (HIBA)
High-conductivity buffer, 619
High-peak capacities, 508
High-performance RPLC–CZE, 589
High resolution, 507
High-sample throughput, 623
High selectivity, 890
High-sensitivity detection, 740, 926
High-speed CZE, 601
High-speed separations, 620
High-throughput DNA sequencing, 779
Histones, 8
Hollow fibers, 518
Homeostasis, 879
Homogeneous reactions, 927
Homogeneous solutions, 948
Homooligonucleotide, 198
Homozygous, CGE separation, 209
Hormone tryptic peptides, 646
Horse heart cytochrome c, 591, 603, 605, 649, 651, 655, 656
Horse heart myoglobin, 959
HPCE instruments, advantages, 732
HPCE–MS coupling, carbohydrate analysis, 735
HPLC combined with CE, 559, 561
HPLC method validations, 558, 559, 562, 565, 568, 569, 574, 577
HPMC, 492, 501, 505. See also Hydroxypropyl methylcellulose
Human chorionic gonadotropin, 691, 704
Human erythrocytes, 902, 942, 953
Human genome project, 473, 641, 664, 767, 786
Human interleukin-3, 667
Human K-*ras* oncogene, CGE separation of, 200
Human plasma, 2D separation, 600
Human recombinant factor, 692, 704
Human recombinant tissue plasminogen activator, 692, 704, 713
Human serum, 593
Human serum albumin, 964
Hyaluronic acid, separation, 752, 759
Hydrocarbon micelles, 93
Hydrochlorothiazide, 566
Hydrodynamic drag, 950, 951, 963

Hydrodynamic flow, 153
Hydrodynamic flow velocity, 452
  Poisseuille, 452
Hydrodynamic injection, 451–454, 477
  CGE, 195
  pressure difference, 451–456
Hydrodynamic mobilization, 225
Hydrodynamic radius, 978
Hydrogen bonding:
  bridge, 497
  CGE, 198
  donors and acceptors, 94, 1007, 1010
  nonaqueous CE, 536
Hydrolysis, glycoforms, 732, 759
Hydrophobic alkyl group, 509
Hydrophobic alkylpolysiloxane coatings, 509
Hydrophobic coating, 229, 490
Hydrophobic interaction, 87, 494, 1001
Hydrophobicity, 1001. *See also* Lipophilicity
Hydrophobic peptides, 649
Hydrophobic polymer, 190
Hydrophobic surface, 509
Hydrostatic injection, 831
Hydroxyalkyl cellulose (HAC), 484, 490, 496, 508, 775
Hydroxybutyric acid (HIBA), 11, 837, 838, 839
Hydroxyethylated polyethyleneimine, 515
Hydroxyisobutyric acid, 837
Hydroxylic coatings, 496
Hydroxylic neutral polymers, 498, 501, 505
Hydroxylic oligomers, 514
Hydroxylic polyethylene glycol–propylene glycol, 518
Hydroxylic surface modifier, 517
Hydroxylic surfactants, 490, 512
Hydroxymethyl cellulose, 208
Hydroxypropyl cellulose, 55
Hydroxypropyl methylcellulose, 208, 484, 694, 697, 698, 709. *See also* HPMC
Hyphenated detection methods, 341
  peptide mapping, 341

IEF, 10, 223
  Joule heating, 225
  mechanism, 225
IgG, 957–959
Imaging detector, 771
Imidazole, 828
Imidopeptides, 666
Iminodiacetic acid metal, 891

Immobilization:
  of antibody, 883, 886–888, 889
  of concanavalin A, 893
  of enzymes, 360, 892, 927
  of lectins, 891, 893
  of ligand, 948
Immobilized enzyme microreactor, 913
Immunoabsorbents, 883
  concentrator, 884, 888
Immunoaffinity capillary electrophoresis, (IACE) 880
  specific examples, 888
Immunoaffinity techniques, on-capillary, 906
Immunoassay, 7
  particle counting, 907
Immunogeneity, 215
Immunoglobulin, 957
Immunoglobulin G (IgG), 753, 757, 758
Immunology, 706
Impregnated membrane, 883
In-column detection, 284
Incubation time, enzyme analysis by EMMA, 927, 928, 932
Indirect chiral separation, 813
Indirect detection, 10, 316, 317, 376
  absorption, 381
  applications, 399, 836
  background electrolyte, 376, 378, 388, 395–397
  calibration, 389
  concentration overload, 390
  electrochemical, 383
  fluorescence, 381
  fronting, 390
  noise, 393
  peak triangulation, 390
  schematic, 377
  selection, 395
  system zones, 386–388
  tailing, 390
  thermal dissipation, 394
  UV absorption, 382
Indirect fluorescence detection:
  carbohydrates, 734
  ion analysis, 844
Indirect UV detection, 866, 868
  limit of detection equation, 830
  molar absorptivity of co-ion, 830
  transfer ratio equation, 830
Inductively coupled plasma–MS (ICP–MS), 844

Infinite dilutions, 28
Inflammatory cytokines, 656
Injection:
    amount, 466
    artifacts, 456
    electrokinetic, 451–456, 464
    error, 456
    gravity, 568
    hydrodynamic, 451–454, 477
    length, 450–452, 454
    microchip devices, 617, 618
    precision, 566, 574
    precision in validation, 566
    pressure, 568
    reproducibility, 619
    spontaneous, 456
    technique, 831
    vacuum, 568
Injection stacking, 464, 477
    field-amplified, 464–468, 474, 477
Inorganic anions, 835. *See also* Anion determination, CE
    nonaqueous CE, 547
Inorganic cations, CE separations, 838, 839
Inorganic ions:
    alkali and alkaline earth metals, 840
    amperometric detection, 845
    analytical techniques, 825
    anion determination, 826, 831, 841
    cation determination, 837, 843
    CE determination, 825–851
    co-ion condition, 828
    complexing agents, 838
    conductivity detection, 845
    detection limits, 830, 837, 841
    direct UV detection, 841
    electrolyte co-ion, 827
    ICP–MS, 844
    indirect fluorescence detection, 844
    indirect UV detection, 826
    pharmaceuticals, 865–868
    potentiometric detection, 845
    visualization reagent, 827
In situ polymerization, 512
Instrumentation, glycoprotein separation, 732
Insulating substrates, microchips, 615
Insulin, 707, 716
    dimerization, 961
Integrated devices, 614
Integrated microchip, 629

Interface designs, CE–MS, 408
    liquid-junction interface, 409, 412, 413
    microchips, 439
    microdialysis junction, 439
    sheath-flow interface, 409, 410
    sheathless interface, 414, 422
Interfacing schemes in 2D separation, 588
Intermicellar diffusion, 146
    plate height, 151, 152
Internal rotor, 593
Internal standards, 858–862, 867, 983
    CGE, 196
    cIEF, 229, 238
Intracellular amino acids, 903
Intracellular concentration, 901
Intramolecular interaction, 498
In vitro derivatization, 902
In vivo biochemical events, 909
In vivo derivatization, 901
In vivo microdialysis, CE–MS, 437
In vivo screening, 880
Ion analysis. *See* Inorganic ions
Ion-depleted zone, 456
Ionic block copolymer, 110
Ionic interactions, 497, 498
Ionic polymers, 83
Ionic repulsion, 505
Ionic strength, 688, 704
    chiral separation, 807
    effects in peptide separation, 645–647, 658
Ionizable compounds, MEKC, 119
Ion pairing, 687, 688, 692, 716
    glycoform separation, 755
    reagents, 857
Ion-pair reagents, peptide separation, 650–651
    butanesulfonate, 650, 651
    octanesulfonic acid, 651
    pentanesulfonate, 650, 651
    phytic acid, 651
    sodium butane, 650
    tetrabutylammonium hydroxide, 650
    trifluoroacetic acid, 650
Ion-selective microelectrodes, 361
Isocratic separations, 583
Isoelectric focusing (IEF), 223, 707, 711–713
    determination of pK and mobility, 974, 982
    glycoform separation, 753
    protein separations, 684
Isoelectric point, pI, 223, 980–984
    determination by CE and IEF, 241, 245

Isoelectrostatic, CGE, 191
Isoenzymes, 942
Isorheric separation mode, CGE, 192
Isotachophoresis (ITP), 457–461, 477, 478, 974, 978, 984
   counterions, 259
   criteria for zone stability, 257
   electrolyte system, 256
   principles, 252
   stacking, 845
Isothermal separations, 583
Isotopic compounds, separation of, 55
Isotopic oxygen benzoic acid:
   determination of pK_a, 979
   optimum pH, 979
Isozymes of carbonic anhydrase, 959
ITP. See Isotachophoresis, principles
ITP–CZE, 260, 418, 457
   dispersion of zones, 268
   standardization method, 268
ITP stacking, ion analysis, 845

Joule heating, 4, 688
   bubble formation in CEC, 282
   CGE, 195
   in CEC, 286
   MEKC, 148, 154, 164

Kilobase pair, 191
Kinase A peptide substrates, 649
Kinetic constants, 894
Kinetic parameters of binding, 963
Kohlrausch-regulating function, 48, 255, 384, 457

Lability of covalent coating, 229
Laboratory-on-a-chip, 615
Lactate dehydrogenase (LDH), 906
Lactate dehydrogenase, EMMA determination, 936
Lactoglobulin, 940
Laminar flow, 37, 278, 468
   Hagen–Poiseuille equation, 47
Laser beam waist, 770
Laser-induced fluorescence (LIF), 7, 356, 539, 706, 707
   application in EMMA, 936
   carbohydrate detection, 733, 734, 754, 757
   microchip, 629
Lasers, 312
   argon ion, 325
   diode, 325
   HeCd, 325
   HeNe, 325
LC–MS, 16
Leading electrolyte, 252, 457
Lectin affinity capillary electrophoresis (LACE), 892
Lectin–carbohydrates complex, 883
Lectin chromatography, 754–755, 891
Lens, 313
   ball, 313
   conventional, 313
   quartz, 313
   sapphire, 313
Leucine enkephalin, 662
LIF. See Laser-induced fluorescence (LIF)
Ligand–receptor binding using ACE, 967
Light scattering, 322, 323
Light scattering detection, 343
   amino acids, 345
   condensation nucleation, 343, 344
   laser-based particle-counting, 345, 346
   laser-light, 345
   multiplexed, 773
   peptides, 345
Limit of detection, 826, 830. See also Detection limit
Limit of quantitation, 574
Linear free-energy relationships (LFER), 1001
Linear pH gradient, 229, 235, 236
Linear polyacrylamide gels, 15, 198, 205, 211, 512
Linear polymer, CGE, 189, 205, 211
Linear response, validation, 561, 562
Linear solvation-energy relationships (LSER), 87, 1007
LiPFOS (lithium perfluorooctane sulfonate), 81
Lipophilicity, 1000. See also Hydrophobicity
Lipoproteins, 705–706
Liposomes, 1004
Liquid chromatography (HPLC), 708
   glycoconjugates, 730
   glycopeptide separation, 735, 754
Liquid-junction interface, CE–MS, 409
   schematics, 413
Living cells, 346
Living PEG, 508
Log k vs log P_{ow} relationships:
   CTAB micelles, 1006

LiPFOS micelles, 1007
SC micelles, 1006
SDS micelles, 1005
Log P_{ow}:
   determination, 999
   relationship with retention, 1000
Long-chain surfactants and bile salts, 81
Longitudinal diffusion, 39
   MEKC, 146
Low conductivity buffers, 5
Low-conductivity sample, 619
LSER (linear solvation-energy relationships), 87, 1007
Lysing cells, 901
Lysozyme, 685, 690, 692

Main component assay:
   pharmaceuticals, 858
   validation, 561
Make-up flow, 409
MALDI–MS, 407
   glycoconjugates, 736
Maltodextrin, 759
Maltooligosaccharide, separation of, 6
Mammalian cell lysates, 904
Mass detection limit, 355, 356, 475
Mass sensitivity in CE, 733
Mass spectrometric detection, 16, 538–539
   advantages, 407
   array detectors, 422
   buffer choice, 414
   carbohydrate detection, 733, 735
   CF–FAB, 408
   collision-induced dissociation, 719
   coupled with CE, 733, 735
   disadvantages, 414
   electrospray ionization, 408, 716, 717, 719
   Fourier-transform ion cyclotron resonance, 719
   high-conductivity buffers, 415
   instrumentation, 407
   ion storage analyzers, 423
   MALDI–MS, 714, 716–717
   MEKC, 130, 428
   non-volatile buffers, 414
   selected ion monitoring, 423
   time-of-flight, 424, 427, 439, 717
Mass spectrometry, instrumentation, 407
   interface design, 408
   ionization techniques, 407

Mass transfer rate, 155
   analyte-micelle, 155
   analyte-wall, 155
Material transport on microchip, 617
Matrices in CGE, 775
Matrix-assisted laser desorption ionization (MALDI), 407
Maximum power, CGE, 195
Measurement of diffusion coefficients, 178
MECC, 78. *See also* Micellar electrokinetic chromatography (MEKC)
Melting temperature, determination, 992
Membrane preconcentration, 656
   cartridge, 881, 882
Membrane proteins, 210
Mercaptopropyltrialkoxysilane, 514
Mesh size, CGE, 191
Metal additives, peptide separation, 649
   Ca^{2+}, 649
   Cu^{2+}, 649
   Zn^{2+}, 649
Metal cation analysis, 837, 840. *See also* Cation determinations, CE
Metal electrodes, 360
Metal ion analysis, pharmaceuticals, 866, 873
Metal ions, nonaqueous CE, 541
Metal-binding peptides, 650
Metallic ligand, 891
Metalloprotein, 649
Metamphetamine, determination in urine, 888, 889
Methacryloylpropyl, 512
Methanol, 527, 529–531, 536, 539, 541, 546–550, 571
Method accuracy, validation, 559, 560
Method development strategies for peptide separation, 657–658
Method repeatability, validation, 566
Method validation:
   requirements, 558
   in CE, 557–579
Methycellulose, 14, 229, 693, 775
   CGE, 205
Methylene chloride, 514
Methylenebisacrylamide, 189
Methylformamide, 527, 529, 531–533, 536, 538–539, 541, 550–551
Methyl-1-phenyl-2-pyrazolin-5-one (PMP) as UV-Vis tag, 737, 738
Micellar diffusion coefficient, 147

Micellar electrokinetic chromatography
    (MEKC), 10, 76, 471, 478, 715, 999
  applications, 128
  band broadening, 87
  charged solutes, 85
  on chip, 176
  elution window, 84
  estimation of lipophilicity, 1000
  field amplification, 144
  fluorocarbon–hydrocarbon anionic surfactants, 652
  Joule heating, 148
  MEKC–ESI-MS, 130, 429
  microchip, 627, 631
  microfabricated devices, 614, 626, 627
  migration, ionizable solutes, 119
  migration parameters, 84
  migration scheme, 78, 83
  modifiers, 113
  nonequilibrium dispersion, 148
  nonionic surfactant, 651, 658
  optimum retention, 87
  peak capacity, 87
  of peptides, 651–653
  pharmaceuticals, 856, 860–861, 864, 868
  polysorbate, 652
  prediction of retention, 1011
  resolution, 85
  retention factor, 84
  retention–log P_{ow} relationship, 1000, 1002, 1005–1007
  role of pseudostationary phase, 1005
  selectivity, 89
  stacking, 144
  surface treatment, 509
  taurodeoxycholate, 652
  temperature effect, 125, 157
  Tween 20, 652
  wall adsorption, 148
  zone sharpening, 125
  zwitterionic detergent, 652, 658
Micellar liquid chromatography, 1001
Microanalysis, 926
Microbioanalysis, 899–923
Microcapillary, 891
Microchip(s), 614
  coupling to MS, 631
  integrated, 629
  MEKC, 176
  MS, 439
  precolumn and postcolumn reaction, 614
  solvent programming, 631
  substrate-fused quartz, 615
Microchip column:
  cross microchip, 623
  cyclic format, 622
  free-flow, 624
  linear, 628
  parallel architecture, 625, 631
  separation channel, 628
  serpentine geometry, 621, 630
Microchip components:
  channel, 614, 615
  cross, 614, 615
  detector cell, 626
  frit, 626
  tee, 614, 615, 619
Microchip devices, 20
  fabrication techniques, 615
  fluid manipulation, 617
  injection, 617
  system design, 614
Microchip electrophoresis, 620, 621, 631, 917
  DNA separation, 623
  high-efficiency, 620, 622
  high-speed separation, 620, 623
Microchip fabrication:
  direct bonding, 616
  etching, 616
  photomask, 616
Microchip substrates:
  BK7, 617
  fused quartz, 617
  glass, 617
  polymer, 624
  silicon, 622
Microchip valve, 614
  "double-tee" injector, 620, 621
  fixed-volume valve, 618, 619
  spatial confinement, 617
  split injector, 626
  variable volume valve, 618, 619, 629
Microcolumn in 2D separations, 591
Microcolumn SEC–CZE, 591
Microdialysis–CE–LIF, 910
Microdialysis junction, 439
Microdisk, 359
Microelectrode, 360
Microemulsion electrokinetic chromatography, 1004

INDEX   1035

Microemulsions, 146, 154
Microfabricated devices, 20, 613–633. *See also* Microchip devices
Microheterogeneity, 151
Micromachining, 617
Micromanipulator, 358
Microparticle-packed capillaries, 517
Micropreparative separation, 733
Micropreparative technique, 660
Microreactor–CE–MS peptide mapping, 918–919
Microreactors, 666, 881, 888
Microscale analysis, 927
Microscope objective, 771
Microsphere electrophoretic mobility, 514
Migration in CGE:
  effect of electric field, 187, 188
  effect of molecular weight, 187, 188
  hydrodynamic radius, 188
  matrix pore size, 188
Migration parameters, 84
Migration rate, 390–392, 587
Migration scheme, ionizable solutes, 119
Migration time, 35, 62
  precision in method validation, 566
Migration time in CGE:
  precision, 196
  reproducibility, 200, 214
Migration time, micelle, 79
Miniaturization, 614
Minimal detectable concentration (MDC), 308
Minimal detectable quantity (MDQ), 308
Mixed media in MEKC, 163, 172, 176
Mixed micelles, 106
Mixed polymers, 776
Mobile phase in CEC, 289
Mobility, 27
  absolute, 27
  actual, 27
  concentration, 27
  dimension, 27
  effect of ionic strength, 28, 29
  effect of pH, 31, 254, 258
  effective, 29, 254
  electrophoretic, 454, 457
  reduced, 62, 63
  sign, 27, 62
  temperature effects, 46
Mobility difference, 148
Mobility match, 50, 827

Mobility of micelles, 147
Mobility of peptides:
  determination, 985
  relationship to charge/mass, 985–987
Mobilization in cIEF, 226, 227, 229, 232
  cathodic and anodic, 234
  voltage and vacuum, 236
Mode of electrochemical detection, 359
Modification of proteins, 963–964
Modified capillary, 482
Modified electrodes, Nafion membrane, 358
Molar absorptivity, 573
Molecular imprinted polymer, 297
Molecular mass, 974
Molecular sieving, 188
Molecular weight determination, 408
  glycoconjugates, 736
Monoclonal antibodies, 883, 884
  immobilization, 906
Monoclonal IgG, 964
Monolithic columns, 289, 297
Monosaccharides:
  electropherogram, 740
  enantiomers, 748
Moving-boundary electrophoresis (MBE), 974
Moving-boundary EMMA, 931
Multicapillary electrophoresis, 771
Multidimensional fluorescence, 327–329
  time-resolved fluorescence, 329
Multimodal interaction, 494, 505
Multimolecular layers, 498
Multiple capillary bundles, 884, 886
Multiple capillary scheme, 770
Multiple enzyme analysis, 935
Multiplexed capillary electrophoresis, 767, 768, 769, 785
  capillary gel electrophoresis, 768
  DNA sequencing, 768
  Joule heating, 769
Multiply charged ions, ESI–MS, 408
Multipoint adsorption, 492
Multipoint interaction, 496, 498
Multivalent proteins, 891
Myoglobin tryptic peptides, 656, 667

NANA. *See* N-Acetylneuraminic acid
Nanoliter, 880
NaOH rinsing, 488
Naproxen, 575
Narrow-bore capillary, 508

Native fluorescence, 329
  bilirubins, 329
  of glycoproteins, 735
  peptides, 327
  pharmaceutical compounds, 329
  porphyrins, 329
  vitamin B_6 metabolites, 329
Natural glycoconjugates, 745
Negative charge concentration, 505
Nernst–Einstein equation, 41
Neurohypophyseal peptides, 660, 662
Neuron, 667
Neutral coating, 8, 229, 232
Neutral marker, 59
Neutral molecules:
  separation on microchip, 626
  stacking, 471
Neutral solutes:
  MEKC migration, 78
  nonaqueous CE, 547
Nitroaromatic compounds, MEKC separation, 98
Nitrophenol-releasing substrates, 936
NMR, 342, 343
Noise:
  absorption detection, 382, 393
  background, 378
  baseline, 378
  dark current, 382
  indirect detection, 393
  shot, 382
  validation, 573
Nonaqueous CE, 525–555
  addition of cyclodextrins, 529, 550–552
  apparent pH, 538
  autoprotolysis constant, 533–535
  detection, 538–539
  dielectric constant, 527, 531
  dissociation, 535
  electroosmotic flow, 527–533
  electrophoretic mobility, 533–535
  heteroconjugation, 536
  Joule heating, 529
  Ohm's law plot, 528
  pharmaceuticals, 873
  physical properties of solvents, 527
  pK_a, 534–538
  sample stacking, 528
  selectivity, 533–538
  separation of charged solutes, 540–547

  separation of chiral solutes, 550–553
  separation of uncharged solutes, 547–550
  solubility, 540
  transfer activity coefficient, 534
  viscosity, 527, 531
  volatility, 540
  zeta potential, 527, 531, 533
Nonaqueous solvents for CE, physical properties, 527
Noncovalent coating, 518
Noncovalent modification of proteins, 964
Non-cross-linked polyacrylamide, 229
Non-cross-linked polymers, 189
Nondenaturing condition in CGE, 190
Nondenaturing gel, CGE, 198
Nonequilibrium dispersion, MEKC, 148
Nonequilibrium theory, 151
Nonequilibrium zone broadening, 486
Nonhydrogen bond, 94
Nonionic detergents, 8
Nonionic hydroxyalkyl methylcellulose, 500
Nonionic surfactants, 81, 106
Nonspecific preconcentration, 881
Nonsteroidal anti-inflammatory drugs, 507
Nonsulfated forms of peptide, 645
Normalized areas, 565
Nucleic acids, 767
  melting point determination, 992
  T_m values, 992–995
Number of components in sample, 584
Number of theoretical plates, 583

Octadecylsilane, 287
Octanol–water partition coefficient, 1000
Octapeptide, 644
ODS. See Octadecylsilane
Off-column detection, 356
Off-column reactions, potential problems, 927
Off-line interaction, 888, 891
Off-line sample pretreatment, 881
Offord equation, 985
Ogston theory, 188
Ohm's law, 658
  aqueous vs. nonaqueous, 528
Oligoethylene glycols, 514
Oligonucleotides:
  base number, 198
  CE–MS, 432, 437
  CGE separation of, 203
  migration order in CGE, 198

secondary structure, 198
Oligosaccharides, 215, 730
  borate complexation, 748
  chiral analysis, 749
  cleaving enzyme, 936
  library, 756
  mapping, 732, 756, 757
  separation using CGE, 759
On-capillary binding assays, 905
On-capillary enzymatic reactions, 904
On-capillary immunoaffinity, 906
On-chip reactors, 917
On-column concentration, 127
On-column detection, 283
On-column frit, 358
On-column, homogeneous reaction-based analysis, 927
On-column protein digestion and peptide mapping, 940
On-column stacking, 457, 474
  moving boundary, 457
  stationary boundary, 457
One-dimensional separation, 585
On-line analyte concentrator, 883
On-line assays, 904
On-line capillary microreactor–CE systems, 909
On-line chemical reaction to CE, 926
On-line digestion and mapping of tRNA, electropherogram, 915
On-line immunoaffinity, 879
On-line lectin affinity capillary electrophoresis (LACE), 892, 893
On-line peptide mapping, 916, 927
On-line preconcentration, 891
  nonspecific, 881
  specific, 883
On-line preconcentration, CE–MS, 416, 419
On-line reaction, 901
On-line sampling, 900
On-line microreactors, 899
Open-channel electrochromatography, 614, 626, 627
Open-tubular precapillary, 891
Opioid peptides, 644
Optical fibers, 770
Optical gating, 601, 602
Optimization in peptide separation, 657–658
  buffer concentration, 658
  pH, 657
  voltage, 658

Optimization, MEKC, 117
Optimum CD concentration, 809
Organelles, 926
Organic additives, peptide separation, 647–649, 653, 658
  acetonitrile, 647
  Brij 35, 649
  chitosan, 649
  cyclodextrin, 649, 652
  1,3-diaminopropane, 649
  glycerol, 649
  hexadecyltrimethylammonium bromide, 649, 652
  hexamethonium bromide, 649
  hexanesulfonic acid, 649, 651
  methanol, 649, 653
  polyethyleneimine, 667
  sodium dodecyl sulfate, 649, 652, 653
  2,2,2-trifluoroethanol, 667
Organic modifiers in MEKC separations, 116
Organic polymer capillaries, 517
Organic solvents, 113
  in capillary electrophoresis, 525–555
Organized media concentration, effect on efficiency, 158
Organosilane, 512
Orthogonality, 586–587
Orthogonal separation technique, 587
Ovalbumin, 685, 704, 753
  diffusion coefficient, 991
Overload, 390
Oxidoreductase, 936
Oxytocin, 662

P_{mw}, 79
PAAEE. See Poly(N-acrylaminoethoxyethanol) gel
Packed-bed liquid chromatography, microfabricated devices, 614, 626
Packed capillary, 279
PAH separations, 473, 474
  CEC separation, 290, 293
  MEKC separation, 111
Parabolic flow, 7
Parallel laser-excited fluorescence detectors, 771, 772
Parallel measurement, 614
Parallel sequencing, 769
Partial capillary filling, 177
Particle-counting immunoassay, 907

Partition coefficient, 1000
  micelle-water, 79
Pathlength, 949
PCR (polymerase chain reaction), 776
PCR in CGE analysis, 204
PCR, microchip:
  multiplexed, 631
  silicon thermal cycler, 631
Peak area precision, validation, 566
Peak area reproducibility, CGE, 196
Peak asymmetry, 828, 829
Peak-broadening effects, 35
  contributing factors, 53
Peak capacity, 583, 584, 586
  MEKC, 87
Peak deformation, 50
Peak dispersion, 35. *See also* Band broadening
  contribution of EOF, 63
  sources, 35
Peak efficiency, 574
  CGE, 191
Peak identification, 319
Peak leading edge, 50
Peak purity, 319, 328
  validation, 562
Peak shape, 50
Peak tailing:
  in CEC, 280
  in MEKC, 174
Peak tracking, 318, 328
Peak trailing edge, 50
Peak triangulation, 48, 390
Peak volume, 590
PEG, 492, 496. *See also* Poly(ethylene glycol) (PEG)
  covalent, 514
  living, 508
PEI, 496
Pepsin, 753, 940
Pepsinogen A, C, 669
Peptidases, 935
Peptide analysis, 638-671
Peptide mapping, 8, 591, 754, 881
  on-column, 940
  on-line, 927
Peptides:
  absolute mobility, 984-988
  large, 652
  relationship of mobility to charge/mass, 985-987

Peptides and proteins, CE-MS, 416, 420, 424, 430, 437
Peptide separation:
  effects of parameters, 639-650
  method development strategies, 657-658
  nonaqueous CE, 541, 544
Peptide separation, effect of:
  applied voltage, 640
  buffer additive, 647-650
  capillary inner diameter, 640
  capillary length, 639
  capillary temperature, 642
  ionic strength, 645
  pH, 643
Peptidyl ligands, 957
Performance of analytical separation, 507
Permanent coating, 483, 489, 491
  alkyl chain length, 494
  biogenic amines, 495
  commercial use, 492
  development, 516
PGA, 498
  adhesion, 505
pH:
  chiral separation, 807
  effect in MEKC, 120
  effects in nonaqueous CE, 536-538
  effects in peptide separation, 643-645, 653, 657
  extremes, 229
  mobility curves, 975-980
  mobilization in cIEF, 234
  optimum, 978-980
Pharmaceutical analysis, 853-877
  application areas, 854-855
  chiral analysis, 862-855
  dissolution analysis, 868
  excipient analysis, 868
  impurity determination, 855-858
  inorganic contaminants, 868
  inorganic ion determination, 865-868
  intercompany collaboration, 867
  intercompany method-transfer, 861
  main-component assay, 858-862
  micellar electrokinetic chromatography (MEKC), 856, 860-861, 864, 868
  raw material analysis, 868
  validation, 858
Pharmaceutical formulations, CGE assay, 200
Pharmaceuticals, CE-MS, 430

Pharmacological effects, 880
Phenoxyacid herbicides, 474
Phenyl diisothiocyanate (DITC), 884, 886
Phosphatases, 935
Phosphopeptide, 662
Phosphoric acid, 232
Phosphorylated and deamidated peptides, 645
Phosphothioate oligonucleotides, CGE separation, 198, 202
Photobleach, 603
Photobleaching fluorophores, 734
Photodiode, 311
  multiwavelength, 311, 318, 319
Photomultiplier tube, 311
Physical gels, 189
Physicochemical measurements for peptides, 663–665
  binding constant, 664, 665
  electrophoretic, 664, 665
  kinetics, 664, 665
  stoichiometries, 664, 665
Physicochemical parameters, 973–998
Physiologically derived proteins, 518
Physisorbed modifiers, 495
Physisorption mechanism, 497
pI, 980–984
  proteins, 982–984
  determination by CE and IEF, 242, 245
Picoliter sampling capability, 900
Pixels, 771
pK values:
  amphoteric compounds, 980–981
  silanols, 981–982
  weak electrolytes, 975–980
pK_a, 30
  amino groups, 691
  effect of solvent, 32, 534–538
  imidazole, 701
  phylate, 692
  shift, 34
Plasmid DNA, 917
Plateau product profile in EMMA, 931
Plate count(s), 827
  EOF, 501
Plate height, 36, 143
  dependence on voltage in MEKC, 149
Plate height equations, 53
  coiling, 39
  electromigration dispersion, 48, 65
  EOF, 64

injection, 37
longitudinal diffusion, 48, 64
thermal effects, 47, 65
wall adsorption, 51
Plate number, 40–42, 143
  influence of pH, 44
Plate number-injection, 450
Plug flow, 7, 278
  profile, 35
Plug–plug electrophoretic mixing (also transient engagement), 929
Plug size, 143
Pneumatic injection, 37
POE. *See* Poly(ethylene oxide) (POE)
Point mutation analysis, CGE, 190, 195, 196
Polarity switching, 467–473
Polyacrylamide gel:
  cross-linked, 189
  for DNA sequencing, 201
  separation of DNA, 205
  separation of proteins and peptides, 210
  use in EMMA, 932
  use in IEF, 225
Polyacrylamide (PAA), 484, 713, 768, 775, 974
  coating, 690, 699–700
  glycosaminoglycan separation, 751
  polysaccharide separation, 751
Poly(*N*-acrylaminoethoxyethanol) gel (PAAEE), 195
Polyamide resin, reactive, 517
Polyanionic cyclodextrin, 814
Polyaromatic hydrocarbons, MEKC separation, 111
Polybrene, 490, 709
  chemisorption, 505
Polyclonal antibodies, 885
Polycyclic aromatic hydrocarbons (PAHs), nonaqueous CE, 550, 552
Polydispersity, 152
Poly(ethylene glycol) (PEG), 492, 709
  CGE, 205, 211
Poly(ethylene oxide) (POE), 190, 518, 695, 696, 713, 775, 778, 779
Polyimide coating, 144
  in CEC, 281, 283
Polyion, 187
Polymer additives, CGE, 205
Polymerase chain reaction (PCR) 20, 190, 776
  microfabricated devices, 614
Polymer capillaries, 517

1040　　　　　　　　　　　　　INDEX

Polymer-filled capillaries, 12
Polymeric additives, neutral, 489
Polymeric layers, covalently bonded, 512
Polymeric modifier, 496
Polymeric pseudostationary phases, 83, 110
Polymerization, *in situ*, 512
Polymerized micelles, 83
　chiral separation, 83, 173, 176, 805, 806
Polymer networks, 188
Polymers, 548
　cationic, 691, 701, 709
　cellulose acetates, 694–695
　chitosan, 691
　dextrans, 695, 701, 713
　hydroxyethyl cellulose, 695, 696
　hydroxypropyl methyl cellulose, 695, 696
　polybrene, 709
　poly(ethylene oxide), 695, 696
　poly(vinyl alcohol), 694
　polyvinylimidazole, 701
　pullulan, 713
　sieving effects, 713
　Tween 20, 710
Polymers in CGE:
　separation of dsDNA, 205
　separation of proteins and peptides, 192, 211
Polymer tubing, 518
Polymethylsiloxane OV1, 509
Polymorphism, 208
Poly(oxyethylene) (POE), 550
Polypeptides, 651
Polypropylene hollow fibers, 518
Polysaccharides:
　borate complexation, 745–749
　buffer additives, 745
　electromigration, 745
　fluorescent tags, 759
　HPCE analysis, 757
　separation using polyacrylamide, 751
　separation using polymers, 751
　separation using pulsed-field HPCE, 761
Polysaccharides, chiral selectors:
　anionic, 799, 800
　dextrins and dextrans, 799
　neutral, 799
Polysorbed, 489
Poly(tetrafluoroethylene) (PTFE), 231, 492, 982
Poly(vinyl acetate), 514
Poly(9-vinyladenine), 198

Poly(vinyl alcohol) (PVA), 484, 492, 496, 505, 775
　coatings, 518
　covalently bonded, 514
　immobilized modifier, 501
　shielding layer, 501
Polyvinylimidazole, 701
Polyvinylpyrrolidone, 701
Pores in gels, structure and size, 189
Pore structure, entangled polymers, 205
Porous glass beads, 893
Porous graphite tubing, 358
Positional isomers, optimum pH, 980
Postcolumn reactions, 927
　chemiluminescence detection, 937–938
Postcolumn reactor, microchip, 614, 615, 631
Potential sweep, 361
Potentiometric detection, 845
Potentiometry, 361
Power, 5
Power-generated, 45
Precision:
　internal standard, 561
　validation, 566
Precolumn-performing chemical reactions and CE, 627
Precolumn reactor(s), 927
Precolumn reactor, microchip, 614, 615, 627
　biochemical reaction and analysis, 629
　electrophoretic sizing, 629
　kinetics, 629
　restriction digestion, 629
　schematic, 628, 630
　variation of product formation with time, 628
Preconcentration cartridge, 881
Preconcentration CE–MS, 419
Preconcentration, on-line:
　nonspecific, 881
　selectivity, 890
　specific, 883
Preconcentrator, 655–656, 668
Prediction of retention in MEKC, 1011
Preinjection, 456
　of water in CGE, 196
Preparative isoelectric focusing, 225
Pressure-driven pumping, 278
Pressure injections in CGE, 190
Pressure limitation, 279
Pressurized injection, 37
Pretreatment of capillary, 487

Priminox, 32, 492
Probability of resolving single peak, 584
Product profiles in EMMA:
  plateau, 930
  triangular, 931
Proline-rich peptide, 660, 665
Propanol, 539
Propylene carbonate, 527
Proteases, 935
Protein(s):
  absolute mobility, 984–988
  basic proteins, 690, 696, 698, 701
  CE separations, 683–727
  denatured proteins, 685
  electrophoretic mobility, 692
  folding, 711
  food proteins, 710
  glycoforms, 691, 692, 704, 713
  isoelectric point, 685, 696, 712
  pH, 690–691, 704, 716
  pI determination, 982
  recovery, 686
  serum proteins, 703, 710
  wall adsorption, 685
Protein(s), chiral selectors, 804
  α_1-acid glycoprotein, 804, 812, 818
  bovine serum albumin, 804, 818
  ovomucoid, 804
Protein adsorption, 494, 496
  control, 514
Protein binding to two ligands, 957
Protein CGE separation:
  complexes with SDS, 187, 189, 210
  glycosylated, 214
  migration in polymer networks, 188
  molecular weight determination, 190
  SDS–PAGE, 210
  using chemical gels, 211
  using coated capillaries, 213
  using dextrans and POEs, 211
  using physical gels, 211
Protein charge ladder, 963
Protein–DNA complex, 968
Protein effective charge, measurement, 963
Protein electrostatic interactions, 968
Protein expression, 10, 11
Protein fingerprinting, 591
Protein folding, 215
Protein G, 887, 891
Protein glycosylation, 731

Protein hydrolysates, 927
Protein–ligand binding constant, 950, 951
Protein–ligand interactions, 948
Protein molecular weight determination, 213, 214
Protein precipitation, 230, 232
Protein–protein interactions, 951, 961
Protein samples, on-line preconcentration, 891
Protein separation, 8, 16
  artifacts, 685
  conventional electrophoresis, 684
  electrophoresis in fused-silica capillaries, 684
  surface adsorption reduction, 509
Protein wall adsorption, 685, 968
Proteoglycan, 757
Proteolysis, glycoforms, 732
Proteolytic enzymes, 927
Protolysis, 30
  constant, 34
  equilibrium, 32
  silanol groups, 58
Protolytic equilibria, 975
Pseudo ITP stacking, 467, 469
Pseudostationary phase, effects, log P_{ow} estimation, 1005
Pseudostationary phase, types, 79, 97
Pulse amperometric detection, 360
Pumping mechanisms:
  electrically driven, 278
  electroosmotic pumping, 278
  pressure driven, 278
Purity check, peptide, 660
Pyrazoline, as UV-Vis tag, 737
Pyromellitic acid, 834

Quantitative structure–activity relationship (QSAR), 999
Quantum yield, 773
Quartz glass surface, 514
Quartz substrates, 617
Quartz zeta potential, 514
Quaternary amines, 494
Quaternary amine surfactants, 827

Racemic acidic drugs, 507
Radial potential, 482
Radical catalysts, 508
Radionuclide, 338–340
Raman, 341, 342
  scattering, 322

Random pixel addressing, 773
Random walk, 152
Rate constant, 963
  of desorption, 51
Ratiogram, 782, 784
Reactive polyamide resin, 517
Reactor capillary, 927
Reagent-filled capillary, 928
Rear boundary, 619
Receptor–ligand interactions, 215
Receptor binding of neutral ligands, 955
Recombinant human erythropoietin, 8
Recombinant human insulin-like growth factor, 660
Recombinant protein(s), 708–709
Recombinant protein A, 883
Recorder time constant, 145
Recovery, validation:
  extraction, 564
  spiking, 564–565, 570
Redox reactions, catalysis, 360
Reduced mobility, 63
Reductive amination, 215
Refractive index, 333–335
  concentration gradient, 335
  off-axis, 334
  on-axis, 334
Refractive index detection, 844
  imaging in cIEF, 239, 241
Regeneration of wall surface, 487, 496
Relative binding affinity of isozymes, 959
Removal of heat, 4
Remoxipride, 573
Repeatability:
  EOF, 488
  validation, 566, 568
Replaceable gels in CGE, 189, 203, 204
Replaceable polymer matrix, 190
Reproducibility, 696
  in CGE, 200, 214
Reptation migration model, 188, 210
Resistance to mass transfer, 51, 146, 486
  intercolumn, 148
  intermicellar, 148
  micellar, 148
Resolution, 53, 485, 574, 575, 828
  effect of EOF, 68
  efficiency, 67
  enhancements, 691, 692, 703, 704, 709
  MEKC, 85
  optimization in CZE, 56
  selectivity, 67
  sieving limits, 713
Response factor, validation, 565
Response time, 306
Restriction enzyme digest analysis, microchip, 626
Restriction enzyme fragments, CGE analysis, 204, 206, 207
Restriction fragment digest, 776
Retardation and selectivity, 503
Retardation coefficient, 187
Retention factor, 84
  charged solutes, 85
  uncharged solutes, 84
Retention vs. log P_{ow} relationship, 1002
  CTAB micelles, 1006
  LiPFOS micelles, 1006
  SC micelles, 1006
  SDS micelles, 1005
Reversal of charge, 491
Reversal of EOF, 490
Reverse charge polarity, 505
Reversed phase LC, 79, 583
  column, 693
  estimation of lipophilicity, 1000
  retention vs. log P_{ow}, 1000
Rigid gels, 225
Rinsing, 488
  of capillary, 487
  NaOH, 488
  strong acid or base, 487
Rinsing procedure, 495
RNase A, 685, 690, 704
RNase B, 753, 754
Robustness, validation, 569

Saccharides, 730
Salt mobilization in cIEF, 234
Sample capacity, 528
Sample cleanup, 623
Sample concentration, CE–MS:
  cIEF, 438
  isotachophoresis, 416
  precolumn and membranes, 416, 438
  sample stacking, 415
Sample handling, 881
Sample-induced isotachophoresis, 267, 269
Sample injection, types, 37
Sample introduction. *See* Injection

Sample loading, 573
  microchip, 617
Sample matrix, 487
Sample plug length, 900
Sample plug size, 143
Sample preparation, glycobiology, 731
Sample pretreatment, 881
Sample self-stacking, 272
Sample stacking, 258, 619, 831
  CE–MS, 416
  nonaqueous CE, 527
Sample volume, 880, 881
Sample zone length, 452–466
Sampling, on-line, 900
Sampling bias, CGE, 196
Sampling efficiency, 900
Sampling frequency in 2D separation, 588
Sampling single cells, 900
Sampling with on-line reaction, 901
Sanger reaction, 770, 774, 777, 778, 780
Scanning electrochemistry, 361
Scatchard analysis, 950, 953
Schiff-base mechanism, 737, 741
SDS (sodium dodecyl sulfate), 81, 494
  adsorption, 509
  as denaturant, 190
SDS–PAGE, 10, 12, 210, 974
  protein separations, 684
SDS–protein complex, 187, 189, 210
SDS–protein mixtures, 211–215
SDS–protein separation, effect of temperature, 192
Secondary chemical equilibria, 117
Selected ion monitoring, 407, 423
Selective preconcentration, 891
Selectivity, 54, 533–538, 687, 694, 700, 715–716
Selectivity coefficient, 48, 55
Self-sharpening, 251, 254
Self-stacking, 269, 271
Semiconductor, 615, 616
Semipermanent coating, 494
Semipermanent surface modification, 494
Sensitivity:
  of microchip devices, 619
  MS detection, 415, 422
  validation, 571, 573–574
Sensitivity enhancement, peptide analysis, 654–656
  preconcentrator, 655–656, 668

stacking, 654, 658
Separation:
  of acidic peptides, 515
  of anions, 541–547
  of basic compounds, 505
  of basic drugs, 515, 517
  of basic proteins, 496, 515
  of biomolecules, 490
  of cationic and anionic small molecules, 501
  of cations, 540–541
  of chiral solutes, 550–543
  of drugs, 541
  of metal ions, 541
  of peptides, 541
  of pharmaceuticals, 517
  of polymers, 508
  of proteins, 517
  of small molecules, 517
Separation assessment, 503
Separation capillary, 927
Separation channel, microchip, 617
Separation current, 358
Separation mechanism, 587
Separation selectivity, chiral, 793
Separation time, CGE, 191
Sequencing, glycoconjugates, 736
Serpentine column geometry, 621
Serum, 888, 891, 892
Serum IgE, electropherogram, 892
Shake-flask method, 1000
Sheath flow, 16, 409, 770
  CE–MS interface, 410, 411
  cell, 306, 323–325
Sheathless interface, CE–MS, 409, 412, 414, 422
Shielding, PVA layer, 501
Shot noise, 311, 312
Sialic acids, 741, 745, 753
Sickle cell anemia, 708
Sieving effect, CGE, 191, 205
Sieving electrophoresis, 518
Sieving matrices, 189, 775, 974
Sieving mechanism, 13, 213
Sieving media, microchip:
  hydroxyethyl cellulose, 624
  linear polyacrylamide, 624, 625
Sieving medium, carbohydrate separation, 751
Signal-to-noise ratio, 382, 835
Silanization, 489, 491
  protein separation, 515

Silanization (*Contd.*)
  reagent, 508
  sublayer, 512
Silanol groups, 228, 229, 482, 483, 487, 491, 685
  determination of pK_a, 981–982
  ionization, 58, 689
Silica surface, 483
Siloxane, 229, 512
Siloxy linkage, 229
Silver staining, 225
Silylated glass beads, 884
Silylation reagent, 512
Simultaneous determination of relative binding affinities, 957, 959
Single-base resolution, 774
Single cell, 359, 593, 900, 926, 941
Single-cell analysis, 906
Single-enzyme molecule detection, 904
Single-molecule detection, 7, 321, 631
Single-stranded DNA, 190
Single-stranded oligonucleotides, 189
Single-stranded oligonucleotides, CGE separation, 196
Sintered glass frits, 884
Six-port valve in LC-CZE, 589–591
Size-based separations, 12
Size-dependent retardation, 187
Size exclusion chromatography, 583
Size exclusion in CGE, 187, 198
Size of elution window in MEKC, 84, 85, 108
Size-selective separation, 494, 512
Size selectivity, 205
Skewed peaks, 157
Slab-gel electrophoresis, 3, 10, 186, 768
Sleep-inducing peptides, 662
Slew-up time, 596–597
Slurry packed columns, 280
  sonication, 281
Smoluchowski equation, 527
Snail neuron, 600
Snake venom, 662
Sodium cholate (SC), 81
Sodium deoxycholate (SDC), 81, 693
Sodium dodecyl sulfate, (SDS), 10, 81, 494, 693, 696, 698, 700, 713, 715
Sodium hydroxide, 232
Soil adsorption, 1001
Solder glass, 358
Solgel technology, 287, 518

Solid phase chemistry, microfabricated devices, 614
Solid phase extraction, 656
Solid supports, 885
Solute–micelle interaction, 87, 1001
Solvation, 34
Solvatochromic comparison methods, 88
Solvent degassing, 282
Solvent programming, microchip, 631
Solvents for CE, physical properties, 527
Somatostatin, 662, 668
Soybean trypsin inhibitor, 959, 962
Specific preconcentrator, 883
Spermine, 495, 688, 689, 691
Splitter tee, 607
Spontaneous injection, 456
ssDNA. *See* Single-stranded DNA
Stacking, 144, 251, 258, 267, 451–474, 654, 658
  applications, 471
  band broadening, 463
  injection, 464, 477
  neutral molecules, 471
  on-column, 457, 477
  sample zone length, 452–466
Starburst dendrimers, 110, 547–548
Static coating, 508
Stationary phase in CEC:
  α-acid glycoprotein (AGP), 288
  chiral, 288
  hydroxy-β-cyclodextrin, 288
  octadecylsilane, 288
Statistical model of overlap, 583, 584
Steady-state boundaries, 254
Stepwise field strength gradient, 191
Stern double layer, 482, 489, 501
Stern–Gouy–Chapman theory, 58, 285
Stoichiometry of binding, determination, 964
Stokes radius, 693, 694
Stray light rejection, 770
Streptavidin, 964
Structural analysis, of glycoconjugates, 736
Structure–retention relationship, 87
STS (sodium tetradecyl sulfate), 81
Sublayer formation, 512
Submicron particles, 296
Substrates:
  enzymatic assays, 926, 928
  enzymatic determination by EMMA, 938
Sudan III, 84, 151
Sugar-binding activities, 892

Sugar derivatization, 733
Sugar-specific enzymes, 732
Sulfatases, 935
Sulfobetain, SB-12, 81
Sulfonamide ligand, 955
Sulfonamides, 976
Sulfonated aminonaphthalenes, fluorescence tag, 741
Sumatriptan, 566
Suppressed conductivity detection, 845
Suppression of analyte–wall interactions, 483, 517
Suppression of EOF, 484
Surface adsorption, 500
Surface-bound mediators, 360
Surface contamination, 487
Surface copolymerization, 512
Surface deactivation, 490
Surface generation, bound dextran, 514
Surface modification, 497, 686–688, 690, 692, 693, 696
  dynamic coating, 483
  MEKC, 509
  modifiers, 487, 497
  permanent coating, 483
  procedures, 488
Surface wetting:
  aqueous polymer, 514
  two-phase system, 514
Surfactant concentration, MEKC, 96
Surfactants:
  anionic, 81, 693
  effects on MEKC separations, 89–95
  nonionic, 81, 689
  pluronic, 81, 693
  sodium dodecyl sulfate, 81, 693, 696, 698, 700, 713, 715
  Triton X-100, 693
Symmetrical peaks, 828, 830
Synchronized cyclic electrophoresis, microfabricated devices, 613
Synthetic branched and multiple-antigen peptides, 645
Synthetic peptides, 660, 645, 651, 662
System zones, 386

Tagging agents:
  2-aminopyridine, 737, 748
  aminopyridyl, 737, 757
  ANDSA, 741
  ANTS, 741, 742, 748, 753, 757, 759, 760
  benzo-derivatives, 737
  in carbohydrate analysis, 737
  3-(4-carboxybenzoyl)-2-quinoline carboxaldehyde (CBQCA), 739–741, 745, 747, 757
  3-methyl-1-phenyl-2-pyrazolin-5-one (PMP), 737, 738
  pyrazoline, 737
  sulfonated aminonaphthalenes, 741
Tagging procedures, 737
Tailing peak, 688, 828
  indirect detection, 390
Tandem column, 587
Tandem MS, 735
Taylor–Aris theory, 991
Taylor dispersion, 37
Temperature control, 6
Temperature difference or gradient, 4, 45, 151, 155, 191, 583
Temperature effects:
  DNA separation in CGE, 191
  limitations, 191
  MEKC, 125, 157
  mutation studies in CGE, 195
  protein separation in CGE, 192, 211
Temperature profile, 45, 155
Temporal resolution, 769
Terminating electrolyte, 252, 457
Tetracycline antibiotics, nonaqueous CE, 530
Tetradecyltrimethylammonium bromide, 491
Tetramethylethylenediamine, 698, 699
Tetramethylrhodamine isothiocyanate, 597
Tetrapeptide, 655
Theobromine, 509
Theophylline, 509
Theoretical plates equation, 485. *See* Plate number
Therapeutic drugs, 880
Thermal conductivity, 45
Thermal dispersion coefficient, 48
Thermal effects, 45
Thermal gradient capillary electrophoresis (TGCE), 993–995
Thermal treatment, 496
Thermodynamics potential, 360
Thermooptical, 336–338
  amino acids, 336, 338
  probe beam, 336
  pump beam, 336

Thioation failures, CGE, 200
Three-dimensional separation, 589, 605–607
  coupled columns, 605
  peptides, 661
Tight-binding ligands, 957
Time constant, detector and recorder, 144–145
T_m values, Nucleic acids, 992–995
T_{mc}, migration time, micelle, 79
  marker, 84
Toluenesulfonic acid, 562
Toxicological effects, 880
Trace enrichment, 474
Transfer activity coefficient, 32, 534
Transference numbers, 59
Transfer ratio, 377, 383
  equation, 830
  expressions, 384
  rules, 388–389
Transient engagement (also plug–plug electrophoretic mixing), 929
Transient isotachophoresis (tITP), 267, 655, 881
  tITP–CZE, 457
  tITP–MS, 416
Translational stage, 769
Transverse-flow-gating interface, 598
Tridecapeptide pheromone, 668
Triethyleneamine, 515
Tris(hydroxymethyl)aminomethane, 488
Triton X-100, 708
Tryptic derivatized human plasma, 600
Tryptic digest:
  of ovalbumin, 607
  of porcine thyroglobulin, 600
Tryptic peptide mapping, 912, 914
Tryptophan, 566
Tryptophan-containing peptides, 652
Tubing material, 487
Turbulent mixing, 927
Turnover number of enzymes, 931
Tween 20 [polyoxyethylene (20) sorbitan monolaurate], 81, 490, 553, 710
Two-color sequencing, 781
Two-dimensional confinement, 631
Two-dimensional separations, 581–611
  criteria, 586
  high-speed LC–CZE, 601–605
  micro RPLC–CE, 597–601
  micro SEC–CZE, 591–597
  peptides, 660, 667
  RPLC–CE, 589–591

UV detection, 609
Two-phase systems, 514

Ubiquitin, 662
Ultrafiltration of sample, CGE, 196
Ultrasonic bath, 282
Unbound form, 955
Uncharged solutes:
  MEKC migration, 78
  nonaqueous CE, 547
Universal detection, 10, 316, 333
  far UV, 316
  indirect, 316, 317
Unretained solute, 79
Unstable EOF, 496
Urea, 653, 658
  addition in MEKC, 115
  as denaturant, 190
Ureic acid, 509
Urine sample, 600
  determination of drugs and metabolites, 885, 893
  direct injection, 7
UV-active co-ions, inorganic ion analysis, 835
UV spectra, validation, 562
UV-transparent polymer matrices, 211
UV-Vis absorbance, 304–320
  detection cell, 305
  detection limits, 308, 310, 311, 313
  focusing optics, 313
  light sources, 311, 313
  photodetectors, 311
  sensitivity, 310, 313, 315
  universal detection, 316, 333
  wavelength selector, 312
UV-Vis tags, 737
  aminopyridine, 737, 748
  aminopyridyl, 737, 757
  benzo-derivatives, 737
  carbohydrates, 737
  methyl-1-phenyl-2-pyrazolin-5-one (PMP), 737, 748

Validation:
  accuracy (cross-correlation), 559
  assay, 559
  bioassay, 560
  buffer depletion effects, 561
  chiral purity determination, 560
  difference CE and HPLC, 558

identity confirmation, 558
inorganic ion contents, 560
instrument-to-instrument transfer, 558
linearity, 561
main component assay, 558, 561
peak purity, 561
pharmaceuticals, 858
purity determination, 558
reagents supply, 562
recovery, 564
related impurities, 560
repeatability, 566
requirements, 558
response factors, 565
robustness, 569
sample diluent, 570
selectivity, 570
sensitivity, 571
solution stability, 574
system stability, 574
training, 575
transfer of methods between laboratories, 575
Valve-loop interface, 593
Van Deemter equation, 286
Van der Waals interaction, 491, 512
Vancomycin, 668, 957
Variance additivity, 36
detector, 145
Velocity, 453
dependence on polymer, concentration in CGE, 187
electrophoretic, 453, 460, 466
equation in CGE, 187
hydrodynamic, 452
total, net, apparent, 62
Vertical electric field, 501
Vinyl acetate, 514
Vinylated silica surface, 512
Viscosity measurements, 992
Viscosity of buffer, 500

Visualizing ion, 376, 393–397, 827
Voltage-driven transport, 685
Voltage effects on DNA separation, 191
Voltage gradients in CGE, 191

Walden rule, 31
Wall adsorption, 50, 486
  in ACE, 951, 952
  distribution coefficient, 51
  MEKC, 148, 155
  proteins, 8
Wall coating, 190. *See also* Coated capillaries
Wall contamination, 487
Wall interaction, 190, 229, 230, 486. *See also* Wall adsorption
Wavelength selection, 318
Weak acids and bases, 29
Weak electrolytes:
  determination of pK_a, 981
  silica surface density, 981
Whole-cell injection, 901, 902
Whole-column detection in cIEF, 238, 240

Yoctomole, 880

Z-cell, 313–315
Zeptomole, 7, 880
Zero net charge, 223, 225, 242
Zero-potential mode (EMMA), 931
Zeta potential, 482, 685, 686, 982
Zone broadening, sources of, 53. *See also* Band broadening
Zone electrophoresis (ZE), 974
Zone length. *See* Injection length
Zone sharpening, MEKC, 124–127
Zone stability, 257
Zwitterionic buffers, 8, 952
Zwitterionic surfactants, 81, 106
Zwitterions, 5, 223